The Practice of Statistics for Business and Economics

FIFTH EDITION

Layth C. Alwan
University of Wisconsin–Milwaukee

Bruce A. Craig
Purdue University

George P. McCabe
Purdue University

Austin • Boston • New York • Plymouth

Vice President, STEM: Daryl Fox
Program Director: Andrew Dunaway
Program Manager: Sarah Seymour
Senior Marketing Manager: Nancy Bradshaw
Marketing Assistant: Madeleine Inskeep
Executive Development Editor: Katrina Mangold
Development Editor: Leslie Lahr
Executive Media Editor: Catriona Kaplan
Associate Editor: Andy Newton
Assistant Editor: Justin Jones
Director of Content Management Enhancement: Tracey Kuehn
Senior Managing Editor: Lisa Kinne
Senior Content Project Manager: Edgar Doolan
Project Manager: Valerie Brandenburg, Lumina Datamatics, Ltd.
Director of Design, Content Management: Diana Blume
Design Services Manager: Natasha Wolfe
Cover Design Manager: John Callahan
Text Designer: Blake Logan
Director of Digital Production: Keri deManigold
Senior Media Project Manager: Elton Carter
Senior Workflow Manager: Paul Rohloff
Executive Permissions Editor: Cecilia Varas
Composition: Lumina Datamatics, Ltd.
Printing and Binding: LSC Communications
Cover Image: Suriyapong Thongsawang/Getty Images

Library of Congress Control Number: 2019947671

Student Edition Hardcover:
ISBN-13: 978-1-319-10900-4
ISBN-10: 1-319-10900-4

Student Edition Loose-leaf:
ISBN-13: 978-1-319-27266-1
ISBN-10: 1-319-27266-5

© 2020, 2016, 2011, 2009 by W. H. Freeman and Company
All rights reserved.
1 2 3 4 5 6   24 23 22 21 20

Macmillan Learning
One New York Plaza
Suite 4600
New York, NY 10004-1562
www.macmillanlearning.com

In 1946, William Freeman founded W. H. Freeman and Company and published Linus Pauling's *General Chemistry*, which revolutionized the chemistry curriculum and established the prototype for a Freeman text. W. H. Freeman quickly became a publishing house where leading researchers can make significant contributions to mathematics and science. In 1996, W. H. Freeman joined Macmillan and we have since proudly continued the legacy of providing revolutionary, quality educational tools for teaching and learning in STEM.

Brief Contents

CHAPTER 1	Examining Distributions	1
CHAPTER 2	Examining Relationships	63
CHAPTER 3	Producing Data	123
CHAPTER 4	Probability: The Study of Randomness	175
CHAPTER 5	Random Variables and Probability Distributions	219
CHAPTER 6	Sampling Distributions	293
CHAPTER 7	Introduction to Inference	337
CHAPTER 8	Inference for Means	395
CHAPTER 9	One-Way Analysis of Variance	457
CHAPTER 10	Inference for Proportions	505
CHAPTER 11	Inference for Categorical Data	541
CHAPTER 12	Inference for Regression	569
CHAPTER 13	Multiple Regression	617
CHAPTER 14	Time Series Forecasting	675

The Core book includes Chapters 1–14. Chapters 15–18 are individual optional Companion Chapters and can be found at www.macmillanlearning.com/psbe5e.

CHAPTER 15	Statistics for Quality: Control and Capability	15-1
CHAPTER 16	Two-Way Analysis of Variance	16-1
CHAPTER 17	Nonparametric Tests	17-1
CHAPTER 18	Logistic Regression	18-1

Contents

To Instructors: About This Book xii
SaplingPlus for Statistics xx
To Students: What Is Statistics? xxv
Index of Cases xxviii
Index of Data Tables xxix
Beyond the Basics Index xxxi
About the Authors xxxii

CHAPTER 1 Examining Distributions 1

Introduction 1

1.1 Data 2
  SECTION 1.1 SUMMARY 6
  SECTION 1.1 EXERCISES 7

1.2 Displaying Distributions with Graphs 8
  Categorical variables: bar graphs and pie charts 8
  Quantitative variables: histograms 13
  CASE 1.1 Treasury Bills 13
  Quantitative variables: stemplots 16
  Interpreting histograms and stemplots 18
  Time plots 20
  SECTION 1.2 SUMMARY 22
  SECTION 1.2 EXERCISES 22

1.3 Describing Distributions with Numbers 24
  CASE 1.2 Time to Start a Business 24
  Measuring center: the mean 25
  Measuring center: the median 26
  Comparing the mean and the median 27
  Measuring spread: the quartiles 28
  The five-number summary and boxplots 30
  The 1.5 × IQR rule for outliers 31
  Measuring spread: the standard deviation 33
  Choosing measures of center and spread 34
  BEYOND THE BASICS: Risk and return 35
  SECTION 1.3 SUMMARY 36
  SECTION 1.3 EXERCISES 37

1.4 Density Curves and the Normal Distributions 40
  Density curves 40
  The median and mean of a density curve 41
  Normal distributions 44
  The 68–95–99.7 rule 45
  The standard Normal distribution 46
  Normal distribution calculations 48
  Using the standard Normal table 50
  Inverse Normal calculations 51
  Assessing the Normality of data 52
  BEYOND THE BASICS: Density estimation 56
  SECTION 1.4 SUMMARY 57
  SECTION 1.4 EXERCISES 58

CHAPTER 1 REVIEW EXERCISES 60

CHAPTER 2 Examining Relationships 63

Introduction 63

2.1 Scatterplots 64
  CASE 2.1 Education Expenditures and Population: Benchmarking 66
  Interpreting scatterplots 68
  The log transformation 69
  Adding categorical variables to scatterplots 71
  SECTION 2.1 SUMMARY 72
  SECTION 2.1 EXERCISES 72

2.2 Correlation 74
  The correlation r 75
  Facts about correlation 76
  SECTION 2.2 SUMMARY 78
  SECTION 2.2 EXERCISES 79

2.3 Least-Squares Regression 81
  The least-squares regression line 82
  Facts about least-squares regression 87
  Interpretation of r^2 88
  Residuals 89
  The distribution of the residuals 93
  Influential observations 93
  SECTION 2.3 SUMMARY 95
  SECTION 2.3 EXERCISES 96

2.4 Cautions about Correlation and Regression 99
  Extrapolation 99
  Lurking variables 100

Correlation does not imply causation 102
BEYOND THE BASICS: Big data 103
SECTION 2.4 SUMMARY 103
SECTION 2.4 EXERCISES 104

2.5 Data Analysis for Two-Way Tables 105
CASE 2.2 Does the Right Music Sell the Product? 105
Marginal distributions 106
Conditional distributions 108
Mosaic plots and software output 110
Simpson's paradox 111
SECTION 2.5 SUMMARY 114
SECTION 2.5 EXERCISES 114
CHAPTER 2 REVIEW EXERCISES 117

CHAPTER 3 Producing Data 123

Introduction 123

3.1 Sources of Data 124
Anecdotal data 124
Available data 125
Sample surveys and experiments 127
Other sources and uses of data 128
SECTION 3.1 SUMMARY 129
SECTION 3.1 EXERCISES 129

3.2 Designing Samples 130
Simple random samples 133
Stratified samples 136
Multistage samples 137
Cautions about sample surveys 138
BEYOND THE BASICS: Capture-recapture sampling 140
SECTION 3.2 SUMMARY 141
SECTION 3.2 EXERCISES 141

3.3 Designing Experiments 143
Comparative experiments 146
Randomized comparative experiments 147
Completely randomized designs 148
How to randomize 149
The logic of randomized comparative experiments 152
Cautions about experimentation 154
Matched pairs designs 156
Block designs 156
SECTION 3.3 SUMMARY 158
SECTION 3.3 EXERCISES 159

3.4 Data Ethics 161
Institutional review boards 162
Informed consent 163
Confidentiality 163
Clinical trials 165
Behavioral and social science experiments 167
SECTION 3.4 SUMMARY 168
SECTION 3.4 EXERCISES 168
CHAPTER 3 REVIEW EXERCISES 170

CHAPTER 4 Probability: The Study of Randomness 175

Introduction 175

4.1 Randomness 176
The language of probability 177
Thinking about randomness and probability 178
SECTION 4.1 SUMMARY 179
SECTION 4.1 EXERCISES 179

4.2 Probability Models 182
Sample spaces 182
Probability rules 185
Assigning probabilities: Finite number of outcomes 187
CASE 4.1 Uncovering Fraud by Digital Analysis 187
Assigning probabilities: Equally likely outcomes 189
Independence and the multiplication rule 190
Applying the probability rules 192
SECTION 4.2 SUMMARY 194
SECTION 4.2 EXERCISES 194

4.3 General Probability Rules 197
General addition rules 198
Two-Way Table of Counts to Probabilities 200
Conditional probability 201
General multiplication rule 205
Tree diagrams 206
Bayes' rule 208
Independence again 211
SECTION 4.3 SUMMARY 212
SECTION 4.3 EXERCISES 213
CHAPTER 4 REVIEW EXERCISES 216

CHAPTER 5 Random Variables and Probability Distributions 219

Introduction 219

5.1 Random Variables 220
Discrete random variables 221
CASE 5.1 Tracking Perishable Demand 222
Continuous random variables 224
SECTION 5.1 SUMMARY 226
SECTION 5.1 EXERCISES 227

5.2 Means and Variances of Random Variables 228
The mean of a random variable 229
Mean and the law of large numbers 231
Thinking about the law of large numbers 233
Rules for means 234
CASE 5.2 Portfolio Analysis 234
The variance of a random variable 238
Rules for variances and standard deviations 240
SECTION 5.2 SUMMARY 245
SECTION 5.2 EXERCISES 246

5.3 Common Discrete Distributions 249
Binomial distributions 249
The binomial distributions for sample counts 250
The binomial distributions for statistical sampling 251
CASE 5.3 Inspecting a Supplier's Products 252
Finding binomial probabilities 253
Binomial formula 256
Binomial mean and standard deviation 258
Assessing the binomial assumptions with data 260
Poisson Distributions 261
The Poisson setting 262
The Poisson model 264
Poisson approximation of the binomial 264
Assessing the Poisson assumptions with data 265
SECTION 5.3 SUMMARY 267
SECTION 5.3 EXERCISES 268

5.4 Common Continuous Distributions 272
Uniform distributions 273
Revisiting Normal distributions 275
Normal approximation for binomial distribution 277
The continuity correction 280
Normal approximation for Poisson distribution 281
Exponential distributions 282
SECTION 5.4 SUMMARY 284
SECTION 5.4 EXERCISES 285
CHAPTER 5 REVIEW EXERCISES 287

CHAPTER 6 Sampling Distributions 293

Introduction 293

6.1 Toward Statistical Inference 294
Sampling variability 296
Sampling distributions 296
Bias and variability in estimation 300
Sampling from large populations 303
Why randomize? 304
SECTION 6.1 SUMMARY 304
SECTION 6.1 EXERCISES 305

6.2 The Sampling Distribution of the Sample Mean 307
The mean and standard deviation of $\bar{x}$ 311
The central limit theorem 312
How large is large enough? 315
Two more facts 320
SECTION 6.2 SUMMARY 321
SECTION 6.2 EXERCISES 322

6.3 The Sampling Distribution of the Sample Proportion 324
Sample proportion mean and standard deviation 325
Normal approximation for proportions 328
SECTION 6.3 SUMMARY 331
SECTION 6.3 EXERCISES 331
CHAPTER 6 REVIEW EXERCISES 333

CHAPTER 7 Introduction to Inference 337

Introduction 337
Overview of inference 338

7.1 Estimating with Confidence 339
Statistical confidence 340
Confidence intervals 342
Confidence interval for a population mean 344
CASE 7.1 Bankruptcy Attorney Fees 345
How confidence intervals behave 348
Some cautions 350
SECTION 7.1 SUMMARY 351
SECTION 7.1 EXERCISES 352

7.2 Tests of Significance 355
The reasoning of significance tests 355
CASE 7.2 Fill the Bottles 356
Step 1: Stating the hypotheses 358
Step 2: Calculating the value of a test statistic 360
Step 3: Finding the *P*-value 362
Step 4: Stating a conclusion 364
Summary of the *z* test for one population mean 366
Two-sided significance tests and confidence intervals 368
Assessing significance with *P*-values versus critical values 371
SECTION 7.2 SUMMARY 372
SECTION 7.2 EXERCISES 373

7.3 Use and Abuse of Tests 376
Choosing a level of significance 376
Statistical significance does not imply practical significance 378
Statistical inference is not valid for all sets of data 379
Beware of searching for significance 380
SECTION 7.3 SUMMARY 381
SECTION 7.3 EXERCISES 381

7.4 Prediction Intervals 383
Concept of random deviations 384
Prediction of a single observation 384
SECTION 7.4 SUMMARY 388
SECTION 7.4 EXERCISES 388
CHAPTER 7 REVIEW EXERCISES 389

CHAPTER 8 Inference for Means 395
Introduction 395

8.1 Inference for the Mean of a Population 396
t distributions 396
The one-sample *t* confidence interval 398
CASE 8.1 Battery Life of a Smartphone 399
The one-sample *t* test 400
Using software 403
Matched pairs *t* procedures 406
Robustness of the one-sample *t* procedures 409
Inference for non-Normal populations 410
BEYOND THE BASICS: The bootstrap 412
SECTION 8.1 SUMMARY 414
SECTION 8.1 EXERCISES 415

8.2 Comparing Two Means 419
The two-sample *t* statistic 420
The two-sample *t* confidence interval 421
The two-sample *t* significance test 423
Robustness of the two-sample procedures 424
Inference for small samples 426
The pooled two-sample *t* procedures 428
CASE 8.2 Active versus Failed Retail Companies 429
SECTION 8.2 SUMMARY 432
SECTION 8.2 EXERCISES 433

8.3 Additional Topics on Inference 437
Sample size for confidence intervals 438
Power of a significance test 441
Inference as a decision 445
SECTION 8.3 SUMMARY 449
SECTION 8.3 EXERCISES 449
CHAPTER 8 REVIEW EXERCISES 451

CHAPTER 9 One-Way Analysis of Variance 457
Introduction 457

9.1 One-Way Analysis of Variance 458
The ANOVA setting 458
Comparing means 459
Revisiting the pooled two-sample *t* statistic 460
An overview of ANOVA 461
CASE 9.1 Tip of the Hat and Wag of the Finger? 461
The ANOVA model 464
Estimates of population parameters 466
Testing hypotheses in one-way ANOVA 469
The ANOVA table 470
The *F* test 473
Using software 476
BEYOND THE BASICS: Testing the equality of spread 478
SECTION 9.1 SUMMARY 479
SECTION 9.1 EXERCISES 480

9.2 Additional Comparisons of Group Means 483
Contrasts 483
CASE 9.2 Evaluation of a New Educational Product 483
Multiple comparisons 489
Simultaneous confidence intervals 492

Assessing the power of the ANOVA *F* test 493
SECTION 9.2 SUMMARY 495
SECTION 9.2 EXERCISES 496
CHAPTER 9 REVIEW EXERCISES 497

CHAPTER 10 Inference for Proportions 505

Introduction 505

10.1 Inference for a Single Proportion 506
CASE 10.1 Trends in the Workplace 506
Large-sample confidence interval for a single proportion 507
BEYOND THE BASICS: Plus four confidence interval for a single proportion 510
Significance test for a single proportion 511
Choosing a sample size for a confidence interval 514
CASE 10.2 Marketing Christmas Trees 515
Choosing a sample size for a significance test 517
SECTION 10.1 SUMMARY 518
SECTION 10.1 EXERCISES 519

10.2 Comparing Two Proportions 523
Large-sample confidence intervals for a difference in proportions 524
CASE 10.3 Social Media in the Supply Chain 525
BEYOND THE BASICS: Plus four confidence intervals for a difference in proportions 527
Significance tests 527
Choosing a sample size for two sample proportions 531
BEYOND THE BASICS: Relative risk 534
SECTION 10.2 SUMMARY 535
SECTION 10.2 EXERCISES 536
CHAPTER 10 REVIEW EXERCISES 537

CHAPTER 11 Inference for Categorical Data 541

Introduction 541

11.1 Inference for Two-Way Tables 542
Two-way tables 542
CASE 11.1 Are Flexible Companies More Competitive? 543
Describing relations in two-way tables 545
The null hypothesis: no association 548
Expected cell counts 548
The chi-square test 549
The chi-square test and the z test 551
Models for two-way tables 552
BEYOND THE BASICS: Meta-analysis 554
SECTION 11.1 SUMMARY 556
SECTION 11.1 EXERCISES 557

11.2 Goodness of Fit 559
SECTION 11.2 SUMMARY 563
SECTION 11.2 EXERCISES 563
CHAPTER 11 REVIEW EXERCISES 564

CHAPTER 12 Inference for Regression 569

Introduction 569

12.1 Inference about the Regression Model 570
Statistical model for simple linear regression 571
From data analysis to inference 572
CASE 12.1 The Relationship between Income and Education for Entrepreneurs 572
Estimating the regression parameters 577
Conditions for regression inference 581
Confidence intervals and significance tests 582
The word "regression" 586
Inference about correlation 587
SECTION 12.1 SUMMARY 589
SECTION 12.1 EXERCISES 590

12.2 Using the Regression Line 596
Confidence and prediction intervals 597
BEYOND THE BASICS: Nonlinear regression 601
SECTION 12.2 SUMMARY 602
SECTION 12.2 EXERCISES 602

12.3 Some Details of Regression Inference 604
Standard errors 604
Analysis of variance for regression 607
SECTION 12.3 SUMMARY 610
SECTION 12.3 EXERCISES 611
CHAPTER 12 REVIEW EXERCISES 612

CHAPTER 13 Multiple Regression 617

Introduction 617

13.1 Data Analysis for Multiple Regression 618
Using a linear model with multiple variables 618

CASE 13.1 The Inclusive Development Index (IDI) 620
Data for multiple regression 621
Preliminary data analysis for multiple regression 623
Estimating the multiple regression coefficients 625
Regression residuals 627
The regression standard error 629
SECTION 13.1 SUMMARY 630
SECTION 13.1 EXERCISES 631

13.2 Inference for Multiple Regression 633
Multiple linear regression model 634
CASE 13.2 Predicting Movie Revenue 635
Estimating the parameters of the model 635
Inference about the regression coefficients 636
Inference about prediction 639
ANOVA table for multiple regression 639
Squared multiple correlation R^2 641
Inference for a collection of regression coefficients 643
SECTION 13.2 SUMMARY 645
SECTION 13.2 EXERCISES 646

13.3 Multiple Regression Model Building 650
CASE 13.3 Prices of Homes 650
Models for curved relationships 653
Models with categorical explanatory variables 655
More elaborate models 659
Variable selection methods 662
BEYOND THE BASICS: Regression trees 665
SECTION 13.3 SUMMARY 666
SECTION 13.3 EXERCISES 667

CHAPTER 13 REVIEW EXERCISES 669

CHAPTER 14 Time Series Forecasting 675

Introduction 675

Overview of Time Series Forecasting 676

14.1 Assessing Time Series Behavior 676
CASE 14.1 Adidas Stock Price Returns 678
Random process model 679
Nonrandom processes 679
CASE 14.2 Amazon Sales 681
Runs test 682
Autocorrelation function 685
Forecasts of a random process 690

SECTION 14.1 SUMMARY 692
SECTION 14.1 EXERCISES 692

14.2 Random Walks 693
Price changes versus returns 698
Deterministic and stochastic trends 702
BEYOND THE BASICS: Dickey-Fuller tests 705
SECTION 14.2 SUMMARY 707
SECTION 14.2 EXERCISES 707

14.3 Basic Smoothing Models 709
Moving-average models 710
CASE 14.3 Lake Michigan Water Levels 711
Forecasting accuracy 714
Moving average and seasonal indexes 717
Exponential smoothing models 725
SECTION 14.3 SUMMARY 731
SECTION 14.3 EXERCISES 732

14.4 Regression-Based Forecasting Models 734
Modeling deterministic trends 734
Modeling seasonality 740
Residual checking 747
Modeling with lagged variables 749
BEYOND THE BASICS: ARCH models 758
SECTION 14.4 SUMMARY 761
SECTION 14.4 EXERCISES 761

CHAPTER 14 REVIEW EXERCISES 763

Tables T-1
Answers to Odd-Numbered Exercises S-1
Index I-1
Notes and Data Sources (offered online at www.macmillanlearning.com/psbe5e)

The following optional Companion Chapters can be found online at www.macmillanlearning.com/psbe5e.

CHAPTER 15 Statistics for Quality: Control and Capability 15-1

Introduction 15-1
Quality overview 15-2
Systematic approach to process improvement 15-3
Process improvement toolkit 15-4

15.1 Statistical Process Control 15-7
SECTION 15.1 SUMMARY 15-9
SECTION 15.1 EXERCISES 15-9

15.2 Variable Control Charts 15-10
$\bar{x}$ and R charts 15-12

CASE 15.1 Turnaround Time for Lab
    Results 15-14
CASE 15.2 O-Ring Diameters 15-18
$\bar{x}$ and s charts 15-22
Assumptions underlying subgroup charts 15-24
Charts for individual observations 15-28
SECTION 15.2 SUMMARY 15-34
SECTION 15.2 EXERCISES 15-35

15.3 Process Capability Indices 15-38
SECTION 15.3 SUMMARY 15-44
SECTION 15.3 EXERCISES 15-45

15.4 Attribute Control Charts 15-46
Control charts for sample proportions 15-46
CASE 15.3 Reducing Absenteeism 15-47
Control charts for counts per unit of
    measure 15-51
SECTION 15.4 SUMMARY 15-53
SECTION 15.4 EXERCISES 15-53
CHAPTER 15 REVIEW EXERCISES 15-54

CHAPTER 16 Two-Way Analysis of Variance 16-1

Introduction 16-1

16.1 The Two-Way ANOVA Model 16-2
Advantages of two-way ANOVA 16-2
The two-way ANOVA model 16-5
Main effects and interactions 16-7
SECTION 16.1 SUMMARY 16-12
SECTION 16.1 EXERCISES 16-13

16.2 Inference for Two-Way ANOVA 16-14
The ANOVA table for two-way ANOVA 16-14
Carrying out a two-way ANOVA 16-15
CASE 16.1 Discounts and Expected Prices 16-15
CASE 16.2 Expected Prices, Continued 16-17
SECTION 16.2 SUMMARY 16-21
SECTION 16.2 EXERCISES 16-21
CHAPTER 16 REVIEW EXERCISES 16-22

CHAPTER 17 Nonparametric Tests 17-1

Introduction 17-1

17.1 The Wilcoxon Rank Sum Test 17-3
CASE 17.1 Price Discrimination? 17-3
The rank transformation 17-5
The Wilcoxon rank sum test 17-6
The Normal approximation 17-8
What hypotheses does Wilcoxon test? 17-10
Ties 17-11
CASE 17.2 Consumer Perceptions of
    Food Safety 17-11
Rank versus t tests 17-13
SECTION 17.1 SUMMARY 17-14
SECTION 17.1 EXERCISES 17-14

17.2 The Wilcoxon Signed Rank Test 17-17
The Normal approximation 17-20
Ties 17-22
SECTION 17.2 SUMMARY 17-23
SECTION 17.2 EXERCISES 17-24

17.3 The Kruskal-Wallis Test 17-26
Hypotheses and assumptions 17-27
The Kruskal-Wallis test 17-27
SECTION 17.3 SUMMARY 17-30
SECTION 17.3 EXERCISES 17-30
CHAPTER 17 REVIEW EXERCISES 17-31

CHAPTER 18 Logistic Regression 18-1

Introduction 18-1

18.1 The Logistic Regression Model 18-2
CASE 18.1 Clothing Color and Tipping 18-2
Binomial distributions and odds 18-3
Model for logistic regression 18-4
Fitting and interpreting the logistic regression
    model 18-6
SECTION 18.1 SUMMARY 18-8
SECTION 18.1 EXERCISES 18-9

18.2 Inference for Logistic Regression 18-10
Examples of logistic regression analyses 18-13
SECTION 18.2 SUMMARY 18-16
SECTION 18.2 EXERCISES 18-17

18.3 Multiple Logistic Regression 18-18
SECTION 18.3 SUMMARY 18-20
SECTION 18.3 EXERCISES 18-21
CHAPTER 18 REVIEW EXERCISES 18-22

To Instructors: About This Book

Statistics is the science of data. ***The Practice of Statistics for Business and Economics (PSBE)*** is an introduction to statistics for students of business and economics based on this principle. We present methods of basic statistics in a way that emphasizes working with data and mastering statistical reasoning. *PSBE* is elementary in mathematical level but conceptually rich in statistical ideas. After completing a course based on our text, we would like students to be able to think objectively about conclusions drawn from data and use statistical methods in their own work.

In *PSBE* we combine attention to basic statistical concepts with a comprehensive presentation of the elementary statistical methods that students will find useful in their work. We believe that you will enjoy using *PSBE* for several reasons:

1. *PSBE* examines the nature of modern statistical practice at a level suitable for beginners. We focus not only on the analysis of data and the traditional topics of probability and inference but also on the sources and production of data.

2. *PSBE* emphasizes data production and data analysis, while inference is treated as a tool that helps us to draw conclusions from data in an appropriate way.

3. *PSBE* presents data analysis as more than a collection of techniques for exploring data. We emphasize systematic ways of thinking about data. Simple principles guide the analysis: always plot your data; look for overall patterns and deviations from them; when looking at the overall pattern of a distribution for one variable, consider shape, center, and spread; for relations between two variables, consider form, direction, and strength; always ask whether a relationship between variables is influenced by other variables lurking in the background. We warn students about pitfalls in clear, cautionary discussions.

4. *PSBE* uses real examples and exercises from business and economics to illustrate and enforce key ideas. Students learn the technique of least-squares regression and how to interpret the regression slope. But they also learn the conceptual ties between regression and correlation, the importance of looking for influential observations.

5. *PSBE* is aware of current developments both in statistical science and in teaching statistics. Brief optional "Beyond the Basics" sections give quick overviews of topics such as density estimation, the bootstrap, big data, relative risk, regression trees, and meta-analysis.

Themes of this Book

Look at your data is a consistent theme in *PSBE*. Rushing to inference without first exploring the data is the most common source of statistical errors that we see in working with users from many fields. A second theme is that *where the data come from matters.* When we do statistical inference, we are acting as if the data come from a properly randomized sample or experimental design. A basic understanding of these designs helps students grasp how inference works. The distinction between observational and experimental data helps students understand the truth of the mantra "association does not imply causation." Moreover, managers need to understand the use of sample surveys for market research and customer satisfaction as well as the use of statistically designed experiments for the development and improvement of products and processes.

Another theme that runs through *PSBE* is that data lead to decisions in a specific setting. A calculation or graph or "reject H_0" is not the conclusion of an exercise in statistics. We encourage students to state a conclusion in the context of a specific problem and we hope that you will require them to do so.

Finally, we think that a first course in statistics should equip students to use statistics (and learn more statistics as needed) by presenting the major concepts and most-used tools of the discipline.

What's New in the Fifth Edition

- **Learning objectives.** We have added learning objectives to the beginning of each section in every chapter. These are designed to give students and instructors an overview of what the students are expected to be able to do when they complete their study of the section. They are not only expected to learn basic concepts; they are expected to acquire the skills needed to apply these concepts in realistic situations.

- **Exercises and examples.** Over 60% of the exercises and examples are new or revised. We have placed additional emphasis on making the business or economics relevance of the exercises clear to the reader.

- **Examining distributions.** Similarities and differences in methods for categorical data and quantitative data are given additional emphasis, and concerns of reviewers on particular issues have been addressed (Chapter 1).

- **Examining relationships.** Material in the Introduction and the first section of Chapter 2 have been reorganized and a brief treatment of covariance has been added.

- **Producing data.** Chapter 3 has been revised to strengthen connections with the material on examining distributions in the first chapter.

- **Probability.** We have reorganized the sections on probability models and general probability rules so that they are now self-contained in this one chapter (Chapter 4).

- **Random variables.** We combined the sections on random variables that previously appeared across two chapters into this one chapter now devoted to random variables and common distributions for random variables (Chapter 5).

- **Common distributions.** Given our reorganization of random variable topics into a single chapter (Chapter 5), we now include a section on common discrete distributions and a section on common continuous distributions. The section on discrete distribution covers the binomial and Poisson distributions. The section on continuous distributions covers the uniform distribution, a revisiting of the Normal distribution, and the exponential distribution.

- **Introduction to inference.** We have reorganized our introduction to inference into two shorter chapters. The first chapter (Chapter 6) focuses solely on sampling distributions. The second chapter (Chapter 7) focuses on inference for one mean with the population standard deviation known. This includes new subsections more clearly delineating the basic steps of hypothesis testing as well as expanded discussion of the uses and abuses of tests. Finally, this chapter introduces a new section on prediction intervals for a single variable, a very relevant topic when considering predictive analytics.

- **Inference for means.** The introductory material on power is now combined with the material on sample size calculations for means in the last section of this chapter (Chapter 8). In addition, the material on inference as a decision, which includes discussion of Type I and Type II errors and their relationship with power, appears in this section.

- **One-way analysis of variance.** This chapter (Chapter 9) has been moved to follow inference for means so the book's flow is now methods to compare one, two, and then more than two means followed by one, two, and then more than two proportions/two-way tables.

- **Sample size determination using software for means and proportions.** Additional material on choosing sample sizes for one, two, and more than two means as well as one and two proportions is included in Chapters 8, 9, and 10, respectively.

- **Inference for proportions.** This material has been moved to Chapter 10. Some potentially confusing issues related to percents and proportions have been clarified and the plus four procedures have been moved to Beyond the Basics items.

- **Inference for categorical data.** This material has been moved to Chapter 11. An introduction has been added and the exercise structure has been changed so that exercises appear after each section with chapter exercises at the end. Enhanced connections between the concepts in this chapter and those in Chapter 2 (descriptive statistics for two-way tables) and Chapter 4 (probability models) have been added.

- **Time Series.** Chapter 14 has been reorganized so that smoothing methods (moving-average and exponential smoothing) are introduced before regression-based forecasting methods. We have added the topic on the computation seasonal additive indexes along with the computation of seasonal ratios. There is new detailed discussion on the contrasting of random processes versus nonrandom processes. We have added two new Beyond the Basics sections, one on Dickey-Fuller tests and the other on ARCH models.

- **Quality Control.** Chapter 15 was reorganized so that there is a standalone section on process capability. The chapter includes new discussion on the underlying assumptions of control charts.

- **Increased emphasis on software.** We have increased our emphasis on graphical displays of data. Software displays have been updated and are given additional prominence.

- **Software output.** Carefully selected instances of software output shown in the text can be switched to match whatever software the student is using when working in our SaplingPlus e-book.

- **Video technology manuals.** In addition, links to video technology manuals are supplied at various places in each chapter so students can see how specific analyses are performed using Excel, JMP, Minitab, SPSS, CrunchIt!, TI calculators, and R.

Content and Style

PSBE adapts to the business and economics statistics setting the approach to introductory instruction that was inaugurated and proved successful in the best-selling general statistics text *Introduction to the Practice of Statistics* (ninth edition, Freeman 2017). *PSBE* features use of real data in examples and exercises and emphasizes statistical thinking as well as mastery of techniques. As the continuing revolution in computing automates most tiresome details, an emphasis on statistical concepts and on insight from data becomes both more practical for students and teachers and more important for users who must supply what is not automated.

Chapters 1 and 2 present the methods and unifying ideas of data analysis. Students appreciate the usefulness of data analysis, and that they can actually do it relieves a bit of their anxiety about statistics. We hope that they will grow accustomed to examining data and will continue to do so even when formal inference to answer a specific question is the ultimate goal. Note in particular that Chapter 2 gives an extended treatment of correlation and regression as descriptive tools, with attention to issues such as influential observations and the dangers posed by lurking variables. These ideas and tools have wider scope than an emphasis on inference (Chapters 12 and 13) allows. We think that a full discussion of data analysis for both one and several variables before students meet inference in these settings both reflects statistical practice and is pedagogically helpful.

Teachers will notice some nonstandard ideas in these chapters, particularly regarding the Normal distributions—we capitalize "Normal" to avoid suggesting that these distributions are "normal" in the usual sense of the word. We introduce density curves and Normal distributions in Chapter 1 as models for the overall pattern of some sets of data. Only later (Chapter 4) do we see that the same tools can describe probability distributions. Although unusual, this presentation reflects the historical origin of Normal distributions and also helps break up the mass of probability that is so often a barrier that students fail to surmount.

We use the notation $N(\mu, \sigma)$ rather than $N(\mu, \sigma^2)$ for Normal distributions. The traditional notation is in fact indefensible other than as inherited tradition. The standard deviation, not the variance, is the natural measure of scale in Normal distributions, visible on the density curve, used in standardization, and so on. We want students to think in terms of mean and standard deviation, so we talk in these terms.

In Chapter 3, we discuss random sampling and randomized comparative experiments. The exposition pays attention to practical difficulties, such as nonresponse in sample surveys, that can greatly reduce the value of data. We think that an understanding of such broader issues is particularly important for managers who must use data but do not themselves produce data. Discussion of statistics in practice alongside more technical material is part of our emphasis on data leading to practical decisions. We include a section on data ethics, a topic of increasing importance for business managers. Chapter 4 then presents probability. Chapter 4 contains only the probability material that is needed to understand statistical inference, and this material is presented quite informally. The sections on probability models and general probability rules variables have been reorganized so that they are self-contained in this chapter. Chapter 5 now focuses on random variables and common distributions (discrete and continuous). Chapter 6 introduces the concepts of parameters and statistics, sampling distributions, and bias and precision. This chapter provides a nice lead-in to Chapter 7, which provides the reasoning of inference.

The remaining chapters present statistical inference, still encouraging students to ask where the data come from and to look at the data rather than quickly choosing a statistical test from an Excel menu. Chapter 7, which describes the reasoning of inference, is the cornerstone. Chapter 8 presents methods for analyzing means from one and two populations, while Chapter 9 treats the more general case of several means by introducing analysis of variance (ANOVA). In a similar way, Chapter 10 presents methods for analyzing proportions from one and two populations while Chapter 11 treats the more general case of two-way tables of counts. Chapter 12 discusses inference for regression with a single explanatory variable while Chapter 13 treats the more general case of several explanatory variables. Chapter 14 discusses time series, a very important topic for many business and economics applications. Instructors who wish to customize a single-semester course or to add a second semester will find a wide choice of additional advanced topics in the Companion Chapters that extend *PSBE*. These chapters are:

Chapter 15 Statistics for Quality: Control and Capability
Chapter 16 Two-Way Analysis of Variance
Chapter 17 Nonparametric Tests
Chapter 18 Logistic Regression

Companion Chapters can be found on the book's website: www.macmillanlearning.com/psbe5e.

Accessible Technology

Any mention of the current state of statistical practice reminds us that quick, cheap, and easy computation has changed the field. Procedures such as our recommended two-sample *t* and logistic regression depend on software. Even the mantra "look at your data" depends in practice on software, as making multiple plots by hand is too tedious when quick decisions are required. Moreover, automating calculations and graphs increases students' ability to complete problems, reduces their frustration, and helps them concentrate on ideas and problem recognition rather than mechanics.

We therefore strongly recommend that a course based on PSBE be accompanied by software of your choice.

Instructors will find using software easier because all data sets for *PSBE* can be found in numerous common formats on the website www.macmillanlearning.com/psbe5e. Students will find additional support for the various software programs in their SaplingPlus e-book.

Carefully selected instances of output can be switched to match whatever software the student is using.

Additionally, video links are supplied at various places in each chapter so students can see how specific analyses are performed using Excel, JMP, Minitab, SPSS, CrunchIt!, TI calculators, and R. Technology Appendices for each chapter, available on the *PSBE* website, give detailed instructions on how to perform key statistical procedures in various statistical software packages including Excel, JMP, R, Minitab, TI 83/84 calculators, SPSS, and CrunchIt!.

Microsoft Excel is frequently used for statistical analysis in business. Our displays of output therefore include Excel, though output from several other programs also appears. *PSBE* is not tied to specific software. Even so, one of our emphases is that a student who has mastered the basics of, say, regression can interpret and use regression output from almost any software.

We are well aware that Excel lacks many advanced statistical procedures. More seriously, Excel's statistical procedures have been found to be inaccurate, and they lack adequate warnings for users when they encounter data for which they may give incorrect answers. There is good reason for people whose profession requires continual use of statistical analysis to avoid Excel. But there are also good practical reasons why managers, whose life is not statistical, prefer a program that they regularly use for other purposes. Excel appears to be adequate for simpler analyses of the kind that occur in many business applications.

Technology can be used to assist learning statistics as well as doing statistics. The design of good software for learning is often quite different from that of software for doing. We want to call particular attention to the set of statistical applets available in SaplingPlus and on the *PSBE* website: www.macmillanlearning.com/psbe5e. These interactive graphical programs provide a very effective way to help students grasp the sensitivity of correlation and regression to outliers, the idea of a confidence interval, the way ANOVA responds to both within-group and among-group variation, and many other statistics fundamentals. Exercises using these applets appear throughout the text, marked by a distinctive icon. We urge you to assign some of these, and we suggest that if your classroom is suitably equipped, you use these very helpful tools for classroom presentations as well.

Carefully Structured Pedagogy

Few students find statistics easy. An emphasis on real data and real problems helps maintain motivation, and there is no substitute for clear writing. Beginning with data analysis builds confidence and gives students a chance to become familiar with software before the statistical content becomes intimidating. We have adopted several structural devices to aid students. Major settings that drive the exposition are presented as cases with more background information than other examples. A distinctive icon ties together examples and exercises based on a case.

At key points in the text, Reminder margin notes direct the reader to the first explanation of a topic, providing page numbers for easy reference.

The exercises are structured with particular care. Short "Apply Your Knowledge" sections pose straightforward problems immediately after each major new idea. These give students stopping points (in itself a great help to beginners) and also tell them that "you should be able to do these things right now." Each numbered section in the text ends with a substantial set of exercises, and more appear as review exercises at the end of each chapter.

Acknowledgments

We are indebted to Karen Carson whose editorial vision and leadership helped to keep the *PSBE* project relevant, interesting, and useful to instructors and readers alike.

We offer our sincere thanks to the following instructors who offered their time and expertise to insure accuracy across the project and complete the ancillary package for this fifth edition: Tadd Colver, Utah State University, accuracy reviewed the fifth edition text; James Adcock, University of Waterloo, authored Clicker questions, lecture slides, and test bank; Terri Rizzo, Lakehead University, accuracy reviewed Clicker questions and test bank; Patricia Humphrey, Georgia Southern University, authored the instructor's guide and technology appendices; Edward Hoang, University of Colorado, Colorado Springs, authored the solutions and back of book answers; Andrew Stephenson, Georgia Gwinnett College, accuracy reviewed exercise solutions and back of book answers; Mark McKibben, West Chester University, authored the practice quizzes; Jean-Marie Magnier, Springfield Technical Community College, accuracy reviewed the practice quizzes.

We are also grateful to the many colleagues and students who have provided helpful comments about *PSBE*. In particular, we would like to thank the following instructors who, as reviewers, offered specific comments on the Fifth Edition:

Worku Aberra, *Dawson College*
Imam Alam, *University of Northern Iowa*
Shafkat Ali, *York University*
Afshin Amiraslany, *University of Saskatchewan*
Wendine R. Bolon, *Monmouth College*
Jennifer A. Bossard, *Doane University*
Ba Chu, *Carleton University*
Marcus D. Casey, *University of Illinois at Chicago*
Paolo Catasti, *Virginia Commonwealth University*
Laurel Chiappetta, *University of Pittsburgh*
Clare Chua, *Ryerson University*
Richard Cox, *Arizona State University*
Joan Donohue, *University of South Carolina*
Andrew Flostrand, *Beedie School of Business*
Mark A. Gebert, *University of Kentucky*
Alex J. Gialanella, *Fordham University*
Matthew Gnagey, *Weber State University*
Elkafi Hassini, *McMaster University*
Susan Kay Herring, *Sonoma State University*
Edward Hoang, *University of Colorado, Colorado Springs*
Aloyce R. Kaliba, *Southern University and A&M College*
William Kinney, *Bethel University*
Warren Kriesel, *University of Georgia*
Moshe Lander, *Concordia University*

Robert Lantis, *Ball State University*
Sally A. Lesik, *Central Connecticut State University*
Tung Liu, *Ball State University*
Janelle Mann, *University of Manitoba*
Edward Markowski, *Old Dominion University*
Ata Mazaheri, *University of Toronto Scarborough*
David E McFeely, *San Jose State University*
Mikhail I. Melnik, *Kennesaw State University*
Kendra Brown Mhoon, *University of Houston Downtown*
Stein Monteiro, *York University*
Bryan Nelson, *University of Pittsburgh*
Joseph R. Nolan, *Northern Kentucky University*
Joe Nowakowski, *Muskingum University*
Judith Palm, *Vancouver Island University*
Nitin V. Paranjpe, *Wayne State University*
Hilde Patron Boenheim, *University of West Georgia*
Courtney Pham, *Missouri State University*
Simcha Pollack, *St. John's University*
Wei Qi, *McGill University*
Mike Racer, *University of Memphis*
Luminita Razaila, *University of North Florida*
James Reimer, *Lethbridge College*
Terri Rizzo, *Lakehead University*

Valerie Rochester, *Carleton University*
P. Wesley Routon, *Georgia Gwinnett College*
Deborah Rumsey, *Ohio State University*
Andrei Semenov, *York University*
Rafael Solis, *California State University, Fresno*
Andrew Stephenson, *Georgia Gwinnett College*
Constant Tra, *University of Nevada, Las Vegas*

Richard J. Volpe III, *California Polytechnic State University*
Yuehua Wu, *York University*
Cleusa S. Yamamoto, *Douglas College*
Yuri Yatsenko, *Houston Baptist University, Dunham College of Business*

SaplingPlus for Statistics

Assessment—The "Office Hours" experience, while doing Homework

Tutorial-Style Formative Assessment

For select questions in our formative assessment environment, students' incorrect answers receive full solutions and feedback to guide their study. Many exercises are also designed to deliver error-specific feedback based on their common misconceptions about topics/learning objectives. This socratic feedback mechanism emulates the office hours experience, encouraging students to think critically about their identified misconception. Students are also provided with the fully-worked solutions to reinforce concepts and provide an in-product study guide for every problem.

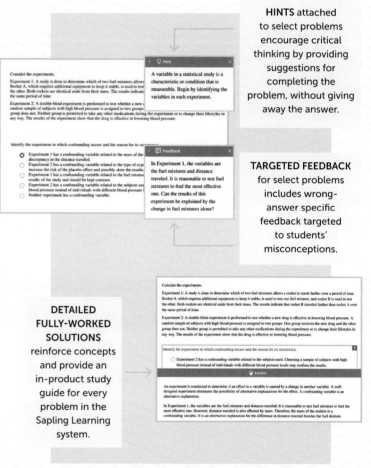

HINTS attached to select problems encourage critical thinking by providing suggestions for completing the problem, without giving away the answer.

TARGETED FEEDBACK for select problems includes wrong-answer specific feedback targeted to students' misconceptions.

DETAILED FULLY-WORKED SOLUTIONS reinforce concepts and provide an in-product study guide for every problem in the Sapling Learning system.

Adaptive Assessment Focused on Fundamental Concepts

LEARNINGCURVE

LearningCurve's adaptive quizzing encourages students to learn through practice. Through gamified elements, students are challenged to gain points by submitting correct answers to meet or exceed an instructor-determined score. Faculty and students are then provided detailed analytics on their performance via their own personal study plan with pathways to additional learning tools and insights. This tool focuses on getting students prepared for their upcoming class.

Analytics

SAPLING GRADEBOOK

STUDENT ANALYSIS

Through our heatmap dashboard instructors can get a quick visual representation of student performance on a given question. They can then dive deeper into each individual student to view and evaluate every submission they've made on a problem, getting to see a student's "work" on the question.

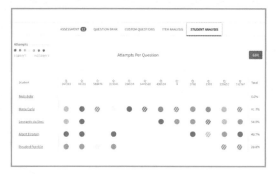

ITEM ANALYSIS

Instructors can also evaluate an aggregate view of their class's performance on a problem (item) through our **"Item Analysis"** tab. They can quickly see what percentage of students got this question correct, incorrect, or unanswered through the progress bar at the top. They can also see all of the student responses rolled up; giving them immediate insight into what their students' most common misconceptions are on this problem.

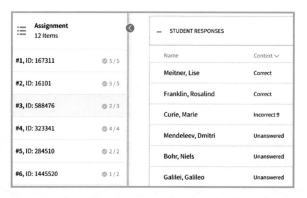

Interactive e-book

The **e-book** provides powerful study tools for students, multimedia content, and easy customization for instructors. Students can search, highlight, and bookmark specific information, making it easier to study and access key content. Media assets such as data sets, glossary terms, select exercise answers, and videos are linked within the e-book in **SaplingPlus,** allowing ease-of-use for students.

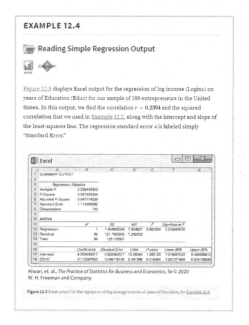

Integration

With **Deep Asset Linking,** instructors can organize individual pieces of Macmillan content into LMS Content folders. Students complete single-sign-on through the link to connect their Macmillan and LMS accounts. Once this occurs, students will have seamless access to the Macmillan content through their campus LMS using only the LMS username and password.

- **Multiple gradebook columns** generate automatically for each Macmillan assignment deployed to the LMS, allowing for assignments to be combined into gradebook categories within the LMS for gradebook calculations like dropping lowest scores or weighted percentages.

- **Automatic grade sync** occurs with Deep integration.

Media/Learning Objects

NEW PSBE-SPECIFIC RESOURCES

IN-CLASS LEARNING ACTIVITIES

Based on case studies from the textbook, in-class learning activities include guided tasks and problem-solving activities designed to promote critical thinking and collaboration. In these activities, students may formulate questions, analyze data, derive conclusions, and reflect on their learning.

R MARKDOWN FILES demonstrate how to solve selected data-driven **PSBE** exercises using **R**.

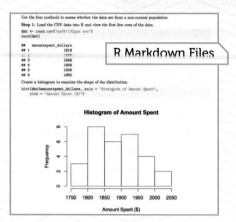
R Markdown Files

VIDEO TECHNOLOGY MANUALS available for TI-83/84 calculators, Minitab, Excel, JMP, SPSS, R, RStudio, Rcmdr, and CrunchIt!® provide brief instructions for using specific statistical software. Icons in the book and e-book link to specific video technology manuals at relevant locations within the text.

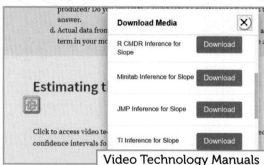
Video Technology Manuals

SOFTWARE OUTPUT

Carefully selected instances of **Software Output** in **PSBE** can be switched in the e-book to match the software that the student is using. Icons in the book and e-book clearly indicate where different software images are available.

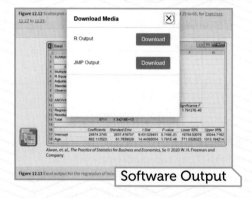
Software Output

Data Tools

DATA FILES are available in JMP, ASCII, Excel, TI, Minitab, SPSS, R, and CSV formats.

CRUNCHIT!® is Macmillan Learning's own web-based statistical software that allows users to perform all the statistical operations and graphing needed for an introductory statistics course and more. It saves users time by automatically loading data from PSBE, and it provides the flexibility to edit and import additional data.

JMP STUDENT EDITION (developed by SAS) is easy to learn and contains all the capabilities required for introductory statistics. JMP is the leading commercial data analysis software of choice for scientists, engineers and analysts at companies throughout the globe (for Windows and Mac).

Paired with Assessment:

PSBE-SPECIFIC ASSESSMENT

PRE-BUILT CHAPTER HOMEWORK ASSIGNMENTS are curated assignments containing questions directly from the book, with select questions containing error-specific feedback to guide students to concept mastery.

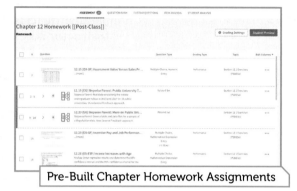
Pre-Built Chapter Homework Assignments

TRIED AND TRUE

STATTUTOR TUTORIALS offer multimedia tutorials that explore important concepts and procedures in a presentation that combines video, audio, and interactive features. The newly revised format includes built-in, assignable assessments and a bright new interface.

STATISTICAL APPLETS give students hands-on opportunities to familiarize themselves with important statistical concepts and procedures, in an interactive setting that allows them to manipulate variables and see the results graphically. Icons in the textbook indicate when an applet is available for the material being covered.

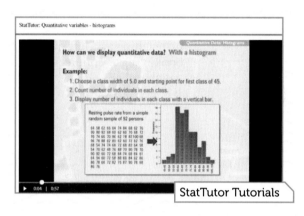
StatTutor Tutorials

STATBOARDS VIDEOS are brief whiteboard videos that illustrate difficult topics through additional examples, written and explained by a select group of statistics educators.

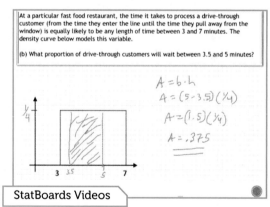
StatBoards Videos

STATISTICAL VIDEO SERIES consists of StatClips, StatClips Examples, and Statistically Speaking "Snapshots." View animated lecture videos, whiteboard lessons, and documentary-style footage that illustrate key statistical concepts and help students visualize statistics in real-world scenarios.

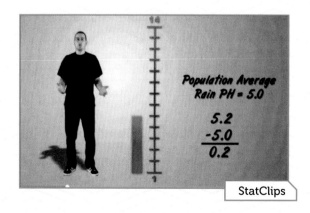
StatClips

Additional Resources Available with *PSBE*:

Companion Website www.macmillanlearning.com/psbe5e This open-access website includes statistical applets, data set files, and the technology appendices. The website also offers four optional companion chapters covering statistics for quality control and capability, two-way analysis of variance, nonparametric tests, and logistic regression.

Instructor access to the Companion Website requires user registration as an instructor and features all the open-access student web materials, plus:

- **Instructor's Solutions Manual**
- **Instructor's Guide**
- **Lecture PowerPoint Slides**
- **Clicker Questions**
- **Image PowerPoint Slides**
- **Test Bank**

iClicker is a two-way radio-frequency classroom response solution developed by educators for educators. Each step of iClicker's development has been informed by teaching and learning. To learn more about packaging iClicker with this textbook, please contact your local Macmillan sales representative or visit www.iclicker.com.

To Students: What Is Statistics?

Statistics is the science of collecting, organizing, and interpreting numerical facts, which we call *data*. We are bombarded by data in our everyday lives. The news mentions movie box-office sales, the latest poll of the president's popularity, and the average high temperature for today's date. Advertisements claim that data show the superiority of the advertiser's product. All sides in public debates about economics, education, and social policy argue from data. A knowledge of statistics helps separate sense from nonsense in this flood of data.

The study and collection of data are also important in the work of many professions, so training in the science of statistics is valuable preparation for a variety of careers. Each month, for example, government statistical offices release the latest numerical information on unemployment and inflation. Economists and financial advisors, as well as policymakers in government and business, study these data to make informed decisions. The sports industry uses data to help make decisions regarding player performance, team rosters, and fan experience. Doctors must understand the origin and trustworthiness of the data that appear in medical journals. Politicians rely on data from polls of public opinion. Business decisions are based on market research data that reveal consumer tastes and preferences. Engineers gather data on the quality and reliability of manufactured products. Most areas of academic study make use of numbers, and therefore also make use of the methods of statistics. This means it is extremely likely that your undergraduate research projects will involve, at some level, the use of statistics.

Learning from Data

The goal of statistics is to learn from data. To learn, we often perform calculations or make graphs based on a set of numbers. But to learn from data, we must do more than calculate and plot, because data are not just numbers; they are numbers that have some context that helps us understand them.

When you do statistical problems, even straightforward textbook problems, don't just graph or calculate. Think about the context and state your conclusions in the specific setting of the problem. As you are learning how to do statistical calculations and graphs, remember that the goal of statistics is not calculation for its own sake, but gaining understanding from numbers. The calculations and graphs can be automated by a calculator or software, but you must supply the understanding. This book presents only the most common specific procedures for statistical analysis. A thorough grasp of the principles of statistics will enable you to quickly learn more advanced methods as needed. On the other hand, a fancy computer analysis carried out without attention to basic principles will often produce elaborate nonsense. As you read, seek to understand the principles as well as the necessary details of methods and recipes.

The Rise of Statistics

Historically, the ideas and methods of statistics developed gradually as society grew interested in collecting and using data for a variety of applications. The earliest origins of statistics lie in the desire of rulers to count the number of inhabitants or measure the value of taxable land in their domains. As the physical sciences developed in the seventeenth and eighteenth centuries, the importance of careful measurements of weights, distances, and other physical quantities grew. Astronomers and surveyors striving for exactness had to deal with variation in their measurements. Many measurements should be better than a single measurement, even though they vary among themselves. How can we best combine many varying observations? Statistical methods that are still important were invented in order to analyze scientific measurements.

By the nineteenth century, the agricultural, life, and behavioral sciences also began to rely on data to answer fundamental questions. How are the heights of parents and children related? Does a new variety of wheat produce higher yields than the old, and under what conditions of rainfall and fertilizer? Can a person's mental ability and behavior be measured just as we measure height and reaction time? Effective methods for dealing with such questions developed slowly and with much debate.

As methods for producing and understanding data grew in number and sophistication, the new discipline of statistics took shape in the twentieth century. Ideas and techniques that originated in the collection of government data, in the study of astronomical or biological measurements, and in the attempt to understand heredity or intelligence came together to form a unified "science of data." That science of data—statistics—is the topic of this text.

Business Analytics

The business landscape has become increasingly dominated with the terms of "business analytics," "predictive analytics," "data science," and "big data." These terms refer to the skills, technologies, and practices in the exploration of business performance data. Companies (for-profit and nonprofit) are increasingly making use of data and statistical analysis to discover meaningful patterns to drive decision making in all functional areas, including accounting, finance, human resources, marketing, and operations. The demand for business managers with statistical and analytic skills has been growing rapidly and is projected to continue for many years to come. Our goal with this text is to provide you with a solid foundation on a variety of statistical methods and the way to think critically about data. These skills will serve you well in this data-driven business world.

The Organization of this Book

The first three chapters of this book concern data analysis and data production. The first two chapters deal with statistical methods for organizing and describing data. These chapters progress from simpler to more complex data. Chapter 1 examines data on a single variable; and Chapter 2 is devoted to relationships among two or more variables. You will learn both how to examine data produced by others and how to organize and summarize your own data. These summaries will be first graphical, then numerical, and then, when appropriate, in the form of a mathematical model that gives a compact description of the overall pattern of the data. Chapter 3 outlines arrangements (called designs) for producing data that answer specific questions. The principles presented in this chapter will help you to design proper samples and experiments for your research projects, and to evaluate other such investigations in your field of study.

Chapters 4 to 7 introduce statistical inference—formal methods for drawing conclusions from properly produced data. Statistical inference uses the language of probability to describe how reliable its conclusions are, so some basic facts about probability are needed to understand inference. Probability is the subject of Chapters 4, 5, and 6. Chapter 7, perhaps the most important chapter in the text, introduces the reasoning of statistical inference. Effective inference is based on good procedures for producing data (Chapter 3), careful examination of the data (Chapters 1 and 2), and an understanding of the nature of statistical inference, as discussed in Chapter 6.

Chapters 8 to 13 describe some of the most common specific methods of inference. Chapters 8 and 9 describe methods for drawing conclusions about means from one, two, and three or more samples. Chapters 10 and 11 present similar material for proportions and categorical data. Chapters 12 and 13 present statistical inference methods for predicting a single outcome variable from one or more explanatory variables.

The remaining five chapters introduce somewhat more advanced methods of inference, dealing with time series, statistics for quality, two-way analysis of variance, nonparametric tests, and logistic regression.

What Lies Ahead

The Practice of Statistics for Business and Economics presents data from a variety of different areas. Many exercises ask you to express the understanding gained from your study of the data. In practice, you would know much more about the background of the data and about the questions the data hope to answer. Nonetheless, it is important to form the habit of asking "What do the data tell me?" rather than just concentrating on making graphs and doing calculations or blindly accepting what someone else tells you they mean.

When doing exercises, make sure to use the help provided by technology to make graphs and to perform calculations. This allows you more time to think critically about the conclusions. There are many kinds of statistical software, from spreadsheets to large programs for advanced users of statistics.

Because graphing and calculating are automated in statistical practice, the most important assets you can gain from the study of statistics are an understanding of the big ideas and the beginnings of good judgment in working with data. Ideas and judgment can't (at least yet) be automated. They guide you in telling the computer what to do and in interpreting its output. This book tries to explain the most important ideas of statistics, not just teach methods. Some examples of big ideas that you will meet are "always plot your data," "randomized comparative experiments," and "statistical significance."

You learn statistics by doing statistical problems. "Practice, practice, practice." Be prepared to work problems. The basic principle of learning is persistence. Being organized and persistent is more helpful in reading this book than knowing lots of math. The main ideas of statistics, like the main ideas of any important subject, took a long time to discover and take some time to master. The gain will be worth the pain.

Index of Cases

CASE 1.1	Treasury Bills 13
CASE 1.2	Time to Start a Business 24
CASE 2.1	Education Expenditures and Population: Benchmarking 66
CASE 2.2	Does the Right Music Sell the Product? 105
CASE 4.1	Uncovering Fraud by Digital Analysis 187
CASE 5.1	Tracking Perishable Demand 222
CASE 5.2	Portfolio Analysis 234
CASE 5.3	Inspecting a Supplier's Products 252
CASE 7.1	Bankruptcy Attorney Fees 345
CASE 7.2	Fill the Bottles 356
CASE 8.1	Battery Life of a Smartphone 399
CASE 8.2	Active versus Failed Retail Companies 429
CASE 9.1	Tip of the Hat and Wag of the Finger? 461
CASE 9.2	Evaluation of a New Educational Product 483
CASE 10.1	Trends in the Workplace 506
CASE 10.2	Marketing Christmas Trees 515
CASE 10.3	Social Media in the Supply Chain 525
CASE 11.1	Are Flexible Companies More Competitive? 543
CASE 12.1	The Relationship between Income and Education for Entrepreneurs 572
CASE 13.1	The Inclusive Development Index (IDI) 620
CASE 13.2	Predicting Movie Revenue 635
CASE 13.3	Prices of Homes 650
CASE 14.1	Adidas Stock Price Returns 678
CASE 14.2	Amazon Sales 681
CASE 14.3	Lake Michigan Water Levels 711
CASE 15.1	Turnaround Time for Lab Results 15-14
CASE 15.2	O-Ring Diameters 15-18
CASE 15.3	Reducing Absenteeism 15-47
CASE 16.1	Discounts and Expected Prices 16-15
CASE 16.2	Expected Prices, Continued 16-17
CASE 17.1	Price Discrimination? 17-3
CASE 17.2	Consumer Perceptions of Food Safety 17-11
CASE 18.1	Clothing Color and Tipping 18-2

Index of Data Tables

TABLE 1.1	Service times (seconds) for calls to a customer service center	15
TABLE 6.1	Duration (in Seconds) of 35 Bike Trips	310
TABLE 8.1	Monthly rates of return on a portfolio (percent)	404
TABLE 8.2	Parts measurements using optical software	406
TABLE 8.3	Cash flow margins of active and failed retail firms	429
TABLE 9.1	Price promotion data	501
TABLE 12.1	Return on Treasury bills and rate of inflation	585
TABLE 12.2	Sales price and assessed value (in thousands of $) of 35 homes in a midwestern county	591
TABLE 12.3	In-state tuition and fees (in dollars) for 33 public universities	591
TABLE 12.4	Net new money (millions of $) flowing into stock and bond mutual funds	595
TABLE 12.5	Selling price and size of homes	595
TABLE 12.6	Annual number of tornadoes in the United States between 1953 and 2017	615
TABLE 13.1	Countries with advanced economies: IDI, percent employment, and median income	621
TABLE 13.2	Insured commercial banks by state or other area	622
TABLE 13.3	Regression coefficients and t statistics for Exercise 13.57	648
TABLE 13.4	Regression coefficients and t statistics for Exercise 13.59	649
TABLE 13.5	Homes for sale in zip code 47904	653
TABLE 13.6	Expected price data	670
TABLE 15.1	Control chart constants	15-13
TABLE 15.2	Thirty control chart subgroups of lab testing turnaround times (in minutes)	15-15
TABLE 15.3	Twenty-five control chart subgroups of O-ring measurements (in inches)	15-19
TABLE 15.4	Hospital losses for 15 samples of joint replacement patients	15-24
TABLE 15.5	Points per minute scored by LeBron James each game played during the 2017–2018 regular season (read left to right)	15-30
TABLE 15.6	Daily calibration subgroups for a Lunar bone densitometer	15-35
TABLE 15.7	Aluminum percent measurements	15-36
TABLE 15.8	Accelerated lifetime measurements (hrs) for an electronic component	15-37
TABLE 15.9	Proportions of workers absent during four weeks	15-48

TABLE 15.10	Proportions of ambulatory surgery patients of Bellin Health System likely to recommend Bellin for ambulatory surgery 15-50	
TABLE 15.11	Counts of OSHA reportable injuries per month for 24 consecutive months 15-52	
TABLE 15.12	$\bar{x}$ and s for samples of film thickness 15-54	
TABLE 16.1	Tool diameter data 16-27	
TABLE 16.2	Expected price data 16-28	
TABLE 17.1	Calories and sodium in three types of hot dogs 17-32	

Beyond the Basics Index

CHAPTER 1	Risk and return	35
CHAPTER 1	Density estimation	56
CHAPTER 2	Big data	103
CHAPTER 3	Capture-recapture sampling	140
CHAPTER 8	The bootstrap	412
CHAPTER 9	Testing the equality of spread	478
CHAPTER 10	Plus four confidence interval for a single proportion	510
CHAPTER 10	Plus four confidence intervals for a difference in proportions	527
CHAPTER 10	Relative risk	534
CHAPTER 11	Meta-analysis	554
CHAPTER 12	Nonlinear regression	601
CHAPTER 13	Regression trees	665
CHAPTER 14	Dickey-Fuller tests	705
CHAPTER 14	ARCH models	758

About the Authors

Layth C. Alwan is an Associate Professor of Supply Chain, Operations Management, and Business Statistics, Sheldon B. Lubar School of Business, University of Wisconsin-Milwaukee. He received an AB in mathematics, SB in statistics, MBA, and PhD in business statistics/operations management, all from the University of Chicago, and an MS in computer science from DePaul University. Professor Alwan is an author of many research articles related to operations management, business forecasting, and statistical process control. He has consulted for many leading companies on statistical issues related to operations, forecasting, and quality applications. On the teaching front, he focuses on engaging and motivating business students on how statistical thinking and data analysis methods have practical importance in business. He is the recipient of several teaching awards, including Business School Teacher of the Year (three times) and Executive MBA Outstanding Teacher of the Year.

Bruce A. Craig is Professor of Statistics and Director of the Statistical Consulting Service at Purdue University. He received his BS in mathematics and economics from Washington University in St. Louis and his PhD in statistics from the University of Wisconsin-Madison. He is an elected fellow of the American Association for the Advancement of Science and of the American Statistical Association, where he served as chair of its section on Statistical Consulting in 2009. He is also an active member of the International Biometrics Society and was elected by the voting membership to the Eastern North American Region's Regional Committee between 2003 and 2006. Professor Craig has served on the editorial board of several statistical journals and has been a member of numerous data and safety monitoring boards, including Purdue's institutional review board. He is an author or coauthor of over 120 peer-reviewed journal articles. His research interests focus on the development of novel statistical methodology to address research questions, primarily in the life sciences. Areas of current interest are diagnostic testing and assessment, measurement error models, and animal abundance estimation.

George P. McCabe is a Professor of Statistics at Purdue University. In 1966 he received a BS in mathematics from Providence College and in 1970 a PhD in mathematical statistics from Columbia University. His entire professional career has been spent at Purdue with sabbaticals at Princeton; the Commonwealth Scientific and Industrial Research Organization (CSIRO) in Melbourne (Australia); the University of Berne (Switzerland); the National Institute of Standards and Technology (NIST) in Boulder, Colorado; and the National University of Ireland in Galway. Professor McCabe is an elected fellow of the American Association for the Advancement of Science and of the American Statistical Association; he was the 1998 Chair of its section on Statistical Consulting. In 2008–2010, he served on the Institute of Medicine Committee on Nutrition Standards for the National School Lunch and Breakfast Programs. He has served on the editorial boards of several statistics journals.

He has consulted with many major corporations and has testified as an expert witness on the use of statistics in several cases. Professor McCabe's research interests have focused on applications of statistics. Much of his recent work has focused on problems in nutrition including nutrient requirements, calcium metabolism, and bone health. He is author or coauthor of over 200 publications in many different journals.

Flying Colours Ltd/Getty Images

CHAPTER 1

Examining Distributions

Introduction

Statistics is the science of learning from data. Data are numerical or qualitative descriptions of the objects that we want to study. In this chapter, we will master the art of examining data.

Data are used to inform decisions in business and economics in many different settings.

- How does *Trader Joe's* decide where to open a new store?
- Which programs for training workers will be needed in the next decade?
- How will changes in the tax code affect donations to your nonprofit organization?

We begin with some basic ideas about data. We learn about the different types of data that are collected and how data sets are organized.

Next, we start our process of learning from data by looking at graphs. These visual displays can help us to see the overall patterns in a set of data. To effectively use graphical software, however, we need practice. Careful judgment is required to choose the type of graph to use as well as to select from the many options that are available to us. Using graphs to learn about data is a process that generally requires refining our initial attempt. Graphs are also important as aids when we communicate the results of our analysis. Frequently, the graphs that we use to learn about our data are not the same as the graphs that we use to communicate what we have learned to others.

Our process of learning from data continues by computing numerical summaries. These sets of numbers describe key characteristics of the patterns that we see in our graphical summaries.

CHAPTER OUTLINE

1.1 Data

1.2 Displaying Distributions with Graphs

1.3 Describing Distributions with Numbers

1.4 Density Curves and the Normal Distributions

A statistical model is an idealized framework that helps us understand data and the relationships between the different descriptions of the objects of interest. In the final section of this chapter, we introduce the idea of a density curve as a way to describe, or model, the distribution of numerical data. The most important statistical model is the Normal distribution, which is introduced here. Normal distributions are used to statistically model many sets of data. They also play a fundamental role in the methods that we will use to draw conclusions from data.

1.1 Data

When you complete this section, you will be able to:

- Give examples of cases in a data set.
- Identify the variables in a data set and demonstrate how a label can be used as a variable in a data set.
- Identify the values of a variable and classify variables as categorical or quantitative.
- Describe the key characteristics of a set of data.
- Explain how a rate can be used to give a more meaningful measure than a count.

data set   A statistical analysis starts with a **data set.** We construct a data set by first deciding which *cases* or units we want to study. For each case, the data set records information about characteristics that we call *variables*.

CASES, LABELS, VARIABLES, AND VALUES

Cases are the objects described by a set of data. Cases may be customers, companies, subjects in a study, or a stock.

A **label** is a special variable used in some data sets to distinguish the different cases.

A **variable** is a characteristic of a case.

Different cases can have different **values** for the variables.

EXAMPLE 1.1

BITCOIN

Bitcoins Bitcoin is a decentralized digital currency that functions without a central bank. It was created by a person or persons using the name Satoshi Nakamoto and was first released as open-source software in 2009. Since then, many of these so called crypto currencies have raised money through Initial Coin Offerings (ICOs).[1] Figure 1.1 gives information for the nine ICOs that have raised more than $100 million; these are the cases. Data for each

	A	B	C	D
1	ID	Name	Location	Amount
2	1	Filecoin	North America	257
3	2	Tezos	Europe	236
4	3	EOS	North America	200
5	4	Paragon	North America	183
6	5	The DAO	Stateless/Unknown	168
7	6	Bancor	Middle East	153
8	7	Polkadot	Europe	121
9	8	QASH	Asia	112
10	9	Status	Europe	109

FIGURE 1.1 Bitcoin data, Example 1.1.

ICO are listed in a different row. ID, a label, has the ICOs numbered from 1 to 9. The next columns give the name of the ICO, the location, and the amount of the ICO in millions of U.S. dollars. ∎

Some variables, such as Location, simply place ICOs into categories. Other variables, like Amount, take on numerical values that we can use to do arithmetic. It makes sense to give an average of the ICO amounts, but it does not make sense to give an "average" location.

observation

We sometimes use the term **observation** to describe the data for a particular case. So, in our Bitcoin example, we have nine observations. Each observation consists of the data entered into the four columns of Figure 1.1.

In this and the following two chapters, we focus on relatively simple situations where the structure of the data is straightforward. In other situations, some judgment may be needed to define the characteristics of the data. For example, suppose we study the daily prices of a sample of 500 stocks over a period of a year. On one hand, we could define the cases as the stocks and the daily prices as individual variables. On the other hand, we could use the combination of a stock and a particular day as the case. The choice will often depend on the particular analysis being performed.

CATEGORICAL AND QUANTITATIVE VARIABLES

A **categorical variable** places a case into one of several groups or categories.

A **quantitative variable** takes numerical values for which arithmetic operations, such as adding and averaging, make sense.

Some summaries that we will study differ for these two types of variables. However, the overall goal is the same: use graphical and numerical summaries to describe the distribution of the values of a variable.

EXAMPLE 1.2

BITCOIN

Categorical and Quantitative Variables for Bitcoins The Bitcoin file has four variables: ID, name, location, and amount. ID, name, and location are categorical variables. Amount is a quantitative variable. ∎

You should choose the label for your cases carefully. In our Bitcoin example, Location would not be a good choice for a label because, for example, Filecoin and EOS are both located in North America. In contrast, the variable ID has different values, 1 and 3, for these cases and is the label variable that we use for this data set. Do you think that the variable Name could be used as a label for this data set? Explain why or why not.

In practice, any set of data is accompanied by background information that helps us understand it. When you plan a statistical study or explore data from someone else's work, ask yourself the following questions:

1. **Who?** Which **cases** do the data describe? **How many** cases appear in the data set?

2. **What?** How many **variables** do the data contain? What are the **exact definitions** of these variables? In which **unit of measurement** is each variable recorded?

3. **Why? What purpose** do the data have? Do we hope to answer some specific questions? Do we want to draw conclusions about cases other than the ones we actually have data for? Are the variables that are recorded suitable for the intended purpose?

spreadsheet   The display in Figure 1.1 is from an Excel **spreadsheet.** Spreadsheets are very useful for doing the kind of simple computations that you will undertake in Exercise 1.2. You can type in a formula and have the same computation performed for each row.

APPLY YOUR KNOWLEDGE

1.1 Read the spreadsheet. Refer to Figure 1.1. Give the location and the amount of the ICO named Polkadot.

1.2 Convert the dollars to euros. Refer to Example 1.1. Add another column to the Bitcoin spreadsheet that gives the value of the ICO in millions of euros. Assume that 1 dollar equals 0.81 euro. Explain how you computed the entries in this column. Does the new column contain values for a categorical variable or for a quantitative variable? Explain your answer. BITCOIN

1.3 Who, what, and why questions for the Bitcoin data. What cases do the data in Figure 1.1 describe? How many cases are there? How many variables are there? What are their definitions and units of measurement? What purpose do you think the data could have?

Note that the names we have chosen for the variables in our spreadsheet do not have spaces. Someone else might have chosen the name "ICO Proceeds" for the amount of the ICO rather than Amount. In some statistical software packages, however, spaces are not allowed in variable names. If you are creating spreadsheets for eventual use with statistical software that has this constraint, you need to avoid spaces in variable names. Another convention is to use an underscore (_) where you would normally use a space. For our data set, we could have used ICO_Proceeds for the amount of the ICO.

EXAMPLE 1.3

Accounting Class Data Suppose that you are a teaching assistant for an accounting class and one of your jobs is to keep track of the grades for students in two sections of the course. Grades are earned on weekly homework assignments, a midterm exam, and a final exam. Each of these components is given a numerical score, and the components are added to get a total score that can range from 0 to 1000. Cutoffs of 900, 800, 700, and so on, are used to assign letter grades of A, B, C, and so on.

You decide to use a spreadsheet for this job. The cases are the students in the class. Each case will be a row in your spreadsheet and have six variables:

- an identifier for each student
- the number of points earned for homework
- the number of points earned for the midterm exam
- the number of points earned for the final exam
- the total number of points earned
- the letter grade earned

The total number of points earned and the letter grades are computed from the homework and exam scores. There are no units of measurement for the student identifier and the letter grade, as they are categorical variables. The student identifier is also a label that we would not analyze. The other variables are measured in "points." Because we can do arithmetic with their values, these variables are quantitative variables. ■

EXAMPLE 1.4

Accounting Class Data for a Different Purpose Suppose the data for the students in the accounting class were also to be used to study relationships between student characteristics and performance in the course. For this purpose, we might want to use a data set that includes additional variables such as Sex, PrevAcct (whether the student has previously taken an accounting course in high school), and Year (student classification as first, second, third, or fourth year). Each of these additional variables is categorical. ■

ordered categorical variable

In our examples of accounting class data, the possible values for the grade variable are A, B, C, D, and F. This is an **ordered categorical variable** because A is better than B, which is better than C, and so on. However, when computing grade-point averages, many colleges and universities translate these letter grades into numbers using the system A = 4, B = 3, C = 2, D = 1, and F = 0. The transformed variable with numeric values is considered to be quantitative because we can average the numerical values across different courses to obtain a grade-point average.

Sometimes, experts argue about these kinds of numerical scales. They ask whether the difference between an A and a B is the same as the difference between a D and an F. Similarly, many questionnaires ask people to respond on a 1 to 5 scale, with 1 representing strongly agree, 2 representing agree, and so on. Again, we could ask whether the five possible values for this scale are equally spaced in some sense. From a practical point of view, the averages that can be computed when we convert categorical scales such as these to numerical values frequently provide a very useful way to summarize data.

nominal variable

Not all categorical variables are ordered. Some statistical software uses the term **nominal** for a categorical variable that is not ordered. When studying a categorical variable, it is very important to know if it is ordered.

APPLY YOUR KNOWLEDGE

1.4 Apartment rentals for students. A data set lists apartments available for students to rent. Information provided includes the zip code, the monthly rent, whether a fitness center is provided at the apartment complex, whether pets are allowed, the number of bedrooms, and the distance to campus. Describe the cases in the data set, give the number of variables, and specify whether each variable is categorical or quantitative. Discuss the basis for your choice to specify the number of bathrooms as categorical or quantitative.

Knowledge of the context of data includes an understanding of the variables that are recorded. Often, the variables in a statistical study are easy to understand: height in centimeters, study time in minutes, and so on. But different areas of application can also have their own special variables. A marketing research department measures consumer behavior using a scale developed for its customers. A health food store combines various types of data into a single measure that it will use to determine whether to put a new store in a particular location. These kinds of variables are measured with special **instruments.** Part of mastering your field of work is learning which variables are important and how they are best measured.

instrument

rate

Be sure that each variable really does measure what you want it to measure. A poor choice of variables can lead to misleading conclusions. Often, for example, the **rate** at which something occurs is a more meaningful measure than a simple count of occurrences.

EXAMPLE 1.5

Comparing Colleges Based on Graduates Think about comparing colleges based on the numbers of graduates. This view tells you something about the relative sizes of different colleges. However, if you are interested in how well colleges succeed at graduating students whom they admit, it would be better to use a rate. For example, you can find data on the Internet on the six-year graduation rates of different colleges. These rates are computed by examining the progress of first-year students who enroll in a given year. Suppose that at College A there were 1000 first-year students in a particular year, and 800 graduated within six years. The graduation rate is

$$\frac{800}{1000} = 0.80$$

or 80%. College B had 2000 first-year students who entered in the same year, and 1200 graduated within six years. The graduation rate is

$$\frac{1200}{2000} = 0.60$$

or 60%. How do we compare these two colleges? College B has more graduates, but College A has a better graduation rate. ■

APPLY YOUR KNOWLEDGE

1.5 Which variable would you choose? Refer to the previous example on colleges and their graduates.

(a) Give a setting in which you would prefer to evaluate the colleges based on their numbers of graduates. Give a reason for your choice.

(b) Give a setting in which you would prefer to evaluate the colleges based on their graduation rates. Give a reason for your choice.

In Example 1.5, when we computed the graduation rate, we used the total number of students to adjust the number of graduates. We constructed a new variable by dividing the number of graduates by the total number of students. Computing a rate is just one of several ways of **adjusting one variable to create another.** We often divide one variable by another to compute a more meaningful variable to study.

adjusting one variable to create another

Exercise 1.5 illustrates an important point about presenting the results of your statistical calculations. *Always consider how to best communicate your results to a general audience.* For example, the numbers produced by your calculator or by statistical software frequently contain more digits than are needed. Be sure that you do not include extra information generated by software that will distract from a clear explanation of what you have found.

SECTION 1.1 SUMMARY

- A data set contains information on a number of **cases.** Cases may be customers, companies, subjects in a study, units in an experiment, or other objects.

- For each case, the data give values for one or more **variables.** A variable describes some characteristic of a case, such as a person's height, gender, or salary. Variables can have different **values** for different cases.

- A **label** is a special variable used to identify cases in a data set.

- Some variables are **categorical** and others are **quantitative.** A categorical variable places each individual into a category, such as male or female. A quantitative variable has numerical values that measure some characteristic of each case, such as height in centimeters or annual salary in dollars.

- The **key characteristics** of a data set answer the questions Who?, What?, and Why?.

- A **rate** is sometimes a more meaningful measure than a count.

SECTION 1.1 EXERCISES

For Exercises 1.1 to 1.3, see page 4; for 1.4, see page 5; and for 1.5, see page 6.

1.6 Summer jobs. You are collecting information about summer jobs that are available for college students in your area. Describe a data set that you could use to organize the information that you collect.

(a) What are the cases?

(b) Identify the variables and their possible values.

(c) Classify each variable as categorical or quantitative. Be sure to include at least one of each.

(d) Use a label and explain how you chose it.

(e) Summarize the key characteristics of your data set.

1.7 Employee application data. The personnel department keeps records on all employees in a company. Here is the information kept in one of the data files: employee identification number, last name, first name, middle initial, department, number of years with the company, salary, education (coded as high school, some college, or college degree), and age.

(a) What are the cases for this data set?

(b) Identify each item in the data file as a label, a quantitative variable, or a categorical variable.

(c) Set up a spreadsheet that could be used to record the data. Give appropriate column headings, and include three sample cases.

1.8 Where should you locate your business? You are interested in choosing a new location for your business. Create a list of criteria that you would use to rank cities. Include at least six variables, and give reasons for your choices. Will you use a label? Classify each variable as quantitative or categorical.

1.9 Survey of customers. A survey of customers of a restaurant near your campus wanted to obtain data for the following variables: (a) sex of the customer, coded as female or male; (b) age of the customer, in years; (c) satisfaction with the cleanliness of the restaurant; (d) satisfaction with the food; (e) satisfaction with the service; (f) whether the respondent is a college student; and (g) whether the respondent ate there at least once a week. Responses for items (c), (d), and (e) are given on a scale of 1 (very dissatisfied) to 5 (very satisfied). Classify each of the survey variables as categorical or quantitative, and give reasons for your answers.

1.10 Your survey of customers. Refer to the previous exercise. Make up your own customer survey with at least six questions. Include at least two categorical variables and at least two quantitative variables. Identify which variables are categorical and which are quantitative. Give reasons for your answers.

1.11 Study habits of students. You are planning a survey to collect information about the study habits of college students. Describe two categorical variables and two quantitative variables that you might measure for each student. Give the units of measurement for the quantitative variables.

1.12 How would you rate colleges? Popular magazines rank colleges and universities on their "academic quality" in serving undergraduate students. Describe five variables that you would like to see measured for each college if you were choosing where to study. Give reasons for each of your choices.

1.13 Attending college in your state or in another state. The U.S. Census Bureau collects a large amount of information concerning higher education.[2] For example, the Census Bureau provides a table that includes the following variables: state, number of students from the state who attend college, and number of students who attend college in their home state.

(a) What are the cases for this set of data?

(b) Is there a label variable? If yes, what is it?

(c) Identify each variable as categorical or quantitative.

(d) Consider a variable computed as the number of students in each state who attend college in the state divided by the total number of students from the state who attend college. Explain how you would use this variable to describe something about the states.

1.14 How should you express the change? Between the first exam and the second exam in your statistics course, you increased the amount of time that you spent working the exercises. Which of the following three ways would you choose to express the results of your increased work: (a) give the grades on the two exams, (b) give the ratio of the grade on the second exam divided by the grade on the first exam, or (c) take the difference between the grade on the second exam and the grade on the first exam, and express it as a percent of the grade on the first exam. Give reasons for your answer.

1.2 Displaying Distributions with Graphs

When you complete this section, you will be able to:

- Use a bar graph, a pie chart, and a Pareto chart to describe the distribution of a categorical variable.
- Use a stemplot and a histogram to describe the distribution of a quantitative variable.
- Examine the distribution of a quantitative variable with respect to the overall pattern of the data and deviations from that pattern and identify the shape, center, and spread of its distribution.
- Choose an appropriate graphical summary for a given set of data.
- Identify and describe any outliers in the distribution of a quantitative variable.
- Use a time plot to describe the trend of a quantitative variable that is measured over time.

distribution

For each variable, the cases generally will have different values for the variables. The **distribution** of a variable describes how the values of a variable vary from case to case. We can use graphical and numerical descriptions for a distribution. In this section we start with graphical summaries; we consider numerical summaries in the following section.

exploratory data analysis

Statistical tools and ideas help us examine data to describe their main features. This examination is called **exploratory data analysis.** Like an explorer crossing unknown lands, we want first to simply describe what we see. Here are two basic strategies that help us organize our exploration of a set of data:

- Begin by examining each variable by itself. Then move on to study the relationships among the variables.
- Begin with a graph or graphs. Then add numerical summaries of specific aspects of the data.

We follow these principles in organizing our learning. The rest of this chapter presents methods for describing a single variable. We study relationships among two or more variables in Chapter 2. Within each chapter, we begin with graphical displays, then add numerical summaries for a more complete description.

When we perform an exploratory data analysis, our focus is on a careful description of the values of the variables in our data set. In many applications, particularly in business and economics, our real interest is in using our descriptions to predict something in the future. We use the term **predictive analytics** to describe data used in this way.

predictive analytics

For example, if *Trader Joe's* wanted to use data to decide where to open a new store, the company might analyze data from its current stores with a focus on characteristics of stores that are very successful. If managers can find a new location with characteristics that are similar they would *predict* that a new store in that location will be successful.

Categorical variables: bar graphs and pie charts

distribution of a categorical variable

The values of a categorical variable are labels for the categories, such as "Yes" and "No." The **distribution of a categorical variable** lists the categories and gives either the count or the percent of cases that fall in each category.

EXAMPLE 1.6

CARDS

Choosing a Credit Card In a study, 1659 U.S. adults were asked about their reasons for choosing a credit card. Here are the most important reasons that they gave for making their choice.³

Reason	Count (n)
Cash back	680
Rewards	448
Interest rate	200
Easy to get	149
Brand	133
Other	49
Total	1659

Reason is the categorical variable in this example, and the values are the six reasons given for the choice. ■

Note that the last value of the variable Reason is "Other," which includes all other most important reasons not reported here. For data sets that have a large number of values for a categorical variable, we often create a category such as this that combines categories with relatively small counts or percents. *Careful judgment is needed when doing this.* You don't want to cover up some important piece of information contained in the data by combining data in this way.

EXAMPLE 1.7

CARDS

Favorites as Percents When we look at the reasons for choosing a credit card, we see that cash back is the clear winner. As shown in the data set, 680 adults reported cash back as their most important reason for choosing a credit card. To interpret this number, we need to know that the total number of adults surveyed was 1659. When we say that cash back is the winner, we can describe this win by saying that 41% (680 divided by 1659) of the adults reported cash back as their most important reason. Here is a table of the percents for the different reasons:

Reason	Percent (%)
Cash back	41
Rewards	27
Interest rate	12
Easy to get	9
Brand	8
Other	3
Total	100

■

The use of graphical methods allows us to see this information and other characteristics of the data easily. We now examine two types of graphs. The first type is a **bar graph**.

bar graph

EXAMPLE 1.8

CARDS

Bar Graph for the Credit Card Choices Data Figure 1.2 displays the credit card choices data using a **bar graph**. The heights of the six bars show the percents of adults who reported each of the reasons for choosing a credit card as their most important factor. ■

FIGURE 1.2 Bar graph for the credit card choices data in Example 1.8.

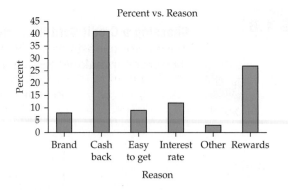

pie chart

The categories in a bar graph or pie chart can be put in any order. In Figures 1.2 and 1.3, we ordered the favorite choices alphabetically. You should always consider the best way to order the values of the categorical variable in a bar graph. Choose an ordering that will be useful to you. If you have difficulty deciding, ask a friend if your choice communicates the message that you expect it to convey. You could also use counts in place of percents. The second type of graph is a **pie chart.**

EXAMPLE 1.9

Pie Chart for the Credit Card Choices Data The **pie chart** in Figure 1.3 helps us see which part of the whole each group forms. Here it is very easy to see that cash back is the most important reason for about 41% of the adults. ∎

FIGURE 1.3 Pie chart for the credit card choices data in Example 1.9.

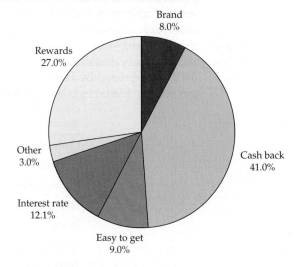

APPLY YOUR KNOWLEDGE

1.15 Compare the bar graph with the pie chart. Refer to the bar graph in Figure 1.2 and the pie chart in Figure 1.3 for the credit card choices data. Which graphical display does a better job of describing the data? Give reasons for your answer.

Bar graphs and pie charts are often very useful tools for describing the distribution of categorical variables in a data set. When they are used to describe the distribution of a variable common to a future setting, they can be powerful predictive analytics tools. Here is an example.

EXAMPLE 1.10

BCOSTS

Analyze the Costs of Your Business Businesses' budgets include many items that are often organized into cost centers. Data for a company with 10 different cost centers is summarized in Figure 1.4. Cost center is a categorical variable with 10 possible values: salaries, maintenance, and research, plus seven other cost centers. Annual cost is a quantitative variable that gives the sum of the amounts spent in each cost center.[4] The distribution of costs in this data set could be used to predict costs for the following year. ■

FIGURE 1.4 Business cost center data, Example 1.10.

Cost Analysis

Cost center	Annual cost	Percent of total	Cumulative percent
Parts and materials	$1,325,000.00	31.17%	31.17%
Manufacturing equipment	$900,500.00	21.19%	52.36%
Salaries	$575,000.00	13.53%	65.89%
Maintenance	$395,000.00	9.29%	75.18%
Office lease	$295,000.00	6.94%	82.12%
Warehouse lease	$250,000.00	5.88%	88.00%
Insurance	$180,000.00	4.23%	92.24%
Benefits and pensions	$130,000.00	3.06%	95.29%
Vehicles	$125,000.00	2.94%	98.24%
Research	$75,000.00	1.76%	100.00%
Total	$4,250,500.00	100.00%	

We have discussed two tools to make a graphical summary for these data—pie charts and bar charts. Let's consider possible uses of the data to help us to choose a useful graph. Which cost centers are generating large annual costs? Notice that the display of the data in Figure 1.4 is organized to help us answer this question. The cost centers are ordered by the annual cost, largest to smallest. The data display also gives the annual cost as a percent of the total. We see that parts and materials have an annual cost of $1,325,000, which is 31% of $4,250,500, the total cost.

The last column in the display gives the cumulative percent, which is the sum of the percents for the cost center in the given row and all above it. We see that the three largest cost centers—parts and materials, manufacturing equipment, and salaries—account for 66% of the total annual costs.

APPLY YOUR KNOWLEDGE

1.16 Rounding in the cost analysis. Refer to Figure 1.4 and the discussion following. In the discussion, we rounded the percents given in the figure. Do you think this is a good idea? Explain why or why not.

1.17 Focus on the 80%. Many analyses using data such as those given in Figure 1.4 focus on the items that make up the top 80% of the total cost. Which items make up the top 80% in our cost analysis data? (Note that you will not be able to answer this question for exactly 80%, so use the closest percent above or below.) Be sure to explain your choice, and give a reason for it.

Pareto chart

A bar graph like Figure 1.2 whose categories are ordered from most frequent to least frequent is called a **Pareto chart**.[5] Pareto charts are frequently used in quality control settings. There, the purpose is often to identify common types of defects in a manufactured product. Deciding upon strategies for corrective action can then be based on what would be most cost effective. Chapter 15 gives more examples of settings in which Pareto charts are used.

Let's use a Pareto chart to look at our cost analysis data.

EXAMPLE 1.11

BCOSTS

Pareto Chart for Cost Analysis Figure 1.5 displays the Pareto chart for the cost analysis data. Here it is easy to see that the parts and materials cost center has the highest annual cost. Research is the cost center with the lowest cost, representing less than 2% of the total. Notice the red curve that is superimposed on the graph. (It is actually a smoothed curve joined at the midpoints of the positions of the bars on the x axis.) This curve gives the cumulative percent of total cost as we move from left to right in the figure. The data display in Figure 1.4 and the graph in Figure 1.5 both describe the same data. Which do you prefer? Why? ■

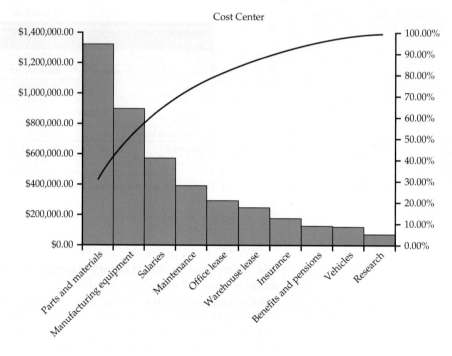

FIGURE 1.5 Pareto chart of business cost center data, Example 1.11.

APPLY YOUR KNOWLEDGE

1.18 Population of Canadian provinces and territories. Here are the 2017 population estimates of the 13 Canadian provinces and territories based on the 2011 Census:[6] CANADAP

Province or territory	Population
Newfoundland and Labrador	528,817
Prince Edward Island	152,021
Nova Scotia	953,869
New Brunswick	759,655
Quebec	8,394,034
Ontario	14,193,384
Manitoba	1,338,109
Saskatchewan	1,163,925
Alberta	4,286,134
British Columbia	4,817,160
Yukon	38,459
Northwest Territories	44,520
Nunavut	37,996

Display these data in a bar graph using the alphabetical order of provinces and territories in the table. You can choose to use counts or percents in your bar graph. Give a reason for your choice.

1.19 Try a Pareto chart. Refer to the previous exercise. CANADAP

(a) Use a Pareto chart to display these data.

(b) Compare the bar graph from the previous exercise with your Pareto chart. Which do you prefer? Why?

Bar graphs, pie charts, and Pareto charts can help you see characteristics of a categorical variable distribution quickly. We now examine quantitative variables, for which graphs are essential tools in analysis.

Quantitative variables: histograms

Because quantitative variables often take many values, a graph of the distribution is clearer if nearby values are grouped together. This type of graph, called a **histogram**, is the most common graph of the distribution of a quantitative variable where we group near values into **classes**.

histogram

Treasury Bills Treasury bills, also known as T-bills, are bonds issued by the U.S. Department of the Treasury. You buy them at a discount from their face value, and they mature in a fixed period of time. For example, you might buy a $1000 T-bill for $980. When it matures, six months later, you would receive $1000—your original $980 investment plus $20 interest. This interest rate is $20 divided by $980, which is 2.04% for six months. Interest is usually reported as a rate per year, so for this example the interest rate would be 4.08%. Rates are determined by an auction that is held every four weeks. The data set, TBILL, contains the interest rates for T-bills for each auction from December 12, 1958, to March 30, 2018.[7] ∎

CASE 1.1

TBILL

Our data set contains 3095 cases. The two variables in the data set are the date of the auction and the interest rate. Here, the cases are defined by the first of these variables. To learn something about T-bill interest rates, we begin with a histogram.

EXAMPLE 1.12

CASE 1.1

A Histogram of T-Bill Interest Rates To make a histogram of the T-bill interest rates, we proceed as follows.

Step 1. Divide the range of the interest rates into **classes** of equal width. The T-bill interest rates range from 0.03% to 15.76%, so we choose as our classes

TBILL

$$0.00 \leq \text{rate} < 1.00$$
$$1.00 \leq \text{rate} < 2.00$$
$$\vdots$$
$$15.00 \leq \text{rate} < 16.00$$

classes

Be sure to specify the classes precisely so that each case falls into *exactly one* class. This is done by using a "less than or equal to" for one boundary of the range, and a "less than" for the other boundary of the range. In our example, an interest rate of 1.98% would fall into the second class, but 2.00% would fall into the third.

Step 2. Count the number of cases in each class. Here are the counts for the first four classes:

Class	Count	Class	Count
$0.00 \leq \text{rate} < 1.00$	463	$2.00 \leq \text{rate} < 3.00$	198
$1.00 \leq \text{rate} < 2.00$	203	$3.00 \leq \text{rate} < 4.00$	377

Step 3. Draw the histogram using counts or percents. Mark on the horizontal axis the scale for the variable whose distribution you are displaying. The variable is Rate in this example. The boundaries run from 0 to 16 to span the data. The vertical axis contains the scale of counts. Each bar represents a class. The base of the bar covers the class, and the bar height is the class count. There is no horizontal space between the bars unless a class is empty, so that its bar has height zero. Figure 1.6 is our histogram. ∎

FIGURE 1.6 Histogram for T-bill interest rates, Example 1.12.

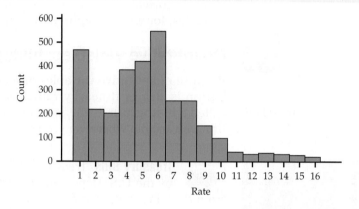

Although histograms resemble bar graphs, their details and uses are distinct. A histogram shows the distribution of counts or percents among the values of a single quantitative variable. A bar graph compares the counts of different values of a categorical variable. The horizontal axis of a bar graph need not have any measurement scale but simply identifies the different values of the categorical variable. We often draw bar graphs with blank space between the bars to separate the items being compared. Draw histograms with no space to indicate that all values of the variable are covered. *Some spreadsheet programs, which are not primarily intended for statistics, will draw histograms as if they were bar graphs, with space between the bars.* Often, you can tell the software to eliminate the extra space to produce a proper histogram.

Our eyes respond to the *area* of the bars in a histogram.[8] Because the classes are all the same width, area is determined by height and all classes are fairly represented. There is no one right choice for the number of classes in a histogram. Choosing too few classes will often smooth over important features of a distribution. Conversely, choosing too many classes may produce a graph that is not very useful with most classes having one or no observations. Neither choice will give a good picture of the shape of the distribution. *Always use your judgment in choosing classes to display the shape.* Statistical software will choose the classes for you, but you usually have options available for changing them.

The histogram function in the *One-Variable Statistical Calculator* applet on the textbook website allows you to change the number of classes by dragging with the mouse so that it is easy to see how the choice of classes affects the histogram. The next example illustrates a situation where the wrong choice of classes will cause you to miss a very important characteristic of the data set.

EXAMPLE 1.13

CC80

Calls to a Customer Service Center Many businesses operate call centers to serve customers who want to place an order or make an inquiry. Customers want their requests handled thoroughly. Businesses want to treat customers well, but they also want to avoid wasted time on the phone. For this reason they monitor the length of calls and encourage their representatives to keep calls short.

We have data on the length of all 31,492 calls made to the customer service center of a small bank in a month. Table 1.1 displays the lengths of the first 80 calls.[9]

1.2 Displaying Distributions with Graphs

TABLE 1.1 Service times (seconds) for calls to a customer service center

77	289	128	59	19	148	157	203
126	118	104	141	290	48	3	2
372	140	438	56	44	274	479	211
179	1	68	386	2631	90	30	57
89	116	225	700	40	73	75	51
148	9	115	19	76	138	178	76
67	102	35	80	143	951	106	55
4	54	137	367	277	201	52	9
700	182	73	199	325	75	103	64
121	11	9	88	1148	2	465	25

Take a look at the data in Table 1.1. In this data set, the *cases* are calls made to the bank's call center. The *variable* is the length of each call. The *units of measurement* are seconds. We see that the call lengths vary a great deal. The longest call lasted 2631 seconds, almost 44 minutes. More striking is that 8 of these 80 calls lasted less than 10 seconds. What's going on? ∎

We started our study of the customer service center data by examining a few cases, the ones displayed in Table 1.1. It would be very difficult to examine all 31,492 cases in this way. We need a better method. Let's try a histogram.

EXAMPLE 1.14

CC

Histogram for Customer Service Center Call Lengths Figure 1.7 is a histogram of the lengths of all 31,492 calls. Based on our inspection of the data in Table 1.1, we chose classes in such a way that we can see the calls that last a very short time. We did not plot the few lengths greater than 1200 seconds (20 minutes). As expected, the graph shows that most calls last less than 300 seconds (5 minutes), with some lasting much longer when customers have complicated problems. More striking is the fact that 7.6% of all calls are no more than 10 seconds long. It turns out that the bank penalized representatives whose average call length was too long—so some representatives just hung up on customers in an effort to bring their average length down. Neither the customers nor the bank was happy about this. The bank changed its policy, and later data showed that calls less than 10 seconds had almost disappeared. ∎

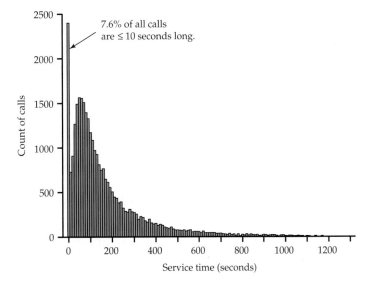

FIGURE 1.7 The distribution of call lengths for 31,492 calls to a bank's customer service center, Example 1.14. The data show a surprising number of very short calls. These are mostly due to representatives deliberately hanging up to bring down their average call length.

EXAMPLE 1.15

Another Histogram for Customer Service Center Call Lengths Figure 1.8 is a histogram of the lengths of all 31,492 calls with class boundaries of 0, 100, 200, ... seconds. Statistical software made this choice as the default option. Notice that the spike representing the very brief calls that appears in Figure 1.7 is covered up in the 0 to 100 seconds class in Figure 1.8. ∎

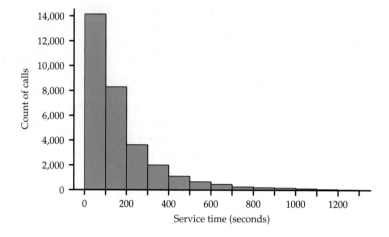

FIGURE 1.8 The default histogram produced by software for the call lengths, Example 1.15. This choice of classes hides the large number of very short calls that is revealed by the histogram of the same data in Figure 1.7.

If we let software choose the classes, we would miss one of the most important features of the data, the calls of very short duration. We were alerted to this unexpected characteristic of the data by our examination of the 80 cases displayed in Table 1.1. *Beware of letting statistical software do your thinking for you.* Example 1.15 illustrates the danger of doing this. To analyze data effectively, we often need to look at data in more than one way. For histograms, examining several choices of classes will help to determine a good choice.

APPLY YOUR KNOWLEDGE

1.20 Exam grades in an accounting course. The following table summarizes the exam scores of students in an accounting course. Use the summary to sketch a histogram that shows the distribution of scores.

Class	Count
$60 \leq$ score < 70	3
$70 \leq$ score < 80	12
$80 \leq$ score < 90	25
$90 \leq$ score < 100	15

1.21 Suppose some students scored 100. No students earned a perfect score of 100 on the exam described in the previous exercise. Note that the last class included only scores that were greater than or equal to 90 and *less than* 100. Explain how you would change the class definitions for a similar exam on which some students earned a perfect score.

Quantitative variables: stemplots

Histograms are not the only graphical display of distributions of quantitative variables. For small data sets, a *stemplot* is quicker to make and presents more detailed information. It is sometimes referred to as a **back-of-the-envelope** technique. Popularized by the statistician John Tukey, this visual display was

back-of-the-envelope

designed to give a quick and informative look at the distribution of a quantitative variable. A stemplot was originally designed to be made by hand, although many statistical software packages include this capability.

> **STEMPLOT**
>
> To make a **stemplot:**
>
> 1. Separate each observation into a **stem,** consisting of all but the final (rightmost) digit, and a **leaf,** the final digit. Stems may have as many digits as needed, but each leaf contains only a single digit.
>
> 2. Write the stems in a vertical column with the smallest at the top, and draw a vertical line at the right of this column.
>
> 3. Write each leaf in the row to the right of its stem, in increasing order out from the stem.

EXAMPLE 1.16

TBILL30

CASE 1.1 **A Stemplot of T-Bill Interest Rates** The histogram that we produced in Example 1.12 to examine the T-bill interest rates used all 3095 cases in the data set. To illustrate the idea of a stemplot, we take a simple random sample of size 30 from this data set. We learn more about how to take such samples in Chapter 3. Here are the data:

3.4	4.9	1.8	12.3	2.6	0.1	6.5	6.5	6.1	3.6
9.0	4.7	0.2	7.0	8.0	3.4	3.7	5.1	0.1	4.5
10.0	0.2	2.3	5.6	5.0	0.5	4.9	4.5	3.9	4.8

The original data set gave the interest rates with two digits after the decimal point. To make the job of preparing our stemplot easier, we first rounded the values to one place following the decimal.

Figure 1.9 illustrates the key steps in constructing the stemplot for these data. How does the stemplot for this sample of size 30 compare with the histogram based on all 3095 interest rates that we examined in Figure 1.6 (page 14)? ∎

FIGURE 1.9 Steps in creating a stemplot for the sample of 30 T-bill interest rates, Example 1.16. (a) Write the stems in a column, from smallest to largest, and draw a vertical line to their right. (b) Add each leaf to the right of its stem. (c) Arrange each leaf in increasing order out from its stem.

```
 0              0 | 1 2 1 2 5       0 | 1 1 2 2 5
 1              1 | 8               1 | 8
 2              2 | 6 3             2 | 3 6
 3              3 | 4 6 4 7 9       3 | 4 4 6 7 9
 4              4 | 9 7 5 9 5 8     4 | 5 5 7 8 9 9
 5              5 | 1 6 0           5 | 0 1 6
 6              6 | 5 5 1           6 | 1 5 5
 7              7 | 0               7 | 0
 8              8 | 0               8 | 0
 9              9 | 0               9 | 0
10             10 | 0              10 | 0
11             11 |                11 |
12             12 | 3              12 | 3
   (a)              (b)                 (c)
```

rounding

You can choose the classes in a histogram. In contrast, the classes (the stems) of a stemplot are given to you. When the observed values have many digits, it is often best to **round** the numbers to just a few digits before making a stemplot, as we did in Example 1.16.

splitting stems

You can also **split stems** to double the number of stems when all the leaves would otherwise fall on just a few stems. Each stem then appears twice. Leaves 0 to 4 go on the upper stem, and leaves 5 to 9 go on the lower stem. Rounding and splitting stems are matters for judgment, like choosing the classes in a histogram.

Stemplots work well for small sets of data. When there are more than 100 observations, a histogram is almost always a better choice.

Stemplots can also be used to compare two distributions. This type of plot is called a **back-to-back stemplot.** In this type of stemplot, we draw vertical lines to the left and right of the stems and put the leaves for one group on the right of the stem and the leaves for the other group on the left. Here is an example.

back-to-back stemplot

EXAMPLE 1.17

TBILLJJ

CASE 1.1 **A Back-to-Back Stemplot of T-Bill Interest Rates in January and July** For this back-to-back stemplot, we took a sample of 31 January T-bill interest rates and another sample of 31 July T-bill interest rates. We then rounded the rates to one digit after the decimal. The plot is shown in Figure 1.10. The stems with the largest number of entries are 4 for January and 7 for July. The rates for July appear to be somewhat larger than those for January. In the next section, we learn how to calculate numerical summaries that help us make this kind of comparison. ■

FIGURE 1.10 Back-to-back stemplot used to compare T-bill interest rates in January and July, Example 1.17.

```
      5 3 3 2 | 0 | 1 3
            8 | 1 | 1 6 9
      9 6 6 6 | 2 | 5
        9 7 1 | 3 |
  9 9 8 7 5 3 3 | 4 | 5 9
        5 3 3 1 | 5 | 0 4 4 4 7 7 9
            2 1 | 6 | 0 0 4 5
            7 0 | 7 | 0 1 1 2 4 5 6 6 7 7
                | 8 | 0
              6 | 9 |
                |10 |
              9 |11 |
              3 |12 | 1
         January       July
```

Special considerations apply for very large data sets. It is often useful to take a sample and examine it in detail using the numerical and graphical measures that we studied in this chapter as a first step. This is what we did in Example 1.13. Sampling can be done in many different ways. A company with a very large number of customer records, for example, might look at those from a particular region or country for an initial analysis.

Interpreting histograms and stemplots

Making a statistical graph is not an end in itself. The purpose of the graph is to help you understand the data. After you make a graph, always ask, "What do I see?" Once you have displayed a distribution, you can see its important features.

EXAMINING A DISTRIBUTION

In any graph of data, look for the **overall pattern** and for striking **deviations** from that pattern.

You can describe the overall pattern of a histogram by its **shape, center,** and **spread.**

An important kind of deviation is an **outlier,** an individual value that falls outside the overall pattern.

We will learn how to describe the distribution's center and spread numerically in Section 1.3. For now, we can describe the center of a distribution by its *midpoint,* the value with roughly half the observations taking smaller values and half taking larger values. We can describe the spread of a distribution by giving the *smallest and largest values*.

EXAMPLE 1.18

TBILL

CASE 1.1 **The Distribution of T-Bill Interest Rates** Let's look again at the histogram in Figure 1.6 (page 14) and the TBILL data file. The distribution has two *peaks,* one at the 0–1 class and another at the 5–6 class. The distribution is somewhat *right-skewed*—that is, the right tail extends farther from the center of the distribution than does the left tail.

This data set includes some relatively large interest rates. The largest is 15.76%. What do we think about this value? Is it so extreme relative to the other values that we would call it an *outlier?* To qualify for this status, an observation should stand apart from the other observations, being either alone or with very few other cases. A careful examination of the data indicates that this 15.76% rate does not qualify for outlier status. There are interest rates of 15.72%, 15.68%, and 15.58%. In fact, there are 15 interest rates of 15% or higher. ■

When you describe a distribution, concentrate on the main features. Look for major peaks, not for minor ups and downs in the bars of the histogram. Look for clear outliers, not just for the smallest and largest observations. Look for rough *symmetry* or clear *skewness*.

SYMMETRIC AND SKEWED DISTRIBUTIONS

A distribution is **symmetric** if the right and left sides of the histogram are approximately mirror images of each other.

A distribution is **skewed to the right** if the right side of the histogram (containing the half of the observations with larger values) extends much farther out than the left side. It is **skewed to the left** if the left side of the histogram extends much farther out than the right side. We also use the terms **"positively skewed"** for distributions that are skewed to the right and **"negatively skewed"** for distributions that are skewed to the left. Positive skewness is the most common type of skewness seen in real data.

EXAMPLE 1.19

IQ

IQ Scores of Fifth-Grade Students Figure 1.11 displays a histogram of the IQ scores of 60 fifth-grade students. There is a single peak around 110, and the distribution is approximately symmetric. The tails decrease smoothly as we move away from the peak. Measures like IQ scores are usually constructed so that they have distributions like the one shown in Figure 1.11. ■

FIGURE 1.11 Histogram of the IQ scores of 60 fifth-grade students, Example 1.19.

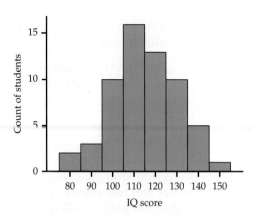

The overall shape of a distribution provides important information about a variable. Some types of data regularly produce distributions that are symmetric or skewed. For example, data on the diameters of ball bearings produced by a manufacturing process tend to be symmetric. Data on incomes (whether of individuals, companies, or nations) are usually strongly skewed to the right. In other words, there are many lower to moderate incomes, some moderate to higher incomes, and a few very high incomes. Distributions can have many different shapes. By performing a careful analysis of data using the techniques in this chapter, you will be able to uncover interesting patterns in distributions that will lead you to ask about the stories behind those patterns.

APPLY YOUR KNOWLEDGE

1.22 **Make a stemplot.** Make a stemplot for a distribution that has a single peak, and is approximately symmetric with one high and two low outliers.

1.23 **Make another stemplot.** Make a stemplot of a distribution that is skewed toward large values.

Time plots

Many variables are measured at intervals over time. We might, for example, measure the cost of raw materials for a manufacturing process each month or the price of a stock at the end of each day. In these examples, our main interest is change over time. To display change over time, make a *time plot*.

> **TIME PLOT**
>
> A **time plot** of a variable plots each observation against the time at which it was measured. Always put time on the horizontal scale of your plot and the variable you are measuring on the vertical scale. Connecting the data points by lines helps emphasize any change over time.

More details about how to analyze data that vary over time are given in Chapter 14. For now, we examine how a time plot can reveal some additional important information about T-bill interest rates.

Histograms and stemplots can be very useful tools for describing the distribution of quantitative variables in a data set. When they are used to describe the distribution of a variable in some similar or future setting, these graphics can be powerful predictive analytics tools. However, caution must be used when making these predictions, particularly when time is an important component of understanding the distribution of a variable. Here is an example.

EXAMPLE 1.20

CASE 1.1 **A Time Plot for T-Bill Interest Rates** The website of the Federal Reserve Bank of St. Louis provided a very interesting graph of T-bill interest rates,[10] shown in Figure 1.12. A time plot shows us the relationship between two variables, in this case interest rate and the auctions that occurred at four-week intervals. Notice how the Federal Reserve Bank included information about a third variable in this plot. The third variable is a categorical variable that indicates whether the United States was in a recession. It is indicated by the light colored areas in the plot. ∎

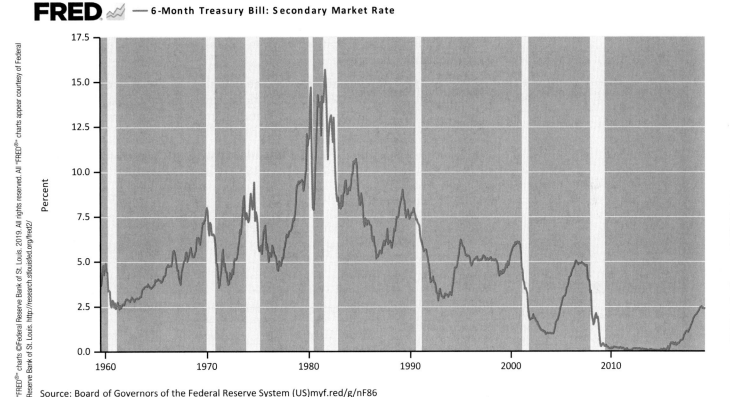

FIGURE 1.12 Time plot for the T-bill interest rates, Example 1.20.

APPLY YOUR KNOWLEDGE

CASE 1.1 **1.24 What does the time plot show?** Carefully examine the time plot in Figure 1.12.

(a) How did the T-bill rates change from 1960 to 1980?

(b) How did the T-bill rates change from 1980 to 2005?

(c) How did the T-bill rates change from 2005 to the end of the plot?

(d) What can you say about the relationship between the rates and the recession periods?

In Example 1.12 (page 13), we examined the distribution of T-bill interest rates for the period December 12, 1958, to March 30, 2018. The histogram in Figure 1.6 showed us the shape of the distribution. By looking at the time plot in Figure 1.12, we now see that there is more to this data set than is revealed by the histogram. This scenario illustrates the types of steps used in an effective statistical analysis of data. We are rarely able to completely plan our analysis in advance, set up the appropriate steps to be taken, and then click on the appropriate buttons in a software package to obtain useful results. Instead, an effective analysis requires that we proceed in an organized way, use a variety of analytical tools as we proceed, and exercise careful judgment at each step in the process.

SECTION 1.2 SUMMARY

- **Exploratory data analysis** uses graphs and numerical summaries to describe the variables in a data set and the relations among them.

- The **distribution** of a variable describes which values the variable takes and how often it takes these values.

- To describe a distribution, begin with a graph. **Bar graphs** and **pie charts** describe the distribution of a categorical variable, and **Pareto charts** identify the most important categories for a categorical variable. **Histograms** and **stemplots** graph the distributions of quantitative variables.

- When examining any graph, look for an **overall pattern** and for notable **deviations** from the pattern.

- **Shape, center,** and **spread** describe the overall pattern of the distribution of a quantitative variable. Some distributions have simple shapes, such as **symmetric** and **skewed.** Not all distributions have a simple overall shape, especially when there are few observations.

- **Outliers** are observations that lie outside the overall pattern of a distribution. Always look for outliers and try to explain them.

- When observations on a variable are taken over time, make a **time plot** that graphs time horizontally and the values of the variable vertically. A time plot can reveal interesting patterns in a set of data.

SECTION 1.2 EXERCISES

For Exercise 1.15, see page 10; for 1.16 and 1.17, see page 11; for 1.18 and 1.19, see pages 12–13; for 1.20 and 1.21, see page 16; for 1.22 and 1.23, see page 20; and for 1.24, see page 21.

1.25 Which graphical display should you use? For each of the following scenarios, decide which graphical display (pie chart, bar graph, Pareto chart, stemplot, or histogram) you would use to describe the distribution of the variable. Give a reason for your choice and if there is an alternative choice that would also be reasonable, explain why your choice was better than the alternative.

(a) The number of sales for your online company on each of the seven days in the past week.

(b) The amounts of each of the sales on Monday of the past week.

(c) The number of items bought by each of the customers who made a purchase on Monday of the past week.

(d) The total amount of sales during the past year for each of the customers who made a purchase on Monday of the past week.

1.26 Garbage is big business. The formal name for garbage is "municipal solid waste"(MSW). In the United States, approximately 250 million tons of garbage is generated in a year. Following is a breakdown of the materials that made up American MSW in a recent year.[11] GARBAGE

Material	Weight (million tons)	Percent of total
Food scraps	38.4	14.7
Glass	11.5	4.4
Metals	23.3	8.9
Paper, paperboard	68.6	26.3
Plastics	33.3	12.8
Rubber, leather	8.2	3.1
Textiles	16.2	6.2
Wood	16.1	6.2
Yard trimmings	34.5	13.2
Other	10.5	4.0

(a) Add the percents. The sum is not exactly equal to 100%. Why?

(b) Make a bar graph of the percents. The graph gives a clearer picture of the main contributors to garbage if you order the bars from tallest to shortest.

(c) Also make a pie chart of the percents. Comparing the two graphs, notice that it is easier to see the small differences among "Food scraps," "Plastics," and "Yard trimmings" in the bar graph.

1.27 Market share for desktop and laptop browsers. The following table gives the market share for the major browsers.[12] SEARCH

Browser	Market share	Browser	Market share
Chrome	60.2%	Edge	4.1%
Internet Explorer	12.5%	Safari	3.6%
Firefox	12.0%	Other	7.6%

(a) Use a bar graph to display the market shares.

(b) Summarize what the graph tells you about market shares for desktop browsers.

1.28 Reliability of household appliances. You are writing an article for a consumer magazine based on a survey of the magazine's readers. Of 12,376 readers who reported owning Brand A dishwashers, 2932 required a service call during the past year. Only 190 service calls were reported by the 475 readers who owned Brand B dishwashers.

(a) Why is the count of service calls (2932 versus 190) not a good measure of the reliability of these two brands of dishwashers?

(b) Use the information given to calculate a suitable measure of reliability. What do you conclude about the reliability of Brand A and Brand B?

1.29 Your Facebook app can generate a million dollars a month. A report on Facebook suggests that Facebook apps can generate large amounts of money, as much as $1 million per month.[13] The following table gives the numbers of active Facebook users by country for the top 10 countries based on the number of users in July 2017.[14] FACEBK

Country	Facebook users (in millions)
India	241
United States	240
Brazil	139
Indonesia	126
Mexico	85
Philippines	69
Vietnam	64
Thailand	57
Turkey	56
United Kingdom	44

(a) Use a bar graph to describe the numbers of users in these countries.

(b) Describe the major features of your graph in a short paragraph.

1.30 Facebook use increases by country. Refer to the previous exercise. The article giving these data was titled "India overtakes the USA to become Facebook's #1 country." The numbers of users, however, do not tell the whole story. The population of India is 1282 million and the population of the United States is 327 million. FACEBK

(a) Use these data along with the data in the previous exercise to calculate the Facebook market penetration—the number of users divided by the population—for the two countries.

(b) Use a graphical display to show the market penetrations for the two countries.

(c) Write a short paragraph about Facebook in India and the United States based on your analysis.

1.31 Products for senior citizens. The market for products designed for senior citizens in the United States is expanding. Here is a stemplot of the percents of residents aged 65 and older in the 50 states for 2012 as estimated by the U.S. Census Bureau American Community Survey.[15] The stems are whole percents, and the leaves are tenths of a percent. Describe the shape, center, and spread of this distribution. US65

9	4
10	0
11	5
12	379
13	68889
14	01233333457778
15	01223334557888899
16	1577
17	08
18	2
19	1

1.32 The Canadian market. Refer to Exercise 1.31. Here are similar data for the 13 Canadian provinces and territories:[16] CANADAP

Province/territory	Percent older than age 65
Newfoundland and Labrador	19.8
Prince Edward Island	19.0
Nova Scotia	19.8
New Brunswick	20.1
Quebec	18.5
Ontario	16.7
Manitoba	15.2
Saskatchewan	15.0
Alberta	12.4
British Columbia	18.3
Yukon	12.5
Northwest Territories	7.4
Nunavut	3.9

(a) Display the data graphically, and describe the major features of your plot.

(b) Explain why you chose the particular format for your graphical display. Which other types of graph could you have used? What are the strengths and weaknesses of each option for displaying this set of data?

1.3 Describing Distributions with Numbers

When you complete this section, you will be able to:

- Describe the center of a distribution by using the mean or the median.
- Describe the spread of a distribution by using the interquartile range (IQR) or the standard deviation.
- Describe a distribution by using the five-number summary.
- Describe a distribution or compare data sets measured on the same variable by using boxplots.
- Identify outliers by using the 1.5 × *IQR* rule.
- Choose measures of center and spread for a particular set of data.

In the previous section, we used the shape, center, and spread as ways to describe the overall pattern of any distribution for a quantitative variable. In this section, we will learn specific ways to use numbers to measure the center and spread of a distribution. The numbers, like the graphs of Section 1.1, are aids to understanding the data, not "the answer" in themselves.

TTS24

CASE 1.2 Time to Start a Business An entrepreneur faces many bureaucratic and legal hurdles when starting a new business. The World Bank collects information about starting businesses throughout the world. It has determined the time, in days, to complete all of the procedures required to start a business.[17] Data for 187 countries are included in the data set, TTS. For this section, we examine data, rounded to integers, for a sample of 24 of these countries. Here are the data:

19	17	43	7	12	27	67	49	6	6	29	12
12	9	17	23	1	12	14	18	6	7	9	31

EXAMPLE 1.21

TTS24

CASE 1.2 The Distribution of Business Start Times The stemplot in Figure 1.13 shows us the *shape, center,* and *spread* of the business start times. The stems are tens of days, and the leaves are days. Note that there is a stem at 5 but it has no leaves because there are no times in the 50s. The distribution is skewed to the right, with a very long tail of high values. All but seven of the times are less than 20 days. The center appears to be about 12 days, and the values range from 1 day to 67 days. There do not appear to be any outliers. ∎

```
0 | 1 6 6 6 7 9 9
1 | 2 2 2 2 4 7 7 8 9
2 | 3 7 9
3 | 1
4 | 3 9
5 |
6 | 7
```

FIGURE 1.13 Stemplot for sample of 24 business start times, Example 1.21.

Measuring center: the mean

A description of a distribution almost always includes a measure of its center. The most common measure of center is the ordinary arithmetic average, or *mean*.

THE MEAN $\bar{x}$

To find the **mean** of a set of observations, add their values and divide by the number of observations. If the n observations are $x_1, x_2, \ldots, x_n$, their mean is

$$\bar{x} = \frac{x_1 + x_2 + \cdots + x_n}{n}$$

or, in more compact notation,

$$\bar{x} = \frac{1}{n} \sum x_i$$

The $\sum$ (capital Greek sigma) in the formula for the mean is short for "add them all up." The subscripts on the observations x_i are just a way of keeping the n observations distinct. They do not necessarily indicate order or any other special facts about the data. The bar over the x indicates the mean of all the x-values. Pronounce the mean $\bar{x}$ as "x-bar." This notation is very common. When writers who are discussing data use $\bar{x}$ or $\bar{y}$, they are talking about a mean.

EXAMPLE 1.22

TTS24

CASE 1.2 **Mean Time to Start a Business** The mean time to start a business is

$$\bar{x} = \frac{x_1 + x_2 + \cdots + x_n}{n}$$
$$= \frac{19 + 17 + \cdots + 31}{24}$$
$$= \frac{453}{24} = 18.875$$

The mean time to start a business for the 24 countries in our data set is 19 days. Note that we have rounded the answer. Our goal in using the mean to describe the center of a distribution is not to demonstrate that we can compute this point with great accuracy. The additional digits do not provide any additional useful information. In fact, they distract our attention from the important digits that are truly meaningful. Do you think it would be better to report the mean as 18.9 days? ■

In practice, you can enter the data into statistical software or a calculator to find the mean. You don't have to actually add and divide.

APPLY YOUR KNOWLEDGE

CASE 1.2 **1.33 Include the outlier.** For Case 1.2, a random sample of 24 countries was selected from a data set that included 187 countries. The South American country of Venezuela, where the start time is 230 days, was not included in the random sample. Consider the effect of adding Venezuela to the original set. Show that the mean for the new sample of 25 countries has increased to 27 days. (This is a rounded number. You should report the mean with two digits after the decimal to show that you have performed this calculation.) TTS25

1.34 Find the mean of the accounting exam scores. Here are the scores on the first exam in an accounting course for 10 students:

75 87 94 85 74 98 93 52 80 91

Find the mean first-exam score for these students. ACCT

1.35 Calls to a customer service center. The service times for 80 calls to a customer service center are given in Table 1.1 (page 15). Use these data to compute the mean service time. CC80

Exercise 1.33 illustrates an important fact about the mean as a measure of center: it is sensitive to the influence of one or more extreme observations. These may be outliers, but a skewed distribution that has no outliers will also pull the mean toward its long tail. Because the mean cannot resist the influence of extreme observations, we say that it is *not* a **resistant measure** of center.

Measuring center: the median

In Section 1.1, we used the midpoint of a distribution as an informal measure of center. The *median* is the formal version of the midpoint, with a specific rule for calculation.

> **THE MEDIAN M**
>
> The **median M** is the midpoint of a distribution, the number such that half the observations are smaller and the other half are larger. To find the median of a distribution:
>
> 1. Arrange all observations in order of size, from smallest to largest.
>
> 2. If the number of observations n is odd, the median M is the middle observation in the ordered list. Find the location of the median by counting $(n + 1)/2$ observations up from the bottom of the list.
>
> 3. If the number of observations n is even, the median M is the mean of the two middle observations in the ordered list. The location of the median is again $(n + 1)/2$ from the bottom of the list.

Note that the formula $(n + 1)/2$ *does not give the median, just the location of the median in the ordered list.* Medians require little arithmetic, so they are easy to find by hand for small sets of data. Arranging even a moderate number of observations in order is very tedious, however, so that finding the median by hand for larger sets of data is unpleasant. Even simple calculators have an $\bar{x}$ button, but you will need software or a graphing calculator to automate finding the median.

EXAMPLE 1.23

Median Time to Start a Business To find the median time to start a business for our 24 countries, we first arrange the data in order from smallest to largest:

1   6   6   6   7   7   9   9   12 12 12 12
14 17 17 18 19 23 27 29 31 43 49 67

1.3 Describing Distributions with Numbers

The count of observations $n = 24$ is even. The median, then, is the average of the two center observations in the ordered list. To find the location of the center observations, we first compute

$$\text{location of } M = \frac{n+1}{2} = \frac{25}{2} = 12.5$$

Therefore, the center observations are the 12th and 13th observations in the ordered list. The median is

$$M = \frac{12 + 14}{2} = 13 \quad \blacksquare$$

Note that you can use the stemplot directly to compute the median. In the stemplot the cases are already ordered, and you simply need to count from the top or the bottom to the desired location.

APPLY YOUR KNOWLEDGE

1.36 Find the median of the accounting exam scores. Here are the scores on the first exam in an accounting course for 10 students:

75 87 94 85 74 98 93 52 80 91

Find the median first-exam score for these students. ACCT

1.37 Calls to a customer service center. The service times for 80 calls to a customer service center are given in Table 1.1 (page 15). Use these data to compute the median service time. CC80

CASE 1.2 **1.38 Include the outlier.** Include Venezuela, where the start time is 230 days, in the data set, and show that the median is 14 days. Note that with this case included, the sample size is now 25 and the median is the 13th observation in the ordered list. Write out the ordered list and circle the outlier. Describe the effect of the outlier on the median for this set of data. TTS25

Comparing the mean and the median

Exercises 1.33 (page 25) and 1.38 (page 27) illustrate an important difference between the mean and the median. Including the data for Venezuela pulls the mean time to start a business up from 19 days to 27 days. By comparison, the increase in the median is very small, from 13 days to 14 days.

This result shows that the median is more *resistant* than the mean. If the largest starting time in the data set was 1200 days, the median for all 25 countries would still be 14 days. The largest observation just counts as one observation above the center, no matter how far above the center it lies. By contrast, the mean uses the actual value of each observation, so it will chase a single large observation upward.

The best way to compare the response of the mean and median to extreme observations is to use an interactive applet that allows you to place points on a line and then drag them with your computer's mouse. Exercises 1.60 to 1.62 use the *Mean and Median* applet on the website for this textbook to compare the mean and the median.

The mean and median of a symmetric distribution are close together. If the distribution is exactly symmetric, the mean and median are exactly the same. In a skewed distribution, the mean is farther out in the long tail than is the median.

Consider the prices of new homes in the United States.[18] The mean price in 2017 was $384,900, while the median was $323,100. This distribution is strongly skewed to the right, because there are many moderately priced houses and only a few very expensive mansions. The few expensive houses pull the mean up but do not affect the median.

Reports about house prices, incomes, and other strongly skewed distributions usually give the median ("midpoint") rather than the mean ("arithmetic average"). However, if you are a tax assessor interested in the total value of houses in your area, use the mean. The total is the mean times the number of houses, but it has no connection with the median. The mean and median measure center in different ways, and both are useful.

APPLY YOUR KNOWLEDGE

1.39 Gross domestic product. The success of companies expanding to developing regions of the world depends in part on the prosperity of the countries in those regions. Here are World Bank data on the average growth of gross domestic product (percent per year) for the period 2000 to 2016 for 12 countries in Asia:[19] GDPA

Country	Growth
Bangladesh	6.0
China	9.9
India	7.5
Indonesia	5.5
Japan	0.7
Korea (South)	6.6
Malaysia	4.9
Pakistan	4.2
Philippines	6.2
Singapore	5.8
Thailand	3.9
Vietnam	6.4

(a) Make a stemplot of the data.

(b) Describe the distribution.

(c) Find the mean growth rate. Do you think that the mean gives a good description of these data? Explain your answer.

(d) Find the median growth rate. Do you think that the median gives a good description of these data? Explain your answer.

Measuring spread: the quartiles

A measure of center alone can be misleading. Two nations with the same median household income are very different if one has extremes of wealth and poverty and the other has little variation among households. A drug with the correct mean concentration of active ingredient is dangerous if some batches are much too high and others much too low. In both these examples, we are interested in the *spread* or *variability* of the values of incomes and drug potencies as well as their centers. In fact, *the simplest useful numerical description of a distribution consists of both a measure of center and a measure of spread.*

One way to measure spread is to give the smallest and largest observations. For example, the times to start a business in our data set that included Venezuela ranged from 1 to 230 days. Without Venezuela, the range is 1 to 67 days. These largest and smallest observations show the full spread of the data and are highly influenced by outliers.

We can improve our description of spread by also giving several percentiles. The **pth percentile** of a distribution is the value such that p percent of the observations fall at or below it. The median is just the 50th percentile, so the use of percentiles to report spread is particularly appropriate when the median is the measure of center.

pth percentile

The most commonly used percentiles other than the median are the *quartiles*. The first quartile is the 25th percentile, and the third quartile is the 75th percentile. That is, the first and third quartiles show the spread of the middle half of the data. (The second quartile is the median itself.) To calculate a percentile, arrange the observations in increasing order, and count up the required percent from the bottom of the list. Our definition of percentiles is a bit inexact because there is not always a value with exactly p percent of the data at or below it. We are content to take the nearest observation for most percentiles, but the quartiles are important enough to require an exact recipe. Our rule for calculating the quartiles uses the rule for the median. The exact rule for these computations varies with the software used.

THE QUARTILES Q_1 AND Q_3

To calculate the **quartiles:**

1. Arrange the observations in increasing order, and locate the median M in the ordered list of observations.

2. The **first quartile Q_1** is the median of the observations whose position in the ordered list is to the left of the location of the overall median.

3. The **third quartile Q_3** is the median of the observations whose position in the ordered list is to the right of the location of the overall median.

Here is an example that shows how the rules for the quartiles work for even numbers of observations.

EXAMPLE 1.24

TTS24

Finding the Quartiles Here is the ordered list of the times to start a business in our sample of 24 countries:

```
 1    6    6    6    7    7    9    9   12   12   12   12
14   17   17   18   19   23   27   29   31   43   49   67
```

The count of observations $n = 24$ is even, so the median is at position $(24 + 1)/2 = 12.5$; that is, it is between the 12th and the 13th observation in the ordered list. There are 12 cases above this position and 12 below it. The first quartile is the median of the first 12 observations, and the third quartile is the median of the last 12 observations. Check that $Q_1 = 8$ and $Q_3 = 25$. ∎

Notice that the quartiles are resistant. For example, Q_3 would have the same value if the highest start time was 670 days rather than 67 days.

There are slight differences in the methods used by software to compute percentiles. However, the results will generally be quite similar, except in cases where the sample sizes are very small.

Be careful when several observations take the same numerical value. Write down all the observations, and apply the rules just as if they all had distinct values.

The five-number summary and boxplots

The smallest and largest observations tell us little about the distribution as a whole, but they give information about the tails of the distribution that is missing if we know only Q_1, M, and Q_3. To get a quick summary of both center and spread, combine all five numbers. The result is the *five-number summary* and a graph based on it.

THE FIVE-NUMBER SUMMARY AND BOXPLOTS

The **five-number summary** of a distribution consists of the smallest observation, the first quartile, the median, the third quartile, and the largest observation, written in order from smallest to largest. In symbols, the five-number summary is

$$\text{Minimum} \quad Q_1 \quad M \quad Q_3 \quad \text{Maximum}$$

A **boxplot** is a graph of the five-number summary.

- A central box spans the quartiles.
- A line in the box marks the median.
- Lines, called whiskers, extend from the box to the minimum and maximum values, or, for modified boxplots, to some other boundaries beyond which individual observations are plotted.

Boxplots are very useful for the side-by-side comparison of several distributions.

You can draw boxplots either horizontally or vertically. Be sure to include a numerical scale in the graph. When you look at a boxplot, first locate the median, which marks the center of the distribution. Then look at the spread. The quartiles show the spread of the middle half of the data, and the extremes (the smallest and largest observations) show the spread of the entire data set. We now have the tools for a preliminary examination of the customer service center call lengths.

EXAMPLE 1.25

CC80

Service Center Call Lengths Table 1.1 (page 15) displays the customer service center call lengths for a random sample of 80 calls that we discussed in Example 1.13 (page 14). The five-number summary for these data is 1.0, 54.4, 103.5, 200, 2631. The distribution is highly skewed. The mean is 197 seconds, a value that is very close to the third quartile. The boxplot is displayed in Figure 1.14. The skewness of the distribution is the major feature that we see in this plot. Note that the mean is marked with a "+" and appears very close to the upper edge of the box. ∎

FIGURE 1.14 Boxplot for sample of 80 service center call lengths, Example 1.25.

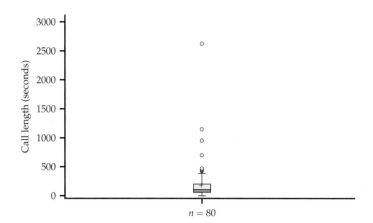

The 1.5 × *IQR* rule for outliers

If we look at the data in Table 1.1 (page 15), we can spot a clear outlier, a call lasting 2631 seconds, more than twice the length of any other call. How can we describe the spread of this distribution? The smallest and largest observations are extremes that do not describe the spread of the majority of the data. The distance between the quartiles (the range of the center half of the data) is a more resistant measure of spread than the range. This distance is called the *interquartile range*.

THE INTERQUARTILE RANGE *IQR*

The **interquartile range *IQR*** is the distance between the first and third quartiles:

$$IQR = Q_3 - Q_1$$

EXAMPLE 1.26

IQR for Service Center Call Lengths. The five-number summary for our data on service center call lengths is 1.0, 54.5, 103.5, 200, and 2631. Therefore, we calculate

$$IQR = Q_3 - Q_1$$
$$IQR = 200 - 54.5$$
$$= 145.5 \quad \blacksquare$$

The quartiles and the *IQR* are not affected by changes in either tail of the distribution. They are resistant, therefore, because changes in a few data points have no further effect once these points move outside the quartiles.

However, *no single numerical measure of spread, such as IQR, is very useful for describing skewed distributions*. The two sides of a skewed distribution have different spreads, so one number can't summarize them. We can often detect skewness from the five-number summary by comparing how far the first quartile and the minimum are from the median (left tail) with how far the third quartile and the maximum are from the median (right tail). The interquartile range can be used as the basis for a rule of thumb for identifying outliers.

THE 1.5 × *IQR* RULE FOR OUTLIERS

Call an observation a possible outlier if it falls more than 1.5 × *IQR* above the third quartile or below the first quartile.

EXAMPLE 1.27

CC80

Outliers for Call Length Data. For the call length data in Table 1.1 (page 15),

$$1.5 \times IQR = 1.5 \times 145.5 = 218.25$$

Any values below $54.5 - 218.25 = -163.75$ or above $200 + 218.25 = 418.25$ are flagged as possible outliers. There are no low outliers, but the eight longest calls are flagged as possible high outliers. Their lengths are

438   465   479   700   700   951   1148   2631

It is difficult to imagine calls lasting this long. ∎

The $1.5 \times IQR$ rule can be incorporated into a boxplot. In this version, the whiskers are terminated at $1.5 \times IQR$. Points beyond the whiskers are plotted individually and are possible outliers according to the $1.5 \times IQR$ rule. This is what was done in Figure 1.15.

side-by-side boxplots   **Side-by-side boxplots** can be very useful to compare two or more distributions. Here is an example.

EXAMPLE 1.28

TBILLJJ

Compare the T-Bill Rates in January and July In Example 1.17 (page 18) we used a back-to-back stemplot to compare the T-bill rates for the months of January and June. Figure 1.15 gives side-by-side boxplots for the two months generated with JMP. Notice that this software plots the individual observations as dots in addition to the modified boxplots as default options. ∎

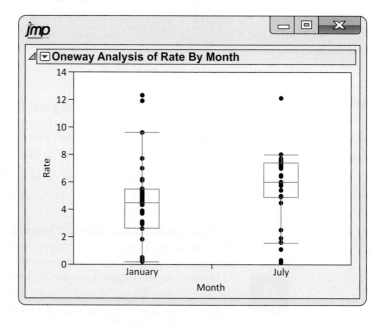

FIGURE 1.15 Side-by-side modified boxplots with observations to compare T-bill rates in January and July from JMP, Example 1.28.

APPLY YOUR KNOWLEDGE

1.40 Stemplots or boxplots for comparing T-bill rates. The T-bill rates for January and July are graphically compared using a back-to-back stemplot in Figure 1.10 (page 18) and using side-by-side boxplots in Figure 1.15. Which graphical display do you prefer for these data? Give reasons for your answer.

CASE 1.2 1.41 Time to start a business. Refer to the data on times to start a business in 24 countries described in Case 1.2 on page 24. Use a boxplot to display the distribution. Discuss the features of the data that you see in the boxplot, and compare this boxplot with the stemplot in Figure 1.13 (page 24). Which do you prefer? Give reasons for your answer. TTS24

1.42 Accounting exam scores. Here are the scores on the first exam in an accounting course for 10 students:

$$75 \quad 87 \quad 94 \quad 85 \quad 74 \quad 98 \quad 93 \quad 52 \quad 80 \quad 91$$

Display the distribution with a boxplot. Discuss whether a stemplot would provide a better way to look at this distribution. ACCT

Measuring spread: the standard deviation

The five-number summary is not the most common numerical description of a distribution. That distinction belongs to the combination of the mean to measure center and the *standard deviation* to measure spread. The standard deviation measures spread by looking at how far the observations are from their mean.

THE STANDARD DEVIATION s

The **variance** s^2 of a set of observations is essentially the average of the squares of the deviations of the observations from their mean. In symbols, the variance of n observations $x_1, x_2, \ldots, x_n$ is

$$s^2 = \frac{(x_1 - \bar{x})^2 + (x_2 - \bar{x})^2 + \cdots + (x_n - \bar{x})^2}{n - 1}$$

or, more compactly,

$$s^2 = \frac{1}{n - 1} \sum (x_i - \bar{x})^2$$

The **standard deviation** s is the square root of the variance s^2:

$$s = \sqrt{\frac{1}{n - 1} \sum (x_i - \bar{x})^2}$$

Notice that the "average" in the variance s^2 divides the sum by 1 less than the number of observations, that is, $n - 1$ rather than n. The reason is that the deviations $x_i - \bar{x}$ always sum to exactly 0, so that knowing $n - 1$ of them determines the last one. Only $n - 1$ of the squared deviations can vary freely, and we average by dividing the total by $n - 1$. The number $n - 1$ is called the **degrees of freedom** of the variance or standard deviation.

degrees of freedom

In practice, use software to obtain the standard deviation. However, doing an example step-by-step will help you understand how the variance and standard deviation work.

EXAMPLE 1.29

Standard Deviation for Time to Start a Business In Example 1.22 (page 25), we found that the mean time to start a business for the 24 countries in our data set was 19 days. Here, we keep an extra three digits ($\bar{x} = 18.875$) to make sure that our intermediate calculations are accurate. We organize the arithmetic in a table.

TTS24

Observations x_i	Deviations $x_i - \bar{x}$		Squared deviations $(x_i - \bar{x})^2$	
19	$19 - 18.875 =$	0.125	$(0.125)^2 =$	0.00015625
17	$17 - 18.875 =$	-1.875	$(-1.875)^2 =$	3.51562500
...	... =	...	... =	...
31	$31 - 18.875 =$	12.125	$(12.125)^2 =$	147.01562500
	sum =	0.000	sum =	5652.625

Note that the sum of the deviations is not exactly zero in our calculations because of round-off error. The variance is the sum of the squared deviations divided by 1 less than the number of observations:

$$s^2 = \frac{5652.625}{23} = 254.766$$

The standard deviation is the square root of the variance:

$$s = \sqrt{254.766} = 15.7 \text{ days} \ \blacksquare$$

More important than the details of hand calculation are the properties that determine the usefulness of the standard deviation:

- s measures spread about the mean and should be used only when the mean is chosen as the measure of center.

- $s = 0$ only when there is *no spread*. This happens only when all observations have the same value. Otherwise, s is greater than zero. As the observations become more spread out about their mean, s gets larger.

- s has the same units of measurement as the original observations. For example, if you measure wages in dollars per hour, s is also in dollars per hour.

- Like the mean $\bar{x}$, s is not resistant. Strong skewness or a few outliers can greatly increase s.

Numerical summaries can be very useful tools for describing the distribution of quantitative variables in a data set. When you use them to describe the numerical summary of a variable in some similar or future setting, you will find that they can be powerful predictive analytics tools. For the time to start a business data, it would be reasonable to consider using the time *for a particular country* in the data set as a predictor of the time in a year that is not too far in the future because the same circumstances that influence the time to start a business would be expected to be present then. We noted, however, that Venezuela is an outlier in the distribution. This may be due to a particular set of circumstances that we hope would be corrected in the near future. If so, the entry for Venezuela in the data set would not be a good predictor. The next exercise provides another view of the effect of Venezuela on a numerical summary.

APPLY YOUR KNOWLEDGE

CASE 1.2 **1.43 Time to start a business.** Verify the statement in the last bullet about the standard deviation using the data on the time to start a business. First, use the 24 cases from Case 1.2 (page 24) to calculate a standard deviation. Next, include the country Venezuela, where the time to start a business is 230 days. Show that the inclusion of this single outlier increases the standard deviation from 15.7 to 44.9. TTS24 TTS25

You may rightly feel that the importance of the standard deviation is not yet clear. We will see in the next section that the standard deviation is the natural measure of spread for an important class of symmetric distributions, the Normal distributions. The usefulness of many statistical procedures is tied to distributions with particular shapes, and that is certainly true of the standard deviation.

Choosing measures of center and spread

How do we choose between the five-number summary and $\bar{x}$ and s to describe the center and spread of a distribution? Because the two sides of a strongly skewed distribution have different spreads, no single number such as s describes the spread well. The five-number summary, with its two quartiles and two extremes, does a better job.

1.3 Describing Distributions with Numbers

CHOOSING A SUMMARY

The five-number summary is usually better than the mean and standard deviation for describing a skewed distribution or a distribution with extreme outliers. Use $\bar{x}$ and s only for reasonably symmetric distributions that are free of outliers.

APPLY YOUR KNOWLEDGE

1.44 Accounting exam scores. Following are the scores on the first exam in an accounting course for 10 students. We found the mean of these scores in Exercise 1.34 (page 26) and the median in Exercise 1.36 (page 27). ACCT

75   87   94   85   74   98   93   52   80   91

(a) Make a stemplot of these data.

(b) Compute the standard deviation.

(c) Are the mean and the standard deviation effective in describing the distribution of these scores? Explain your answer.

1.45 Calls to a customer service center. We displayed the distribution of the lengths of 80 calls to a customer service center in Figure 1.14 (page 31). CC80

(a) Compute the mean and the standard deviation for these 80 calls (the data are given in Table 1.1, page 15).

(b) Find the five-number summary.

(c) Which summary does a better job of describing the distribution of these calls? Give reasons for your answer.

BEYOND THE BASICS

Risk and return

A central principle in the study of investments is that taking bigger risks is rewarded by higher returns, at least on the average, over long periods of time. It is usual practice in finance to measure risk by the standard deviation of returns, on the grounds that investments whose returns show a large spread from year to year are less predictable and, therefore, more risky than those whose returns have a small spread. Compare, for example, the approximate mean and standard deviation of the annual percent returns on American common stocks and U.S. Treasury bills over a 50-year period starting in 1950:

Investment	Mean return	Standard deviation
Common stocks	14.0%	16.9%
Treasury bills	5.2%	2.9%

Stocks are risky. They went up 14% per year on the average during this period. The large standard deviation reflects the fact that stocks have produced both large gains and large losses. When you buy a Treasury bill, by contrast, you are lending money to the government for one year. You know that the government will pay you back with interest. Because the standard deviation for Treasury bills is smaller than the standard deviation for common stocks, Treasury bills are much less risky than buying stocks, so (on the average) you get a smaller return.

```
 0 | 9
 1 | 2 5 5 6 6 8
 2 | 1 5 7 7 9
 3 | 0 1 1 5 5 8 9 9
 4 | 2 4 7 7 8
 5 | 1 1 2 2 2 5 6 6 8
 6 | 2 1 5 6 8
 7 | 2 7 8
 8 | 0 4 8
 9 | 8
10 | 4 5
11 | 3
12 |
13 |
14 | 7
```

FIGURE 1.16(a) Stemplot of the annual returns on Treasury bills for 50 years.

```
-2 | 8
-1 | 9 1 1 0 0
-0 | 9 6 4 3
 0 | 0 0 0 1 2 3 8 9 9
 1 | 1 3 3 4 4 6 6 6 7 8
 2 | 0 1 1 2 3 4 4 4 5 7 7 9 9
 3 | 0 1 1 3 4 6 7
 4 | 5
 5 | 0
```

FIGURE 1.16(b) Stemplot of the annual returns on common stocks for 50 years.

Are $\bar{x}$ and s good summaries for distributions of investment returns? Figures 1.16(a) and 1.16(b) display stemplots of the annual returns for both investments. You see that returns on Treasury bills have a right-skewed distribution. Convention in the financial world calls for use of $\bar{x}$ and s because some parts of investment theory use them. For describing this right-skewed distribution, however, the five-number summary would be more informative.

Remember that a graph gives the best overall picture of a distribution. Numerical measures of center and spread report specific facts about a distribution, but they do not describe its entire shape. Numerical summaries do not disclose the presence of multiple peaks or gaps, for example. **Always plot your data.**

SECTION 1.3 SUMMARY

• A numerical summary of a distribution should report its **center** and its **spread** or **variability**.

• The **mean** $\bar{x}$ and the **median** M describe the center of a distribution in different ways. The mean is the arithmetic average of the observations, and the median is the midpoint of the values.

• When you use the median to indicate the center of the distribution, describe its spread by giving the **quartiles.** The **first quartile** Q_1 has one-fourth of the observations below it, and the **third quartile** Q_3 has three-fourths of the observations below it.

• The **five-number summary**—consisting of the median, the quartiles, and the high and low extremes—provides a quick overall description of a distribution. The median describes the center, and the quartiles and extremes show the spread.

• **Boxplots** based on the five-number summary are useful for comparing several distributions. The box spans the quartiles and shows the spread of the central half of the distribution. The median is marked within the box. Lines, called whiskers, extend from the box to the extremes or to some other boundary beyond which individual observations are plotted.

• The **variance** s^2 and, especially, its square root, the **standard deviation** s, are common measures of spread about the mean as center. The standard deviation s is zero when there is no spread and gets larger as the spread increases.

1.3 Describing Distributions with Numbers

- A **resistant measure** of any aspect of a distribution is relatively unaffected by changes in the numerical value of a small proportion of the total number of observations, no matter how large these changes are. The median and quartiles are resistant, but the mean and the standard deviation are not.

- The mean and standard deviation are good descriptions for symmetric distributions without outliers. They are most useful for the Normal distributions, introduced in the next section. The five-number summary is a better exploratory summary for skewed distributions and distributions with outliers.

SECTION 1.3 EXERCISES

For Exercise 1.33, see page 25; for 1.34 and 1.35, see page 26; for 1.36 to 1.38, see page 27; for 1.39, see page 28; for 1.40 to 1.42, see pages 32–33; for 1.43, see page 34; and for 1.44 and 1.45, see page 35.

1.46 Gross domestic product for 197 countries. The gross domestic product (GDP) of a country is the total value of all goods and services produced in the country. It is an important measure of the health of a country's economy. For this exercise, you will analyze the 2016 GDP for 197 countries. The values are given in billions of U.S. dollars.[20] GDP

(a) Compute the mean and the standard deviation.

(b) Which countries do you think are outliers? Identify them by name and explain why you consider them to be outliers.

(c) Recompute the mean and the standard deviation without your outliers. Explain how the mean and standard deviation changed when you deleted the outliers.

1.47 Use the resistant measures for GDP. Repeat parts (a) and (c) of the previous exercise using the median and the quartiles. Summarize your results and compare them with those of the previous exercise. GDP

1.48 *Forbes* rankings of best countries for business. The *Forbes* website ranks countries based on their characteristics that are favorable for business.[21] One of the characteristics it uses for its rankings is trade balance, defined as the difference between the value of a country's exports and its imports. A negative trade balance occurs when a country imports more than it exports. Similarly, the trade balance will be positive for a country that exports more than it imports. Data related to the rankings are given for 151 countries. BESTBUS

(a) Describe the distribution of trade balance using the mean and the standard deviation.

(b) Do the same using the median and the quartiles.

(c) Using only information from parts (a) and (b), give a description of the data. *Do not* look at any graphical summaries or other numerical summaries for this part of the exercise.

1.49 What do the trade balance graphical summaries show? Refer to the previous exercise. BESTBUS

(a) Use graphical summaries to describe the distribution of the trade balance for these countries.

(b) Give the names of the countries that correspond to extreme values in this distribution.

(c) Reanalyze the data without the outliers.

(d) Summarize what you have learned about the distribution of the trade balance for these countries. Include appropriate graphical and numerical summaries as well as comments about the outliers.

1.50 GDP growth for 151 countries. Refer to the previous two exercises. Another variable that *Forbes* uses to rank countries is growth in GDP per capita, the GDP divided by the population size. BESTBUS

(a) Use graphical summaries to describe the distribution of the growth in GDP for these countries.

(b) Give the names of the countries that correspond to extreme values in this distribution.

(c) Reanalyze the data without the outliers.

(d) Summarize what you have learned about the distribution of the growth in GDP for these countries. Include appropriate graphical and numerical summaries as well as comments about the outliers.

1.51 Create a data set. Create a data set that illustrates the idea that an extreme observation can have a large effect on the mean but not on the median.

1.52 Variability of an agricultural product. A quality product is one that is consistent and has very little variability in its characteristics. Controlling variability can be more difficult with agricultural products than with those that are manufactured. The following table gives the individual weights, in ounces, of the 25 potatoes sold in a 10-pound bag. POTATO

| 7.8 | 7.9 | 8.2 | 7.3 | 6.7 | 7.9 | 7.9 | 7.9 | 7.6 | 7.8 | 7.0 | 4.7 | 7.6 |
| 6.3 | 4.7 | 4.7 | 4.7 | 6.3 | 6.0 | 5.3 | 4.3 | 7.9 | 5.2 | 6.0 | 3.7 | |

(a) Summarize the data graphically and numerically. Give reasons for the methods you chose to use in your summaries.

(b) Do you think that your numerical summaries do an effective job of describing these data? Why or why not?

(c) There appear to be two distinct clusters of weights for these potatoes. Divide the sample into two subsamples based on the clustering. Give the mean and standard deviation for each subsample. Do you think that this way of summarizing these data is better than a numerical summary that uses all the data as a single sample? Give a reason for your answer.

1.53 Apple is the number one brand. A brand is a symbol or images that are associated with a company. An effective brand identifies the company and its products.

Using a variety of measures, dollar values for brands can be calculated.[22] The most valuable brand is Apple, with a value of $170 billion. Apple is followed by Google, at $101.8 billion; Microsoft, at $87 billion; Facebook, at $73.5 billion; and Coca-Cola, at $56.4 million. For this exercise, you will use the brand values, reported in billions of dollars, for the top 100 brands. BRANDS

(a) Graphically display the distribution of the values of these brands.

(b) Use numerical measures to summarize the distribution.

(c) Write a short paragraph discussing the dollar values of the top 100 brands. Include the results of your analysis.

1.54 Advertising for best brands. Refer to the previous exercise. To calculate the value of a brand, the *Forbes* website uses several variables, including the amount the company spent for advertising. For this exercise, you will analyze the amounts of these companies spent on advertising, reported in billions of dollars. BRANDS

(a) Graphically display the distribution of the dollars spent on advertising by these companies.

(b) Use numerical measures to summarize the distribution.

(c) Write a short paragraph discussing the advertising expenditures of the top 100 brands. Include the results of your analysis.

1.55 Gosset's data on double stout sales. William Sealy Gosset worked at the Guinness Brewery in Dublin and made substantial contributions to the practice of statistics.[23] In his work at the brewery, he collected and analyzed a great deal of data. Archives with Gosset's handwritten tables, graphs, and notes have been preserved at the Guinness Storehouse in Dublin.[24] In one study, Gosset examined the change in the double stout market before and after World War I (1914–1918). For various regions in England and Scotland, he calculated the ratio of sales in 1925, after the war, as a percent of sales in 1913, before the war. Here are the data: STOUT

Bristol	94	Glasgow	66
Cardiff	112	Liverpool	140
English Agents	78	London	428
English O	68	Manchester	190
English P	46	Newcastle-on-Tyne	118
English R	111	Scottish	24

(a) Compute the mean for these data.

(b) Compute the median for these data.

(c) Which measure do you prefer for describing the center of this distribution? Explain your answer. (You may include a graphical summary as part of your explanation.)

1.56 Measures of spread for the double stout data. Refer to the previous exercise. STOUT

(a) Compute the standard deviation for these data.

(b) Compute the quartiles for these data.

(c) Which measure do you prefer for describing the spread of this distribution? Explain your answer. (You may include a graphical summary as part of your explanation.)

1.57 Are there outliers in the double stout data? Refer to the previous two exercises. STOUT

(a) Find the *IQR* for these data.

(b) Use the $1.5 \times IQR$ rule to identify and name any outliers.

(c) Make a boxplot for these data and describe the distribution using only the information in the boxplot.

(d) Make a modified boxplot for these data and describe the distribution using only the information in the boxplot.

(e) Make a stemplot for these data.

(f) Compare the boxplot, the modified boxplot, and the stemplot. Evaluate the advantages and disadvantages of each graphical summary for describing the distribution of the double stout data.

1.58 The alcohol content of beer. Brewing beer involves a variety of steps that can affect the alcohol content. A website gives the percent alcohol for 175 domestic brands of beer.[25] BEER

(a) Use graphical and numerical summaries of your choice to describe the data. Give reasons for your choice.

(b) The data set contains an outlier. Explain why this particular beer is unusual.

(c) For the outlier, give a short description of how you think this particular beer should be marketed.

1.59 Outlier for alcohol content of beer. Refer to the previous exercise. BEER

(a) Calculate the mean with and without the outlier. Do the same for the median. Explain how these values change when the outlier is excluded.

(b) Calculate the standard deviation with and without the outlier. Do the same for the quartiles. Explain how these values change when the outlier is excluded.

(c) Write a short paragraph summarizing what you have learned in this exercise.

1.60 Calories in beer. Refer to the previous two exercises. The data set also lists calories per 12 ounces of beverage. BEER

(a) Analyze the data and summarize the distribution of calories for these 175 brands of beer.

(b) In Exercise 1.58, you identified one brand of beer as an outlier. To what extent is this brand an outlier in the distribution of calories? Explain your answer.

(c) Does the distribution of calories suggest marketing strategies for this brand of beer? Describe some marketing strategies.

1.61 Outliers for older residents. The stemplot in Exercise 1.31 (page 23) displays the distribution of the percents of residents aged 65 and older in the 50 states. Stemplots help you find the five-number summary because they arrange the observations in increasing order. US65

(a) Give the five-number summary of this distribution.

(b) Does the $1.5 \times IQR$ rule identify any outliers? If yes, give the names of the states with the percents of the population older than age 65.

The following three exercises use the *Mean and Median* applet available at the textbook website to explore the behavior of the mean and median.

1.62 Mean = median? Place two observations on the line by clicking below it. Why does only one arrow appear?

1.63 Extreme observations. Place three observations on the line by clicking below it—two close together near the center of the line and one somewhat to the right of these two.

(a) Pull the rightmost observation out to the right. (Place the cursor on the point, hold down a mouse button, and drag the point.) How does the mean behave? How does the median behave? Explain briefly why each measure acts as it does.

(b) Now drag the rightmost point to the left as far as you can. What happens to the mean? What happens to the median as you drag this point past the other two? Watch carefully.

1.64 Don't change the median. Place five observations on the line by clicking below it.

(a) Add one additional observation *without changing the median*. Where is your new point?

(b) Use the applet to convince yourself that when you add yet another observation (there are now seven in all), the median does not change no matter where you put the seventh point. Explain why this must be true.

1.65 $\bar{x}$ and s are not enough. The mean $\bar{x}$ and standard deviation s measure center and spread but are not a complete description of a distribution. Data sets with different shapes can have the same mean and standard deviation. To demonstrate this fact, find $\bar{x}$ and s for these two small data sets. Then make a stemplot of each and comment on the shape of each distribution. ABDATA

Data A:	9.14	8.14	8.74	8.77	9.26	8.10
	6.13	3.10	9.13	7.26	4.74	
Data B:	6.58	5.76	7.71	8.84	8.47	7.04
	5.25	5.56	7.91	6.89	12.50	

CASE 1.1 **1.66 Returns on Treasury bills.** Figure 1.16(a) (page 36) is a stemplot of the annual returns on U.S. Treasury bills for 50 years. (The entries are rounded to the nearest tenth of a percent.)

(a) Use the stemplot to find the five-number summary of T-bill returns.

(b) The mean of these returns is about 5.19%. Explain from the shape of the distribution why the mean return is larger than the median return.

1.67 A standard deviation contest. You must choose four numbers from the whole numbers 10 to 20, with repeats allowed.

(a) Choose four numbers that have the smallest possible standard deviation.

(b) Choose four numbers that have the largest possible standard deviation.

(c) Is more than one choice possible in (a)? In (b)? Explain.

1.68 Imputation. Various problems with data collection can cause some observations to be missing. Suppose a data set has 20 cases. Here are the values of the variable x for 10 of these cases: IMPUTE

27   16   6   11   28   21   6   15   14   20

The values for the other 10 cases are missing. One way to deal with missing data is called **imputation**. The basic idea is that missing values are replaced, or imputed, with values that are based on an analysis of the data that are not missing. For a data set with a single variable, the usual choice of a value for imputation is the mean of the values that are not missing.

(a) Find the mean and the standard deviation for these data.

(b) Create a new data set with 20 cases by setting the values for the 10 missing cases to 15. Compute the mean and standard deviation for this data set.

(c) Summarize what you have learned about the possible effects of this type of imputation on the mean and the standard deviation.

1.69 A different type of mean. The **trimmed mean** is a measure of center that is more resistant than the mean but uses more of the available information than the median. To compute the 5% trimmed mean, discard the highest 5% and the lowest 5% of the observations, and compute the mean of the remaining 90%. Trimming eliminates the effect of a small number of outliers. Use the data on the values of the top 100 brands that we studied in Exercise 1.53 (page 37) to find the 5% trimmed mean. Compare this result with the value of the mean computed in the usual way. BRANDS

1.4 Density Curves and the Normal Distributions

When you complete this section, you will be able to:

- Compare the mean and the median for symmetric and skewed distributions.
- Sketch a Normal distribution for any given mean and standard deviation.
- Apply the 68–95–99.7 rule to find proportions of observations within one, two, and three standard deviations of the mean for any Normal distribution.
- Transform values of a variable from a general Normal distribution to the standard Normal distribution.
- Compute areas under a Normal curve using software or Table A.
- Perform inverse Normal calculations to find values of a Normal variable corresponding to various areas.
- Assess the extent to which the distribution of a set of data can be approximated by a Normal distribution.

quantitative variable, p. 3

We now have a kit of graphical and numerical tools for describing distributions. What is more, we have a clear strategy for exploring data on a single quantitative variable:

1. Plot your data: make a graph, usually a histogram or a stemplot.
2. Look for the overall pattern (shape, center, spread) and for striking deviations such as outliers.
3. Calculate numerical summaries to briefly describe center and spread.

Here is one more step to add to this strategy:

4. Sometimes the overall pattern of a large number of observations is so regular that we can describe it by a smooth curve that just depends on the mean and the standard deviation.

Density curves

mathematical model

A density curve is a **mathematical model** for the distribution of a quantitative variable. Mathematical models are idealized descriptions. They allow us to easily make many statements in an idealized world. The statements are useful when the idealized world is similar to the real world. The density curves that we study give a compact picture of the overall pattern of data. They ignore minor irregularities as well as outliers. For some situations, we are able to capture all of the essential characteristics of a distribution with a density curve. For other situations, our idealized model misses some important characteristics. As with so many things in statistics, your careful judgment is needed to decide what is important and how close is good enough.

EXAMPLE 1.30

ASM

Manufacturing Workers Figure 1.17 is a histogram of the numbers of manufacturing workers in the 50 states of the United States based on the Annual Survey of Manufactures (ASM) conducted by the U.S. Census Bureau.[26] The histogram shows that about half (27 out of 50) the states employ 200,000 or fewer workers. The distribution is skewed to the right and there is one outlier with a very large number of manufacturing workers. The density curve in Figure 1.17 fits the distribution described by the histogram fairly well. ∎

FIGURE 1.17 Histogram of manufacturing workers in the 50 states of the United States in 2016, Example 1.30.

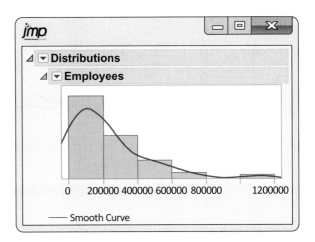

rate, p. 5

The outlier in the manufacturing workers distribution is California, with 1,119,896 workers in this classification. This represents about 10% of the manufacturing workers in the United States. Certainly, we would not want to ignore California in a discussion of U.S. manufacturing. Identification of California as an outlier here leads to other questions that can be answered with data. California has 14% of the U.S. population. Should we analyze these data using a rate? Should we divide the number of manufacturing workers by the population of the states to compute a rate? Or should we divide by the number of workers in the states?

Here are some details about density curves. We need these basic ideas to understand the rest of this chapter.

DENSITY CURVE

A **density curve** is a curve that

- is always on or above the horizontal axis
- has area exactly 1 underneath it.

A density curve describes the overall pattern of a distribution. The area under the curve and within any range of values is the proportion of all observations that fall in that range.

The median and mean of a density curve

Our measures of center and spread apply to density curves as well as to actual sets of observations. The median and quartiles are easy. Areas under a density curve represent proportions of the total number of observations. The median is the point with half the observations on either side. So **the median of a density curve is the equal-areas point**—the point with half the area under the curve to its left and the remaining half of the area to its right. The quartiles divide the area under the curve into quarters. One-fourth of the area under the curve is to the left of the first quartile, and three-fourths of the area is to the left of the third quartile. You can roughly locate the median and quartiles of any density curve by eye by dividing the area under the curve into four equal parts.

EXAMPLE 1.31

Symmetric Density Curves Because density curves are idealized patterns, a symmetric density curve is exactly symmetric. The median of a symmetric density curve is, therefore, at its center. Figure 1.18(a) shows the median of a symmetric curve. ∎

FIGURE 1.18(a) The median and mean for a symmetric density curve, Example 1.31.

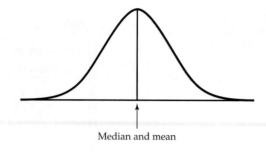

Median and mean

The situation is different for skewed density curves. Here is an example.

EXAMPLE 1.32

Skewed Density Curves It isn't so easy to spot the equal-areas point on a skewed curve. There are mathematical ways of finding the median for any density curve. We did that to mark the median on the skewed curve in Figure 1.18(b). ∎

FIGURE 1.18(b) The median and mean for a right-skewed density curve, Example 1.32.

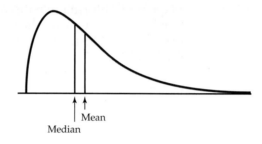

Mean
Median

APPLY YOUR KNOWLEDGE

1.70 Another skewed curve. Sketch a curve similar to Figure 1.18(b) for a left-skewed density curve.

What about the mean? The mean of a set of observations is its arithmetic average. If we think of the observations as weights strung out along a thin rod, the mean is the point at which the rod would balance. This fact is also true of density curves. **The mean is the point at which the curve would balance if the area under the curve were made of solid material.**

> **MEDIAN AND MEAN OF A DENSITY CURVE**
>
> The **median** of a density curve is the equal-areas point, the point that divides the area under the curve in half.
>
> The **mean** of a density curve is the balance point, at which the curve would balance if made of solid material.
>
> The median and mean are the same for a symmetric density curve. They both lie at the center of the curve. The mean of a skewed curve is pulled away from the median in the direction of the long tail.

When necessary, we can once again call on more advanced mathematics to find the standard deviation. The study of mathematical methods for doing calculations with density curves is part of theoretical statistics. Although we are concentrating on statistical practice in this book, we often make use of the results of mathematical study.

EXAMPLE 1.33

Mean and Median Figure 1.19 illustrates this fact about the mean: a symmetric curve balances at its center because the two sides are identical. **The mean and median of a symmetric density curve are equal,** as in Figure 1.18(a). Figure 1.18(b) shows how the mean of this skewed density curve is pulled toward the long tail more than is the median. There are mathematical ways of calculating the mean for any density curve, so we are able to mark the mean as well as the median in Figure 1.18(b). ■

FIGURE 1.19 The mean is the balance point of a density curve (Example 1.33).

Because a density curve is an idealized description of the distribution of data, we need to distinguish between the mean and standard deviation of the density curve and the mean $\bar{x}$ and standard deviation s computed from the actual observations. The usual notation for the mean of an idealized distribution is μ (the Greek letter mu). We write the standard deviation of a density curve as σ (the Greek letter sigma).

mean μ
standard deviation σ

The mean, the median, and the standard deviation are three measures that can be used to describe a density curve. However, there are a wide variety of density curves with very different shapes that are not very well described by these measures. Furthermore, there are density curves for which these quantities do not exist. Fortunately, many of the density curves that we see in practice are similar to the curves in Figures 1.18(a) and 1.18(b).

APPLY YOUR KNOWLEDGE

1.71 A symmetric curve. Sketch a density curve that is symmetric but has a shape different from that of the curve in Figure 1.18(a) (page 42).

1.72 A uniform distribution. Figure 1.20 displays the density curve of a **uniform distribution.** This curve takes the constant value 1 over the interval from 0 to 1 and is 0 outside that range of values. This means that the data described by this distribution have values that are uniformly spread between 0 and 1. Use areas under this density curve to answer the following questions.

uniform distribution

(a) What percent of the observations lie below 0.7?

(b) What percent of the observations lie above 0.4?

(c) What percent of the observations lie between 0.4 and 0.7?

(d) Why is the total area under this curve equal to 1?

(e) What is the mean μ of this distribution?

FIGURE 1.20 The density curve of a uniform distribution, Exercise 1.72.

1.73 Three curves. Figure 1.21 displays three density curves, each with three points marked. At which of these points on each curve do the mean and the median fall?

FIGURE 1.21 Three density curves, Exercise 1.73.

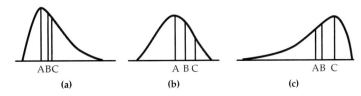

Normal distributions

Normal distributions

One particularly important class of density curves has already appeared in Figure 1.18(a). These density curves are symmetric, single-peaked, and bell-shaped. They are called *Normal curves,* and they describe **Normal distributions.** All Normal distributions have the same overall shape. The exact density curve for a particular Normal distribution is described by giving its mean μ and its standard deviation σ. The mean is located at the center of the symmetric curve and is the same as the median. Changing μ without changing σ moves the Normal curve along the horizontal axis without changing its spread. The standard deviation σ controls the spread of a Normal curve. Figure 1.22 shows two Normal curves with different values of σ. The curve with the larger standard deviation is more spread out.

The standard deviation σ is the natural measure of spread for Normal distributions. Not only do μ and σ completely determine the shape of a Normal curve, but we can also locate σ by eye on the curve. Here's how. Imagine that you are skiing down a mountain that has the shape of a Normal curve. At first, you descend at an ever-steeper angle as you go out from the peak:

Fortunately, before you find yourself going straight down, the slope begins to grow flatter rather than steeper as you go out and down:

The points at which this change of curvature takes place are located along the horizontal axis at distance σ on either side of the mean μ. Remember that μ and σ alone do not specify the shape of most distributions and that the shape of density curves in general does not reveal σ. These are special properties of Normal distributions.

Why are the Normal distributions important in statistics? Here are three reasons.

1. Normal distributions are good descriptions for some distributions of *real data*. Distributions that are often close to Normal include scores on tests taken by many people (such as GMAT exams), repeated careful measurements of the same quantity (such as measurements taken from a production process), and characteristics of biological populations (such as yields of corn).

2. Normal distributions are good approximations to the results of many kinds of *chance outcomes,* such as tossing a coin many times.

3. Many of the *statistical inference* procedures that we study in later chapters are based on Normal distributions.

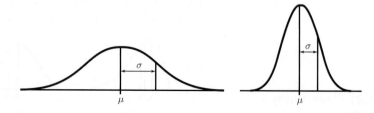

FIGURE 1.22 Two Normal curves, showing the mean μ and the standard deviation σ.

The 68–95–99.7 rule

Although many Normal curves are possible, they all have common properties. In particular, all Normal distributions obey the following rule.

THE 68–95–99.7 RULE

In the Normal distribution with mean μ and standard deviation σ:

- **68%** of the observations fall within σ of the mean μ.
- **95%** of the observations fall within 2σ of μ.
- **99.7%** of the observations fall within 3σ of μ.

Figure 1.23 illustrates the 68–95–99.7 rule. By remembering these three numbers, you can think about Normal distributions without constantly making detailed calculations.

FIGURE 1.23 The 68–95–99.7 rule for Normal distributions.

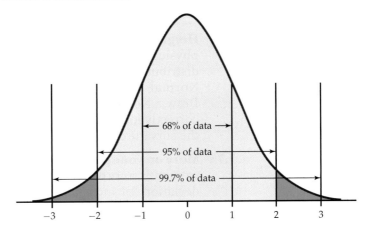

EXAMPLE 1.34

Using the 68–95–99.7 Rule The distribution of weights of 9-ounce bags of a particular brand of potato chips is approximately Normal with mean $\mu = 9.12$ ounces and standard deviation $\sigma = 0.15$ ounce. Figure 1.24 shows what the 68–95–99.7 rule says about this distribution.

FIGURE 1.24 The 68–95–99.7 rule applied to the distribution of weights of bags of potato chips, Example 1.34.

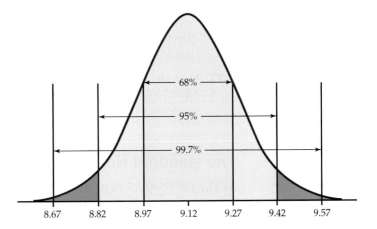

Two standard deviations is 0.3 ounce for this distribution. The 95 part of the 68–95–99.7 rule says that the middle 95% of 9-ounce bags weigh between $9.12 - 0.3 = 8.82$ ounces and $9.12 + 0.3 = 9.42$ ounces, that is, between

8.82 ounces and 9.42 ounces. This fact is exactly true for an exactly Normal distribution. It is approximately true for the weights of 9-ounce bags of chips because the distribution of these weights is approximately Normal.

The other 5% of bags have weights outside the range from 8.82 to 9.42 ounces. Because the Normal distributions are symmetric, half of these bags are on the heavy side. Thus, the heaviest 2.5% of 9-ounce bags are heavier than 9.42 ounces.

The 99.7 part of the 68–95–99.7 rule says that almost all bags (99.7% of them) have weights between $\mu - 3\sigma$ and $\mu + 3\sigma$. This range of weights is 8.67 to 9.57 ounces. ■

Because we will mention Normal distributions often, a short notation is helpful. We abbreviate the Normal distribution with mean μ and standard deviation σ as $N(\mu,\sigma)$. For example, the distribution of weights in the previous example is $N(9.12, 0.15)$.

APPLY YOUR KNOWLEDGE

1.74 **Heights of young women.** Product designers often must consider physical characteristics of their target population. For example, the distribution of heights of women aged 20 to 29 years is approximately Normal with mean 64 inches and standard deviation 2.7 inches. Draw a Normal curve on which this mean and standard deviation are correctly located. (*Hint:* Draw the curve first, locate the points where the curvature changes, then mark the horizontal axis.)

1.75 **More on young women's heights.** The distribution of heights of young women is approximately Normal with mean 64 inches and standard deviation 2.7 inches. Use the 68–95–99.7 rule to answer the following questions.

(a) What percent of these women are taller than 66.7 inches?

(b) Between what heights do the middle 95% of young women fall?

(c) What percent of young women are shorter than 58.6 inches?

1.76 **Test scores.** Many states have programs for assessing the skills of students in various grades. The National Assessment of Educational Progress (NAEP) is one such program.[27] In a recent year the mean mathematics score for forth-graders was 242, and the standard deviation was 29. Assuming that these scores are approximately Normally distributed, $N(242, 29)$, use the 68–95–99.7 rule to give a range of scores that includes 95% of these students.

1.77 **Use the 68–95–99.7 rule.** Refer to the previous exercise. Use the 68–95–99.7 rule to give a range of scores that includes 99.7% of these students.

The standard Normal distribution

As the 68–95–99.7 rule suggests, all Normal distributions share many common properties. In fact, all Normal distributions are the same if we measure in units of size σ about the mean μ as center. Changing to these units is called *standardizing*. To standardize a value, subtract the mean of the distribution and then divide by the standard deviation.

STANDARDIZING AND z-SCORES

If x is an observation from a distribution that has mean μ and standard deviation σ, the **standardized value** of x is

$$z = \frac{x - \mu}{\sigma}$$

A standardized value is often called a **z-score.**

A z-score tells us how many standard deviations the original observation falls away from the mean, and in which direction. Observations larger than the mean are positive when standardized, and observations smaller than the mean are negative when standardized.

EXAMPLE 1.35

Standardizing Potato Chip Bag Weights The weights of 9-ounce potato chip bags are approximately Normal with $\mu = 9.12$ ounces and $\sigma = 0.15$ ounce. The standardized weight is

$$z = \frac{\text{weight} - 9.12}{0.15}$$

A bag's standardized weight is the number of standard deviations by which its weight differs from the mean weight of all bags. A bag weighing 9.3 ounces, for example, has *standardized* weight

$$z = \frac{9.3 - 9.12}{0.15} = 1.2$$

or 1.2 standard deviations above the mean. Similarly, a bag weighing 8.7 ounces has standardized weight

$$z = \frac{8.7 - 9.12}{0.15} = -2.8$$

or 2.8 standard deviations below the mean bag weight. ∎

If the variable we standardize has a Normal distribution, standardizing does more than give a common scale. That is, it transforms all Normal distributions into a single distribution, and this distribution is still Normal. Standardizing a variable that has any Normal distribution produces a new variable that has the *standard Normal distribution*.

STANDARD NORMAL DISTRIBUTION

The **standard Normal distribution** is the Normal distribution $N(0,1)$ with mean 0 and standard deviation 1.

If a variable x has any Normal distribution $N(\mu,\sigma)$ with mean μ and standard deviation σ, then the standardized variable

$$z = \frac{x - \mu}{\sigma}$$

has the standard Normal distribution.

APPLY YOUR KNOWLEDGE

1.78 SAT versus ACT. Emily scores 700 on the Mathematics part of the SAT. The distribution of SAT scores in a reference population is Normal, with mean 500 and standard deviation 100. Michael takes the American College Testing (ACT) Mathematics test and scores 29. ACT scores are Normally distributed with mean 18 and standard deviation 6. Find the standardized scores for both students. Assuming that both tests measure the same kind of ability, who has the higher score?

Normal distribution calculations

Areas under a Normal curve represent proportions of observations from that Normal distribution. There is no easy formula by which to find areas under a Normal curve. To find areas of interest, either software that calculates areas or a table of areas can be used. The table and most software calculate one kind of area: **cumulative proportions.** A cumulative proportion is the proportion of observations in a distribution that lie at or below a given value. When the distribution is given by a density curve, the cumulative proportion is the area under the curve to the left of a given value. Figure 1.25 shows this idea more clearly than words do.

cumulative proportion

The key to calculating Normal proportions is to match the area you want with areas that represent cumulative proportions and then get areas for the cumulative proportions. The following examples illustrate the methods.

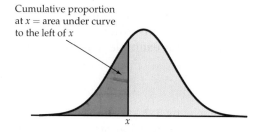

FIGURE 1.25 The cumulative proportion for a value x is the proportion of all observations from the distribution that are less than or equal to x. This is the area to the left of x under the Normal curve.

EXAMPLE 1.36

The NCAA Standard for SAT Scores The National Collegiate Athletic Association (NCAA) requires Division I athletes to get a combined score of at least 820 on the SAT Mathematics and Verbal tests to compete in their first college year. (Higher scores are required for students with poor high school grades.) The scores of the 1.4 million students who took the SATs were approximately Normal with mean 1026 and standard deviation 209. What proportion of all students had SAT scores of at least 820?

Here is the calculation in pictures: the proportion of scores above 820 is the area under the curve to the right of 820. That's the total area under the curve (which is always 1) minus the cumulative proportion up to 820. Note that we have used software for these calculations that is more accurate. ∎

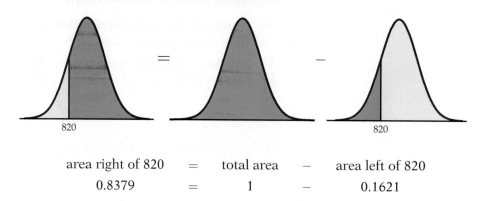

area right of 820   =   total area   −   area left of 820
0.8379   =   1   −   0.1621

Thus, the proportion of all SAT test-takers who would be NCAA qualifiers is 0.8379, or about 84%. ∎

There is *no* area under a smooth curve and exactly over the point 820. Consequently, the area to the right of 820 (the proportion of scores > 820) is the same as the area at or to the right of this point (the proportion of scores ≥ 820). The actual data may contain a student who scored exactly 820 on the SAT. That the proportion of scores exactly equal to 820 is 0 for a Normal distribution is a consequence of the idealized smoothing of Normal distributions for data.

EXAMPLE 1.37

NCAA Partial Qualifiers The NCAA considers a student to be a "partial qualifier"—eligible to practice and receive an athletic scholarship, but not to compete—if the combined SAT score is at least 720. What proportion of all students who take the SAT would be partial qualifiers? That is, what proportion have scores between 720 and 820? Here are the pictures:

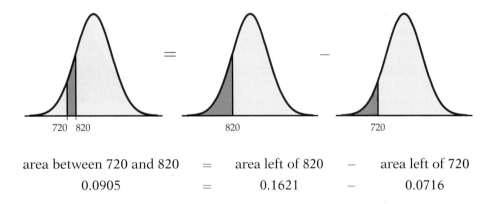

area between 720 and 820   =   area left of 820   −   area left of 720
0.0905   =   0.1621   −   0.0716

About 9% of all students who take the SAT have scores between 720 and 820. ∎

It is easy to use software to find the numerical values of the areas in Examples 1.36 and 1.37. In Excel, you can use the norm.dist function. To find the area under a Normal curve with mean μ and standard deviation σ to the left of x, use =NORM.DIST(x, μ, σ, TRUE). So, to find the area for Example 1.37, we would use

=NORM.DIST(820,1026,209,TRUE)-NORM.DIST(720,1026,209,TRUE)

The value that appears in the cell with this entry is 0.090572157. Your software may refer to these areas as "cumulative probabilities." We will learn in Chapter 4 why the language of probability fits. If you make a sketch of the area you want, you will rarely go wrong.

You can use the *Normal Curve* applet on the text website to find Normal proportions. This applet is more flexible than most software—it will find any Normal proportion, not just cumulative proportions. The applet is an excellent way to understand Normal curves.

If you are not using software, you can find cumulative proportions for Normal curves from a table. That requires an extra step, as we now explain.

Using the standard Normal table

The extra step in finding cumulative proportions from a table is that we must first standardize to express the problem in the standard scale of z-scores. This allows us to get by with just one table, a table of *standard Normal cumulative proportions*. Table A in the back of the book gives cumulative proportions for the standard Normal distribution. Table A also appears on the inside front cover. The pictures at the top of the table remind us that the entries are cumulative proportions, areas under the curve to the left of a value z.

EXAMPLE 1.38

Find the Proportion from z What proportion of observations on a standard Normal variable Z take values less than $z = 1.47$?

To find the area to the left of 1.47, locate 1.4 in the left-hand column of Table A, then locate the remaining digit 7 as 0.07 in the top row. The entry opposite 1.4 and under 0.07 is 0.9292. This is the cumulative proportion we seek. Figure 1.26 illustrates this area. ∎

FIGURE 1.26 The area under the standard Normal curve to the left of the point $z = 1.47$ is 0.9292 (Example 1.38).

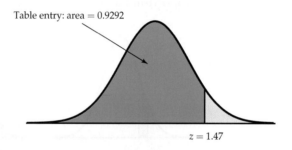

Now that you see how Table A works, let's redo the NCAA Examples 1.36 and 1.37 using the table.

EXAMPLE 1.39

Find the Proportion from x What proportion of all students who take the SAT have scores of at least 820? The picture that leads to the answer is exactly the same as in Example 1.36 (page 46). The extra step is that we first standardize so as to read cumulative proportions from Table A. If X is SAT score, we want the proportion of students for whom $X \geq 820$.

Step 1. *Standardize.* Subtract the mean, then divide by the standard deviation, to transform the problem about X into a problem about a standard Normal Z:

$$X \geq 820$$
$$\frac{X - 1026}{209} \geq \frac{820 - 1026}{209}$$
$$Z \geq -0.99$$

Step 2. *Use the table.* Look at the pictures in Example 1.36. From Table A, we see that the proportion of observations less than -0.99 is 0.1611. The area to the right of -0.99 is, therefore, $1 - 0.1611 = 0.8389$. This is about 84%. ∎

The area from the table in Example 1.37 (0.8389) is slightly less accurate than the area from software in Example 1.36 (0.8379) because we must round z to two places when we use Table A. The difference is rarely important in practice.

EXAMPLE 1.40

Proportion of Partial Qualifiers What proportion of all students who take the SAT would be partial qualifiers in the eyes of the NCAA? That is, what proportion of students have SAT scores between 720 and 820? First, sketch the areas, exactly as in Example 1.37. We again use X as shorthand for an SAT score.

Step 1. *Standardize.*

$$720 \leq X < 820$$
$$\frac{720 - 1026}{209} \leq \frac{X - 1026}{209} < \frac{820 - 1026}{209}$$
$$-1.46 \leq Z < -0.99$$

Step 2. *Use the table.*

area between -1.46 and -0.99 = (area left of -0.99) − (area left of -1.46)
$$= 0.1611 - 0.0721 = 0.0890$$

As in Example 1.37, about 9% of students would be partial qualifiers. ∎

Sometimes we encounter a value of z more extreme than those appearing in Table A. For example, the area to the left of $z = -4$ is not given directly in the table. The z-values in Table A leave only area 0.0002 in each tail unaccounted for. For practical purposes, we can act as if there is zero area outside the range of Table A.

APPLY YOUR KNOWLEDGE

1.79 Find the proportion. Use the fact that the NAEP math scores from Exercise 1.76 (page 46) are approximately Normal, $N(242,29)$. Find the proportion of students who have scores less than 200. Find the proportion of students who have scores greater than or equal to 200. Sketch the relationship between these two calculations using pictures of Normal curves similar to the ones given in Example 1.36 (page 48).

1.80 Find another proportion. Use the fact that the NAEP math scores are approximately Normal, $N(242,29)$. Find the proportion of students who have scores between 200 and 300. Use pictures of Normal curves similar to the ones given in Example 1.37 (page 49) to illustrate your calculations.

Inverse Normal calculations

Examples 1.36 through 1.39 illustrate the use of Normal distributions to find the proportion of observations in a given event, such as "SAT score between 720 and 820." We may, instead, want to find the observed value corresponding to a given proportion.

Statistical software will do this directly. Without software, use Table A backward, finding the desired proportion in the body of the table and then reading the corresponding z from the left column and top row.

EXAMPLE 1.41

How High for the Top 10%? Scores on the SAT Verbal test in recent years follow approximately the $N(505,110)$ distribution. How high must a student score be to place in the top 10% of all students taking the SAT?

Again, the key to the problem is to draw a picture. Figure 1.27 shows that we want the score x with area above it to be 0.10. That's the same as area below x equal to 0.90.

FIGURE 1.27 Locating the point on a Normal curve with area 0.10 to its right, Example 1.41. The result is $x = 646$, or $z = 1.28$ in the standard scale.

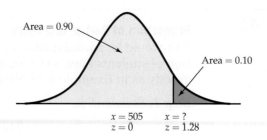

Statistical software has a function that will give you the x for any cumulative proportion you specify. The function often has a name such as "inverse cumulative probability." Plug in mean 505, standard deviation 110, and cumulative proportion 0.9. The software tells you that $x = 645.97$. We see that a student must score at least 646 to place in the highest 10%. ■

Without software, first find the standard score z with cumulative proportion 0.9, then "unstandardize" to find x. Here is the two-step process:

1. *Use the table.* Look in the body of Table A for the entry closest to 0.9. It is 0.8997. This is the entry corresponding to $z = 1.28$. So $z = 1.28$ is the standardized value with area 0.9 to its left.

2. *Unstandardize* to transform the solution from z back to the original x scale. We know that the standardized value of the unknown x is $z = 1.28$. So x itself satisfies

$$\frac{x - 505}{110} = 1.28$$

Solving this equation for x gives

$$x = 505 + (1.28)(110) = 645.8$$

This equation should make sense: it finds the x that lies 1.28 standard deviations above the mean on this particular Normal curve. That is the "unstandardized" meaning of $z = 1.28$. The general rule for unstandardizing a z-score is

$$x = \mu + z\sigma$$

Software makes the calculation easy. In Excel, use the norm.inv function. To find the value of x that corresponds to an area A to the left of x for a Normal curve with mean μ and standard deviation σ, use =NORM.INV(A, μ, σ.) So, to find the value of x for Example 1.41, we would use

=NORM.INV(0.9,505,110)

The value that appears in the cell with this entry is 645.9706722.

APPLY YOUR KNOWLEDGE

1.81 What score is needed to be in the top 25%? Consider the NAEP math scores, which are approximately Normal, $N(242, 29)$. How high a score is needed to be in the top 25% of students who take this exam?

1.82 Find the score that 70% of students will exceed. Consider the NAEP math scores, which are approximately Normal, $N(242, 29)$. Seventy percent of the students will score above x on this exam. Find x.

Assessing the Normality of data

The Normal distributions provide good models for some distributions of real data. Examples include the miles per gallon ratings of vehicles, average payrolls of Major League Baseball teams, and statewide unemployment rates.

The distributions of some other common variables are usually skewed and, therefore, distinctly non-Normal. Examples include personal income, gross sales of business firms, and the service lifetime of mechanical or electronic components. While experience can suggest whether a Normal model is plausible in a particular case, it is risky to assume that a distribution is Normal without actually inspecting the data.

The decision to describe a distribution by a Normal model may determine the later steps in our analysis of the data. Calculations of proportions, as we have done earlier, and statistical inference based on such calculations follow from the choice of a model. How can we judge whether data are approximately Normal?

A histogram or stemplot can reveal distinctly non-Normal features of a distribution, such as outliers, pronounced skewness, or gaps and clusters. If the stemplot or histogram appears roughly symmetric and single-peaked, however, we need a more sensitive way to judge the adequacy of a Normal model. The most useful tool for assessing Normality is another graph, the **Normal quantile plot**.[28]

Here is the idea of a simple version of a Normal quantile plot. It is not feasible to make Normal quantile plots by hand, but software makes them for us, using more sophisticated versions of this basic idea.

1. Arrange the observed data values from smallest to largest. Record what percentile of the data each value occupies. For example, the smallest observation in a set of 20 is at the 5% point, the second smallest is at the 10% point, and so on.

2. Find the same percentiles for the Normal distribution using Table A or statistical software. Percentiles of the standard Normal distribution are often called **Normal scores**. For example, $z = -1.645$ is the 5% point of the standard Normal distribution, and $z = -1.282$ is the 10% point.

3. Plot each data point x against the corresponding Normal score z. If the data distribution is close to standard Normal, the plotted points will lie close to the 45-degree line $x = z$. If the data distribution is close to any Normal distribution, the plotted points will lie close to some straight line.

Any Normal distribution produces a straight line on the plot because standardizing turns any Normal distribution into a standard Normal distribution. Standardizing is a transformation that can change the slope and intercept of the line in our plot but cannot turn a line into a curved pattern.

USE OF NORMAL QUANTILE PLOTS

If the points on a Normal quantile plot lie close to a straight line, the plot indicates that the data are Normal. Systematic deviations from a straight line indicate a non-Normal distribution. Outliers appear as points that are far away from the overall pattern of the plot.

Figures 1.28 through 1.30 (pages 54 through 55) are Normal quantile plots for data we have met earlier. The data x are plotted vertically against the corresponding Normal scores z plotted horizontally. For small data sets, the x axis extends from -3 to 3 because almost all of a standard Normal curve lies between these values.

EXAMPLE 1.42

IQ Scores Are Normal In Example 1.19 (page 19) we examined the distribution of IQ scores for a sample of 60 fifth-grade students. Figure 1.28 gives a Normal quantile plot for these data. Notice that the points have a pattern that is fairly close to a straight line. This pattern indicates that the distribution is approximately Normal. When we constructed a histogram of the data in Figure 1.11 (page 20), we noted that the distribution has a single peak, is approximately symmetric, and has tails that decrease in a smooth way. We can now add to that description by stating that the distribution is approximately Normal. ∎

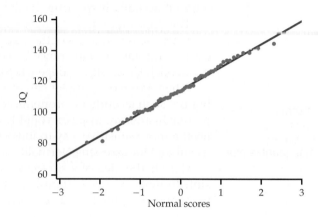

FIGURE 1.28 Normal quantile plot for the IQ data, Example 1.42. This pattern indicates that the data are approximately Normal.

Figure 1.28 does, of course, show some deviation from a straight line. Real data almost always show some departure from the theoretical Normal model. It is important to confine your examination of a Normal quantile plot to searching for shapes that show *clear departures from Normality*. Don't overreact to minor wiggles in the plot. When we discuss statistical methods that are based on the Normal model, we will pay attention to the sensitivity of each method to departures from Normality. Many common methods work well as long as the data are reasonably symmetric and outliers are not present.

EXAMPLE 1.43

TBILL

CASE 1.1 **T-Bill Interest Rates Are Not Normal** We made a histogram for the distribution of interest rates for T-bills in Example 1.12 (page 13). A Normal quantile plot for these data is shown in Figure 1.29. This plot shows some interesting features of the distribution. First, in the central part, from about $z = -1.5$ to $z = 2$, the points fall approximately on a straight line. This suggests that the distribution is approximately Normal in this range. In both the lower and the upper extremes, the points flatten out. This occurs at an interest rate of nearly zero for the lower tail and at 15% for the upper tail. ∎

FIGURE 1.29 Normal quantile plot for the T-bill interest rates, Example 1.43. These data are not approximately Normal.

1.4 Density Curves and the Normal Distributions

The idea that distributions are approximately Normal within a range of values is an old tradition. The remark "All distributions are approximately Normal in the middle" has been attributed to the statistician Charlie Winsor.[29]

APPLY YOUR KNOWLEDGE

CASE 1.2 **1.83 Length of time to start a business.** In Exercise 1.33, we noted that the sample of times to start a business from 25 countries contained an outlier. For Venezuela, the reported time is 230 days. This case is the most extreme in the entire data set. Figure 1.30 shows the Normal quantile plot for all 187 countries, including Venezuela. TTS

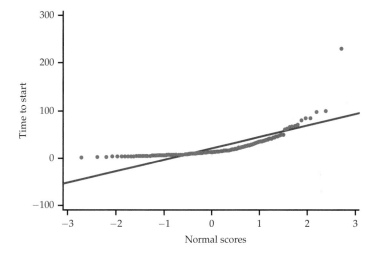

FIGURE 1.30 Normal quantile plot for the length of time required to start a business, Exercise 1.83.

(a) These data are skewed to the right. How does this feature appear in the Normal quantile plot?

(b) Find the point for Venezuela on the plot. Do you think that Venezuela is truly an outlier, or is it part of a very long tail in this distribution? Explain your answer.

1.84 Compare the upper tails. Refer to the previous exercise. Compare the shape of the upper portion of this Normal quantile plot with the upper portion of the plot for the T-bill interest rates in Figure 1.29, and with the upper portion of the plot for the IQ scores in Figure 1.28. Make a general statement about what the shape of the upper portion of a Normal quantile plot tells you about the upper tail of a distribution.

There are several variations on the way that diagnostic plots are used to assess Normality. We have chosen to plot the data on the y axis and the Normal scores on the x axis, but some software packages switch the axes. These plots are sometimes called "Q-Q Plots." Other plots transform the data and the Normal scores into cumulative probabilities and are called "P-P Plots." The basic idea behind all these plots is the same. Plots with points that lie close to a straight line indicate that the data are approximately Normal. *When using these diagnostic plots, you should always look at a histogram or other graphical summary of the distribution to help you interpret the plot.*

Normal distributions and the associated proportions that we can compute can be very useful tools for describing the quantitative variables in a data set. When they are used to describe a variable in some similar or future setting, however, we need to be sure that the mean and standard deviation are reasonably close to what we expect in the new setting. If so, they can be powerful predictive analytics tools.

BEYOND THE BASICS

Density estimation

A density curve gives a compact summary of the overall shape of a distribution. Many distributions do not have the Normal shape. Some other families of density curves are used as mathematical models for these kinds of distribution shapes.

density estimation

Modern software offers a more flexible option: **density estimation.** A density estimator does not start with any specific shape, such as the Normal shape. Instead, it looks at the data and draws a density curve that describes the overall shape of the data.

EXAMPLE 1.44

IQ

IQ Scores In Example 1.42 we observed that the points in the Normal quantile plot for the IQ data were very close to a straight line. This suggests that a Normal distribution is a good fit for these data. Figure 1.31 provides another way to look at this issue. Here we see the histogram with a density estimate, the red curve, along with the best-fitting Normal density curve, the green curve. Because the two curves are approximately the same, we are confident in any further analysis of these data based on the assumption that the data are approximately Normal. ■

FIGURE 1.31 Histogram of IQ scores, with a density estimate and a Normal curve, for Example 1.44. The IQ scores are approximately Normal.

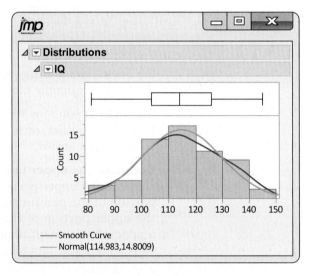

Here is another example where the we see a different picture.

EXAMPLE 1.45

TTS

Times to Start a Business In Exercise 1.83, we examined the Normal quantile plot for the time to start a business data. Figure 1.32 shows the histogram for these data along with a density estimate and the best-fitting Normal distribution. The two density curves are very different, and we conclude that a Normal distribution does not give a good fit for these data. Not only are the data strongly skewed, but there is also a clear outlier. We should be very cautious about using a statistical analysis based on an assumption that the data are approximately Normal in this case. ■

FIGURE 1.32 Histogram of the length of time required to start a business, with a density estimate and a Normal curve, for Example 1.45. The Normal distribution is not a good fit for these data.

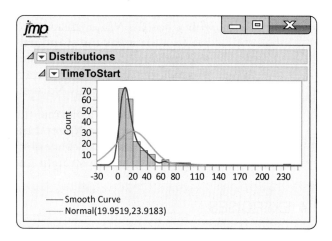

SECTION 1.4 SUMMARY

- We can sometimes describe the overall pattern of a distribution by a **density curve.** A density curve has total area 1 underneath it. An area under a density curve gives the proportion of observations that fall in a range of values.

- A density curve is an idealized description of the overall pattern of a distribution that smooths out the irregularities in the actual data. We write the mean of a density curve as μ and the standard deviation of a density curve as σ to distinguish them from the mean $\bar{x}$ and standard deviation s of the actual data.

- The mean, the median, and the quartiles of a density curve can be located by eye. The **mean μ** is the balance point of the curve. The **median** divides the area under the curve in half. The **quartiles** and the median divide the area under the curve into quarters. The **standard deviation σ** cannot be located by eye on most density curves.

- The mean and median are equal for symmetric density curves. The mean of a skewed curve is located farther toward the long tail than is the median.

- The **Normal distributions** are described by a special family of bell-shaped, symmetric density curves, called **Normal curves.** The mean μ and standard deviation σ completely specify a Normal distribution $N(\mu, \sigma)$. The mean is the center of the curve, and σ is the distance from μ to the change-of-curvature points on either side.

- To **standardize** any observation x, subtract the mean of the distribution and then divide by the standard deviation. The resulting *z*-score

$$z = \frac{x - \mu}{\sigma}$$

says how many standard deviations x lies from the distribution mean.

- All Normal distributions are the same when measurements are transformed to the standardized scale. In particular, all Normal distributions satisfy the **68–95–99.7 rule,** which describes what percent of observations lie within one, two, and three standard deviations of the mean.

- If x has the $N(\mu,\sigma)$ distribution, then the **standardized variable** $z = (x - \mu)/\sigma$ has the **standard Normal distribution $N(0,1)$** with mean 0 and standard deviation 1. Table A gives the proportions of standard Normal observations that are less than z for many values of z. By standardizing, we can use Table A for any Normal distribution.

- The adequacy of a Normal model for describing a distribution of data is best assessed by a **Normal quantile plot,** which is available in most statistical software packages. A pattern on such a plot that deviates substantially from a straight line indicates that the data are not Normal.

SECTION 1.4 EXERCISES

For Exercise 1.70, see page 42; for 1.71 to 1.73, see page 43; for 1.74 to 1.77, see page 46; for 1.78, see page 48; for 1.79 and 1.80, see page 51; for 1.81 and 1.82, see page 52; and for 1.83 and 1.84, see page 55.

1.85 Find the error. Each of the following statements contains an error. Describe the error and then correct the statement.

(a) The area under the curve for a density curve is equal to 2 for some distributions.

(b) A density curve is a mathematical model for the relationship between two variables.

(c) The density curve for a distribution that is skewed to the right cannot have more than one peak.

1.86 Find the error. Each of the following statements contains an error. Describe the error and then correct the statement.

(a) A normal distribution will usually have outliers.

(b) For a symmetric distribution, the mean will be smaller than the median.

(c) The 68–95–99.7 rule applies to skewed distributions.

1.87 Sketch some Normal curves.

(a) Sketch a Normal curve that has mean 20 and standard deviation 5.

(b) On the same x axis, sketch a Normal curve that has mean 25 and standard deviation 5.

(c) How does the Normal curve change when the mean is varied but the standard deviation stays the same?

1.88 The effect of changing the standard deviation.

(a) Sketch a Normal curve that has mean 30 and standard deviation 10.

(b) On the same x axis, sketch a Normal curve that has mean 30 and standard deviation 5.

(c) How does the Normal curve change when the standard deviation is varied but the mean stays the same?

1.89 Know your density. Sketch density curves that might describe distributions with the following shapes.

(a) Symmetric, but with two peaks (that is, two strong clusters of observations).

(b) Single peak and skewed to the left.

1.90 Gross domestic product. Refer to Exercise 1.46 (page 37), where we examined the gross domestic product of 197 countries. GDP

(a) Compute the mean and the standard deviation.

(b) Apply the 68–95–99.7 rule to this distribution.

(c) Compare the results of the rule with the actual percents within one, two, and three standard deviations of the mean.

(d) Summarize your conclusions.

1.91 Do women talk more? Conventional wisdom suggests that women are more talkative than men. One study designed to examine this stereotype collected data on the speech of 42 women and 37 men in the United States.[30]

(a) The mean number of words spoken per day by the women was 14,297 with a standard deviation of 9065. Use the 68–95–99.7 rule to describe this distribution.

(b) Do you think that applying the rule in this situation is reasonable? Explain your answer.

(c) The men averaged 14,060 words per day with a standard deviation of 9056. Answer the questions in parts (a) and (b) for the men.

(d) Do you think that the data support the conventional wisdom? Explain your answer. Note that in Section 7.2, we will learn formal statistical methods to answer this type of question.

1.92 Data from Mexico. Refer to the previous exercise. A similar study in Mexico was conducted with 31 women and 20 men. The women averaged 14,704 words per day with a standard deviation of 6215. For men, the mean was 15,022 and the standard deviation was 7864.

(a) Answer the questions from the previous exercise for the Mexican study.

(b) The means for both men and women are higher for the Mexican study than for the U.S. study. What conclusions can you draw from this observation?

1.93 Total scores for accounting course. Following are the total scores of 10 students in an accounting course:

75 87 94 85 74 98 93 52 80 91

Previous experience with this course suggests that these scores should come from a distribution that is approximately Normal with mean 80 and standard deviation 10. ACCT

(a) Using these values for μ and σ, standardize the scores of these 10 students.

(b) If the grading policy is to give a grade of A to the top 15% of scores based on the Normal distribution with mean 80 and standard deviation 10, what is the cutoff for an A in terms of a standardized score?

(c) Which students earned an A for this course?

1.94 Assign more grades. Refer to the previous exercise. The grading policy says that the cutoffs for the other grades correspond to the following: the bottom 5% receive an F, the next 15% receive a D, the next 35% receive a C, and the next 30% receive a B. These cutoffs are based on the $N(80, 10)$ distribution.

(a) Give the cutoffs for the grades in terms of standardized scores.

(b) Give the cutoffs in terms of actual scores.

(c) Do you think that this method of assigning grades is a good one? Give reasons for your answer.

1.95 Visualizing the standard deviation. Figure 1.33 shows two Normal curves, both with mean 0. Approximately what is the standard deviation of each of these curves?

1.96 Exploring Normal quantile plots.

(a) Create three data sets: one that is clearly skewed to the right, one that is clearly skewed to the left, and one that is clearly symmetric and mound-shaped. (As an alternative to creating data sets, you can look through this chapter and find an example of each type of data set requested.)

(b) Using statistical software, obtain Normal quantile plots or density estimates for each of your three data sets.

(c) Clearly describe the pattern of each data set based on your results in part (b).

1.97 Length of pregnancies. The length of human pregnancies from conception to birth varies according to a distribution that is approximately Normal with mean 266 days and standard deviation 16 days. Use the 68–95–99.7 rule to answer the following questions.

(a) Between what values do the lengths of the middle 95% of all pregnancies fall?

(b) How short are the shortest 2.5% of all pregnancies?

1.98 Uniform random numbers. Use software to generate 150 observations from the distribution described in Exercise 1.72 (page 43). (The software will probably call this a "uniform distribution.") Make a histogram of these observations. How does the histogram compare with the density curve in Figure 1.20? Make a Normal quantile plot of your data. According to this plot, how does the uniform distribution deviate from Normality?

1.99 Use Table A or software. Use Table A or software to find the proportion of observations from a standard Normal distribution that falls in each of the following regions. In each case, sketch a standard Normal curve and shade the area representing the region.

(a) $z \leq -2.05$

(b) $z \geq -2.05$

(c) $z > 1.70$

(d) $-2.05 < z < 1.70$

1.100 Use Table A or software. Use Table A or software to find the value of z for each of the following situations. In each case, sketch a standard Normal curve and shade the area representing the region.

(a) Fifteen percent of the values of a standard Normal distribution are greater than z.

(b) Fifteen percent of the values of a standard Normal distribution are greater than or equal to z.

(c) Fifteen percent of the values of a standard Normal distribution are less than z.

(d) Fifty percent of the values of a standard Normal distribution are less than z.

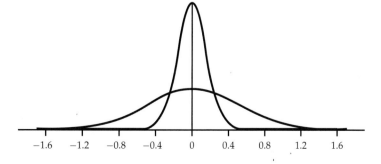

FIGURE 1.33 Two Normal curves with the same mean but different standard deviations (Exercise 1.95).

1.101 Use Table A or software. Consider a Normal distribution with mean 250 and standard deviation 25.

(a) Find the proportion of the distribution with values between 175 and 200. Illustrate your calculation with a sketch.

(b) Find the value of x such that the proportion of the distribution with values between $250 - x$ and $250 + x$ is 0.90. Illustrate your calculation with a sketch.

1.102 Length of pregnancies. The length of human pregnancies from conception to birth varies according to a distribution that is approximately Normal with mean 266 days and standard deviation 16 days.

(a) What percent of pregnancies last fewer than 255 days (that is about 8.5 months)?

(b) What percent of pregnancies last between 240 and 270 days (roughly between 8 and 9 months)?

(c) How long do the longest 25% of pregnancies last?

1.103 Quartiles of Normal distributions. The median of any Normal distribution is the same as its mean. We can use Normal calculations to find the quartiles for Normal distributions.

(a) What is the area under the standard Normal curve to the left of the first quartile? Use this to find the value of the first quartile for a standard Normal distribution. Find the third quartile similarly.

(b) Your work in part (a) gives the Normal scores z for the quartiles of any Normal distribution. What are the quartiles for the lengths of human pregnancies? (Use the distribution given in the previous exercise.)

1.104 Deciles of Normal distributions. The **deciles** of any distribution are the 10th, 20th, ..., 90th percentiles. The first and last deciles are the 10th and 90th percentiles, respectively.

(a) What are the first and last deciles of the standard Normal distribution?

(b) The weights of 9-ounce potato chip bags are approximately Normal with a mean of 9.12 ounces and a standard deviation of 0.15 ounce. What are the first and last deciles of this distribution?

1.105 Normal random numbers. Use software to generate 150 observations from the standard Normal distribution. Make a histogram of these observations. How does the shape of the histogram compare with a Normal density curve? Make a Normal quantile plot of the data. Does the plot suggest any important deviations from Normality? (Repeating this exercise several times is a good way to become familiar with how Normal quantile plots look when data are actually close to Normal.)

1.106 Trade balance. Refer to Exercises 1.48 and 1.49 (page 37) where you examined the distribution of trade balance for 151 countries in the best countries for business data set. Generate a histogram and a normal quantile plot for these data. Describe the shape of the distribution and indicate whether the normal quantile plot suggests that this distribution is Normal. BESTBUS

1.107 Gross domestic product per capita. Refer to the previous exercise. The data set also contains the GDP per capita, calculated by dividing the GDP by the size of the population for each country. BESTBUS

(a) Generate a histogram and a normal quantile plot for these data.

(b) Describe the shape of the distribution and indicate whether the normal quantile plot suggests that this distribution is Normal.

(c) Explain why GDP per capita might be a better variable to use than GDP for assessing how favorable a country is for business.

CHAPTER 1 REVIEW EXERCISES

1.108 Jobs for business majors. Which types of jobs are available for students who graduate with a business degree? The website **careerbuilder.com** lists job opportunities classified in a variety of ways. A recent posting had 25,120 jobs. The following table gives types of jobs and the numbers of postings listed under the classification "business administration" on a recent day.[31] BUSJOBS

Type	number	Type	number
Management	10916	Finance	2339
Sales	5981	Health care	2231
Information technology	4605	Accounting	2175
Customer service	4116	Human resources	1685
Marketing	3821		

Describe these data using the methods you learned in this chapter, and write a short summary describing jobs that are available for those graduates who have a business degree. Include comments on the limitations that should be kept in mind when interpreting this particular set of data.

1.109 Flopping in the 2014 World Cup. Soccer players are often accused of spending an excessive amount of time dramatically falling to the ground, followed by other activities suggesting that a possible injury is very serious. It has been suggested that these tactics are often designed to influence the call of a referee or to take extra time off the clock. Recordings of the first 32 games of the 2014 World Cup were analyzed, and there were 302 times when the referee interrupted the match because of a possible injury. The number of injuries and the total time, in minutes, spent flopping for each of the 32 teams that participated in these matches was recorded.[32] Here are the data: FLOPS

Country	Injuries	Time	Country	Injuries	Time
Brazil	17	3.30	Uruguay	9	4.12
Chile	16	6.97	Greece	9	2.65
Honduras	15	7.67	Cameroon	8	3.15
Nigeria	15	6.42	Germany	8	1.97
Mexico	15	3.97	Spain	8	1.82
Costa Rica	13	3.80	Belgium	7	3.38
USA	12	6.40	Japan	7	2.08
Ecuador	12	4.55	Italy	7	1.60
France	10	7.32	Switzerland	7	1.35
South Korea	10	4.52	England	7	3.13
Algeria	10	4.05	Argentina	6	2.80
Iran	9	5.43	Ghana	6	1.85
Russia	9	5.27	Australia	6	1.83
Ivory Coast	9	4.63	Portugal	4	1.82
Croatia	9	4.32	Netherlands	4	1.65
Colombia	9	4.32	Bosnia and Herzegovina	2	0.40

Describe these data using the methods you learned in this chapter, and write a short summary describing flopping in the 2014 World Cup based on your analysis.

1.110 Another look at T-bill rates. Refer to Example 1.12 with the histogram in Figure 1.6 (page 14), Example 1.20 with the time plot in Figure 1.12 (page 21), and Example 1.43 with the normal quantile plot in Figure 1.29 (page 54). These examples tells us something about the distribution of T-bill rates and how they vary over time. For this exercise, we will focus on very small rates. TBILL

(a) How do the very small rates appear in each of these plots?

(b) Make a histogram that improves upon Figure 1.6 in terms of focusing on these small rates.

1.111 Another look at marketing products for seniors in Canada. In Exercise 1.32 (page 23), you analyzed data on the percent of the population older than age 65 in the 13 Canadian provinces and territories. Those regions with relatively large percents might be good prospects for marketing products for seniors. In addition, you might want to examine the change in this population over time and then focus your marketing on provinces and territories where this segment of the population is increasing. CANADAP

(a) For 2011 and for 2017, describe the total population, the population older than age 65, and the percent of the population older than age 65 for each of the 13 Canadian provinces and territories. (Note that you will need to compute some of these quantities from the information given in the data set.)

(b) Write a brief marketing proposal for targeting seniors based on your analysis.

1.112 Best brands variables. Refer to Exercises 1.53 and 1.54 (pages 37–38). The data set, brands, contains values for seven variables: (1) rank, a number between 1 and 100 with 1 being the best brand; (2) company name; (3) value of the brand, in millions of dollars; (4) change, difference between last year's rank and current rank; (5) revenue, in US$ billions; (6) company advertising, in US$ millions; and (7) industry. BRANDS

(a) Identify each of these variables as categorical or quantitative.

(b) Is there a label variable in the data set? If yes, identify it.

(c) What are the cases? How many are there?

1.113 Best brands industry. Refer to the previous exercise. Describe the distribution of the variable industry using the methods you have learned in this chapter. BRANDS

1.114 Best brands revenue. Refer to the Exercise 1.112. Describe the distribution of the variable revenue using the methods you have learned in this chapter. Your summary should include information about this characteristic of these data. BRANDS

1.115 Beer variables. Refer to Exercises 1.58 through 1.60 (pages 38–39). The data set, beer, contains values for five variables: (1) brand; (2) brewery; (3) percent alcohol; (4) calories per 12 ounces; and (5) carbohydrates in grams. BEER

(a) Identify each of these variables as categorical or quantitative.

(b) Is there a label variable in the data set? If yes, identify it.

(c) What are the cases? How many are there?

1.116 Beer carbohydrates. Refer to the previous exercise. Describe the distribution of the variable carbohydrates using the methods you have learned in this chapter. Note that some cases have missing values for this variable. Your summary should include information about this characteristic of these data. BEER

1.117 Beer breweries. Refer to Exercise 1.115. Describe the distribution of the variable brewery using the methods you have learned in this chapter. BEER

1.118 Companies of the world. The Word Bank collects large amounts of data related to business issues from different countries. One set of data records the number of companies that are incorporated in each country and that are listed on the country's stock exchange at the end of the year.[33] Examine the numbers of companies for 2017 using the methods that you learned in this chapter. INCCOM

1.119 Companies of the world. Refer to the previous exercise. Examine the data for 2007, and compare your results with what you found in the previous exercise. INCCOM

1.120 Which colors sell? Customers' preference for vehicle colors can influence their purchase choice. Here are data on the most popular colors for SUVs, minivans, and light trucks in the United States according to the Kelley Blue Book.[34] VCOLOR

Color	(percent)
White	19.3
Silver	18.0
Black	12.4
Medium dark blue	11.4
Medium dark gray	7.5
Medium red	7.1
Medium dark green	6.7
Light brown	5.1
Bright red	4.5
Gold	1.8
Other	6.2

Use the methods you learned in this chapter to describe these vehicle color preferences. How would you use this information for marketing vehicles in North America?

1.121 Identify the histograms. A survey of a large college class asked the following questions:

1. Are you female or male? (In the data, male = 0, female = 1.)

2. Are you right-handed or left-handed? (In the data, right = 0, left = 1.)

3. What is your height in inches?

4. How many minutes do you study on a typical week night?

Figure 1.34 shows histograms of the student responses, in scrambled order and without scale markings. Which histogram goes with each variable? Explain your reasoning.

1.122 Grading managers. Some companies "grade on a bell curve" to compare the performance of their managers. This forces the use of some low performance ratings so that not all managers are graded "above average." A company decides to give A's to the managers and professional workers who score in the top 15% on their performance reviews, C's to those who score in the bottom 15%, and B's to the rest. Suppose that a company's performance scores are Normally distributed. This year, managers with scores less than 25 received C's, and those with scores above 475 received A's. What are the mean and standard deviation of the scores?

1.123 What influences buying? Product preference depends in part on the age, income, and gender of the consumer. A market researcher selects a large sample of potential car buyers. For each consumer, she records gender, age, household income, and automobile preference. Which of these variables are categorical and which are quantitative?

1.124 Simulated observations. Most statistical software packages have routines for simulating values having specified distributions. Use your statistical software to generate 30 observations from the $N(25, 4)$ distribution. Compute the mean and standard deviation $\bar{x}$ and s of the 30 values you obtain. How close are $\bar{x}$ and s to the μ and σ of the distribution from which the observations were drawn?

Repeat 24 more times the process of generating 25 observations from the $N(25, 4)$ distribution and recording $\bar{x}$ and s. Make a stemplot of the 25 values of $\bar{x}$ and another stemplot of the 25 values of s. Make Normal quantile plots of both sets of data. Briefly describe each of these distributions. Are they symmetric or skewed? Are they roughly Normal? Where are their centers? (The distributions of measures like $\bar{x}$ and s when repeated sets of observations are made from the same theoretical distribution will be very important in later chapters.)

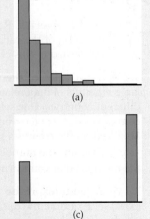
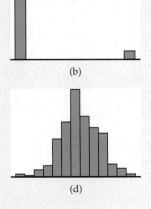

FIGURE 1.34 Match each histogram with its variable (Exercise 1.121).

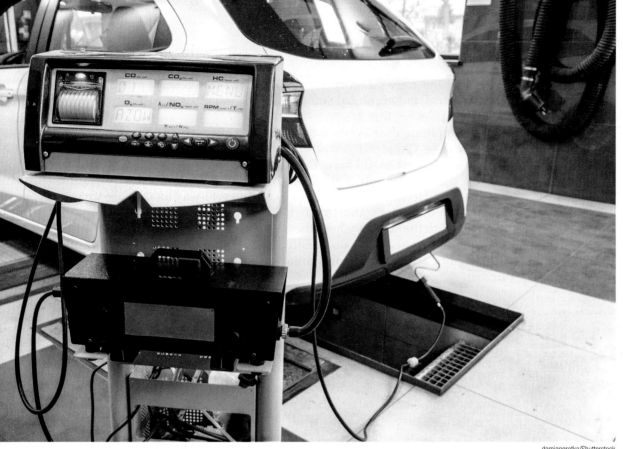
damiangretka/Shutterstock

CHAPTER 2

Examining Relationships

Introduction

Our topic in this chapter is relationships between a pair of variables. Statistical summaries of relationships are used to inform decisions in business and economics in many different settings. Often, we take the view that one of the variables explains or influences the other.

- A realty company wants to know how well the number of square feet in a home predicts its selling price.

- A supermarket wants to predict the impact of digital discount coupons on the shopping preferences of its customers.

- Walmart is exploring the relationship between its catalog sales and its online sales for clothing targeted to different age groups. This relationship will help the company allocate advertising dollars to these two media.

In all these examples, the two variables, commonly labeled y and x, are quantitative variables. The first four sections of this chapter focus on this setting. We start with the study of a graphical summary, the scatterplot, in Section 2.1. Then, we explore the use of a numerical summary, the correlation, in Section 2.2. The least-squares regression line, a numerical summary that is also used to enhance our graphical summary, the scatterplot, is the subject of Section 2.3. Finally, Section 2.4 discusses cautions to keep in mind when using these methods.

CHAPTER OUTLINE

2.1 Scatterplots

2.2 Correlation

2.3 Least-Squares Regression

2.4 Cautions about Correlation and Regression

2.5 Data Analysis for Two-Way Tables

When *y* and *x* are categorical variables, the same general principles apply. Use numerical and graphical summaries to analyze the data. These methods are presented in Section 2.5.

The remaining case is when one variable is quantitative and the other is categorical. In Chapter 1, we studied some basic numerical and graphical summaries that are useful here. For each value of the categorical variable, we describe the distribution of the quantitative variable. More advanced methods that apply to this setting are one-way and two-way analysis of variance, discussed in Chapters 9 and 16. Logistic regression, discussed in Chapter 18, is restricted to the setting where the categorical variable is the response variable and it has only two values. Situations where there are more than two values can be treated by using similar methods that are beyond the scope of our text.

2.1 Scatterplots

When you complete this section, you will be able to:

- Use software to make a scatterplot to examine a relationship between two quantitative variables.
- Describe the overall pattern in a scatterplot and any striking deviations from that pattern.
- Use a scatterplot to describe the form, direction, and strength of a relationship and to identify outliers.
- Explain the effect of a change of units on a scatterplot.
- Use a log transformation to change a curved relationship into a linear relationship.
- Use different plotting symbols to include information about a categorical variable in a scatterplot.

Most statistical studies examine data for more than one variable. Fortunately, statistical analysis of several-variable data builds on the tools we used to examine individual variables. The principles that guide our work also remain the same:

- First, plot the data; then, add numerical summaries.
- Look for overall patterns and deviations from those patterns.
- When the overall pattern is quite regular, use a compact mathematical model to describe it.

In many situations, such as the ones described in the Introduction, we take the view that one variable *explains* the other. We use the terms *response variable* and *explanatory variable* to describe the variables in these settings.

RESPONSE VARIABLE AND EXPLANATORY VARIABLE

A **response variable** measures an outcome of a study. An **explanatory variable** explains or influences changes in a response variable.

independent variable
dependent variable

You will often find explanatory variables called **independent variables** and response variables called **dependent variables.** The idea behind this language is that the response variable *depends* on the explanatory variable. Because the words "independent" and "dependent" have other meanings in statistics that are unrelated to the explanatory–response distinction, we prefer to avoid these words.

It is easiest to identify explanatory and response variables when we actually control the values of one variable to see how it affects another variable.

EXAMPLE 2.1

The Best Price? Price is important to consumers and, therefore, to retailers. Sales of an item typically increase as its price falls. Rising sales can lead to increased profits. However, if the price is reduced too much, profits will suffer. A retail chain introduces a new TV that can respond to voice commands at several different price points and monitors sales. The chain wants to discover the price at which its profits are greatest. Price is the explanatory variable, and total profit from sales of the TV is the response variable. ∎

The distinction between explanatory and response variables is not appropriate for all studies involving two variables. Sometimes we just want to examine the relationship between the two variables.

EXAMPLE 2.2

Inventory and Sales Emily is a district manager for a retail chain. She wants to know how the average monthly inventory and monthly sales for the stores in her district are related to each other. Emily doesn't think that either inventory level or sales explains the other. She has two related variables, and neither is an explanatory variable.

Zachary manages another district for the same chain. He asks, "Can I predict a store's monthly sales if I know its inventory level?" Zachary is treating the inventory level as the explanatory variable and the monthly sales as the response variable. ∎

In Example 2.1, price differences actually *cause* differences in profits by impacting the sales of TVs. In contrast, there is no cause-and-effect relationship between inventory levels and sales in Example 2.2. Because inventory and sales are closely related, we can nonetheless use a store's inventory level to predict its monthly sales. We will learn how to make such a prediction in Section 2.3. Prediction requires that we identify an explanatory variable and a response variable. Some other statistical techniques ignore this distinction. *Remember: Calling one variable "explanatory" and the other "response" doesn't necessarily mean that changes made to one will cause changes in the other.*

APPLY YOUR KNOWLEDGE

label variable, p. 2

2.1 Relationship between job satisfaction and sleep. A study is designed to examine the relationship between employees' job satisfaction and the amount of sleep they get. Think about creating a data set for this study.

(a) What are the cases?

(b) Would your data set have a label variable? If yes, describe it.

(c) What are the variables? Are they quantitative or categorical?

(d) Is there an explanatory variable and a response variable? Explain your answer.

66 Chapter 2 Examining Relationships

2.2 Price versus size. You visit a local Starbucks to buy a Triple Mocha Frappuccino. The barista explains that this blended coffee beverage comes in three sizes and asks if you want a Tall, a Grande, or a Venti. The prices are $4.75, $5.25, and $5.75, respectively.

(a) What are the variables and cases?

(b) Which variable is the explanatory variable? Which is the response variable? Explain your answers.

(c) The Tall contains 12 ounces of beverage, the Grande contains 16 ounces, and the Venti contains 20 ounces. Answer parts (a) and (b) with ounces in place of the names for the sizes.

 Education Expenditures and Population: Benchmarking The data file, EDSPEND, provides the following information for each of the 50 states in the United States:

- The state name
- State spending on education ($ billion)
- Local government spending on education ($ billion)
- Spending (total of state and local) on education ($ billion)
- Gross state product ($ billion)
- Growth in gross state product (percent)
- Population (million)

We expect states with larger populations to spend more on education than states with smaller populations.[1] What is the nature of this relationship? Can we use this relationship to evaluate whether some states are spending more than we expect or less than we expect? In this type of exercise, which is called **benchmarking,** the basic idea is to compare processes or procedures of an organization with those of similar organizations. ∎

APPLY YOUR KNOWLEDGE

2.3 Classify the variables. Use the EDSPEND data set for this exercise. Classify each variable as categorical or quantitative. Is there a label variable in the data set? If there is, identify it. EDSPEND

2.4 Describe the variables. Refer to the previous exercise.

(a) Use graphical and numerical summaries to describe the distribution of total spending. EDSPEND

(b) Do the same for population.

(c) Write a short paragraph summarizing your work in parts (a) and (b).

The most common way to visually display the relationship between two quantitative variables is a *scatterplot*.

SCATTERPLOT

A **scatterplot** shows the relationship between two quantitative variables measured for the same cases. The values of one variable appear on the horizontal axis, and the corresponding values of the other variable appear on the vertical axis. Each case in the data set appears as the point in the plot fixed by the values of both variables for that case.

EXAMPLE 2.3

EDSPEND

Spending and Population A state with a larger number of people needs to spend more money on education. Therefore, we can think of population as an explanatory variable and spending on education as a response variable. We begin our study of this relationship with a graphical display of the two variables.

Figure 2.1 is a scatterplot that displays the relationship between the response variable, spending, and the explanatory variable, population. The data appear to cluster around a line with relatively small variation about this pattern. The relationship is positive: states with larger populations generally spend more on education than states with smaller populations. There are three or four states that are somewhat extreme in terms of both population and spending on education, but their values still appear to be consistent with the overall pattern. ■

FIGURE 2.1 Scatterplot of spending on education (in billions of dollars) versus population (in millions), Example 2.3.

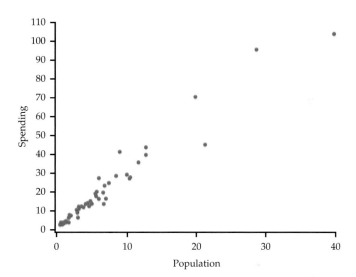

Always plot the explanatory variable, if there is one, on the horizontal axis (the x axis) of a scatterplot. As a reminder, we usually call the explanatory variable x and the response variable y. If there is no explanatory–response distinction, either variable can go on the horizontal axis. The time plots in Section 1.2 (page 20) are special scatterplots where the explanatory variable x is a measure of time.

APPLY YOUR KNOWLEDGE

2.5 Make a scatterplot. EDSPEND

(a) Make a scatterplot similar to Figure 2.1 for the education spending data.

(b) Label the four points with high population and high spending with the names of those states.

2.6 Change the units. EDSPEND

(a) Create a spreadsheet for the education spending data, with education spending expressed in tens of billions of dollars and population in tens of millions. In other words, divide education spending by 10 and divide population by 10.

(b) Make a scatterplot for the data coded in this way.

(c) Describe how this scatterplot differs from Figure 2.1.

examining a distribution, p. 8

Interpreting scatterplots

To interpret a scatterplot, apply the strategies of data analysis learned in Chapter 1.

EXAMINING A SCATTERPLOT

Look for the **overall pattern** and for striking **deviations** from that pattern.

You can describe the overall pattern of a scatterplot by the **form, direction,** and **strength** of the relationship.

An important kind of deviation is an **outlier,** an individual value that falls outside the overall pattern of the relationship.

linear relationship

In this section, our primary focus is on the relationship *form* illustrated by the scatterplot in Figure 2.1. This *form* is clear; the data lie in a roughly straight-line, or linear pattern. To help us see this **linear relationship,** we can use software to put a straight line through the data. We will show how the equation of this line is determined in Section 2.3.

EXAMPLE 2.4

EDSPEND

Scatterplot with a Straight Line Figure 2.2 plots the education spending data along with a fitted straight line. This plot confirms our initial impression about these data. The overall pattern is approximately linear, and a few states have relatively high values for both variables.

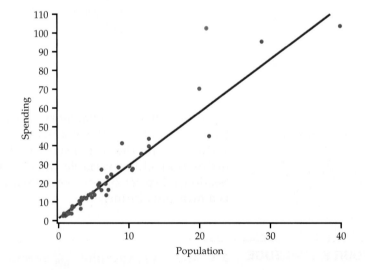

FIGURE 2.2 Scatterplot of spending on education (in billions of dollars) versus population (in millions) with a fitted straight line, Example 2.4.

The relationship in Figure 2.2 also has a clear *direction:* states with higher populations spend more on education than states with smaller populations. This is a *positive association* between the two variables. ■

POSITIVE ASSOCIATION AND NEGATIVE ASSOCIATION

Two variables are **positively associated** when above-average values of one tend to accompany above-average values of the other variable, and when below-average values also tend to occur together.

Two variables are **negatively associated** when above-average values of one variable tend to accompany below-average values of the other variable, and vice versa.

The *strength* of a relationship in a scatterplot is determined by how closely the points follow a clear form. The strength of the linear relationship in Figure 2.1 is fairly strong.

Software is a powerful tool that can help us to see the pattern in a set of data. Many statistical packages have procedures for fitting smooth curves to data measured on a pair of quantitative variables. Here is an example.

EXAMPLE 2.5

EDSPEND

smoothing parameter

Smooth Relationship for Education Spending Figure 2.3 is a scatterplot of the population versus education spending for the 50 states in the United States with a smooth curve generated by software. The smooth curve follows the data too closely. If we increase the extent to which the relationship is smoothed, Figure 2.4 is the result. Here we see the smooth curve is very close to our plot with the line in Figure 2.2. In this way, we have confirmed our view that we can summarize this relationship with a line. Adjusting the level of smoothness is done by changing the **smoothing parameter**. ∎

FIGURE 2.3 Scatterplot of spending on education (in billions of dollars) versus population (in millions) with a smooth curve, Example 2.5. This smooth curve fits the data too well and does not provide a good summary of the relationship.

FIGURE 2.4 Scatterplot of spending on education (in billions of dollars) versus population (in millions) with a better smooth curve, Example 2.5. This smooth curve fits the data well and provides a good summary of the relationship. It shows that the relationship is approximately linear.

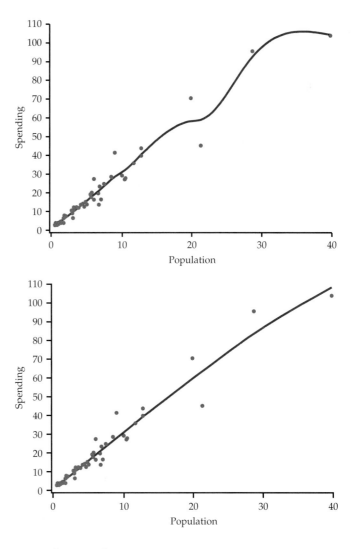

The log transformation

In many business and economic studies, we deal with quantitative variables that take only positive values and are skewed toward high values. In Example 2.4 (page 68), you observed this situation for spending and population size in our education spending data set. One way to make skewed distributions more symmetric is to transform the data in some way.

log transformation

The most important transformation that we will use is the **log transformation.** This transformation can be used only for variables that have positive values. Occasionally, we use it when the data set includes zeros, but, in this case, we first replace the zero values by some small value, often one-half of the smallest positive value in the data set.

You have probably encountered logarithms in one of your mathematics courses as a way to do certain kinds of arithmetic. Usually, these are base 10 logarithms. Performing a log transformation allows us to apply some special interpretations that we will discuss later. Logarithms are a lot more fun when used in statistical analyses than in mathematics. For our statistical applications, **natural logarithms** we will use **natural logarithms**. Statistical software and statistical calculators generally provide easy ways to perform this transformation.

APPLY YOUR KNOWLEDGE

2.7 Transform education spending and population. In Exercise 2.4 (page 66), we examined the distributions of education spending and population. Transform the education spending and population variables using logs, and describe the distributions of the transformed variables. Compare these distributions with those described in Exercise 2.4. EDSPEND

Here, we are concerned with relationships between pairs of quantitative variables. There is no requirement that either or both of these variables should be Normal. For now, though, let's examine the effect of the transformations on the relationship between education spending and population.

EXAMPLE 2.6

EDSPEND

Education Spending and Population with Logarithms Figure 2.5 is a scatterplot of the log of education spending versus the log of population for the 50 states in the United States. The line on the plot fits the data well, and we conclude that the relationship is linear in the transformed variables.

Notice how the data are more evenly spread throughout the range of the possible values. The three or four high values no longer appear to be extreme. Instead, we now see them as the high end of a distribution. ■

FIGURE 2.5 Scatterplot of log spending on education versus log population with a fitted straight line, Example 2.6.

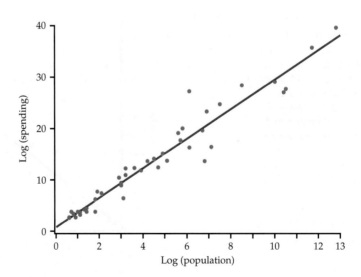

The transformations of the two quantitative variables maintained the linearity of the relationship that we saw in Figure 2.2 (page 68). Sometimes, however, we transform one of the variables to change a nonlinear relationship into a linear one.

The interpretation of scatterplots, including knowing to use transformations, is an art that requires judgment and knowledge about the variables that we are studying.

Always ask yourself if the relationship that you see makes sense. If it does not, then additional analyses are needed to understand the data.

Many statistical procedures work very well with data that are Normal and relationships that are linear. However, there is no requirement that we must have Normal data and linear relationships for everything that we do. In fact, with advances in statistical software, we now have many statistical techniques that work well in a wide range of settings. See Chapters 17 and 18 for examples.

Adding categorical variables to scatterplots

Natural Resources Canada measures fuel efficiency and other characteristics of vehicles.[2] The data file called canfuel gives data for 1045 model-year 2018 conventional vehicles. Variables include fuel efficiency measured in miles per imperial gallon (MPG), carbon dioxide (CO_2) emissions, and the type of fuel used. Four types of vehicles are given:

- X, regular gasoline
- Z, premium gasoline
- D, diesel
- E, ethanol

Although much of our focus in this chapter is on linear relationships, many interesting relationships are more complicated. Our fuel efficiency data provide us with an example.

EXAMPLE 2.7

CANFUEL

Fuel Efficiency and CO_2 Let's look at the relationship between MPG and CO_2 emissions, two quantitative variables, while also taking into account the type of fuel, a categorical variable. The JMP statistical software was used to produce the plot in Figure 2.6. As we can see, there is a negative relationship between the two quantitative variables: better (higher) MPG is associated with lower CO_2 emissions. The relationship is curved, however—not linear.

The legend on the right side of the figure identifies the colors used to plot the four types of fuel, our categorical variable. The vehicles that use regular gasoline (green) and premium gasoline (purple) appear to be mixed together. The diesel-burning vehicles (blue) are close to the gasoline-burning vehicles, but they tend to have higher values for both MPG and emissions. Why do you think that the vehicles that burn ethanol (red) are clearly separated from the other vehicles? ∎

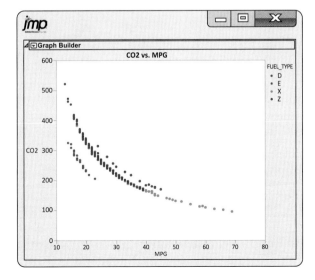

FIGURE 2.6 Scatterplot of CO_2 emissions versus MPG for 1045 vehicles for the model year 2018 using JMP software. Colors correspond to the type of fuel used: blue for diesel, red for ethanol, green for regular gasoline, and purple for premium gasoline, Example 2.7.

Careful judgment is needed in applying this graphical method. Don't be discouraged if your first attempt is not very successful. *To discover interesting things in your data, you will often produce several plots before you find the one that is most effective in describing the data.*[3]

SECTION 2.1 SUMMARY

- To study relationships between variables, we must measure the variables for the same cases.

- If we think that a variable x may explain or even cause changes in another variable y, we call x an **explanatory variable** and y a **response variable.**

- A **scatterplot** displays the relationship between two quantitative variables measured on the same cases. Plot the data for each case as a point on the graph.

- Always plot the explanatory variable, if there is one, on the x axis of a scatterplot. Plot the response variable on the y axis.

- Plot points with different colors or symbols to see the effect of a categorical variable in a scatterplot.

- In examining a scatterplot, look for an overall pattern showing the **form, direction,** and **strength** of the relationship and then for **outliers** or other deviations from this pattern.

- **Linear relationships,** where the points show a straight-line pattern, are an important **form** of relationship between two variables. Curved relationships and clusters are other forms to watch for.

- If the relationship has a clear **direction,** we speak of either **positive association** (high values of the two variables tend to occur together) or **negative association** (high values of one variable tend to occur with low values of the other variable).

- The **strength** of a relationship is determined by how close the points in the scatterplot follow a clear form such as a line.

- The **log transformation** is frequently used in business applications of statistics. It can help us to better see relationships between variables in a scatterplot. It also tends to make skewed distributions more symmetric.

SECTION 2.1 EXERCISES

For Exercises 2.1 and 2.2, see pages 65–66; for 2.3 and 2.4, see page 66; for 2.5 and 2.6, see page 67; and for 2.7, see page 70.

2.8 What's wrong? Explain what is wrong with each of the following statements:

(a) A boxplot can be used to examine the relationship between two variables.

(b) In a scatterplot, we put the quantitative variable on the x axis and the categorical variable on the y axis.

(c) If two variables are positively associated, then high values of one variable are associated with low values of the other variable.

2.9 Make some sketches. For each of the following situations, make a scatterplot that illustrates the given relationship between two variables.

(a) A strong negative linear relationship.

(b) A weak positive relationship that is not linear.

(c) A more complicated relationship. Explain the relationship.

(d) No apparent relationship.

2.10 Companies of the world. In Exercise 1.118 (page 61), you examined data collected by the World Bank on the number of companies that are incorporated and were listed in their country's stock exchange at the end of the year for 2017. In Exercise 1.119, you did the same for the year 2007.[4] In this exercise, you will examine the relationship between the numbers for these two years. INCCOM

(a) Which variable would you choose as the explanatory variable, and which would you choose as the response variable? Give reasons for your answers.

(b) Make a scatterplot of the data.

(c) Describe the form, the direction, and the strength of the relationship.

(d) Are there any outliers? If yes, identify them by name.

2.11 Companies of the world. Refer to the previous exercise. Using the questions there as a guide, describe the relationship between the numbers for 2017 and 1997. Do you expect this relationship to be stronger or weaker than the one you described in the previous exercise? Give a reason for your answer.

2.12 Brand-to-brand variation in a product. Beer100.com advertises itself as "Your Place for All Things Beer." One of its "things" is a list of 175 domestic beer brands with the percent alcohol, calories per 12 ounces, and carbohydrates (in grams).[5] In Exercises 1.58 through 1.60 (page 38), you examined the distribution of alcohol content and the distribution of calories for these beers. BEER

(a) Give a brief summary of what you learned about these variables in those exercises. (If you did not do these exercises when you studied Chapter 1, do them now.)

(b) Make a scatterplot of calories versus percent alcohol.

(c) Describe the form, direction, and strength of the relationship.

(d) Are there any outliers? If yes, identify them by name.

2.13 More beer. Refer to the previous exercise. Repeat the exercise for the relationship between carbohydrates and percent alcohol. Be sure to include summaries of the distributions of the two variables you are studying. BEER

2.14 Marketing in Canada. Many consumer items are marketed to particular age groups in a population. To plan such marketing strategies, it is helpful to know the demographic profile for different areas. Statistics Canada provides a great deal of demographic data organized in different ways.[6] CANADAP

(a) Make a scatterplot of the percent of the population older than age 65 versus the percent of the population younger than age 15.

(b) Describe the form, direction, and strength of the relationship.

2.15 Compare the provinces with the territories. Refer to the previous exercise. The three Canadian territories are the Northwest Territories, Nunavut, and the Yukon Territories. All of the other entries in the data set are provinces. CANADAP

(a) Generate a scatterplot of the Canadian demographic data similar to the one that you made in the previous exercise, but with the points labeled "P" for provinces and "T" for territories (or some other way, if that is easier to do with your software).

(b) Use your new scatterplot to write a new summary of the demographics for the 13 Canadian provinces and territories.

2.16 Sales and time spent on web pages. You have collected data on 1000 customers who visited the web pages of your company last week. For each customer, you recorded the time spent on your pages and the total amount of the customer's purchases during the visit. You want to explore the relationship between these two variables.

(a) What is the explanatory variable? What is the response variable? Explain your answers.

(b) Are these variables categorical or quantitative?

(c) Do you expect a positive or negative association between these variables? Why?

(d) How strong do you expect the relationship to be? Give reasons for your answer.

2.17 A product for lab experiments. Barium-137m is a radioactive form of the element barium that decays very rapidly. It is easy and safe to use for lab experiments in schools and colleges.[7] In a typical experiment, the radioactivity of a sample of barium-137m is measured for one minute. It is then measured for three additional one-minute periods, separated by two minutes. Thus, data are recorded at one, three, five, and seven minutes after the start of the first counting period. The measurement units are counts. Here are the data for one of these experiments:[8] DECAY

Time	1	3	5	7
Count	578	317	203	118

(a) Make a scatterplot of these data. Give reasons for the choice of which variables to use on the x and y axes.

(b) Describe the overall pattern in the scatterplot.

(c) Describe the form, direction, and strength of the relationship.

(d) Identify any outliers.

(e) Is the relationship approximately linear? Explain your answer.

2.18 Use a log for the radioactive decay. Refer to the previous exercise. Transform the counts using a log transformation. Then repeat parts (a) through (e) for the transformed data, and compare your results with those from the previous exercise. DECAY

2.19 Time to start a business. Case 1.2 (page 24) uses the World Bank data on the time required to start a business in different countries. For Example 1.21 and several other examples that followed, we used data for a subset of the countries for 2018. Data are also available for times to start a business in 2008. Let's look at the data for all 187 countries to examine the relationship between the times to start a business in 2018 and the times to start a business in 2008. TTS

(a) Why should you use the time for 2008 as the explanatory variable and the time for 2018 as the response variable?

(b) Make a scatterplot of the two variables.

(c) How many points are in your plot? Explain why there are not 189 points.

(d) Describe the form, direction, and strength of the relationship.

(e) Identify any outliers.

(f) Is the relationship approximately linear? Explain your answer.

2.20 Use logs. Refer to the previous exercise. Rerun your analysis using logs for both variables. Summarize your results and contrast your findings what you reported in the previous exercise. TTS

2.21 Fuel efficiency and CO_2 emissions. Refer to Example 2.7 (page 71), where we examined the relationship between CO_2 emissions and MPG for 1045 vehicles for the model year 2018. In that example, we used MPG as the explanatory variable and CO_2 as the response variable. Fuel efficiency expressed as MPG in our data set is a composite of 55% city fuel efficiency and 45% highway fuel efficiency. The data set also includes variables that give fuel efficiency in terms of fuel consumption expressed as liters per 100 kilometers for both city and highway driving. Using Example 2.7 as a guide, analyze the relationship between city fuel efficiency and highway fuel efficiency and write a short summary of your results. CANFUEL

2.22 Add the type of fuel to the plots. Refer to the previous exercise. As we did in Figure 2.6 (page 71), add the categorical variable, type of fuel, to your plots. Summarize what you have found in this exercise. CANFUEL

2.2 Correlation

When you complete this section, you will be able to:

- Use a correlation to describe the direction and strength of a linear relationship between two quantitative variables.
- Identify situations in which the correlation is not a good measure of association between two quantitative variables.
- For describing the relationship between two quantitative variables, identify the roles of the correlation (a numerical summary) and the scatterplot (a graphical summary).

A scatterplot displays the form, direction, and strength of the relationship between two quantitative variables. Linear relationships are particularly important because a straight line is a simple pattern that is quite common. We say a linear relationship is strong if the points lie close to a straight line; we say it is weak if they are widely scattered about a line. Our eyes are not good judges of how strong a linear relationship is.

The two scatterplots in Figure 2.7 depict exactly the same data, but the lower plot is drawn with an expanded range for both y and x. The lower plot seems to show a stronger linear relationship. *Our eyes are often fooled by changing the plotting scales or the amount of white space around the cloud of points in a scatterplot.*[9] We need to follow our strategy for data analysis by using a numerical measure to supplement the graph. *Correlation* is the measure we use.

FIGURE 2.7 Two scatterplots of the same data. The straight-line pattern in the lower plot appears stronger because of the surrounding open space.

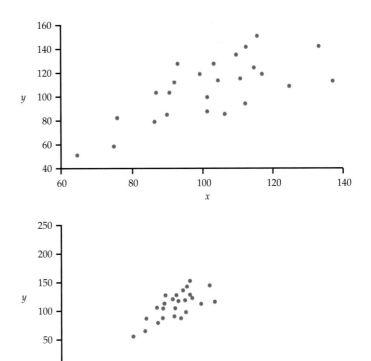

The correlation *r*

CORRELATION

The **correlation** measures the direction and strength of the linear relationship between two quantitative variables. Correlation is usually written as *r*.

Suppose that we have data on variables x and y for n cases. The values for the first case are x_1 and y_1, the values for the second case are x_2 and y_2, and so on. The means and standard deviations of the two variables are $\bar{x}$ and s_x for the x-values, and $\bar{y}$ and s_y for the y-values. The correlation r between x and y is

$$r = \frac{1}{n-1} \sum \left(\frac{x_i - \bar{x}}{s_x} \right) \left(\frac{y_i - \bar{y}}{s_y} \right)$$

As always, the summation sign Σ means "add these terms for all cases." The formula for the correlation r is a bit complex. It helps us to see what correlation is, but just as in making a scatterplot, in practice you should use software that finds r from keyed-in values of the two variables x and y.

standardizing, p. 47

The formula for r begins by standardizing the data. Suppose, for example, that x is height in centimeters and y is weight in kilograms and that we have height and weight measurements for n people. Then $\bar{x}$ and s_x are the mean and standard deviation of the n heights, both in centimeters. The value

$$\frac{x_i - \bar{x}}{s_x}$$

is the standardized height of the ith person. The standardized height says how many standard deviations above or below the mean a person's height lies. Standardized values have no units—in this example, they are no longer measured in centimeters. Similarly, the standardized weights obtained by subtracting $\bar{y}$ and dividing by s_y are no longer measured in kilograms. The correlation r is an average of the products of the standardized height and the standardized weight for the n people.

APPLY YOUR KNOWLEDGE

CASE 2.1 **2.23 Spending on education.** In Example 2.3 (page 67), we examined the relationship between spending on education and population for the 50 states in the United States. Compute the correlation between these two variables. EDSPEND

CASE 2.1 **2.24 Change the units.** Refer to Exercise 2.6 (page 67), where you changed the units to millions of dollars for education spending and to thousands for population. EDSPEND

(a) Find the correlation between spending on education and population using the new units.

(b) Compare this correlation with the one that you computed in the previous exercise.

(c) Generally speaking, what effect, if any, did changing the units in this way have on the correlation?

covariance

A quantity closely related to the correlation is the **covariance**. It is calculated using the following formula:

$$s_{xy} = \frac{1}{n-1} \sum (x_i - \bar{x})(y_i - \bar{y})$$

Note that it is closely related to the correlation

$$r = \frac{s_{xy}}{s_x s_y}$$

so the correlation is a kind of standardized covariance.

Facts about correlation

The formula for correlation helps us see that r is positive when there is a positive association between the variables. Height and weight, for example, have a positive association. People who are above average in height tend to be above average in weight. Both the standardized height and the standardized weight are positive. People who are below average in height tend to have below-average weight, in which case both standardized height and standardized weight are negative. In both cases, the products in the formula for r are mostly positive, so r is positive. In the same way, we can see that r is negative when the association between x and y is negative. A more detailed study of the formula reveals more detailed properties of r. Here is what you need to know to interpret correlation.

1. Correlation makes no distinction between explanatory and response variables. Thus, it makes no difference which variable you call x and which variable you call y when calculating the correlation.

2. Correlation requires that both variables be quantitative, so that it makes sense to do the arithmetic indicated by the formula for r.

We cannot calculate a correlation between the incomes of a group of people and what city they live in because city is a categorical variable.

3. Because r uses the standardized values of the data, r does not change when we change the units of measurement of x, y, or both. Measuring height in inches rather than centimeters, and weight in pounds rather than kilograms, does not change the correlation between height and weight. The correlation r itself has no unit of measurement; it is just a number.

4. Positive r indicates a positive association between the variables, and negative r indicates a negative association.

5. The correlation r is always a number between -1 and 1. Values of r near 0 indicate a very weak linear relationship. The strength of the linear relationship increases as r moves away from 0 toward either -1 or 1. Values of r close to -1 or 1 indicate that the points in a scatterplot lie close to a straight line. The extreme values $r = -1$ and $r = 1$ occur only in the case of a perfect linear relationship, when the points lie exactly along a straight line.

6. Correlation measures the strength of only a *linear relationship* between two variables. Correlation does not describe a curved relationship between variables, no matter how strong that relationship is.

7. Like the mean and standard deviation, the correlation is not resistant: r is strongly affected by a few outlying observations. Use r with caution when outliers appear in the scatterplot.

resistant, p. 26

Note that if all the values of x are the same, the correlation cannot be computed because s_x is zero. The same is true for y. What does your software report when $s_x = 0$?

The scatterplots in Figure 2.8 illustrate how values of r closer to 1 or -1 correspond to stronger linear relationships. To make the meaning of r clearer, the standard deviations of both variables in these plots are equal, and the horizontal and vertical scales are the same. In general, it is not so easy to guess the value of r from the appearance of a scatterplot. Although changing the plotting scales in a scatterplot may mislead our eyes, remember that it does not change the correlation.

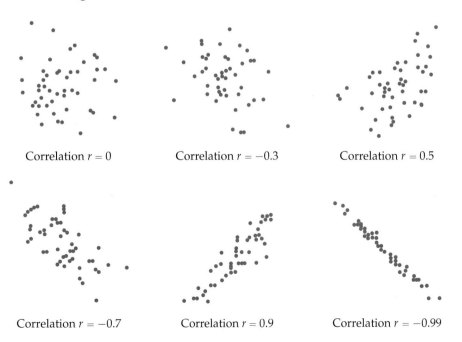

FIGURE 2.8 How the correlation measures the strength of a linear relationship. Patterns closest to a straight line have correlations closer to 1 or −1.

Correlation is not a complete description of two-variable data, even when the relationship between the variables is linear. For this reason, you should give the means and standard deviations of both x and y along with the correlation. (Because the formula for correlation uses the means and standard deviations, these measures are the proper choice to accompany a correlation.) Conclusions based on correlations alone may require rethinking the relationships in the light of a more complete description of the data.

EXAMPLE 2.8

Forecasting Earnings Stock analysts regularly forecast the earnings per share (EPS) of companies they follow. EPS is calculated by dividing a company's net income for a given time period by the number of common stock shares outstanding. We have two analysts' EPS forecasts for a computer manufacturer for the next six quarters. How well do the two forecasts agree? The correlation between them is $r = 0.9$, but the mean of the first analyst's forecasts is $3 per share lower than the second analyst's mean.

These facts do not contradict each other; instead, they are simply different kinds of information. The means show that the first analyst predicts lower EPS than the second. But because the first analyst's EPS predictions are about $3 per share lower than the second analyst's *for every quarter*, the correlation remains high. Adding or subtracting the same number to all values of either x or y does not change the correlation. The two analysts agree on which quarters will see higher EPS values. The high r shows this agreement, despite the fact that the actual predicted values differ by $3 per share. ■

APPLY YOUR KNOWLEDGE

2.25 Strong association but no correlation. Here is a data set that illustrates an important point about correlation: CORR

x	20	30	40	50	60
y	10	30	50	30	10

(a) Make a scatterplot of y versus x.

(b) Describe the relationship between y and x. Is it weak or strong? Is it linear?

(c) Find the correlation between y and x.

(d) What important point about correlation does this exercise illustrate?

2.26 Brand names and generic products.

(a) If a store always prices its generic "store brand" products at exactly 80% of the brand-name products' prices, what would be the correlation between these two prices? (*Hint:* Draw a scatterplot for several prices.)

(b) If the store always prices its generic products $4 less than the corresponding brand-name products, then what would be the correlation between the prices of the brand-name products and the store-brand products?

SECTION 2.2 SUMMARY

- The **correlation** r measures the strength and direction of the linear association between two quantitative variables x and y. Although you can calculate a correlation for any scatterplot, r measures only straight-line relationships.

- Correlation indicates the direction of a linear relationship by its sign: $r > 0$ for a positive association and $r < 0$ for a negative association.

- Correlation always satisfies $-1 \leq r \leq 1$ and indicates the strength of a relationship by how close it is to -1 or 1. Perfect correlation, $r = \pm 1$, occurs only when the points on a scatterplot lie exactly on a straight line.

- Correlation ignores the distinction between explanatory and response variables. The value of r is not affected by changes in the unit of measurement of either variable. Correlation is not resistant, so outliers can greatly change the value of r.

SECTION 2.2 EXERCISES

For Exercises 2.23 and 2.24, see page 76; and for 2.25 and 2.26, see page 78.

2.27 Companies of the world. Refer to Exercise 1.118 (page 61), where we examined data collected by the World Bank on the numbers of companies that are incorporated and are listed on their country's stock exchange at the end of the year. In Exercise 2.10 (page 72), you examined the relationship between these numbers for 2017 and 2007. INCCOM

(a) Find the correlation between these two variables.

(b) Do you think that the correlation you computed gives a good numerical summary of the strength of the relationship between these two variables? Explain your answer.

2.28 Companies of the world. Refer to the previous exercise and to Exercise 2.11 (page 73). Answer parts (a) and (b) for 2017 and 1997. Compare the correlation you found in the previous exercise with the one you found in this exercise. Why do they differ in this way? INCCOM

2.29 A product for lab experiments. In Exercise 2.17 (page 73), you described the relationship between time and count for an experiment examining the decay of barium. DECAY

(a) Is the relationship between these two variables strong? Explain your answer.

(b) Find the correlation.

(c) Do you think that the correlation you computed gives a good numerical summary of the strength of the relationship between these two variables? Explain your answer.

2.30 Use a log for the radioactive decay. Refer to the previous exercise and to Exercise 2.18 (page 73), where you transformed the counts with a logarithm. DECAY

(a) Is the relationship between time and the log of the counts strong? Explain your answer.

(b) Find the correlation between time and the log of the counts.

(c) Do you think that the correlation you computed gives a good numerical summary of the strength of the relationship between these two variables? Explain your answer.

(d) Compare your results here with those you found in the previous exercise. Was the correlation useful in explaining the relationship before the transformation? After? Explain your answers.

(e) Using your answer in part (d), write a short explanation of what these analyses show about the use of a correlation to explain the strength of a relationship.

2.31 Brand-to-brand variation in a product. In Exercise 2.12, you examined the relationship between percent alcohol and calories per 12 ounces for 175 domestic brands of beer. BEER

(a) Compute the correlation between these two variables.

(b) Do you think that the correlation you computed gives a good numerical summary of the strength of the relationship between these two variables? Explain your answer.

2.32 Alcohol and carbohydrates in beer revisited. Refer to the previous exercise. Delete any outliers that you identified in Exercise 2.12. BEER

(a) Recompute the correlation without the outliers.

(b) Write a short paragraph about the possible effects of outliers on the correlation, using this example to illustrate your ideas.

2.33 Marketing in Canada. In Exercise 2.14 (page 73), you examined the relationship between the percent of the population older than age 65 and the percent younger than age 15 for the 13 Canadian provinces and territories. CANADAP

(a) Make a scatterplot of the two variables if you do not have your work from Exercise 2.14.

(b) Find the value of the correlation r.

(c) Does this numerical summary give a good indication of the strength of the relationship between these two variables? Explain your answer.

2.34 Nunavut. Refer to the previous exercise. CANADAP

(a) Do you think that Nunavut is an outlier? Explain your answer.

(b) Find the correlation without Nunavut. Using your work from the previous exercise, summarize the effect of Nunavut on the correlation.

2.35 Education spending and population with logs. In Example 2.3 (page 67), we examined the relationship between spending on education and population, and in Exercise 2.23 (page 76), you found the correlation between these two variables. In Example 2.6 (page 70), we examined the relationship between the variables transformed by logs. **EDSPEND**

(a) Compute the correlation between the variables expressed as logs.

(b) How does this correlation compare with the one you computed in Exercise 2.23? Discuss this result.

2.36 Are they outliers? Refer to the previous exercise. Delete the four states with high values. **EDSPEND**

(a) Find the correlation between spending on education and population for the remaining 46 states.

(b) Do the same for these variables expressed as logs.

(c) Compare your results in parts (a) and (b) with the correlations that you computed with the full data set in Exercise 2.23 and in the previous exercise. Discuss these results.

2.37 Fuel efficiency and CO_2 emissions. In Example 2.7 (page 71), we examined the relationship between highway MPG and CO_2 emissions for 1045 vehicles for the model year 2018. Let's examine the relationship between the two measures of fuel efficiency in the data set, highway MPG and city MPG. **CANFUEL**

(a) Make a scatterplot with city fuel efficiency on the x axis and highway fuel efficiency on the y axis.

(b) Describe the relationship.

(c) Calculate the correlation.

(d) Does this numerical summary give a good indication of the strength of the relationship between these two variables? Explain your answer.

2.38 Consider the fuel type. Refer to the previous exercise and to Figure 2.6 (page 71), where different colors are used to distinguish four different types of fuels used by these vehicles. **CANFUEL**

(a) Make a scatterplot of highway fuel efficiency and city fuel efficiency similar to Figure 2.6 that allows us to see the categorical variable, type of fuel, in the scatterplot. If your software does not have this capability, make different scatterplots for each fuel type.

(b) Discuss the relationship between highway fuel efficiency and city fuel efficiency, taking into account the type of fuel. Compare this view with what you found in the previous exercise where you did not make this distinction.

(c) Find the correlation between highway fuel efficiency and city fuel efficiency for each type of fuel. Write a short summary of what you have found.

2.39 Match the correlation. The *Correlation and Regression* applet at the text website allows you to create a scatterplot by clicking and dragging with the mouse. The applet calculates and displays the correlation as you change the plot. You will use this applet to make scatterplots with 12 points that have a correlation close to 0.8. The lesson is that many patterns can have the same correlation. Always plot your data before you trust a correlation.

(a) Stop after adding the first two points. What is the value of the correlation? Why does it have this value?

(b) Make a lower-left to upper-right pattern of 12 points with correlation about $r = 0.8$. (You can drag points up or down to adjust r after you have 12 points.) Make a rough sketch of your scatterplot.

(c) Make another scatterplot with eight points in a vertical stack at the right of the plot. Add one point far to the left and move it until the correlation is close to 0.8. Make a rough sketch of your scatterplot.

(d) Make yet another scatterplot with 12 points in a curved pattern that starts at the lower left, rises to the right, and then falls again at the far right. Adjust the points up or down until you have a smooth curve with correlation close to 0.8. Make a rough sketch of this scatterplot as well.

2.40 Stretching a scatterplot. Changing the units of measurement can greatly alter the appearance of a scatterplot. Consider the following data: **STRETCH**

x	−4	−4	−3	3	4	4
y	0.5	−0.6	−0.5	0.5	0.5	−0.6

(a) Draw x and y axes, each extending from −6 to 6. Plot the data on these axes.

(b) Calculate the values of new variables $x^* = x/12$ and $y^* = 12y$, starting from the values of x and y. Plot y^* against x^* on the same axes using a different plotting symbol. The two plots are very different in appearance.

(c) Find the correlation between x and y. Then find the correlation between x^* and y^*. How are the two correlations related? Explain why this outcome isn't surprising.

2.41 CEO compensation and stock market performance. An academic study concludes, "The evidence indicates that the correlation between the compensation of corporate CEOs and the performance of their company's stock is close to zero." A business magazine reports this as follows: "A new study shows that companies that pay their CEOs highly tend to perform poorly in the stock market, and vice versa." Explain why the magazine's report is wrong. Write a statement in plain language (don't use the word "correlation") to explain the study's conclusion.

2.42 Investment reports and correlations. Investment reports often include correlations. Following a table of correlations among mutual funds, a report adds, "Two funds can have perfect correlation, yet different levels of risk. For example, Fund A and Fund B may be perfectly correlated, yet Fund A moves 20% whenever Fund B moves 10%." Write a brief explanation, for someone who does not know statistics, of how this can happen. Include a sketch to illustrate your explanation.

2.43 Sloppy writing about correlation. Each of the following statements contains a blunder. Explain in each case what is wrong.

(a) "There is a high correlation between the color of a car and the age of its owner."

(b) "There is a large negative correlation $(r = -1.2)$ between the premium you would pay for a standard automobile insurance policy and the number of years since your last accident."

(c) "The correlation between y and x is $r = 0.5$, but the correlation between $-x$ and $-y$ is $r = -0.5$."

2.3 Least-Squares Regression

When you complete this section, you will be able to:

- Find the equation of the least-squares regression line and draw it on a scatterplot of a set of data.
- Predict a value of the response variable y for a given value of the explanatory variable x using a regression equation.
- Explain the meaning of the term "least squares."
- Calculate the equation of a least-squares regression line from the means and standard deviations of the explanatory and response variables and their correlation.
- Read the output of statistical software to find the equation of the least-squares regression line and the value of r^2.
- Explain the meaning of r^2 in the regression setting.
- Use a residual plot to examine the fit of a regression line to a set of data and to identify influential observations.

Correlation measures the direction and strength of the straight-line (linear) relationship between two quantitative variables. If a scatterplot shows a linear relationship, we would like to summarize this overall pattern by drawing a line on the scatterplot. A *regression line* summarizes the relationship between two variables, but only in a specific setting: when one of the variables helps explain or predict the other. That is, regression describes a relationship between an explanatory variable and a response variable.

REGRESSION LINE

A **regression line** is a straight line that describes how a response variable y changes as an explanatory variable x changes. We often use a regression line to predict the value of y for a given value of x.

EXAMPLE 2.9

IDI

World Economic Forum The World Economic Forum studies data on many variables related to financial development in the countries of the world. In 2017, this organization introduced a new metric, the Inclusive Development Index (IDI), that measures the impact of a country's economic policy and growth on all its citizens.[10] One of the variables used in the metric is median per capita daily income (MI), measured in U.S. dollars. Here are the data for 15 countries that ranked high on the IDI:

Country	IDI	MI	Country	IDI	MI	Country	IDI	MI
Australia	5.36	44.4	Iceland	6.07	43.4	Korea Rep	5.09	34.2
Belgium	5.14	43.8	Ireland	5.44	38.0	Netherlands	5.61	43.3
Canada	5.06	49.2	Israel	4.51	25.8	Norway	6.08	63.8
Czech Republic	5.09	24.3	Italy	4.31	34.3	Portugal	3.97	21.2
Estonia	4.74	22.1	Japan	4.53	34.8	United Kingdom	4.89	39.4

How well does MI predict the IDI? Figure 2.9 is a scatterplot of the data. The correlation is $r = 0.74$. The scatterplot includes a regression line drawn through the points. ■

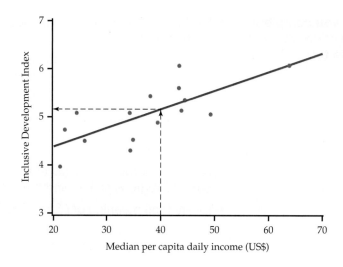

FIGURE 2.9 Scatterplot of the Inclusive Development Index (IDI) and median per capita daily income for 15 countries that rank high on financial development, Example 2.9. The dashed line indicates how to use the regression line to predict the IDI for a country with a median per capita daily income of $40.

prediction

Suppose we want to use this relationship between the IDI and MI $40. To **predict** the IDI, first locate 40 on the x axis. Then go "up and across" as in Figure 2.9 to find the corresponding IDI. We predict that a country with an MI of $40 will have an IDI of about 5.16.

The least-squares regression line

Different people might draw different lines by eye on a scatterplot. To deal with this inconsistency, we need a way to draw a regression line that doesn't depend on our guess as to where the line should be. We will use the line to predict y from x, so the prediction errors we make are errors in y, the vertical direction in the scatterplot. If we predict an IDI of 5.08 and the actual IDI is 5.44, our prediction error is

$$\text{error} = \text{observed } y - \text{predicted } y$$
$$= 5.44 - 5.08 = 0.36$$

APPLY YOUR KNOWLEDGE

2.44 Find a prediction error. Use Figure 2.9 to estimate the IDI for a country that has a MI of $30. If the actual IDI of this country is 4.1, find the prediction error.

2.45 Positive and negative prediction errors. Examine Figure 2.9 carefully. How many of the prediction errors are positive? How many are negative?

EXAMPLE 2.10

The Least-Squares Idea Figure 2.10 illustrates the idea. This plot shows some of the data, along with a line. The vertical distances of the data points from the line appear as vertical line segments. ■

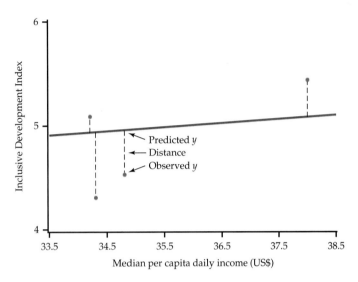

FIGURE 2.10 The least-squares idea. For each observation, find the vertical distance of each point from a regression line. The least-squares regression line makes the sum of the squares of these distances as small as possible.

No line will pass exactly through all the points in the scatterplot. Even so, we want the *vertical* distances of the points from the line to be as small as possible.

There are several ways to make the collection of vertical distances "as small as possible." The most common is the *least-squares* method.

LEAST-SQUARES REGRESSION LINE

The **least-squares regression line** of y on x is the line that makes the sum of the squares of the vertical distances of the data points from the line as small as possible.

One reason for the popularity of the least-squares regression line is that the problem of finding the line has a simple solution. We can give the recipe for the least-squares line in terms of the means and standard deviations of the two variables and their correlation.

EQUATION OF THE LEAST-SQUARES REGRESSION LINE

We have data on an explanatory variable x and a response variable y for n cases. From the data, calculate the means $\bar{x}$ and $\bar{y}$ and the standard deviations s_x and s_y of the two variables and their correlation r. The least-squares regression line is the line

$$\hat{y} = b_0 + b_1 x$$

with **slope**

$$b_1 = r \frac{s_y}{s_x}$$

and **intercept**

$$b_0 = \bar{y} - b_1 \bar{x}$$

We write $\hat{y}$ (read "y hat") in the equation of the regression line to emphasize that the line gives a *predicted* response $\hat{y}$ for any x. In some applications, we use the equation to predict y for values of x that may or may not be in our original set of data. This use of regression is sometimes called predictive analytics or simply *analytics*. Because of the scatter of points about the line, the predicted response will usually not be exactly the same as the actually *observed* response y. In practice, you don't need to calculate the means, standard deviations, and correlation first.

predictive analytics, p. 8

Statistical software or your calculator will give the slope b_1 and intercept b_0 of the least-squares line from keyed-in values of the variables x and y. You can then concentrate on understanding and using the regression line. Be careful, though—different software packages and calculators label the slope and intercept differently in their output, so remember that the slope is the value that multiplies x in the equation.

EXAMPLE 2.11

IDI

The Equation for Predicting IDI The line in Figure 2.9 is in fact the least-squares regression line for predicting the IDI from the MI. The equation of this line is

$$\hat{y} = 3.609 + 0.03872x$$

This line is estimated using data from 15 countries that ranked high on the IDI. ■

slope

The **slope** of a regression line is almost always important for interpreting the data. The slope is the rate of change, the amount of change in $\hat{y}$ when x increases by 1. The slope $b_1 = 0.03872$ in this example says that each additional dollar of median per capita daily income is associated with an additional 0.03872 unit for the IDI.

intercept

The **intercept** of the regression line is the value of $\hat{y}$ when $x = 0$. Although we need the value of the intercept to draw the line, it is statistically meaningful only when x can actually take values close to zero. In our example, $x = 0$ occurs when a country has a zero MI. Such a situation would be very unusual, and we would not include it within the framework of our analysis

prediction

The equation of the regression line makes **prediction** easy. Just substitute a value of x into the equation.

EXAMPLE 2.12

Predict the IDI To predict the net assets per capita for a country that has a MI of $40, we use $x = 40$:

$$\begin{aligned}\hat{y} &= 3.609 + 0.03872x \\ &= 3.609 + (0.03872)(40) \\ &= 3.609 + 1.5488 = 5.16\end{aligned}$$

The predicted IDI is 5.16. See Figure 2.9 (page 82) for a graphical illustration of this calculation. ■

plotting a line

To **plot the line** on the scatterplot, you can use the equation to find $\hat{y}$ for two values of x, one near each end of the range of x in the data. Plot each $\hat{y}$ above its x, and draw the line through the two points. *As a check, it is a good idea to compute y for a third value of x and verify that this point is on your line.*

APPLY YOUR KNOWLEDGE

2.46 A regression line. A regression equation is $y = 10 + 20x$.

(a) What is the slope of the regression line?

(b) What is the intercept of the regression line?

(c) Find the predicted values of y for $x = 0$, for $x = 20$, and for $x = 40$.

(d) Plot the regression line for values of x between 0 and 40.

EXAMPLE 2.13

IDI coefficient

MI and IDI Results Using Software Figure 2.11 displays the selected regression output for the World Economic Forum data from Minitab, Excel, and JMP. The complete outputs contain many other items that we will study in Chapter 12.

Let's look at the Minitab output first. A table, labeled "Coefficients," gives the estimated regression intercept and slope under the column heading "Coef." **Coefficient** is a generic term that refers to the quantities that define a regression equation. Note that the intercept is labeled "Constant," and the slope is labeled with the name of the explanatory variable. In the table, Minitab reports the intercept as 3.609 and the slope as 0.03872, followed by the regression equation.

Excel provides the same information in a slightly different format. Here the intercept is reported as 3.608585784, and the slope is reported as 0.03872102. Check the JMP output to see how the regression coefficients are reported there. ■

FIGURE 2.11 Selected least-squares regression output for the world financial markets data. (a) Minitab. (b) Excel. (c) JMP.

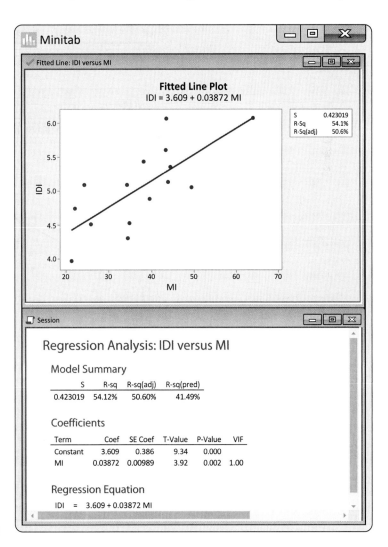

(a)

FIGURE 2.11 Continued

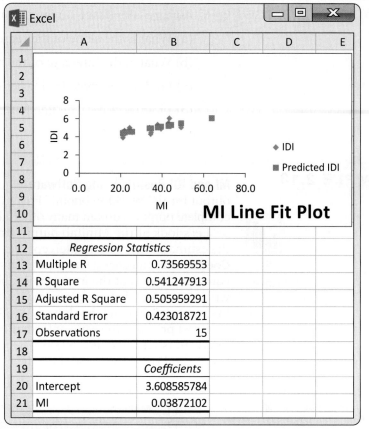

(b)

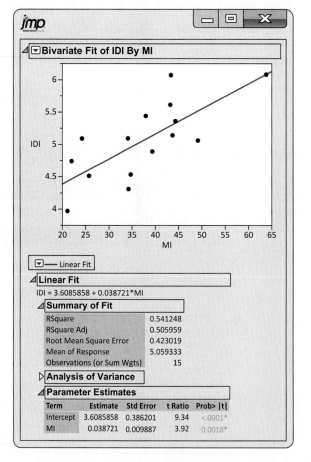

(c)

How many digits should we keep when reporting the results of statistical calculations? The answer depends on how the results will be used. For example, if we are giving a description of the equation, then rounding the coefficients and reporting the equation as $\hat{y} = 3.6 + 0.039x$ would be fine. If we will use the equation to calculate predicted values, we should keep a few more digits and then round the resulting calculation as we did in Example 2.12.

APPLY YOUR KNOWLEDGE

2.47 Predicted values and residuals for MI and IDI. Refer to the World Economic Forum data in Example 2.9. IDI

(a) Use software to compute the coefficients of the regression equation. Indicate where to find the slope and the intercept on the output, and report these values.

(b) Make a scatterplot of the data with the least-squares line.

(c) Find the predicted value of assets for each country.

(d) Find the difference between the actual value and the predicted value for each country.

Facts about least-squares regression

The use of regression to describe the relationship between a response variable and an explanatory variable is one of the most commonly encountered statistical methods, and least squares is the most commonly used technique for fitting a regression line to data. Here are some facts about least-squares regression lines.

Fact 1. There is a close connection between correlation and the slope of the least-squares line. The slope is

$$b_1 = r \frac{s_y}{s_x}$$

This equation says that along the regression line, **a change of one standard deviation in x corresponds to a change of r standard deviations in y.** When the variables are perfectly correlated ($r = 1$ or $r = -1$), the change in the predicted response $\hat{y}$ is the same (in standard deviation units) as the change in x. Otherwise, because $-1 \leq r \leq 1$, the change in $\hat{y}$ is less than the change in x. As the correlation becomes weaker, the prediction $\hat{y}$ moves less strongly in response to changes in x.

Fact 2. The least-squares regression line always passes through the point $(\bar{x}, \bar{y})$ on the graph of y against x. So the least-squares regression line of y on x is the line with slope rs_y/s_x that passes through the point $(\bar{x}, \bar{y})$. We can describe regression entirely in terms of the basic descriptive measures $\bar{x}$, s_x, $\bar{y}$, s_y, and r.

Fact 3. The distinction between explanatory and response variables is essential in regression. Least-squares regression looks at the distances of the data points from the line only in the y direction. If we reverse the roles of the two variables, we get a different least-squares regression line.

EXAMPLE 2.14

Education Spending and Population Figure 2.12 is a scatterplot of the education spending data described in Case 2.1 (page 66). There is a positive linear relationship.

FIGURE 2.12 Scatterplot of spending on education versus the population. The two lines are the least-squares regression lines: using population to predict spending on education (solid) and using spending on education to predict population (dashed), Example 2.14.

EDSPEND

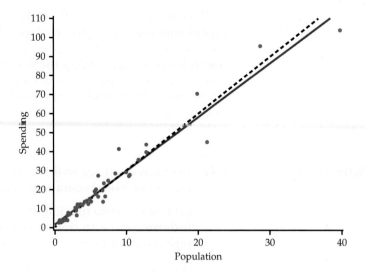

The two lines on the plot are the two least-squares regression lines. The regression line for using population to predict education spending is solid, while the regression line for using education spending to predict population is dashed. *The two regressions give different lines.* In the regression setting, you must choose one variable to be the explanatory variable. ∎

Interpretation of r^2

The square of the correlation r describes the strength of a straight-line relationship. Here is the basic idea. Think about trying to predict a new value of y. With no other information than our sample of values of y, a reasonable choice is $\bar{y}$.

Now consider how your prediction would change if you had an explanatory variable. If we use the regression equation for the prediction, we would use the equation $\hat{y} = b_0 + b_1 x$. This prediction takes into account the value of the explanatory variable x.

Let's compare our two choices for predicting y. With the explanatory variable x, we use $\hat{y}$; without this information, we use $\bar{y}$, the sample of the response variable. How can we compare these two choices? When we use $\bar{y}$ to make a prediction, our prediction error is $y - \bar{y}$. If, instead, we use $\hat{y}$, our prediction error is $y - \hat{y}$. The use of x in our prediction changes our prediction error from $y - \bar{y}$ to $y - \hat{y}$; thus the difference is $\hat{y} - \bar{y}$. Our comparison uses the sums of squares of these differences $\Sigma(y - \bar{y})^2$ and $\Sigma(\hat{y} - \bar{y})^2$. The ratio of these two quantities is the square of the correlation:

$$r^2 = \frac{\Sigma(\hat{y} - \bar{y})^2}{\Sigma(y - \bar{y})^2}$$

The numerator represents the variation in y that is explained by x, and the denominator represents the total variation in y. In Chapter 13, where $\hat{y}$ is a combination of several explanatory variables, we use R^2 to denote this quantity.

> **PERCENT OF VARIATION EXPLAINED BY THE LEAST-SQUARES EQUATION**
>
> To find the percent of variation explained by the least-squares equation, square the value of the correlation and express the result as a percent.

EXAMPLE 2.15

Using r^2 The correlation between the MI and the IDI in Example 2.12 (page 84) is $r = 0.73570$, so $r^2 = 0.54125$. In other words, median per capita daily income explains about 54% of the variability in the Inclusive Development Index. ∎

When you report a regression, give r^2 as a measure of how successful the regression was in explaining the response. The software outputs in Figure 2.11 (pages 85–86) include r^2, either in decimal form or as a percent. *When you see a correlation (often listed as R or Multiple R in outputs), square it to get a better feel for the strength of the association.*

APPLY YOUR KNOWLEDGE

2.48 The "January effect." Some people think that the behavior of the stock market in January predicts its behavior for the rest of the year. Take the explanatory variable x to be the percent change in a stock market index in January and the response variable y to be the change in the index for the entire year. Calculation based on 38 years of data gives

$$\bar{x} = 1.75\% \qquad s_x = 5.36\% \qquad r = 0.596$$
$$\bar{y} = 9.07\% \qquad s_y = 15.35\%$$

(a) What percent of the observed variation in yearly changes in the index is explained by a straight-line relationship with the change during January?

(b) What is the equation of the least-squares line for predicting the full-year change from the January change?

(c) The mean change in January is $\bar{x} = 1.75\%$. Use your regression line to predict the change in the index in a year in which the index rises 1.75% in January.

2.49 Is regression useful? In Exercise 2.39 (page 80), you used the *Correlation and Regression* applet to create three scatterplots having correlation about $r = 0.8$ between the horizontal variable x and the vertical variable y. Create three similar scatterplots again, after clicking the "Show least-squares line" box to display the regression line. Correlation $r = 0.8$ is considered reasonably strong in many areas of work. Because there is a reasonably strong correlation, we might use a regression line to predict y from x. In which of your three scatterplots does it make sense to use a straight line for prediction?

Residuals

A regression line is a mathematical model for the overall pattern of a linear relationship between an explanatory variable and a response variable. Deviations from the overall pattern are also important. In the regression setting, we identify deviations by looking at the scatter of the data points about the regression line. The vertical distances from the points to the least-squares regression line are as small as possible in the sense that they have the smallest possible sum of squares. Because they represent "leftover" variation in the response after fitting the regression line, these distances are called *residuals*.

RESIDUALS

A **residual** is the difference between an observed value of the response variable and the value predicted by the regression line. That is,

$$\text{residual} = \text{observed } y - \text{predicted } y$$
$$= y - \hat{y}$$

EXAMPLE 2.16

EDSPEND

Education Spending and Population Figure 2.13 is a scatterplot showing education spending versus the population for the 50 states that we studied in Case 2.1 (page 66). Included on the scatterplot is the least-squares line. The points for the states with large values for both variables—California, Texas, Florida, and New York—are marked individually.

The equation of the least-squares line is $\hat{y} = 1.38017 + 2.82654x$, where $\hat{y}$ represents education spending and x represents the population of the state.

Let's look carefully at the data for California, $y = 103.7$ and $x = 39.8$. The predicted education spending for a state with 39.8 million people is

$$\hat{y} = 1.38017 + 2.82654(39.8)$$
$$= 113.88$$

The residual for California is the difference between the observed spending (y) and this predicted value.

$$\text{residual} = y - \hat{y}$$
$$= 103.70 - 113.88$$
$$= -10.18 \blacksquare$$

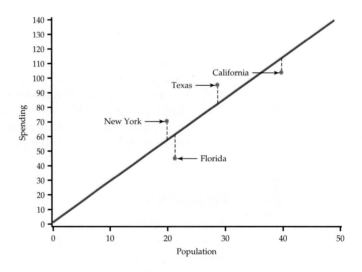

FIGURE 2.13 Scatterplot of spending on education versus the population for 50 states, with the least-squares line and selected points labeled, Example 2.16.

California spends $10.18 million less on education than the least-squares regression line predicts. On the scatterplot, the residual for California is shown as a dashed vertical line between the actual spending and the least-squares line. It appears below the line because the residual is negative.

APPLY YOUR KNOWLEDGE

2.50 Residual for Texas. Refer to Example 2.16. Texas spent $95.3 million on education and has a population of 28.7 million people.

(a) Find the predicted education spending for Texas.

(b) Find the residual for Texas.

(c) Which state, California or Texas, has a greater deviation from the regression line?

There is a residual for each data point. Finding the residuals with a calculator is a bit unpleasant, because you must first find the predicted response for every x. Statistical software gives you the residuals all at once.

Because the residuals show how far the data fall from our regression line, examining the residuals helps us assess how well the line describes the data. Although residuals can be calculated from any model fitted to data, the residuals from the least-squares line have a special property: **the mean of the least-squares residuals is always zero.** When we compute with real data, the residuals are generally rounded off, so the sum will be very small but not always exactly zero.

APPLY YOUR KNOWLEDGE

2.51 Sum the education spending residuals. In addition to the variables described in Case 2.1 (page 66), the EDSPEND data set contains the residuals rounded to two places after the decimal. Find the sum of these residuals. Is the sum exactly zero? If not, explain why. EDSPEND

As usual, when we perform statistical calculations, we prefer to display the results graphically. We can do this for the residuals.

RESIDUAL PLOTS

A **residual plot** is a scatterplot of the regression residuals against the explanatory variable. Residual plots help us assess the fit of a regression line.

EXAMPLE 2.17

Residual Plot for Education Spending Figure 2.14 gives the residual plot for the education spending data. The horizontal line at zero in the plot helps orient us. ∎

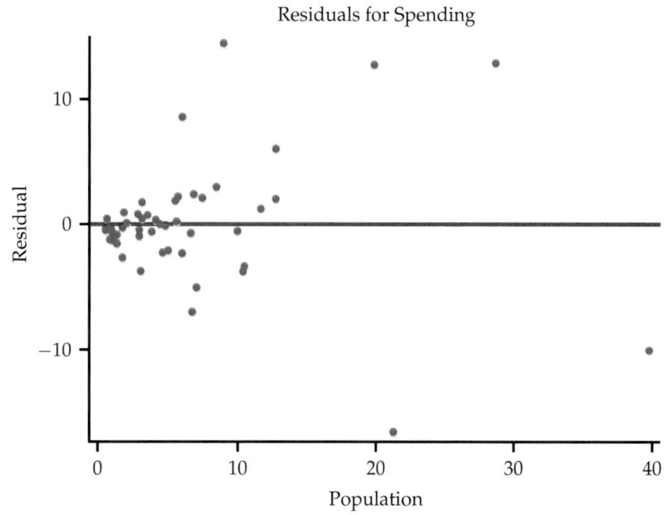

FIGURE 2.14 Residual plot for the education spending data, Example 2.17.

APPLY YOUR KNOWLEDGE

2.52 Identify the four states. In Figure 2.13, four states are identified by name: California, Texas, Florida, and New York. The dashed lines in the plot represent the residuals. **EDSPEND**

(a) Sketch a version of Figure 2.14 or generate your own plot using the EDSPEND data file. Write in the names of the states California, Texas, Florida, and New York on your plot.

(b) Explain how you were able to identify these four points on your sketch.

If the regression line captures the overall relationship between x and y, the residuals should have no systematic pattern. The residual plot will look something like the pattern in Figure 2.15(a). That plot shows a scatter of points about the fitted line, with no unusual individual observations or systematic change as x increases. Here are some things to look for when you examine a residual plot:

- **A curved pattern,** which shows that the relationship is not linear. Figure 2.15(b) is a simplified example. A straight line is not a good summary for such data.

- **Increasing or decreasing spread about the line** as x increases. Figure 2.15(c) is a simplified example. Prediction of y will be less precise for larger x in that example.

- **Individual points with large residuals,** which are outliers in the vertical (y) direction because they lie far from the line that describes the overall pattern.

- **Individual points that are extreme in the x direction,** like California in Figure 2.14. Such points may or may not have large residuals, but they can be very important. We address such points next.

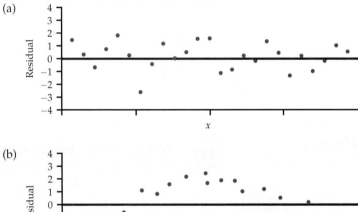
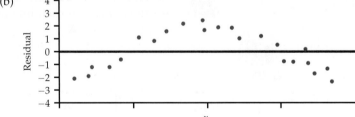
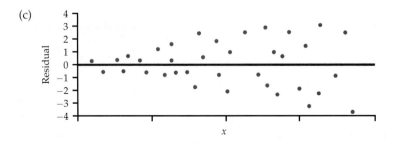

FIGURE 2.15 Idealized patterns in plots of least-squares residuals. Plot (a) indicates that the regression line fits the data well. The data in plot (b) have a curved pattern, so a straight line fits poorly. The response variable y in plot (c) has more spread for larger values of the explanatory variable x, so predictions will be less accurate when x is large.

The distribution of the residuals

When we compute the residuals, we are creating a new quantitative variable for our data set. Each case has a value for this variable. It is natural to ask about the distribution of this variable. We already know that the mean is zero, but we can use the methods we learned in Chapter 1 to examine other characteristics of the distribution. As we will see in Chapter 10, a question of interest with respect to residuals is whether they are approximately Normal. Recall that we used Normal quantile plots to address this issue.

Normal quantile plots, p. 53

EXAMPLE 2.18

EDSPEND

Are the Residuals Approximately Normal? Figure 2.16 gives the Normal quantile plot for the residuals in our education spending example. The distribution of the residuals is not Normal. Most of the points are close to a line in the center of the plot, but there appear to be four outliers—one with a negative residual and three with positive residuals. ■

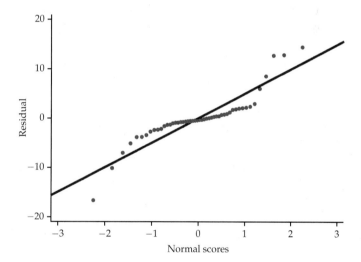

FIGURE 2.16 Normal quantile plot of the residuals for the education spending regression, Example 2.18.

Take a look at the plot of the data with the least-squares line in Figure 2.2 (page 68). Note that you can see the same four points in this plot. If we eliminated these states from our data set, the remaining residuals would be approximately Normal. However, there is nothing wrong with the data for these four states. To account for them, a complete analysis of the data should include a statement that they are somewhat extreme relative to the distribution of the other states.

Influential observations

In the scatterplot of spending on education versus population in Figure 2.12 (page 88), California, Texas, Florida, and New York have somewhat higher values for both variables than the other 46 states. This could be of concern if these cases distort the least-squares regression line. A case that has a big effect on a numerical summary is called **influential.**

influential

EXAMPLE 2.19

Is California Influential? To answer this question, we compare the regression lines with and without California. The result is shown in Figure 2.17. The two lines are very close, so we conclude that California is not influential with respect to the least-squares slope and intercept. ∎

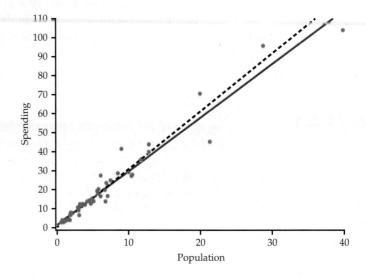

FIGURE 2.17 Two least-squares lines for the education spending data, Example 2.19. The solid line is calculated using all the data. The dashed line leaves out the data for California. The two lines are very similar, so we conclude that California is not influential.

Let's think about a situation in which California would be influential in regard to the least-squares regression line. California's spending on education is $103.7 million. This case is close to both least-squares regression lines in Figure 2.17. Suppose California's spending was much less than $103.7 million. Would this case then become influential?

EXAMPLE 2.20

Suppose California Spent Half as Much? What would happen if California spent about half of what this state actually spent on education—say, $50 million. Figure 2.18 shows the two regression lines, with and without California. Here we see that the regression line changes substantially when California is removed. Therefore, in this setting we would conclude that California is very influential. ∎

FIGURE 2.18 Two least-squares lines for the education spending data, with the California education spending changed to $50 million, Example 2.20. The solid line is calculated using all the data. The dashed line leaves out the data for California, which is influential here. California pulls the least-squares regression line toward it.

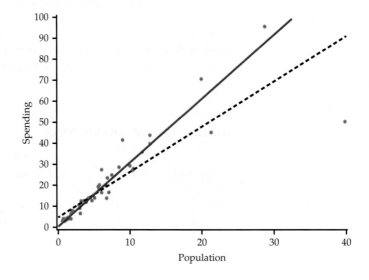

OUTLIERS AND INFLUENTIAL CASES IN REGRESSION

An **outlier** is an observation that lies outside the overall pattern of the other observations. Points that are outliers in the *y* direction of a scatterplot have large regression residuals, but other outliers need not have large residuals.

A case is **influential** for a statistical calculation if removing it would markedly change the result of the calculation. Points that are extreme in the *x* direction of a scatterplot are often influential for the least-squares regression line.

APPLY YOUR KNOWLEDGE

CASE 2.1 **2.53 The influence of Texas.** Make a plot similar to Figure 2.19 giving regression lines with and without Texas. Summarize what this plot describes. EDSPEND

California, Texas, Florida, and New York are somewhat unusual and might be considered outliers. However, these cases are not influential with respect to the least-squares regression line.

Influential cases may have small residuals because they pull the regression line toward themselves. That is, you can't always rely on residuals to point out influential observations. Influential observations can change the interpretation of data. For a linear regression, we compute a slope, an intercept, and a correlation. An individual observation can be influential for one of more of these quantities.

EXAMPLE 2.21

Effects on the Correlation The correlation between the spending on education and population for the 50 states is $r = 0.973$. If we drop California from the data set, the correlation decreases to 0.966. We conclude that California is not influential for the correlation. ∎

The best way to grasp the important idea of influence is to use an interactive animation that allows you to move points on a scatterplot and observe how the correlation and regression respond to these changes. The *Correlation and Regression* applet on the text website allows you to do this. Exercises 2.73 and 2.74 later in this chapter guide the use of this applet.

SECTION 2.3 SUMMARY

- A **regression line** is a straight line that describes how a response variable *y* changes as an explanatory variable *x* changes.

- The most common method of fitting a line to a scatterplot is least squares. The **least-squares regression line** is the straight line $\hat{y} = b_0 + b_1 x$ that minimizes the sum of the squares of the vertical distances of the observed points from the line.

- You can use a regression line to **predict** the value of *y* for any value of *x* by substituting this *x* into the equation of the line.

- The **slope** b_1 of a regression line $\hat{y} = b_0 + b_1 x$ is the rate at which the predicted response $\hat{y}$ changes along the line as the explanatory variable *x* changes. Specifically, b_1 is the change in $\hat{y}$ when *x* increases by 1.

- The **intercept** b_0 of a regression line $\hat{y} = b_0 + b_1 x$ is the predicted response $\hat{y}$ when the explanatory variable $x = 0$. This prediction is of no statistical use unless x can actually take values near 0.

- The least-squares regression line of y on x is the line with slope $b_1 = rs_y/s_x$ and intercept $b_0 = \bar{y} - b_1\bar{x}$. This line always passes through the point $(\bar{x}, \bar{y})$.

- **Correlation and regression** are closely connected. The correlation r is the slope of the least-squares regression line when we measure both x and y in standardized units.

- The square of the correlation r^2 is the fraction of the variability of the response variable that is explained by the explanatory variable using least-squares regression.

- You can examine the fit of a regression line by studying the **residuals,** which are the differences between the observed and predicted values of y. Be on the lookout for outlying points with unusually large residuals, as well as for nonlinear patterns and uneven variation about the line.

- Also look for **influential observations,** individual points that substantially change the regression line. Influential observations are often outliers in the x direction, but they need not have large residuals.

SECTION 2.3 EXERCISES

For Exercises 2.47 and 2.48, see pages 87–89; for 2.49, see page 89; for 2.50, see page 91; for 2.51 and 2.52, see pages 91–92; and for 2.53, see page 95.

2.54 What is the equation for the selling price? You buy items at a cost of x and sell them for y. Assume that your selling price includes a profit of 15% plus a fixed cost of $40.00. Give an equation that can be used to determine y from x.

2.55 Production costs for cell phone batteries. A company manufactures batteries for cell phones. The overhead expenses of keeping the factory operational for a month—even if no batteries are made—total $800,000. Batteries are manufactured in lots (1000 batteries per lot), with each lot costing $9000 to make. In this scenario, $800,000 is the *fixed* cost associated with producing cell phone batteries and $9000 is the *marginal* (or *variable*) cost of producing each lot of batteries. The total monthly cost y of producing x lots of cell phone batteries is given by the equation

$$y = 800{,}000 + 9000x$$

(a) Draw a graph of this equation. (Choose two values of x, such as 0 and 20, to draw the line and a third for a check. Compute the corresponding values of y from the equation. Plot these two points on graph paper and draw the straight line joining them.)

(b) What will it cost to produce 12 lots of batteries (12,000 batteries)?

(c) If each lot cost $12,000 instead of $9000 to produce, what is the equation that describes total monthly cost for x lots produced?

2.56 Inventory of Blu-Ray players. A local consumer electronics store sells exactly 10 Blu-Ray players of a particular model each week. The store expects no more shipments of this particular model, and it has 140 such units in its current inventory.

(a) Give an equation for the number of Blu-Ray players of this particular model in inventory after x weeks. What is the slope of this line?

(b) Draw a graph of this line between now (Week 0) and Week 14.

(c) Would you be willing to use this line to predict the inventory after 20 weeks? Calculate the prediction and think about the reasonableness of the result.

2.57 Compare cell phone payment plans. A cellular telephone company offers two plans. Plan A charges $40 per month for up to 120 minutes of airtime and $0.55 per minute after 120 minutes. Plan B charges $45 per month for up to 220 minutes and $0.50 per minute after 200 minutes.

(a) Draw a graph of the Plan A charge against minutes used from 0 to 250 minutes.

(b) How many minutes a month must the user talk for Plan B to be less expensive than Plan A?

2.58 Companies of the world. Refer to Exercise 1.118 (page 61), where we examined data collected by the World Bank on the numbers of companies that are incorporated and listed on their country's stock exchange at the end of the year. In Exercise 2.10, you examined the relationship between these numbers for 2017 and 2007, and in Exercise 2.27, you found the correlation between these two variables. INCCOM

(a) Find the least-squares regression equation for predicting the 2017 numbers using the 2007 numbers.

(b) France had 465 companies in 2017 and 707 companies in 2007. Use the least-squares regression equation to find the predicted number of companies in 2017 for France.

(c) Find the residual for France.

2.59 Companies of the world. Refer to the previous exercise and to Exercise 2.11 (page 73). France had 740 companies in 1997. Answer parts (a), (b), and (c) of the previous exercise for 2017 and 1997. Compare the results you found in the previous exercise with the ones you found in this exercise. Explain your findings in a short paragraph. INCCOM

2.60 A product for lab experiments. In Exercise 2.17 (page 73), you described the relationship between time and count for an experiment examining the decay of barium. In Exercise 2.29 (page 79), you found the correlation between these two variables. DECAY

(a) Find the least-squares regression equation for predicting count from time.

(b) Use the equation to predict the count at one, three, five, and seven minutes.

(c) Find the residuals for one, three, five, and seven minutes.

(d) Plot the residuals versus time.

(e) What does this plot tell you about the model you used to describe this relationship?

2.61 Use a log for the radioactive decay. Refer to the previous exercise. Also see Exercise 2.18 (page 73), where you transformed the counts with a logarithm, and Exercise 2.30 (page 79), where you found the correlation between time and the log of the counts. Answer parts (a) to (d) of the previous exercise for the transformed counts, and compare the results with those you found in the previous exercise. DECAY

2.62 Fuel efficiency and CO_2 emissions. In Exercise 2.37 (page 80), you examined the relationship between highway MPG and city MPG for 1045 vehicles for the model year 2018. CANFUEL

(a) Use the city MPG to predict the highway MPG. Give the equation of the least-squares regression line.

(b) Find the city MPG and the highway MPG for the Lexus 450h AWD. Use your equation to find the predicted highway MPG for this vehicle.

(c) Find the residual.

2.63 Fuel efficiency and CO_2 emissions. Refer to the previous exercise. CANFUEL

(a) Make a scatterplot of the data with highway MPG as the response variable and city MPG as the explanatory variable. Include the least-squares regression line on the plot. There is an unusual pattern for the vehicles with high city MPG. Describe it.

(b) Make a plot of the residuals versus city MPG. Describe the major features of this plot. How does the unusual pattern noted in part (a) appear in this plot?

(c) The Lexus 450h AWD that you examined in parts (b) and (c) of the previous exercise is in the group of unusual cases mentioned in parts (a) and (b) of this exercise. This hybrid vehicle uses a conventional engine and a electric motor that is powered by a battery that can recharge itself when the vehicle is driven. The conventional engine also turns off when the vehicle is stopped in traffic. As a result of these features, hybrid vehicles are unusually efficient during city driving, but they do not have a similar advantage when driven at higher speeds on the highway. How do these facts explain the residual for this vehicle?

(d) Several Toyota vehicles are also hybrids. Use the residuals to suggest which vehicles are in this category.

2.64 Consider the fuel type. Refer to the previous two exercises and to Figure 2.6 (page 71), where different colors are used to distinguish four different types of fuels used by these vehicles. In Exercise 2.38, you examined the relationship between highway MPG and city MPG for each of the four different fuel types used by these vehicles. Using the previous two exercises as a guide, analyze these data separately for each of the four fuel types. Write a summary of your findings. CANFUEL

2.65 Predict one characteristic of a product using another characteristic. In Exercise 2.12 (page 73), you used a scatterplot to examine the relationship between calories per 12 ounces and percent alcohol in 175 domestic brands of beer. In Exercise 2.31, you calculated the correlation between these two variables. BEER

(a) Find the equation of the least-squares regression line for these data.

(b) Make a scatterplot of the data with the least-squares regression line.

2.66 Predicted values and residuals. Refer to the previous exercise. BEER

(a) Sierra Nevada India Pale Ale is 6.9% alcohol and has 231 calories per 12 ounces. Find the predicted calories for Sierra Nevada India Pale Ale.

(b) Find the residual for Sierra Nevada India Pale Ale.

2.67 Predicted values and residuals. Refer to the previous two exercises. BEER

(a) Make a plot of the residuals versus percent alcohol.

(b) Interpret the plot. Do you see any systematic pattern? Explain your answer.

(c) Examine the plot carefully and determine the approximate location of New Belgium Fat Tire. Is there anything unusual about this case? Explain why or why not.

2.68 Carbohydrates and alcohol in beer revisited. Refer to Exercise 2.65. The data that you used to compute the least-squares regression line includes a beer with a very low alcohol content that might be considered to be an outlier. BEER

(a) Remove this case and recompute the least-squares regression line.

(b) Make a graph of the regression lines with and without this case.

(c) Do you think that this case is influential? Explain your answer.

2.69 Monitoring the water quality near a manufacturing plant. Manufacturing companies (and the Environmental Protection Agency) monitor the quality of the water near manufacturing plants. Measurements of pollutants in water are indirect—a typical analysis involves forming a dye by a chemical reaction with the dissolved pollutant, then passing light through the solution and measuring its "absorbance." To calibrate such measurements, the laboratory measures known standard solutions and uses regression to relate absorbance to pollutant concentration. This is usually done every day. Here is one series of data on the absorbance for different levels of nitrates. Nitrates are measured in milligrams per liter of water.[11] NRATES

Nitrates	Absorbance	Nitrates	Absorbance
50	7.0	800	93.0
50	7.5	1200	138.0
100	12.8	1600	183.0
200	24.0	2000	230.0
400	47.0	2000	226.0

(a) Chemical theory says that these data should lie on a straight line. If the correlation is not at least 0.997, something went wrong and the calibration procedure is repeated. Plot the data and find the correlation. Must the calibration be done again?

(b) What is the equation of the least-squares line for predicting absorbance from concentration? If the lab analyzed a specimen with 500 milligrams of nitrates per liter, what do you expect the absorbance to be? Based on your plot and the correlation, do you expect your predicted absorbance to be very accurate?

2.70 Data generated by software. The following 20 observations on y and x were generated by a computer program. GENDATA

y	x	y	x
34.38	22.06	27.07	17.75
30.38	19.88	31.17	19.96
26.13	18.83	27.74	17.87
31.85	22.09	30.01	20.20
26.77	17.19	29.61	20.65
29.00	20.72	31.78	20.32
28.92	18.10	32.93	21.37
26.30	18.01	30.29	17.31
29.49	18.69	28.57	23.50
31.36	18.05	29.80	22.02

(a) Make a scatterplot and describe the relationship between y and x.

(b) Find the equation of the least-squares regression line and add the line to your plot.

(c) Plot the residuals versus x.

(d) What percent of the variability in y is explained by x?

(e) Summarize your analysis of these data in a short paragraph.

2.71 Add an outlier. Refer to the previous exercise. Add an additional case with $y = 65$ and $x = 35$ to the data set. Repeat the analysis that you performed in the previous exercise and summarize your results, paying particular attention to the effect of this outlier. GENDATB

2.72 Add a different outlier. Refer to the previous two exercises. Add an additional case with $y = 65$ and $x = 19$ to the original data set. GENDATC

(a) Repeat the analysis that you performed in Exercise 2.70 and summarize your results, paying particular attention to the effect of this outlier.

(b) In this exercise and in the previous one, you added an outlier to the original data set and reanalyzed the data. Write a short summary of the changes in correlations that can result from different kinds of outliers.

2.73 Influence on correlation. The *Correlation and Regression* applet at the text website allows you to create a scatterplot and to move points by dragging with the mouse. Click to create a group of 10 points in the lower-left corner of the scatterplot with a strong straight-line pattern (correlation about 0.9).

(a) Add one point at the upper right that is in line with the first 10. How does the correlation change?

(b) Drag this last point down until it is opposite the group of 10 points. How small can you make the correlation? Can you make the correlation negative? Notice how a single outlier can greatly strengthen or weaken a correlation. Always plot your data to check for outlying points.

2.74 Influence in regression. As in the previous exercise, create a group of 10 points in the lower-left corner of the scatterplot with a strong straight-line pattern (correlation at least 0.9). Click the "Show least-squares line" box to display the regression line.

(a) Add one point at the upper right that is far from the other 10 points but exactly on the regression line. Why does this outlier have no effect on the line even though it changes the correlation?

(b) Now drag this last point down until it is opposite the group of 10 points. You see that one end of the least-squares line chases this single point, while the other end remains near the middle of the original group of 10. What aspect of the last point makes it so influential?

2.75 Employee absenteeism and raises. Data on number of days of work missed and annual salary increase for a company's employees show that, in general, employees who missed more days of work during the year received smaller raises than those who missed fewer days. Number of days missed explained 36% of the variation in salary increases. What is the numerical value of the correlation between number of days missed and salary increase?

2.76 Always plot your data! Four sets of data prepared by the statistician Frank Anscombe illustrate the dangers of calculating without first plotting the data.[12] ANSDATA

(a) Without making scatterplots, find the correlation and the least-squares regression line for all four data sets. What do you notice? Use the regression line to predict y for $x = 10$.

(b) Make a scatterplot for each of the data sets, and add the regression line to each plot.

(c) In which of the four cases would you be willing to use the regression line to describe the dependence of y on x? Explain your answer in each case.

2.4 Cautions about Correlation and Regression

When you complete this section, you will be able to:
- Identify and describe situations where extrapolation should be avoided.
- Identify lurking variables that can influence the interpretation of relationships between two variables.
- Explain the difference between association or correlation and causality when interpreting the relationship between two variables.

Correlation and regression are powerful tools for describing the relationship between two variables. When you use these tools, you must be aware of their limitations, beginning with the fact that **correlation and regression describe only linear relationships.** Also remember that **the correlation r and the least-squares regression line are not resistant.** One influential observation or incorrectly entered data point can greatly change these measures. For this reason, you should always plot your data before interpreting regression or correlation. Here are some other cautions to keep in mind when you apply correlation and regression or read accounts of their use.

Extrapolation

Associations for variables can be trusted only for the range of values for which data have been collected. Even a very strong relationship may not hold outside the data's range.

EXAMPLE 2.22

TARGET

Predicting the Number of Target Stores Here are data on the number of Target stores in operation at the end of each year in the early 1990s, in 2008, in 2014, and in 2018:[13]

Year (x)	1990	1991	1992	1993	2008	2014	2018
Stores (y)	420	463	506	554	1682	1916	1829

A plot of these data is given in Figure 2.19. The data for 1990 through 1993 lie almost exactly on a straight line, which we calculated using only the data from 1990 to 1993. The equation of this line is $y = -88{,}136 + 44.5x$ and $r^2 = 0.9992$. We know that 99.92% of the variation in the number of stores is explained by year for these years. The equation predicts 1220 stores for 2008, but the actual number of stores is much higher, 1682. It predicts 1487 for 2014 and 1665 for 2018. These predictions are very poor because the very strong linear trend evident in the 1990 to 1993 data did not continue to the years 2008, 2014, and 2018. ∎

FIGURE 2.19 Plot of the number of Target stores versus year with the least-squares regression line calculated using data from 1990, 1991, 1992, and 1993, Example 2.22. The poor fits to the number of stores in 2008, 2014, and 2018 illustrate the dangers of extrapolation.

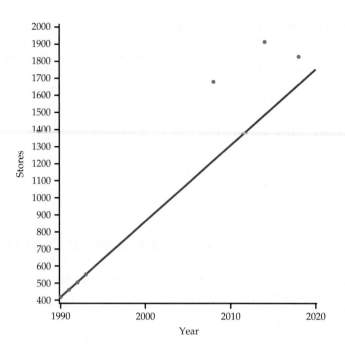

Predictions made far beyond the range for which data have been collected can't be trusted. Few relationships are linear for *all* values of x. It is risky to stray far from the range of x-values that actually appear in your data.

EXTRAPOLATION

Extrapolation is the use of a regression line for prediction far outside the range of values of the explanatory variable x that you used to obtain the line. Such predictions are often not accurate.

In general, extrapolation involves using a mathematical relationship beyond the range of the data that were used to estimate the relationship. The scenario described in the previous example is typical: we try to use a least-squares relationship to make predictions for values of the explanatory variable that are much larger than the values in the data that we have. We can encounter the same difficulty when we attempt to make predictions for values of the explanatory variable that are much smaller than the values in the data that we have.

Careful judgment is needed when making predictions. If the prediction is for values that are within the range of the data that you have, or for values that are not too far above or below the data, then your prediction may be reasonably accurate. Beyond that, you are in danger of making an inaccurate prediction.

Lurking variables

Correlation and regression describe the relationship between two variables. Often, the relationship between two variables is strongly influenced by other variables. We try to measure such potentially influential variables. We can then use more advanced statistical methods to examine all the relationships revealed by our data. Sometimes, however, the relationship between two variables is influenced by other variables that we did not measure or even think about. Variables lurking in the background—measured or not—often help explain statistical associations.

2.4 Cautions about Correlation and Regression

LURKING VARIABLE

A **lurking variable** is a variable that is not among the explanatory or response variables in a study, yet may influence the interpretation of relationships among those variables.

A lurking variable can falsely suggest a strong relationship between x and y, or it can hide a relationship that is really there. Here is an example of a negative correlation that is due to a lurking variable.

EXAMPLE 2.23

Gas and Electricity Bills A single-family household receives bills for gas and electricity each month. The 12 observations for a recent year are plotted with the least-squares regression line in Figure 2.20. We have arbitrarily chosen to put the electricity bill on the x axis and the gas bill on the y axis. There is a clear negative association. Does this mean that a high electricity bill causes the gas bill to be low, and vice versa?

To understand the association in this example, we need to know a little more about the two variables. In this household, heating is done by gas and cooling by electricity. Therefore, in the winter months, the gas bill will be relatively high and the electricity bill will be relatively low. The pattern is reversed in the summer months. The association that we see in this example is due to a lurking variable: time of year. ∎

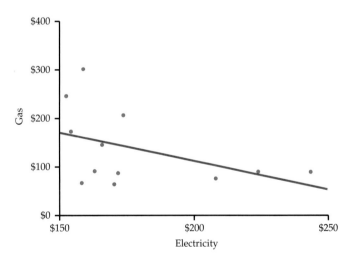

FIGURE 2.20 Scatterplot with the least-squares regression line for predicting monthly charges for gas using monthly charges for electricity for a household, Example 2.23.

APPLY YOUR KNOWLEDGE

2.77 Education and income. There is a strong positive correlation between years of education and income for economists employed by business firms. In particular, economists with a doctorate earn more than economists with only a bachelor's degree. There is also a strong positive correlation between years of education and income for economists employed by colleges and universities. But when all economists are considered, there is a *negative* correlation between education and income. The explanation for this relationship is that business pays high salaries and employs mostly economists with bachelor's degrees, while colleges pay lower salaries and employ mostly economists with doctorates. Sketch a scatterplot with two groups of cases (business and academic) illustrating how a strong positive correlation within each group and a negative overall correlation can occur together.

Correlation does not imply causation

When we study the relationship between two variables, we often hope to show that changes in the explanatory variable *cause* changes in the response variable. But a strong association or correlation between two variables is not enough to draw conclusions about cause and effect. Sometimes, an observed association really does reflect cause and effect. Natural gas consumption in a household that uses natural gas for heating will be higher in colder months because cold weather requires burning more gas to stay warm. In other cases, an association is explained by lurking variables, and the conclusion that x causes y is either wrong or not proved. Here is an example.

EXAMPLE 2.24

Does Television Extend Life? Measure the number of television sets per person x and the average life expectancy y for the world's nations. There is a high positive correlation: nations with many TV sets have higher life expectancies.

The basic meaning of causation is that by changing x, we can bring about a change in y. Could we lengthen the lives of people in Rwanda by shipping them TV sets? No. Rich nations have more TV sets than poor nations. Rich nations also have longer life expectancies because they offer better nutrition, clean water, and better health care. There is no cause-and-effect tie between TV sets and length of life. ■

Correlations such as that in Example 2.24 are sometimes called "nonsense correlations." The correlation is real. What is nonsense is the conclusion that changing one of the variables causes changes in the other. A lurking variable—such as national wealth—that influences both x and y can create a high correlation, even though there is no direct connection between x and y.

APPLY YOUR KNOWLEDGE

2.78 **How's your self-esteem?** People who do well tend to feel good about themselves. Perhaps helping people feel good about themselves will help them do better in their jobs and in life. For a time, raising self-esteem became a goal in many schools and companies. Can you think of explanations for the association between high self-esteem and good performance other than "Self-esteem causes better work"?

2.79 **Are big hospitals bad for you?** A study shows that there is a positive correlation between the size of a hospital (measured by its number of beds x) and the median number of days y that patients remain in the hospital. Does this mean that you can shorten a hospital stay by choosing a small hospital? Why?

2.80 **Do firefighters make fires worse?** Someone says, "There is a strong positive correlation between the number of firefighters at a fire and the amount of damage the fire does. So sending lots of firefighters just causes more damage." Explain why this reasoning is wrong.

These and other examples lead us to the most important caution about correlation, regression, and statistical association between variables in general.

> **ASSOCIATION DOES NOT IMPLY CAUSATION**
>
> An association between an explanatory variable x and a response variable y—even if it is very strong—is not, by itself, good evidence that changes in x actually cause changes in y.

experiment   The best way to get good evidence that x causes y is to do an **experiment** in which we change x and keep lurking variables under control. We will discuss experiments in Chapter 3. When experiments cannot be done, finding the explanation for an observed association is often difficult and controversial. Many of the sharpest disputes in which statistics plays a role involve questions of causation that cannot be settled by experiment. Does gun control reduce violent crime? Does cell phone usage cause brain tumors? Has increased free trade widened the gap between the incomes of more-educated and less-educated American workers? All of these questions have become public issues. All concern associations among variables. And all have this in common: they try to pinpoint cause and effect in a setting involving complex relations among many interacting variables.

BEYOND THE BASICS

Big data

Chapters 1 and 2 of this book are devoted to the important aspect of statistics called *exploratory data analysis* (EDA). We use graphs and numerical summaries to examine data, searching for patterns and paying attention to striking deviations from the patterns we find. In discussing regression, we advanced to using the pattern we find (in this case, a linear pattern) for prediction.

Today we have the capability of collecting very large amounts of data. We use the term *big data* to describe such data sets.[14] For example, we might have all stock market transactions in a day, a week, or a year; or all purchases recorded by the scanners of our retail food chain during the past week. Although many of the basic ideas of data analysis remain the same, the size of big data poses many new challenges.

- When you have 100 terabytes of data, even straightforward calculations and graphics can become very time-consuming. So, efficient algorithms are very important.

- The structure of the database and the process of storing the data, perhaps by unifying data scattered across many departments of a large corporation, require careful thought.

All of these features point to the need for sophisticated computer science. Many statistical ideas and tools, such as those for dealing with many variables in a data set, are very helpful.

SECTION 2.4 SUMMARY

- Correlation and regression must be **interpreted with caution. Plot the data** to be sure the relationship is roughly linear and to detect outliers and influential observations.

- Avoid **extrapolation,** the use of a regression line for making predictions at values of the explanatory variable far outside the range of the data from which the line was calculated.

- **Lurking variables** that you did not measure may explain the relations between the variables you did measure. Correlation and regression can be misleading if you ignore important lurking variables.

- Be careful not to conclude that there is a cause-and-effect relationship between two variables just because they are strongly associated. **High correlation does not imply causation.** The best evidence that an association is due to causation comes from an **experiment** in which the explanatory variable is directly changed and other influences on the response are controlled.

SECTION 2.4 EXERCISES

For Exercise 2.77, see page 101; and for 2.78 and 2.79, see page 102.

2.81 What's wrong? Each of the following statements contains an error. Describe each error and explain why the statement is wrong.

(a) A lurking variable is always a qualitative variable.

(b) If the residuals are all positive, then there will be a positive relationship between the response variable and the explanatory variable.

(c) A negative relationship is never due to causation.

2.82 What's wrong? Each of the following statements contains an error. Describe each error and explain why the statement is wrong.

(a) If we have data at values of x equal to 10, 20, 30, 40, and 50, and we try to predict the value of y at $x = 25$ using a least-squares regression line, we are extrapolating.

(b) In an experiment, the response variable is directly changed.

(c) An outlier will never have a large residual.

2.83 Predict the sales. You analyzed the past 10 years of sales data for your company, and the data fit a straight line very well. Do you think the equation you found would be useful for predicting next year's sales? Would your answer change if the prediction was for sales five years from now? Give reasons for your answers.

2.84 Older workers and income. The effect of a lurking variable can be surprising when cases are divided into groups. Explain how, as a nation's population grows older, mean income can go down for workers in each age group but still go up for all workers.

2.85 Marital status and income. Data show that married, divorced, and widowed men earn quite a bit more than men the same age who have never been married. This does not mean that a man can raise his income by getting married, because men who have never been married are different from married men in many ways other than marital status. Suggest several lurking variables that might help explain the association between marital status and income.

2.86 Sales at a farmers' market. You sell fruits and vegetables at your local farmers' market, and you keep track of your weekly sales. A plot of the data from May through August suggests a increase over time that is approximately linear, so you calculate the least-squares regression line. Your partner likes the plot and the line and suggests that you use it to estimate sales for the rest of the year. Explain why this is probably a very bad idea.

2.87 Does your product have an undesirable side effect? People who use artificial sweeteners in place of sugar tend to be heavier than people who use sugar. Does this mean that artificial sweeteners cause weight gain? Give a more plausible explanation for this association.

2.88 Does your product help nursing-home residents? A group of college students believes that herbal tea has remarkable powers. To test this belief, they make weekly visits to a local nursing home, where they visit with the residents and serve them herbal tea. The nursing-home staff reports that, after several months, many of the residents are healthier and more cheerful. We should commend the students for their good deeds, but doubt that herbal tea helped the residents. Identify the explanatory and response variables in this informal study. Then explain which lurking variables account for the observed association.

2.89 Education and income. There is a strong positive correlation between years of schooling completed x and lifetime earnings y for American men. One possible reason for this association is causation: more education leads to higher-paying jobs. But lurking variables may explain some of the correlation. Suggest some lurking variables that would explain why men with more education earn more.

2.90 Do power lines cause cancer? It has been suggested that electromagnetic fields of the kind present near power lines can cause leukemia in children. Experiments with children and power lines are not ethical. Careful studies have found no association between exposure to electromagnetic fields and childhood leukemia.[15] Suggest several lurking variables for which you would want information to investigate the claim that living near power lines is associated with cancer.

2.5 Data Analysis for Two-Way Tables

When you complete this section, you will be able to:

- Identify the row variable, the column variable, and the cells in a two-way table.
- Interpret the marginal distributions in a two-way table.
- Use the conditional distributions to describe the relationship displayed in a two-way table.
- Determine the marginal distributions and the conditional distributions in a two-way table from software output.
- Interpret examples of Simpson's paradox.

So far, we have concentrated on relationships in which at least the response variable is quantitative. Now we will shift to describing relationships between two or more categorical variables. Some variables—such as sex, race, and occupation—are categorical by nature. Other categorical variables are created by grouping values of a quantitative variable into classes. Published data often appear in grouped form to save space. To analyze categorical data, we use the *counts* or *percents* of cases that fall into various categories.

CASE 2.2 Does the Right Music Sell the Product? Market researchers know that background music can influence the mood and the purchasing behavior of customers. One study in a supermarket in Northern Ireland compared three treatments: no music, French accordion music, and Italian string music. Under each condition, the researchers recorded the numbers of bottles of French, Italian, and other wine purchased.[16] Here is the two-way table that summarizes the data:

WINE

Wine	Music			Total
	None	French	Italian	
French	30	39	30	99
Italian	11	1	19	31
Other	43	35	35	113
Total	84	75	84	243

two-way table
row and column variables

The data table for Case 2.2 is a **two-way table** because it describes two categorical variables. The type of wine is the **row variable** because each row in the table describes the data for one type of wine. The type of music played is the **column variable** because each column describes the data for one type of music. Each combination of the value of a row variable and the value of a column variable defines a **cell**. There are $3 \times 3 = 9$ cells in our wine and music table. The entries in the cells are the counts of bottles of wine of the particular type sold while the given type of music was playing. The two variables in this example, wine and music, are both categorical variables.

cell

This two-way table is a 3×3 table, to which we have added the totals obtained by summing across rows and columns. For example, the first-row total is $30 + 39 + 30 = 99$. The grand total, the number of bottles of wine in the study, can be computed by summing the row totals, $99 + 31 + 113 = 243$, or the column totals, $84 + 75 + 84 = 243$. It is a good idea to do both as a check on your arithmetic.

Marginal distributions

How can we best grasp the information contained in the wine and music table? First, *look at the distribution of each variable separately*. The distribution of a categorical variable says how often each outcome occurred. The "Total" column at the right margin of the table contains the totals for each of the rows. These are called **marginal row totals.** They give the numbers of bottles of wine sold by the type of wine: 99 bottles of French wine, 31 bottles of Italian wine, and 113 bottles of other types of wine. Similarly, the **marginal column totals** are given in the "Total" row at the bottom margin of the table. These are the numbers of bottles of wine that were sold while different types of music were being played: 84 bottles when no music was playing, 75 bottles when French music was playing, and 84 bottles when Italian music was playing.

Percents are often more informative than counts. We can calculate the distribution of wine type in percents by dividing each row total by the table total. This distribution is called the **marginal distribution** of wine type.

> **MARGINAL DISTRIBUTIONS**
>
> To find the marginal distribution for the row variable in a two-way table, divide each row total by the total number of entries in the table. Similarly, to find the marginal distribution for the column variable in a two-way table, divide each column total by the total number of entries in the table.

Although the usual definition of a distribution is in terms of proportions, we often multiply these proportions by 100 to convert them to percents. You can describe a distribution either way as long as you clearly indicate which format you are using.

EXAMPLE 2.25

WINE

Calculating a Marginal Distribution Let's find the marginal distribution for the types of wine sold. The counts that we need for these calculations appear in the margin at the right of the table:

Wine	Total
French	99
Italian	31
Other	113
Total	243

The percent of bottles of French wine sold is

$$\frac{\text{bottles of French wine sold}}{\text{total sold}} = \frac{99}{243} = 0.4074 = 40.74\%$$

Similar calculations for Italian wine and other wine give the following distribution in percents:

Wine	French	Italian	Other
Percent	40.74	12.76	46.50

The total should be 100% because each bottle of wine sold is classified into exactly one of these three categories. In this case, the total is exactly 100%. Small deviations from 100% can occur due to roundoff error. ∎

As usual, we prefer to display numerical summaries using a graph. Figure 2.21 is a bar graph of the distribution of wine type sold. In a two-way table, we have two marginal distributions, one for each of the variables that defines the table.

FIGURE 2.21 Marginal distribution of type of wine sold, Example 2.25.

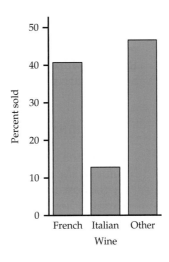

APPLY YOUR KNOWLEDGE

CASE 2.2 **2.91 Marginal distribution for type of music.** Find the marginal distribution for the type of music. Display the distribution using a graph. WINE

In working with two-way tables, you must calculate lots of percents. Here's a tip to help you decide which fraction will give the percent you want. Ask, "Which group represents the total that I want a percent of?" The count for that group is the denominator of the fraction that leads to the percent. In Example 2.25, we wanted percents of "bottles of the different types of wine sold," so the table total is the denominator.

APPLY YOUR KNOWLEDGE

2.92 Construct a two-way table. Construct your own 2 × 3 table. Add the marginal totals and find the two marginal distributions.

2.93 Fields of study for college students. The following table gives the number of students (in thousands) graduating from college with degrees in several fields of study for seven countries:[17] FOS

Field of study	Canada	France	Germany	Italy	Japan	U.K.	U.S.
Social sciences, business, law	64	153	66	125	259	152	878
Science, mathematics, engineering	35	111	66	80	136	128	355
Arts and humanities	27	74	33	42	123	105	397
Education	20	45	18	16	39	14	167
Other	30	289	35	58	97	76	272

(a) Calculate the marginal totals, and add them to the table.

(b) Find the marginal distribution for the country data, and give a graphical display of the distribution.

(c) Do the same for the marginal distribution for the field of study data.

Conditional distributions

The 3 × 3 table for Case 2.2 (page 105) contains much more information than the two marginal distributions. We need to do a little more work to describe the relationship between the type of music playing and the type of wine purchased. **Relationships among categorical variables are described by calculating appropriate percents from the counts given.**

> ### CONDITIONAL DISTRIBUTIONS
>
> To find the conditional distribution of the column variable for a particular value of the row variable in a two-way table, divide each count in the row by the row total. Similarly, to find the conditional distribution of the row variable for a particular value of the column variable in a two-way table, divide each count in the column by the column total.

EXAMPLE 2.26

WINE

CASE 2.2 Wine Purchased When No Music Was Playing Which types of wine were purchased when no music was playing? To answer this question, we find the marginal distribution of wine type for the value of music equal to none. The counts we need appear in the first column of our table:

	Music
Wine	None
French	30
Italian	11
Other	43
Total	84

What percent of French wine was sold when no music was playing? To answer this question, we divide the number of bottles of French wine sold when no music was playing by the total number of bottles of wine sold when no music was playing:

$$\frac{30}{84} = 0.3571 = 35.71\%$$

In the same way, we calculate the percents for Italian and other types of wine. Here are the results:

Wine type:	French	Italian	Other
Percent when no music is playing:	35.7	13.1	51.2

Other wine was the most popular choice when no music was playing, but French wine has a reasonably large share. Notice that these percents sum to 100%; that is, there is no roundoff error here. The distribution is displayed in Figure 2.22. ∎

FIGURE 2.22 Conditional distribution of types of wine sold when no music is playing, Example 2.26.

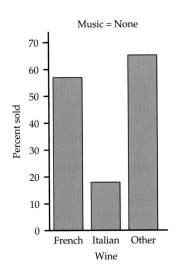

APPLY YOUR KNOWLEDGE

CASE 2.2 **2.94 Conditional distribution when French music was playing.** WINE

(a) Write down the column of counts that you need to compute the conditional distribution of the type of wine sold when French music was playing.

(b) Compute this conditional distribution.

(c) Display this distribution graphically.

(d) Compare this distribution with the one in Example 2.26. Was there an increase in sales of French wine when French music was playing rather than no music?

CASE 2.2 **2.95 Conditional distribution when Italian music was playing.** WINE

(a) Write down the column of counts that you need to compute the conditional distribution of the type of wine sold when Italian music was playing.

(b) Compute this conditional distribution.

(c) Display this distribution graphically.

(d) Compare this distribution with the one in Example 2.26. Was there an increase in sales of Italian wine when Italian music was playing rather than no music?

CASE 2.2 **2.96 Compare the conditional distributions.** In Example 2.26, we found the distribution of sales by wine type when no music was playing. In Exercise 2.94, you found the distribution when French music was playing, and in Exercise 2.95, you found the distribution when Italian music was playing. Examine these three conditional distributions carefully, and write a paragraph summarizing the relationship between sales of different types of wine and the music played. WINE

For Case 2.2, we examined the relationship between sales of different types of wine and the music that was played by studying the three conditional distributions of type of wine sold, one for each music condition. For these computations, we used the counts from the 3 × 3 table, one column at a time. We could also have computed conditional distributions using the counts for each row. The result would be the three conditional distributions of the type of music played for each of the three wine types. For this example, we think

110 Chapter 2 Examining Relationships

that conditioning on the type of music played gives us the most useful data summary. Comparing conditional distributions can be particularly useful when the column variable is an explanatory variable.

The choice of which conditional distribution to use depends on the nature of the data and the questions that you want to ask. Sometimes you will prefer to condition on the column variable, and sometimes you will prefer to condition on the row variable. Occasionally, both sets of conditional distributions will be useful. Statistical software will calculate all of these quantities. *You need to select the parts of the output that are needed for your particular questions. Don't let computer software make this choice for you.*

APPLY YOUR KNOWLEDGE

2.97 Fields of study by country for college students. In Exercise 2.93, you examined data on fields of study for graduating college students from seven countries. FOS

(a) Find the seven conditional distributions giving the distribution of graduates in the different fields of study for each country.

(b) Display the conditional distributions graphically.

(c) Write a paragraph summarizing the relationship between field of study and country.

2.98 Countries by fields of study for college students. Refer to the previous exercise. Answer the same questions for the conditional distribution of countries for each field of study. FOS

2.99 Compare the two analytical approaches. In the previous two exercises, you examined the relationship between country and field of study in two different ways. FOS

(a) Compare these two approaches.

(b) Which do you prefer? Give a reason for your answer.

(c) What kinds of questions are most easily answered by each of the two approaches? Explain your answer.

We have now studied the concepts of marginal and conditional distributions using numerical and graphical summaries based on data. In Chapter 4, we will revisit these ideas expressed in the language of probability.

Mosaic plots and software output

Statistical software will compute all of the quantities that we have discussed in this section. Included in some output is a very useful graphical summary called a **mosaic plot.** Here is an example.

mosaic plot

EXAMPLE 2.27

WINE

Software Output for Wine and Music Output from JMP statistical software for the wine and music data is given in Figure 2.23. The mosaic plot appears in the top part of the display. Here, we think of music as the explanatory variable and wine as the response variable, so music is displayed across the x axis in the plot. The conditional distributions of wine for each type of music are displayed in the three columns. Note that when French music is playing, 52% of the wine sold is French wine. The red bars display the percents of French wine sold when each type of music is playing. Similarly, the green and blue bars display the correspondences for Italian wine and other wine, respectively. The widths of the three sets of bars display the marginal distribution of music. We can see that the proportions are approximately equal, but a little less French music was played compared with the other two music categories. The marginal distribution of wine is shown as a separate column on the right. The contingency table output summarizes the percents we computed for the marginal and conditional distributions. Can you find them? ∎

FIGURE 2.23 Output from JMP for the wine and music data, Example 2.27.

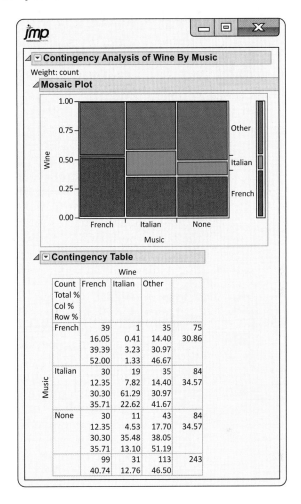

Simpson's paradox

As is the case with quantitative variables, the effects of lurking variables can change or even reverse relationships between two categorical variables. Here is an example that demonstrates the surprises that can await the unsuspecting user of data.

EXAMPLE 2.28

CSERV

Which Customer Service Representative Is Better? A customer service center has a goal of resolving customer questions in 10 minutes or less. Here are the records for two representatives:

	Representative	
Goal met	Ashley	Joshua
Yes	172	118
No	28	82
Total	200	200

Ashley has met the goal 172 times out of 200, a success rate of 86%. For Joshua, the success rate is 118 out of 200, or 59%. Ashley clearly has the better success rate. ∎

EXAMPLE 2.29

CSERV

Let's Look at the Data More Carefully Here are the counts broken down by week:

Goal met	Week 1		Week 2	
	Ashley	Joshua	Ashley	Joshua
Yes	162	19	10	99
No	18	1	10	81
Total	180	20	20	180

For Week 1, Ashley met the goal 90% of the time (162/180), while Joshua met the goal 95% of the time (19/20). Joshua had the better performance in Week 1. What about Week 2? Here, Ashley met the goal 50% of the time (10/20), while the success rate for Joshua was 55% (99/180). Joshua again had the better performance. How does this analysis compare with the analysis that combined the counts for the two weeks? That analysis clearly showed that Ashley had the better performance, 86% versus 59%. ∎

These results can be explained by a *lurking variable* related to week. The first week was during a period when the product had been in use for several months. Most of the calls to the customer service center concerned problems that had been encountered before. The representatives were trained to answer these questions and usually had no trouble in meeting the goal of resolving the problems quickly. By contrast, the second week fell shortly after the release of a new version of the product. Most of the calls during this week concerned new problems that the representatives had not yet encountered. Many more of these questions took longer than the 10-minute goal to resolve.

Look at the total in the bottom row of the detailed table. During the first week, when calls were easy to resolve, Ashley handled 180 calls and Joshua handled 20. The situation was exactly the opposite during the second week, when calls were difficult to resolve. There were 20 calls for Ashley and 180 for Joshua.

The original two-way table, which did not take the week into account, was misleading. This example illustrates *Simpson's paradox*.

SIMPSON'S PARADOX

An association or comparison that holds for all of several groups can reverse direction when the data are combined to form a single group. This reversal is called **Simpson's paradox.**

The lurking variables in Simpson's paradox are categorical. That is, they break the cases into groups, as when calls are classified by week. Simpson's paradox is just an extreme form of the fact that observed associations can be misleading when there are lurking variables.

2.5 Data Analysis for Two-Way Tables

APPLY YOUR KNOWLEDGE

2.100 Which hospital is safer? Insurance companies and consumers are interested in the performance of hospitals. The government releases data about patient outcomes in hospitals that can be useful in making informed health care decisions. Here is a two-way table of data on the survival of patients after surgery in two hospitals. All patients undergoing surgery in a recent time period are included. "Survived" means that the patient lived at least six weeks following surgery. HOSP

	Hospital A	Hospital B
Died	63	16
Survived	2037	784
Total	2100	800

What percent of patients treated in Hospital A died? What percent of patients treated in Hospital B died? These are the numbers one might see reported in the media.

2.101 Patients in "poor" or "good" condition. Not all surgery cases are equally serious, however. Patients are classified as being in either "poor" or "good" condition before surgery. Here are the data broken down by patient condition. Check that the entries in the original two-way table are just the sums of the "poor" and "good" entries in this pair of tables. HOSP

Good condition	Hospital A	Hospital B
Died	6	8
Survived	594	592
Total	600	600

Poor condition	Hospital A	Hospital B
Died	57	8
Survived	1443	192
Total	1500	200

(a) Find the percent of patients treated in Hospital A who died and had been classified as being in "poor" condition before surgery. Do the same for patients treated in Hospital B. In which hospital do patients in "poor" condition fare better?

(b) Repeat part (a) for patients classified as being in "good" condition before surgery.

(c) What is your recommendation to someone facing surgery and choosing between these two hospitals?

(d) How can Hospital A do better treating both groups, yet do worse overall? Look at the data and carefully explain how this outcome can happen.

three-way table

aggregation

The data in Example 2.28 can be given in a **three-way table** that reports counts for each combination of three categorical variables: week, representative, and whether the goal was met. In Example 2.29, we constructed two two-way tables for representative by goal, one for each week. The original table, which we showed in Example 2.28, can be obtained by adding the corresponding counts for the two tables in Example 2.29. This process is called **aggregating** the data. When we aggregated data in Example 2.28, we ignored the variable week, which then became a lurking variable. *Conclusions that seem obvious when we look only at aggregated data can become quite different when the data are examined in more detail.*

SECTION 2.5 SUMMARY

- A **two-way table** of counts organizes counts of data classified by two categorical variables. Values of the **row variable** label the rows that run across the table, and values of the **column variable** label the columns that run down the table. Two-way tables are often used to summarize large amounts of information by grouping outcomes into categories.

- The **row totals** and **column totals** in a two-way table give the **marginal distributions** of the two individual variables. It is clearer to present these distributions as percents of the table total. Marginal distributions tell us nothing about the relationship between the variables.

- To find the **conditional distribution** of the row variable for one specific value of the column variable, look only at that one column in the table. Divide each entry in the column by the column total.

- There is a conditional distribution of the row variable for each column in the table. Comparing these conditional distributions is one way to describe the association between the row variables and the column variables. It is particularly useful when the column variable is the explanatory variable.

- **Bar graphs** are a flexible means of presenting categorical data. There is no single best way to describe an association between two categorical variables.

- **Mosaic plots** are effective graphical displays for two-way tables, particularly when the column variable is an explanatory variable.

- A comparison between two variables that holds for each individual value of a third variable can be changed or even reversed when the data for all values of the third variable are combined. This is **Simpson's paradox.** Simpson's paradox is an example of the effect of lurking variables on an observed association.

SECTION 2.5 EXERCISES

For Exercise 2.91, see page 107; for 2.92 and 2.93, see page 107; for 2.94 to 2.96, see page 109; for 2.97 to 2.99, see page 110; and for 2.100 and 2.101, see page 113.

2.102 Facebook users and age. A survey of U.S. adults ages 18 and older asked about their use of social media. One analysis examined the relationship between Facebook usage and age.[18] Here are the data: FACEAGE

Facebook use	Age			
	18–29	30–49	50–64	65 and older
Yes	285	412	354	217
No	67	116	190	312

(a) Find the distribution of age for the Facebook users.

(b) Do the same for those individuals who do not use Facebook.

(c) If you have the appropriate software, use a mosaic plot to illustrate the marginal distribution of age and your results in parts (a) and (b).

(d) Summarize the relationship between age and Facebook usage using the results of parts (a) and (b).

2.103 YouTube users and age. Refer to the previous exercise. The survey asked about usage of other social media. Here are the data for YouTube: TUBEAGE

YouTube use	Age			
	18–29	30–49	50–64	65 and older
Yes	320	449	370	212
No	32	79	174	317

(a) Find the distribution of age for the YouTube users.

(b) Do the same for those individuals who do not use YouTube.

(c) If you have the appropriate software, use a mosaic plot to illustrate the marginal distribution of age and your results in parts (a) and (b).

(d) Summarize the relationship between age and YouTube usage using the results of parts (a) and (b).

2.104 Compare Facebook with YouTube. Refer to the previous two exercises. Use the results that you found there to compare the age distributions of Facebook and YouTube users.

2.5 Data Analysis for Two-Way Tables

2.105 Trust and honesty in the workplace. One of the questions in a survey of high school students asked whether they thought trust and honesty were essential in business and the workplace.[19] Here are the counts classified by sex: TRUST

Trust and honesty are essential	Sex	
	Male	Female
Agree	9,097	10,935
Disagree	685	423

(a) Add the marginal totals to the table.

(b) Calculate appropriate percents to describe the results of this question.

(c) Summarize your findings in a short paragraph.

2.106 Exercise and adequate sleep. A survey of 656 boys and girls, ages 13 to 18, asked about adequate sleep and other health-related behaviors. The recommended amount of sleep is six to eight hours per night.[20] In the survey, 54% of the respondents reported that they got less than this amount of sleep on school nights. The researchers also developed an exercise scale that was used to classify the students as above or below the median in how much they exercised. Here is the 2 × 2 table of counts with students classified as getting or not getting adequate sleep and by the exercise variable: SLEEP

Enough sleep	Exercise	
	High	Low
Yes	151	115
No	148	242

(a) Find the distribution of adequate sleep for the high exercisers.

(b) Do the same for the low exercisers.

(c) If you have the appropriate software, use a mosaic plot to illustrate the marginal distribution of exercise and your results in parts (a) and (b).

(d) Summarize the relationship between adequate sleep and exercise using the results of parts (a) and (b).

2.107 Adequate sleep and exercise. Refer to the previous exercise. SLEEP

(a) Find the distribution of exercise for those respondents who get adequate sleep.

(b) Do the same for those respondents who do not get adequate sleep.

(c) Write a short summary of the relationship between adequate sleep and exercise using the results of parts (a) and (b).

(d) Compare this summary with the summary that you obtained in part (c) of the previous exercise. Which do you prefer? Give a reason for your answer.

2.108 Full-time and part-time college students. The Census Bureau provides estimates of numbers of people in the United States classified in various ways.[21] Let's look at college students. The following table gives us data with which to examine the relationship between age and full-time or part-time status. The numbers in the table are expressed as thousands of U.S. college students. COLSTUD

Age	Status	
	Full-time	Part-time
15–19	3388	389
20–24	5238	1164
25–34	1703	1699
35 and older	762	2045

(a) Find the distribution of age for full-time students.

(b) Do the same for the part-time students.

(c) Use the summaries in parts (a) and (b) to describe the relationship between full- or part-time status and age. Write a brief summary of your conclusions.

2.109 Condition on age. Refer to the previous exercise. COLSTUD

(a) For each age group, compute the percent of students who are full-time and the percent of students who are part-time.

(b) Make a graphical display of the results that you found in part (a).

(c) If you have the appropriate software, make a mosaic plot.

(d) In a short paragraph, describe the relationship between age and full- or part-time status using your numerical and graphical summaries.

(e) Explain why you need only the percents of students who are full-time for your summary in part (b).

(f) Compare this way of summarizing the relationship between these two variables with what you presented in part (c) of the previous exercise.

2.110 Class size and course level. College courses taught at lower levels often have larger class sizes. The following table gives the number of classes classified by course level and class size.[22] For example, there were 202 first-year level courses with between one and nine students. CSIZE

Course level	Class size						
	1–9	10–19	20–29	30–39	40–49	50–99	100 or more
1	202	659	917	241	70	99	123
2	190	370	486	307	84	109	134
3	150	387	314	115	96	186	53
4	146	256	190	83	67	64	17

(a) Fill in the marginal totals in the table.

(b) Find the marginal distribution for the variable course level.

(c) Do the same for the variable class size.

(d) For each course level, find the conditional distribution of class size.

(e) Summarize your findings in a short paragraph.

2.111 Hiring practices. A company has been accused of age discrimination in hiring for operator positions. Lawyers for both sides look at data on applicants for the past three years. They compare hiring rates for applicants younger than 40 years and those 40 years or older. HIRING

Age	Hired	Not hired
Younger than 40	84	1140
or older	4	178

(a) Find the two conditional distributions of hired/not hired—one for applicants who are less than 40 years old and one for applicants who are not less than 40 years old.

(b) Based on your calculations, make a graph to show the differences in distribution for the two age categories.

(c) Describe the company's hiring record in words. Does the company appear to discriminate on the basis of age?

(d) Which lurking variables might be involved here?

2.112 Nonresponse in a survey of companies. A business school conducted a survey of companies in its state. It mailed a questionnaire to 200 small companies, 200 medium-sized companies, and 200 large companies. The rate of nonresponse is important in deciding how reliable survey results are. Here are the data on response to this survey: NRESP

	Small	Medium	Large
Response	139	70	46
No response	81	130	174
Total	220	220	220

(a) What was the overall percent of nonresponse?

(b) Describe how nonresponse is related to the size of the business. (Use percents to make your statements precise.)

(c) Draw a bar graph to compare the nonresponse percents for the three size categories.

2.113 Demographics and new products. Companies planning to introduce a new product to the market must define the "target" for the product. The key question they ask is, "Who do we hope to attract with our new product?" Age and sex are two of the most important demographic variables. The following two-way table describes the age and marital status of American women.[23] The table entries are in thousands of women. AGEGEN

Age (years)	Marital status			
	Never married	Married	Widowed	Divorced
18 to 24	12,112	2,171	23	164
25 to 39	9,472	18,219	177	2,499
40 to 64	5,224	35,021	2,463	8,674
≥ 65	984	9,688	8,699	2,412

(a) Find the sum of the entries for each column.

(b) Find the marginal distributions.

(c) Find the conditional distributions.

(d) If you have the appropriate software, make a mosaic plot.

(e) Write a short description of the relationship between marital status and age for women.

2.114 Demographics, continued. AGEGEN

(a) Using the data in the previous exercise, compare the conditional distributions of marital status for women ages 18 to 24 and women ages 40 to 64. Briefly describe the most important differences between the two groups of women, and back up your description with percents.

(b) Your company is planning to launch a magazine aimed at women who have never been married. Find the conditional distribution of age among never-married women, and display it in a bar graph. Which age group or groups should your magazine aim to attract?

2.115 Demographics and new products—men. Refer to Exercises 2.113 and 2.114. Here are the corresponding counts for men: AGEGEN

Age (years)	Marital status			
	Never married	Married	Widowed	Divorced
18 to 24	13,509	1,245	6	63
25 to 39	12,685	16,029	78	1,790
40 to 64	6,869	34,650	760	6,647
≥ 65	685	12,514	2,124	1,464

Answer the questions from Exercises 2.113 and 2.114 for these counts.

2.116 Discrimination? Wabash Tech has two professional schools, business and law. Here are two-way tables of applicants to both schools, categorized by sex and admission decision. (Although these data are made up, similar situations occur in reality.) DISC

	Business	
	Admit	Deny
Male	480	120
Female	180	20

	Law	
	Admit	Deny
Male	10	90
Female	100	200

(a) Make a two-way table of sex by admission decision for the two professional schools together by summing the entries in these tables.

(b) From the two-way table, calculate the percent of male applicants who are admitted and the percent of female applicants who are admitted. Wabash admits a higher percent of male applicants.

(c) Now compute separately the percents of male and female applicants admitted by the business school and by the law school. Each school admits a higher percent of female applicants.

(d) This is Simpson's paradox: both schools admit a higher percent of the women who apply, but overall, Wabash admits a lower percent of female applicants than of male applicants. Explain carefully, as if speaking to a skeptical reporter, how it can happen that Wabash appears to favor males when each school individually favors females.

2.117 Obesity and health. Recent studies have shown that earlier reports underestimated the health risks associated with being overweight. The error was due to lurking variables. In particular, smoking tends both to reduce weight and to lead to earlier death. Illustrate Simpson's paradox by presenting a simplified version of this situation. That is, make up tables of overweight (yes or no) by early death (yes or no) by smoker (yes or no) such that

- Overweight smokers and overweight nonsmokers both tend to die earlier than those individuals who are not overweight.

- But when smokers and nonsmokers are combined into a two-way table of overweight by early death, persons who are not overweight tend to die earlier.

2.118 Find the table. Here are the row and column totals for a two-way table with two rows and two columns:

a	b	60
c	d	60
70	50	120

Find *two different* sets of counts a, b, c, and d for the body of the table that give these same totals. This exercise shows that the relationship between two variables cannot be obtained from the two individual distributions of the variables.

CHAPTER 2 REVIEW EXERCISES

2.119 Companies of the world with logs. In Exercises 2.10 (page 72), 2.27 (page 79), and 2.58 (page 96), you examined the relationship between the numbers of companies that are incorporated and listed on their country's stock exchange at the end of the year using data collected by the World Bank.[24] In this exercise, you will explore the relationship between the numbers for 2017 and 2007 using logs. Start by computing the logs for the two variables. INCCOM

(a) Which variable do you choose to be the explanatory variable, and which do you choose to be the response variable? Explain your answer.

(b) Plot the data with the least-squares regression line. Summarize the major features of your plot.

(c) Give the equation of the least-squares regression line.

(d) Find the predicted value and the residual for France.

(e) Find the correlation between the two variables.

(f) Compare the results found in this exercise with those you found in Exercises 2.10, 2.27, and 2.58. Do you prefer the analysis with the original data or the analysis using logs? Give reasons for your answer.

2.120 Residuals for companies of the world with logs. Refer to the previous exercise. INCCOM

(a) Use a histogram to examine the distribution of the residuals.

(b) Make a Normal quantile plot of the residuals.

(c) Summarize the distribution of the residuals using the graphical displays that you created in parts (a) and (b).

(d) Repeat parts (a), (b), and (c) for the original data, and compare these results with those you found in parts (a), (b), and (c). Which do you prefer? Give reasons for your answer.

2.121 Dwelling permits and sales for 21 European countries. The Organization for Economic Cooperation and Development (OECD) collects data on Main Economic Indicators (MEIs) for many countries. Each variable is recorded as an index, with the year 2000 serving as a base year. In other words, the variable for each year is reported as a ratio of the value for the year divided by the value for 2000. Using indices in this way makes it easier to compare values for different countries.[25] MEIS

(a) Make a scatterplot with sales as the response variable and permits issued for new dwellings as the explanatory variable. Describe the relationship. Are there any outliers or influential observations?

(b) Find the least-squares regression line and add it to your plot.

(c) What is the predicted value of sales for a country that has an index of 160 for dwelling permits?

(d) The Netherlands has an index of 160 for dwelling permits. Find the residual for this country.

(e) What percent of the variation in sales is explained by dwelling permits?

2.122 Dwelling permits and production. Refer to the previous exercise. MEIS

(a) Make a scatterplot with production as the response variable and permits issued for new dwellings as the explanatory variable. Describe the relationship. Are there any outliers or influential observations?

(b) Find the least-squares regression line and add it to your plot.

(c) What is the predicted value of production for a country that has an index of 160 for dwelling permits?

(d) The Netherlands has an index of 160 for dwelling permits. Find the residual for this country.

(e) What percent of the variation in production is explained by dwelling permits? How does this value compare with the value you found in the previous exercise for the percent of variation in sales that is explained by building permits?

2.123 Sales and production. Refer to the previous two exercises. MEIS

(a) Make a scatterplot with sales as the response variable and production as the explanatory variable. Describe the relationship. Are there any outliers or influential observations?

(b) Find the least-squares regression line and add it to your plot.

(c) What is the predicted value of sales for a country that has an index of 125 for production?

(d) Finland has an index of 125 for production. Find the residual for this country.

(e) What percent of the variation in sales is explained by production? How does this value compare with the percents of variation that you calculated in the two previous exercises?

2.124 Salaries and raises. For this exercise, we consider a hypothetical employee who starts working in Year 1 at a salary of $50,000. Each year, her salary increases by approximately 5%. By Year 20, she is earning $126,000. The following table gives her salary for each year (in thousands of dollars): RAISES

Year	Salary	Year	Salary	Year	Salary	Year	Salary
1	50	6	63	11	81	16	104
2	53	7	67	12	85	17	109
3	56	8	70	13	90	18	114
4	58	9	74	14	93	19	120
5	61	10	78	15	99	20	126

(a) Figure 2.24 is a scatterplot of salary versus year with the least-squares regression line. Describe the relationship between salary and year for this person.

(b) The value of r^2 for these data is 0.9832. What percent of the variation in salary is explained by year? Would you say that this is an indication of a strong linear relationship? Explain your answer.

2.125 Look at the residuals. Refer to the previous exercise. Figure 2.25 is a plot of the residuals versus year. RAISES

(a) Interpret the residual plot.

(b) Explain how this plot highlights the deviations from the least-squares regression line that you can see in Figure 2.24.

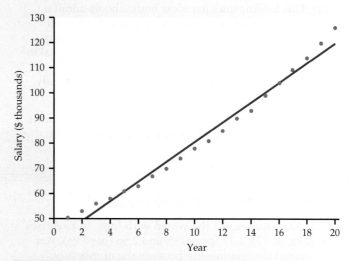

FIGURE 2.24 Plot of salary versus year, with the least-squares regression line, for an individual who receives approximately a 5% raise each year for 20 years, Exercise 2.124.

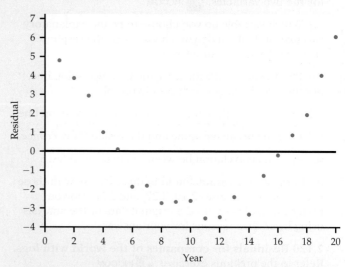

FIGURE 2.25 Plot of residuals versus year for an individual who receives approximately a 5% raise each year for 20 years, Exercise 2.125.

2.126 Try logs. Refer to the previous two exercises. Figure 2.26 is a scatterplot with the least-squares regression line for log salary versus year. For this model, $r^2 = 0.9995$. RAISES

(a) Compare this plot with Figure 2.24. Write a short summary of the similarities and the differences.

(b) Figure 2.27 is a plot of the residuals for the model using year to predict log salary. Compare this plot with Figure 2.25 and summarize your findings.

2.127 Predict some salaries. The individual whose salary we have been studying in Exercises 2.124 through 2.126 wants to do some financial planning. Specifically, she would like to predict her salary five years into the future—that is, for Year 25. She is willing to assume that her employment situation will be stable for the next five years and that it will be similar to the last 20 years. RAISES

(a) Use the least-squares regression equation constructed to predict salary from year to predict her salary for Year 25.

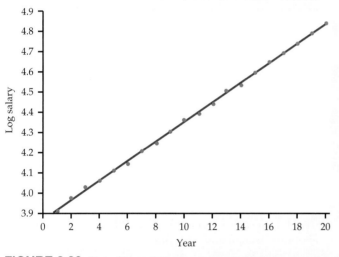

FIGURE 2.26 Plot of log salary versus year, with the least-squares regression line, for an individual who receives approximately a 5% raise each year for 20 years, Exercise 2.126.

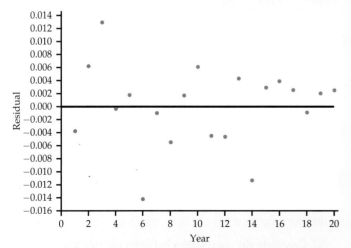

FIGURE 2.27 Plot of residuals, based on log salary, versus year for an individual who receives approximately a 5% raise each year for 20 years, Exercise 2.126.

(b) Use the least-squares regression equation constructed to predict log salary from year to predict her salary for Year 25. Note that you will need to convert the predicted log salary back to the predicted salary. Many calculators have a function that will perform this operation.

(c) Which prediction do you prefer? Explain your answer.

(d) Someone looking at the numerical summaries, and not at the plots, for these analyses says that because both models have very high values of r^2, they should perform equally well in making this prediction. Write a response to this comment.

(e) Write a short paragraph about the value of graphical summaries and the problems of extrapolation using what you have learned from studying these salary data.

2.128 Faculty salaries. Data on the salaries of a sample of professors in a business department at a large university are given in the following table. The salaries are for the academic years 2017–2018 and 2018–2019. FACSAL

2017–2018 salary ($)	2018–2019 salary ($)	2017–2018 salary ($)	2018–2019 salary ($)
148,700	150,700	139,650	141,650
115,700	117,660	135,160	137,150
112,200	114,400	77,290	79,590
103,800	104,900	77,500	80,000
115,000	116,000	86,000	88,400
114,790	116,800	144,850	146,830
106,500	108,700	125,500	127,510
152,000	153,900	116,100	120,100

(a) Construct a scatterplot with the 2018–2019 salaries on the vertical axis and the 2017–2018 salaries on the horizontal axis.

(b) Comment on the form, direction, and strength of the relationship in your scatterplot.

(c) What proportion of the variation in the 2018–2019 salaries is explained by the 2017–2018 salaries?

2.129 Find the line and examine the residuals. Refer to the previous exercise. FACSAL

(a) Find the least-squares regression line for predicting the 2018–2019 salaries from the 2017–2018 salaries.

(b) Analyze the residuals, paying attention to any outliers or influential observations. Write a summary of your findings.

2.130 Bigger raises for those earning less. Refer to the previous two exercises. The 2017–2018 salaries do an excellent job of predicting the 2018–2019 salaries. Is there anything more that we can learn from these data? In this department, there is a tradition of giving higher-than-average percent raises to those professors whose salaries are lower than those of their peers. Let's see if we can find evidence to support this idea in the data. FACSAL

(a) Compute the percent raise for each faculty member. Take the difference between the 2018–2019 salary and the 2017–2018 salary, divide by the 2017–2018 salary, and then multiply by 100. Make a scatterplot with the raise as the response variable and the 2017–2018 salary as the explanatory variable. Describe the relationship that you see in your plot.

(b) Find the least-squares regression line and add it to your plot.

(c) Analyze the residuals. Are there any outliers or influential cases? Make a graphical display and include it in a short summary stating what you conclude.

(d) Is there evidence in the data to support the idea that greater percentage raises are given to those with lower salaries? Summarize your findings and include numerical and graphical summaries to support your conclusion.

2.131 Marketing your college. Colleges compete for students, and many students do careful research when choosing a college. One source of information is the rankings compiled by *U.S. News & World Report*. One of the factors used to evaluate undergraduate programs is the proportion of incoming students who graduate. This quantity, called the graduation rate, can be predicted by other variables such as the SAT or ACT scores and the high school records of the incoming students. One of the components in *U.S. News & World Report* rankings is the difference between the actual graduation rate and the rate predicted by a regression equation.[26] In this chapter, we call this quantity the residual. Explain why the residual is a better measure to evaluate college graduation rates than the raw graduation rate.

2.132 Planning for a new product. The editor of a statistics text would like to plan for the next edition. A key variable is the number of pages that will be in the final version of the new edition. Text files are prepared by the authors using a word processor called LaTeX, and separate files contain figures and tables. For the previous edition of the text, the number of pages in the LaTeX files can easily be determined, as well as the number of pages in the final version of the text. Here are the data: TPAGES

Chapter	1	2	3	4	5	6	7	8	9	10	11	12	13
LaTeX pages	77	73	59	80	45	66	81	45	47	43	31	46	26
Text pages	99	89	61	82	47	68	87	45	53	50	36	52	19

(a) Plot the data and describe the overall pattern.

(b) Find the equation of the least-squares regression line, and add the line to your plot.

(c) Find the predicted number of pages for the next edition if the number of LaTeX pages for a chapter is 62.

(d) Write a short report for the editor explaining to her how you constructed the regression equation and how she could use it to estimate the number of pages in the next edition of the text.

2.133 Points scored in women's basketball games. Perform an Internet search to find the scores for the past season's women's basketball team at a college of your choice. Is there a relationship between the points scored by your chosen team and the points scored by their opponents? Summarize the data and write a report on your findings.

2.134 Look at the data for men. Refer to the previous exercise. Analyze the data for the men's team from the same college, and compare your results with those for the women's team.

2.135 Circular saws. The following table gives the weight (in pounds) and amps for 19 circular saws. Saws with higher amp ratings tend to also be heavier than saws with lower amp ratings. We can quantify this fact using regression. CIRCSAW

Weight	Amps	Weight	Amps	Weight	Amps
11	15	9	10	11	13
12	15	11	15	13	14
11	15	12	15	10	12
11	15	12	14	11	12
12	15	10	10	11	12
11	15	12	13	10	12
13	15				

(a) We will use amps as the explanatory variable and weight as the response variable. Give a reason for this choice.

(b) Make a scatterplot of the data. What do you notice about the weight and amp values?

(c) Report the equation of the least-squares regression line along with the value of r^2.

(d) Interpret the value of the estimated slope.

(e) How much of an increase in amps would you expect to correspond to a one-pound increase in the weight of a saw, on average, when comparing two saws?

(f) Create a residual plot for the model in part (b). Does the model indicate curvature in the data?

2.136 Circular saws. The table in the previous exercise gives the weight (in pounds) and amps for 19 circular saws. The data contain only five different amp ratings among the 19 saws. CIRCSAW

(a) Calculate the correlation between the weights and the amps of the 19 saws.

(b) Calculate the average weight of the saws for each of the five amp ratings.

(c) Calculate the correlation between the average weights and the amps. Is the correlation between average weights and amps greater than, less than, or equal to the correlation between individual weights and amps?

2.137 What correlation does and doesn't say. Construct a set of data with two variables that have different means and correlation equal to one. Use your example to illustrate what correlation does and doesn't say.

2.138 Simpson's paradox and regression. Simpson's paradox occurs when a relationship between variables within groups of observations reverses when all of the data are combined. This phenomenon is usually discussed in terms of categorical variables, but it also occurs in other settings. Here is an example: SIMREG

y	x	Group	y	x	Group
10.1	1	1	18.3	6	2
8.9	2	1	17.1	7	2
8.0	3	1	16.2	8	2
6.9	4	1	15.1	9	2
6.1	5	1	14.3	10	2

(a) Make a scatterplot of the data for Group 1. Find the least-squares regression line and add it to your plot. Describe the relationship between y and x for Group 1.

(b) Do the same for Group 2.

(c) Make a scatterplot using all 10 observations. Find the least-squares line and add it to your plot.

(d) Make a plot with all of the data using different symbols for the two groups. Include the three regression lines on the plot. Write a paragraph about Simpson's paradox for regression using this graphical display to illustrate your description.

2.139 Wood products. A wood product manufacturer is interested in replacing solid-wood building materials with less-expensive products made from wood flakes.[27] The company collected the following data to examine the relationship between the length (in inches) and the strength (in pounds per square inch) of beams made from wood flakes: WOOD

Length:	5	6	7	8	9	10	11	12	13	14
Strength:	446	371	334	296	249	254	244	246	239	234

(a) Make a scatterplot that shows how the length of a beam affects its strength.

(b) Describe the overall pattern of the plot. Are there any outliers?

(c) Fit a least-squares line to the entire set of data. Graph the line on your scatterplot. Does a straight line adequately describe these data?

(d) The scatterplot suggests that the relationship between length and strength can be described by *two* straight lines: one for lengths of 5 to 9 inches and another for lengths of 9 to 14 inches. Fit least-squares lines to these two subsets of the data, and draw the lines on your plot. Do they describe the data adequately? What question would you now ask the wood experts?

2.140 Aspirin and heart attacks. Does taking aspirin regularly help prevent heart attacks? "Nearly five decades of research now link aspirin to the prevention of stroke and heart attacks," according to the Bayer Aspirin website (**bayeraspirin.com**). The most important evidence for this claim comes from the Physicians' Health Study. The subjects in this study were 22,071 healthy male doctors at least 40 years old. Half the subjects, chosen at random, took aspirin every other day. The other half took a placebo, a dummy pill that looked and tasted like aspirin. The results are shown in the following table.[28] (The row for "None of these" is left out of the two-way table.)

	Aspirin group	Placebo group
Fatal heart attacks	10	26
Other heart attacks	129	213
Strokes	119	98
Total	11,037	11,034

What do these data show about the association between taking aspirin and having a heart attack or stroke? Use percents to make your statements precise. Include a mosaic plot if you have access to the needed software. Do you think the study provides evidence that aspirin actually reduces heart attacks (cause and effect)? ASPIRIN

2.141 More smokers live at least 20 more years! You can see the headline: "More smokers than nonsmokers live at least 20 more years after being contacted for study!" A medical study contacted randomly chosen people in a district in England. Here are data on the 1314 women contacted who were either current smokers or who had never smoked. The tables classify these women by their smoking status and age at the time of the survey and whether they were still alive 20 years later.[29] SMOKERS

	Age 18 to 44		Age 45 to 64		Age 65+	
	Smoker	Not	Smoker	Not	Smoker	Not
Dead	19	13	78	52	42	165
Alive	269	327	167	147	7	28

(a) From these data, make a two-way table of smoking (yes or no) by dead or alive. What percent of the smokers stayed alive for 20 years? What percent of the nonsmokers survived? It seems surprising that a higher percent of smokers stayed alive.

(b) The age of the women at the time of the study is a lurking variable. Show that within each of the three age groups in the data, a higher percent of nonsmokers remained alive 20 years later. This is another example of Simpson's paradox.

(c) The study authors give this explanation: "Few of the older women (over 65 at the original survey) were smokers, but many of them had died by the time of follow-up." Compare the percent of smokers in the three age groups to verify the explanation.

2.142 Recycled product quality. Recycling is supposed to save resources. Some people think recycled products are lower in quality than other products, a fact that makes recycling less practical. People who actually use a recycled product may have different opinions from those who don't use it. Here are data on attitudes toward coffee filters made of recycled paper among people who do and don't buy these filters:[30] RECYCLE

Filter buyers	Think the quality of the recycled product is:		
	Higher	The same	Lower
Buyers	20	7	9
Nonbuyers	29	25	43

(a) Find the marginal distribution of opinion about quality. Assuming that these people represent all users of coffee filters, what does this distribution tell us?

(b) How do the opinions of buyers and nonbuyers differ? Use conditional distributions as a basis for your answer. Include a mosaic plot if you have access to the needed software. Can you conclude that using recycled filters causes more favorable opinions? If so, giving away samples might increase sales.

FatCamera/Getty Images

CHAPTER 3

Producing Data

Introduction

Reliable data are needed to make business decisions. Here are some examples where carefully collected data are essential:

- How does Samsung decide which features to highlight in its advertising for the latest Galaxy phone?
- How does Starbucks choose the location for a new store?
- How does Toyota decide how many Prius automobiles to produce in January of next year?

In Chapters 1 and 2, we learned some basic tools of *data analysis*. We used graphs and numbers to describe data. When we do exploratory data analysis, we rely heavily on plotting the data. We look for patterns that suggest interesting conclusions or questions for further study. However, *exploratory analysis alone can rarely provide convincing evidence for its conclusions because striking patterns we find in data can arise from many sources.*

The validity of the conclusions that we draw from an analysis of data depends not only on the use of the best methods to perform the analysis, but also on the quality of the data. Therefore, Section 3.1 begins this chapter with a short overview on sources of data. The two main sources for quality data are designed samples and designed experiments. We study these two sources in Sections 3.2 and 3.3, respectively.

Should an experiment or sample survey that could possibly provide interesting and important information always be performed? How can we

CHAPTER OUTLINE

3.1 Sources of Data

3.2 Designing Samples

3.3 Designing Experiments

3.4 Data Ethics

exploratory data analysis, p. 8

safeguard the privacy of subjects in a sample survey? What constitutes the mistreatment of the people or animals that are studied in an experiment? These are questions of **ethics.** In Section 3.4, we address ethical issues related to the design of studies and the analysis of data.

3.1 Sources of Data

When you complete this section, you will be able to:

- Describe anecdotal data and, using specific examples, explain why they have limited value.
- Identify available data and explain how they can be used in specific examples.
- Identify data collected from sample surveys and explain how they can be used in specific examples.
- Identify data collected from experiments and explain how they can be used in specific examples.
- Distinguish data that are from experiments, from observational studies that are sample surveys, and from observational studies that are not sample surveys.
- Describe the treatment in an experiment.

There are many sources of data. Some data are very easy to collect, but they may not be very useful. Other data require careful planning and must be gathered by professional staff. These data can be much more useful. Whatever the source, a good statistical analysis will start with a careful study of the source of the data.

Anecdotal data

It is tempting to simply draw conclusions from our own experience, making no use of more broadly representative data. An advertisement for a Pilates class says that men need this form of exercise even more than women. The ad describes the benefits that two men received from taking Pilates classes. A newspaper ad states that a particular brand of windows is "considered to be the best" and says that "now is the best time to replace your windows and doors." These types of stories, or *anecdotes*, sometimes provide quantitative data. However, such data do not give us a sound basis for drawing conclusions.

> **ANECDOTAL EVIDENCE**
>
> **Anecdotal evidence** is based on haphazardly selected cases, which often come to our attention because they are striking in some way. These cases need not be representative of any larger group of cases.

APPLY YOUR KNOWLEDGE

3.1 Is this good market research? You and your friends are big fans of Ariana Grande, an American singer and actress. To what extent do you think you can generalize your preference for this performer to all students at your college?

3.2 Should you invest in stocks? You have just accepted a new job and are offered several options for your retirement account. One of these invests approximately 75% of your employer's contribution in stocks. You talk to a friend who joined the company several years ago; he says that after he chose that option, the value of the stocks decreased substantially. He strongly recommends that you choose a different option. Comment on the value of your friend's advice.

3.3 Preference for cold weather. Desiree is a serious long-distance runner who prefers to run in cold, wet weather. Explain why Desiree's preference may not be shared by large numbers of long-distance runners.

3.4 Reliability of a product. A friend has driven a Toyota Camry for more than 200,000 miles with only the usual service maintenance expenses. Explain why not all Camry owners can expect this kind of performance.

Available data

available data

Occasionally, data are collected for a particular purpose but can also serve as the basis for drawing sound conclusions about other research questions. We use the term **available data** for this type of data.

> **AVAILABLE DATA**
>
> **Available data** are data that were produced in the past for some other purpose but that may help answer a present question.

The library and the Internet can be good sources of available data. Because producing new data is expensive, we all use available data whenever possible. Here are two examples.

EXAMPLE 3.1

Unemployment per Job Opening If you visit the U.S. Bureau of Labor Statistics website, **bls.gov,** you can find many interesting sets of data and statistical summaries. One recent study posting described the number of unemployed people per job opening from March 2003 to March 2018.[1] ∎

EXAMPLE 3.2

Can Our Workforce Compete in a Global Economy? To compete in the global economy, many U.S. students need to improve their mathematics.[2] At the website of the National Center for Education Statistics, **nces.ed.gov/nationsreportcard,** you will find full details about the math skills of schoolchildren in the latest National Assessment of Educational Progress. Figure 3.1 shows one of the pages that reports on the changes in eighth-grade mathematics scores from 1990 to 2017.[3] ∎

FIGURE 3.1 The websites of government statistical offices are prime sources of data. Here is a page from the National Assessment of Educational Progress, Example 3.2.

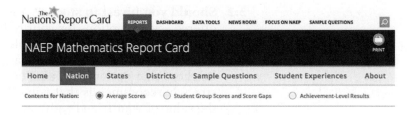
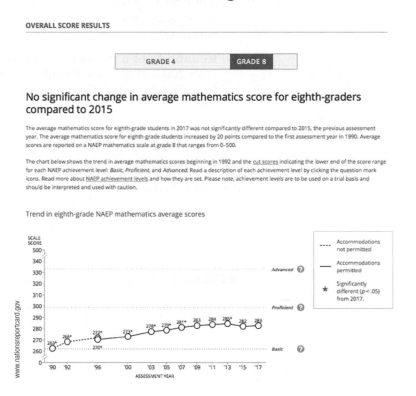

Many nations have a single national statistical office, such as Statistics Canada (**statcan.gc.ca**) and Mexico's INEGI (**inegi.org.mx**). Many different U.S. agencies collect data. You can reach most of them through the government's site (**usa.gov/statistics**).

APPLY YOUR KNOWLEDGE

3.5 Check out the Bureau of Labor Statistics website. Visit the Bureau of Labor Statistics website, **bls.gov**. Find a set of data that interests you. Explain how the data were collected and which questions the study was designed to answer.

Although available data can be very useful for many situations, we often find that clear answers to important questions require that data be produced to answer those specific questions. Are your customers likely to buy a product from a competitor if you raise your price? Is the expected return from a proposed advertising campaign sufficient to justify the cost? The validity of our conclusions from the analysis of data collected to address these issues rests on a foundation of carefully collected data. In this chapter, we learn how to produce trustworthy data and to judge the quality of data produced by others. The techniques for producing data that we study require no formulas, but they are among the most important ideas in statistics. Statistical designs for producing data rely on either *sampling* or *experiments*.

Sample surveys and experiments

How have the attitudes of Americans, on issues ranging from shopping online to satisfaction with work, changed over time? Sample surveys are the usual tool for answering questions like these. A **sample survey** collects data from a sample of cases that represent some larger population of cases.

sample survey

EXAMPLE 3.3

Confidence in Banks and Companies One of the most important sample surveys is the General Social Survey (GSS) conducted by the National Opinion Research Center (NORC), a national organization for research and computing affiliated with the University of Chicago.[4] The GSS interviews about 3000 adult residents of the United States every second year. The survey includes questions about how much confidence people have in banks and companies. ∎

The GSS selects a *sample* of adults to represent the larger population of all English-speaking adults living in the United States. The idea of *sampling* is to study a part to gain information about the whole. Data are often produced by sampling a *population* of people or things. Opinion polls, for example, report the views of the entire country based on interviews with a sample of about 1000 people. Government reports on employment and unemployment are produced from a monthly sample of about 60,000 households. The quality of manufactured items is monitored by inspecting small samples each hour or each shift.

APPLY YOUR KNOWLEDGE

3.6 Are Millennials loyal customers? A website claims that Millennial-generation consumers are very loyal to the brands that they prefer. Other sites disagree. Which additional information do you need to evaluate this claim?

In all our examples, the expense of examining every item in the population makes sampling a practical necessity. Timeliness is another reason for preferring a sample to a **census,** which is an attempt to contact every case in the entire population. We want information on current unemployment and public opinion next week, not next year. Moreover, a carefully conducted sample is often more accurate than a census. Accountants, for example, sample a firm's inventory to verify the accuracy of the records. Counting every item in a warehouse can be both expensive and inaccurate. After all, bored people might not count carefully.

A sample survey collects information about a population by selecting and measuring a sample from the population. It is one kind of *observational study*. The goal is to obtain a picture of the population, disturbed as little as possible by the act of gathering information. This requires a sound design for selecting the sample.

An observational study, even one based on a statistical sample, is a poor way to determine what will happen if we change something. The best way to see the effects of a change is to do an **intervention**—where we actually impose the change. The change imposed is called a **treatment.** When our goal is to understand cause and effect, *experiments* are the only source of fully convincing data. In an experiment, a treatment is imposed and the responses are recorded. Experiments usually require a sound design for assigning the treatments to cases.

census

intervention
treatment

> **OBSERVATION VERSUS EXPERIMENT**
>
> In an **observational study,** we observe cases and measure variables of interest but do not attempt to influence the responses.
>
> In an **experiment,** we deliberately impose some treatment on cases and observe their responses.

explanatory and response variables, p. 64

In an experiment, the treatment is the explanatory variable and the response is the response variable. In an observational study, there can be several response variables and there may not be any explanatory variables.

APPLY YOUR KNOWLEDGE

3.7 **Market share for energy drinks.** A website reports that Red Bull is the top energy drink brand in the United States with 26.5% of the market. Monster Energy is second, with 14.2% of the U.S. market.[5] Do you think that this report is based on an observational study or an experiment? Explain your answer.

3.8 **An advertising agency chooses an ESPN television ad.** An advertising agency developed two versions of an ad that will be shown during a college sporting event on ESPN, but must choose only one to air. The agency recruited 100 college students and divided them into two groups of 50. Each group viewed one of the versions of the ad and then answered a collection of questions about their reactions to the ad. Is the advertising agency using an observational study or an experiment to help make its decision? Give reasons for your answer.

Other sources and uses of data

In the next two sections, we concentrate on data from surveys and experiments. However, it is important to note that some other sources of data can be useful in a variety of applications. In some studies, we analyze one set of data to generate a set of results, for example, to develop a model to predict something. This kind of data set is called a **training data set.** The ability of the model to predict the outcome of interest is then evaluated using a **validation data set,** where the predictions are made and then compared with the actual values. A training data set is typically constructed by randomly selecting half, or a larger fraction, of a given data set and using the remaining observations as the validation data set. A version of this approach is used in **machine learning.** Other applications are described by the term cross-validation.

Data can also take the form of **text.** A survey of customers may include questions that ask for descriptions of why a particular product is liked or not. These responses can be analyzed by software that looks for key words or phrases, which can then be studied with statistical methods.

The purpose of a **meta-analysis** is to summarize results from a collection of studies to generate a conclusion about a common issue addressed by the studies. A meta-analysis data set would contain the summary results from each study with characteristics of the individual studies.

The Research Electronic Data Capture (**REDCap**)[6] is a data collection platform that can be used to collect, organize, and store data. It provides the capability to generate customized data sets that can be used for statistical analyses. Microsoft Access is a tool for organizing data in a **database** from which data sets for statistical analysis can be extracted. The term **data warehouse** is sometimes used to describe a system for organizing, storing, and analyzing complex data.

SECTION 3.1 SUMMARY

- **Anecdotal data** come from stories or reports about cases that do not necessarily represent a larger group of cases.
- **Available data** are data that were produced for some other purpose but that may help answer a question of interest.
- A **sample survey** collects data from a sample of cases that represent some larger population of cases.
- A **census** collects data from all cases in the population of interest.
- A **sample survey** is one type of an **observational study.**
- In an **experiment,** a **treatment** is imposed and the responses are recorded.

SECTION 3.1 EXERCISES

For Exercises 3.1 to 3.4, see pages 124–125; for 3.5, see page 126; for 3.6, see page 127; and for 3.7 and 3.8, see page 128.

In several of the following exercises, you are asked to identify the type of data that are described. Possible answers include anecdotal data, available data, observational data that are from sample surveys, observational data that are not from sample surveys, and experiments. It is possible for some data to be classified in more than one category.

3.9 Software for teaching creative writing. An educational software company wants to compare the effectiveness of its computer animation for teaching creative writing with that of a textbook presentation. The company tests the creative-writing skills of a number of second-year college students and then randomly divides them into two groups. One group uses the animation, and the other studies the text. The company retests all the students and compares the increase in creative-writing skills in the two groups. Is this an experiment? Why or why not? Explain why this is an experiment, making sure to define the treatment and the response variable.

3.10 Apples or apple juice? Food rheologists study different forms of foods and how the form of a food affects how full we feel when we eat it. One study prepared samples of apple juice and samples of apples with the same number of calories. Half of the subjects were fed apples on one day, followed by apple juice on a later day; the other half received the apple juice, followed by the apples. After each eating session, the subjects were asked about how full they felt. Explain why this is an experiment, making sure to define the treatment and the response variable.

3.11 A dissatisfied customer. You like to eat homemade tuna sandwiches. Recently, you noticed that there does not seem to be as much tuna as you expected when you opened a can. Identify the type of data that this represents, and describe how it can or cannot be used to reach a conclusion about the amount of tuna in cans.

3.12 Claims settled for $3,300,000! According to a story in *Consumer Reports*, three major producers of canned tuna agreed to pay $3,300,000 to settle claims in California that the amount of tuna in their cans was less than the amount printed on the label.[7] What kind of data do you think was used in this situation to convince the producers to pay this amount of money to settle the claims? Explain your answer fully.

3.13 Marketing milk. An advertising campaign was developed to promote the consumption of milk by adolescents. Part of the campaign was based on a study conducted to determine the effect of additional milk in the diet of adolescents over a period of 18 months. A control group received no extra milk. Growth rates of total body bone mineral content (TBBMC) over the study period were calculated for each subject. Data for the group who received the extra milk and the control group were then used to examine the relationship between growth rate of TBBMC and age.

(a) How would you classify the data used to evaluate the effect of the additional milk in the diet? Explain your answer.

(b) How would you classify the control group data on growth rate of TBBMC and age for the study of this relationship? Explain your answer.

(c) Can you classify the growth rate of TBBMC variables and age as explanatory or response? If so, which is the explanatory variable? Give reasons for your answer.

3.14 What's wrong? Explain what is wrong in each of the following statements.

(a) Available data are always anecdotal.

(b) A census collects information on a subset of the population of interest.

(c) A sample survey usually involves a treatment.

3.15 Can you market echinacea for the common cold? In a study designed to evaluate the benefits of taking echinacea when you have a cold, 719 patients were randomly divided into four groups. The groups were (1) no pills, (2) pills that had no echinacea, (3) pills that had echinacea but the subjects did not know whether the pills contained echinacea, and (4) pills that had echinacea and the bottle containing the pills stated that the contents included echinacea. The outcome was a measure of the severity of the cold.[8]

(a) Identify the type of data collected in this study. Give reasons for your answer.

(b) Is this an experiment? If yes, what is the treatment?

3.16 Satisfaction with allocation of concert tickets. Your college sponsored a concert that sold out.

(a) After the concert, an article in the student newspaper reported interviews with three students who were unable to get tickets and were very upset with that fact. What kind of data does this represent? Explain your answer.

(b) A week later, the student organization that sponsored the concert set up a website where students could rank their satisfaction with the way that the tickets were allocated using a 5-point scale with values "very satisfied," "satisfied," "neither satisfied nor unsatisfied," "dissatisfied," and "very dissatisfied." The website was open to any students who chose to provide their opinion. How would you classify these data? Give reasons for your answer.

(c) Suppose that the website in part (b) was changed so that only a sample of students from the college were invited by text message to respond, and those who did not respond within three days were sent an additional text message reminding them to respond. How would your answer to part (b) change, if at all?

(d) Write a short summary contrasting the different types of data using your answers to parts (a), (b), and (c) of this exercise.

3.2 Designing Samples

When you complete this section, you will be able to:

- Distinguish between a population and a sample.
- Use the response rate to evaluate a survey.
- Use software or Table B to generate a simple random sample (SRS).
- Construct a stratified random sample using Table B or software to select the samples from the strata.
- Identify voluntary response samples, simple random samples, stratified random samples, and multistage random samples.
- Identify characteristics of samples that limit their usefulness, including undercoverage, nonresponse, response bias, and the wording of questions.

Samsung and O2 want to know how much time smartphone users spend on their smartphones. An automaker hires a market research firm to learn what percent of adults aged 18 to 35 recall seeing television advertisements for a new sport-utility vehicle. Government economists inquire about average household income. In all these cases, we want to gather information about a large group of people. We will not, as in an experiment, impose a treatment and then try to observe the response. Also, time, cost, and inconvenience forbid contacting every person. Instead, in such cases, we gather information about only part of the group—a *sample*—and then seek to draw conclusions about the whole, the *population*. Sample surveys are an important kind of observational study.

> **POPULATION AND SAMPLE**
>
> The entire group of cases that we want to study is called the **population.**
>
> A **sample** is a subset of the population for which we collect data.

Notice that "population" is defined in terms of our desire for knowledge. If we wish to draw conclusions about all U.S. college students, then that group is our population—even if only local students are available for questioning.

3.2 Designing Samples

sample design

The sample is the part from which we draw conclusions about the whole. The **design of a sample** refers to the method used to choose the sample from the population.

EXAMPLE 3.4

Can We Compete Globally? A lack of reading skills has been cited as one factor that limits the United States' ability to compete in the global economy. Various efforts have been made to improve this situation, including the Reading Recovery (RR) program. RR has specially trained teachers who work one-on-one with at-risk first-grade students to help them learn to read.

A study was designed to examine the relationship between the RR teachers' beliefs about their ability to motivate students and the progress of the students whom they teach.[9] The researchers were interested in the opinions of 6112 RR teachers. The researchers send a questionnaire to a random sample of 200 of these teachers. The population consists of all 6112 RR teachers, and the sample is the 200 who were randomly selected. ■

sampling frame

Unfortunately, our idealized framework of population and sample does not exactly correspond to the situations that we face in many cases. In Example 3.4, the list of teachers was prepared at a particular time in the past. It is very likely that some of the teachers on the list are no longer working as RR teachers today. New teachers, who have been trained in updated RR methods, may not have been added to the list. A list of items to be sampled is often called a **sampling frame.** For our example, we view this list as the population. Even so, we may have out-of-date addresses for some individuals who are still working as RR teachers, and some teachers may choose not to respond to our survey questions.

response rate

In reporting the results of a sample survey, it is important to include all details regarding the procedures used. The proportion of the original sample who actually provide usable data is called the **response rate** and should be reported for all surveys. If only 150 of the teachers who were sent questionnaires provided usable data, the response rate would be 150/200, or 75%. Follow-up mailings or phone calls to those who do not initially respond can help increase the response rate.

APPLY YOUR KNOWLEDGE

3.17 Taxes and forestland usage. A study was designed to assess the impact of taxes on forestland usage in part of the Upper Wabash River Watershed in Indiana.[10] A survey was sent to 772 forest owners from this region, and 348 were returned. Consider the population, the sample, and the response rate for this study. Describe these based on the information given, and indicate any additional information that you would need to give a complete answer.

3.18 Job satisfaction. A research team wanted to examine the relationship between employee participation in decision making and job satisfaction in a company. They are planning to randomly select 300 employees from a list of 2500 employees in the company. The Job Descriptive Index (JDI) will be used to measure job satisfaction, and the Conway Adaptation of the Alutto-Belasco Decisional Participation Scale will be used to measure decision participation. Describe the population and the sample for this study. Can you determine the response rate?

Poor sample designs can produce misleading conclusions. Here is an example.

EXAMPLE 3.5

Sampling Product in a Steel Mill A mill produces large coils of thin steel for use in manufacturing home appliances. The quality engineer wants to submit a sample of 5-centimeter squares to detailed laboratory examination. She asks a technician to cut a sample of 10 such squares. Wanting to provide "good" pieces of steel, the technician carefully avoids the visible defects in the coil material when cutting the sample. The laboratory results are wonderful, but the manufacturers complain about the material they are receiving. ∎

In this example, the samples were selected in a manner that guaranteed that they would not be representative of the entire population. In other words, this sampling scheme displays *bias*, or systematic error, in favoring some parts of the population over others.

BIAS

The design of a study is **biased** if it systematically favors certain outcomes.

Online opinion polls are particularly vulnerable to bias because the sample who respond are not representative of the population at large. Online polls use *voluntary response samples*, a particularly common form of biased sample.

VOLUNTARY RESPONSE SAMPLE

A **voluntary response sample** consists of people who choose themselves by responding to a general appeal. Voluntary response samples are biased because people with strong opinions, especially negative opinions, are most likely to respond.

The remedy for bias in choosing a sample is to allow impersonal chance to do the choosing so that there is neither favoritism by the sampler nor the use of voluntary response. Random selection of a sample, as in Example 3.4 (page 131), eliminates bias by giving all cases an equal chance to be chosen.

Voluntary response is one common type of bad sample design. Another is **convenience sampling**, which chooses the cases easiest to reach. Here is an example of convenience sampling.

convenience sampling

EXAMPLE 3.6

Interviewing Customers at the Mall Manufacturers and advertising agencies often use interviews at shopping malls to gather information about the habits of consumers and the effectiveness of ads. Obtaining a sample of mall customers is fast and cheap. But people contacted at shopping malls are not representative of the entire U.S. population. They are richer, for example, and more likely to be teenagers or retired. Moreover, mall interviewers tend to select neat, safe-looking subjects from the stream of customers. Decisions based on mall interviews may not reflect the preferences of other consumers, such as those who shop online. ∎

Both voluntary response samples and convenience samples produce samples that are almost guaranteed not to represent the entire population. These sampling methods display *bias* in favoring some parts of the population over others.

Big data

Big data involves extracting useful information from large and complex data sets. The exciting developments in this field have opened up many opportunities for new uses of data. Some critics, however, have suggested that the results obtained from some big data sets are vulnerable to bias.[11] Here is an example:

EXAMPLE 3.7

Bias and Big Data A study used Twitter and Foursquare data on coffee, food, nightlife, and shopping activity to describe the disruptive effects of Hurricane Sandy.[12] However, the data were dominated by tweets and smartphone activity from Manhattan. Relatively few data were from areas such as Breezy Point, a neighborhood in Queens, where the effects of the hurricane were most severe. ■

APPLY YOUR KNOWLEDGE

3.19 What is the population? For each of the following sampling situations, identify the population as exactly as possible. That is, which kinds of cases are in the population? If the information given is not sufficient, complete the description of the population in a reasonable way.

(a) The Annual Survey of Manufacturers (ASM) is conducted every year, except for years ending in 2 and 7. It is designed to provide statistics on employment, payroll, and other variables related to manufacturing for all manufacturers that have one of more paid employees.[13]

(b) An investigation of a subsidized housing facility with 320 apartments suggested that the owners were fraudulently charging the city by requesting payment for vacant apartments and other irregularities. Records of a sample of 30 apartments were selected for a careful audit.

3.20 Market segmentation and movie ratings. You wonder if that new "blockbuster" movie is really any good. Some of your friends like the movie, but you decide to check the Internet Movie Database (**imdb.com**) to see others' ratings. You find that 1897 people chose to rate this movie, with an average rating of only 4.6 out of 10. You are surprised that most of your friends liked the movie, while many people gave low ratings to the movie online. Are you convinced that a majority of those who saw the movie would give it a low rating? What type of sample are your friends? What type of sample are the raters on the Internet Movie Database? Discuss this example in terms of market segmentation.

Simple random samples

The simplest sampling design amounts to placing names (the population) in a hat and drawing out a handful (the sample). This is *simple random sampling*.

SIMPLE RANDOM SAMPLE

A **simple random sample (SRS)** of size n consists of n cases from the population chosen in such a way that every set of n cases has an equal chance to be the sample actually selected.

We select an SRS by labeling all the cases in the population and using software or a table of random digits to select a sample of the desired size. Notice that an SRS not only gives each case an equal chance to be chosen (thus avoiding bias in the choice), but also gives every possible sample an equal chance to be chosen. Some other random sampling designs give each case, but not each sample, an equal chance. One such design, *systematic random sampling*, is described later in Exercise 3.36 (page 142).

Thinking about random digits helps you to understand randomization even if you will use software in practice. Table B at the back of the book is a table of random digits.

RANDOM DIGITS

A **table of random digits** is a list of the digits 0, 1, 2, 3, 4, 5, 6, 7, 8, 9 that has the following properties:

1. The digit in any position in the list has the same chance of being any one of 0, 1, 2, 3, 4, 5, 6, 7, 8, 9.

2. The digits in different positions are independent in the sense that the value of one has no influence on the value of any other.

You can think of Table B as the result of asking an assistant (or a computer) to mix the digits 0 to 9 in a hat, draw one, then replace the digit drawn, mix again, draw a second digit, and so on. The assistant's mixing and drawing saves us the work of mixing and drawing when we need to randomize. Table B begins with the digits 19223950340575628713. To make the table easier to read, the digits appear in groups of five and in numbered rows. The groups and rows have no meaning—the table is just a long list of digits having Properties 1 and 2 described in the preceding box.

Our goal is to use random digits to select random samples. We need the following facts about random digits, which are consequences of the following basic properties:

- Any *pair* of random digits has the same chance of being any of the 100 possible pairs: 00, 01, 02, . . . , 98, 99.

- Any *triple* of random digits has the same chance of being any of the 1000 possible triples: 000, 001, 002, . . . , 998, 999.

- . . . and so on for groups of four or more random digits.

EXAMPLE 3.8

BRANDS

Brands A brand is a symbol or an image that is associated with a company. An effective brand identifies the company and its products. Using a variety of measures, dollar values for brands can be calculated. In Exercise 1.53 (page 37), you examined the distribution of the values of the top 100 brands.

Suppose that you want to write a research report on some of the characteristics of the companies in this elite group. You decide to look carefully at the websites of 10 companies from the list. One way to select the companies is to use a simple random sample. To create such a sample by using Table B, we start with a list of the companies with the top 100 brands. This is given in the data file, BRANDS. Next, we need to label the companies. In the data file, they are listed with their ranks, 1 to 100. Let's assign the labels 01 to 99 to the first 99 companies and 00 to the company with rank 100. With these labels, we can use Table B to select the SRS.

Let's start with line 156 of Table B. This line has the entries 55494 67690 88131 81800 11188 28552 25752 21953. These are grouped in sets of five digits, but we need to use sets of two digits for our randomization. Here is line 156 organized into sets of two digits: 55 49 46 76 90 88 13 18 18 00 11 18 82 85 52 25 75 22 19 53.

Using these random digits, we select eBay (55), Siemens (49), CVS (46), LEGO (76), Philips (90), Huawei (88), IBM (13), and Oracle (18); we then skip the second 18 because we have already selected Oracle to be in our SRS; and finally we select Allianz (00, recoded from rank 100), and BMW (11). ∎

Most statistical software will select an SRS for you, eliminating the need for Table B. The *Simple Random Sample* applet on the text website is another convenient way to automate this task.

Excel and other spreadsheet software can do the job. There are four steps:

1. Create a data set with all the elements of the population in the first column.
2. Assign a random number to each element of the population; put these in the second column.
3. Sort the data set by the random number column.
4. The simple random sample is obtained by taking elements in order from the sorted list until the desired sample size is reached.

We illustrate this procedure with a simplified version of Example 3.8.

EXAMPLE 3.9

BRANDS

Select a Random Sample Figure 3.2(a) gives the spreadsheet with the company names in column B. Only the first 12 of the 100 companies in the top 100 brands list are shown.

The random numbers generated by the RAND() function are given in the next column in Figure 3.2(b). The sorted data set is given in Figure 3.2(c). The 10 brands were selected for our random sample are Siemens, Hershey, Microsoft, Heineken, Marlboro, eBay, Mercedes-Benz, IKEA, J.P. Morgan, and Home Depot. ∎

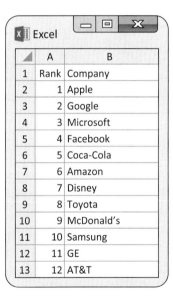

FIGURE 3.2 Selection of a simple random sample of brands using Excel, Example 3.9: (a) labels; (b) random numbers; (c) randomly sorted labels.

APPLY YOUR KNOWLEDGE

3.21 Ringtones for cell phones. You decide to change the ringtones for your cell phone by choosing two from a list of the 10 most popular ringtones.[14] Here is the list: RINGS

| Shake It Off | All about That Bass | Burnin' It Down | Black Widow | Don't Tell 'Em |
| Stay with Me | Lifestyle | Happy | Bang Bang | Royals |

Select your two ringtones using a simple random sample.

3.22 Listen to three songs. The walk to your statistics class takes about 10 minutes, about the amount of time needed to listen to three songs on your iPod. You decide to take a simple random sample of songs from the top 10 songs listed on the *Billboard* Hot 100 chart.[15] Here is the list: SONGS

This Is America	Nice for What	God's Plan	Psycho	Meant to Be
The Middle	No Tears Left to Cry	Look Alive	Never Be the Same	Perfect

Select the three songs for your iPod using a simple random sample.

Stratified samples

The general framework for designs that use chance to choose a sample is a *probability sample*.

> **PROBABILITY SAMPLE**
>
> A **probability sample** is a sample chosen by chance. We must know what samples are possible and what chance, or probability, each possible sample has.

An SRS gives each possible sample of a given size an equal chance of being selected. As a result, each member of the population is equally likely to be selected. Other probability sampling designs also give each member of the population an *equal* chance to be selected. This may not be true in more elaborate sampling designs. In every case, however, the use of chance to select the sample is the essential principle of statistical sampling.

Designs for sampling from large populations spread out over a wide area are usually more complex than an SRS. For example, it is common to sample important groups within the population separately, then combine these samples. This is the idea of a *stratified sample*.

> **STRATIFIED RANDOM SAMPLE**
>
> To select a **stratified random sample,** first divide the population into groups of similar cases, called **strata.** Then choose a separate SRS in each stratum and combine these SRSs to form the full sample.

Choose the strata based on facts known before the sample is taken. For example, a population of election districts might be divided into urban, suburban, and rural strata. A stratified design can produce more exact information than an SRS of the same size by taking advantage of the fact that cases in the same stratum are similar to one another. Think of the extreme case in which all cases in each stratum are identical: just one case from each stratum is then enough to completely describe the population.

EXAMPLE 3.10

Fraud against Insurance Companies A dentist is suspected of defrauding insurance companies by describing some dental procedures incorrectly on claim forms and overcharging for them. An investigation begins by examining a sample of his bills for the past three years. Because there are five suspicious types of procedures, the investigators take a stratified sample. That is, they randomly select bills for each of the five types of procedures separately. ∎

Multistage samples

Another common means of restricting random selection is to choose the sample in stages. This is common practice for national samples of households or people. For example, data on employment and unemployment are gathered by the government's Current Population Survey, which conducts interviews in about 60,000 households each month. The cost of sending interviewers to the widely scattered households in an SRS would be too high. Moreover, the government wants data broken down by states and large cities. The Current Population Survey, therefore, uses a **multistage sampling design.** The final sample consists of clusters of nearby households that an interviewer can easily visit. Most opinion polls and other national samples are also multistage, though interviewing in most national samples today is done by telephone rather than in person, eliminating the economic need for clustering. The Current Population Survey sampling design is roughly as follows:[16]

Stage 1. Divide the United States into 2007 geographical areas called primary sampling units (PSUs). PSUs do not cross state lines. Select a sample of 754 PSUs. This sample includes the 428 PSUs with the largest populations and a stratified sample of 326 of the others.

Stage 2. Divide each PSU selected into smaller areas called "blocks." Stratify the blocks using ethnic and other information, and take a stratified sample of the blocks in each PSU.

Stage 3. Sort the housing units in each block into clusters of four nearby units. Interview the households in a probability sample of these clusters.

Analysis of data from sampling designs more complex than an SRS takes us beyond basic statistics. But the SRS is the building block of more elaborate designs, and analysis of other designs differs more in complexity of detail than in fundamental concepts.

APPLY YOUR KNOWLEDGE

3.23 Who goes to the market research workshop? A small advertising firm has 30 junior associates and 10 senior associates. The junior associates are WSHOP

Abrams	Dawson	Heathcote	Marks	Seeger
Armstrong	Dowell	Howe	Mills	Sloan
Bell	Enos	Johnson	Perkins	Swihart
Boyer	Fletcher	Kim	Perez	Turner
Cantrell	Gastineau	Knapp	Reed	Xie
Colver	Haas	Lewicki	Rollins	Zhang

The senior associates are

| Cline | Boudreaux | Eastern | Eifert | Haarms |
| Hardin | Kiesler | Oden | Traylor | Whilby |

The firm will send four junior associates and two senior associates to a workshop on current trends in market research. It decides to choose those who will go by random selection. Use Table B or software to choose a stratified random sample of four junior associates and two senior associates. If you use Table B, start at line 121 to choose your sample.

3.24 **Sampling by accountants.** Accountants use stratified samples during audits to verify a company's records of such things as accounts receivable. The stratification is based on the dollar amount of the item and often includes 100% sampling of the largest items. One company reports 3000 accounts receivable. Of these, 20 are in amounts greater than $50,000; 380 are in amounts between $1000 and $50,000; and the remaining 2600 are in amounts less than $1000. Using these groups as strata, you decide to verify all of the largest accounts and to sample 5% of the midsize accounts and 1% of the small accounts. How would you label the two strata from which you will sample? Use software or Table B, starting at line 125, to select the accounts to be audited.

Cautions about sample surveys

Random selection eliminates bias in the choice of a sample from a list of the population. Sample surveys of large human populations, however, require much more than a good sampling design. To begin, we need an accurate and complete list of the population. Because such a list is rarely available, most samples suffer from some degree of *undercoverage*. A sample survey of households, for example, will miss not only homeless people, but also prison inmates and students in dormitories. An opinion poll conducted by telephone will miss the 6% of American households without residential phones. Thus, results of national sample surveys have some bias if the people not covered—who most often are poor people—differ from the rest of the population.

A more serious source of bias in most sample surveys is *nonresponse*, which occurs when a selected case cannot be contacted or refuses to cooperate. The nonresponse rate for sample surveys often reaches 50% or more, even with careful planning and several callbacks. Because nonresponse is higher in urban areas, most sample surveys substitute other people in the same area to avoid favoring rural areas in the final sample. If the people contacted differ from those who are rarely at home or who refuse to answer questions, some bias remains.

> **UNDERCOVERAGE AND NONRESPONSE**
>
> **Undercoverage** occurs when some groups in the population are left out of the process of choosing the sample.
>
> **Nonresponse** occurs when a case chosen for the sample cannot be contacted or does not cooperate.

EXAMPLE 3.11

Nonresponse in the Current Population Survey How bad is nonresponse? The Current Population Survey (CPS) uses a probability sample of approximately 60,000 occupied households to study a variety of topics concerning the labor market (see Figure 3.3). It has a low nonresponse rate relative to other polls. Only about 10% of the households in the CPS sample refuse to take part, and another 3% can't be contacted.[17] ∎

FIGURE 3.3 The General Social Survey (GSS) assesses attitudes on a variety of topics, Example 3.11.

What about polls done by the media and by market research and opinion-polling firms? We don't know their rates of nonresponse because they won't say. That in itself is a bad sign.

EXAMPLE 3.12

Change in Nonresponse in Pew Surveys The Pew Research Center conducts research using surveys on a variety of issues, attitudes, and trends. A study by the center examined the decline in the response rates to its surveys over time. After a long period of decline, the response rates now appear to have stabilized. Here are some data that describe the trend.[18]

Year	1997	2000	2003	2006	2009	2012	2016
Nonresponse rate	64%	72%	75%	79%	85%	91%	91%

The center is devising alternative methods that show some promise of improving the response rates of its surveys. ∎

Most sample surveys, and almost all opinion polls, are now carried out by telephone or online. This and other details of the interview method can affect the results. When presented with several options for a reply—such as completely agree, mostly agree, mostly disagree, and completely disagree—people tend to be a little more likely to respond to the first one or two options presented.

The behavior of the respondent or of the interviewer can also cause **response bias** in sample results. Respondents may lie, especially if asked about illegal or unpopular behavior. The race or gender of the interviewer can influence responses to questions about race relations or attitudes toward feminism. Answers to questions that ask respondents to recall past events are often inaccurate because of faulty memory.

response bias

wording of questions

The **wording of questions** is the most important influence on the answers given to a sample survey. Confusing or leading questions can introduce strong bias, and even minor changes in wording can change a survey's outcome. Here is an example.

EXAMPLE 3.13

The Form of the Question Is Important Do you exercise regularly and eat a healthy diet? This is a *double-barreled* question. A "yes" answer should mean that the person does both things: exercises regularly and eats a healthy diet.

Many respondents would do only one of these things and would be confused by the question. A good practice is to avoid double-barreled questions and ask two separate questions to obtain the desired information. ∎

APPLY YOUR KNOWLEDGE

3.25 Random digit dialing. Ideally, the sampling frame should include every case in the population, but in practice this is often difficult.

(a) Suppose that a sample of households in a community is selected at random from a published directory available at **dexpages.com.** Which households are omitted from this frame? Which types of people do you think are likely to live in these households? These people will probably be underrepresented in the sample.

(b) It is usual in telephone surveys to use random digit dialing equipment that selects the last four digits of a telephone number at random after being given the exchange (the first three digits). Which of the households that you mentioned in your answer to part (a) will be included in the sampling frame by random digit dialing?

The statistical design of sample surveys is a science, but this science is only part of the art of sampling. Because of nonresponse, response bias, and the difficulty of posing clear and neutral questions, you should hesitate to fully trust reports about complicated issues based on surveys of large human populations. *Insist on knowing the exact questions asked, the rate of nonresponse, and the date and method of the survey before you trust a poll result.*

BEYOND THE BASICS

Capture-recapture sampling

Pacific salmon return to reproduce in the river where they were hatched three or four years earlier. How many salmon made it back this year? The answer will help determine quotas for commercial fishing on the west coast of Canada and the United States. Biologists estimate the size of animal populations with a special kind of repeated sampling, called **capture-recapture sampling.** More recently, capture-recapture methods have been used on human populations.

capture-recapture sampling

EXAMPLE 3.14

Sampling for a Major Industry in British Columbia The old method of counting returning salmon involved placing a "counting fence" in a stream and counting all the fish caught by the fence. This is expensive and difficult. For example, fences are often damaged by high water.

Repeat sampling using small nets is more practical. During this year's spawning run in the Chase River in British Columbia, Canada, you net 200 coho salmon, tag the fish, and release them. Later in the week, your nets capture 120 coho salmon in the river, of which 12 have tags.

The proportion of your second sample that have tags should estimate the proportion in the entire population of returning salmon that are tagged. So if N is the unknown number of coho salmon in the Chase River this year, we should have approximately

proportion tagged in sample = proportion tagged in population

$$\frac{12}{120} = \frac{200}{N}$$

Solving for N estimates that the total number of salmon in this year's spawning run in the Chase River is approximately

$$N = 200 \times \frac{120}{12} = 2000 \ \blacksquare$$

The capture-recapture idea extends the use of a sample proportion to estimate a population proportion. The idea works well if both samples are SRSs from the population and the population remains unchanged between samples. In practice, complications arise. For example, some tagged fish might be caught by bears or otherwise die between the first and second samples.

Variations on capture-recapture samples are widely used in wildlife studies and are now finding other applications. One way to estimate the census undercount in a district is to consider the census as "capturing and marking" the households that respond. Census workers then visit the district, take an SRS of households, and see how many of those counted by the census show up in the sample. Capture-recapture estimates the total count of households in the district. As with estimating wildlife populations, there are many practical pitfalls with this sampling technique.

SECTION 3.2 SUMMARY

- A sample survey selects a **sample** from the **population** that is the object of our study. We base conclusions about the population on data collected from the sample.

- The **sample design** of a sample refers to the method used to select the sample from the population. **Probability sampling designs** use chance to select a sample.

- The basic probability sample is a **simple random sample (SRS).** An SRS gives every possible sample of a given size the same chance to be chosen.

- Choose an SRS by labeling the members of the population and using a **table of random digits** to select the sample. Software can automate this process.

- To choose a **stratified random sample,** divide the population into **strata,** or groups of cases that are similar in some way that is important to the response. Then choose a separate SRS from each stratum, and combine all the SRSs to form the full sample.

- **Multistage samples** select successively smaller groups within the population in stages, resulting in a sample consisting of clusters of cases. Each stage may employ an SRS, a stratified sample, or another type of sample.

- Failure to use probability sampling often results in **bias,** or systematic errors in the way the sample represents the population. **Voluntary response** samples, in which the respondents choose themselves, are particularly prone to large bias.

- In human populations, even probability samples can suffer from bias due to **undercoverage** or **nonresponse,** from **response bias** due to the behavior of the interviewer or the respondent, or from misleading results due to **poorly worded questions.**

SECTION 3.2 EXERCISES

For Exercises 3.17 and 3.18, see page 131; for 3.19 and 3.20, see page 133; for 3.21 and 3.22, see pages 135–136; for 3.23 and 3.24, see pages 137–138; and for 3.25, see page 140.

3.26 What's wrong? Explain what is wrong in each of the following statements.

(a) A simple random population is one way to randomly select cases from a sample.

(b) Random digits cannot be used to select a sample from a population that has fewer than 10 cases.

(c) When you select a simple random sample, the sample will always have more cases than the population.

3.27 What's wrong? Explain what is wrong with each of the following random selection procedures, and explain how you would do the randomization correctly.

(a) To determine the reading level of an introductory statistics text, you evaluate all of the written material in the first chapter.

(b) You want to sample student opinions about a proposed change in procedures for changing majors. You hand out questionnaires to 50 students as they arrive for class at 7:30 A.M.

(c) A population of subjects is put in alphabetical order, and a simple random sample of size 20 is taken by selecting the first 20 subjects in the list.

3.28 Importance of students as customers. A committee on community relations in a college town plans to survey local businesses about the importance of students as customers. From telephone book listings, the committee chooses 50 businesses at random. Of these, 28 return the questionnaire mailed by the committee. What is the population for this sample survey? What is the sample? What is the rate (percent) of nonresponse?

3.29 How many text messages? You would like to know something about how many text messages you will receive in the next 100 days. Counting the number for each of the 100 days would take more time than you would like to spend on this project, so you randomly select 10 days from the hundred to count.

(a) Describe the population for this setting.

(b) What is the sample?

3.30 Identify the populations. For each of the following sampling situations, identify the population as exactly as possible. That is, which cases are in the population? If the information given is not complete, complete the description of the population in a reasonable way.

(a) A college has changed its core curriculum and wants to obtain detailed feedback information from the students during each of the first 12 weeks of the coming semester. Each week, a random sample of five students will be selected to be interviewed.

(b) The American Community Survey (ACS) replaced the census "long form" starting with the 2010 census. The main part of the ACS contacts 250,000 addresses by mail each month, with follow-up by phone and in person if there is no response. Each household answers questions about their housing, economic, and social status.

(c) An opinion poll contacts 1161 adults and asks them, "Which political party do you think has better ideas for leading the country in the twenty-first century?"

3.31 Interview potential customers. You have been hired by a company that is planning to build a new apartment complex for students in a college town. The company wants you to collect information about preferences of potential customers for its complex. Most of the college students who live in apartments live in one of 33 complexes. You decide to select five apartment complexes at random for in-depth interviews with residents. Select a simple random sample of five of the following apartment complexes. If you use Table B, start at line 127. RESID

Ashley Oaks	Country View	Mayfair Village
Bay Pointe	Country Villa	Nobb Hill
Beau Jardin	Crestview	Pemberly Courts
Bluffs	Del-Lynn	Peppermill
Brandon Place	Fairington	Pheasant Run
Briarwood	Fairway Knolls	Richfield
Brownstone	Fowler	Sagamore Ridge
Burberry	Franklin Park	Salem Courthouse
Cambridge	Georgetown	Village Manor
Chauncey Village	Greenacres	Waterford Court
Country Squire	Lahr House	Williamsburg

3.32 Using GIS to identify mint field conditions. A geographic information system (GIS) is to be used to distinguish different conditions in mint fields. Ground observations will be used to classify regions of each field as either healthy mint, diseased mint, or weed-infested mint. The GIS divides mint-growing areas into regions called pixels. An experimental area contains 200 pixels. For a random sample of 20 pixels, ground measurements will be made to determine the status of the mint, and these observations will be compared with information obtained by the GIS. Select the random sample. If you use Table B, start at line 122 and choose only the first eight pixels in the sample.

3.33 Select a simple random sample. After you have labeled the cases in a population, the *Simple Random Sample* applet automates the task of choosing an SRS. Use the applet to choose the sample in the previous exercise.

3.34 Select a simple random sample. There are approximately 256 active telephone area codes covering the United States.[19] You want to choose an SRS of 25 of these area codes for a study of available telephone numbers. Use software or the *Simple Random Sample* applet to choose your sample. ACODES

3.35 Repeated use of Table B. In using Table B repeatedly to choose samples, you should not always begin at the same place, such as line 101. Why not?

3.36 Systematic random samples. Systematic random samples are often used to choose a sample of apartments in a large building or dwelling units in a block at the last stage of a multistage sample. An example will illustrate the idea of a systematic sample. Suppose that we must choose four addresses out of 100. Because 100/4 = 25, we can think of the list as four lists of 25 addresses. Choose one of the first 25 at random, using Table B. The sample contains this address and the addresses 25, 50, and 75 places down the list from it. If 13 is chosen, for example, then the systematic random sample consists of the addresses numbered 13, 38, 63, and 88.

(a) A study of dating among college students wanted a sample of 40 of the 5000 single male students on campus. The sample consisted of every 125th name from a list of the 5000 students. Explain why the survey chooses every 125th name.

(b) Use software or Table B at line 145 to choose the starting point for this systematic sample.

3.37 Systematic random samples versus simple random samples. The previous exercise introduces systematic random samples. Explain carefully why a systematic random sample *does* give every case the same chance to be chosen but is *not* a simple random sample.

3.38 Random digit telephone dialing for market research. A market research firm in California uses random digit dialing to choose telephone numbers at random. Numbers are selected separately within each California area code. The size of the sample in each area code is proportional to the population living there.

(a) What is the name for this kind of sampling design?

(b) California area codes, in rough order from north to south, are

209	213	310	323	408	415	510	530	559	562
619	626	650	661	707	714	760	805	818	831
858	909	916	925	949					

Another California survey does not call numbers in all area codes but starts with an SRS of ten area codes. Choose such an SRS. If you use Table B, start at line 112. CACODES

3.39 Select employees for an awards committee. A department has 30 hourly workers and 10 salaried workers. The hourly workers are

Abdelbaki	Gougeon	Nolan	Schruth	Wark
Bean	Herring	Parker	Shook	Webster
Crissinger	Humphrey	Poliak	Solis	Werner
Draper	Jorgensen	Rizzo	Song	Whitten
Evans	Manley	Robertson	Stokes	Wu
Gebert	Miller	Samons	Swartz	Yu

and the salaried workers are

Crowe	Fatemi	Hilde	Rumsey	Sklar
Drougas	Gnagey	Kim	Schroeder	Zhang

The committee will have six hourly workers and two salaried workers. Random selection will be used to select the committee members. Select a stratified random sample of six hourly workers and two salaried workers. CMEMB

3.40 When do you ask? When observations are taken over time, it is important to check for patterns that may be important for the interpretation of the data. In Section 1.1, we learned to use a time plot for this purpose. Describe and discuss a sample survey question where you would expect to have variation over time (answers would be different at different times) for the following situations:

(a) Data are taken at each hour of the day from 8 A.M. to 6 P.M.

(b) Data are taken on each of the seven days of the week.

(c) Data are taken during each of the 12 months of the year.

3.41 Survey questions. Comment on each of the following as a potential sample survey question. Is the question clear? Is it slanted toward a desired response?

(a) "Some cell phone users have developed brain cancer. Should all cell phones come with a warning label explaining the danger of using cell phones?"

(b) "Do you agree that a national system of health insurance should be favored because it would provide health insurance for everyone and would reduce administrative costs?"

(c) "In view of escalating environmental degradation and incipient resource depletion, would you favor economic incentives for recycling of resource-intensive consumer goods?"

3.3 Designing Experiments

When you complete this section, you will be able to:

- Identify experimental units, subjects, treatments, and outcomes for a comparative experiment.
- Describe a placebo effect in an experiment.
- Identify bias in an experiment.
- Explain the need for a control group in an experiment.
- Explain the need for randomization in an experiment.
- When carrying out an experiment, apply the basic principles of experimental design: compare, randomize, and replicate.
- Use software or a table of random digits to randomly assign experimental units to treatments in an experiment.
- Identify whether an experiment is run using a completely randomized design, a matched pairs design, or a block design.

A study is an experiment when we actually do something to people, animals, or objects so that we can then observe the response. Here is the basic vocabulary of experiments.

> **TREATMENTS, EXPERIMENTAL UNITS, AND SUBJECTS**
>
> The conditions studied in an experiment are called **treatments**. The **experimental units** are the cases that are assigned to treatments. When the experimental units are human beings, they are called **subjects**.

explanatory and response variables, p. 64

factors

level of a factor

Because the purpose of an experiment is to reveal the response of one variable to changes in other variables, the distinction between explanatory and response variables is important. The explanatory variables in an experiment are often called **factors**. Many experiments study the joint effects of several factors. In such an experiment, each treatment is formed by combining a specific value (often called a **level**) of each of the factors.

EXAMPLE 3.15

confounded

Is the Cost Justified? The increased costs for teacher salaries and facilities associated with smaller class sizes can be substantial. Are smaller classes really better? We might do an observational study that compares students who happened to be in smaller and larger classes in their early school years. Small classes are expensive, so they are more common in schools that serve richer communities. Students in small classes tend to also have other advantages: their schools have more resources, their parents are better educated, and so on. The size of the classes is **confounded** with other characteristics of the students, making it impossible to isolate the effects of small classes.

The Tennessee STAR program was an experiment on the effects of class size. The *subjects* were 6385 students who were beginning kindergarten. Each student was assigned to one of three *treatments*: regular class (22 to 25 students) with one teacher, regular class with a teacher and a full-time teacher's aide, and small class (13 to 17 students) with one teacher. These treatments are levels of a single *factor*: the type of class. The students stayed in the same type of class for four years, then all returned to regular classes. In later years, students from the small classes had higher scores on standard tests, were less likely to fail a grade, had better high school grades, and so on. The benefits of small classes were greatest for minority students.[20] ■

Example 3.15 illustrates the big advantage of experiments over observational studies: *In principle, experiments can give good evidence for causation.* In an experiment, we study the specific factors we are interested in, while controlling the effects of lurking variables. All the students in the Tennessee STAR program followed the usual curriculum at their schools. Because students were randomly assigned to different class types within their schools, school resources and family backgrounds were not confounded with class type. The only systematic difference was the type of class. When students from the small classes did better than those in the other two types, we can be confident that class size made the difference.

EXAMPLE 3.16

Effects of TV Advertising What are the effects of repeated exposure to an advertising message? The answer may depend both on the length of the ad and on how often it is repeated. An experiment investigates this question using undergraduate students as subjects. All subjects view a 40-minute television program that includes ads for a digital camera. Some subjects see a 30-second commercial; others, a 90-second version. The same commercial is repeated one, three, or five times during the program. After viewing, all of the subjects answer questions about their recall of the ad, their attitude toward the camera, and their intention to purchase it. These are the response variables.[21]

This experiment has two factors: length of the commercial, with two levels; and repetitions, with three levels. All possible combinations of the 2×3 factor levels form six treatment combinations. Figure 3.4 shows the layout of these treatments. ■

FIGURE 3.4 The treatments in the experimental design of Example 3.16. Combinations of levels of the two factors form six treatments.

	Factor B Repetitions		
	1 time	3 times	5 times
Factor A Length — 30 seconds	1	2	3
90 seconds	4	5	6

Experimentation allows us to study the effects of the specific treatments we are interested in. Moreover, we can control the environment of the subjects to hold constant the factors that are of no interest to us, such as the specific product advertised in Example 3.16. On the one hand, the ideal case is a laboratory experiment in which we control all lurking variables and so see only the effect of the treatments on the response. On the other hand, the effects of being in an artificial environment such as a laboratory may affect the outcomes. *The balance between control and realism is an important consideration in the design of experiments.*

interaction

Another advantage of experiments is that we can study the combined effects of several factors simultaneously. The **interaction** of several factors can produce effects that could not be predicted from looking at the effect of each factor alone. Perhaps longer commercials increase interest in a product, and more commercials also increase interest, but if we make a commercial longer *and* show it more often, viewers get annoyed and their interest in the product drops. The two-factor experiment in Example 3.16 will help us find out. In chapter 16 we study this type of design in detail.

APPLY YOUR KNOWLEDGE

3.42 Radiation and storage time for food products. Storing food for long periods of time is a major challenge for those planning for human space travel beyond the moon. One problem is that exposure to radiation decreases the length of time that food can be stored. One experiment examined the effects of nine different levels of radiation on a particular type of fat, or lipid.[22] The amount of oxidation of the lipid is the measure of the extent of the damage due to the radiation. Three samples are exposed to each radiation level. Give the experimental units, the treatments, and the response variable. Describe the factor and its levels. There are many different types of lipids. To what extent do you think the results of this experiment can be generalized to other lipids?

3.43 **Can they use the Web?** A course in computer graphics technology requires students to learn multiview drawing concepts. This topic is traditionally taught using supplementary material printed on paper. The instructor for the course believes that a web-based interactive drawing program will be more effective in increasing the drawing skills of the students.[23] The 50 students who are enrolled in the course will be randomly assigned to either the paper-based instruction or the web-based instruction. A standardized drawing test will be given before and after the instruction. Explain why this study is an experiment, and give the experimental units, the treatments, and the response variable. Describe the factor and its levels. To what extent do you think the results of this experiment can be generalized to other settings?

3.44 **Is the packaging convenient for the customer?** A manufacturer of food products uses package liners that are sealed by applying heated jaws after the package is filled. The customer peels the sealed pieces apart to open the package. What effect does the temperature of the jaws have on the force needed to peel the liner? To answer this question, engineers prepare 20 package liners. They seal four liners at each of five different temperatures: 250°F, 275°F, 300°F, 325°F, and 350°F. Then they measure the force needed to peel each seal.

(a) What are the experimental units studied?

(b) There is one factor (explanatory variable). What is it, and what are its levels?

(c) What is the response variable?

Comparative experiments

Many experiments have a simple design with only a single treatment, which is applied to all experimental units. The design of such an experiment can be outlined as

$$\text{Treatment} \longrightarrow \text{Observe response}$$

EXAMPLE 3.17

Increase the Sales Force A company has increased its sales force in the hope that its sales will increase. The company compares sales before the sales force increase with sales after the increase. Sales are up, so the manager who suggested the change gets a bonus.

$$\text{Increase the sales force} \longrightarrow \text{Observe sales change} \blacksquare$$

The sales experiment described in Exercise 3.17 was poorly designed to evaluate the effect of increasing the sales force. Perhaps sales would have increased because of seasonal variation in demand or other factors affecting the business.

placebo effect

In medical settings, an improvement in condition is sometimes due to a phenomenon called the **placebo effect.** A placebo is a dummy or fake treatment, such as a sugar pill. Many participants, regardless of treatment, respond favorably due to the personal attention or to the expectation that the treatment will help them.

For the sales force study, we don't know whether the increase in sales was due to increasing the sales force or due to other factors. The experiment gave inconclusive results because the effect of increasing the sales force was confounded with other factors that could have had an effect on sales. The best way to avoid confounding is to do a **comparative experiment.** Think about a study where the sales force is increased in half of the regions where the product is sold and is not changed in the other regions. A comparison of sales from the two sets of regions would provide an evaluation of the effect of the increasing the sales force.

In medical settings, it is standard practice to randomly assign patients to either a **control group** or a **treatment group.** All patients are treated the same in every way except that the treatment group receives the treatment that is being evaluated. In the setting of our comparative sales experiment, we would randomly divide the regions into two groups. One group will have the sales force increased and the other group will not.

Uncontrolled experiments in medicine and the behavioral sciences can be dominated by such influences as the details of the experimental arrangement, the selection of subjects, and the placebo effect. The result is often *bias*.

bias, p. 132

An uncontrolled study of a new medical therapy, for example, is biased in favor of finding the treatment effective because of the placebo effect. It should not surprise you to learn that uncontrolled studies in medicine give new therapies a much higher success rate than proper comparative experiments do. Well-designed experiments usually compare several treatments.

APPLY YOUR KNOWLEDGE

3.45 Does using statistical software improve exam scores? An instructor in an elementary statistics course wants to know if using a new statistical software package will improve students' final-exam scores. He asks for volunteers, and approximately half of the class agrees to work with the new software. He compares the final-exam scores of the students who used the new software with the scores of those who did not. Discuss possible sources of bias in this study.

Randomized comparative experiments

The **design of an experiment** first describes the response variables, the factors (explanatory variables), and the layout of the treatments, with *comparison* as the leading principle. The second aspect of design is the rule used to assign the subjects to the treatments. Comparison of the effects of several treatments is valid only when all treatments are applied to similar groups of subjects. If one corn variety is planted on more fertile ground, or if one cancer drug is given to less seriously ill patients, comparisons among treatments are biased. How can we assign cases to treatments in a way that is fair to all the treatments?

Our answer is the same as in sampling: let chance make the assignment. The use of chance to divide subjects into groups is called **randomization.** Groups formed by randomization don't depend on any characteristic of the subjects or on the judgment of the experimenter. An experiment that uses both comparison and randomization is a **randomized comparative experiment.** Here is an example.

EXAMPLE 3.18

Testing a Breakfast Food A food company assesses the nutritional quality of a new "instant breakfast" product by feeding it to newly weaned male white rats. The response variable is a rat's weight gain over a 28-day period. A control group of rats eats a standard diet but otherwise receives exactly the same treatment as the experimental group.

This experiment has one factor (the diet) with two levels. The researchers use 30 rats for the experiment and so divide them into two groups of 15. To do this in an unbiased fashion, put the cage numbers of the 30 rats in a hat, mix them up, and draw 15. These rats form the experimental group and the remaining 15 make up the control group. *Each group is an SRS of the available rats.* Figure 3.5 outlines the design of this experiment. ■

FIGURE 3.5 Outline of a randomized comparative experiment, Example 3.18.

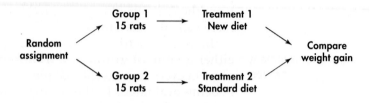

APPLY YOUR KNOWLEDGE

3.46 Diagram the food storage experiment. Refer to Exercise 3.42 (page 145). Draw a diagram similar to Figure 3.5 that describes the food for space travel experiment.

3.47 Diagram the Web use. Refer to Exercise 3.43 (page 146). Draw a diagram similar to Figure 3.5 that describes the computer graphics drawing experiment.

Completely randomized designs

The design in Figure 3.5 combines comparison and randomization to arrive at the simplest statistical design for an experiment. This "flowchart" outline presents all the essentials: randomization, the sizes of the groups and which treatment they receive, and the response variable. There are, as we will see later, statistical reasons for generally using treatment groups that are approximately equal in size. We call designs like that in Figure 3.5 *completely randomized*.

COMPLETELY RANDOMIZED DESIGN

In a **completely randomized** experimental design, all the subjects are allocated at random among all the treatments.

Completely randomized designs can compare any number of treatments. Here is an example that compares three treatments.

EXAMPLE 3.19

Utility Companies and Energy Conservation Many utility companies have introduced programs to encourage energy conservation among their customers. An electric company considers placing electronic meters in households to show what the cost would be if the electricity use at that moment continued for a month. Will these meters reduce electricity use? Would cheaper methods work almost as well? The company decides to design an experiment.

One cheaper approach is to give customers a chart and information about monitoring their electricity use. The experiment compares these two approaches (meter, chart) and also a control. The control group of customers receives information about energy conservation but no help in monitoring electricity use. The response variable is total electricity used in a year. The company finds 60 single-family residences in the same city willing to participate, so it assigns 20 residences at random to each of the three treatments. Figure 3.6 outlines the design. ■

FIGURE 3.6 Outline of a completely randomized design comparing three treatments, Example 3.19.

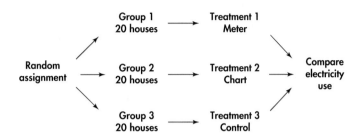

How to randomize

The idea of randomization is to assign experimental units to treatments by drawing names from a hat. In practice, experimenters use software to carry out randomization. In Example 3.19, we have 60 residences that need to be randomly assigned to three treatments. Most statistical software will be able to do the randomization required.

We prefer to use software for randomizing. However, if you do not have that option available to you, you can use a *table of random digits,* such as Table B. If you use software, the steps are similar to what we used to select an SRS in Example 3.9 (page 135). Here are the steps needed:

Step 1: Label. Give each experimental unit a unique label. For privacy reasons, we might want to use a numerical label and keep a file that identifies the experimental units with the numbers in a separate place.

Step 2: Use the computer. Once we have the labels, we create a data file with the labels and generate a random number for each label. In Excel, this can be done with the RAND() function. Finally, we sort the entire data set based on the random numbers. Groups are formed by selecting units in order from the sorted list.

EXAMPLE 3.20

Do the Randomization for the Utility Company Experiment Using Excel In the utility company experiment of Example 3.19, we must assign 60 residences to three treatments. First, we generate the labels. Let's use numerical labels and keep a separate file that gives the residence address for each number. So for Step 1, we will use these labels, 1 to 60:

$$1, 2, 3, \ldots, 59, 60$$

To illustrate Step 2, we will show several Excel files. To see what we are doing, it will be easier if we reduce the number of residences to be randomized. So, let's randomize 12 residences to the three treatments. Our labels are

$$1, 2, 3, 4, 5, 6, 7, 8, 9, 10, 11, 12$$

For the first part of Step 2, we create an Excel file with the numbers 1 to 12 in the first column. This file is shown in Figure 3.7(a). Next, we use the RAND() function in Excel to generate 12 random numbers in the second column. The result is shown in Figure 3.7(b). We then sort the file based in the random numbers. We create a third column with the following treatments: "Meter" for the first four, "Chart" for the next four, and "Control" for the last four. The result is displayed in Figure 3.7(c). ■

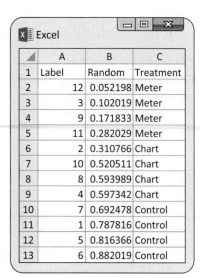

FIGURE 3.7 Randomization of 12 experimental units to three treatments using Excel, Example 3.20: (a) labels; (b) random numbers; (c) randomly sorted labels with treatment assignments.

If software is not available, you can use the random digits in Table B to do the randomization. The method is similar to the one we used to select an SRS in Example 3.8 (page 137). Here are the steps that you need:

Step 1: Label. Give each experimental unit a numerical label. Each label must contain the same number of digits. So, for example, if you are randomizing 10 experimental units, you could use the labels, 0, 1, ..., 8, 9; or 01, 02, ..., 10. Note that with the first choice you need only one digit, but for the second choice, you need two.

Step 2: Table. Start anywhere in Table B and read digits in groups corresponding to one-digit or two-digit groups. (You really do not want to use Table B for more than 100 experimental units. Software is needed here.)

EXAMPLE 3.21

Do the Randomization for the Utility Company Experiment Using Random Digits
As we did in Example 3.20, we will illustrate the method by randomizing 12 residences to three treatments. For Step 1, we assign the 12 residences the following labels:

01, 02, 03, 04, 05, 06, 07, 08, 09, 10, 11, 12

Compare these labels with the ones we used in Example 3.20. Here, we need the same number of digits for each label, so we put a zero as the first digit for the first nine labels.

For Step 2, we will use Table B starting at line 115. Here are the table entries for that line:

61041 77684 94322 24709 73698 14526 31893 32591

To make our work a little easier, we rewrite these digits in pairs:

61 **04** 17 76 84 94 32 22 47 **09** 73 69 81 45 26 31 89 33 25 91

We now select four labels for the first treatment, "Meter." Reading pairs of digits from left to right and ignoring pairs that do not correspond to any of our labels, we see the labels 04 and 09. The corresponding residences will receive the "Meter" treatment. We will continue the process of finding labels with values between 01 and 12 on line 116 of Table B. Here, we find 03, 10, 06, and 11. The residences with labels 03 and 10 will receive the "Meter" treatment. The residences with labels 06 and 11 will receive the "Chart" treatment. On line 117, we find 06, but this label has already been selected. Similarly, on line 118, we find labels 03 and 04, but these labels have already been used. We also find 02 and 07, which we assign to "Chart." The remaining labels 01, 05, 08, and 12 are assigned to "Control." In summary, 03, 04, 09, and 10 are assigned to "Meter," 02, 06, 07, and 11 are assigned to "Chart," and 01, 05, 08, and 12 are assigned to "Control." ■

As Example 3.21 illustrates, randomization requires two steps: assign labels to the experimental units and then use Table B to select labels at random. *Be sure that all labels are the same length so that all have the same chance to be chosen.* You can read digits from Table B in any order—along a row, down a column, and so on—because the table has no order. As an easy standard practice, we recommend reading along rows. In Example 3.21, we needed four lines of random digits to complete the randomization. If we wanted to reduce this amount, we could use more than one label for each residence. For example, we could use labels 01, 21, 41, 61, and 81 for the first residence; 02, 22, 42, 62, and 82 for the second residence; and so forth.

Examples 3.18 and 3.19 describe completely randomized designs that compare levels of a single factor. In Example 3.18, the factor is the diet fed to the rats. In Example 3.19, it is the method used to encourage energy conservation. Completely randomized designs can have more than one factor. The advertising experiment of Example 3.16 has two factors: the length and the number of repetitions of a television commercial. Their combinations form the six treatments outlined in Figure 3.4 (page 145). A completely randomized design assigns subjects at random to these six treatments. Once the layout of treatments is set, the randomization needed for a completely randomized design is tedious but straightforward.

APPLY YOUR KNOWLEDGE

3.48 **Does child care help recruit employees?** Will providing child care for employees make a company more attractive to women? You are designing an experiment to answer this question. You prepare recruiting material for two fictitious companies, both in similar businesses in the same location. Company A's brochure does not mention child care. There are two versions of Company B's brochure. One is identical to Company A's brochure. The other is also the same, but also includes a description of the company's onsite child care facility. Your subjects are 30 women who are college seniors seeking employment. Each subject will read recruiting material for Company A and one of the versions of the recruiting material for Company B. You will give each version of Company B's brochure to half the women. After reading the material for both companies, each subject chooses the one she would prefer to work for. You expect that a higher percent of those who read the description that includes child care will choose Company B. CCARE

(a) Outline an appropriate design for the experiment.

(b) The names of the subjects appear below. Use software or Table B, beginning at line 152, to do the randomization required by your design. List the subjects who will read the version that mentions child care.

Abrams	Danielson	Gutierrez	Lippman	Rosen
Afifi	Edwards	Hwang	McNeill	Thompson
Brown	Fluharty	Iselin	Morse	Travers
Cansico	Garcia	Janle	Ng	Turing
Chen	Gerson	Kaplan	Quinones	Ullmann
Curzakis	Gupta	Lattimore	Roberts	Wong

3.49 Sealing food packages. Use a diagram to describe a completely randomized experimental design for the package liner experiment of Exercise 3.44. (Show the size of the groups, the treatment each group receives, and the response variable. Figures 3.5 and 3.6 are models to follow.) Use software or Table B, starting at line 130, to do the randomization required by your design.

The logic of randomized comparative experiments

Randomized comparative experiments are designed to give good evidence that differences in the treatments actually *cause* the differences we see in the response. The logic is as follows:

- Random assignment of subjects forms groups that should be similar in all respects before the treatments are applied.

- Comparative design ensures that influences other than the experimental treatments operate equally on all groups.

- Therefore, differences in average response must be due either to the treatments or to the play of chance in the random assignment of subjects to the treatments.

That "either-or" deserves more thought. In Example 3.18, we cannot say that *any* difference in the average weight gains of rats fed the two diets must be caused by a difference between the diets. There would be some difference even if both groups received the same diet because some rats grow faster than others. If chance assigns the faster-growing rats to one group or the other, this creates a chance difference between the groups. We would not trust an experiment with just one rat in each group, for example. The results would depend on which group got lucky and received the faster-growing rat. If we assign many rats to each diet, however, the effects of chance will average out, and there will be little difference in the average weight gains in the two groups unless the diets themselves cause a difference. "Use enough subjects to reduce chance variation" is the third big idea of statistical design of experiments.

> **PRINCIPLES OF EXPERIMENTAL DESIGN**
>
> 1. **Compare** two or more treatments. This will control the effects of lurking variables on the response.
>
> 2. **Randomize**—use chance to assign subjects to treatments.
>
> 3. **Replicate** each treatment on enough subjects to reduce chance variation in the results.

EXAMPLE 3.22

Cell Phones and Driving Does talking on a hands-free cell phone distract drivers? Undergraduate students "drove" in a high-fidelity driving simulator equipped with a hands-free cell phone. The car ahead brakes: how quickly does the subject respond? Twenty students (the control group) simply drove. Another 20 (the experimental group) talked on the cell phone while driving. The simulator gave the same driving conditions to both groups.

This experimental design has good control because the only difference in the conditions for the two groups is the use of the cell phone. Students are randomized to the two groups, so we satisfy the second principle. Based on past experience with the simulators, the length of the drive and the number of subjects were judged to provide sufficient information to make the comparison. We will learn more about choosing sample sizes for experiments starting in Chapter 8. ∎

We hope to see a difference in the responses so large that it is unlikely to happen just because of chance variation. We can use the laws of probability, which give a mathematical description of chance behavior, to learn if the treatment effects are larger than we would expect to see if only chance were operating. If they are, we call them *statistically significant*.

STATISTICAL SIGNIFICANCE

An observed effect so large that it would rarely occur by chance is called **statistically significant.**

If we observe statistically significant differences among the groups in a comparative randomized experiment, we have good evidence that the treatments actually caused these differences. You will often see the phrase "statistically significant" in reports of investigations in many fields of study. The great advantage of randomized comparative experiments is that they can produce data that give good evidence for a cause-and-effect relationship between the explanatory and response variables. We know that, in general, a strong association does not imply causation. A statistically significant association in data from a well-designed experiment does imply causation. Statistical significance and related issues regarding statistical inference are discussed in Chapter 7.

APPLY YOUR KNOWLEDGE

3.50 Utility companies. Example 3.19 (page 148) describes an experiment to learn whether providing households with electronic meters or charts will reduce their electricity consumption. An executive of the utility company objects to including a control group. He says, "It would be simpler to just compare electricity use last year (before the meter or chart was provided) with consumption in the same period this year. If households use less electricity this year, the meter or chart must be working." Explain clearly why this design is inferior to that in Example 3.19.

3.51 Statistical significance. The financial aid office of a university asks a sample of students about their employment and earnings. The report says that "for academic year earnings, a significant difference was found between the sexes, with men earning more on the average. In addition, no significant difference was found between the earnings of black and white students." Explain the meaning of "a significant difference" and "no significant difference" in plain language.

Completely randomized designs can compare any number of treatments. The treatments can be formed by levels of a single factor or by more than one factor. Here is an example with two factors.

EXAMPLE 3.23

Randomization for the TV Commercial Experiment Figure 3.4 (page 145) displays six treatments formed by the two factors in an experiment on response to a TV commercial. Suppose that we have 150 students who are willing to serve as subjects. We must assign 25 students at random to each group. Figure 3.8 outlines the completely randomized design.

FIGURE 3.8 Outline of a completely randomized design for comparing six treatments, Example 3.23.

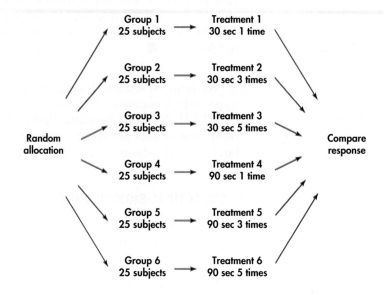

To carry out the random assignment, label the 150 students 001 to 150. (Three digits are needed to label 150 subjects.) Enter Table B and read three-digit groups until you have selected 25 students to receive Treatment 1 (a 30-second ad shown once). If you start at line 140, the first few labels for Treatment 1 subjects are 129, 048, and 003.

Continue in Table B to select 25 more students to receive Treatment 2 (a 30-second ad shown three times). Then select another 25 for Treatment 3 and so on until you have assigned 125 of the 150 students to Treatments 1 through 5. The 25 students who remain get Treatment 6. The randomization is straightforward but very tedious to do by hand. We recommend using software such as the *Simple Random Sample* applet. Exercise 3.63 asks you to use the applet to do the randomization for this example. ■

APPLY YOUR KNOWLEDGE

3.52 Do the randomization. Use Excel to carry out the randomization in Example 3.23.

Cautions about experimentation

The logic of a randomized comparative experiment depends on our ability to treat all the subjects identically in every way except for the actual treatments being compared. Good experiments therefore require careful attention to details.

Many—perhaps most—experiments have some weaknesses in detail. The environment of an experiment can influence the outcomes in unexpected ways. Although experiments are the gold standard for evidence of cause and effect, really convincing evidence usually requires that a number of studies in different places with different details produce similar results. The most serious potential weakness of experiments is **lack of realism.** The subjects or treatments or setting of an experiment may not realistically duplicate the conditions we really want to study. Here are two examples.

lack of realism

EXAMPLE 3.24

Layoffs and Feeling Bad How do layoffs at a workplace affect the workers who remain on the job? Psychologists asked student subjects to proofread text for extra course credit, then "let go" some of the workers (who were actually accomplices of the experimenters). Some subjects were told that those let go had performed poorly (Treatment 1). Others were told that not all could be kept and that it was just luck that they were kept and others let go (Treatment 2). We can't be sure that the reactions of the students are the same as those of workers who survive a layoff in which other workers lose their jobs. Many behavioral science experiments use student subjects in a campus setting. Do the conclusions apply to the real world? ■

EXAMPLE 3.25

Does the Regulation Make the Product Safer? Do those high center brake lights, required on all cars sold in the United States since 1986, really reduce rear-end collisions? Randomized comparative experiments with fleets of rental and business cars, done before the lights were required, showed that the third brake light reduced rear-end collisions by as much as 50%. Unfortunately, requiring the third light in all cars led to only a 5% drop.

What happened? Most cars did not have the extra brake light when the experiments were carried out, so it caught the eye of following drivers. Now that almost all cars have the third light, they no longer capture attention. ■

Lack of realism can limit our ability to apply the conclusions of an experiment to the settings of greatest interest. Most experimenters want to generalize their conclusions to some setting wider than that of the actual experiment. Statistical analysis of the original experiment cannot tell us how far the results will generalize. Nonetheless, the randomized comparative experiment, because of its ability to give convincing evidence for causation, is one of the most important ideas in statistics.

APPLY YOUR KNOWLEDGE

3.53 **Managers and stress.** Some companies employ consultants to train their managers in meditation in the hope that this practice will relieve stress and make the managers more effective on the job. An experiment that claimed to show that meditation reduces anxiety proceeded as follows. The experimenter interviewed the subjects and rated their level of anxiety. Then the subjects were randomly assigned to two groups. The experimenter taught one group how to meditate, and they meditated daily for a month. The other group was simply told to relax more. At the end of the month, the experimenter interviewed all the subjects again and rated their anxiety level. The meditation group now had less anxiety. Psychologists said that the results were suspect because the ratings were not blind—that is, the experimenter knew which treatment each subject received. Explain what this means and how lack of blindness could bias the reported results.

3.54 **Frustration and teamwork.** A psychologist wants to study the effects of failure and frustration on the relationships among members of a work team. She forms a team of students, brings them to the psychology laboratory, and has them play a game that requires teamwork. The game is rigged so that they lose regularly. The psychologist observes the students through a one-way window and notes the changes in their behavior during an evening of game playing. Why is it doubtful that the findings of this study tell us much about the effect of working for months developing a new product that never works right and is finally abandoned by your company?

Matched pairs designs

Completely randomized designs are the simplest statistical designs for experiments. They illustrate clearly the principles of control, randomization, and replication of treatments on a number of subjects. However, completely randomized designs are often inferior to more elaborate statistical designs. In particular, matching the subjects in various ways can produce more precise results than simple randomization.

matched pairs design

One common design that combines matching with randomization is the **matched pairs design.** A matched pairs design compares just two treatments. Choose pairs of subjects that are as closely matched as possible. Assign one of the treatments to each subject in a pair by tossing a coin or reading odd and even digits from Table B. We also use the term "matched pairs" for a design where each subject receives both treatments. Here, we would randomize the order in which the treatments are given or we could randomly assign half of the subjects to one order (for example, A, then B) and give the other half the reverse order (B, then A).

EXAMPLE 3.26

Matched Pairs for the Cell Phone Experiment Example 3.22 (page 152) describes an experiment on the effects of talking on a cell phone while driving. The experiment compared two treatments: driving in a simulator and driving in a simulator while talking on a hands-free cell phone. The response variable is the time the driver takes to apply the brake when the car in front brakes suddenly. In Example 3.22, 40 student subjects were assigned at random, 20 students to each treatment. Subjects differed in driving skill and reaction times. The completely randomized design relies on chance to create two similar groups of subjects.

In fact, the experimenters used a matched pairs design in which all subjects drove under both conditions. They compared each subject's reaction times with and without the phone. If all subjects drove first with the phone and then without it, the effect of talking on the cell phone would be confounded with the fact that this is the first run in the simulator. The proper procedure requires that all subjects first be trained in using the simulator, that the *order* in which a subject drives with and without the phone be random, and that the two drives take place on separate days to reduce the chance that the results of the second treatment will be affected by the first treatment. ■

The completely randomized design uses chance to decide which 20 subjects will drive with the cell phone. The other 20 drive without it. The matched pairs design uses chance to decide which 20 subjects will drive first with and then without the cell phone. The other 20 drive first without and then with the phone.

Block designs

Matched pairs designs apply the principles of comparison of treatments, randomization, and replication. However, the randomization is not complete—we do not randomly assign all the subjects at once to the two treatments. Instead, we only randomize within each matched pair. This allows matching to reduce the effect of variation among the subjects. Matched pairs are an example of *block designs.*

BLOCK DESIGN

A **block** is a group of subjects that are known before the experiment to be similar in some way expected to affect the response to the treatments. In a **block design,** the random assignment of individuals to treatments is carried out separately within each block.

A block design combines the idea of creating equivalent treatment groups by matching with the principle of forming treatment groups at random. Here is a typical example of a block design.

EXAMPLE 3.27

Men, Women, and Advertising An experiment to compare the effectiveness of three television commercials for the same product will want to look separately at the reactions of men and women, as well as assess the overall response to the ads.

A completely randomized design considers all subjects, both men and women, as a single pool. The randomization assigns subjects to three treatment groups without regard to their gender. This ignores the differences between men and women. A better design considers women and men separately. Randomly assign the women to three groups, one to view each commercial. Then separately assign the men at random to three groups. Figure 3.9 outlines this improved design. ∎

FIGURE 3.9 Outline of a block design, Example 3.27.

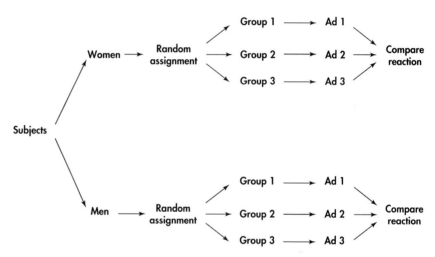

A block is a group of subjects formed before an experiment starts. We reserve the word "treatment" for a condition that we impose on the subjects. We don't speak of six treatments in Example 3.27, even though we can compare the responses of six groups of subjects formed by the two blocks (men, women) and the three commercials. Block designs are similar to stratified samples. Blocks and strata both group similar individuals together. We use two different names only because the idea developed separately for sampling and experiments.

Blocks are another form of *control*. They control the effects of some outside variables by bringing those variables into the experiment to form the blocks. The advantages of block designs are the same as the advantages of stratified samples, in that blocks allow us to draw separate conclusions about each block—for example, about men and women in the advertising study in Example 3.27. Blocking also allows more precise overall conclusions because the systematic differences between men and women can be removed when we study the overall effects of the three commercials when there is no interaction between sex and treatment.

The idea of blocking is an important additional principle of statistical design of experiments. A wise experimenter will form blocks based on the most important unavoidable sources of variability among the experimental subjects. Randomization will then average out the effects of the remaining variation and allow an unbiased comparison of the treatments.

Like the design of samples, the design of complex experiments is a job for experts. Now that we have seen a bit of what is involved, we will usually just act as if most experiments were completely randomized.

APPLY YOUR KNOWLEDGE

3.55 Does charting help investors? Some investment advisers believe that charts of past trends in the prices of securities can help predict future prices. Most economists disagree. In an experiment to examine the effects of using charts, business students trade (hypothetically) a foreign currency at computer screens. There are 20 student subjects available, named for convenience A, B, C, . . . , T. Their goal is to make as much money as possible, and the best performances are rewarded with small prizes. The student traders have the price history of the foreign currency in dollars in their computers. They may or may not also have software that highlights trends. Describe two designs for this experiment—a completely randomized design and a matched pairs design in which each student serves as his or her own control. In both cases, carry out the randomization required by the design.

SECTION 3.3 SUMMARY

- In an experiment, we impose one or more **treatments** on the **experimental units** or **subjects.** Each treatment is a combination of levels of the explanatory variables, which we call **factors.**

- The **design** of an experiment describes the choice of treatments and the manner in which the subjects are assigned to the treatments.

- The basic principles of statistical design of experiments are **control, randomization,** and **replication.**

- The simplest form of control is **comparison.** Experiments should compare two or more treatments to avoid **confounding** of the effect of a treatment with other influences, such as lurking variables.

- **Randomization** uses chance to assign subjects to the treatments. It creates treatment groups that are similar (except for chance variation) before the treatments are applied. Randomization and comparison together prevent **bias,** or systematic favoritism, in experiments.

- You can carry out randomization by giving numerical labels to the subjects and using software or a **table of random digits** to choose treatment groups.

- **Replication** of each treatment using many subjects reduces the role of chance variation and makes the experiment more sensitive to differences among the treatments.

- Good experiments require attention to detail as well as good statistical design. **Lack of realism** in an experiment can prevent us from generalizing its results.

- In addition to comparison, a second form of control is to restrict randomization by forming **blocks** of subjects that are similar in some way that is important to the response. Randomization is then carried out separately within each block.

- **Matched pairs** are a common form of blocking for comparing just two treatments. In some matched pairs designs, each subject receives both treatments in a random order. In others, the subjects are matched in pairs as closely as possible, and one subject in each pair receives each treatment.

SECTION 3.3 EXERCISES

For Exercises 3.42 to 3.44, see pages 145–146; for 3.45, see page 147; for 3.46 and 3.47, see page 148; for 3.48 and 3.49, see pages 151–152; for 3.50 and 3.51, see page 153; for 3.52, see page 155; for 3.53 and 3.54, see page 155; and for 3.55, see page 158.

3.56 What is needed? Explain what is deficient in each of the following proposed experiments, and explain how you would improve the experiment.

(a) Your company wants to compare two versions of a new granola cereal. For a one-month period, the first product is sold at 10 stores in southern California and the second is sold at 10 stores in a suburb of New York City. At the end of the one-month period, the sales are compared.

(b) Real-time feedback uses shopping carts equipped with scanners to give feedback regarding the costs of customers' purchases. You want to compare the total purchases of shoppers using feedback with those who use conventional carts. Shopping carts equipped with scanners were used by all shoppers in a grocery store on a Monday morning. The average spending of these shoppers was compared with the average spending of all shoppers in the same store during the previous month.

(c) You want to compare the times taken to assemble a machine using two different sets of instructions. Twenty volunteers will assemble the machines. In a morning session, they will use the first set of instructions. After a free lunch, they will use the second set.

3.57 What is wrong? Explain what is wrong with each of the following randomization procedures, and describe how you would do the randomization correctly.

(a) Ten subjects are to be assigned to two treatments, five to each. For each subject, a random digit is generated. If the digit is 0, 1 2, 3, or 4, the subject is assigned to the first treatment; if the digit is 5, 6, 7, 8, or 9, the subject is assigned to the second treatment.

(b) An experiment will assign 60 rats to three different treatment conditions. The rats arrive from the supplier in batches of 20, and the treatment lasts two weeks. The first batch of 20 rats is randomly assigned to one of the three treatments, and data for these rats are collected. After a one-week break, another batch of 20 rats arrives and is assigned to one of the two remaining treatments. The process continues until the last batch of rats is given the treatment that has not been assigned to the three previous batches.

(c) A list of 40 subjects is entered into a computer file and then sorted by first name. The subjects are assigned to four treatments by taking the first 10 subjects for Treatment 1, the next 10 subjects for Treatment 2, and so forth.

3.58 Evaluate a new method for training new employees. A new method for training new employees is to be evaluated by randomly assigning new employees to either the current training program or the new method. A questionnaire will be used to evaluate the satisfaction of the new employees with the training. Explain how this experiment should be done.

3.59 Can you change attitudes of workers about teamwork? You will conduct an experiment designed to change attitudes of workers about teamwork. Discuss some variables that you might use if you were to use a block design for this experiment.

3.60 An experiment for a new product. Compost tea is rich in microorganisms that help plants grow. It is made by soaking compost in water.[24] Design a comparative experiment that will provide evidence about whether compost tea works for a particular type of plant that interests you. Be sure to provide all details regarding your experiment, including the response variable or variables that you will measure. Assuming that the experiment shows positive results, write a short description about how you would use the results in a marketing campaign for compost tea.

3.61 Marketing your training materials. Water quality of streams and lakes is an issue of concern to the public. Although trained professionals typically are used to take reliable measurements, many volunteer groups are gathering and distributing information based on data that they collect.[25] You are part of a team to train volunteers to collect accurate water quality data. Design an experiment to evaluate the effectiveness of the training. Write a summary of your proposed design to present to your team. Be sure to include all the details that they will need to evaluate your proposal. How would you use the results of the experiment to market your training materials?

3.62 Randomly assign the subjects. You can use the *Simple Random Sample* applet to choose a treatment group at random once you have labeled the subjects. Example 3.22 (page 152) describes an experiment in which 20 students are chosen from a group of 40 for the treatment group in a study of the effect of cell phones on driving. Use the applet to choose the 20 students for the experimental group. Which students did you choose? The remaining 20 students make up the control group.

3.63 Randomly assign the subjects. The *Simple Random Sample* applet allows you to randomly assign experimental units to more than two groups without difficulty. Example 3.23 (page 154) describes a randomized comparative experiment in which 150 students are randomly assigned to six groups of 25.

(a) Use the applet to randomly choose 25 out of 150 students to form the first group. Which students are in this group?

(b) The "population hopper" now contains the 125 students who were not chosen, in scrambled order. Click "Sample" again to choose 25 of these remaining students to make up the second group. Which students were chosen?

(c) Click "Sample" three more times to choose the third, fourth, and fifth groups. Don't take the time to write down these groups. Check that there are only 25 students remaining in the "population hopper." These subjects get Treatment 6. Which students are they?

3.64 Random digits. Table B is a table of random digits. Which of the following statements are true of a table of random digits, and which are false? Explain your answers.

(a) Each pair of digits has chance 1/11 of being 11.

(b) There are no more than five 5s in each row of 40 digits.

(c) The digits 000 can never appear as a group, because this pattern is not random.

3.65 Health benefits of bee pollen. "Bee pollen is effective for combating fatigue, depression, cancer, and colon disorders." So says a website that offers the pollen for sale. We wonder if bee pollen really does prevent colon disorders. Here are two ways to study this question. Explain why the first design will produce more trustworthy data.

(a) Find 400 women who do not have colon disorders. Randomly assign 200 to take bee pollen capsules and the other 200 to take placebo capsules that are identical in appearance. Follow both groups for five years.

(b) Find 200 women who take bee pollen regularly. Match each with a woman of the same age, race, and occupation who does not take bee pollen. Follow both groups for five years.

3.66 Price cuts on athletic shoes. Stores advertise price reductions to attract customers. What type of price cut is most attractive? Market researchers prepared ads for athletic shoes announcing different levels of discounts (20%, 40%, or 60%). The student subjects who read the ads were also given "inside information" about the fraction of shoes on sale (50% or 100%). Each subject then rated the attractiveness of the sale on a scale of 1 to 7.

(a) There are two factors. Make a sketch like Figure 3.4 (page 145) that displays the treatments formed by all combinations of levels of the factors.

(b) Outline a completely randomized design using 50 student subjects. Use software or Table B at line 121 to choose the subjects for the first treatment.

3.67 Effects of price promotions. A researcher is studying the effect of price promotions on consumers' expectations. She makes up a history of the store price of a hypothetical brand of laundry detergent for the past year. Students in a marketing course view the price history on a computer. Some students see a steady price, while others see regular promotions that temporarily cut the price. Then the students are asked what price they would expect to pay for the detergent.

(a) Is this study an experiment? Explain your answer.

(b) What are the explanatory and response variables?

3.68 Aspirin and heart attacks. "Nearly five decades of research now link aspirin to the prevention of stroke and heart attacks." So says the Bayer Aspirin website, **bayeraspirin.com.** The most important evidence for this claim comes from the Physicians' Health Study, a large medical experiment involving 22,000 male physicians. One group of about 11,000 physicians took an aspirin every second day, while the rest took a placebo. After several years, the study found that subjects in the aspirin group had significantly fewer heart attacks than subjects in the placebo group.

(a) Identify the experimental subjects, the factor and its levels, and the response variable in the Physicians' Health Study.

(b) Use a diagram to outline a completely randomized design for the Physicians' Health Study.

(c) What does it mean to say that the aspirin group had "significantly fewer heart attacks"?

3.69 Marketing to children. If children are given more choices within a class of products, will they tend to prefer that product to a competing product that offers fewer choices? Marketers want to know. An experiment prepared three sets of beverages. Set 1 contained two milk drinks and two fruit drinks. Set 2 had two fruit drinks and four milk drinks. Set 3 contained four fruit drinks but only two milk drinks. The researchers divided 120 children aged 4 to 12 years into three groups at random. They offered each group one of the sets. As each child chose a beverage to drink from the set presented, the researchers noted whether the choice was a milk drink or a fruit drink.

(a) What are the experimental subjects?

(b) What is the factor and what are its levels? What is the response variable?

(c) Use a diagram to outline a completely randomized design for the study.

(d) Explain how you would assign labels to the subjects. Use software to do the randomization or Table B at line 145 to choose the first five subjects assigned to the first treatment.

3.70 Effects of TV advertising. You decide to use a completely randomized design in the two-factor experiment on response to advertising described in Example 3.16 (page 145). The 30 students named below will serve as subjects. Outline the design. Then use software or Table B at line 110 to randomly assign the subjects to the six treatments. TVADS

Alomar	Denman	Han	Liang	Padilla	Valasco
Asihiro	Durr	Howard	Maldonado	Plochman	Vaughn
Bennett	Edwards	Hruska	Marsden	Rosen	Wei
Chao	Fleming	James	O'Brian	Trujillo	Willis
Clemente	George	Kaplan	Ogle	Tullock	Zhang

3.71 Temperature and work performance. An expert on worker performance is interested in the effect of room temperature on the performance of tasks requiring manual dexterity. She chooses temperatures of 20°C (68°F), 25°C (77°F), and 30°C (86°F) as treatments. The response variable is the number of correct insertions, during a 30-minute period, in a peg-and-hole apparatus that requires the use of both hands simultaneously. Each subject is trained on the apparatus and then is asked to make as many insertions as possible in 30 minutes of continuous effort.

(a) Outline a completely randomized design to compare dexterity at 20°C, 25°C, and 30°C using 45 subjects.

(b) Because people differ greatly in dexterity, the wide variation in individual scores may hide the systematic effect of temperature unless each group includes many subjects. Describe in detail the design of an experiment in which each subject is tested on each of the three temperatures.

3.4 Data Ethics

When you complete this section, you will be able to:

- Describe the purpose of an institutional review board and describe what kinds of expertise its members require.
- Describe informed consent and evaluate whether it has been given in specific examples.
- Determine when data have been kept confidential in a study.
- Evaluate a clinical trial from the viewpoint of ethics.

The production and use of data often involve ethical questions. We won't discuss the telemarketer who begins a telephone sales pitch with "I'm conducting a survey." Such deception is clearly unethical. It enrages legitimate survey organizations, which find the public less willing to talk with them. Neither will we discuss those few researchers who, in the pursuit of professional advancement, publish fake data. There is no ethical question here—faking data to advance your career is just wrong. But just how honest must researchers be about real, unfaked data? Here is an example that suggests the answer is "More honest than they often are."

EXAMPLE 3.28

Provide All the Critical Information Papers reporting scientific research are supposed to be short, with no extra baggage. Brevity, however, can allow researchers to avoid complete honesty about their data. Did they choose their subjects in a biased way? Did they report data on only some of their subjects? Did they try several statistical analyses and report only the ones that looked best? The statistician John Bailar screened more than 4000 medical papers in more than a decade as consultant to the *New England Journal of Medicine*. He says, "When it came to the statistical review, it was often clear that critical information was lacking, and the gaps nearly always had the practical effect of making the authors' conclusions look stronger than they should have."[26] The situation is no doubt worse in fields that screen published work less carefully. ∎

The most complex issues of data ethics arise when we collect data from people. Ethical difficulties are easier to identify in experiments that impose a treatment on people. However, sample surveys can also result in harm, particularly when they concern sensitive issues. Here are some basic standards of data ethics that must be obeyed by any study that gathers data from human subjects, whether it is a sample survey or an experiment.

BASIC DATA ETHICS

The organization that carries out the study must have an **institutional review board** that reviews all planned studies in advance to protect the subjects from possible harm.

All subjects in a study must give their **informed consent** before data are collected.

All subject data must be kept **confidential.** Only statistical summaries for groups of subjects may be made public.

The law requires that studies carried out or funded by the federal government obey these principles.[27] But neither the law nor the consensus of experts is completely clear about the details of their application.

Institutional review boards

The purpose of an institutional review board (IRB) is not to decide whether a proposed study will produce valuable information or whether it is statistically sound. Instead, the board's purpose is, in the words of one university's board, "to protect the rights and welfare of human subjects (including patients) recruited to participate in research activities." The board reviews the plan of the study and can require changes in it. It reviews the consent form to ensure that subjects are informed about the nature of the study and about any potential risks. Once research begins, the board monitors the study's progress at least once a year.

The most pressing issue concerning IRBs is whether their workload has become so large that their effectiveness in protecting subjects drops. When the government temporarily stopped human-subject research at Duke University Medical Center in 1999 due to inadequate protection of subjects, more than 2000 studies were going on. That's a lot of review work. There are shorter review procedures for projects that involve only minimal risks to subjects, such as most sample surveys. When a board is overloaded, however, there is a temptation to put more proposals in the minimal-risk category to speed the work.

In 2018, the U.S. Department of Health and Human Services, Office for Human Research Protections issued revisions to the requirements for IRBs. Changes to the "Common Rule" and details regarding the implementation of the changes are available at **hhs.gov/ohrp,** along with other information about the work of Office for Human Research Protections.

APPLY YOUR KNOWLEDGE   *The exercises in this section on ethics are designed to help you think about the issues that we are discussing and to formulate some opinions. In general, there are no wrong or right answers, but you need to give reasons for your answers.*

3.72 **Who should be on an institutional review board?** Government regulations require that IRBs consist of at least five people, including at least one scientist, one nonscientist, and one person from outside the institution. Most boards are larger, but many contain just one outsider.

(a) Why should review boards contain people who are not scientists?

(b) Do you think that one outside member is enough? How would you choose that member? (For example, would you prefer a medical doctor? A member of the clergy? An activist for patients' rights?)

3.73 **Do these proposals involve minimal risk?** You are a member of your college's IRB. You must decide whether several research proposals qualify for lighter review because they involve only minimal risk to subjects. Federal regulations say that "minimal risk" means the risks are no greater than "those ordinarily encountered in daily life or during the performance of routine physical or psychological examinations tests." That's vague. Which of these do you think qualifies as "minimal risk"?

(a) Draw a drop of blood by pricking a finger so as to measure blood sugar.

(b) Draw blood from the arm for a full set of blood tests.

(c) Insert a tube that remains in the arm, so that blood can be drawn regularly.

Informed consent

Both words in the phrase "informed consent" are important, and both can be controversial. Subjects must be *informed* in advance about the nature of a study and any risk of harm it may bring. In the case of a sample survey, physical harm is not possible. For such a study, the subjects should be told what kinds of questions the survey will ask and about how much of their time it will take. Experimenters must tell subjects the nature and purpose of the study and outline possible risks. Subjects must then give their *consent* in writing.

EXAMPLE 3.29

Who Can Give Informed Consent? Are there some subjects who can't give informed consent? It was once common, for example, to test new vaccines on prison inmates who gave their consent in return for good-behavior credit. Now, we worry that prisoners are not really free to refuse, and the law forbids almost all medical research in prisons.

Children can't give fully informed consent, so the usual procedure is to ask their parents. A study of new ways to teach reading is about to start at a local elementary school, so the study team sends consent forms home to parents. Many parents don't return the forms. Can their children take part in the study because the parents did not say "No," or should we allow only children whose parents returned the form and said "Yes"?

What about research into new medical treatments for people with mental disorders? What about studies of new ways to help emergency room patients who may be unconscious? In most cases, there is not time to get the consent of the family. Does the principle of informed consent bar realistic trials of new treatments for unconscious patients?

These are questions without clear answers. Reasonable people differ strongly on all of them. There is nothing simple about informed consent.[28] ■

The difficulties of informed consent do not vanish even for capable subjects. Some researchers, especially in medical trials, regard consent as a barrier to getting patients to participate in research. To "get around" this barrier, they may not explain all possible risks; they may not point out that other therapies might be better than those being studied; and they may be too optimistic in talking with patients even when the consent form has all the right details. Of course, mentioning every possible risk leads to very long consent forms that really are barriers. "They are like rental car contracts," one lawyer said. Some subjects don't read forms that run five or six printed pages. Others are frightened by the large number of possible (but unlikely) disasters that might happen and so refuse to participate. Of course, unlikely disasters sometimes happen. When they do, lawsuits follow—and the consent forms become yet longer and more detailed.

Confidentiality

Ethical problems do not disappear once a study has been cleared by the review board and has actually collected data about the subjects. It is important to protect the subjects' privacy by keeping all data about subjects confidential. The report of an opinion poll may say what percent of the 1200 respondents felt that legal immigration should be reduced, but it should not state what *you* said about this or any other issue.

anonymity   Confidentiality is not the same as **anonymity.** Anonymity means that subjects are anonymous—their names are not known even to the director of the study. Anonymity is rare in statistical studies. Even where it is possible (mainly in surveys conducted by mail), anonymity prevents any follow-up to improve nonresponse or inform subjects of results.

Any breach of confidentiality is a serious violation of data ethics. The best practice is to separate the identity of the subjects from the rest of the data at once. Sample surveys, for example, use the identification only to check on who did or did not respond. In an era of advanced technology, however, it is no longer enough to be sure that each set of data protects people's privacy. The government, for example, maintains a vast amount of information about citizens in many separate databases—census responses, tax returns, Social Security information, data from surveys such as the Current Population Survey, and so on. Many of these databases can be searched by computers for statistical studies. A clever computer search of several databases might be able, by combining information, to identify you and learn a great deal about you even if your name and other identification have been removed from the data available for search. A colleague from Germany once remarked that "female full professor of statistics with PhD from the United States" was enough to identify her among all the 83 million residents of Germany. Privacy and confidentiality of data are hot issues among statisticians in the digital age.

EXAMPLE 3.30

Data Collected by the Government Citizens are required to give information to the government. Think of tax returns and Social Security contributions. The government needs these data for administrative purposes—to see if we paid the right amount of tax and how large a Social Security benefit we are owed when we retire. Some people believe that people should be able to forbid any other use of their data, even when all identification of those data is removed. This would prevent the use of government records to study, say, the ages, incomes, and household sizes of Social Security recipients. Such a study could well be vital to debates on reforming Social Security. ■

APPLY YOUR KNOWLEDGE

3.74 Should we allow this personal information to be collected? In which of the following circumstances would you allow collecting personal information without the subjects' consent?

(a) A government agency takes a random sample of income tax returns to obtain information on the average income of people in different occupations. Only the incomes and occupations are recorded from the returns, not the names.

(b) A social psychologist attends public meetings of a religious group to study the behavior patterns of members.

(c) A social psychologist pretends to be converted to membership in a religious group and attends private meetings to study the behavior patterns of members.

3.75 How can we obtain informed consent? A researcher suspects that traditional religious beliefs tend to be associated with an authoritarian personality. She prepares a questionnaire that measures authoritarian tendencies and also asks many religious questions. Write a description of the purpose of this research to be read by subjects as part of the informed consent. You must balance the conflicting goals of not deceiving the subjects about what the questionnaire will reveal about them and of not biasing the sample by scaring off religious people.

Clinical trials

Clinical trials are experiments that study the effectiveness of medical treatments on actual patients. Medical treatments can harm as well as heal, so clinical trials spotlight the ethical problems of experiments with human subjects. Here are the starting points for a discussion:

- Randomized comparative experiments are the only way to see the true effects of new treatments. Without them, risky treatments that are no more effective than placebos will become common.

- Clinical trials produce great benefits, but most of these benefits go to future patients. The trials also pose risks, and these risks are borne by the subjects of the trial. So we must balance future benefits against present risks.

- Both medical ethics and international human rights standards say that "the interests of the subject must always prevail over the interests of science and society."

The quoted words are from the 1964 Helsinki Declaration of the World Medical Association, the most respected international standard. The most outrageous examples of unethical experiments are those that ignore the interests of the subjects.

EXAMPLE 3.31

The Tuskegee Study In the 1930s, syphilis was common among black men in the rural South, a group that had almost no access to medical care. The Public Health Service Tuskegee study recruited 399 poor black sharecroppers with syphilis and 201 others without the disease to observe how syphilis progressed when no treatment was given. Beginning in 1943, penicillin became available to treat syphilis. The study subjects were not treated. In fact, the Public Health Service prevented any treatment until word leaked out and forced an end to the study in the 1970s.

The Tuskegee study is an extreme example of investigators following their own interests and ignoring the well-being of their subjects. A 1996 review said, "It has come to symbolize racism in medicine, ethical misconduct in human research, paternalism by physicians, and government abuse of vulnerable people." In 1997, President Bill Clinton formally apologized to the surviving participants in a White House ceremony.[29] ∎

Because "the interests of the subject must always prevail," medical treatments can be tested in clinical trials only when there is reason to hope that they will help the patients who are subjects in the trials. Future benefits aren't enough to justify experiments with human subjects. Of course, if strong evidence already shows that a treatment works and is safe, it is unethical *not* to give it. Here are the words of Dr. Charles Hennekens of the Harvard Medical School, who directed the large clinical trial that showed that aspirin reduces the risk of heart attacks:

> ... there must be sufficient belief in the agent's potential to justify exposing half the subjects to it. On the other hand, there must be sufficient doubt about its efficacy to justify withholding it from the other half of subjects who might be assigned to placebos.[30]

Why is it ethical to give a control group of patients a placebo? Well, we know that placebos often work. Moreover, placebos have no harmful side effects. So in the state of balanced doubt described by Dr. Hennekens, the placebo group may be getting a better treatment than the drug group. If we *knew* which treatment was better, we would give it to everyone. When we don't know, it is ethical to try both and compare them.

The idea of using a control or a placebo is a fundamental principle to be considered in designing experiments. In many situations, deciding what to use as an appropriate control requires some careful thought. The choice of the control can have a substantial impact on the conclusions drawn from an experiment. Here is an example.

EXAMPLE 3.32

Was the Claim Misleading? The manufacturer of a breakfast cereal designed for children claims that eating this cereal has been clinically shown to improve attentiveness by nearly 20%. The study used two groups of children who were tested before and after breakfast. One group received the cereal for breakfast, while breakfast for the control group was water. The results of tests taken three hours after breakfast were used to make the claim.

The Federal Trade Commission investigated the marketing of this product. They charged that the claim was false and violated federal law. The charges were settled, and the company agreed to not use misleading claims in its advertising.[31] ∎

It is not sufficient to obtain appropriate controls. The data from all groups must be collected and analyzed in the same way. Here is an example of this type of flawed design.

EXAMPLE 3.33

The Product Doesn't Work! Two scientists published a paper claiming to have developed a very exciting new method to detect ovarian cancer using blood samples. The potential market for such a procedure is substantial, and there is no specific screening test currently available. When other scientists were unable to reproduce the results in different labs, the original work was examined more carefully. The original study used blood samples from women with ovarian cancer and from healthy controls. The blood samples were all analyzed using a mass spectrometer. The control samples were analyzed on one day and the cancer samples were analyzed on the next day. This design was flawed because it could not control for changes over time in the measuring instrument.[32] ∎

APPLY YOUR KNOWLEDGE

3.76 Should the treatments be given to everyone? Effective drugs for treating AIDS are very expensive, so most African nations cannot afford to give them to large numbers of people. Yet AIDS is more common in parts of Africa than anywhere else. Several clinical trials being conducted in Africa are looking at ways to prevent pregnant mothers infected with HIV (the virus that causes AIDS) from passing the infection to their unborn children, a major source of HIV infections in Africa. Some people say these trials are unethical because they do not give effective AIDS drugs to their subjects, as would be required in rich nations. Others reply that the trials are looking for treatments that can work in the real world in Africa and that they promise benefits at least to the children of their subjects. What do you think?

3.77 Is this study ethical? Researchers on aging proposed to investigate the effect of supplemental health services on the quality of life of older people. Eligible patients of a large medical clinic were to be randomly assigned to treatment and control groups. The treatment group would be offered hearing aids, dentures, transportation, and other services not available without charge to the control group. The review board indicated that providing these services to some but not other persons in the same institution raised ethical questions. Do you agree?

Behavioral and social science experiments

When we move from medicine to the behavioral and social sciences, the direct risks to experimental subjects are less acute, but so are the possible benefits to the subjects. Consider, for example, the experiments conducted by psychologists in their study of human behavior.

EXAMPLE 3.34

Personal Space Psychologists observe that people have a "personal space" and are uneasy if others come too close to them. We don't like strangers to sit at our table in a coffee shop if other tables are available, and we see people move apart in elevators if there is room to do so. Americans tend to require more personal space than people in most other cultures. Can violations of personal space have physical, as well as emotional, effects?

Investigators set up shop in a men's public restroom. They blocked off urinals to force men walking in to use either a urinal next to an experimenter (treatment group) or a urinal separated from the experimenter (control group). Another experimenter, using a periscope from a toilet stall, measured how long the subject took to start urinating and how long he continued.[33] ∎

This personal space experiment illustrates the difficulties facing those who plan and review behavioral studies:

- There is no risk of harm to the subjects, although they would certainly object to being watched through a periscope. Even when physical harm is unlikely, do other types of harm need to be considered? Emotional harm? Undignified situations? Invasion of privacy?

- What about informed consent? The subjects did not even know they were participating in an experiment. Many behavioral experiments rely on hiding the true purpose of the study. The subjects would change their behavior if told in advance what the investigators were studying. Subjects are asked to consent on the basis of vague information. They receive full information only after the experiment.

The "Ethical Principles" of the American Psychological Association require consent unless a study merely observes behavior in a public place. They allow deception only when it is necessary to the study, does not hide information that might influence a subject's willingness to participate, and is explained to subjects as soon as possible. The personal space study (from the 1970s) does not meet current ethical standards.

As this discussion suggests, the basic requirement for informed consent is understood differently in medicine and psychology. Here is an example of another setting with yet another interpretation of what is ethical. The subjects get no information and give no consent. They don't even know that an experiment may be sending them to jail for the night.

EXAMPLE 3.35

Reducing Domestic Violence How should police respond to domestic-violence calls? In the past, the usual practice was to remove the offender and order the offender to stay out of the household overnight. Police were reluctant to make arrests because the victims rarely pressed charges. Women's groups argued that arresting offenders would help prevent future violence even if no charges were filed. Is there evidence that arrest will reduce future offenses? That's a question that experiments have tried to answer.

A typical domestic-violence experiment compares two treatments: arrest the suspect and hold the suspect overnight or warn and release the suspect. When police officers reach the scene of a domestic-violence call, they calm the participants and investigate. Weapons or death threats require an arrest. If the facts permit an arrest but do not require it, an officer radios headquarters for instructions. The person on duty opens the next envelope in a file prepared in advance by a statistician. The envelopes contain the treatments in random order. The police either make an arrest or warn and release, depending on the contents of the envelope. The researchers then watch police records and visit the victim to see if the domestic violence recurs.

Such experiments show that arresting domestic-violence suspects does reduce their future violent behavior.[34] As a result of this evidence, arrest has become the common police response to domestic violence. ∎

The domestic-violence experiments shed light on an important issue of public policy. Because there is no informed consent, the ethical rules that govern clinical trials and most social science studies would forbid these experiments. They were cleared by review boards because, in the words of one domestic-violence researcher, "These people became subjects by committing acts that allow the police to arrest them. You don't need consent to arrest someone."

SECTION 3.4 SUMMARY

- The purpose of an **institutional review board** is to protect the rights and welfare of the human subjects in a study. Institutional review boards review **informed consent** forms that subjects will sign before participating in a study.

- Information about subjects in a study must be kept **confidential,** but statistical summaries of groups of subjects may be made public.

- **Clinical trials** are experiments that study the effectiveness of medical treatments on actual patients.

- Some studies in the **behavioral** and **social sciences** are observational, while others are designed experiments.

SECTION 3.4 EXERCISES

For Exercises 3.72 and 3.73, see pages 162–163; for 3.74 and 3.75, see page 164; and for 3.76 and 3.77, see page 166.

Most of these exercises pose issues for discussion. There are no right or wrong answers, but there are more and less thoughtful answers.

3.78 How should the samples be analyzed? Refer to the ovarian cancer diagnostic test study in Example 3.33 (page 166). Describe how you would process the samples through the mass spectrometer.

3.79 The Vytorin controversy. Vytorin is a combination pill designed to lower cholesterol. The combination consists of a relatively inexpensive and widely used drug, Zocor, and a newer drug called Zetia. Early study results suggested that Vytorin was no more effective than Zetia. Critics claimed that the makers of the drugs tried to change the response variable for the study, and two congressional panels investigated why there was a two-year delay in the release of the results. Use the Web to search for more information about this controversy, and write a report of what you find. Include an evaluation in the framework of ethical use of experiments and data. A good place to start your search would be to look for the phrase "Vytorin's shortcomings."

3.80 What's wrong? Explain what is wrong in each of the following scenarios.

(a) An institutional review board should determine if a proposed study is statistically sound.

(b) It is sufficient for informed consent to be obtained through a telephone call if it is too costly to have the subjects consent in writing.

(c) Clinical trials that use a placebo are always ethical.

3.81 Anonymity and confidentiality in mail surveys. Some common practices may appear to offer anonymity while actually delivering only confidentiality. Market researchers often use mail surveys that do not ask the respondent's identity but contain hidden codes on the questionnaire that identify the respondent. A false claim of anonymity is clearly unethical. If only confidentiality is promised, is it also unethical to say nothing about the identifying code, perhaps causing respondents to believe their replies are anonymous?

3.82 Studying your blood. Long ago, doctors drew a blood specimen from you when you were treated for anemia. Unknown to you, the sample was stored. Now researchers plan to use stored samples from you and many other people to look for genetic factors that may influence anemia. It is no longer possible to ask your consent. Modern technology can read your entire genetic makeup from the blood sample.

(a) Do you think it violates the principle of informed consent to use your blood sample if your name is on it but you were not told that it might be saved and studied later?

(b) Suppose that your identity is not attached. The blood sample is known only to come from (say) "a 20-year-old white female being treated for anemia." Is it now ethical to use the sample for research?

(c) Perhaps we should use biological materials such as blood samples only from patients who have agreed to allow the material to be stored for later use in research. It isn't possible to say in advance what kind of research, so this falls short of the usual standard for informed consent. Is this practice acceptable, given complete confidentiality and the fact that using the sample can't physically harm the patient?

3.83 Anonymous? Confidential? One of the most important nongovernment surveys in the United States is the National Opinion Research Center's General Social Survey. The GSS regularly monitors public opinion on a wide variety of political and social issues. Interviews are conducted in person in the subject's home. Are a subject's responses to GSS questions anonymous, confidential, or both? Explain your answer.

3.84 Anonymous? Confidential? Texas A&M, like many universities, offers free screening for HIV, the virus that causes AIDS. The announcement says, "Persons who sign up for the HIV screening will be assigned a number so that they do not have to give their name." They can learn the results of the test by telephone, still without giving their name. Does this practice offer *anonymity* or just *confidentiality*?

3.85 Political polls. Candidates for public office hire polling organizations to take sample surveys to find out what the voters think about the issues. What information should the pollsters be required to disclose?

(a) What does the standard of informed consent require the pollsters to tell potential respondents?

(b) Should polling organizations be required to give respondents the name and address of the organization that carries out the poll?

(c) The polling organization usually has a professional name such as "Samples Incorporated," so respondents don't know that the poll is being paid for by a political party or candidate. Would revealing the sponsor to respondents bias the poll? Should the sponsor always be announced whenever poll results are made public?

3.86 Making poll results public. Some people think that the law should require that all political poll results be made public. Otherwise, the possessors of poll results can use the information to their own advantage. They can act on the information, release only selected parts of it, or time the release for best effect. A candidate's organization replies that it is paying for the poll in an effort to gain information for its own use, not to amuse the public. Do you favor requiring complete disclosure of political poll results? What about other private surveys, such as market research surveys of consumer tastes?

3.87 Student subjects. Students taking Psychology 001 are required to serve as experimental subjects. Students in Psychology 002 are not required to serve, but they are given extra credit if they do so. Students in Psychology 003 are required either to sign up as subjects or to write a term paper. Serving as an experimental subject may be educational, but current ethical standards frown on using "dependent subjects" such as prisoners or charity medical patients. Students are certainly somewhat dependent on their teachers. Do you object to any of these course policies? If so, which ones, and why?

3.88 How many have HIV? Researchers from Yale, working with medical teams in Tanzania, wanted to know how common infection with HIV (the virus that causes AIDS) is among pregnant women in that African country. To do this, they planned to test blood samples drawn from pregnant women.

Yale's institutional review board insisted that the researchers get the informed consent of each woman and tell her the results of the test. This is the usual procedure in developed nations. The Tanzanian government did not want to tell the women why blood was drawn or tell them the test results. The government feared panic if many people turned out to have an incurable disease for which the country's medical system could not provide care. The study was canceled. Do you think that Yale was right to apply its usual standards for protecting subjects?

3.89 AIDS trials in Africa. One of the most important goals of AIDS research is to find a vaccine that will protect against HIV infection. Because AIDS is so common in parts of Africa, that is the easiest place to test a vaccine. It is likely, however, that a vaccine would be so expensive that it could not (at least at first) be widely used in Africa. Is it ethical to test in Africa if the benefits go mainly to rich countries? The treatment group of subjects would get the vaccine, and the placebo group would later be given the vaccine if it proved effective. So the actual subjects would benefit and the future benefits then would go elsewhere. What do you think?

3.90 Asking teens about sex. The Centers for Disease Control and Prevention, in a survey of teenagers, asked the subjects if they were sexually active. Those who said "Yes" were then asked, "How old were you when you had sexual intercourse for the first time?" Should consent of parents be required to ask minors about sex, drugs, and other such issues, or is consent of the minors themselves enough? Give reasons for your opinion.

3.91 Deceiving subjects. Students sign up to be subjects in a psychology experiment. When they arrive, they are told that interviews are running late and are taken to a waiting room. The experimenters then stage a theft of a valuable object left in the waiting room. Some subjects are alone with the thief, and others are in pairs—these are the treatments being compared. Will the subject report the theft? The students had agreed to take part in an unspecified study, and the true nature of the experiment is explained to them afterward. Do you think this study is ethically OK?

3.92 Deceiving subjects. A psychologist conducts the following experiment: she measures the attitude of subjects toward cheating, then has them play a game rigged so that winning without cheating is impossible. The computer that organizes the game also records—unknown to the subjects—whether they cheat. Then attitude toward cheating is retested.

Subjects who cheat tend to change their attitudes to find cheating more acceptable. Those who resist the temptation to cheat tend to condemn cheating more strongly on the second test of attitude. These results confirm the psychologist's theory.

This experiment tempts subjects to cheat. The subjects are led to believe that they can cheat secretly when, in fact, they are observed. Is this experiment ethically objectionable? Explain your position.

CHAPTER 3 REVIEW EXERCISES

3.93 Price promotions and consumer behavior. A researcher is studying the effect of price promotions on consumer behavior. Subjects are asked to choose between purchasing a single case of a soft drink for $4.00 or three cases of the same soft drink for $10.00. Is this study an experiment? Why? What are the explanatory and response variables?

3.94 What type of study? What is the best way to answer each of the following questions: an experiment, a sample survey, or an observational study that is not a sample survey? Explain your choices.

(a) Do students learn statistics more effectively in a traditional setting or in a flipped classroom?

(b) What is the average length of time for calls to a customer service center?

(c) What proportion of first-year college students use their cell phones for more than an hour per day?

3.95 Choose the type of study. Give an example of a question about your customers, their behavior, or their opinions that would best be answered by

(a) a sample survey.

(b) an observational study that is not a sample survey.

(c) an experiment.

3.96 Compare McDonald's fries with Burger King's. Do consumers prefer fries from McDonald's or Burger King?

(a) Discuss how you might make this a blind test in which neither source of the fries is identified. Do you think that your blinding will be successful for all subjects?

(b) Describe briefly the design of a matched pairs experiment to investigate this question. How will you use randomization?

3.97 Coupons and customer expectations. A researcher studying the effect of coupons on consumers' expectations makes up two different series of ads for a hypothetical brand of cola for the past year. Students in a family science course view one or the other sequence of ads on a computer. Some students see a sequence of ads with no coupon offered on the cola, while others see regular coupon offerings that effectively lower the price of the cola temporarily. Next, the students are asked what price they would expect to pay for the cola.

(a) Is this study an experiment? Why?

(b) What are the explanatory and response variables?

3.98 How many online purchases? An opinion poll calls 1000 randomly chosen residential telephone numbers, and then asks to speak with an adult member of the household. The interviewer asks, "How many online purchases have you made in the past 2 months?"

(a) What population do you think the poll has in mind?

(b) In all, 437 people respond. What is the rate (percent) of nonresponse?

(c) For the question asked, what source of response error is likely present?

(d) Write a variation on this question that would reduce the associated response error.

3.99 Marketing a dietary supplement. Your company produces a dietary supplement that contains a significant amount of calcium as one of its ingredients. The company would like to be able to market this fact successfully to one of the target groups for the supplement: men with high blood pressure. To this end, you must design an experiment to demonstrate that added calcium in the diet reduces blood pressure. You have available 30 men with high blood pressure who are willing to serve as subjects. CALSUPP

(a) Outline an appropriate design for the experiment, taking the placebo effect into account.

(b) The names of the subjects appear below. Do the randomization required by your design, and list the subjects to whom you will give the drug. (If you use Table B, enter the table at line 136.)

Alomar	Denman	Han	Liang	Rosen
Asihiro	Durr	Howard	Maldonado	Solomon
Bikalis	Farouk	Imrani	Moore	Townsend
Chen	Fratianna	James	O'Brian	Tullock
Cranston	Green	Krushchev	Plochman	Willis
Curtis	Guillen	Lawless	Rodriguez	Zhang

3.100 A hot fund. A large mutual funds group assigns a young securities analyst to manage its small biotechnology stock fund. The fund's share value increases an impressive 43% during the first year under the new manager. Explain why this performance does not necessarily establish the manager's ability.

3.101 Employee meditation. You see a news report of an experiment that claims to show that a meditation technique increased job satisfaction of employees. The experimenter interviewed the employees and assessed their levels of job satisfaction. The subjects then learned how to meditate and did so regularly for a month. The experimenter reinterviewed them at the end of the month and assessed their job satisfaction levels again.

(a) There was no control group in this experiment. Why is this a blunder? What lurking variables might be confounded with the effect of meditation?

(b) The experimenter who diagnosed the effect of the treatment knew that the subjects had been meditating. Explain how this knowledge could bias the experimental conclusions.

(c) Briefly discuss a proper experimental design, with controls and blind diagnosis, to assess the effect of meditation on job satisfaction.

3.102 Executives and exercise. A study of the relationship between physical fitness and leadership uses as subjects middle-aged executives who have volunteered for an exercise program. The executives are divided into a low-fitness group and a high-fitness group on the basis of a physical examination. All subjects then take a psychological test designed to measure leadership, and the results for the two groups are compared. Is this an observational study or an experiment? Explain your answer.

3.103 Does the new pizza taste better? You need to help your company to decide whether to bring a new variety of pizza to market. You plan a study where pizza eaters are asked to taste the new pizza as well as the current pizza that your company sells. Is this an observational study or an experiment? Why?

3.104 Questions about attitudes. Write two questions about an attitude that concerns you for use in a sample survey. Create the first question so that it is biased in one direction, and make the second question biased in the opposite direction. Explain why your questions are biased, and then write a third question that has little or no bias.

3.105 Will the regulation make the product safer? Canada requires that cars be equipped with "daytime running lights," headlights that automatically come on at a low level when the car is started. Some manufacturers are now equipping cars sold in the United States with running lights. Will running lights reduce accidents by making cars more visible?

(a) Briefly discuss the design of an experiment to help answer this question. In particular, which response variables will you examine?

(b) Example 3.25 (page 155) discusses center brake lights. What cautions do you draw from that example that apply to an experiment on the effects of running lights?

3.106 Learning about markets. Your economics professor wonders if playing market games online will help students understand how markets set prices. You suggest an experiment: have some students use the online games, while others discuss markets in recitation sections. The course has two lectures, at 10:30 A.M. and 3:30 P.M. Six recitation sections are attached to each lecture, and the students are already assigned to recitations. For practical reasons, all students in each recitation must follow the same program.

(a) The professor says, "Let's just have the 10:30 group do online work in recitation and the 3:30 group do discussion." Why is this a bad idea?

(b) Outline the design of an experiment with the 12 recitation sections as cases. Carry out your randomization, and include in your outline the recitation numbers assigned to each treatment.

3.107 How much do students earn? A university's financial aid office wants to know how much it can expect students to earn from summer employment. This information will be used to set the level of financial aid. The population consists of 2640 students who have completed at least one year of study but have not yet graduated. The university will send a questionnaire to an SRS of 500 of these students, drawn from an alphabetized list.

(a) Describe how you will label the students to select the sample.

(b) Use software to select the students in the sample.

3.108 Attitudes toward collective bargaining. A labor organization wants to study the attitudes of college faculty members toward collective bargaining. These attitudes appear to be different depending on the type of college. The American Association of University Professors classifies colleges as follows:

- Class I: offer doctorate degrees and award at least 15 per year.
- Class IIA: award degrees above the bachelor's but are not in Class I.
- Class IIB: award no degrees beyond the bachelor's.
- Class III: two-year colleges.

Discuss the design of a sample of faculty from colleges in your state, with total sample size about 200.

3.109 Student attitudes concerning labor practices. You want to investigate the attitudes of students at your school about the labor practices of factories that make college-brand apparel. You have a grant that will pay the costs of contacting about 500 students.

(a) Specify the exact population for your study. For example, will you include part-time students?

(b) Describe your sample design. Will you use a stratified sample?

(c) Briefly discuss the practical difficulties that you anticipate. For example, how will you contact the students in your sample?

3.110 Opioid addiction. Opioid addiction is a major problem in the United States today. Suggest three different strategies that could be used to address this problem and outline the design of an experiment to compare their effectiveness. Be sure to specify the response variables you will measure.

3.111 Experiments and surveys for business. Write a short report describing the differences and similarities between experiments and surveys that would be used in business. Include a discussion of the advantages and disadvantages of each.

3.112 The product should not be discolored. Few people want to eat discolored french fries. Potatoes are kept refrigerated before being cut for french fries to prevent spoiling and preserve flavor. But immediate processing of cold potatoes causes discoloring due to complex chemical reactions. The potatoes must, therefore, be brought to room temperature before processing. Fast-food chains and other sellers of french fries must understand potato behavior. Design an experiment in which tasters will rate the color and flavor of french fries prepared from several groups of potatoes. The potatoes will be freshly harvested, or stored for a month at room temperature, or stored for a month refrigerated. They will then be sliced and cooked either immediately or after an hour at room temperature.

(a) What are the factors and their levels, the treatments, and the response variables?

(b) Describe and outline the design of this experiment.

(c) It is efficient to have each taster rate fries from all treatments. How will you use randomization in presenting fries to the tasters?

3.113 Quality of service. Statistical studies can often help service providers assess the quality of their service. The U.S. Postal Service is one such provider of services. We wonder if the number of days a letter takes to reach another city is affected by the time of day it is mailed and whether the zip code is used. Describe briefly the design of a two-factor experiment to investigate this question. Be sure to specify the treatments exactly and to tell how you will handle lurking variables such as the day of the week on which the letter is mailed.

3.114 iPhone versus Android. Many people hold very strong opinions about the superiority of the products that they use. Design an experiment to compare customer satisfaction with the iPhone versus the Android smartphone operating system. Consider whether you will block on the type of phone currently being used. Write a summary of your study design, including your reasons for the choices you make. Be sure to include the question or questions that you will use to measure customer satisfaction.

3.115 Design your own experiment. The previous two exercises illustrate the use of statistically designed experiments to answer questions of interest to consumers as well as to businesses. Select a question of interest to you that an experiment might answer, and briefly discuss the design of an appropriate experiment.

3.116 Randomization for testing a breakfast food. To demonstrate how randomization reduces confounding, return to the breakfast food testing experiment described in Example 3.18 (page 147). Label the 30 rats 01 to 30. Suppose that, unbeknownst to the experimenter, the 10 rats labeled 01 to 10 have a genetic defect that will cause them to grow more slowly than normal rats. If the experimenter simply puts rats 01 to 15 in the experimental group and rats 16 to 30 in the control group, this lurking variable will bias the experiment against the new food product.

Use software or Table B to assign 15 rats at random to the experimental group as in Example 3.18. Record how many of the 10 rats with genetic defects are placed in the experimental group and how many are in the control group. Repeat the randomization using different lines in Table B until you have done five random assignments. What is the mean number of genetically defective rats in the experimental and control groups in your five repetitions?

3.117 Two ways to ask sensitive questions. Sample survey questions are usually read from a computer screen. In a computer-aided personal interview (CAPI), the interviewer reads the questions and enters the responses. In a computer-aided self-interview (CASI), the interviewer stands aside and the respondent reads the questions and enters responses. One method almost always shows a higher percent of subjects admitting use of illegal drugs. Which method? Explain why.

3.118 Your institutional review board. Your college or university has an institutional review board that screens all studies that use human subjects. Get a copy of the document that describes this board (you can probably find it online).

(a) According to this document, what are the duties of the board?

(b) How are members of the board chosen? How many members are not scientists? How many members are not employees of the college? Do these members have some special expertise, or are they simply members of the "general public"?

3.119 Online behavioral advertising. The Federal Trade Commission's report "Self-Regulatory Principles for Online Behavioral Advertising" defines behavioral advertising as "the tracking of a consumer's online activities over time—including the searches the consumer has conducted, the web pages visited and the content viewed—to deliver advertising targeted to the individual consumer's interests." The report suggests four governing concepts for online behavioral advertising:

1. Transparency and control: when companies collect information from consumers for advertising, they should tell the consumers about how the data will be collected, and consumers should be given a choice about whether to allow the data to be collected.

2. Security and data retention: data should be kept secure and should be retained only as long as needed.

3. Privacy: before data are used in a way that differs from how the companies originally said they would use the information, companies should obtain consent from consumers.

4. Sensitive data: consent should be obtained before using any sensitive data.[35]

Write a report discussing your opinions concerning online behavioral advertising and the four governing concepts. Pay particular attention to issues related to the ethical collection and use of statistical data.

3.120 Confidentiality at NORC. The National Opinion Research Center conducts a large number of surveys and has established procedures for protecting the confidentiality of its survey participants. For its Survey of Consumer Finances, it provides a pledge to participants regarding confidentiality. This pledge is available at **scf.norc.org/confidentiality.html**. Review the pledge and summarize its key parts. Do you think that the pledge adequately addresses issues related to the ethical collection and use of data? Explain your answer.

3.121 What's wrong? Explain what is wrong in each of the following statements. Give reasons for your answers.

(a) Software should never be used for a simple random sample when the population is smaller than 25.

(b) Matched pairs designs and block designs are always inferior to randomized comparative experiments.

(c) Multistage samples can be used for assigning subjects to treatments.

CHAPTER 4

Probability: The Study of Randomness

Introduction

In this chapter, we study the basic concepts of probability. So far, we have spent the first two chapters focused on exploring and describing data in hand. We then learned in Chapter 3 how to produce quality data that can be used to reliably infer conclusions about a larger population.

You might then ask yourself, "Where does the study of probability fit in our data journey? Isn't the next step learning statistical methods?" The answers to these questions lie in recognizing that the reasoning of statistical inference rests on asking, "How often would this method give a correct answer if I used it very many times?" When we produce data by random sampling or randomized comparative experiments, the laws of probability answer this question. As such, *probability* can be viewed as the backbone of statistical inference.

The importance of probability ideas for statistical inference is reason enough to delve into this and the following chapter. However, the concepts of probability also come into play in many scenarios in business and economics. Here are just two examples:

- As a business student, there is a good chance you are pursuing an accounting major with the hope of becoming a certified public accountant (CPA). Did you know that CPAs can boost their earnings potential by an additional 10% to 25% by adding a certification for fraud detection? Certified fraud accountants must have in their toolkit a probability distribution, known as the Benford distribution, that we study in this chapter. Liberty Mutual Insurance, Citibank, MasterCard, Deloitte, and the FBI are just a few of the organizations that employ fraud accountants.

CHAPTER OUTLINE

4.1 Randomness

4.2 Probability Models

4.3 General Probability Rules

- There are around 200 for-profit marketing research firms in the United States.[1] Marketing research firms conduct surveys and analyze purchases in different market segments to estimate various probabilities of consumer behavior. For example, a retailer might want to know the probabilities that consumers will use a mobile wallet (e.g., Apple Pay, Samsung Pay, or Google Pay) *given* various demographic profiles that include sex and age. Determining these probabilities can help companies develop effective marketing strategies.

We begin our exploration by introducing the concept of probability. This is followed by a section on the basic probability model. The chapter then concludes with a section on important rules we need for our discussion of random variables in Chapter 5 and methods of statistical inference in the remainder of the textbook.

4.1 Randomness

When you complete this section, you will be able to:
- Recognize a random phenomenon.
- Describe the *probability* of an outcome in reference to many trials.
- Identify trials as independent or not.

simple random sample (SRS), p. 133

When you toss a coin, sometimes you get a head, at other times you get a tail. When you pick an SRS, the observed sample may be one of many possibilities. In either of these cases, the result cannot be predicted with certainty in advance because it varies across tosses or draws. But a regular pattern does emerge in the collection of results, and it becomes clearer the more the repetitions. This remarkable fact is the basis for the idea of probability.

EXAMPLE 4.1

Auditing Financial Records An accounting colleague tells you that the proportion of 1s in the leftmost digit ("first digit") of financial numbers is likely to be much greater than $1/9 = 0.111$. You are very skeptical of this claim and decide to conduct an audit of randomly selected financial reimbursement requests from company employees. With each selected reimbursement request amount, you record whether the leftmost digit is 1. The first randomly selected reimbursement request amount is $543.94, so the leftmost digit is 5. This is not a 1, so the proportion of 1s starts at 0. On the next eight reimbursement request amounts, you observe the leftmost digit values of 7, 1, 2, 9, 4, 3, 6, and 8. A value of 1 was only observed on the third record, so the proportion of 1s jumps from 0 to 1/3 and then steadily decreases down to 1/9, as seen in the following table.

Financial Record	Leftmost Digit	Proportion of 1s
1	5	0/1
2	7	0/2
3	1	1/3
4	2	1/4
5	9	1/5
6	4	1/6
7	3	1/7
8	6	1/8
9	8	1/9

At this moment, you pause and notice that one out of the nine financial numbers has a leftmost digit of 1. Your skepticism seems to be confirmed and you are ready to correct your colleague's misguided thinking. However, you have time on your hands and decide to continue the audit. As the new records accumulate, you notice that the proportion of 1s increases and starts settling down to a number around 0.3, as shown in Figure 4.1. If you were to get more and more records, the proportion would converge to 0.301. We say that 0.301 is the **probability** that the leftmost digit of a financial record is 1. The probability 0.301 appears as a horizontal line on the graph. ∎

probability

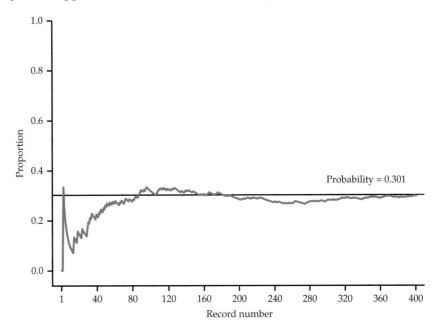

FIGURE 4.1 The proportion of randomly selected financial numbers with the leftmost digit being 1 as more financial records are being audited. Eventually, the proportion approaches 0.301.

Example 4.1 suggests one common trap when it comes to assessing the probability of a phenomenon. Namely, after nine observed outcomes, there is a temptation to conclude that the probability of the leftmost digit being 1 is 0.111. However, as we learned from the continuation of the experiment, the proportion in a small or moderate number of outcomes can be far from this probability. *Many people prematurely assess the probability of a phenomenon based only on short-term outcomes.* Probability describes what happens in the long run.

The language of probability

"Random" in statistics is not a synonym for "haphazard," but rather a description of a kind of order that emerges only in the long run. We often encounter the unpredictable side of randomness in our everyday experience, but we rarely see enough repetitions of the same *random* phenomenon to observe the long-term regularity that probability describes. You can see that regularity emerging in Figure 4.1. In the very long run, the proportion of financial records that give a leftmost digit of 1 is 0.301. This is the intuitive idea of probability. Probability 0.301 means that it "occurs 30.1% of the time in a very large number of trials." In statistics, the term **trial** represents a single performance of a well-defined experiment such as drawing a single financial record or a single toss of a specific coin.

trial

empirical

In this sense, the idea of probability is **empirical**; that is, it is based on observation. Probability describes what happens in very many trials, and we must actually observe many trials to pin down a probability. Perhaps the most iconic application of the notion of probability emerging in the long run is the experiment of tossing a coin. Some diligent people have, in fact, made thousands of tosses.

EXAMPLE 4.2

Some Coin Tossers The French naturalist Count Buffon (1707–1788) tossed a coin 4040 times. Result: 2048 heads, or proportion 2048/4040 = 0.5069 for heads.

Around 1900, the English statistician Karl Pearson heroically tossed a coin 24,000 times. Result: 12,012 heads, a proportion of 0.5005.

While imprisoned by the Germans during World War II, the South African mathematician John Kerrich tossed a coin 10,000 times. Result: 5067 heads, a proportion of 0.5067. ∎

The experiments of these individuals did not just result in heads: they also observed the other possible outcome of tails. Pearson, for example, found the proportion of tails to be 0.4995. Their experiments revealed the long-term regularity across all the possible outcomes. In other words, they were able to pin down the distribution of outcomes. In the next section, we will indeed go beyond the study of the probability of getting a 1 as the leftmost digit to the consideration of the distribution of all possible leftmost digits for financial numbers.

distribution, p. 8

fair

The *Probability* applet available on the text website allows you to animate experiments with two possible outcomes. For example, it allows you to choose the probability of a head and simulate any number of tosses of a coin with that probability. A coin is called **fair** if the probability of a head is 0.5. Try the applet for the scenario of a fair coin toss and see the value to which the proportion converges over many tosses. If you want to basically simulate the scenario of Example 4.1, then equate a head outcome with getting a 1 for a leftmost digit and set the probability to 0.30.

RANDOMNESS AND PROBABILITY

We call a phenomenon **random** if individual outcomes are uncertain but there is, nonetheless, a regular distribution of outcomes in a large number of repetitions.

The **probability** of any outcome of a random phenomenon is the proportion of times the outcome would occur in a very long series of repetitions.

APPLY YOUR KNOWLEDGE

4.1 Other digits. Refer to Example 4.1 (pages 176-177) and the table showing the sequence of the first nine randomly selected financial records.

(a) Using the same run of financial records, what would be the sequence of proportion values if our interest is with the leftmost digit being 3?

(b) Using the same run of financial records, what would be the sequence of proportion values if our interest is with the leftmost digit being 5?

(c) Using the same run of financial records, what would be the sequence of proportion values if our interest is with the leftmost digit *not* being 1?

Thinking about randomness and probability

Randomness is everywhere. In our personal lives, we observe randomness with varying outdoor temperatures, our blood pressure readings, our commuting times to school or work, and the scores of our favorite sports team. Businesses exist in a world of randomness in the forms of varying dimensions on manufactured parts, customers' waiting times, demand for products or services, prices of a company's stock, injuries in the workplace, and customers' abilities to pay off a loan.

Probability theory is the branch of mathematics that describes random behavior; its advanced study entails high-level mathematics. However, as we will discover, many of the key ideas are basic. Managers who assimilate these key ideas are better able to cope with the stark realities of randomness. They become better decision makers.

Of course, we never observe a probability exactly. We could always continue tossing the coin, for example. Mathematical probability is an idealization based on imagining what would happen in an indefinitely long series of trials. The best way to understand randomness is to observe random behavior—not only the long-run regularity, but the unpredictable results of short runs. You can do this with physical devices such as coins and dice, but computer simulations of random behavior enable faster exploration.

As you explore randomness, remember:

independent
- You must have a long series of **independent** trials. That is, the probability of each outcome does not change based on the outcomes of previous trials. Imagine a crooked gambling house where the operator of a roulette wheel does not allow there to be two reds in a row. The trials of the roulette spins would then not be independent.

- The idea of probability is empirical. Computer simulations start with given probabilities and imitate random behavior, but we can estimate a real-world probability only by actually observing many trials.

- Nonetheless, computer simulations are very useful because we need long runs of trials. In situations such as coin tossing, the proportion of an outcome often requires several hundred trials to settle down to the probability of that outcome. The kinds of physical techniques suggested in the exercises are too slow for this. Short runs give only rough estimates of a probability.

SECTION 4.1 SUMMARY

- A **random phenomenon** has outcomes that we cannot predict with certainty but that, nonetheless, have a regular distribution in very many repetitions.

- The **probability** of an outcome is the proportion of times the outcome occurs in many repeated trials of a random phenomenon.

- Trials are **independent** if the outcome of one trial does not influence the outcome of any other trial.

SECTION 4.1 EXERCISES

For Exercise 4.1, see page 178.

4.2 Trials and probability. Consider the coin-tossing experiments noted in Example 4.2.

(a) Based on the definition of a *trial* (page 177), how many trials were conducted by Buffon, Pearson, and Kerrich, respectively?

(b) Do you think it is reasonable to combine the results by calculating $(2048 + 12{,}012 + 5067)/(4040 + 24{,}000 + 10{,}000) = 0.5028$? Explain your answer.

4.3 Not just coins. With Example 4.2, we noted the most recognizable experiment of chance, the coin toss. The coin has two random outcomes, heads and tails. Provide two examples of business scenarios in which there are two distinct but uncertain outcomes.

4.4 Common misconception. We often hear the phrase "law of averages," which is used to suggest certain outcomes of independent trials are "due" to happen to compensate for short-term results. For example, if a baseball player with a 0.250 batting average has a slump of 10 strikeouts in a row, a sports commentator might naïvely state, "He is due for a hit because of the law of averages." Or, in a gambling context, we might hear someone say, "The roulette wheel has landed on red in four consecutive spins. Bet on black because of the law of averages." The common usage of this phrase is an erroneous generalization of what we expect in the long term applied to the short term. For this exercise, consider the tossing of a fair coin.

(a) Suppose you toss the coin five times in front of a friend and observe HHHHH. Your friend states that "a tail is surely coming up next." How do you respond to your friend?

(b) Amazingly, suppose you get 10 heads on your first 10 tosses in front of your friend. You ask your friend, "If I were to toss this coin 90 more times and given the results of the first 10 tosses, how many heads do we expect in the first 100 tosses?" Your friend responds, "50 heads since the coin is fair." Explain how your friend's incorrect answer most likely relates to the common interpretation of the law of averages.

(c) Continuing with part (b), how many heads should you expect in the 100 tosses? (*Hint:* The 10 heads on the first 10 tosses are given. Think only about the remaining 90 tosses and how many heads are expected. Then combine these facts.)

4.5 Financial fraud. It has been estimated that one in six fraud victims knew the perpetrator as a friend or an acquaintance. Financial fraud includes crimes such as unauthorized credit card charges, withdrawal of money from a savings or checking account, and opening an account in someone else's name. Suppose you want to simulate the outcome that a fraud victim knew the perpetrator versus the outcome that the fraud victim did not know the perpetrator.

(a) If you were to roll a single die, explain how you would match the outcomes of the die with the fraud scenario.

(b) If you were to roll a pair of dice, explain how you would match the outcomes of the pair of dice with the fraud scenario.

(c) If you had software that randomly generated numbers between 0 and 1, explain how you would match the outcomes of the random number simulator with the fraud scenario.

4.6 Credit monitoring. In a recent study of consumers, 25% reported purchasing a credit-monitoring product that alerts them to any activity on their credit report. Suppose you want to use a physical device to simulate the outcome of a consumer purchasing the credit-monitoring product versus the outcome of the consumer not purchasing the product. Describe how you could use two fair coins to conduct a simulation experiment to mimic consumer behavior. In particular, what outcomes of the two flipped coins would you associate with purchasing the product and what outcomes would you associate with not purchasing the product?

4.7 Random digits. As discussed in Chapter 3, generation of random numbers is one approach for obtaining a simple random sample (SRS). If we were to look at the random generation of digits, the mechanism should give each digit probability 0.1. Consider the digit "0" in particular.

(a) The table of random digits (Table B) was produced by a random mechanism that gives each digit probability 0.1 of being a 0. What proportion of the first 200 digits in the table are 0s? Is this estimate, based on 200 repetitions, close to the true value of 0.1? Would you expect the proportion of the next 200 repetitions to be closer? What about the combined proportion of 400 repetitions? Explain your answer.

(b) Now use software assigned by your instructor to generate 1000 repetitions:

- *Excel users:* Enter the formula = **RANDBETWEEN(0, 9)** in cell A1. Now, drag and copy the contents of cell A1 into cells A2:A1000. You will find 1000 random digits appear. Any attempt to copy these digits for sorting purposes will result in the digits changing. You will need to "freeze" the generated values. To do so, highlight column A and copy the contents and then **Paste Special as Values** the contents into the same column. The values will now not change. To get a count of the 0s, enter the formula = **COUNTIF(A1:A1000,0)** in cell B1.

- *JMP users:* With a new data table, right-click on the header of Column 1 and choose **Column Info.** In the drag-down dialog box named **Initialize Data,** pick **Random** option. Choose the bullet option of **Random Integer,** and set **Minimum/Maximum** to 0 and 9. Input the value of 1000 into the **Number of rows** box, and then click **OK.** To get a count of the 0s, go to **Summary** under the **Table** tab. Then click in column 1 in the **Group** box and then click **OK.**

- *Minitab users:* Do the following pull-down sequence: **Calc → Random Data → Integer**. Enter "1000" in the **Number of rows of data to generate** box, type "c1" in the **Store in column(s)** box, enter "0" in the **Minimum value** box, and enter "9" in the **Maximum** box, and then click **OK**. To get a count of the 0s, do the following pull-down sequence: **Stat → Basic Statistics → Display Descriptive Statistics**. Then click the **Statistics** button and clear all the boxes, leaving only **N nonmissing** with a check mark. Click **OK**. Enter "c1" into both the boxes of **Variables** and **By variables** and then click **OK**.

- *R users:* Type following at the R prompt:

    X <- sample(0:9, 1000, replace = T)

Use the "table(X)" command to find the count of the 0s.

Based on the software you used, what proportion of the 1000 randomly generated digits are 0s? Is this proportion close to 0.1?

4.8 Are Visa's prices independent? Over time, stock prices are always on the move. Consider a time series of 1064 consecutive daily prices of Visa's stock from the beginning of January 2014 to near the end of March 2018.[2]

(a) Using software, plot the prices over time. Are the prices constant over time? Describe the nature of the price movement over time.

(b) Now consider the relationship between price on any given day with the price on the prior day. The previous day's price is sometimes referred to as the *lag* price. You will want to put the lagged prices in another column or variable:

- **Excel users:** Highlight and copy the price values, and paste them in a new column shifted down by one row.

- **JMP users:** Click on the price column header name to highlight the column of price values. Copy the highlighted values. Now click anywhere on the nearest empty column, resulting in the column being filled with missing values. Double-click on the cell in row 2 of the newly formed column. With row 2 cell open, paste the price values to create a column of lagged prices. (*Note:* A column of lagged values can also be created with JMP's **Lag** function found in the **Formula** option of the column.)

- **Minitab users: Stat → Time Series → Lag.**

- **R users:** There are convenient ways to create a lag variable in R based on downloading specialized time-series packages. However, without a special package, the following commands will create a lag variable of the price variable:

y <- read.csv("ex04-08Visa.csv",header=T)

Price = y$Price

lagPrice <-c(NA,Price[seq_along(Price)-1])

Referring back to Chapter 2 and scatterplots (page 66), create a scatterplot of Visa's price on a given day versus the price on the previous day. Does the scatterplot suggest that the price series behaves as a series of independent trials? Explain why or why not.

4.9 Are Visa's price changes independent? Refer to the daily price series of Visa's stock in Exercise 4.8. Instead of looking at the prices themselves, consider now the daily *changes* in prices found in the provided data file. One way to create a column of differences is to subtract the lag price variable from the original price variable. **VISA**

(a) Using software, plot the price changes over time. Describe the nature of the price changes over time.

(b) Now consider the relationship between a given price change and the previous price change. Create a lag of price changes by following the steps of Exercise 4.8(b). Create a scatterplot of price change versus the previous price change. Does the scatterplot seem to suggest that the price-change series behaves essentially as a series of independent trials? Explain why or why not.

(c) This exercise explored the relationship or lack of it between price changes of successive days. If you want to feel more confident about a conclusion of independence of price changes over time, which additional scatterplots might you consider creating?

4.10 Use the *Probability* applet. The idea of probability is that the *proportion* of heads in many tosses of a balanced coin eventually gets close to 0.5. But does the actual *count* of heads get close to one-half the number of tosses? Let's find out. Set the "Probability of Heads" in the *Probability* applet to 0.5 and the number of tosses to 50. You can extend the number of tosses by clicking "Toss" again to get 50 more. Don't click "Reset" during this exercise.

(a) After 50 tosses, what is the proportion of heads? What is the count of heads? What is the difference between the count of heads and 25 (one-half the number of tosses)?

(b) Keep going to 150 tosses. Again record the proportion and count of heads and the difference between the count and 75 (half the number of tosses).

(c) Keep going. Stop at 300 tosses and again at 600 tosses to record the same facts. Although it may take a long time, the laws of probability say that the proportion of heads will always get close to 0.5 and also that the difference between the count of heads and half the number of tosses will always grow without limit.

4.11 A question about dice. In the seventeenth century, here is a question that a French gambler asked mathematicians Fermat and Pascal: what is the probability of getting at least one 6 in rolling four dice? The *Law of Large Numbers* applet allows you to roll several dice and watch the outcomes. (Ignore the title of the applet for now.) Because simulation—just like real random phenomena—often takes very many trials to estimate a probability accurately, let's simplify the question: is this probability clearly greater than 0.5, clearly less than 0.5, or quite close to 0.5? Use the applet to roll four dice until you can confidently answer this question. You will have to set "Rolls" to 1 so that you have time to look at the four up-faces. Keep clicking "Roll dice" to roll again and again. How many times did you roll four dice? What percent of your rolls produced at least one 6?

4.12 Proportions of Visa's price changes. Continue the study of daily price changes of Visa's stock from Exercise 4.9. Consider three possible outcomes: (1) positive price change, (2) no price change, and (3) negative price change. **VISA**

(a) Find the proportions of each of these outcomes. This is most easily done by sorting the price change data into another column of the software and then counting the number of negative, zero, and positive values.

(b) Would the proportions found in part (a) serve as reasonable estimates for the true probabilities of price changes being positive, zero, or negative? Explain your answer in the context of a repeated experiment.

4.13 Thinking about probability statements. Probability is a measure of how likely an event is to occur. Match one of the probabilities that follow with each statement of likelihood given. (The probability is usually a more exact measure of likelihood than is the verbal statement.)

0   0.01   0.3   0.6   0.99   1

(a) This event is impossible. It can never occur.

(b) This event is certain. It will occur on every trial.

(c) This event is very unlikely, but it will occur once in a while in a long sequence of trials.

(d) This event is not extremely likely, but it occurs more often than not.

4.2 Probability Models

When you complete this section, you will be able to:
- Describe a sample space from a description of a random phenomenon.
- Identify the correct assignment of probabilities to a set of events.
- Apply the addition rule for disjoint events.
- Apply the complement rule.
- Apply the multiplication rule to independent events.

probability model

The idea of probability as the proportion of a particular outcome in very many repeated trials guides our intuition, but is hard to express in mathematical form. A description of a random phenomenon in the language of mathematics is called a **probability model**. To see how to proceed, think first about a very simple random phenomenon, tossing a coin once. When we toss a coin, we cannot know the outcome in advance. What do we know? We are willing to say that the outcome will be either heads or tails. Because the coin appears to be balanced, we believe that each of these outcomes has probability 1/2. This description of coin tossing has two parts:

- a list of possible outcomes
- a probability for each outcome

This two-part description is the starting point for a probability model. We begin by describing the outcomes of a random phenomenon and then learn how to assign these probabilities ourselves.

Sample spaces

A probability model first tells us what outcomes are possible.

> **SAMPLE SPACE**
>
> The **sample space** S of a random phenomenon is the set of all distinct possible outcomes.

The name "sample space" is natural in random sampling, where each possible outcome is a sample and the sample space contains all possible samples. To specify S, we must state what constitutes an individual outcome and then state which outcomes can occur. We often have some freedom in defining the sample space, so the choice of S is a matter of convenience as well as correctness. The idea of a sample space, and the freedom we may have in specifying it, are best illustrated by examples.

EXAMPLE 4.3

Sample Space for Tossing a Coin Toss a coin. There are only two possible outcomes, and the sample space is

$$S = \{\text{heads, tails}\}$$

or, more briefly, $S = \{H, T\}$. ∎

EXAMPLE 4.4

Sample Space for Random Digits Type "= RANDBETWEEN(0,9)" into any Excel cell and press enter. Record the value of the digit that appears in the cell. The possible outcomes are

$$S = \{0, 1, 2, 3, 4, 5, 6, 7, 8, 9\}$$ ∎

EXAMPLE 4.5

Sample Space for Tossing a Coin Four Times Toss a coin four times and record the results. That's a bit vague. To be exact, record the results of each of the four tosses in order. A possible outcome is then HTTH. Counting shows that there are 16 possible outcomes. The sample space S is the set of all these outcomes.

$$S = \{\text{HHHH, THHH, HTHH, HHTH, HHHT, TTHH, THTH, THHT,}$$
$$\text{HTTH, HTHT, HHTT, TTTH, TTHT, THTT, HTTT, TTTT}\}$$

Suppose that our only interest is the number of heads in four tosses. Now we can be exact in a simpler fashion. The random phenomenon is to toss a coin four times and count the number of heads. The sample space contains only five outcomes:

$$S = \{0, 1, 2, 3, 4\}$$

This example illustrates the importance of carefully specifying what constitutes an individual outcome. ■

Although these examples seem remote from the practice of statistics in business, the connection is surprisingly close. Suppose that in conducting a marketing survey, you select four people at random from a large population and ask each if he or she has used a given product. The answers are Yes or No. The possible outcomes—the sample space—are exactly as in Example 4.5 if we replace heads by Yes and tails by No. Similarly, the possible outcomes of an SRS of 1500 people are the same in principle as the possible outcomes of tossing a coin 1500 times. One of the great advantages of mathematics is that the essential features of quite different phenomena can be described by the same mathematical model, which, in our case, is the probability model.

The sample spaces considered so far correspond to situations in which there is a finite list of all possible values. In other sample spaces, theoretically, the list of outcomes is infinite.

EXAMPLE 4.6

Using Software Most statistical software has a function that will generate a random number between 0 and 1. The sample space is

$$S = \{\text{all numbers between 0 and 1}\}$$

This S is a mathematical idealization with an infinite number of outcomes. In reality, any specific random number generator produces numbers with some limited number of decimal places so that, strictly speaking, not all numbers between 0 and 1 are possible outcomes. For example, in default mode, Excel reports random numbers like 0.798249, with six decimal places. The entire interval from 0 to 1 is easier to think about. It also has the advantage of being a suitable sample space for different software systems that produce random numbers with different numbers of digits. ■

APPLY YOUR KNOWLEDGE

4.14 Describing sample spaces. For each of the following questions, define a sample space S for the associated random phenomenon. In some cases, you have some freedom in your choice of S.

(a) Will a randomly selected Silicon Valley tech startup be in existence in five years?

(b) What is final grade of a randomly selected business statistics student?

(c) What is the primary major of a randomly selected student at your business school?

4.15 Describing sample spaces. In each of the following situations, describe a sample space S for the random phenomenon. Explain why, *theoretically*, a list of all possible outcomes is not finite.

(a) You record the number of tosses of a die until you observe a six.

(b) You record the number of tweets per week that a randomly selected student makes.

A sample space S lists the possible outcomes of a random phenomenon. To complete a mathematical description of the random phenomenon, we must also give the probabilities with which these outcomes occur.

The true long-term proportion of any outcome—say, "exactly 2 heads in 4 tosses of a coin"—can be found only empirically, and then only approximately. How then can we describe probability mathematically? Rather than immediately attempting to give "correct" probabilities, let's confront the easier task of laying down rules that any assignment of probabilities must satisfy. We need to assign probabilities not only to single outcomes but also to sets of outcomes.

EVENT

An **event** is an outcome or a set of outcomes of a random phenomenon. That is, an event is a subset of the sample space.

EXAMPLE 4.7

Exactly Two Heads in Four Tosses Take the sample space S for four tosses of a coin to be the 16 possible outcomes in the form HTHH. Then "exactly 2 heads" is an event. Call this event A. The event A expressed as a set of outcomes is

$$A = \{\text{TTHH, THTH, THHT, HTTH, HTHT, HHTT}\}\ \blacksquare$$

In a probability model, events have probabilities. What properties must any assignment of probabilities to events have? Here are some basic facts about any probability model. These facts follow from the idea of probability as "the long-run proportion of repetitions in which an event occurs."

1. **Any probability is a number between 0 and 1.** Any proportion is a number between 0 and 1, so any probability is also a number between 0 and 1. An event with probability 0 never occurs, and an event with probability 1 occurs in every trial. An event with probability 0.5 occurs in half the trials in the long run.

2. **All possible outcomes of the sample space together must have probability 1.** Because every trial will produce an outcome, the sum of the probabilities for all possible outcomes must be exactly 1.

3. **If two events have no outcomes in common, the probability that one or the other occurs is the sum of their individual probabilities.** If one event occurs in 40% of all trials, a different event occurs in 25% of all trials, and the two can never occur together, then one or the other occurs in 65% of all trials because 40% + 25% = 65%.

4. **The probability that an event does not occur is 1 minus the probability that the event does occur.** If an event occurs in 70% of all trials, it fails to occur in the other 30%. The probability that an event occurs and the probability that it does not occur always add to 100%, or 1.

Probability rules

Formal probability uses mathematical notation to state Facts 1 to 4 more concisely. We use capital letters near the beginning of the alphabet to denote events. If A is any event, we write its probability as $P(A)$. Here are our probability facts in formal language. As you apply these rules, remember that they are just another form of intuitively true facts about long-run proportions.

PROBABILITY RULES

Rule 1. The probability $P(A)$ of any event A satisfies $0 \leq P(A) \leq 1$.

Rule 2. If S is the sample space in a probability model, then $P(S) = 1$.

Rule 3. Two events A and B are **disjoint (also called mutually exclusive)** if they have no outcomes in common and so can never occur together. If A and B are disjoint,

$$P(A \text{ or } B) = P(A) + P(B)$$

This is the **addition rule for disjoint events.**

Rule 4. The **complement** of any event A is the event that A does not occur, written as A^c. The **complement rule** states that

$$P(A^c) = 1 - P(A)$$

You may find it helpful to draw a picture to remind yourself of the meaning of disjoint events and complements. A picture like Figure 4.2 that shows the sample space S as a rectangular area and events as areas within S is called a **Venn diagram.** The events A and B in Figure 4.2 are disjoint because they do not overlap. As Figure 4.3 shows, the complement A^c contains exactly the outcomes that are not in A.

Venn diagram

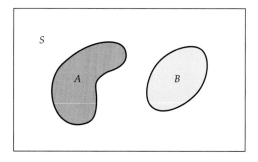

FIGURE 4.2 Venn diagram showing disjoint events A and B.

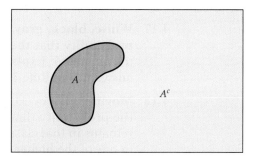

FIGURE 4.3 Venn diagram showing the complement A^c of an event A. The complement consists of all outcomes that are not in A.

EXAMPLE 4.8

Favorite Vehicle Colors What is your favorite color for a vehicle? Our preferences can be related to our personality, our moods, or particular objects. Here is a probability model for color preferences in North America.[3]

Color	White	Black	Gray	Silver
Probability	0.25	0.21	0.16	0.11

Color	Red	Blue	Brown	Other
Probability	0.10	0.08	0.04	0.05

Each probability is between 0 and 1. The probabilities add to 1 because these outcomes together make up the sample space S. Our probability model corresponds to selecting a person at random and asking him or her about a favorite color. ∎

Let's use the probability Rules 3 and 4 to find some probabilities for favorite vehicle colors.

EXAMPLE 4.9

Black or Silver? What is the probability that a person's favorite vehicle color is black or silver? If the favorite is black, it cannot be silver, so these two events are disjoint. Using Rule 3, we find

$$P(\text{black or silver}) = P(\text{black}) + P(\text{silver})$$
$$= 0.21 + 0.11 = 0.32 \quad \blacksquare$$

There is a 32% chance that a randomly selected person will choose black or silver as his or her favorite color. Suppose that we want to find the probability that the favorite color is not blue.

EXAMPLE 4.10

Use the Complement Rule To solve this problem, we could use Rule 3 and add the probabilities for white, black, silver, gray, red, brown, and other. However, it is easier to use the probability that we have for blue and Rule 4. The event that the favorite is not blue is the complement of the event that the favorite is blue. Using our notation for events, we have

$$P(\text{not blue}) = 1 - P(\text{blue})$$
$$= 1 - 0.08 = 0.92 \quad \blacksquare$$

We see that 92% of people have a favorite vehicle color that is not blue.

APPLY YOUR KNOWLEDGE

4.16 **Red or brown.** Refer to Example 4.8, and find the probability that the favorite color is red or brown.

4.17 **White, black, gray, silver, or red.** Refer to Example 4.8, and find the probability that the favorite color is white, black, gray, silver, or red using Rule 4. Explain why this calculation is easier than finding the answer using Rule 3.

4.18 **Moving up.** An economist studying economic class mobility finds that the probability that the son of a father in the lowest economic class remains in that class is 0.46. What is the probability that the son moves to one of the higher classes?

4.19 **Occupational deaths.** Government data on job-related deaths assign a single occupation for each such death that occurs in the United States. The data on occupational deaths in 2016 show that the probability is 0.187 that a randomly chosen death was a construction worker and 0.267 that it was transportation operator. What is the probability that a randomly chosen death was either construction related or transportation related? What is the probability that the death was related to some other occupation?

4.20 Grading Canadian health care. Each year, the Canadian Medical Association uses the marketing research firm Ipsos to measure public opinion with respect to the Canadian health care system. Between July 22 and July 26 of 2016, Ipsos interviewed a random sample of 1286 adults.[4] The people in the sample were asked to grade the overall quality of health care services as an A, B, C, or F, where an A is the highest grade and an F is a failing grade. Here are the results:

Outcome	Probability
A	0.37
B	0.36
C	?
F	0.05

These proportions are probabilities for choosing an adult at random and asking the person's opinion on the Canadian health care system.

(a) What is the probability that a person chosen at random gives a grade of C? Why?

(b) If a "positive" grade is defined as A or B, what is the probability of a positive grade?

Assigning probabilities: Finite number of outcomes

The individual outcomes of a random phenomenon are always disjoint. So, the addition rule provides a way to assign probabilities to events with more than one outcome: start with probabilities for individual outcomes and add to get probabilities for events. This idea works well when there are only a finite (fixed and limited) number of outcomes.

PROBABILITIES IN A FINITE SAMPLE SPACE

Assign a probability to each individual outcome. These probabilities must be numbers between 0 and 1 and must sum to 1.

The probability of any event is the sum of the probabilities of the outcomes making up the event.

CASE 4.1 **Uncovering Fraud by Digital Analysis** In Example 4.1 (pages 176–177), we explored the concept of probability by observing the proportion of times the leftmost digit ("first digit") of a multi-digit financial number is 1 over many financial records. As shown with Figure 4.1 (page 177), this proportion gets closer to 0.301, not 1/9 as many of us would assume. Surprisingly, a probability of 1/9 for any leftmost digit value (1 or otherwise) is often not the case for legitimately reported financial numbers. It is a striking fact that the first digits of numbers in legitimate records often follow a distribution known as *Benford's law*. Here it is (note that the first digit can't be 0):

First digit	1	2	3	4	5	6	7	8	9
Proportion	0.301	0.176	0.125	0.097	0.079	0.067	0.058	0.051	0.046

It is a regrettable fact that financial fraud permeates business and governmental sectors. In a recent study, the Association of Certified Fraud Examiners (ACFE) estimated that a typical organization loses 5% of revenues each year to fraud.[5] ACFE estimated the median loss to a company per fraudulent case at $150,000, with the global fraud loss amounting to nearly $4 trillion. Common examples of business fraud include:

- Corporate financial statement fraud: including reporting fictitious revenues, understating expenses, and artificially inflating reported assets.

- Personal expense fraud: employee reimbursement claims for fictitious or inflated business expenses (e.g., personal travel, meals).

- Billing fraud: submission of inflated invoices or invoices for fictitious goods or services to be paid to an employee-created shell company.

- Cash register fraud: making false entries on a cash register for fraudulent removal of cash.

In all these situations, the individuals committing fraud need to "invent" fake financial entry numbers. No matter how the invented numbers are created, the first digits of the fictitious numbers will most likely not follow the probabilities given by Benford's law. As such, Benford's law serves as an important "digital analysis" tool of auditors, typically CPAs, trained to look for fraudulent behavior. ■

Of course, not all sets of data follow Benford's law. Numbers that are assigned, such as Social Security numbers, do not. Nor do data with a fixed maximum, such as deductible contributions to individual retirement accounts (IRAs). Nor, of course, do random numbers. But a given remarkable number of financial-related data sets do closely obey Benford's law, so its role in auditing of financial and accounting statement cannot be ignored.

EXAMPLE 4.11

Find Some Probabilities for Benford's Law Consider the events

$$A = \{\text{first digit is 5}\}$$
$$B = \{\text{first digit is 3 or less}\}$$

From the table of probabilities in Case 4.1,

$$P(A) = P(5) = 0.079$$
$$P(B) = P(1) + P(2) + P(3)$$
$$= 0.301 + 0.176 + 0.125 = 0.602 \quad ■$$

Note that $P(B)$ is not the same as the probability that a first digit is strictly less than 3. The probability $P(3)$ that a first digit is 3 is included in "3 or less" but not in "less than 3."

APPLY YOUR KNOWLEDGE

4.21 Household space heating. Draw a U.S. household at random and record the primary source of energy for household use. "At random" means that we give every household the same chance to be chosen. That is, we choose an SRS of size 1. For 2018, here is the distribution of primary sources for U.S. households[6]:

Primary Source	Probability
Natural gas	0.436
Electricity	0.426
Distillate fuel oil	0.045
Propane	0.040
Wood	0.034
Nonmarketable (e.g., solar, wind)	0.019

(a) Show that this is a legitimate probability model.

(b) What is the probability that a randomly chosen U.S. household uses natural gas or electricity as its primary source of energy?

CASE 4.1 **4.22 Benford's law.** Using the probabilities for Benford's law, find the probability that the first digit is anything other than 4.

CASE 4.1 **4.23 Use the addition rule.** Use the addition rule (page 185) with the probabilities for the events A and B from Example 4.11 to find $P(A \text{ or } B)$.

EXAMPLE 4.12

Find More Probabilities for Benford's Law Check that the probability of the event C that a first digit is even is

$$P(C) = P(2) + P(4) + P(6) + P(8) = 0.391$$

Consider again event B from Example 4.11 (page 188), which had an associated probability of 0.602. The probability

$$P(B \text{ or } C) = P(1) + P(2) + P(3) + P(4) + P(6) + P(8) = 0.817$$

is *not* the sum of $P(B)$ and $P(C)$ because events B and C are not disjoint. The outcome of 2 is common to both events. *Be careful to apply the addition rule only to disjoint events.* In Section 4.3, we expand upon the probability given in this section to handle the case of nondisjoint events. ■

Assigning probabilities: Equally likely outcomes

Assigning correct probabilities to individual outcomes often requires long observation of the random phenomenon. In some circumstances, however, we are willing to assume that individual outcomes are equally likely because of some balance in the phenomenon. For example, ordinary coins have a physical balance that should make heads and tails very close to being equally likely. Software algorithms can also be designed to generate random digits that are equally likely for all practical purposes.

EXAMPLE 4.13

First Digits That Are Equally Likely You might think that first digits are distributed with equal chance among the digits 1 to 9 in business records. The nine possible outcomes would then be equally likely. The sample space for a single digit is

$$S = \{1, 2, 3, 4, 5, 6, 7, 8, 9\}$$

Because the total probability must be 1, the probability of each of the nine outcomes must be 1/9. That is, the assignment of probabilities to outcomes is

First digit	1	2	3	4	5	6	7	8	9
Probability	1/9	1/9	1/9	1/9	1/9	1/9	1/9	1/9	1/9

The probability of the event B that a randomly chosen first digit is 3 or less is

$$P(B) = P(1) + P(2) + P(3)$$
$$= \frac{1}{9} + \frac{1}{9} + \frac{1}{9} = \frac{3}{9} = 0.333$$

Compare this with the Benford's law probability in Case 4.1 (pages 187–188). A crook who fakes financial numbers in business records by picking digits "at random" will end up with too few first digits that are 3 or less. ∎

In Example 4.13, all outcomes have the same probability. Because there are nine equally likely outcomes, each must have probability 1/9. Because exactly three of the nine equally likely outcomes are 3 or less, the probability of this event is 3/9. In the special situation in which all outcomes are equally likely, we have a simple rule for assigning probabilities to events.

EQUALLY LIKELY OUTCOMES

If a random phenomenon has k possible outcomes, all equally likely, then each individual outcome has probability $1/k$. The probability of any event A is

$$P(A) = \frac{\text{count of outcomes in } A}{\text{count of outcomes in } S}$$
$$= \frac{\text{count of outcomes in } A}{k}$$

Most random phenomena do not have equally likely outcomes, so the general rule for finite sample spaces (page 187) is more important than the special rule for equally likely outcomes.

APPLY YOUR KNOWLEDGE

4.24 Possible outcomes for rolling a die. A die has six sides with one to six spots on the sides. Give the probability distribution for the six possible outcomes that can result when a perfect die is rolled.

Independence and the multiplication rule

Rule 3, the addition rule for disjoint events, describes the probability that *one or the other* of two events A and B occurs when A and B cannot occur together. Now we describe the probability that *both* events A and B occur, again only in a special situation. More general rules appear in Section 4.3.

Suppose that you toss a balanced coin twice. You are counting heads, so two events of interest are

$$A = \{\text{first toss is a head}\}$$
$$B = \{\text{second toss is a head}\}$$

The events A and B are not disjoint. They occur together whenever both tosses give heads. We want to compute the probability of the event $\{A \text{ and } B\}$ that *both* tosses are heads. The Venn diagram in Figure 4.4 illustrates the event $\{A \text{ and } B\}$ as the overlapping area that is common to both A and B.

The coin tossing of Buffon, Pearson, and Kerrich described in Example 4.2 (page 178) makes us willing to assign probability 1/2 to a head when we toss a coin. So,

$$P(A) = 0.5$$
$$P(B) = 0.5$$

FIGURE 4.4 Venn diagram showing the events A and B that are not disjoint. The event {A and B} consists of outcomes common to A and B.

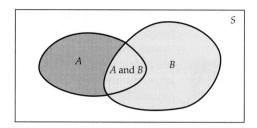

What is $P(A \text{ and } B)$? Our common sense says that it is 1/4. The first toss will give a head half the time and then the second will give a head on half of those trials, so two tosses will give heads on $1/2 \times 1/2 = 1/4$ of all trials in the long run. This reasoning assumes that the second toss still has probability 1/2 of a head after the first has given a head. This is true—we can verify it by tossing a coin twice many times and observing the proportion of heads on the second toss after the first toss has produced a head. We say that the events "head on the first toss" and "head on the second toss" are *independent*. Here is our final probability rule.

MULTIPLICATION RULE FOR INDEPENDENT EVENTS

Rule 5. Two events A and B are **independent** if knowing that one occurs does not change the probability that the other occurs. If A and B are independent,

$$P(A \text{ and } B) = P(A)P(B)$$

This is the **multiplication rule for independent events.**

Our definition of independence is rather informal here, but we will make this informal idea precise in Section 4.3. In practice, though, we rarely need a precise definition of independence, because independence is usually *assumed* as part of a probability model when we want to describe random phenomena that seem to be physically unrelated to each other.

EXAMPLE 4.14

Determining Independence Using the Multiplication Rule Consider a manufacturer that uses two suppliers for an identical part that enters the production line. Sixty percent of the parts come from one supplier (Supplier 1), while the remaining 40% come from the other supplier. Internal quality audits find that there is a 1% chance that a randomly chosen part from the production line is defective. External supplier audits reveal that two parts per thousand are defective from Supplier 1. Are the events of a part coming from a particular supplier—say, Supplier 1—and a part being defective independent?

Define the two events as follows:

$S1$ = A randomly chosen part comes from Supplier 1

D = A randomly chosen part is defective

We have $P(S1) = 0.60$ and $P(D) = 0.01$. The product of these probabilities is

$$P(S1)P(D) = (0.60)(0.01) = 0.006$$

However, supplier audits of Supplier 1 indicate that $P(S1 \text{ and } D) = 0.002$. Given that $P(S1 \text{ and } D) \neq P(S1)P(D)$, we conclude that the supplier and defective part events are not independent. ■

The multiplication rule $P(A \text{ and } B) = P(A)P(B)$ holds if A and B are *independent* but not otherwise. The addition rule $P(A \text{ or } B) = P(A) + P(B)$ holds if A and B are *disjoint* but not otherwise. Resist the temptation to use these simple rules when the circumstances that justify them are not present. *You must also be careful not to confuse disjointness and independence. Disjoint events cannot be independent.* If A and B are disjoint, then the fact that A occurs tells us that B cannot occur—look back at Figure 4.2 (page 185). Thus, disjoint events are not independent. Unlike disjointness, picturing independence with a Venn diagram is not obvious. As we will show later in Example 4.27, the mosaic plot (introduced in Chapter 2) provides a better way to visualize independence or lack of it.

mosaic plot, p. 110

APPLY YOUR KNOWLEDGE

4.25 High school rank. Select a first-year college student at random and ask what his or her academic rank was in high school. Here are the probabilities, based on proportions from a large sample survey of first-year students:

Rank	Top 20%	Second 20%	Third 20%	Fourth 20%	Lowest 20%
Probability	0.41	0.23	0.29	0.06	0.01

(a) Choose two first-year college students at random. Why is it reasonable to assume that their high school ranks are independent?

(b) What is the probability that both were in the top 20% of their high school classes?

(c) What is the probability that the first was in the top 20% and the second was in the lowest 20%?

4.26 College-educated part-time workers? For people aged 25 years or older, government data show that 39.7% of employed people have at least four years of college and that 16.6% of employed people work part-time. Can you conclude that because $(0.397)(0.166) = 0.066$, about 6.6% of employed people aged 25 years or older are college-educated part-time workers? Explain your answer.

Applying the probability rules

If two events A and B are independent, then their complements A^c and B^c are also independent and A^c is independent of B. Suppose, for example, that 75% of all registered voters in a suburban district are Republicans. If an opinion poll interviews two voters chosen independently, the probability that the first is a Republican and the second is not a Republican is $(0.75)(1 - 0.75) = 0.1875$.

The multiplication rule also extends to collections of more than two events, provided that all are independent. Independence of events A, B, and C means that no information about any one or any two events can change the probability of the remaining events. The formal definition is a bit messy, but fortunately independence is usually assumed in setting up a probability model. We can then use the multiplication rule freely.

By combining the rules we have learned, we can compute probabilities for rather complex events. Here is an example.

EXAMPLE 4.15

False-Positives in Job Drug Testing In 2017, Quest Diagnostics reported that drug use in the American workforce reached the highest positivity rate in more than a decade. Job applicants in both the public sector and the private sector are often finding that pre-employment drug testing is a requirement.

An estimated 56% of U.S. companies require drug testing of new job applicants, and 29% of companies randomly test hired employees.[7] From an applicant's or employee's perspective, one primary concern with drug testing is a "false-positive" result — that is, an indication of drug use when the individual has not actually used drugs. If a job applicant tests positive, some companies allow the applicant to pay for a retest. For existing employees, a positive result is sometimes followed up with a more sophisticated and expensive test. Beyond cost considerations, there are issues of defamation, wrongful discharge, and emotional distress.

Tests based on urine samples are mostly commonly used because they are fast and inexpensive. When applied to people who are free of illegal drugs, these tests have been reported to have false-positive rates ranging from 0.2% to 2.5%. If 150 employees are tested and all 150 are free of illegal drugs, what is the probability that at least one false-positive will occur, assuming the lower-end false-positive rate of 0.2%?

It is reasonable to assume as part of the probability model that the test results for different individuals are independent. The probability that the test is positive for a single person is 0.2%, or 0.002, so the probability of a negative result is $1 - 0.002 = 0.998$ by the complement rule. The probability of at least one false-positive among the 150 people tested is, therefore,

$$\begin{aligned} P(\text{at least 1 positive}) &= 1 - P(\text{no positives}) \\ &= 1 - P(150 \text{ negatives}) \\ &= 1 - 0.998^{150} \\ &= 1 - 0.741 = 0.259 \end{aligned}$$

The probability is greater than 1/4 that at least one of the 150 people will test positive for illegal drugs, even though no one has taken such drugs. ∎

APPLY YOUR KNOWLEDGE

4.27 Misleading résumés. For more than two decades, Jude Werra, president of an executive recruiting firm, has tracked executive résumés to determine the rate of misrepresenting education credentials and/or employment information. On a biannual basis, Werra reports a now nationally recognized statistic known as the "Liars Index." In a recent report, Werra found that 24.39% of executive job applicants lied on their résumés.[8]

(a) Suppose five résumés are randomly selected from an executive job applicant pool. What is the probability that all of the résumés are truthful?

(b) What is the probability that at least one of five randomly selected résumés has a misrepresentation?

4.28 Failing to detect drug use. In Example 4.15, we considered how drug tests can indicate illegal drug use when no illegal drugs were actually used. Consider now another type of false test result. Suppose an employee is suspected of having used an illegal drug and is given two tests that operate independently of each other. Test A has probability 0.9 of being positive if the illegal drug has been used. Test B has probability 0.8 of being positive if the illegal drug has been used. What is the probability that *neither* test is positive if the illegal drug has been used?

4.29 Bright lights? A string of holiday lights contains 20 lights. The lights are wired in series, so that if any light fails, the whole string will go dark.

Each light has probability 0.02 of failing during a three-year period. Assume that the failure of each light is independent.

(a) What is the probability that the string of lights will remain bright for a three-year period?

(b) How does the probability that the string of lights will remain bright for a three-year period change if the string of lights had fewer lights or more lights?

SECTION 4.2 SUMMARY

- A **probability model** for a random phenomenon consists of a sample space S and an assignment of probabilities P.

- The **sample space** S is the set of all possible outcomes of the random phenomenon. Sets of outcomes are called **events**. P assigns a number $P(A)$ to an event A as its probability.

- The **complement** A^c of an event A consists of exactly the outcomes that are not in A.

- Events A and B are **disjoint or mutually exclusive** if they have no outcomes in common.

- Events A and B are **independent** if knowing that one event occurs does not change the probability we would assign to the other event.

- Any assignment of probability must obey the rules that state the basic properties of probability:

Rule 1. $0 \leq P(A) \leq 1$ for any event A.

Rule 2. $P(S) = 1$.

Rule 3. Addition rule: If events A and B are **disjoint,** then $P(A \text{ or } B) = P(A) + P(B)$.

Rule 4. Complement rule: For any event A, $P(A^c) = 1 - P(A)$.

Rule 5. Multiplication rule: If events A and B are **independent,** then $P(A \text{ and } B) = P(A)P(B)$.

SECTION 4.2 EXERCISES

For Exercises 4.14 and 4.15, see pages 183–184; for 4.16 to 4.20, see pages 186–187; for 4.21 to 4.23, see pages 188–189; for 4.24, see page 190; for 4.25 and 4.26, see page 192; and for 4.27 to 4.29, see pages 193–194.

4.30 Wearable trackers. An Ipsos 2017 survey interviewed a random sample of 1003 adult Canadians on use of digital devices as it pertains to health. Here are the results on the use of wearable trackers (e.g., Fitbit and Apple Watch):[9]

	Yes, currently using	Have used before, but not currently	Heard of it, but never used	Never heard of this
Probability	0.20	0.16	?	0.13

(a) What probability should replace "?" in the distribution?

(b) What is the probability that a randomly chosen adult is using or has used a wearable tracker?

(c) How many adults in the survey have never heard of a wearable tracker?

4.31 Confidence in institutions. A Gallup Poll (June 7–11, 2017) interviewed a random sample of 1009 adults (18 years or older). The people in the sample were asked about their level of confidence in a variety of institutions in the United States. Here are the results for their confidence in small and big businesses:[10]

	Great deal	Quite a lot	Some	Very little	None	No opinion
Small business	0.33	0.37	0.23	0.07	0.00	0.00
Big business	0.09	0.12	0.38	0.36	0.04	0.01

(a) What is the probability that a randomly chosen person has either no opinion, no confidence, or very little confidence in small businesses? Find the similar probability for big businesses.

(b) Using your answer from part (a), determine the probability that a randomly chosen person has at least some confidence in small businesses. Again based on part (a), find the similar probability for big businesses.

4.32 Immigrant demographics—language. Choose an immigrant living in Montreal, Canada, at random and ask, "What is your mother tongue?" Here is the distribution of responses:[11]

Language	Arabic	Spanish	Italian
Probability	0.180	0.129	?

Language	Creole	Mandarin	Other
Probability	0.065	0.042	0.476

(a) What probability should replace "?" in the distribution?

(b) Only English and French are considered official languages in Canada. With the information provided, we cannot determine precisely the probability that a randomly chosen immigrant's mother tongue is an official language. What is the most that we can say about this probability?

4.33 Resident survey. Toronto Community Housing Corporation (TCHC) is the second-largest income-based housing provider in North America. Each year, TCHC conducts a resident survey. Based on a random sample of 3214 residents in 2017, here is the distribution of responses on overall satisfaction with delivery of TCHC services (e.g., repairs and maintenance):[12]

	Very satisfied	Somewhat satisfied	Neither satisfied nor dissatisfied
Probability	0.32	0.35	0.13

	Somewhat dissatisfied	Very dissatisfied
Probability	0.07	0.13

(a) Show that this is a legitimate probability distribution.

(b) What is the probability that a randomly chosen resident feels somewhat or very satisfied?

4.34 World Internet usage. Approximately 40.4% of the world's population uses the Internet (as of December 2017).[13] Furthermore, a randomly chosen Internet user has the following probabilities of being from the given country of the world:

Region	China	India	U.S.	Japan
Probability	0.186	0.111	0.075	0.029

(a) What is the probability that a randomly chosen Internet user does not live in one of the four countries listed in this table?

(b) What is the probability that a randomly chosen Internet user does not live in the United States?

(c) At *least* what proportion of Internet users are from Asia?

4.35 Modes of transportation. Governments (local and national) find it important to gather data on modes of transportation for commercial and workplace movement. Such information is useful for policymaking as it pertains to infrastructure (such as roads and railways), urban development, energy use, and pollution. Based on 2016 Canadian and 2015 U.S. government data, here are the distributions of the primary means of transportation to work for employees working outside the home:[14]

	Car (self or pool)	Public transport	Bicycle or motorcycle	Walk	Other
Canada	?	0.124	0.014	0.055	0.011
U.S.	?	0.052	0.006	0.028	0.012

(a) What is the probability that a randomly chosen Canadian employee who works outside the home uses a car to get to work? What is the probability that a randomly chosen U.S. employee who works outside the home uses a car?

(b) Transportation systems primarily based on cars are regarded as unsustainable because of the excessive energy consumption and the effects on the health of populations. The Canadian government includes public transit, walking, and cycles as "sustainable" modes of transportation. For both countries, determine the probability that a randomly chosen employee who works outside the home uses sustainable transportation. How do you assess the relative status of sustainable transportation for these two countries?

4.36 Car colors. Choose a new car at random and note its color. Here are the probabilities of the most popular colors for cars purchased in China in 2016:[15]

Color	White	Black	Brown	Silver	Yellow	Gray
Probability	0.57	0.18	0.08	0.05	0.05	0.02

(a) What is the probability that a randomly chosen car is either silver or white?

(b) In North America, the probability of a new car being red is 0.10. What can you say about the probability of a new car in China being red?

4.37 Land in Iowa. Choose an acre of land in Iowa at random. The probability is 0.92 that it is farmland and 0.01 that it is forest.

(a) What is the probability that the acre chosen is not farmland?

(b) What is the probability that it is either farmland or forest?

(c) What is the probability that a randomly chosen acre in Iowa is something other than farmland or forest?

4.38 Stock market movements. You note the price of the Dow Jones Industrial Index price (DJI) for Monday, and then you watch the prices of the DJI for Tuesday and Wednesday. Give a sample space for each of the following random phenomena.

(a) For days of Tuesday and Wednesday, you record the sequence of up-days, down-days, and no-change days.

(b) For days of Tuesday and Wednesday, you record the number of up-days.

4.39 Colors of M&M'S. The colors of candies such as M&M'S are carefully chosen to match consumer preferences. The color of an M&M drawn at random from a bag has a probability distribution determined by the proportions of colors among all M&M'S of that type.

(a) Here is the distribution for plain M&M'S:

Color	Blue	Orange	Green	Brown	Yellow	Red
Probability	0.24	0.20	0.16	0.14	0.14	?

What must be the probability of drawing a red candy?

(b) What is the probability that a plain M&M is any of orange, green, or yellow?

4.40 Almond M&M'S. Exercise 4.39 gives the probabilities that an M&M candy is each of blue, orange, green, brown, yellow, and red. If "Almond" M&M'S are equally likely to be any of these colors, what is the probability of drawing a blue Almond M&M?

4.41 Legitimate probabilities? In each of the following situations, state whether the given assignment of probabilities to individual outcomes is legitimate—that is, satisfies the rules of probability. If not, give specific reasons for your answer.

(a) When a coin is spun, $P(H) = 0.55$ and $P(T) = 0.45$.

(b) When a coin is flipped twice, $P(HH) = 0.4$, $P(HT) = 0.4$, $P(TH) = 0.4$, and $P(TT) = 0.4$.

(c) Plain M&M'S have not always had the mixture of colors given in Exercise 4.39. In the past, there were no red candies and no blue candies. Tan had probability 0.10, and the other four colors had the same probabilities given in Exercise 4.39.

4.42 Who goes to Paris? Abby, Deborah, Sam, Tonya, and Roberto work in a firm's public relations office. Their employer must choose two of them to attend a conference in Paris. To avoid unfairness, the choice will be made by drawing two names from a hat. (This is an SRS of size 2.)

(a) Write down all possible choices of two of the five names. This is the sample space.

(b) The random drawing makes all choices equally likely. What is the probability of each choice?

(c) What is the probability that Tonya is chosen?

(d) What is the probability that neither of the two men (Sam and Roberto) is chosen?

4.43 Equally likely events. For each of the following situations, explain why you think that the events are equally likely or not. Explain your answers.

(a) The outcome of the next tennis match for Simona Halep is either a win or a loss. (You might want to check the Internet for information about this tennis player.)

(b) You draw a king or a two from a shuffled deck of 52 cards.

(c) You are observing turns at an intersection. You classify each turn as a right turn or a left turn.

(d) For college basketball games, you record the times that the home team wins and the number of times that the home team loses.

4.44 Using Internet sources. Internet sites often vanish or move, which means that references to them can't be followed. In fact, 13% of Internet sites referenced in major scientific journals are lost within two years after publication.

(a) If a paper contains seven Internet references, what is the probability that all seven are still good two years later?

(b) What specific assumptions did you make to calculate this probability?

4.45 Everyone gets audited. Wallen Accounting Services specializes in tax preparation for individual tax returns. Data collected from past records reveal that 9% of the returns prepared by Wallen have been selected for audit by the Internal Revenue Service. Today, Wallen has six new customers. Assume the chances of these six customers being audited are independent.

(a) What is the probability that all six new customers will be selected for audit?

(b) What is the probability that none of the six new customers will be selected for audit?

(c) What is the probability that exactly one of the six new customers will be selected for audit?

4.46 Hiring strategy. A chief executive officer (CEO) has resources to hire one vice president or three managers. He believes that he has probability 0.6 of successfully recruiting the vice president candidate and probability 0.8 of successfully recruiting each of the manager candidates. The three candidates for manager will make their decisions independently of each other. The CEO must successfully recruit either the vice president or all three managers to consider his hiring strategy a success. Which strategy should he choose?

4.47 A random walk on Wall Street? The "random walk" theory of securities prices holds that price movements in disjoint time periods are independent of each other. Suppose that we record only whether the price is up or down each year and that the probability that our portfolio rises in price in any one year is 0.65. (This probability is approximately correct for a portfolio containing equal dollar amounts of all common stocks listed on the New York Stock Exchange.)

(a) What is the probability that our portfolio goes up for three consecutive years?

(b) If you know that the portfolio has risen in price two years in a row, what probability do you assign to the event that it will go down next year?

(c) What is the probability that the portfolio's value moves in the same direction in both of the next two years?

4.48 The multiplication rule for independent events. The probability that a randomly selected person prefers the vehicle color white is 0.24. Can you apply the multiplication rule for independent events in the situations described in parts (a) and (b)? If your answer is Yes, apply the rule.

(a) Two people are chosen at random from the population. What is the probability that both prefer white?

(b) Two people who are sisters are chosen. What is the probability that both prefer white?

(c) Write a short summary about the multiplication rule for independent events using your answers to parts (a) and (b) to illustrate the basic idea.

4.49 What's wrong? In each of the following scenarios, there is something wrong. Describe what is wrong and give a reason for your answer.

(a) If two events are disjoint, we can multiply their probabilities to determine the probability that they will both occur.

(b) If the probability of A is 0.6 and the probability of B is 0.5, the probability of both A and B happening is 1.1.

(c) If the probability of A is 0.35, then the probability of the complement of A is -0.35.

4.50 What's wrong? In each of the following scenarios, there is something wrong. Describe what is wrong and give a reason for your answer.

(a) If the sample space consists of two outcomes, then each outcome has probability 0.5.

(b) If we select a digit at random, then the probability of selecting a 2 is 0.2.

(c) If the probability of A is 0.2, the probability of B is 0.3, and the probability of A and B is 0.5, then A and B are independent.

4.51 Playing the lottery. An instant lottery game gives you probability 0.02 of winning on any one play. Plays are independent of each other. If you play five times, what is the probability that you win at least once?

4.52 Axioms of probability. Show that any assignment of probabilities to events that obeys Rules 2 and 3 on page 185 automatically obeys the complement rule (Rule 4). This implies that a mathematical treatment of probability can start from just Rules 1, 2, and 3. These rules are sometimes called *axioms* of probability.

4.53 Independence of complements. Show that if events A and B obey the multiplication rule, $P(A \text{ and } B) = P(A)P(B)$, then A and the complement B^c of B also obey the multiplication rule, $P(A \text{ and } B^c) = P(A)P(B^c)$. That is, if events A and B are independent, then A and B^c are also independent. (*Hint:* Start by drawing a Venn diagram and noticing that the events "A and B" and "A and B^c" are disjoint.)

4.3 General Probability Rules

When you complete this section, you will be able to:
- Apply the general addition rule for union of two events.
- Convert a two-way table of counts to a table of joint and marginal probabilities.
- Compute conditional probabilities given joint and marginal probabilities.
- Apply the general multiplication rule.
- Use a tree diagram to find probabilities.
- Use Bayes' rule to find probabilities.
- Determine whether two events that both have positive probability are independent.

In the previous section, we encountered and used five basic rules of probability (page 194). To lay the groundwork for probability, we considered simplified settings such as dealing with only one or two events or making assumptions that the events are disjoint or independent. In this section, we learn more general laws that govern the assignment of probabilities. These more general laws of probability allow us to apply probability models to more complex random phenomena.

General addition rules

Probability has the property that if A and B are disjoint events, then $P(A \text{ or } B) = P(A) + P(B)$. But what if there are more than two events or the events are not disjoint? These circumstances are covered by more general addition rules for probability.

> **UNION**
>
> The **union** of any collection of events is the event that at least one of the collection occurs.

For two events A and B, the union is the event $\{A \text{ or } B\}$ that A or B or both occur. From the addition rule for two disjoint events, we can obtain rules for more general unions. Suppose first that we have several events—say A, B, and C—that are disjoint in pairs. That is, no two can occur simultaneously. The Venn diagram in Figure 4.5 illustrates three disjoint events. The addition rule for two disjoint events extends to the following law.

FIGURE 4.5 The addition rule for disjoint events: $P(A \text{ or } B \text{ or } C) = P(A) + P(B) + P(C)$ when events A, B, and C are disjoint.

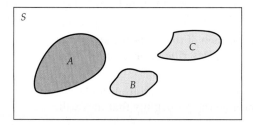

> **ADDITION RULE FOR DISJOINT EVENTS**
>
> If events A, B, and C are disjoint in the sense that no two have any outcomes in common, then
>
> $$P(A \text{ or } B \text{ or } C) = P(A) + P(B) + P(C)$$
>
> This rule extends to any number of disjoint events.

EXAMPLE 4.16

Disjoint Events Generate a random integer in the range of 10 to 59. What is the probability that the ten's digit will be odd? The event that the ten's digit is odd is the union of three disjoint events:

$$A = \{10, 11, \ldots, 19\}$$
$$B = \{30, 31, \ldots, 39\}$$
$$C = \{50, 51, \ldots, 59\}$$

In each of these events, there are 10 outcomes out of the 50 possible outcomes. This implies $P(A) = P(B) = P(C) = 0.2$. As a result, the probability that the ten's digit is odd is

$$P(A \text{ or } B \text{ or } C) = P(A) + P(B) + P(C)$$
$$= 0.2 + 0.2 + 0.2 = 0.6 \blacksquare$$

APPLY YOUR KNOWLEDGE

4.54 Probability that you roll an even number. Suppose you roll a die. What is the probability that your roll is an even number?

If events A and B are not disjoint, they can occur simultaneously. The probability of their union is then *less* than the sum of their probabilities. As Figure 4.6 suggests, the outcomes common to both are counted twice when we add probabilities, so we must subtract this probability once. Here is the addition rule for the union of any two events, disjoint or not.

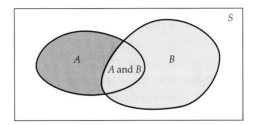

FIGURE 4.6 The general addition rule: $P(A \text{ or } B) = P(A) + P(B) - P(A \text{ and } B)$ for any events A and B.

GENERAL ADDITION RULE FOR UNIONS OF TWO EVENTS

For any two events A and B,

$$P(A \text{ or } B) = P(A) + P(B) - P(A \text{ and } B)$$

If A and B are disjoint, the event $\{A \text{ and } B\}$ that both occur has no outcomes in it. This *empty event* is the complement of the sample space S and must have probability 0. So the general addition rule includes Rule 3, the addition rule for disjoint events.

EXAMPLE 4.17

Making Partner Vanessa and Gabriel are anxiously awaiting word on whether they have been made partners of their law firm. Vanessa guesses that her probability of making partner is 0.7 and that Gabriel's probability is 0.5. (These are personal probabilities reflecting Vanessa's assessment of chance.) This assignment of probabilities does not give us enough information to compute the probability that at least one of the two is promoted. In particular, adding the individual probabilities of promotion gives the impossible result 1.2. If Vanessa also guesses that the probability that *both* she and Gabriel are made partners is 0.3, then by the general addition rule

$$P(\text{at least one is promoted}) = 0.7 + 0.5 - 0.3 = 0.9$$

The probability that *neither* is promoted is then 0.1 by the complement rule. ∎

APPLY YOUR KNOWLEDGE

4.55 Probability that the sum of dice is even or greater than 8. Suppose you roll a pair of dice and record the sum of the dice. What is the probability that the sum is even or greater than 8?

Venn diagrams are a great help in finding probabilities because you can just think of adding and subtracting areas. Figure 4.7 shows some events and their probabilities for Example 4.17. What is the probability that Vanessa is promoted and Gabriel is not?

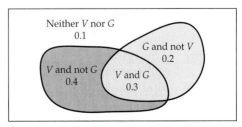

FIGURE 4.7 Venn diagram and probabilities for Example 4.17.

V = Vanessa is made partner
G = Gabriel is made partner

The Venn diagram shows that this is the probability that Vanessa is promoted minus the probability that both are promoted, $0.7 - 0.3 = 0.4$. Similarly, the probability that Gabriel is promoted and Vanessa is not is $0.5 - 0.3 = 0.2$. The four probabilities that appear in the figure add to 1 because they refer to four disjoint events that make up the entire sample space.

Two-Way Table of Counts to Probabilities

two-way table, p. 105

In Section 2.5 (page 105), we discussed several basic concepts and terminologies in the study of the relationship between categorical variables. A two-way table of counts was introduced as a means of organizing data about two categorical variables. In this two-way table setting, we discussed the ideas of distribution, such as a marginal distribution that gives the percents of counts of a row (column) variable relative to the total number of counts in the table. We now revisit the two-way table presentation while using the language of probabilities.

EXAMPLE 4.18

Two-Way Table of P2P Lending Counts Lending Club is the largest peer-to-peer (P2P) lending company in the United States. The company provides a fully online lending platform directly connecting borrower and investors. By cutting out banks from the process, borrowers typically get lower rates on loans. In applying for a loan, borrowers state the purpose of their loan request. The most common purposes are paying off credit cards (CC), debt consolidation (DC), and funding a home improvement project (HI). Among the information gathered is the borrower's dwelling status: home paid in full (Own), has a mortgage (Mortgage), and renting an apartment (Rent). Following are data on nearly 120,000 approved loans for the fourth quarter of 2017 organized as a two-way table of counts:[16]

	Dwelling status			
Purpose	Own	Mortgage	Rent	Total
Credit card (CC)	3265	11,396	10,645	25,306
Debt consolidation (DC)	7245	30,476	24,731	62,452
Home improvement (HI)	1432	6430	816	8678
Other	3358	8758	10,094	22,210
Total	15,300	57,060	46,286	118,646

Randomly choose a borrower from the 118,646 borrowers. What is the probability that a borrower sought a loan for a home improvement project? Because "choose at random" gives all 118,646 borrowers the same chance, the probability is just the following proportion:

$$P(\text{HI}) = \frac{8678}{118,646} = 0.07314$$

By using the total count in the HI row, this calculation does not assume anything about the borrower's dwelling status. Such a probability is referred to as a **marginal probability.** Each row (column) has an associated marginal probability computed by taking the marginal row (column) total divided by the total number of entries in the table. The collection of marginal probabilities for the rows (columns) defines the row (column) marginal distribution.

marginal probability

marginal distribution, p. 106

Now let's consider the probability that a randomly chosen borrower has a mortgage *and* sought a loan for a home improvement project. To find this probability, we grab the count of 6430 from the table cell associated with home improvement and mortgage. The desired probability is then the following proportion:

$$P(\text{Mortgage and HI}) = \frac{6430}{118,646} = 0.05419$$

joint probability This probability is known as a **joint probability**. In general, a joint probability is the probability of the co-occurrence of two or more events. ■

In the previous example, we illustrated the computation of only one marginal probability and only one joint probability. Let's now consider the entire two-way table.

EXAMPLE 4.19

P2P Lending Probabilities By doing similar computations on all the row and column counts and on all the cell counts, we obtain the following table of joint and marginal probabilities (rounded to five decimal places):

	Dwelling Status			
Purpose	Own	Mortgage	Rent	Probability
Credit card (CC)	0.02752	0.09605	0.08972	0.21329
Debt consolidation (DC)	0.06106	0.25686	0.20844	0.52637
Home improvement (HI)	0.01207	0.05419	0.00688	0.07314
Other	0.02830	0.07382	0.08508	0.18720
Probability	0.12896	0.48093	0.39012	1

joint probability table The inner table of cell probabilities is referred to as a **joint probability table.** These are the probabilities associated with events that involve *both* loan purpose and dwelling status. In Example 4.18, we found the marginal probability of a borrower seeking a home improvement project loan by taking the row count of 8678 and dividing it by the total count of 118,646. With a joint probability table, we can find the same marginal probability by adding the joint probabilities in that row:

$$P(\text{HI}) = 0.01207 + 0.05419 + 0.00688 = 0.07314$$ ■

APPLY YOUR KNOWLEDGE

4.56 YouTube users. Refer to Exercise 2.103 (page 114) to find a two-way table of YouTube usage versus age groups. Based on these respondents, construct a joint probability table along with the marginal probabilities.

4.57 P2P lending. Refer to Example 4.19.

(a) Determine the probability that a randomly chosen borrower took out a loan for either debt consolidation or home improvement.

(b) Determine the probability that a randomly chosen borrower either is a renter or took out a debt consolidation loan.

Conditional probability

In Examples 4.18 and 4.19, the probabilities were computed from the total count of 118,646 borrowers. We now consider how the probability we assign to an event can change (or not) if we know that some other event has occurred. A probability of an event given information that another event occurred is known as a **conditional probability**. The idea of conditional probabilities is key to many applications of probability.

EXAMPLE 4.20

Calculating a Conditional Probability In Examples 4.18 and 4.19, we found that the probability that a randomly chosen borrower sought a loan for a home improvement project is the marginal probability of 0.07314. Suppose now we are told that the borrower has a mortgage. From the two-way table shown in Example 4.18, we find that there are 57,060 borrowers who have a mortgage. Out of these 57,060 borrowers, 6430 borrowers are doing a home improvement project. As such, the probability that the borrower is doing a home improvement project, *given the information that the borrower has a mortgage*, is

$$P(\text{HI}|\text{Mortgage}) = \frac{6430}{57{,}060} = 0.11268 \ \blacksquare$$

The new notation of $P(B\,|\,A)$ for a conditional probability can be read as "the probability that B occurs given the information that A occurs" or, more simply, "the probability of B given A." As can be seen, the additional information of the borrower having a mortgage adjusted the probability that a borrower is doing a home improvement project from 0.07314 to 0.11268. As we will note at the end of this section, this inequality in probabilities is suggestive that the events are not independent.

APPLY YOUR KNOWLEDGE

4.58 Facebook users. Refer to Exercise 2.102 (page 114) to find a two-way table of Facebook usage versus age groups.

(a) For the survey respondents, what is the probability that a randomly chosen respondent is a Facebook user?

(b) What is the conditional probability a randomly chosen respondent is a Facebook user, given the person's age is between 30 and 49 years?

(c) What is the conditional probability a randomly chosen respondent is between 30 and 49 years old, given that the person is a Facebook user?

4.59 Two-way table. Consider the following incomplete table of probabilities:

	A_1	A_2	Probability
B_1			0.54
B_2		0.24	
Probability	0.37		

(a) Determine all the missing probability entries.

(b) Suppose the probability table was determined from a two-way table with a total count of 200. Based on part (a), reconstruct the two-way table.

(c) Determine $P(B_1\,|\,A_1)$ and $P(A_1\,|\,B_1)$.

Do not confuse the probabilities of $P(B\,|\,A)$ and $P(A \text{ and } B)$. They are generally not equal. Consider, for example, that the computed probability of 0.11268 from Example 4.20 is *not* the probability that a randomly selected borrower is doing a home improvement project and has a mortgage. Even though these probabilities are different, they are connected in a special way. Consider first the probability that a borrower has a mortgage. Then, for the mortgage population, find the probability that a borrower is doing a home improvement project. Multiply these two probabilities that are found in Examples 4.19 and 4.20:

$$P(\text{Mortgage and HI}) = P(\text{Mortgage}) \times P(\text{HI}|\text{Mortgage})$$
$$= (0.48093)(0.11268) = 0.05419$$

This probability can be found in the joint probability table of Example 4.19. We have just discovered the general multiplication rule of probability.

MULTIPLICATION RULE

The probability that both of two events A and B happen together can be found by

$$P(A \text{ and } B) = P(A)P(B \mid A)$$

Here $P(B \mid A)$ is the conditional probability that B occurs, given the information that A occurs.

EXAMPLE 4.21

Downloading Music from the Internet The multiplication rule is just common sense made formal. For example, suppose that 29% of Internet users download music files, and 67% of downloaders say they don't care if the music is copyrighted. So the percent of Internet users who download music (event A) *and* don't care about copyright (event B) is 67% of the 29% who download, or

$$(0.67)(0.29) = 0.1943 = 19.43\%$$

The multiplication rule expresses this as

$$\begin{aligned} P(A \text{ and } B) &= P(A) \times P(B \mid A) \\ &= (0.29)(0.67) = 0.1943 \end{aligned}$$ ∎

APPLY YOUR KNOWLEDGE

4.60 Focus group probabilities. A focus group of 11 consumers has been selected to view a new TV commercial. Even though all of the participants will provide their opinion, two members of the focus group will be randomly selected and asked to answer even more detailed questions about the commercial. The group contains six men and five women. What is the probability that the two participants chosen to answer questions will both be women? (*Hint:* Choosing two women is selecting a woman and *then* selecting another woman.)

4.61 Buying from Japan. Functional Robotics Corporation buys electrical controllers from a Japanese supplier. The company's treasurer thinks that there is probability 0.4 that the dollar will fall in value against the Japanese yen in the next month. The treasurer also believes that *if* the dollar falls, there is probability 0.8 that the supplier will demand renegotiation of the contract. What probability has the treasurer assigned to the event that the dollar falls and the supplier demands renegotiation?

If $P(A)$ and $P(A \text{ and } B)$ are given, we can rearrange the multiplication rule to produce a *definition* of the conditional probability $P(B \mid A)$ in terms of unconditional probabilities.

DEFINITION OF CONDITIONAL PROBABILITY

When $P(A) > 0$, the **conditional probability** of B given A is

$$P(B \mid A) = \frac{P(A \text{ and } B)}{P(A)}$$

Keep in mind the distinct roles in $P(B \mid A)$ of the event B whose probability we are computing and the event A that represents the information we are given. The conditional probability $P(B \mid A)$ makes no sense if the event A can never

occur, so we require that $P(A) > 0$ whenever we talk about $P(B \mid A)$. Finally, remember that $P(A \text{ and } B) = P(B \text{ and } A)$, which implies that the numerator of the conditional probability can be $P(A \text{ and } B)$ or $P(B \text{ and } A)$.

EXAMPLE 4.22

Personalized Advertising Awareness To what extent are Internet users aware that personalized advertising exists? To explore this question, a U.K. study of Internet users asked respondents, "If someone in the same country as you visits the same website or app at the same time as you, which one of these things applies to any advertising shown?"[17] Here are the joint probabilities classified by socioeconomic status and question response:

Socioeconomic Status	See Different Advertisements	See Same Advertisements	Unsure
AB	0.1856	0.0638	0.0406
C1	0.1682	0.0725	0.0493
C2	0.1029	0.0567	0.0504
DE	0.1008	0.0735	0.0357

The status codes represent "Upper or middle managerial occupations" (AB), "Supervisor or clerical occupations" (C1), "Skilled manual occupations" (C2), and "Unskilled occupations" (DE). Let's compute the probability that an Internet user is aware that websites/apps show different advertisements, given that the user is of DE status. We know that the probability that an Internet user is of DE status *and* is aware of personalized advertising is 0.1008 from the table of probabilities. But what we want here is a conditional probability, given that an Internet user is of DE status. Rather than asking about awareness of personalized advertising among all users, we restrict our attention to the subpopulation of Internet users of DE status. Let

DE = the Internet user is a DE status

PA = the Internet user is aware of personalized advertising

Our formula is

$$P(PA \mid DE) = \frac{P(DE \text{ and } PA)}{P(DE)}$$

We read $P(DE \text{ and } PA) = 0.1008$ from the table, as mentioned previously. What about $P(DE)$? This is the probability that a user is of DE status. Notice that there are three responses in our table that fit this status. To find the probability needed, we add the entries:

$$P(DE) = 0.1008 + 0.0735 + 0.0357 = 0.21$$

We are now ready to complete the calculation of the conditional probability:

$$P(PA \mid DE) = \frac{P(DE \text{ and } PA)}{P(DE)}$$
$$= \frac{0.1008}{0.21}$$
$$= 0.48$$

Given the Internet user has an unskilled occupation, the probability that an Internet user is aware of personalized advertising is 48%. ∎

Here is another way to give the information in the last sentence of this example: 48% of Internet users with unskilled occupations are aware of personalized users. Which way do you prefer?

APPLY YOUR KNOWLEDGE

4.62 Find the conditional probability. Refer to Example 4.22.

(a) What is the probability that an Internet user with an upper or middle managerial position (AB) is aware of personalized advertising? Compare this probability with the conditional probability found in Example 4.22. What insights do you gain from this comparison?

(b) What is the probability that an Internet user has an unskilled occupation (DE), given that the user is aware of personalized advertising? Explain in your own words the difference between this calculation and the one that we did in Example 4.22.

4.63 What rule did we use? In Example 4.22, we calculated $P(DE)$. What rule did we use for this calculation? Explain why this rule applies in this setting.

General multiplication rule

The definition of conditional probability reminds us that, in principle, all probabilities—including conditional probabilities—can be found from the assignment of probabilities to events that describe random phenomena. More often, however, conditional probabilities are part of the information given to us in a probability model, and the multiplication rule is used to compute $P(A \text{ and } B)$. This rule extends to more than two events.

The union of a collection of events is the event that *any* of them occur. The corresponding term for the event that *all* of them occur is *intersection*.

> **INTERSECTION**
>
> The **intersection** of any collection of events is the event that *all* the events occur.

To extend the multiplication rule to the probability that all of several events occur, the key is to condition each event on the occurrence of *all* the preceding events. For example, the intersection of three events A, B, and C has probability

$$P(A \text{ and } B \text{ and } C) = P(A)P(B \mid A)P(C \mid A \text{ and } B)$$

Notice how the multiplication rule builds upon the preceding events. First, determine the probability of A; then, find the probability of B *given* that A occurred; and finally, find the probability of C *given* that both A and B occurred. The desired probability is the multiplication of these probabilities.

EXAMPLE 4.23

Career in Big Business: NFL Worldwide, the sports industry has become synonymous with big business. It has been estimated by the United Nations that sports account for nearly 3% of global economic activity. The most profitable sport in the world is professional football under the management of the National Football League (NFL).[18] With multimillion-dollar signing contracts, the economic appeal of pursuing a career as a professional sports athlete is unquestionably strong. But what are the realities? Only 6.8% of high school football players go on to play at the college level. Of these, only 1.5% will play in the NFL.[19] About 40% of the NFL players have a career of more than three years. Define these events for the sport of football:

$$\text{Coll} = \{\text{competes in college}\}$$
$$\text{NFL} = \{\text{competes in the NFL}\}$$
$$\text{ThreePlus} = \{\text{has an NFL career longer than 3 years}\}$$

What is the probability that a high school football player competes in college and then goes on to have an NFL career of more than three years? We know that

$$P(\text{Coll}) = 0.068$$
$$P(\text{NFL} \mid \text{Coll}) = 0.015$$
$$P(\text{ThreePlus} \mid \text{Coll and NFL}) = 0.4$$

The probability we want is, therefore,

$$P(\text{Coll and NFL and ThreePlus}) = P(\text{Coll})P(\text{NFL} \mid \text{Coll})P(\text{ThreePlus} \mid \text{Coll and NFL})$$
$$= 0.068 \times 0.015 \times 0.40 = 0.00041$$

Only about four of every 10,000 high school football players can expect to compete in college and have an NFL career of more than three years. High school football players would be wise to concentrate on studies rather than unrealistic hopes of fortune from pro football. ■

Tree diagrams

In Example 4.23, we investigated the likelihood of a high school football player going on to play collegiately and then have an NFL career of more than three years. The sports of football and basketball are unique in that players are prohibited from going straight into professional ranks from high school. Baseball, however, has no such restriction. Some baseball players might make the professional ranks through the college route, while others might ultimately make it coming out of high school, albeit often with a journey through the minor leagues.

Determining the probability that a baseball player becomes a professional player involves more elaborate calculations than the football scenario. We illustrate with our next example how the use of a **tree diagram** can help organize our thinking.

tree diagram

EXAMPLE 4.24

How Many Go to MLB? For baseball, 7.1% of high school players go on to play at the college level. Of these, 9.1% will play in Major League Baseball (MLB).[20] Define the event A as playing professionally in the MLB. The probability of a high school player ultimately playing professionally is $P(A)$. To find $P(A)$, consider the tree diagram shown in Figure 4.8.

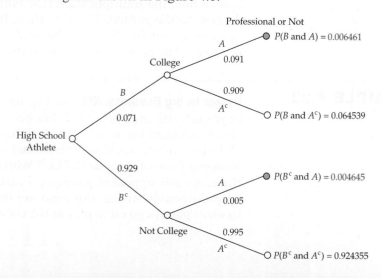

FIGURE 4.8 Tree diagram and probabilities for Example 4.24.

Each segment in the tree is one stage of the problem. Each complete branch shows a path that a player can take. Starting at the left, high school baseball players either do (B) or do not (B^c) compete in college. We know that the probability of competing in college is $P(B) = 0.071$, so the probability of not competing is $P(B^c) = 0.929$ by the complement rule. These probabilities mark the leftmost branches in the tree.

Thereafter, the probability written on each segment is the *conditional probability* that a player follows that segment given that he has reached the point from which it branches. Conditional on competing in college, the probability of playing in MLB is $P(A \mid B) = 0.091$. So the conditional probability of *not* playing in MLB is

$$P(A^c \mid B) = 1 - P(A \mid B) = 1 - 0.091 = 0.909$$

These conditional probabilities mark the paths branching out from B in Figure 4.8.

The lower half of the tree diagram describes players who do not compete in college (B^c). For baseball, in years past, the majority of destined professional players did not take the route through college. However, nowadays it is relatively unusual for players to go straight from high school to MLB.[21] It is reported that the conditional probability that a high school athlete reaches MLB, given that he does not compete in college, is $P(A \mid B^c) = 0.005$. We can now mark the two paths branching from B^c in Figure 4.8.

The probability of reaching A through college (top half of the tree) is

$$P(B \text{ and } A) = P(B)P(A \mid B)$$
$$= 0.071 \times 0.091 = 0.006461$$

The probability of reaching B without college is

$$P(B^c \text{ and } A) = P(B^c)P(A \mid B^c)$$
$$= 0.929 \times 0.005 = 0.004645$$

Because these two paths to A (MLB play) are disjoint, $P(A)$ is simply the sum of their probabilities by the addition rule. Combining all these facts, the final result is then

$$P(A) = P(B \text{ and } A) + P(B^c \text{ and } A)$$
$$= P(B)P(A \mid B) + P(B^c)P(A \mid B^c)$$
$$= 0.006461 + 0.004645$$
$$= 0.0111$$

About eleven high school baseball players out of 1000 will play professionally. Even though this probability is quite small, it is comparatively much greater than the chances of making it to the professional ranks in basketball and football. ∎

Tree diagrams combine the addition and multiplication rules. Given the paths to the branch ends are disjoint, the addition rule for disjoint events (page 198) allows us to simply add the probabilities of these events. The multiplication rule says that the probability of reaching the end of any complete branch is the product of the probabilities written on the segments leading to the end of that branch. Also note that since the disjoint events at the branch ends represent *all* possible outcomes in the sample space, the sum of their probabilities must be 1.

APPLY YOUR KNOWLEDGE

4.64 MLB. Refer to Figure 4.8 to answer the following questions:

(a) Confirm that 1 is the sum of the probabilities for the branch ends.

(b) What values on the tree need to be added to determine the probability that a high school athlete will *not* make it to the MLB?

Bayes' rule

There is another kind of probability question that we might ask in the context of studies of athletes. Our earlier calculations look forward toward professional sports as the final stage of an athlete's career. Now let's concentrate on professional athletes and look back at their earlier careers. In terms of conditional probabilities, we utilized $P(A \mid B)$ in Example 4.24. Now we are interested in reversing the conditioning to consider $P(B \mid A)$.

EXAMPLE 4.25

Professional Athletes' Pasts What proportion of professional athletes competed in college? In the notation of Example 4.24, this is the conditional probability $P(B \mid A)$. Before we compute this probability, let's take stock of a few facts. First, the multiplication rule tells us

$$P(B \text{ and } A) = P(B)P(A \mid B)$$
$$P(B^c \text{ and } A) = P(B^c)P(A \mid B^c)$$

We know the probabilities $P(B)$ and $P(B^c)$ that a high school baseball player does and does not compete in college. We also know the conditional probabilities $P(A \mid B)$ and $P(A \mid B^c)$ that a player from each group reaches the MLB ranks. Example 4.24 shows how to use this information to calculate $P(A)$. The method can be summarized in a single expression that adds the probabilities of the two paths to A in the tree diagram:

$$P(A) = P(B)P(A \mid B) + P(B^c)P(A \mid B^c)$$

Combining these facts, we can now make the following computation:

$$P(B \mid A) = \frac{P(B \text{ and } A)}{P(A)}$$
$$= \frac{P(B)P(A \mid B)}{P(B)P(A \mid B) + P(B^c)P(A \mid B^c)}$$
$$= \frac{0.071 \times 0.091}{0.071 \times 0.091 + 0.929 \times 0.005}$$
$$= 0.582$$

About 58% of MLB players competed in college. ■

In calculating the "reverse" conditional probability of Example 4.25, we had two disjoint events in B and B^c whose probabilities add to exactly 1. We also had the conditional probabilities of event A given each of the disjoint events. More generally, in some applications we may have more than two disjoint events whose probabilities add up to 1. For example, suppose the sample space is divided into three disjoint events B_1, B_2, and B_3 as depicted in Figure 4.9(a). Notice that event A (shaded area event) is made up by the disjoint pieces of A that overlap the B's. As such, we have

$$P(A) = P(B_1 \text{ and } A) + P(B_2 \text{ and } A) + P(B_3 \text{ and } A)$$
$$= P(B_1)P(A \mid B_1) + P(B_2)P(A \mid B_2) + P(B_3)P(A \mid B_3)$$

The first line of the equation states that if A is divided into three disjoint pieces, the $P(A)$ is simply the sum of the probabilities of those pieces; remember that this is the application of the addition rule for disjoint events. The second line of the equation employs the multiplication rule to rewrite the probabilities of the pieces in terms of conditional probabilities. This second line is often referred to as the **law of total probability.** As an alternative visualization, Figure 4.9(b) shows the tree representation where event A is made

FIGURE 4.9 Event A divided into three disjoint pieces as shown by (a) a Venn diagram (event A is the shaded area) and (b) a tree diagram.

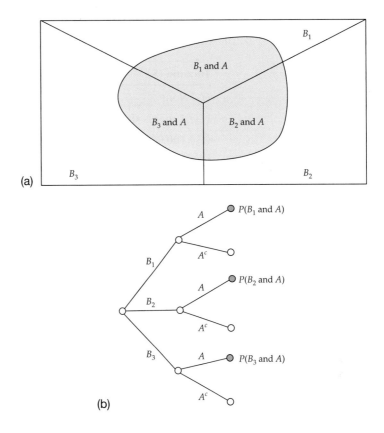

up of the three highlighted branches. $P(A)$ is then the sum of the probabilities of these highlighted branches.

The law of total probability can be extended into any number of events as long as they are disjoint and cover all of the sample space. Put in general notation, we have another probability law.

BAYES' RULE

Suppose that $B_1, B_2, \ldots, B_k$ are disjoint events that divide up the whole sample space so that their probabilities add to exactly 1. That is, any outcome is in exactly one of these events. Then, if A is any other event whose probability is not 0,

$$P(B_i \mid A) = \frac{P(B_i \text{ and } A)}{P(A)}$$
$$= \frac{P(A \mid B_i)P(B_i)}{P(A \mid B_1)P(B_1) + P(A \mid B_2)P(B_2) + \cdots + P(A \mid B_k)P(B_k)}$$

The numerator in Bayes' rule is always one of the terms in the sum that makes up the denominator. This rule is named after Reverend Thomas Bayes, who wrestled with the idea of calculating an "inverse probability" based on outcomes from some event A back to the B_i in an essay published in 1763 after his death. Our next example utilizes Bayes' rule with several disjoint events.

EXAMPLE 4.26

IRS Audits The IRS audits some tax returns in greater detail to verify that the tax reported is correct. The audit rates vary depending on the size of the individual's adjusted gross income. In 2016, the IRS reported the percentages of total returns by adjusted gross income categories and the percentages of returns that were audited within a given income category:[22]

Income ($)	Returns Filed (%)	Audit (%)
(I_1) None	1.70	3.25
(I_2) 1 to less than 100,000	82.38	0.62
(I_3) 100,000 to less than 200,000	11.72	0.62
(I_4) 200,000 to less than 500,000	3.38	1.01
(I_5) 500,000 to less than 1 million	0.54	2.06
(I_6) 1 million to less than 5 million	0.25	4.60
(I_7) 5 million or more	0.03	13.24

Two facts can be drawn from this table. First, the income categories ($I_1, I_2, I_3, I_4, I_5, I_6, I_7$) are disjoint and cover all possible incomes. Thus, the percentages of returns filed add up to 100%. Second, the audit values represent conditional probabilities. For example, 3.25% of individuals with no reported income have been audited. Notionally, this implies $P(\text{Audited} \mid I_1) = 0.0325$.

Suppose a return is randomly selected and it was audited. What is the probability that the individual's adjusted gross income falls in the range of $1 to $100,000 ($I_2$)? Because income categories are disjoint and their probabilities add to 1, we can employ Bayes' rule. It will be convenient to present the calculations of the terms in Bayes' rule as a table.

Income	$P(\text{Income})$	$P(\text{Audit} \mid \text{Income})$	$P(\text{Audit} \mid \text{Income})P(\text{Income})$
I_1	0.0170	0.0325	$(0.0325)(0.0170) = 0.00055$
I_2	0.8238	0.0062	$(0.0062)(0.8238) = 0.00511$
I_3	0.1172	0.0062	$(0.0062)(0.1172) = 0.00073$
I_4	0.0338	0.0101	$(0.0101)(0.0338) = 0.00034$
I_5	0.0054	0.0206	$(0.0206)(0.0054) = 0.00011$
I_6	0.0025	0.0460	$(0.0460)(0.0025) = 0.00012$
I_7	0.0003	0.1324	$(0.1324)(0.0003) = 0.00004$

Here is the computation of the desired probability using Bayes' rule along with the preceding computed values:

$$P(I_2 \mid \text{Audit}) = \frac{P(\text{Audit} \mid I_2)P(I_2)}{P(\text{Audit})}$$

$$= \frac{0.00511}{0.00055 + 0.00511 + 0.00073 + 0.00034 + 0.00011 + 0.00012 + 0.00004}$$

$$= \frac{0.00511}{0.007} = 0.73$$

prior probability

posterior probability

Without any information on the return being audited, $P(I_2) = 0.8238$. This probability can be referred to as a **prior probability.** Bayes' rule allows us to adjust this probability based on the information of the return being audited to come up with $P(I_2 \mid \text{Audit}) = 0.73$. This adjusted probability is referred to as a **posterior probability.** In this case, the posterior probability is less than the prior probability. Intuitively, this downward adjustment is due to the fact that the probability of an audit for category I_2 (0.0062) is less than the overall probability of an audit (0.007) as found above in the denominator of our Bayes' computation. ■

4.3 General Probability Rules

It turns out that the concepts of conditional probability and Bayes' rule serve as a basis for a classification method used in predictive analytics. A classification method attempts to identify which category from a set of categories a new observation belongs to. An example is a spam filter that designates a given email as "spam" or "not spam" based on features of the email message. One classification method used in practice is known as a naïve Bayes classifier. Given certain simplifying assumptions about reality, a naïve Bayes classifier is an application of Bayes' rule to estimate the posterior probability that a new observation belongs to a given category. These posterior probabilities are then used to ultimately classify the new observation. There are many other classification methods, including logistic regression (the focus of Chapter 18).

APPLY YOUR KNOWLEDGE

4.65 IRS audits. Refer to Example 4.26.

(a) Without speaking in terms of conditional probability, explain in words what the probability of 0.00511 represents.

(b) Given the information that a return was audited, what is the probability that the individual's income is in the range of $100,000 to $200,000?

Independence again

The conditional probability $P(B \mid A)$ is generally not equal to the unconditional probability $P(B)$. The difference arises because the occurrence of event A generally gives us some additional information about whether event B occurs. If knowing that A occurs gives no additional information about B, then A and B are independent events. The formal definition of independence is expressed in terms of conditional probability.

> **INDEPENDENT EVENTS**
>
> Two events A and B that both have positive probability are **independent** if
> $$P(B \mid A) = P(B)$$

The alternative term for unconditional probability is marginal probability (page 200). Thus, with independence, the preceding definition states that the conditional probability equals the marginal probability.

With this formal definition of independence, we now see that the multiplication rule for independent events, $P(A \text{ and } B) = P(A)P(B)$, provided in Section 4.2 (page 191) is a special case of the general multiplication rule, $P(A \text{ and } B) = P(A)P(B \mid A)$ (page 203). Recalling that $P(A \text{ and } B)$ is called a joint probability (page 201), the multiplication rule for independence states that the joint probability is the product of the marginal probabilities.

EXAMPLE 4.27

P2P Lending In Example 4.18 (pages 200–201) and Example 4.19 (page 201), we found that the probability a borrower does a home improvement project is $P(\text{HI}) = 0.07314$. However, in Example 4.20 (page 202), given the information that the borrower has a mortgage, we found that the probability a borrower does a home improvement project is $P(\text{HI} \mid \text{Mortgage}) = 0.11268$. This large difference in probabilities (marginal versus conditional) is suggestive of the lack of independence between the events. This is just one comparison between marginal and conditional probabilities extracted from the table of Example 4.18.

mosaic plot, p. 110

We can, of course, compute all the other probabilities and compare the values. However, a nice alternative is to visualize the conditional probabilities in relationship to the marginal probabilities with a mosaic plot.

Figure 4.10 shows a JMP-produced mosaic plot displaying the conditional and marginal probabilities. The marginal probabilities for loan purpose are displayed to the right as a single strip with different colors by loan purpose. The lengths of the colors are associated with the probability values. To the left of the marginal probability plot, we find the conditional probabilities displayed in the three columns each labeled by dwelling status. It appears that the conditional probabilities for paying off credit cards (CC) are nearly equal to each other and, in turn, nearly equal to the marginal probability of CC. However, the conditional probabilities for "other" loans and home improvement loans vary greater across dwelling statuses and differ substantially from their respective marginal probabilities. As such, the mosaic plot gives a strong indication that loan purpose and dwelling status are not independent events. ∎

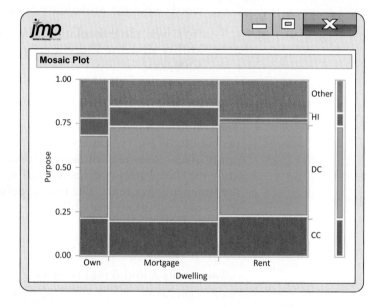

FIGURE 4.10 Mosaic plot from JMP displaying the conditional and marginal probabilities between loan purpose and dwelling status, Example 4.27.

Example 4.27 demonstrates how a mosaic plot provides a nice visualization technique for informally assessing independence versus lack of independence between two categorical variables. In Chapter 11, we will supplement the mosaic plot with formal testing procedures dedicated to the question of whether data on two categorical variables suggest their independence.

SECTION 4.3 SUMMARY

- The **complement** A^c of an event A contains all outcomes that are not in A. The **union** {A or B} of events A and B contains all outcomes in A, in B, and in both A and B. The **intersection** {A and B} contains all outcomes that are in both A and B, but not outcomes in A alone or B alone.

- The **joint probability** $P(A \text{ and } B)$ is the probability of the co-occurrence of events A and B.

- The **conditional probability** $P(B \mid A)$ of an event B, given an event A, is defined by

$$P(B \mid A) = \frac{P(A \text{ and } B)}{P(A)}$$

when $P(A) > 0$. In practice, conditional probabilities are most often found from directly available information.

- The essential general rules of elementary probability are

    Legitimate values: $0 \le P(A) \le 1$ for any event A

    Total probability 1: $P(S) = 1$

    Complement rule: $P(A^c) = 1 - P(A)$

    Addition rule: $P(A \text{ or } B) = P(A) + P(B) - P(A \text{ and } B)$

    Multiplication rule: $P(A \text{ and } B) = P(A)P(B \mid A)$

- If A and B are **disjoint,** then $P(A \text{ and } B) = 0$. The general addition rule for unions then becomes the special addition rule, $P(A \text{ or } B) = P(A) + P(B)$.

- If events A, B, and C are disjoint in the sense that no two have any outcomes in common, then

$$P(A \text{ or } B \text{ or } C) = P(A) + P(B) + P(C)$$

This rule extends to any number of disjoint events.

- A and B are **independent** when $P(B \mid A) = P(B)$; that is, these events are independent when the conditional probability equals the **marginal probability.** The multiplication rule for intersections then becomes $P(A \text{ and } B) = P(A)P(B)$.

- In problems with several stages, draw a **tree diagram** to organize use of the multiplication and addition rules.

- Suppose that $B_1, B_2, \ldots, B_k$ are disjoint events that divide up the whole sample space so that their probabilities add to exactly 1—that is, any outcome is in exactly one of these events. Then, if A is any other event whose probability is not 0,

$$P(B_i \mid A) = \frac{P(A \mid B_i)P(B_i)}{P(A \mid B_1)P(B_1) + P(A \mid B_2)P(B_2) + \cdots + P(A \mid B_k)P(B_k)}$$

Here, the denominator expression of $P(A)$ is called the **law of total probability.** In the context of **Bayes' rule,** $P(B_i)$ is a **prior probability** while the adjusted probability of $P(B_i \mid A)$ is a **posterior probability.**

SECTION 4.3 EXERCISES

For Exercise 4.54, see page 198; for 4.55, see page 199; for 4.56 and 4.57, see page 201; for 4.58 and 4.59, see page 202; for 4.60 and 4.61, see page 203; for 4.62 and 4.63, see page 205; for 4.64, see page 207; and for 4.65, see page 211.

4.66 Find and explain some probabilities.

(a) Can we have an event A that has negative probability? Explain your answer.

(b) Suppose $P(A) = 0.1$ and $P(B) = 0.3$. Explain what it means for A and B to be disjoint. Assuming that they are disjoint, find the probability that A or B occurs.

(c) Explain in your own words the meaning of the rule $P(S) = 1$.

(d) Consider an event A. What is the name for the event that A does not occur? If $P(A) = 0.25$, what is the probability that A does not occur?

(e) Suppose that A and B are independent and that $P(A) = 0.7$ and $P(B) = 0.4$. Explain the meaning of the event $\{A \text{ and } B\}$, and find its probability.

4.67 Unions.

(a) Assume that $P(A) = 0.3$, $P(B) = 0.2$, and $P(C) = 0.1$. If the events A, B, and C are disjoint, find the probability that the union of these events occurs.

(b) Draw a Venn diagram to illustrate your answer to part (a).

(c) Find the probability of the complement of the union of A, B, and C.

4.68 Conditional probabilities. Suppose that $P(A) = 0.6$, $P(B) = 0.3$, and $P(B \mid A) = 0.2$.

(a) Find the probability that both A and B occur.

(b) Use a Venn diagram to explain your calculation.

(c) What is the probability of the event that B occurs and A does not?

4.69 Find the probabilities. Suppose that the probability that A occurs is 0.6 and the probability that A and B occur is 0.4.

(a) Find the probability that B occurs given that A occurs.

(b) Illustrate your calculations in part (a) using a Venn diagram.

4.70 What's wrong? In each of the following scenarios, there is something wrong. Describe what is wrong and give a reason for your answer.

(a) $P(A \text{ or } B)$ is always equal to the sum of $P(A)$ and $P(B)$.

(b) The probability of an event minus the probability of its complement is always equal to 1.

(c) Two events are disjoint if $P(B \mid A) = P(B)$.

4.71 Attendance at two-year and four-year colleges. In a large national population of college students, 72% attend four-year institutions and the rest attend two-year institutions. Males make up 45% of the students in the four-year institutions and 47% of the students in the two-year institutions.

(a) Find the four probabilities for each combination of gender and type of institution in the following table. Be sure that your probabilities sum to 1.

	Men	Women
Four-year institution		
Two-year institution		

(b) Consider randomly selecting a female student from this population. What is the probability that she attends a four-year institution?

4.72 Draw a tree diagram. Refer to the previous exercise. Draw a tree diagram to illustrate the probabilities in a situation in which you first identify the type of institution attended and then identify the gender of the student. Label the tree by school type or gender and label segments by appropriate probabilities.

4.73 Draw a different tree diagram for the same setting. Refer to the previous two exercises. Draw a tree diagram to illustrate the probabilities in a situation in which you first identify the gender of the student and then identify the type of institution attended. Label the tree by school type or gender and label segments by appropriate probabilities.

(a) Explain why the segment probabilities in this tree diagram are different from those that you used in the previous exercise.

(b) Explain why the probabilities of reaching the ends of the branches from this tree are the same as reaching the ends of the branches from the previous exercise.

4.74 Education and income. Call a household prosperous if its income exceeds $100,000. Call the household educated if one of the householders completed college. Select an American household at random, and let A be the event that the selected household is prosperous and B be the event that it is educated. According to the Current Population Survey, $P(A) = 0.138$, $P(B) = 0.261$, and the probability that a household is both prosperous and educated is $P(A \text{ and } B) = 0.082$. What is the probability $P(A \text{ or } B)$ that the household selected is either prosperous or educated?

4.75 Find a conditional probability. In the setting of the previous exercise, what is the conditional probability that a household is prosperous, given that it is educated? Explain why your result shows that events A and B are not independent.

4.76 Draw a Venn diagram. Draw a Venn diagram that shows the relation between the events A and B in Exercise 4.74. Indicate each of the following events on your diagram and use the information in Exercise 4.74 to calculate the probability of each event. Finally, describe in words what each event is.

(a) $\{A \text{ and } B\}$

(b) $\{A^c \text{ and } B\}$

(c) $\{A \text{ and } B^c\}$

(d) $\{A^c \text{ and } B^c\}$

4.77 Sales of cars and light trucks. Motor vehicles sold to individuals are classified as either cars or light trucks (including SUVs) and as either domestic or imported. In a recent year, 69% of vehicles sold were light trucks, 78% were domestic, and 55% were domestic light trucks. Let A be the event that a vehicle is a car and B be the event that it is imported. Write each of the following events in set notation and give its probability.

(a) The vehicle is a light truck.

(b) The vehicle is an imported car.

4.78 Conditional probabilities and independence. Using the information in Exercise 4.77, answer these questions.

(a) Given that a vehicle is imported, what is the conditional probability that it is a light truck?

(b) Are the events "vehicle is a light truck" and "vehicle is imported" independent? Justify your answer.

4.79 Driverless cars. Do men and women differ in their attitudes toward the development of driverless cars? A Pew Research study of a sample of men and women revealed the following results on being enthusiastic (somewhat or very) versus worried (somewhat or very):[23]

	Sex	
Response	Male	Female
Enthusiastic	941	710
Worried	1105	1379

(a) Give an example of a joint event.

(b) Is the probability that a randomly chosen respondent is enthusiastic about the development of driverless cars a conditional probability? Give an explanation for your answer.

4.80 Driverless cars. For a randomly selected respondent, use the table from the previous exercise to find the following:

(a) What is the probability that the respondent is enthusiastic?

(b) What is the probability that the respondent is female *and* enthusiastic about the development of driverless cars?

(c) What is the probability that the respondent is female *or* enthusiastic about the development of driverless cars?

(d) Given that the respondent is female, what is the probability that the respondent is enthusiastic about the development of driverless cars? Compare this probability with part (a) and explain what can be inferred.

4.81 P2P lending. In Example 4.27 (pages 211–212), comparison of $P(\text{HI})$ and $P(\text{HI} \mid \text{Mortgage})$ was suggestive of lack of independence between the events. Consider another comparison. Refer to Example 4.19 (page 201) and obtain the values for $P(\text{HI})$, $P(\text{Mortgage})$, and $P(\text{Mortgage and HI})$. Without computing a conditional probability, how can you use these three values to check for independence? (*Hint:* Refer to the multiplication rule for independent events as stated in Section 4.2.)

4.82 Loan officer decision. A loan officer is considering a loan request from a customer of the bank. Based on data collected from the bank's records over many years, there is an 8% chance that a customer who has overdrawn an account will default on the loan. However, there is only a 0.6% chance that a customer who has never overdrawn an account will default on the loan. Based on the customer's credit history, the loan officer believes there is a 40% chance that this customer will overdraw his account. Let D be the event that the customer defaults on the loan, and let O be the event that the customer overdraws his account.

(a) Express the three probabilities given in the problem in the notation of probability and conditional probability.

(b) What is the probability that the customer will default on the loan?

4.83 Loan officer decision. Considering the information provided in the previous exercise, calculate $P(O \mid D)$. Show your work. Also, express this probability in words in the context of the loan officer's decision. If new information about the customer becomes available before the loan officer makes her decision and if this information indicates that there is only a 25% chance that this customer will overdraw his account, rather than a 40% chance, how does this change $P(O \mid D)$?

4.84 High school football players. Using the information in Example 4.23 (pages 205–206), determine the proportion of high school football players expected to play professionally in the NFL.

4.85 High school baseball players. It is estimated that 56% of MLB players have careers of three or more years. Using the information in Example 4.24 (pages 206–207), determine the proportion of high school players expected to play three or more years in MLB.

4.86 Telemarketing. A telemarketing company calls telephone numbers chosen at random. It finds that 70% of calls are not completed (the party does not answer or refuses to talk), that 20% result in talking to a woman, and that 10% result in talking to a man. After that point, 30% of the women and 20% of the men actually buy something. What percent of calls result in a sale? (Draw a tree diagram.)

4.87 Preparing for the GMAT. A company that offers courses to prepare would-be MBA students for the GMAT examination finds that 40% of its customers are currently undergraduate students and 60% are college graduates. After completing the course, 50% of the undergraduates and 70% of the graduates achieve scores of at least 600 on the GMAT. Use a tree diagram to organize this information.

(a) What percent of customers are undergraduates *and* score at least 600? What percent of customers are graduates *and* score at least 600?

(b) What percent of all customers score at least 600 on the GMAT?

4.88 Sales to women. In the setting of Exercise 4.86, what percent of sales are made to women? (Write this as a conditional probability.)

4.89 Success on the GMAT. In the setting of Exercise 4.87, what percent of the customers who score at least 600 on the GMAT are undergraduates? (Write this as a conditional probability.)

4.90 Successful bids. Consolidated Builders has bid on two large construction projects. The company president believes that the probability of winning the first contract (event A) is 0.6, that the probability of winning the second (event B) is 0.5, and that the probability of winning both jobs (event $\{A \text{ and } B\}$) is 0.3. What is the probability of the event $\{A \text{ or } B\}$—that Consolidated will win at least one of the jobs?

4.91 Independence? In the setting of the previous exercise, are events A and B independent? Do a calculation that proves your answer.

4.92 Successful bids, continued. Draw a Venn diagram that illustrates the relation between events A and B in Exercise 4.90. Write each of the following events in terms of A, B, A^c, and B^c. Indicate the events on your diagram and use the information in Exercise 4.90 to calculate the probability of each.

(a) Consolidated wins both jobs.

(b) Consolidated wins the first job, but not the second.

(c) Consolidated does not win the first job, but does win the second.

(d) Consolidated does not win either job.

4.93 Credit card defaults. The credit manager for a local department store is interested in customers who default (ultimately failed to pay their entire balance). Of those customers who default, 88% were late (by a week or more) with two or more monthly payments. This prompts the manager to suggest that future credit be denied to any customer who is late with two monthly payments. Further study shows that 3% of all credit customers default on their payments and 40% of those who have not defaulted have had at least two late monthly payments in the past.

(a) What is the probability that a customer who has two or more late payments will default?

(b) Under the credit manager's policy, in a group of 100 customers who have their future credit denied, how many would we expect *not* to default on their payments?

(c) Does the credit manager's policy seem reasonable? Explain your response.

4.94 Podcasting. The medium of podcasting is evolving and engaging a growing U.S. audience. As such, advertisers are interested in understanding the demographics of the podcast audience. Based on a Nielsen study of smartphone usage along with smartphone demographic data, it can be determined that 16% of IOS-based smartphone owners (I) listen to podcasts, 10.2% of Android-based smartphone owners (A) listen to podcasts, and 7.7% of other operating system–based smartphone owners (O) listen to podcasts. In the United States, the proportions of IOS-based phone owners, Android phone owners, and other phone owners are 0.42, 0.51, and 0.07, respectively.[24]

(a) Use Bayes' rule to find the probability that a randomly selected smartphone podcast listener owns an IOS-based smartphone.

(b) Use Bayes' rule to find the probability that a randomly selected smartphone podcast listener owns an Android-based smartphone.

(c) What is the value of the denominator in your Bayes' computation for parts (a) and (b)? In terms of the context of this exercise, what is the interpretation of this denominator value?

4.95 Supplier quality. A manufacturer of an assembled product uses three different suppliers for a particular component. By means of supplier audits, the manufacturer estimates the following percentages of defective parts by supplier:

Supplier	1	2	3
Percent defective	0.4%	0.3%	0.6%

Shipments from the suppliers are continually streaming to the manufacturer in small lots from each of the suppliers. As a result, the inventory of parts held by the manufacturer is a mix of parts representing the relative supplier rate from each supplier. In the current inventory, there are 423 parts from Supplier A, 367 parts from Supplier B, and 205 parts from Supplier C. Suppose a part is randomly chosen from inventory. Define "S1" as the event the part came from Supplier 1, "S2" as the event the part came from Supplier 2, and "S3" as the event the part came from Supplier 3. Also, define "D" as the event the part is defective.

(a) Based on the inventory mix, determine $P(S1)$, $P(S2)$, and $P(S3)$.

(b) If the part is found to be defective, use Bayes' rule to determine the probability that it came from Supplier 3.

CHAPTER 4 REVIEW EXERCISES

4.96 Why not? Suppose that $P(A) = 0.4$. Explain why $P(A \text{ and } B)$ cannot be 0.6.

4.97 Using probability rules. Let $P(A) = 0.8$, $P(B) = 0.4$, and $P(C) = 0.1$.

(a) Explain why it is not possible that events A and B can be disjoint.

(b) What is the smallest possible value for $P(A \text{ and } B)$? What is the largest possible value for $P(A \text{ and } B)$? It might be helpful to draw a Venn diagram.

(c) If events A and C are independent, what is $P(A \text{ or } C)$?

4.98 Roll a pair of dice two times. Consider rolling a pair of fair dice two times. For a given roll, consider the total on the up-faces. For each of the following pairs of events, tell whether they are disjoint, independent, or neither.

(a) $A = 2$ on the first roll, $B = 8$ or more on the first roll.

(b) $A = 2$ on the first roll, $B = 8$ or more on the second roll.

(c) $A = 5$ or less on the second roll, $B = 4$ or less on the first roll.

(d) $A = 5$ or less on the second roll, $B = 4$ or less on the second roll.

4.99 Find the probabilities. Refer to the previous exercise. Find the probabilities for each event.

4.100 How possible? Suppose $P(A) > 0$ and $P(B \mid A) = 0$. What must be true about the events A and B?

4.101 Wine tasters. Two wine tasters rate each wine that they taste on a scale of 1 to 5. From data on their ratings of a large number of wines, we obtain the following probabilities for both tasters' ratings of a randomly chosen wine:

Taster 1	Taster 2				
	1	2	3	4	5
1	0.03	0.02	0.01	0.00	0.00
2	0.02	0.07	0.06	0.02	0.01
3	0.01	0.05	0.25	0.05	0.01
4	0.00	0.02	0.05	0.20	0.02
5	0.00	0.01	0.01	0.02	0.06

(a) Why is this a legitimate assignment of probabilities to outcomes?

(b) What is the probability that the tasters agree when rating a wine?

(c) What is the probability that Taster 1 rates a wine higher than 3? What is the probability that Taster 2 rates a wine higher than 3?

4.102 Full-time versus part-time students. For 2017, the following table gives the counts (in thousands) of enrolled students in the United States classified as being of full-time or part-time status and by type of school attended.[25]

Status	Type of school		
	Two-year college	Four-year college	Graduate school
Full-time	2781	9335	2304
Part-time	1564	1816	1394

(a) What is the probability that a randomly chosen enrolled student attends a four-year college?

(b) What is the probability that a randomly chosen enrolled student is a part-time student?

(c) If a randomly chosen enrolled student is in a four-year school, what is the probability that the student is part-time student? Compare this probability with the value from part (b). What does your comparison suggest?

(d) Draw a tree diagram beginning with school type, followed by status (full-time and part-time). Label the tree by school type or status, and label segments by appropriate probabilities (see Figure 4.8 as an example).

(e) Show how the tree can be used to find the probability that a randomly chosen enrolled student is a full-time student.

4.103 Wine tasting. In the setting of Exercise 4.101, Taster 1's rating for a wine is 3. What is the conditional probability that Taster 2's rating is higher than 3?

4.104 An interesting case of independence. Independence of events is not always obvious. Toss two balanced coins independently. The four possible combinations of heads and tails in order each have probability 0.25. The events

A = head on the first toss

B = both tosses have the same outcome

may seem intuitively related. Show that $P(B \mid A) = P(B)$, so that A and B are, in fact, independent.

4.105 Three events. For any three events A, B, and C, the extended addition rule is

$$P(A \text{ or } B \text{ or } C) = P(A) + P(B) + P(C) - P(A \text{ and } B) \\ - P(A \text{ and } C) - P(B \text{ and } C) \\ + P(A \text{ and } B \text{ and } C)$$

Draw a Venn diagram with all three events overlapping. Then use the Venn diagram to explain with words why the extended addition rule for three events is correct.

4.106 Sample surveys for sensitive issues. It is difficult to conduct sample surveys on sensitive issues because many people will not answer questions if the answers might embarrass them. **Randomized response** is an effective way to guarantee anonymity while collecting information on topics such as student cheating or sexual behavior. Here is the idea. To ask a sample of students whether they have plagiarized a term paper while in college, have each student toss a coin in private. If the coin lands heads *and* they have not plagiarized, they are to answer No. Otherwise, they are to give Yes as their answer. Only the student knows whether the answer reflects the truth or just the coin toss, but the researchers can use a proper random sample with follow-up for nonresponse and other good sampling practices.

Suppose that, in fact, the probability is 0.3 that a randomly chosen student has plagiarized a paper. Draw a tree diagram in which the first stage is tossing the coin and the second stage is the truth about plagiarism. The outcome at the end of each branch is the answer given to the randomized-response question. What is the probability of a No answer in the randomized-response poll? If the probability of plagiarism were 0.2, what would be the probability of a No response on the poll? Now suppose that you get 39% No answers in a randomized-response poll of a large sample of students at your college. What do you estimate to be the percent of the population who have plagiarized a paper?

guruXOX/Shutterstock

Random Variables and Probability Distributions

Introduction

In Chapter 4, we studied the basic concepts of probability, including sample spaces, events, and the assignment of probabilities to the outcomes of a random phenomenon. In this chapter, we focus our attention on the assignment of probabilities to *numerical* outcomes of a random phenomenon. Specifically, we introduce the idea of a random variable. You will learn that random variables can be of two types: discrete or continuous. Given that most statistical applications are associated with numerical outcomes, the concept of a random variable serves as an important transition from general probability to statistical inference and modeling techniques.

From the definition of a random variable, you will learn how probability distributions are associated with random variables depending on their type (discrete versus continuous). You will also be introduced to the concepts of mean and variance of random variables, along with various rules of means and variances.

A number of discrete probability distributions have wide applications in business and are of great theoretical importance. In this chapter, we introduce the two most commonly used discrete probability distributions: the binomial and Poisson distributions. Our discussion of continuous random variables will both revisit the Normal distributions and explore two other common distributions

CHAPTER OUTLINE

5.1 Random Variables

5.2 Means and Variances of Random Variables

5.3 Common Discrete Distributions

5.4 Common Continuous Distributions

(uniform and exponential). Here are just a few scenarios where the concepts of random variables come into play:

- Financial advisers at wealth management firms such as Wells Fargo, Fidelity Investments, and J.P. Morgan Chase routinely provide advice to their clients on investments. Which ones (e.g., stocks, mutual funds, bonds) should their clients buy? How much should their clients invest in each possible instrument? We will learn that their advice is guided by concepts studied in this chapter.

- Online bookseller Amazon.com serves its U.S. customers with inventory consolidated in only a handful of warehouses. Each Amazon warehouse pools demand over a large geographic area, which leads to maintaining a lower total inventory versus having many smaller warehouses. We will discover the underlying principle that explains why this strategy provides Amazon with a competitive edge.

- If you follow soccer, you have undoubtedly heard of Manchester United, Arsenal, and Chelsea. It is fascinating to learn that the number of goals scored in a game by these teams are well described by the Poisson distribution! Sports analytics is sweeping across all facets of the sports industry. Many sports teams (baseball, basketball, football, hockey, and soccer) use data to drive decisions on player acquisition, game strategy, and training methods.

- When customers arrive to a business to receive a service, businesses strive to design the most effective system to serve them. To design effective systems (e.g., determining staffing levels), businesses need an understanding of arrival behavior, such as the times between customer arrivals. The exponential distribution is often used to model time-related applications.

5.1 Random Variables

When you complete this section, you will be able to:

- Describe the meaning of a random variable.
- Distinguish between a discrete and a continuous random variable.
- Describe the probability distribution of a discrete random variable.
- Use the probability distribution of a discrete random variable to calculate probabilities of events.
- Describe the connection between density curves and probabilities for continuous random variables.

Sample spaces need not consist of numbers. When we toss a coin four times, we can record the outcome as a string of heads and tails, such as HTTH. In statistics, however, we are most often interested in numerical outcomes such as the count of heads in the four tosses. It is convenient to use a shorthand notation: Let X be the number of heads. If our outcome is HTTH, then $X = 2$. If the next outcome is TTTH, the value of X changes to 1. The possible values of X are 0, 1, 2, 3, and 4. Tossing a coin four times will give X one of these possible values. Tossing the coin four more times will give X another and probably different value. We call X a *random variable* because its numeric value varies over sets of four coin tosses.

> **RANDOM VARIABLE**
>
> A **random variable** is a variable whose value is a numerical outcome of a random phenomenon.

In our coin-tossing example, the random variable is the number of heads in the four tosses.

We usually denote random variables by capital letters found near the end of the alphabet, such as X or Y. Of course, the random variables of greatest interest to us are outcomes such as the mean $\bar{x}$ of a random sample, for which we will keep the familiar notation.[1] As we progress from general rules of probability toward statistical inference, we will concentrate on random variables.

To build a probability model for a random variable, we need to define the sample space and the probabilities associated with the various events. For a random variable, the sample space S consists of the possible values of the random variable. There are two main ways of assigning probabilities to the events of a random variable and this depends on which type of random variable it is.

Discrete random variables

In Chapter 4, we learned several rules of probability, but only one method of assigning probabilities: state the probabilities of the individual outcomes and assign probabilities to events by summing over the outcomes. These outcome probabilities must all be between 0 and 1 and sum to 1. When the outcomes are numerical, they are values of a random variable. We now attach a name to random variables having probability assigned in this way.

DISCRETE RANDOM VARIABLE

A **discrete random variable** X has possible values that can be given in an ordered list. The **probability distribution** of X lists the values and their probabilities:

Value of X	x_1	x_2	x_3	$\cdots$
Probability	p_1	p_2	p_3	$\cdots$

The probabilities p_i must satisfy two requirements:

1. Every probability p_i is a number between 0 and 1.
2. The sum of the probabilities is 1; $p_1 + p_2 + \cdots = \sum p_i = 1$.

Find the probability of any event by adding the probabilities p_i of the particular values x_i that make up the event.

Without explicitly stating so, we experienced discrete random variable scenarios in Chapter 4. For example, consider the distribution of the first digit for data that follow Benford's law, as introduced in Case 4.1 (page 187). If we define X as the first digit, then this variable is indeed a discrete random variable. For many situations that we will study, the number of possible values is a finite number, k. In the case of Benford's law, $k = 9$, with X taking the possible values of 1, 2, 3, 4, 5, 6, 7, 8, and 9.

In other settings, the number of possible values can be infinite. Think about counting the number of tosses of a coin until you get a head. In this case, the set of possible values for X is given by $\{1, 2, 3, \ldots\}$. As another example, suppose X represents the number of customers who make complaints to a retail store during a certain time period. Now, the set of possible values for X is given by $\{0, 1, 2, \ldots\}$. In both of these examples, we say that there is a **countably infinite** number of possible values. Simply defined, *countably infinite* means that we can match each possible outcome with the counting or natural numbers of $\{0, 1, 2, \ldots\}$.

countably infinite

In summary, a discrete random variable has either a finite number of possible values or a countably infinite number of possible values.

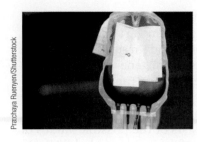

CASE 5.1 Tracking Perishable Demand

Whether a business is in manufacturing, retailing, or service, it inevitably needs to hold inventory to meet the demand for the items it keeps in stock. One of the most basic decisions in the control of an inventory management system is the decision of how many items should be ordered to be stocked. Ordering too much leads to unnecessary inventory costs, while ordering too little risks the organization experiencing stock-out situations. Hospitals have a unique challenge in the inventory management of blood. Blood is a perishable product, so blood inventory management involves a trade-off between shortage and wastage. The demand for blood and its components fluctuates. Hospitals routinely track daily blood demand to estimate rates of usage so that they can manage their blood inventory.

For this case, we consider the daily usage of red blood cells (RBC) O+ transfusion blood bags collected from a Midwest hospital.[2] These transfusion data are categorized as "new-aged" blood cells, which are used for the most critical patients, such as patients with cancer or immune deficiencies. If these blood cells are unused by day's end, then they are downgraded to the category of medium-aged blood cells. Here is the distribution of the number of bags X used in a day:

Bags used	0	1	2	3	4	5	6
Probability	0.202	0.159	0.201	0.125	0.088	0.087	0.056

Bags used	7	8	9	10	11	12
Probability	0.025	0.022	0.018	0.008	0.006	0.003

probability histogram

We can use histograms to show probability distributions as well as distributions of data. Figure 5.1 displays the **probability histogram** of the blood bag probabilities. The height of each bar shows the probability of the outcome at its base. Because the heights are probabilities, they add to 1. As usual, all the bars in a histogram have the same width. Thus, the bar areas also display the relative assignment of probability to outcomes. For the blood bag distribution, we can visually see that more than 50% of the distribution is less than or equal to two bags and the distribution is generally skewed to the right. ■

Probability histograms are used to visualize a discrete probability model in the same way a histogram is used to visualize a sample of data. They also make it easy to quickly compare two distributions. For example, Figure 5.2 compares the probability model for equally likely random digits (Example 4.13; page 189) with the model given by Benford's law (Case 4.1; page 187).

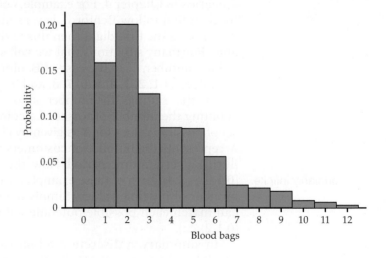

FIGURE 5.1 Probability histogram for blood bag demand probabilities. The height of each bar shows the probability assigned to a single outcome.

FIGURE 5.2 Probability histograms for (a) equally likely random digits 1 to 9 and (b) Benford's law.

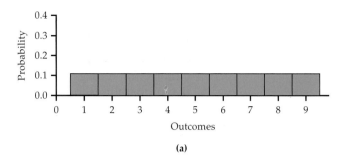

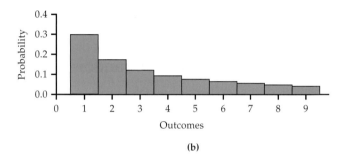

Let's now use the probability distribution of Case 5.1 to compute the probability of an event.

EXAMPLE 5.1

 Demand for at Least One Bag? Consider the event that daily demand is at least one bag. In the language of random variables,

$$P(X \geq 1) = P(X = 1) + P(X = 2) + \cdots + P(X = 11) + P(X = 12)$$
$$= 0.159 + 0.201 + \cdots + 0.006 + 0.003 = 0.798$$

However, adding together 12 probabilities is a bit of a tedious affair. There is a much easier way to get at the ultimate probability when we think about the complement rule. The probability of at least one bag being required is more simply found, as follows:

$$P(X \geq 1) = 1 - P(X = 0)$$
$$= 1 - 0.202 = 0.798 \blacksquare$$

APPLY YOUR KNOWLEDGE

5.1 How many cars? Choose an American household at random and let the random variable X be the number of cars (including SUVs and light trucks) they own. Here is the probability model if we ignore the few households that own more than five cars:

Number of cars X	0	1	2	3	4	5
Probability	0.09	0.36	0.35	0.13	0.05	0.02

(a) Verify that this is a legitimate discrete distribution. Display the distribution in a probability histogram.

(b) Say in words what the event $\{X \geq 1\}$ is. Find $P(X \geq 1)$.

(c) Your company builds houses that include a two-car garage. What percent of households have more cars than the garage can hold?

CASE 5.1 **5.2 High demand.** Refer to Case 5.1 (page 222) for the probability distribution of daily demand for blood transfusion bags.

(a) What is the probability that the hospital will face a high demand of either 11 or 12 bags? Compute this probability directly using the respective probabilities for 11 and 12 bags.

(b) Now show how the complement rule would be used to find the same probability of part (a).

(c) Consider the calculations performed in parts (a) and (b) and the calculations performed in Example 5.1 (page 223). Explain under what circumstances does the use of the complement rule ease computations.

Continuous random variables

When we use the table of random digits to select a digit between 0 and 9, the result is a discrete random variable. The probability model assigns probability 1/10 to each of the 10 possible outcomes. Suppose that we want to choose a number at random between 0 and 1, allowing *any* number between 0 and 1 as the outcome. Software random number generators will perform this operation for us.

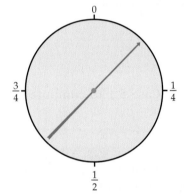

FIGURE 5.3 A spinner that generates a random number between 0 and 1.

You can visualize such a random number by thinking of a spinner (Figure 5.3) that turns freely on its axis and slowly comes to a stop. The pointer can come to rest anywhere on a circle that is marked from 0 to 1. The sample space is now an interval of numbers:

$$S = \{\text{all numbers } x \text{ such that } 0 \leq x \leq 1\}$$

How can we assign probabilities to events such as $\{0.3 \leq X \leq 0.7\}$? As in the case of selecting a random digit, we would like all possible outcomes to be equally likely. But we cannot assign probabilities to each individual value of X and then sum them, because there are infinitely many possible values.

Earlier, we noted that discrete random variables can sometimes take on an infinite number of possible values corresponding to the set of counting numbers $\{0, 1, 2, \ldots\}$. However, the infinity associated with the spinner's possible outcomes is a different infinity. There is no way to match the infinite number of decimal values in the range from 0 to 1 to the counting numbers. We are dealing with the possible outcomes being associated with the *real numbers* as opposed to the counting numbers. As such, we say here that this situation involves an **uncountably infinite** number of possible values.

uncountably infinite

density curve, p. 41

In light of these facts, we need to use a new way of assigning probabilities directly to events—as *areas under a density curve*. The idea of a density curve was first introduced in Section 1.4 as a mathematical model for the distribution of a quantitative variable. As mentioned in Section 1.4 (page 41), any density curve has area exactly 1 underneath it, corresponding to total probability of 1.

EXAMPLE 5.2

uniform distribution

Uniform Random Numbers A random number generator will spread its output uniformly across the entire interval from 0 to 1 similar to the spinner in Figure 5.3. The results of a long sequence of generated numbers are represented by the density curve of a **uniform distribution**.

This density curve appears in red in Figure 5.4. It has height 1 over the interval from 0 to 1, and height 0 everywhere else. The area under the density curve is 1: the area of a square with base 1 and height 1. The probability of any event is the area under the density curve and above the event in question.

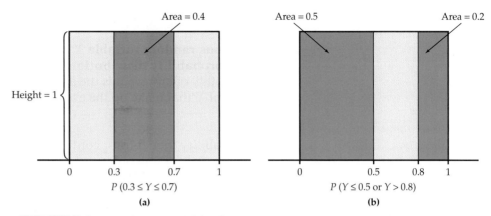

FIGURE 5.4 Assigning probabilities for generating a random number between 0 and 1, for Example 5.2. The probability of any interval of numbers is the area above the interval and under the density curve.

As Figure 5.4(a) illustrates, the probability that the random number generator produces a number X between 0.3 and 0.7 is

$$P(0.3 \leq X \leq 0.7) = 0.4$$

because the area under the density curve and above the interval from 0.3 to 0.7 is 0.4. The height of the density curve is 1, and the area of a rectangle is the product of height and length, so the probability of any interval of outcomes is just the length of the interval.

Similarly,

$$P(X \leq 0.5) = 0.5$$
$$P(X > 0.8) = 0.2$$
$$P(X \leq 0.5 \text{ or } X > 0.8) = 0.7$$

Notice that the last event consists of two nonoverlapping intervals, so the total area above the event is found by adding two areas, as illustrated by Figure 5.4(b). This assignment of probabilities obeys all of our rules for probability. ■

Example 5.2 illustrated for the uniform distribution the important concept of assigning probabilities to events. Figure 5.5 illustrates this idea in general form. We call X in Example 5.2 a *continuous random variable* because its values are not isolated numbers, but rather an interval of numbers.

FIGURE 5.5 The probability distribution of a continuous random variable assigns probabilities as areas under a density curve. The total area under any density curve is 1.

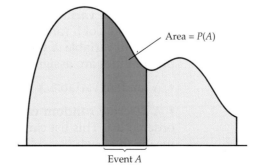

CONTINUOUS RANDOM VARIABLE

A **continuous random variable** X takes all values in an interval of numbers. The **probability distribution** of X is described by a density curve. The probability of any event is the area under the density curve and above the values of X that make up the event.

The probability model for a continuous random variable assigns probabilities to intervals of outcomes rather than to individual outcomes. In fact, **all continuous probability distributions assign probability 0 to every individual outcome.** Only intervals of values have positive probability. To see that this is true, consider a specific outcome such as $P(X = 0.8)$ in the context of Example 5.2. For the uniform distribution of Example 5.2, we noted that the probability of any interval is, in the end, the same as its length. The point 0.8 on the x axis is an interval with no length, so its probability is 0.

Although this fact may seem odd, it makes intuitive, as well as mathematical, sense. The random number generator produces a number between 0.79 and 0.81 with probability 0.02. An outcome between 0.799 and 0.801 has probability 0.002. A result between 0.799999 and 0.800001 has probability 0.000002. You see that as we approach 0.8, the probability gets closer to 0.

To be consistent, the probability of an outcome *exactly* equal to 0.8 must be 0. Because there is no probability exactly at $X = 0.8$, the two events $\{X > 0.8\}$ and $\{X \geq 0.8\}$ have the same probability. In general, we can ignore the distinction between $>$ and $\geq$ when finding probabilities for continuous random variables. Similarly, we can ignore the distinction between $<$ and $\leq$ in the continuous case. *However, when dealing with discrete random variables, we cannot ignore these distinctions. Thus, it is important to recognize whether you are dealing with continuous or discrete random variables when doing probability calculations.* In Section 5.4, we will revisit continuous random variables in the context of a few common continuous distributions.

APPLY YOUR KNOWLEDGE

5.3 Uniform probability. For the uniform distribution described in Example 5.2, find the probabilities that

(a) X is between 0.34 and 0.78.

(b) X is 0.6.

(c) X is less than 0.2 or greater than 0.5.

(d) X is greater than 0.

(e) X is greater than 1.

SECTION 5.1 SUMMARY

- A **random variable** is a variable taking numerical values determined by the outcome of a random phenomenon. The **probability distribution** of a random variable X tells us what the possible values of X are and how probabilities are assigned to those values.

- A random variable X and its distribution can be **discrete** or **continuous.**

- A **discrete random variable** has possible values that can be given in an ordered list. This list can be finite or countable infinite. Countably infinite means that each possible outcome corresponds to one of the counting numbers $\{0, 1, 2, \ldots\}$.

- For a discrete random variable, the probability distribution assigns each of these values a probability between 0 and 1 such that the sum of all the probabilities is 1. The probability of any event is the sum of the probabilities of all the values that make up the event.

- A **continuous random variable** takes all values in some interval of numbers. For continuous random variables, there is an uncountable infinite number of possible values. A **density curve** describes the probability distribution of a continuous random variable. The probability of any event is the area under the curve and above the values that make up the event.

- You can picture a probability distribution by drawing a **probability histogram** in the discrete case or by graphing the density curve in the continuous case.

SECTION 5.1 EXERCISES

For Exercises 5.1 and 5.2, see pages 223–224; for 5.3, see page 226.

CASE 5.1 **5.4 Daily demand.** Refer to Case 5.1 (page 222) and the distribution for the number of blood bags X in a day.

(a) What is $P(X \geq 6)$?

(b) What is $P(X > 6)$?

(c) What is $P(1 < X \leq 4)$?

(d) What is $P(1 < X < 4)$?

(e) What is $P(X \neq 1)$?

5.5 How many courses? At a small liberal arts college, students can register for one to six courses. Let X be the number of courses taken in the fall by a randomly selected student from this college. In a typical fall semester, 5% of students take one course, 5% take two courses, 13% take three courses, 26% take four courses, 36% take five courses, and 15% take six courses. Let X be the number of courses taken in the fall by a randomly selected student from this college. Describe the probability distribution of this random variable.

5.6 Make a graphical display. Refer to the previous exercise. Use a probability histogram to provide a graphical description of the distribution of X.

5.7 Find some probabilities. Refer to Exercise 5.5.

(a) Find the probability that a randomly selected student takes three or fewer courses.

(b) Find the probability that a randomly selected student takes four or five courses.

(c) Find the probability that a randomly selected student takes eight courses.

5.8 Texas hold 'em. The game of Texas hold 'em starts with each player receiving two cards. Here is the probability distribution for the number of aces in two-card hands:

Number of aces	0	1	2
Probability	0.8507	0.1448	0.0045

(a) Verify that this assignment of probabilities satisfies the requirement that the sum of the probabilities for a discrete distribution must be 1.

(b) Make a probability histogram for this distribution.

(c) What is the probability that a hand contains at least one ace? Show two different ways to calculate this probability.

5.9 How large are households? Choose an American household at random, and let X be the number of persons living in the household. If we ignore the few households with more than seven inhabitants, the probability model for X is as follows:

Household size X	1	2	3	4	5	6	7
Probability	0.27	0.33	0.16	0.14	0.06	0.03	0.01

(a) Verify that this is a legitimate probability distribution.

(b) What is $P(X \geq 5)$?

(c) What is $P(X > 5)$?

(d) What is $P(2 < X \leq 4)$?

(e) What is $P(X \neq 1)$?

(f) Write the event that a randomly chosen household contains more than two persons in terms of X. What is the probability of this event?

CASE 5.1 **5.10 Two-day demand.** Refer to the distribution of daily demand for blood bags X in Case 5.1 (page 222). Let Y be the total demand over two days. Assume that demand is independent from day to day.

(a) List the possible values for Y.

(b) From the distribution of daily demand, we find that the probability that no bags are demanded on a given day is 0.202. In that light, suppose a hospital manager states, "The chances that no bags are demanded over two consecutive days is two times 0.202 or 0.404." Make a simple argument to the manager explaining the mistake in probability conclusion. (*Hint:* Use the manager's logic to determine the chance of no bags demanded over many more than two consecutive days to show the error in the manager's thinking.)

(c) What is $P(Y = 0)$? (*Hint:* Refer to the multiplication rule for independent events on page 191.)

5.11 Discrete or continuous. In each of the following situations, decide whether the random variable is discrete or continuous, and give a reason for your answer.

(a) Your web page has five different links, and a user can click on one of the links or can leave the page. You record the length of time that a user spends on the web page before clicking one of the links or leaving the page.

(b) The number of hits on your web page.

(c) The yearly income of a visitor to your web page.

5.12 Find the probabilities. Let the random variable X be a random number with the uniform density curve in Figure 5.4 (page 225). Find the following probabilities:

(a) $P(X \geq 0.30)$.

(b) $P(X = 0.30)$.

(c) $P(0.30 < X < 1.30)$.

(d) $P(0.20 \leq X \leq 0.25 \text{ or } 0.7 \leq X \leq 0.9)$.

(e) X is not in the interval 0.4 to 0.7.

5.13 Spell-checking software. Spell-checking software catches "nonword errors," which are strings of letters that are not words, as when "the" is typed as "eth." When undergraduates are asked to write a 250-word essay (without spell checking), the number of nonword errors X has the following distribution:

Value of X	0	1	2	3	4
Probability	0.1	0.3	0.3	0.2	0.1

(a) Sketch the probability distribution for this random variable.

(b) Write the event "at least one nonword error" in terms of X. What is the probability of this event?

(c) Describe the event $X \leq 2$ in words. What is its probability? What is the probability that $X < 2$?

CASE 5.1 **5.14 How much to order?** Faced with the demand for the perishable product of blood, hospital managers need to establish an ordering policy that deals with the trade-off between shortage and wastage. As it turns out, this scenario, referred to as a single-period inventory problem, is well known in the area of operations management, and there is an optimal policy. What we need to know is the per-item cost of being short (C_S) and the per-item cost of being in excess (C_E). In terms of the blood example, the hospital estimates that for every bag short, there is a cost of $80 per bag, which includes expediting and emergency delivery costs. Any transfusion blood bags left in excess at day's end are associated with a $20 per bag cost, which includes the original cost of purchase along with end-of-day handling costs. With the objective of minimizing long-term average costs, the following critical ratio (CR) needs to be computed:

$$CR = \frac{C_S}{C_S + C_E}$$

Recognize that CR will always be in the range of 0 to 1. It turns out that the optimal number of items to order is the *smallest* value of k such that $P(X \leq k)$ is at least the CR value. Refer to the distribution of daily demand for blood bags X in Case 5.1 (page 222).

(a) Based on the given values of C_S and C_E, what is the value of CR?

(b) Given the CR found in part (a), determine the optimal order quantity of blood bags per day.

(c) Keeping C_E at $20, for what range of values of C_S does the hospital order three bags?

5.2 Means and Variances of Random Variables

When you complete this section, you will be able to:

- Use a probability distribution to find the mean, variance, and standard deviation of a discrete random variable.
- Apply the law of large numbers to describe the behavior of the sample mean as the sample size increases.
- Find means of linear transformations, sums, and differences of random variables.
- Find variances and standard deviations of linear transformations, sums, and differences of random variables.

The probability histograms and density curves that picture the probability distributions of random variables resemble our earlier pictures of distributions of data. In describing data, we moved from graphs to numerical measures, such as means and standard deviations. Now we make the same move to expand our descriptions of the distributions of random variables. We can speak of the mean winnings in a game of chance or the standard deviation of the randomly varying number of calls a travel agency receives in an hour. In this section, we learn more about how to compute these descriptive measures and about the laws they obey.

The mean of a random variable

In Chapter 1 (page 25), we learned that the mean $\bar{x}$ is the average of the observations in a *sample*. Recall that a random variable X is a numerical outcome of a random process. Think about repeating this random process many times and recording the resulting values of X. You can think of the mean of a random variable as the average of a very large sample where the relative frequencies of the values are the same as their probabilities.

If we think of the random process as corresponding to the population, then the mean of the random variable is a parameter of this population. Here is an example.

EXAMPLE 5.3

The Tri-State Pick 3 Lottery Most states and Canadian provinces have government-sponsored lotteries. Here is a simple lottery wager from the Tri-State Pick 3 game that New Hampshire shares with Maine and Vermont. You choose a three-digit number from 000 to 999. The state chooses a three-digit winning number at random and pays you $500 if your number is chosen.

Because there are 1000 three-digit numbers, you have probability $1/1000 = 0.001$ of winning. Taking X to be the amount your ticket pays you, the probability distribution of X is

Payoff X	$0	$500
Probability	0.999	0.001

The random process consists of drawing a three-digit number. The population consists of the numbers 000 to 999. Each of these possible outcomes is equally likely in this example. In the setting of sampling in Chapter 3 (page 133), we can view the random process as selecting an SRS of size 1 from the population. The random variable X is 1 if the selected number is equal to the one that you chose and is 0 if it is not.

What is your average payoff from many tickets? The ordinary average of the two possible outcomes $0 and $500 is $250, but that makes no sense as the average because $500 is much less likely than $0. In the long run, you receive $500 once in every 1000 tickets and $0 on the remaining 999 of 1000 tickets. The long-run average payoff is

$$\$500 \frac{1}{1000} + \$0 \frac{999}{1000} = \$0.50$$

or 50 cents. This number is the mean of the random variable X. (Tickets cost $1, so in the long run, the state keeps half the money you wager.) ■

If you play Tri-State Pick 3 several times, we would, as usual, call the mean of the actual amounts you win $\bar{x}$. The mean in Example 5.3 is a different quantity—it is the long-run average winnings you expect if you play a very large number of times.

APPLY YOUR KNOWLEDGE

5.15 Find the mean of the probability distribution. You toss a fair coin. If the outcome is heads, you win $5.00; if the outcome is tails, you win nothing. Let X be the amount that you win in a single toss of a coin. Find the probability distribution of this random variable and its mean.

Just as probabilities are an idealized description of long-run proportions, the mean of a probability distribution describes the long-run average outcome. We can't call this mean $\bar{x}$, so we need a different symbol. The symbol for the **mean of a probability distribution** is μ, the Greek letter mu. We used μ in Chapter 1 for the mean of a Normal distribution, so this is not a new notation. We will often be interested in several random variables, each having a different probability distribution with a different mean.

mean μ

To remind ourselves that we are talking about the mean of X, we often write μ_X rather than simply μ. In Example 5.3, $\mu_X = \$0.50$. Notice that, as often happens, the mean is not a possible value of X. You will often find the mean of a random variable X called the **expected value** of X. This term can be misleading because we don't necessarily expect an observation on X to equal its expected value.

expected value

The mean of any discrete random variable is found just as described in Example 5.3. It is an average of the possible outcomes, but a weighted average in which each outcome is weighted by its probability. Because the probabilities add to 1, we have total weight 1 to distribute among the outcomes. An outcome that occurs half the time has probability one-half and gets one-half the weight in calculating the mean. Here is the general definition.

MEAN OF A DISCRETE RANDOM VARIABLE

Suppose that X is a **discrete random variable** whose distribution is

Value of X	x_1	x_2	x_3	$\cdots$
Probability	p_1	p_2	p_3	$\cdots$

To find the **mean** of X, multiply each possible value by its probability, then add all the products:

$$\mu_X = x_1 p_1 + x_2 p_2 + \cdots$$
$$= \sum x_i p_i$$

EXAMPLE 5.4

The Mean of Equally Likely First Digits If the first digits in a set of data all have the same probability (see Figure 5.2(a), page 223), the probability distribution of the first digit X is

First digit X	1	2	3	4	5	6	7	8	9
Probability	1/9	1/9	1/9	1/9	1/9	1/9	1/9	1/9	1/9

The mean of this distribution is

$$\mu_X = 1 \times \frac{1}{9} + 2 \times \frac{1}{9} + 3 \times \frac{1}{9} + 4 \times \frac{1}{9} + 5 \times \frac{1}{9} + 6 \times \frac{1}{9} + 7 \times \frac{1}{9} + 8 \times \frac{1}{9} + 9 \times \frac{1}{9}$$
$$= 45 \times \frac{1}{9} = 5 \quad \blacksquare$$

Suppose that the random digits had a different probability distribution. In Case 4.1 (page 187) and Figure 5.2(b) (page 223), we described Benford's law as a probability distribution that describes first digits of numbers in many real situations. Let's calculate the mean for Benford's law.

EXAMPLE 5.5

The Mean of First Digits That Follow Benford's Law The following distribution of the first digit applies for data that follow Benford's law. We use the letter V for this random variable to distinguish it from the one that we studied in Example 5.4. The distribution of V is

First digit V	1	2	3	4	5	6	7	8	9
Probability	0.301	0.176	0.125	0.097	0.079	0.067	0.058	0.051	0.046

The mean of V is

$$\mu_V = (1)(0.301) + (2)(0.176) + (3)(0.125) + (4)(0.097) + (5)(0.079) + (6)(0.067) +$$
$$(7)(0.058) + (8)(0.051) + (9)(0.046)$$
$$= 3.441$$

The mean reflects the greater probability of smaller first digits under Benford's law than when each of the first digits 1 to 9 is equally likely. ∎

Figure 5.6 locates the means of X and V on the two probability histograms. Because the discrete uniform distribution of Figure 5.6(a) is symmetric, the mean lies at the center. We can't locate the mean of the right-skewed distribution of Figure 5.6(b) by eye; instead, calculation is needed.

What about continuous random variables? The probability distribution of a continuous random variable is described by a density curve. Chapter 1 showed how to find the mean of the distribution: it is the point at which the area under the density curve would balance if it were made out of solid material. The mean lies at the center of symmetric density curves such as Normal curves.[3] Exact calculation of the mean of a distribution with a skewed density curve requires advanced mathematics.[4] The idea that the mean is the balance point of the distribution applies to discrete random variables as well, but in the discrete case, we have an arithmetic formula (page 230) that gives us this point.

mean as balance point, p. 42

Mean and the law of large numbers

With probabilities in hand, we have shown that, for discrete random variables, the mean of the distribution (μ) can be determined by computing a weighted average in which each possible value of the random variable is weighted by its

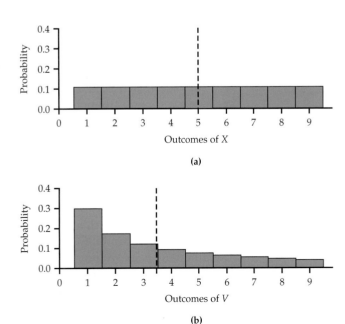

FIGURE 5.6 Locating the mean of a discrete random variable on the probability histogram for (a) digits between 1 and 9 chosen at random and (b) digits between 1 and 9 chosen from records that obey Benford's law.

probability. For example, in Example 5.5, we found that the mean of the first digit of numbers obeying Benford's law is 3.441.

Now suppose that we are unaware of the probabilities of Benford's law but we still want to determine the mean of the distribution. To do so, we choose an SRS of financial statements and record the first digits of entries known to follow Benford's law. We then calculate the sample mean $\bar{x}$ to estimate the unknown population mean μ. In the vocabulary of statistics, μ is referred to as a *parameter* and $\bar{x}$ is called a *statistic*.

It seems reasonable to use $\bar{x}$ to estimate μ. An SRS should fairly represent the population, so the mean $\bar{x}$ of the sample should be somewhere near the mean μ of the population. Of course, we don't expect $\bar{x}$ to be exactly equal to μ, and we realize that if we choose another SRS, the luck of the draw will probably produce a different $\bar{x}$. How can we control the variability of the sample means? The answer is to increase the sample size. If we keep adding observations to our random sample, the statistic $\bar{x}$ is *guaranteed* to get as close as we wish to the parameter μ and then stay that close. We have the comfort of knowing that if we gather up more financial statements and keep recording more first digits, eventually we will estimate the mean value of the first digit very accurately. This remarkable fact is called the *law of large numbers*. It is remarkable because it holds for *any* population, not just for some special class such as Normal distributions.

LAW OF LARGE NUMBERS

Draw independent observations at random from any population with finite mean μ. As the number of observations drawn increases, the mean $\bar{x}$ of the observed values becomes progressively closer to the population mean μ.

The behavior of $\bar{x}$ is similar to the idea of probability. In the long run, the *proportion* of outcomes taking any value gets close to the *probability* of that value, and the *average outcome* gets close to the distribution *mean*. Figure 4.1 (page 177) shows how proportions approach probability in one example. Here is an example of how sample means approach the distribution mean.

EXAMPLE 5.6

Applying the Law of Large Numbers to Benford's Law Distribution

CASE 4.1 Suppose that the first randomly drawn financial statement entry has an 8 as its first digit. Thus, our initial sample mean is 8. We proceed to select a second financial statement entry, and find the first digit to be 3, so for $n = 2$ the mean is now

$$\bar{x} = \frac{8+3}{2} = 5.5$$

As this stage, we might be tempted to think that digits are equally likely because we have observed a large digit and a small digit. The flaw in this thinking is obvious: we are believing that short-run results accurately reflect long-run behavior. With a clear mind, we proceed to collect more observations and continue to update the sample mean. Figure 5.7 shows that the sample mean changes as we increase the sample size. Notice that the first point is 8 and the second point is the calculated mean of 5.5. More importantly, notice that the mean of the observations gets close to the distribution mean $\mu = 3.441$ and settles down to that value. The law of large numbers says that this *always* happens. ■

FIGURE 5.7 The law of large numbers in action. As we take more observations, the sample mean always approaches the mean (μ) of the population.

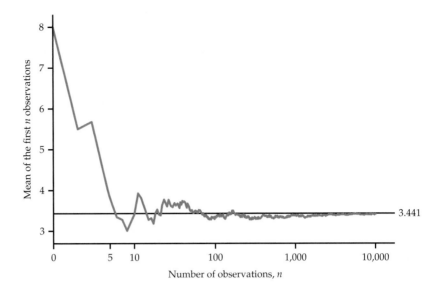

APPLY YOUR KNOWLEDGE

5.16 **Use the *Law of Large Numbers* applet.** The *Law of Large Numbers* applet animates a graph like Figure 5.7 for rolling dice and getting a specific number like 6. Use it to better understand the law of large numbers by creating a similar graph.

The mean μ of a random variable is the average value of the variable in two senses. By definition, μ is the average of the possible values, weighted by their probability of occurring. The law of large numbers says that μ is also the long-run average of many independent observations on the variable. The law of large numbers can be proved mathematically starting from the basic laws of probability.

Thinking about the law of large numbers

The law of large numbers says broadly that the average result of many independent observations is stable and predictable. The gamblers in a casino may win or lose, but the casino will win in the long run because the law of large numbers says what the average outcome of many thousands of bets will be. An insurance company deciding how much to charge for life insurance and a fast-food restaurant deciding how many beef patties to prepare also rely on the fact that averaging over many individuals produces a stable result. It is worth the effort to think a bit more closely about this important fact.

The "law of small numbers"

Both the rules of probability and the law of large numbers describe the regular behavior of chance phenomena *in the long run*. Psychologists have discovered that our intuitive understanding of randomness is quite different from the true laws of chance.[5] For example, most people believe in an incorrect "law of small numbers." That is, we expect even short sequences of random events to show the kind of average behavior that, in fact, appears only in the long run.

Some teachers of statistics begin a course by asking students to toss a coin 50 times and bring the sequence of heads and tails to the next class. The teacher then announces which students just wrote down a random-looking sequence rather than actually tossing a coin. The faked tosses don't have enough "runs" of consecutive heads or consecutive tails. Runs of the same outcome don't look random to us but are, in fact, common. For example, the probability of a run of three or more consecutive heads or tails in just 10 tosses is greater than 0.8.[6] The runs of consecutive heads or consecutive tails that appear in real coin tossing (and that are predicted by the mathematics of probability) seem surprising to us. Because we don't expect to see long runs, we may conclude that the coin tosses are not independent or that some influence is disturbing the random behavior of the coin.

EXAMPLE 5.7

The "Hot Hand" in Basketball Belief in the law of small numbers influences behavior. If a basketball player makes several consecutive shots, both the fans and her teammates believe that she has a "hot hand" and is more likely to make the next shot. This is doubtful.

Careful study suggests that runs of baskets made or missed are no more frequent in basketball than would be expected if each shot were independent of the player's previous shots. Baskets made or missed are just like heads and tails in tossing a coin. (Of course, some players make 30% of their shots in the long run and others make 50%, so a coin-toss model for basketball must allow coins with different probabilities of a head.) Our perception of hot or cold streaks simply shows that we don't perceive random behavior very well.[7] ■

Our intuition doesn't do a good job of distinguishing random behavior from systematic influences. This is also true when we look at data. We need statistical inference to supplement exploratory analysis of data because probability calculations can help verify that what we see in the data is more than a random pattern.

How large is a large number?

The law of large numbers says that the actual mean outcome of many trials gets close to the distribution mean μ as more trials are made. But it doesn't say how many trials are needed to guarantee a mean outcome close to μ—that depends on the *variability* of the random outcomes. The more variable the outcomes, the more trials that are needed to ensure that the mean outcome $\bar{x}$ is close to the distribution mean μ. Casinos well understand this relationship: the outcomes of games of chance are variable enough to hold the interest of gamblers. Only the casino plays often enough to rely on the law of large numbers. Gamblers get entertainment; the casino has a business.

Rules for means

Imagine you are a financial adviser who must advise clients regarding how to distribute their assets among different investments such as individual stocks, mutual funds, bonds, and real estate. With data available on all these financial instruments, you are able to gather a variety of insights, such as the proportion of the time a particular stock outperformed the market index, the average performance of the different investments, the consistency or inconsistency of the performance of the different investments, and relationships among the investments. In other words, you are gathering measures of probability, mean, standard deviation, and correlation. In general, the discipline of finance relies heavily on a solid understanding of probability and statistics. In the next case, we explore how the concepts of this chapter play a fundamental role in constructing an investment portfolio.

CASE 5.2 **Portfolio Analysis** One fundamental measure of performance of an investment is its *rate of return*. For a stock, the rate of return of an investment over a time period is basically the percent change in the share price during the time period. However, corporate actions such as dividend payments and stock splits can complicate the calculation. A stock's closing price can be amended to include any distributions and corporate actions to give us an adjusted closing price. The percent change of adjusted closing prices can then serve as a reasonable calculation of the actual return.

For example, the closing adjusted price of the well-known S&P 500 market index was $2713.83 for January 2018 and was $2605.00 for February 2018. So, the index's *monthly* rate of return for that time period was

$$\frac{\text{change in price}}{\text{starting price}} = \frac{2605.00 - 2713.83}{2713.83} = -0.0401, \text{ or } -4.01\%$$

Investors want high positive returns, but they also want safety. During the period from 2000 to early 2018, the S&P 500's monthly returns fell as low as −17% and climbed as high as +11%. The variability of returns, called **volatility** in finance, is a measure of the risk of an investment. A highly volatile stock, which may go either up or down by large percents, is more risky than a Treasury bill, whose return is very predictable.

A *portfolio* is a collection of investments held by an individual or an institution. **Portfolio analysis** involves studying the risk and return of a portfolio based on the risk and return of the individual investments it contains. That's where statistics comes in: the return on an investment over some period of time is a random variable. We are interested in the *mean* return, and we measure volatility by the *standard deviation* of returns. Indeed, investment firms will report online the historical mean and standard deviation of returns of individual stocks or funds.[8]

Suppose that we are interested in building a simple portfolio based on allocating funds into one of two investments. Let's take one of the investments to be the commonly chosen S&P 500 index. The key now is to pick another investment that does *not* have a high positive correlation with the market index. Investing in two investments that have very high positive correlation with each other is tantamount to investing in just one.

Possible choices against the S&P 500 index are different asset classes like real estate, gold, energy, and utilities. For example, suppose we build a portfolio with 70% of funds invested in the S&P 500 index and 30% in a well-known utilities sector fund (XLU). If X is the monthly return on the S&P 500 index and Y is the monthly return on the utilities fund, the portfolio rate of return is

$$R = 0.7X + 0.3Y$$

How can we find the mean and standard deviation of the portfolio return R starting from information about X and Y? We must now develop the machinery to do this. ∎

We will first start with the mean and then move on to the standard deviation. Think first not about investments, but about making refrigerators. You are studying flaws in the painted finish of refrigerators made by your firm. Dimples and paint sags (runny or drippy appearance) are two kinds of surface flaws. Not all refrigerators have the same number of dimples: many have none, some have one, some two, and so on. You ask for the average number of imperfections on a refrigerator. Your inspectors report finding an average of 0.7 dimple and 1.4 sags per refrigerator. How many total imperfections of both kinds (on the average) are present on a refrigerator? That's easy: if the average number of dimples is 0.7 and the average number of sags is 1.4, then counting both gives an average of $0.7 + 1.4 = 2.1$ flaws.

In more formal language, the number of dimples on a refrigerator is a random variable X that varies as we inspect one refrigerator after another. We know only that the mean number of dimples is $\mu_X = 0.7$. The number of paint sags is a second random variable Y having mean $\mu_Y = 1.4$. (As usual, the subscripts keep straight which variable we are talking about.) The total number of both dimples and sags is another random variable, the sum $X + Y$. Its mean μ_{X+Y} is the average number of dimples and sags together. It is just the sum of the individual means μ_X and μ_Y. That's an important rule for how means of random variables behave.

Here's another rule. A large lot of plastic coffee-can lids has a mean diameter of 4.2 inches. What is the mean in centimeters? There are 2.54 centimeters in an inch, so the diameter in centimeters of any lid is 2.54 times its diameter in inches. If we multiply every observation by 2.54, we also multiply their average by 2.54. The mean in centimeters must be 2.54 × 4.2, or about 10.7 centimeters. More formally, the diameter in inches of a lid chosen at random from the lot is a random variable X with mean μ_X. The diameter in centimeters is $2.54X$, and this new random variable has mean $2.54\mu_X$.

The point of these examples is that means behave like averages. Here are the rules we need.

RULES FOR MEANS OF LINEAR TRANSFORMATIONS, SUMS, AND DIFFERENCES

Rule 1. If X is a random variable and a and b are fixed numbers, then

$$\mu_{a+bX} = a + b\mu_X$$

Rule 2. If X and Y are random variables, then

$$\mu_{X+Y} = \mu_X + \mu_Y$$

Rule 3. If X and Y are random variables, then

$$\mu_{X-Y} = \mu_X - \mu_Y$$

EXAMPLE 5.8

Aggregating Demand in a Supply Chain To remain competitive, companies worldwide are increasingly recognizing the need to effectively manage their supply chains. Let us consider a simple but realistic supply chain scenario. ElectroWorks is a company that manufactures and distributes electronic parts to various regions in the United States. To serve the Chicago–Milwaukee region, the company has one warehouse in Milwaukee and another in Chicago. Because the company produces thousands of parts, it is considering an alternative strategy of locating a single, centralized warehouse between the two markets—say, in Kenosha, Wisconsin—that will serve all customer orders. Delivery time, referred to as *lead time,* from manufacturing to warehouse(s) and ultimately to customers is unaffected by the new strategy.

To illustrate the implications of the centralized warehouse, let us focus on one specific part: SurgeArrester. The lead time for this part from manufacturing to warehouses is one week. Based on historical data, the lead time demands for the part in each of the markets are Normally distributed with

X = Milwaukee warehouse $\quad \mu_X = 415$ units $\quad \sigma_X = 48$ units
Y = Chicago warehouse $\quad \mu_Y = 2689$ units $\quad \sigma_Y = 272$ units

If the company were to centralize, what would be the mean of the total aggregated lead time demand $X + Y$? Using Rule 2, we can easily find the mean overall lead time demand:

$$\mu_{X+Y} = \mu_X + \mu_Y = 415 + 2689 = 3104 \blacksquare$$

At this stage, we have only part of the picture on the aggregated demand random variable—namely, its mean value. In Example 5.14 (page 242), we continue our study of aggregated demand to include the variability dimension that, in turn, will reveal operational benefits from the proposed strategy of centralizing. Let's now return to the portfolio scenario of Case 5.2 to demonstrate the use of a combination of the mean rules.

EXAMPLE 5.9

Portfolio Analysis The past behavior of the two securities in the portfolio is pictured in Figure 5.8, which plots the monthly returns for the utility sector index (XLU) against the S&P 500 market index from January 2000 to February 2018. As we can see, the returns on the two indices have a moderate level of positive correlation. This fact will be used later to obtain a complete assessment of the expected performance of the portfolio. For now, we can calculate mean returns from the 219 data points shown on the plot:[9]

$$X = \text{monthly return for S\&P 500 index} \quad \mu_X = 0.351\%$$
$$Y = \text{monthly return for Utility index} \quad \mu_Y = 0.660\%$$

By combining Rules 1 and 2, we find the mean return on the portfolio based on a 70/30 mix of S&P index shares and utility shares:

$$R = 0.7X + 0.3Y$$
$$\mu_R = 0.7\mu_X + 0.3\mu_Y$$
$$= (0.7)(0.351) + (0.3)(0.660) = 0.444\%$$

expected return

This calculation uses historical data on returns. Next month may, of course, be very different. In finance, the term **expected return** is typically used instead of mean return. ∎

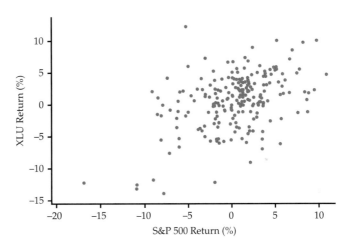

FIGURE 5.8 Monthly returns on the Utilities Sector index versus the S&P 500 index (January 2000 to February 2018), for Example 5.9.

APPLY YOUR KNOWLEDGE

5.17 Find μ_Y. The random variable X has mean $\mu_X = 6$. If $Y = 28 - 2X$, find μ_Y.

5.18 Find μ_W. The random variable U has mean $\mu_U = 21$, and the random variable V has mean $\mu_V = 12$. If $W = 2U + 3V$, find μ_W.

5.19 Managing a new-product development process. Managers often have to oversee a series of related activities directed to a desired goal or output. As a new-product development manager, you are responsible for two sequential steps of the product development process—namely, the development of product specifications followed by the design of the manufacturing process. Let X be the number of weeks required to complete the development of product specifications, and let Y be the number of weeks required to complete the design of the manufacturing process. Based on experience, you estimate the following probability distribution for the first step:

Weeks (X)	1	2	3
Probability	0.3	0.5	0.2

For the second step, your estimated distribution is

Weeks (Y)	1	2	3	4	5
Probability	0.1	0.15	0.4	0.3	0.05

(a) Calculate μ_X and μ_Y.

(b) The cost per week for the activity of developing product specifications is \$8000, while the cost per week for the activity of designing the manufacturing process is \$30,000. Calculate the mean cost for each step.

(c) Calculate the mean completion time and mean cost for the two steps combined.

CASE 5.2 **5.20 Mean return on portfolio.** The addition rule for means extends to sums of any number of random variables. Let's look at a portfolio containing three mutual funds from three different industrial sectors: biotechnology, information services, and defense. The monthly returns on Fidelity Select Biotechnology fund (FBIOX), Fidelity Software and IT Services fund (FSCSX), and Fidelity Select Defense and Aerospace fund (FSDAX) for 10 years ending in February 2018 had approximately these means:[10]

X = Biotechnology monthly return $\mu_X = 1.592\%$

Y = Software and IT Services monthly return $\mu_Y = 1.444\%$

Z = Defense and Aerospace monthly return $\mu_Z = 1.178\%$

What is the mean monthly return for a portfolio consisting of 50% Biotechnology, 30% Software and IT Services, and 20% Defense and Aerospace?

The variance of a random variable

The mean is a measure of the center of a distribution. Another important characteristic of a distribution is its spread. The variance and the standard deviation are the standard measures of spread that accompany the choice of the mean to measure the center. Just as for the mean, we need a distinct symbol to distinguish the variance of a random variable from the variance s^2 of a data set. We write the variance of a random variable X as σ_X^2. Once again, the subscript reminds us which variable we have in mind. The definition of the variance σ_X^2 of a random variable is similar to the definition of the sample variance s^2 given in Chapter 1 (page 33). Namely, the variance is an average value of the squared deviations of the variable X from its mean μ_X—that is, the average value of $(X - \mu_X)^2$.

As with the computation of the mean of a random variable, we use a weighted average of these squared deviations based on the *probability* of each outcome. Calculating this weighted average is straightforward for discrete random variables but requires advanced mathematics in the continuous case. Here is the definition.

VARIANCE OF DISCRETE RANDOM VARIABLE

Suppose that X is a **discrete random variable** whose distribution is

Value of X	x_1	x_2	x_3	...
Probability	p_1	p_2	p_3	...

and that μ_X is the mean of X. The **variance** of X is

$$\sigma_X^2 = (x_1 - \mu_X)^2 p_1 + (x_2 - \mu_X)^2 p_2 + \cdots$$
$$= \sum (x_i - \mu_X)^2 p_i$$

The **standard deviation** σ_X of X is the square root of the variance.

EXAMPLE 5.10

Find the Mean and the Variance In Case 5.1, we saw that the distribution of the daily demand X of transfusion blood bags is

Bags used	0	1	2	3	4	5	6
Probability	0.202	0.159	0.201	0.125	0.088	0.087	0.056

Bags used	7	8	9	10	11	12
Probability	0.025	0.022	0.018	0.008	0.006	0.003

We can find the mean and variance of X by arranging the calculation in the form of a table. Both μ_X and σ_X^2 are the sums of columns in this table.

x_i	p_i	$x_i p_i$	$(x_i - \mu_X)^2 p_i$		
0	0.202	0.00	$(0 - 2.754)^2 (0.202)$	=	1.53207
1	0.159	0.159	$(1 - 2.754)^2 (0.159)$	=	0.48917
2	0.201	0.402	$(2 - 2.754)^2 (0.201)$	=	0.11427
3	0.125	0.375	$(3 - 2.754)^2 (0.125)$	=	0.00756
4	0.088	0.352	$(4 - 2.754)^2 (0.088)$	=	0.13662
5	0.087	0.435	$(5 - 2.754)^2 (0.087)$	=	0.43887
6	0.056	0.336	$(6 - 2.754)^2 (0.056)$	=	0.59004
7	0.025	0.175	$(7 - 2.754)^2 (0.025)$	=	0.45071
8	0.022	0.176	$(8 - 2.754)^2 (0.022)$	=	0.60545
9	0.018	0.162	$(9 - 2.754)^2 (0.018)$	=	0.70223
10	0.008	0.080	$(10 - 2.754)^2 (0.008)$	=	0.42004
11	0.006	0.066	$(11 - 2.754)^2 (0.006)$	=	0.40798
12	0.003	0.036	$(12 - 2.754)^2 (0.003)$	=	0.25647
	$\mu_X = 2.754$			$\sigma_X^2 =$	6.151

We see that $\sigma_X^2 = 6.151$. The standard deviation of X is $\sigma_X = \sqrt{6.151} = 2.48$. This standard deviation is a measure of the variability of the daily demand of blood bags. As in the case of distributions for data, the connection of standard deviation to probability is easiest to understand for Normal distributions (e.g., the 68–95–99.7 rule). For general distributions, we are content to understand that the standard deviation provides us with a basic measure of variability. ■

68–95–99.7 rule, p. 45

The presented formula for σ_X^2 clearly shows that the variance is a weighted average of the squared deviations of the variable X from its mean μ_X. An alternative formula involves less arithmetic operations and is often easier to use, but we do not present it here in the main text because it does not obviously communicate the concept of variance. This alternative formula is shown in Exercise 5.120 (page 288).

APPLY YOUR KNOWLEDGE

5.21 Managing new-product development process. Exercise 5.19 (page 237) gives the distribution of time to complete two steps in the new-product development process.

(a) Calculate the variance and the standard deviation of the number of weeks X to complete the development of product specifications.

(b) Calculate σ_Y^2 and σ_Y for the design of the manufacturing-process step.

Rules for variances and standard deviations

What are the facts for variances that parallel Rules 1, 2, and 3 for means? *The mean of a sum of random variables is always the sum of their means, but this addition rule is true for variances only in special situations.* To understand why, take X to be the percent of a family's after-tax income that is spent, and take Y to be the percent that is saved. When X increases, Y decreases by the same amount. Although X and Y may vary widely from year to year, their sum $X + Y$ is always 100% and does not vary at all. It is the association between the variables X and Y that prevents their variances from adding.

If random variables are independent, this kind of association between their values is ruled out and their variances do add. As defined in Chapter 4 for general events A and B (page 191), two random variables X and Y are independent if knowing that any event involving X alone did or did not occur tells us nothing about the occurrence of any event involving Y alone.

Probability models often assume independence when the random variable outcomes appear unrelated to each other. *You should ask in each instance whether the assumption of independence seems reasonable.*

correlation, p. 75

When random variables are not independent, the variance of their sum depends on the *correlation* between them as well as on their individual variances. In Chapter 2, we examined the correlation r between two observed variables measured on the same individuals. We defined the correlation r as an average of the products of the standardized x and y observations. The correlation between two random variables is defined in the same way, once again using a weighted average with probabilities as weights. We won't give the details—it is enough to know that the correlation between two random variables has the same basic properties as the correlation r calculated from data. We use ρ, the Greek letter rho, for the correlation between two random variables. The correlation ρ is a number between -1 and 1 that measures the direction and strength of the linear relationship between two variables. **The correlation between two independent random variables is zero.**

Returning to the family finances example, if X is the percent of a family's after-tax income that is spent and Y is the percent that is saved, then $Y = 100 - X$. This is a perfect linear relationship with a negative slope, so the correlation between X and Y is $\rho = -1$. With the correlation at hand, we can state the rules for variances.

> **RULES FOR VARIANCES AND STANDARD DEVIATIONS OF LINEAR TRANSFORMATIONS, SUMS, AND DIFFERENCES**
>
> **Rule 1.** If X is a random variable and a and b are fixed numbers, then
> $$\sigma_{a+bX}^2 = b^2 \sigma_X^2$$
>
> **Rule 2.** If X and Y are independent random variables, then
> $$\sigma_{X+Y}^2 = \sigma_X^2 + \sigma_Y^2$$
> $$\sigma_{X-Y}^2 = \sigma_X^2 + \sigma_Y^2$$

This is the **addition rule for variances of independent random variables.**

Rule 3. If X and Y have correlation ρ, then

$$\sigma^2_{X+Y} = \sigma^2_X + \sigma^2_Y + 2\rho\sigma_X\sigma_Y$$

$$\sigma^2_{X-Y} = \sigma^2_X + \sigma^2_Y - 2\rho\sigma_X\sigma_Y$$

This is the **general addition rule for variances of random variables.**

To find the standard deviation of linear transformations, sums, or differences, apply the square root operation *after* the desired variance has been determined.

Adding a constant a to a random variable changes its mean but does not change its variability. Thus, as seen with Rule 1, the constant a is not part of the variance formula for the transformed random variable. Because a variance is the average of squared deviations from the mean, multiplying X by a constant b multiplies σ^2_X by the square of the constant. For example, the variances of $2X$ and $-2X$ are both equal to $4\sigma^2_X$. Since we find the standard deviation *after* finding the variance, this implies that the standard deviations of $2X$ and $-2X$ are both equal to $2\sigma_X$; remember that standard deviation cannot be negative, so we always take the positive square root.

Because the square of -1 is 1, the addition rule says that the variance of a difference between independent random variables is the *sum* of the variances. For independent random variables, the difference $X - Y$ is more variable than either X or Y alone because variations in both X and Y contribute to variation in their difference.

Let's take a closer look at Rule 2 to underscore the need to take the square root after the variance has been determined. Suppose X and Y are independent random variables with $\sigma_X = 3$ and $\sigma_Y = 4$. It is tempting to believe that σ_{X+Y} is the sum of these two standard deviations and equal to 7. *But, this result is wrong.* Instead, the variance of the sum of the random variables should be calculated first and *then* the square root taken. These steps are combined here:

$$\sigma_{X+Y} = \sqrt{\sigma^2_{X+Y}} = \sqrt{\sigma^2_X + \sigma^2_Y} = \sqrt{3^2 + 4^2} = 5$$

EXAMPLE 5.11

Payoff in the Tri-State Pick 3 Lottery The payoff X of a $1 ticket in the Tri-State Pick 3 game is $500 with probability 1/1000 and 0 the rest of the time. Here is the combined calculation of mean and variance:

x_i	p_i	$x_i p_i$	$(x_i - \mu_X)^2 p_i$		
0	0.999	0	$(0 - 0.5)^2 (0.999)$	=	0.24975
500	0.001	0.5	$(500 - 0.5)^2 (0.001)$	=	249.50025
		$\mu_X = 0.5$	σ^2_X	=	249.75

The mean payoff is 50 cents. The standard deviation is $\sigma_X = \sqrt{249.75} = \$15.80$. It is usual for games of chance to have large standard deviations because large variability makes gambling exciting. ∎

If you buy a Pick 3 ticket, your winnings are $W = X - 1$ because the dollar you paid for the ticket must be subtracted from the payoff. Let's find the mean and variance for this random variable.

EXAMPLE 5.12

Winnings in the Tri-State Pick 3 Lottery By the rules for means, the mean amount you win is

$$\mu_W = \mu_X - 1 = -\$0.50$$

That is, you lose an average of 50 cents on a ticket. The rules for variances remind us that the variance and standard deviation of the winnings $W = X - 1$ are the same as those of X. Subtracting a fixed number changes the mean but not the variance. ∎

Suppose now that you buy a $1 ticket on each of two different days. The payoffs X and Y on the two tickets are independent because separate drawings are held each day. Your total payoff is $X + Y$. Let's find the mean and standard deviation for this payoff.

EXAMPLE 5.13

Two Tickets The mean for the payoff for the two tickets is

$$\mu_{X+Y} = \mu_X + \mu_Y = \$0.50 + \$0.50 = \$1.00$$

Because X and Y are independent, the variance of $X + Y$ is

$$\sigma^2_{X+Y} = \sigma^2_X + \sigma^2_Y = 249.75 + 249.75 = 499.5$$

The standard deviation of the total payoff is

$$\sigma_{X+Y} = \sqrt{499.5} = \$22.35$$

This is not the same as the sum of the individual standard deviations, which is $15.80 + $15.80 = $31.60. Variances of independent random variables add together; standard deviations do not. ∎

When we add random variables that are correlated, we need to use the correlation for the calculation of the variance, but not for the calculation of the mean. Here are two examples.

EXAMPLE 5.14

Aggregating Demand in a Supply Chain In Example 5.8 (page 236), we learned that the lead time demands for SurgeArresters in two markets are Normally distributed with

X = Milwaukee warehouse $\mu_X = 415$ units $\sigma_X = 48$ units
Y = Chicago warehouse $\mu_Y = 2689$ units $\sigma_Y = 272$ units

Based on the given means, we found that the mean aggregated demand μ_{X+Y} is 3104. The variance and standard deviation of the aggregated *cannot be computed* from the information given so far. Not surprisingly, demands in the two markets are not independent because of the proximity of the regions. Therefore, Rule 2 for variances does not apply. We need to know ρ, the correlation between X and Y, to apply Rule 3. Historically, the correlation between Milwaukee demand and Chicago demand is about $\rho = 0.52$. To find the variance of the overall demand, we use Rule 3:

$$\sigma^2_{X+Y} = \sigma^2_X + \sigma^2_Y + 2\rho\sigma_X\sigma_Y$$
$$= (48)^2 + (272)^2 + (2)(0.52)(48)(272)$$
$$= 89,866.24$$

The variance of the sum $X + Y$ is greater than the sum of the variances $\sigma_X^2 + \sigma_Y^2$ because of the positive correlation between the two markets. We find the standard deviation from the variance:

$$\sigma_{X+Y} = \sqrt{89{,}866.24} = 299.78$$

Notice that even though the variance of the sum is greater than the sum of the variances, the standard deviation of the sum is less than the sum of the standard deviations. Here lies the potential benefit of a centralized warehouse. To protect against stock-outs, ElectroWorks maintains safety stock for a given product at each warehouse. Safety stock is extra stock in hand over and above the mean demand. For example, if ElectroWorks has a policy of holding two standard deviations of safety stock, then the amount of safety stock (rounded to the nearest integer) at warehouses would be

Location	Safety Stock
Milwaukee warehouse	2(48) = 96 units
Chicago warehouse	2(272) = 544 units
Centralized warehouse	2(299.78) = 600 units

The combined safety stock for the Milwaukee and Chicago warehouses is 640 units, which is 40 more units required than if distribution was operated out of a centralized warehouse. Now imagine the implication for safety stock when you take into consideration not just one part but *thousands* of parts that need to be stored.

risk pooling

This example illustrates the important supply chain concept known as **risk pooling**. Many companies such as Walmart and e-commerce retailer Amazon take advantage of the benefits of risk pooling as illustrated by this example. ■

EXAMPLE 5.15

CASE 5.2 **Portfolio Analysis** With the rules for variances, we can complete our analysis of the portfolio constructed on a 70/30 mix of S&P 500 index shares and utility sector shares. Based on monthly returns between 2000 and 2018, we have

$X =$ monthly return for S&P 500 index $\mu_X = 0.351\%$ $\sigma_X = 4.195\%$
$Y =$ monthly return for utilities index $\mu_Y = 0.660\%$ $\sigma_Y = 4.287\%$
Correlation between X and Y: $\rho = 0.449$

In Example 5.9 (page 237), we found that the mean return R is 0.444%. To find the variance of the portfolio return, combine Rules 1 and 3:

$$\begin{aligned}\sigma_R^2 &= \sigma_{0.7X}^2 + \sigma_{0.3Y}^2 + 2\rho\sigma_{0.7X}\sigma_{0.3Y} \\ &= (0.7)^2\sigma_X^2 + (0.3)^2\sigma_Y^2 + 2\rho(0.7 \times \sigma_X)(0.3 \times \sigma_Y) \\ &= (0.7)^2(4.195)^2 + (0.3)^2(4.287)^2 + (2)(0.449)(0.7 \times 4.195)(0.3 \times 4.287) \\ &= 13.67 \\ \sigma_R &= \sqrt{13.67} = 3.697\%\end{aligned}$$

We see that the portfolio has a smaller mean return than investing all of our funds in the utilities index. However, what is gained is that the portfolio has less variability (or risk) than investing all of our funds in one or the other index. ■

Example 5.15 illustrates the first step in modern finance, using the mean and standard deviation to describe the behavior of a particular portfolio. We illustrated a 70/30 mix in this example, but what is needed is an exploration of different combinations to seek the best construction of the portfolio.

EXAMPLE 5.16

Portfolio Analysis By performing the mean computations of Example 5.9 (page 237) and the standard deviation computations of Example 5.15 for different mixes, we find the following values.

S&P 500 Proportion	μ_R	σ_R
0.0	0.660	4.287
0.1	0.629	4.064
0.2	0.598	3.879
0.3	0.567	3.739
0.4	0.536	3.648
0.5	0.506	3.610
0.6	0.475	3.627
0.7	0.444	3.697
0.8	0.413	3.819
0.9	0.382	3.986
1.0	0.351	4.195

minimum variance portfolio

From Figure 5.9, we see that the plot of the portfolio mean returns against the corresponding standard deviations forms a parabola. The point on the parabola where the portfolio standard deviation is lowest is the **minimum variance portfolio** (MVP). From the preceding table, we see that the MVP is somewhere near a 50/50 allocation between the two investments. The solid curve of the parabola provides the preferable options where the expected return is, for a given level of risk (i.e., standard deviation), higher than the dashed line option. ■

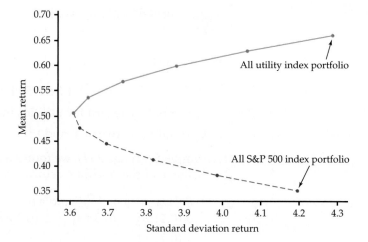

FIGURE 5.9 Mean return of portfolio versus standard deviation of portfolio, for Example 5.16.

APPLY YOUR KNOWLEDGE

5.22 Comparing sales. Tamara and Derek are sales associates in a large electronics and appliance store. Their store tracks each associate's daily sales in dollars. Tamara's sales total X varies from day to day with mean and standard deviation

$$\mu_X = \$1100 \quad \text{and} \quad \sigma_X = \$100$$

Derek's sales total Y also varies, with

$$\mu_Y = \$1000 \quad \text{and} \quad \sigma_Y = \$80$$

Because the store is large and Tamara and Derek work in different departments, we might assume that their daily sales totals vary independently of each other. What are the mean and standard deviation of the difference $X - Y$ between Tamara's daily sales and Derek's daily sales? Tamara sells more on the average. Do you think she sells more every day? Why?

5.23 Comparing sales. It is unlikely that the daily sales of Tamara and Derek in the previous problem are uncorrelated. They will both sell more during the weekends, for example. Suppose that the correlation between their sales is $\rho = 0.4$. Now what are the mean and standard deviation of the difference $X - Y$? Can you explain conceptually why a positive correlation between two variables reduces the variability of the difference between them?

5.24 Managing a new-product development process. Exercise 5.19 (page 237) gives the distributions of X, the number of weeks to complete the development of product specifications, and Y, the number of weeks to complete the design of the manufacturing process. You did some useful variance calculations in Exercise 5.21 (page 240). The cost per week for developing product specifications is $8000, while the cost per week for designing the manufacturing process is $30,000.

(a) Calculate the standard deviation of the cost for each of the two activities using Rule 1 for variances (page 240).

(b) Assuming the activity times are independent, calculate the standard deviation for the total cost of both activities combined.

(c) Assuming $\rho = 0.8$, calculate the standard deviation for the total cost of both activities combined.

(d) Assuming $\rho = 0$, calculate the standard deviation for the total cost of both activities combined. How does this compare with your result in part (b)? In part (c)?

(e) Assuming $\rho = -0.8$, calculate the standard deviation for the total cost of both activities combined. How does this compare with your result in part (b)? In part (c)? In part (d)?

SECTION 5.2 SUMMARY

- The probability distribution of a random variable X, like a distribution of data, has a **mean** μ_X and a **variance** σ_X^2.

- The **law of large numbers** says that the average of the values of X observed in many trials must approach μ.

- The **mean** μ is the balance point of the probability histogram or density curve. If X is **discrete** with possible values x_i having probabilities p_i, the mean is the average of the values of X, each weighted by its probability:

$$\mu_X = x_1 p_1 + x_2 p_2 + \cdots = \sum x_i p_i$$

- The **variance** σ_X^2 is the average squared deviation of the values of the variable from their mean. For a discrete random variable,

$$\sigma_X^2 = (x_1 - \mu_X)^2 p_1 + (x_2 - \mu_X)^2 p_2 + \cdots = \sum (x_i - \mu_X)^2 p_i$$

- The **standard deviation** σ_X is the square root of the variance. The standard deviation measures the variability of the distribution around the mean. It is easiest to interpret for Normal distributions.

- The **mean and variance of a continuous random variable** can be computed from the density curve, but to do so requires more advanced mathematics.

- If a and b are fixed numbers, then

$$\mu_{a+bX} = a + b\mu_X$$
$$\sigma^2_{a+bX} = b^2\sigma^2_X$$

- If X and Y are any two random variables, then

$$\mu_{X+Y} = \mu_X + \mu_Y$$
$$\mu_{X-Y} = \mu_X - \mu_Y$$

- If X and Y are any two random variables having correlation ρ, then

$$\sigma^2_{X+Y} = \sigma^2_X + \sigma^2_Y + 2\rho\sigma_X\sigma_Y$$
$$\sigma^2_{X-Y} = \sigma^2_X + \sigma^2_Y - 2\rho\sigma_X\sigma_Y$$

- If X and Y are **independent,** then $\rho = 0$. In this case,

$$\sigma^2_{X+Y} = \sigma^2_X + \sigma^2_Y$$
$$\sigma^2_{X-Y} = \sigma^2_X + \sigma^2_Y$$

- To find the standard deviation of linear transformations, sums, or differences, apply the square root operation *after* the desired variance has been determined.

SECTION 5.2 EXERCISES

For Exercise 5.15, see page 230; for 5.16, see page 233; for 5.17 to 5.20, see pages 237–238; for 5.21, see page 240; and for 5.22 to 5.24, see pages 244–245.

CASE 5.2 **5.25 Portfolio analysis.** Show that if 20% of the portfolio is based on the S&P 500 index, then the mean and standard deviation of the portfolio are indeed the values given in Example 5.16 (page 244).

5.26 Find some means. Suppose that X is a random variable with mean 20 and standard deviation 4. Also suppose that Y is a random variable with mean 40 and standard deviation 7. Find the mean of the random variable Z for each of the following cases. Be sure to show your work.

(a) $Z = 40 - 5X$

(b) $Z = 15X - 20$

(c) $Z = X + Y$

(d) $Z = X - Y$

(e) $Z = -2X + 3Y$

5.27 Find the variance and the standard deviation. A random variable X has the following distribution.

X	−1	0	1	2
Probability	0.1	0.4	0.3	0.2

Find the variance and the standard deviation for this random variable. Show your work.

5.28 Find some variances and standard deviations. Suppose that X is a random variable with mean 20 and standard deviation 4. Also suppose that Y is a random variable with mean 40 and standard deviation 7. Assume that X and Y are independent. Find the variance and the standard deviation of the random variable Z for each of the following cases. Be sure to show your work.

(a) $Z = 40 - 5X$

(b) $Z = 15X - 20$

(c) $Z = X + Y$

(d) $Z = X - Y$

(e) $Z = -2X + 3Y$

5.29 What happens if the correlation is not zero? Suppose that X is a random variable with mean 20 and standard deviation 4. Also suppose that Y is a random variable with mean 40 and standard deviation 7. Assume that the correlation between X and Y is 0.5. Find the variance and standard deviation of the random variable Z for each of the following cases. Be sure to show your work.

(a) $Z = X + Y$

(b) $Z = X - Y$

(c) $Z = -2X + 3Y$

5.30 What's wrong? In each of the following scenarios, there is something wrong. Describe what is wrong, and give a reason for your answer.

(a) If you toss a fair coin three times and get heads all three times, then the probability of getting a tail on the next toss is much greater than one-half.

(b) If you multiply a random variable by 10, then the mean is multiplied by 10 and the variance is multiplied by 10.

(c) When finding the mean of the sum of two random variables, you need to know the correlation between them.

5.31 Difference between heads and tails. Suppose a fair coin is tossed three times.

(a) Using the labels of "H" and "T," list all the possible outcomes in the sample space.

(b) For each outcome in the sample space, define the random variable D as the number of heads minus the number of tails observed. Use the fact that all outcomes of part (a) are equally likely to find the probability distribution of D.

(c) Use the probability distribution found in (b) to find the mean and standard deviation of D.

5.32 Mean of the distribution for the number of aces. In Exercise 5.8 (page 227), you examined the probability distribution for the number of aces when you are dealt two cards in the game of Texas hold 'em. Let X represent the number of aces in a randomly selected deal of two cards in this game. Here is the probability distribution for the random variable X:

Value of X	0	1	2
Probability	0.8507	0.1448	0.0045

Find μ_X, the mean of the probability distribution of X.

5.33 Standard deviation of the number of aces. Refer to the previous exercise. Find the standard deviation of the number of aces.

5.34 Difference between heads and tails. In Exercise 5.31, you computed the mean and standard deviation directly from the probability distribution of random variable D. Now define X as the number of heads in the three flips, and define Y as the number of tails in the three flips.

(a) Find the probability distribution for X along with the mean μ_X and standard deviation σ_X.

(b) Find the probability distribution for Y along with the mean μ_Y and standard deviation σ_Y.

(c) Explain why the correlation ρ between X and Y is -1.

(d) Define D as $X - Y$. Use the rules of means and variances along with $\rho = 1$ to find the mean and standard deviation of D. Confirm the values are the same as those found in Exercise 5.31.

5.35 Pick 3 and the law of large numbers. In Example 5.3 (page 229), the mean payoff for the Tri-State Pick 3 lottery was found to be $0.50. In our discussion of the law of large numbers, we learned that the mean of a probability distribution describes the long-run average outcome. In this exercise, you will explore this concept using technology.

- *Excel users:* Input the values "0" and "500" in the first two rows of column A. Now input the corresponding probabilities of 0.999 and 0.001 in the first two rows of column B. Next, choose "Random Number Generation" from the **Data Analysis** menu box. Enter "1" in the **Number of Variables** box, enter "30000" in the **Number of Random Numbers** box, choose "Discrete" for the **Distribution** option, enter the cell range of the values and their probabilities ($A$1:$B$2) in **Value and Probability Input Range** box, and finally enter "C1" for the **Output Range**. Click **OK** to find 30,000 random values outputted in the worksheet. In any cell, enter =AVERAGE(c1:c30000) to find the average of the 30,000 values.

- *JMP users:* With a new data table, right-click on header of column 1 and choose **Column Info**. In the drag-down dialog box named **Initialize Data**, pick **Random** option. Choose the bullet option of **Random Indicator**. Input the values of "0" and "500" in the first two **Value** dialog boxes, and input the values of 0.999 and 0.001 in the corresponding **Proportion** dialog boxes. Input "30000" in the **Number of rows** box, and then click **OK** to find 30,000 random values outputted in the worksheet. Find the average of the 30,000 values.

- *Minitab users:* Input the values "0" and "500" in the first two rows of column 1 (c1) of a worksheet. Now input the corresponding probabilities of 0.999 and 0.001 in the first two rows of column 2 (c2). Do the following pull-down sequence: Calc → Random Data → Discrete. Enter "30000" in the **Number of rows of data to generate** box, type "c3" in the **Store in column(s)** box, click-in "c1" in the **Values in** box, and click-in "c2" in the **Probabilities in** box. Click **OK** to find 30,000 random values outputted in the worksheet. Find the average of the 30,000 values.

- *R users:* Type following command at the R prompt:

X<-sample(c(0,500), 30000,
prob=c(0.999,0.001), replace=T)

Use the "mean(X)" command to find the average of the 30,000 random values.

Whether you used Excel, JMP, Minitab, or R, how does the average value of the 30,000 X-values compare with the mean reported in Example 5.3?

CASE 5.2 **5.36 Households and families in government data.** In U.S. government data, a household consists of all occupants of a dwelling unit, while a family consists of two or more persons who live together and are related by blood or marriage. So all families form households, but some households are not families. Here are the distributions of household size and of family size in the United States:

Number of persons	1	2	3	4	5	6	7
Household probability	0.27	0.33	0.16	0.14	0.06	0.03	0.01
Family probability	0.00	0.44	0.22	0.20	0.09	0.03	0.02

Compare the two distributions using probability histograms on the same scale. Also compare the two distributions using means and standard deviations.

Write a summary of your comparisons using your calculations to back up your statements.

CASE 5.2 **5.37 Perfectly negatively correlated investments.** Consider the following quote from an online site providing investment guidance: "Perfectly negatively correlated investments would provide 100% diversification, as they would form a portfolio with zero variance, which translates to zero risk." Consider a portfolio based on two investments (X and Y) with standard deviations of σ_X and σ_Y. In line with the quote, assume that the two investments are perfectly negatively correlated ($\rho = -1$).

(a) Suppose $\sigma_X = 4$, $\sigma_Y = 2$, and the portfolio mix is 70/30 of X to Y. What is the standard deviation of the portfolio? Does the portfolio have zero risk?

(b) Suppose $\sigma_X = 4$, $\sigma_Y = 2$, and the portfolio mix is 50/50. What is the standard deviation of the portfolio? Does the portfolio have zero risk?

(c) Suppose $\sigma_X = 4$, $\sigma_Y = 4$, and the portfolio mix is 50/50. What is the standard deviation of the portfolio? Does the portfolio have zero risk?

(d) Is the online quote a universally true statement? If not, how would you modify it so that it can be stated that the portfolio has zero risk?

5.38 What happens when the correlation is 1? We know that variances add if the random variables involved are uncorrelated ($\rho = 0$), but not otherwise. The opposite extreme is perfect positive correlation ($\rho = 1$). Show by using the general addition rule for variances that in this case the standard deviations add together— that is, $\sigma_{X+Y} = \sigma_X + \sigma_Y$ if $\rho_{XY} = 1$.

5.39 Making glassware. In a process for manufacturing glassware, glass stems are sealed by heating them in a flame. The temperature of the flame varies. Here is the distribution of the temperature X measured in degrees Celsius:

Temperature	540°	545°	550°	555°	560°
Probability	0.1	0.25	0.3	0.25	0.1

(a) Find the mean temperature μ_X and the standard deviation σ_X.

(b) The target temperature is 550°C. Use the rules for means and variances to find the mean and standard deviation of the number of degrees off target, $X - 550$.

(c) A manager asks for results in degrees Fahrenheit. The conversion of X into degrees Fahrenheit is given by

$$Y = \frac{9}{5}X + 32$$

What are the mean μ_Y and standard deviation σ_Y of the temperature of the flame in the Fahrenheit scale?

 Portfolio analysis. Here are the means, standard deviations, and correlations for the monthly returns from three Fidelity mutual funds for 10 years ending in February 2018. Because there are three random variables, there are three correlations. We use subscripts to indicate the pair of random variables to which a correlation refers.

X = Biotechnology monthly return $\quad \mu_X = 1.592\% \quad \sigma_X = 7.204\%$

Y = Software and IT Services monthly return $\quad \mu_Y = 1.444\% \quad \sigma_Y = 5.675\%$

Z = Defense and Aerospace monthly return $\quad \mu_Z = 1.178\% \quad \sigma_Z = 5.594\%$

Correlations
$\rho_{XY} = 0.535 \quad \rho_{XZ} = 0.564 \quad \rho_{YZ} = 0.753$

Exercises 5.40 through 5.42 make use of these historical data.

CASE 5.2 **5.40 Diversification.** Currently, Michael is exclusively invested in the Fidelity Biotechnology fund. Even though the mean return for this fund is quite high, it comes with greater volatility and risk. So, Michael decides to diversify his portfolio by constructing a portfolio of 80% Biotechnology fund and 20% Software and IT Services fund. Based on the provided historical performance, what is the expected return and standard deviation of this portfolio? Relative to his original investment scheme, what is the percentage reduction in his risk level (as measured by standard deviation) associated with this particular portfolio?

CASE 5.2 **5.41 More on diversification.** Continuing with the previous exercise, suppose Michael's primary goal is to seek a portfolio mix of the Biotechnology and Software and IT Services funds that will give him *minimal* risk as measured by standard deviation of the portfolio. Compute the standard deviations for portfolios based on the proportion of Biotechnology fund in the portfolio ranging from 0 to 1 in increments of 0.1. You may wish to do these calculations in Excel. What is your recommended mix of Biotechnology and Software and IT Services funds for Michael? What is the standard deviation for your recommended portfolio?

CASE 5.2 **5.42 Larger portfolios.** Portfolios often contain more than two investments. The rules for means and variances continue to apply, though the arithmetic gets messier. A portfolio containing proportions a of Biotechnology fund, b of Software and IT Services fund, and c of Defense and Aerospace fund has a return $R = aX + bY + cZ$. Because $a, b,$ and c are the proportions invested in the three funds, $a + b + c = 1$. The mean and variance of the portfolio return R are

$$\mu_R = a\mu_X + b\mu_Y + c\mu_Z$$
$$\sigma_R^2 = a^2\sigma_X^2 + b^2\sigma_Y^2 + c^2\sigma_Z^2 + 2ab\rho_{XY}\sigma_X\sigma_Y + 2ac\rho_{XZ}\sigma_X\sigma_Z + 2bc\rho_{YZ}\sigma_Y\sigma_Z$$

Having seen the advantages of diversification, Michael decides to invest his funds 20% in Biotechnology, 35% in Software and IT Services, and 45% in Defense and Aerospace. What are the (historical) mean and standard deviation of the monthly returns for this portfolio?

5.3 Common Discrete Distributions

When you complete this section, you will be able to:

- Identify the required conditions such that a count X can be modeled using the binomial distribution.
- Determine when the binomial distribution can be used as an adequate approximation for the study of counts from an SRS.
- Compute binomial probabilities using a formula or software.
- Calculate the mean and standard deviation of a binomial random variable.
- Identify the required conditions such that a count X can be modeled using the Poisson distribution.
- Compute Poisson probabilities using a formula or software.
- Determine when the Poisson distribution can be used as an adequate approximation of the binomial distribution.
- Check the adequacy of the binomial or Poisson models using the observed data.

In our general discussions about discrete random variables in Sections 5.1 and 5.2, we were introduced to the following concepts:

- The probability distribution (page 221), which gives us the values the discrete random variable takes, and the probabilities of those values.
- The mean of a discrete random variable (page 230).
- The variance (standard deviation) of a discrete random variable (pages 238–239).

In this section, we investigate two important discrete distributions that have wide applications in real-world settings. These distributions both relate to the study of counts but under a different set of conditions.

Binomial distributions

Consider the following example.

EXAMPLE 5.17

Cola Wars A blind taste test of two diet colas (labeled "A" and "B") asks 200 randomly chosen consumers which cola is preferred. We would like to view the responses of these consumers as representative of a larger population of consumers who hold similar preferences. That is, we will view the responses of the sampled consumers as an SRS from a population. ∎

When there are only two possible outcomes for a random variable, we can summarize the results by giving the count for one of the possible outcomes. We let n represent the sample size, and we use X to represent the random variable that gives the count for the outcome of interest.

EXAMPLE 5.18

The Random Variable of Interest In our marketing study of consumers, $n = 200$. We will ask each consumer in our study whether he or she prefers Cola A or Cola B. The variable X is the number of consumers who prefer Cola A. Suppose that we observe $X = 138$. ∎

In our example, we chose the random variable X to be the number of consumers who prefer Cola A over Cola B. We could have chosen X to be the number of consumers who prefer Cola B over Cola A. The choice is yours. Often, we make the choice based on how we would like to describe the results in a written summary.

The binomial distributions for sample counts

The distribution of a count X depends on how the data are produced. Here is a simple but common situation.

THE BINOMIAL SETTING

1. There is a fixed number of observations n.
2. The n observations are all **independent.** That is, knowing the result of one observation tells you nothing about the other observations.
3. Each observation falls into one of just two categories, which, for convenience, we call "success" and "failure."
4. The probability of a success, call it p, is the same for each observation.

Think of tossing a coin n times as an example of the binomial setting. Each toss, or observation, gives either a heads or a tails, and the outcomes of successive tosses are independent. If we call heads a success, then p is the probability of a head and remains the same as long as we toss the same coin. The number of heads we count is a random variable X. The distribution of X, and more generally the distribution of the count of successes in any binomial setting, is completely determined by the number of observations n and the success probability p.

BINOMIAL DISTRIBUTION

The distribution of the count X of successes in the binomial setting is the **binomial distribution** with parameters n and p. The parameter n is the number of observations, and p is the probability of a success on any one observation. The possible values of X are the whole numbers from 0 to n. As an abbreviation, we say that X is $B(n, p)$.

The binomial distributions are an important family of discrete probability distributions. That said, *the most important skill when using binomial distributions is the ability to recognize situations to which they do and don't apply.* This can be done by checking all the facets of the binomial setting.

EXAMPLE 5.19

Binomial Examples? (a) Analysis of the 50 years of weekly S&P 500 price changes reveals that the directions are independent of each other, with the probability of a positive price change being 0.56. Defining a "success" as a positive price change, let X be the number of successes over the next year—that is, over the next 52 weeks. Given the independence of trials, it is reasonable to assume that X has the $B(52, 0.56)$ distribution.

(b) Engineers define reliability as the probability that an item will perform its function under specific conditions for a specific period of time. Replacement heart valves made of animal tissue, for example, have probability 0.77 of performing well for 15 years.[11] The probability of failure within 15 years is, therefore, 0.23. It is reasonable to assume that valves in different patients fail (or not) independently of each other. The number of patients in a group of 500 who will need another valve replacement within 15 years has the $B(500, 0.23)$ distribution.

(c) Deal 10 cards from a shuffled standard deck of 52 cards (26 red cards and 26 black cards) and count the number X of red cards. There are 10 observations, and each gives either a red or a black card. A "success" is a red card. But the observations are *not* independent. If the first card is black, the second is more likely to be red because there are more red cards than black cards left in the deck. The count X does *not* have a binomial distribution. ■

APPLY YOUR KNOWLEDGE

In each of Exercises 5.43 to 5.46, X is a count. Does X have a binomial distribution? If so, give the distribution of X. If not, give your reasons as to why not.

5.43 Toss a coin. Toss a fair coin 20 times. Let X be the number of heads that you observe.

5.44 Card dealing. Define X as the number of red cards observed in the following card-dealing scenarios:

(a) Deal one card from a standard 52-card deck.

(b) Deal one card from a standard 52-card deck, record its color, return it to the deck, and shuffle the cards. Repeat this experiment 10 times.

5.45 Customer satisfaction calls. The service department of an automobile dealership follows up each service encounter with a customer satisfaction survey by means of a phone call. On a given day, let X be the number of customers a service representative has to call until a customer is willing to participate in the survey.

5.46 Teaching office software. A company uses a computer-based system to teach clerical employees new office software. After a lesson, the computer presents 10 exercises. The student solves each exercise and enters the answer. The computer gives additional instruction between exercises if the answer is wrong. The count X is the number of exercises that the student gets right.

The binomial distributions for statistical sampling

What is the estimated proportion of consumers who use smartphones to research a product prior to going to a store? What is the estimated proportion of unemployed adults in a given region? What is the estimated proportion of registered voters who favor a particular candidate? These questions all relate to making inferences about the proportion p of "successes" in a population. In each of these cases, a survey or a poll based on an SRS from the population would be drawn and the obtained data then used for estimation purposes. As will be noted in Chapters 6 and 10, the binomial distribution underlies the development of the most common inferential procedures about p. With this point in mind, let us consider the specific example of drawing an SRS from a population.

Inspecting a Supplier's Products A manufacturing firm purchases components for its products from suppliers. Good practice calls for suppliers to manage their production processes to ensure good quality. You can find some discussion of statistical methods for managing and improving quality in Chapter 15 (online). Unfortunately, there have been quality lapses in the switches supplied by a regular vendor. While working with the supplier to improve its processes, the manufacturing firm temporarily institutes an *acceptance sampling* plan to assess the quality of shipments of switches. If a random sample from a shipment contains too many switches that don't conform to specifications, the firm will not accept the shipment.

A quality engineer at the firm chooses an SRS of 150 switches from a shipment of 10,000 switches (the population). Suppose that (unknown to the engineer) 8% of the switches in the shipment are nonconforming. The engineer counts the number X of nonconforming switches in the sample. Is the count X of nonconforming switches in the sample a binomial random variable? ∎

It turns out the setting described in Case 5.3 is not quite a binomial setting. Just as removing one card in Example 5.19(c) changed the makeup of the deck, removing one switch changes the proportion of nonconforming switches remaining in the shipment. If there are initially 800 nonconforming switches, the proportion remaining is $800/9999 = 0.080008$ if the first switch drawn is okay and $799/9999 = 0.079908$ if the first switch fails inspection. That is, the state of the second switch chosen is not independent of the first. Nevertheless, these proportions are so close to 0.08 that, for practical purposes, we can act as if removing one switch has no effect on the proportion of nonconforming switches remaining. We act as if the count X of nonconforming switches in the sample has the binomial distribution $B(150, 0.08)$.

In general, *choosing an SRS from a population is never quite a binomial setting.* This is the case for the sampling of components in a shipment as illustrated in Case 5.3—but it is also the case when conducting a survey or poll of people to make inferences about the population proportion. Exercises 5.74 and 5.75 (pages 270–271) explore the exact distribution, known as a **hypergeometric distribution**, that the binomial distribution is approximating in the case of an SRS.

hypergeometric distribution

Fortunately, the binomial distribution is suitable for practical applications when the population size is large relative to the sample size, as typically is the case in nearly all real-world applications.

DISTRIBUTION OF COUNT OF SUCCESSES IN AN SRS

A population contains proportion p of successes. If the population is much larger than the sample, the count X of successes in an SRS of size n has approximately the binomial distribution $B(n, p)$.

The accuracy of this approximation improves as the size of the population increases relative to the size of the sample. As a rule of thumb, we use the binomial distribution for counts when the population is at least 20 times as large as the sample.

In Chapters 6 and 10, the binomial distribution will be used in the development of inference-based methods for proportions. The discussion here explains why we will encounter conditional statements such as "It is approximately correct for an SRS from a large population" or "Choose an SRS of size n from a large population."

Finding binomial probabilities

Later, we give a formula for the probability that a binomial random variable takes any of its values. Even though it is useful to know and work with the formula for fundamental understanding of the binomial distribution, you will seldom have to use this formula for calculations in practical applications. Some calculators and most statistical software packages calculate binomial probabilities.

EXAMPLE 5.20

The Probability of Nonconforming Switches The quality engineer in Case 5.3 inspects an SRS of 150 switches from a large shipment in which 8% fail to conform to specifications. What is the probability that exactly 10 switches in the sample fail inspection? What is the probability that the quality engineer finds no more than 10 nonconforming switches? Figure 5.10 shows the Minitab output for the desired probabilities. You see from the output that the count X has the $B(150, 0.08)$ distribution

$$P(X = 10) = 0.106959$$
$$P(X \leq 10) = 0.338427$$

It is easy to repeat these calculations in other software. The following are examples from Excel and R:

- For Excel, we use the "BINOM.DIST()" function. This function has four arguments: the first is the value of the number of successes (k), the second is n, the third is p, and the fourth is either "0" (reports individual probability) or "1" (reports cumulative probability). In any empty cell, enter the formula =BINOM.DIST(10,150,0.08,0) to find the value 0.106959.

- Using the R software, the probabilities are obtained using the first three arguments in two different functions.

```
dbinom(10,150,0.08)
[1] 0.1069587
pbinom(10,150,0.08)
[1] 0.3384272
```

Typically, the output supplies more decimal places than we need and sometimes uses labels that may not be helpful (e.g., "Probability Density Function" when the distribution is discrete, not continuous). But, as usual with software, we can ignore these distractions and find the results we need. ∎

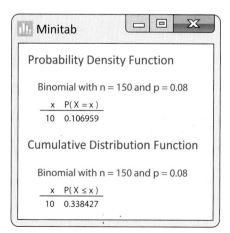

FIGURE 5.10 Binomial probabilities for Example 5.20; output from Minitab software.

If you do not have suitable computing facilities, you can still shorten the work of calculating binomial probabilities for some values of n and p by looking up probabilities in Table C in the back of this book. The entries in the table are the probabilities $P(X=k)$ for a binomial random variable X.

EXAMPLE 5.21

The Probability Histogram Suppose that the quality engineer chooses just 15 switches for inspection. What is the probability that no more than one of the 15 is nonconforming? The count X of nonconforming switches in the sample has approximately the $B(15, 0.08)$ distribution. Figure 5.11 is a probability histogram for this distribution. The distribution is strongly skewed. Although X can take any whole-number value from 0 to 15, the probabilities of values larger than 5 are so small that they do not appear in the histogram.

We want to calculate

$$P(X \leq 1) = P(X=0) + P(X=1)$$

when X has the $B(15, 0.08)$ distribution. To use Table C for this calculation, look opposite $n = 15$ and under $p = 0.08$. This part of the table appears at the left. The entry opposite each k is $P(X = k)$. Blank entries are 0 to four decimal places, so we have omitted most of them here. You see that

$$P(X \leq 1) = P(X = 0) + P(X = 1)$$
$$= 0.2863 + 0.3734 = 0.6597$$

		p
n	k	.08
15	0	.2863
	1	.3734
	2	.2273
	3	.0857
	4	.0223
	5	.0043
	6	.0006
	7	.0001
	8	
	9	

Approximately two-thirds of all samples will contain no more than one nonconforming switch. In fact, almost 29% of the samples will contain no bad switches. A sample of size 15 cannot be trusted to provide adequate evidence about the presence of nonconforming items in this population. In contrast, for a sample of size 50, there is only a 1.5% risk that no bad switch will be revealed in a sample from this population. Calculations such as these can be used to design appropriate acceptance sampling schemes. ∎

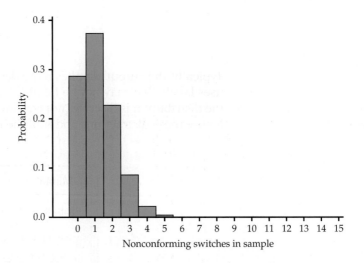

FIGURE 5.11 Probability histogram for the binomial distribution with $n = 15$ and $p = 0.08$, for Example 5.21.

The values of p that appear in Table C are all 0.5 or smaller. When the probability of a success is greater than 0.5, restate the problem in terms of the number of failures. The probability of a failure is less than 0.5 when the probability of a success exceeds 0.5. When using the table, always stop to ask whether you must count successes or failures.

EXAMPLE 5.22

		p
n	k	.07
15	0	.3367
	1	.3801
	2	.2003
	3	.0653
	4	.0148
	5	.0024
	6	.0003
	7	
	8	
	9	

Priority Mail The U.S. Postal Service (USPS) provides a two-day priority shipping service for letters, documents, or merchandise. In 2018, USPS reported that the nationwide percent of on-time two-day deliveries was 93%.[12] Suppose a small business uses USPS's two-day service for fulfilling 15 customer orders and later learns that four orders did not arrive on time. This news left the small business questioning its loyalty to USPS. Is it unusual for USPS to perform so poorly?

To answer this question, we can ask one of two equivalent questions: (1) What is the probability of observing 11 or fewer on time deliveries? or (2) What is the probability of observing four or more late deliveries? In either case, we assume that deliveries are independent with probability of being on time equal to 0.93. Software can compute either probability directly. However, because the probability of being on time is greater than 0.5, we count late deliveries so that we can use Table C. The probability of a late delivery is $1 - 0.93$, or 0.07. The number X of late deliveries in 15 attempts has the binomial distribution with $n = 15$ and $p = 0.07$.

We want the probability of four or more late deliveries. From Table C, this is

$$P(X \geq 4) = P(X = 4) + P(X = 5) + P(X = 6) + \cdots + P(X = 15)$$
$$= 0.0148 + 0.0024 + 0.0003 + \cdots + 0.0000 = 0.0175$$

We would expect to find four or more late deliveries out of 15 deliveries less than 2% of the time. This is a fairly unusual outcome and falls outside the range of unusual chance variation in USPS delivery performance. ∎

APPLY YOUR KNOWLEDGE

5.47 Find the probabilities.

(a) Suppose that X has the $B(6, 0.25)$ distribution. Use software or Table C to find $P(X = 0)$ and $P(X \geq 4)$.

(b) Suppose that X has the $B(6, 0.75)$ distribution. Use software or Table C to find $P(X = 6)$ and $P(X \leq 2)$.

(c) Explain the relationship between your answers to parts (a) and (b) of this exercise.

5.48 Restaurant survey. You operate a restaurant. You read that a sample survey by the National Restaurant Association shows that 60% of adults are more likely to visit a restaurant that offers locally sourced food items. To help plan your menu, you decide to conduct a sample survey in your own area. You will use random digit dialing to contact an SRS of 20 households by telephone.

(a) If the national result holds in your area, it is reasonable to use the binomial distribution with $n = 20$ and $p = 0.60$ to describe the count X of respondents who favor restaurants offering locally sourced food items. Explain why.

(b) Fourteen of the 20 respondents (70% of the sample) say they favor restaurants that offer locally sourced food items. Is this reason enough to believe that the percent in your area is higher than the national percent of 60%? To answer this question, use software to find the probability that X is 14 or greater if $p = 0.6$ is true. If this probability is very small, that is reason to think that p is actually greater than 0.6 in your area.

5.49 Do our athletes graduate? A university claims that at least 80% of its basketball players get degrees. An investigation examines the fate of 20 players who entered the program over a period of several years that ended six years ago. Of these players, 11 graduated and the remaining nine are no longer in school. If the university's claim is true, the number of players who graduate among the 20 should have the binomial distribution with $n = 20$ and p at least equal to 0.8.

(a) Use software to find the probability that 11 or fewer players graduate using $p = 0.8$.

(b) Without doing any computation, explain what the probability that 11 or fewer players graduate would be in comparison to the result from part (a) if $p > 0.8$.

(c) What does the probability you found in part (a) suggest about the university's claim?

Binomial formula

We can find a formula that generates the binomial probabilities from software or by using Table C. Finding the formula for the probability that a binomial random variable takes a particular value entails adding probabilities for the different ways of getting exactly that many successes in n observations. An example will guide us toward the formula we want.

EXAMPLE 5.23

Determining Consumer Preferences Suppose that market research shows that your product is preferred over competitors' products by 25% of all consumers. If X is the count of the number of consumers who prefer your product in a group of five consumers, then X has a binomial distribution with $n = 5$ and $p = 0.25$, provided the five consumers make choices independently. What is the probability that exactly two consumers in the group prefer your product? We are seeking $P(X = 2)$.

Because the method doesn't depend on the specific example, we will use "S" for success and "F" for failure. Here, "S" stands for a consumer preferring your product over the competitors' products. We do the work in two steps.

Step 1. Find the probability that a specific two of the five consumers—say, the first and the third—give successes. This is the outcome SFSFF. Because choices are independent, the multiplication rule for independent events applies. The probability we want is

$$P(\text{SFSFF}) = P(S)P(F)P(S)P(F)P(F)$$
$$= (0.25)(0.75)(0.25)(0.75)(0.75)$$
$$= (0.25)^2(0.75)^3$$

Step 2. Observe that the probability of *any one* arrangement of two S's and three F's has this same probability. This is true because we multiply together 0.25 twice and 0.75 three times whenever we have two S's and three F's. The probability that $X = 2$ is the probability of getting two S's and three F's in any arrangement whatsoever. Here are all the possible arrangements:

SSFFF SFSFF SFFSF SFFFS FSSFF
FSFSF FSFFS FFSSF FFSFS FFFSS

There are 10 of them, all with the same probability. The overall probability of two successes is therefore

$$P(X = 2) = 10(0.25)^2(0.75)^3 = 0.2637$$

Approximately 26% of the time, samples of five independent consumers will produce exactly two who prefer your product over competitors' products. ■

The steps of this calculation in Example 5.23 work for any binomial probability. To combine them, we must count the number of arrangements of k successes in n observations. We use the following fact to do this counting without actually listing all the arrangements.

BINOMIAL COEFFICIENT

The number of ways of arranging k successes among n observations is given by the **binomial coefficient**

$$\binom{n}{k} = \frac{n!}{k!(n-k)!}$$

for $k = 0, 1, 2, \ldots, n$.

factorial

The formula for binomial coefficients uses **factorial** notation. The factorial $n!$ for any positive whole number n is

$$n! = n \times (n-1) \times (n-2) \times \cdots \times 3 \times 2 \times 1$$

Also, $0! = 1$. Notice that the larger of the two factorials in the denominator of a binomial coefficient will cancel much of the $n!$ in the numerator. For example, the binomial coefficient we need for Example 5.23 is

$$\binom{5}{2} = \frac{5!}{2!3!}$$
$$= \frac{(5)(4)(3)(2)(1)}{(2)(1) \times (3)(2)(1)}$$
$$= \frac{(5)(4)}{(2)(1)} = \frac{20}{2} = 10$$

This agrees with our previous calculation.

The notation $\binom{n}{k}$ is not related to the fraction $\frac{n}{k}$. A helpful way to remember its meaning is to read it as "binomial coefficient n choose k." Binomial coefficients have many uses in mathematics, but we are interested in them only as an aid to finding binomial probabilities. The binomial coefficient $\binom{n}{k}$ counts the number of ways in which k successes can be distributed among n observations. The binomial probability $P(X = k)$ is this count multiplied by the probability of any specific arrangement of the k successes. Here is the formula we seek.

BINOMIAL PROBABILITY

If X has the binomial distribution $B(n, p)$, with n observations and probability p of success on each observation, the possible values of X are $0, 1, 2, \ldots, n$. If k is any one of these values, the **binomial probability** is

$$P(X = k) = \binom{n}{k} p^k (1-p)^{n-k}$$

Here is an example of the use of the binomial probability formula.

EXAMPLE 5.24

Inspecting Switches Consider the scenario of Example 5.21 (page 254), in which the number X of switches that fail inspection closely follows the binomial distribution with $n = 15$ and $p = 0.08$.

The probability that no more than one switch fails is

$$P(X \leq 1) = P(X = 0) + P(X = 1)$$
$$= \binom{15}{0}(0.08)^0(0.92)^{15} + \binom{15}{1}(0.08)^1(0.92)^{14}$$
$$= \frac{15!}{0!15!}(1)(0.2863) + \frac{15!}{1!14!}(0.08)(0.3112)$$
$$= (1)(1)(0.2863) + (15)(0.08)(0.3112)$$
$$= 0.2863 + 0.3734 = 0.6597$$

This calculation used the facts that $0! = 1$ and that $a^0 = 1$ for any number $a \neq 0$. The result agrees with that obtained from Table C in Example 5.21. ■

APPLY YOUR KNOWLEDGE

5.50 Hispanic representation. A factory employs several thousand workers, of whom 30% are Hispanic. If the 10 members of the union executive committee were chosen from the workers at random, the number of Hispanics on the committee X would have the binomial distribution with $n = 10$ and $p = 0.3$.

(a) Use the binomial formula to find $P(X = 3)$.

(b) Use the binomial formula to find $P(X \leq 3)$.

5.51 Misleading résumés. In Exercise 4.27 (page 193), it was stated that 24.39% of executive job applicants lied on their résumés. Suppose an executive job hunter randomly selects five résumés from an executive job applicant pool. Let X be the number of misleading résumés found in the sample.

(a) What are the possible values of X?

(b) Use the binomial formula to find $P(X = 2)$.

(c) Use the binomial formula to find the probability of at least one misleading résumé being included in the sample.

Binomial mean and standard deviation

If a count X has the $B(n, p)$ distribution, what are the mean μ_X and the standard deviation σ_X? We can guess the mean. If a basketball player makes 75% of her free throws, the mean number made in 12 tries should be 75% of 12, or 9. That's μ_X when X has the $B(12, 0.75)$ distribution.

Intuition suggests more generally that the mean of the $B(n, p)$ distribution should be np. Can we show that this is correct and also obtain a short formula for the standard deviation? Because binomial distributions are discrete probability distributions, we could find the mean and variance by using the binomial probabilities along with general formula for computing the mean and variance given in Section 5.2. But, there is an easier way.

A binomial random variable X is the count of successes in n independent observations that each have the same probability p of success. Let the random variable S_i indicate whether the ith observation is a success or failure by taking the values $S_i = 1$ if a success occurs and $S_i = 0$ if the outcome is a failure.

The S_i are independent because the observations are independent, and each S_i has the same simple distribution:

Outcome	1	0
Probability	p	$1-p$

From the definition of the mean of a discrete random variable (page 230), we know that the mean of each S_i is

$$\mu_S = (1)(p) + (0)(1-p) = p$$

Using this mean in the definition of the variance of a discrete random variable (pages 238–239), we find the variance of S_i as follows:

$$\begin{aligned}\sigma_S^2 &= (1-\mu_S)^2 p + (0-\mu_S)^2(1-p) \\ &= (1-p)^2 p + (0-p)^2(1-p) \\ &= (1-p)[(1-p)p + p^2] \\ &= (1-p)(p - p^2 + p^2) \\ &= p(1-p)\end{aligned}$$

Because each S_i is 1 for a success and 0 for a failure, to find the total number of successes X we add the S_i's:

$$X = S_1 + S_2 + \cdots + S_n$$

Apply the addition rules for means and variances to this sum. To find the mean of X, we add the means of the S_i's:

$$\begin{aligned}\mu_X &= \mu_S + \mu_S + \cdots + \mu_S \\ &= n\mu_S = np\end{aligned}$$

Similarly, because the S_i's are independent, the variance of X is the sum of the variances of S_i and, thus, is $\sigma_X^2 = np(1-p)$. The standard deviation σ_X is the square root of the variance. Here is the result.

BINOMIAL MEAN AND STANDARD DEVIATION

If a count X has the binomial distribution $B(n, p)$, then

$$\mu_X = np$$
$$\sigma_X = \sqrt{np(1-p)}$$

EXAMPLE 5.25

Inspecting Switches Continuing Case 5.3 (page 252), the count X of nonconforming switches is binomial with $n = 150$ and $p = 0.08$. The mean and standard deviation of this binomial distribution are

$$\begin{aligned}\mu_X &= np \\ &= (150)(0.08) = 12 \\ \sigma_X &= \sqrt{np(1-p)} \\ &= \sqrt{(150)(0.08)(0.92)} = \sqrt{11.04} = 3.3226\end{aligned}$$ ■

APPLY YOUR KNOWLEDGE

5.52 Hispanic representation. Refer to the information provided in Exercise 5.50 (see page 258).

(a) What is the mean number of Hispanics on randomly chosen committees of 10 workers?

(b) What is the standard deviation σ of the count X of Hispanic members?

(c) Suppose now that 10% of the factory workers are Hispanic. Then $p = 0.1$. What is σ in this case? What is σ if $p = 0.01$? What does your work show about the behavior of the standard deviation of a binomial distribution as the probability of a success gets closer to 0?

5.53 **Do our athletes graduate?** Refer to the information provided in Exercise 5.49 (see page 256).

(a) Find the mean number of graduates out of 20 players if 80% of players graduate.

(b) Find the standard deviation σ of the count X if 80% of players graduate.

(c) Suppose now that the 20 players came from a population of which $p = 0.9$ graduated. What is the standard deviation σ of the count of graduates? If $p = 0.99$, what is σ? What does your work show about the behavior of the standard deviation of a binomial distribution as the probability p of success gets closer to 1?

Assessing the binomial assumptions with data

In the examples of this section, the probability calculations rest on the assumption that the count random variable X is well described by the binomial distribution. Our confidence in this assumption depends to a certain extent on the strength of our belief that the conditions of the binomial setting are at play. But ultimately we should allow the data to judge the validity of our beliefs. In Chapter 1, we used the Normal quantile plot to check the compatibility of the data with the Normal distribution. The binomial distribution has its own unique features that we can check against the data. Let's explore the applicability of the binomial distribution with the following example.

Normal quantile plot, p. 53

EXAMPLE 5.26

INJECT

Checking for Binomial Compatibility Consider an application in which $n = 200$ manufactured fuel injectors are sampled periodically to check for compliance with specifications. Figure 5.12 shows the counts of defective injectors found in 40 consecutive samples. The counts appear to be behaving randomly over time. Summing over the 40 samples, we find the total number of observed defects to be 210 out of the 8000 total number of injectors inspected. This is associated with a proportion defective of 0.02625. Assuming that the random variable X of the defect counts for each sample follows the $B(200, 0.02625)$ distribution, the standard deviation of X is:

$$\sigma_{\hat{p}} = \sqrt{np(1-p)}$$
$$= \sqrt{200(0.02625)(0.97375)} = 2.26$$

sample variance, p. 33

In terms of variance, the variance of the counts is expected to be around 2.26^2 or 5.11. Computing the sample variance s^2 on the observed counts, we find a variance of 9.47. The observed variance of the counts is nearly twice what would be expected if the counts were truly following the binomial distribution. It appears that the binomial model does not fully account for the overall variation of the counts.

The statistical software JMP provides a nice option of superimposing a binomial distribution fit on the observed counts. Figure 5.13 shows the

FIGURE 5.12 Sequence plot of counts of fuel injector defects per 200 inspected over 40 samples, for Example 5.26.

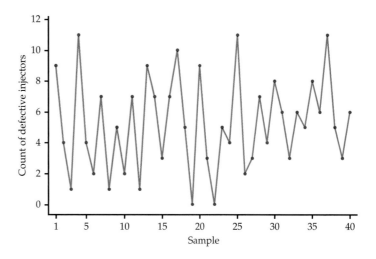

FIGURE 5.13 Binomial distribution fit to fuel injector defect count data, for Example 5.26.

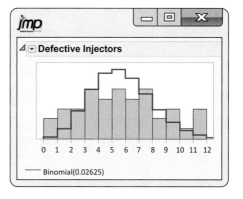

$B(200, 0.02625)$ distribution overlaid on the histogram of the count data. The mismatch between the binomial distribution fit and the observed counts is clear. The observed counts are spread out more than expected by the binomial distribution, with a greater number of counts found at both the lower and upper ends of the histogram. These count data are showing **overdispersion**—that is, the counts have greater variability than would be expected from the assumed count distribution. One likely explanation for the extra variability is change in the probability of defects between production runs due to adjustments in machinery, changes in the quality of incoming raw material, and even changes in personnel. As it currently stands, it would be ill advised to base probability computations on the binomial distribution. ∎

overdispersion

Poisson Distributions

A count X has a binomial distribution when it is produced under the binomial setting. If one or more facets of this setting do not hold, the count X will have a different distribution. In this section, we discuss one of these distributions.

Frequently, we meet counts that are open-ended (that is, are not based on a fixed number of n observations): the number of customers at a popular cafe between 12:00 P.M. and 1:00 P.M.; the number of finish defects in the sheet metal of a car; the number of workplace injuries during a given month; the number of impurities in a liter of water. These are all counts that could be 0, 1, 2, 3, and so on indefinitely. Recall that when count values potentially go on indefinitely, they are said to be *countably infinite* (page 221).

The Poisson setting

The Poisson distribution is another model for a count and can often be used in these open-ended situations. With this distribution, the count represents the number of events (call them "successes") that occur in some fixed unit of measure such as an interval of time, a region of area, or a region of space. The Poisson distribution is appropriate under the following conditions.

THE POISSON SETTING

1. The number of successes that occur in two nonoverlapping units of measure are **independent.**
2. The probability that a success will occur in a unit of measure is the same for all units of equal size and is proportional to the size of the unit.
3. The probability that more than one event occurs in a unit of measure is negligible for very small-sized units. In other words, the events occur one at a time.

For binomial distributions, the important quantities were n, the fixed number of observations, and p, the probability of success on any given observation. For Poisson distributions, the only important quantity is the mean number of successes μ occurring per unit of measure.

POISSON DISTRIBUTION

The distribution of the count X of successes in the Poisson setting is the **Poisson distribution** with **mean** μ. The parameter μ is the mean number of successes per unit of measure. The possible values of X are the whole numbers 0, 1, 2, 3, ... If k is any whole number, then[*]

$$P(X = k) = \frac{e^{-\mu}\mu^k}{k!}$$

The **standard deviation** of the distribution is $\sqrt{\mu}$.

EXAMPLE 5.27

Number of Wi-Fi Interruptions Suppose that the number of Wi-Fi interruptions (e.g., interference from a neighbor's network, significant slowdowns with no disconnection, or disconnections) on your home network varies, with an average of 0.9 interruption per day. If we assume that the Poisson setting is reasonable for this situation, we can model the daily count of interruptions X using the Poisson distribution with $\mu = 0.9$. What is the probability of having no more than two interruptions tomorrow?

We can calculate $P(X \leq 2)$ either using software or the Poisson probability formula. Using the probability formula:

$$P(X \leq 2) = P(X = 0) + P(X = 1) + P(X = 2)$$
$$= \frac{e^{-0.9}(0.9)^0}{0!} + \frac{e^{-0.9}(0.9)^1}{1!} + \frac{e^{-0.9}(0.9)^2}{2!}$$
$$= 0.4066 + 0.3659 + 0.1647$$
$$= 0.9372$$

---

[*] The e in the Poisson probability formula is a mathematical constant equal to 2.71828 to five decimal places. Many calculators have an e^x function.

Using Excel, we can apply the "POISSON.DIST()" function to find the individual probabilities or cumulative probability. This function has three arguments: the first is the value of k, the second is the mean value μ, and the third is either "0" (reports individual probability) or "1" (reports cumulative probability). Here we can see what Excel reports for the individual probabilities:

	A	B
1	k	P(X = k)
2	0	0.40657
3	1	0.365913
4	2	0.164661
5	Sum	0.937143

The reported value of 0.937143 was obtained by using Excel's SUM function. Using the R software, the probability can be found as follows:

```
dpois(0,0.9) + dpois(1,0.9) + dpois(2,0.9)
[1] 0.9371431
```

Excel's and R's answers and the preceding hand-computed answer differ slightly due to roundoff error in the hand calculation. There is roughly a 94% chance that you will have no more than two Wi-Fi interruptions tomorrow. ■

As when using the binomial distribution, Poisson probability calculations are rarely done by hand if the event includes numerous possible values for X. Most software provides functions to calculate $P(X = k)$ and the cumulative probabilities of the form $P(X \leq k)$. These cumulative probability calculations make solving many problems less tedious. Here's an example.

EXAMPLE 5.28

Counting ATM Customers Suppose the number of persons using an ATM in any given hour between 9 A.M. and 5 P.M. can be modeled by a Poisson distribution with $\mu = 8.5$. What is the probability that more than 10 persons will use the ATM between 3 P.M. and 4 P.M.?

Calculating this probability requires two steps:

1. Write $P(X > 10)$ as an expression involving a cumulative probability:

$$P(X > 10) = 1 - P(X \leq 10)$$

2. Calculate $P(X \leq 10)$ and subtract the value from 1.

- Using Excel, we again employ the "POISSON.DIST()" function to compute a cumulative probability, with the third argument being "1" in this case. To get the desired probability, enter the formula =1-POISSON.DIST(10,8.5,1) in any cell. The formula returns a probability of 0.236638.
- Using R, we can get the desired probability as follows:

```
1-ppois(10,8.5)
[1] 0.236638
```

The probability that more than 10 persons will use the ATM between 3 P.M. and 4 P.M. is about 0.24. Under the Poisson setting, this probability of 0.24 applies not only to the 3–4 P.M. hour but to any hour during the day period of 9 A.M. to 5 P.M.

Relying on software to get the cumulative probability is much quicker and less prone to error than the method described in Example 5.27. For this case, that method would involve determining 11 probabilities and then summing their values. ■

APPLY YOUR KNOWLEDGE

5.54 ATM customers. Refer to Example 5.28. Use the Poisson model to compute the probability that four or fewer customers will use the ATM during any given hour between 9 A.M. and 5 P.M.

5.55 Number of Wi-Fi interruptions. Refer to Example 5.27. What is the probability of having at least one Wi-Fi interruption on any given day?

The Poisson model

If we add counts from two nonoverlapping areas, we are just counting the successes in a larger area. That count still meets the conditions of the Poisson setting. If the individual areas are equal in size, our unit of measure doubles, resulting in the mean of the new count being twice as large. In general, if X is a Poisson random variable with mean μ_X and Y is a Poisson random variable with mean μ_Y, and if Y is independent of X, then $X + Y$ is a Poisson random variable with mean $\mu_X + \mu_Y$. This fact means that we can combine areas or look at a portion of an area and still use Poisson distributions to model the count.

EXAMPLE 5.29

Paint Finish Flaws Auto bodies are painted during manufacture by robots programmed to move in such a way that the paint is uniform in thickness and quality. You are testing a newly programmed robot by counting paint sags caused by small areas receiving too much paint. Sags are more common on vertical surfaces. Suppose that counts of sags on the roof follow the Poisson model with mean 0.7 sag per square yard and that counts on the side panels of the auto body follow the Poisson model with mean 1.4 sags per square yard. Counts in nonoverlapping areas are independent. Then

- The number of sags in two square yards of roof is a Poisson random variable with mean $0.7 + 0.7 = 1.4$.

- The total roof area of the auto body is 4.8 square yards. The number of paint sags on a roof is a Poisson random variable with mean $4.8 \times 0.7 = 3.36$.

- A square foot is 1/9 square yard. The number of paint sags in a square foot of roof is a Poisson random variable with mean $1/9 \times 0.7 = 0.078$.

- If we examine one square yard of roof and one square yard of side panel, the number of sags is a Poisson random variable with mean $0.7 + 1.4 = 2.1$. ■

Poisson approximation of the binomial

Although the settings of the binomial and Poisson are different, there are times when the Poisson distribution can be used to approximate a binomial distribution calculation. Specifically in cases where n is large but p is so small that $np < 10$, the Poisson distribution with $\mu = np$ yields very accurate results.

EXAMPLE 5.30

Low-Defect Manufacturing In some manufacturing industries, particularly electronics, defective items produced are measured in parts per million (ppm). Consider a computer chip manufacturer that has achieved 30 ppm defective chips produced. In a manufacturing run of 100,000 chips, what is the probability that there are exactly two defective chips?

From a binomial setup, we define X as the number of defective chips with $n = 100{,}000$ and $p = 30/1{,}000{,}000 = 0.00003$. Using the binomial probability formula, we have

$$P(X = 2) = \binom{100000}{2}(0.00003)^2(0.99997)^{99998}$$

Even though this calculation is doable, it is a bit tedious to perform with a calculator. When the calculation is complete, the computed binomial probability is 0.224041. Now, consider the Poisson approximation with $\mu = np = 100{,}000(0.00003) = 3$. Using the Poisson probability formula, we have

$$P(X = 2) = \frac{e^{-3}(3)^2}{2!}$$

Obviously, this computation is much simpler. The approximation gives a value of 0.224042, which is off by only 0.000001 from the exact result. This very accurate approximation is due to the fact that n is very large along with $np < 10$. ∎

APPLY YOUR KNOWLEDGE

5.56 A safety initiative. A large manufacturing plant has averaged six "reportable accidents" per month. Suppose that these accident counts over time follow a Poisson distribution. A "safety culture change" initiative attempts to reduce the number of accidents at the plant. After the initiative, there were 50 reportable accidents during the year.

(a) Based on an average of six accidents per month, what is the distribution of the number of reportable accidents in a year?

(b) Based on an average of six accidents per month, use software to find the probability of 50 or fewer accidents in a year. Does the computed probability provide evidence that the initiative did reduce the accident rate? Explain why or why not.

5.57 Low-defect manufacturing. Refer to the computer chip manufacturer described in Example 5.30. Without aid of software, use the Poisson approximation to find:

(a) The probability of at least one defective chip.

(b) The probability of no more than three defective chips.

Assessing the Poisson assumptions with data

Similar to the binomial distribution, the applicability of the Poisson distribution requires that certain specific conditions are met. In particular, we model counts with the Poisson distribution if we are confident that the counts arise from a Poisson setting (page 262). Let's consider a couple of examples to see if the Poisson model reasonably applies.

EXAMPLE 5.31

EPL

English Premier League Goals Consider data on the total number of goals scored per soccer game in the English Premier League (EPL) for the 2016–2017 regular season.[13] Over the 380 games played in the season, the average number of goals per game was 2.8.

The Poisson distribution has a unique characteristic: the standard deviation of the Poisson random variable is equal to the square root of the mean. In turn, this implies that the mean of a Poisson random variable X equals its variance; that is, $\sigma_X^2 = \mu$. This fact provides us with a very convenient quick check for Poisson compatibility—namely, compare the mean observed count

FIGURE 5.14 Poisson distribution fit to EPL goals per game, for Example 5.31.

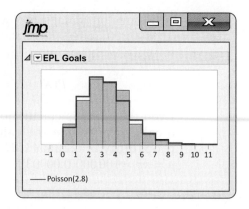

with the observed variance. For the goal data, we would find that the sample variance of the counts is 2.79, which is remarkably close to the mean of 2.8. This suggests that the Poisson distribution might serve as a reasonable model for counts on EPL goals per game.

Figure 5.14 shows a JMP-produced graph of a Poisson distribution with $\mu = 2.8$ overlaid on the count data. The Poisson distribution and observed counts show quite a good match. It would be reasonable to assume that the variability in goals scored in EPL games is well accounted for by the Poisson distribution. ∎

The next example shows a different story.

EXAMPLE 5.32

SHAREH

zero inflation

Shareholder Proposals The U.S. Securities and Exchange Commission (SEC) has established regulations that allow shareholders of a public company who own at least $2000 in market value of a company's outstanding stock to submit shareholder proposals. A shareholder proposal is a resolution put forward by a shareholder, or group of shareholders, to be voted on at the company's annual meeting. Shareholder proposals serve as a means for investor activists to effect changes in corporate governance and activities. Proposals can range from executive compensation to corporate social responsibility issues, such as companies' stances on human rights, labor relations, and global warming. The SEC requires companies to disclose shareholder proposals on the company's proxy statement. Proxy statements are publicly available.

In a study of 1532 companies, data were gathered on the counts of shareholder proposals per year.[14] The mean number of shareholder proposals was found to be 0.5157 per year. We would find that observed variance of the counts is 1.1748, which is more than twice the mean value. This implies that the counts are varying to a greater degree than expected by the Poisson model. As noted in Example 5.26 (page 260), this phenomenon is known as overdispersion. Figure 5.15 shows a Minitab-produced graph of the observed number of counts along with the expected number of counts for a Poisson distribution with mean 0.5157. The figure shows the incompatibility of the Poisson model with the observed count data. We find that there are more zero counts than expected, along with more higher counts than expected. The extra abundance of zeroes in the data is known as the **zero inflation** phenomenon. The researchers who conducted this study hypothesize that the increased count of zeroes is due to many companies choosing to privately resolve shareholder concerns so as to protect their corporate image. ∎

FIGURE 5.15 Poisson distribution fit to counts on shareholder proposals, for Example 5.32.

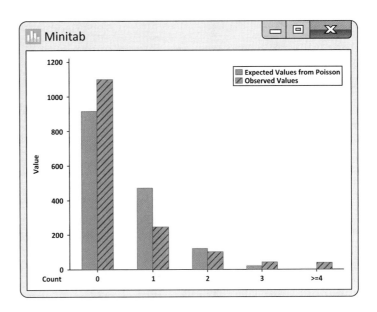

SECTION 5.3 SUMMARY

- A **count** X of successes has the **binomial distribution** $B(n, p)$ in the **binomial setting:** there are n trials, all independent, each resulting in a success or a failure, and each having the same probability p of a success.

- For a population that contains proportion p of successes, the count X of successes in an SRS of size n does not exactly follow a binomial distribution. However, the binomial distribution $B(n, p)$ is an adequate approximation when the population is at least 20 times as large as the sample. The accuracy of this approximation improves as the size of the population increases relative to the size of the sample.

- If X has a binomial distribution with parameters n and p, the possible values of X are the whole numbers $0, 1, 2, \ldots, n$. The **binomial probability** that X takes any value is

$$P(X = k) = \binom{n}{k} p^k (1-p)^{n-k}$$

Binomial probabilities are most easily found by software, but this formula is practical for calculations when n is small. Table C contains binomial probabilities for some values of n and p.

- The **binomial coefficient**

$$\binom{n}{k} = \frac{n!}{k!(n-k)!}$$

counts the number of ways k successes can be arranged among n observations. Here, the **factorial** $n!$ is

$$n! = n \times (n-1) \times (n-2) \times \cdots \times 3 \times 2 \times 1$$

for positive whole numbers n, and $0! = 1$.

- The mean and standard deviation of a **binomial count** X are

$$\mu_X = np \qquad \sigma_X = \sqrt{np(1-p)}$$

- A simple check for the adequacy of the binomial model is to compare the binomial-based standard deviation (or variance) with the observed count standard deviation (or variance). In addition, some software can fit the binomial model on the observed histogram to assess compatibility.

- A count X of successes has a **Poisson distribution** in the **Poisson setting**: the number of successes that occur in two nonoverlapping units of measure are independent; the probability that a success will occur in a unit of measure is the same for all units of equal size and is proportional to the size of the unit; the probability that more than one event occurs in a unit of measure is negligible for very small-sized units. In other words, the events occur one at a time.

- If X has the Poisson distribution with mean μ, then the standard deviation of X is $\sqrt{\mu}$, and the possible values of X are the whole numbers 0, 1, 2, 3, and so on.

- The **Poisson probability** that X takes any of these values is

$$P(X = k) = \frac{e^{-\mu}\mu^k}{k!} \quad k = 0,1,2,3,\cdots$$

- Sums of independent Poisson random variables also have the Poisson distribution. For example, in a Poisson model with mean μ per unit of measure, the count of successes in a units is a Poisson random variable with mean $a\mu$.

- When n is large and p is so small that $np < 10$, the Poisson distribution with $\mu = np$ provides an accurate approximation of the binomial distribution.

- A simple check for the adequacy of the Poisson model is to compare the closeness of the observed mean count with the observed variance of the counts. In addition, some software can fit the Poisson model on the observed histogram to assess compatibility.

SECTION 5.3 EXERCISES

For Exercises 5.43 to 5.46, see page 251; for 5.47 to 5.49, see pages 255–256; for 5.50 and 5.51, see page 258; for 5.52 and 5.53, see pages 259–260; for 5.54 and 5.55, see page 264; and for 5.56 and 5.57, see page 265.

5.58 What's wrong? Explain what is wrong in each of the following scenarios.

(a) In the binomial setting, X is a proportion.

(b) The variance for a binomial count is $\sqrt{p(1-p)/n}$.

(c) For $n = 20$, a binomial count of 25 was observed.

(d) The probability of observing k successes in n trials is $\frac{n!}{k!(n-k)!}p^k$.

5.59 What's wrong? Explain what is wrong in each of the following scenarios.

(a) If you toss a fair coin four times and a head appears each time, then the next toss is more likely to be a tail than a head.

(b) If you toss a fair coin four times and observe the pattern HTHT, then the next toss is more likely to be a head than a tail.

(c) The quantity $\hat{p}$ is one of the parameters for a binomial distribution.

(d) The binomial distribution can be used to model the daily number of pedestrian/cyclist near-crash events on campus.

5.60 Should you use the binomial distribution? In each of the following situations, is it reasonable to use a binomial distribution for the random variable X? Give reasons for your answer in each case. If a binomial distribution applies, give the values of n and p.

(a) In a random sample of 20 students in a fitness study, X is the mean daily exercise time of the sample.

(b) A manufacturer of running shoes picks a random sample of 20 shoes from the production of shoes each day for a detailed inspection. X is the number of pairs of shoes with a defect.

(c) A college tutoring center chooses an SRS of 50 students. The students are asked whether they have used the tutoring center for any sort of tutoring help. X is the number who say that they have.

(d) X is the number of days during the school year when you skip a class.

5.61 Should you use the binomial distribution? In each of the following situations, is it reasonable to use a binomial distribution for the random variable X? Give reasons for your answer in each case. If a binomial distribution applies, give the values of n and p.

(a) A poll of 200 college students asks whether they usually feel irritable in the morning. X is the number who reply that they do usually feel irritable in the morning.

(b) You toss a fair coin until a head appears. X is the count of the number of tosses that you make.

(c) Most calls made at random by sample surveys don't succeed in talking with a person. Of such calls made in New York City, only one-twelfth succeed. A survey calls 500 randomly selected numbers in New York City. X is the number of times that a person is reached.

(d) You deal 10 cards from a shuffled deck of standard playing cards and count the number X of black cards.

5.62 Digital-self checkers. Consumers are increasingly keeping track on their digital presence. A 2017 Bank of America survey found that 11% of Generation Z smartphone users (born in the year 2000 and later) perform a Web search of themselves daily.[15] Suppose you were to draw a random sample of 10 Generation Z smartphone owners.

(a) The number in your sample who are daily self-checkers has a binomial distribution. What are n and p?

(b) Use the binomial formula to find the probability that exactly two of the 10 are daily self-checkers in your sample.

(c) Use the binomial formula to find the probability that two or fewer are daily self-checkers in your sample.

(d) What is the mean number of owners in such samples who are daily self-checkers? What is the standard deviation?

5.63 Random stock prices. As noted in Example 5.19(a) (page 250), the S&P 500 index has a probability 0.56 of increasing in any week. Moreover, the change in the index in any given week is not influenced by whether it rose or fell in earlier weeks. Let X be the number of weeks among the next five weeks in which the index rises.

(a) X has a binomial distribution. What are n and p?

(b) What are the possible values that X can take?

(c) Use the binomial formula to find the probability of each value of X. Draw a probability histogram for the distribution of X.

(d) What are the mean and standard deviation of this distribution?

5.64 Public transit. Many U.S. cities encourage a shift of commuters toward the use of public transportation for commuting. In New York City, 57% of commuters use public transportation.[16]

(a) If you choose 10 NYC commuters at random, what is the probability that exactly four commuters use public transportation for their daily commute? Determine this probability by hand using the binomial probability formula.

(b) If you choose 10 NYC commuters at random, what is the probability that more than half use public transportation? Determine this probability by hand using the binomial probability formula.

(c) If you choose 100 NYC commuters at random, what is the probability that 50 or fewer in your sample use public transportation? Use software to determine your answer.

5.65 Uber female drivers. In a 2018 study of nearly 2 million Uber drivers, researchers found that the nationwide proportion of female Uber drivers was 27.3%.[17] Use this proportion to compute the following probabilities.

(a) If you were to take five Uber trips, what is the probability that exactly one of the the trips is with a female driver? Determine this probability by hand using the binomial probability formula.

(b) If you were to take five Uber trips, what is the probability that two or more of the trips are with female drivers? Determine this probability by hand using the binomial probability formula.

(c) Suppose you were to track the number of female drivers in your next 30 Uber trips. What is the probability of having 10 or more female drivers? Use software to determine your answer.

5.66 Internet video postings. Suppose (as is roughly true) that 30% of all adult Internet users have posted videos online. A survey interviews an SRS of 1555 adult Internet users.

(a) What is the distribution of the number X in the sample who have posted videos online?

(b) Use software to find the exact probability that 450 or fewer of the people in the sample have posted videos online.

5.67 Random digits. Each entry in a table of random digits like Table B has probability 0.1 of being a 0, and digits are independent of each other.

(a) Suppose you want to determine the probability of getting at least one 0 in a group of five digits. Explain what is wrong with the logic of computing it as $0.1 + 0.1 + 0.1 + 0.1 + 0.1$ or 0.5.

(b) Find the probability that a group of five digits from the table will contain at least one 0.

(c) In Table B, there are 40 digits on any given line. What is the mean number of zeros in lines 40 digits long?

5.68 Are we shipping on time? Your mail-order company advertises that it ships 90% of its orders within three working days. You select an SRS of 100 of the 5000 orders received in the past week for an audit. The audit reveals that 86 of these orders were shipped on time.

(a) If the company really ships 90% of its orders on time, what is the probability that 86 or fewer in an SRS of 100 orders are shipped on time? Use software to find the probability.

(b) A critic says, "Aha! You claim 90%, but in your sample the on-time percent is only 86%. So the 90% claim is wrong." Explain in simple language why your probability calculation in part (a) shows that the result of the sample does not refute the 90% claim.

5.69 Checking for survey errors. One way of checking for undercoverage, nonresponse, and other sources of error in a sample survey is to compare the sample with known facts about the population. Approximately 13% of American adults are black. The number X of blacks in a random sample of 1500 adults should, therefore, vary with the binomial ($n = 1500$, $p = 0.13$) distribution.

(a) What are the mean and standard deviation of X?

(b) Use software to find the probability that the sample will contain 150 or fewer blacks.

(c) If 150 or fewer blacks are found in a survey of 1500 adults, what does this suggest about the representativeness of the sample? Explain in simple language why your probability calculation in part (c) supports your argument.

5.70 Show that these facts are true. Use the definition of binomial coefficients to show that each of the following facts is true. Then restate each fact in words in terms of the number of ways that k successes can be distributed among n observations.

(a) $\binom{n}{n} = 1$ for any whole number $n \geq 1$.

(b) $\binom{n}{0} = 1$ for any whole number $n \geq 1$.

(c) $\binom{n}{n-1} = n$ for any whole number $n \geq 1$.

(d) $\binom{n}{k} = \binom{n}{n-k}$ for any n and k with $k \leq n$.

5.71 Does your vote matter? Consider a common situation in which a vote takes place among a group of people and the winning result is associated with having one vote greater than the losing result. For example, if a management board of 11 members votes yes or no on a particular issue, then minimally a 6-to-5 vote is needed to decide the issue either way. Your vote would have mattered if the other members voted 5-to-5.

(a) You are on this committee of 11 members. Assume that there is a 50% chance that each of the other members will vote yes and assume that the members are voting independently of each other. What is the probability that your vote will matter?

(b) There is a closely contested election between two candidates for town mayor in a town with 523 eligible voters. Assume that all eligible voters will vote, with a 50% chance that a voter will vote for a particular candidate. What is the probability that your vote will matter?

5.72 Tossing a die. You are tossing a balanced die that has probability 1/6 of coming up 1 on each toss. Tosses are independent. You are interested in how long you must wait to get the first 1.

(a) The probability of getting a 1 on the first toss is 1/6. What is the probability that the first toss is not a 1 and the second toss is a 1?

(b) What is the probability that the first two tosses are not 1s and the third toss is a 1? This is the probability that the first 1 occurs on the third toss.

(c) Now you see the pattern. What is the probability that the first 1 occurs on the fourth toss? On the fifth toss?

5.73 The geometric distribution. Generalize your work in Exercise 5.72. You have independent trials, each resulting in a success or a failure. The probability of a success is p for each trial. The binomial distribution describes the count of successes in a fixed number of trials. Now, the number of trials is not fixed; instead, continue until you get a success. The random variable Y is the number of the trial in which the first success occurs. What are the possible values of Y? What is the probability $P(Y = k)$ for any of these values? (*Comment:* The distribution of the number of trials to the first success is called a **geometric distribution**.)

5.74 The hypergeometric distribution. The discussion following Case 5.3 (page 252) pointed out that when sampling an SRS from a population, the binomial distribution is a good *approximation* for modeling the probability of success when the population is at least 20 times as large as the sample. In the case of drawing an SRS, the distribution that the binomial is approximating is known as a hypergeometric distribution. In a hypergeometric setting, there is a finite population of size N. In this population, K elements (e.g., individuals or objects) are successes and $N - K$ elements are failures. A sample of n individuals (or objects) is selected without replacement in such a way that every possible sample of size n has the same chance to be chosen. Sampling without replacement implies that an individual selected is not available to be selected again. The hypergeometric distribution gives the probability of observing k successes in the sample:

$$P(X = k) = \frac{\binom{K}{k}\binom{N-K}{n-k}}{\binom{N}{n}}$$

For parts (a) and (b), without the aid of software, use the hypergeometric distribution formula to find the probabilities in question.

(a) In a small town, there are 60 registered female voters and 55 registered male voters. An SRS of 10 voters is drawn. What is the probability that exactly six of the sampled voters are female?

(b) An inventory parts rack has 15 similar parts from two different suppliers (A and B). Six parts are from Supplier A and nine parts are from Supplier B. Suppose five parts are randomly selected from the rack. What is the probability that at least one part in the five parts selected is from Supplier A?

5.75 Binomial approximation of the hypergeometric distribution. In Case 5.3 (page 252), an SRS of 150 switches was drawn from a shipment of 10,000 switches (population) in which 8% of the switches fail to conform to specifications; that is, there are 800 nonconforming switches. Given that the population size 10,000 is more than 20 times the sample size of 150, the binomial distribution was used as a suitable approximation for the distribution of the count of nonconforming switches in the sample. In Example 5.20 (page 253), using the $B(150, 0.08)$ distribution, it was found that the probability of observing 10 nonconforming switches in a sample of 150 switches is 0.106959. The exact distribution is the hypergeometric distribution shown in the previous exercise. In this exercise, you will explore how well the binomial distribution approximates the hypergeometric distribution for different population sizes N. Assume that the sample size is always $n = 150$. As with Case 5.3, assume that 8% of the switches in the population are nonconforming. In terms of the notation of the previous exercise, this implies $K = 0.08N$.

Since the hypergeometric calculations can be unwieldy to perform by calculator, you should use software. In Excel, use the "HYPGEOM.DIST()" function for computing $P(X = k)$. This function has five arguments: the first is the value of the number of successes (k), the second is K, the third is n, the fourth is N, and the fifth should be set to "0" (reports individual probability). Using Excel (or other software), determine the missing values below:

N	$P(X = 10)$
1000	
10,000	
100,000	
1,000,000	
10,000,000	

What do you observe about the hypergeometric probabilities as N increases?

5.76 How many calls? Calls to the customer service department of a cable TV provider are made randomly and independently at a rate of 11 calls per minute. The company has a staff of 20 customer service specialists who handle all the calls. Assume that none of the specialists are on a call at this moment and that a Poisson model is appropriate for the number of incoming calls per minute.

(a) Use the Poisson probability formula to compute the probability of the customer service department receiving exactly 20 calls in the next minute.

(b) Use software to find the probability of the customer service department receiving more than 20 calls in the next minute.

(c) Use software to find the probability of the customer service department receiving fewer than 11 calls in the next minute.

5.77 EPL goals. Refer to Example 5.31 (page 265), in which we found that the total number of goals scored in an EPL game is well modeled by the Poisson distribution. Compute the following probabilities without the aid of software.

(a) What is the probability that a game will end in a 0–0 tie?

(b) What is the probability that three or more goals are scored in a game?

5.78 Email. Suppose the average number of emails received by a particular employee at your company is seven emails per hour. Suppose the count of emails received can be adequately modeled as a Poisson random variable. Compute the following probabilities without the aid of software.

(a) What is the probability of this employee receiving exactly seven emails in any given hour?

(b) What is the probability of receiving fewer than seven emails in any given hour?

(c) What is the probability of receiving at least one email in any given hour?

(d) What is the probability of receiving at least one email in any given 30-minute span?

5.79 Traffic model. The number of vehicles passing a particular mile marker during 15-minute units of time can be modeled as a Poisson random variable. Counting devices show that the average number of vehicles passing the mile marker every 15 minutes is 48.7.

(a) Use software to find the probability of 50 or more vehicles passing the marker during a 15-minute time period.

(b) What is the mean number of vehicles passing the marker in a 30-minute time period?

(c) Use software to find the probability of 100 or more vehicles passing the marker during a 30-minute time period.

5.80 Flaws in carpets. Flaws in carpet material follow the Poisson model with mean 0.6 flaw per square yard. Suppose an inspector examines a sample of carpeting measuring 1.5 yards by 1.5 yards.

(a) What is the distribution for the number of flaws in the sample carpeting?

(b) Use the Poisson probability formula to compute the probability that the total number of flaws the inspector finds is exactly five.

(c) Use the Poisson probability formula to compute the probability that the total number of flaws the inspector finds is two or less.

5.81 Email, continued. Refer to Exercise 5.78, where we learned that a particular employee at your company receives an average of seven emails per hour.

(a) What is the distribution of the number of emails over the course of an eight-hour workday?

(b) What is the standard deviation of the number of emails during an eight-hour workday?

(c) Use software to find the probability of receiving 50 or more emails during an eight-hour workday.

5.82 Combining distributions. Suppose the number of people arriving for treatment at a small hospital emergency room during the night shift follows a Poisson model with mean 8.6. Historically, the proportion of people coming to the emergency room needing to be transferred to a hospital room is 0.4. What is the probability that six people arrive to the emergency room *and* three or more of these people need to be transferred to hospital rooms? (*Hint:* View this question in terms of two events A and B. Recognize that you are finding $P(A \text{ and } B)$. From Chapter 4, this joint probability is $P(A)P(B \mid A)$; use each of the distributions discussed in this section—Poisson and binomial—to find these two probabilities separately and then multiply them.)

5.83 How many zeroes expected? Refer to Example 5.32 (page 266). We would find that 1099 of the observed counts have a value of 0. Based on the provided information in the example, how many more observed zeroes are there in the data set than the number that the best-fitting Poisson model would expect?

5.84 Mishandled baggage. In the airline industry, the term "mishandled baggage" refers to baggage that was lost, delayed, damaged, or stolen. In 2017, American Airlines had an average of 3.38 mishandled bags per 1000 passengers.[18] Consider an incoming Boeing 787 American Airlines flight carrying 260 passengers. Let X be the number of mishandled bags.

(a) Use the binomial distribution to find the probability that there will be at least one mishandled bag.

(b) Use the Poisson approximation to find the probability of part (a).

5.85 Calculator convenience. Suppose that X follows a Poisson distribution with mean μ.

(a) Show that $P(X = 0) = e^{-\mu}$.

(b) Show that $P(X = k) = \frac{\mu}{k} P(X = k - 1)$ for any whole number $k \geq 1$.

(c) Suppose $\mu = 3$. Use part (a) to compute $P(X = 0)$.

(d) Part (b) gives us a nice calculator convenience that allows us to multiply a given Poisson probability by a factor to get the next Poisson probability. What would you multiply the probability from part (a) to get $P(X = 1)$? What would you then multiply $P(X = 1)$ by to get $P(X = 2)$?

5.86 Baseball runs scored. We found in Example 5.31 (page 265) that, in soccer, goal scoring is well described by the Poisson model. It will be interesting to investigate if that phenomenon carries over to other sports. Consider data on the number of runs scored per game by the Cleveland Indians for the 2017 season. Parts (a) through (e) can be done with any software. CLE

(a) Produce a histogram of the runs. Describe the distribution.

(b) What is the mean number of runs scored by the Indians? What is the sample variance of the runs scored?

(c) What do your answers from part (b) tell you about the applicability of the Poisson model for these data?

(d) If you were to use the Poisson model, in how many games in a 162-game season would you expect the Indians not to score?

(e) Sort the runs scored column and count the actual number of games in which the Indians did not score. Compare this count with your answer in part (d) and respond.

(f) *JMP and Minitab users only:* Provide output of a Poisson fit along with the observed runs. Discuss what you see.

5.4 Common Continuous Distributions

When you complete this section, you will be able to:

- Find probabilities from a general uniform distribution defined over an interval from *a* to *b*.
- Find the mean and standard deviation of a general uniform distribution.
- Use the Normal distribution (with and without continuity correction) for approximating binomial probabilities.
- Use the Normal distribution for approximating Poisson probabilities.
- Explain the connection between a Poisson distribution and an exponential distribution.
- Determine the mean and standard deviation of an exponential distribution having a rate parameter λ, and find probabilities from an exponential distribution using software.

In Sections 5.1 and 5.2, we encountered a variety of special-case discrete distributions (e.g., Benford distribution) that are particularly applicable to certain scenarios. In these two sections, our focus was to use the illustrated discrete distributions to introduce the concepts of means and variances of random variables. In Section 5.3, we turned our attention to two common discrete distributions—binomial and Poisson—that have widespread applicability in many business settings.

Recall that for a discrete random variable, the number of possible outcomes is finite or the possible outcomes can be listed in an infinite sequence corresponding to the counting numbers. In contrast, a continuous random variable has outcomes associated with the real numbers, not the counting numbers. For this reason, the means by which we compute probabilities for discrete and continuous random variables are very different. For discrete random variables, the probabilities for the possible outcome values are either directly provided or computed from a probability distribution formula (e.g., binomial and Poisson) to find the probability for each possible outcome value. However, as first introduced in Section 1.4 (page 41) and as noted earlier in this chapter (page 224), with continuous random variables, we encounter density curves. Probability corresponds to the *area* under the curve, *not* the height of the curve.

This section reinforces the idea of finding probabilities from continuous distributions by considering three common continuous distributions: generalized uniform distributions, Normal distributions, and exponential distributions. For this section, we focus on these three distributions for the following reasons:

- The uniform distribution is associated with the simplest of density curves. As such, the uniform distribution allows us to connect the idea of density curves to probabilities by means of simple calculations.

- The Normal distribution was introduced in Section 1.4 (page 44). Our coverage of the Normal distribution in this section is not a repeat of the discussion in Section 1.4. Instead, we will revisit the Normal distribution for two reasons. First, the Normal distribution has an important connection to the discrete distributions (binomial and Poisson) covered in Section 5.3. Second, a refresher on the Normal distribution is particularly useful at this juncture, given the Normal distribution's role in the statistical inference carried out in Chapters 6 and 7. It will also be used for inference of proportions in Chapter 10.

- The exponential distribution is an important distribution for the study of certain time-related scenarios. Our highlighting of this distribution is particularly motivated by the fact that it has a direct connection to the Poisson distribution covered in Section 5.3.

You will encounter Normal distributions after this chapter as we examine certain inference-related topics. But, as you will learn, several other continuous distributions that are not covered here are essential to know for statistical inference—most notably, t distributions, F distributions, and chi-square distributions. The density curves for the variety of continuous distributions you will encounter, in this section and in later chapters, are all different in shape and form. Nevertheless, they all share a common link: the areas under the curves correspond to probabilities.

Uniform distributions

Example 5.2 (page 224) considered one of the simplest continuous distributions—namely, the uniform distribution. The motivation for Example 5.2 was understanding the distribution associated with random number generators that attempt to produce numbers "at random" between 0 and 1. The ability of random number generators to produce uniform distributed numbers between

FIGURE 5.16 General uniform distribution over the interval of a to b.

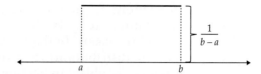

0 and 1 is critical for performing simulation studies. It turns out that a variety of techniques can be used to transform uniform distributed random numbers into random numbers from a variety of other distributions, including the Normal distribution.

The uniform distribution is not restricted to the interval from 0 to 1. In general, we can have a uniform distribution over an interval from a to b, as illustrated in Figure 5.16. The density curve is plotted as the horizontal line over the interval a to b at a height of $1/(b-a)$. Outside the interval of a to b, the density curve has height 0, implying that no values outside the interval are possible. Because of the look of a uniform distribution, it is also known as a **rectangular distribution**. Remember that with all density curves, the total area under the curve is 1. This fact explains why the height of the uniform density curve is indeed $1/(b-a)$ since

rectangular distribution

$$\text{Total area} = (\text{base})(\text{height}) = (b-a)\frac{1}{b-a} = 1$$

What is the value of the mean of the uniform distribution? Your intuition will lead you to the correct conclusion: it is at the midpoint of the a to b interval.

$$\mu = \frac{a+b}{2}$$

Finding the standard deviation of a uniform random variable X requires advanced mathematics not covered in this text. Its formula is

$$\sigma = \sqrt{\frac{(b-a)^2}{12}}$$

EXAMPLE 5.33

How Long to Wait? Historically, aside from unusual events (e.g., severe weather or power outages), the arrival time X of a commuter train to a particular stop is a random variable following a uniform distribution between 8:00 A.M. and 8:05 A.M. Suppose a commuter arrives to the stop at 8:01 A.M.. What is the probability that the commuter misses the train?

Define X as the train's arrival time (in minutes) relative to 8:00 A.M. This implies that 8:05 A.M. is associated with the value of 5. The values of 0 and 5 therefore define the endpoints of the interval—that is, $a = 0$ and $b = 5$. The density curve for the range of X from 0 to 5 is placed at $1/(5-0) = 1/5$. The probability of missing a train if a commuter arrives at 8:01 a.m. is depicted by the shaded area of Figure 5.17. This area is simply

$$P(X < 1) = (\text{base})(\text{height}) = (1-0)\frac{1}{5} = \frac{1}{5} = 0.2 \blacksquare$$

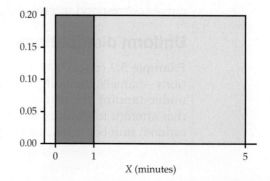

FIGURE 5.17 Uniform distribution over the interval of 0 to 5. The shaded area is $P(X < 1)$, calculated in Example 5.33.

APPLY YOUR KNOWLEDGE

5.87 Commuter train. Refer to Example 5.33. With this problem, consider the time format of hr:mm:ss (hours/minutes/seconds).

(a) Suppose you arrive to the stop at 7:57:10 A.M. What is the probability that you will have to wait for five or more minutes for the train?

(b) Suppose you arrive to the stop at 8:03:28 A.M. What is the probability that you will have to wait for five or more minutes for the train?

(c) What is the mean arrival time written in terms of hr:mm:ss?

Revisiting Normal distributions

The density curves that are most familiar to us are the Normal curves, as first introduced in Section 1.4 (page 44). Following are some key points pertaining to the Normal distribution:

- The Normal density curve is specified by a particular equation. The height of the density curve at a given point x is

$$\frac{1}{\sigma\sqrt{2\pi}} e^{-\frac{1}{2}\left(\frac{x-\mu}{\sigma}\right)^2}$$

We will not make direct use of this fact. We simply recognize it as the basis of mathematical work with Normal distributions. When drawn, Normal density curves are characterized by being bell-shaped, symmetric, and unimodal.

- As can be seen from the density curve equation, two parameters specify a Normal distribution: the mean μ and the standard deviation σ.

- Stating that a random variable X has the $N(\mu,\sigma)$ distribution is a brief way of saying that a random variable X is Normally distributed with mean μ and standard deviation σ.

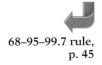

68–95–99.7 rule, p. 45

- A useful rule to know for all Normal distributions is the 68–95–99.7 rule. It states that for a Normal distribution, about 68% of the values fall within 1σ of the mean, about 95% of the values fall within 2σ of the mean, and 99.7% of the values fall within 3σ of the mean.

- If X has the $N(\mu,\sigma)$ distribution, then the standardized variable $Z = (X - \mu)/\sigma$ has a standard Normal distribution $N(0,1)$.

Let's turn to a couple of examples to review basic calculations with the Normal distribution. Refer to Section 1.4 for additional examples.

EXAMPLE 5.34

Find the Probability from the Normal Distribution The actual tread life X of a 40,000-mile automobile tire has a Normal probability distribution with $\mu = 50,000$ miles and $\sigma = 5500$ miles. We say that the random variable X has an $N(50,000, 5500)$ distribution. From a manufacturer's perspective, it would be useful to know the probability that a tire fails to meet the guaranteed wear life of 40,000 miles. Figure 5.18 shows this probability as an area under a Normal density curve. To find the desired probability using Table A, we standardize the variable X,

$$P(X < 40,000) = P\left(\frac{X - 50,000}{5500} < \frac{40,000 - 50,000}{5500}\right)$$
$$= P(Z < -1.82)$$
$$= 0.0344$$

In Excel, we can find the probability more precisely by entering the formula =NORM.DIST(40000,50000,5500,1) in any cell. The last input of "1" is to

FIGURE 5.18 The Normal distribution with $\mu = 50{,}000$ and $= 5500$. The shaded area is $P(X < 40{,}000)$, calculated in Example 5.34.

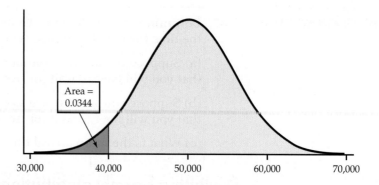

find the probability to the left of 40,000. The formula returns a probability of 0.034518. The manufacturer should expect to incur warranty costs for approximately 3.5% of its tires. ∎

As noted in Section 5.1 (page 226), the probability that a continuous variable is equal to a specific value is 0. This implies that for Example 5.34, $P(X < 40{,}000)$ and $P(X \leq 40{,}000)$ are equal.

EXAMPLE 5.35

Inverse Normal Calculation Rather than finding a probability as done in the previous example, consider a reverse calculation. In particular, what is the tread life value that represents the 80th percentile? In other words, we want to find the tread life value associated with an area of 0.80 below it. Without software, first find the standard score z with cumulative probability 0.80, then unstandardize to find x. In Table A, the closest entry to 0.80 is $z = 0.84$. This implies that the unknown x value satisfies this equation

$$\frac{x - 50{,}000}{5500} = 0.84$$

Solving this equation for x gives

$$x = 50{,}000 + (0.84)(5500) = 54{,}620$$

In Excel, we can find the percentile point more precisely by entering the formula `=NORM.INV(0.8,50000,5500)` in any cell. This formula returns a more accurate value of 54,628.92. ∎

APPLY YOUR KNOWLEDGE

5.88 Normal probabilities using a table. Suppose that X has a Normal distribution with $\mu = 40$ and $\sigma = 7$. Using Table A, what are the following probabilities?

(a) $X < 30$

(b) $X > 45$

(c) $25 < X < 50$

5.89 Normal probabilities using software. Refer to the previous exercise and now find the probabilities for parts (a)–(c) using software.

5.90 Inverse Normal calculations using a table. Suppose that X has a Normal distribution with $\mu = 100$ and $\sigma = 25$. Using Table A to find the following percentile values:

(a) 10th percentile.

(b) 95th percentile.

5.91 Inverse Normal calculations using software. Refer to the previous exercise and now find the percentiles for parts (a) and (b) using software.

5.92 Tread life. Example 5.34 gives the Normal distribution $N(50,000, 5500)$ for the tread life X of a type of tire (in miles). Apply the 68–95–99.7 rule to this distribution. Give the ranges of tread life values that are within one, two, and three standard deviations of the mean.

Now that we have reviewed the Normal distribution, we next consider how we can use the Normal distribution to approximate probabilities for the two common discrete distributions (binomial and Poisson) introduced in Section 5.3. Similar to the Poisson approximation of the binomial (page 264), these approximations are done to simplify computation.

Normal approximation for binomial distribution

Recall from Section 5.3 that a count X of successes has the binomial distribution $B(n, p)$ in a binomial setting. The binomial probability formula (page 257) provides us with a means of finding the exact probability of observing k successes in n trials. For large n, manual calculation of this formula, even with the aid of a calculator, is impractical or even impossible.

EXAMPLE 5.36

Priority Mail In Example 5.22 (page 255), we found the probability of observing four or more late USPS priority mail deliveries out of $n = 15$ deliveries, with a probability of $p = 0.07$ of any given delivery being late. To find the desired probability, we turned to the pre-calculated binomial probabilities given in Table C. Alternatively, we could have manually calculated the desired probability by using the binomial probability formula.

Now imagine this question of late deliveries plays out on a much larger scale. The USPS delivers nearly 500 million pieces of mail each day. Consider 10 million priority mail deliveries. What is the probability that 701,000 or more of these deliveries will be late? The probability of having exactly 701,000 late deliveries is

$$\frac{10,000,000!}{701,000! \, 9,299,000!} (0.07)^{701,000} (0.93)^{9,299,000}$$

This is not a computation you can perform with your calculator. But, the real problem is actually much worse. We need to find the probability of 701,000 or more deliveries being late. This implies having to also find the individual probabilities for 701,001, 701,002, and so on up to 10,000,000.

Fortunately, the binomial functions available in software can handle very large n values, including 10,000,000. However, for large enough n, even software is unable to use the binomial probability formula directly because the factorial term becomes astronomical. In such situations, software uses a variety of computational algorithms to give nearly exact results. Even with that being said, software can still be pushed to its limits, leaving it unable to compute binomial probabilities. ■

From the preceding example, for very large n, we seem to be left without any hope of finding binomial probabilities without software. In actuality, this is not quite true. Amazingly, there is a connection between the Normal distribution and the binomial distribution that allows us to *estimate* quite accurately binomial probabilities for large enough n.

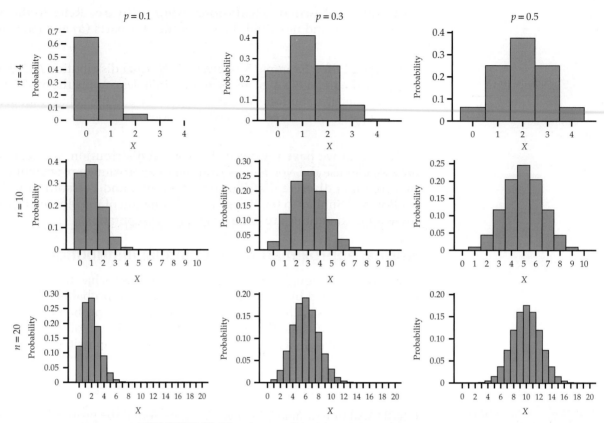

FIGURE 5.19 The shapes of the binomial distribution for different values of n and p.

Figure 5.19 shows the binomial distributions for different values of p (columns) and n (rows). From these graphs, we see that, for a given p, the shape of the binomial distribution becomes more symmetrical as n gets larger. In particular, *as the number of trials n gets larger, the binomial distribution gets closer to a Normal distribution.* We can also see from Figure 5.19 that, for a given n, the binomial distribution is more symmetrical as p approaches 0.5. The upshot is that the accuracy of the Normal approximation depends on the values of both n and p. Check out this relationship yourself with the *Normal Approximation to Binomial* applet. This applet allows you to change n or p and see the effect on the binomial probability histogram and the Normal curve that approximates it.

NORMAL APPROXIMATION FOR BINOMIAL DISTRIBUTION

Draw an SRS of size n from a large population having population proportion p of successes. Let X be the count of successes in the sample. When n is large, the distribution of the count statistic is approximately Normal:

$$X \text{ is approximately } N\left(np, \sqrt{np(1-p)}\right)$$

As a rule of thumb, we use this approximation for values of n and p that satisfy $np \geq 10$ and $n(1-p) \geq 10$.

This Normal approximation is easy to remember because it says that a binomial random variable X is approximately Normal, with its usual mean and standard deviation (see page 259). Recall earlier that we described a Poisson approximation for the binomial (page 264) but that focused on situations when $np < 10$.

In this chapter, we have made a clear distinction between discrete random variables and continuous random variables. The intersection of these two types of random variables with the discrete binomial distribution and the continuous Normal distribution might strike you as being strange and unexpected. In the next chapter, you will learn why it occurs. The Normal distribution finding its way into a binomial setting here serves a preview of why it plays a key role for many *statistical inference* procedures.

EXAMPLE 5.37

CASE 5.3

Using the Normal Approximation As described in Case 5.3 (page 252), a quality engineer inspects an SRS of 150 switches from a large shipment, in which 8% of the units fail to meet specifications. The count X of nonconforming switches in the sample was therefore assumed to be the $B(150, 0.08)$ distribution. In Example 5.25 (page 259), we found $\mu_X = 12$ and $\sigma_X = 3.3226$.

The Normal approximation for the probability of no more than 10 nonconforming switches is the area to the left of $X = 10$ under the Normal curve. Using Table A,

$$P(X \leq 10) = P\left(\frac{X - 12}{3.3226} \leq \frac{10 - 12}{3.3226}\right)$$
$$\doteq P(Z \leq -0.60) = 0.2743$$

In Example 5.20 (page 253), software told us that the actual binomial probability that there are no more than 10 nonconforming switches in the sample is $P(X \leq 10) = 0.3384$. The Normal approximation is only roughly accurate. One of our rule-of-thumb criteria for using the Normal approximation is that $np \geq 10$. Because $np = 12$ here, this combination of n and p is close to the border of the values for which we are willing to use the approximation. Given that this rule-of-thumb criterion is barely met, it is not a surprise that the approximation is only roughly accurate. ∎

The distribution of the count of nonconforming switches in a sample of 15 is distinctly non-Normal, as Figure 5.11 (page 254) showed. When we increase the sample size to 150, however, the shape of the binomial distribution becomes roughly Normal. Figure 5.20 displays the probability histogram of the binomial distribution with the density curve of the approximating Normal distribution superimposed on it. Both distributions have the same mean and standard deviation. The Normal curve fits the histogram reasonably well. But look closer: the histogram is slightly skewed to the right—a property that the symmetric Normal curve can't quite match for this particular combination of n and p.

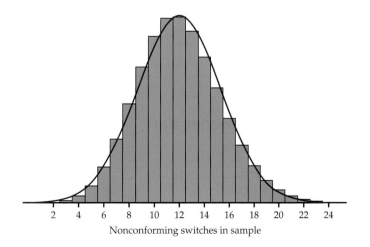

FIGURE 5.20 Probability histogram and Normal approximation for the binomial distribution with $n = 150$ and $p = 0.08$, for Example 5.37.

APPLY YOUR KNOWLEDGE

5.93 Restaurant survey. Return to the survey described in Exercise 5.48 (page 255) in which X is the count of respondents who favor restaurants offering locally sourced food items. You plan to use random digit dialing to contact an SRS of 200 households by telephone, rather than just 20. Suppose you find that 140 of the 200 respondents (70% of the sample) favor restaurants that offer locally sourced food items. We wonder if this is enough evidence to conclude that the percent in your area is higher that the national level of 60%.

(a) What are the mean and standard deviation of the number of people who favor restaurants that offer locally sourced food items in your sample if $p = 0.6$ is true?

(b) Are the rule-of-thumb criteria met for using the Normal approximation?

(c) Based on the Normal approximation, what is the probability that X is greater than or equal to 140?

(d) The probability of part (c) answers the question, "How likely is a sample with at least 70% successes when the population has 60% successes?" Given the probability you found in part (c), are you leaning toward concluding that p is equal to or greater than the national proportion? Explain your reasoning.

The continuity correction

Figure 5.21 illustrates an idea that greatly improves the accuracy of the Normal approximation to binomial probabilities. The binomial probability $P(X \leq 10)$ is the area of the histogram bars for values 0 to 10. The bar for $X = 10$ actually extends between 9.5 to 10.5. Because the discrete binomial distribution puts probability only on whole numbers, the probabilities $P(X \leq 10)$ and $P(X \leq 10.5)$ are the same. The Normal distribution spreads probability continuously, so these two Normal probabilities are different. The Normal approximation is more accurate if we consider $X = 10$ to extend from 9.5 to 10.5, matching the bar in the probability histogram.

The event $\{X \leq 10\}$ includes the outcome $X = 10$. Figure 5.21 shades the area under the Normal curve that matches all the histogram bars for outcomes 0 to 10, bounded on the right not by 10, but by 10.5. So $P(X \leq 10)$ is calculated as $P(X \leq 10.5)$. By contrast, $P(X < 10)$ excludes the outcome $X = 10$, so we exclude the entire interval from 9.5 to 10.5 and calculate $P(X \leq 9.5)$ from

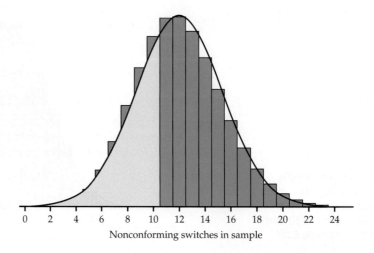

FIGURE 5.21 Area under the Normal approximation curve for the probability in Example 5.37.

the Normal table. Here is the result of the Normal calculation in Example 5.37 (page 279) improved in this way:

$$P(X \leq 10) = P(X \leq 10.5)$$
$$= P\left(\frac{X-12}{3.3226} \leq \frac{10.5-12}{3.3226}\right)$$
$$\doteq P(Z \leq -0.45) = 0.3264$$

The improved approximation misses the exact binomial probability value of 0.3384 by only 0.012. Acting as though a whole number occupies the interval from 0.5 below to 0.5 above the number is called the **continuity correction** to the Normal approximation.

continuity correction

When n is very large, the difference between the Normal approximation with or without continuity correction is marginal. In general, if you need accurate values for binomial probabilities, try to first use software to do exact calculations. If software is not available, use the continuity correction especially when n is not very large. However, because most statistical problems do not require extremely accurate probability calculations, the use of the continuity correction can be viewed as optional.

APPLY YOUR KNOWLEDGE

5.94 Nonconforming switches. Refer to Example 5.37 (page 279). Use the Normal approximation with continuity correction to find:

(a) $P(X \geq 10)$.

(b) $P(X > 11)$.

(c) $P(5 \leq X \leq 9)$.

Normal approximation for Poisson distribution

Under certain circumstances, as we showed earlier, the Poisson distribution can serve as an approximation for the binomial distribution. Given the connections between the Normal distribution and the binomial distribution, it should come as no surprise that the Normal distribution can be used to approximate the Poisson distribution. Indeed, this is the case. As a rule of thumb, when $\mu > 10$, Poisson probabilities can be approximated using the Normal distribution with mean μ and standard deviation $\sqrt{\mu}$. Here is an example.

EXAMPLE 5.38

Number of 311 Calls The government 311 hotline service is a non-emergency phone number that people can call in many U.S. and Canadian cities to find information about services, make complaints, or report problems like trash collection issues and potholes. Most cities perform analytics on call volumes to their call centers for purposes of maintaining adequate staffing and timely services. For the city of Chicago, the average number of 311 calls in the summer months is about 9.7 per hour for the night shift (11:00 P.M. to 7:00 A.M.).[19] Suppose that the number of calls per hour follows a Poisson distribution. What is the probability that over the night shift, the call center receives more than 100 calls?

Since there are eight hours in the shift, the mean is $8 \times 9.7 = 77.6$. To find this probability manually by calculator, we would need to perform the following calculation:

$$P(X > 100) = 1 - P(X \leq 100)$$
$$= 1 - \left[\frac{e^{-77.6}(77.6)^0}{0!} + \frac{e^{-77.6}(77.6)^1}{1!} + \cdots + \frac{e^{-77.6}(77.6)^{100}}{100!}\right]$$

Not only is this calculation tedious but, depending on your calculator, you may not even be able to enter large factorial terms found in the denominators. We can, however, turn to software. In Excel, we can enter the formula =1-POISSON.DIST(100,77.6,1) in any cell. Excel returns a value of 0.00615. This result implies that there is slightly more than an 0.6% chance of receiving more than 100 calls. For the Normal approximation, we compute

$$P(X > 100) = P\left(\frac{X - 77.6}{\sqrt{77.6}} > \frac{100 - 77.6}{\sqrt{77.6}}\right)$$
$$= P(Z > 2.54)$$
$$= 1 - P(Z < 2.54)$$
$$= 1 - 0.9945 = 0.0055$$

The approximation is quite accurate, differing from the actual probability by only 0.00065. ■

The Normal approximation for the Poisson distribution can be further improved by applying the continuity correction idea shown for approximating the binomial distribution. In the end, although the Normal approximation is adequate for many practical purposes, we recommend using statistical software when possible so you can obtain exact Poisson probabilities.

Exponential distributions

As we have learned, one important application of the Poisson distribution is in the modeling of independent events that occur over time. These events might be visits to an emergency room, visits to a particular website, customer calls received by a call center, or customer arrivals to a repair center. Consider now another perspective on these time-oriented Poisson settings. In particular, consider the *times* between the occurrences of these events. For example, a call center might be interested in studying the times between customer calls. Notice that in this Poisson setting, our perspective shifts from the number of calls (a discrete random variable) to the time between calls (a continuous random variable).

exponential distribution   For a Poisson setting, the random variable X of time between events is a continuous random variable following an **exponential distribution**. The exponential distribution is defined by a parameter λ (with $\lambda > 0$) and the height of the density curve at a given point $x \geq 0$ is

$$\lambda e^{-\lambda x}$$

The exponential distribution, as stated earlier, has a direct relationship with the Poisson distribution. Indeed, λ represents the average number of events per unit of time. This parameter is what we called μ in the Poisson setting. The exponential parameter λ is often called the *rate parameter*. The mean and standard deviation of exponential distribution are both equal to $1/\lambda$. In some software, the mean of the exponential distribution may be referred to as the *scale parameter*. Depending on the specific software, you may enter the rate parameter λ or the mean (scale) parameter $1/\lambda$.

Figure 5.22 shows the exponential density curves for a couple of values of λ. The exponential distribution is strongly right-skewed.

EXAMPLE 5.39   **Time between ATM Customer Arrivals** In the scenario described in Example 5.28 (page 263), the mean number of customers arriving at an ATM was 8.5 customers per hour. What is the probability that a customer will arrive at the ATM in less than one minute after the arrival of the previous customer? The arrival rate in Example 5.28 is based on hour time units. The question is stated in terms of

FIGURE 5.22 Exponential density curves for different λ.

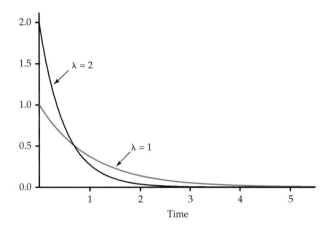

minute time units. We can either convert minutes to hours or convert hours to minutes; let's do the latter. Thus, the arrival rate is $\lambda = 8.5/60 = 0.1417$ customers per minute.

We now want to find $P(X < 1)$, which is the same as finding $P(X \leq 1)$ since, for a continuous random variable, $P(X = 1) = 0$. Figure 5.23 displays the probability we wish to find. Using Excel, we can use the "EXPON.DIST()" function to find the cumulative probability. This function has three arguments: the first is the value of x, the second is the rate parameter λ, and the third is either "0" (reports the height of the density curve) or "1" (reports cumulative probability). Thus, we would enter in any Excel cell =EXPON.DIST(1,0.1417,1). The formula returns a value of 0.132118. ∎

In this ATM setting, there is roughly a 13% chance that a new customer will arrive within one minute of the previous customer. If the previous customer has not finished doing the ATM transactions, the following customer will have to wait in line. Waiting lines are referred to as *queues*. In the business world, many processes create waiting lines or queues. For example, we often have to wait in line to check in baggage at an airport, enter a theater, or get service at a FedEx counter. Not only must people wait for service, but broken machines must wait to be repaired, ships have to wait in harbors to be loaded or unloaded, and airplanes have to circle airports to be cleared for landing. In fact, these events are so common that a special branch of probability known as **queuing theory** is devoted to the study of such waiting time scenarios. The Poisson and exponential distributions serve as fundamental distributions to this branch of study.

queuing theory

Beyond waiting line applications, exponential distributions are useful for the study of the working life of products and equipment, as illustrated in Example 5.40.

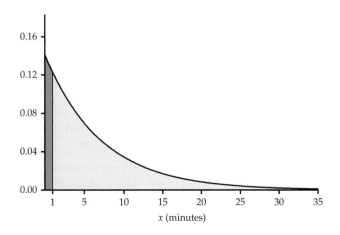

FIGURE 5.23 Exponential distribution with $\lambda = 0.1417$. The shaded area is $P(X < 1)$, calculated in Example 5.39.

EXAMPLE 5.40

iPhone Battery Life In a study of the iPhone X conducted by CNET, researchers found that under normal working conditions the phone's battery lasts 11.5 hours on average. What is the probability that a fully charged iPhone X will have a battery life of more than 13 hours? Unlike Example 5.39, this problem is not set up as arrivals of some event dictated by a Poisson distribution. Here, we are told that 11.5 hours is the mean of the exponential distribution. As noted earlier, the mean of the exponential distribution is $1/\lambda$. Setting $1/\lambda = 11.5$, we find $\lambda = 0.087$. In Excel, to find $P(X > 13)$, enter the formula `=1-EXPON.DIST(13,0.087,1)` in any cell. The formula returns a value of 0.32271. ∎

APPLY YOUR KNOWLEDGE

5.95 Time between ATM customer arrivals. Refer to Example 5.39 (page 282). It was noted in the example that we can either convert minutes to hours or convert hours to minutes; the latter was done. Suppose that we take λ to be 8.5 customers per hour. If we are interested in determining the same probability of a customer arriving less than one minute after the arrival of the previous customer, what specifically should we enter into an Excel cell to find the desired result?

5.96 Time between 311 calls. Refer to Example 5.38 (page 281), in which it is reported that the average number of number of 311 calls made in the city of Chicago during the night shift is 9.7 calls per hour. Assume that the number of calls per time unit follows the Poisson distribution.

(a) What is the average time *between* successive calls in minutes?

(b) Use software to find the probability that the time between successive calls is less than two minutes.

(c) Use software to find the probability that the time between successive calls is greater than 15 minutes.

(d) Use software to find the probability that the time between successive calls is between 5 and 10 minutes.

SECTION 5.4 SUMMARY

- A general **uniform distribution** over an interval from a to b has a density curve that is plotted as the horizontal line over the interval a to b at a height of $1/(b-a)$ and has height 0 outside the interval.

- The mean and standard deviation of a uniform random variable X are

$$\mu = \frac{a+b}{2}$$

$$\sigma = \sqrt{\frac{(b-a)^2}{12}}$$

- The **Normal density curves** are specified by two parameters: the mean μ and the standard deviation σ. Normal density curves are characterized by being bell-shaped, symmetric, and unimodal.

- Stating that a random variable X has the $N(\mu, \sigma)$ distribution is a brief way of saying that a random variable X is Normally distributed with mean μ and standard deviation σ.

- If X has the $N(\mu, \sigma)$ distribution, then the standardized variable $Z = (X - \mu)/\sigma$ has a **standard Normal distribution** $N(0,1)$. Probabilities (and percentiles) for any Normal distribution can be calculated by software or from the standard Normal table (Table A).

- The **Normal approximation to the binomial distribution** says that if X is a count having the $B(n,p)$ distribution, then when n is large,

$$X \text{ is approximately } N\left(np, \sqrt{np(1-p)}\right)$$

We will use this approximation when $np \geq 10$ and $n(1-p) \geq 10$.

- The **continuity correction** improves the accuracy of the Normal approximations.

- The **Normal approximation to the Poisson distribution** says that if X is a count having the Poisson distribution with mean μ, then when μ is large,

$$X \text{ is approximately } N\left(\mu, \sqrt{\mu}\right)$$

We will use this approximation when $\mu > 10$.

- If the number of occurrences of some event over time follows a Poisson distribution with mean μ, then the time between these occurrences is a continuous random variable that follows an **exponential distribution** with rate parameter λ, which is equal to the Poisson distribution mean μ.

- The exponential distribution is right-skewed with mean and standard deviation both equal to $1/\lambda$. The mean parameter $1/\lambda$ is also referred to as a **scale** parameter.

SECTION 5.4 EXERCISES

For Exercise 5.87, see page 275; for 5.88 to 5.92, see pages 276–277; for 5.93, see page 280; for 5.94, see page 281; and for 5.95 and 5.96, see page 284.

5.97 What's wrong? Explain what is wrong in each of the following scenarios.

(a) For any random variable X (discrete or continuous), $P(X \leq a) \neq P(X < a)$.

(b) The heights of density curves are the probabilities of a continuous random variable.

(c) The uniform distribution has a skewed density curve.

(d) The Normal distribution and the uniform distribution share a common property in that their density curves lie above the horizontal axis for all values of X.

5.98 Uniform distribution. Consider a uniform random variable X over the interval 4 to 9.

(a) Find the mean and standard deviation of this distribution.

(b) Find $P(X \leq 5.4)$.

(c) Find $P(X > 7.1)$.

(d) Find $P(4.5 \geq X \leq 8.2)$.

(e) Find $P(X \neq 5)$.

(f) Find the 80th percentile.

5.99 Uniform distribution. Consider a uniform random variable X over the interval -3 to 5.

(a) Find the mean and standard deviation of this distribution.

(b) Find $P(X \leq 3.5)$.

(c) Find $P(X > -1)$.

(d) Find $P(-2.7 \geq X \leq -0.55)$.

(e) Find the 90th percentile.

(f) Find the 10th percentile.

5.100 The sum of two uniform random numbers. Generate *two* random numbers between 0 and 1 and take Y to be their sum. Then Y is a continuous random variable that can take any value between 0 and 2. The density curve of Y is the triangle shown in Figure 5.24.

(a) Verify by geometry that the area under this curve is 1.

(b) What is the probability that Y is less than 1? (Sketch the density curve, shade the area that represents the probability, and then find that area. Do this for part (c) also.)

(c) What is the probability that Y is greater than 0.6?

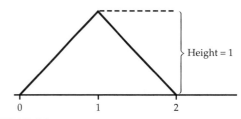

FIGURE 5.24 The density curve for the sum of two random numbers, for Exercise 5.100. This density curve spreads probability between 0 and 2.

5.101 Normal distribution. Suppose that X has the $N(230, 40)$ distribution. Using software, find the following:

(a) $P(X < 180)$.

(b) $P(X > 280)$.

(c) $P(160 < X < 200)$.

(d) 80th percentile.

(e) 5th percentile.

5.102 Transforming a uniform random variable. Many software programs (including Excel) provide a function to randomly generate numbers uniformly over the interval 0 to 1. Suppose you wish to randomly generate uniform numbers over a general interval of a to b. Define X as a uniform random variable over the interval 0 to 1 and define Y as a uniform random variable over the interval a to b. It turns out that Y is a simple linear transformation of X:

$$Y = a + cX$$

The purpose of this exercise is to determine c.

(a) Find μ_X, the mean of distribution of X.

(b) Using the rules for means (page 236) and the mean value from part (a), express μ_Y as a function of a and c.

(c) Refer to page 274 and express μ_Y in terms of a and b.

(d) Equate the two expressions from parts (b) and (c) and solve for c. Rewrite $Y = a + cX$ based on what you found c to be.

(e) Based on part (d), what is the linear transformation of X to generate uniform random numbers over the interval 7 to 11?

(f) Based on part (d), what is the linear transformation of X to generate uniform random numbers over the interval -10 to 4?

5.103 Hybrid learning. Higher education institutions are increasingly introducing courses based on hybrid instruction, a combination of conventional face-to-face and online instruction. In a recent nationwide study by EDUCAUSE Center for Analysis and Research, it was found that 77% of students believe that hybrid learning is the most effective learning environment for them.[20] You decide to poll a sample of the incoming students at your institution to gauge their attitude toward hybrid courses. In an SRS of 200 incoming students, you find that 170 prefer this type of course delivery.

(a) What are the mean and standard deviation of the number of students who prefer hybrid learning in your sample if your institution's student population follows the same proportion as the nationwide student population?

(b) Using the 68–95–99.7 rule, if you had drawn an SRS from the national body of students, you would expect the count of students who prefer hybrid learning to fall between what two values about 95% of the time?

(c) Based on your observation that 170 students prefer hybrid learning, do you think that the incoming students at your institution prefer this type of instruction more, less, or about the same as students nationally? Explain your answer.

5.104 Public transit. Refer to Exercise 5.64 (page 269), where it is reported that 57% of New York City commuters use public transportation. Suppose you choose 500 NYC commuters at random.

(a) Are the rule-of-thumb criteria met for using the Normal approximation?

(b) What is the probability that 275 or fewer commuters in your sample use public transportation? Use the Normal approximation *without* continuity correction to determine your answer.

(c) What is the probability that 300 or more commuters in your sample use public transportation? Use the Normal approximation *without* continuity correction to determine your answer.

(d) Repeat part (b) using continuity correction.

(e) Repeat part (c) using continuity correction.

5.105 Uber female drivers. Refer to Exercise 5.65 (page 269), where it is reported that the nationwide proportion of female Uber drivers is 27.3%. Use this proportion to compute the desired probabilities. Suppose you were to track the number of female drivers X in your next 100 Uber trips.

(a) Are the rule-of-thumb criteria met for using the Normal approximation?

(b) What is $P(20 \leq X \leq 30)$? Use the Normal approximation *without* continuity correction.

(c) Repeat part (b) using continuity correction.

5.106 Finding $P(X = k)$. In Example 5.20 (page 253), we found $P(X = 10) = 0.106959$ when X has a $B(150, 0.08)$ distribution. Suppose we wish to find $P(X = 10)$ using the Normal approximation.

(a) What is the value for $P(X = 10)$ if the Normal approximation is used *without* continuity correction?

(b) What is the value for $P(X = 10)$ if the Normal approximation is used now *with* continuity correction?

5.107 Internet video postings. Refer to Exercise 5.66 (page 269), where it was stated that approximately 30% of all adult Internet users have posted videos online. A sample survey interviews an SRS of 1555 Internet users.

(a) What is the mean number of adult Internet users who post videos online? What is the standard deviation?

(b) Use the Normal approximation *without* continuity correction to find the probability that 450 or fewer of the people in the sample have posted videos online.

(c) Repeat part (c) using continuity correction.

5.108 Website visits. In 2017, Amazon's website averaged 4560 unique visitors (worldwide) per minute. Assume that the visits are well modeled by the Poisson distribution. Some software packages will have trouble calculating Poisson probabilities with such a large value of μ.

(a) Use Excel's or R's Poisson function to calculate the probability of 4700 or more unique visitors during any given minute.

(b) Find the probability of part (a) using the Normal approximation.

(c) Do this part if you are a Minitab user. Try calculating the probability of part (a) using Minitab's Poisson option. Did you get an answer? If not, how did the software respond? What is the largest value of μ that Minitab can handle?

5.109 Website visits, continued. Refer to the previous exercise and consider the number of unique visitors to Amazon's website in one hour. Use the Normal distribution and the 68–95–99.7 rule to find the range in which we would expect 95% of the number of visitors to fall.

5.110 Initial public offerings. The number of companies making their initial public offering of stock (IPO) can be modeled by a Poisson distribution with a mean of 20 per month. Use the Normal approximation to find the probability of 250 or more IPOs occurring in a year.

5.111 iPhone battery life. Refer to Example 5.40 (page 284). Use software to find the desired probabilities.

(a) What is the probability that the battery life of a fully charged iPhone X will last no more than 2 hours?

(b) What is the probability that the battery life of a fully charged iPhone X will last between 5 and 15 hours?

5.112 Calculating exponential distribution probabilities. In Examples 5.39 (page 282) and 5.40 (page 284), we showed how software can be used to find the desired probabilities from the exponential distribution. However, for the exponential distribution, it turns out that there is a straightforward equation for finding $P(X \leq x)$ when $x \geq 0$:

$$P(X \leq x) = 1 - e^{-\lambda x}$$

(a) Refer to Example 5.39, in which $P(X < 1)$ was found. Using a calculator, show that the same probability value is found by means of the preceding equation.

(b) In general, what would be the equation for finding $P(X > x)$?

(c) Refer to Example 5.40, in which $P(X > 13)$ was found. Using a calculator, show that the same probability value is found by means of the equation found in part (b).

5.113 Percentiles on an exponential distribution. As shown in Example 5.35 (page 276), Excel has a formula for doing inverse calculations to find percentile points on the Normal distribution. However, unlike most statistical software, Excel does not provide a formula for performing inverse calculations on exponential distributions. Fortunately, there is a way to find percentiles for an exponential distribution using the equation of the previous exercise.

Refer to Example 5.39 (page 282), where $\lambda = 0.1417$. The exponential distribution for this example is shown in Figure 5.23. In this exercise, you will find the 90th percentile point on this distribution.

(a) Set the probability formula shown in the previous exercise to 0.90 with $\lambda = 0.1417$. Now, solve for x. To do so, rearrange the equation so that $e^{-\lambda x}$ is on one side of the equal sign. Then use the fact that the natural log is the inverse of the exponential function. In particular, $log(e^a) = a$, where log is the natural log function. What value of x (i.e., the 90th percentile point) do you find?

(b) Suppose you want to confirm in Excel that the value of part (a) is indeed the 90th percentile point. What specifically should you enter into an Excel cell to confirm your result?

5.114 Probability of being less than the mean. For any Normal distribution, the probability that X is less than the mean is always 0.5—that is, $P(X \leq \mu) = 0.5$. This fact is, of course, due to the symmetry of the Normal distribution. For any exponential distributions, it turns out that the probability that X is less than the mean is always equal to the same value. Assume that X has an exponential distribution with rate parameter λ.

(a) In terms of λ, what is μ, the mean of the exponential distribution?

(b) Use part (a) along with the equation given in Exercise 5.112 to find the mathematical formula for $P(X \leq \mu)$. What is the formula and what is the value of $P(X \leq \mu)$?

(c) Consider now two different values for λ—namely, 2 and 10. What are the means for these two different exponential distributions? In Excel, what specifically should be entered in Excel cells to find $P(X \leq \mu)$ for each of these λ values? Refer to Example 5.39 (page 282) for an explanation of the Excel's exponential function. Enter these functions in Excel. What values does Excel report? Compare them with the general result from part (b).

CHAPTER 5 REVIEW EXERCISES

5.115 Repeat an experiment many times. Here is a probability distribution for a random variable X:

Value of X	−6	7
Probability	0.4	0.6

A single experiment generates a random value from this distribution. Suppose this experiment is repeated many times.

(a) What will be the approximate proportion of times that the value of −6 is observed? Give a reason for your answer.

(b) What will be the approximate value of the mean of these random values? Give a reason for your answer.

5.116 Work with a transformation. Here is a probability distribution for a random variable X:

Value of X	2	7
Probability	0.8	0.2

(a) Find the mean and the standard deviation of this distribution.

(b) Let $Y = 4X - 1$. Use the rules for means and variances to find the mean and the standard deviation of the distribution of Y.

(c) For part (b), give the rules that you used to find your answer.

5.117 A different transformation. Refer to the previous exercise. Now let $Y = 4X^2 - 1$.

(a) Find the distribution of Y.

(b) Find the mean and standard deviation for the distribution of Y.

(c) Explain why the rules that you used for part (b) of the previous exercise do not work for this transformation.

5.118 Some probability distributions. Here is a probability distribution for a random variable X:

Value of X	2	3	4
Probability	0.3	0.3	0.4

(a) Find the mean and standard deviation for this distribution.

(b) Construct a different probability distribution with the same possible values, the same mean, and a larger standard deviation. Show your work and report the standard deviation of your new distribution.

(c) Construct a different probability distribution with the same possible values, the same mean, and a smaller standard deviation. Show your work and report the standard deviation of your new distribution.

5.119 Find some conditional probabilities. Choose a point at random in the square with sides $0 \leq x \leq 1$ and $0 \leq y \leq 1$; thus, the probability that the point falls in any region within the square is the area of that region. Let X be the x coordinate and Y be the y coordinate of the point chosen. Find the conditional probability $P(Y < 1/3 \mid Y > X)$. (*Hint:* Sketch the square and the events $Y < 1/3$ and $Y > X$.)

5.120 Alternative computation of variance. By algebraically rewriting the formula for variance (page 239), it can be shown that the variance can be computed by the following formula.

$$\sigma_X^2 = \mu_{X^2} - \mu_X^2$$

Said in words, the variance of X equals the mean of X^2 minus the square of the mean of X. Refer to Example 5.10 (page 239) to find the distribution along with the computed mean and variance for daily demand for transfusion blood bags X.

(a) Find μ_{X^2}.

(b) Use the value found in part (a) in the alternative computation for variance to confirm its value is the same as reported in Example 5.10.

CASE 5.1 **5.121 Blood bag demand.** Refer to the distribution of daily demand for blood bags X in Case 5.1 (page 222). Assume that demand is independent from day to day.

(a) What is the probability that at least one bag will be demanded every day of a given month? Assume 30 days in the month.

(b) What is the interpretation of one minus the probability found part (a)?

(c) What is the probability that the blood bank will go a whole year (365 days) without experiencing a demand of 12 bags on a given day?

5.122 Risk pooling in a supply chain. Example 5.14 (page 242) compares a decentralized versus a centralized inventory system as it ultimately relates to the amount of safety stock (extra inventory over and above mean demand) held in the system. Suppose that the CEO of ElectroWorks requires a 99% customer service level. This means that the probability of satisfying customer demand during the lead time is 0.99. Assume that lead time demands for the Milwaukee warehouse, Chicago warehouse, and centralized warehouse are Normally distributed with the means and standard deviations found in the example.

(a) For a 99% service level, how much safety stock of the SurgeArrester part does the Milwaukee warehouse need to hold? Round your answer to the nearest integer.

(b) For a 99% service level, how much safety stock of the SurgeArrester part does the Chicago warehouse need to hold? Round your answer to the nearest integer.

(c) For a 99% service level, how much safety stock of the SurgeArrester part does the centralized warehouse need to hold? Round your answer to the nearest integer. How many more units of the part need to be held in the decentralized system than in the centralized system?

5.123 Life insurance. Assume that a 25-year-old man has these probabilities of dying during the next five years:

Age at death	25	26	27	28	29
Probability	0.00039	0.00044	0.00051	0.00057	0.00060

(a) What is the probability that the man does not die in the next five years?

(b) An online insurance site offers a term insurance policy that will pay $100,000 if a 25-year-old man dies within the next five years. The cost is $175 per year. Thus, the insurance company will take in $875 from this policy if the man does not die within five years. If he does die, the company must pay $100,000. Its loss depends on how many premiums the man paid, as follows:

Age at death	25	26	27	28	29
Loss	$99,825	$99,650	$99,475	$99,300	$99,125

What is the insurance company's mean cash intake (income) from such policies?

5.124 Risk for one versus many life insurance policies. It would be quite risky for an insurance company to insure the life of only one 25-year-old man under the terms of Exercise 5.123. There is a high probability that person would live and the company would gain $875 in premiums. But if he were to die, the company would lose almost $100,000. We have seen that the risk of an investment is often measured by the standard deviation of the return on the investment. The more variable the return is (the larger σ is), the riskier the investment is.

(a) Suppose only one person's life is insured. Compute the standard deviation of the income X that the insurer will receive. Find σ_X, using the distribution and mean you found in Exercise 5.123.

(b) Suppose that the insurance company insures two men. Define the total income as $T = X_1 + X_2$, where X_i is the income made from man i. Find the mean and standard deviation of T.

(c) You should have found that the standard deviation computed in part (b) is greater than that found in part (a). But this does not necessarily imply that insuring two people is riskier than insuring one person. What needs to be recognized is that the mean income has also increased. So, to measure the riskiness of each scenario, we need to scale the standard deviation values relative to the mean values. This is simply done by computing σ / μ, which is called the **coefficient of variation** (CV). Compute the coefficients of variation for insuring one person and for insuring two people. What do the CV values suggest about the relative riskiness of the two scenarios?

(d) Compute the mean total income, the standard deviation of total income, and the CV of total income when 30 people are insured.

(e) Compute the mean total income, the standard deviation of total income, and the CV of total income when 1000 people are insured.

(f) As you will learn in Chapter 6, a remarkable result in probability theory states that the sum of a large number of independent random variables follows approximately the Normal distribution even if the random variables themselves are not Normal. In most cases, 30 is sufficiently "large." Given this fact, use the mean and standard deviation from part (d) to compute the probability that the insurance company will lose money from insuring 30 people—that is, compute $P(T < 0)$. Compute now the probability of a loss to the company if 1000 people are insured. What did you learn from these probability computations?

5.125 Benford's law. In Chapter 4, we learned about the striking fact that the first digits of numbers in legitimate records often follow a distribution known as Benford's law. Here it is:

First digit	1	2	3	4	5	6	7	8	9
Proportion	0.301	0.176	0.125	0.097	0.079	0.067	0.058	0.051	0.046

(a) What is the probability that a randomly chosen invoice has a first digit of 1, 2, or 3?

(b) Suppose 10 invoices are randomly chosen. What is the probability that four or more of the invoices will have a first digit of 1, 2, or 3? Use the binomial formula.

(c) Now do a larger study, examining a random sample of 1000 invoices. Use software to find the exact probability that 620 or more of the invoices have first digits of 1, 2, or 3.

(d) Using Table A and no software, use the Normal approximation with continuity correction to find the probability of part (b).

5.126 Wi-Fi interruptions. Refer to Example 5.27 (page 262), in which we were told that the mean number of Wi-Fi interruptions per day is 0.9. We also found in Example 5.27 that the probability of no interruptions on a given day is 0.4066.

(a) Treating each day as a trial in a binomial setting, use the binomial formula to compute the probability of no interruptions in a week.

(b) Now, instead of using the binomial model, let's use the Poisson distribution exclusively. What is the mean number of Wi-Fi interruptions during a week?

(c) Based on the Poisson mean of part (b), use the Poisson distribution to compute the probability of no interruptions in a week. Confirm that this probability is the same as that found part (a). Explain in words why the two ways of computing no interruptions in a week give the same result.

(d) Explain why using the binomial distribution to compute the probability that only one day in the week will not be interruption free would *not* give the same probability had we used the Poisson distribution to compute the probability that only one interruption occurs during the week.

5.127 Benford's law, continued. Benford's law suggests that the proportion of legitimate invoices with a first digit of 1, 2, or 3 is much greater than if the digits were distributed as equally likely outcomes. As a fraud investigator, you would be suspicious of some potential wrongdoings if the count of invoices with a first digit of 1, 2, or 3 is too low. You decide that if the count is in the lower 5% of counts expected by Benford's law, then you will call for a detailed investigation for fraud. Assuming the expected proportion of invoices with a first digit of 1, 2, or 3 given by Benford's law, use software and the binomial distribution to find that cutoff point for becoming suspicious of fraud. Specifically, find the largest number m out of $n = 1000$ invoices such that $P(X \leq m)$ is less than 0.05.

5.128 Environmental credits. An opinion poll asks an SRS of 300 adults whether they favor tax credits for companies that demonstrate a commitment to preserving the environment. Suppose that, in fact, 45% of the population favor this idea.

(a) Use software to find the exact probability that more than half of the sample are in favor.

(b) Use the Normal approximation (with continuity correction) to find the probability that more than half of the sample are in favor.

5.129 Leaking gas tanks. Leakage from underground gasoline tanks at service stations can damage the environment. It is estimated that 25% of these tanks leak. You examine 20 tanks chosen at random, independently of each other.

(a) What is the mean number of leaking tanks in such samples of 20?

(b) What is the probability that 10 or more of the 20 tanks leak? Compute this probability manually using the binomial formula.

(c) Now you do a larger study, examining a random sample of 2000 tanks nationally. Use software to find the probability that at least 550 of these tanks are leaking.

5.130 Is this coin balanced? While he was a prisoner of the Germans during World War II, John Kerrich tossed a coin 10,000 times. He got 5067 heads. Take Kerrich's tosses to be an SRS from the population of all possible tosses of his coin. If the coin is perfectly balanced, $p = 0.5$. Is there reason to think that Kerrich's coin gave too many heads to be balanced? To answer this question, find the probability that a balanced coin would give 5067 or more heads in 10,000 tosses. What do you conclude?

5.131 Six Sigma. Six Sigma is a quality improvement strategy that strives to identify and remove the causes of defects. Processes that operate with Six Sigma quality produce defects at a level of 3.4 defects per million. Suppose 10,000 independent items are produced from a Six Sigma process.

(a) Using the binomial distribution, find the probability that at least one defect is produced.

(b) Use the Poisson distribution to approximate the probability. Comment on how well the approximation did.

(c) Explain why the Normal distribution would not provide a good approximation in this setting.

5.132 Binomial distribution? Suppose a manufacturing colleague tells you that 1% of items produced in the first shift are defective, while 1.5% in the second shift are defective and 2% in the third shift are defective. He notes that the number of items produced is approximately the same from shift to shift, which implies an average defective rate of 1.5%. He further states that because the items produced are independent of each other, the binomial distribution with p of 0.015 will represent the number of defective items in an SRS of items taken in any given day. What is your reaction?

5.133 Poisson distribution? Suppose you find in your spam folder an average of two spam emails every 10 minutes. Furthermore, you find that the rate of spam email from midnight to 6 a.m. is twice the rate during other parts of the day. Explain whether the Poisson distribution is an appropriate model for the spam process.

5.134 Airline overbooking. Airlines regularly overbook flights to compensate for no-show passengers. In doing so, airlines are balancing the risk of having to compensate bumped passengers against lost revenue associated with empty seats. Historically, no-show rates in the airline industry range from 10% to 15%. Assuming a no-show rate of 12.5%, what is the probability that no passenger will be bumped if an airline books 215 passengers on a 200-seat plane?

5.135 Inventory control. OfficeShop experiences a one-week order time to restock its HP printer cartridges. During this reorder time, also known as *lead time*, OfficeShop wants to ensure a high level of customer service by not running out of cartridges. Suppose the average lead time demand for a particular HP cartridge is 15 cartridges. OfficeShop makes a restocking order when there are 18 cartridges on the shelf. Assuming the Poisson distribution models the lead time demand process, what is the probability that OfficeShop will be short of cartridges during the lead time?

5.136 More about inventory control. Refer to the previous exercise. In practice, the amount of inventory held on the shelf during the lead time is known as the *reorder point*. Firms use the term *service level* to indicate the percentage of the time that the amount of inventory is sufficient to meet demand during the reorder period. Use software and the Poisson distribution to determine the reorder points so that the service level is minimally

(a) 90%.

(b) 95%.

(c) 99%.

5.137 How many close friends? How many close friends do you have? Suppose that the number of close friends adults claim to have varies from person to person, with mean $\mu = 9$ and standard deviation $\sigma = 2.4$. An opinion poll asks this question of an SRS of 1100 adults. We see in Chapter 6 that, in this situation, the sample mean response $\bar{x}$ has approximately the Normal distribution with mean 9 and standard deviation 0.0724. What is $P(8 \leq \bar{x} \leq 10)$, the probability that the statistic $\bar{x}$ estimates the parameter μ to within ± 1?

5.138 Normal approximation for a sample proportion. A sample survey contacted an SRS of 700 registered voters in Oregon shortly after an election and asked respondents whether they had voted. Voter records show that 56% of registered voters had actually voted. We will see in Chapter 6 that, in this situation. the proportion $\hat{p}$ of the sample who voted has approximately the Normal distribution with mean $\mu = 0.56$ and standard deviation $\sigma = 0.019$.

(a) If the respondents answer truthfully, what is $P(0.52 \leq \hat{p} \leq 0.60)$? This is the probability that the statistic $\hat{p}$ estimates the parameter 0.56 within plus or minus 0.04.

(b) In fact, 72% of the respondents said they had voted ($\hat{p} = 0.72$). If respondents answer truthfully, what is $P(\hat{p} \geq 0.72)$? This probability is so small that it is good evidence that some people who did not vote claimed that they did vote.

5.139 Morning coffee. Suppose, in any given 10 minute interval, the number of arrivals at a Starbucks drive-up window during the morning hours can be modeled by a Poisson distribution with $\mu = 16.25$. Define X as the time (in minutes) between two successive arrivals at the drive-up window.

(a) What is the mean of the arrival count per minute?

(b) What is the value of the rate parameter λ?

(c) What is the mean time between successive arrivals?

(d) What is the standard deviation of the time between successive arrivals?

(e) What is $P(X \leq 0.5)$?

(f) What is $P(X > 1)$?

5.140 LED light bulbs. Philips markets LED post bulbs, which are used for lighting urban parks and for street lighting. The bulbs are reported to have a mean life of 50,000 hours. If a light is operated 12 hours per day, this translates into a mean life of 11.42 years. Assume that the life of a bulb X follows the exponential distribution and that the bulbs are operated for 12 hours per day.

(a) What is the probability that a bulb will fail within one year?

(b) What is the probability that a bulb will last longer than 15 years?

(c) Suppose a small plaza area has four lamp posts, each with one bulb. Assuming that the bulbs are independent of each other, what is the probability that all will fail within five years?

maximkabb/Getty Images

CHAPTER 6

Sampling Distributions

Introduction

In Chapters 1 and 2, we learned how to use graphical and numerical methods to summarize the key features of *sample* data. Chapter 1 also introduced us to the class of Normal distributions, which plays a fundamental role in many methods of statistical analysis. In Chapter 3, we learned that any data summary is only as good as the means by which the data were produced and that generalizations are most secure when we produce data by random sampling or randomized experimentation. In Chapter 4, we studied the basic concepts of probability. Chapter 5 introduced the idea of random variables as well as the binomial distribution and its connection to the Normal distribution.

The topics of all these previous chapters set us up for an important juncture in this book as we continue to learn how the area of statistics is the science of learning from data. Namely, we are now ready for a formal transition into the topic of statistical inference. Statistical inference draws conclusions about a population or process from data. It emphasizes substantiating these conclusions via probability calculations because probability allows us to take chance variation into account.

The use of probability concepts allows us to introduce the idea of a *sampling distribution* of a statistic, the theme of this chapter. A sampling distribution shows how the statistic would vary in identical repeated data collections. This probability distribution answers the question, "What would happen if we did this experiment or sampling many times?" Such distributions provide the necessary link between probability and the data in your sample or from your experiment. They are key to understanding statistical inference.

CHAPTER OUTLINE

6.1 Toward Statistical Inference

6.2 The Sampling Distribution of the Sample Mean

6.3 The Sampling Distribution of the Sample Proportion

The last two sections of this chapter study the sampling distributions of two common statistics: the sample mean (for quantitative data) and the sample proportion (for categorical data). Here are just two examples of the need to understand the behaviors of these two statistics:

- Retailers seek insights into consumer spending behavior to make strategic decisions ranging from the amount of inventory to hold by product type, the nature and timing of sales promotions, and modes of delivery (e.g., in-store pickup, free shipping) to consumers. To gain these insights, retailers often hire consulting firms to perform consumer-focused analytics. For example, in a Deloitte study of 5085 consumers, researchers found that GenZ consumers spent an average of $1200 during the holiday season while baby boomer consumers spent an average of $1274. How different might these averages be if another 5085 consumers were studied? For the populations of GenZ and baby boomer consumers, can we conclude that the average baby boomer's holiday spending amount is greater than the average GenZ consumer's holiday spending amount? To answer these questions, there must be an understanding of how sample means vary from sample to sample.

- Procter & Gamble (P&G) has identified "customer understanding" as one of its five core strengths. P&G invests hundreds of millions of dollars annually to conduct thousands of marketing research studies to determine customers' preferences, which are typically translated into proportions. For example, in a P&G Consumer Cleaning Insights Survey involving 1008 randomly selected consumers, it was reported that a "majority (51 percent) agree that if the restaurant is really clean, they are more likely to overlook poor service." Would this result of the "majority" hold up if P&G selected another 1008 consumers? To answer this question, marketing analysts at P&G need a base understanding of how proportions vary from sample to sample.

The general framework for constructing a sampling distribution is the same for all statistics, so we initially focus on these two statistics (means and proportions) given their prevalent use in inference. As part of our study of sampling distributions, we revisit the Normal distribution and make use of our background knowledge of the binomial distribution.

6.1 Toward Statistical Inference

When you complete this section, you will be able to:

- Identify populations, parameters, samples, sample statistics, and the relationships among these terms.
- Define sampling distributions in terms of probability distributions.
- Interpret a sampling distribution to describe properties of a sample statistic.
- Define what estimators are and how they relate to sample statistics.
- Define the meaning of bias of an estimator.
- Identify ways to reduce bias and variability of an estimator.
- Indicate how large a population should be so that variability of a sample statistic depends little on the size of the population.

population and sample, p. 130

As we proceed into the key ideas of this chapter, it is important to recall the fundamental distinction between populations and samples:

- A population is the collection of all possible cases that we want to study.
- A sample is a subset of the population for which we collect data.

The National Retail Federation (NRF) is the world's largest retail trade association, representing the full spectrum of retail stores from department and discount stores to Internet retailers. NRF routinely conducts surveys to take the pulse of consumer behavior and spending. In one such survey, NRF interviewed a random *sample* of 2067 households with children between the ages of 6 and 17. The sample was obtained to see how households would shop for clothing, electronics, shoes, and school supplies for the upcoming school year (K–12).[1] Unlike Christmas or other gift-giving holidays (e.g., Mother's Day and Valentine's Day), back-to-school spending is needs-driven and not discretionary. As it turns out, back-to-school spending is the second biggest spending season for retailers behind winter holiday spending. In the survey, NRF found that the average back-to-school spending was $687.72. That's the truth about the 2067 households in the sample. But what is the truth about the millions of back-to-college spending households who make up this *population* in question? In particular, what might be the average back-to-school expenditure in the population of back-to-school spending households?

As there is a distinction between populations and samples, there is a distinction in the vocabulary that we use for numerical descriptions of population and sample.

PARAMETERS AND STATISTICS

A **parameter** is a numerical value that describes the **population.** The value of a parameter is a fixed (unchanging) number. Since we rarely have complete knowledge about a population, its value is typically unknown.

A **statistic** is a numerical value that describes a **sample.** Different samples will generally give different values for the statistic. We often use a statistic to estimate an unknown parameter.

EXAMPLE 6.1

Back-to-School Spending: Statistic versus Parameter In the back-to-school survey, NRF reports that the average expenditure is $687.72 and that 21% of households shop 1–2 weeks before school starts.

The **sample mean** $\bar{x} = \$687.72$ is a *statistic*. The corresponding *parameter* is the **average** (call it μ) of all back-to-school households. This parameter is referred to as the **population mean.**

sample proportion

Similarly, the **sample proportion** that shops 1–2 weeks before school starts,

$$\hat{p} = \frac{434}{2067} = 0.21 = 21\%$$

is a *statistic*. The corresponding *parameter* is the **proportion** (call it p) of all back-to-school spending households that shop 1–2 weeks before school starts. This parameter is referred to as the **population proportion.**

population proportion

We don't know the values of the parameters μ and p, so we use the statistics $\bar{x}$ and $\hat{p}$, respectively, to estimate them. ∎

Chapter 1 introduced a variety of statistics (e.g., sample mean, sample median, and sample standard deviation) as descriptive summaries of a sample. Summarizing a sample with numerical values along with visual aids (graphs) is an important first step of any statistical study. But, our use of statistics most often does not end with a summary of the sample results. After summarizing a sample, we then use the statistics to draw conclusions about the population from which the sample was selected.

estimate Consider again the NRF survey of 2067 randomly selected households. Because the sample was chosen at random, it's reasonable to believe that these 2067 households represent the entire population fairly well. As such, NRF researchers can use the observed value of $687.72 for the *sample mean statistic* as an **estimate** of the average back-to-school expenditure in the population of back-to-school spending households—that is, as an estimate of the *population mean parameter*.

statistical inference This is a basic idea in statistics: use a fact about a sample to estimate the truth about the whole population. We call this process **statistical inference** because we infer conclusions about the entire population from data on selected individuals.

APPLY YOUR KNOWLEDGE

6.1 **Workplace bullying.** In a 2017 study commissioned by Workplace Bullying Institute, a random sample of 1008 employees taken nationwide were asked if they experienced any of the following types of mistreatment: abusive conduct that is threatening, intimidating, humiliating, work sabotage, or verbal abuse. The study found that 19% of respondents directly experienced abusive conduct at work.[2] Describe the statistic, population, and population parameter for this setting.

Sampling variability

If the NRF took a second random sample of 2067 households, the new sample would have different households in it. It is almost certain that the sample mean $\bar{x}$ would not again be $687.72. Likewise, we would not expect there to be exactly 434 households that shop 1–2 weeks before school starts. In other words, the value of a statistic will vary from sample to sample. This basic fact *sampling variability* is called **sampling variability:** the value of a statistic varies in repeated random sampling.

Random sampling eliminates any preferences or favoritism from the act of choosing a sample, but we still might have low confidence in making judgments about the population due to the *variability* that results from random sampling. For example, what if a second sample of 2067 households resulted in 10% of households shopping 1–2 weeks before school starts? Do these results, 21% and 10%, leave you confident about the value of the true population proportion? When sampling variability is too great, we can't trust the results of any one sample.

bias, p. 132 We assess the level of sampling variability by using the second advantage of random sampling (the first advantage is the elimination of *sampling bias*). Namely, if we take many random samples of the same size from the same population, then the variation from sample to sample will follow a predictable pattern. **All statistical inference is based on one idea: to see how trustworthy a procedure is, ask what would happen if we repeated it many times.**

Sampling distributions

With random sampling, the values of a statistic vary randomly from sample to sample and are unknown prior to the samples being selected. Viewed from this perspective, we see that statistics are *random variables*. As you learned from Chapter 5, random variables are associated with *probability distributions*. So, as with any other random variable, a statistic has a probability distribution that provides the probabilities of its possible values. When dealing with a statistic, we reserve a special name for its probability distribution. In particular, *sampling distribution* a probability distribution of a statistic is called a **sampling distribution.**

A natural question is, How do we go about finding the sampling distribution? Recall the coin-tossing experiments described in Example 4.2 (page 178). To determine the probabilities of heads and tails, diligent people tossed a coin thousands of times. This approach to pinning down the distribution of outcomes is empirical. The same technique can be applied to any statistic; namely, we can record the outcomes of a statistic over many repeated samples. In particular, we might consider the following:

- Take a large number of samples from the same population.
- Calculate the statistic for each sample.
- Make a histogram of the values of the statistic.
- Examine the distribution displayed in the histogram for shape, center, and spread, as well as outliers or other deviations.

In practice, it is too expensive or impractical to take many samples from a large population such as all back-to-school households in the United States. But we can gain insights about sampling distribution behavior by using computer software. In particular, we can assume some population with known parameter values. We can then imitate taking many samples using software from this specified population to emulate the chance behavior of the statistic. This process is called **simulation**.

simulation

EXAMPLE 6.2

Take Many Samples Let's simulate drawing simple random samples (SRSs) of size 100 from the population of back-to-school households. Suppose the true population proportion of households (p) that shop 1–2 weeks before school starts is 0.2. (Of course, we would not sample in practice if we already knew that $p = 0.2$. We are sampling here to understand how the statistic $\hat{p}$ behaves.)

Suppose the sample proportion from our first SRS is $\hat{p} = 0.25$. That is more than the population proportion. Take another SRS of size 100. The sample proportion for this sample is $\hat{p} = 0.18$. That's less than the population proportion. If we take more samples of size 100, we will get different values of $\hat{p}$ and a pattern will emerge. Figure 6.1 is a histogram of the sample proportions for 1000 samples, each of size 100. We find that the sample proportions range from a value as low as 0.08 to as high as 0.36. This suggests that a survey based on this sample size is prone to give us a sample proportion considerably off from the population proportion.

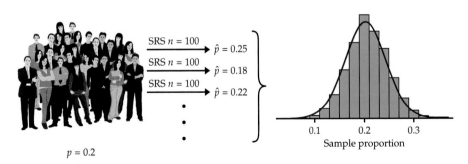

FIGURE 6.1 The results of many SRSs have a regular pattern, Example 6.2. Here we draw 1000 SRSs of size 100 from the same populations. The population parameter is $p = 0.2$. The histogram shows the distribution of 1000 sample proportions.

However, NRF samples 2067 households, not just 100. Figure 6.2 shows a histogram of 1000 sample proportions, each based on an SRS of 2067.

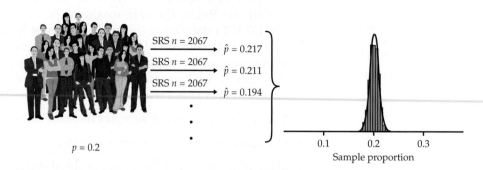

FIGURE 6.2 The distribution of the sample proportion for 1000 SRSs of size 2067 drawn from the same population as in Figure 6.1. The two histograms have the same scale. The statistic from the larger sample is less variable.

Figures 6.1 and 6.2 are drawn on the same scale. Comparing them shows what happens when we increase the size of our samples from 100 to 2067. As seen, the sample proportion based on the larger sample is less variable. These histograms display the *sampling distribution* of the statistic $\hat{p}$ for the two sample sizes. ∎

Strictly speaking, the sampling distribution is the ideal pattern that would emerge if we looked at all possible samples of size n (whether 100 or 2067) from our population. A distribution obtained from a fixed number of trials, like the 1000 trials in Figures 6.1 and 6.2, is only an approximation to the sampling distribution. Here is the general definition.

SAMPLING DISTRIBUTION

The **sampling distribution** of a statistic is the distribution of values taken by the statistic in all possible samples of the same size from the same population.

By being a distribution of values of the statistic in *all* possible samples of the same sample size, the above definition is reminding us that a sampling distribution is a probability distribution.

As it turns out, for many statistics, including $\hat{p}$ and $\bar{x}$, we can use probability theory, mathematics of chance behavior, to determine sampling distributions exactly (or approximately) *without* the use of simulation. However, there are some settings in which we are not so lucky to have results directly obtained from probability theory. In Section 8.1 (pages 412–413), we mention a simulation-based method known as **bootstrapping** that can provide approximate sampling distributions when theory cannot help or we do not have the luxury of sampling many times over from the population. In all cases, the interpretation of a sampling distribution is the same, whether we obtain it by simulation or by the mathematics of probability.

APPLY YOUR KNOWLEDGE

6.2 Average wait time. As one effort to improve student satisfaction with the college advising office, the wait time (between checking in and entering the adviser's office) is recorded for each student. Each day these wait times are averaged to get a daily average. Many daily averages are gathered and summarized with a histogram. Can this histogram be viewed as an approximation of the sampling distribution of the sample mean statistic? Explain your answer.

FIGURE 6.3 Normal quantile plot of the sample proportions in Figure 6.1. The distribution is close to Normal except for a very slight skewness and some clustering due to the fact that the sample proportions from a sample size 100 can take only values that are a multiple of 0.01.

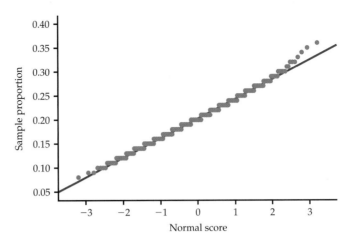

With a sampling distribution in hand (by simulation or theory), we focus our attention on certain basic features of the distribution. Let's use the tools of data analysis to describe the sampling distributions of Figures 6.1 and 6.2.

- **Shape:** Both histograms look Normal. However, there is a slight skewness (to the right) for the values of $\hat{p}$ for our samples of 100. Figure 6.3 is a Normal quantile plot of these values. With only a slight bend, it shows that the distribution of Figure 6.1 is quite close to Normal. The 1000 values for the samples of size 2067 in Figure 6.2 are even closer to Normal. The Normal curves superimposed on the histograms describe the overall shape quite well.

- **Center:** In both cases, the value of the sample proportion $\hat{p}$ varies from sample to sample, but the values are centered around the value of 0.20. Recall that $p = 0.2$ is the true population parameter. Some samples have $\hat{p}$ less than 0.2 and some greater, but there is no tendency to be always low or always high. That is, $\hat{p}$ has no *bias* as an estimator of p. This is true for both large and small samples.

- **Spread:** The values of $\hat{p}$ from samples of size 2067 are much less spread out than the values from samples of size 100.

Although these results describe just two sets of simulations for the sample proportion, they reflect facts that are generally true whenever we use random sampling.

APPLY YOUR KNOWLEDGE

6.3 Which is it? Suppose 1000 SRSs of size $n = 20$ and 1000 SRSs of size $n = 100$ are drawn from a population in which the true proportion is $p = 0.7$. For each sample, the sample proportion $\hat{p}$ is computed and recorded. For each sample size, separate histograms of the sample proportions are then constructed.

(a) For each histogram, where will the values of the sample proportions be centered?

(b) One of the histograms appears slightly skewed. Explain which histogram that would likely be.

(c) For one histogram, about 95% of the sample proportions lie between 0.6 and 0.8. For the other histogram, about 95% of the sample proportions lie between 0.5 and 0.9. Explain which histogram is associated with which sample size.

Bias and variability in estimation

We introduced the term "estimate" (page 296) as a value of a statistic used to infer the value of a population parameter. When a statistic is used to estimate a population parameter, it is called an **estimator.** Thus, sampling distributions describe the behavior of estimators.

estimator

The sampling distribution in Figure 6.2 (page 298) shows that a sample of size 2067 will almost always give an estimate $\hat{p}$ that is close to the truth about the population. Figure 6.2 illustrates this fact for just one value of the population proportion ($p = 0.2$), but it is true for any proportion. In contrast, as seen from Figure 6.1 (page 297), samples of size 100 might give an estimate of 10% or 30% when the truth is 20%.

Thinking about Figures 6.1 and 6.2 helps us restate the idea of bias when we use a statistic like $\hat{p}$ to estimate a parameter like p. It also reminds us that variability matters as much as bias.

BIAS AND VARIABILITY OF AN ESTIMATOR

Bias of an estimator concerns the center of the sampling distribution in comparison to the true value of the parameter being estimated. An estimator is said to be **unbiased** if the mean of its sampling distribution is equal to the true value of the parameter being estimated: that is, the bias is zero.

The **variability of a estimator** is described by the spread of its sampling distribution. This spread is determined by the sampling design and the sample size n.

Shooting arrows at a target with a bull's-eye is a nice way to think in general about bias and variability of *any* estimator, not just the sample proportion. We can think of the true value of the population parameter as the bull's-eye on a target and of the estimator as an arrow shot at the bull's-eye. Where an arrow lands is an estimate. Bias and variability describe what happens when an archer shoots many arrows at the target. *Bias* means that the aim is off; the arrows land consistently off the bull's-eye in the same direction. As a consequence, the estimates do not center on the population value. Large *variability* means that the shots are widely scattered on the target. In other words, there is a lack of precision, or consistency, among the estimates. Figure 6.4 illustrates the target analogy of these two types of errors.

(a) Large bias, small variability

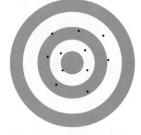

(b) Small bias, large variability

(c) Large bias, large variability

(d) Small bias, small variability

FIGURE 6.4 Bias and variability in shooting arrows at a target. Bias means the archer systematically misses in the same direction. Variability means that the arrows are scattered.

Notice that small variability (repeated shots that are close together) can accompany large bias (the arrows are consistently away from the bull's-eye in one direction). Likewise, small bias (the arrows center on the bull's-eye) can accompany large variability (repeated shots that are widely scattered). A good sampling scheme, like a good archer, must have both small bias and small variability.

Two primary culprits may cause an estimator to be biased:

1. The sampling scheme systematically favors certain outcomes.

2. The method of the calculation of the estimate lends itself to inherent bias. We illustrate this type of bias in Example 6.3.

The first source of bias is a sampling bias associated with not obtaining a representative sample of the population. For example, conducting a consumer survey by calling only landline phones will not be representative, since it fails to include consumers who have only cell phones. But even if we include both landline and cell phones, we may still not have a representative sample. What we need is randomization, such as that associated with obtaining an SRS. We will further emphasize the importance of randomization at the end of this section.

MANAGING SAMPLING BIAS

To reduce sampling bias, use random sampling. When we start with a list of the entire population, simple random sampling produces unbiased estimates—the values of an estimator computed from an SRS neither consistently overestimate nor consistently underestimate the value of the population parameter.

Even with a well-designed study based on randomization (such as taking an SRS) to reduce sampling bias, an estimator can still be biased based on how it is calculated.

EXAMPLE 6.3

Estimating Population Variance In Section 1.3, the sample variance s^2 was introduced as

$$s^2 = \frac{\sum(x_i - \bar{x})^2}{n-1}$$

sample variance, p. 33

The reason for dividing by $n - 1$ was due to the fact that only $n - 1$ of the squared deviations can vary freely. The number $n - 1$ is known as the degrees of freedom of the variance estimator. Notwithstanding this explanation, you might still wonder: what would happen if we were to use the seemingly "natural" choice of n in the denominator? To this end, let's consider an alternative estimator:

$$\hat{\sigma}^2 = \frac{\sum(x_i - \bar{x})^2}{n}$$

To explore the behavior of $\hat{\sigma}^2$, we will assume that the population is Normal with mean $\mu = 0$ and variance $\sigma^2 = 100$. Suppose that the sample size is $n = 10$. Even though the sampling distribution of $\hat{\sigma}^2$ can be exactly determined mathematically based on probability theory, let's use simulation instead to gain insights. Specifically, we will draw an SRS of 10 observations from the Normal distribution with mean of 0 and variance of 100 and compute

$$\hat{\sigma}^2 = \frac{\sum(x_i - \bar{x})^2}{10}$$

Figure 6.5 shows the simulated sampling distribution of $\hat{\sigma}^2$ based on 1000 samples. The sampling distribution for the estimator $\hat{\sigma}^2$ is distinctly right skewed. But, where are the values centered at? The mean of the 1000 values is 90.002, which rounds to 90. If $\hat{\sigma}^2$ is an unbiased estimate of σ^2, then we should find the mean of the simulated values to be close to 100. Our simulation is suggesting that $\hat{\sigma}^2$ is a *biased* estimator of σ^2.

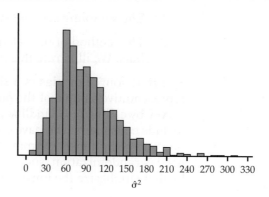

FIGURE 6.5 The histogram shows the distribution of 1000 sample variance estimates computed with *n* in the denominator rather than $n - 1$. The underlying population has variance $\sigma^2 = 100$.

What if we had simulated s^2 instead? The answer to this question doesn't require another simulation, but rather can be found by doing some simple algebra. Convince yourself the following is true:

$$s^2 = \frac{n}{n-1} \hat{\sigma}^2$$

For $n = 10$, this implies $s^2 = (10/9)\hat{\sigma}^2$. We found that the mean of the 1000 $\hat{\sigma}^2$ values is 90. This implies that the mean of 1000 s^2 values would be $(10/9)90 = 100$. By means of simulation, we see the key reason for dividing by the degrees of freedom of $n - 1$: so that the associated estimator will then be *unbiased*. As you move forward, you will encounter variance estimators in different settings in which the degrees of freedom may not be $n - 1$. But, the principle remains the same. Namely, by dividing by the appropriate degrees of freedom, we ensure that the associated variance estimator will be unbiased. ∎

Again referring to Figure 6.4 (page 300), in addition to wanting estimators with small bias, we desire estimators that have small variability. Here's how we can achieve this.

MANAGING VARIABILITY

To reduce the variability of an estimator from an SRS, use a larger sample. You can make the variability as small as you want by taking a large enough sample.

In practice, the NRF takes only one sample of consumers. We don't know how close to the truth an estimate from this one sample is because we don't know the truth about the population. But *large random samples almost always give an estimate that is close to the truth*. Looking at the sampling distribution of Figure 6.2 (page 298) shows that we can trust the result of one sample based on the sample size of $n = 2067$. Similarly, the Current Population Survey's sample of about 60,000 households estimates the U.S. national unemployment rate very accurately.

margin of error

The size of the sample then determines how close to the population truth the sample result is likely to be. Results from a sample survey usually come with a **margin of error** that sets bounds on the size of the likely error.

> **MARGIN OF ERROR**
>
> The **margin of error** is a numerical measure of the spread of a sampling distribution. It can be used to set bounds on the size of the likely error when using the statistic as an estimator of a population parameter.

The margin of error directly reflects the variability of the estimator, so it is smaller for larger samples. Example 6.8 (page 314) gives you a preview of how margin of error is computed. We will provide more details on margin of error and its connection to sample size in the next chapter. It will play a critical role in subsequent chapters thereafter.

APPLY YOUR KNOWLEDGE

6.4 Twitter polls. Twitter provides the option for users to conduct polls. Twitter promotes this feature as follows: "If you want the public's opinion on anything—what to name your dog, who will win tonight's game, which election issue people care most about—there's no better place to get answers than on Twitter. For poll creators, it's a new way to engage with Twitter's massive audience and understand exactly what people think." Depending on the user conducting the poll and the use of hashtags, Twitter polls can be considerably larger than opinion polls that incorporate probability sampling. Does a larger sample size for a Twitter poll mean more trustworthy results? Explain your answer.

Sampling from large populations

NRF's sample of 2067 households focuses on those that have children between the ages of 6 and 17; this translates to sampling about 1 out of every 16,000 such households in the United States. Had the NRF sampled 2067 households in Pennsylvania, this would translate to sampling about 1 out every 1000 households in Pennsylvania. Does this necessarily imply that the Pennsylvania sample would give more precise estimates than the United States sample? The answer is no.

> **LARGE POPULATIONS DO NOT REQUIRE LARGE SAMPLES**
>
> The variability of a statistic from a random sample depends little on the size of the population, as long as the population is at least 20 times larger than the sample.

Why does the size of the population have little influence on the behavior of statistics from random samples? To see why this is plausible, imagine sampling harvested corn by thrusting a scoop into a lot of corn kernels. The scoop doesn't know whether it is surrounded by a bag of corn or by an entire truckload. As long as the corn is well mixed (so that the scoop selects a random sample), the variability of the result depends only on the size of the scoop.

For sufficiently large populations, variability is controlled by the size of the sample, *not* the relative size of the sample to the population size. An SRS of size 2000 from the 250 million adults of the United States gives results as precise as an SRS of size 2000 from the roughly 740,000 adults of Austin, Texas. This is good news for designers of national samples. An SRS of size 2000 from 250 million adults translates to sampling 1 out of every 125,000 U.S. adults. It is *mistaken* thinking to conclude that sampling 1 in 125,000 Austin adults would give the same precision of results as the national survey. Randomly sampling 1 in 125,000 Austin adults would imply an SRS of size 6. To obtain equally trustworthy results from an SRS, designers of samples (national or local) must use the same sample size.

Why randomize?

The act of randomizing guarantees that the results of analyzing our data are subject to the laws of probability. The behavior of statistics is described by a sampling distribution. The form of the sampling distribution is known and, in many cases, is approximately Normal. Often, the center of the distribution lies at the true parameter value so that the notion that randomization eliminates bias is made more precise. The spread of the distribution describes the variability of the statistic and can be made as small as we wish by choosing a large enough sample. In a randomized experiment, we can reduce variability by choosing larger groups of subjects for each treatment.

These facts are at the heart of formal statistical inference. The remainder of this chapter and the following chapters have much to say in more technical language about sampling distributions and the way statistical conclusions are based on them. What any user of statistics must understand is that all the technical talk has its basis in a simple question: *what would happen if the sample or the experiment were repeated many times?* This reasoning applies not only to an SRS, but also to the complex sampling designs actually used by opinion polls and other national sample surveys. The same conclusions hold as well for randomized experimental designs. The details vary with the design, but the basic facts are true whenever randomization is used to produce data.

As discussed in Section 3.2 (page 138), even with a well-designed sampling plan, survey samples can suffer from problems of undercoverage and nonresponse. The sampling distribution shows only how a statistic varies due to the operation of chance in randomization. *It reveals nothing about possible bias due to undercoverage or nonresponse in a sample or to lack of realism in an experiment.* The actual error in estimating a parameter by a statistic can be much larger than the sampling distribution suggests. What is worse, there is no way to say how large the added error is. The real world is considerably less orderly than statistics textbooks imply.

SECTION 6.1 SUMMARY

- A numerical value that describes a population is a **parameter.** A numerical value that is be computed from the data and describes a sample is a **statistic.**

- The purpose of sampling or experimentation is usually **inference:** the use of sample statistics to make statements and draw conclusions about unknown population parameters.

- With random sampling, the values of a statistic vary randomly from sample to sample; thus, a statistic is a **random variable.**

- The probability distribution of a statistic is called a **sampling distribution.** The sampling distribution answers the question, "What would happen if we repeated the sample or experiment many times?" Formal statistical inference is based on the sampling distributions of statistics.

- A statistic that is used to estimate a population parameter is called an **estimator.**

- An estimator of a parameter may suffer from **bias** and/or from high **variability.** Bias means that the center of the sampling distribution is not equal to the true value of the parameter. The variability of an estimator is described by the spread of its sampling distribution. Variability is usually reported by giving a **margin of error** for conclusions based on sample results.

- Properly chosen statistics from randomized data production designs have no bias resulting from the way the sample is selected or the way the experimental units are assigned to treatments. We can reduce the variability of the statistic by increasing the size of the sample or the size of the experimental groups.

- As long as the population is at least 20 times larger than the sample, the size of the population has little influence on the behavior of statistics from random samples. Instead, variability is controlled by the size of the sample.

SECTION 6.1 EXERCISES

For Exercise 6.1, see page 296; for 6.2, see page 298; for 6.3, see page 299; and for 6.4, see page 303.

6.5 Describe the population and sample. For each of the following situations, describe the population, the sample, and statistic.

(a) A random sample of 100 graduates of your university's business school found that 82% found a job within six months of graduating.

(b) A national survey of 32,585 currently enrolled college students reported that the average number of times a student has visited the career services office is 1.9.

(c) A random sample of 200 shipping documents reveals that 2% have an understatement of sales due to shipments made but not recorded as sales.

6.6 What population and sample? Fifty employees from a firm of 3100 employees are randomly selected to be on a committee to evaluate how to implement sensitivity training. Currently, training is done in person, but a proposal has been made to implement the required training online. Each of the committee members is asked to vote Yes or No on the proposal.

(a) Describe the population for this setting.

(b) What is the sample?

(c) Describe the statistic and how it would be calculated.

(d) What is the population parameter?

6.7 What is the bias value? In Example 6.3 (page 301), we learned that an estimator for variance ($\hat{\sigma}^2$) based on dividing by n is biased, whereas an estimator (s^2) based on dividing by $n-1$ is unbiased. In statistics, bias is measured as the mean of the estimator minus the true value of the parameter being estimated. Answer the following questions in the context of estimating the population variance.

(a) For the setting of Example 6.3, what is the bias value of $\hat{\sigma}^2$?

(b) Suppose that the population variance is 200 and $n = 50$. What is the bias value of $\hat{\sigma}^2$?

(c) An estimator is said to have downward bias if the bias is negative and upward bias if the bias is positive. Which type of bias does $\hat{\sigma}^2$ have?

6.8 What's wrong? State what is wrong in each of the following scenarios.

(a) A parameter describes a sample.

(b) Bias and variability are two names for the same thing.

(c) Large samples are always better than small samples.

6.9 Pop-up website survey. When visiting the website of a company, you have no doubt encountered a pop-up survey requesting your participation. These surveys give two options: "No, thanks" and "Yes, I'll give feedback." Suppose a company reports that the 91% of respondents are pleased with the ease of its website. In light of the discussions of this section, what concerns might you have with this reported statistic?

6.10 Is it unbiased? A statistic has a sampling distribution that is somewhat skewed. The median is 10 and the quartiles are 4 and 20. The mean is 16.

(a) If the population parameter is 10, is the estimator unbiased?

(b) If the population parameter is 16, is the estimator unbiased?

(c) If the population parameter is 12, is the estimator unbiased?

6.11 Simulate a sampling distribution. In Exercise 1.72 (page 43) and Example 5.2 (page 224), you examined the density curve for a uniform distribution ranging from 0 to 1. The population mean for this uniform distribution is 0.5 and the population variance is 1/12. Let's simulate taking samples of size 2 from this distribution and compute the sample mean for each sample.

- **Excel users:** Enter the formula **=RAND()** into cells A1 and B1. Now, drag and copy the contents of cells A1 and B1 down to the row 1000. You will find 1000 random digits appear. Any attempt to copy these digits for sorting purposes will result in the digits changing. To "freeze" the generated values, highlight columns A and B and copy the contents; then **Paste Special as Values** the contents into the same columns. Now define column C as the average of columns A and B.

- **JMP users:** With a new data table, right-click on the header of Column 1 and choose **Column Info.** In the drag-down dialog box titled **Initialize Data,** pick the **Random** option. Click the **Random Uniform** option and enter "1000" into the **Number of rows** and click OK. Repeat the procedure for Column 2. For Column 3, pick the **Formula** option from the drag-down dialog box titled **Column Properties.** In the Formula dialog box, use the calculator buttons to enter the formula of adding Columns 1 and 2 and then dividing by 2. Click OK twice to find the sample means in Column 3 of the data table.

- **Minitab users:** **Calc → Random Data → Uniform.** Enter 1000 in the **Number of row of data to generate** dialog box, type "c1" and "c2" in the **Store in column(s)** dialog box, enter 1000 in the **Number of trials** dialog box, and click OK. Now use **Calculator** to define another column in the worksheet as the average of columns c1 and c2.

- **R users:** Type following at the R prompt:
```
x1<-runif(1000)
x2<-runif(1000)
xbar<-(x1+x2)/2
```

(a) The theoretical mean for this sampling distribution is the mean of the population that we sample from. How close is your simulation estimate to this parameter value?

(b) The theoretical standard deviation for this sampling distribution is the square root of 1/24. How close is your simulation estimate to this parameter value?

(c) For one of the columns (variables), obtain a histogram and a Normal quantile plot for individual observations. Additionally, obtain a histogram and a Normal quantile plot for the simulated sampling distribution. Compare the plots and write a short summary of your findings.

6.12 Bias and variability. Figure 6.6 shows histograms of four sampling distributions of statistics intended to estimate the same parameter. Label each distribution relative to the others as high or low bias and as high or low variability.

6.13 Appropriate estimate? Bathroom exhaust fans have been found to cause many residential fires,

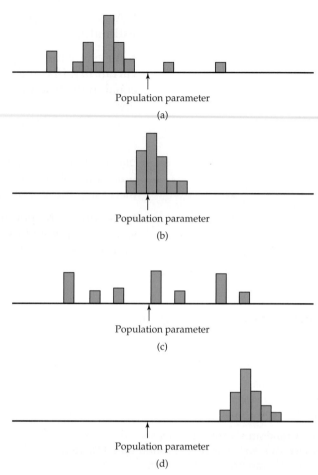

FIGURE 6.6 Determine which of these sampling distributions displays high or low bias and high or low variability, Exercise 6.12.

resulting in millions of dollars of property damage nationwide. Causes of fires include the igniting of built-up lint in the fan's motor and melting of copper wire connections. An exhaust fan manufacturer wishes to determine the lifetimes of a model released in a particular year that was sold to millions of consumers. From fire departments and law firms, the manufacturer has data on the date when an exhaust fan caused a fire. Based on 107 incidents, the sample mean time from purchase to a fire incident is $\bar{x} = 8.3$ years. Do you think this value serves as an appropriate estimate of the mean lifetime of the model? Explain your answer.

6.14 Biased variance estimator and sample size. Refer to Example 6.3 (page 301) and consider again the estimator for variance ($\hat{\sigma}^2$) based on dividing by n.

(a) In general, for a population with variance σ^2 and a sample size of n, what is the mean of the sampling distribution of $\hat{\sigma}^2$?

(b) Refer to Exercise 6.7 for the definition of bias of an estimator. Based on part (a), what is the bias of $\hat{\sigma}^2$ in general?

(c) Based on parts (a) and (b), discuss what occurs as the sample size gets larger and larger.

6.2 The Sampling Distribution of the Sample Mean

When you complete this section, you will be able to:

- Explain the difference between the sampling distribution of $\bar{x}$ and the population distribution.
- Find the mean and standard deviation of $\bar{x}$ for an SRS of size n from a population with mean μ and standard deviation σ.
- Determine how much larger n has to be for an SRS to reduce the standard deviation of $\bar{x}$ by a certain factor.
- Explain the basic result of the central limit theorem and describe how it relates to n.
- Utilize the central limit theorem to approximate the sampling distribution of $\bar{x}$ and perform probability calculations based on this approximation.
- Find probabilities for a linear combination of independent Normal random variables.

A variety of statistics are used to describe quantitative data. The sample mean, median, variance, and standard deviation are all examples of statistics based on quantitative data. In Example 6.3 (page 301), we used simulation to study the behaviors of two competing estimators of population variance. Statistical theory describes the sampling distributions of these statistics. In this section, we will examine the sample mean in terms of its behavior as a random variable. Because sample means are just averages of observations, they are among the most frequently used statistics.

Suppose that you plan to survey 1000 students at your university about their sleeping habits. The sampling distribution of the average hours of sleep per night describes what this average would be if many simple random samples of 1000 students were drawn from the population of students at your university. In other words, it gives you an idea of the results that you are likely to see from your survey. It tells you whether you should expect this average to be near the population mean and whether the variation of the statistic is roughly ±2 hours or ±2 minutes.

When we talk about distributions, it is important to understand the distinction between a sampling distribution and a population distribution. Imagine choosing an *individual* case at random from a population and measuring a quantity. The quantities obtained from repeated draws of an individual case from a population have a probability distribution that is the distribution of the population. In contrast, a sampling distribution describes how the value of a *statistic* varies in many samples taken from the population.

EXAMPLE 6.4

Total Sleep Time of College Students A study describes the distribution of total sleep time among college students as approximately Normal with a mean of 7.4 hours and a standard deviation of 0.7 hour.[3] Suppose that we select a college student at random and obtain his or her sleep time. This result is a random variable X because, prior to the random sampling, we don't know the sleep time. We do know, however, that in repeated sampling, X will have the same approximate $N(7.4, 0.7)$ distribution that describes the pattern of sleep time in the entire population. We call $N(7.4, 0.7)$ the *population distribution*. ■

POPULATION DISTRIBUTION

The **population distribution** of a variable is the distribution of its values for all cases of the population. The population distribution is also the probability distribution of the variable when we choose one case at random from the population.

In this example, the population of all college students actually exists, so we can, in principle, draw an SRS of students from it. At other times, however, our population of interest does not actually exist. For example, suppose that we are interested in studying final-exam scores in a statistics course, and we have the scores of the 34 students who took the course last semester. For the purposes of statistical inference, we might want to consider these 34 students as part of a hypothetical population of similar students who would take this course. In this sense, these 34 students represent not only themselves but also a larger population of similar students. The key idea is to think of the observations that you have as coming from a population with a probability distribution.

APPLY YOUR KNOWLEDGE

6.15 Number of steps per day. In a large-scale study of nearly 400,000 U.S. Apple iPhone smartphone users with the *Azumio Argus* app, researchers gathered data on the number of steps per day taken by users.[4] Researchers found an average of 4774 steps and a standard deviation of 2659.

(a) State the population that this study describes and the statistic.

(b) On the basis of the 68–95–99.7 rule (page 45), explain why the population distribution is not likely to be Normal. What would you suspect the shape of the population distribution to be?

Using simulation, we studied the sampling distribution of the sample proportion $\hat{p}$ in Example 6.2 (page 297) and the sampling distribution of the sample variance s^2 in Example 6.3 (page 301). Because the general framework for constructing a sampling distribution is the same for all statistics, let's do the same here to understand the sampling distribution of $\bar{x}$.

EXAMPLE 6.5

BIKE

Sample Means Are Approximately Normal In urban settings worldwide, bicycle-sharing systems have been implemented that allow individuals to use a bike for a very short term for a price. The systems work on the basis of allowing an individual to borrow a bike from a dock and return it to another dock in the city. In the United States, the largest privately owned public bike-sharing systems are Citi Bike in New York and Miami; Capital Bikeshare in Washington, D.C.; Divvy in Chicago; and Hubway in Boston.

For Citi Bike in New York, in October 2017, there were nearly 1.9 million bike rides. Among these rides are three customer segments: (1) single rides, (2) day passes, and (3) annual members. Annual members are allowed unlimited 45-minute rides, whereas other segments are restricted to 30-minute rides. For the annual member segment, there were more than 930,000 bike rides.[5] Due to a steep charge for each additonal 15 minutes past the 45-minute limit, nearly all the duration times are less than 120 minutes (99.999% of the recorded times). Figure 6.7(a) displays the distribution of bike-trip duration times (in seconds).

(We omitted a few extreme outliers that likely reflect failure to return bikes.) The distribution is clearly very different from the Normal distribution: it is extremely skewed to the right and very spread out. The right tail is actually even longer (extending beyond 7000 seconds) than what appears in the figure because there are too few high duration times for the histogram bars to be visible on this scale. The population mean is $\mu = 747.918$ seconds.

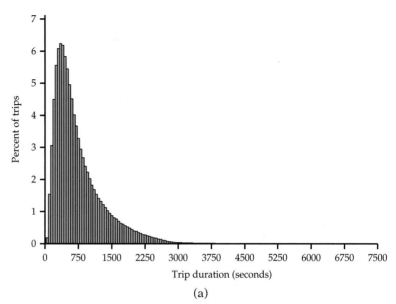

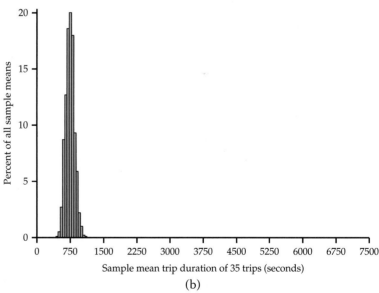

FIGURE 6.7 (a) The distribution of bike-trip duration times for more than 930,000 trips, Example 6.5. (b) The distribution of sample means $\bar{x}$ for 1000 random samples of size 35 from this population. Both histograms have the same scales and histogram classes to allow for direct comparison.

Table 6.1 contains the duration times of a random sample of $n = 35$ bike trips from this population. The mean of these 35 trips is $\bar{x} = 864.29$ seconds. If we were to take another sample of size 35, we would likely get a different value of $\bar{x}$, because this new sample would contain a different set of bike trips. To find the sampling distribution of $\bar{x}$, we take many SRSs of size 35 and calculate $\bar{x}$ for each sample. Figure 6.7(b) is the distribution of the values of $\bar{x}$ for 1000 random samples. The scales and choice of classes are exactly the same as in Figure 6.7(a), so we can make a direct comparison. Notice something remarkable: even though the distribution of the individual duration times is strongly skewed and very spread out, the distribution of the sample means is quite symmetric and much less spread out.

TABLE 6.1 Duration (in Seconds) of 35 Bike Trips

421	675	464	579	509	400	2633
375	1070	1019	501	249	236	173
1500	384	351	2361	1393	336	561
927	1570	156	234	305	1888	1066
318	3802	118	1456	329	1552	339

Figure 6.8(a) is the distribution of the $\bar{x}$ values on a scale that more clearly shows its shape. We can see that the distribution of sample means is close to the Normal distribution. The Normal quantile plot of Figure 6.8(b) further confirms the compatibility of the distribution of sample means with the Normal distribution. Furthermore, the histogram in Figure 6.8(a) appears to be essentially centered on the population mean μ value. Specifically, the mean of the 1000 sample means is 743.273, which is nearly equal to the μ-value of 747.918. ∎

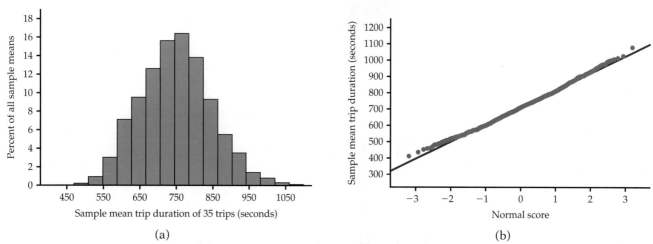

FIGURE 6.8 (a) The distribution of sample means $\bar{x}$ from Figure 6.7(b) in more detail. (b) Normal quantile plot of these 1000 sample means. The distribution is very close to Normal.

This example illustrates three important points discussed in this section.

FACTS ABOUT SAMPLE MEANS

1. The distribution of sample means (for $n > 1$) has smaller variance than does the population distribution.
2. The distribution of sample means is centered on the population mean.
3. For sufficiently large n, the distribution of sample means will be approximately Normal, even if the population distribution is not.

These three facts contribute to the popularity of sample means in statistical inference of the population mean.

The mean and standard deviation of $\bar{x}$

The sample mean $\bar{x}$ from a sample or an experiment is an estimate of the mean μ of the underlying population. The sampling distribution of $\bar{x}$ is determined by the design used to produce the data, the sample size n, and the population distribution.

Select an SRS of size n from a population, and measure a variable X on each individual case in the sample. The n measurements are values of n random variables $X_1, X_2, \ldots, X_n$. A single X_i is a measurement on one individual case selected at random from the population and, therefore, has the distribution of the population. If the population is large relative to the sample, we can consider $X_1, X_2, \ldots, X_n$ to be independent random variables, each having the same distribution. This is our probability model for measurements on each individual case in an SRS.

The sample mean of an SRS of size n is

$$\bar{x} = \frac{1}{n}(X_1 + X_2 + \cdots + X_n)$$

rules for means, p. 236

If the population has mean μ, then μ is the mean of the distribution of each observation X_i. To get the mean of $\bar{x}$, we use the rules for means of random variables. Specifically,

$$\mu_{\bar{x}} = \frac{1}{n}(\mu_{X_1} + \mu_{X_2} + \cdots + \mu_{X_n})$$
$$= \frac{1}{n}(\mu + \mu + \cdots + \mu) = \mu$$

unbiased estimator, p. 300

That is, *the mean of $\bar{x}$ is the same as the mean of the population*. The sample mean $\bar{x}$ is therefore an unbiased estimator of the unknown population mean μ.

The observations are independent, so the addition rule for variances also applies:

rules for variances, pp. 240–241

$$\sigma_{\bar{x}}^2 = \left(\frac{1}{n}\right)^2 (\sigma_{X_1}^2 + \sigma_{X_2}^2 + \cdots + \sigma_{X_n}^2)$$
$$= \left(\frac{1}{n}\right)^2 (\sigma^2 + \sigma^2 + \cdots + \sigma^2)$$
$$= \frac{\sigma^2}{n}$$

With n in the denominator, the variability of $\bar{x}$ about its mean decreases as the sample size grows. Thus, a sample mean from a large sample will usually be very close to the true population mean μ. Here is a summary of these facts.

MEAN AND STANDARD DEVIATION OF A SAMPLE MEAN

Let $\bar{x}$ be the mean of an SRS of size n from a population having mean μ and standard deviation σ. The mean and the standard deviation of $\bar{x}$ are

$$\mu_{\bar{x}} = \mu$$
$$\sigma_{\bar{x}} = \frac{\sigma}{\sqrt{n}}$$

How precisely does a sample mean $\bar{x}$ estimate a population mean μ? Because the values of $\bar{x}$ vary from sample to sample, we must give an answer in terms of the sampling distribution. We know that $\bar{x}$ is an unbiased estimator of μ, so its values in repeated samples are not systematically too high or too low. Most samples will give an $\bar{x}$-value close to μ if the sampling distribution is concentrated close to its mean μ. Thus, the precision of estimation depends on the spread of the sampling distribution.

Because the standard deviation of $\bar{x}$ is $\sigma/\sqrt{n}$, the standard deviation of the statistic decreases in proportion to the square root of the sample size. This implies, for example, that a sample size must be multiplied by 4 to divide the statistic's standard deviation in half. By comparison, a sample size must be multiplied by 100 to reduce the standard deviation by a factor of 10.

EXAMPLE 6.6

Standard Deviations for Sample Means of Trip Durations The standard deviation of the population of bike-trip durations in Figure 6.7(a) (page 309) is $\sigma = 570.469$ seconds. The duration of any single bike trip will often be far from the population mean. If we choose an SRS of 35 trips, the standard deviation of their mean duration is

$$\sigma_{\bar{x}} = \frac{570.469}{\sqrt{35}} = 96.427 \text{ seconds}$$

Averaging over more bike trips reduces the variability and makes it more likely that $\bar{x}$ is close to μ. For example, if we were to sample 140 (4×35) trips, the standard deviation will be half as large:

$$\sigma_{\bar{x}} = \frac{570.469}{\sqrt{140}} = 48.213 \text{ seconds} \quad \blacksquare$$

APPLY YOUR KNOWLEDGE

6.16 Find the mean and the standard deviation of the sampling distribution. Compute the mean and standard deviation of the sampling distribution of the sample mean when you plan to take an SRS of size 49 from a population with mean 420 and standard deviation 21.

6.17 The effect of increasing the sample size. In the setting of the previous exercise, repeat the calculations for a sample size of 441. Explain the effect of the sample size increase on the mean and standard deviation of the sampling distribution.

The central limit theorem

We have described the center and spread of the probability distribution of a sample mean $\bar{x}$, but not its shape. The shape of the distribution of $\bar{x}$ depends on the shape of the population distribution. Here is one important case: if the population distribution is Normal, then so is the distribution of the sample mean.

SAMPLING DISTRIBUTION OF A SAMPLE MEAN

If a population has the $N(\mu, \sigma)$ distribution, then the sample mean $\bar{x}$ of n independent observations has the $N(\mu, \sigma/\sqrt{n})$ distribution.

6.2 The Sampling Distribution of the Sample Mean

This is a somewhat special result. Many population distributions are not Normal. The bike-trip duration times in Figure 6.7(a) (page 309), for example, are *extremely* skewed. Yet Figures 6.7(b) (page 309) and 6.8 (page 310) show that the means of samples of size 35 are close to Normal.

One of the most famous facts of probability theory says that, for large sample sizes, the distribution of $\bar{x}$ is close to a Normal distribution. This is true no matter what shape the population distribution has, as long as the population has a finite standard deviation σ. This is the **central limit theorem**. It is much more useful than the fact that the distribution of $\bar{x}$ is exactly Normal if the population is exactly Normal.

central limit theorem

CENTRAL LIMIT THEOREM

Draw an SRS of size n from any population with mean μ and finite standard deviation σ. When n is large, the sampling distribution of the sample mean $\bar{x}$ is approximately Normal:

$$\bar{x} \text{ is approximately } N\left(\mu, \frac{\sigma}{\sqrt{n}}\right)$$

In general, the approximation improves as n gets larger.

Example 6.5 (page 308) showed the remarkable central limit theorem effect for a population distribution that was highly skewed but with a well-behaved pattern. With this said, it is natural to wonder whether the central limit theorem truly stands up to its claim for highly irregular population distributions. To answer this question, it is well worth trying another example.

EXAMPLE 6.7

LOAN

P2P Lending In Example 4.18 (page 200), we studied data on loan types versus dwelling status from Lending Club, which is a peer-to-peer (P2P) lending company. With this example, let's consider data on loan amounts. Figure 6.9(a) displays the distribution of 118,646 loan amounts (in dollars). Both the distribution displayed in Figure 6.7(a), on page 309, and the distribution displayed in Figure 6.9(a) are clearly not Normal. However, the natures of the non-Normality are very different. In Figure 6.7(a), the distribution is highly skewed but has a smooth pattern. In contrast, the distribution shown in Figure 6.9(a) is choppy and irregular. It is counterintuitive to believe that regularity in sampling distribution can emerge from such a population as claimed by the central limit theorem.

Following the same procedure as was used in Example 6.5, we take many SRSs and calculate $\bar{x}$ for each sample. In Example 6.5, the sample means were based on sample sizes of $n = 35$. Let's challenge the central limit theorem by using a smaller sample size of $n = 25$. Figure 6.9(b) is the distribution of the values of $\bar{x}$ for 1000 random samples. As expected from the central limit theorem, the distribution of the sample means is essentially the Normal distribution. Even for professionally trained statisticians, the central limit theorem effect never ceases to be a source of amazement. ∎

FIGURE 6.9 (a) The distribution of loan amounts for 118,646 loans, Example 6.7. (b) The distribution of sample means $\bar{x}$ for 1000 random samples of size 25 from this population.

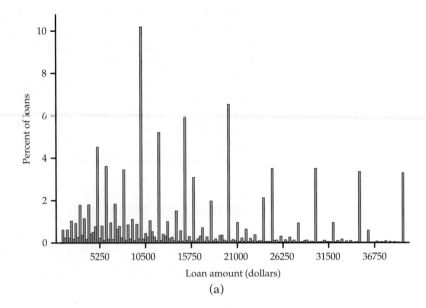

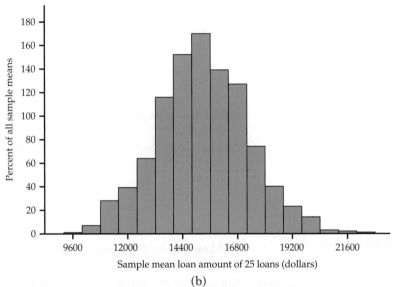

Now that we know that the Normal distribution can serve as the basis for describing the behavior of sample means, we can better describe how precisely a random sample of observations estimates the mean of the underlying population.

EXAMPLE 6.8

68–95–99.7 rule, p. 45

How Close Will the Sample Mean Be to the Population Mean? Returning to the bike-sharing application, with the Normal distribution to work with, we can better describe how precisely a random sample of 35 bike trips estimates the mean duration of all the bike trips in the population. The population standard deviation for the more than 930,000 trips in the population of Figure 6.7(a) (page 309) is $\sigma = 570.469$ seconds. From Example 6.6 (page 312), we know $\sigma_{\bar{x}} = 96.427$ seconds. By the 95 part of the 68–95–99.7 rule, about 95% of all samples will have a sample mean $\bar{x}$ within two standard deviations of μ—that is, within ± 192.854 seconds of μ. This 192.854 value is what we earlier referred to as a margin of error (page 303). ∎

If we view the sample mean based on $n = 35$ as not sufficiently precise for estimation purposes, then we must consider a larger sample size to reduce the standard deviation of $\bar{x}$. Referring to Example 6.6 (page 312), we must bear in mind that reduction in the standard deviation of $\bar{x}$ occurs in proportion to the square root of the sample size. Thus, to cut in half the ± 192.854 accuracy bounds, we need to take a sample size of 140. If we wish to cut these bounds in half once more, we need to take a sample size of 560. In light of the increased cost and time associated with larger sample sizes, practitioners must consider the basic trade-off between estimation accuracy and sample size choice.

The main point of Example 6.8 is that the central limit theorem allows us to use Normal probability calculations to answer questions about sample means, even when the population distribution is not Normal.

APPLY YOUR KNOWLEDGE

6.18 Use the 68–95–99.7 rule. You take an SRS of size 49 from a population with mean 185 and standard deviation 84.

(a) According to the central limit theorem, what is the approximate sampling distribution of the sample mean?

(b) Use the 95 part of the 68–95–99.7 rule to compute the margin of error.

(c) Explain how the margin of error found in part (a) is interpreted in terms of the variability of the sample mean.

(d) Use the 99.7 part of the 68–95–99.7 rule to compute the margin of error. Compare this computed margin of error with the one computed in part (b).

6.19 The effect of changing the sample size. You take an SRS of size n from a population with mean μ and standard deviation σ.

(a) Define m_1 as the margin of error based on the 95 part of the 68–95–99.7 rule. What is the equation for m_1 when $n = 200$?

(b) Define m_2 as the margin of error based on the 95 part of the 68–95–99.7 rule. What is the equation for m_2 when $n = 50$?

(c) By what factor is m_2 greater than m_1? Explain how value of this factor is related to the sample sizes of parts (a) and (b).

How large is large enough?

Our statement of the central limit theorem (page 313) has the condition "when n is large." In practical applications, it can be difficult to know when n is large enough. This difficulty stems from that fact that the approximation depends on the shape of the underlying population distribution.

One often hears $n \geq 30$ as a "magical" rule of thumb for all underlying populations. However, for the population distribution of loan amounts displayed in Figure 6.9(a) (page 314), we saw that a sample size of 25 worked quite well in producing Normality for the sampling distribution of the sample mean. For the population distribution of bike-trip durations displayed in Figure 6.7(a) (page 309), we produced Normality in the sampling distribution using a sample size of 35. As it turns out, if we had used a sample size of 25 for the bike-trip duration population, then the associated sampling distribution would be clearly not Normal. It might seem surprising that Normality was achieved with a smaller sample size for the loan amount population than for the bike-trip duration population. The reason that this result occurs is because the loan amount population is more evenly spread out. In other words, the extreme skewness in the bike-trip duration population presented more of a challenge to the central limit theorem.

If the population distribution is fairly symmetric, then the Normal approximation can be quite good for n much less than 30. Indeed, in quality control applications, it is common to use the Normal distribution as a baseline model for monitoring sample means based on sample sizes of four or five, even when underlying population distribution may not be quite Normal. *Nevertheless, there are population distributions (such as extreme skewness) for which n needs to be considerably larger than 30.*

Here is a detailed study of another skewed distribution.

EXAMPLE 6.9

The Central Limit Theorem in Action Figure 6.10 shows the central limit theorem in action for another very non-Normal population. Figure 6.10(a) displays the density curve of a single observation from the population. The distribution is strongly right-skewed, and the most probable outcomes are near 0. The mean μ of the shown distribution is 1, and its standard deviation σ is also 1. This particular continuous distribution is called an exponential distribution. The exponential distribution was introduced in Chapter 5 for modeling time between events and modeling the working life of products and equipment.

exponential distribution, p. 282

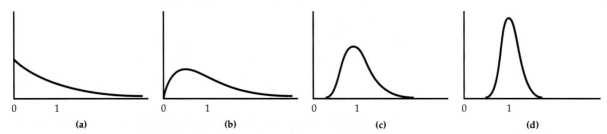

FIGURE 6.10 The central limit theorem in action: the distribution of sample means $\bar{x}$ from a strongly non-Normal population becomes more Normal as the sample size increases. (a) The distribution of 1 observation. (b) The distribution of $\bar{x}$ for 2 observations. (c) The distribution of $\bar{x}$ for 10 observations. (d) The distribution of $\bar{x}$ for 25 observations.

Figures 6.10(b), (c), and (d) are the density curves of the sample means of 2, 10, and 25 observations from this population. As n increases, the shape becomes more Normal. The mean remains at $\mu = 1$, but the standard deviation decreases, taking the value $1/\sqrt{n}$. The density curve for 10 observations is still somewhat skewed to the right but already resembles a Normal curve, having $\mu = 1$ and $\sigma = 1/\sqrt{10} = 0.32$. The density curve for $n = 25$ is yet more Normal. The contrast between the shape of the population distribution and the shape of the distribution of the mean of 10 or 25 observations is striking. ■

You can also use the *Central Limit Theorem* applet to study the sampling distribution of $\bar{x}$. From one of three population distributions, 10,000 SRSs of a user-specified sample size n are generated, and a histogram of the sample means is constructed. You can then compare this estimated sampling distribution with the Normal curve that is based on the central limit theorem.

EXAMPLE 6.10

Using the *Central Limit Theorem* Applet In Example 6.9, we considered sample sizes of $n = 2$, 10, and 25 from an exponential distribution. Figure 6.11 shows a screenshot of the *Central Limit Theorem* applet for the exponential distribution when $n = 10$. The mean and standard deviation of this sampling distribution are 1 and $1/\sqrt{10} = 0.316$, respectively. From the 10,000 SRSs, the mean is estimated to be 1.002, and the estimated standard deviation is 0.319. These results are both quite close to the true values. In Figure 6.10(c), we saw that the density curve for 10 observations is still somewhat skewed to the right. We can see this same behavior in Figure 6.11 when we compare the histogram with the Normal curve based on the central limit theorem. ∎

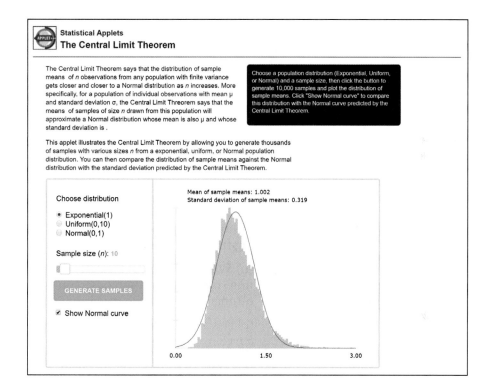

FIGURE 6.11 Screenshot of the *Central Limit Theorem* applet for the exponential distribution when $n = 10$, Example 6.10.

Try using the applet for the other sample sizes mentioned in Example 6.9. You should get histograms shaped like the density curves shown in Figure 6.10. You can also consider other sample sizes by sliding n from 1 to 100. As you increase n, the shape of the histogram moves closer to the Normal curve that is based on the central limit theorem.

In statistical practice, the question of how large a sample size we might need ultimately rests on the inferential procedures that we implement. For example, Chapter 8 introduces inference procedures called t procedures that are based on the use of the sample mean statistic. These procedures are used in the study of the mean(s) of population distribution(s) for quantitative variables (e.g., dollars, minutes, meters). These t procedures provide exact results when the population distribution is Normal. When the population is not Normal, t procedures provide approximately correct results for large n; the central limit theorem, in part, explains why this is the case. In the presence of strong skewness, t procedures for inference about a single population mean provide good performance when the sample is large, roughly $n \geq 40$. Chapter 8 will offer general practical guidelines for the implementation of t procedures.

quantitative variable, p. 3

Our illustrations of the central limit theorem with the bike-trip duration, loan amount, and exponential populations, along with our discussion of t procedures, have centered on population distributions of quantitative variables. However, many important applications are associated with categorical variables. Most notably, consider the situation where we have only two categories (e.g., success and failure). As done with our study of the binomial distribution (page 258), we can assign a value of 1 to a success and a value of 0 to a failure. In a sample of n observations based on this assignment, the sum of the observations is simply the count of the number of successes in the n observations. We can also find the sample mean of the observations, which is the sample proportion of successes, $\hat{p}$, as introduced in Example 6.1 (page 295). Section 6.3 is devoted to the sampling distribution of the sample proportion. For now, it should come as no surprise that sample proportions, being sample means, are approximately Normal for large n due to the central limit theorem. However, given the extreme nature of the individual observations as being only one of two values (0 or 1), the sample size requirement for approximate Normality may be considerably larger than it is in most cases for quantitative variables.

categorical variable, p. 3

APPLY YOUR KNOWLEDGE

6.20 Use the *Central Limit Theorem* applet. Let's consider the uniform distribution between 0 and 10. For this distribution, all intervals of the same length between 0 and 10 are equally likely. This distribution has a mean of 5 and standard deviation of 2.89.

(a) Approximate the population distribution by setting $n = 1$ and clicking the "Generate Samples" button.

(b) What are your estimates of the population mean and the population standard deviation based on the 10,000 SRSs? Are these population estimates close to the true values?

(c) Describe the shape of the histogram and compare it with the Normal curve.

6.21 Use the *Central Limit Theorem* applet again. Refer to the previous exercise. In the setting of Example 6.9 (page 316), let's approximate the sampling distribution for samples of size $n = 2, 10,$ and 25 observations.

(a) For each sample size, compute the mean and standard deviation of $\bar{x}$.

(b) For each sample size, use the applet to approximate the sampling distribution. Report the estimated mean and standard deviation. Are they close to the true values calculated in part (a)?

(c) For each sample size, compare the shape of the sampling distribution with the Normal curve based on the central limit theorem.

(d) For this population distribution, what sample size do you think is needed to make you feel comfortable using the central limit theorem to approximate the sampling distribution of $\bar{x}$? Explain your answer.

Now that we know that the sampling distribution of the sample mean $\bar{x}$ is approximately Normal for a sufficiently large n, let's consider some probability calculations.

EXAMPLE 6.11

Time between 311 Calls In Example 5.38 (page 281), it was reported that, for the city of Chicago, the average number of 311 calls in the summer months is about 9.7 per hour for the night shift. In terms of minutes, this translates to 0.162 call per minute. As in Example 5.38, suppose that the number of calls per time unit follows a Poisson distribution. This implies that the time X between calls is governed by the exponential distribution. As noted in Chapter 5 (page 282), the mean and the standard deviation of the exponential distribution are both equal to the inverse of the rate of arrivals (in this case, rate of incoming 311 calls). This implies that the time between calls has an exponential distribution with mean $\mu = (1/0.162) = 6.173$ minutes and standard deviation $\sigma = 6.173$ minutes. You record the next 50 times between 311 calls. What is the probability that their average exceeds 5 minutes?

The central limit theorem says that the sample mean time $\bar{x}$ (in minutes) between 311 calls has approximately the Normal distribution with mean equal to the population mean $\mu = 6.173$ minutes and standard deviation

$$\frac{\sigma}{\sqrt{50}} = \frac{6.173}{\sqrt{50}} = 0.873 \text{ minute}$$

The sampling distribution of $\bar{x}$ is, therefore, approximately $N(6.173, 0.873)$. Figure 6.12 shows this Normal curve (solid) as well as the actual density curve of $\bar{x}$ (dashed).[6]

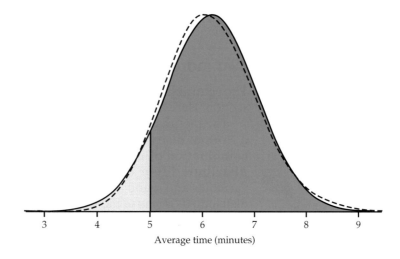

FIGURE 6.12 The exact distribution (dashed) and the Normal approximation from the central limit theorem (solid) for the average time between 311 calls, Example 6.11.

The probability we want is $P(\bar{x} > 5)$. This is the area to the right of 5 under the solid Normal curve in Figure 6.12. Using the Normal table, the Normal distribution calculation gives

$$P(\bar{x} > 5) = P\left(\frac{\bar{x} - 6.173}{0.873} > \frac{5 - 6.173}{0.873}\right)$$
$$= P(Z > -1.34) = 0.9099$$

The exactly correct probability is the area under the dashed density curve in the figure: 0.9180. The central limit theorem Normal approximation is off by only about 0.0081. ∎

APPLY YOUR KNOWLEDGE

6.22 Find a probability. Refer to Example 6.11. Find the probability that the mean time between 311 calls is less than 7 minutes. The exact probability is 0.8302. Compare your answer with the exact one.

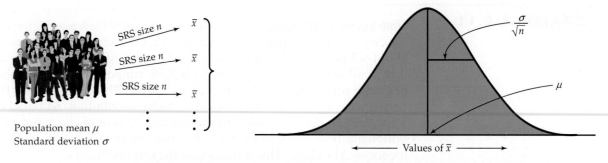

FIGURE 6.13 The sampling distribution of a sample mean $\bar{x}$ has mean μ and standard deviation $\sigma/\sqrt{n}$. The sampling distribution is Normal if the population distribution is Normal; it is approximately Normal for large samples in any case.

Figure 6.13 summarizes the facts about the sampling distribution of $\bar{x}$ in a way that emphasizes the big idea of a sampling distribution. The general framework for constructing the sampling distribution of $\bar{x}$ is shown on the left.

- Take many random samples of size n from a population with mean μ and standard deviation σ.
- Find the sample mean $\bar{x}$ for each sample.
- Collect all the $\bar{x}$'s and display their distribution.

The sampling distribution of $\bar{x}$ is shown on the right. Keep this figure in mind as you go forward.

Two more facts

The central limit theorem is the big fact of probability theory applied in this section. Here are a couple of related facts that are also useful to know.

The fact that the sample mean of an SRS from a Normal population has a Normal distribution is a special case of a more general fact: **any linear combination of independent Normal random variables is also Normally distributed.** For example, if X and Y are independent Normal random variables and a and b are any constant numbers, then $aX + bY$ is Normally distributed. Because the constants of a and b can be positive or negative, this fact about a linear combination applies to sums and differences of random variables. The mean and standard deviation of $aX + bY$ are found using the rules for means and variances presented in Chapter 5.

rules for means, p. 236

rules for variances, pp. 240–241

EXAMPLE 6.12

Difference in Packaged Contents Consider the content weight of a randomly selected prepackaged 8-ounce bag of salad mix. Based on historical data, the distribution of the content amount is the $N(8.12, 0.02)$ distribution. The same salad mix is also available in a 16-ounce container. The salad content in containers varies according to the $N(16.19, 0.05)$ distribution. If the content in bags and containers varies independently, what is the probability that the total content in two randomly selected bags exceeds the content in one container?

First, we can define the content in two bags X by $B_1 + B_2$, where both B_1 and B_2 follow the $N(8.12, 0.02)$ distribution. As a note of caution, it would be incorrect to define X as $2B$ (see Exercise 6.27). The two-bag content amount is Normally distributed, with a mean and variance of

$$\mu_{B_1+B_2} = \mu_{B_1} + \mu_{B_2} = 8.12 + 8.12 = 16.24$$

$$\sigma^2_{B_1+B_2} = \sigma^2_{B_1} + \sigma^2_{B_2} = 0.02^2 + 0.02^2 = 0.0008$$

The standard deviation is then $\sqrt{0.0008} = 0.028284$. Defining Y as the content in a 16-ounce container, the difference in content (two bags minus container) is given by $X - Y$. This random variable of $X - Y$ is Normally distributed, with a mean and variance of

$$\mu_{X-Y} = \mu_X - \mu_Y = 16.24 - 16.19 = 0.05$$
$$\sigma^2_{X-Y} = \sigma^2_X + \sigma^2_Y = 0.028284^2 + 0.05^2 = 0.0033$$

Because $\sqrt{0.0033} = 0.0574$, $X - Y$ has the $N(0.05, 0.0574)$ distribution. The desired probability is

$$P(X > Y) = P(X - Y > 0)$$
$$= P\left(\frac{(X-Y) - 0.05}{0.0574} > \frac{0 - 0.05}{0.0574}\right)$$
$$= P(Z > -0.87) = 0.8078$$

From another perspective, we can conclude that although, on average, two bags have more salad content than one container, the probability is 0.1922 ($= 1 - 0.8078$) that a container will have more content than two bags. ∎

Another useful fact is that **more general versions of the central limit theorem say that the distribution of a sum or average of many small random quantities is close to Normal.** This is true even if the quantities are not independent (as long as they are not too highly correlated) and even if they have different distributions (as long as no single random quantity is so large that it dominates the others). These more general versions of the central limit theorem suggest why the Normal distributions are common models for observed data. Any variable that is a sum of many small random influences will have approximately a Normal distribution.

SECTION 6.2 SUMMARY

- The **sample mean** $\bar{x}$ of an SRS of size n drawn from a large population with mean μ and standard deviation σ has a sampling distribution with mean and standard deviation

$$\mu_{\bar{x}} = \mu$$
$$\sigma_{\bar{x}} = \frac{\sigma}{\sqrt{n}}$$

- The sample mean $\bar{x}$ is an unbiased estimator of the population mean μ. When $n > 1$, the sample mean is less variable than a single observation. The standard deviation decreases in proportion to the square root of the sample size n. This implies that to reduce the standard deviation by a factor of K, we need to increase the sample size by a factor of K^2.

- When the population distribution is Normal, the sampling distribution of $\bar{x}$ is exactly Normal.

- The **central limit theorem** states that for large n, the sampling distribution of $\bar{x}$ is approximately $N(\mu, \sigma/\sqrt{n})$ for any population with mean μ and finite standard deviation σ. The approximation improves as n gets larger. This allows us to approximate probability calculations of $\bar{x}$ using the Normal distribution.

- Linear combinations of independent Normal random variables have Normal distributions. In particular, if the population has a Normal distribution, so does $\bar{x}$.

SECTION 6.2 EXERCISES

For Exercise 6.15, see page 308; for 6.16 and 6.17, see page 312; for 6.18 and 6.19, see page 315; for 6.20 and 6.21, see page 318; and for 6.22, see page 319.

6.23 What's wrong? Explain what is wrong in each of the following statements.

(a) If the population standard deviation is 30, then the standard deviation of $\bar{x}$ for an SRS of 10 observations will be 30/10 = 3.

(b) When taking SRSs from a large population, larger sample sizes will result in larger standard deviations of $\bar{x}$.

(c) For an SRS from a large population, both the mean and the standard deviation of $\bar{x}$ depend on the sample size n.

(d) The 68–95–99.7 rule says that about 95% of the observed values will fall within $\mu \pm 2\sigma/\sqrt{n}$.

6.24 What's wrong? Explain what is wrong in each of the following statements.

(a) The central limit theorem states that for large n, the population mean μ is approximately Normal.

(b) For large n, the distribution of observed values will be approximately Normal.

(c) For sufficiently large n, the 68–95–99.7 rule says that $\bar{x}$ should be within $\mu \pm 2\sigma$ about 95% of the time.

6.25 Number of employees. There are nearly 6 million firms in the United States with paid employees. The mean number of employees in these firms is about **21**. A university selects a random sample of 100 firms in Colorado and finds that they average about **12** employees. Is each of the bold numbers a parameter or a statistic?

6.26 Number of companies friended on social networks. Based on a sample of 1535 Internet users, a University of Southern California–sponsored study reported Internet users followed an average of 9.9 companies on social network sites such as Facebook.[7] State the population for this survey, the statistic, and some likely values from the population distribution.

6.27 Sum of random variables. In Example 6.12 (page 320), the total content of two randomly selected prepackaged 8-ounce bags of salad mix was defined by $X = B_1 + B_2$ and the variance of X was found to be 0.0008. Suppose we were define X as $2B$, where B follows the $N(8.12, 0.02)$ distribution.

(a) Explain why X defined as $2B$ is incorrect for the setting of the problem. What is $2B$ representing, as opposed to $B_1 + B_2$?

(b) Find the variance of $2B$ and compare it with the variance of $B_1 + B_2$ found in Example 6.12.

6.28 Total sleep time of college students. In Example 6.4 (page 307), the total sleep time per night among college students was approximately Normally distributed with mean $\mu = 7.4$ hours and standard deviation $\sigma = 0.7$ hour. You plan to take an SRS of size $n = 50$ and compute the average total sleep time.

(a) What is the standard deviation for the average time?

(b) Use the 95 part of the 68–95–99.7 rule to describe the variability of this sample mean.

(c) What is the probability that your average will be greater than 7.5 hours?

6.29 Determining sample size. Refer to the previous exercise. Now you want to use a sample size such that about 95% of the averages fall within ±10 minutes (0.17 hour) of the true mean $\mu = 7.4$.

(a) Based on your answer to part (b) in Exercise 6.28, should the sample size be larger or smaller than 50? Explain.

(b) What standard deviation of $\bar{x}$ do you need so that 95% of all samples will have a mean within 10 minutes of μ?

(c) Using the standard deviation you calculated in part (b), determine the number of students you need to sample.

6.30 Number of friends on Facebook. Among adult active Facebook users, it has been found that the average user has about 340 friends. This distribution takes only integer values, so it is certainly not Normal. It is also highly skewed to the right, with a median of 200 friends.[8] Suppose that $\sigma = 300$ and you take an SRS of 80 Facebook users.

(a) For your sample, what are the mean and standard deviation of $\bar{x}$, the mean number of friends per user?

(b) Use the central limit theorem to find the probability that the average number of friends for 80 Facebook users is less than 300.

(c) What are the mean and standard deviation of the total number of friends in your sample? (*Hint*: Use the rules for means and variances for a sum of independent random variables found in Section 5.2.)

(d) What is the probability that the total number of friends among your sample of 80 Facebook users is greater than 25,000?

6.31 Generating a sampling distribution. Let's illustrate the idea of a sampling distribution in the case of a very small sample from a very small population. The population consists of 10 medium-sized businesses. The variable of interest is the company size, measured in terms of the number of employees. For convenience, the 10 companies have been labeled with the integers 1 to 10.

Company	1	2	3	4	5	6	7	8	9	10
Size	82	62	80	58	72	73	65	66	74	62

The parameter of interest is the mean size μ in this population. The sample is an SRS of size $n = 3$ drawn from the population. Software can be used to generate an SRS.

(a) Find the mean of the 10 sizes in the population. This is the population mean μ.

(b) Now use software to make an SRS of size 3.

- **▣ Excel users:** A simple way to draw a random sample is to enter "=RANDBETWEEN(1,10)" in any cell. Take note of the number that represents the company and record in another column the corresponding size. Press the F9 key to change the random entry. If you get a repeat, press F9 again. Do this until you get three distinct values.

- *jmp* **JMP users:** Enter the size values in a data table. Do the following pull-down sequence: **Tables → Subset.** In the drag-down dialog box titled **Initialize Data,** pick **Random** option. Choose the bullet option of **Random-sample size,** enter "3" in its dialog box, and then click OK. You will find an SRS of three company sizes in a new data table.

- *Minitab users:* Enter the size values in column one (c1) of the worksheet. Do the following pull-down sequence: **Calc → Random Data → Sample from Samples.** Enter "3" in the **Number of rows to sample,** type "c1" in the **From columns** box, and type "c2" in the **Store samples in** box, and then click OK. You will find an SRS of three company sizes in c2.

- **R users:** Type the following at the R prompt to find an SRS of three company sizes in numeric vector s1:
  `x<-c(82,62,80,58,72,73,65,66,74,62)`
  `s1<-sample(x,3,replace=FALSE)`

With your SRS, calculate the sample mean $\bar{x}$. This statistic is an estimate of μ.

(c) Repeat this process nine more times. Make a histogram of the 10 values of $\bar{x}$. You are constructing the sampling distribution of $\bar{x}$. Is the center of your histogram close to μ?

6.32 ACT scores of high school seniors. The scores of your state's high school seniors on the ACT college entrance examination in a recent year had mean $\mu = 22.3$ and standard deviation $\sigma = 6.2$. The distribution of scores is only roughly Normal.

(a) What is the approximate probability that a single student randomly chosen from all those taking the test scores 27 or higher?

(b) Now consider an SRS of 16 students who took the test. What are the mean and standard deviation of the sample mean score $\bar{x}$ of these 16 students?

(c) What is the approximate probability that the mean score $\bar{x}$ of these 16 students is 27 or higher?

(d) Which of your two Normal probability calculations in parts (a) and (c) is more accurate? Why?

6.33 Safe flying weight. In response to the increasing weight of airline passengers, the Federal Aviation Administration told airlines to assume that passengers average 190 pounds in the summer, including clothing and carry-on baggage. But passengers vary: the FAA gave a mean but not a standard deviation. A reasonable standard deviation is 35 pounds. Weights are not Normally distributed, especially when the population includes both men and women, but they are not very non-Normal. A commuter plane carries 19 passengers. What is the approximate probability that the total weight of the passengers exceeds 4000 pounds? (*Hint:* To apply the central limit theorem, restate the problem in terms of the mean weight.)

6.34 Grades in an accounting course. Indiana University posts the grade distributions for its courses online.[9] Students in four sections of Introduction to Accounting (all sections are taught by the same instructor) in the spring 2018 semester received 23.6% A's, 58.1% B's, 17.7% C's, 0.6% D's, and 0% F's.

(a) Using the common scale A = 4, B = 3, C = 2, D = 1, F = 0, take X to be the grade of a randomly chosen Intro to Accounting student. Use the definitions of the mean (page 230) and standard deviation (pages 238–239) for discrete random variables to find the mean μ and the standard deviation σ of grades in this course.

(b) With nearly 700 students in the four sections, the population of spring semester students is a large enough course that we can take the grades of an SRS of 25 students to be independent of each other. If $\bar{x}$ is the average of these 25 grades, what are the mean and standard deviation of $\bar{x}$?

(c) What is the probability that a randomly chosen Intro to Accounting student gets a B or better, $P(X \geq 3)$?

(d) What is the approximate probability $P(\bar{x} \geq 3)$ that the grade-point average for 25 randomly chosen Intro to Accounting students is B or better?

6.35 Increasing sample size. Heights of adults are well approximated by the Normal distribution. Suppose that the population of adult U.S. males has mean of 69 inches and standard deviation of 2.8 inches.

(a) What is the probability that a randomly chosen male adult is taller than 6 feet?

(b) What is the probability that the sample mean of two randomly chosen male adults is greater than 6 feet?

(c) What is the probability that the sample mean of five randomly chosen male adults is greater than 6 feet?

(d) Provide an intuitive argument as to why the probability of the sample mean being greater than 6 feet decreases as n gets larger.

6.36 Supplier delivery times. Supplier on-time delivery performance is critical to enabling the buyer's organization to meet its customer service commitments. Therefore, monitoring supplier delivery times is also critical. Based on a great deal of historical data, a manufacturer of personal computers finds for one of its just-in-time suppliers that the delivery times are random and well approximated by the Normal distribution with mean 51.7 minutes and standard deviation 9.5 minutes.

(a) What is the probability that a particular delivery will exceed one hour?

(b) Based on part (a), what is the probability that a particular delivery arrives in less than one hour?

(c) What is the probability that the mean time of five deliveries will exceed one hour?

6.3 The Sampling Distribution of the Sample Proportion

When you complete this section, you will be able to:

- Explain the connection between a count X and a sample proportion $\hat{p}$.
- Determine the mean and standard deviation of the sample proportion $\hat{p}$.
- Explain how the sample proportion $\hat{p}$ can be viewed as a special case of the sample mean statistic.
- Determine when one can approximate the sampling distribution of the sample proportion using the Normal distribution.
- Use the Normal approximation to perform probability calculations for the sample proportion.

In Example 6.2 (page 297), we discussed the use of simulation to study the sampling distribution of the sample proportion. In this section, we will use probability theory to more precisely describe the sampling distribution of the sample proportion. Let's start with an example.

EXAMPLE 6.13

Mobile Activities Before Shopping A Nielsen sample survey asked 2009 Canadian adult consumers about their smartphone activities prior to shopping at a physical store.[10] One question asked was if the consumer used the smartphone to research the product by means of a search engine prior to going to the store. We would like to view the response of these respondents as a representative of a larger population of Canadian consumers. ∎

We let n represent the sample size and the random variable X represent the count for the outcome of interest.

EXAMPLE 6.14

Count Random Variable In this survey of Canadian consumers, $n = 2009$. The variable X is the number of consumers who responded that they use smartphones to research products prior to going to the store. In this case, $X = 1386$. ∎

binomial distribution and SRS, p. 252

If you studied Section 5.3, you will recognize that the binomial distribution can be used as an adequate approximation for the study of counts from an SRS when the population is large relative to the sample size. The rule of thumb for the applicability of the binomial distribution with an SRS is that the population be at least 20 times as large as the sample, as clearly is the case here. Thus, the random variable X follows an approximate binomial distribution with $n = 2009$ and some underlying probability of success p, which in this case is the probability of a respondent using a smartphone for product research prior to shopping. When a random variable has only two possible outcomes, it is more common to use the sample proportion $\hat{p} = X/n$ as the summary rather than the count X.

EXAMPLE 6.15

The Sample Proportion For the survey result of Example 6.14, the sample proportion of Canadian consumers who use smartphones to research products prior to going to the store is

$$\hat{p} = \frac{1386}{2009} = 0.69$$ ∎

Notice that this summary takes into account the sample size n. We need to know n to properly interpret the meaning of the random variable X. For example, the conclusion we would draw about consumers' behavior would be quite different if we had observed $X = 1386$ from a sample twice as large. *Be careful not to directly compare counts when the sample sizes are different.* Instead, to allow for direct comparison, divide the counts by their associated sample sizes to obtain sample proportions.

APPLY YOUR KNOWLEDGE

6.37 Seniors waived out of a math prerequisite. In a random sample of 300 business students who are in or have taken business statistics, 7.1% reported that they had been waived out of taking the college math prerequisite for business statistics due to their AP calculus examination score. Give n, X, and $\hat{p}$ for this setting.

6.38 OTT advertising. Over-the-top (OTT) is a term describing the delivery of video and audio content via the Internet, without requiring users to subscribe to a traditional cable or satellite pay-TV service. A recent survey of 19,867 adult online consumers asked whether they were willing to pay to remove advertising with OTT services. Of the participants, 7152 answered Yes. The other 12,715 answered No.[11]

(a) What is n?

(b) Choose one of the two possible outcomes to define the random variable, X. Give a reason for your choice.

(c) What is the value of X?

(d) Find the sample proportion, $\hat{p}$.

Sample proportion mean and standard deviation

population proportion

The sample proportion of 0.69 found in Example 6.15 is an estimate of the **population proportion** of Canadian consumers who use smartphones to research products prior going to the store. In business applications, we often want to estimate the proportion p of "successes" in a population. Our estimator is the sample proportion of successes:

$$\hat{p} = \frac{\text{count of successes in sample}}{\text{size of sample}}$$
$$= \frac{X}{n}$$

Be sure to distinguish between the proportion $\hat{p}$ and the count X. The count takes whole-number values anywhere in the range from 0 to n, but a proportion is always a number in the range of 0 to 1.

In the binomial setting, the count X has a binomial distribution. However, the proportion $\hat{p}$ does *not* have a binomial distribution. We can, however, do probability calculations for $\hat{p}$ by restating them in terms of the count X and using binomial methods.

EXAMPLE 6.16

Mobile Activities before Shopping In Example 6.15, we learned that a Nielsen survey reveals that 69% of Canadian adult respondents use smartphones to research products prior to going to the store. However, smartphone usage varies based on age. For example, baby boomers are less likely to use smartphones than younger respondents. You decide to take a nationwide random sample of 2500 Canadian college students and ask if they use smartphones to research products prior to

going to the store. Suppose that 75% of *all* Canadian college students indicate that they do so. Given this population proportion of $p = 0.75$, what is the probability that the sample proportion from your survey is at least 73%?

Define X as the count of college students who use smartphones for product research prior to shopping. The binomial distribution $B(2500, 0.75)$ can be used to model the random variable X. However, the sample proportion $\hat{p} = X/2500$ does *not* have a binomial distribution because it is not a count. But we can translate any question about a sample proportion $\hat{p}$ into a question about the count X. Because 73% of 2500 is 1825,

$$P(\hat{p} \geq 0.73) = P(X \geq 1825)$$
$$= P(X = 1825) + P(X = 1826) + \cdots + P(X = 2500)$$

This is a rather tedious calculation, because we must add 676 binomial probabilities. By adding these probabilities, software tells us that $P(\hat{p} \geq 0.73) = 0.9897$. But it turns out that we can arrive at a value quite close to this probability without taking such a sum. ∎

binomial mean and standard deviation, p. 259

As a first step, we need to find the mean and standard deviation of a sample proportion. For the binomial distribution, we know the mean and standard deviation of a sample count variable X are $\mu_X = np$ and $\sigma_X = \sqrt{np(1-p)}$, respectively. To get the mean of $\hat{p}$, we apply the rule for means (page 236) associated with multiplying a random variable by a fixed number:

$$\mu_{\hat{p}} = \frac{1}{n}\mu_X = \frac{1}{n}(np) = p$$

Since the standard deviation of X is $\sqrt{np(1-p)}$, the variance of X is then $\sigma_X^2 = np(1-p)$. We can now apply the rule for variances (pages 240–241) associated with multiplying a random variable by a fixed number:

$$\sigma_{\hat{p}}^2 = \left(\frac{1}{n}\right)^2 \sigma_X^2 = \left(\frac{1}{n}\right)^2 np(1-p) = \frac{p(1-p)}{n}$$

Taking the square root of this quantity gives the standard deviation for $\hat{p}$:

$$\sigma_{\hat{p}} = \sqrt{\frac{p(1-p)}{n}}$$

Here is the summary of our results along with a rule of thumb of applicability.

MEAN AND STANDARD DEVIATION OF A SAMPLE PROPORTION

Let $\hat{p}$ be the sample proportion of successes in an SRS of size n drawn from a large population having population proportion p of successes. The mean and standard deviation of $\hat{p}$ are

$$\mu_{\hat{p}} = p$$

$$\sigma_{\hat{p}} = \sqrt{\frac{p(1-p)}{n}}$$

The formula for $\sigma_{\hat{p}}$ is exactly correct in the binomial setting. It is approximately correct for an SRS from a large population. We will use it when the population is at least 20 times as large as the sample.

The fact that the mean of $\hat{p}$ is p indicates that it is an *unbiased* estimator of p. We observed this fact empirically with Figures 6.1 (page 297) and 6.2 (page 298) when we saw that the simulated sampling distributions of the sample proportions were centered on the true population proportion of $p = 0.2$, regardless of the value of n. What we saw then empirically is now verified by the laws of probability.

We can also see from the standard deviation of $\hat{p}$ that its variability around its mean gets smaller as the sample size increases. So, a sample proportion from a large sample will usually lie quite close to the population proportion p. We also observed this fact with the simulated sampling distributions of Figures 6.1 and 6.2. Now we have discovered exactly how the standard deviation decreases. With $\sqrt{n}$ in the denominator of the standard deviation, the sample size must be multiplied by 4 if we wish to the divide the standard deviation in half. This "discovery" is actually not too surprising in light of our study of the sample mean statistic $\bar{x}$ that has the same $\sqrt{n}$ factor in its standard deviation. As pointed out earlier (page 318), the sample proportion is simply the average of the 0 (failure) and 1 (success) values; thus, it is a special case of the sample mean.

Let's now use these formulas to calculate the mean and standard deviation for Example 6.16.

EXAMPLE 6.17

The Mean and the Standard Deviation For the setting of Example 6.16, the mean and standard deviation of the sample proportion of the Canadian college students who use smartphones for product research prior to shopping are

$$\mu_{\hat{p}} = p = 0.75$$

$$\sigma_{\hat{p}} = \sqrt{\frac{p(1-p)}{n}} = \sqrt{\frac{(0.75)(0.25)}{2500}} = 0.00866 \;\blacksquare$$

APPLY YOUR KNOWLEDGE

6.39 Find the mean and the standard deviation. Past audits reveal that 4% of a company's shipping documents are inaccurate, understating the sales due to shipments made but not recorded as sales. Since there are more than 10,000 shipping documents per fiscal year, audits are based on random samples. Suppose an audit of 300 randomly selected shipping documents will be conducted. Assume that the underlying proportion of inaccurate shipping documents is 4%.

(a) Find the mean of the sample proportion of inaccurate documents.

(b) Can we closely approximate the standard deviation of the sample proportion when $n = 300$? Explain your answer.

(c) Find the standard deviation of the sample proportion of inaccurate documents.

(d) Are your answers to parts (a) and (c) the same as the mean and the standard deviation of the sample proportion of accurate documents? Explain your answer.

6.40 Mean and standard deviation comparison. Refer to Example 6.2 (page 297), in which we showed the simulated sampling distributions of $\hat{p}$ for $n = 100$ and $n = 2067$ with $p = 0.2$ in both cases. PHAT

(a) Based on theory, what is the mean of the sampling distribution of $\hat{p}$ for $n = 100$ and $n = 2067$?

(b) Based on theory, what is the standard deviation of the sampling distribution of $\hat{p}$ for $n = 100$ and $n = 2067$?

(c) The 1000 simulated $\hat{p}$ values for each sample size are provided. Find the means and standard deviations of the simulated sample proportions, and compare their values with what you found in parts (a) and (b).

Normal approximation for proportions

Using simulation, we discovered in Section 6.1 that the sampling distribution of a sample proportion $\hat{p}$ is close to Normal, see Figures 6.1 (page 297) and 6.2 (page 298). Now we know that the distribution of $\hat{p}$ is that of a binomial count divided by the sample size n. This seems at first to be a contradiction. However, we should quickly remind ourselves that the sample proportion is a special case of the sample mean. So, when we think about the behavior of the sample proportion, we should think related to our lessons about the sample mean in the context of the remarkable central limit theorem.

In Chapter 5 (page 259), to find the mean and variance of a binomial random variable X, we defined the count X as a sum

$$X = S_1 + S_2 + \cdots + S_n$$

of independent random variables S_i that take the value 1 if a success occurs on the ith trial and the value 0 otherwise. The proportion of successes $\hat{p} = X/n$ can then be calculated as simply the sample mean of the S_i values:

$$\hat{p} = \frac{X}{n} = \frac{S_1 + S_2 + \cdots + S_n}{n}$$

Also, as we have learned in Section 6.2, the central limit theorem applies to *sums and averages* of independent random variables. From the perspective of a sum, this now makes it clear why we learned in Chapter 5 that for large n, the binomial distribution approaches the Normal distribution. From the perspective of an average, we now see that $\hat{p} = X/n$ is also approximated by the Normal distribution for large n.

Normal approximation of binomial, p. 278

Let's see this relation in action. Figure 6.14 displays the probability histogram of the *exact* distribution of the sample proportion of Canadian college students who use smartphones for product research prior to shopping, based on the binomial distribution $B(2500, 0.75)$. There are hundreds of narrow bars, one for each of the 2501 possible values of $\hat{p}$. It would be a mess to try to show all these probabilities on the graph. *The key takeaway from the figure is that the probability histogram looks very Normal!* Once again, this is the central limit theorem at work.

Combining the key facts we have learned, we can make the following statement about the sampling distribution of $\hat{p}$.

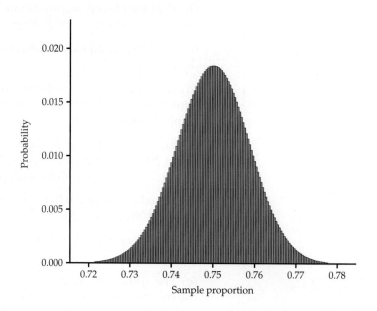

FIGURE 6.14 Probability histogram of the sample proportion $\hat{p}$ based on a binomial count with $n = 2500$ and $p = 0.75$. The distribution is very close to Normal.

SAMPLING DISTRIBUTION OF $\hat{p}$

Draw an SRS of size n from a large population having population proportion p of successes. Let $\hat{p}$ be the sample proportion of successes. When n is large, the sampling distribution of $\hat{p}$ is approximately

$$N\left(p, \sqrt{\frac{p(1-p)}{n}}\right)$$

As a rule of thumb, this approximation is adequate for values of n and p that satisfy $np \geq 10$ and $n(1-p) \geq 10$.

This rule of thumb is directly borrowed from the Normal approximation for the binomial distribution (page 278).

EXAMPLE 6.18

Compare the Normal Approximation with the Exact Calculation Let's compare the Normal approximation for the probability question of Example 6.16 (page 325) with the exact calculation from software. We want to calculate $P(\hat{p} \geq 0.73)$ when the sample size is $n = 2500$ and the population proportion is $p = 0.75$. Example 6.17 (page 327) shows that

$$\mu_{\hat{p}} = p = 0.75$$

$$\sigma_{\hat{p}} = \sqrt{\frac{p(1-p)}{n}} = 0.00866$$

We see from Figure 6.14 that the Normal approximation will no doubt be satisfactory. Let's put that figure out of mind and check our rule of thumb. We have $np = 2500(0.75) = 1875$ and $n(1-p) = 2500(0.25) = 625$. Both these values are much greater than the value of 10. So, we can be confident in proceeding with the Normal approximation. Namely, we act as if $\hat{p}$ were Normal with mean 0.75 and standard deviation 0.008866. The approximate probability, as illustrated in Figure 6.15, is

$$P(\hat{p} \geq 0.73) = P\left(\frac{\hat{p} - 0.75}{0.00866} \geq \frac{0.73 - 0.75}{0.00866}\right)$$

$$\doteq P(Z \geq -2.31) = 0.9896$$

That is, about 99% of all samples have a sample proportion that is at least 0.73. Because the sample was large, this Normal approximation is quite accurate. It misses the exact value of 0.9897 reported by software by only 0.0001. ∎

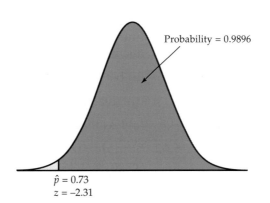

FIGURE 6.15 The Normal probability calculation for Example 6.18.

Let's take a closer look at the rule of thumb that was originally introduced with the Normal approximation of the binomial distribution. Recall that $\hat{p}$ is just a binomial random variable divided by n, so the use of the same rule of thumb is not unexpected. Referring back to Figure 5.19 (page 278), we see that the closer p is to 0.5, the smaller n needs to be for approximate Normality. For $p = 0.5$, we have $n(0.5) \geq 10$ and $n(1 - 0.5) \geq 10$, which implies that n needs to be only at least 20. However, as p moves closer to 0 or 1, the recommended minimum n increases. For example, for $p = 0.1$, the rule of thumb suggests that $n \geq 100$.

The rule of thumb for the minimum sample size presumes that we know p. But with statistical inference, we do not know the population parameter p. So, can this rule of thumb be of any use? Yes! We have yet to cover the details of performing inference, but here are two snapshots illustrating how this rule of thumb will come into play.

- Hypothesis testing: With hypothesis testing, we typically make a statement about population parameter(s). We then develop statistical procedures to measure how well the data and the statement agree, leading us to decide to either refute or not refute the statement. For example, we might make a statement ("hypothesis") that the population proportion p is some value p_0. The statistical procedures to test such a statement about p are developed under the assumption that p is indeed equal to p_0. If we wish to use the Normal distribution in the implementation of these procedures, we would need to select a sufficiently large sample size n. With the assumption $p = p_0$ along with our rule of thumb, the recommendation is to choose n such that $np_0 \geq 10$ and $n(1 - p_0) \geq 10$.

- Confidence intervals: A confidence interval provides a range of values to serve as an estimate for an unknown parameter. With confidence intervals for the population proportion, there is no assumption about the value of p. So, to use the Normal approximation as the basis for constructing confidence intervals, we use $\hat{p}$ in place of p in the rule of thumb. This gives a rule of thumb of $n\hat{p} \geq 10$ and $n(1 - \hat{p}) \geq 10$. But, $n\hat{p}$ is the observed number of successes in the sample and $n(1 - \hat{p})$ is the observed number of failures in the sample. This leaves us with a recommendation for using a Normal-based confidence interval for the population proportion when the number of successes and the number of failures are both at least 10.

Chapter 10 is devoted to the details of inference for population proportions. You will indeed encounter the just-described rules of thumb for hypothesis testing and confidence intervals in that chapter.

APPLY YOUR KNOWLEDGE

6.41 Use the Normal approximation. Refer to the setting of Exercise 6.39 (page 327). For a sample size of 400, use the Normal approximation to find the probability that the sample proportion of inaccurate shipping documents is

(a) between 0.02 and 0.04.

(b) greater than 0.045.

(c) less than 0.01.

6.42 Sample size determination. Refer to the setting of Exercise 6.39 (page 327). Based on the discussed rule of thumb, what is the minimal sample size one should use to obtain an adequate Normal approximation of the sampling distribution of $\hat{p}$?

SECTION 6.3 SUMMARY

- In a binomial setting, for a **count** X of successes in n trials, the **sample proportion** $\hat{p}$ is an estimator of the population proportion p.

- The mean and standard deviation of $\hat{p}$ are

$$\mu_{\hat{p}} = p$$

$$\sigma_{\hat{p}} = \sqrt{\frac{p(1-p)}{n}}$$

The sample proportion $\hat{p}$ is, therefore, an unbiased estimator of the population proportion p. The formula for $\sigma_{\hat{p}}$ is exactly correct in the binomial setting. It is approximately correct for an SRS from a large population—in particular, when the population is at least 20 times as large as the sample.

- The sample proportion $\hat{p}$ is a special case of the sample mean statistic where the observations are 1s (successes) and 0s (failures).

- When n is large, the **sampling distribution** of $\hat{p}$ is approximately

$$N\left(p, \sqrt{\frac{p(1-p)}{n}}\right)$$

As a rule of thumb, this approximation is adequate for values of n and p that satisfy $np \geq 10$ and $n(1-p) \geq 10$.

SECTION 6.3 EXERCISES

For Exercises 6.37 and 6.38, see page 325; for 6.39 and 6.40, see page 327; and for 6.41 and 6.42, see page 330.

6.43 AP statistics. Suppose the population proportion of students at a university who have taken an AP statistics course in high school is 0.35.

(a) If you were to take an SRS of 40 students to survey them on whether they have taken an AP statistics course, what is the standard deviation of $\hat{p}$?

(b) Using the 68–95–99.7 rule, we would expect $\hat{p}$ to fall between what two percents about 95% of the time?

(c) Suppose the following data represent the responses (1 for Yes and 0 for No):

0	0	0	1	0	1	0	1	0	1
0	1	0	0	0	0	0	0	1	0
1	0	1	1	1	0	0	1	1	0
1	1	1	0	1	0	0	1	0	1

What is the value of $\hat{p}$ for this sample?

(d) Your friend tells you that your observed sample proportion is very unusual. Do you agree? Respond to your friend.

6.44 Hybrid learning. Higher education institutions are increasingly introducing courses based on hybrid instruction, a combination of conventional face-to-face and online instruction. In a recent nationwide study by EDUCAUSE Center for Analysis and Research, researchers found that 77% of students believe that hybrid learning is the most effective learning environment for them.[12] You decide to poll a sample of the incoming students at your institution to gauge their attitude toward hybrid courses. Suppose you take an SRS 200 of incoming students.

(a) What are the mean and standard deviation of $\hat{p}$ if $p = 0.77$?

(b) Using the 68–95–99.7 rule, we would expect $\hat{p}$ to fall between what two percents about 95% of the time?

(c) What is the probability that at least 80% of the incoming students believe hybrid learning is the most effective learning environment?

6.45 Working or studying? As an entity of the U.S. Department of Education, the National Center for Education Statistics (NCES) collects and analyzes data related to education. Based on U.S. Census data, the NCES reports that for 2016, 17% of 20- to 24-year-olds were neither enrolled in school nor working.[13] You decide to survey 20- to 24-year-olds in your county. In an SRS of four hundred 20- to 24-year-olds of your county, you find 44 are neither enrolled in school nor working.

(a) What is the sample proportion of 20- to 24-year-olds in your county who are neither enrolled in school nor working?

(b) Assume that the population proportion of all 20- to 24-year-olds is 17%. Assuming that this is true for your county as well, what is the standard deviation of $\hat{p}$?

(c) Using the 68–95–99.7 rule, you would expect $\hat{p}$ to fall between what two percents about 95% of the time?

(d) Based on your result in part (c), do you think that the proportion of 20- to 24-year-olds from your county as a whole who are neither enrolled in school nor working is more than, less than, or about the same as the national level? Explain your answer.

6.46 LASIK surgery. A Food and Drug Administration study found that after six months, 41% of patients having Lasik corrective eye surgery experience visual aberrations.[14] Suppose an SRS of 150 Lasik patients who had the surgery at least six months ago is taken. Assume that $p = 0.41$.

(a) What are the mean and standard deviation of $\hat{p}$?

(b) What is the probability that the percent of patients experiencing visual aberrations is less than 30%?

(c) What is the probability that the percent of patients experiencing visual aberrations is greater than 50%?

(d) Using the 68–95–99.7 rule, we would expect $\hat{p}$ to fall between what two percents about 95% of the time?

6.47 Approximations. The methods associated with sample proportion are often approximations rather than exact probability results. We have given rules of thumb for the safe use of these approximations.

(a) In a survey of nearly 70,000 students, the International Center for Academic Integrity found that nearly 40% of undergraduates admitted to cheating on a test.[15] You are interested in the rate of cheating among 80 members of a fraternity. Suppose that, in fact, 40% of the 80 members would say Yes to cheating. Explain why you *cannot* safely proceed with the sampling distribution of $\hat{p}$ presented in this section.

(b) In many high-volume electronics industries (e.g., microchip manufacturing), it is not unusual to have defect rates of 0.01% (that's 0.0001 as a decimal fraction). For such manufacturing scenarios, explain why you *cannot* safely use the Normal approximation for the sample proportion of defects for an SRS of 1000 manufactured items.

6.48 Simulate a sampling distribution for $\hat{p}$. In this exercise, you will use software to first generate binomial counts from the $B(n, p)$ distribution. We can use this fact to simulate the sampling distribution for $\hat{p}$. In this exercise, you will generate 1000 sample proportions for $p = 0.70$ and $n = 100$.

- *Excel users:* Choose "Random Number Generation" from the **Data Analysis** menu box. Enter "1" in the **Number of Variables** box, enter "1000" in the **Number of Random Numbers** box, choose "Binomial" for the **Distribution** option, and finally enter "A1" for the **Output Range.** Click **OK** to find 1000 random values outputted in the worksheet. In cell B1, enter "=A1/100" and drag the formula down the column to cell B1000 to obtain 1000 simulated sample proportions.

- *JMP users:* With a new data table, right-click on the header of Column 1 and choose **Column Info.** Enter 1000 in the **Number of rows** dialog box. In the drag-down dialog box titled **Column Properties,** pick the **Formula** option. You will then encounter a Formula dialog box. Open the **Random** formula group and click **Random Binomial** so that it appears in the formula dialog region. Give the values of 100 for n and 0.7 for p. Click the division symbol found on the calculator pad, and divide the binomial function by 100. Click OK twice to return to the data table. You will find 1000 simulated sample proportions generated.

- *Minitab users:* **Calc → Random Data → Binomial.** Enter 1000 in the **Number of row of data to generate** dialog box, type "c1" in the **Store in column(s)** dialog box, enter 100 in the **Number of trials** dialog box, and enter 0.7 in the **Event probability** dialog box. Click OK to find the random binomial counts in column c1 of the worksheet. Now use **Calculator** to define another column as the binomial counts divided by 100 to obtain 1000 simulated sample proportions.

- *R users:* Type the following at the R prompt:

```
X<-rbinom(1000,100,0.7)
phat<-X/100
```

(a) Produce a histogram and a Normal quantile plot of the randomly generated sample proportions and describe its shape.

(b) What is the sample mean of the 1000 proportions? How close is this simulation estimate to the parameter value?

(c) What is the sample standard deviation of the 1000 proportions? How close is this simulation estimate to the theoretical standard deviation?

6.49 Shooting free throws. At the college level of men's basketball, based on the average of millions of free throws attempted over nearly 50 seasons, it has been found that roughly 69% of free throw attempts are successful. The season percents of free throws made fluctuate from a low of about 67% to a high of 70%.[16] In any given season, the number of free throw attempts is around 300,000. Assume that 300,000 free throws will be attempted in the upcoming season.

(a) What are the mean and standard deviation of $\hat{p}$ (sample proportion of free throws made in the season) if the population proportion is $p = 0.69$?

(b) Using the 68–95–99.7 rule, we would expect $\hat{p}$ to fall between what two percents about 95% of the time?

(c) Given the width of the interval in part (b) and the historical range of season percents, do you think that it is reasonable to assume that the population proportion has been the same over the last 50 seasons? Explain your answer.

6.50 Multiple-choice tests. Here is a simple probability model for multiple-choice tests. Suppose that each student has probability p of correctly answering a question chosen at random from a universe of possible questions. (A strong student has a higher p than a weak student.) The correctness of an answer to a question is independent of the correctness of answers to other questions. Emily is a good student for whom $p = 0.88$.

(a) Use the Normal approximation to find the probability that Emily scores 85% or lower on a 100-question test.

(b) If the test contains 250 questions, what is the probability that Emily will score 85% or lower?

(c) How many questions must the test contain to reduce the standard deviation of Emily's proportion of correct answers to half its value for a 100-item test?

(d) Diane is a weaker student for whom $p = 0.72$. Does the answer you gave in part (c) for the standard deviation of Emily's score apply to Diane's standard deviation as well?

CHAPTER 6 REVIEW EXERCISES

6.51 Survey about tax cuts. To gauge the attitudes of a state's citizens about corporate tax cuts, the office of the governor randomly surveys several chambers of commerce across the state. The office finds the sample proportion in favor of tax cuts to be 87% with a margin of error of ±4 percentage points. Would the governor be justified in applying these survey results to the total population of the state's citizens? Explain why or why not.

6.52 Survey about media-use habits. The nonprofit organization Common Sense Media, along with researchers from Northwestern University, conducted a survey of 1786 parents of children age 8 to 18 living in the United States to study parents' electronic-use habits.[17] The methodology section notes that the report is based on a nationally representative survey and that parents were

> . . . recruited using probability-based methods such as address-based sampling and random-digit-dial telephone calls. Households that were not already online were provided with a notebook computer and Internet access for the purpose of participating in surveys.

It was reported that the average time a parent spent with screen media (including TV, video, computers, smartphones, and tablets) was 9 hours and 22 minutes per day.

(a) Describe the population for this setting. What is the population parameter?

(b) What is the value of the sample mean strictly in units of hours?

(c) In terms of the discussion and terminology of Section 6.1, what were the researchers attempting to reduce by providing notebook computers and Internet access to households that weren't already online?

(d) If the researchers didn't provide notebook computers and Internet access to households that weren't already online, in what direction do you suspect the value of the sample average would change relative to the reported average? Explain your reasoning.

6.53 What population and sample? Fifty employees from a firm of 3100 employees are randomly selected to participate in a committee to evaluate how to implement sensitivity training. Currently, training is done in person, but a proposal has been made to implement the required training online. Each committee member is asked to vote Yes or No on the proposal.

(a) Describe the population for this setting.

(b) What is the sample?

(c) Describe the statistic and explain how it would be calculated.

(d) What is the population parameter?

6.54 What's wrong? Explain what is wrong in each of the following statements.

(a) The standard deviation of $\hat{p}$ increases with larger sample sizes.

(b) The central limit theorem states that for a sufficiently large sample size, the shape of the sampling distribution is approximately the shape of the population distribution.

(c) The sample proportion computed from an SRS taken from a population having a proportion p has a standard deviation exactly equal to $\sqrt{p(1-p)/n}$.

6.55 What is the real sample mean? Many companies have specially designated budgets for the awarding of gift cards to employees as part of their incentive and recognition programs. In 2018, Incentive Research Foundation (IRF) conducted a survey of medium- (annual revenue between $100 million and $1 billion) and large-sized (annual revenue exceeding $1 billion) companies to study how much money is budgeted toward gift cards.[18] IRF sampled 300 companies, of which 48% were medium-sized and 52% were large-sized. It reported that 69% of medium-sized companies had a budget for gift cards and 61% of large-sized companies had a budget for gift cards. For those companies *having* a budget for gift cards, the average annual spending for gift cards was $450,000 and $1,010,000 for medium- and large-sized companies, respectively.

(a) Explain why these averages would be biased estimates of average spending by the general population of medium- and large-sized companies. Relative to the true mean spending in the general populations of medium- and large-sized companies, would the reported IRF averages be biased low or high?

(b) Based on the provided information, determine the sample mean spending for all medium-sized companies found in the survey. Also, determine the sample mean spending for all large-sized companies found in the survey.

6.56 Public transit riders. The American Public Transportation Association performed a nationwide study of nearly 700,000 public transit users to study various demographics of passengers.[19] It was found that 7% of public transit users are students.

(a) Explain why it is incorrect to conclude that 7% of students are public transit users.

(b) Describe the population for the setting of the survey.

(c) Suppose you wish to study the demographics of public transit users in your region. You take an SRS of 200 public transit users. Assuming that the regional proportion of transit users who are students is the same as the national proportion, what is the standard deviation of $\hat{p}$?

(d) Continuing with part (c), using the 68–95–99.7 rule, you would expect $\hat{p}$ to fall between what two percents about 95% of the time?

6.57 Standard deviation comparison. Refer to Example 6.5 (page 308), in which we showed the simulated sampling distribution of $\bar{x}$ for $n = 35$ from the distribution of bike-trip durations. In Example 6.6 (page 312), we found the theoretical standard deviation of $\bar{x}$ to be 96.427 seconds. The 1000 simulated $\bar{x}$ values are provided. Find the standard deviation of simulated sample means and compare its value with the theoretical value. SIMBIKE

6.58 Number of steps per day. Refer to the setting of Exercise 6.15 (page 308). Suppose you randomly chose 50 U.S. iPhone users with the *Azumio Argus* app and ask them to report the number of steps they took on a given day to find the average number of steps.

(a) What is the standard deviation of the average number of steps?

(b) Using the 95 part of the 68–95–99.7 rule, describe the variability of this sample mean.

(c) What is the probability that your average will be greater than 5000 steps?

6.59 Mean and standard deviation comparison. Refer to Example 6.7 (page 313), in which we showed the distribution of 118,646 Lending Club loan amounts in dollars. For that distribution, the mean (μ) is $15,317.30 and the standard deviation (σ) is $9955.04. In Example 6.7, we also showed the simulated sampling distribution of $\bar{x}$ for $n = 25$ from the loan amount distribution. The 1000 simulated $\bar{x}$ values for each sample size are provided. SIMLOAN

(a) What are the theoretical mean and the standard deviation of the sampling distribution of $\bar{x}$?

(b) Find the mean and the standard deviation of 1000 simulated sample means, and compare their values with the theoretical values of part (a).

(c) Obtain a Normal quantile plot of the simulated sample means. What do you conclude? Relate your conclusion in the context of the central limit theorem.

6.60 Sample size? In a Deloitte global survey of more than 51,000 mobile users on the issue of consumer privacy and security, researchers found that 97% of users of ages 18 to 34 accepted mobile app terms and conditions without reading them.[20] Suppose you want to conduct your own survey on this issue in your region. Assuming $p = 0.97$, given the rules of thumb for the safe use of the Normal approximation of the $\hat{p}$, what minimum sample size for your SRS should you choose?

6.61 Sample median. In Chapter 1 (page 27), you learned that the mean and median will differ for skewed distributions. Clearly, the median estimator would be a biased estimator of the population mean when the population is skewed. But consider the case in which the population is symmetric and, therefore, the population mean equals the population median. Would it matter if we were to use the sample median or mean for estimating the population mean for symmetric distributions? You will explore this question in this exercise. The data file provides 1000 simulated sample means and sample medians based on $n = 50$ sampled from a Normal distribution with $\mu = 300$ and $\sigma = 60$. SIMMED

(a) Find the averages of the 1000 simulated sample means and sample medians. Are these averages close to the population mean of 300? In terms of estimating the population mean, what can you say about the sample mean and sample median estimators?

(b) Based on theory, what is the standard deviation of the sampling distribution of $\bar{x}$?

(c) Find the standard deviation of the simulated sample means. How does this value compare with the value found in part (b)?

(d) Find the standard deviation of the simulated sample medians and compare it with the standard deviation of the simulated sample means from part (c). Refer to the four bull's-eyes of Figure 6.4 (page 300). Based on what you found in part (a) along with the standard deviations of sample means and medians, which of the four bull's-eyes would you associate with the sample mean and with the sample median as estimators for μ?

6.62 A random walk. The random walk hypothesis is a financial theory related to the efficient-market hypothesis that suggests stock price changes over time have the same distribution and are independent of each other—and, therefore, cannot be predicted. For any given period, a positive (negative) price change in the next period implies that the stock went up (down) from the current period's price. Define X_i as the price change of a given stock i periods into the future. The change in a stock's price after n periods is the sum of these random movements:

$$Y = X_1 + X_2 + \cdots + X_n$$

Consider the weekly price changes of Facebook stock from May 2012 to June 2018. A time-series analysis of the weekly price changes would show them to behave randomly over time, with a distribution having a mean of 0.501 and a standard deviation of 3.664. Assume these mean and standard deviation values are μ and σ for the underlying distribution of weekly price changes.

(a) What is the mean and standard deviation of the change Y in stock price after 52 weeks into the future?

(b) Use the central limit theorem to find the approximate probability that Facebook's stock price 52 weeks into the future will be less than the current price.

(c) Using the 68–95–99.7 rule, we would expect the change in Facebook's stock price after 52 weeks to fall between what two values with probability about 95%?

6.63 Special cases. Consider the following special case scenarios.

(a) What are the values of $\bar{x}$ when $n = 1$? Explain what must be true for the sampling distribution of $\bar{x}$ to be Normal in the case of $n = 1$.

(b) For a population with mean μ and $\sigma = 0$, what will be true about the sampling distribution of $\bar{x}$ for any value of n?

(c) Discuss the nature of the sampling distribution of $\hat{p}$ when $p = 1$ or $p = 0$.

(d) For a given sample size n, what value of p will result in the largest standard deviation for $\hat{p}$? *Hint:* Systematically try different values of p from 0 to 1.

6.64 Sample mean distribution. Consider the following distribution for a discrete random variable X:

k	−2	−1	0	1
$P(X = k)$	1/4	1/4	1/4	1/4

Imagine a simple experiment of randomly generating a value for X, recording it, and then repeating this process a second time. Recognize that it is possible to get the same result in both trials. Finally, take the average of the two observed values.

(a) Manually draw the probability distribution of X.

(b) Find $P(X < 0)$ for either of the trials.

(c) Find the probability that X is less than 0 for both trials.

(d) List out all possible outcomes of the experiment. Find all possible values of $\bar{x}$, and determine the probability distribution for the possible sample mean values.

(e) Based on the probabilities found in part (d), manually draw the probability distribution distribution for the sample mean statistic. Describe the shape of this probability in relation to the probability distribution of part (a). Which phenomenon discussed in this chapter is taking place?

(f) Find the probability that the sample mean statistic is less 0. Explain why this probability is not the same as that found in part (c).

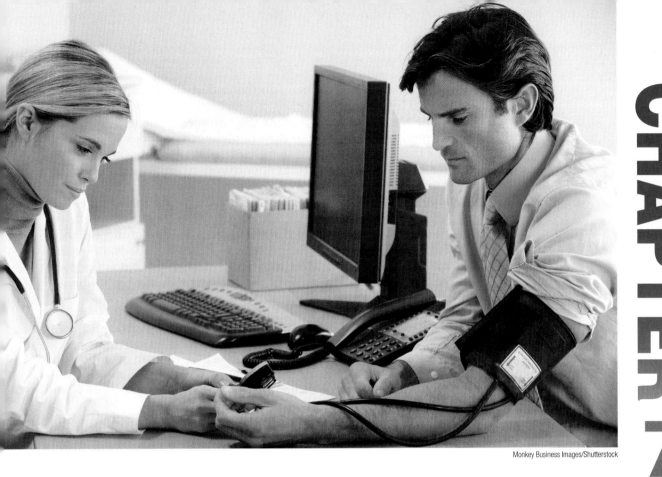

CHAPTER 7

Introduction to Inference

Introduction

Data-driven companies—both in manufacturing and service—gather data on various aspects of their businesses to draw conclusions about their own performance and about their markets.

- When Coca-Cola or Pepsi are filling millions of two-liter bottles, how can these companies be sure that the average fill amount remains on target at 2 liters?

- In response to customer complaints, AT&T attempts to improve customers' waiting times with its call centers. How can AT&T be confident that its efforts have reduced average wait time?

- Kaplan claims that its GMAT test prep courses will increase the average GMAT score of their students. Do before and after scores of Kaplan prep course students support this claim?

These are all examples in which statistical inference—namely, drawing conclusions about a population or process from sample data—could be used. Inference takes into account the natural variability in the sample data, thereby providing a statement of how much confidence we can place in the conclusions. Although there are numerous methods for inference, there are only a few general types of statistical inference. This chapter introduces the two most common types: *confidence intervals* and *tests of significance*.

Because the underlying reasoning for these two types of inference remains the same across different settings, this chapter considers just one simple setting: inference about the mean of a large population whose standard deviation is known. Although, in practice, the population standard deviation is rarely

CHAPTER OUTLINE

7.1 Estimating with Confidence

7.2 Tests of Significance

7.3 Use and Abuse of Tests

7.4 Prediction Intervals

known, this setting allows us to focus on the underlying rationale of these types of statistical inference with a class of distributions that we are familiar with: namely, the Normal distributions. As you will see, the material of Chapter 6 on the sampling distribution of the sample mean provides a necessary foundation for the developments of this chapter. Likewise, our coverage of confidence intervals provides a natural opportunity to introduce the idea of prediction intervals in Section 7.4.

Later chapters will present inference methods to use in more realistic settings that we encounter when learning to explore data. In fact, there are libraries—both of books and of computer software—full of more elaborate statistical techniques. Informed use of any of these methods, however, requires a firm understanding of the underlying reasoning. That is the goal of this chapter. A computer or calculator will do the arithmetic, but *you must still exercise sound judgment based on understanding.*

Overview of inference

The purpose of statistical inference is to draw conclusions from data. Formal inference emphasizes substantiating our conclusions via probability calculations. Probability allows us to take chance variation into account. Here is an example.

EXAMPLE 7.1

VZVOL

Pattern in Trading Volumes Investors and analysts are continually seeking ways to understand price movements in stocks. One factor often considered is trading volume, that is, the quantity of shares that change owners. In studying the relationship between trading volumes and price movements, an understanding of the trading-volume process becomes a source of interest. Are trading volumes random from trading period to trading period? Or, is there some pattern in trading volumes over time?

Figure 7.1 is a time plot of the daily trading volumes for Verizon Communications stock from January 2 to June 25, 2018, with the mean trading volume indicated by a horizontal black line.[1] Do the daily volumes appear to be randomly fluctuating around this mean? More specifically, could you simulate the trading-volume process with the process of flipping a fair coin, with heads

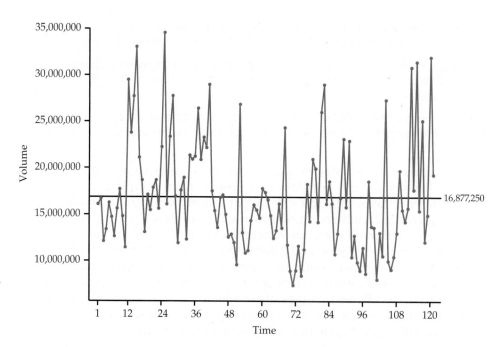

FIGURE 7.1 Time plot of daily trading volumes of Verizon Communications stock (January 2, 2018 to June 25, 2018), with the mean trading volume indicated.

being an observation below the mean and tails being an observation above the mean? Or, does it appear that the volumes move in patterns around the mean level? In particular, are there "strings" (runs) of observations either all below or all above the mean?

As we will learn in Chapter 14, a runs test indicates that observing a more extreme number of runs than those shown in Figure 7.1 would occur only 3.7% of the time if, in fact, trading volumes are truly random over time. Because this chance is fairly small, we are inclined to believe that trading volumes are *not* purely random over time. This probability calculation helps us to distinguish between patterns that are consistent or inconsistent with random behavior over time. ■

In this chapter, we introduce the two most frequently used types of statistical inference. Section 7.1 concerns *confidence intervals* for estimating the value of a population parameter. Section 7.2 presents *tests of significance* which assess the evidence for a claim or baseline scenario. Both types of inference are based on the sampling distributions of statistics. That is, both report probabilities that state *what would happen if we used the inference method many times*. That kind of probability statement is characteristic of standard statistical inference. The reporting of 3.7% in Example 7.1 provides an example of a probability statement (known as *P*-value) used in statistical inference. Our choice of a type of analysis unfamiliar to you in Example 7.1 was quite purposeful. Specifically, our purpose is to hint to the fact that the concepts and basic reasoning of inference developed in this chapter carry through the remainder of the book.

sampling distribution, p. 298

There are a wide variety of statistics used to summarize data. In this chapter, we concentrate on the sample mean. Because sample means are just averages of observations, they are among the most frequently used statistics. From Section 6.2, we learned that the sampling distribution of the sample mean based on n observations has a standard deviation of $\sigma/\sqrt{n}$, where σ is the population standard deviation. In practice, we rarely know sigma and must estimate it.

standard deviation of sample mean, p. 311

By making the assumption of σ known, the procedures set forth in this chapter will be based on the standard Normal distribution. In the next chapter, there will be no assumption made about σ so we estimate it with the sample standard deviation s. When the standard deviation is estimated, a family of distributions known as the *t distributions* is used in place of the standard Normal distribution. However, for large sample sizes the *t* distribution is practically indistinguishable from the standard Normal distribution. This implies a certain practicality to the methods of this chapter. Namely, we are safe to consider the Normal-based procedures of this chapter even if we use the sample standard deviation in place of the unknown σ when the sample size is large.

With this said, the goal of this chapter is not to provide methods of inference for general use. Instead, view this chapter as providing you with the common concepts, terminologies, and basic reasoning of inference methods which, in turn, will ease your transition to the remaining chapters on statistical methods.

7.1 Estimating with Confidence

When you complete this section, you will be able to:

- Explain the purpose of a confidence interval and describe its common form in terms of an estimate and its margin of error.
- Explain the meaning of confidence level.

- Construct a confidence interval for μ based on a random sample of size n from a population having known standard deviation σ.
- Explain how a change in sample size or confidence level changes the margin of error.
- Determine the sample size needed to obtain a specified margin of error.
- Identify situations where the inference about μ using the presented confidence interval may be suspect.

The SAT is a widely used measure of readiness for college study. It consists of two required sections, one for mathematical reasoning ability (SATM) and one for reading and writing (SATRW). Possible scores on each section range from 200 to 800, for a total range of 400 to 1600. Since 1995, section scores have been *recentered* so that the mean is approximately 500 with a standard deviation of 100 in a large "standardized group." This scale has been maintained so that scores have a constant interpretation.

EXAMPLE 7.2

Estimating the Mean SATM Score for Seniors in California Suppose that you want to estimate the mean SATM score for the 492,835 high school seniors in California.[2] You know better than to trust data from the students who choose to take the SAT. Only about 38% of California students typically take the SAT. These self-selected students are planning to attend college and are not representative of all California seniors. At considerable effort and expense, you give the test to a simple random sample (SRS) of 500 California high school seniors. The mean score for your sample is $\bar{x} = 485$. What can you say about the mean score μ in the population of all 492,835 seniors? ∎

unbiased estimator, p. 300

law of large numbers, p. 232

The sample mean $\bar{x}$ is the natural estimator of the unknown population mean μ. We know that $\bar{x}$ is an unbiased estimator of μ. More important, the law of large numbers says that the sample mean must approach the population mean as the size of the sample grows. The value $\bar{x} = 485$, therefore, appears to be a reasonable estimate of the mean score μ that all 492,835 students would achieve if they took the test.

But how reliable is this estimate? A second sample of 500 students would surely not give a sample mean of 485 again. Unbiasedness says only that there is no systematic tendency to underestimate or overestimate the truth. Could we plausibly get a sample mean of 465 or 510 in repeated samples? *An estimate without an indication of its variability is of little value.*

Statistical confidence

central limit theorem, p. 313

The unbiasedness of an estimator concerns the center of its sampling distribution, but questions about variation are answered by looking at its spread. From the central limit theorem, we know that if the entire population of SATM scores has mean μ and standard deviation σ, then in repeated samples of size 500 the sample mean $\bar{x}$ is approximately $N(\mu, \sigma/\sqrt{500})$. Let us suppose that we know that the standard deviation σ of SATM scores in our California population is $\sigma = 100$. This means that, in repeated sampling, the sampling distribution of the sample mean $\bar{x}$ is approximately Normal, centered at the unknown population mean μ with a standard deviation of

$$\sigma_{\bar{x}} = \frac{100}{\sqrt{500}} = 4.5$$

FIGURE 7.2 Distribution of the sample mean for Example 7.2. $\bar{x}$ lies within ±9 points of μ in 95% of all samples. This also means that μ is within ±9 points of $\bar{x}$ in those samples.

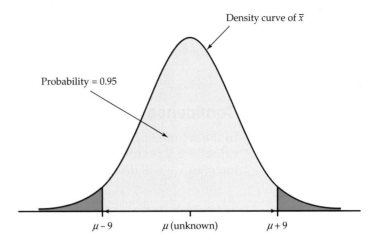

Now we are ready to proceed. Consider this line of thought, which is illustrated by Figure 7.2:

- The 68–95–99.7 rule says that the probability is about 0.95 that $\bar{x}$ will be within nine points (that is, two standard deviations of $\bar{x}$) of the population mean score μ.

- To say that $\bar{x}$ lies within nine points of μ is the same as saying that μ is within nine points of $\bar{x}$.

- So about 95% of all samples will contain the true μ in the interval from $\bar{x} - 9$ to $\bar{x} + 9$.

We have simply restated a fact about the sampling distribution of $\bar{x}$. *The language of statistical inference uses this fact about what would happen in the long run to express our confidence in the results of any one sample.* Our sample gave $\bar{x} = 485$. We say that we are *95% confident* that the unknown mean score for all California seniors lies between

$$\bar{x} - 9 = 485 - 9 = 476$$

and

$$\bar{x} + 9 = 485 + 9 = 494$$

Be sure you understand the grounds for our confidence. There are only two possibilities for our SRS:

1. The interval between 476 and 494 contains the true μ.
2. The interval between 476 and 494 does not contain the true μ.

We cannot know whether our sample is one of the 95% for which the interval $\bar{x} \pm 9$ contains μ or one of the unlucky 5% for which it does not contain μ. The statement that we are 95% confident is shorthand for saying, "We arrived at these numbers by a method that gives correct results 95% of the time."

APPLY YOUR KNOWLEDGE

7.1 Company invoices. The mean amount μ for all the invoices for your company last month is not known. Based on your past experience, you are willing to assume that the standard deviation of invoice amounts is about $320. If you take a random sample of 100 invoices, what is the value of the standard deviation for $\bar{x}$?

7.2 Use the 68–95–99.7 rule. In the setting of the previous exercise, the 68–95–99.7 rule says that the probability is about 0.95 that $\bar{x}$ is within _____ of the population mean μ. Fill in the blank.

7.3 An interval for 95% of the sample means. In the setting of the previous two exercises, about 95% of all samples will capture the true mean of all the invoices in the interval $\bar{x}$ plus or minus _____. Fill in the blank.

Confidence intervals

In the setting of Example 7.2 (page 340), the interval of numbers between the values $\bar{x} \pm 9$ is called a *95% confidence interval* for μ. Like most confidence intervals we will discuss, this one has the form

$$\text{estimate} \pm \text{margin of error}$$

margin of error

The estimate ($\bar{x} = 485$ in this case) is our guess for the value of the unknown parameter. The **margin of error** (9 here) reflects how accurate we believe our guess is, based on the variability of the estimate and how confident we are that the procedure will produce an interval that will contain the true population mean μ.

Figure 7.3 illustrates the behavior of 95% confidence intervals in repeated sampling from a Normal distribution with mean μ. The center of each interval (marked by a dot) is at $\bar{x}$ and varies from sample to sample. The sampling distribution of $\bar{x}$ (also Normal) appears at the top of the figure to show the long-term pattern of this variation.

The 95% confidence intervals, $\bar{x} \pm$ margin of error, from 25 SRSs appear below the sampling distribution. The arrows on either side of the dot ($\bar{x}$) span the confidence interval. All except one of the 25 intervals contain the true value of μ. In those intervals that contain μ, sometimes μ is near the middle of the interval and sometimes it is closer to one of the ends. This again reflects the variation of $\bar{x}$. In practice, we don't know the value of μ, but we have a method such that, in a very large number of samples, 95% of the confidence intervals will contain μ.

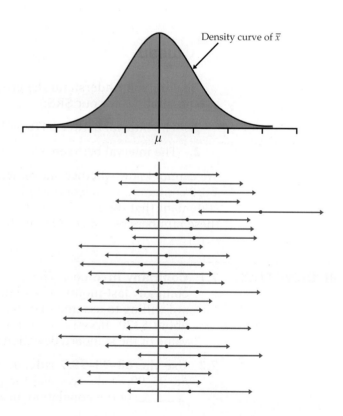

FIGURE 7.3 Twenty-five samples from the same population gave these 95% confidence intervals. In the long run, 95% of all samples give an interval that covers μ.

Statisticians have constructed confidence intervals for many different parameters based on a variety of designs for data collection. We will meet a number of these in later chapters. Two important things about a confidence interval are common to all settings:

1. It is an interval of the form (a, b), where a and b are numbers computed from the sample data.

confidence level

2. It has a property called a **confidence level** that gives the probability of producing an interval that contains the unknown parameter.

Users can choose the confidence level, but 95% is the standard for most situations. Occasionally, 90% or 99% is used. We will use C to stand for the confidence level in decimal form. For example, a 95% confidence level corresponds to $C = 0.95$.

CONFIDENCE INTERVAL

A level C **confidence interval** for a parameter is an interval computed from sample data by a method that has probability C of producing an interval containing the true value of the parameter.

With the *Confidence Interval* applet, you can construct diagrams similar to the one displayed in Figure 7.3. The only difference is that the applet displays the Normal population distribution at the top rather than the Normal sampling distribution of $\bar{x}$. You choose the confidence level C, the sample size n, and whether you want to generate 1 or 25 intervals at a time. A running total (and percent) of the number of intervals that contain μ is displayed so you can study more than 25 intervals.

When generating a single interval, the data for the SRS are shown below the confidence interval. The spread in these data reflects the spread of the population distribution. This spread is assumed known, and it does not change with sample size. What does change, as you vary n, is the margin of error because it reflects the uncertainty in the estimate of μ. As you increase n, you'll find that the span of the confidence interval gets smaller and smaller.

APPLY YOUR KNOWLEDGE

7.4 Generating a single confidence interval. Using the default settings in the *Confidence Interval* applet (95% confidence level and $n = 20$), click "Sample" to choose an SRS and display its confidence interval.

(a) Is the spread in the data, shown as yellow dots below the confidence interval, larger than the span of the confidence interval? Explain why this would typically be the case.

(b) For the same data set, you can compare the span of the confidence interval for different values of C by sliding the confidence level to a new value. For the SRS you generated in part (a), what happens to the span of the interval when you move C to 99%? What about 90%? Describe the relationship you find between the confidence level C and the span of the confidence interval.

7.5 80% confidence intervals. The idea of an 80% confidence interval is that the interval captures the true parameter value in 80% of all samples. That's not high enough confidence for practical use, but 80% hits and 20% misses make it easy to see how a confidence interval behaves in repeated samples from the same population.

(a) Set the confidence level in the *Confidence Interval* applet to 80%. Click "Sample 25" to choose 25 SRSs and display their confidence intervals. How many of the 25 intervals contain the true mean μ? What proportion contain the true mean?

(b) We can't determine whether a new SRS will result in an interval that contains μ or not. The confidence level only tells us what percent will contain μ in the long run. Click "Sample 25" again to get the confidence intervals from 50 SRSs. What proportion hit? Keep clicking "Sample 25" and record the proportion of hits among 100, 200, 300, 400, and 500 SRSs. As the number of samples increases, we expect the percent of captures to get closer to the confidence level, 80%. Do you find this pattern in your results?

Confidence interval for a population mean

We will now construct a level C confidence interval for the mean μ of a population when the data are an SRS of size n. The construction is based on the sampling distribution of the sample mean $\bar{x}$. This distribution is exactly $N(\mu, \sigma/\sqrt{n})$ when the population has the $N(\mu, \sigma)$ distribution. The central limit theorem says that this same sampling distribution is approximately correct for large samples. We will assume we are in one of these two situations.

central limit theorem, p. 313

Our construction of a 95% confidence interval for the mean SATM score began by noting that any Normal distribution has probability of about 0.95 within ± 2 standard deviations of its mean. To construct a level C confidence interval, we first catch the central C area under a Normal curve. That is, we must find the number z^* such that any Normal distribution has probability C within $\pm z^*$ standard deviations of its mean.

Because all Normal distributions have the same standardized form, we can obtain everything we need from the standard Normal curve. Figure 7.4 shows how C and z^* are related. Figure 7.4 reminds us that any Normal curve has probability C between the point z^* standard deviations below the mean and the point z^* standard deviations above the mean. The sample mean $\bar{x}$ has the Normal distribution with mean μ and standard deviation $\sigma/\sqrt{n}$, so there is probability C that $\bar{x}$ lies between

$$\mu - z^* \frac{\sigma}{\sqrt{n}} \quad \text{and} \quad \mu + z^* \frac{\sigma}{\sqrt{n}}$$

This is exactly the same as saying that the unknown population mean μ lies between

$$\bar{x} - z^* \frac{\sigma}{\sqrt{n}} \quad \text{and} \quad \bar{x} + z^* \frac{\sigma}{\sqrt{n}}$$

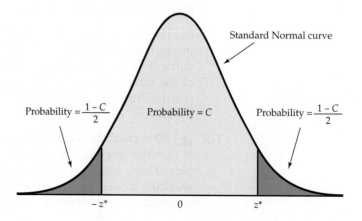

FIGURE 7.4 To construct a level C confidence interval, we must find the number z^*. The area between the values $-z^*$ and z^* under the standard Normal curve is C.

That is, there is probability C that the interval $\bar{x} \pm z^*\sigma/\sqrt{n}$ contains μ. This is our confidence interval. The estimate of the unknown μ is $\bar{x}$, and the margin of error is $z^*\sigma/\sqrt{n}$.

CONFIDENCE INTERVAL FOR A POPULATION MEAN

Choose an SRS of size n from a population having unknown mean μ and known standard deviation σ. The **margin of error** for a level C confidence interval for μ is

$$m = z^* \frac{\sigma}{\sqrt{n}}$$

Here, z^* is the value on the standard Normal curve with area C between the points $-z^*$ and z^*. The level C **confidence interval** for μ is

$$\bar{x} \pm m$$

The confidence level of this interval is exactly C when the population distribution is Normal and is approximately C when n is large in other cases.

The interval $\bar{x} \pm z^*\sigma/\sqrt{n}$ assumes that the standard deviation σ of the population is known. In practice, σ is usually not known, which seems to leave this interval of little use. In the next chapter, however, you will learn how to construct the confidence interval when the sample standard deviation s is used in place of σ. It turns out that when n is large, the interval $\bar{x} \pm z^*s/\sqrt{n}$ is a reasonably accurate approximation. This confidence interval for μ is often called the "large-sample confidence interval for a population mean" and is often sufficiently accurate when $n \geq 30$. However, if the data seem to arise from an extremely skewed distribution, a safer threshold is $n \geq 40$. Even though the interval $\bar{x} \pm z^*s/\sqrt{n}$ is a reasonable construction when n is large, software will *always* report the more accurate construction outlined in Chapter 8. Remember the primary goal of this chapter is to provide you with the general concepts, terminologies, and basic reasoning of inference methods.

Values of z^* for many choices of C appear in the row labeled z^* at the bottom of Table D. Here are the most important entries from that row:

z^*	1.645	1.960	2.576
C	90%	95%	99%

Notice that for 95% confidence, the value 2 obtained from the 68–95–99.7 rule is replaced with the more precise 1.96. In Exercise 7.28 (page 354), you will explore how software can be used to find values of z^* for any choice of C.

CASE 7.1 **Bankruptcy Attorney Fees** For individuals or businesses, bankruptcy is the process of legally declaring the inability to pay outstanding debts. Nearly a million people in the United States file for bankruptcy each year. For individuals, bankruptcies are filed under either Chapter 7 or Chapter 13 bankruptcy. Simply stated, Chapter 7 bankruptcy quickly discharges a debtor's unsecured debts (e.g., medical bills, credit cards, and personal loans) by surrendering their nonexempt assets (e.g., property that is not the primary home, jewelry, newer vehicles with equity, and artwork). Chapter 13 does not force liquidation of nonexempt assets but rather requires the debtor to establish a repayment plan (typically, three to five years) to pay back all or a portion of the debt before being discharged. Bankruptcy attorneys help individuals

decide between the two bankruptcy options and thereafter file and represent the debtor in the bankruptcy case.

Due to the differences between the two bankruptcy options, attorneys charge different fee amounts. To estimate attorney fees for Chapters 7 and 13 on individual bankruptcy cases, researchers conducted a nationwide random sample of individuals who had bankruptcy filings.[3] ■

Often with surveys, as with this bankruptcy study, there are nonrespondents. *Nonresponse should always be considered as a source of bias.* The researchers of the bankruptcy study appropriately conducted an extensive study of key characteristics (e.g., known financial variables) of the debtors and found no significant differences between those who responded and the nonrespondents. As such, the researchers believed any nonresponse to be an ignorable source of bias and proceeded to use the available samples for statistical inference.

EXAMPLE 7.3

Confidence Interval for Chapter 13 Attorney Fees For the Chapter 13 filings involving an attorney, the sample size n is 1511 and the average attorney fee is $3243.57. The researchers of the study found the sample distribution of the fees to be fairly symmetric with thicker tails and a higher peak than a Normal distribution. Such a distribution is said to have positive excess **kurtosis**. Nevertheless, because the sample size is so large, we can rely on the central limit theorem to assure us that the confidence interval based on the Normal distribution will be a very good approximation.

Let's compute the approximate 95% confidence interval for the mean attorney fee among all Chapter 13 debtors who used an attorney. We assume that the standard deviation for the population of attorney fees is $973.21 as found from the study; the very large sample size of 1511 allows to safely proceed with the Normal-based procedure. For 95% confidence, we see from Table D that $z^* = 1.960$. The margin of error for the 95% confidence interval for μ is, therefore,

$$m = z^* \frac{\sigma}{\sqrt{n}}$$
$$= 1.960 \frac{973.21}{\sqrt{1511}}$$
$$= 49.07$$

With the computed margin of error, the 95% confidence interval is

$$\bar{x} \pm m = 3243.57 \pm 49.07$$
$$= (3194.50, 3292.64)$$

We are 95% confident that the average attorney fee among all Chapter 13 debtors who used an attorney is between $3194.50 and $3292.64. In this application, the extra digits in the confidence interval provide little additional useful information. So, we might simply round the values and report the confidence interval as approximately being (3195, 3293). ■

Suppose that the researchers who designed this study had used a different sample size. How would this affect the confidence interval? We can answer this question by changing the sample size in our calculations and keeping the confidence level C, sample mean ($\bar{x}$), and population standard deviation (σ) the same.

EXAMPLE 7.4

How Sample Size Affects the Confidence Interval As in Example 7.3, the sample mean of the attorney fee is $3243.57 and the population standard deviation is $973.21. Suppose that the sample size is only 168 but still large enough for us to rely on the central limit theorem. In this case, the margin of error for 95% confidence is

$$m = z^* \frac{\sigma}{\sqrt{n}}$$
$$= 1.960 \frac{973.21}{\sqrt{168}}$$
$$= 147.17$$

and the approximate 95% confidence interval is

$$\bar{x} \pm m = 3243.57 \pm 147.17$$
$$= (3096.40, 3390.74) \blacksquare$$

Notice that the margin of error (147.17) for this example is about three times as large as the margin of error (49.07) that we computed in Example 7.3. The only change that we made was to assume a sample size of 168, which is about one-ninth of the original 1511. We triple the margin of error when we reduce the sample size to one-ninth of the original value. Or, from the other perspective, the margin of error reduced by a factor of 3 when the sample size increased by a factor of $3^2 = 9$. Figure 7.5 illustrates the effect in terms of the intervals.

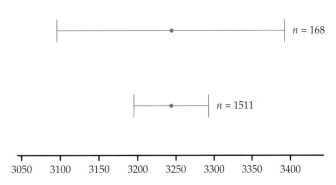

FIGURE 7.5 Confidence intervals for $n = 1511$ and $n = 168$, Examples 7.3 and 7.4. A sample size nine times as large results in a confidence interval that is one-third as wide.

In general, for a given confidence level C and population standard deviation σ, margin of error decreases in proportion to the square root of the sample size n. This implies that to reduce the margin of error by a factor of K, we need to increase the sample size by a factor of K^2.

APPLY YOUR KNOWLEDGE

CASE 7.1 7.6 Average attorney fee for Chapter 7 cases. Refer to Example 7.3 (page 346). The average attorney fee for $n = 2834$ Chapter 7 bankruptcy cases was $1152.37. If the population standard deviation is $642.10, give the 95% confidence interval for μ, the average attorney fee from the population of Chapter 7 bankruptcy cases.

CASE 7.1 7.7 Changing the sample size. In the setting of the previous exercise, would the margin of error for 95% confidence be roughly doubled or halved if the sample size were changed to $n = 709$? Verify your answer by performing the calculations.

CASE 7.1 7.8 Changing the confidence level. In the setting of Exercise 7.6, would the margin of error for 90% confidence be larger or smaller? Verify your answer by performing the calculations.

The argument leading to the form of confidence intervals for the population mean μ rested on the fact that the statistic $\bar{x}$ used to estimate μ has a Normal distribution. Because many sample estimates have Normal distributions (at least approximately), it is useful to notice that the confidence interval has the form

$$\text{estimate} \pm z^*\sigma_{\text{estimate}}$$

The estimate based on the sample is the center of the confidence interval. The margin of error is $z^*\sigma_{\text{estimate}}$. The desired confidence level determines z^* from Table D. The standard deviation of the estimate is found from knowledge of the sampling distribution in a particular case. When the estimate is $\bar{x}$ from an SRS, the standard deviation of the estimate is $\sigma_{\text{estimate}} = \sigma/\sqrt{n}$. We will return to this general form numerous times in the following chapters.

How confidence intervals behave

The margin of error $z^*\sigma/\sqrt{n}$ for the mean of a Normal population illustrates several important properties that are shared by all confidence intervals in common use. The user chooses the confidence level, and the margin of error follows from this choice.

Both high confidence and a small margin of error are desirable characteristics of a confidence interval. High confidence says that our method almost always gives correct answers. A small margin of error says that we have pinned down the parameter quite precisely.

Suppose that in planning a study you calculate the margin of error and decide that it is too large. Here are your choices to reduce it:

- Use a lower level of confidence (smaller C).
- Choose a larger sample size (larger n).
- Reduce σ.

For most problems, you would choose a confidence level of 90%, 95%, or 99%, so z^* will be 1.645, 1.960, or 2.576, respectively. Figure 7.4 (page 344) shows that z^* will be smaller for lower confidence (smaller C). The bottom row of Table D also shows this. If n and σ are unchanged, a smaller z^* leads to a smaller margin of error.

EXAMPLE 7.5

CASE 7.1

How the Confidence Level Affects the Confidence Interval Suppose that for the attorney fee data in Example 7.3 (page 346), we wanted 99% confidence. Table D tells us that for 99% confidence, $z^* = 2.576$. The margin of error for 99% confidence based on 1511 observations is

$$m = z^* \frac{\sigma}{\sqrt{n}}$$

$$= 2.576 \frac{973.21}{\sqrt{1511}}$$

$$= 64.49$$

and the 99% confidence interval is

$$\bar{x} \pm m = 3243.57 \pm 64.49$$

$$= (3179.08, 3308.06)$$

Requiring 99%, rather than 95%, confidence has increased the margin of error from 49.07 to 64.49. Figure 7.6 compares the two intervals. ∎

FIGURE 7.6 Confidence intervals, Examples 7.3 and 7.5. The larger the value of C, the wider the interval.

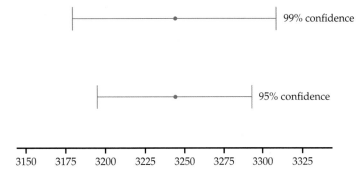

Similarly, choosing a larger sample size n reduces the margin of error for any fixed confidence level. The square root in the formula implies that we must multiply the number of observations by 4 in order to cut the margin of error in half. If we want to reduce the margin of error by a factor of 4, we must take a sample 16 times as large.

As seen from its calculation, the margin of error is directly related to size of the standard deviation σ, the measure of population variation. You can think of the variation among individuals in the population as noise that obscures the average value μ. It is harder to pin down the mean μ of a highly variable population; that is why the margin of error of a confidence interval increases with σ. In practice, we can sometimes reduce σ by carefully controlling the measurement process. We also might change the mean of interest by restricting our attention to only part of a large population. Focusing on a subpopulation (e.g., men vs. women, urban vs. rural) will often result in a smaller σ.

Example 7.4 (page 347) shows the impact of changing sample size n on the margin of error. Given the interplay between sample size and margin of error, a natural question arises. How large of a sample do we need to ensure that the margin of error is no more than some specified value? Let's look again at the margin of error:

$$m = z^* \frac{\sigma}{\sqrt{n}}$$

By rearranging the above formula, we can find the sample size n that gives a desired margin of error. Here is the result.

SAMPLE SIZE FOR SPECIFIED MARGIN OF ERROR

The confidence interval for a population mean will have a specified margin of error m when the sample size is

$$n = \left(\frac{z^*\sigma}{m}\right)^2$$

If the computed n is a fractional number, always round up to the nearest integer value.

This formula does not account for sampling costs. In practice, obtaining observations costs time and money. The required sample size may be prohibitively expensive. In such situations, you might consider a larger margin of error and/or a lower confidence level to find a workable sample size.

EXAMPLE 7.6

 How Many Debtors Should We Survey? Suppose that we are planning to survey Chapter 13 debtors as described in Example 7.3 (page 346). If we want the margin of error for the average attorney fee to be $35 with 95% confidence, what sample size n do we need? For 95% confidence, Table D gives $z^* = 1.960$. For σ, we will use the value from the previous study, $973.21. If the margin of error is $35, we have

$$n = \left(\frac{z^*\sigma}{m}\right)^2 = \left(\frac{1.96 \times 973.21}{35}\right)^2 = 2970.22$$

Because 2970 observations will give a slightly wider interval than desired, we round up to 2971 observations, which will give a slightly narrower interval. Thus, we should choose $n = 2971$. ■

In Section 8.3, we will revisit the question of sample size determination when the population standard deviation σ is unknown.

APPLY YOUR KNOWLEDGE

7.9 Starting salaries. You are planning a survey of starting salaries for recent business majors. In the 2018 survey by the National Association of Colleges and Employers, the average starting salary was reported to be $56,720.[4] If you assume that the standard deviation is $11,500, what sample size do you need to have a margin of error equal to $1000 with 95% confidence?

7.10 Changes in sample size. Suppose that, in the setting of the previous exercise, you have the resources to contact 600 recent graduates. If all respond, will your margin of error be larger or smaller than $1000? What if only 50% respond? Verify your answers by performing the calculations.

Some cautions

We have already seen that small margins of error and high confidence can require large numbers of observations. You should also be keenly aware that *any formula for inference is correct only in specific circumstances*. Our formula $\bar{x} \pm z^*\sigma/\sqrt{n}$ for estimating a population mean comes with the following list of warnings for the user:

- There is no correct method for inference from data haphazardly collected with bias of unknown size. Fancy formulas cannot rescue badly produced data. The data should be an SRS from the population. We are safe if we actually did a randomization and drew an SRS. We are not in great danger if the data can plausibly be thought of as independent observations from a population. That is the case in Examples 7.3 through 7.5, where we redefine our population to correspond to survey respondents.

- The formula is not applicable for probability sampling designs more complex than an SRS. Correct methods for other designs are available. We will not discuss confidence intervals based on multistage or stratified samples (pages 136–137). If you plan such samples, be sure that you (or your statistical consultant) know how to carry out the inference you desire.

resistant measure, p. 26

- Because $\bar{x}$ is not a resistant measure, outliers can have a large effect on the confidence interval. *You should search for outliers and try to correct them or justify their removal before computing the interval*. If the outliers cannot be removed, ask your statistical consultant about procedures that are not sensitive to outliers.

- If the sample size is small and the population is not Normal, the true confidence level will be different from the value C used in computing the interval. *Prior to any calculations, examine your data carefully for skewness and other signs of non-Normality.* Remember, though, that the interval relies only on the distribution of $\bar{x}$, which, even for quite small sample sizes, is much closer to Normal than is the distribution of the individual observations. When $n \geq 15$, the confidence level is not greatly disturbed by non-Normal populations unless extreme outliers or quite strong skewness are present. Our attorney data in Example 7.3 are not Normal, but because of the very large sample size of 1511, we are confident that the distribution of the sample mean will be nearly Normal.

- Finally, you should understand what statistical confidence does *not* say. Based on our SRS in Example 7.2, we are 95% confident that the mean SATM score for the California students lies between 476 and 494 (page 341). This says that this interval was calculated by a method that gives correct results in 95% of all possible samples. It does *not* say that the probability is 0.95 that the true mean falls between 476 and 494. *No randomness remains after we draw a particular sample and compute the interval.* The true mean either is or is not between 476 and 494. The probability calculations of standard statistical inference describe how often the *method*, not a particular sample, gives correct answers.

Every inference procedure that we will meet has its own list of warnings. Because many of the warnings are similar to those we have mentioned, we do not print the full warning label each time. It is easy to state (from the mathematics of probability) conditions under which a method of inference is exactly correct. These conditions are *never* fully met in practice. For example, no population is exactly Normal. *Deciding when a statistical procedure should be used in practice often requires judgment assisted by exploratory analysis of the data.* Mathematical facts are, therefore, only a part of statistics. The difference between statistics and mathematics can be stated in this way: mathematical theorems are true; statistical methods are often effective when used with skill and good judgment.

APPLY YOUR KNOWLEDGE

7.11 Nonresponse in a survey. The Kaiser Family Foundation (Kaiser) and the Health Research & Educational Trust (HRET) conduct an annual national survey of firms on a variety of health coverage issues, including premiums and employee contributions. In the survey methodology section of the 2016 survey report, the following is stated:

> *2016 Annual Employer Health Benefits Survey (Kaiser/HRET) reports findings from a telephone survey of 1,933 randomly selected public and private employers with three or more workers.... In 2016, the overall response rate is 40%, which includes firms that offer and do not offer health benefits.*

(a) How many firms were initially contacted for this study?

(b) Provide a couple of reasons a selected firm might not provide information. Based on these reasons, do you think the reported averages of annual firm and worker premium contributions are biased? Similarly, are the margins of error based just on the reported firms good measures of precision for confidence interval construction?

SECTION 7.1 SUMMARY

- The purpose of a **confidence interval** is to estimate an unknown parameter with an indication of how accurate the estimate is and of how confident we are that the result is correct. Any confidence interval has two parts: an interval computed from the data and a confidence level. The interval often has the form

$$\text{estimate} \pm \text{margin of error}$$

- The **confidence level** states the probability that the method will give a correct answer. That is, if you use 95% confidence intervals, in the long run 95% of your intervals will contain the true parameter value. When you apply the method once (that is, to a single sample), you do not know if your interval gave a correct answer (this happens 95% of the time) or not (this happens 5% of the time).

- The **margin of error** for a level C confidence interval for the mean μ of a Normal population with known standard deviation σ, based on an SRS of size n, is given by

$$m = z^* \frac{\sigma}{\sqrt{n}}$$

Here, z^* is obtained from the row labeled z^* at the bottom of Table D. The probability is C that a standard Normal random variable takes a value between $-z^*$ and z^*. The confidence interval is

$$\bar{x} \pm m$$

If the population is not Normal and n is large, the confidence level of this interval is approximately correct.

- Other things being equal, the margin of error of a confidence interval decreases as
  - the confidence level C decreases,
  - the sample size n increases, and
  - the population standard deviation σ decreases.

- For a given confidence level C and population standard deviation σ, to reduce the margin of error by a factor of K, we need to increase the sample size by a factor of K^2.

- The sample size n required to obtain a confidence interval of specified margin of error m for a population mean is

$$n = \left(\frac{z^* \sigma}{m}\right)^2$$

If the computed n is a fractional number, always round up to the nearest integer value.

- A specific confidence interval formula is correct only under specific conditions. The most important conditions concern the method used to produce the data. Other factors such as the form of the population distribution may also be important. These conditions should be investigated *prior* to any calculations.

SECTION 7.1 EXERCISES

For Exercises 7.1 to 7.3, see pages 341–342; for Exercises 7.4 and 7.5, see pages 343–344; for 7.6 to 7.8, see page 347; for 7.9 and 7.10, see page 350; and for 7.11, see page 351.

7.12 Margin of error and the confidence interval. A study based on a sample of size 30 reported a mean of 78 with a margin of error of 9 for 99% confidence. The margin of error was computed based on the population standard deviation being known.

(a) Give the 99% confidence interval.

(b) What is the value of the population standard deviation?

(c) If you wanted 95% confidence for the same study, would your margin of error be greater than, equal to, or less than 9? Explain your answer.

7.13 Change the sample size. Suppose that the sample mean is 50 and the population standard deviation is 7. Make a diagram similar to Figure 7.5 (page 347) that illustrates the effect of sample size on the width of a 95% interval. Use the following sample sizes: 10, 20, 40, and 80. Summarize what the diagram shows.

7.14 Change the confidence. Consider the setting of the previous exercise. Suppose that the sample mean is still 50, the sample size is 30, and the population standard deviation is 7. Make a diagram similar to Figure 7.6 (page 349) that illustrates the effect of the confidence level on the width of the interval. Use 80%, 90%, 95%, and 99%. Summarize what the diagram shows.

7.15 Populations sampled and margins of error. Consider the following two scenarios. (A) Take a simple random sample of 100 sophomore students at your college or university. (B) Take a simple random sample of 100 sophomore students in your major at your college or university. For each of these samples, you will record the amount spent on textbooks used for classes during the fall semester. Which sample should have the smaller margin of error for 95% confidence? Explain your answer.

7.16 Reporting margins of error. An Associated Press article of June 19, 2018, reported Commerce Department estimates of changes in the construction industry:

A surge of construction in the Midwest drove U.S. housing starts up 5 percent in May from the prior month. The Commerce Department said Wednesday that housing starts rose to a seasonally adjusted annual rate of 1.35 million, the strongest pace since July 2007.

If we turn to the original Commerce Department report (released on June 19, 2018), we read:

Privately-owned housing starts in May were at a seasonally adjusted annual rate of 1,350,000. This is 5.0 percent (±10.2 percent) above the revised April estimate of 1,286,000.

(a) The reported 10.2% is the margin of error based on a 90% level of confidence. Given that fact, what is the 90% confidence interval for the percent change in housing starts from April to May?

(b) Explain why a credible media report should state: "The Commerce Department has no evidence that privately-owned housing starts rose or fell in May from the previous month."

7.17 Confidence interval mistakes and misunderstandings. Suppose that 500 randomly selected alumni of the University of Okoboji were asked to rate the university's academic advising services on a 1 to 10 scale. The sample mean $\bar{x}$ was found to be 8.6. Assume that the population standard deviation is known to be $\sigma = 2.2$.

(a) Ima Bitlost computes the 95% confidence interval for the average satisfaction score as $8.6 \pm 1.96(2.2)$. What is her mistake?

(b) After correcting her mistake in part (a), she states, "I am 95% confident that the sample mean falls between 8.4 and 8.8." What is wrong with this statement?

(c) She quickly realizes her mistake in part (b) and instead states, "The probability that the true mean is between 8.4 and 8.8 is 0.95." What misinterpretation is she making now?

(d) Finally, in her defense for using the Normal distribution to determine the confidence interval, she says, "Because the sample size is quite large, the population of alumni ratings will be approximately Normal." Explain to Ima her misunderstanding and correct this statement.

7.18 More confidence interval mistakes and misunderstandings. Suppose that 100 randomly selected members of the Karaoke Channel were asked how much time they typically spend on the site during the week.[5] The sample mean $\bar{x}$ was found to be 3.8 hours. Assume that the population standard deviation is known to be $\sigma = 2.9$.

(a) Cary Oakey computes the 95% confidence interval for the average time on the site as $3.8 \pm 1.96(2.9/100)$. What is his mistake?

(b) He corrects this mistake and then states that "95% of the members spend between 3.23 and 4.37 hours a week on the site." What is wrong with his interpretation of this interval?

(c) The margin of error is slightly larger than half an hour. To reduce this to roughly 15 minutes, Cary says that the sample size needs to be doubled to 200. What is wrong with this statement?

7.19 In the extremes. As suggested in our discussions, 90%, 95%, and 99% are probably the most common confidence levels chosen in practice.

(a) In general, what would be a 100% confidence interval for the mean μ? Explain why such an interval is of no practical use.

(b) What would be a 0% confidence interval? Explain why it makes sense that the resulting interval provides you with 0% confidence.

7.20 Average starting salary. The University of Texas at Austin McCombs School of Business performs and reports an annual survey of starting salaries for recent bachelor's in business administration graduates.[6] For 2017, there were a total of 598 respondents.

(a) Respondents who were finance majors were 41.42% of the total responses. Rounding to the nearest integer, what is n for the finance major sample?

(b) For the sample of finance majors, the average salary is $68,145 with a standard deviation of $13,489. What is the 90% confidence interval for average starting salaries for finance majors?

7.21 Survey response and margin of error. Suppose that a business conducts a marketing survey. As is often done, the survey is conducted by telephone. As it turns out, the business was only able to elicit responses from less than 10% of the randomly chosen customers. The low response rate is attributable to many factors, including caller ID screening. Undaunted, the marketing manager was pleased with the sample results because the margin of error was quite small, and thus the manager felt that the business had a good sense of the customers' perceptions on various issues. Do you think the small margin of error is a good measure of the accuracy of the survey's results? Explain.

7.22 Fuel efficiency. Computers in some vehicles calculate various quantities related to performance. One of these is the fuel efficiency, or gas mileage, usually expressed as miles per gallon (mpg). For one vehicle equipped in this way, the car was set to 60 miles per hour by cruise control, and the mpg were recorded at random times.[7] Here are the mpg values from the experiment:

📊 MILEAGE

37.2 21.0 17.4 24.9 27.0 36.9 38.8 35.3 32.3 23.9
19.0 26.1 25.8 41.4 34.4 32.5 25.3 26.5 28.2 22.1

Suppose that the standard deviation of the population of mpg readings of this vehicle is known to be $\sigma = 6.5$ mpg.

(a) What is $\sigma_{\bar{x}}$, the standard deviation of $\bar{x}$?

(b) Based on a 95% confidence level, what is the margin of error for the mean estimate?

(c) Given the margin of error computed in part (b), give a 95% confidence interval for μ, the mean highway mpg for this vehicle. The vehicle sticker information for the vehicle states a highway average of 27 mpg. Are the results of this experiment consistent with the vehicle sticker?

7.23 Fuel efficiency in metric units. In the previous exercise, you found an estimate with a margin of error for the average miles per gallon. Convert your estimate and margin of error to the metric units kilometers per liter (kpl). To change mpg to kpl, use the fact that 1 mile = 1.609 kilometers and 1 gallon = 3.785 liters.

7.24 Confidence intervals for average annual income. Based on a 2015 survey, the National Statistics Office of the Republic of the Philippines released a report on various estimates related to family income and expenditures in Philippine pesos. With respect to annual family income, we would find the following reported:[8]

	Estimate	Standard error	Lower	Upper
Average annual income	266,962	2,836	?	272,524

The "Lower" and "Upper" headers signify lower and upper confidence interval limits. As will be noted in Chapter 8, the "standard error" for estimating the mean is $s/\sqrt{n}$. But because the sample sizes of the national survey are large, $\sigma/\sqrt{n}$ is taken as $s/\sqrt{n}$.

(a) What is the value of the lower confidence limit?

(b) What is the value of the margin of error?

(c) Determine the level of confidence C used.

7.25 What is the cost? In Exercise 7.22, you found an estimate with a margin of error for the fuel efficiency expressed in miles per gallon. Suppose that fuel costs $3.80 per gallon. Find the estimate and margin of error for fuel efficiency in terms of miles per dollar. To convert miles per gallon to miles per dollar, divide miles per gallon by the cost in dollars per gallon.

7.26 More than one confidence interval. As we prepare to take a sample and compute a 95% confidence interval, we know that the probability that the interval we compute will cover the parameter is 0.95. That's the meaning of 95% confidence. If we plan to use several such intervals, however, our confidence that *all* of them will give correct results is less than 95%. Suppose that we plan to take independent samples each month for five months and report a 95% confidence interval for each set of data.

(a) What is the probability that all five intervals will cover the true means? This probability (expressed as a percent) is our overall confidence level for the five simultaneous statements.

(b) Suppose that we wish to have an overall confidence level of 95% for the five simultaneous statements. About what confidence level should we pick for the construction of the individual intervals?

7.27 State of well-being. The Gallup-Healthways Well-Being Index is a single metric on a 0 to 100 scale based on the five elements of purpose, social, financial, community, and physical health. In 2017, the estimate for the index on the national level is 61.5 points. The results were based on telephone interviews conducted January 2 to December 30, 2017, with a random sample of 160,498 U.S. adults. Material provided with the results of the poll noted:

For results based on the total sample of national adults, the margin of sampling error for the Well-Being Index score is 0.15 points at the 95% confidence level.[9]

The poll uses a complex multistage sample design, but the specialized sample estimator for the mean has approximately a Normal sampling distribution.

(a) The announced poll result was 61.5 ± 0.15 points. Can we be certain that the true population mean falls in this interval? Explain your answer.

(b) Explain to someone who knows no statistics what the announced result 61.5 ± 0.15 points means.

(c) This confidence interval has the same form we have met earlier:

$$\text{estimate} \pm z^* \sigma_{\text{estimate}}$$

What is the standard deviation σ_{estimate} of the estimated points?

(d) Does the announced margin of error include errors due to practical problems such as nonresponse? Explain your answer.

7.28 Finding z^* from software. On page 345, we gave the values of z^* for $C = 90\%$, 95%, and 99%. These values were obtained from Table D. In this exercise, you will use software to find these z^* values. Software is typically set up to give a z value from the standard Normal distribution with a specified probability to the *left* of it. For example, looking at Figure 7.4 (page 344) with $C = 95\%$, the probability to the left of z^* is 0.975.

- **Excel** *users:* Enter the values of 0.95, 0.975, and 0.995 into cells A1, A2, and A3, respectively. In cell B1, enter the formula = **NORM.S.INV(A1)**. Now, drag and copy the contents of cells B1 down to B3. You will find the z^* values.

- **JMP** *users:* Enter the values of 0.95, 0.975, and 0.995 into the first three cells of column 1. Right-click on the header of column 2 and choose **New Columns.** In the drag-down dialog box named **Column Properties,** pick the **Formula** option. Open the **Probability** formula group and click **Normal Quantile** so that it appears in the formula dialog region. In the empty box of the Normal Quantile function, click in column 1. Click OK twice to return to the data table to find the z^* values.

- **Minitab** *users:* Enter the values of 0.95, 0.975, and 0.995 into the first three cells of column c1 of a worksheet. Do the following selection: **Calc → Probability Distributions → Normal.** Choose the **Inverse cumulative probability** option. Type "c1" in the **Input column** dialog box. Click OK to find the z^* values reported in the Session window.

- **R** *users:* Type the following at the R prompt:
  ```
  p <- c(0.95, 0.975, 0.995)
  qnorm(p)
  ```

(a) Use the software of your choice to find values of z^* for $C = 90\%$, 95%, and 99%. Compare these software values with the Table D values noted on page 345.

(b) Use the software of your choice to find values of z^* for $C = 85\%$ and 97%.

7.29 Excel's margin of error function. Using Excel, we can employ the "CONFIDENCE.NORM()" function to find the margin of error. The function has three arguments. The first argument is a value for α, which is equal to $1 - C$. The second argument is the value of σ, and the third argument is the sample size n. Use this Excel function to find the margins of error computed in Example 7.3 (page 346) and Example 7.5 (page 348). Report your results to at least four decimal places.

7.30 Sample size determination. Refer to Example 6.6 (page 312) to find the standard deviation of duration times for shared Citi Bikes is $\sigma = 570.469$ seconds.

(a) Use the sample size formula (page 349) to determine what sample size you need to have a margin of error equal to two minutes with 90% confidence. Explain why you must always round up to the next higher whole number when using the formula for n.

(b) What minimum sample size do you need to have a margin of error no greater than one minute with 95% confidence?

7.2 Tests of Significance

When you complete this section, you will be able to:
- Explain the basic reasoning underlying statistical tests and outline the four steps common to all tests of significance.
- Perform a z test (one- or two-sided) for one population mean.
- Describe the relationship between a level α two-sided significance test for μ and the $1 - \alpha$ confidence interval.
- Explain the advantages of reporting the P-value versus the statistical signfiicance of a statistical test.

The confidence interval is appropriate when our goal is to estimate population parameters. The second common type of inference is directed at a quite different goal: to assess the evidence provided by the data in favor of some claim about the population parameters.

The reasoning of significance tests

A significance test is a formal procedure for comparing observed data with a hypothesis whose truth we want to assess. The hypothesis is a statement about the parameters in a population or model. The results of a test are expressed in terms of a probability that measures how well the data and the hypothesis agree. We use the following case study and subsequent examples to illustrate these ideas.

Fill the Bottles Perhaps one of the most common applications of hypothesis testing of the mean is the quality control problem of assessing whether the underlying population mean is on "target." Consider the case of Bestea Bottlers. One of Bestea's most popular products is the 16-ounce, or 473-milliliter (ml), bottle of sweetened green iced tea. Annual production at any of its given facilities is in the millions of bottles. There is some variation from bottle to bottle because the filling machinery is not perfectly precise. Bestea has two concerns: whether there is a problem of underfilling (customers are then being shortchanged, which is a form of false advertising) or whether there is a problem of overfilling (resulting in unnecessary cost to the bottler). ∎

Notice that in Case 7.2, there is an intimate understanding of what is important to be discovered. In particular, is the population mean too high or too low relative to a desired level? With an understanding of what role the data play in the discovery process, we are able to formulate appropriate hypotheses. If the bottler were concerned only about the possible underfilling of bottles, then the hypotheses of interest would change. Let us proceed with the question of whether the bottling process is either underfilling or overfilling bottles.

EXAMPLE 7.7

TEA1

Are the Bottles Being Filled as Advertised? The filling process is not new to Bestea. Data on past production show that the distribution of the contents is close to Normal, with standard deviation $\sigma = 2$ ml. To assess the state of the bottling process, 20 bottles were randomly selected from a facility's streaming high-volume production line. The sample mean content ($\bar{x}$) is found to be 474.54 ml. Is a sample mean of 474.54 ml convincing evidence that the mean fill of all bottles produced by the current process differs from the desired level of 473 ml?

Without proper statistical thinking, we might knee jerk one of two possible conclusions:

• "The mean of the bottles sampled is not 473 ml so the process is not filling the bottles at a mean level of 473 ml."

• "The difference of 1.54 ml is small relative to the 473 ml baseline so there is nothing unusual going on here."

Both responses fail to consider the underlying variability of the population, which ultimately implies a failure to consider the sampling variability of the mean statistic.

So, what is the conclusion? One way to answer this question is to compute the probability of observing a sample mean at least as far from 473 ml as 1.54 ml, *assuming*, in fact, the underlying process mean is equal to 473 ml. Taking into account sampling variability, the answer is 0.00058. (You learn how to find this probability in Example 7.12.) Because this probability is so small, we are led to the conclusion that the underlying bottling process does not have a mean of $\mu = 473$ ml. The estimated average overfilling amount of 1.54 ml per bottle may seem fairly inconsequential. But, when it is put in the context of the high-volume production bottling environment and the potential cumulative waste across many bottles, correcting the potential overfilling is of great *practical* importance. ∎

What are the key steps in the above example?

• We started with a question about the underlying mean of the current filling process. We then asked whether the data from the process are compatible with a mean fill of 473 ml.

- Next, we compared the mean given by the data, $\bar{x} = 474.54$ ml, with the value assumed in the question, 473 ml.
- The result of the comparison takes the form of a probability, 0.00058.

The probability is quite small. Something that happens with probability 0.00058 occurs only about 6 times out of 10,000. In this case, we have two possible explanations:

1. The underlying process mean equals 473 ml but we just happened to have observed something that is very unusual.

2. The assumption that underlies the calculation (the underlying process mean equals 473 ml) is not true.

Because this probability is so small, we prefer the second conclusion: the process mean is not 473 ml. It should be emphasized that to "conclude" does not mean we know the truth or that we are right. There is always a chance that *our conclusion is wrong*. Always bear in mind that when dealing with data, there are no guarantees. We now turn to an example in which the data suggest a different conclusion.

EXAMPLE 7.8

TEA2

Is It Right Now? In Example 7.7, sample evidence suggested that the mean fill amount was not at the desired target of 473 ml. In particular, it appeared that the process was overfilling the bottles on average. In response, Bestea's production staff made adjustments to the facility's process and collected a sample of 20 bottles from the "corrected" process. For this sample, we find $\bar{x} = 472.56$ ml. In this case, the sample mean is less than 473 ml—to be exact, 0.44 ml less than 473 ml.

Did the production staff overreact and adjust the mean level too low? We need to ask a similar question as in Example 7.7. In particular, what is the probability that the mean of a sample of size $n = 20$ from a Normal population with mean $\mu = 473$ and standard deviation $\sigma = 2$ is as far away or farther away from 473 ml as 0.44 ml? The answer is about 0.327. A sample result this far from 473 ml would happen just by chance in 32.7% of samples from a population having a true mean of 473 ml. An outcome that could so easily happen just by chance is not convincing evidence that the population mean differs from 473 ml.

At this moment, Bestea does not have strong evidence to further tamper with the process settings. But, with this said, no decision is static or necessarily correct. Considering that the cost of underfilling in terms of disgruntled customers is potentially greater than the waste cost of overfilling, Bestea personnel might be well served to gather more data if there is any suspicion that the process mean fill amount is too low. In Section 8.3, we discuss sample size considerations for detecting departures from the null hypothesis that are considered important given a specified probability of detection. ■

The probabilities in Examples 7.7 and 7.8 are measures of the compatibility of the data (sample means of 474.54 and 472.56) with the *null hypothesis* that $\mu = 473$. Figure 7.7 compares the two results graphically. The Normal curve is the sampling distribution of $\bar{x}$ when $\mu = 473$. You can see that we are not particularly surprised to observe $\bar{x} = 472.56$, which is a departure of 0.44 from 473. However, $\bar{x} = 474.54$, which is a departure of 1.54 from 473, is clearly an unusual data result. Equally so, $\bar{x} = 471.46 = 473 - 1.54$ would be viewed as an unusual data result.

FIGURE 7.7 The mean fill amount for a sample of 20 bottles will have this sampling distribution if the mean for all bottles is $\mu = 473$ ml. A sample mean $\bar{x} = 474.54$ ml is so far out on the curve that it would rarely happen just by chance.

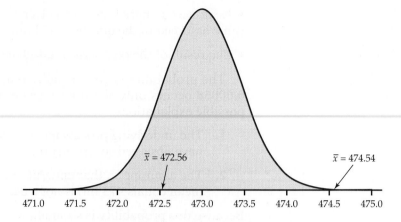

Herein lies the core reasoning of statistical tests: *a data result that is extreme if a hypothesis were true is evidence that the hypothesis may not be true.* In the following four subsections, we delineate the four common steps of all tests of significance.

Step 1: Stating the hypotheses

In Examples 7.7 and 7.8, we asked whether the fill data are plausible if, in fact, the true mean fill amount for all bottles (μ) is 473 ml. That is, we ask if the data provide evidence *against* the claim that the population mean is 473. The first step in a test of significance is to state a claim that we will try to find evidence *against*.

> **NULL HYPOTHESIS H_0**
>
> The **null hypothesis** is the statement of claim that is assumed initially true. The test of significance is designed to assess the strength of the evidence against the null hypothesis. We abbreviate "null hypothesis" as H_0.

A null hypothesis is a statement about the one or more population or process parameters. For example, the null hypothesis for Examples 7.7 and 7.8 is

$$H_0: \mu = 473$$

Here the statement is that the process mean (μ) of all filled bottles, including those we do not have data on, has a specific value of 473. In the context of the bottle-filling operation, this null hypothesis of $\mu = 473$ can be thought of as the null hypothesis of the "status quo"; that is, the manufacturing process is operating normally with its mean at the desired target.

In many applications, the null hypothesis is a statement of "no effect" or "no difference." For example, suppose you want to address the question of whether the average starting salaries of finance majors and accounting majors are the same. Suppose that μ_1 represents the mean starting salary of the population of all finance majors and μ_2 represents the mean starting salary of the population of all accounting majors. In this scenario, the null hypothesis is a statement of "no difference" and would be written as:

$$H_0: \mu_1 - \mu_2 = 0$$

alternative hypothesis

It is convenient also to give a name to the statement that we hope or suspect is true instead of H_0. This is called the **alternative hypothesis** and is abbreviated as H_a. The alternative hypothesis can be thought of as a statement contradictory to H_0.

EXAMPLE 7.9

Bottle Fill Amount: The Hypotheses As noted in the setting of Case 7.2 (page 356), the manufacturer is concerned with the problem of underfilling and overfilling. As such, the alternative hypothesis is that the mean fill amount is not 473. The competing hypotheses are then given as follows:

$$H_0: \mu = 473$$
$$H_a: \mu \neq 473$$

The estimate of μ is the sample mean $\bar{x}$. Values of $\bar{x}$ far from 473 on either the low or the high side count as evidence against the null hypothesis. ∎

Hypotheses are always set up in relationship to some population, process, or model. For this reason, we always state H_0 and H_a in terms of parameters. Because H_a expresses the effect that we hope to find evidence *for*, we will sometimes begin with H_a and then set up H_0 as the statement that the hoped-for effect is not present. Stating H_a, however, is often the more difficult task. It is not always clear, in particular, whether H_a should be **one-sided** or **two-sided,** which refers to whether a parameter differs from its null hypothesis value in a specific direction or in either direction.

one-sided or two-sided alternatives

In Example 7.9, the choice of the alternative is clear. Namely, the alternative $H_a: \mu \neq 473$ in the bottle-filling example is two-sided because we are looking to see if there is evidence that the mean fill amount is off target, either below or above target. Here, the alternative is not a good situation in the sense that the process mean is off target. Thus, it is not our *hope* that the alternative is true. However, it is our hope that we can detect when the process has gone off target so that corrective actions can be taken. Here is a setting in which a one-sided alternative is appropriate.

EXAMPLE 7.10

Have We Reduced Processing Time? Your company hopes to reduce the mean time μ required to process customer orders. At present, this mean is 3.8 days. You study the process and eliminate some unnecessary steps. Did you succeed in decreasing the average processing time? You hope to show that the mean is now less than 3.8 days, so the alternative hypothesis is one-sided, $H_a: \mu < 3.8$. The null hypothesis is the "no-change" value, $H_0: \mu = 3.8$ days. ∎

The alternative hypothesis should express the hopes or suspicions that we bring to the data. *It is cheating to first look at the data and then frame H_a to fit what the data show.* If you do not have a specific direction firmly in mind in advance, you must use a two-sided alternative. Moreover, some users of statistics argue that we should always use a two-sided alternative.

The choice of the hypotheses in Example 7.10 as

$$H_0: \mu = 3.8$$
$$H_a: \mu < 3.8$$

deserves a final comment. We do not expect that the elimination of steps in order processing would actually increase the processing time. However, we can allow for an increase by including this case in the null hypothesis. Then we would write

$$H_0: \mu \geq 3.8$$
$$H_a: \mu < 3.8$$

This statement is logically satisfying because the hypotheses account for all possible values of μ. However, only the parameter value in H_0 that is closest to H_a influences the form of the test in all common significance-testing situations. Think of it this way: if the data lead us away from $\mu = 3.8$ and toward believing that $\mu < 3.8$, then the data would certainly lead us away from believing that $\mu > 3.8$ because this involves values of μ that are in the opposite direction to that which the data are pointing. Moving forward, we take H_0 to be the simpler statement that the parameter *equals* a specific value, in this case $H_0: \mu = 3.8$.

APPLY YOUR KNOWLEDGE

7.31 Customer feedback. Feedback from your customers shows that many think it takes too long to fill out the online order form for your products. You redesign the form and plan a survey of customers to determine whether they think that the new form is actually an improvement. Sampled customers will respond using a 5-point scale: −2 if the new form takes much less time than the old form; −1 if the new form takes a little less time; 0 if the new form takes about the same time; +1 if the new form takes a little more time; and +2 if the new form takes much more time. The mean response from the sample is $\bar{x}$, and the mean response for all of your customers is μ. State null and alternative hypotheses that provide a framework for examining whether the new form is an improvement.

7.32 Laboratory quality control. Hospital laboratories routinely check their diagnostic equipment to ensure that patient lab test results are accurate. To check if the equipment is well calibrated, lab technicians make several measurements on a control substance known to have a certain quantity of the chemistry being measured. Suppose that a vial of controlled material has 4.1 nanomoles per L (mmol/L) of potassium. The technician runs the lab equipment on the control material 10 times and compares the sample mean reading $\bar{x}$ with the target mean μ using a significance test. State the null and alternative hypotheses for this test.

Step 2: Calculating the value of a test statistic

We learn the form of significance tests in a number of common situations. Here are some principles that apply to most tests and help in understanding the form of tests:

- The test is based on a statistic (estimator) that estimates the parameter that appears in the hypotheses. Usually, this is the same estimate we would use in a confidence interval for the parameter. When H_0 is true, we expect the estimate to take a value near the parameter value specified by H_0. We call this specified value the hypothesized value.

- Values of the estimate far from the hypothesized value give evidence against H_0. The alternative hypothesis determines which directions count against H_0.

- To assess how far the estimate is from the hypothesized value, standardize the estimate. In many common situations, the **test statistic** has the form

$$\frac{\text{estimate} - \text{hypothesized value}}{\text{standard deviation of the estimate}}$$

The language of the denominator above is a bit informal. More formally, the denominator represents the standard deviation of the statistic as given by its sampling distribution. For example, if the sample mean is used as the estimate, the denominator would be the standard deviation of the sample mean, which is $\sigma/\sqrt{n}$.

Many test statistics will have this form, but not all. All test statistics, however, regardless of the form, serve a similar purpose. Namely, a test statistic measures how compatible the observed data are with the null hypothesis. Furthermore, due to the fact that the statistic used in the calculation of a test statistic is a random variable, it should be recognized that a test statistic itself is a *random variable*. We use the probability distribution of the test statistic for making a probability calculation that we need for our test of significance.

Let's return to our bottle-filling example and calculate the test statistic.

EXAMPLE 7.11

Bottle Fill Amount: The Test Statistic For Example 7.9 (page 359), the null hypothesis is $H_0: \mu = 473$. In Example 7.7 (page 356), the observations from an SRS of size $n = 20$ from a population of bottles with $\sigma = 2$. The observed average fill amount is $\bar{x} = 474.54$. The test statistic, which we call z for this problem, is the standardized version of $\bar{x}$:

$$z = \frac{\bar{x} - \mu}{\sigma/\sqrt{n}}$$

This z statistic is the distance between the sample mean and the hypothesized population mean in the standard scale of z-scores. In this example,

$$z = \frac{474.54 - 473}{2/\sqrt{20}} = 3.44 \ \blacksquare$$

As stated in Example 7.7, past production shows that the fill amounts of the individual bottles are not too far from the Normal distribution. In that light, we can be confident enough that, with a sample size of 20, the distribution of the sample $\bar{x}$ is close enough to the Normal for working purposes. In turn, the standardized test statistic z will have approximately the $N(0,1)$ distribution. Even without a formal probability calculation, by simply recalling the 68–95–99.7 rule for the Normal, we realize that a z-score of 3.44 is an unusual value in that it falls outside the range of where 99.7% of the standard Normal distribution lies. This suggests incompatibility of the observed sample result with the null hypothesis. We now use the Normal distribution to arrive at a formal probability statement.

68–95–99.7 rule, p. 45

APPLY YOUR KNOWLEDGE

7.33 Departure from null hypothesis and z statistic. Suppose the null hypothesis is $H_0: \mu = 70$ and $\sigma = 5$. For parts (a)–(e) below, assume that the sample size is 36.

(a) Suppose $\bar{x} = 72$. What is the value of the z statistic?

(b) Suppose $\bar{x} = 74$. What is the value of the z statistic?

(c) Suppose $\bar{x} = 68$. What is the value of the z statistic?

(d) Suppose $\bar{x} = 66$. What is the value of the z statistic?

(e) In terms of direction and amount, explain the relationship between the departure of $\bar{x}$ from the null hypothesis value of μ and the z statistic value.

7.34 Sample size and z statistic. Suppose the null hypothesis is $H_0: \mu = 50$ and $\sigma = 20$.

(a) Suppose $\bar{x} = 60$. What is the value of the z statistic if $n = 15$?

(b) Suppose $\bar{x} = 60$. What is the value of the z statistic if $n = 20$?

(c) Suppose $\bar{x} = 40$. What is the value of the z statistic if $n = 15$?

(d) Suppose $\bar{x} = 40$. What is the value of the z statistic if $n = 20$?

(e) For a given departure of $\bar{x}$ from the null hypothesis value of μ, explain the effect of changing n on the z statistic value.

Step 3: Finding the *P*-value

If all test statistics were Normal, we could base our conclusions on the value of the z test statistic. In fact, the Supreme Court of the United States has said that "two or three standard deviations" ($z = 2$ or 3) is its criterion for rejecting H_0 (see Exercise 7.39 on page 366), and this is the criterion used in most applications involving the law. But because not all test statistics are Normal, we use the language of probability to express the meaning of a test statistic.

A test of significance finds the probability of getting an outcome *as extreme or more extreme than the actually observed outcome*. "Extreme" means "far from what we would expect if H_0 were true." The direction or directions that count as "far from what we would expect" are determined by the competing hypothesis.

P-VALUE

The probability, computed assuming that H_0 is true, that the test statistic would take a value as extreme or more extreme than that actually observed is called the ***P*-value** of the test. The smaller the *P*-value, the stronger the evidence against H_0 provided by the data.

The key to calculating the *P*-value is the sampling distribution of the test statistic. For the problems we consider in this chapter, we need only the standard Normal distribution for the test statistic z.

EXAMPLE 7.12

CASE 7.2 **Bottle Fill Amount: The *P*-Value** In Example 7.11 (page 361), we found that the test statistic for testing $H_0: \mu = 473$ versus $H_a: \mu \neq 473$ is $z = 3.44$. If H_0 is true, then z is a single observation from the standard Normal, $N(0,1)$, distribution. Since H_a is a two-sided alternative, values of z away from zero *in either direction* count against the null hypothesis. The *P*-value is the probability of observing a value of Z at least as extreme as the one that we observed, $z = 3.44$. This implies that we must consider departures of at least 3.44 standard deviations below and above zero. Figure 7.8

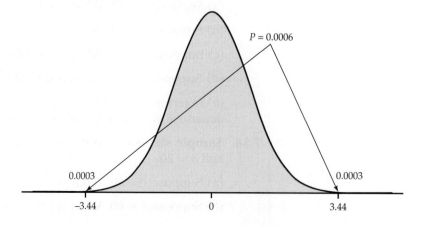

FIGURE 7.8 The *P*-value for Example 7.12. The two-sided *P*-value is the probability (when H_0 is true) that $\bar{x}$ takes a value as extreme or more extreme than the actual observed value, $z = 3.44$. Because the alternative hypothesis is two-sided, we use both tails of the distribution.

illustrates this calculation. From Table A, our table of standard Normal probabilities, we find

$$P(Z \geq 3.44) = 1 - 0.9997 = 0.0003$$

Because the Normal distribution is symmetric, the probability for being extreme in the negative direction is the same:

$$P(Z \leq -3.44) = 0.0003$$

So the P-value is

$$P(Z \leq -3.44 \text{ or } Z \geq 3.44) = P(Z \leq -3.44) + P(Z \geq 3.44)$$
$$= 2P(Z \geq 3.44)$$
$$= 2(0.0003) = 0.0006$$

In Example 7.7, we reported that a probability of 0.00058 was obtained from software. The value of 0.0006 found from the tables is essentially the same. ∎

It is worth emphasizing that we would have arrived at the same P-value of 0.0006 if we observed $z = -3.44$. For a two-sided alternative, it is the absolute value $|z|$ that matters, not whether z is positive or negative. As will be illustrated in Example 7.14, for a one-sided alternative, the P-value calculation is based on only one of the two tails of the standard Normal distribution.

When using tables, we may not always be able to report the exact P-value. Suppose that for Example 7.12, the test statistic value was 4.5. The P-value is then $2P(Z \geq 4.5)$. In Table A, the largest entry we have is $z = 3.49$ with $P(Z \leq 3.49) = 0.9998$. So, $P(Z > 3.49) = 1 - 0.9998 = 0.0002$. Therefore, we can only conclude that $P(Z \geq 4.5) < 0.0002$, which implies that $2P(Z \geq 4.5) < 0.0004$. In the end, we would report the P-value as $P < 0.0004$. It is also important to be aware that software vary in their reporting of P-values. Here are some examples of how software report in *default* mode:

- Excel and R attempt to report P-values as exact as possible. For very small P-values, both use scientific (E) notation. For example, Excel reports $2P(Z \geq 4.5)$ as 6.79535E-06, which is shorthand for 0.00000679535.

- JMP and SAS report P-values rounded to the ten-thousandths place so P-values less than 0.0001 are reported as $< .0001$. Thus, both JMP and SAS report $2P(Z \geq 4.5)$, as $< .0001$.

- Minitab reports P-values rounded to the thousandths place. This means the P-value of 0.0006 of Example 7.12 would be reported as 0.001. If, however, the P-value is less than 0.0005, Minitab will report the P-value as 0.000.

All these software platforms allow the user to format the reported P-values to more or fewer decimal places than given in default mode.

The calculation of a P-value in Example 7.12 provides an opportunity to make an important caution about one of the most common misinterpretations of P-values. *A P-value does not measure the probability that the null hypothesis is true.* For example, the P-value of 0.0006 is not saying that the probability is 0.0006 that the null hypothesis of H_0: $\mu = 473$ is true, nor does it say that the probability is 0.9994 that the alternative hypothesis of H_a: $\mu \neq 473$ is true. It is, however, correct to interpret a P-value as a statistical summary of the compatibility of the observed data with the specified null hypothesis. For Example 7.12, the P-value of 0.0006 is the probability of observing a test statistic (z) as large or larger than 3.44 (in magnitude) under the assumption that the null hypothesis of $\mu = 473$ is true. This small P-value is reflecting an incompatibility between the observed data and the null hypothesis and, as such, provides evidence against the null hypothesis.

APPLY YOUR KNOWLEDGE

7.35 Spending on housing. The Census Bureau reports that households spend an average of 31% of their total spending on housing. A homebuilders association in Cleveland wonders if the national finding applies in its area. It interviews a sample of 40 households in the Cleveland metropolitan area to learn what percent of their spending goes toward housing. Take μ to be the mean percent of spending devoted to housing among all Cleveland households. We want to test the hypotheses

$$H_0: \mu = 31\%$$
$$H_a: \mu \neq 31\%$$

The population standard deviation is $\sigma = 9.6\%$.

(a) The study finds $\bar{x} = 28.6\%$ for the 40 households in the sample. What is the value of the test statistic z? Sketch a standard Normal curve, and mark z on the axis. Shade the area under the curve that represents the P-value.

(b) Calculate the P-value. Are you convinced that Cleveland differs from the national average?

Step 4: Stating a conclusion

We started our discussion of significance tests with the statement of null and alternative hypotheses. We then learned that a test statistic is the tool used to examine the compatibility of the observed data with the null hypothesis. Finally, we translated the test statistic into a P-value to quantify the evidence against H_0. One important final step is needed: to state our conclusion.

It is common to compare the P-value we calculated with a fixed value that we regard as decisive. This amounts to announcing in advance how much evidence against H_0 we will require to reject H_0. The decisive value is called the **significance level.** It is commonly denoted by α (the Greek letter alpha). If we choose $\alpha = 0.05$, we are requiring that the data give evidence against H_0 so strong that it would happen no more than 5% of the time (1 time in 20) when H_0 is true. If we choose $\alpha = 0.01$, we are insisting on stronger evidence against H_0, evidence so strong that it would appear only 1% of the time (1 time in 100) if H_0 is, in fact, true.

signicance level

> **STATISTICAL SIGNIFICANCE**
>
> If the P-value is as small or smaller than α, we say that the test result is **statistically significant at level α.**

"Significant" in the statistical sense does not mean "important." The original meaning of the word is "signifying something." In statistics, the term is used to indicate only that the evidence against the null hypothesis has reached the standard set by α. A P-value is more informative than a statement simply declaring the test result as being "statistically significant" or as being "not statistically significant." The P-value allows us to go beyond a statement of significance (or not) at some fixed level α to assess significance at any level we choose. For example, a result with $P = 0.03$ is significant at the $\alpha = 0.05$ level but is not significant at the $\alpha = 0.01$ level. We discuss this in more detail at the end of this section.

EXAMPLE 7.13

Bottle Fill Amount: The Conclusion In Example 7.12 (page 362), we found that the P-value is

$$P = 2P(Z \geq 3.44) = 2(0.0003) = 0.0006$$

If the underlying process mean is truly 473 ml, there is only a 6 in 10,000 chance of observing a sample mean deviating as extreme as 1.54 ml (in either direction) away from this hypothesized mean. Because this P-value is *smaller* than the $\alpha = 0.05$ significance level, we conclude that our test result is significant. We could report the result as "the data show strong evidence that the underlying process mean filling amount is not at the desired value of 473 ml ($z = 3.44, P = 0.0006$)." ∎

When drawing any conclusion from a statistical test, be careful not to assume absolute certainty about your conclusion. In Example 7.13, the statement of having "strong evidence that the underlying mean is not at the desired value of 473 ml" does not imply that we know for sure the mean is not 473 ml. When using an $\alpha = 0.05$ significance level, the probability is 0.05 of rejecting H_0 when it is true. As will be discussed in Section 8.3, this mistaken conclusion is referred to as a **Type I error** and, in general, α is the probability of committing this error. Thus, there is a chance to get an extreme ("unlucky") result even when the null hypothesis is true. When faced with an extreme result, a more plausible explanation is that the alternative hypothesis is true as opposed to the null hypothesis is true and we just happened to have gotten an unlucky result. But, we still cannot be certain! In the scientific community, it is considered good practice to test additional samples from the population in question with the same methods to provide supporting or contradictory evidence with regard to the initial findings. This is known as performing **replication studies.**

Examples 7.9 through 7.13 in sequence showed us that a test of significance is a systematic set of steps for assessing the significance of the evidence provided by the data against a null hypothesis. These steps provide the general template for *all* tests of significance. Here is a general summary of the four common steps we covered.

TESTS OF SIGNIFICANCE: FOUR COMMON STEPS

1. State the *null hypothesis* H_0 and the *alternative hypothesis* H_a. The test is designed to assess the strength of the evidence against H_0; H_a is the statement that we accept if the evidence enables us to reject H_0.

2. Calculate the value of the *test statistic* on which the test will be based. This statistic measures how compatible the observed data are with the null hypothesis H_0.

3. Find the *P-value* for the observed test statistic. This is the probability, calculated assuming that H_0 is true, that the test statistic will weigh against H_0 at least as strongly as it does for these data.

4. State a conclusion. This is done by choosing a *significance level* α, how much evidence against H_0 you regard as decisive. If the P-value is less than or equal to α, you reject the null hypothesis in favor of the alternative hypothesis. If the P-value is greater than α, you conclude that the data do not provide sufficient evidence to reject the null hypothesis. Your conclusion is a sentence or two that summarizes what you have found by using a test of significance.

We will learn the details of many tests of significance in the following chapters. The proper test statistic is determined by the hypotheses and the data collection design. We use computer software or a calculator to find its numerical value and the P-value. The computer will not formulate your hypotheses for you, however. Nor will it decide if significance testing is appropriate or help you to interpret the P-value that it presents to you. These steps require judgment based on a sound understanding of this type of inference.

APPLY YOUR KNOWLEDGE

7.36 A new supplier. A new supplier offers a good price on a catalyst used in your production process. You compare the purity of this catalyst with that from your current supplier. The P-value for a test of "no difference" is 0.43. Can you be confident that the purity of the new product is the same as the purity of the product that you have been using? Discuss.

7.37 P-value and significance level. The P-value for a significance test is 0.034.

(a) Do you reject the null hypothesis at level $\alpha = 0.05$?

(b) Do you reject the null hypothesis at level $\alpha = 0.01$?

(c) Explain how you determined your answers in parts (a) and (b).

7.38 More on P-value and significance level. The P-value for a significance test is 0.067.

(a) Do you reject the null hypothesis at level $\alpha = 0.05$?

(b) Do you reject the null hypothesis at level $\alpha = 0.01$?

(c) Explain how you determined your answers in parts (a) and (b).

7.39 The Supreme Court speaks. Court cases in such areas as employment discrimination often involve statistical evidence. The Supreme Court has said that z-scores beyond $z^* = 2$ or 3 are generally convincing statistical evidence. For a two-sided test, what significance level corresponds to $z^* = 2$? To $z^* = 3$?

Summary of the z test for one population mean

We have noted the four steps common to *all* tests of significance. Using the bottle-filling scenario of Case 7.2, we have introduced you to these steps in the context of testing one population mean when the population standard deviation σ is known. Here is a summary of the test for one population mean.

z TEST FOR A POPULATION MEAN

To test the hypothesis $H_0: \mu = \mu_0$ based on an SRS of size n from a population with unknown mean μ and known standard deviation σ, compute the **test statistic**

$$z = \frac{\bar{x} - \mu_0}{\sigma/\sqrt{n}}$$

In terms of a standard Normal random variable Z, the P-value for a test of H_0 against

$H_a: \mu > \mu_0$ is $P(Z \geq z)$

$H_a: \mu < \mu_0$ is $P(Z \leq z)$

$H_a: \mu \neq \mu_0$ is $2P(Z \geq |z|)$

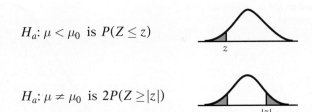

These *P*-values are exact if the population distribution is Normal and are approximately correct for large *n* in other cases. For a given significance level α, if the *P*-value is less than or equal to α, the null hypothesis is rejected in favor of the alternative hypothesis. If the *P*-value is greater than α, there is not sufficient evidence to reject the null hypothesis.

Similar to the Normal-based confidence interval methods (page 345), if the sample size is large and the sample standard deviation *s* is used in place of σ, then the Normal-based testing procedure provides sufficiently accurate *P*-values.

The bottle-filling scenario of Case 7.2 provided us with an example of performing a two-sided test of the population mean. Let's now consider a one-sided test.

EXAMPLE 7.14

Blood Pressures of Executives The medical director of a large company is concerned about the effects of stress on the company's younger executives. According to the National Center for Health Statistics, the mean systolic blood pressure for males 35 to 44 years of age is 128, and the standard deviation in this population is 15. The medical director examines the records of 72 executives in this age group and finds that their mean systolic blood pressure is $\bar{x} = 129.93$. Is this evidence that the mean blood pressure for all the company's young male executives is higher than the national average? To proceed, we make the assumption that the population standard deviation is known—in this case, that executives have the same $\sigma = 15$ as the general population.

Step 1: hypotheses. The hypotheses about the unknown mean μ of the executive population are

$$H_0: \mu = 128$$
$$H_a: \mu > 128$$

Step 2: test statistic. The *z* test requires that the 72 executives in the sample are an SRS from the population of the company's young male executives. We must ask how the data were produced. If records are available only for executives with recent medical problems, for example, the data are of little value for our purpose. It turns out that all executives are given a free annual medical exam and that the medical director selected 72 exam results at random. The *z* statistic is

$$z = \frac{\bar{x} - \mu_0}{\sigma/\sqrt{n}} = \frac{129.93 - 128}{15/\sqrt{72}}$$
$$= 1.09$$

Step 3: P-value. Draw a picture to help find the *P*-value. Figure 7.9 shows that the *P*-value is the probability that a standard Normal variable *Z* takes a value of 1.09 or greater. From Table A we find that this probability is

$$P = P(Z \geq 1.09) = 1 - 0.8621 = 0.1379$$

FIGURE 7.9 The P-value for the one-sided test, Example 7.14.

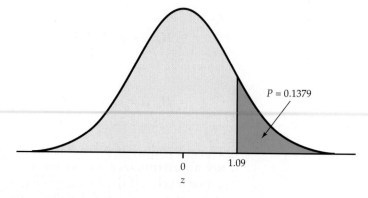

Step 4: conclusion. Given that a P-value of 0.1379 is fairly large, we report the result as "the data fail to provide evidence that would lead us to conclude that the mean blood pressure of the company's young male executives is higher than that of the general population of men of the same age group ($z = 1.09, P = 0.1379$)." ■

The reported statement does not imply that we conclude that the null hypothesis is true, only that the level of evidence we require to reject the null hypothesis is not met. Tests of significance assess *evidence* against H_0. If the evidence is strong, we can confidently reject H_0 in favor of the alternative. Failing to find evidence against H_0 means only that the data are reasonably consistent with H_0, not that we have clear evidence that H_0 is true. This requires a different test of significance that is discussed in the next chapter.

APPLY YOUR KNOWLEDGE

7.40 Testing a random number generator. Statistical software has a "random number generator" that is supposed to produce numbers uniformly distributed between 0 and 1. If this is true, the numbers generated come from a population with $\mu = 0.5$. A command to generate 100 random numbers gives outcomes with mean $\bar{x} = 0.531$ and $s = 0.294$. Because the sample is reasonably large, take the population standard deviation also to be $\sigma = 0.294$. Do we have evidence that the mean of all numbers produced by this software is not 0.5?

7.41 Computing the test statistic and P-value. You will perform a significance test of $H_0: \mu = 20$ based on an SRS of $n = 36$. Assume that $\sigma = 12$.

(a) If $\bar{x} = 25$, what is the test statistic z?

(b) What is the P-value if $H_a: \mu > 20$?

(c) What is the P-value if $H_a: \mu \neq 20$?

7.42 One-sided and two-sided P-values. The P-value for a two-sided z test is 0.09.

(a) State the P-values for the two one-sided tests.

(b) What additional information do you need to properly assign these P-values to the > and < (one-sided) alternatives?

Two-sided significance tests and confidence intervals

Recall the basic idea of a confidence interval, discussed in Section 7.1. We constructed an interval that would include the true value of μ with a specified probability C. Suppose that we use a 95% confidence interval ($C = 0.95$). Then the values of μ_0 that are not in our interval would seem to be incompatible

EXAMPLE 7.15

IPO Initial Returns The decision to go public is clearly one of the most significant decisions to be made by a privately owned company. Such a decision is typically driven by the company's desire to raise capital and expand its operations. The first sale of stock to the public by a private company is referred to as an initial public offering (IPO). One of the important measurables for an IPO is the first-day (initial) return, which is defined as:

$$\text{IPO initial return} = \frac{\text{first day closing price} - \text{offer price}}{\text{offer price}}$$

The first-day closing price represents what market investors are willing to pay for the company's shares. If the offer price is lower than the first-day closing price, the IPO is said to be underpriced. In terms of the IPO initial return, an underpriced IPO is associated with a positive initial return. Similarly, an overpriced IPO is associated with a negative initial return.

Numerous studies in the finance literature consistently report that IPOs, on average, are underpriced in the United States and in international markets. The underpricing phenomena represents a perplexing puzzle in finance circles because it seems to contradict the assumption of market efficiency. In a study of the Indian market, researchers gathered data on 464 IPOs and found the mean initial return to be 25.21%, and the standard deviation of the returns was found to be 68.65%.[10] A question that might be asked is if the Indian IPO initial returns are showing a mean return different than 0—that is, neither a tendency toward underpricing nor overpricing. This calls for a test of the hypotheses

$$H_0: \mu = 0$$
$$H_a: \mu \neq 0$$

We carry out the test twice, first with the usual significance test using $\alpha = 0.01$ and then with a 99% confidence interval. ∎

First, let's do the test. The mean of the sample is $\bar{x} = 25.21$. Given the large sample size of $n = 464$, it is fairly safe to use the reported standard deviation of 68.65% as σ. The test statistic is

$$z = \frac{\bar{x} - \mu_0}{\sigma/\sqrt{n}} = \frac{25.21 - 0}{68.65/\sqrt{464}} = 7.91$$

Because the alternative is two-sided, the P-value is

$$P = 2P(Z \geq |7.91|)$$

The largest value of z in Table A is 3.49. All we can say from Table A is that the P-value is less than $2P(Z \geq 3.49) = 2(1 - 0.9998) = 0.0004$. Software can be used to find a more accurate value of the P-value. However, because the P-value is clearly less than 0.01, we reject H_0. There is strong evidence to conclude that the mean initial return for the Indian IPO population is not 0.

For 99% confidence, Table D gives $z^* = 2.576$. The confidence interval is then

$$\bar{x} \pm z^* \frac{\sigma}{\sqrt{n}} = 25.21 \pm 2.576 \frac{68.65}{\sqrt{464}}$$
$$= 25.21 \pm 8.21$$
$$= (17.00, 33.42)$$

FIGURE 7.10 The link between two-sided significance tests and confidence intervals. For Example 7.15, values of μ falling outside a 99% confidence interval can be rejected at the 1% level; values falling inside the interval cannot be rejected. This holds for any significance level α and $(1 - \alpha)$ confidence interval.

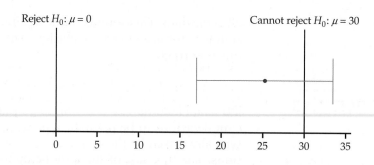

The hypothesized value $\mu_0 = 0$ falls well outside this confidence interval. In other words, it is in the region we are 99% confident μ is *not* in. Thus, we can reject

$$H_0: \mu = 0$$

at the 1% significance level. This is the same conclusion arrived upon by the significance test. However, we might want to test the Indian market against other markets. Initial mean returns vary by country, by industry sector, and by the size of companies. Suppose that we wish to test the Indian market against a market that has a μ value of 30. In Exercise 7.45 (page 372), you are asked to draw conclusions using a test of significance. Alternatively, because the value of 30 lies inside the 99% confidence interval for μ, we cannot reject

$$H_0: \mu = 30$$

Figure 7.10 illustrates both cases.

For the setting of Example 7.15, the identical conclusion of rejecting $H_0: \mu = 0$ by means of a 1% significance test and by means of a 99% confidence interval is not accidental. The calculations of the test statistic and the confidence interval both involve $\bar{x}$ and $\sigma/\sqrt{n}$. Furthermore, the fact that the P-value is less than 0.01 corresponds directly to the fact that $|z| > 2.576$. So, in the end, we have similar calculations, just different perspectives. In general, a two-sided test at significance level α can be carried out directly from a confidence interval with confidence level $C = 1 - \alpha$.

TWO-SIDED SIGNIFICANCE TESTS AND CONFIDENCE INTERVALS

A level α two-sided significance test that rejects a hypothesis $H_0: \mu = \mu_0$ exactly corresponds to the value μ_0 falling outside a level $1 - \alpha$ confidence interval for μ.

APPLY YOUR KNOWLEDGE

7.43 Does the confidence interval include μ_0? The P-value for a two-sided test of the null hypothesis $H_0: \mu = 70$ is 0.021.

(a) Does the 95% confidence interval include the value 70? Explain.

(b) Does the 99% confidence interval include the value 70? Explain.

7.44 Can you reject the null hypothesis? A 95% confidence interval for a population mean is (132,142).

(a) Can you reject the null hypothesis that $\mu = 147$ at the 5% significance level? Why?

(b) Can you reject the null hypothesis that $\mu = 134$ at the 5% significance level? Why?

Assessing significance with P-values versus critical values

We can find from Table A or the bottom of Table D that a value of $z^* = 1.96$ gives us a point on the standard Normal distribution such that 5% of the distribution is beyond ± 1.96. For a two-sided test, by comparing the observed z with the values of ± 1.96, we can develop a decision rule to reject or not reject H_0 at the 5% significance level. In particular, if $z \leq -1.96$ or $z \geq 1.96$, we would reject the null hypothesis for $\alpha = 0.05$. The range of values beyond ± 1.96 is referred to as the **rejection region.** As an example, suppose we are conducting a two-sided test and find the observed z to be 2.41. Since 2.41 falls beyond 1.96, we would reject the null hypothesis at the 5% significance level. Similarly, if we observed $z = -2.41$, we would reject the null hypothesis. Thus, a more compact form of giving the decision rule for a two-sided test at the 5% significance level is to reject H_0 if $|z| \geq 1.96$.

In the case of one-sided tests, the boundary between the do not reject and reject regions is the z-value associated with all of α in one tail (lower tail for < alternative, upper tail for > alternative). The use of a rejection region to implement the significance testing is typically called the **critical value approach.**

The critical value approach is straightforward: find the observed z, then compare it with a critical value associated with a fixed α, and draw a conclusion. Without the aid of a computer, assessing significance using the critical value approach is easier than the P-value approach because no probability calculation is required. You may ask yourself why wasn't this approach introduced earlier in place of the P-value approach?

Imagine again that we are conducting a two-sided test and our observed $z = 2.41$. Suppose that we use the critical value approach and then report our results as follows: "The data lead us to reject the null hypothesis at the 5% level of significance." Our statement is lacking. Beyond being significant at the 5% level of significance, the reader is left with no sense of the extent of the evidence. The statement also implicitly imposes upon the reader the 5% level of significance as his or her threshold, which may not be the case. Even if we were to report $z = 2.41$, this is still unsatisfactory in that we would be *forcing* the reader to find out for themselves the conclusion at another significance level (e.g., 1% level). Alternatively, we can compute the P-value

$$P = 2P(Z \geq |2.41|) = 0.016$$

Notice how much more informative and convenient for others it is if we report the following: "The data lead us to reject the null hypothesis ($z = 2.41, P = 0.016$) at the 5% level of significance." Namely, the reader of this statement sees that the result is significant at the $\alpha = 0.05$ level because $0.016 < 0.05$. But it is not significant at the $\alpha = 0.01$ level because the P-value is greater than 0.01. From Figure 7.11, we see that *the P-value is the smallest level α at which the test statistic is significant.* Knowing the P-value allows us, or anyone else, to assess significance at any level with ease. With this said, the P-value is not the "answer all" of a statistical study. As will be emphasized in

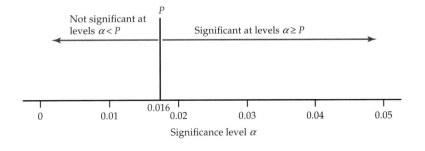

FIGURE 7.11 Link between the P-value and the significance level α. An outcome with P-value P is significant at all levels α at or above P and is not significant at smaller levels α.

APPLY YOUR KNOWLEDGE

7.45 IPO initial returns. Refer to Example 7.15 (page 369), where it was reported that the mean initial return is 25.21% and the standard deviation of the returns is 68.65% based on $n = 464$. Suppose that we wish to test $H_0: \mu = 30$ against $H_a: \mu \neq 30$.

(a) What is the value of the z statistic?

(b) For $\alpha = 0.01$, what is the rejection region for the z test?

(c) Based on the critical value approach and $\alpha = 0.01$, what do you conclude? Explain.

7.46 Blood pressures of executives. Refer to Example 7.14 (page 367), where the competing hypotheses are $H_0: \mu = 128$ versus $H_a: \mu > 128$.

(a) For $\alpha = 0.05$, what is the rejection region for the z test?

(b) In Example 7.14, the z-statistic was found to be 1.09. Based on the critical value approach and $\alpha = 0.05$, what do you conclude? Explain.

(c) What values of $\bar{x}$ correspond to rejection of the null hypothesis for $\alpha = 0.05$?

SECTION 7.2 SUMMARY

- **Tests of significance** have a common template of four steps: (1) stating the hypotheses; (2) calculating the value of a test statistic; (3) finding the P-value; and (4) stating a conclusion.

- A test of significance assesses the evidence provided by data against a **null hypothesis H_0** and in favor of an **alternative hypothesis H_a**.

- The hypotheses are stated in terms of population parameters. H_a says that a parameter differs from the parameter value of H_0 in a specific direction (**one-sided alternative**) or in either direction (**two-sided alternative**).

- The test is based on a **test statistic**. The **P-value** is the probability, computed assuming that H_0 is true, that the test statistic will take a value at least as extreme as that actually observed. Small P-values indicate strong evidence against H_0. Calculating P-values requires knowledge of the sampling distribution of the test statistic when H_0 is true.

- If the P-value is as small or smaller than a specified value α, the test result is **statistically significant** at significance level α.

- Significance tests for the hypothesis $H_0: \mu = \mu_0$ concerning the unknown mean μ of a population are based on the z **statistic**:

$$z = \frac{\bar{x} - \mu_0}{\sigma/\sqrt{n}}$$

The z test assumes an SRS of size n, a known population standard deviation σ, and either a Normal population or a large sample. P-values are computed from the Normal distribution (Table A).

- The **critical value approach** to significance testing rejects H_0 if the test statistic falls in a **rejection region.** For the z test, this approach uses **standard Normal critical values** such as those found in the z^* row of Table D.

SECTION 7.2 EXERCISES

For Exercises 7.31 and 7.32, see page 360; for 7.33 and 7.34, see pages 361–362; for 7.35, see page 364; for 7.36 to 7.39, see page 366; for 7.40 to 7.42, see page 368; for 7.43 and 7.44, see page 370; and for 7.45 and 7.46, see page 372.

7.47 What's wrong? Here are several situations in which there is an incorrect application of the ideas presented in this section. Write a short explanation of what is wrong in each situation and why it is wrong.

(a) A manager wants to test the null hypothesis that average weekly demand is not equal to 150 units.

(b) A random sample of size 25 is taken from a population that is assumed to have a standard deviation of 10. The standard deviation of the sample mean is $10/25 = 0.4$.

(c) A researcher tests the following null hypothesis: H_0: $\bar{x} = 17$.

7.48 What's wrong? Here are several situations in which there is an incorrect application of the ideas presented in this section. Write a short explanation of what is wrong in each situation and why it is wrong.

(a) A report says that the alternative hypothesis is rejected at level $\alpha = 0.01$ because the P-value is 0.003.

(b) A significance test rejected the null hypothesis that the sample mean is 90.

(c) A report on a study says that the results are statistically significant and the P-value is 0.83.

(d) The z statistic had a value of 0.024, and the null hypothesis was rejected at the 5% level because $0.024 < 0.05$.

(e) It is stated that the null hypothesis is not rejected because the large P-value of 2.31 is greater than 0.05.

7.49 What's wrong? Here are several situations in which there is an incorrect application of the ideas presented in this section. Write a short explanation of what is wrong in each situation and why it is wrong.

(a) The z statistic had a value of -2.3 for a two-sided test. The null hypothesis is not rejected for $\alpha = 0.05$ because $-2.3 < 1.96$.

(b) A two-sided test is conducted to test H_0: $\mu = 10$, and the observed sample mean is $\bar{x} = 19$. The null hypothesis is rejected because $19 \neq 10$.

(c) The z statistic had a value of 1.2 for a two-sided test. The P-value was calculated as $2P(Z \leq 1.2)$.

(d) The observed sample mean $\bar{x}$ is 5 for a sample size $n > 1$. The population standard deviation is 2. For testing the null hypothesis mean of μ_0, a z statistic of $(5 - \mu_0)/2$ is calculated.

7.50 Interpreting P-value. The reporting of P-values is standard practice in statistics. Unfortunately, misinterpretations of P-values by producers and readers of statistical reports are common. The previous two exercises dealt with a few incorrect applications of the P-value. This exercise explores the P-value a bit further.

(a) Suppose that the P-value is 0.028. Explain what is wrong with stating, "The probability that the null hypothesis is true is 0.028."

(b) Suppose that the P-value is 0.07. Explain what is wrong with stating, "The probability that the alternative hypothesis is true is 0.93."

(c) In terms of a probability language, a P-value is a conditional probability. Define the event D as "observing a test statistic as extreme or more extreme than actually observed." Consider two conditional probabilities: $P(H_0 \text{ is true} \mid D)$ versus $P(D \mid H_0 \text{ is true})$. Explain which of these two conditional probabilities represents a P-value.

7.51 Hypotheses. Each of the following situations requires a significance test about a population mean μ. State the appropriate null hypothesis H_0 and alternative hypothesis H_a in each case.

(a) David's car averages 28 miles per gallon on the highway. He now switches to a new motor oil that is advertised as increasing gas mileage. After driving 2500 highway miles with the new oil, he wants to determine if his gas mileage actually has increased.

(b) The diameter of a spindle in a small motor is supposed to be 4 millimeters. If the spindle is either too small or too large, the motor will not perform properly. The manufacturer measures the diameter in a sample of motors to determine whether the mean diameter has moved away from the target.

(c) Many studies have shown that many herbal supplement pills are not filled with the content that is advertised but rather have significant amounts of contaminants and fillers.[11] The percents of real produce versus filler vary by company. A consumer advocacy group randomly selects bottles and tests each pill for its percent of ginseng. The group is testing the pills to see if there is evidence that the percent of ginseng is less than 90%.

7.52 Hypotheses. In each of the following situations, a significance test for a population mean μ is called for. State the null hypothesis H_0 and the alternative hypothesis H_a in each case.

(a) A university gives credit in French language courses to students who pass a placement test. The language department wants to know if students who get credit in this way differ in their understanding of spoken French from students who actually take the French courses. Experience has shown that the mean score of students in the courses on a standard listening test is 26. The language department gives the same listening test to a sample of 35 students who passed the credit examination to see if their performance is different.

(b) Experiments on learning in animals sometimes measure how long it takes a mouse to find its way through a maze. The mean time is 22 seconds for one particular maze. A researcher thinks that a loud noise will cause the mice to complete the maze faster. She measures how long each of 12 mice takes with a noise as stimulus.

(c) The examinations in a large accounting class are scaled after grading so that the mean score is 75. A self-confident teaching assistant thinks that his students have a higher mean score than the class as a whole. His students this semester can be considered a sample from the population of all students he might teach, so he compares their mean score with 75.

7.53 Hypotheses. In each of the following situations, state an appropriate null hypothesis H_0 and alternative hypothesis H_a. Be sure to identify the parameters that you use to state the hypotheses. (We have not yet learned how to test these hypotheses.)

(a) A sociologist asks a large sample of high school students which academic subject they like best. She suspects that a higher percent of males than of females will name economics as their favorite subject.

(b) An education researcher randomly divides sixth-grade students into two groups for physical education class. He teaches both groups basketball skills, using the same methods of instruction in both classes. He encourages Group A with compliments and other positive behaviors, but acts cool and neutral toward Group B. He hopes to show that positive teacher attitudes result in a higher mean score on a test of basketball skills than do neutral attitudes.

(c) An economist believes that among employed young adults, there is a positive correlation between income and the percent of disposable income that is saved. To test this, she gathers income and savings data from a sample of employed persons in her city aged 25 to 34.

7.54 Hypotheses. Translate each of the following research questions into appropriate H_0 and H_a.

(a) Census Bureau data show that the mean household income in the area served by a shopping mall is $62,500 per year. A market research firm questions shoppers at the mall to find out whether the mean household income of mall shoppers is higher than that of the general population.

(b) Last year, your company's service technicians took an average of 2.6 hours to respond to trouble calls from business customers who had purchased service contracts. Do this year's data show a different average response time?

7.55 Exercise and statistics exams. A study examined whether exercise affects how students perform on their final exam in statistics. The P-value was given as 0.68.

(a) State null and alternative hypotheses that could be used for this study. (Note that there is more than one correct answer.)

(b) Do you reject the null hypothesis? State your conclusion in plain language.

(c) What other facts about the study would you like to know for a proper interpretation of the results?

7.56 Financial aid. The financial aid office of a university asks a sample of students about their employment and earnings. The report says that "for academic year earnings, a significant difference $(P = 0.038)$ was found between the sexes, with men earning more on the average. No difference $(P = 0.476)$ was found between the earnings of black and white students."[12] Explain both of these conclusions, for the effects of sex and of race on mean earnings, in language understandable to someone who knows no statistics.

7.57 Computing the P-value. A test of the null hypothesis H_0: $\mu = \mu_0$ gives a test statistic $z = 1.87$.

(a) What is the P-value if the alternative is H_a: $\mu > \mu_0$?

(b) What is the P-value if the alternative is H_a: $\mu < \mu_0$?

(c) What is the P-value if the alternative is H_a: $\mu \neq \mu_0$?

7.58 More on computing the P-value. A test of the null hypothesis H_0: $\mu = \mu_0$ gives a test statistic $z = -1.27$.

(a) What is the P-value if the alternative is H_a: $\mu > \mu_0$?

(b) What is the P-value if the alternative is H_a: $\mu < \mu_0$?

(c) What is the P-value if the alternative is H_a: $\mu \neq \mu_0$?

7.59 Who is the author? Statistics can help decide the authorship of literary works. Sonnets by a certain Elizabethan poet are known to contain an average of $\mu = 6.9$ new words (words not used in the poet's other works). The standard deviation of the number of new words is $\sigma = 2.7$. Now a manuscript with five new sonnets has come to light, and scholars are debating whether it is the poet's work. The new sonnets contain an average of $\bar{x} = 11.2$ words not used in the poet's known works. We expect poems by another author to contain more new words, so to see if we have evidence that the new sonnets are not by our poet, we test

$$H_0: \mu = 6.9$$
$$H_a: \mu > 6.9$$

Give the z test statistic and its P-value. What do you conclude about the authorship of the new poems?

7.60 Study habits. The Survey of Study Habits and Attitudes (SSHA) is a psychological test that measures the motivation, attitude toward school, and study habits of students. Scores range from 0 to 200. The mean score for U.S. college students is about 115, and the standard deviation is about 30. A teacher who suspects that older students have better attitudes toward school gives the SSHA to 25 students who are at least 30 years of age. Their mean score is $\bar{x} = 133.2$.

(a) Assuming that $\sigma = 30$ for the population of older students, carry out a test of

$$H_0: \mu = 115$$
$$H_a: \mu > 115$$

Report the *P*-value of your test, draw a sketch illustrating the *P*-value, and state your conclusion clearly.

(b) Your test in part (a) required two important assumptions in addition to the assumption that the value of σ is known. What are they? Which of these assumptions is most important to the validity of your conclusion in part (a)?

7.61 Corn yield. The 10-year historical average yield of corn in the United States is about 160 bushels per acre. A survey of 50 farmers this year gives a sample mean yield of $\bar{x} = 158.4$ bushels per acre. We want to know whether this is good evidence that the national mean this year is not 160 bushels per acre. Assume that the farmers surveyed are an SRS from the population of all commercial corn growers and that the standard deviation of the yield in this population is $\sigma = 5$ bushels per acre. Report the value of the test statistic z, give a sketch illustrating the *P*-value, and report the *P*-value for the test of

$$H_0: \mu = 160$$
$$H_a: \mu \neq 160$$

Are you convinced that the population mean is not 160 bushels per acre? Is your conclusion correct if the distribution of corn yields is somewhat non-Normal? Why?

7.62 Student survey. Peru State College in Nebraska conducts an annual survey ("Campus Climate, Safety, and Sexual Assault") of on-campus students to gauge the campus climate.[13] All survey questions are based upon a 7-point scale, ranging from 1 (strongly disagree) to 7 (strongly agree). Higher scores indicate more favorable perceptions of the dimension being questioned. Administration has a benchmark goal of 5.5 for any given factor. Any evidence of not meeting or exceeding this benchmark for a given factor suggests an area on which the college should focus its attention for improvement purposes. Thus, the competing hypotheses are

$$H_0: \mu = 5.5$$
$$H_a: \mu < 5.5$$

Test the above competing hypotheses for each of the following factors studied. In each case, assuming the reported standard deviation to be σ, report the test statistic, *P*-value, and your recommendation to administration.

(a) On the dimension of visibility of diversity, there were 152 responses with the sample mean being 5.12 and a standard deviation of 1.29.

(b) On the dimension of co-curricular environment, there were 148 responses with the sample mean being 5.58 and a standard deviation of 1.12.

(c) On the dimension of administrative leadership qualities, there were 151 responses with the sample mean being 5.39 and a standard deviation of 1.38.

7.63 Academic probation and TV watching. There are other z statistics that we have not yet met. We can use Table D to assess the significance of any z statistic. A study compares the habits of students who are on academic probation with students whose grades are satisfactory. One variable measured is the hours spent watching television last week. The null hypothesis is "no difference" between the means for the two populations. The alternative hypothesis is two-sided. The value of the test statistic is $z = -1.38$.

(a) Is the result significant at the 5% level?

(b) Is the result significant at the 1% level?

7.64 Impact of $\bar{x}$ on significance. The *Statistical Significance* applet illustrates statistical tests with a fixed level of significance for Normally distributed data with known standard deviation. Open the applet and keep the default settings for the null ($\mu = 0$) and the alternative ($\mu > 0$) hypotheses, the sample size ($n = 10$), the standard deviation ($\sigma = 1$), and the significance level ($\alpha = 0.05$). In the "I have data, and the observed $\bar{x}$ is $\bar{x} =$" box, enter the value 1. Is the difference between $\bar{x}$ and μ_0 significant at the 5% level? Repeat for $\bar{x}$ equal to 0.1, 0.2, 0.3, 0.4, 0.5, 0.6, 0.7, 0.8, 0.9. Make a table giving $\bar{x}$ and the results of the significance tests. What do you conclude?

7.65 Effect of changing α on significance. Repeat the previous exercise with significance level $\alpha = 0.01$. How does the choice of α affect which values of $\bar{x}$ are far enough away from μ_0 to be statistically significant?

7.66 Changing to a two-sided alternative. Repeat the previous exercise, but with the two-sided alternative hypothesis. How does this change affect which values of $\bar{x}$ are far enough away from μ_0 to be statistically significant at the 0.01 level?

7.67 Changing the sample size. Refer to Exercise 7.64. Suppose that you increase the sample size n from 10 to 40. Again make a table giving $\bar{x}$ and the results of the significance tests at the 0.05 significance level. What do you conclude?

7.68 Impact of $\bar{x}$ on the *P*-value. We can also study the *P*-value using the *Statistical Significance* applet. Reset the applet to the default settings for the null ($\mu = 0$) and the alternative ($\mu > 0$) hypotheses, the sample size ($n = 10$), the standard deviation ($\sigma = 1$), and the significance level ($\alpha = 0.05$). In the "I have data, and the observed $\bar{x}$ is $\bar{x} =$" box, enter the value 1. What is the *P*-value? It is shown at the top of the blue vertical line. Repeat for $\bar{x}$ equal to 0.1, 0.2, 0.3, 0.4, 0.5, 0.6, 0.7, 0.8, 0.9. Make a table giving $\bar{x}$ and *P*-values. How does the *P*-value change as $\bar{x}$ moves farther away from μ_0?

7.69 Changing to a two-sided alternative, continued. Repeat the previous exercise but with the two-sided alternative hypothesis. How does this change affect the P-values associated with each $\bar{x}$? Explain why the P-values change in this way.

7.70 Other changes and the P-value. Refer to the previous exercise.

(a) What happens to the P-values when you change the significance level α to 0.01? Explain the result.

(b) What happens to the P-values when you change the sample size n from 10 to 40? Explain the result.

7.71 Why is it significant at the 5% level? Explain in plain language why a significance test that is significant at the 1% level must always be significant at the 5% level.

7.72 Finding a P-value. You have performed a two-sided test of significance and obtained a value of $z = 3.1$.

(a) Use Table A to find the P-value for this test.

(b) Use software to find the P-value even more accurately.

7.73 Test statistic and levels of significance. Consider a significance test for a null hypothesis versus a two-sided alternative. Give a value of z that will give a result significant at the 1% level but not at the 0.5% level.

7.74 Finding a P-value. You have performed a one-sided test of significance for greater-than alternative and obtained a value of $z = -0.382$.

(a) Use Table A to find the approximate P-value for this test.

(b) Use software to find the P-value even more accurately.

7.3 Use and Abuse of Tests

When you complete this section, you will be able to:

- Explain the general considerations when choosing a significance level.
- Discriminate between practical significance and statistical significance.
- Identify poorly designed studies where formal statistical inference is suspect.
- Explain the consequences of searching solely for statistical significance, whether through the investigation of multiple tests or by identifying and testing using the same data set.

Carrying out a test of significance is often quite simple, especially if the P-value is given effortlessly by a computer. Using tests wisely is not so simple. Each test is valid only in certain circumstances, with properly produced data being particularly important.

The z test, for example, should bear the same warning labels that were attached in Section 7.1 to the corresponding confidence interval (pages 350–351). Similar warnings accompany the other tests that we will learn. There are additional caveats that concern tests more than confidence intervals, enough to warrant this separate section. Some hesitation about the unthinking use of significance tests is a sign of statistical maturity.

Choosing a level of significance

The goal of a test of significance is to give a clear statement of the degree of evidence provided by the sample against the null hypothesis. The P-value does this. It is common practice to report P-values and to describe results as statistically significant whenever $P \leq \alpha$, where α is commonly chosen to be 0.05. However, there is no sharp border between "significant" and "not significant," only increasingly strong evidence as the P-value decreases.

EXAMPLE 7.16

Information Provided by the P-Value Suppose that the test statistic for a two-sided significance test for a population mean is $z = 1.952$. We can calculate the P-value to be

$$P = 2P(Z \geq |1.952|) = 0.051$$

What if, however, our test statistic is $z = 1.969$? We would then find the P-value to be

$$P = 2P(Z \geq |1.969|) = 0.049$$

Fixated on $\alpha = 0.05$, in one case ($P = 0.051$) we fail to reject the null hypothesis, but in the other case ($P = 0.049$) we reject the null hypothesis. Clearly, to implement a decision rule such that P-values of 0.051 and 0.049 give such contrasting conclusions defies common sense. A more sensible perspective is to view these two P-values no differently than each other in that they are both suggestive but not conclusive evidence against the null hypothesis. In either case, if the effect in question is interesting and potentially important, we might want to design another study with a larger sample to investigate it further. ■

Here is another example in which the P-value leads us to a more conclusive decision.

EXAMPLE 7.17

More on Information Provided by the P-Value We have a test statistic of $z = -4.75$ for a two-sided significance test on a population mean. Software tells us that the P-value is 0.000002. This means that there are two chances in 1,000,000 of observing a sample mean this far or farther away from the null hypothesized value of μ. This kind of event is virtually impossible if the null hypothesis is true. There is no ambiguity in the result; we can clearly reject the null hypothesis. ■

One reason for the common use of $\alpha = 0.05$ is the great influence of Sir R. A. Fisher, the inventor of formal statistical methods for analyzing experimental data. Here is his opinion on choosing this level of significance: "Personally, the writer prefers to set a low standard of significance at the 5 percent point, and ignore entirely all results which fail to reach this level. A scientific fact should be regarded as experimentally established only if a properly designed experiment *rarely fails* to give this level of significance."[14] Fisher's opinion was made nearly a century ago. To this day, many researchers continue to adhere to the 0.05 cutoff value for reporting significant results. In addition, by default, statistical software will often flag results associated with $P \leq 0.05$; for example, JMP displays P-values less than or equal to 0.05 in a colored typeface as opposed to a black typeface for P-values greater than 0.05.

In this book, we often illustrate methods and discuss data results in the context of the 5% level of significance due to its common usage. However, do not mistake our illustrations of methods using $\alpha = 0.05$ as advocacy for its general use in practice. There is nothing sacred about $\alpha = 0.05$. The choice of α can vary from application to application and is often dependent on the costs of making errors in decisions; in Section 8.3, we discuss in detail the two fundamental types of errors made in hypothesis testing.

Recently, in certain scientific communities (including economics) where the statistical threshold for significance has historically been 0.05, researchers have been proposing that the threshold be lowered to 0.005.[15] By lowering the threshold to $\alpha = 0.005$, it is argued that an initial discovery of a significant result has a greater chance of being replicated (confirmed) in follow-up studies. The hope is then that more results affecting policy changes, such as those from economic studies, are true results. With this said, other scientific communities may have even more stringent requirements. For example, researchers in high-energy physics use a threshold of $\alpha = 0.003$ for "evidence of a new particle," but use a threshold of $\alpha = 0.0000003$ as the standard for a "discovery."[16]

In the end, remember that a P-value is a measure of the strength of evidence against the null hypothesis on a continuous scale. Aside from specific levels of significance dictated by an area (e.g., high-energy physics), we can *roughly* state that small P-values (e.g., less than 0.01) provide convincing evidence against the null hypothesis, while large values (e.g., greater than 0.10) do not provide convincing evidence. In-between P-values fall in a more vague region of moderate ("suggestive") evidence to inconclusive evidence against the null hypothesis. In the end, one should approach hypothesis testing holistically and with common sense and good judgment.

Statistical significance does not imply practical significance

When a null hypothesis ("no effect" or "no difference") can be rejected because of a small P-value, there is good evidence that an effect is present. That effect, however, can be extremely small. *When large samples are available, even tiny deviations from the null hypothesis will be significant.*

EXAMPLE 7.18

It's Significant, but Is It Important? Suppose that we are testing the hypothesis of no correlation between two variables. Figure 7.12 is an example of 1000 (x, y) pairs that have a small observed correlation of -0.071. For these data, the P-value for testing the null hypothesis of no correlation is 0.024. The small P-value does *not* mean that there is a strong association, only that there is strong evidence of *some* association. The proportion of the variability in one of the variables explained by the other is only $r^2 = 0.005$, or 0.5%. ■

big data, p. 132

For practical purposes, we might well decide to ignore this association. *Statistical significance is not the same as practical significance.* You may ask yourself, "Why is it that such a small correlation that is not even visible in a plot turns out to be statistically significant?" The answer stems from the size of the sample ($n = 1000$), which is fairly large. Any effect, no matter how small, can be "teased" out as being statistically significant if the sample size is large enough. Now with the advent of big data, the importance of distinguishing statistical significance from practical significance is particularly pertinent. With data sets having millions or billions of observations, the most minuscule of effects are brought to the surface as being statistically significant with extraordinarily small P-values. It is incumbent upon analysts of big data sets to assess whether a statistically significant effect

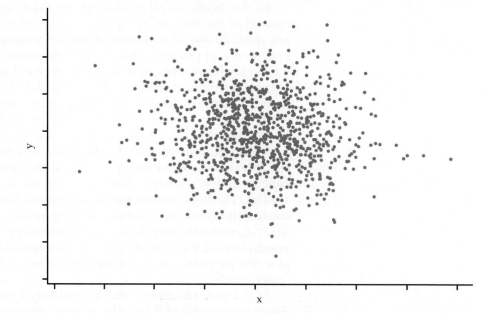

FIGURE 7.12 Scatterplot of $n = 1000$ observations with an observed correlation of -0.071, Example 7.18. There is not a strong association between the two variables, even though there is significant evidence ($P = 0.024$) that the population correlation is not zero.

is of practical significance: for example, in terms of scientific or economic significance.

There is a corollary to large data sets bringing out statistically significant effects that are not practically significant. Namely, it is quite possible for large and meaningful effects to be associated with insignificant results; for example, a P-value failing to attain some standard such as the 5% level of significance. The inability to identify these large effects may be due to the sample size being too small. If you wish to detect an effect of a certain size with high probability, you can plan your study sample size accordingly. This probability is the *power* of the test. Power calculations are discussed in Section 8.3.

The remedy for attaching too much importance to statistical significance is to pay attention to the actual experimental results as well as to the P-value. Plot your data and examine them carefully. Beware of outliers. A few outlying observations can produce significant results if you blindly apply common tests of significance. Outliers can also destroy the significance of otherwise convincing data. *The user of statistics who feeds the data to a computer without exploratory analysis will often be embarrassed.* It is usually wise to give a confidence interval for the parameter in which you are interested along with the P-value. Understanding and background knowledge of the practical application will guide you to assess whether the estimated effect size reflected by the confidence interval is important enough for action. Confidence intervals are not used as often as they should be, while tests of significance are perhaps overused.

APPLY YOUR KNOWLEDGE

7.75 Is it significant? More than 200,000 people worldwide take the GMAT examination each year when they apply for MBA programs. Their scores vary Normally with mean about $\mu = 525$ and standard deviation about $\sigma = 100$. One hundred students go through a rigorous training program designed to raise their GMAT scores. Test the following hypotheses about the training program

$$H_0 : \mu = 525$$
$$H_a : \mu > 525$$

in each of the following situations.

(a) The students' average score is $\bar{x} = 541.4$. What is the P-value? Is this result significant at the 5% level?

(b) Now suppose that the average score is $\bar{x} = 541.5$. What is the P-value? Is this result significant at the 5% level?

(c) Explain how you would reconcile the differences in conclusions between parts (a) and (b), especially if any increase greater than 15 points is considered a success.

Statistical inference is not valid for all sets of data

design of experiments, p. 143

In Chapter 3, we learned that badly designed surveys or experiments often produce invalid results. *Formal statistical inference cannot correct basic flaws in the design.*

Tests of significance and confidence intervals are based on the laws of probability. Randomization in sampling or experimentation ensures that these laws apply. But we must often analyze data that do not arise from randomized samples or experiments. To apply statistical inference to such data, we must have confidence in a probability model for the data. For example, the diameters of successive holes bored in auto engine blocks during production may or may not behave like independent observations from a Normal distribution.

We can check these assumptions by examining the data. If the assumptions appear to be satisfied, we can apply the methods of this chapter to make inference about the process mean diameter μ. Do ask how the data were produced, and don't be too impressed by reported P-values until you are confident that the data were analyzed properly.

APPLY YOUR KNOWLEDGE

7.76 Student satisfaction. Each year *Forbes* publishes its rankings of 650 American colleges. The category of student satisfaction carries a weight of 25% toward the overall score of a college. The major component of the student satisfaction measure is based on student evaluations from RateMyProfessor for the college. Explain why inference about the satisfaction levels of a given college are suspect with this approach.

Beware of searching for significance

Statistical significance is an outcome much desired by researchers and data analysts. It means (or ought to mean) that you have found an effect that you were looking for. *The reasoning behind statistical significance works well if you decide what effect you are seeking beforehand, design an experiment or sample to search for it, and use a test of significance to weigh the evidence you get.* Because a successful search for a new scientific phenomenon often ends with statistical significance, it is all too tempting to make significance itself the object of the search. Such a temptation manifests itself when researchers conduct multiple analyses on a data set, often without prior established hypotheses, and report only significant P-values to then attach hypotheses postanalysis. Such cherry picking of significant results is pejoratively termed "P-value hacking" or "data dredging" in the scientific community and is unacceptable. There can also be unintentional but naïve practices leading to questionable conclusions.

EXAMPLE 7.19

Cell Phones and Brain Cancer Might the radiation from cell phones be harmful to users? Many studies have found little or no connection between using cell phones and various illnesses. Here is part of a news account of one study:

> *A hospital study that compared brain cancer patients and a similar group without brain cancer found no statistically significant association between cell phone use and a group of brain cancers known as gliomas. But when 20 types of glioma were considered separately, an association was found between phone use and one rare form. Puzzlingly, however, this risk appeared to decrease rather than increase with greater mobile phone use.*[17]

Think for a moment: Suppose that the 20 null hypotheses of no connection between cell phone radiation and a particular cancer type are all true. Then each test has a 5% chance of being significant at the 5% level. That's what $\alpha = 0.05$ means—results this extreme occur only 5% of the time just by chance when the null hypothesis is true. Because 5% is 1/20, we expect about one of 20 tests to give a significant result just by chance. Running one test and reaching the $\alpha = 0.05$ level is reasonably good evidence that you have found something; running 20 tests and reaching that level only once is not. ∎

The peril of multiple testing is increased now that a few simple commands will set software to work performing a slew of complicated tests and operations on your data. The dangers of unbridled multiple testing are never more evident than with the big data movement sweeping through the corporate world and the scientific communities. With big data, corporate analytics professionals are combing through massive data sets across multiple variables on consumer behavior with the hope of finding significant relationships that can be leveraged for competitive advantage. By searching through mega data sets and thousands of variables, it is not hard to imagine that significant relationships are bound to be identified. However, these significant relationships are often caused entirely by chance and have no real predictive power. Such relationships, in the end, are commonly referred to as **false-positives.**

false-positives

Our warnings are not to suggest that searching data for patterns is not legitimate. It certainly is. Many important discoveries, scientific and business related, have been made by accident rather than by design. Exploratory analysis of data is an essential part of statistics. We do mean that the usual reasoning of statistical inference does not apply when the search for a pattern is successful, but made absent of a prior established hypothesis. *It is not legitimate practice to test a hypothesis on the same data that first suggested that hypothesis.* A more credible approach is clear. Once you have a hypothesis, then design a study to search specifically for the effect you now think is there. If the result of this study is statistically significant, you have real evidence.

SECTION 7.3 SUMMARY

- *P*-values are more informative than simply a reject-or-not result of a fixed-level α test. Sensible interpretation of *P*-values should not include having sharp borders between significance and insignificance. Beware of placing too much weight on traditional values of α, such as $\alpha = 0.05$.

- Very small effects can be highly significant (small *P*-value), especially when a test is based on a large sample. A statistically significant effect need not be practically important. Plot the data to display the effect you are seeking, and use confidence intervals to estimate the actual value of parameters.

- Significance tests are not always valid. Faulty data collection or substantial violations of underlying assumptions can invalidate a test.

- Many tests run at once will probably produce some significant results by chance alone, even if all the null hypotheses are true.

SECTION 7.3 EXERCISES

For Exercise 7.75, see page 379; for 7.76, see page 380.

7.77 Your role on a team. You are the statistical expert on a team that is planning a study. After you have made a careful presentation of the mechanics of significance testing, one of the team members suggests using $\alpha = 0.20$ for the study because you would be more likely to obtain statistically significant results with this choice. Explain in simple terms why this would not be a good use of statistical methods.

7.78 What do you know? A research report described two results that both achieved statistical significance at the 5% level. The *P*-value for the first is 0.049; for the second it is 0.00002. Do the *P*-values add any useful information beyond that conveyed by the statement that both results are statistically significant? Write a short paragraph explaining your views on this question.

7.79 Find some journal articles. Find two journal articles that report results with statistical analyses. For each article, summarize how the results are reported and write a critique of the presentation. Be sure to include details regarding use of significance testing at a particular level of significance, *P*-values, and confidence intervals.

7.80 Vitamin C and colds. In a study of the suggestion that taking vitamin C will prevent colds, 400 subjects are assigned at random to one of two groups. The experimental group takes a vitamin C tablet daily, while the control group takes a placebo. At the end of the experiment, the researchers calculate the difference between the percents of subjects in the two groups who were free of colds. This difference is statistically significant ($P = 0.03$) in favor of the vitamin C group. Can we conclude that vitamin C has a strong effect in preventing colds? Explain your answer.

7.81 How far do rich parents take us? How much education children get is strongly associated with the wealth and social status of their parents, termed "socioeconomic status," or SES. The SES of parents, however, has little influence on whether children who have graduated from college continue their education. One study looked at whether college graduates took the graduate admissions tests for business, law, and other graduate programs. The effects of the parents' SES on taking the LSAT test for law school were "both statistically insignificant and small."

(a) What does "statistically insignificant" mean?

(b) Why is it important that the effects were small in size as well as statistically insignificant?

7.82 Do you agree? State whether or not you agree with each of the following statements and provide a short summary of the reasons for your answers.

(a) If the P-value is larger than 0.05, the null hypothesis is true.

(b) Practical significance is not the same as statistical significance.

(c) We can perform a statistical analysis using any set of data.

(d) If you find an interesting pattern in a set of data, it is appropriate to then use a significance test to determine its significance.

(e) It's always better to use a significance level of $\alpha = 0.05$ than to use $\alpha = 0.01$ because it is easier to find statistical significance.

7.83 Turning insignificance into significance. Every user of statistics should understand the distinction between statistical significance and practical importance. A sufficiently large sample will declare very small effects statistically significant. Consider the following *randomly* generated digits used to form (x, y) observation pairs:

SIGNIF

x	1	7	9	4	6	4	6	5	0	1
y	0	0	4	3	7	5	5	2	4	5

Read the 10 ordered pair values into statistical software. We will want to test the significance of the observed correlation.

(a) Make a scatterplot of the data and describe what you see.

(b) Compute and report the sample correlation. Software will report the P-value for testing the null hypothesis that the true population correlation is 0. What is the P-value? Is it consistent with what you observed in part (a)?

(c) Copy and paste the 10 ordered pair values into the same two columns to create two replicates of the original data set. Your sample size is now $n = 20$. Produce a scatterplot and compare it with part (a). Has the sample correlation changed? What is the P-value now?

(d) Add more replicates to the two columns so that you can get P-values for $n = 30, 40, 50,$ and 60. Using these values along with what was found in parts (b) and (c), make a table of the P-values versus n. Describe what is happening with the P-values as n increases. Has the correlation changed with the increase in n?

(e) Keep replicating until the P-value becomes less than 0.05. What is the value of n?

(f) Briefly discuss the general lesson learned with this exercise.

7.84 Predicting success of trainees. What distinguishes managerial trainees who eventually become executives from those who, after expensive training, don't succeed and leave the company? We have abundant data on past trainees—data on their personalities and goals, their college preparation and performance, even their family backgrounds and their hobbies. Statistical software makes it easy to perform dozens of significance tests on these dozens of variables to see which ones best predict later success. From running such tests, we find that future executives are significantly more likely than washouts to have an urban or suburban upbringing and an undergraduate degree in a technical field.

Explain clearly why using these "significant" variables to select future trainees is not wise. Then suggest a follow-up study using this year's trainees as subjects that should clarify the importance of the variables identified by the first study.

7.85 More than one test. A P-value based on a single test is misleading if you perform several tests. The **Bonferroni procedure** gives a significance level for several tests together. Level α then means that if *all* the null hypotheses are true, the probability is α that *any* of the tests rejects its null hypothesis.

If you perform two tests and want to use the $\alpha = 5\%$ significance level, Bonferroni says to require a P-value of $0.05/2 = 0.025$ to declare either one of the tests significant. In general, if you perform k tests and want protection at level α, use α/k as your cutoff for statistical significance for each test.

You perform six tests and obtain individual P-values of 0.376, 0.037, 0.009, 0.007, 0.004, and <0.001. Which of these are statistically significant using the Bonferroni procedure with $\alpha = 0.05$?

7.86 More than one test. Refer to the previous exercise. A researcher has performed 12 tests of significance and wants to apply the Bonferroni procedure with $\alpha = 0.05$. The calculated P-values are 0.039, 0.549, 0.003, 0.316, 0.001, 0.006, 0.251, 0.031, 0.778, 0.012, 0.002, and <0.001. Which of these tests reject their null hypotheses with this procedure?

7.87 More than one test and critical value. Suppose that you are performing 12 two-sided tests of significance using the Bonferroni procedure with $\alpha = 0.05$.

(a) If you were to perform the testing procedure using a critical value z^*, what would be z^*?

(b) As the number of test increases, what will happen to z^*?

7.88 False-positive rate. With the big data movement, companies are searching through thousands of variables to find patterns in the data to make better predictions on key business variables. For example, Walmart used predictive analytics and found that sales of strawberry

Pop-Tarts increased significantly when the surrounding region was threatened with an impending hurricane.[18] Imagine yourself in a business analytics position at a company and that you are trying to find variables that significantly correlate with company sales y. Among the variables you are going to compare y against are 80 variables that are truly unrelated to y. In other words, for each of these 80 variables, the null hypothesis is true that the correlation between y and the variables is 0. You are unaware of this fact. Suppose that the 80 variables are independent of each other and that you perform correlation tests between y and each of the variables at the 5% level of significance.

(a) What is the probability that you find at least one of the 80 variables to be significant with y? This probability is referred to as a **false-positive rate.** If you had done only one comparison, what would be the false-positive rate?

(b) Refer to Exercise 7.85 to apply the Bonferroni procedure with $\alpha = 0.05$. What is now the probability that you find at least one of the 80 variables to be significant with y? What do you find this false-positive rate to be close to?

(c) For the significant correlations you do find in your current data, explain how you can use new data on the variables in question to feel more confident about actually using the discovered variables for company purposes.

7.89 False-positives. Refer to the setting of the previous problem. Define X as the number of false-positives occurring among the 80 correlation tests.

(a) What is the distribution of the number X of tests that are significant?

(b) Find the probability that two or more of the tests are significant.

7.4 Prediction Intervals

When you complete this section, you will be able to:

- Explain how the random deviations ε_i can be used in defining observations y_i coming from a Normal distribution with mean μ and standard deviation σ.
- Define what a point prediction is and specify what criterion leads to the sample mean being the point prediction.
- Construct and interpret a level C prediction interval for a new observation based on a random sample of n observations from a Normally distributed population with standard deviation σ.
- Explain the similarities and differences between confidence and prediction intervals.

The predictive analytics movement has resulted in more and more companies using prediction techniques to support decision making. An operations manager, for example, has to have some idea about likely demand when deciding how much inventory to hold. A financial manager can better prepare a budget plan with a prediction of future revenue. When a quality control manager decides to continue the operations of a manufacturing process, there is often a statistical prediction of the process remaining in control. When a hospital administrator decides on ER staffing, an essential part of the decision is a prediction of incoming patients by day of the week and by time of the day. Finally, a realtor can better advise a client when aided with a prediction of the client's home selling price.

Because of variability in the population, predictions are subject to variability. As such, we cannot expect to predict with certainty the value of a future outcome. The strategy then is to provide a best guess along with an interval that likely contains the future outcome. In this section, we discuss the construction of such an interval for prediction of a future observation from a Normally distributed population with known standard deviation σ. Because the construction of such an interval entails the estimation of the population mean, such an interval is an *inference method*, as is a confidence interval.

In fact, you might consider why not use the confidence interval for the mean, presented in this chapter, to be such an interval. *Before moving forward, we must caution against this common temptation.* A confidence interval only captures the uncertainty of our estimate of the population mean μ. A confidence interval for the mean refers to a range of values that is likely to contain the value of the unknown population parameter μ. Our goal here is different. We are seeking a range that is likely to contain a future outcome. As we will show, it is necessary to capture the uncertainty of a future outcome drawn from the population in addition to the uncertainty of our estimate of μ.

Concept of random deviations

When making a prediction of a variable of interest, we refer to the variable as the response variable. As noted in Chapter 2 (page 67), we denote a response variable as y. Using this notation, we use $y_1, y_2, \ldots, y_n$ to denote our random sample of size n from a Normal population with mean μ and standard deviation σ.

To reflect the fact that these observations vary, we can express the observations by

$$y_i = \mu + \varepsilon_i, \qquad i = 1, \ldots, n$$

random deviation where the ε_i are referred to as **random deviations.** As the random deviations vary, so do the observations of the response variable y. We assume that these deviations ε_i are independent and Normally distributed with mean 0 and standard deviation σ.

We can view the response variable y as being made up of two separate pieces: (1) a fixed component (μ) and (2) a random component (ε). From a broader perspective, the component μ can be regarded as the underlying model, while the deviations ε represent the "noise" variation in y that obscures our ability to see the exact model. This breakdown between model and random deviations is a fundamental concept in regression analysis and will be revisited in Chapters 12 and 13.

Prediction of a single observation

Consider now our desire to predict a new (future) outcome of the response variable y given a set of n outcomes $y_1, \ldots, y_n$. Because of random deviations ε, the value of μ is unknown and, thus, we need to estimate it from our sample. In Section 2.3 (pages 83–84), the notation of $\hat{y}$ was introduced to represent the prediction of y based on an estimated model.

For our scenario, the question now is "What is a reasonable choice for $\hat{y}$?" Given that the sample mean is an unbiased estimator of μ, a reasonable and intuitive choice is $\hat{y} = \bar{y}$. It can also be shown that the least-squares idea of minimizing the sum of squared deviations between the y values and the predicted $\hat{y}$ values also leads to the choice of $\hat{y} = \bar{y}$.

least-squares idea p. 83
point prediction

A single best guess of a new outcome of y is called a **point prediction** and can be denoted by $\hat{y}_{\text{new}}$. For example, we may be interested in next year's annual precipitation in Chicago. If we use the observed data of annual precipitation for the last 150 years, a sensible point prediction is the mean of the sample, about 34.6 inches. This point estimate is only a best guess. It does not convey anything about uncertainty. An examination of the precipitation process would show it to be stable and well approximated by the Normal distribution with a standard deviation of 6.2 inches. Using the 68–95–99.7 rule, we can make a reasonable range of values for our prediction of next year's annual precipitation. Namely, using the 95 part of the rule, we are 95% confident that next year's precipitation will lie in the interval of $34.6 \pm 2(6.2) = (22.2, 47.0)$. Let us now consider a more formal approach in the construction of such intervals.

Suppose again we have the random sample $y_1, y_2, \ldots, y_n$. We wish to predict a new observation y_{new}, which will be subject to a new random deviation ε_{new}

$$y_{\text{new}} = \mu + \varepsilon_{\text{new}}$$

The difference between the actual value of y_{new} and its point prediction ($\hat{y}_{\text{new}}$) is a *prediction error* given by

$$\begin{aligned}\text{prediction error} &= y_{\text{new}} - \hat{y}_{\text{new}} \\ &= \mu + \varepsilon_{\text{new}} - \hat{y}_{\text{new}} \\ &= \mu + \varepsilon_{\text{new}} - \bar{y} \\ &= \varepsilon_{\text{new}} + (\mu - \bar{y})\end{aligned}$$

We see that prediction error is decomposed into two sources of error. The first term relates to the contribution of the new random deviation to the prediction error. The second term, $\mu - \bar{y}$, can be thought of as *estimation error*; namely, it reflects how off our estimate of μ is from its actual value.

Since the new random deviation ε_{new} is independent of the past y observations, it is also independent of $\bar{y}$. Applying the rule for the variance of a sum of independent random variables, the variance of prediction error is

rules for variance, pp. 240–241

$$\begin{aligned}\sigma^2_{\text{pred}} &= \sigma^2_{\varepsilon_{\text{new}}} + \sigma^2_{\mu - \bar{y}} \\ &= \sigma^2_{\varepsilon_{\text{new}}} + \sigma^2_{\bar{y}} \\ &= \sigma^2 + \frac{\sigma^2}{n} \\ &= \sigma^2\left(1 + \frac{1}{n}\right)\end{aligned}$$

The standard deviation of prediction error is accordingly

$$\sigma_{\text{pred}} = \sigma\sqrt{1 + \frac{1}{n}}$$

prediction interval

Putting the pieces together in a manner similar to that of a confidence interval, a **prediction interval** for a new observation of y is

$$\bar{y} \pm z^*\sigma\sqrt{1 + \frac{1}{n}}$$

It is instructive to compare the prediction interval with the confidence interval for the mean μ. Recognizing that $\sigma/\sqrt{n} = \sigma\sqrt{1/n}$, we can write the confidence interval as

$$\bar{y} \pm z^*\sigma\sqrt{\frac{1}{n}}$$

The difference in the intervals is the factor $\sqrt{1 + 1/n}$ for the prediction interval versus the factor of $\sqrt{1/n}$ for the confidence interval. Since the prediction interval factor is larger than the confidence interval factor, the prediction interval is wider than the confidence interval. Conceptually, the prediction interval is wider because it must take into account the uncertainty of a new outcome *and* the uncertainty about μ.

PREDICTION INTERVAL FOR A SINGLE OBSERVATION

For a sample of n independent observations from a Normally distributed population having unknown mean μ and known standard deviation σ, a level C **prediction interval for a single new observation** of y is

$$\bar{y} \pm z^*\sigma\sqrt{1 + \frac{1}{n}}$$

It cannot be overemphasized that the applicability of the presented prediction interval relies on the Normality of the individual observations. In contrast, the confidence interval for the mean relies on the Normality of the sample mean statistic, which can be Normal even when the individual observations are not, due to the central limit theorem effect.

A prediction interval should not be interpreted to imply that there is a specified probability that a new observation will fall in the interval. As an example, for a 95% prediction interval, it would be incorrect to state that the probability is 0.95 that a new observation will fall within the interval. Consider a simple scenario where the response variable y has the $N(100,10)$ distribution. Knowing μ and σ, it *would be correct* to state that the probability is 0.95 that a new observation will fall within

$$\mu \pm 1.96\sigma = 100 \pm 1.960(10) = (80.4, 119.6)$$

However, now assume that we do not know μ and have to estimate it with $\bar{y}$. Suppose the sample size $n = 20$. The 95% prediction interval is then

$$\bar{y} \pm 1.960\,\sigma\sqrt{1 + \frac{1}{n}} = \bar{y} \pm 1.960(10)\sqrt{1 + \frac{1}{20}}$$
$$= \bar{y} \pm 20.08$$

If the sample mean is $\bar{y} = 103$, the 95% prediction interval is

$$\bar{y} \pm 20.08 = 103 \pm 20.08 = (82.92, 123.08)$$

Using the $N(100,10)$ distribution, the probability that a new observation falls in the above interval is 0.9457, a probability less than 0.95. Had $\bar{y} = 102$, the probability that a new observation falls in the 95% prediction interval would be 0.9511, a probability greater than 0.95. Remember that we do not know $\mu = 100$ so these probabilities are not known to us in reality. Even though we cannot specify the actual probability that a particular prediction interval will include a new observation, we can interpret a prediction interval in much the same way as a confidence interval. In particular, 95% prediction intervals, like 95% confidence intervals, are correct 95% of the time in repeated use.

EXAMPLE 7.20

HOMES

Milwaukee Home Sales For the year 2017, consider a sample of $n = 182$ home sales in district 2 of the city of Milwaukee.[19] For this sample, the average selling price is \$89,550.846. The standard deviation of selling prices is \$23,363.969, which we will assume to be the σ for the population of homes.

Given that n is sufficiently large, we can rely on the central limit theorem to assure us that the confidence interval for the mean is a good approximation. The approximate 95% confidence interval for the mean is

$$\bar{y} \pm z^* \frac{\sigma}{\sqrt{n}} = 89550.846 \pm 1.960 \frac{23363.969}{\sqrt{182}}$$
$$= (86156, 92945)$$

We are 95% confident that the population mean selling price is between \$86,156 and \$92,945. Consider now a different question. What is the prediction interval for the selling price of an individual home that has yet to be observed? Before constructing such an interval, we must check for the Normality of the individual observations. Figure 7.13 is a Normal quantile plot of the selling prices. Because the plot looks fairly straight, we can proceed confidently with the construction of the approximate 95% prediction interval:

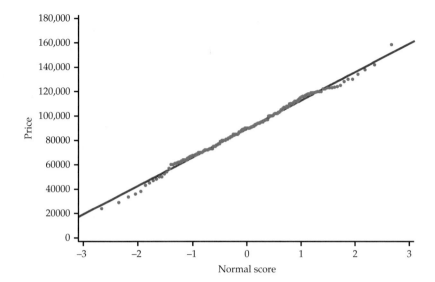

FIGURE 7.13 Normal quantile plot of 182 Milwaukee home selling prices, Example 7.20.

$$\bar{y} \pm z^*\sigma\sqrt{1+\frac{1}{n}} = 89550.846 \pm 1.960(23363.969)\sqrt{1+\frac{1}{182}}$$
$$= (43632, 135470)$$

The prediction interval is considerably wider than the corresponding confidence interval. As noted, the confidence interval only takes into account the uncertainty about μ, while the prediction interval takes into account the uncertainty about μ along with the variability of a single new outcome. The prediction interval indicates that we are 95% *confident* that the future home selling price will fall between \$43,632 and \$135,470. ■

Some software has the option of reporting a prediction interval for a new observation based only on the sample mean $\bar{y}$ as being the predicted value. Figure 7.14 shows JMP summary statistics and prediction limits output for the data of Example 7.20. We find that the software-reported limits rounded to the nearest dollar are \$43,324 and \$135,778. This interval is slightly wider than that found in Example 7.20. In particular, the software-reported limits are about \$300 farther out than the computed values of Example 7.20. The differences are due to the fact that the software is taking into account that the standard deviation of 23,363.969 is an estimate of the population standard deviation rather than viewing it as σ.

As will be discussed in the next chapter, a distribution known as the t distribution is used in place of the standard Normal distribution for determining critical values when σ is not known. Using a t distribution, the critical value of $z^* = 1.960$ would be replaced with a value of about 1.973. This slight change in the critical value accounts for the slightly wider software-reported prediction interval.

Whether we consider the prediction interval in Example 7.20 or the slightly more precise interval of Figure 7.14, a more pertinent concern is the practicality of the interval. We have a point prediction of \$89,551 (rounded to the nearest dollar) for the selling price of a new observation. However, the prediction interval

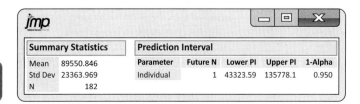

FIGURE 7.14 JMP output of summary statistics and 95% prediction interval limits for 182 Milwaukee home selling prices, Example 7.20.

is rather wide. The problem is that we are relying solely on the distribution of selling prices with our predictive model being $\hat{y} = \bar{y}$ coupled with the fact that distribution has a large standard deviation. To shrink the width of the prediction interval, we need to incorporate more information into our predictive model, which will result in our predictive model $\hat{y}$ being more "sophisticated" than simply being $\bar{y}$. A prediction interval for the selling price of a home will surely be tighter if we were to consider factors such as the size of the home, number of bedrooms, number of bathrooms, exterior type, and lot size. This notion of incorporating more information into a predictive model for more precise predictions is the key idea underlying a variety of topics including regression models (Chapters 12 and 13) and time series forecasting methods (Chapter 14).

SECTION 7.4 SUMMARY

- To say that $y_1, y_2, \ldots, y_n$ is a random sample from a Normally distributed population with mean μ and standard deviation σ, we can express the observations by

$$y_i = \mu + \varepsilon_i, \quad i = 1, \ldots, n$$

where ε_i is a **random deviation.** Random deviations are assumed to be independent of each other and have the Normal distribution with mean 0 and standard deviation σ.

- A single best guess of a new outcome of y is called a **point prediction** and denoted by $\hat{y}$. Based on the least-squares criterion, for a sample of $y_1, y_2, \ldots, y_n$, the point predictor $\hat{y}$ is simply given by $\bar{y}$.

- For a sample of n independent observations from a Normally distributed population having unknown mean μ and known standard deviation σ, a level C **prediction interval** for a single new observation of y is

$$\bar{y} \pm z^* \sigma \sqrt{1 + \frac{1}{n}}$$

where z^* is obtained from the row labeled z at the bottom of Table D. The probability is C that a standard Normal random variable takes a value between $-z$ and z^*.

- The interpretation of the level C for a prediction interval is similar to that of a confidence interval. Namely, if you use 95% prediction intervals, in the long run 95% of your intervals will include the predicted new observation.

- The prediction interval presented in this section is correct under specific conditions of independence and Normality. These conditions should be investigated *prior* to any calculations.

SECTION 7.4 EXERCISES

CASE 7.2 **7.90 Fill amount prediction.** Refer to the setting of Example 7.7 (page 356) in which the sample mean of 20 bottles was found to be 474.54 ml. Suppose that the process is unchanged and a random bottle is selected.

(a) What is the point prediction for the fill amount in that bottle?

(b) What is a 95% prediction interval for the fill amount in that bottle?

(c) What was stated in Example 7.7 that supports the prediction interval found in part (b) as being appropriate to interpret?

CASE 7.2 **7.91 Fill amount prediction, continued.** Refer to the setting of Example 7.8 (page 357) in which production staff made process adjustments and found the sample mean of 20 bottles to be 472.56 ml. Suppose the process is unchanged and a random bottle is selected.

(a) What is the point prediction for the fill amount in that bottle?

(b) What is a 90% prediction interval for the fill amount in that bottle?

7.92 What's wrong? Here are several situations where there is an incorrect application of the ideas presented in this section. Briefly explain what is wrong in each situation and why it is wrong.

(a) Suppose that the 95% confidence interval for the mean is (120, 150). This implies we are 95% confident a new observation will fall between 120 and 150.

(b) Suppose that a population has a Normal distribution and the computed 90% prediction interval is (200, 300). This implies that the probability is 0.90 that a new observation will fall between 200 and 300.

(c) Since the 95% prediction interval of (30, 70) includes the value of 65, we cannot reject the null hypothesis of $\mu = 65$ at the 5% level of significance.

(d) For predicting next year's global temperature index, we should use $\hat{y} = \bar{y}$ as a point prediction, where $\bar{y}$ is the average temperature of the last 100 years.

7.93 Interval width. In Section 7.1, for a given confidence level C, you learned that the margin of error of a confidence interval changes as the sample size n changes.

(a) What happens to the width of a confidence interval as the sample size n increases? Theoretically, the width of the interval can be pushed closer and closer to what value as the sample size n increases?

(b) What is the theoretical limit of the width of a prediction interval as the sample size n increases?

(c) Suppose $\sigma = 5$. What is the narrowest width of a 95% prediction interval can be?

7.94 Giannis Antetokounmpo rebounds. Sports analytics have permeated nearly all sports at all levels. Predictive analytic methods are used to study and monitor a variety of statistics on teams and players.

Consider the provided data on the number of rebounds that NBA superstar Giannis Antetokounmpo made for each of the 75 consecutive 2017–2018 regular season games that he played in.[20] An analysis of the consecutive rebounds would show them to behave randomly over time with no patterns. GIANNIS

(a) Obtain a histogram and Normal quantile plot of the rebounds data. Even though technically the distribution cannot exactly be Normal due to the discreteness of the data values, do the plots suggest that Normality is a reasonable approximation? Explain.

(b) Use software to find the values of $\bar{y}$ and the sample standard deviation s.

(c) Assuming that Giannis's rebounding process continues in the same manner and using the standard deviation value found in part (b) as σ, what is a 90% prediction interval for the number of rebounds in his next game? Determine the limits by hand computation, not by software. Report the prediction interval as computed in decimal form.

(d) In general, when dealing with predicting integer outcomes, as with rebounds, how would you suggest converting the lower and upper limit values to integers so that the resulting interval is minimally of the specified C level? Be careful when thinking about the conversion of a lower limit value versus an upper limit value. Explain.

(e) Following your approach of part (d), what is the conservative 90% prediction interval with integer endpoints for the prediction interval found in part (c)?

(f) If you were to expand your predictive model to narrow the prediction interval width, list at least two factors or pieces of information that you would gather.

CHAPTER 7 REVIEW EXERCISES

7.95 Deconstructing a confidence interval. A 95% confidence interval for a population mean is (220, 234). The population standard deviation is known to be $\sigma = 50$.

(a) What is the value of $\bar{x}$?

(b) What is the sample size?

7.96 Census Bureau. In the official training manual of the U.S. Census Bureau, it is stated that "A (margin of error) MOE is a measure of the possible variation of the estimate around the population value."[21] In the same Census Bureau documentation, MOE is defined as $1.645 \times SE$, where SE is the square root of the variance of the estimator. This definition is found after a statement that the Census Bureau Standard is the 90% percent confidence level.

(a) In the training manual, the Census Bureau denotes its standard as $MOE_{90\%}$ and proceeds to show the general formula for converting its standard to margins of error based on other confidence levels. Provide a general formula for converting $MOE_{90\%}$ to $MOE_{95\%}$.

(b) In the 2016 Census Bureau report titled "Health Insurance Coverage in the United States," the estimated number of uninsured (in thousands) is 28,052 with a margin of error of 512. What is the 90% confidence interval for the number of uninsured in the United States?[22]

(c) Using your conversion formula from part (a), what is $MOE_{95\%}$?

(d) Based on parts (b) and (c), what is the 95% confidence interval for the number of uninsured in the United States?

7.97 Deconstructing a confidence interval. A 99% confidence interval for a population mean is (70, 80). The sample size is 32.

(a) What is the value of $\bar{x}$?

(b) What is the value of the population standard deviation σ?

7.98 What's wrong? Provide a brief explanation of what is wrong with each of the following statements.

(a) As the sample size increases, the margin of error for a 95% confidence interval for μ increases.

(b) As the confidence level C increases, the sample size required to obtain a confidence interval of specified margin of error decreases.

(c) As the population standard deviation decreases, the margin of error for a 95% confidence interval for μ increases.

(d) When σ is known and n is large, the margin of error is a random variable that follows a Normal distribution.

7.99 Coverage percent of 95% confidence interval. For this exercise, use the *Confidence Interval* applet. Set the confidence level at 95%, and click the "Sample" button 10 times to simulate 10 confidence intervals. Record the percent hit (that is, percent of intervals including the population mean). Simulate another 10 intervals by clicking another 10 times (do not click the "Reset" button). Record the percent hit for your 20 intervals. Repeat the process of simulating 10 additional intervals and recording the results until you have a total of 200 intervals. Plot your results and write a summary of what you have found.

7.100 Coverage percent of 90% confidence interval. Refer to the previous exercise. Do the simulations and report the results for 90% confidence.

7.101 Change the confidence level. Refer to Example 7.15 (page 369) and construct a 95% confidence interval for the mean initial return for the population of Indian IPO firms.

7.102 Change the confidence level. Refer to Example 7.15 (page 369) and construct a 90% confidence interval for the mean initial return for the population of Indian IPO firms.

7.103 What is the standard deviation? In a 2018 CMO Survey of top marketing managers from 340 firms, respondents were asked to rate their optimism about the U.S. economy on a scale from 0 to 100, with 0 being the least optimistic.[23] The mean optimism score was 68.94, and the reported 95% confidence interval for the mean was (67.40, 70.47). What is the value of the standard deviation of the individual responses? Show your calculations.

7.104 Survey of Prosthodontists. In a 2017 survey of prosthodontists conducted by the American College of Prosthodontists, a number of confidence intervals are reported in the form of $\bar{x} \pm m$, where m is the margin of error.[24]

(a) In the report, the following is stated:

Based on the responses, the average age of respondent prosthodontists was calculated to be 51.20 years with a standard deviation of 14.12 years and based on 603 prosthodontist respondents.

Determine the margin of error based on a 95% level of confidence. Report it to the nearest hundredths place.

(b) Based on part (a), express the 95% confidence interval in the form of $\bar{x} \pm m$.

(c) The report states the following:

This means that the 95% confidence interval estimate of the age of all prosthodontists in the U.S. using the responses from the 2017 survey ranges from: 51.20 years +/−2.26 years, or 48.94 years to 53.46 years

Respond to the above quote in light of your answer to part (b). What does the value of 2.26 actually represent?

7.105 Very small P-value. For Example 7.15 (page 369), we noted that the P-value for testing the null hypothesis of $\mu = 0$ is $2P(Z \geq 7.91)$. Based on Table A, we noted that the P-value is obviously less than 0.0004.

(a) Just how small is the P-value? Excel will actually report very small probabilities. Use the NORM.DIST function to find the probability.

(b) Relate the extremely small probability found in part (a) to a friend with the small probability event of winning the multistate Powerball lottery, which has a probability of 1 in 175 million.

7.106 Supply chain practices. In a Stanford University study of supply chain practices, researchers gathered data on numerous companies and computed the correlations between various managerial practices and metrics on social responsibility.[25] In the report, the researchers only report correlations that meet the following criteria: correlation value ≥ 0.2 and P-value ≤ 0.05. Why do you think the researchers are not reporting statistically significant correlations that are less than 0.2?

7.107 Wine. Many food products contain small quantities of substances that would give an undesirable taste or smell if they were present in large amounts. An example is the "off odors" caused by sulfur compounds in wine. Oenologists (wine experts) have determined the odor threshold, that is, the lowest concentration of a compound that the human nose can detect. For example, the odor threshold for dimethyl sulfide (DMS) is given in the oenology literature as 25 micrograms per liter of wine (μg/l). Untrained noses may be less sensitive, however. Here are the DMS odor thresholds for 10 beginning students of oenology:

31 31 43 36 23 34 32 30 20 24

Assume (this is not realistic) that the standard deviation of the odor threshold for untrained noses is known to be $\sigma = 7$ μg/l. ODOR

(a) Make a stemplot to verify that the distribution is roughly symmetric, with no outliers. (A Normal quantile plot confirms that there are no systematic departures from Normality.)

(b) Give a 95% confidence interval for the mean DMS odor threshold among all beginning oenology students.

(c) Are you convinced that the mean odor threshold for beginning students is higher than the published threshold, 25 μg/l? Carry out a significance test to justify your answer.

7.108 Too much cellulose to be profitable? Excess cellulose in alfalfa reduces the "relative feed value" of the product that will be fed to dairy cows. If the cellulose content is too high, the price will be lower and the producer will have less profit. An agronomist examines the cellulose content of one type of alfalfa hay. Suppose that the cellulose content in the population has standard deviation $\sigma = 8$ milligrams per gram (mg/g). A sample of 15 cuttings has mean cellulose content $\bar{x} = 145$ mg/g.

(a) Give a 90% confidence interval for the mean cellulose content in the population.

(b) A previous study claimed that the mean cellulose content was $\mu = 140$ mg/g, but the agronomist believes that the mean is higher than that figure. State H_0 and H_a and carry out a significance test to see if the new data support this belief.

(c) The statistical procedures used in parts (a) and (b) are valid when several assumptions are met. What are these assumptions?

7.109 Where do you buy? Consumers can purchase nonprescription medications at food stores, mass merchandise stores such as Kmart and Walmart, or pharmacies. About 45% of consumers make such purchases at pharmacies. What accounts for the popularity of pharmacies, which often charge higher prices?

A study examined consumers' perceptions of overall performance of the three types of store using a long questionnaire that asked about such things as "neat and attractive store," "knowledgeable staff," and "assistance in choosing among various types of nonprescription medication." A performance score was based on 27 such questions. The subjects were 201 people chosen at random from the Indianapolis telephone directory. Here are the means and standard deviations of the performance scores for the sample:[26]

Store type	$\bar{x}$	s
Food stores	18.67	24.95
Mass merchandisers	32.38	33.37
Pharmacies	48.60	35.62

We do not know the population standard deviations, but a sample standard deviation s from so large a sample is usually close to σ. Use s in place of the unknown σ in this exercise.

(a) What population do you think the authors of the study want to draw conclusions about? What population are you certain they can draw conclusions about?

(b) Give 95% confidence intervals for the mean performance for each type of store.

(c) Based on these confidence intervals, are you convinced that consumers think that pharmacies offer higher performance than the other types of stores? In Chapter 9, we study a statistical method for comparing the means of several groups.

7.110 Confidence level? The Kaiser Family Foundation (Kaiser) and the Health Research & Educational Trust (HRET) conduct an annual national survey of firms on a number of health coverage issues.[27] The survey presents and compares sample means and proportions across a variety of categories (e.g., firm size, insurance plan type) to determine if there are significant statistical differences between these categories. In the methodology section of the survey report, the following is stated

All statistical tests are performed at the .05 confidence level, unless otherwise noted.

Is the term "confidence level" used in the survey report appropriate terminology? If not, what term would you suggest in its place?

7.111 Using software on a data set. Refer to Exercise 7.107 and the DMS odor threshold data. As noted in the exercise, assume $\sigma = 7$ µg/l. Read the data into statistical software, and obtain the 95% confidence interval for the mean DMS. From a set of data, standard Excel does not provide an option for determining a confidence interval for the mean when σ is known. Provide output from the software used. ODOR

- *JMP users:* With data in a data table, select the data in the **Distribution** platform to get the histogram and other summary statistics. With the red arrow option pull-down, go to **Confidence Interval** and then select **Other.** You will then find an option to provide a known sigma.

- *Minitab users:* With data in a worksheet, do the following pull-down sequence: **Stat → Basic Statistics → 1–Sample Z.** You will then find an option to provide a known sigma.

- *R users:* There is no option for when σ is known in the base package of R. However, there are a number of packages that can easily be installed that have the capability. For example, here are the commands to get an 80% confidence interval for a data set of three observations with $\sigma = 2$:

```
x <- c(4,10,1)
install.packages("TeachingDemos")
library(TeachingDemos)
z.test(x,sd=2,conf.level=0.80)
```

7.112 Using software with summary measures. Most statistical software packages provide an option of finding confidence interval limits by inputting the sample mean, sample size, population standard deviation, and desired confidence level. With summary measures, standard Excel does not provide a direct option for reporting confidence interval limits for the mean when σ is known. Excel does, however, have a function for computing the margin of error component of the confidence interval (see Exercise 7.29, page 355).

- *JMP users:* Do the following pull-down sequence: **Help → Sample Data** and then select **Confidence Interval for One Mean** found in the **Calculators** group.

- *Minitab users:* Do the following pull-down sequence: **Stat → Basic Statistics → 1 Sample Z** and select **Summarized data** option.

- **R users:** There is no option for when σ is known in the base package of R. However, there are a number of packages that can easily be installed that have the capability. For example, here are the commands to get an 80% confidence interval for a data set of three observations with $\bar{x} = 5$ and $\sigma = 2$:

  ```
  install.packages("TeachingDemos")
  library(TeachingDemos)
  z.test(5,sd=2,n=3,conf.level=0.80)
  ```

(a) Have software find the 95% confidence interval for the mean when $\bar{x} = 20$, $n = 27$, and $\sigma = 4$. Provide output from the software used.

(b) Have software find the 93.5% confidence interval using the information of part (a). Provide output from the software used.

CASE 7.2 **7.113 Is it right now?** Refer to the setting of Example 7.8 (page 357) in which it is noted that Bestea's production staff made adjustments to the process and collected a sample of 20 bottles from the "corrected" process. The sample mean was found to be $\bar{x} = 472.56$ ml. A probability value of 0.327 was also reported.

(a) What does the 0.327 value represent?

(b) Using Table A, show how the 0.327 value is determined.

(c) Without actually constructing a 95% confidence interval for the population mean, explain how the 0.327 value will tell us whether or not the interval will include the value of $\mu = 473$.

7.114 CEO pay. A study of the pay of corporate chief executive officers (CEOs) examined the increase in cash compensation of the CEOs of 104 companies, adjusted for inflation, in a recent year. The mean increase in real compensation was $\bar{x} = 6.9\%$, and the standard deviation of the increases was $s = 55\%$. Is this good evidence that the mean real compensation μ of all CEOs increased that year? The hypotheses are

$$H_0: \mu = 0 \quad \text{(no increase)}$$
$$H_a: \mu > 0 \quad \text{(an increase)}$$

Because the sample size is large, the sample s is close to the population σ, so take $\sigma = 55\%$.

(a) Sketch the Normal curve for the sampling distribution of $\bar{x}$ when H_0 is true. Shade the area that represents the P-value for the observed outcome $\bar{x} = 6.9\%$.

(b) Calculate the P-value.

(c) Is the result significant at the $\alpha = 0.05$ level? Do you think the study gives strong evidence that the mean compensation of all CEOs went up?

7.115 Are Verizon daily trading volumes random over time? In Example 7.1 (page 338), we looked at a time plot of daily trading volumes over the course of 121 consecutive days. The question is whether the observed data suggest that the underlying volume process is random or not over time. Thus, the null hypothesis is that the underlying process is random and the alternative hypothesis is that the underlying process is nonrandom ("patterned"). One way to test these competing hypotheses is to count the observed number of runs (sequences of consecutive observations all below or above the sample mean). For the data shown in Figure 7.1, we find 48 observed runs. If the null hypothesis is true, the expected number of runs (μ_R) is 58.92 with a standard deviation of 5.24. It can be theoretically shown that the observed runs statistic is approximately Normal. Significantly too many runs or too few runs is evidence against the null hypothesis.

(a) Find the value of z test statistic.

(b) Find the P-value. Show your computations. Verify that this value was reported in Example 7.1. What do you conclude based on a 5% level of significance?

7.116 Big data effect. In this exercise, you will explore the effect of increasing the sample size n to very large values often seen in big data settings. Assume that the population being sampled from has standard deviation $\sigma = 10$ and that $\bar{x} = 100.01$ for all sample sizes considered. Determine all the values for z and P-value when conducting a two-sided test of $H_0: \mu = 100$. Suppose further that a departure of 0.01 from $\mu = 100$ is of no meaningful consequence for the associated application.

n	Test statistic z	P-value
10		
1000		
100,000		
10,000,000		
1,000,000,000		

Discuss the lesson learned with this exercise in the context of the information found in Section 7.3.

7.117 Do you agree? State whether or not you agree with each of the following statements, and provide a short summary of the reasons for your answers.

(a) The one-sided P-value is 0.03. This implies that the two-sided P-value is 0.06.

(b) If the z statistic is 1.8, the two-sided P-value is $2P(Z = |1.8|)$.

(c) The one-sided P-value can't be greater than 0.5.

(d) The z statistic has a value between ± 1.96. This implies that the two-sided P-value is greater than 0.05.

7.118 Verizon daily returns. In Example 7.1 (page 338), we looked at a time plot of daily trading volumes from January 2 to June 25, 2018. Now consider the daily returns of Verizon stock over the same period of time. Daily returns are simply the percent change in stock prices. Unlike the daily volumes, an analysis of the consecutive returns would show them behaving randomly over time with no patterns. (*Note*: There are 121 trading days from January 2 to June 25, 2018. However, because the first observed return is computed after the second day in the sequence, there are 120 returns.) VZRET

(a) Obtain a histogram and Normal quantile plot of the returns data. Do the plots suggest that Normality is a reasonable approximation? Explain.

(b) Use software to find the values of $\bar{y}$ and the sample standard deviation s.

(c) Assuming that the daily process continues in the same manner and using the standard deviation value found in part (b) as σ, what is a 95% prediction interval for next daily return? Determine the limits by hand computation, not by software. Report the prediction interval as computed in decimal form.

7.119 Welfare reform. A study compares two groups of mothers with young children who were on welfare two years ago. One group attended a voluntary training program offered free of charge at a local vocational school and advertised in the local news media. The other group did not choose to attend the training program. The study finds a significant difference $(P < 0.01)$ between the proportions of the mothers in the two groups who are still on welfare. The difference is not only significant, but quite large. The report says that with 95% confidence the percent of the nonattending group still on welfare is 21% ± 4% higher than that of the group who attended the program. You are on the staff of a member of Congress who is interested in the plight of welfare mothers and who asks you about the report.

(a) Explain briefly, and in nontechnical language, what "a significant difference $(P < 0.01)$" means.

(b) Explain clearly and briefly what "95% confidence" means.

(c) Is this study good evidence that requiring job training of all welfare mothers would greatly reduce the percent who remain on welfare for several years?

7.120 Have we reduced processing time? Refer to Example 7.10 (page 359), where the competing hypotheses are H_0: $\mu = 3.8$ versus H_a: $\mu < 3.8$. Suppose that you observe a large sample of process customer orders, in particular, $n = 100$. You find that that average processing time is 3.4 days with $s = 1.32$. Given the large sample size, you proceed with a z test of the mean.

(a) What is the value of the z statistic?

(b) For $\alpha = 0.01$, what is the rejection region for the z test?

(c) Based on the critical value approach and $\alpha = 0.01$, what do you conclude? Explain.

(d) What is the P-value?

CHAPTER 8

Inference for Means

Introduction

We began our study of data analysis in Chapter 1 by learning graphical and numerical tools for describing the distribution of a single variable and for comparing several distributions. Our study of the practice of statistical inference begins in the same way, with inference about a single distribution and comparison of two distributions.

Two important aspects of any distribution are its center and spread. If the distribution is Normal, we describe its center by the mean μ and its spread by the standard deviation σ. In this chapter, we will consider confidence intervals and significance tests for inference about a population mean μ and the difference between population means $\mu_1 - \mu_2$. These methods allow us to answer questions such as these:

- Product testing provides consumers with in-use information on products. If a product tester measures the battery life of the latest Samsung smartphone by continuously surfing the Web until the phone dies, do you think your typical phone use would result in a longer or shorter battery life? How would you go about answering this question?

- A smart shopping cart includes a scanner, which reports the total cost of the goods in the cart. Do you think its use would influence your spending? If so, would you spend more or less? Grocery store chains, such as Safeway and Kroger, are interested in understanding these preferences and spending effects.

- Do you expect to be treated rudely by salespeople of high-end retail such as Gucci and Burberry? If yes, why? There are some who argue that this rudeness adds value to the goods being sold. Do you agree? If so, would rudeness add value even at a mass market store, such as Old Navy or Kohl's?

CHAPTER OUTLINE

8.1 Inference for the Mean of a Population

8.2 Comparing Two Means

8.3 Additional Topics on Inference

Chapter 8 Inference for Means

Chapter 7 emphasized the reasoning of significance tests and confidence intervals; now we emphasize statistical practice and no longer assume that population standard deviations are known. As a result, we replace the standard Normal sampling distribution with the family of t distributions. The t procedures for inference about means are among the most commonly used statistical methods in business and economics.

8.1 Inference for the Mean of a Population

When you complete this section, you will be able to:

- Construct a confidence interval for the population mean μ from a simple random sample (SRS).
- Describe a confidence interval for μ in terms of an estimate and its margin of error.
- Perform a one-sample t significance test for μ from an SRS and summarize the results.
- Identify when the matched pairs t procedures should be used instead of the two-sample t procedures.
- Check conditions to assess whether the one-sample t procedures can be useful for non-Normal data.

Both confidence intervals and tests of significance for the mean μ of a Normal population are based on the sample mean $\bar{x}$, which estimates the unknown μ. The sampling distribution of $\bar{x}$ depends on the standard deviation σ. This fact causes no difficulty when σ is known. When σ is unknown, we must estimate σ even though we are primarily interested in μ.

In this section, we meet the sampling distribution of the standardized version of $\bar{x}$ when we use the sample standard deviation s as our estimate of the standard deviation σ. We then use this sampling distribution in our discussion of both confidence intervals and significance tests for inference about the mean μ.

t distributions

sampling distribution of $\bar{x}$, p. 312

Suppose that we have a simple random sample (SRS) of size n from a Normally distributed population with mean μ and standard deviation σ. The sample mean $\bar{x}$ is then Normally distributed with mean μ and standard deviation $\sigma/\sqrt{n}$. When σ is not known, we estimate it with the sample standard deviation s, and then we estimate the standard deviation of $\bar{x}$ by $s/\sqrt{n}$. This quantity is called the *standard error* of the sample mean, and we denote it by $\text{SE}_{\bar{x}}$.

> **STANDARD ERROR**
>
> When the standard deviation of a statistic is estimated from the data, the result is called the **standard error** of the statistic. The standard error of the sample mean is
>
> $$\text{SE}_{\bar{x}} = \frac{s}{\sqrt{n}}$$

The term "standard error" is sometimes used for the actual standard deviation of a statistic. The estimated value is then called the "estimated standard error." In this book, we use the term "standard error" *only* when the standard deviation of a statistic is estimated from the data. The term has this meaning in the output of many statistical software packages and in reports of research in many fields that apply statistical methods.

In the previous chapter, the standardized sample mean, or one-sample z statistic,

$$z = \frac{\bar{x} - \mu}{\sigma/\sqrt{n}}$$

was used to introduce us to the procedures for inference about μ. This statistic has the standard Normal distribution $N(0,1)$. However, when we substitute the standard error $s/\sqrt{n}$ for the standard deviation of $\bar{x}$, this statistic no longer has a Normal distribution. It has a distribution that is new to us, called a *t distribution*.

THE *t* DISTRIBUTIONS

Suppose that an SRS of size n is drawn from an $N(\mu, \sigma)$ population. Then the **one-sample *t* statistic**

$$t = \frac{\bar{x} - \mu}{s/\sqrt{n}}$$

has the ***t* distribution** with $n - 1$ **degrees of freedom.**

degrees of freedom, p. 33

A particular *t* distribution is specified by its *degrees of freedom*. We use $t(k)$ to stand for the *t* distribution with k degrees of freedom. The degrees of freedom for this *t* statistic come from the sample standard deviation s in the denominator of *t*. We saw in Chapter 1 that s has $n - 1$ degrees of freedom. We will meet many other *t* statistics with different degrees of freedom in this and later chapters.

The *t* distributions were discovered in 1908 by William S. Gosset. Gosset was a statistician employed by the Guinness brewing company, which prohibited its employees from publishing their discoveries that were brewing related. In this case, the company let him publish under the pen name "Student" using an example that did not involve brewing. The *t* distributions are often called "Student's *t*" in his honor.

The density curves of the $t(k)$ distributions are similar in shape to the standard Normal curve. That is, they are symmetric about 0 and are bell-shaped. Figure 8.1 compares the density curves of the standard Normal distribution and the *t* distributions with 5 and 15 degrees of freedom. The similarity in shape is apparent, as is the fact that the *t* distributions have more probability in the tails and less in the center.

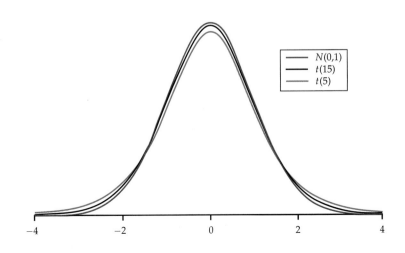

FIGURE 8.1 Density curves for the standard Normal (red), $t(5)$ (blue), and $t(15)$ (black) distributions. All are symmetric with center 0. The *t* distributions have more probability in the tails than the standard Normal distribution.

This greater spread is due to the extra variability caused by substituting the random variable s for the fixed parameter σ. Comparing the two t curves, we see that as the degrees of freedom k increase, the $t(k)$ density curve gets closer to the $N(0,1)$ curve. This reflects the fact that s will generally be closer to σ as the sample size increases.

APPLY YOUR KNOWLEDGE

8.1 One-bedroom rental apartment. Realtor.com contains several hundred postings for one-bedroom apartments in your city. You choose 16 at random and calculate a mean monthly rent of $879 and a standard deviation of $168.

(a) What is the standard error of the mean?

(b) What are the degrees of freedom for the one-sample t statistic?

8.2 Changing the sample size. Refer to the previous exercise. Suppose that you drew a new SRS, this time containing 20 postings.

(a) Would you expect the standard error of the mean to be larger or smaller in this case? Explain your answer.

(b) State why you can't be certain that the standard error for this new SRS will be larger or smaller.

With the t distributions to help us, we can analyze an SRS from a Normal population with unknown σ or a large sample from a non-Normal population with unknown σ. We usually rely on software to do the calculations involving t distributions. When software is not available, Table D in the back of the book can be used. It provides the critical values t^* for the t distributions. For convenience, the table entries are labeled both by the value of the upper-tail probability p needed for significance tests and by the confidence level C (in percent) required for confidence intervals. The standard Normal critical values in the bottom row of entries are labeled z^*.

The one-sample t confidence interval

z confidence interval, p. 345

The one-sample t confidence interval is similar in both reasoning and computational detail to the z procedures of Chapter 7. There, the margin of error for the population mean was $z^*\sigma/\sqrt{n}$. When σ is unknown, we replace it with its estimate s and switch from z^* to t^*. This means that the margin of error for the population mean when we use the data to estimate σ is $t^*s/\sqrt{n}$.

> **ONE-SAMPLE t CONFIDENCE INTERVAL**
>
> Suppose that an SRS of size n is drawn from a population having unknown mean μ. A level C **confidence interval** for μ is
>
> $$\bar{x} \pm m$$
>
> In this formula, the **margin of error** is
>
> $$m = t^* \text{SE}_{\bar{x}} = t^* \frac{s}{\sqrt{n}}$$
>
> where t^* is the value for the $t(n-1)$ density curve with area C between $-t^*$ and t^*. This interval is exact when the population distribution is Normal and is approximately correct for large n in other cases.

CASE 8.1 **Battery Life of a Smartphone** These days, the battery life of a smartphone is more important than ever. Demanding apps, bigger screens, and an increased reliance on smartphones have all contributed to the need to charge a phone at least once daily. To assess battery life, product-testing groups put smartphones through a stress test and record the time until the phone dies. *Consumer Reports*, for example, tests batteries using a robotic finger that simulates what a person might do with a phone during a typical day.[1] Other groups will set the phone to play a video on continuous loop or continuously surf the Web until it is out of juice.

Suppose a product-testing group that uses the Web-surfing test reports your smartphone has a battery life of 684 minutes (11.4 hours). Because you use your phone intermittently and for more than just surfing, you expect your battery to have a longer life and decide to investigate this by starting each weekday on a full charge and seeing how long your phone lasts while doing your normal activities. ■

EXAMPLE 8.1

BLIFE

CASE 8.1 **Estimating the Average Battery Life of a Smartphone** The following data are the times in minutes for eight weekdays:

$$787 \quad 721 \quad 685 \quad 727 \quad 673 \quad 751 \quad 769 \quad 643$$

We want to find a 95% confidence interval for μ, the average number of minutes before your smartphone runs out of juice.

The sample mean is

$$\bar{x} = \frac{787 + 721 + \cdots + 643}{8} = 719.5$$

and the sample standard deviation is

$$s = \sqrt{\frac{(787 - 719.5)^2 + (721 - 719.5)^2 + \cdots + (643 - 719.5)^2}{8 - 1}} = 49.66$$

with degrees of freedom $n - 1 = 7$. The standard error of $\bar{x}$ is

$$\mathrm{SE}_{\bar{x}} = \frac{s}{\sqrt{n}} = \frac{49.66}{\sqrt{8}} = 17.56$$

df = 7		
t^*	1.895	2.365
C	90%	95%

From Table D, we find $t^* = 2.365$. In Excel, this value can be obtained by entering the formula `=T.INV(0.975,7)` into a cell. The first input is 1 minus the upper-tail probability p and the second value is the degrees of freedom.

The margin of error is then

$$m = 2.365 \times \mathrm{SE}_{\bar{x}} = (2.365)(17.56) = 41.5$$

and the 95% confidence interval is

$$\bar{x} \pm m = 719.5 \pm 41.5$$
$$= (678.0, 761.0)$$

Thus, we are 95% confident that the average battery life of your smartphone is between 678.0 and 761.0 minutes. ■

In this example, we have given the actual interval as our answer. Sometimes, we prefer to report the mean and margin of error: the average amount of time is 719.5 minutes with a margin of error of 41.5 minutes.

The use of the *t* confidence interval in Example 8.1 rests on a couple assumptions. First, we assume that our sample of weekdays is an SRS from the population of all possible weekdays, a population that does not actually exist. Because you randomly chose the weekdays from among the 20 weekdays of the month, this is likely reasonable. Whether we can generalize these results beyond this month depends on how similar your phone use is over time. Second, because our sample size is not large, we assume that the distribution of times is Normal. With only eight observations, this assumption cannot be effectively checked. We can, however, check if the data suggest a severe departure from Normality. Figure 8.2 shows the Normal quantile plot, and we can clearly see there are no outliers or severe skewness. Deciding whether to use the *t* confidence interval for inference about μ is often a judgment call. We provide some practical guidelines to assist in this decision later in this section.

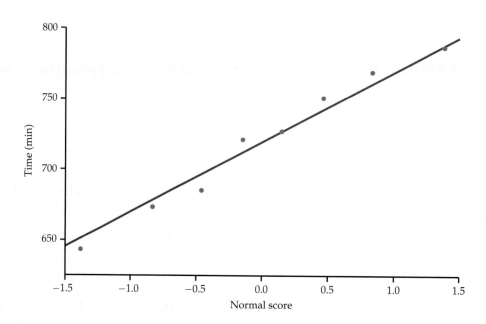

FIGURE 8.2 Normal quantile plot of the data, Example 8.1.

APPLY YOUR KNOWLEDGE

8.3 More on apartment rents. Refer to Exercise 8.1 (page 398). Assuming the rents in your city are Normally distributed, construct a 95% confidence interval for the mean monthly rent of all advertised one-bedroom apartments.

8.4 90% versus 95% confidence interval. If you chose 90%, rather than 95%, confidence in the previous exercise, would your margin of error be larger or smaller? Explain your answer.

The one-sample *t* test

z test for a population mean, p. 366–367

Significance tests of the mean μ using the standard error of $\bar{x}$ are also very similar to the *z* test described in the previous chapter. We still carry out the four steps required to do a significance test (page 365), but because we use *s* in place of σ, the distribution we use to find the *P*-value changes from the standard Normal to a *t* distribution. Here are the details.

ONE-SAMPLE t TEST

Suppose that an SRS of size n is drawn from a population having unknown mean μ. To test the hypothesis $H_0: \mu = \mu_0$, compute the **one-sample t statistic**

$$t = \frac{\bar{x} - \mu_0}{s/\sqrt{n}}$$

In terms of a random variable T having the $t(n-1)$ distribution, the P-value for a test of H_0 against

$H_a: \mu > \mu_0$ is $P(T \geq t)$

$H_a: \mu < \mu_0$ is $P(T \leq t)$

$H_a: \mu \neq \mu_0$ is $2P(T \geq |t|)$

These P-values are exact if the population distribution is Normal and are approximately correct for large n in other cases.

EXAMPLE 8.2

CASE 8.1 **Does Your Phone's Average Battery Life Differ from the Reported Value?** Can the Web-surfing test result be used as an estimate of your smartphone's average battery life? To help answer this, we can use the sample of times in Example 8.1 to test whether the average battery life of your phone differs from that reported by the testing group. Specifically, we want to test

BLIFE

$$H_0: \mu = 684$$
$$H_a: \mu \neq 684$$

at the 0.05 significance level. Recall that $n = 8$, $\bar{x} = 719.5$, and $s = 49.66$. The t statistic is

$$t = \frac{\bar{x} - \mu_0}{s/\sqrt{n}} = \frac{719.5 - 684}{49.66/\sqrt{8}}$$
$$= 2.022$$

This means that the sample mean $\bar{x} = 719.5$ is slightly more than two standard errors above the null hypothesized value of 684.

Because the degrees of freedom are $n - 1 = 7$, this t statistic has the $t(7)$ distribution. Figure 8.3 shows that the P-value is $2P(T \geq |2.022|)$, where T has the $t(7)$ distribution.

Using Table D, we see that $P(T \geq 1.895) = 0.05$ and $P(T \geq 2.365) = 0.025$. Thus, we only know that the P-value is between $2 \times 0.025 = 0.05$ and $2 \times 0.05 = 0.10$. In Excel, the exact P-value for the two-sided alternative can be obtained using the function =T.DIST.2T(2.022,7), where the first input is the absolute value of the test statistic and the second input is the degrees of freedom. It gives $P = 0.0829$.

	df = 7	
p	0.05	0.025
t^*	1.895	2.365

FIGURE 8.3 Sketch of the P-value calculation, Example 8.2.

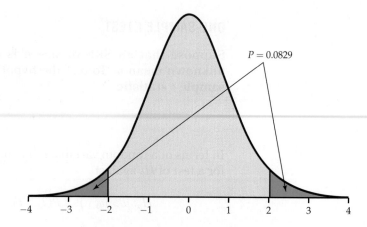

These data are compatible with an average of $\mu = 684$ minutes. Under H_0, a difference this large or larger would occur more than one time in twenty simply due to chance. There is not enough evidence to reject the null hypothesis at the 0.05 level. ■

In this example, we tested the null hypothesis $\mu = 684$ against the two-sided alternative $\mu \neq 684$. Because we had suspected that your average time would be larger, we could have used a one-sided test.

EXAMPLE 8.3

BLIFE

CASE 8.1 **One-sided Test for Average Battery Life of a Smartphone** To test whether the average battery life is larger than the testing group's mean, our hypotheses are

$$H_0: \mu = 684$$
$$H_a: \mu > 684$$

The t test statistic does not change: $t = 2.022$. As Figure 8.4 illustrates, however, the P-value is now $P(T \geq 2.022)$, half of the value in Example 8.2. From Table D, we can determine that $0.025 < P < 0.05$; again rounding to four decimal places, Excel gives $P = 0.0414$, using the function =T.DIST.RT(2.022,7). There is now enough evidence to reject the null hypothesis at the 0.05 significance level and conclude that average battery life of your phone is greater than that reported by the testing group. ■

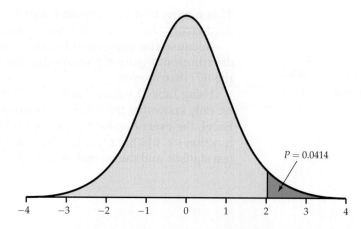

FIGURE 8.4 Sketch of the P-value calculation, Example 8.3.

8.1 Inference for the Mean of a Population

For these data, our conclusion depended on the choice between a one-sided and a two-sided alternative hypothesis. Because this can happen, the choice of alternative hypothesis needs to be made prior to analysis. *It is wrong to examine the data first and then decide to do a one-sided test in the direction indicated by the data.* If in doubt, always use a two-sided test. This is the alternative hypothesis to use when there is no prior suspicion that the mean is larger or smaller.

APPLY YOUR KNOWLEDGE

8.5 Apartment rents. Refer to Exercise 8.1 (page 398). Assuming the rents in your city are Normally distributed, do these data give good reason to believe that the average rent for all advertised one-bedroom apartments in your city is different from $800 per month? Make sure to state the hypotheses; find the t statistic, degrees of freedom, and P-value; and state your conclusion using the 5% significance level.

8.6 Significant? A test of a null hypothesis versus the greater than alternative gives $t = 1.81$.

(a) The sample size is 13. Is the test result significant at the 5% level? Explain how you obtained your answer.

(b) The sample size is 9. Is the test result significant at the 5% level?

(c) Sketch the two t distributions to illustrate your answers.

8.7 Average quarterly return. A stockbroker predicts the short-run direction of the market using the average quarterly return of stock mutual funds. He believes the market direction will be up when the average return is greater than 1.5%. He will get complete quarterly return information soon, but right now he has data from a random sample of 25 stock funds. The mean quarterly return of the sample is 2.3%, and the standard deviation is 2.6%. Based on this sample, can the broker state the market direction is up?

(a) State appropriate null and alternative hypotheses. Explain how you decided between a one- or two-sided alternative.

(b) Find the t statistic and its degrees of freedom. State your conclusion using the $\alpha = 0.05$ significance level.

Using software

For small data sets, such as the one in Example 8.1, it is easy to perform the computations for confidence intervals and significance tests with an ordinary calculator and Table D. For larger data sets, however, software or a statistical calculator eases our work.

EXAMPLE 8.4

DIVRS

Diversify or Be Sued An investor sued his broker and brokerage firm, claiming a lack of diversification in his stock portfolio led to poor performance. Table 8.1 gives the rates of return for the 39 months that the account was managed by the broker. An arbitration panel compared these returns with the average of the Standard & Poor's 500-stock index for the same period.[2] Let's do the same using the one-sample t test.

Consider the 39 monthly returns as a random sample from the population of possible monthly returns that the brokerage firm would generate during this time period. Are these returns compatible with a population mean of $\mu = 1.38\%$, the average S&P 500 return during that period? Our hypotheses are

$$H_0: \mu = 1.38$$
$$H_a: \mu \neq 1.38$$

TABLE 8.1 Monthly rates of return on a portfolio (percent)

−8.36	1.63	−2.27	−2.93	−2.70	−2.93	−9.14	−2.64
6.82	−2.35	−3.58	6.13	7.00	−15.25	−8.66	−1.03
−9.16	−1.25	−1.22	−10.27	−5.11	−0.80	−1.44	1.28
−0.65	4.34	12.22	−7.21	−0.09	7.34	5.04	−7.24
−2.14	−1.01	−1.41	12.03	−2.56	4.33	2.35	

Figure 8.5 gives a histogram and summary statistics for these data using JMP. There are no outliers, and the distribution shows no strong skewness. Thus, we are reasonably confident that the distribution of $\bar{x}$ is approximately Normal, and can proceed with our inference based on Normal theory. Minitab, SPSS, and Excel outputs appear in Figure 8.6.

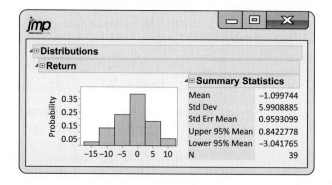

FIGURE 8.5 Histogram and summary statistics of monthly rates of return for a stock portfolio using JMP, Example 8.4.

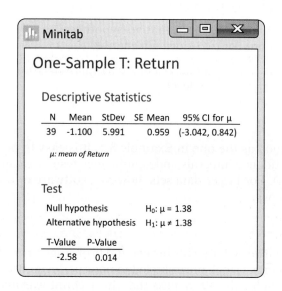
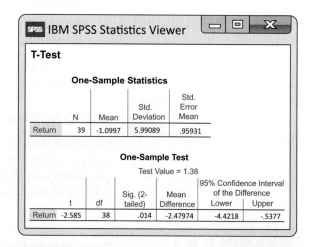

FIGURE 8.6 Minitab, SPSS, and Excel outputs, Examples 8.4 and 8.5.

Here is one way to report the conclusion: the mean monthly return on investment for this client's account was $\bar{x} = -1.1\%$. This differs significantly from 1.38%, the performance of the S&P 500 for the same period ($t = -2.58$, df = 38, $P = 0.014$). ■

The hypothesis test in Example 8.4 leads us to conclude that the mean return on the client's account differs from that of the stock index. Now let's assess the return on the client's account with a confidence interval.

EXAMPLE 8.5

DIVRS

Estimating the Mean Monthly Return The mean monthly return on the client's portfolio was $\bar{x} = -1.1\%$, and the standard deviation was $s = 5.99\%$. Figures 8.5 and 8.6 all give output for a 95% confidence interval for the population mean μ. Note that Excel gives the margin of error next to the label "Confidence Level(95.0%)" rather than the actual confidence interval. We see that the 95% confidence interval is $(-3.04, 0.84)$, or (from Excel) -1.0997 ± 1.9420. ■

Because the S&P 500 return, 1.38%, falls outside the confidence interval, we know that μ differs significantly from 1.38% at the $\alpha = 0.05$ level. The interval suggests that the broker's management of this account had a long-term mean somewhere between a loss of 3.04% and a gain of 0.84% per month. However, we are interested not in the actual mean but in the difference between the broker's process and the diversified S&P 500 index.

EXAMPLE 8.6

DIVRS

Estimating the Difference from a Standard The difference between the mean of the investor's account and the S&P 500 is $\bar{x} - \mu = -1.10 - 1.38 = -2.48\%$. In Example 8.5, we found that the 95% confidence interval for the investor's account was $(-3.04, 0.84)$. To obtain the corresponding interval for the difference, subtract 1.38 from each of the endpoints. The resulting interval is $(-3.04 - 1.38, 0.84 - 1.38)$, or $(-4.42, -0.54)$. This interval is presented in the SPSS output of Figure 8.6. We conclude with 95% confidence that the underperformance was between -4.42% and -0.54%. This estimate helps to set the compensation owed to the investor. ■

The assumption that these 39 monthly returns represent an SRS from the population of monthly returns from that brokerage during this time period is certainly debatable. One could also argue that we should not consider the S&P 500 monthly average return as a constant standard. Its return would also vary from month to month. If those monthly returns were available, an alternative analysis would be to compare the average difference between each monthly return for this account and for the S&P 500. This type of comparison is discussed next.

APPLY YOUR KNOWLEDGE

CASE 8.1 **8.8 Using software to compute a confidence interval.** In Example 8.1 (page 399), we calculated the 95% confidence interval for the average battery life of a smartphone. Use software to compute this interval, and verify that you obtain the same interval. BLIFE

CASE 8.1 **8.9 Using software to perform a significance test.** In Example 8.2 (page 401), we tested whether the average battery life of your smartphone was different from a reported value. Use software to perform this test and obtain the exact P-value. BLIFE

Matched pairs *t* procedures

confounding, p. 144

matched pairs design, p. 156

The battery life problem of Case 8.1 concerns only a single population. We know that comparative studies are usually preferred to single-sample investigations because of the protection they offer against confounding. For that reason, inference about a parameter of a single distribution is less common than comparative inference.

One common comparative design, however, makes use of single-sample procedures. In a matched pairs study, subjects are matched in pairs and the outcomes are compared within each matched pair. For example, an experiment to compare two marketing campaigns might use pairs of subjects that are the same age, sex, and income level. The experimenter could toss a coin to assign the two campaigns to the two subjects in each pair. The idea is that matched subjects are more similar than unmatched subjects, so comparing outcomes within each pair is more efficient (i.e., reduces the standard deviation of the estimated difference of treatment means). Matched pairs are also common when randomization is not possible. For example, before-and-after observations on the same subjects call for a matched pairs analysis.

EXAMPLE 8.7

GEPRT

The Effect of Altering a Software Parameter The MeasureMind® 3D MultiSensor metrology software is used by various companies to measure complex machine parts. As part of a technical review of the software, researchers at GE Healthcare discovered that unchecking one option reduced measurement time by 10%. This time reduction would help the company's productivity provided the option has no impact on the measurement outcome. To investigate this, the researchers measured 76 parts using the software both with and without this option checked.[3]

Table 8.2 gives the measurements (in microns) for the first 10 parts. For analysis, we subtract the measurement with the option on from the measurement with the option off. These differences form a single sample and appear in the "Diff" column.

TABLE 8.2 Parts measurements using optical software

Part	OptionOn	OptionOff	Diff	Part	OptionOn	OptionOff	Diff
1	118.63	119.01	0.38	6	117.78	118.04	0.26
2	117.34	118.51	1.17	7	119.29	119.25	−0.04
3	119.30	119.50	0.20	8	120.26	118.84	−1.42
4	119.46	118.65	−0.81	9	118.42	117.78	−0.64
5	118.12	118.06	−0.06	10	119.49	119.66	0.17

To assess whether there is a difference between the measurements with and without this option, we test

$$H_0: \mu = 0$$
$$H_a: \mu \neq 0$$

Here μ is the mean difference for the entire population of parts. The null hypothesis says that there is no difference, and H_a says that there is a difference, but does not specify a direction.

The 76 differences have

$$\bar{x} = 0.027 \quad \text{and} \quad s = 0.607$$

Figure 8.7 shows a histogram of the differences. It is reasonably symmetric with no outliers, so based on the guidelines we discuss next (pages 409–410), we can comfortably use the one-sample t procedures. *Remember to always check assumptions before proceeding with statistical inference.*

FIGURE 8.7 Histogram of the differences in measurements (option off minus option on), Example 8.7.

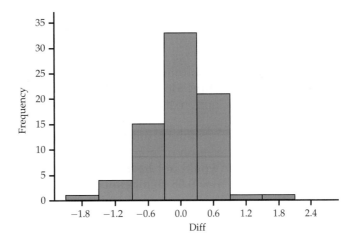

The one-sample t statistic is

$$t = \frac{\bar{x} - 0}{s/\sqrt{n}} = \frac{0.027}{0.607/\sqrt{76}} = 0.39$$

The P-value is found using the $t(75)$ distribution because the degrees of freedom are 1 less than the sample size.

Table D does not provide a row for 75 degrees of freedom, but for both $t(60)$ and $t(80)$, $t = 0.39$ lies to the left of the first column entry. This means the P-value is greater than $2(0.25) = 0.50$. Using software for all the calculations gives the exact value $P = 0.6967$. There is little evidence to suggest this option has an impact on the measurements. When reporting results, it is usual to omit the details of routine statistical procedures; our test would be reported in the form: "The difference in measurements was not statistically significant ($t = 0.39$, df $= 75$, $P = 0.70$)." ■

equivalence testing

This result, however, does not fully address the goal of their study. *A lack of statistical significance does not prove the null hypothesis is true.* If that were the case, we would simply design poor experiments whenever we wanted to prove the null hypothesis. The more appropriate method of inference in this setting is to consider **equivalence testing.** With this approach, we try to prove that the mean difference is within some acceptable region around 0. This can be done using a confidence interval.

EXAMPLE 8.8

GEPRT

Are the Two Means Equivalent? Suppose the GE Healthcare researchers state that a mean difference less than 0.25 microns is not important. To see if the data support a mean difference within 0.00 ± 0.25 microns, we construct a 90% confidence interval for the mean difference.

The standard error is

$$\text{SE}_{\bar{x}} = \frac{s}{\sqrt{n}} = \frac{0.607}{\sqrt{76}} = 0.070$$

so the margin of error is

$$m = t^* \times \text{SE}_{\bar{x}} = (1.665)(0.070) = 0.116$$

where the value $t^* = 1.665$ comes from the Excel cell entry =T.INV(0.95,75). The confidence interval is

$$\bar{x} \pm m = 0.027 \pm 0.116$$
$$= (-0.089, 0.143)$$

This interval is entirely within the 0.00 ± 0.25-micron region that the researchers state is not important. Thus, we can conclude at the 5% significance level that the two means are equivalent. The company can turn this option off to save time obtaining measurements. ∎

If the resulting 90% confidence interval would have been outside the stated region or contained values both within and outside the stated region, we would not have been able to conclude that the means are equivalent at the $\alpha = 0.05$ significance level.

ONE-SAMPLE TEST OF EQUIVALENCE

Suppose that an SRS of size n is drawn from a population having unknown mean μ. To test, at significance level α, if μ is within a range of equivalency to μ_0, specified by the interval $\mu_0 \pm \delta$:

1. Compute the confidence interval with $C = 1 - 2\alpha$.

2. Compare this interval with the range of equivalency.

If the confidence interval falls entirely within $\mu_0 \pm \delta$, conclude that μ is equivalent to μ_0. If the confidence interval is outside the equivalency range or contains values both within and outside the range, then there is not enough evidence to conclude μ is equivalent to μ_0.

This approach is exact when the population is Normally distributed and approximately correct for large n in other cases.

APPLY YOUR KNOWLEDGE

8.10 Cold plasma technology. Cold plasma (CP) technology has been shown to be an effective tool for shelf-life extension of food.[4] For it to become widely used, however, CP's effects on food quality must be understood. Consider the following study that compared the taste of fruit juice with and without CP treatment. Each juice was rated, by a set of taste experts, using a 0 to 100 scale, with 100 being the highest rating. For each expert, a coin was tossed to see which juice was tasted first. JUICE

	Expert									
	1	2	3	4	5	6	7	8	9	10
With CP:	93	65	37	78	89	49	88	55	63	62
Without CP:	92	67	47	79	84	52	80	52	67	69

Is there a difference in taste? State the appropriate hypotheses, and carry out a matched pairs t test using $\alpha = 0.05$.

8.11 Are the means equivalent? Refer to the previous exercise. Other research has suggested a difference of ±4 on this taste scale is indistinguishable. Use the data to get a 90% confidence interval for the mean difference in taste scores between CP and non-CP treated fruit juice. Is there enough evidence to claim the two means are equivalent?

Robustness of the one-sample *t* procedures

The matched pairs *t* procedures use one-sample *t* confidence intervals and significance tests for differences. They are, therefore, based on an assumption that the *population of differences* has a Normal distribution. For the histogram of the 76 differences in Example 8.7 shown in Figure 8.7 (page 407), the data appear to be slightly skewed. Does this non-Normality suggest that we should not use the *t* procedures for these data?

All inference procedures are based on some conditions, such as Normality. Procedures that are not strongly affected by violations of a condition are called *robust*. Robust procedures are very useful in statistical practice because they can be used over a wide range of conditions with good performance.

ROBUST PROCEDURES

A statistical inference procedure is called **robust** if the probability calculations required are insensitive to violations of the conditions that usually justify the procedure.

resistant measure, p. 26

central limit theorem, p. 313

The condition that the population be Normal rules out outliers, so the presence of outliers shows that this condition is not fulfilled. The *t* procedures are not robust against outliers because these statistics depend on $\bar{x}$ and s, and they are not resistant to outliers.

Fortunately, the *t* procedures are quite robust against non-Normality of the population, particularly when the sample size is large. The *t* procedures rely only on the Normality of the sample mean $\bar{x}$. This condition is satisfied when the population is Normal, but the central limit theorem tells us that a mean $\bar{x}$ from a large sample follows a Normal distribution closely even when individual observations are not Normally distributed.

To convince yourself of this fact, use the *Distribution of the One-Sample t Statistic* applet to study the sampling distribution of the one-sample *t* statistic. From one of three population distributions, 10,000 SRSs of a user-specified sample size *n* are generated, and a histogram of the *t* statistics is constructed. You can then compare this estimated sampling distribution with the $t(n-1)$ distribution. When the population distribution is Normal, the sampling distribution is always *t* distributed. For the other two distributions, you should see that as *n* increases, the histogram looks more like the $t(n-1)$ distribution.

To assess whether the *t* procedures can be used in practice, Normal quantile plots, stemplots, histograms, and boxplots are all good tools for checking for skewness and outliers. For most purposes, the one-sample *t* procedures can be safely used when $n \geq 15$ unless an outlier or clearly marked skewness is present. In fact, the condition that the data are an SRS from the population of interest is the more crucial assumption, except in the case of small samples. Here are practical guidelines, based on the sample size and plots of the data, for inference on a single mean:[5]

- *Sample size less than 15:* Use *t* procedures if the data are close to Normal. If the data are clearly non-Normal or if outliers are present, do not use *t*.

- *Sample size at least 15:* The t procedures can be used except in the presence of outliers or strong skewness.

- *Large samples:* The t procedures can be used even for clearly skewed distributions when the sample is large, roughly $n \geq 40$.

For the battery life study in Example 8.1, there are only $n = 8$ observations. We are comfortable that the t procedures give approximately correct results because the data are close to Normal. On the other hand, the measurement study in Example 8.7 has $n = 76$ observations. With that large a sample, our only concern is the presence of outliers.

APPLY YOUR KNOWLEDGE

8.12 Significance test for the cold plasma technology study? Consider the paired differences in Exercise 8.10. Construct a Normal quantile plot or histogram to assess Normality. Do you feel comfortable that the t procedures will give approximately correct results? Explain your answer.

8.13 Significance test for the average T-bill interest rate? Consider data on the T-bill interest rate presented in Figure 1.29 (page 54). Would you feel comfortable applying the t procedures in this case? Explain your answer.

Inference for non-Normal populations

So what do we do when our populations are clearly non-Normal and we *do not* think that the sample size is large enough to rely on the robustness of the t procedures? If you face this problem, you should consult an expert. Three general strategies are available:

- In some cases, a distribution other than a Normal distribution describes the data well. There are many non-Normal models for data, and inference procedures for these models are available.

- Because skewness is the chief barrier to the use of t procedures on data without outliers, you can attempt to transform skewed data so that the distribution is symmetric and as close to Normal as possible. Confidence levels and P-values from the t procedures applied to the transformed data will be quite accurate for even moderate sample sizes. Methods are generally available for transforming the results back to the original scale.

distribution-free procedures

nonparametric procedures

- The third strategy is to use a **distribution-free** inference procedure. Such procedures do not assume that the population distribution has any specific form, such as Normal. Distribution-free procedures are often called **nonparametric procedures.** Chapter 17 discusses many of the traditional nonparametric procedures.

Each of these strategies quickly takes us beyond the basic practice of statistics. We emphasize procedures based on Normal distributions because they are the most common in practice, because their robustness makes them widely useful, and (most importantly) because we are first of all concerned with understanding the principles of inference.

Distribution-free significance tests do not require that the data follow any specific type of distribution such as Normal. This gain in generality isn't free: if the data really are close to Normal, distribution-free tests do not detect true alternatives as often as t tests. They also don't quite answer the same question. The t tests concern the population *mean*. Distribution-free tests ask about the population *median,* as is natural for distributions that are skewed.

The sign test

sign test The simplest distribution-free test, and one of the most useful, is the **sign test.** The test gets its name from the fact that we look only at the signs of the differences, not their actual values. The following example illustrates this test.

EXAMPLE 8.9

GEPRT

Does Altering a Software Parameter Affect the Median? In Example 8.7 (page 406), we used the matched pairs t test to compare measurement data from two software algorithms. The slight skewness in the distribution of differences, however, may make the P-value only roughly correct. The sign test is based on the following simple observation: of the 76 parts measured, 43 had a larger measurement with the option off and 33 had a larger measurement with the option on.

To perform a significance test based on these counts, let p be the probability that a randomly chosen part would have a larger measurement with the option turned off. The null hypothesis of "no effect" says that these two measurements are just repeat measurements, so the measurement with the option off is equally likely to be larger or smaller than the measurement with the option on. Therefore, we want to test

$$H_0: p = 1/2$$
$$H_a: p \neq 1/2$$

The 76 parts are independent trials, so the number that had larger measurements with the option off has the binomial distribution $B(76, 1/2)$ if H_0 is true. The P-value for the observed count 43 is, therefore, $2P(X \geq 43)$, where X has the $B(76, 1/2)$ distribution. You can compute this probability with software or the Normal approximation to the binomial:

Normal approximation to the binomial, p. 277

$$2P(X \geq 43) = 2P\left(Z \geq \frac{43 - 76(0.5)}{\sqrt{76(0.5)(1 - 0.5)}}\right)$$
$$= 2P\left(Z \geq \frac{43 - 38}{\sqrt{19}}\right)$$
$$= 2P(Z \geq 1.147)$$
$$= 2(0.1251)$$
$$= 0.2502$$

As in Example 8.7, there is not strong evidence that the two measurements are different. ■

There are several varieties of sign test, all based on counts and the binomial distribution. The sign test for matched pairs is the most useful. The null hypothesis of "no effect" is then always $H_0: p = 1/2$. The alternative can be one-sided in either direction or two-sided, depending on the type of change we are considering.

SIGN TEST FOR MATCHED PAIRS

Ignore pairs with difference 0; the number of trials n is the count of the remaining pairs. The test statistic is the count X of pairs with a positive difference. P-values for X are based on the binomial $B(n, 1/2)$ distribution.

The matched pairs *t* test in Example 8.7 tested the hypothesis that the mean of the distribution of differences is 0. The sign test in Example 8.9 is, in fact, testing the hypothesis that the *median* of the differences is 0. If p is the probability that a difference is positive, then $p = 1/2$ when the median is 0. This is true because the median of the distribution is the point with probability 1/2 lying to its right. As Figure 8.8 illustrates, $p > 1/2$ when the median is greater than 0, again because the probability to the right of the median is always 1/2. The sign test of H_0: $p = 1/2$ against H_a: $p > 1/2$ is a test of

$$H_0: \text{population median} = 0$$

$$H_a: \text{population median} > 0$$

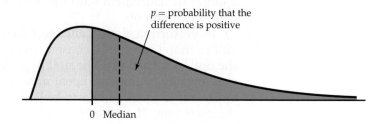

FIGURE 8.8 Why the sign test tests the median difference: when the median is greater than 0, the probability p of a positive difference is greater than 1/2, and vice versa.

The sign test in Example 8.9 makes no use of the actual scores—it just counts how many parts had a larger measurement with the option off. Any parts that did not have different measurements would be ignored altogether. Because the sign test uses so little of the available information, it does not detect true alternatives as often as the *t* test when the data are close to Normal. Chapter 17 describes other distribution-free tests that use more of the available information.

APPLY YOUR KNOWLEDGE

8.14 Sign test for the cold-plasma study. Exercise 8.10 (page 408) gives data on the taste of fruit juice with and without CP treatment. Is there evidence that the median difference in taste is nonzero? State the hypotheses, carry out the sign test, and report your conclusion.

BEYOND THE BASICS

The bootstrap

A modern computer-intensive nonparametric procedure that is especially useful for confidence intervals is the *bootstrap*. Confidence intervals are based on sampling distributions. To construct the *t* confidence intervals, we've used the facts that

- The sampling distribution of $\bar{x}$ is $N(\mu, \sigma/\sqrt{n})$ when the data are an SRS from an $N(\mu, \sigma)$ population.

- If the data are non-Normal and n large, the central limit theorem tells us that the distribution of $\bar{x}$ is approximately $N(\mu, \sigma/\sqrt{n})$, provided the distribution of the data is not strongly skewed and there are no outliers.

What if the population does not appear to be Normal and we have only a small sample? Then we do not know what the sampling distribution of $\bar{x}$ looks like. The bootstrap is a procedure for approximating sampling distributions when theory cannot tell us their shape.[6] It is an example of how the use of fast and easy computing is changing the way we do statistics.

resample

The basic idea is to act as if our sample were the population. We take many samples from it. Each of these is called a **resample.** We calculate the mean $\bar{x}$ for each resample. We get different results from different resamples because we sample *with replacement*. Thus, an observation in the original sample can appear more than once in a resample. We treat the resulting distribution of $\bar{x}$'s as if it were the sampling distribution and use it to perform inference. If we want a 95% confidence interval, for example, we could use the middle 95% of this distribution.

EXAMPLE 8.10

BLIFE

Bootstrap Confidence Interval. Consider the eight battery life measurements (in minutes) in Example 8.1:

787 721 685 727 673 751 769 643

We defended the use of the one-sample t confidence interval for an earlier analysis. Let's now compare those results with the confidence interval constructed using the bootstrap.

This can be done using JMP by right clicking on the mean value in the Summary Statistics produced from the Distribution platform and selecting bootstrap. We decide to collect the $\bar{x}$'s from 1000 resamples of size $n = 8$. Figure 8.9 shows the histogram of these 1000 $\bar{x}$'s. One resample was

787 721 721 721 643 721 787 685

with $\bar{x} = 723.25$. The middle 95% of our 1000 $\bar{x}$'s runs from 690.3 to 752.7. We repeat the procedure and get the interval (687.3, 752.5). The two bootstrap intervals are relatively close to each other and are more narrow than the one-sample t confidence interval (678.0, 761.0). ∎

FIGURE 8.9 Histogram of 1000 $\bar{x}$'s based on resamples of size $n = 8$ using JMP. The middle 95% of these $\bar{x}$'s represents a 95% bootstrap confidence interval.

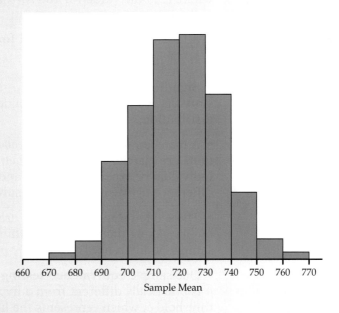

Using the middle 95% of the resampled $\bar{x}$'s is just one of many methods to form the 95% confidence interval from the resamples. While conceptually straightforward, the resulting intervals tend to be too narrow when the sample size is small. Other methods do a better job with coverage but require more computation. As a result, the bootstrap is practical only when you can use a computer to take and store a large number of samples quickly.

SECTION 8.1 SUMMARY

- Significance tests and confidence intervals for the mean μ of a Normal population are based on the sample mean $\bar{x}$ of an SRS. Because of the central limit theorem, the resulting procedures are approximately correct for other population distributions when the sample is large.

- The **standard error** of the sample mean is

$$\text{SE}_{\bar{x}} = \frac{s}{\sqrt{n}}$$

- The standardized sample mean, or **one-sample z statistic**,

$$z = \frac{\bar{x} - \mu}{\sigma/\sqrt{n}}$$

has the $N(0,1)$ distribution. If the standard deviation of $\bar{x}$ is replaced by the standard error, the **one-sample t statistic**

$$t = \frac{\bar{x} - \mu}{s/\sqrt{n}}$$

has the t **distribution** with $n-1$ degrees of freedom.

- There is a t distribution for every positive **degrees of freedom** k. All are symmetric distributions similar in shape to Normal distributions. The $t(k)$ distribution approaches the $N(0,1)$ distribution as k increases.

- The **margin of error** for level C confidence is

$$m = t^* \times \text{SE}_{\bar{x}} = t^* \frac{s}{\sqrt{n}}$$

where t^* is the value for the $t(n-1)$ density curve with area C between $-t^*$ and t^*.

- A level C **confidence interval for the mean** μ of a Normal population is

$$\bar{x} \pm m$$

- Significance tests for H_0: $\mu = \mu_0$ are based on the one-sample t statistic. P-values and critical values are computed from the $t(n-1)$ distribution.

- A matched pairs analysis is needed when subjects or experimental units are matched in pairs or when there are two measurements on each individual or experimental unit and the question of interest concerns the difference between the two measurements.

- These one-sample procedures are used to analyze **matched pairs** data by first taking the difference within each matched pair to produce a single sample of differences.

- One-sample **equivalence testing** assesses whether a population mean μ is practically different from a hypothesized mean μ_0. This test requires a threshold δ, which represents the largest difference between μ and μ_0 such that the means are considered equivalent.

- The t procedures are relatively **robust** against lack of Normality, especially for larger sample sizes. The t procedures are useful for non-Normal data when $n \geq 15$ unless the data show outliers or strong skewness.

- The **sign test** is a **distribution-free test** because it uses probability calculations that are correct for a wide range of population distributions.

- The sign test for "no treatment effect" in matched pairs counts the number of positive differences. The P-value is computed from the $B(n, 1/2)$ distribution, where n is the number of nonzero differences. Because the sign test uses limited information, it does not detect true alternatives as often as the t test in cases where use of the t test is justified.

SECTION 8.1 EXERCISES

For Exercises 8.1 and 8.2, see page 398; for 8.3 and 8.4, see page 400; for 8.5 to 8.7, see page 403; for 8.8 and 8.9, see page 405; for 8.10 and 8.11, see pages 408–409; for 8.12 and 8.13, see page 410; and for 8.14 see page 412.

8.15 What's wrong? For each of the following statements, explain what is wrong and why.

(a) As the degrees of freedom k increase, the $t(k)$ density curve moves away from the $N(0,1)$ curve.

(b) The 95% margin of error of the sample mean is $t^* s^2 / n$, with t^* coming from the $t(n-1)$ distribution.

(c) A researcher wants to test $H_0: \bar{x} = 1000$ versus the one-sided alternative $H_a: \bar{x} < 1000$.

(d) One-sample t procedures can safely be used in practice when $n \geq 15$.

8.16 Finding the t^*-values. What value t^* from software, or Table D, should be used to calculate the margin of error for a confidence interval for the mean of the population in each of the following situations?

(a) A 95% confidence interval based on $n = 15$ observations.

(b) A 90% confidence interval from an SRS of 24 observations.

(c) A 95% confidence interval from a sample of size 24.

(d) These cases illustrate how the size of the margin of error depends on the confidence level and on the sample size. Summarize the relationships illustrated.

8.17 A one-sample t test. The one-sample t statistic for testing

$$H_0: \mu = 10$$
$$H_a: \mu > 10$$

from a sample of $n = 28$ observations has the value $t = 2.13$.

(a) What are the degrees of freedom for this statistic?

(b) Give the two critical values t^* from Table D that bracket t.

(c) What are the right-tail probabilities p for these two entries?

(d) Between what two values does the P-value of the test fall?

(e) Is the value $t = 2.13$ significant at the 5% level? Is it significant at the 1% level?

(f) If you have software available, find the exact P-value.

8.18 Another one-sample t test. The one-sample t statistic for testing

$$H_0: \mu = 60$$
$$H_a: \mu \neq 60$$

from a sample of $n = 30$ observations has the value $t = -2.09$.

(a) What are the degrees of freedom for t?

(b) Locate the two critical values t^* from Table D that bracket t. What are the right-tail probabilities p for these two values?

(c) How would you report the P-value for this test?

(d) Is the value $t = -2.09$ statistically significant at the 5% level? At the 1% level?

(e) If you have software available, find the exact P-value.

8.19 A final one-sample t test. The one-sample t statistic for testing

$$H_0: \mu = 20$$
$$H_a: \mu < 20$$

based on $n = 200$ observations has the value $t = -2.74$.

(a) What are the degrees of freedom for this statistic?

(b) How would you report the P-value based on Table D?

(c) If you have software available, find the exact P-value.

8.20 Business bankruptcies in Canada. Business bankruptcies in Canada are monitored by the Office of the Superintendent of Bankruptcy Canada (OSB).[7] Included in each report are the assets and liabilities the company declared at the time of the bankruptcy filing. A study is based on a random sample of 60 reports from the province of Ontario. The average debt (liabilities minus assets) is $2.68 million with a standard deviation of $4.85 million.

(a) Construct a 95% one-sample t confidence interval for the average debt of these companies at the time of filing.

(b) For this population of companies, debt is positive. Because the sample standard deviation is larger than the sample mean, this debt distribution is right skewed. Provide a defense for using the t confidence interval in this case.

8.21 Fuel economy. In 2017, the Environmental Protection Agency (EPA) changed how it calculated window-sticker gas mileage. For many models, this meant a 1–2 miles per gallon (mpg) reduction in combined highway/city mileage.[8] Here are some combined mpg test values for the 2018 Toyota RAV4 LE: MILEAGE

25.9	25.4	24.7	26.1	24.7	25.7	25.2	26.2
26.1	25.3	24.3	24.6	26.5	25.8	25.8	25.6

(a) Create a histogram or Normal quantile plot of these values. Is it appropriate to analyze these data using the t procedures? Briefly explain your answer.

(b) Construct a 95% confidence interval for μ, the average combined mpg for this type of car.

8.22 Is the average mpg now lower? Refer to the previous exercise. The average combined mpg for the 2016 Toyota RAV4 LE was 26.0 mpg. Is the average combined mpg for the 2018 model lower? MILEAGE

(a) Perform a significance test using the 0.05 significance level. Be sure to specify the hypotheses, the test statistic, the P-value, and your conclusion.

(b) Explain why is it difficult to conclude that a statistically significant change is due solely to the new EPA calculation method.

(c) Propose an experiment that would allow you to test if the average mpg change is due to the new EPA methodology.

8.23 The return-trip effect. We often feel that the return trip from a destination takes less time than the trip to the destination even though the distance traveled is usually identical. To better understand this effect, a group of researchers ran a series of experiments.[9] In one experiment, they surveyed 69 participants who had just returned from a day trip by bus. Each was asked to rate how long the return trip had taken, compared with the initial trip, on an 11-point scale from $-5 =$ a lot shorter to $5 =$ a lot longer. The sample mean was -0.55, and the sample standard deviation was 2.16.

(a) These data are integer values representing an ordered set of opinions. Do you think it's reasonable to use t-based methods on this type of data? Explain your answer.

(b) Is there evidence that the mean rating is different from zero? Regardless of your opinion in (a), carry out the significance test using $\alpha = 0.05$ and summarize the results.

(c) Construct a 95% confidence interval for the mean. Do you think that this average distance from 0 on this Likert scale is of practical importance? To help answer this question, compute the standardized difference from 0 (i.e., the effect size).

8.24 Food costs. The Consumer Expenditure Survey provides information on the buying habits of U.S. consumers.[10] In the latest report, the average annual amount a person under the age of 25 spent on food was $4759 with a standard error of $289.

(a) Assuming a sample size of $n = 2000$, calculate a 90% confidence interval for the average annual amount a person under the age of 25 spends on food.

(b) Will this interval capture 90% of all annual food expenditures by persons under the age of 25? Explain your answer.

8.25 Statistical process control. An iodized salt manufacturer must maintain strict control over the amount of iodine in the product. An SRS of 25 servings of salt (1.5 grams each) was collected as part of a Six Sigma quality control effort within the company. Here are the amounts of iodine (micrograms per serving) for this sample: IODINE

69.55	70.22	71.18	70.10	69.24
69.41	69.59	70.16	70.22	70.48
69.98	69.94	69.83	70.67	68.78
69.26	70.66	70.47	70.81	69.67
70.71	70.72	69.86	70.38	69.88

(a) Create a histogram, boxplot, and Normal quantile plot of these amounts.

(b) Write a careful description of the distribution. Make sure to note any outliers, and comment on the skewness or Normality of the data.

(c) Based on your observations in part (b), is it appropriate to analyze these data using the t procedures? Briefly explain your response.

8.26 Estimating the average amount of iodine. Refer to the previous exercise. IODINE

(a) Find the mean, the standard deviation, and the standard error of the mean for this sample.

(b) If you were to calculate the margin of error for the average amount of iodine at 90% and 95% confidence, which would be smaller? Briefly explain your reasoning without doing the calculations.

(c) Calculate the 90% and 95% confidence intervals for the mean amount of iodine per serving.

(d) Compare the widths of these two intervals. Does this comparison support your answer to part (b)? Explain.

8.27 Significance test for the average amount of iodine. Refer to the previous two exercises. IODINE

(a) Do these data provide evidence that the average amount of iodine in a serving is 70.0 micrograms? Using a significance level of 5%, state your hypotheses, the P-value, and your conclusion.

(b) Do these data provide evidence that the average amount of iodine in a serving is 70.5 micrograms? Using a significance level of 5%, state your hypotheses, the P-value, and your conclusion.

(c) Explain the relationship between your conclusions to parts (a) and (b) and the 95% confidence interval calculated in the previous exercise.

8.28 Investigating the endowment effect. Consider an ice-cold glass of lemonade on a hot July day. What is the maximum price you'd be willing to pay for it? What is the minimum price at which you'd be willing to sell it? For most people, the maximum buying price will be less than the minimum selling price. In behavioral economics, this occurrence is called the endowment effect. People seem to add value to products, regardless of attachment, just because they own them.

As part of a series of studies, a group of researchers recruited 40 students from a graduate marketing course and obtained from each of them their maximum (to pay) and minimum (to sell) prices for a Vosges Woolloomooloo gourmet chocolate bar made with milk chocolate and coconut.[11] Test the null hypothesis that there is no difference between the two prices. Also construct a 95% confidence interval of the endowment effect. ENDOW

8.29 Plant capacity. A leading company chemically treats its product before packaging. The company monitors the weight of product per hour that each machine treats. An SRS of 90 hours of production data for a particular machine is collected. The measured variable is in pounds. PRDWGT

(a) Use graphical methods to describe the distribution of pounds treated. Is it appropriate to analyze these data using t distribution methods? Explain.

(b) Calculate the mean, standard deviation, standard error, and margin of error for 90% confidence.

(c) Report the 90% confidence interval for the mean pounds treated per hour by this particular machine.

(d) Test whether these data provide evidence that the mean pounds of product treated in one hour is greater than 33,000. Use a significance level of 5%, and state your hypotheses, the P-value, and your conclusion.

8.30 Credit card fees. A bank wonders whether offering a higher cash back percent would increase the amount charged on its credit card. The bank makes this offer to an SRS of 85 of its existing credit card customers. It then compares how much these customers charge this year with the amount that they charged last year. The mean is $612, and the standard deviation is $1327.

(a) Is there significant evidence at the 1% level that the mean amount charged increases under the higher cash back offer? State H_0 and H_a and carry out a t test.

(b) Give a 95% confidence interval for the mean amount of the increase.

(c) The distributions of the amount charged are skewed to the right, but outliers are prevented by the credit limit that the bank enforces on each card. Use of the t procedures is justified in this case even though the population distribution is not Normal. Explain why.

(d) A critic points out that the customers would probably have charged more this year than last even without the new offer because the economy is more prosperous and interest rates are lower. Briefly describe the design of an experiment to study the effect of the increased cash back offer that would avoid this criticism.

8.31 Supermarket shoppers. A marketing consultant observed 40 consecutive shoppers at a supermarket. One variable of interest was how much each shopper spent in the store. Here are the data (in dollars), arranged in increasing order: SHOPRS

5.32	8.88	9.26	10.81	12.69	15.23	15.62	17.00
17.35	18.43	19.50	19.54	20.59	22.22	23.04	24.47
25.13	26.24	26.26	27.65	28.08	28.38	32.03	34.98
37.37	38.64	39.16	41.02	42.97	44.67	45.40	46.69
49.39	52.75	54.80	59.07	60.22	84.36	85.77	94.38

(a) Display the data using a stemplot. Make a Normal quantile plot if your software allows. The data are clearly non-Normal. In what way? Because $n = 40$, the t procedures remain quite accurate.

(b) Calculate the mean, the standard deviation, and the standard error of the mean.

(c) Find a 95% t confidence interval for the mean spending for all shoppers at this store.

8.32 The influence of big shoppers. Eliminate the three largest observations and redo parts (a), (b), and (c) of the previous exercise. Do these observations have a large influence on the results? SHOPRS

8.33 Sell in May and go away? August through October has historically been a challenging time for stock market investors. Because other months have historically had above average returns, one simple investment strategy is to be in cash during these challenging months and be in stocks otherwise. Two researchers compared the Sell-in-May (SIM) strategy with the Buy-and-Hold (B&H) strategy using the S&P 500 index fund from 1993 through 2016. They show that the SIM strategy over this time period would have resulted in almost a doubling of wealth.[12]

But is this increase statistically significant? The following table summarizes the log return under each strategy for each of the 23 years. We'll consider these $n = 23$ years to be an SRS from the population of all possible years. Comparing total returns is comparable to comparing the average log returns. SIMAGA

(a) Explain why a matched pairs t test is appropriate for this analysis.

(b) Create a histogram or Normal quantile plot of the differences. Write a careful description of the distribution, making sure to note any outliers and/or skewness.

(c) Is it appropriate to analyze the data using the t procedures? Explain your answer.

8.34 Sell in May, continued. Refer to the previous exercise. Let's compare the two strategies with and without any outliers to see if they alter the conclusions. SIMAGA

(a) Perform the analysis using all $n = 23$ years and the 0.05 significance level. Be sure to specify the hypotheses, the test statistic, the P-value, and your conclusion.

(b) Repeat part (a) after removing any outliers.

(c) Compare the results from the two analyses. How would you answer the question of whether the two strategies have different average log returns?

8.35 Rudeness and its effect on onlookers. Many believe that an uncivil environment has a negative effect on people. A pair of researchers performed a series of experiments to test whether witnessing rudeness and disrespect affects task performance.[13] In one study, 34 participants met in small groups and witnessed the group organizer being rude to a "participant" who showed up late for the group meeting. After the exchange, each participant performed an individual brainstorming task in which he or she was asked to produce as many uses for a brick as possible in five minutes. The mean number of uses was 7.88 with a standard deviation of 2.35.

(a) Suppose that prior research has shown that the average number of uses a person can produce in five minutes under normal conditions is 10. Given that the researchers hypothesize that witnessing this rudeness will decrease performance, state the appropriate null and alternative hypotheses.

(b) The data for this analysis are integers. Do you think that the t procedures are still appropriate? Explain why or why not.

(c) Carry out the significance test using a significance level of 0.05. Give the P-value and state your conclusion.

8.36 Design of controls. The design of controls and instruments has a large effect on how easily people can use them. A student project investigated this effect by asking 25 right-handed students to turn a knob (with their right hands) that moved an indicator by screw action. There were two identical instruments, one with a right-hand thread (the knob turns clockwise) and the other with a left-hand thread (the knob turns counterclockwise). The following table gives the times required (in seconds) to move the indicator a fixed distance:[14] CNTROLS

Subject	Right thread	Left thread	Subject	Right thread	Left thread
1	113	137	14	107	87
2	105	105	15	118	166
3	130	133	16	103	146
4	101	108	17	111	123
5	138	115	18	104	135
6	118	170	19	111	112
7	87	103	20	89	93
8	116	145	21	78	76
9	75	78	22	100	116
10	96	107	23	89	78
11	122	84	24	85	101
12	103	148	25	88	123
13	116	147			

(a) Each of the 25 students used both instruments. Discuss briefly how the experiment should be arranged and how randomization should be used.

(b) The project hoped to show that right-handed people find right-hand threads easier to use. State the appropriate H_0 and H_a about the mean time required to complete the task.

(c) Carry out a test of your hypotheses. Give the P-value and report your conclusions.

8.37 Is the difference important? Give a 90% confidence interval for the mean time advantage of right-hand over left-hand threads in the setting of the previous exercise. Do you think that the time saved would be of practical importance if the task were performed many times—for example, by an assembly-line worker? To help answer this question, find the mean time for right-hand threads as a percent of the mean time for left-hand threads. CNTROLS

8.38 Confidence interval? As CEO, you obtain the salaries of all 31 individuals working in your marketing department. You feed these salaries into your statistical software package, and the output produced includes a confidence interval. Is this a valid confidence interval? Explain your answer.

8.39 A field trial. An agricultural field trial compares the yields of two varieties of tomatoes for commercial use. The researchers divide in half each of eight small plots of land in different locations and plant each tomato variety on one half of each plot. After harvest, they compare the yields in pounds per plant at each location. The eight differences (Variety A − Variety B) give the following statistics: $\bar{x} = -0.52$ and $s = 0.62$. Is there a difference between the yields of these two varieties? Write a summary paragraph to answer this question. Make sure to include H_0, H_a, and the P-value with degrees of freedom.

8.40 Sign test comparing SIM and B&H strategies. Because of the outlier and skewness of the data in Exercise 8.34, let's assess whether the median difference of log returns under SIM and B&H is different from 0. SIMAGA

(a) State the hypotheses two ways, in terms of a population median and in terms of the probability of the SIM log return being larger than the B&H log return.

(b) Carry out the sign test using software or by hand, finding the approximate P-value using the Normal approximation to the binomial distributions. Report your conclusion.

8.41 Design of controls, continued. Apply the sign test to the data in Exercise 8.36 to assess whether the subjects can complete a task with a right-hand thread significantly faster than with a left-hand thread. CNTROLS

(a) State the hypotheses two ways, in terms of a population median and in terms of the probability of completing the task faster with a right-hand thread.

(b) Carry out the sign test. Find the approximate P-value using the Normal approximation to the binomial distributions, and report your conclusion.

8.2 Comparing Two Means

When you complete this section, you will be able to:

- Describe a confidence interval for the difference between two population means in terms of an estimate and its margin of error.
- Construct a confidence interval for the difference between two population means $\mu_1 - \mu_2$ from two SRSs, one from each population.
- Perform a two-sample t significance test and summarize the results.
- Explain when the two-sample t procedures can be useful for non-Normal data.

How do retail companies that fail differ from those that succeed? An accounting professor compares two samples of retail companies: one sample of failed retail companies and one of retail companies that are still active. Which of two incentive packages will lead to higher use of a bank's credit cards? The bank designs an experiment where credit card customers are assigned at random to receive one or the other incentive package. *Two-sample problems* such as these are among the most common situations encountered in statistical practice.

TWO-SAMPLE PROBLEMS

- The goal of inference is to compare the means of the response variable in two groups.
- Each group is considered to be a sample from a distinct population.
- The responses in each group are independent of each other and those in the other group.

You must carefully distinguish two-sample problems from the matched pairs designs studied earlier. In two-sample problems, there is no matching of the units in the two samples, and the two samples may be of different sizes. As a result, inference procedures for two-sample data differ from those for matched pairs.

We can present two-sample data graphically with a back-to-back stemplot for small samples (page 18) or with side-by-side boxplots for larger samples (page 32). Now we will apply the ideas of formal inference in this setting. When both population distributions are symmetric, and especially when they are at least approximately Normal, a comparison of the mean responses in the two populations is most often the goal of inference.

We have two independent samples, from two distinct populations (such as failed companies and active companies). We measure the same quantitative response variable (such as the cash flow margin) in both samples. We will call the variable x_1 in the first population and x_2 in the second because the variable may have different distributions in the two populations. Here is the notation that we will use to describe the two populations:

Population	Variable	Mean	Standard deviation
1	x_1	μ_1	σ_1
2	x_2	μ_2	σ_2

We want to compare the two population means, either by giving a confidence interval for $\mu_1 - \mu_2$ or by testing the hypothesis of no difference, $H_0: \mu_1 = \mu_2$. We base inference on two independent SRSs, one from each population. Here is the notation that describes the samples:

Population	Sample size	Sample mean	Sample standard deviation
1	n_1	$\bar{x}_1$	s_1
2	n_2	$\bar{x}_2$	s_2

Throughout this section, the subscripts 1 and 2 show the population to which a parameter or a sample statistic refers.

The two-sample t statistic

The natural estimator of the difference $\mu_1 - \mu_2$ is the difference between the sample means, $\bar{x}_1 - \bar{x}_2$. If we are to base inference on this statistic, we must know its distribution. Here are some facts:

rules for means, p. 236

- The mean of the difference $\bar{x}_1 - \bar{x}_2$ is the difference of the means $\mu_1 - \mu_2$. This follows from the addition rule for means and the fact that the mean of any $\bar{x}$ is the same as the mean μ of the population.
- The variance of the difference $\bar{x}_1 - \bar{x}_2$ is

$$\frac{\sigma_1^2}{n_1} + \frac{\sigma_2^2}{n_2}$$

rules for variances, p. 240–241

Because the samples are independent, their sample means $\bar{x}_1$ and $\bar{x}_2$ are independent random variables. The addition rule for variances says that the variance of the difference of two independent random variables is the sum of their variances.

linear combination of Normals, p. 320

- If the two population distributions are both Normal, then the distribution of $\bar{x}_1 - \bar{x}_2$ is also Normal. This is true because each sample mean alone is Normally distributed and a difference of Normal random variables is also Normal.

Because any Normal random variable has the $N(0,1)$ distribution when standardized, we have arrived at a new z statistic. The *two-sample z statistic*

$$z = \frac{(\bar{x}_1 - \bar{x}_2) - (\mu_1 - \mu_2)}{\sqrt{\dfrac{\sigma_1^2}{n_1} + \dfrac{\sigma_2^2}{n_2}}}$$

has the standard Normal $N(0,1)$ sampling distribution and would be used in inference when the two population standard deviations σ_1 and σ_2 are known.

In practice, however, σ_1 and σ_2 are not known. We estimate them by the sample standard deviations s_1 and s_2 from our two samples. Following the pattern of the one-sample case, we substitute the standard errors for the standard deviations in the two-sample z statistic. The result is the **two-sample t statistic:**

two-sample t statistic

$$t = \frac{(\bar{x}_1 - \bar{x}_2) - (\mu_1 - \mu_2)}{\sqrt{\dfrac{s_1^2}{n_1} + \dfrac{s_2^2}{n_2}}}$$

Unfortunately, this statistic does *not* have a t distribution. A t distribution replaces an $N(0,1)$ distribution only when a *single* standard deviation is replaced by an estimate. In this case, we replace *two* standard deviations (σ_1 and σ_2) by their estimates (s_1 and s_2).

df approximation

Nonetheless, we can approximate the distribution of the two-sample t statistic by using the $t(k)$ distribution with an **approximation for the degrees of freedom k.** We use these approximations to find approximate values of t^* for confidence intervals and to find approximate P-values for significance tests. The choice of approximation rarely makes a difference in the conclusion.

Satterthwaite approximation

Most statistical software uses the **Satterthwaite approximation** to approximate the $t(k)$ distribution unless the user requests another method. Use of this approximation without software is a bit complicated.[15] In general, the resulting k will not be an integer.

conservative

If you cannot access software, we recommend using degrees of freedom k equal to the smaller of $n_1 - 1$ and $n_2 - 1$. This approximation is appealing because it is **conservative**.[16] That is, margins of error for confidence intervals are larger than they need to be, so the true confidence level is larger than C. Likewise, P-values for significance tests will be larger, making it more difficult to reject H_0.

The two-sample t confidence interval

We now apply the basic ideas about t procedures to the problem of comparing two means when the standard deviations are unknown. Similar to the one-sample case, we start with confidence intervals.

TWO-SAMPLE t CONFIDENCE INTERVAL

Draw an SRS of size n_1 from a population with unknown mean μ_1 and an independent SRS of size n_2 from another population with unknown mean μ_2. The level C **confidence interval for $\mu_1 - \mu_2$** is

$$(\bar{x}_1 - \bar{x}_2) \pm t^* \sqrt{\frac{s_1^2}{n_1} + \frac{s_2^2}{n_2}}$$

The **margin of error** is

$$t^* \sqrt{\frac{s_1^2}{n_1} + \frac{s_2^2}{n_2}}$$

where t^* is the value for the $t(k)$ density curve with area C between $-t^*$ and t^*. The value of the degrees of freedom k is usually approximated by software. The smaller of $n_1 - 1$ and $n_2 - 1$ is recommended only when software is not available. When the populations are Normal, this interval has confidence level at least C and is approximately correct in other cases when $n_1 + n_2$ is large.

EXAMPLE 8.11

SMART

Smart Shopping Carts and Spending Smart shopping carts are shopping carts equipped with scanners that track the total cost of the items in the cart. While both consumers and retailers have expressed interest in the use of this technology, actual implementation has been slow. One reason for this is uncertainty in how real-time spending feedback affects shopping behavior. Retailers do not want to adopt a technology that is going to lower sales.

To help understand the smart shopping cart's influence on spending behavior, a group of researchers designed a study to compare spending with and without this real-time feedback. Each participant was asked to shop at an online grocery store for items on a common grocery list. The goal was to keep total spending around $35. Half the participants were randomly assigned to receive real-time feedback—specifically, the names of the products currently in their cart and the total cost. The other participants only saw the total cost when they completed their shopping.

Figure 8.10 shows side-by-side boxplots of the data.[17] There appears to be a slight skewness in the total cost, but no extreme outliers in either group. Given these results and the large sample sizes, we feel confident in using the t procedures. Like the one-sample case, deciding whether to use the t procedures is often a judgment call. We provide some guidelines to assist in this determination later in the section.

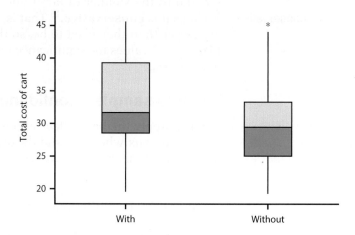

FIGURE 8.10 Side-by-side boxplots of total price for the smart shopping cart study, Example 8.11.

In general, the participants with real-time feedback appear to have spent more than those without feedback. The summary statistics are

Group	n	$\bar{x}$	s
With feedback	49	33.137	6.568
Without feedback	48	30.315	6.846

We'd like to estimate the difference in the two means and provide an estimate of the precision. Plugging in these summary statistics, the 95% confidence interval for the difference in means is

$$(\bar{x}_1 - \bar{x}_2) \pm t^* \sqrt{\frac{s_1^2}{n_1} + \frac{s_2^2}{n_2}} = (33.137 - 30.315) \pm t^* \sqrt{\frac{6.568^2}{49} + \frac{6.846^2}{48}}$$

$$= 2.822 \pm (t^* \times 1.363)$$

Using software, the degrees of freedom are 94.63 and $t^* = 1.985$. This approximation gives

$$2.822 \pm (1.985 \times 1.363) = 2.822 \pm 2.706 = (0.12, 5.53)$$

The conservative approach would use the smaller of

$$n_1 - 1 = 49 - 1 = 48 \quad \text{and} \quad n_2 - 1 = 48 - 1 = 47$$

Using the Excel function =T.INV(0.975,47), $t^* = 2.012$. With this approximation we have

$$2.822 \pm (2.012 \times 1.363) = 2.822 \pm 2.742 = (0.08, 5.56)$$

The conservative approach does give a larger interval than the more accurate approximation used by software. However, the difference is very small (just a few cents at each end). We estimate the mean difference in spending to be $2.82 with a margin of error of slightly more than $2.70 (or $2.74). ∎

APPLY YOUR KNOWLEDGE

8.42 How to complete a task. As assembly-line manager, you decide to compare two approaches that are used to complete a time-consuming task. You randomly assign each worker to one of the approaches and measure the time (in seconds) it takes to complete the task 10 times. Assume that the data are close to Normal with $\bar{x}_1 = 100$, $\bar{x}_2 = 94$, $s_1 = 11$, $s_2 = 8$, $n_1 = 15$, and $n_2 = 15$. Find a 95% confidence interval for the average difference in completion time (using software, $k = 25.57$ and $t^* = 2.060$).

8.43 Another two-sample t confidence interval. Refer to the previous exercise. Suppose instead your study results were $\bar{x}_1 = 100$, $\bar{x}_2 = 94$, $s_1 = 11$, $s_2 = 8$, $n_1 = 25$, and $n_2 = 25$. Find a 95% confidence interval for the average difference in completion time (using software, $k = 43.84$ and $t^* = 2.017$). Compare the width of this interval with that of the one in the previous exercise.

The two-sample t significance test

The same ideas that we used for the two-sample t confidence intervals also apply to *two-sample t significance tests*. We can use either software or the conservative approach with Table D to approximate the P-value.

TWO-SAMPLE t SIGNIFICANCE TEST

Draw an SRS of size n_1 from a population with unknown mean μ_1 and an independent SRS of size n_2 from another population with unknown mean μ_2. To test the hypothesis H_0: $\mu_1 = \mu_2$, compute the **two-sample t statistic**

$$t = \frac{(\bar{x}_1 - \bar{x}_2) - (\mu_1 - \mu_2)}{\sqrt{\dfrac{s_1^2}{n_1} + \dfrac{s_2^2}{n_2}}}$$

and use P-values or critical values for the $t(k)$ distribution, where the degrees of freedom k are approximated by software.

This procedure provides accurate P-values when the populations are Normal or in other cases when $n_1 + n_2$ is large.

EXAMPLE 8.12

SMART

Does Real-time Feedback Influence Spending? For the grocery spending study described in Example 8.11, we want to see if there is a difference in average spending between the group of participants that had real-time feedback and the group that did not. For a formal significance test, the hypotheses are

$$H_0: \mu_1 = \mu_2 \qquad H_0: \mu_1 - \mu_2 = 0$$

or

$$H_a: \mu_1 \neq \mu_2 \qquad H_a: \mu_1 - \mu_2 \neq 0$$

The two-sample t test statistic is

$$t = \frac{(\bar{x}_1 - \bar{x}_2) - 0}{\sqrt{\dfrac{s_1^2}{n_1} + \dfrac{s_2^2}{n_2}}}$$

$$= \frac{33.137 - 30.315}{\sqrt{\dfrac{6.568^2}{49} + \dfrac{6.846^2}{48}}} = 2.07$$

The P-value for the two-sided test is $2P(T \geq |2.07|)$. Software gives the approximate P-value as 0.0410 and uses 94.63 as the degrees of freedom.

Without software, we'd approximate the degrees of freedom k to be 47. Because there is no row for $k = 47$, we consider both $k = 40$ and $k = 50$. In both cases, 2.07 lies between the entries for $p = 0.02$ and $p = 0.025$. Thus, we can conclude that P lies between $2(0.02) = 0.04$ and $2(0.025) = 0.05$. The data do suggest that consumers on a budget will spend more when provided with real-time feedback ($t = 2.07$, df $= 47$, $0.04 < P < 0.05$). ■

	p	
df	0.025	0.02
40	2.021	2.123
50	2.009	2.109

APPLY YOUR KNOWLEDGE

8.44 How to complete a task, continued. Refer to Exercise 8.42. Perform a significance test to see if there is a difference in means between the two approaches using $\alpha = 0.05$. Make sure to specify the hypotheses, the test statistic, and its P-value, and state your conclusion.

8.45 Another two-sample t-test. Refer to Exercise 8.43.

(a) Perform a significance test to see if there is a difference in means between the two approaches using $\alpha = 0.05$.

(b) Describe how you could use the 95% confidence interval you calculated in Exercise 8.43 to determine if there is a difference between the means of the two approaches at significance level 0.05.

Robustness of the two-sample procedures

The two-sample t procedures are more robust than the one-sample t methods. When the sizes of the two samples are equal and the distributions of the two populations being compared have similar shapes, probability values from the t table are quite accurate for a broad range of distributions, even when the sample sizes are as small as $n_1 = n_2 = 5$.[18] When the two population distributions have different shapes, larger samples are needed. The guidelines given on pages 409–410 for the use of one-sample t procedures can be adapted to two-sample procedures by replacing "sample size" with the "sum of the sample sizes" $n_1 + n_2$. Specifically,

- *If $n_1 + n_2$ is less than 15:* Use t procedures if the data are close to Normal. If the data in either sample are clearly non-Normal or if outliers are present, do not use t.

- *If $n_1 + n_2$ is at least 15 and less than 40:* The t procedures can be used except in the presence of outliers or strong skewness.

- *Large samples:* The t procedures can be used even for clearly skewed distributions when the sample is large, roughly $n_1 + n_2 \geq 40$.

These guidelines are rather conservative, especially when the two samples are of equal size. In planning a two-sample study, you should usually choose equal sample sizes. The two-sample t procedures are most robust against non-Normality in this case, and the conservative probability values are most accurate.

Here is another example with large sample sizes that are almost equal. Even though the distributions are likely non-Normal, we are confident that the sample means will be approximately Normal and the two-sample t procedures will provide an accurate P-value.

EXAMPLE 8.13

Airbnb Rent Amounts A real estate company is considering the development of an Airbnb rental in Amsterdam. Two locations in the city are being considered. The Centrum-West region, which is very popular among artists, students, and tourists, and the Zuid region, which is again popular with tourists and contains the Heineken brewery. A random sample of Airbnb rental amounts for double occupancy accommodations were obtained from both regions.[19] Here are the summary statistics, in $US:

Region	n	$\bar{x}$	s
Centrum-West	60	140.83	47.85
Zuid	55	114.15	36.24

The Centrum-West rents are higher on the average. But we have data from only a random sample of Airbnb sites. Can we conclude that the average rents in these two regions are not the same? Or is this difference merely due to random variation?

Because we did not specify a direction for the difference before looking at the data, we choose a two-sided alternative. The hypotheses are

$$H_0: \mu_1 = \mu_2$$
$$H_a: \mu_1 \neq \mu_2$$

Because the samples are moderately large, we can confidently use the t procedures even though we only have the numerical summaries and cannot verify the Normality condition.

The two-sample t statistic is

$$t = \frac{(\bar{x}_1 - \bar{x}_2) - 0}{\sqrt{\frac{s_1^2}{n_1} + \frac{s_2^2}{n_2}}}$$

$$= \frac{140.83 - 114.15}{\sqrt{\frac{47.85^2}{60} + \frac{36.24^2}{55}}}$$

$$= 3.39$$

Our conservative approach finds the P-value by comparing 3.39 to critical values for the $t(54)$ distribution because the smaller sample has 55 observations. We must double the table tail area p because the alternative is two-sided.

Table D does not have entries for 54 degrees of freedom so we look at the rows for $k = 50$ and $k = 60$. Our calculated value of t is larger than the $p = 0.001$ entry and smaller than the $p = 0.0005$ entry in the table for both rows. Doubling each of these, we conclude that the P-value is between 0.001 and 0.002. The data give conclusive evidence that the mean rent in Centrum-West is higher than the mean rent in Zuid ($t = 3.39$, df $= 54$, $0.001 < P < 0.002$). ∎

	p	
df	0.001	0.0005
50	3.261	3.551
60	3.232	3.496

In this example, the P-value is very small because $t = 3.39$ says that the observed mean is more than three standard deviations above the hypothesized mean. The difference in mean rents is not only highly significant but large enough ($26.68 per night) to be important to the real estate company.

In this and other examples, we can choose which population to label 1 and which to label 2. After inspecting the data, we chose Centrum-West as Population 1 because this choice makes the t statistic a positive number. This avoids any possible confusion from reporting a negative value for t. Choosing the population labels is *not* the same as choosing a one-sided alternative after looking at the data. Choosing hypotheses after seeing a result in the data is a violation of sound statistical practice.

Inference for small samples

Small samples require special care. We do not have enough observations to examine the distribution shapes, and only extreme outliers stand out. The margins of error of confidence intervals tend to be large and the ability of significance tests to detect true mean differences (i.e., reject H_0) tends to be low. Despite these difficulties, we can often draw important conclusions from studies with small sample sizes. If the size of an effect is large, as it is in the following example, it can still be evident even if the n's are small.

EXAMPLE 8.14

WHEAT

Wheat Prices The U.S. Department of Agriculture (USDA) uses sample surveys to produce important economic estimates.[20] One pilot study estimated durum wheat prices in July and in January using independent samples of $n = 5$ wheat producers in the two months. The data are

Month	Price of wheat ($/bushel)				
January	5.8976	6.2064	6.1278	6.0468	5.9692
July	6.2335	6.3215	6.4457	6.1488	6.3079

back-to-back stemplot, p. 18

First, let's examine the distributions with a back-to-back stemplot after rounding each price to the nearest cent.

```
   January      July
        70 | 5.9 |
         5 | 6.0 |
         3 | 6.1 | 5
         1 | 6.2 | 3
           | 6.3 | 12
           | 6.4 | 5
```

The pattern is reasonably clear. Although there is variation among prices within each month, the top four prices are all from July and the four lowest prices are from January.

A significance test can confirm that the difference between months is too large to easily arise just by chance. We test

$$H_0: \mu_1 = \mu_2$$
$$H_a: \mu_1 \neq \mu_2$$

The price is higher in July ($t = 3.28$, df $= 7.91$, $P = 0.0114$). The difference in sample means is 24.2 cents. ∎

Figure 8.11 gives outputs for this analysis from four software systems. Although the formats and labels differ, the basic information is the same. All report the sample sizes, the sample means and standard deviations (or variances), the t statistic, and its P-value. All agree that the P-value is very small, though some give more detail than others. Excel and JMP outputs, for example,

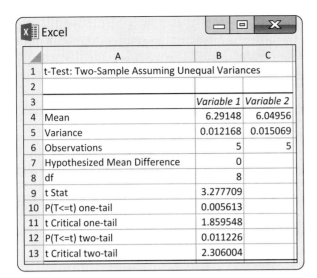
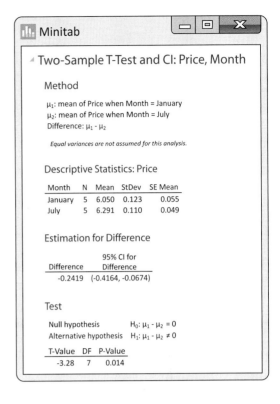
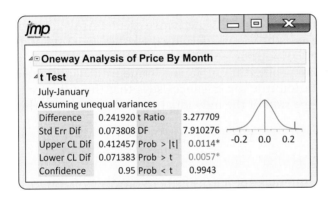
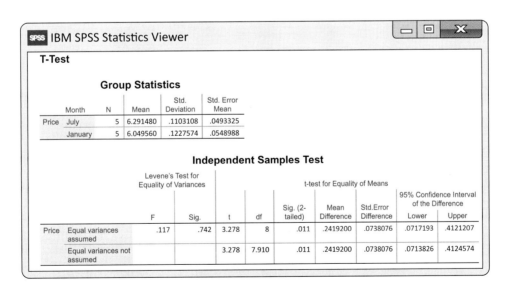

FIGURE 8.11 Excel, Minitab, JMP, and SPSS outputs, Example 8.14.

provide both one-sided and two-sided P-values. Minitab labels the groups in alphabetical order. In this example, January is then the first population and $t = -3.28$, the negative of our result. *Always check the means first and report the statistic (you may need to change the sign) in an appropriate way.* Be sure to also mention the size of the effect you observed, such as "The sample mean price for July was 24.2 cents higher than in January."

The SPSS output reports the results of *two t* procedures: the general two-sample procedure that we have just studied and a special procedure that assumes that the two population variances are equal. The "equal-variances" procedures are most helpful in cases like this when the sample sizes n_1 and n_2 are small and it is reasonable to assume equal variances. When appropriate, these methods result in slightly smaller margins of error and slightly greater ability to detect true mean differences. To understand why this is the case, let's briefly explore these procedures.

The pooled two-sample *t* procedures

Suppose that the two Normal population distributions whose means we want to compare have the *same* standard deviation. How does this additional assumption impact our *t* statistic? Let's investigate! As we did with the general two-sample *t* statistic, we will develop the *z* statistic first and from it the *t* statistic.

Call the common—but still unknown—standard deviation of both populations σ. The addition rule for variances says that the difference $\bar{x}_1 - \bar{x}_2$ has variance equal to the *sum* of the individual variances, which in this case is

$$\frac{\sigma^2}{n_1} + \frac{\sigma^2}{n_2} = \sigma^2\left(\frac{1}{n_1} + \frac{1}{n_2}\right)$$

The standardized difference of means is therefore

$$z = \frac{(\bar{x}_1 - \bar{x}_2) - (\mu_1 - \mu_2)}{\sigma\sqrt{\frac{1}{n_1} + \frac{1}{n_2}}}$$

This is the special two-sample *z* statistic for the case in which the populations have the same σ.

To get to the *t* statistic, we replace the unknown σ with its estimate. Because both sample variances s_1^2 and s_2^2 estimate σ^2, it would make sense to combine them into a single estimate. It turns out the best way to do this is to average them with weights equal to their degrees of freedom. This gives more weight to the information from the larger sample. The resulting estimator of σ^2 is

$$s_p^2 = \frac{(n_1 - 1)s_1^2 + (n_2 - 1)s_2^2}{n_1 + n_2 - 2}$$

This is called the **pooled estimator of σ^2** because it combines the information in both samples.

Because we replace a single standard deviation σ by its estimate s_p, the resulting *t* statistic has a *t* distribution. The degrees of freedom are $n_1 + n_2 - 2$, the sum of the degrees of freedom of the two sample variances. These degrees of freedom are always at least as large as the degrees of freedom for the general two-sample procedure. The larger degrees of freedom and the pooled estimator of variance are the reasons these procedures are helpful. However, to get these gains, we assumed a common variance.

POOLED TWO-SAMPLE t PROCEDURES

Draw an SRS of size n_1 from a population with unknown mean μ_1 and an independent SRS of size n_2 from another population with unknown mean μ_2. Suppose that the two populations have the same unknown standard deviation. A level C **confidence interval** for $\mu_1 - \mu_2$ is

$$(\bar{x}_1 - \bar{x}_2) \pm t^* s_p \sqrt{\frac{1}{n_1} + \frac{1}{n_2}}$$

Here, t^* is the value for the $t(n_1 + n_2 - 2)$ density curve with area C between $-t^*$ and t^*.

To test the hypothesis $H_0: \mu_1 = \mu_2$, compute the **pooled two-sample t statistic**

$$t = \frac{\bar{x}_1 - \bar{x}_2}{s_p \sqrt{\frac{1}{n_1} + \frac{1}{n_2}}}$$

and use P-values from the $t(n_1 + n_2 - 2)$ distribution.

These procedures are exact when the populations are Normal and approximately true for large n_1 and n_2 in other cases.

CMPS

CASE 8.2 Active versus Failed Retail Companies In what ways are companies that fail different from those that continue to do business? To answer this question, one study compared various characteristics of active and failed retail firms.[21] One of the variables was the cash flow margin. Roughly speaking, this is a measure of how efficiently a company converts its sales dollars to cash and is a key profitability measure. The higher the percent, the more profitable the company. The data for 101 companies appear in Table 8.3.

TABLE 8.3 Cash flow margins of active and failed retail firms

Active firms						Failed firms		
−15.57	4.13	−19.37	17.27	32.29	−1.44	23.87	49.07	−7.53
23.43	−8.75	−1.35	34.55	1.70	−0.67	−23.91	7.29	−14.81
3.17	11.62	9.38	13.40	2.20	−22.26	−5.12	−24.34	−38.27
−0.35	−27.78	0.65	−40.82	23.55	24.45	7.71	−28.79	−38.35
−9.65	−16.01	36.31	−27.71	9.73	40.48	9.88	−7.99	−18.91
3.37	5.80	−15.60	−3.58	8.46	8.83	−46.38	−41.30	0.37
40.25	−13.39	15.86	−2.25	12.97	28.21	1.41	−25.56	5.28
11.02	30.00	4.84	30.60	6.57	−20.31	−15.13	8.48	15.72
27.97	3.72	−0.71	−16.46	7.76	−4.20	−11.00	1.27	14.23
13.08	−9.31	20.21	−10.45	21.39				
−22.10	−24.55	28.93	35.83	21.02				
12.28	0.43	22.49	−8.54	−30.46				
−1.89	27.92	32.79	−0.52	6.35				

As usual, we first examine the data. Histograms for the two groups of firms are given in Figure 8.12. The distribution for the active firms looks more

FIGURE 8.12 Histograms of the cash flow margin, Example 8.15.

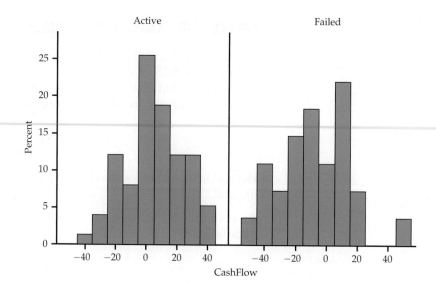

Normal than the distribution for the failed firms. However, there are no outliers or strong departures from Normality that will prevent us from using the t procedures for these data. Let's compare the mean cash flow margin for the two groups of firms using a significance test.

EXAMPLE 8.15

CMPS

CASE 8.2 **Does the Cash Flow Margin Differ?** Take Group 1 to be the firms that were active and Group 2 to be those that failed. The question of interest is whether or not the mean cash flow margin is different for the two groups. We therefore test

$$H_0: \mu_1 = \mu_2$$
$$H_a: \mu_1 \neq \mu_2$$

Here are the summary statistics:

Group	Firms	n	$\bar{x}$	s
1	Active	74	5.42	18.80
2	Failed	27	−7.14	21.67

This difference in sample standard deviations is not particularly unusual even in samples this large. We are willing to assume equal population standard deviations. The pooled sample variance is

$$s_p^2 = \frac{(n_1 - 1)s_1^2 + (n_2 - 1)s_2^2}{n_1 + n_2 - 2}$$

$$= \frac{(73)(18.80)^2 + (26)(21.67)^2}{74 + 27 - 2} = 383.94$$

so that

$$s_p = \sqrt{383.94} = 19.59$$

The pooled two-sample t statistic is

$$t = \frac{\bar{x}_1 - \bar{x}_2}{s_p\sqrt{\dfrac{1}{n_1} + \dfrac{1}{n_2}}}$$

$$= \frac{5.42 - (-7.14)}{19.59\sqrt{\dfrac{1}{74} + \dfrac{1}{27}}} = 2.85$$

The P-value is $2P(T \geq |2.85|)$, where T has the $t(99)$ distribution.

In Table D, we have entries for 80 and 100 degrees of freedom. In both rows, our calculated value of t is between the $p = 0.005$ and $p = 0.0025$ entries in the table. Doubling these, we conclude that the two-sided P-value is between 0.005 and 0.01. Statistical software gives the result $P = 0.005$. There is strong evidence that the average cash flow margins are different. ∎

	p	
df	0.005	0.0025
80	2.639	2.887
100	2.626	2.871

Of course, a P-value is rarely a complete summary of a statistical analysis. To make a judgment regarding the size of the difference between the two groups of firms, we need a confidence interval.

EXAMPLE 8.16

CASE 8.2 **How Different Are Cash Flow Margins?** The difference in mean cash flow margins for active versus failed firms is

$$\bar{x}_1 - \bar{x}_2 = 5.42 - (-7.14) = 12.56$$

For a 95% margin of error, software tells us to use the critical value $t^* = 1.984$ from the $t(99)$ distribution. The margin of error is

$$t^* s_p \sqrt{\frac{1}{n_1} + \frac{1}{n_2}} = (1.984)(19.59)\sqrt{\frac{1}{74} + \frac{1}{27}}$$

$$= 8.74$$

We report that the active firms have current cash flow margins that average 12.56% higher than failed firms, with margin of error 8.74% for 95% confidence. Alternatively, we are 95% confident that the difference is between 3.82% and 21.30%. ∎

The pooled two-sample t procedures are anchored in statistical theory and have long been the standard version of the two-sample t in textbooks. But they require the condition that the two unknown population standard deviations are equal. This condition is hard to verify.

The pooled t procedures are, therefore, a bit risky. They are reasonably robust against both non-Normality and unequal standard deviations when the sample sizes are nearly the same. When the samples are quite different in size, the pooled t procedures become sensitive to unequal standard deviations and should be used with caution unless the samples are large. Unequal standard deviations are very common. In particular, it is common for the spread of data to increase as the mean increases. We recommend regular use of the unpooled t procedures, particularly when software automates the Satterthwaite approximation.

APPLY YOUR KNOWLEDGE

8.46 Using software. Figure 8.11 (page 427) gives the outputs from four software systems for comparing prices received by wheat producers in July and January for small samples of five producers in each month. For all software, the unpooled version uses the Satterthwaite approximation for degrees of freedom, but some round down or to the nearest integer. Summarize what each software does and provide its P-value.

8.47 Airbnb rents revisited. Example 8.13 (page 425) gives summary statistics for Airbnb rent amounts in two regions of Amsterdam. The two sample standard deviations are relatively close, so we may be willing to assume equal population standard deviations. Calculate the pooled t test statistic and its degrees of freedom from the summary statistics. Use Table D to assess significance. How do your results compare with the unpooled analysis in the example?

SECTION 8.2 SUMMARY

- Significance tests and confidence intervals for the difference of the means μ_1 and μ_2 of two Normal populations are based on the difference $\bar{x}_1 - \bar{x}_2$ of the sample means from two independent SRSs. Because of the central limit theorem, the resulting procedures are approximately correct for other population distributions when the sample sizes are large.

- When independent SRSs of sizes n_1 and n_2 are drawn from two Normal populations with parameters μ_1, σ_1 and μ_2, σ_2 the **two-sample z statistic**

$$z = \frac{(\bar{x}_1 - \bar{x}_2) - (\mu_1 - \mu_2)}{\sqrt{\frac{\sigma_1^2}{n_1} + \frac{\sigma_2^2}{n_2}}}$$

has the $N(0,1)$ distribution.

- The **two-sample t statistic**

$$t = \frac{(\bar{x}_1 - \bar{x}_2) - (\mu_1 - \mu_2)}{\sqrt{\frac{s_1^2}{n_1} + \frac{s_2^2}{n_2}}}$$

does *not* have a t distribution. However, software can give accurate P-values and critical values using the **Satterthwaite approximation.**

- **Conservative inference procedures** for comparing μ_1 and μ_2 use the two-sample t statistic and the $t(k)$ distribution with degrees of freedom k equal to the smaller of $n_1 - 1$ and $n_2 - 1$. Use this method only when you are not using software.

- An approximate level C **confidence interval** for $\mu_1 - \mu_2$ is given by

$$(\bar{x}_1 - \bar{x}_2) \pm t^* \sqrt{\frac{s_1^2}{n_1} + \frac{s_2^2}{n_2}}$$

Here, t^* is the value for the $t(k)$ density curve with area C between $-t^*$ and t^*, where k is estimated by software. The **margin of error** is

$$t^* \sqrt{\frac{s_1^2}{n_1} + \frac{s_2^2}{n_2}}$$

- Significance tests for $H_0: \mu_1 = \mu_2$ are based on the **two-sample t statistic**

$$t = \frac{\bar{x}_1 - \bar{x}_2}{\sqrt{\dfrac{s_1^2}{n_1} + \dfrac{s_2^2}{n_2}}}$$

The P-value is approximated using the $t(k)$ distribution, where k is found by software.

- The guidelines for practical use of two-sample t procedures are similar to those for one-sample t procedures but considering $n_1 + n_2$. Equal sample sizes are recommended.

- If we can assume that the two populations have equal variances, **pooled two-sample t procedures** can be used. These are based on the **pooled estimator**

$$s_p^2 = \frac{(n_1 - 1)s_1^2 + (n_2 - 1)s_2^2}{n_1 + n_2 - 2}$$

of the unknown common variance and the $t(n_1 + n_2 - 2)$ distribution.

SECTION 8.2 EXERCISES

For Exercises 8.42 and 8.43, see page 423; for 8.44 and 8.45, see page 424; and for 8.46 and 8.47, see page 432.

In exercises that call for two-sample t procedures, try to use the degrees of freedom approximation provided by software. For exercises involving summarized data, this approximation is provided to you. If you instead use the conservative approximation, the smaller of $n_1 - 1$ and $n_2 - 1$, be sure to clearly state this.

8.48 What's wrong? For each of the following statements, explain what is wrong and why.

(a) A researcher wants to test $H_0: \bar{x}_1 = \bar{x}_2$ versus the two-sided alternative $H_a: \bar{x}_1 \neq \bar{x}_2$.

(b) A study recorded the credit card IQ scores of 90 college freshmen. The scores of the 44 males in the study were compared with the scores of all 90 freshmen using the two-sample methods of this section.

(c) A two-sample t statistic gave a P-value of 0.98. From this, we can reject the null hypothesis with 95% confidence.

(d) A researcher is interested in testing the one-sided alternative $H_a: \mu_1 > \mu_2$. The significance test for $\mu_1 - \mu_2$ gave $t = -2.29$. With a P-value for the two-sided alternative of 0.008, he concluded that his P-value was 0.004.

8.49 Understanding concepts. For each of the following, answer the question and give a short explanation of your reasoning.

(a) A 95% confidence interval for the difference between two means is reported as $(-0.3, 0.7)$. What can you conclude about the results of a level $\alpha = 0.05$ significance test of the null hypothesis that the population means are equal versus the two-sided alternative?

(b) Will larger samples generally give a larger or smaller margin of error for the difference between two sample means?

8.50 Determining significance. For each of the following, answer the question and give a short explanation of your reasoning.

(a) A significance test for comparing two means gave $t = -1.95$ with 13 degrees of freedom. Can you reject the null hypothesis that the μ's are equal versus the two-sided alternative at the 5% significance level?

(b) Answer part (a) for the one-sided alternative that the difference in means is negative.

(c) Answer part (a) for the one-sided alternative that the difference in means is positive.

8.51 The efficacy of digital mindfulness training. There is growing evidence that in-person mindfulness training can reduce stress. Little is known, however, about the efficacy of self-guided digital training. To investigate this, a group of researchers randomized 69 participants to either a digital training group or a control group.[22] For the digital group, participants were asked to complete the first 10 guided meditations using the mindfulness app *Headspace*. For the control group, participants were asked to listen to the 10 excerpts from an audiobook on mindfulness using *Headspace*. The following table summarizes the change from baseline in feelings of stress as measured using the Stress Overload Scale (SOS).

Group	n	$\bar{x}$	s
Digital	41	−5.39	8.36
Control	28	0.10	10.46

(a) Can we conclude that the change from baseline is different across the two groups? Specify the hypotheses, test statistic, P-value, and conclusion using $\alpha = 0.01$ (software gives $k = 49.35$).

(b) Can we conclude that the average stress level was reduced in the digital group? Specify the hypotheses, test statistic, P-value, and conclusion using $\alpha = 0.01$.

8.52 Incomplete follow-up. Refer to the previous exercise. The researchers report that 19 participants ($n = 13$ digital and $n = 6$ control) did not complete training and thus were not included in the analysis. Does this information in any way alter your conclusions in the previous exercise? Explain your answer.

8.53 Trustworthiness and eye color. Why do we naturally tend to trust some strangers more than others? One group of researchers decided to study the relationship between eye color and trustworthiness.[23] In their experiment, the researchers took photographs of 80 students (20 males with brown eyes, 20 males with blue eyes, 20 females with brown eyes, and 20 females with blue eyes), each seated in front of a white background looking directly at the camera with a neutral expression. These photos were cropped so the eyes were horizontal and at the same height in the photo and so the neckline was visible. They then recruited 105 participants to judge the trustworthiness of each student photo. This was done using a 10-point scale, where 1 meant very untrustworthy and 10 very trustworthy. The 80 scores from each participant were then converted to z-scores, and the average z-score of each photo (across all 105 participants) was used for the analysis. Here is a summary of the results:

Eye color	n	$\bar{x}$	s
Brown	40	0.55	1.68
Blue	40	−0.38	1.53

Can we conclude from these data that brown-eyed students appear more trustworthy compared with their blue-eyed counterparts? Test the hypothesis that the average z-scores for the two groups are the same (software gives $k = 77.33$).

8.54 Sadness and spending. The "misery is not miserly" phenomenon refers to a sad person's spending judgment going haywire. In a recent study, 31 young adults were given $10 and randomly assigned to either a sad or a neutral group. The participants in the sad group watched a video about the death of a boy's mentor (from *The Champ*), and those in the neutral group watched a video on the Great Barrier Reef. After the video, each participant was offered the chance to trade $0.50 increments of the $10 for an insulated water bottle.[24] Here are the data: SADNESS

Group	Purchase price ($)
Neutral	0.00 2.00 0.00 1.00 0.50 0.00 0.50
	2.00 1.00 0.00 0.00 0.00 0.00 1.00
Sad	3.00 4.00 0.50 1.00 2.50 2.00 1.50 0.00 1.00
	1.50 1.50 2.50 4.00 3.00 3.50 1.00 3.50

(a) Examine each group's prices graphically. Is use of the t procedures appropriate for these data? Carefully explain your answer.

(b) Make a table with the sample size, mean, and standard deviation for each of the two groups.

(c) State appropriate null and alternative hypotheses for comparing these two groups.

(d) Perform the significance test at the $\alpha = 0.05$ level, making sure to report the test statistic, degrees of freedom, and P-value. What is your conclusion?

(e) Construct a 95% confidence interval for the mean difference in purchase price between the two groups.

8.55 Noise levels in fitness classes. Fitness classes often have very loud music that could affect hearing. One study collected noise levels (decibels) in both high-intensity and low-intensity fitness classes across eight commercial gyms in Sydney, Australia.[25] NOISE

(a) Create a histogram or Normal quantile plot for the high-intensity classes. Do the same for the low-intensity classes. Are the distributions reasonably Normal? Summarize the distributions in words.

(b) Test the equality of means using a two-sided alternative hypothesis and significance level $\alpha = 0.05$.

(c) Are the t procedures appropriate given your observations in part (a)? Explain your answer.

(d) Remove the one low decibel reading for the low-intensity group and redo the significance test. How does this outlier affect the results?

(e) Do you think the results of the significance test from part (b) or (d) should be reported? Explain your answer.

8.56 Noise levels in fitness classes, continued. Refer to the previous exercise. In most countries, the workplace noise standard is 85 db (over eight hours). For every 3-dB increase above that, the amount of exposure time is halved. This means that the exposure time for a dB level of 91 is two hours and for a dB level of 94 it is one hour. NOISE

(a) Construct a 95% confidence interval for the mean dB level in high-intensity classes.

(b) Using the interval in part (a), construct a 95% confidence interval for the number of one-hour classes per day an instructor can teach before possibly risking hearing loss. (*Hint:* This is a linear transformation.)

(c) Repeat parts (a) and (b) for low-intensity classes.

(d) Explain how one might use these intervals to determine the staff size of a new gym.

8.57 Drive-thru speed of service. *QSRMagazine.com* assessed 2011 random drive-thru visits at quick-service restaurants.[26] One benchmark assessed was speed of service. For McDonald's, the average time for 179 visits was 239.03 seconds. For Burger King, the average time based on 171 visits was 189.48 seconds. The sample standard deviations were not provided but assume they were $s_M = 68.3$ and $s_{BK} = 57.8$ seconds. Is there significant evidence that the average drive-thru visit at Burger King is at least 30 seconds shorter than the average visit to McDonald's at the 5% significance level (software gives $k = 343.05$)?

8.58 Dust exposure at work. Exposure to dust at work can lead to lung disease later in life. One study measured the workplace exposure of tunnel construction workers.[27] Part of the study compared 115 drill and blast workers with 220 outdoor concrete workers. Total dust exposure was measured in milligram years per cubic meter (mgy/m^3). The mean exposure for the drill and blast workers was 18.0 mgy/m^3 with a standard deviation of 7.8 mgy/m^3. For the outdoor concrete workers, the corresponding values were 6.5 and 3.4 mgy/m^3, respectively.

(a) The sample included all workers for a tunnel construction company who received medical examinations as part of routine health checkups. Discuss the extent to which you think these results apply to other similar types of workers.

(b) Use a 95% confidence interval to describe the difference in the exposures (software gives $k = 137.07$ and $t^* = 1.977$). Write a sentence that gives the interval and provides the meaning of 95% confidence.

(c) Test the null hypothesis that the exposures for these two types of workers are the same. Justify your choice of a one-sided or two-sided alternative. Report the test statistic, the degrees of freedom, and the *P*-value. Give a short summary of your conclusion.

(d) The authors of the article describing these results note that the distributions are somewhat skewed. Do you think that this fact makes your analysis invalid? Give reasons for your answer.

8.59 Not all dust is the same. Not all dust particles that are in the air around us cause problems for our lungs. Some particles are too large and stick to other areas of our body before they can get to our lungs. Others are so small that we can breathe them in and out and they will not deposit in our lungs. The researchers in the study described in the previous exercise also measured respirable dust. This is dust that deposits in our lungs when we breathe it. For the drill and blast workers, the mean exposure to respirable dust was 6.3 mgy/m^3 with a standard deviation of 2.8 mgy/m^3. The corresponding values for the outdoor concrete workers were 1.4 and 0.7 mgy/m^3, respectively. Analyze these data using the questions in the previous exercise as a guide (software gives $k = 121.50$ and $t^* = 1.980$).

CASE 8.2 **8.60 Active companies versus failed companies.** Examples 8.15 and 8.16 (pages 430–431) compare active and failed companies under the special assumption that the two populations of firms have the same standard deviation. In practice, we prefer not to make this assumption because it is hard to verify, so let's reanalyze the data without making this assumption. Write a short summary describing the differences between these results and the ones in Examples 8.15 and 8.16. CMPS

8.61 When is 30–31 days not equal to a month? Time can be expressed on different levels of scale—days, weeks, months, and years. Can the scale provided influence perception of time? For example, if you placed an order over the phone, would it make a difference if you were told the package would arrive in four weeks or one month? To investigate this, two researchers asked a group of 267 college students to imagine their car needed major repairs and would have to stay at the shop. Depending on the group he or she was randomized to, the student was either told it would take one month or 30–31 days. Each student was then asked to give best- and worst-case estimates of when the car would be ready. The interval between these two estimates (in days) was the response. Here are the results:[28]

Group	n	$\bar{x}$	s
30–31 days	177	20.4	14.3
One month	90	24.8	13.9

(a) Given that the interval cannot be less than 0, the distributions are likely skewed. Comment on the appropriateness of using the *t* procedures.

(b) Test that the average interval is the same for the two groups using the $\alpha = 0.05$ significance level. Report the test statistic, the degrees of freedom, and the *P*-value (software gives $k = 119.90$). Give a short summary of your conclusion.

8.62 When is 52 weeks not equal to a year? Refer to the previous exercise. The researchers also had 60 marketing students read an announcement about a construction project. The expected duration was either one year or 52 weeks. Each student was then asked to state the earliest and latest completion date.

Group	n	$\bar{x}$	s
52 weeks	30	84.1	55.8
1 year	30	139.6	73.1

Test whether the average interval is the same for the two groups using the $\alpha = 0.05$ significance level. Report the test statistic, the degrees of freedom, and the *P*-value (software provides $k = 54.23$). Give a short summary of your conclusion.

8.63 Fitness and ego. Employers sometimes seem to prefer executives who appear physically fit, despite the legal troubles that may result. Employers may also favor certain personality characteristics. Fitness and personality are related. In one study, middle-aged college faculty who had volunteered for a fitness program were divided into low-fitness and high-fitness groups based on a physical examination. After the program, the subjects took the Cattell Sixteen Personality Factor Questionnaire.[29] Here are the data for the "ego strength" personality factor: EGO

Low fitness			High fitness		
4.99	5.53	3.12	6.68	5.93	5.71
4.24	4.12	3.77	6.42	7.08	6.20
4.74	5.10	5.09	7.32	6.37	6.04
4.93	4.47	5.40	6.38	6.53	6.51
4.16	5.30		6.16	6.68	

(a) Is the difference in mean ego strength significant at the 5% level? At the 1% level? Be sure to state H_0 and H_a.

(b) Can you generalize these results to the population of all middle-aged men? Give reasons for your answer.

(c) Can you conclude that increasing fitness *causes* an increase in ego strength? Give reasons for your answer.

8.64 Study design matters! In the previous exercise, you analyzed data on the ego strength of high-fitness and low-fitness participants after a six-month campus fitness program. Suppose that instead you had data on the ego strengths of these men both *before and after* the six-month program.

(a) Explain carefully the statistical procedures you would use to assess whether the low-fitness (or high-fitness) program affected ego strength.

(b) Explain carefully the statistical procedures you would use to compare the average change in ego strength between the low-fitness and high-fitness groups.

(c) If the test in (b) resulted in a statistically larger *increase* in ego strength for the high-fitness group, can you conclude that increasing fitness *causes* an increase in ego strength? Give reasons for your answer.

8.65 Sales of small appliances. A market research firm supplies manufacturers with estimates of the retail sales of their products from samples of retail stores. Marketing managers are prone to look at the estimate and ignore sampling error. Suppose that an SRS of 50 stores this month shows mean sales of 53 units of a small appliance, with standard deviation 12 units. During the same month last year, an SRS of 65 stores gave mean sales of 50 units, with standard deviation 11 units. An increase from 50 to 53 is a rise of 6%. The marketing manager is happy, because sales are up 6%.

(a) Use the two-sample t procedure to give a 95% confidence interval for the difference in mean number of units sold at all retail stores (software gives $k = 100.63$ and $t^* = 1.984$).

(b) Explain in language that the manager can understand why he cannot be confident that sales rose by 6%, and that in fact sales may even have dropped.

8.66 Compare two marketing strategies. A bank compares two proposals to increase the amount that its credit card customers charge on their cards. (The bank earns a percent of the amount charged, paid by the stores that accept the card.) Proposal A offers to eliminate the annual fee for customers who charge $3600 or more during the year. Proposal B offers a small percent of the total amount charged as a cash rebate at the end of the year. The bank offers each proposal to an SRS of 150 of its existing credit card customers. At the end of the year, the total amount charged by each customer is recorded. Here are the summary statistics:

Group	n	$\bar{x}$	s
A	150	$3358	$498
B	150	$3021	$461

(a) Do the data show a significant difference between the mean amounts charged by customers offered the two plans? Give the null and alternative hypotheses, and calculate the two-sample t statistic. Obtain the P-value (either approximately from Table D or more accurately using $k = 296.24$ from software). State your practical conclusions.

(b) The distributions of amounts charged are skewed to the right, but outliers are prevented by the limits that the bank imposes on credit balances. Do you think that skewness threatens the validity of the test that you used in part (a)? Explain your answer.

8.67 More on smart shopping carts. Recall Example 8.11 (page 421). The researchers also had participants, who were not told they were on a budget, go through the same online grocery shopping exercise. SMART1

(a) For this set of participants, construct a table that includes the sample size, mean, and standard deviation of the total cost for the subset of participants with feedback and those without.

(b) Generate histograms or Normal quantile plots for each subset. Comment on the distributions and whether it is appropriate to use the t procedures.

(c) Test whether the average cost of the cart is the same for these two groups using the 0.05 significance level. Write a short summary of your findings. Make sure to compare them with the results in Example 8.11.

8.68 New hybrid tablet and laptop? The purchasing department has suggested your company switch to a new hybrid tablet and laptop. As CEO, you want data to be assured that employees will like these new hybrids over the old laptops. You designate the next 14 employees needing a new laptop to participate in an experiment in which seven will be randomly assigned to receive the standard laptop and the remainder will receive the new hybrid tablet and laptop. After a month of use, these employees will express their satisfaction with their new computers by responding to the statement "I like my new computer" on a scale from 1 to 5, where 1 represents "strongly disagree," 2 is "disagree," 3 is "neutral," 4 is "agree," and 5 is "strongly agree."

(a) The employees with the hybrid computers have an average satisfaction score of 4.3 with standard deviation 0.3. The employees with the standard laptops have an average of 3.5 with standard deviation 0.9. Give a 95% confidence interval for the difference in the mean satisfaction scores for all employees (software provides $k = 7.32$ and $t^* = 2.365$).

(b) Would you reject the null hypothesis that the mean satisfaction for the two types of computers is the same versus the two-sided alternative at significance level 0.05? Use your confidence interval to answer this question. Explain why you do not need to calculate the test statistic.

8.69 Why randomize? Refer to the previous exercise. A coworker suggested that you give the new hybrid computers to the next seven employees who need new computers and the standard laptop to the following seven. Explain why your randomized design is better.

8.70 Pooled procedures. Refer to the previous two exercises. Reanalyze the data using the pooled procedure. Does the conclusion depend on the choice of method? The standard deviations are quite different for these data, so we do not recommend use of the pooled procedures in this case.

8.71 Satterthwaite approximation. The degrees of freedom given by the Satterthwaite approximation are always at least as large as the smaller of $n_1 - 1$ and $n_2 - 1$ and never larger than the sum $n_1 + n_2 - 2$. In Exercise 8.55 (page 434), you were asked to compare the analyses with and without a very low decibel reading in the low-intensity group. Redo those analyses and make a table showing the sample sizes n_1 and n_2, the standard deviations s_1 and s_2, and the Satterthwaite degrees of freedom for each of these analyses. Based on these results, suggest when the Satterthwaite degrees of freedom will be closer to the smaller of $n_1 - 1$ and $n_2 - 1$ and when it will be closer to $n_1 + n_2 - 2$. NOISE

8.72 Pooled equals unpooled? The software outputs in Figure 8.11 (page 427) give the *same value* for the pooled and unpooled t statistics. Do some simple algebra to show that this is always true when the two sample sizes n_1 and n_2 are the same. In other cases, the two t statistics usually differ.

8.73 The advantage of pooling. For the analysis of wheat prices in Example 8.14 (page 437), there are only five observations per month. When sample sizes are small, we have very little information to make a judgment about whether the population standard deviations are equal. The potential gain from pooling is large when the sample sizes are very small. Assume that we will perform a two-sided test using the 5% significance level.

(a) Find the critical value for the unpooled t test statistic that does not assume equal variances using the minimum of $n_1 - 1$ and $n_2 - 1$ for the degrees of freedom.

(b) Find the critical value for the pooled t test statistic.

(c) How does comparing these critical values show an advantage of the pooled test?

8.74 The advantage of pooling. Suppose that in the setting of the previous exercise, you are interested in 95% confidence intervals for the difference rather than significance testing. Find the widths of the intervals for the two procedures (assuming or not assuming equal standard deviations). How do they compare? WHEAT

8.3 Additional Topics on Inference

When you complete this section, you will be able to:

- Use software to compute the sample size needed for a desired margin of error for a mean μ or for a difference in means $\mu_1 - \mu_2$.
- Define what is meant by the power of a test.
- Use software to determine the sample sizes necessary for a one- and two-sample t test to have adequate power to detect an alternative.
- Use these sample-size calculations to determine whether a proposed study should be performed.
- Describe the two types of possible errors (Type I and Type II) when performing a significance test and relate them to the significance level and power of the test.

In this section, we describe two topics that are related to the procedures we have learned for inference about population means. First, we focus on a very important issue when planning a study: choosing the sample size. *A wise user of statistics does not plan for inference without at the same time planning data collection.* While the actual formulas are a bit technical, only a general understanding of the calculations is necessary. We can rely on software to do the heavy lifting.

The second topic pertains to the use of inference as a decision. So far, we have presented tests of significance as methods for assessing the strength of evidence *against the null hypothesis*. Sometimes, we are really concerned about making a decision or choosing an action based on our evaluation of the data. This change in focus alters the reasoning we use when performing these tests.

Sample size for confidence intervals

We can arrange to have both high confidence and a small margin of error by choosing an appropriate sample size. Let's first focus on the one-sample t confidence interval. Its margin of error is

$$m = t^* \text{SE}_{\bar{x}} = t^* \frac{s}{\sqrt{n}}$$

Besides the confidence level C and sample size n, this margin of error depends on the sample standard deviation s. Because we don't know the value of s until we collect the data, we must guess a value to use in our calculations.

We will call this guessed value s_g. We typically use results from a pilot study or from studies published earlier to help us with this guess. If you had an idea of the expected range of the data, you could also use

$$s_g = \frac{\text{range}}{4}$$

following the 68–95–99.7 rule. *It is always better to use a value of the standard deviation that is a little larger than what is expected.* This may result in a sample size that is a little larger than needed, but it helps avoid the situation when the resulting margin of error is larger than desired.

Given the desired margin of error m and a guess for s, we can find the sample size by plugging everything into the margin of error formula and solving for n. In fact, we did this in Chapter 7 (page 349). The one complication here is that t^* depends not only on the confidence level C but also on the sample size n. Here are the details.

> **SAMPLE SIZE FOR DESIRED MARGIN OF ERROR FOR A MEAN μ**
>
> The level C confidence interval for a mean μ will have an expected margin of error less than or equal to a specified value m when the sample size is such that
>
> $$m \geq t^* s_g / \sqrt{n}$$
>
> Here, t^* is the critical value for confidence level C with $n - 1$ degrees of freedom, and s_g is the guessed value for the population standard deviation.

Finding the smallest sample size n that satisfies this requirement can be done using the following iterative search:

1. Get an initial sample size by replacing t^* with z^*. Compute $n = (z^* s_g / m)^2$ and round up to the nearest integer.
2. Use this sample size to obtain t^*, and check if $m \geq t^* s_g / \sqrt{n}$.
3. If the requirement is satisfied, then this n is the needed sample size. If the requirement is not satisfied, increase n by 1 and return to Step 2.

Notice that this method makes no reference to the size of the *population*. It is the size of the *sample* that determines the margin of error. The size of the population does not influence the sample size we need as long as the population is much larger than the sample. Here is an example.

EXAMPLE 8.17

Planning a New Battery-Life Study. In Example 8.1 (page 399), we calculated a 95% confidence interval for the average battery life of your smartphone. The margin of error based on an SRS of $n = 8$ days was 41.5 minutes. Suppose that a new study is being planned and the goal is to have a margin of error of 30 minutes. How many days do you need?

The sample standard deviation in Example 8.1 was 49.66 minutes. To be conservative, we'll guess that the population standard deviation is 55 minutes.

1. To compute an initial n, we replace t^* with z^*. This results in

$$n = \left(\frac{z^* s_g}{m}\right)^2 = \left[\frac{1.96(55)}{30}\right]^2 = 12.91$$

Round up to get $n = 13$.

2. We now check to see if this sample size satisfies the requirement when we switch back to t^*. For $n = 13$, we have $n - 1 = 12$ degrees of freedom and $t^* = 2.179$. Using this value, the expected margin of error is

$$2.179(55)/\sqrt{13} = 33.24$$

This is larger than $m = 30$ minutes, so the requirement is not satisfied.

3. The following table summarizes these calculations for some larger values of n.

n	$t^* s_g / \sqrt{n}$
13	33.24
14	31.76
15	30.46
16	29.31

The requirement is first satisfied when $n = 16$. Thus, we need to double the number of days to $n = 16$ in order for the expected margin of error to be no more than 30 minutes.

Figure 8.13 shows the Minitab input window needed to do these calculations. Because the default confidence level is 95%, only the desired margin of error m and the estimate for s need to be entered. The software does the rest once you click OK. ∎

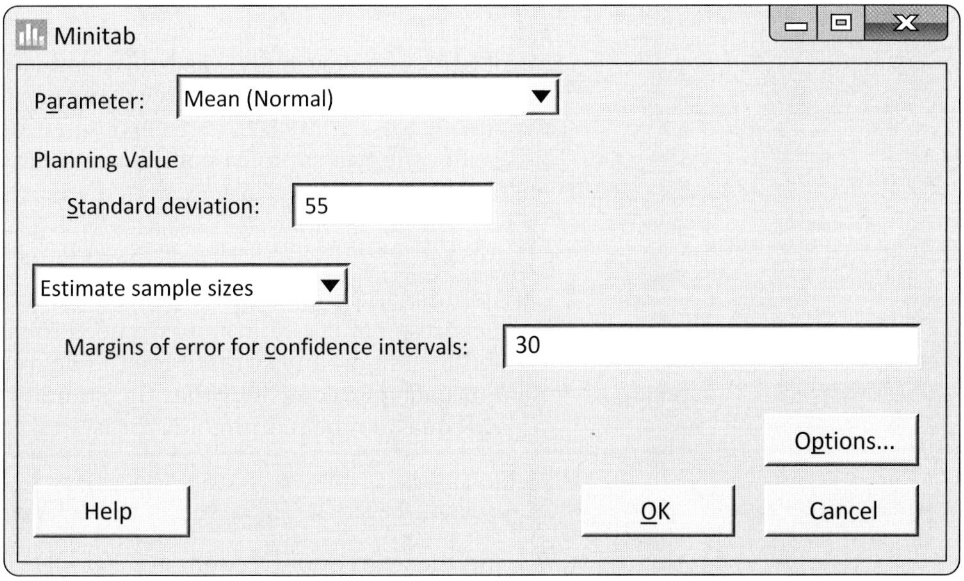

FIGURE 8.13 The Minitab input window for the sample size calculation performed in Example 8.17.

Note that the $n = 16$ refers to the *expected* margin of error being no more than 30 minutes. This does not guarantee that the margin of error for the sample we collect will be less than 30 minutes. That is because the sample standard deviation s varies from sample to sample and these calculations are treating it as a fixed quantity. More advanced sample size procedures ask you to also specify the probability of obtaining a margin of error less than the desired value. For the current approach, this probability is roughly 50%. For a probability closer to 100%, the sample size will need to be larger. For example, suppose we wanted this probability to be roughly 80%. In SAS, we'd perform these calculations using the commands

```
proc power;
  onesamplemeans CI=t stddev=55 halfwidth=30
  probwidth=0.80 ntotal=.;
run;
```

The needed sample size increases from $n = 16$ to $n = 19$.

Unfortunately, the actual number of usable observations is often less than that planned at the beginning of a study. This is particularly true of data collected in surveys or studies that involve a time commitment from the participants. Careful study designers often assume a nonresponse rate or dropout rate that specifies what proportion of the originally planned sample will fail to provide data. We use this information to calculate the sample size to be used at the start of the study. For example, if we need $n = 100$ but expect only 25% of the population to respond to our survey, we would need to start with a sample size that is $4 \times 100 = 400$ to obtain usable information from 100 people.

These sample size calculations also do not account for collection costs. In practice, taking observations costs time and money. There are times when the required sample size may be impossibly expensive. In those situations, one might consider a larger margin of error and/or a lower confidence level to be acceptable.

APPLY YOUR KNOWLEDGE

8.75 How do design choices change the sample size? Refer to Example 8.17. For each of the following changes, state whether the needed sample size will increase or decrease and explain your reasoning.

(a) The desired margin of error is 15 minutes rather than 30 minutes.

(b) A 90% rather than 95% confidence level is used.

(c) We use $s_g = 50$ minutes instead of 55 minutes.

8.76 How many postings to sample? In Exercise 8.1 (page 398), a random sample of $n = 16$ postings from Realtor.com resulted in a standard deviation of $168. If a new study was being planned with $s_g = 170$ and the desired 95% margin of error is $50, how many postings need to be sampled?

For the two-sample t confidence interval, the margin of error is

$$m = t^* \sqrt{\frac{s_1^2}{n_1} + \frac{s_2^2}{n_2}}$$

A similar type of iterative search can be used to determine the sample sizes n_1 and n_2, but now we need to guess both standard deviations and decide on an estimate for the degrees of freedom. An alternative approach is to consider that the standard deviations and sample sizes are equal, so the margin of error is

$$m = t^* s_p \sqrt{\frac{2}{n}}$$

and the degrees of freedom are $2(n-1)$. That is the approach most statistical software take.

EXAMPLE 8.18

SMART

Planning a New Smart Shopping Cart Study As part of Example 8.11 (page 421), we calculated a 95% confidence interval for the mean difference in spending when shopping with and without real-time feedback. The 95% margin of error was roughly $2.70. Suppose that a new study is being planned and the desired margin of error is $1.50. How many shoppers per group do we need?

The sample standard deviations in Example 8.11 were $6.59 and $6.85. To be a bit conservative, we'll guess that the two population standard deviations are both $7.00. To compute an initial n, we replace t^* with z^*. This results in

$$n = \left(\frac{\sqrt{2}z^* s_g}{m}\right)^2 = \left[\frac{\sqrt{2}(1.96)(7)}{1.5}\right]^2 = 167.3$$

We round up to get $n = 168$. The following table summarizes the margin of error for this and some larger values of n.

n	$t^* s_g \sqrt{2/n}$
168	1.502
169	1.498
170	1.493

The requirement is first satisfied when $n = 169$. This sample size is almost 3.5 times the sample size used in Example 8.11. We may not be able to recruit this large a sample. If so, we should consider a larger margin of error or lower confidence level. ∎

In SAS, we can perform these calculations using the commands

```
proc power;
    twosamplemeans CI=diff stddev=7 halfwidth=1.5
    probwidth=0.50 npergroup=.;
run;
```

Similar to the one-sample case, the probability of obtaining a margin of error less than the desired value can be altered by changing the probwidth setting.

APPLY YOUR KNOWLEDGE

8.77 Do we need a larger sample size? Refer to Example 8.18. For each of the following changes, state whether the needed sample size will increase or decrease and explain your reasoning.

(a) The desired margin of error is $2.00 instead of $1.50.

(b) A 99% rather than 95% confidence level is used.

(c) We use $n - 1$ instead of $2(n - 1)$ for the degrees of freedom.

8.78 Planning a new shopping cart study. Refer to the Example 8.18. What is the required sample size if the goal is have the 99% margin of error no more than $2.00? Use $s_g = 7.00$ and $2(n - 1)$ for the degrees of freedom.

Power of a significance test

two-sided tests and confidence intervals, p. 368

Although we prefer to use P-values rather than the reject-or-not view of the level α significance test, the latter view is very important for planning studies and for understanding statistical decision theory. Significance tests are closely related to confidence intervals—in fact, we saw that a two-sided test can be carried out directly from a confidence interval.

The significance level, like the confidence level, says how reliable the method is in repeated use. For example, if we perform level $\alpha = 0.05$ significance tests repeatedly when H_0 is, in fact, true, the conclusion to reject H_0 will occur $100 \times \alpha = 5\%$ of the time and the conclusion to not reject H_0 will occur the remaining 95% of the time.

When an alternative is true, the probability to reject H_0 will deviate away from $100 \times \alpha$. The higher it is, the more sensitive, or responsive, the test is to that alternative. We call this probability the *power* of the test and it is commonly used to determine the sample size for a test of significance.

> **POWER**
>
> The probability that a level α significance test will reject H_0 when a particular alternative value of the parameter is true is called the **power** of the test to detect that alternative.

The power of a statistical test measures its ability to detect deviations from the null hypothesis. Because we usually hope to show that the null hypothesis is false, it is important to design a study with high power. Unfortunately, because of inadequate planning, users frequently fail to find evidence for the effects that they believe to be present. This is often the result of an inadequate sample size. Power calculations, performed prior to running a study, are a way to avoid this pitfall and ensure the sample size is sufficiently large to answer the research question.

Just like the margin of error, the power of a significance test depends on various study-specific factors. Each factor and how it impacts power is as follows:

- **Significance level α.** A significance test at the 5% level will have a greater chance of rejecting H_0 than a test at the 1% level because the strength of evidence required for rejection is less.

- **Population standard deviation(s).** More variation means more uncertainty (less precision) about the mean(s). With more uncertainty, we are less likely to reject H_0.

- **Sample size(s).** More data will provide more information about the mean(s) and thus a better chance of distinguishing the alternative.

- **The alternative.** The further the alternative is from the null hypothesis, the easier it is to distinguish it and thus reject H_0.

We showed earlier that the margin of error depends on the first three factors. Any change in these factors that increases the power of a significance test also results in a smaller margin of error. This is due to the close relationship between significance tests and confidence intervals.

We usually rely on subject-specific knowledge to choose the alternative. It is the smallest departure from H_0 that matters in terms of decision making. For example, if allowing credit card purchases at your stores is only profitable if average monthly sales increase by at least $200, you would not be interested in detecting an average monthly sales increase of $100. On the other hand, you would be very interested in detecting an average monthly sales increase of $400.

There are times, however, when the choice of the alternative is not as clear-cut. In these situations, we commonly consider the **effect size,** which is the departure from H_0 divided by the population standard deviation. This standardizes the departure and puts it on a common scale. Conventional guidelines consider an effect size of 0.2 to be small, 0.5 to be medium, and 0.8 to be large.[30]

CALCULATING POWER

In general, calculating the power of a significance test involves two steps:

1. Write the event, in terms of the test statistic, that the test rejects H_0.
2. Find the probability of this event under the specified alternative.

Computing power can be quite technical. Fortunately, statistical software will do both steps for us. We just need to provide the four study-specific factors. When working with software that does not include effect size, set the alternative equal to the effect size and set the population standard deviation equal to 1.

The power of the one-sample *t* test

Computing the exact power of the one-sample t test against a specific alternative value of the population mean μ requires a new distribution, known as the **noncentral t distribution.** Calculations involving this distribution are not practical by hand but are easy with software that calculates probabilities for this new distribution. Here is an example.

noncentral *t* distribution

EXAMPLE 8.19

Is the Sample Size Large Enough? In Example 8.2 (page 401), we compared the average battery life of your smartphone to the average life stated by a product-testing group. A friend of yours with the same type of smartphone is planning to do the same. She decides that detecting a mean at least 30 minutes larger is useful in practice. Can she rely on a sample of 8 days to detect a difference of this size? Or does she need a larger sample size?

She wishes to compute the power of the t test for

$$H_0: \mu = 684$$
$$H_a: \mu > 684$$

against the alternative that $\mu = 684 + 30 = 714$ when $n = 8$. This gives us most of the information we need to compute the power. The other important piece is a guess of σ. She decides to use the results from your study and therefore rounds up $s = 49.66$ to $s_g = 50$ for her calculations.

Figure 8.14 shows Minitab output for the exact power calculation. It is about 45% and is represented by a dot on the power curve at a difference of 30.

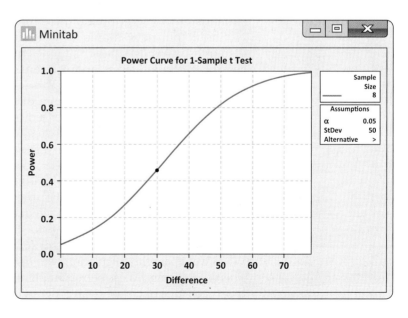

FIGURE 8.14 Minitab output (a power curve) for the one-sample power calculation in Example 8.19.

This curve is very informative. We see that with a sample size of 8, the power is greater than 80% only for differences larger than about 48 minutes. Your friend will definitely want to increase the sample size if she wants adequate power for a difference of 30 minutes. Currently, the chance of detecting this difference is slightly less than getting a head on the flip of a fair coin. ■

APPLY YOUR KNOWLEDGE

8.79 Power for other values of μ. Using the curve in Figure 8.14, what is the power for the alternative

(a) $\mu = 725$?

(b) $\mu = 765$?

(c) $\mu = 684$?

8.80 How is power affected? Refer to Example 8.19.

(a) If your friend used $s_g = 55$ instead of 50, would the power at $\mu = 714$ decrease, increase, or stay the same? Explain your answer.

(b) If your friend used $\alpha = 0.01$ instead of $\alpha = 0.05$, would the power at $\mu = 714$ decrease, increase, or stay the same? Explain your answer.

The power of the two-sample *t* test

Moving from the one-sample to two-sample case is relatively straightforward given access to software. The approach is similar and exact power calculations require the noncentral t distributions. For our discussion, we will consider only the common case where the null hypothesis is "no difference," $\mu_1 - \mu_2 = 0$. We also illustrate the calculation for the pooled two-sample t test. This is the most common test for power calculations because it requires a guess at only one common standard deviation σ, and the degrees of freedom, which are $n_1 + n_2 - 2$, do not need to be approximated.

EXAMPLE 8.20

CASE 8.2

Active versus Failed Companies In Case 8.2, we compared the cash flow margin for 74 active and 27 failed companies. Using the pooled two-sample procedure, the difference was statistically significant ($t = 2.85$, df $= 99$, $P = 0.005$). Because this study is a couple of years old, let's plan a similar study to determine if these findings continue to hold.

Should our new sample have similar numbers of firms? Or could we save resources by using smaller samples and still be able to declare that the successful and failed firms are different? To answer this question, we do some power calculations.

We want to be able to detect a difference in the means that is about the same as the value that we observed in our previous study. So, in our calculations, we will use the alternative $\mu_1 - \mu_2 = 12.00$. For our guess at the common standard deviation, we take the pooled value from our previous study, $s = 19.59$.

This leaves only two pieces of additional information: a significance level α and the sample sizes n_1 and n_2. For the first, we will choose the standard value $\alpha = 0.05$. For the sample sizes, we decide to try several different values but consider only examples where $n_1 = n_2$.

We start with $n_1 = 26$ and $n_2 = 26$, and software gives the power as 0.582. When we repeat the calculation with $n_1 = 41$ and $n_2 = 41$, we get a power of 78%. This result using JMP is shown in Figure 8.15. We need a relatively large sample from each type of company in order to detect a difference of 12%. ■

FIGURE 8.15 JMP input/output window for the two-sample power calculation in Example 8.20.

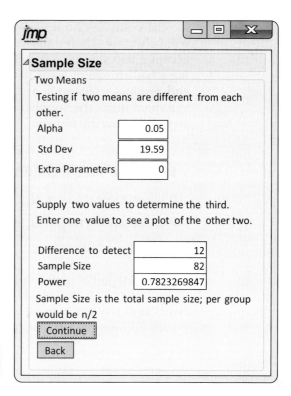

APPLY YOUR KNOWLEDGE

8.81 **Power and $\mu_1 - \mu_2$.** If you repeat the calculations in Example 8.20 for other values of $\mu_1 - \mu_2$ that are smaller than 12, would you expect the power to increase or decrease? Explain your answer.

8.82 **Power and the standard deviation.** If the true population standard deviation were 25 instead of the 19.59 hypothesized in Example 8.20, would the power increase or decrease? Explain your answer.

Inference as a decision

We have presented tests of significance as methods for assessing the strength of evidence *against the null hypothesis*. This assessment is made by the P-value, which is a probability computed under the assumption that H_0 is true. The alternative hypothesis (the statement we seek evidence for) enters the test only to help us see what outcomes count against the null hypothesis and to assess its power.

There is another way to think about these issues. Sometimes, we are really concerned about making a decision or choosing an action based on our evaluation of the data. Consider a company that uses high-quality ball bearings in their product. When a shipment arrives from a supplier, the company inspects a random sample of bearings from the thousands of bearings found in the shipment. On the basis of this sample outcome, the company either accepts or rejects the shipment.

Two types of error

Tests of significance concentrate on H_0, the null hypothesis. If a decision is called for, however, there is no reason to single out H_0. There are simply two hypotheses, and we must accept one and reject the other. We will call the two hypotheses H_0 and H_a, but H_0 no longer has the special status that it had in tests of significance. In the ball-bearing problem, the company must decide between

H_0: the shipment of bearings meets standards

H_a: the shipment does not meet standards

on the basis of a sample of bearings.

The company hopes that the decision will be correct, but sometimes it will be wrong. There are two types of incorrect decisions. The company can accept a bad shipment of bearings, or it can reject a good shipment. Accepting a bad shipment leads to a variety of costs to the company (e.g., injury to end-product users such as skateboarders or bikers), while rejecting a good shipment hurts the supplier. To help distinguish these two types of error, we give them specific names.

TYPE I AND TYPE II ERRORS

If we reject H_0 (accept H_a) when in fact H_0 is true, this is a **Type I error.**

If we accept H_0 (reject H_a) when in fact H_a is true, this is a **Type II error.**

The possibilities are summed up in Figure 8.16. If H_0 is true, our decision either is correct (if we accept H_0) or is a Type I error. If H_a is true, our decision either is correct or is a Type II error. Only one error is possible at one time. Figure 8.17 applies these ideas to the ball-bearing example.

FIGURE 8.16 The two types of error in testing hypotheses.

Decision based on sample	Truth about the population: H_0 true	Truth about the population: H_a true
Reject H_0	Type I error	Correct decision
Accept H_0	Correct decision	Type II error

FIGURE 8.17 The two types of error for the sampling of bearings application.

Decision based on sample	Truth about the lot: Does meet standards	Truth about the lot: Does not meet standards
Reject the lot	Type I error	Correct decision
Accept the lot	Correct decision	Type II error

Error probabilities

We can assess any rule for making decisions in terms of the probabilities of the two types of error. This is in keeping with the idea that statistical inference is based on probability. We cannot (short of inspecting the whole shipment) guarantee that good shipments of bearings will never be rejected and bad shipments will never be accepted. But by random sampling and the laws of probability, we can say what the probabilities of both kinds of error are.

EXAMPLE 8.21

Diameters of Bearings The diameter of a particular precision ball bearing has a target value of 20 millimeters (mm) with tolerance limits of ±0.001 mm around the target. When a shipment of the bearings arrives, the consumer takes an SRS of 25 bearings from the shipment and measures their diameters. The consumer rejects the bearings if the sample mean diameter is significantly different from 20 mm at the 5% significance level.

This is a test of the hypotheses

$$H_0: \mu = 20$$

$$H_a: \mu \neq 20$$

To carry out the test, the company will perform a one-sample t-test and rejects H_0 if

$$t < -2.064 \quad \text{or} \quad t > 2.064$$

A Type I error is to reject H_0 when in fact $\mu = 20$.

What about Type II errors? Because there are many values of μ in H_a, we concentrate on one value. Based on the tolerance limits, the supplier agrees that if there is evidence that the mean of ball bearings in the lot is 0.001 mm away from the desired mean of 20 mm, then the whole shipment should be rejected. So, a particular Type II error is to accept H_0 when in fact $\mu = 20 + 0.001 = 20.001$.

Figure 8.18 shows how the two probabilities of error are obtained from the two sampling distributions of $\bar{x}$, for $\mu = 20$ and for $\mu = 20.001$. When $\mu = 20$, H_0 is true and to reject H_0 is a Type I error. When $\mu = 20.001$, accepting H_0 is a Type II error. ∎

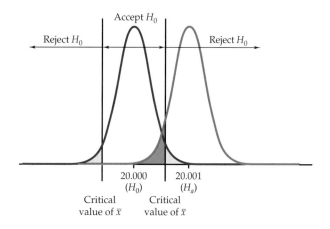

FIGURE 8.18 The two error probabilities, Example 8.21. The probability of a Type I error (yellow area) is the probability of rejecting H_0: $\mu = 20$ when in fact $\mu = 20$. The probability of a Type II error (blue area) is the probability of accepting H_0 when in fact $\mu = 20.001$.

The probability of a Type I error is the probability of rejecting H_0 when it is really true. In Example 8.21, this is the probability that $|t(24)| \geq 2.064$ when $\mu = 20$. But this is exactly the significance level of the test. The critical value

2.064 was chosen to make this probability 0.05, so we do not have to compute it again. The definition of "significant at level 0.05" is that sample outcomes this extreme will occur with probability 0.05 when H_0 is true.

SIGNIFICANCE AND TYPE I ERROR

The significance level α of any fixed level test is the probability of a Type I error. That is, α is the probability that the test will reject the null hypothesis H_0 when H_0 is in fact true.

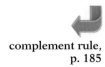

complement rule, p. 185

The probability of a Type II error is more complicated to compute but we can again use the relationship with tests of significance. The probability of a Type II error is the probability that the test will *fail to reject H_0* when μ has an alternative value. The *power* of the test for the alternative value is the probability that the test *does reject H_0* when H_a is true. These are complementary events so we can use the complement rule to compute one given the other.

POWER AND TYPE II ERROR

The power of a fixed level test for a particular alternative is 1 minus the probability of a Type II error for that alternative.

The two types of error and their probabilities give another interpretation of the significance level and power of a test. The distinction between tests of significance and tests as rules for deciding between two hypotheses lies not in the calculations, but in the reasoning that motivates the calculations.

In a test of significance, we focus on a single hypothesis (H_0) and a single probability (the P-value). The goal is to measure the strength of the sample evidence against H_0. Calculations of power are done to check the sensitivity of the test. If we cannot reject H_0, we conclude only that there is not sufficient evidence against H_0, not that H_0 is actually true.

If the same inference problem is thought of as a decision problem, we focus on two hypotheses and give a rule for deciding between them based on the sample evidence. We must, therefore, focus equally on two probabilities—the probabilities of the two types of error. We must choose one or the other hypothesis and cannot abstain on grounds of insufficient evidence.

Such a clear distinction between the two ways of thinking is helpful for understanding. In practice, the two approaches often merge. We continued to call one of the hypotheses in a decision problem H_0. The common practice of *testing hypotheses* mixes the reasoning of significance tests and decision rules as follows:

1. State H_0 and H_a just as in a test of significance.

2. Think of the problem as a decision problem so that the probabilities of Type I and Type II errors are relevant.

3. Because of Step 1, Type I errors are more serious. So choose an α (significance level) and consider only tests with probability of a Type I error no greater than α.

4. Among these tests, select one that makes the probability of a Type II error as small as possible (i.e., power as large as possible). If this probability is too large, you will have to take a larger sample to reduce the chance of an error.

Testing hypotheses may seem to be a hybrid approach. It was, historically, the effective beginning of decision-oriented ideas in statistics. An impressive mathematical theory of hypothesis testing was developed between 1928 and 1938 by Jerzy Neyman and Egon Pearson. The decision-making approach came later (1940s). Because decision theory in its pure form leaves you with two error probabilities and no simple rule on how to balance them, it has been used less often than either tests of significance or tests of hypotheses. Decision ideas have been applied in testing problems mainly by way of the Neyman-Pearson hypothesis-testing theory. That theory asks you first to choose α, and the influence of Fisher has often led users of hypothesis testing comfortably back to $\alpha = 0.05$ or $\alpha = 0.01$. Fisher, who was exceedingly argumentative, violently attacked the Neyman-Pearson decision-oriented ideas, and the argument still continues.

SECTION 8.3 SUMMARY

- The **sample size** required to obtain a confidence interval with an expected margin of error no larger than m for a population mean satisfies the constraint

$$m \geq t^* s_g / \sqrt{n}$$

where t^* is the critical value for the desired level of confidence with $n - 1$ degrees of freedom, and s_g is the guessed value for the population standard deviation.

- The sample sizes necessary for a two-sample confidence interval can be obtained using a similar constraint, but we would now need to guess both standard deviations and decide on an estimate for the degrees of freedom. We suggest using the smaller of $n_1 - 1$ and $n_2 - 1$.

- The **power** of a significance test measures its ability to detect an alternative hypothesis. The power for a specific alternative is calculated as the probability that the test will reject H_0 when that alternative is true. Increasing the size of the sample increases the power.

- The power of the one- and two-sample t tests involves the **noncentral t distributions.** Calculations involving these distributions are not practical by hand but are easy with software.

- To compute power, the researcher must provide the significance level α, sample size(s), guesses of the standard deviation(s), and specific alternative.

- In the case of testing H_0 versus H_a, decision analysis chooses a decision rule on the basis of the probabilities of two types of error. A **Type I error** occurs if H_0 is rejected when it is in fact true. A **Type II error** occurs if H_0 is accepted when in fact H_a is true.

- In a fixed level α significance test, the significance level α is the probability of a Type I error, and the power for a specific alternative is 1 minus the probability of a Type II error for that alternative.

SECTION 8.3 EXERCISES

For Exercises 8.75 and 8.76, see page 440; for 8.77 and 8.78, see page 441; for 8.79 and 8.80, see page 444; and for 8.81 and 8.82, see page 445.

8.83 What's wrong? For each of the following statements, explain what is wrong and why.

(a) To reduce the margin of error in half, the sample size needs to be doubled.

(b) When testing $H_0: \mu = 20$ versus the two-sided alternative, the power at $\mu = 15$ is larger than at $\mu = 25$.

(c) Lowering the significance level from 0.05 to 0.01 will increase the power for any alternative.

(d) Increasing sample size not only increases the power for all alternatives, but it also decreases the probability of a Type I error.

8.84 Apartment rental rates. You hope to rent an unfurnished one-bedroom apartment in Dallas next year. You call a friend who lives there and ask him to give you an estimate of the mean monthly rate. Having recently taken a statistics course, your geeky friend asks about the desired margin of error and confidence level for the estimate. He also tells you that the standard deviation of monthly rents for one-bedrooms is $360.

(a) For 95% confidence and a margin of error of $100, how many apartments should your friend randomly sample from the local newspaper?

(b) Suppose that you instead want the margin of error to be $50. How many apartments should your friend sample?

(c) Explain why in either case your geeky friend's sample may result in a bigger margin of error than desired.

8.85 More on apartment rental rates. Refer to the previous exercise. Will the 95% confidence interval include approximately 95% of the rents of all unfurnished one-bedroom apartments in this area? Explain why or why not.

8.86 Average hours per week on Facebook. The *Student Monitor* surveys 1200 undergraduates from 100 colleges semiannually to understand trends among college students.[31] Recently, the *Student Monitor* reported that Facebook users spend an average 5.6 hours per week on the site. Given the security issues with Facebook, you suspect that this amount of time is too large for your campus and plan a survey.

(a) You feel that a reasonable estimate of the standard deviation is 11.0 hours. What sample size is needed so that the expected margin of error of your estimate is not larger than one hour for 95% confidence?

(b) The distribution of times is likely to be heavily skewed to the right. Do you think that this skewness will invalidate the use of the t confidence interval in this case? Explain your answer.

8.87 Average hours per week listening to the radio. Refer to the previous exercise. The *Student Monitor* also reported that the average amount of time listening to the radio was 11.5 hours.

(a) Given an estimated standard deviation of 5.1 hours, what sample size is needed so that the expected margin of error of your estimate is not larger than one hour for 95% confidence?

(b) If your survey is going to ask about Facebook use and radio use, which of the two calculated sample sizes should you use? Explain your answer.

8.88 Accuracy of a laboratory scale. To assess the accuracy of a laboratory scale, a standard weight known to weigh 10 grams is weighed repeatedly. The scale readings are Normally distributed with unknown mean (this mean is 10 grams if the scale has no bias). The standard deviation of the scale readings in the past has been 0.002 gram. How many measurements must be averaged to get an expected margin of error no more than 0.001 with 98% confidence?

8.89 Credit card fees. The bank in Exercise 8.30 (page 417) tested a new idea on a sample of 125 customers. Suppose that the bank wanted to be quite certain of detecting a mean increase of $\mu = \$300$ in the credit card amount charged at the $\alpha = 0.01$ significance level. Perhaps a sample of only $n = 60$ customers would accomplish this. Find the approximate power of the test with $n = 60$ for the alternative $\mu = \$300$ as follows:

(a) What is the t critical value for the one-sided test with $\alpha = 0.01$ and $n = 60$?

(b) Write the criterion for rejecting H_0: $\mu = 0$ in terms of the t statistic. Then take $s = 1330$ and state the rejection criterion in terms of $\bar{x}$.

(c) Assume that $\mu = 300$ (the given alternative) and that $\sigma = 1330$. The approximate power is the probability of the event you found in part (b), calculated under these assumptions. Find the power. Would you recommend that the bank do a test on 60 customers, or should more customers be included?

8.90 A field trial. The tomato experts who carried out the field trial described in Exercise 8.39 (page 418) suspect that the relative lack of significance there is due to low power. They would like to be able to detect a mean difference in yields of 0.3 pound per plant at the 0.05 significance level. Based on the previous study, use 0.51 as an estimate of both the population σ and the value of s in future samples.

(a) What is the power of the test from Exercise 8.39 with $n = 10$ for the alternative $\mu = 0.3$?

(b) If the sample size is increased to $n = 15$ plots of land, what will be the power for the same alternative?

8.91 Assessing noise levels in fitness classes. In Exercise 8.55 (page 434), you compared the noise levels in both high-intensity and low-intensity fitness classes. Suppose you are concerned with these results and want to see if the noise levels in high-intensity fitness classes in your city are above the "standard" level ($\mu = 85$ dB). You plan to take an SRS of $n = 24$ classes in your neighborhood. Assuming $\sigma = 2.8$, $\alpha = 0.05$, and the alternative mean is $\mu = 86$ dB, what is the approximate power?

8.92 Digital mindfulness training: power. Exercise 8.51 (page 433) summarizes data on the change in stress levels across two different meditation regimens. Suppose that you are designing a new study to compare face-to-face training with digital training. You are hoping to recruit a total sample of just $N = 50$

participants (equal number assigned to each regimen) and would like to identify any difference in the change of SOS score that is more than 4.5 units. Your guess at the standard deviation is $s_g = 8$. If you use a pooled two-sample t test with significance level 0.05, what is the power of the test for this design?

8.93 Power, continued. Repeat the power calculation in the previous exercise for 70, 90, and 110 total participants. Summarize your power study. A graph of the power against sample size will help.

8.94 Approximating power. Refer to Exercise 8.92. When software is not available, one can approximate the power by assuming the standard deviation σ is known and using the Normal distribution. Here are the steps.

(a) Given $N = 50$, state the degrees of freedom for the two-sample pooled t test.

(b) Assuming $\alpha = 0.05$, use Table D to approximate the critical t value, t^*.

(c) This means we reject when

$$\left| \frac{(\bar{x}_1 - \bar{x}_2) - 0}{8\sqrt{2/25}} \right| \geq t^*$$

Rewrite this event so that it is in terms of $\bar{x}_1 - \bar{x}_2$.

(d) Compute the probability of this event assuming $(\bar{x}_1 - \bar{x}_2) \sim N(4.5, 8)$. How does it compare to the calculation in Exercise 8.92?

8.95 Ego strength: power. You want to compare the ego strengths of MBA students who plan to seek work at consulting firms and those who favor manufacturing firms. Based on the data from Exercise 8.63 (page 436), you will use $\sigma = 0.7$ for planning purposes. The pooled two-sample t test with $\alpha = 0.01$ will be used to make the comparison. You judge a difference of 0.5 points to be of interest.

(a) Find the power for the design with 20 MBA students in each group.

(b) The power in part (a) is not acceptable. Redo the calculations for 30 students in each group and $\alpha = 0.05$.

8.96 Starting salaries. In a recent survey by the National Association of Colleges and Employers, the average starting salary for business majors was reported to be $56,720.[32] You are planning to do a survey of starting salaries for recent business majors from your university. Using an estimated standard deviation of $15,500, what sample size do you need to have a margin of error equal to $5000 with 95% confidence?

8.97 Changes in sample size. Suppose that, in the setting of the previous exercise, you have the resources to contact 50 recent graduates. If all respond, will your margin of error be larger or smaller than $5000? What if only 50% respond? Verify your answers by performing the calculations.

CHAPTER 8 REVIEW EXERCISES

In exercises that call for two-sample t procedures, try to use the degrees of freedom approximation provided by software. For exercises involving summarized data, this approximation is provided to you. If you instead use the conservative approximation, the smaller of $n_1 - 1$ and $n_2 - 1$, be sure to clearly state this.

8.98 LSAT scores. The scores of four classmates on the Law School Admission Test are

168   127   147   155

Find the mean, the standard deviation, and the standard error of the mean. Is it appropriate to calculate a confidence interval based on these data? Explain why or why not.

8.99 t is robust. A manufacturer of flash drives employs a market research firm to estimate retail sales of its products. Here are last month's sales of 128-GB flash drives from an SRS of 50 stores in the Midwest sales region: RETAIL

29	31	45	40	32	21	23	28	19	11
35	21	17	23	22	22	33	31	34	15
32	27	33	24	21	28	16	67	21	39
33	56	48	14	40	8	47	21	21	25
53	28	35	16	20	24	45	56	28	23

(a) Make a stemplot of the data to confirm that the distribution is skewed to the right. Even though the data are not Normal, explain why the t procedures can be used to analyze these data.

(b) Let's verify this robustness. Three bootstrap (pages 412–413) simulations, each with 1000 repetitions, give these 95% confidence intervals for mean sales in the entire region: (26.32, 33.10), (26.14, 33.22), and (26.46, 33.20). Find the 95% t confidence interval for the mean. Is it essentially the same as the bootstrap intervals? Explain your answer.

8.100 Number of critical food violations. The results of a major city's restaurant inspections are available through its online newspaper.[33] Critical food violations are those that put patrons at risk of getting sick and must be immediately corrected by the restaurant. An SRS of $n = 300$ inspections from the more than 10,000 inspections since January 2012 had $\bar{x} = 0.90$ violations and $s = 2.35$ violations.

(a) Test the hypothesis, using $\alpha = 0.05$, that the average number of critical violations is less than 1.25. State the two hypotheses, the test statistic, and the P-value.

(b) Construct a 95% confidence interval for the average number of critical violations and summarize your result.

(c) Which of the two summaries (significance test versus confidence interval) do you find more helpful in this case? Explain your answer.

(d) These data are integers ranging from 0 to 14. The data are also skewed to the right, with 70% of the values either a 0 or 1. Given this information, do you feel use of the t procedures is appropriate? Explain your answer.

8.101 Interpreting software output. You use statistical software to perform a significance test of the null hypothesis that two means are equal. The software reports P-values for the two-sided alternative. Your alternative is that the first mean is less than the second mean.

(a) The software reports $t = -1.87$ with a P-value of 0.076. Would you reject H_0 with $\alpha = 0.05$? Explain your answer.

(b) The software reports $t = 1.87$ with a P-value of 0.076. Would you reject H_0 with $\alpha = 0.05$? Explain your answer.

8.102 Which design? The following situations all require inference about a mean or means. Identify each as (1) a single sample, (2) matched pairs, or (3) two independent samples. Explain your answers.

(a) Your customers are college students. You are interested in comparing the interest in a new product that you are developing between those students who live in the dorms and those who live elsewhere.

(b) Your customers are college students. You are interested in finding out which of two new product labels is more appealing.

(c) Your customers are college students. You are interested in assessing their interest in a new product.

8.103 Which design? The following situations all require inference about a mean or means. Identify each as (1) a single sample, (2) matched pairs, or (3) two independent samples. Explain your answers.

(a) You want to estimate the average age of your store's customers.

(b) You do an SRS survey of your customers every year. One of the questions on the survey asks about customer satisfaction on a 7-point scale with the response 1 indicating "very dissatisfied" and 7 indicating "very satisfied." You want to see if the mean customer satisfaction has improved from last year.

(c) You ask an SRS of customers their opinions on each of two new floor plans for your store.

8.104 Does flipping a classroom work? One approach to active learning is a "flipped classroom." This commonly involves students watching video lectures outside of class and working on problem-solving activities in class. Research has primarily focused on comparing teaching approaches using end-of-class outcomes, such as final grades. In a recent project, researchers compared the lasting benefits of a flipped classroom by comparing grades received in the subsequent course of the series. Here are the results:[34]

Group	n	$\bar{x}$	s
Flipped	166	2.45	1.09
Traditional	129	1.59	1.20

(a) Do the average grades received differ across the two delivery styles of the previous course? Test the hypothesis that the average grades of the two groups are the same (software gives $k = 261.40$).

(b) Is this an experiment or an observational study? Explain your answer.

(c) Based on your answers to parts (a) and (b), what are your conclusions?

8.105 Two-sample t test versus matched pairs t test, continued. Refer to the previous exercise. Perhaps an easier way to see the major difference in the two analysis approaches for these data is by computing 95% confidence intervals for the mean difference.

(a) Compute the 95% confidence interval using the two-sample t confidence interval.

(b) Compute the 95% confidence interval using the matched pairs t confidence interval.

(c) Compare the estimates (i.e., the centers of the intervals) and margins of error. What is the major difference between the two approaches for these data?

8.106 Two-sample t test versus matched pairs t test. Consider the following data set. The data were actually collected in pairs, and each row represents a pair.
PAIRED

Group 1	Group 2
48.86	48.88
50.60	52.63
51.02	52.55
47.99	50.94
54.20	53.02
50.66	50.66
45.91	47.78
48.79	48.44
47.76	48.92
51.13	51.63

(a) Suppose that we ignore the fact that the data were collected in pairs and mistakenly treat this as a two-sample problem. Compute the sample mean and variance for each group. Then compute the two-sample t statistic, degrees of freedom, and P-value for the two-sided alternative.

(b) Now analyze the data in the proper way. Compute the sample mean and variance of the differences. Then compute the t statistic, degrees of freedom, and P-value.

(c) Describe the differences in the two test results.

8.107 Corn seed prices. The U.S. Department of Agriculture (USDA) uses sample surveys to obtain important economic estimates. One USDA pilot study estimated from a sample of 30 farms the amount a farmer will pay per planted acre of corn. The mean price was reported as $680.36 with a standard error of $15.49. Give a 95% confidence interval for the amount a farmer will pay per planted acre of corn.[35]

8.108 Does dress affect competence and intelligence ratings? Researchers performed a study to examine whether or not women are perceived as less competent and less intelligent when they dress in a sexy manner versus a business-like manner. Competence was rated from 1 (not at all) to 7 (extremely), and a 1 to 5 scale was used for intelligence. Under each condition, 17 subjects provided data. Here are summary statistics:[36]

Rating	Sexy $\bar{x}$	Sexy s	Business-like $\bar{x}$	Business-like s
Competence	4.13	0.99	5.42	0.85
Intelligence	2.91	0.74	3.50	0.71

Analyze the two variables, and write a report summarizing your work. Be sure to include details regarding the statistical methods you used, your assumptions, and your conclusions (software gives $k = 31.3$ and $k = 31.9$ degrees of freedom for competence and intelligence, respectively).

8.109 Can snobby salespeople boost retail sales? Researchers asked 180 women to read a hypothetical shopping experience where they entered a luxury store (e.g., Louis Vuitton, Gucci, Burberry) and ask a salesperson for directions to the items they seek. For half the women, the salesperson was condescending while doing this. The other half were directed in a neutral manner. After reading the experience, participants were asked various questions, including what price they were willing to pay (in dollars) for a particular product from the brand.[37] Here is a summary of the results.

Chain	n	$\bar{x}$	s
Condescending	90	4.44	3.98
Neutral	90	3.95	2.88

Were the participants who were treated rudely willing to pay more for the product? Analyze the data, and write a report summarizing your work. Be sure to include details regarding the statistical methods you used, your assumptions, and your conclusions. If you use two-sample t procedures, software gives $k = 162.1$ degrees of freedom.

8.110 Evaluate the dress study. Refer to Exercise 8.108. Participants in the study viewed a videotape of a woman described as a 28-year-old senior manager for a Chicago advertising firm who had been working for this firm for seven years. The same woman was used for each of the two conditions, but she wore different clothing each time. For the business-like condition, the woman wore little makeup, black slacks, a turtleneck, a business jacket, and flat shoes. For the sexy condition, the same woman wore a tight knee-length skirt, a low-cut shirt with a cardigan over it, high-heeled shoes, and more makeup, and her hair was tousled. The subjects who evaluated the videotape were male and female undergraduate students who were predominantly Caucasian, from middle- to upper-class backgrounds, and between the ages of 18 and 24. The content of the videotape was identical in both conditions. The woman described her general background, life in college, and hobbies.

(a) Write a critique of this study, with particular emphasis on its limitations and how you would take these into account when drawing conclusions based on the study.

(b) Propose an alternative study that would address a similar question. Be sure to provide details about how your study would be run.

8.111 More on snobby salespeople. Refer to Exercise 8.109. Researchers also asked a different 180 women to read the same hypothetical shopping experience, but now they entered a mass market store (e.g., Gap, American Eagle, H&M). Here are those results (in dollars) for the two conditions.

Chain	n	$\bar{x}$	s
Condescending	90	2.90	3.28
Neutral	90	2.98	3.24

Were the participants who were treated rudely willing to pay more for the product? Analyze the data, and write a report summarizing your work. Be sure to include details regarding the statistical methods you used, your assumptions, and your conclusions. Also compare these results with the ones from Exercise 8.109. If you use two-sample t procedures, software gives $k = 178.0$ degrees of freedom.

8.112 Transforming the response. Refer to Exercises 8.109 and 8.111. The researchers state that they took the natural log of the willingness to pay variable in order to "normalize the distribution" prior to analysis. Thus, their test results are based on log dollar measurements. For the t procedures used in the previous two exercises, do you feel this transformation is necessary? Explain your answer.

8.113 Personalities of hotel managers. Successful hotel managers must have personality characteristics often thought of as feminine (such as "compassionate") as well as those often thought of as masculine (such as "forceful"). The Bem Sex-Role Inventory (BSRI) is a personality test that gives separate ratings for female and male stereotypes, both on a scale of 1 to 7. Here are summary statistics for a sample of 148 male general managers of three-star and four-star hotels.[38] The data come from a comprehensive mailing to these hotels. The response rate was 48%, which is good for mail surveys of this kind. Although nonresponse remains an issue, users of statistics usually act as if they have an SRS when the response rate is "good enough."

Masculinity score	Femininity score
$\bar{x} = 5.91$	$\bar{x} = 5.29$
$s = 0.57$	$s = 0.75$

The mean BSRI masculinity score for the general male population is $\mu = 4.88$. Is there evidence that hotel managers on the average score higher in masculinity than the general male population?

8.114 Another personality trait of hotel managers. Continue your study from the previous exercise. The mean BSRI femininity score in the general male population is $\mu = 5.19$. (It does seem odd that the mean femininity score is higher than the mean masculinity score, but such is the world of personality tests. The two scales are separate.) Is there evidence that hotel managers on the average score higher in femininity than the general male population?

8.115 Alcohol content of wine. The alcohol content of wine depends on the grape variety, the way in which the wine is produced from the grapes, the weather, and other influences. Here are data on the percent of alcohol in wine produced from the same grape variety in the same year by 48 winemakers in the same region of Italy:[39] WINE

12.86	12.88	12.81	12.70	12.51
12.60	12.25	12.53	13.49	12.84
12.93	13.36	13.52	13.62	12.25
13.16	13.88	12.87	13.32	13.08
13.50	12.79	13.11	13.23	12.58
13.17	13.84	12.45	14.34	13.48
12.36	13.69	12.85	12.96	13.78
13.73	13.45	12.82	13.58	13.40
12.20	12.77	14.16	13.71	13.40
13.27	13.17	14.13		

(a) Make a stemplot of the data. The distribution is a bit irregular, but there is no reason to avoid use of t procedures for $n = 48$.

(b) Give a 95% confidence interval for the mean alcohol content of wine of this type.

8.116 Accuracy performance of hockey players. Individual alpha peak frequency (IAPF) is an important indicator in electoencephalogram (EEG) studies and is linked to a variety of cognition functions. Higher IAPF is thought to reflect the trait of cognitive preparedness. Because athletes rely on speedy information processing and shorter reaction times, it is thought their IAPF will be above average and changes in the IAPF can be used to assess psychomotor demand of a sports task.

In one study, researchers recruited 28 university hockey players (15 female, 13 male).[40] Each player had an initial eyes-open EEG reading, then proceeded to do 50 shots on net (accuracy challenge), and then had another eyes-open EEG reading. Two subjects were removed from analysis due to faulty EEG readings. HOCKEY

(a) The number of target hits on the accuracy challenge (PERF) is an integer between 0 and 50. Is the use of t procedures to compare males and females appropriate? Explain your opinion.

(b) Compare the males and females on the accuracy challenge. State the two hypotheses, the test statistic, and its degrees of freedom and P-value.

8.117 Psychomotor demand of the accuracy test. Refer to the previous exercise for a description of the problem. Let's now look at the change AIPF. HOCKEY

(a) Is a two-sample or matched pairs t test appropriate to assess the psychomotor demand of accuracy test among all hockey players? Explain your answer.

(b) Perform the appropriate t test. State the two hypotheses, the test statistic, and its degrees of freedom and P-value. What are your conclusions?

(c) Suppose you wanted to compare psychomotor demand across sexes. Is a two-sample or matched pairs t test appropriate? Explain your answer.

(d) Perform the appropriate t test. State the two hypotheses, the test statistic, and its degrees of freedom and P-value. What are your conclusions?

8.118 Psychomotor demand of the accuracy test, continued. Refer to the previous exercise. HOCKEY

(a) There is a potential outlier when considering the change in IAPF. Identify this hockey player by ID. Is the player male of female?

(b) Repeat the analyses of the previous exercise with this observation removed from the analysis. How does it impact the results and conclusions? Write a short summary report.

8.119 Converting a confidence interval. In Example 8.1 (page 399), we computed a 95% confidence interval for the average battery life of your phone in minutes. Suppose instead, we wanted the confidence interval in hours. BLIFE

(a) Convert each of the eight times into hours by dividing it by 60. Then construct the 95% confidence interval for the average time in hours using these converted times.

(b) Describe how you could have used the interval computed in Example 8.1 to get this interval.

8.120 Durable press and breaking strength. "Durable press" cotton fabrics are treated to improve their recovery from wrinkles after washing. Unfortunately, the treatment also reduces the strength of the fabric. A study compared the breaking strength of fabric treated by two commercial durable press processes. Five specimens of the same fabric were assigned at random to each process. Here are the data, in pounds of pull needed to tear the fabric[41]: BRKSTR

| Permafresh 55 | 29.9 | 30.7 | 30.0 | 29.5 | 27.6 |
| Hylite LF | 28.8 | 23.9 | 27.0 | 22.1 | 24.2 |

Is there good evidence that the two processes result in different mean breaking strengths?

8.121 Find a confidence interval. Continue your work from the previous exercise. A fabric manufacturer wants to know how large a strength advantage fabrics treated by the Permafresh method have over fabrics treated by the Hylite process. Give a 95% confidence interval for the difference in mean breaking strengths. BRKSTR

8.122 Recovery from wrinkles. Of course, the reason for durable press treatment is to reduce wrinkling. "Wrinkle recovery angle" measures how well a fabric recovers from wrinkles. Higher is better. Here are data on the wrinkle recovery angle (in degrees) for the same fabric specimens discussed in the previous two exercises: WRINKLE

| Permafresh 55 | 136 | 135 | 132 | 137 | 134 |
| Hylite LF | 143 | 141 | 146 | 141 | 145 |

Which process has better wrinkle resistance? Is the difference statistically significant?

8.123 Competitive prices? A retailer entered into an exclusive agreement with a supplier who guaranteed to provide all products at competitive prices. The retailer eventually began to purchase supplies from other vendors who offered better prices. The original supplier filed a legal action, claiming violation of the agreement. In defense, the retailer had an audit performed on a random sample of invoices. For each audited invoice, all purchases made from other suppliers were examined and the prices were compared with those offered by the original supplier. For each invoice, the percent of purchases for which the alternate supplier offered a lower price than the original supplier was recorded. Here are the data:[42] CMPPRIC

100	0	0	100	33	45	100	34	78
100	77	33	100	69	100	89	100	100
100	100	100	100	100	100	100		

Report the average of the percents with a 95% margin of error. Do the sample invoices suggest that the original supplier's prices are not competitive on the average?

8.124 Executives learn Spanish. A company contracts with a language institute to provide instruction in Spanish for its executives who will be posted overseas. The following table gives the pretest and posttest scores on the Modern Language Association's listening test in Spanish for 20 executives.[43] SPNISH

Subject	Pretest	Posttest	Subject	Pretest	Posttest
1	30	29	11	30	32
2	28	30	12	29	28
3	31	32	13	31	34
4	26	30	14	29	32
5	20	16	15	34	32
6	30	25	16	20	27
7	34	31	17	26	28
8	15	18	18	25	29
9	28	33	19	31	32
10	20	25	20	29	32

(a) We hope to show that the training improves listening skills. State an appropriate H_0 and H_a. Describe in words the parameters that appear in your hypotheses.

(b) Make a graphical check for outliers or strong skewness in the data that you will use in your statistical test, and report your conclusions on the validity of the test.

(c) Carry out a test. Can you reject H_0 at the 5% significance level? At the 1% significance level?

(d) Give a 90% confidence interval for the mean increase in listening score due to the intensive training.

8.125 Learning Spanish. Refer to the previous exercise. Use the sign test to assess whether the intensive language training improves Spanish listening skills. State the hypotheses, give the P-value using the binomial table (Table C), and report your conclusion.

8.126 Investigating the endowment effect, continued. Refer to Exercise 8.28 (page 416). The group of researchers also asked these same 40 students from a graduate marketing course to consider a Vosges Oaxaca gourmet chocolate bar made with dark chocolate and chili pepper. Test the null hypothesis that there is no difference between the two prices. Also construct a 95% confidence interval of the endowment effect. ENDOW1

8.127 Testing job applicants. The one-hole test is used to test the manipulative skill of job applicants. This test requires subjects to grasp a pin, move it to a hole, insert it, and return for another pin. The score on the test is the number of pins inserted in a fixed time interval. One study compared male college students with experienced female industrial workers. Here are the data for the first minute of the test:[44]

Group	n	$\bar{x}$	s
Students	750	35.12	4.31
Workers	412	37.32	3.83

(a) We expect that the experienced workers will outperform the students, at least during the first minute, before learning occurs. State the hypotheses for a statistical test of this expectation and perform the test. Give a P-value, and state your conclusions (software gives $k = 803.8$ degrees of freedom).

(b) The distribution of scores is slightly skewed to the left. Explain why the procedure you used in part (a) is nonetheless acceptable.

(c) One purpose of the study was to develop performance norms for job applicants. Based on the preceding data, what is the range that covers the middle 95% of experienced workers? (Be careful! This is not the same as a 95% confidence interval for the mean score of experienced workers.)

(d) The five-number summary of the distribution of scores among the workers is

23   33.5   37   40.5   46

for the first minute and

32   39   44   49   59

for the fifteenth minute of the test. Display these summaries graphically, and describe briefly the differences between the distributions of scores in the first and fifteenth minutes.

8.128 Ego strengths of MBA graduates: power. In Exercise 8.95 (page 451), you found the power for a study designed to compare the "ego strengths" of two groups of MBA students. Now you must design a study to compare MBA graduates who reached partner in a large consulting firm with those who joined the firm but failed to become partners.

Assume the same value of $\sigma = 0.7$ and use $\alpha = 0.05$. You are planning to have 20 subjects in each group. Calculate the power of the pooled two-sample t test that compares the mean ego strengths of these two groups of MBA graduates for several values of the true difference. Include values that have a very small chance of being detected and some that are virtually certain to be seen in your sample. Plot the power versus the true difference and write a short summary of what you have found.

8.129 Sign test for the endowment effect. Refer to Exercise 8.28 (page 416) and Exercise 8.126. We can also compare the endowment effects of each chocolate bar. Is there evidence that the median difference in endowment effects (Woolloomooloo minus Oaxaca) is greater than 0? Perform a sign test using the 0.05 significance level.
ENDOW2

mavoimages/Deposit Photos

CHAPTER 9

One-Way Analysis of Variance

Introduction

In the previous chapter, we discussed the tools needed to compare the means of two groups and found that the two-sample t procedures are sufficiently robust to be widely useful. Businesses, however, frequently consider studies that involve more than two groups. For example,

- A new company needs to determine its best digital advertising channel. It decides to compare the company's website traffic under three different channels: search engine advertising on Google, display advertising, and social media advertising on LinkedIn.

- Guayule is a shrub grown in the Southwest that contains a type of rubber that can be used in tires. Cooper Tire & Rubber investigates the tread life of six concept tires made entirely of rubber from guayule.[1]

- Are there ways to make multiblade cartridges last longer? Gillette researchers compare seven different methods to increase cartridge life. These include storing the razor's cartridge in rubbing alcohol, olive oil, or water in between uses.

How should we analyze data from studies with many groups? Based on what we learned in Chapter 8, we might consider a set of two-sample t tests, each comparing a different pair of groups. This approach, however, can involve a large number of t tests (for example, the Gillette study would involve 21 t tests), and each comparison uses only a subset of the available data. A more unified approach would be better.

In this chapter, we discuss such an approach that allows us to compare any number of means. These techniques generalize the two-sample t methods and share its robustness and usefulness. They also introduce us to the method of analysis of variance and the family of F distributions.

CHAPTER OUTLINE

9.1 One-Way Analysis of Variance

9.2 Additional Comparisons of Group Means

9.1 One-Way Analysis of Variance

When you complete this section, you will be able to:

- Describe the underlying idea of the *F* test for ANOVA in terms of the variation among group means and the variation within groups.
- Identify the ANOVA table sources of variation and associated degrees of freedom from the description of a study.
- Use ANOVA table output to obtain the *F* test result and the coefficient of determination.
- Summarize what the ANOVA *F* test can tell you about the group means and what it cannot.
- Use software to generate plots and sample statistics used to check the conditions that are needed for ANOVA.

Which of three digital advertising channels produces the largest website traffic for a new company? Which of six concept tires has the best treadwear performance? When comparing different groups, the data are subject to sampling variability. We would not, for example, expect the same website traffic if the company repeated the study over a different set of weeks. We also would not expect the same tread lifetime data if the tire experiment were repeated under similar conditions.

Because of this variability, we pose the question for inference in terms of the *mean* response. In Section 8.2 (page 419), we met procedures for comparing the means of two populations. We now extend those methods to problems involving the means of more than two populations. The statistical methodology for comparing several means is called **analysis of variance,** or simply ANOVA **ANOVA.** In this and the next section, we examine the basic ideas and conditions that are needed for ANOVA. Although the details differ, many of the concepts are similar to those discussed in the two-sample case.

When there is only one way to classify the populations of interest, we use one-way ANOVA **one-way ANOVA** to analyze the data. We call the categorical explanatory variable that classifies these populations a factor. For example, to compare the three digital advertising channels, we use one-way ANOVA with advertising channel as our factor. This chapter presents the details for one-way ANOVA.

factor, p. 144

In many other comparison studies, there is more than one way to classify the populations. For the tire study, the Cooper R&D team might also consider tread life under three different road temperatures. How much greater will the wear be for each concept tire at a higher temperature? Analyzing the effect two-way ANOVA of concept tire *and* road temperature together requires **two-way ANOVA.** We discuss this technique in Chapter 16.

The ANOVA setting

The data for one-way ANOVA are collected just as they are in the two-sample population case. We draw a simple random sample (SRS) from each population and use the data to test the null hypothesis that the population means are all equal.

Consider the following two examples.

EXAMPLE 9.1

Comparing Produce Displays A grocery store chain wants to compare three different produce displays. It is interested in whether one of the layouts better attracts shoppers and results in more sales. To investigate, the chain randomly assigns 20 stores to each of the three displays and records the amount of produce sold in a one-week period. ∎

EXAMPLE 9.2

Average Age of Customers How do five Starbucks coffeehouses in the same city differ in terms of customer demographics? Are certain coffeehouses more popular among Generation Z? A market research team randomly asks 40 customers of each store to respond to a questionnaire. One variable of interest is the customer's age. ■

These two examples are similar in that

- There is a single quantitative response variable measured on many units; the units are the 60 *stores* in the first example and 200 *customers* in the second. The response variable is the *weekly sales* in the first example and the *age of the customer* in the second.

- The goal is to compare several populations: stores with *three* different produce displays in the first example and customers of the *five* coffeehouses in the second.

- There is a single categorical explanatory variable, or factor, that classifies these populations: *produce display* in the first example and *coffeehouse* in the second.

observation versus experiment, p. 128

bias and variability, p. 300

There is, however, an important difference. Example 9.1 describes an experiment in which stores are randomly assigned to displays. Example 9.2 is an observational study in which customers are selected during a particular time of day, and not all may agree to provide data. This means that the samples in the second example are not truly SRSs. The research team will treat them as such because they believe that the selective sampling and nonresponse are ignorable sources of bias and variability, but others may not agree. *Always consider the sampling design and potential sources of bias in an observational study.*

In both examples, one-way ANOVA is used to compare the mean responses. The same ANOVA methods apply to data from randomized experiments (Example 9.1) and to data from random samples (Example 9.2). *However, it is important to keep the data-production method in mind when interpreting the results.* A strong case for causation is best made by a randomized experiment.

Comparing means

The question we ask in ANOVA is, "Do all groups have the same population mean?" We often use the term "groups" for the populations to be compared in a one-way ANOVA. To answer this question, we compare sample means. Figure 9.1 displays the sample means for the Example 9.1 experiment. It appears that Display 2 has the highest average sales. But are the observed differences among the sample means just the result of chance variation? We would not expect sample means to be equal even if the population means are all identical.

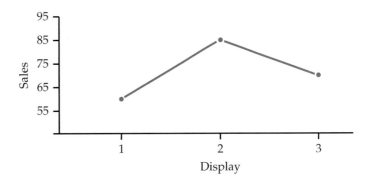

FIGURE 9.1 Mean weekly sales for three different produce displays.

FIGURE 9.2 (a) Side-by-side boxplots for three groups with large within-group variation. The differences among centers may be just chance variation. (b) Side-by-side boxplots for three groups with the same centers as in Figure 9.2(a) but with small within-group variation. The differences among centers are more likely to be significant.

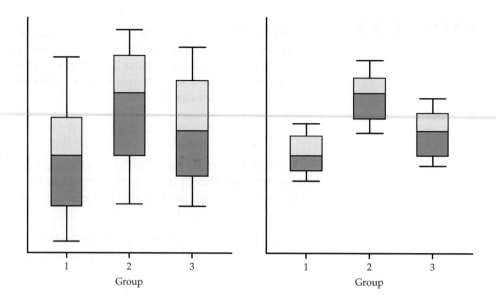

standard deviation of $\bar{x}$, p. 311

The purpose of ANOVA is to assess whether the observed differences among sample means are *statistically significant*, in other words, to test whether the variation in sample means is too large to plausibly be due to chance. This question can't be answered from the sample means alone. Because the standard deviation of a sample mean $\bar{x}$ is the population standard deviation σ divided by $\sqrt{n}$, the answer depends upon both the variation within the groups of observations and the sizes of the samples.

Side-by-side boxplots help us see the within-group variation. Compare Figures 9.2(a) and 9.2(b). The sample medians are the same in both figures and roughly match the means in Figure 9.1. The large variation within the groups in Figure 9.2(a) suggests that the differences among sample medians could be due simply to chance. The data in Figure 9.2(b) are much more convincing evidence that the groups differ.

Even the boxplots omit essential information, however. To assess the observed differences, we must also know how large the samples are. Nonetheless, boxplots are a good preliminary display of ANOVA data. While ANOVA compares means and boxplots display medians, these two measures of center will be close together for distributions that are nearly symmetric. If the distributions are not symmetric, we may consider a transformation prior to displaying and analyzing the data. We could also consider a nonparametric approach described in Chapter 17.

transforming data, p. 69

Revisiting the pooled two-sample t statistic

pooled two-sample t statistics, p. 429

Two-sample t statistics compare the means of two populations. If the two populations are assumed to have equal but unknown standard deviations and the sample sizes are both equal to n, the t statistic is

$$t = \frac{\bar{x}_1 - \bar{x}_2}{s_p\sqrt{\dfrac{1}{n} + \dfrac{1}{n}}} = \frac{\sqrt{\dfrac{n}{2}}(\bar{x}_1 - \bar{x}_2)}{s_p}$$

The square of this t statistic is

$$t^2 = \frac{\dfrac{n}{2}(\bar{x}_1 - \bar{x}_2)^2}{s_p^2}$$

It turns out that if we use ANOVA to compare two populations, the ANOVA test statistic is exactly equal to t^2. Thus, we can learn something about how ANOVA works by looking carefully at the statistic in this form.

between-group variation   The numerator in the t^2 statistic measures the variation **between** the groups in terms of the difference between their sample means $\bar{x}_1$ and $\bar{x}_2$. This is multiplied by a factor for the common sample size n. Thus, the numerator can be large because of a large difference between the sample means or because the common sample size n is large.

within-group variation   The denominator measures the variation **within** groups by s_p^2, the pooled estimator of the common variance. For the same sample means and n, the smaller the within-group variation, the larger the test statistic and, thus, the more significant the result (recall Figure 9.2).

Although the general form of the ANOVA test statistic is more complicated, the idea is the same. To assess whether several populations all have the same mean, we compare the variation *among* the means of several groups with the variation *within* groups. Because we are comparing variation, the method is called *analysis of variance*.

APPLY YOUR KNOWLEDGE

9.1 What's wrong? For each of the following, explain what is wrong and why.

(a) ANOVA tests the null hypothesis that the sample means are all equal.

(b) Within-group variation is the variation in the data due to the differences in the group sample means.

(c) A two-way ANOVA is used when comparing two populations.

(d) A strong case for causation is best made by an observational study.

9.2 Viewing side-by-side boxplots. Consider the coffeehouse setting in Example 9.2 (page 459) and assume all the population means are the same except for the second coffeehouse, which has a lower average age. Using Figure 9.2 as a guide, sketch possible side-by-side boxplots for the following two situations:

(a) The within-group variation is small enough that ANOVA will almost surely detect that the means differ.

(b) The within-group variation is so large that ANOVA likely won't detect this difference in means.

An overview of ANOVA

Now that we have a general idea of how ANOVA works, we will explore the specifics of one-way ANOVA using Case 9.1. Because the computations are more lengthy than those for the t test, we generally use computer software to perform the calculations. Automating the calculations frees us from the burden of arithmetic and allows us to concentrate on interpretation.

CASE 9.1 Tip of the Hat and Wag of the Finger? It seems as if every day we hear of another public figure acting badly. In business, immoral actions by a CEO not only threaten the CEO's professional reputation, but can also affect the company's bottom line. Quite often, however, consumers continue to support the company regardless of how they feel about the CEO's actions. A group of researchers propose this is because consumers engage in moral

MORAL

decoupling.[2] This is a process by which judgments of performance are separated from judgments of morality.

To demonstrate moral decoupling, the researchers performed an experiment involving 121 participants. Each participant was randomly assigned to one of three groups: moral decoupling, moral rationalization, or control. For the first two groups, participants were primed by reading statements either arguing that immoral actions should remain separate from judgments of performance (moral decoupling) or statements arguing that people should not always be at fault for their immoral actions because of situational pressures (moral rationalization). In the control group, participants read about the importance of humor.

All participants then read a scenario about a CEO of a consumer electronics company who had helped his company become a leader in innovative and stylish products over the course of the past decade. Last month, however, he was caught in a scandal and confirmed he supported racist and sexist hiring policies. After reading the scenario, participants were asked to indicate their likelihood of purchasing the company's products in the future by answering several items, each measured on a 0 to 100 scale. Here is a summary of the data:

Group	n	$\bar{x}$	s
Control	41	58.11	22.88
Moral decoupling	43	75.06	16.35
Moral rationalization	37	74.04	18.02

The two moral reasoning groups have higher sample means than the control group, suggesting people in these two groups are more likely to buy this company's products in the future. ANOVA will allow us to determine whether this observed pattern in sample means can be extended to the group means. ■

We should always start an ANOVA with a careful examination of the data using graphical and numerical summaries in order to get an idea of what to expect from the analysis and also to check for unusual patterns in the data. *As with the two-sample t methods, outliers and extreme departures from Normality can invalidate the computed results.*

EXAMPLE 9.3

MORAL

A Graphical Summary of the Data Let's use the labels C, D, and R for the control, moral decoupling, and moral reasoning groups, respectively. Histograms of the three groups are given in Figure 9.3. *Note that the heights of the bars are percents rather than counts. This is commonly done when the group sample sizes vary.* Figure 9.4 gives side-by-side boxplots for these data. These summaries show a lot of overlap in the scores across groups. There also do not appear to be any outliers, and the distributions have some, but not severe, skewness.

Figure 9.4 also plots the three sample means. It appears the control group mean is lower than the other two. However, given the large amount of overlap in the scores across groups, this pattern could just be the result of chance variation. ■

FIGURE 9.3 Histograms for Example 9.3.

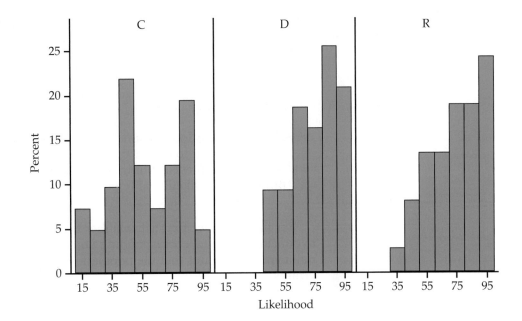

FIGURE 9.4 Side-by-side boxplots for Example 9.3. The three sample means are also shown using connect lines.

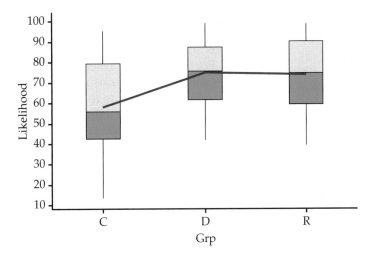

The setting of Case 9.1 is an experiment in which participants were randomly assigned to one of three groups. Each of these groups has a mean, and our inference asks questions about these means. The participants in this study were all recruited through the same university. They also participated in return for financial payment.

Formulating a clear definition of the populations being compared with ANOVA can be difficult. Often, some expert judgment is required, and different consumers of the results may have differing opinions. Whether the samples in this study should be considered as SRSs from the population of consumers at the university or from the population of all consumers in the United States is open to debate. Regardless, we are more confident in generalizing our conclusions to similar populations when the results are clearly significant than when the level of significance just barely passes the standard of $P = 0.05$.

APPLY YOUR KNOWLEDGE

CASE 9.1 **9.3 Defining the population.** The participants in Case 9.1 were undergraduates, staff, and area residents recruited through the University of Pennsylvania. The researchers report that 58% were females and the mean age was 20. Do you think this population can be considered representative of consumers at the university? The Philadelphia area? The United States? Provide an explanation with your answers.

CASE 9.1 **9.4 Randomization.** The participants in Case 9.1 were each randomly assigned to one of the three conditions. Explain why this process is important in establishing a cause-and-effect relationship between condition and the likelihood to purchase.

For inference, we first ask whether there is sufficient evidence in the data to conclude that the corresponding population means are not all equal. Our null hypothesis states that the population mean score is the same for all three groups. The alternative is that they are not all the same. *Rejecting the null hypothesis that the means are all the same is not the same as concluding that all the means are different from one another.* The ANOVA alternative hypothesis can be true in many different ways. It could be true because all of the means are different or simply because one of them differs from the rest.

Because the one-way ANOVA alternative is more complex than the two-population comparison alternative, we typically need to perform further analysis if we reject the null hypothesis. This additional analysis allows us to draw conclusions about which population means differ from others and by how much.

For Case 9.1, the researchers hypothesize that these moral reasoning strategies should allow a consumer to continue to support the company. Therefore, a reasonable question to ask is whether the mean for the control group is smaller than the others. When there are particular versions of the alternative hypothesis that are of interest, we use *contrasts* to examine them. *Note that to use contrasts, it is necessary that the questions of interest be formulated before examining the data.* It is cheating to make up these questions after analyzing the data.

If we have no specific relations among the means in mind before looking at the data, we instead use a *multiple-comparisons* procedure to determine which pairs of population means differ significantly. We explore both contrasts and multiple comparisons in the next section.

The ANOVA model

When analyzing data, it is often helpful to consider the following equation:

$$\text{DATA} = \text{FIT} + \text{RESIDUAL}$$

It reminds us that we look for an overall pattern (FIT) in the data and deviations (RESIDUAL) from it. We will use this approach to describe the statistical models used in ANOVA and later with linear regression. This approach provides a convenient way to summarize the conditions that are the foundation for the analysis. It also gives us the necessary notation to describe the calculations needed.

First, recall the statistical model for a random sample of observations from a single Normal population with mean μ and standard deviation σ. If the observations are

$$x_1, x_2, \ldots, x_n$$

we can describe this model by saying that the x_j are an SRS from the $N(\mu, \sigma)$ distribution. Another way to describe the same model is to think of the x's varying about their population mean. To do this, write each observation x_j as

$$x_j = \mu + \varepsilon_j$$

The ε_j are then an SRS from the $N(0,\sigma)$ distribution. Because μ is unknown, the ε's are unknown as well. This form more closely corresponds to our

$$\text{DATA} = \text{FIT} + \text{RESIDUAL}$$

way of thinking. The FIT part of the model is represented by μ. It is the systematic part of the model. The RESIDUAL part is represented by ε_j. It represents the deviations of the data from the fit and is due to random variation.

There are two unknown parameters in this statistical model: μ and σ. We estimate μ by $\bar{x}$, the sample mean, and σ by s, the sample standard deviation. The differences $e_j = x_j - \bar{x}$ are called the *residuals* and correspond to the ε_j in this statistical model.

The model for one-way ANOVA is very similar. We take random samples from each of I different populations. The sample size is n_i for the ith population. Let x_{ij} represent the jth observation from the ith population. The I population means are the FIT part of the model and are represented by μ_i. The random variation, or RESIDUAL, part of the model is represented by the deviations ε_{ij} of the observations from the means.

THE ONE-WAY ANOVA MODEL

The data for one-way ANOVA are SRSs from each of I populations. The sample from the ith population has n_i observations, $x_{i1}, x_{i2}, \ldots, x_{in_i}$. The **one-way ANOVA model** is

$$x_{ij} = \mu_i + \varepsilon_{ij}$$

for $i = 1, \ldots, I$ and $j = 1, \ldots, n_i$. The ε_{ij} are assumed to be from an $N(0,\sigma)$ distribution. The **parameters of the model** are the I population means $\mu_1, \mu_2, \ldots, \mu_I$ and the common standard deviation σ.

Note that the sample sizes n_i may differ, but the standard deviation σ is assumed to be the same in all the populations. Figure 9.5 pictures this model for $I = 3$. The three population means μ_i are all different, but the spreads of the three Normal distributions are the same, reflecting the condition that all three populations have the same standard deviation.

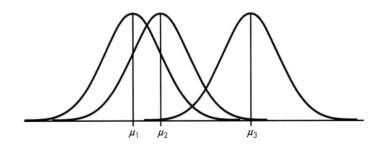

FIGURE 9.5 Model for one-way ANOVA with three groups. The three populations have Normal distributions with the same standard deviation.

EXAMPLE 9.4

CASE 9.1 **ANOVA Model for the Moral Reasoning Study** In Case 9.1, there are three groups that we want to compare, so $I = 3$. The population means μ_1, μ_2, and μ_3 are the mean values for the control group, moral decoupling group, and moral rationalization group, respectively. The sample sizes n_i are 41, 43, and 37. It is common to use numerical subscripts to distinguish the different means, and some software requires that levels of factors in ANOVA be specified as numerical values. An alternative is to use subscripts that suggest the actual groups. Given the labels we used in Example 9.3, we could replace μ_1, μ_2, and μ_3 with μ_C, μ_D, and μ_R, respectively.

level of a factor, p. 144

The observation x_{11} is the likelihood score for the first participant in the control group. Accordingly, the data for the remaining control group participants are denoted by x_{12}, x_{13}, and so on. Similarly, the data for the other two groups have a first subscript indicating the group and a second subscript indicating the participant in that group.

According to our model, the score for the first participant in the control group is $x_{11} = \mu_1 + \varepsilon_{11}$, where μ_1 is the average for the population of *all* consumers under the control condition and ε_{11} is the chance deviation due to this particular participant. Similarly, the score for the last participant assigned to the moral rationalization group is $x_{3,37} = \mu_3 + \varepsilon_{3,37}$, where μ_3 is the average score for all consumers primed with moral rationalization and $\varepsilon_{3,37}$ is the chance deviation due to this participant. ∎

central limit theorem, p. 313

The ANOVA model assumes that the deviations ε_{ij} are independent and Normally distributed with mean 0 and standard deviation σ. For Case 9.1, we have clear evidence that the data are not Normal. The values are numbers that can only range from 0 to 100, and we saw some skewness in the group distributions (Figure 9.3). However, because our inference is based on the sample means, which will be approximately Normal given this study's sample sizes, we are not overly concerned about this violation of model assumptions. We only need the model conditions to be approximately met in order for ANOVA to provide valid results.

APPLY YOUR KNOWLEDGE

9.5 Produce displays. Example 9.1 (page 458) describes a study designed to compare sales based on different produce displays. Write out the ANOVA model for this study. Be sure to give specific values for I and the n_i. List all the parameters of the model.

9.6 Ages of customers at different coffeehouses. In Example 9.2 (page 459), the ages of customers at different coffeehouses are compared. Assuming all who are surveyed respond, write out the ANOVA model for this study. Be sure to give specific values for I and the n_i. List all the parameters of the model.

Estimates of population parameters

The unknown parameters in the statistical model for ANOVA are the I population means μ_i and the common population standard deviation σ. To estimate μ_i we use the sample mean for the ith group:

$$\bar{x}_i = \frac{1}{n_i} \sum_{j=1}^{n_i} x_{ij}$$

residuals   The **residuals** $e_{ij} = x_{ij} - \bar{x}_i$ reflect the variation about the sample means that we see in the data and are used in the calculations of the sample standard deviations

$$s_i = \sqrt{\frac{\sum_{j=1}^{n_i}(x_{ij} - \bar{x}_i)^2}{n_i - 1}} = \sqrt{\frac{\sum_{j=1}^{n_i} e_{ij}^2}{n_i - 1}}$$

In addition to the deviations being Normally distributed, the ANOVA model also states that the population standard deviations are all equal. Before estimating σ, it is important to check this equality assumption using the sample standard deviations. Most computer software provides at least one test for the equality of standard deviations. Unfortunately, many of these tests lack robustness against non-Normality.

Because ANOVA procedures are not extremely sensitive to unequal standard deviations, we do *not* recommend a formal test of equality of standard deviations as a preliminary to the ANOVA. Instead, we use the following rule of thumb.

RULE FOR EXAMINING STANDARD DEVIATIONS IN ANOVA

If the largest sample standard deviation is less than twice the smallest sample standard deviation, we can use methods based on the condition that the population standard deviations are equal and our results will still be approximately correct.[3]

If there is evidence of unequal population standard deviations, we generally try to transform the data so that they are approximately equal. We might, for example, work with $\sqrt{x_{ij}}$ or $\log x_{ij}$. Fortunately, we can often find a transformation that makes the group standard deviations more nearly equal *and also* makes the distributions of observations in each group more nearly Normal (see Exercises 9.81 and 9.82). If the standard deviations are markedly different and cannot be made similar by a transformation, inference requires different methods, such as a nonparametric method (Chapter 17).

EXAMPLE 9.5

MORAL

CASE 9.1 **Are the Standard Deviations Equal?** In the moral reasoning study, there are $I = 3$ groups and the sample standard deviations are $s_1 = 22.88$, $s_2 = 16.35$, and $s_3 = 18.02$. The largest standard deviation (22.88) is not larger than twice the smallest ($2 \times 16.35 = 32.70$), so our rule of thumb indicates we can assume the population standard deviations are equal. ∎

When we assume that the population standard deviations are equal, each sample standard deviation is an estimate of σ. To combine these into a single estimate, we use a generalization of the pooling method introduced in Chapter 8 (page 428).

POOLED ESTIMATOR OF σ

Suppose we have sample variances $s_1^2, s_2^2, \ldots, s_I^2$ from I independent SRSs of sizes $n_1, n_2, \ldots, n_I$ from populations with common variance σ^2. The **pooled sample variance**

$$s_p^2 = \frac{(n_1 - 1)s_1^2 + (n_2 - 1)s_2^2 + \cdots + (n_I - 1)s_I^2}{(n_1 - 1) + (n_2 - 1) + \cdots + (n_I - 1)}$$

is an unbiased estimator of σ^2. The **pooled standard error**

$$s_p = \sqrt{s_p^2}$$

is the estimate of σ.

Pooling gives more weight to groups with larger sample sizes. If the sample sizes are equal, s_p^2 is just the average of the I sample variances. *Note that s_p is not the average of the I sample standard deviations.*

EXAMPLE 9.6

The Common Standard Deviation Estimate In the moral reasoning study, there are $I = 3$ groups and the sample sizes are $n_1 = 41$, $n_2 = 43$, and $n_3 = 37$. The sample standard deviations are $s_1 = 22.88$, $s_2 = 16.35$, and $s_3 = 18.02$, respectively.

The pooled variance estimate is

$$s_p^2 = \frac{(n_1 - 1)s_1^2 + (n_2 - 1)s_2^2 + (n_3 - 1)s_3^2}{(n_1 - 1) + (n_2 - 1) + (n_3 - 1)}$$

$$= \frac{(40)(22.88)^2 + (42)(16.35)^2 + (36)(18.02)^2}{40 + 42 + 36}$$

$$= \frac{43{,}857.26}{118} = 371.67$$

The pooled standard error is

$$s_p = \sqrt{371.67} = 19.28$$

This is our estimate of the common standard deviation σ of the likelihood scores in the three populations of consumers. ∎

APPLY YOUR KNOWLEDGE

9.7 Produce displays. Example 9.1 (page 458) describes a study designed to compare weekly sales based on different produce displays, and in Exercise 9.5 (page 466), you described the ANOVA model for this study. The three displays are designated 1, 2, and 3. The following table summarizes the weekly sales data (in thousands).

Display	$\bar{x}$	s	n
1	60.3	18.8	20
2	85.6	23.4	20
3	69.7	15.2	20

(a) Is it reasonable to pool the standard deviations for these data? Explain your answer.

(b) For each parameter in your model from Exercise 9.5, give the estimate.

9.8 Ages of customers at different coffeehouses. In Example 9.2 (page 459), the ages of customers at different coffeehouses are compared, and you described the ANOVA model for this study in Exercise 9.6 (page 466). It turned out that roughly 25% of those surveyed declined to respond. Here is a summary of the ages of the customers who responded:

Store	$\bar{x}$	s	n
A	29.3	7.9	32
B	45.0	12.0	33
C	31.1	8.9	28
D	35.6	10.0	25
E	28.8	6.9	28

(a) Is it reasonable to pool the standard deviations for these data? Explain your answer.

(b) For each parameter in your model from Exercise 9.6, give the estimate.

Testing hypotheses in one-way ANOVA

Comparison of several means is accomplished by using an F statistic to compare the variation among groups with the variation within groups. We now show how the F statistic expresses this comparison. Calculations are organized in an ANOVA table, which contains the numerical measures of the variation among groups and within groups.

First, we must specify our hypotheses. Now that we've defined the one-way ANOVA model, we can do this in terms of model parameters. As usual, I represents the number of populations to be compared.

HYPOTHESES FOR ONE-WAY ANOVA

The null and alternative hypotheses for one-way ANOVA are

$$H_0: \mu_1 = \mu_2 = \cdots = \mu_I$$

$$H_a: \text{not all of the } \mu_i \text{ are equal}$$

Before proceeding to inference, we should verify the conditions necessary for the ANOVA results to be trusted. There is no point in trying to do inference if the data do not, at least approximately, meet these conditions.

EXAMPLE 9.7

MORAL

Verifying the Conditions for ANOVA We've discussed each of these conditions already. Here is a summary of our assessments for Case 9.1.

SRSs. This is an experiment in which participants were randomly assigned to groups. We clearly have SRSs but defining the population from which we obtained the participants is a challenge. They were all recruited from one university and paid to participate. Can we act as if they were randomly chosen from the university or from the population of all consumers? People may disagree on the answer.

Normality. Are the likelihood scores Normally distributed in each group? We've already argued they are not, but because inference is based on the sample means and we have relatively large sample sizes, we do not need to be concerned about violating this condition. The exception would be if there were outliers or extreme skewness. Figure 9.6 displays the Normal quantile plots for each of the three groups. There are no outliers and no strong departures from Normality.

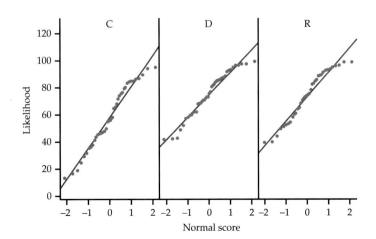

FIGURE 9.6 Normal quantile plots of the purchase intent scores for the three moral reasoning groups in Case 9.1.

Common standard deviation. Are the population standard deviations equal? Because the largest sample standard deviation is not more than twice the smallest sample standard deviation, we proceed assuming the population standard deviations are equal. ∎

Having determined that the data approximately satisfy these three conditions, we proceed with inference.

EXAMPLE 9.8

MORAL

CASE 9.1 **Are the Differences Significant?** The ANOVA results produced by JMP are shown in Figure 9.7. The pooled standard error s_p is reported as 19.28 under the heading "Root Mean Square Error." The calculated value of the F statistic appears under the heading "F Ratio," and its P-value is under the heading "Prob > F." The value of F is 9.93, with a P-value of 0.0001. That is, an F of 9.93 or larger would occur by chance one time in 10,000 when the population means are equal. There is very strong evidence that the three populations of consumers do not all have the same mean score. We can reject the null hypothesis that the three populations have equal means for any common choice of significance level. ∎

FIGURE 9.7 JMP analysis of variance output for the moral reasoning data of Case 9.1.

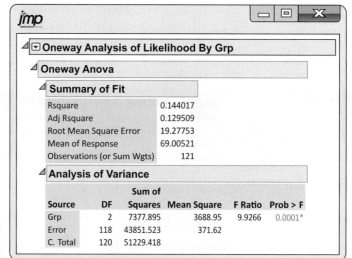

Additional analysis is needed to determine which population means differ from the others or to test if the control group mean is smaller than the other two. These methods are covered in the next section.

APPLY YOUR KNOWLEDGE

CASE 9.1 **9.9 An alternative Normality check.** Figure 9.6 displays Normal quantile plots for each of the three groups. An alternative approach is to make one Normal quantile plot using the residuals $e_{ij} = x_{ij} - \bar{x}_i$ for all three groups. Make this plot and summarize what it shows. MORAL

CASE 9.1 **9.10 An alternative Normality check, continued.** Refer to the previous exercise. Describe the benefits and drawbacks of checking Normality using all the residuals together versus looking at each group sample separately. Which approach do you prefer and why?

The ANOVA table

Software ANOVA output contains more than simply the F test statistic and associated P-value. The additional information shows, among other things, where the test statistic comes from.

9.1 One-Way Analysis of Variance

The information in an analysis of variance is organized in an *ANOVA table*. In the software output in Figure 9.7, the columns of this table are labeled "Source," "DF," "Sum of Squares," "Mean Square," "F Ratio," and "Prob > F." The rows are labeled "Grp," "Error," and "C. Total." These are the three sources of variation in the one-way ANOVA. Other software may use different column and row names, but the rest of the table will be similar.

variation among groups

The "Grp" row in the table corresponds to the FIT term in our DATA = FIT + RESIDUAL way of thinking. It gives information related to the **variation among** group means. "Grp" was the name used in entering the data to distinguish the three populations, so that is why it appears.

variation within groups

The "Error" row in the table corresponds to the RESIDUAL term in DATA = FIT + RESIDUAL. It gives information related to the **variation within** groups. The term "error" is most appropriate for experiments in the physical sciences, where the observations within a group differ because of measurement error. In business and the biological and social sciences, on the other hand, the within-group variation is often due to the fact that not all firms or plants or people are the same. This sort of variation is not due to errors and is better described as *residual*.

Finally, the "C. Total" row in the table corresponds to the DATA term in our DATA = FIT + RESIDUAL framework. So, for analysis of variance,

$$\text{DATA} = \text{FIT} + \text{RESIDUAL}$$

translates into

$$\text{total variation} = \text{variation among groups} + \text{variation within groups}$$

The ANOVA idea is to break the total variation in the responses into two parts: the variation due to differences among the group means and that due to differences within groups.

sum of squares

Variation is expressed by **sums of squares.** We use SSG, SSE, and SST for the sums of squares for groups, error, and total, respectively. Each sum of squares is the sum of the squares of a set of deviations that expresses a source of variation. SST is the sum of squares of $x_{ij} - \bar{x}$, which measure variation of the responses around their overall mean. Variation of the group means around the overall mean $\bar{x}_i - \bar{x}$ is measured by SSG. Finally, SSE is the sum of squares of the deviations $x_{ij} - \bar{x}_i$ of each observation from its group mean.

EXAMPLE 9.9

Sums of Squares for the Three Sources of Variation The "SS" column in Figure 9.7 gives the values for the three sums of squares. They are

$$\text{SSG} = 7377.895$$
$$\text{SSE} = 43{,}851.523$$
$$\text{SST} = 51{,}229.418.$$

Verify that SST = SSG + SSE. In this example, it appears that most of the variation is coming from Error, that is, from within groups. ■

It is always true that SST = SSG + SSE. This is the algebraic version of the ANOVA idea: total variation is the sum of between-group variation and within-group variation.

degrees of freedom, p. 33

Associated with each sum of squares is a quantity called the degrees of freedom. Because SST measures the variation of all N observations around the overall mean, its degrees of freedom are DFT = $N - 1$, the degrees of freedom for the sample variance of the N responses. Similarly, because SSG measures the variation of the I sample means around the overall mean, its degrees of freedom are DFG = $I - 1$. Finally, SSE is the sum of squares of the deviations $x_{ij} - \bar{x}_i$. Here, we have N observations being compared with I sample means and DFE = $N - I$.

EXAMPLE 9.10

Degrees of Freedom for the Three Sources of Variation In Case 9.1, we have $I = 3$ and $N = 121$. Therefore,

$$\text{DFG} = I - 1 = 3 - 1 = 2$$
$$\text{DFE} = N - I = 121 - 3 = 118$$
$$\text{DFT} = N - 1 = 121 - 1 = 120$$

These are the entries in the "DF" column in Figure 9.7. ∎

Note that the degrees of freedom add in the same way that the sums of squares add. That is. DFT = DFG + DFE.

mean square

For each source of variation, the **mean square** is the sum of squares divided by the degrees of freedom. You can verify this by doing the divisions for the values given on the output in Figure 9.7. Generally, the ANOVA table includes mean squares only for the first two sources of variation. The mean square corresponding to the total source is the sample variance that we would calculate assuming that we have one sample from a single population—that is, assuming that the means of the three groups are the same.

> **SUMS OF SQUARES, DEGREES OF FREEDOM, AND MEAN SQUARES**
>
> **Sums of squares** represent variation present in the data. They are calculated by summing squared deviations. In one-way ANOVA, there are three **sources of variation:** groups, error, and total. The sums of squares are related by the formula
>
> $$\text{SST} = \text{SSG} + \text{SSE}$$
>
> Thus, the total variation is composed of two parts, one due to groups and one due to "error" (variation within groups).
>
> **Degrees of freedom** are related to the deviations that are used in the sums of squares. The degrees of freedom are related in the same way as the sums of squares:
>
> $$\text{DFT} = \text{DFG} + \text{DFE}$$
>
> To calculate each **mean square,** divide the corresponding sum of squares by its degrees of freedom.

We can use the mean square for error to find s_p, the pooled estimate of the parameter σ of our model. It is true in general that

$$s_p^2 = \text{MSE} = \frac{\text{SSE}}{\text{DFE}}$$

In other words, the mean square for error is an estimate of the within-group variance, σ^2. The estimate of σ is therefore the square root of this quantity. So,

$$s_p = \sqrt{\text{MSE}}$$

APPLY YOUR KNOWLEDGE

CASE 9.1 **9.11 Total mean square.** The ANOVA table output in Figure 9.7 (page 470) does not give the total mean square MST = SST/DFT. Calculate this quantity. Then find the mean and variance of all 121 observations and verify that MST is the variance of all the responses. MORAL

9.12 Computing the pooled variance estimate. In Example 9.6, we computed the pooled standard error s_p using the population standard deviations and sample sizes. Now use the MSE from the ANOVA table in Figure 9.7 (page 470) to estimate σ. Verify that it is equal to 19.28.

The *F* test

F statistic

If H_0 is true, there are no differences among the group means. In that case, MSG will reflect only chance variation and should be about the same as MSE. The ANOVA ***F* statistic** simply compares these two mean squares, $F = \text{MSG}/\text{MSE}$. Thus, this statistic is near 1 if H_0 is true and tends to be larger if H_a is true. In our example, MSG = 3688.95 and MSE = 371.62, so the ANOVA F statistic is

$$F = \frac{\text{MSG}}{\text{MSE}} = \frac{3688.95}{371.62} = 9.93$$

F distributions

To calculate its P-value, we need to know the sampling distribution of the F statistic when H_0 is true. If the data are Normal, the F statistic follows an F distribution. The ***F* distributions** are a family of distributions with two parameters: the degrees of freedom of the mean square in the numerator and denominator of the F statistic. The F distributions are another of R. A. Fisher's contributions to statistics and are called F in his honor. In fact, Fisher introduced F statistics for comparing several means.

The numerator degrees of freedom are always mentioned first. Interchanging the degrees of freedom changes the distribution, so the order is important. Our brief notation will be $F(j,k)$ for the F distribution with j degrees of freedom in the numerator and k in the denominator. The F distributions are not symmetric, but are right-skewed. The density curve in Figure 9.8 illustrates the general shape. Because variation cannot be negative, the F statistic takes only positive values, and the F distribution has no probability below 0. The peak of the F density curve is near 1 and values much greater than 1 provide evidence against the null hypothesis.

For one-way ANOVA, the degrees of freedom for the numerator are DFG = $I - 1$ and the degrees of freedom for the denominator are DFE = $N - I$. We use the notation $F(I - 1, N - I)$ for this distribution. The *One-Way ANOVA* applet is

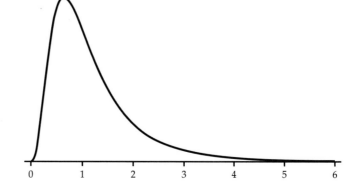

FIGURE 9.8 The density for the $F(9, 10)$ distribution. The F distributions are skewed to the right.

an excellent way to see how the value of the F statistic and its P-value depend on both the variability of the data within the groups and the differences between the group means. Exercises 9.26, 9.27, and 9.58 make use of this applet.

The ANOVA F test shares the robustness of the two-sample t test. It is relatively insensitive to moderate non-Normality and unequal variances, especially when the sample sizes are similar. The constant variance assumption is more important when we compare means using contrasts and multiple comparisons, additional analyses that are generally performed after the ANOVA F test.

THE ANOVA F TEST

To test the null hypothesis in a one-way ANOVA, calculate the **F statistic**

$$F = \frac{\text{MSG}}{\text{MSE}}$$

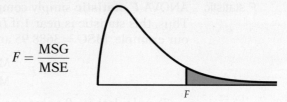

When H_0 is true, the F statistic has the $F(I-1, N-I)$ distribution. When H_a is true, the F statistic tends to be large. We reject H_0 in favor of H_a if the F statistic is sufficiently large.

The **P-value** of the F test is the probability that a random variable having the $F(I-1, N-I)$ distribution is greater than or equal to the calculated value of the F statistic.

Tables of F critical values are available for use when software does not give the P-value or you cannot use a software function, such as F.DIST.RT in Excel. However, tables of F critical values are awkward because a separate table is needed for every pair of degrees of freedom. Table E in the back of the book contains the F critical values for probabilities $p = 0.100, 0.050, 0.025, 0.010$, and 0.001. For one-way ANOVA, we use critical values from the table corresponding to $I-1$ degrees of freedom in the numerator and $N-I$ degrees of freedom in the denominator. *When determining the P-value, remember that the F test is always one-sided because any differences among the group means tend to make F large.*

EXAMPLE 9.11

df = (2, 100)

p	Critical value
0.100	2.36
0.050	3.09
0.025	3.83
0.010	4.82
0.001	7.41

 Comparing the Likelihood of Purchase Means In the moral strategy study, $F = 9.93$. There are three populations, so the degrees of freedom in the numerator are DFG $= I - 1 = 2$. The degrees of freedom in the denominator are DFE $= N - I = 121 - 3 = 118$. To get the P-value in Excel, we would enter F.DIST.RT(9.93,2,118). It provides $P = 0.000103$. In Table E, first find the column corresponding to two degrees of freedom in the numerator. For the degrees of freedom in the denominator, there are entries for 100 and 200. These entries are very close. To be conservative, we use critical values corresponding to 100 degrees of freedom in the denominator because these are slightly larger. Because 9.93 is beyond 7.41, we reject H_0 and conclude that the differences in means are statistically significant, with $P < 0.001$. ∎

one-way ANOVA table

The following display shows the general form of a **one-way ANOVA table** with the F statistic. The formulas in the sum of squares column can be used for calculations in small problems. There are other formulas that are more efficient for hand or calculator use, but ANOVA calculations are usually done by computer software.

Source	Degrees of freedom	Sum of squares	Mean square	F
Groups	$I - 1$	$\sum_{\text{groups}} n_i(\bar{x}_i - \bar{x})^2$	SSG/DFG	MSG/MSE
Error	$N - I$	$\sum_{\text{groups}} (n_i - 1)s_i^2$	SSE/DFE	
Total	$N - 1$	$\sum_{\text{obs}} (x_{ij} - \bar{x})^2$	SST/DFT	

coefficient of determination

One other item given by some software for ANOVA is worth noting. For an analysis of variance, we define the **coefficient of determination** as

$$R^2 = \frac{\text{SSG}}{\text{SST}}$$

interpretation of r^2, p. 88

Like r^2 for regression, the coefficient of determination describes the proportion of the variation of the response variable that is described by the groups. We can easily calculate the value from the ANOVA table entries.

EXAMPLE 9.12

Coefficient of Determination for the Moral Strategy Study The software-generated ANOVA table for Case 9.1 is given in Figure 9.7. From that display, we see that SSG = 7377.895 and SST = 51,229.418. The coefficient of determination is

$$R^2 = \frac{\text{SSG}}{\text{SST}} = \frac{7377.895}{51,229.418} = 0.14 \blacksquare$$

About 14% of the variation in the likelihood scores is explained by the different groups. The other 86% of the variation is due to participant-to-participant variation within each of the groups. We can see this in the histograms of Figure 9.3 (page 463) and the boxplots of Figure 9.4 (page 463). Each of the groups has a large amount of variation, and there is a substantial amount of overlap in the distributions. *The fact that we have strong evidence ($P < 0.001$) against the null hypothesis that the three population means are all the same does not tell us that the distributions of values are far apart.*

APPLY YOUR KNOWLEDGE

9.13 Using Table E or software. An ANOVA is run to compare five groups. There are seven observations per group.

(a) Give the degrees of freedom for the ANOVA F statistic.

(b) How large would the F statistic need to be to have a P-value less than 0.05?

(c) Suppose instead that there were 11 observations per group. How large would the F statistic need to be to have a P-value less than 0.05?

(d) Explain why the answer to part (c) is smaller than what you found for part (b).

9.14 What's wrong? For each of the following, explain what is wrong and why.

(a) The pooled estimate s_p is a parameter of the ANOVA model.

(b) The mean squares in an ANOVA table will add, that is, MST = MSG + MSE.

(c) For an ANOVA F test with $P = 0.41$, we conclude that the group means are the same.

(d) A very small F test P-value implies that the group distributions of responses are far apart.

Using software

We used JMP to illustrate the ANOVA analysis of the moral strategy study. Other statistical software packages give similar outputs, and you should be able to extract all the ANOVA information we have discussed. Here is an example on which to practice this skill.

EXAMPLE 9.13

TIDE

Does the Color Green Impact Perceived Product Efficacy? In recent years, considerable attention has been given to issues related to the environment. This has resulted in an increased interest among consumers for environmentally friendly products. It has not, however, resulted in consumers buying more of these products. Why is this? Are the products considered too expensive? Or are there other perceptions about these products that steer consumers away?

The color green and eco-labels are the two most common packaging cues used to signal environmental friendliness. Do these cues also alter the perception of a product? A group of researchers investigated this through a series of studies. In one study, 121 undergraduate business students from a mid-sized Canadian university were randomized to four conditions.[4] Each participant was shown a standard container of Tide laundry detergent and asked, through a series of 7-point Likert-scale questions, their impressions of the product's efficacy.

The Tide containers were identical across conditions except for the color of the cap and the inclusion of an eco-label. The researchers were interested in assessing whether the use of the color green and/or the use of a certified eco-label could taint impressions about the efficacy of the product. The following table summarizes the responses for the four conditions. Outputs from Excel, JMP, and Minitab are given in Figure 9.9.

Condition	n	$\bar{x}$	s
Blue cap, Without eco-label	36	6.16	0.63
Blue cap, With eco-label	29	5.74	0.76
Green cap, Without eco-label	29	5.67	0.80
Green cap, With eco-label	27	5.93	0.73

There is evidence at the 5% significance level to reject the null hypothesis that the four conditions have equal means ($P = 0.034$). In Exercises 9.47 and 9.48 (page 496), you are asked to perform further inference using contrasts. In Exercise 9.72 (page 500), you are asked to check model conditions. ∎

FIGURE 9.9 Excel, JMP, and Minitab outputs for the environmentally friendly product study of Example 9.13.

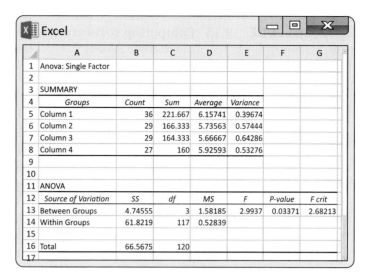

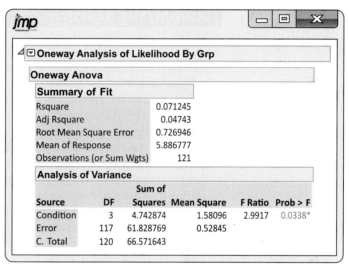

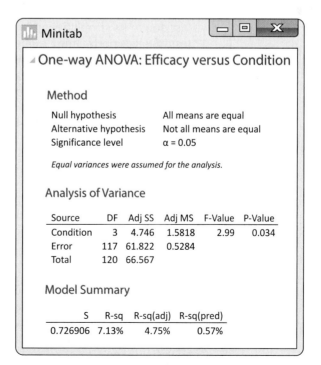

APPLY YOUR KNOWLEDGE

9.15 Comparing software outputs. The pooled standard error for the data in Example 9.13 is $s_p = 0.73$. Look at the software outputs in Figure 9.9.

(a) Excel does not report s_p. How can you find its value from Excel output?

(b) Explain to someone new to ANOVA why JMP labels this quantity as "Root Mean Square Error."

9.16 Comparing software output labels. Refer to the output in Figure 9.9. Different names are given to the sources of variation in the ANOVA tables.

(a) What are the names given to the source we call Groups?

(b) What are the names given to the source we call Error?

(c) What are the names given to the P-value?

BEYOND THE BASICS

Testing the equality of spread

The two most basic descriptive features of a distribution are its center and spread. We have described procedures for inference about population means for Normal populations and found that these procedures are often useful for non-Normal populations as well. It is natural to turn next to inference about spread.

While the standard deviation is a natural measure of spread for Normal distributions, it is not for distributions in general. In fact, because skewed distributions have unequally spread tails, no single numerical measure is adequate to describe the spread of a skewed distribution. Because of this, we recommend caution when testing equal standard deviations and interpreting the results.

Most formal tests for equal standard deviations are extremely sensitive to non-Normal populations. Of the tests commonly available in software packages, we suggest using the *modified Levene's* (or Brown-Forsythe) test due to its simplicity and robustness against non-Normal data.[5] The test involves performing a one-way ANOVA on a transformation of the response variable, constructed to measure the spread in each group. If the populations have the same standard deviation, then the average deviation from the population center should also be the same.

MODIFIED LEVENE'S TEST FOR EQUALITY OF STANDARD DEVIATIONS

To test for the equality of the I population standard deviations, perform a one-way ANOVA using the transformed response

$$y_{ij} = |x_{ij} - M_i|$$

where M_i is the sample median for population i. We reject the assumption of equal spread if the P-value of this test is less than the significance level α.

This test uses a more robust measure of deviation, replacing the mean with the median and replacing squaring with the absolute value. Also, the transformed response variable is straightforward to create, so this test can easily be performed regardless of whether your software specifically has it.

EXAMPLE 9.14

MORAL

Are the Standard Deviations Equal? Figure 9.10 shows output of the modified Levene's test for the moral reasoning study. In JMP, the test is called Brown-Forsythe. The P-value is 0.0377, which is smaller than $\alpha = 0.05$, suggesting the variances are not the same. This is not that surprising a result given the boxplots in Figure 9.4 (page 463). There, we see that the spread in the scores for the control group is much larger than the spread in the other two groups. ∎

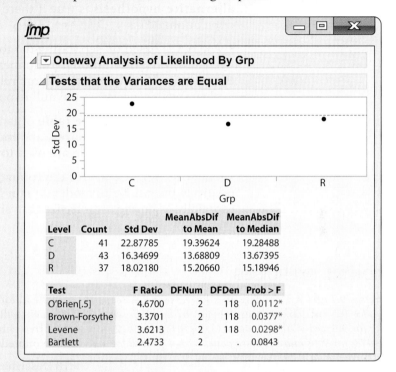

FIGURE 9.10 JMP output for the test that the variances are equal.

Remember that our rule of thumb (page 467) is used to assess whether different standard deviations will impact the ANOVA results. *It's not a formal test that the standard deviations are equal.* In Exercises 9.81 and 9.82 (page 502), we examine a transformation of the likelihood scores that better stabilizes the variances. Part of those exercises will be a comparison of the transformed likelihood score ANOVA results with the results of this section.

SECTION 9.1 SUMMARY

- **One-way analysis of variance (ANOVA)** is used to compare several population means based on independent SRSs from each population. The ANOVA model assumes that the populations are Normal and that, although they may have different means, they have the same standard deviation.

- To do an analysis of variance, first examine the data. The model assumptions only need to be approximately met in order to obtain valid results. Side-by-side boxplots give an overview. Normal quantile plots (either for each group separately or for the residuals) allow you to detect outliers or extreme deviations from Normality.

- In addition to Normality, the ANOVA model assumes equal population standard deviations. Compute the ratio of the largest to the smallest

- sample standard deviation. If this ratio is less than two and the Normality assumption is reasonable, ANOVA can be performed.

- If the data do not support equal standard deviations, consider transforming the response variable. This often makes the group standard deviations more nearly equal and makes the group distributions more Normal.

- The **null hypothesis** is that the population means are *all equal*. The **alternative hypothesis** is true if there are *any* differences among the population means.

- ANOVA is based on separating the total variation observed in the data into two parts: **variation among** group means and **variation within** groups. If the variation among groups is large relative to the variation within groups, we have evidence against the null hypothesis.

- An **analysis of variance table** organizes the ANOVA calculations. **Degrees of freedom, sums of squares,** and **mean squares** appear in the table. The **F statistic** and its **P-value** are used to test the null hypothesis.

- The ANOVA F test shares the **robustness** of the two-sample t test. It is relatively insensitive to moderate non-Normality and unequal variances, especially when the sample sizes are similar.

SECTION 9.1 EXERCISES

For Exercises 9.1 and 9.2, see page 461; for 9.3 and 9.4, see page 464; for 9.5 and 9.6, see page 466; for 9.7 and 9.8, see page 468; for 9.9 and 9.10, see page 470; for 9.11 and 9.12, see page 473; for 9.13 and 9.14, see pages 475–476; and for 9.15 and 9.16, see page 478.

9.17 The ANOVA framework. For each of the following situations, identify the response variable and the populations to be compared, and give I, the n_i, and N.

(a) Last semester, an alcohol awareness program was conducted for three groups of students at an eastern university. Follow-up questionnaires were sent to the participants two months after each presentation. There were 237 responses from students in an elementary statistics course, 147 from a health and safety course, and 86 from a cooperative housing unit. One of the questions was, "Did you discuss the presentation with any of your friends?" The answers were rated on a 7-point scale with 1 corresponding to "not at all" and 7 corresponding to "a great deal."

(b) A researcher is interested in students' opinions regarding an additional annual fee to support non-income-producing varsity sports. Students were asked to rate their acceptance of this fee on a 7-point scale. She received 114 responses, of which 37 were from students who attend varsity football or basketball games only, 18 were from students who also attend other varsity competitions, and 59 were from students who did not attend any varsity games.

(c) A university sandwich shop wants to compare the effects on sales of providing free food with a sandwich order. The experiment will be conducted from 11:00 A.M. to 2:00 P.M. for the next 25 weekdays. On each day, customers will be offered one of the following: a free drink, free chips, a free cookie, or nothing. Each option will be offered five times.

9.18 Describing the ANOVA model. For each of the following situations, identify the response variable and the populations to be compared, and give I, the n_i, and N.

(a) A developer of a virtual-reality (VR) teaching tool for people who are deaf wants to compare the effectiveness of different navigation methods. A total of 30 children are available for the experiment, of which equal numbers were randomly assigned to use a joystick, wand, or gesture-based pinch gloves. The time (in seconds) to complete a designed VR path is recorded for each child.

(b) A waiter designed a study to see the effects of his behaviors on the tip amounts that he received. For some customers, he would tell a joke; for others, he would describe two of the food items as being particularly good that night; for others, he would check on the table every 10 minutes; and for others, he would behave normally. Using a table of random numbers, he assigned equal numbers of his next 28 customers to his different behaviors.

(c) A supermarket wants to compare the effects on overall sales of providing free samples of food. An experiment will be conducted from 5:00 P.M. to 6:00 P.M. for the next 20 weekdays. On each day, customers will be offered one of the following when they enter the store: a small cube of cheese pierced by a toothpick, a soft iced cookie, a cracker with choice of hummus spread, or nothing.

9.19 Provide some details. Refer to Exercise 9.17. For each situation, give the following:

(a) Degrees of freedom for group, for error, and for the total.

(b) Null and alternative hypotheses.

(c) Numerator and denominator degrees of freedom for the F statistic.

9.20 Provide some details. Refer to Exercise 9.18. For each situation, give the following:

(a) Degrees of freedom for group, for error, and for the total.

(b) Null and alternative hypotheses.

(c) Numerator and denominator degrees of freedom for the F statistic.

9.21 How much can you generalize? Refer to Exercise 9.17. For each situation, discuss the method of obtaining the data and how this would affect the extent to which the results can be generalized.

9.22 How much can you generalize? Refer to Exercise 9.18. For each situation, discuss the method of obtaining the data and how this would affect the extent to which the results can be generalized.

9.23 What's wrong? For each of the following, explain what is wrong and why.

(a) You use one-way ANOVA to compare the variances of several populations.

(b) A multiple-comparisons procedure is used to compare a relation among means that was specified prior to looking at the data.

(c) When rejecting the null hypothesis, we can conclude that all the means are different from one another.

(d) The ANOVA F statistic will be large when the within-group variation is much larger than the between-group variation.

9.24 A one-way ANOVA example. A study compared six groups with five observations per group. An F statistic of 3.07 was reported.

(a) Give the degrees of freedom for this statistic and the entries from Table E that correspond to this distribution.

(b) Sketch a picture of this F distribution with the information from the table included.

(c) Based on the table information, how would you report the P-value?

(d) Can you reject the null hypothesis that the means are the same at the $\alpha = 0.05$ significance level? Explain your answer.

9.25 Find the F statistic. For each of the following situations, find the F statistic and the degrees of freedom. Then draw a sketch of the distribution under the null hypothesis and shade in the portion corresponding to the P-value. State how you would report the P-value.

(a) Compare four groups with 11 observations per group, MSE = 50, and MSG = 137.

(b) Compare three groups with 9 observations per group, SSG = 36, and SSE = 100.

9.26 **The effect of within-group variation.** Go to the *One-Way ANOVA* applet. In the applet display, the black dots are the mean responses in three treatment groups. Move these up and down until you get a configuration with P-value about 0.01. Note the value of the F statistic. Now increase the variation within the groups without changing their means by dragging the mark on the standard deviation scale to the right. Describe what happens to the F statistic and the P-value. Explain why this happens.

9.27 **The effect of between-group variation.** Go to the *One-Way ANOVA* applet. Set the standard deviation near the middle of its scale and drag the black dots so that the three group means are approximately equal. Note the value of the F statistic and its P-value. Now increase the variation among the group means: drag the mean of the second group up and the mean of the third group down. Describe the effect on the F statistic and its P-value. Explain why they change in this way.

9.28 Internet banking. A study in Finland looked at consumer perceptions of Internet banking (IB).[6] Data were collected via personal, structured interviews as part of a nationwide consumer study. The sample included 300 active users of IB between 15 and 74 years old. Based on the survey, users were broken down into three groups based on their familiarity with the Internet. For this exercise, we consider the consumer's perception of status or image in the eyes of other consumers. Standardized scores were used for analysis.

Familiarity	Mean	n
Low	0.21	77
Medium	−0.14	133
High	0.03	90

(a) To compare the mean scores across familiarity levels, what are the degrees of freedom for the ANOVA F statistic?

(b) The MSG = 3.12. If $s_p = 1.05$, what is the F statistic?

(c) Give an approximate (from a table) or exact (from software) P-value. What do you conclude?

9.29 Effects of music on imagery training. Music that matches a sports activity's requirements has been shown to enhance sports performance. Very little, however, is known about the effects of music on imagery in the context of sports performance. In one study, 63 novice dart throwers were randomly assigned equally to three groups.[7] During each of 12 dart-throwing imagery sessions, one group listened to unfamiliar relaxing music (URM); another listened to unfamiliar arousing music (UAM); and the other listened to no music (NM). Dart-throwing performance was assessed before and after the imagery sessions using a 40-dart distance score (darts closer to the center of the board were given a higher score). Here are the results for the gain in score (after − before):

Group	$\bar{x}$	s
URM	37.24	5.66
UAM	17.57	5.30
NM	13.19	6.14

(a) Is it reasonable to assume the variance is the same across groups? Explain your answer.

(b) Compute the estimated common standard deviation.

(c) Plot the means versus the imagery group. Do there appear to be differences in the average gain in performance? Use the estimated common standard deviation to explain your answer.

9.30 Effects of music on imagery training, continued. Refer to the previous exercise.

(a) What are the numerator and denominator degrees of freedom for this study's ANOVA F test?

(b) For this study, SSG = 328.17 and SSE = 1956.50. Compute the F statistic.

(c) Using Table E or software, what is the P-value for this study? What are your conclusions?

9.31 Word-of-mouth communications. Consumers often seek opinions on products from other consumers. These word-of-mouth communications are considered valuable because they are thought to be less biased toward the product and more likely to contain negative information. What makes certain opinions with negative information more credible than others? A group of researchers think it may have to do with the use of dispreferred markers. Dispreferred markers indicate that the communicator has just said, or is about to say, something unpleasant or negative. To investigate this, they recruited 257 subjects and randomly assigned them to three groups: positive-only review, balanced review, and balanced review with a dispreferred marker. Each subject read about two friends discussing one of their cars. The positive-only group heard that it has been owned for three years, rides well, and gets good gas mileage. The other two groups also heard that the radio and air conditioner cannot run at the same time.[8] One of the variables measured is the credibility of the friend describing her car. Here is part of the ANOVA table for these data:

Source	Degrees of freedom	Sum of squares	Mean square	F
Groups		183.59		
Error		2643.53		
Total	256			

(a) Fill in the missing entries in the ANOVA table.

(b) State H_0 and H_a for this experiment.

(c) What is the distribution of the F statistic under the assumption that H_0 is true? Using Table E, give an approximate P-value for the ANOVA test. Write a brief conclusion.

(d) What is s_p^2, the estimate of the within-group variance? What is the pooled standard error s_p?

9.32 Word-of-mouth communications, continued. Another variable measured in the experiment described in the previous exercise was the likability of the friend describing her car. Higher values of this score indicate a better opinion. Here is part of the ANOVA table for these data:

Source	Degrees of freedom	Sum of squares	Mean square	F
Groups				9.20
Error			0.93	
Total	256			

(a) Fill in the missing entries in the ANOVA table.

(b) State H_0 and H_a for this experiment.

(c) What is the distribution of the F statistic under the assumption that H_0 is true? Using Table E, give an approximate P-value for the ANOVA test. What do you conclude?

(d) What is s_p^2, the estimate of the within-group variance? What is s_p?

9.33 Are the produce display means the same? Let's now use software to analyze the data related to Example 9.1 (page 458). PRDSALE

(a) Fit the one-way ANOVA model and obtain the residuals. Generate a histogram or Normal quantile to assess Normality. Comment on what you find. You already addressed the constant variance condition in Exercise 9.7 (page 468).

(b) Based on your output, what is the estimate of σ? Compare it to your estimate of σ in Exercise 9.7.

(c) Report the F statistic, its degrees of freedom, and P-value. What do you conclude?

9.34 Does green impact purchase intention? Refer to Example 9.13. In addition to scoring product efficacy, each participant was asked to rate their purchase intentions. TIDE1

(a) Construct a table of group means and standard deviations.

(b) Is it reasonable to pool the standard deviations? Explain your answer.

(c) Fit the one-way ANOVA model and obtain the residuals. Generate a Normal quantile plot or histogram of the residuals. Are the errors approximately Normal? Explain your answer.

(d) Report the F statistic, its degrees of freedom, and P-value. What do you conclude about the four conditions?

9.35 Purchase intention analysis, continued. Refer to the previous exercise. Purchase intent was scored using two 7-point items. This means the response is discrete and thus non-Normal. TIDE1

(a) Explain why one may still be comfortable using ANOVA in this setting.

(b) A transformation of the response can often improve the model conditions necessary for inference. Consider the transformation y^2. Refit the model using this new response and assess both constant variance and Normality. Do you find the conditions better met under this approximation?

(c) Report the F statistic, its degrees of freedom, and P-value for this transformed response.

(d) Compare your results in (c) to the results in the previous exercise. Do any of your conclusions change?

9.2 Additional Comparisons of Group Means

When you complete this section, you will be able to:

- Distinguish between the use of contrasts to examine particular versions of the alternative hypothesis and the use of a multiple-comparisons method to compare pairs of means.
- Construct a confidence interval or perform a t significance test for a contrast and summarize the results.
- Describe the use of a multiple-comparisons method in terms of controlling false rejections.
- Interpret statistical software ANOVA output and draw conclusions regarding differences among population means.
- Use software to determine the power of the ANOVA F test or contrast t test for a given set of population means and group sample size n.

The ANOVA F test gives a general answer to a general question: are the differences among observed group means significant? Unfortunately, a small P-value simply tells us that the group means are not all the same. It does not tell us specifically which means differ from each other. Plotting and inspecting the means give us some indication of where the differences lie, but we would like to supplement inspection with formal inference. This section presents two approaches to the task of comparing group means. It concludes with discussion on the power of the ANOVA F test.

Contrasts

The preferred approach of comparing group means is to pose specific questions before the data are collected. We can answer specific questions of this kind and attach a level of confidence to the answers we give. We now explore these ideas in the setting of Case 9.2.

EDPRD

CASE 9.2 **Evaluation of a New Educational Product** Your company markets educational materials aimed at parents of young children. You are planning a new product that is designed to improve children's reading comprehension. Your product is based on new ideas from educational research, and you would like to claim that children will acquire better reading comprehension skills utilizing these new ideas than with the traditional approach. Marketing plans to include the results of a study conducted to compare two versions of the new approach with the traditional method in its advertising.[9] The standard method is called Basal, and the two variations of the new method are called DRTA and Strat.

For the study, 66 children were randomly divided into three groups of 22. Each group was taught by one of the three methods. The response variable is a measure of reading comprehension called COMP that was obtained by a test taken after the instruction was completed. Can your company claim that the new methods are superior to Basal? ■

We can compare the new with the standard by posing and answering specific questions about the mean responses. First, we present the ANOVA results.

EXAMPLE 9.15

CASE 9.2 **Are the Comprehension Scores Different?** Figure 9.11 gives the summary statistics for COMP computed by JMP. We've labeled the groups as B for Basal, D for DRTA, and S for Strat. We could have just as easily numbered them 1, 2, and 3, respectively, as some software uses only numeric values for the factor. Side-by-side boxplots appear in Figure 9.12. Also included are the group sample means. The ANOVA results are given in Figure 9.13, and a Normal quantile plot of the residuals appears in Figure 9.14. ∎

EDPRD

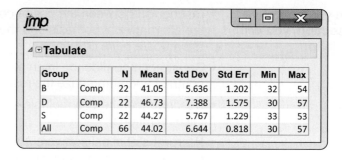

FIGURE 9.11 Summary statistics for the comprehension scores in the three groups for the new-product evaluation study of Case 9.2.

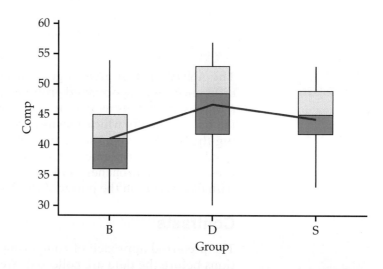

FIGURE 9.12 Side-by-side boxplots of the comprehension scores in the new-product evaluation study of Case 9.2. The sample means are also shown using connect lines.

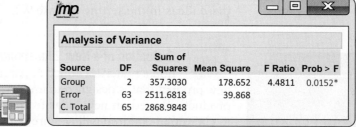

FIGURE 9.13 JMP analysis of variance output for the comprehension scores in the new-product evaluation study of Case 9.2.

The residuals appear Normal (Figure 9.14) and our rule for examining standard deviations indicates we can assume equal population standard deviations (i.e., $7.388 < 2 \times 5.636$). Figure 9.13 shows that $F = 4.48$ with degrees of freedom 2 and 63, and a P-value of 0.015. We have strong evidence against the ANOVA null hypothesis

$$H_0: \mu_B = \mu_D = \mu_S$$

where the subscripts correspond to the group labels.

FIGURE 9.14 Normal quantile plot of the residuals for the comprehension scores in the new-product evaluation study of Case 9.2.

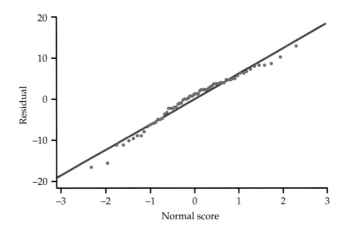

Having evidence that the three population means are not all the same does not tell marketing exactly what they'd like to know. For example, the alternative hypothesis is true if

$$\mu_B < \mu_D = \mu_S$$

or if

$$\mu_B > \mu_D = \mu_S$$

or if

$$\mu_B < \mu_D < \mu_S$$

We would like our analysis to provide us with more specific information.

EXAMPLE 9.16

CASE 9.2 **The Major Question** The two new methods are based on the same ideas. Are they superior to the standard method? We can formulate this question as the null hypothesis

$$H_{01}: \frac{1}{2}(\mu_D + \mu_S) = \mu_B$$

with the alternative

$$H_{a1}: \frac{1}{2}(\mu_D + \mu_S) > \mu_B$$

The hypothesis H_{01} compares the average of the two innovative methods (DRTA and Strat) with the standard method (Basal). The alternative is one-sided because the researchers are interested in demonstrating that the new methods are better than the old. ∎

Because we have a second question regarding the means, we include the additional subscript 1 in H_{01} and H_{a1} in order to identify that these are the null and alternative hypotheses for the first question. We will use the additional subscript 2 for the set of hypotheses associated with our other question.

EXAMPLE 9.17

CASE 9.2 **A Secondary Question** A secondary question involves a comparison of the two new methods. We formulate this as the hypothesis that the methods DRTA and Strat are equally effective,

$$H_{02}: \mu_D = \mu_S$$

versus the alternative

$$H_{a2}: \mu_D \neq \mu_S \blacksquare$$

Each of H_{01} and H_{02} says that a combination of population means is 0. These combinations of means are called contrasts. We use ψ, the Greek letter psi, for contrasts among population means. The two contrasts that arise from our two null hypotheses are

$$\psi_1 = -\mu_B + \frac{1}{2}(\mu_D + \mu_S)$$
$$= (-1)\mu_B + (0.5)\mu_D + (0.5)\mu_S$$

and

$$\psi_2 = \mu_D - \mu_S$$
$$= (0)\mu_B + (1)\mu_D + (-1)\mu_S$$

In each case, the value of the contrast is 0 when H_0 is true. We chose to define the contrasts so that they will be positive when the alternative hypothesis is true. Whenever possible, this is a good idea because it makes some computations easier.

A contrast expresses an effect in the population as a combination of population means. To estimate the contrast, form the corresponding **sample contrast** by using sample means in place of population means. Under the ANOVA assumptions, a sample contrast is a linear combination of independent Normal variables and, therefore, has a Normal distribution. We can obtain the standard error of a contrast by using the rules for variances. Inference is based on t statistics. Here are the details.

sample contrast

rules for variances, p. 240–241

> **CONTRASTS**
>
> A **contrast** is a combination of population means of the form
>
> $$\psi = \sum a_i \mu_i$$
>
> where the sum of the coefficients a_i is 0. The corresponding **sample contrast** is the same combination of sample means,
>
> $$c = \sum a_i \bar{x}_i$$
>
> The **standard error of c** is
>
> $$SE_c = s_p \sqrt{\sum \frac{a_i^2}{n_i}}$$
>
> To test the null hypothesis H_0: $\psi = 0$, use the **t statistic**
>
> $$t = \frac{c}{SE_c}$$
>
> with degrees of freedom DFE that are associated with s_p. The alternative hypothesis can be one-sided or two-sided.
>
> A **level C confidence interval for ψ** is
>
> $$c \pm t^* SE_c$$
>
> where t^* is the value for the $t(DFE)$ density curve with area C between $-t^*$ and t^*.

rules for means, p. 236

Because each $\bar{x}_i$ estimates the corresponding μ_i, the addition rule for means tells us that the mean μ_c of the sample contrast c is ψ. In other words, c is an unbiased estimator of ψ. Testing the hypothesis that a contrast is 0

EXAMPLE 9.18

The Coefficients for the Contrasts In our example, the coefficients in the contrasts are $a_1 = -1$, $a_2 = 0.5$, $a_3 = 0.5$ for ψ_1, and $a_1 = 0$, $a_2 = 1$, $a_3 = -1$ for ψ_2, where the subscripts 1, 2, and 3 correspond to B, D, and S, respectively. In each case the sum of the a_i is 0. ∎

We look at inference for each of these contrasts in turn.

EXAMPLE 9.19

Are the New Methods Better? Refer to Figures 9.11 and 9.13 (page 484). The sample contrast that estimates ψ_1 is

$$c_1 = -\bar{x}_B + \frac{1}{2}(\bar{x}_D + \bar{x}_S)$$

$$= -41.05 + \frac{1}{2}(46.73 + 44.27) = 4.45$$

with standard error

$$SE_{c_1} = 6.314\sqrt{\frac{(-1)^2}{22} + \frac{(0.5)^2}{22} + \frac{(0.5)^2}{22}}$$

$$= 1.65 \quad \blacksquare$$

The t statistic for testing H_{01}: $\psi_1 = 0$ versus H_{a1}: $\psi_1 > 0$ is

$$t = \frac{c_1}{SE_{c_1}}$$

$$= \frac{4.45}{1.65} = 2.70$$

Because s_p has 63 degrees of freedom, software using the $t(63)$ distribution gives the one-sided P-value as 0.0044. If we used Table D, we would conclude that $P < 0.005$. The P-value is small, so there is strong evidence against H_{01}. In this study, the new methods produce higher mean scores than the old.

The size of the improvement can be described by a confidence interval. To find the 95% confidence interval for ψ_1, we combine the estimate with its margin of error:

$$c_1 \pm t^* SE_{c_1} = 4.45 \pm (2.00)(1.65)$$

$$= 4.45 \pm 3.30$$

The interval is (1.15, 7.75). We are 95% confident that the mean improvement obtained by using one of the innovative methods rather than the old method is between 1.15 and 7.75 points.

EXAMPLE 9.20

Comparing the Two New Methods The second sample contrast, which compares the two new methods, is

$$c_2 = 46.73 - 44.27$$

$$= 2.46$$

with standard error

$$SE_{c_2} = 6.314\sqrt{\frac{(1)^2}{22} + \frac{(-1)^2}{22}}$$
$$= 1.90 \blacksquare$$

The t statistic for assessing the significance of this contrast is

$$t = \frac{2.46}{1.90} = 1.29$$

The P-value for the two-sided alternative is 0.2020. We conclude that either the two new methods have the same population means or the sample sizes are not sufficiently large to distinguish them. A confidence interval helps clarify this statement. To find the 95% confidence interval for ψ_2, we combine the estimate with its margin of error:

$$c_2 \pm t^* SE_{c_2} = 2.46 \pm (2.00)(1.90)$$
$$= 2.46 \pm 3.80$$

The interval is $(-1.34, 6.26)$. With 95% confidence, we state that the difference between the population means for the two new methods is between -1.34 and 6.26.

EXAMPLE 9.21

EDPRD

CASE 9.2 **Using Software** Figure 9.15 displays the JMP output for the analysis of these contrasts. The row labeled "t Ratio" gives the t statistics 2.702 and 1.289 for our two contrasts. The P-values are given in the row labeled "Prob > |t|." These are correct for two-sided alternative hypotheses. The values are 0.0088 and 0.202. To convert the computer-generated results to apply to our one-sided alternative concerning ψ_1, simply divide the reported P-value by 2 after checking that the value of c is in the direction of H_a (i.e., that c is positive). $\blacksquare$

FIGURE 9.15 JMP output for contrasts for the comprehension scores in the new-product evaluation study of Case 9.2.

Contrast		
Test Detail		
B	-1	0
D	0.5	1
S	0.5	-1
Estimate	4.4545	2.4545
Std Error	1.6487	1.9038
t Ratio	2.7018	1.2893
Prob>|t|	0.0088	0.202
SS	291.03	66.273

Some statistical software packages report the test statistics associated with contrasts as F statistics rather than t statistics. For example, JMP provides the numerators of the F statistics in the row labeled "SS." The denominator of the

F statistic in both cases is the MSE from the ANOVA table. These F statistics are the squares of the t statistics described in Example 9.21. The associated P-values are always for the two-sided alternative.

Questions about population means are expressed as hypotheses about contrasts. A contrast should express a specific question that we have in mind when designing the study. *When contrasts are formulated before seeing the data, inference about contrasts is valid whether or not the ANOVA H_0 of equality of means is rejected.* Because the F test answers a very general question, it is less powerful than tests for contrasts designed to answer specific questions. Specifying the important questions before the analysis is undertaken enables us to use this powerful statistical technique.

APPLY YOUR KNOWLEDGE

9.36 **Using different coefficients.** Refer to Example 9.19 (page 487). Suppose we had selected the coefficients $a_1 = -2$, $a_2 = 1$, and $a_3 = 1$. Would this choice of coefficients alter our inference? Explain your answer.

9.37 **Why ANOVA with a contrast?** Refer to Example 9.20 (page 487). Because the question involves only two populations, we could have used a pooled two-sample t test for inference. Explain why a contrast has more power than this t test to detect a difference. (*Hint:* How do the two standard errors differ?)

Multiple comparisons

In many studies, specific questions cannot be formulated in advance of the analysis. If H_0 is not rejected, we conclude that the population means are indistinguishable on the basis of the data given. On the other hand, if H_0 is rejected, we would like to know which pairs of means differ. Multiple-comparisons methods address this issue. *It is important to keep in mind that multiple-comparisons methods are commonly used only after rejecting the ANOVA H_0.*

Return once more to the reading comprehension study described in Case 9.2. We found in Example 9.15 that the means were not all the same ($F = 4.48$, df = 2 and 63, $P = 0.015$).

EXAMPLE 9.22

CASE 9.2 **A t Statistic to Compare Two Means** Refer to Figures 9.11 and 9.13 (page 484). There are three pairs of population means. We can compare Groups 1 and 2, Groups 1 and 3, and Groups 2 and 3. For each of these pairs, we can write a t statistic for the difference in means. To compare Basal with DRTA (1 with 2), we compute

$$t_{12} = \frac{\bar{x}_1 - \bar{x}_2}{s_p\sqrt{\dfrac{1}{n_1} + \dfrac{1}{n_2}}}$$

$$= \frac{41.05 - 46.73}{6.31\sqrt{\dfrac{1}{22} + \dfrac{1}{22}}}$$

$$= -2.99$$

The subscripts on t specify which groups are compared. ■

pooled two-sample
t procedures, p. 428

These t statistics are very similar to the pooled two-sample t statistic for comparing two population means. The difference is that we now have more than two populations, so each statistic uses the pooled estimator s_p from all groups rather than the pooled estimator from just the two groups being compared. This additional information about the common σ increases the power of the tests. The degrees of freedom for all of these statistics are DFE = 63, those associated with s_p.

Because we do not have any specific ordering of the means in mind as an alternative to equality, we must use a two-sided approach to the problem of deciding which pairs of means are significantly different.

MULTIPLE COMPARISONS

To perform a **multiple-comparisons procedure,** compute t **statistics** for all pairs of means using the formula

$$t_{ij} = \frac{\bar{x}_i - \bar{x}_j}{s_p \sqrt{\dfrac{1}{n_i} + \dfrac{1}{n_j}}}$$

If

$$|t_{ij}| \geq t^{**}$$

we declare that the population means μ_i and μ_j are different. Otherwise, we conclude that the data do not distinguish between them. The value of t^{**} depends upon which multiple-comparisons procedure we choose.

LSD method

One obvious choice for t^{**} is the upper $\alpha/2$ critical value for the $t(\text{DFE})$ distribution. This choice simply carries out as many separate significance tests of fixed level α as there are pairs of means to be compared. The procedure based on this choice is called the **least-significant differences method,** or simply LSD. *LSD has some undesirable properties, particularly if the number of means being compared is large.* Suppose, for example, that there are $I = 20$ groups and we use LSD with $\alpha = 0.05$. There are 190 different pairs of means. If we perform 190 t tests, each with an error rate of 5%, our overall error rate will be unacceptably large. We would expect about 5% of the 190 to be significant even if the corresponding population means are all equal. Because 5% of 190 is 9.5, we would expect 9 or 10 false rejections.

Because the LSD procedure fixes the probability of a false rejection for each single pair of means being compared, it does not control the overall probability of *some* false rejection among all pairs. Other choices of t^{**} control possible errors in other ways. The choice of t^{**} is, therefore, a complex problem, and a detailed discussion of it is beyond the scope of this text. Many choices for t^{**} are used in practice. One major statistical package allows selection from a list of more than a dozen choices.

Bonferroni method

We discuss only one of these, the **Bonferroni method.** Use of this procedure with $\alpha = 0.05$, for example, guarantees that the probability of *any* false rejection among all comparisons made is no greater than 0.05. This is much stronger protection than controlling the probability of a false rejection at 0.05 for *each separate* comparison.

EXAMPLE 9.23

Which Means Differ? We apply the Bonferroni multiple-comparisons procedure with $\alpha = 0.05$ to the data from the new-product evaluation study in Example 9.15. The value of t^{**} for this procedure (from software or special tables) is 2.46. The t statistic for comparing Basal with DRTA is $t_{12} = -2.99$. Because $|-2.99|$ is greater than 2.46, the value of t^{**}, we conclude that the DRTA method produces higher reading comprehension scores than Basal. ■

Usually, we use software to perform the multiple-comparisons procedure. The formats differ from package to package but they all give the same basic information.

EXAMPLE 9.24

EDPRD

Computer Output for Multiple Comparisons The output from SAS for Bonferroni comparisons appears in Figure 9.16. The first row of numbers gives the results for comparing Basal with DRTA, Groups 1 and 2. The difference between the means is given as -5.682 with a standard error of 1.904. SAS provides a Bonferroni-adjusted P-value for the comparison under the heading "Adj P." The value is 0.0121. Therefore, we can declare that the means for Basal and DRTA are different according to the Bonferroni procedure as long as we are using a value of α that is larger than 0.0121. In particular, these groups are significantly different at the *overall* $\alpha = 0.05$ level. The last two entries in the row give the Bonferroni 95% confidence interval. We will discuss this later. ■

FIGURE 9.16 SAS Bonferroni multiple-comparisons output for the comprehension scores in the new-product evaluation study of Case 9.2.

			Differences of Least Squares Means								
Effect	Group	_Group	Estimate	Standard Error	DF	t Value	Pr > |t|	Adjustment	Adj P	Adj Lower	Adj Upper
Group	B	D	-5.6818	1.9038	63	-2.98	0.0040	Bonferroni	0.0121	-10.3643	-0.9993
Group	B	S	-3.2273	1.9038	63	-1.70	0.0950	Bonferroni	0.2849	-7.9098	1.4552
Group	D	S	2.4545	1.9038	63	1.29	0.2020	Bonferroni	0.6060	-2.2280	7.1370

SAS provides the t statistics for multiple comparisons. If a software does not provide them, simply divide the estimated difference in the means by its standard error to get them. For example, the t statistic when comparing DRTA with Strat is

$$t_{23} = \frac{2.45}{1.90} = 1.29$$

EXAMPLE 9.25

Displaying Multiple-Comparisons Results When there are many groups, the many results of multiple comparisons are difficult to describe. Here is a table of the means and standard deviations for the three treatment groups. To report the results of multiple comparisons, use letters to label the means of pairs of groups that do *not* differ at the overall 0.05 significance level.

Group	$\bar{x}$	s	n
Basal	41.05^A	2.97	22
DRTA	46.73^B	2.65	22
Strat	$44.27^{A,B}$	3.34	22

Label *A* shows that Basal and Strat do not differ. Label *B* shows that DRTA and Strat do not differ. Because Basal and DRTA do not have a common label, they do differ. ■

The display in Example 9.25 shows that, at the overall 0.05 significance level, Basal does not differ from Strat and Strat does not differ from DRTA, yet Basal does differ from DRTA. These conclusions appear to be illogical. If μ_1 is the same as μ_3, and μ_3 is the same as μ_2, doesn't it follow that μ_1 is the same as μ_2? Logically, the answer must be Yes.

This apparent contradiction points out the nature of the conclusions of tests of significance. A careful statement would say that we found significant evidence that Basal differs from DRTA and failed to find evidence that Basal differs from Strat or that Strat differs from DRTA. *Failing to find strong enough evidence that two means differ doesn't say that they are equal.* It is very unlikely that any two methods of teaching reading comprehension would give *exactly* the same population means, but the data can fail to provide good evidence of a difference. This is particularly true in multiple-comparisons methods such as Bonferroni that use a single α for an entire set of comparisons.

Simultaneous confidence intervals

One way to deal with the difficulties of interpretation is to give confidence intervals for the differences. The intervals remind us that the differences are not known exactly. We want to give simultaneous confidence intervals—that is, intervals for all the differences among the population means with, say, 95% confidence that *all the intervals at once* cover the true population differences. Again, there are many competing procedures—in this case, many methods of obtaining simultaneous intervals.

SIMULTANEOUS CONFIDENCE INTERVALS FOR DIFFERENCES BETWEEN MEANS

Simultaneous confidence intervals for all differences $\mu_i - \mu_j$ between population means have the form

$$(\bar{x}_i - \bar{x}_j) \pm t^{**} s_p \sqrt{\frac{1}{n_i} + \frac{1}{n_j}}$$

The critical values t^{**} are the same as those used for the multiple-comparisons procedure chosen.

The confidence intervals generated by a particular choice of t^{**} are closely related to the multiple-comparisons results for that same method. If one of the confidence intervals includes the value 0, then that pair of means will not be declared significantly different, and vice versa.

EXAMPLE 9.26

CASE 9.2 **Software Output for Confidence Intervals** For simultaneous 95% Bonferroni confidence intervals, SAS gives the output in Figure 9.16 for the data in Case 9.2. We are 95% confident that *all three* intervals simultaneously contain the true values of the population mean differences. After rounding the output in the last two columns, the confidence interval for the difference between the mean of the Basal group and the mean of the DRTA group is (−10.36, −1.00). This interval does not include zero, so we conclude that the DRTA method results in higher mean comprehension scores than the Basal method. This is the same conclusion we obtained from the significance test, but the confidence interval provides us with additional information about the size of the difference. ∎

APPLY YOUR KNOWLEDGE

9.38 Why no additional analyses? Explain why it is unnecessary to further analyze the data using a multiple-comparisons procedure when $I = 2$.

9.39 Computing the adjusted Bonferroni P-value. SAS does not provide the Bonferroni t^{**} value. Instead, it adjusts the P-value. Using Figure 9.16 (page 490), compute the three ratios of the adjusted P-value to the unadjusted P-value and compare them to the number of comparisons performed. What do you find?

Assessing the power of the ANOVA F test

power, p. 442

power of the two-sample t test, p. 444

The power of a statistical test is a measure of the test's ability to detect deviations from the null hypothesis. In Chapter 8, we described the use of power calculations to ensure adequate sample size in one- and two-sample studies. We now extend these methods to any number of populations.

Because the one-way ANOVA F test is a generalization of the two-sample t test, it should not be surprising that the procedure for calculating power of the ANOVA F test is quite similar. In both cases, the following four design factors must be specified in order to compute power:

- Significance level α
- Common population standard deviation
- Group sample sizes
- Alternative H_a

noncentral F distribution

The difference comes in the specific calculations. Instead of using a t and noncentral t distribution to compute the probability of rejecting H_0, we now use an F and a **noncentral F distribution.** Calculations involving the noncentral F distributions are again not practical by hand but are quite easy with software.

The last three design factors in the previous list determine the appropriate noncentral F distribution. Most software assumes a constant sample size for the group sizes. For the alternative, some software doesn't request group means but rather the smallest difference between means that is judged important. Here is an example using two software packages that approach specifying H_a differently.

EXAMPLE 9.27

CASE 9.2

The Effect of Fewer Subjects The reading comprehension study described in Case 9.2 had 22 subjects in each group. Suppose a confirmatory study is being planned with only 10 subjects per group. How sensitive is the F test of this study to population means that are similar in size to those observed in the actual study?

Based on the results of the actual study, we will calculate the power for the alternative $\mu_1 = 41$, $\mu_2 = 47$, $\mu_3 = 44$, with $\sigma = 7$. We'll use $n_1 = n_2 = n_3 = 10$ and consider $\alpha = 0.05$. Figure 9.17 shows the power calculation output from JMP and Minitab.

For JMP, we specify the alternative group means, the standard deviation, and the total sample size N. The power is calculated when the "Continue" button is clicked. It is roughly 35%, which is very low.

For Minitab, we specify the common sample size n, the standard deviation, and the smallest difference between means that is deemed important. The largest difference among the population means is $6 = 47 - 41$, so that was entered. In this case, the calculated power is the same as that of JMP. ∎

FIGURE 9.17 JMP and Minitab power calculation outputs for Example 9.27.

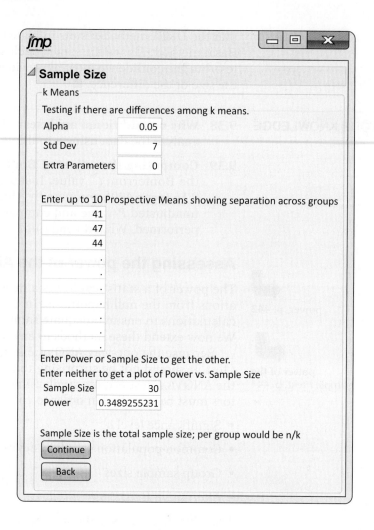

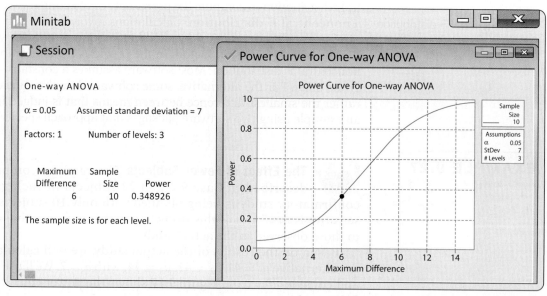

These two H_a specification approaches won't always give the same result. Usually the power is lower when only specifying a difference among the population means. This is because the other population means are not specified and the software considers a worst-case scenario.

If the assumed values of the μ_i in Example 9.27 describe differences among the groups that the company wants to confirm, then it wants to use more than 10 subjects per group in this confirmatory study. Let's now consider other sample sizes that are larger.

EXAMPLE 9.28

CASE 9.2

Choosing the Sample Size for the Confirmatory Study To decide on an appropriate sample size, we repeat the power calculation for different values of n, the number of subjects in each group. Here are the results:

n	DFG	DFE	F^*	Power
20	2	57	3.16	0.65
30	2	87	3.10	0.84
40	2	117	3.07	0.93
50	2	147	3.06	0.97
100	2	297	3.03	≈ 1

Try using JMP to verify these calculations. With $n = 40$, the experimenters have a 93% chance of rejecting H_0 with $\alpha = 0.05$ and thereby demonstrating that the groups have different means. In the long run, 93 out of every 100 such experiments would reject H_0 at the $\alpha = 0.05$ level of significance. Using 50 subjects per group increases the chance of finding significance to 97%. With 100 subjects per group, the experimenters are virtually certain to reject H_0. The exact power for $n = 100$ is 0.99990. In most real-life situations, the additional cost of increasing the sample size from 50 to 100 subjects per group would not be justified by the relatively small increase in the chance of obtaining statistically significant results.

APPLY YOUR KNOWLEDGE

9.40 Understanding power calculations. Refer to Example 9.27. Suppose that the researcher decided to use $\mu_1 = 39$, $\mu_2 = 44$, and $\mu_3 = 49$ in the power calculations. With $n = 10$ and $\sigma = 7$, would the power be larger or smaller than 35%? Explain your answer.

9.41 Understanding power calculations, continued. If all the group means are equal (H_0 is true), what is the power of the F test? Explain your answer.

SECTION 9.2 SUMMARY

• The ANOVA F test does not say which of the group means differ. It is, therefore, usual to add comparisons among the means to basic ANOVA.

• Specific questions formulated before examination of the data can be expressed as **contrasts**. Tests and confidence intervals for contrasts provide answers to these questions.

• If no specific questions are formulated before examination of the data and if the null hypothesis of equality of population means is rejected, **multiple-comparisons methods** are used to assess the statistical significance of the differences between pairs of means. These methods are less powerful than contrasts, so use contrasts whenever a study is designed to answer specific questions.

• The **power** of the F test depends upon the sample sizes, the variation among population means, and the within-group standard deviations. Some software allows easy calculation of power.

SECTION 9.2 EXERCISES

For Exercises 9.36 and 9.37, see page 489; for 9.38 and 9.39, see page 493; and for 9.40 and 9.41, see page 495.

9.42 Define a contrast. An ANOVA was run with five groups. Give the coefficients for the contrast that compares the average of the means of the first four groups with the mean of the last group.

9.43 Find the standard error. Refer to the previous exercise. Suppose that there are 8 observations in each group and that $s_p = 6$. Find the standard error for the contrast.

9.44 Is the contrast significant? Refer to the previous exercise. Suppose that the average of the first four groups minus the average of the last group is 1.2. State an appropriate null hypothesis for this comparison and find the test statistic with its degrees of freedom. Can you draw a conclusion? Or do you need to know the alternative hypothesis?

9.45 Give the confidence interval. Refer to the previous exercise. Give a 95% confidence interval for the difference between the average of the means of the first four groups and the mean of the last group.

CASE 9.1 **9.46 Additional analysis for the moral strategy example.** Refer to Case 9.1 (page 461) for a description of the study and Figure 9.7 (page 470) for the ANOVA results. The researchers hypothesize that the control group would be less likely to continue to buy products because they were not primed with a moral reasoning strategy. MORAL

(a) Because this hypothesis was declared prior to examining the data, can the researchers investigate H_0 regardless of the ANOVA F test result? Explain your answer.

(b) Test the alternative hypothesis that the mean of the control group is less than the average of the other two groups using $\alpha = 0.05$. Make sure to specify the contrast coefficients, the contrast estimate, the contrast standard error, degrees of freedom, and P-value.

9.47 Writing contrasts. Return to the color green study described in Example 9.13 (page 476). Let $\mu_1, \mu_2, \mu_3,$ and μ_4 represent the mean scores for the four conditions as presented. Write a contrast for each of the following mean comparisons of interest.

(a) The average of the blue cap conditions versus the average of the green cap conditions.

(b) The average of the eco-label present conditions versus the average of the eco-label absent conditions.

(c) The difference between the blue cap conditions with and without the eco-label versus the difference between green cap conditions with and without the eco-label.

9.48 Analyzing contrasts. Refer to the previous exercise and Figure 9.9 (page 477). Answer the following questions for the three contrasts that you defined in the previous exercise. This can be done with or without software. TIDE

(a) For each contrast, give H_0 and an appropriate H_a. In choosing the alternatives, you should use information given in the description of the problem, but you may not consider any impressions obtained by inspection of the sample means.

(b) Find the values of the corresponding sample contrasts $c_1, c_2,$ and c_2.

(c) Using the value $s_p = 0.73$, calculate the standard error of each sample contrast.

(d) Give the test statistics and approximate P-values. What do you conclude?

9.49 Analyzing contrasts, continued. Refer to the previous exercise. In addition to product efficacy, the researchers also looked at purchase intentions. Repeat the parts of the previous exercise using purchase intentions as the response. This response was analyzed in Exercise 9.34 (page 482). TIDE1

9.50 Which means differ significantly? Here is a table of means for a one-way ANOVA with four groups:

Group	$\bar{x}$	s	n
Group 1	128.2	7.7	20
Group 2	148.0	8.7	20
Group 3	148.6	10.8	20
Group 4	135.0	8.6	20

Based on this table, the standard error for a comparison of two means is 2.854. Also, according to the LSD and Bonferroni multiple-comparisons procedures with $\alpha = 0.05$, pairs of means are different if the test statistic is larger than $t^{**} = 1.99$ and $t^{**} = 2.71$, respectively.

(a) Using the LSD procedure, mark the means of each pair of groups that do *not* differ significantly with the same letter. Summarize the results.

(b) Using the Bonferroni procedure, mark the means of each pair of groups that do *not* differ significantly with the same letter. Summarize the results.

(c) Do the results vary based on procedure? If so, which do you prefer? Explain your reasoning.

9.51 Power calculations for planning a study. You are planning a study to see if the eye color of the model advertising leave-in conditioner affects a consumer's likelihood to purchase. You plan to photoshop the advertisement so the model has blue, green, or brown eyes. These images will then be randomly assigned to female university students to get their likelihood of purchase.

The likelihood of purchase instrument you plan to use involves a series of 7-point Likert scale questions. Previous research suggests $\sigma = 1.7$. You decide as an alternative the means $\mu_1 = 3.5, \mu_2 = 3.4,$ and $\mu_s = 4.0$.

(a) Pick several values for n (the number of students assigned to each eye color) and calculate the power of the ANOVA F test for each of your choices.

(b) Plot the power versus the sample size. Describe the general shape of the plot.

(c) What choice of n would you choose for this study? Give reasons for your answer.

9.52 Power against a different alternative. Refer to the previous exercise. Suppose we increase μ_3 to 4.2. For each of the choices of n in the previous example, would the power be larger or smaller under this new set of alternative means? Explain your answer.

9.53 Power calculation, revisited. In Example 8.20 (page 444), we computed the power of the two-sample t when $\sigma = 19.59$, $n_1 = n_2 = 41$ and $\mu_1 - \mu_2 = 12$. Repeat this calculation now assuming you'll perform one-way ANOVA. If you need to specify population means, use $\mu_1 = 12$ and $\mu_2 = 0$. Compare your answer to the previous one of 78.23%. What does this tell you about a pooled two-sample t test and one-way ANOVA when $I = 2$?

CHAPTER 9 REVIEW EXERCISES

9.54 How large does the F statistic need to be? For each of the following situations, state how large the F statistic needs to be for rejection of the null hypothesis at the 0.05 level.

(a) Compare five groups with three observations per group.

(b) Compare five groups with five observations per group.

(c) Compare five groups with seven observations per group.

(d) Summarize what you have learned about F distributions from this exercise.

9.55 Use the F statistic. A study compared six groups with five observations per group. An F statistic of 3.13 was reported.

(a) Give the degrees of freedom for this statistic and the entries from Table E that correspond to this distribution.

(b) Sketch a picture of this F distribution with the information from the table included.

(c) Based on the table information, how would you report the P-value?

(d) Can you reject the null hypothesis that the means are the same at the $\alpha = 0.05$ significance level? Explain your answer.

(e) Can you conclude that all pairs of means are different? Explain your answer.

9.56 Visualizing the ANOVA model. For each of the following situations, draw a picture of the ANOVA model similar to Figure 9.5 (page 465). Use numerical values for the μ_i. To sketch the Normal curves, you may want to review the 68–95–99.7 rule on page 45.

(a) $\mu_1 = 13$, $\mu_2 = 18$, $\mu_3 = 19$, and $\sigma = 3$.

(b) $\mu_1 = 14$, $\mu_2 = 15$, $\mu_3 = 20$, $\mu_4 = 18$, and $\sigma = 2$.

(c) $\mu_1 = 12$, $\mu_2 = 16$, $\mu_3 = 20$, and $\sigma = 4$.

9.57 Visualizing the ANOVA model, continued. Refer to the previous exercise. If SRSs of size $n = 10$ were obtained from each of the three populations, under which setting would you most likely obtain a significant ANOVA F test result? Explain your answer.

9.58 The effect of increased sample size. Set the standard deviation for the One-Way ANOVA applet at a middle value and drag the black dots so that the means are roughly 4.00, 4.75, and 5.00.

(a) What are the F statistic, its degrees of freedom, and the P-value?

(b) Slide the sample size bar to the right so $n = 80$. Also drag the black dots back to the values of 4.00, 4.75, and 5.00. What are the F statistic, its degrees of freedom, and the P-value?

(c) Explain why the F statistic and P-value change in this way as n increases.

9.59 Pooling variances. An experiment was run to compare four groups. The sample sizes were 35, 32, 150, and 30, and the corresponding estimated standard deviations were 25, 22, 13, and 23.

(a) Is it reasonable to assume equal standard deviations when we analyze these data using ANOVA? Give a reason for your answer.

(b) Find the pooled variance.

(c) What is the value of the pooled standard error?

(d) Explain why your answer in part (c) is much closer to the standard deviation for the third group than to any of the other standard deviations.

9.60 Public transit use and physical activity. In one study on physical activity, participants used accelerometers and a seven-day travel log to monitor their physical activity.[10] Researchers used the data from each participant to quantify the amount of daily physical activity and to classify each participant as a nontransit user, or a low-, mid-, or high-frequency transit user. Here is a summary of physical activity (in minutes per day) broken down into walking and nonwalking activities. One-way ANOVA was used to compare the groups across each activity.

Physical activity	Nontransit $n = 394$	Low frequency $n = 99$	Mid frequency $n = 73$	High frequency $n = 83$	Overall P-value
Walking	21.8^A	25.8^{A,B}	34.4^{B,C}	36.5^C	< 0.001
Nonwalking	16.0	13.5	11.9	15.2	0.24

(a) Would this be considered an observational study or an experiment? Explain your answer.

(b) What are the numerator and denominator degrees of freedom for the F tests?

(c) State the null and alternative hypotheses associated with each of the overall P-values.

(d) The superscript letters in each row summarize the multiple-comparisons results. Write a short paragraph explaining what these results tell you with regard to walking and nonwalking physical activity.

9.61 Innovative wine packaging. Are wine bottles outdated? A group of researchers randomly assigned 247 German consumers to one of three wine packaging styles. They were a bottle with a screw cap (SC), a wine bag-in-box (BiB), and a four-pack of single-serving StackTek glasses (ST). Each participant was asked to state their degree of acceptance to the statement "I would buy wine in this packaging" using a 7-point Likert scale with 1 corresponding to "totally unacceptable" and 7 corresponding to "totally acceptable." Here are the results:[11]

Group	n	$\bar{x}$	s
SC	145	6.15	1.40
BiB	147	2.93	2.00
ST	135	2.64	1.86

(a) Is this an observational study or an experiment? Explain your answer.

(b) Plot the means versus the packaging group. Does there appear to be a difference in the average purchase intention?

(c) Is it reasonable to assume a common standard deviation? Justify your answer.

(d) The data are integer values representing increasing levels of acceptance to a statement. Do you think it is reasonable to use one-way ANOVA for these data? Explain your answer.

(e) SSG = 543.14. Compute the F statistic. Also state the degrees of freedom and P-value. What do you conclude?

9.62 Age differences across coffeehouses? Recall Example 9.2 (page 459) and Exercise 9.8 (page 468). Use the data set and output shown in Figure 9.18 to answer the following questions. AGECOFF

(a) In addition to the conditions of SRSs from each group and constant variance, the data must be approximately Normal. Do you think it is reasonable to assume that here? Include any plots or numeric summaries you use to answer this.

(b) Write the null and alternative hypotheses associated with the F test in Figure 9.18.

(c) Report the F statistic, its degrees of freedom, and P-value. What are your conclusions?

FIGURE 9.18 Excel output comparing the average age at five coffeehouses, for Exercises 9.62 and 9.63.

9.63 Age differences across coffeehouses, continued. Recall the previous exercise. Using the estimates of the group means and σ in Figure 9.18, compare the groups using the least-significance differences method (LSD). Write a short summary of what you find.

9.64 Time levels of scale. Recall Exercise 8.62 (page 435). This experiment actually involved three groups. The last group was told the construction project would last 12 months. Here is a summary of the interval lengths (in days) between the earliest and latest completion dates. TIMESCL

Group	n	$\bar{x}$	s
1: 52 weeks	30	84.1	55.8
2: 12 months	30	104.6	70.1
3: 1 year	30	139.6	73.1

(a) Is this an observational study or an experiment? Explain your answer.

(b) Use graphical methods to describe the three populations.

(c) Examine the conditions necessary for ANOVA. Summarize your findings.

9.65 Time levels of scale, continued. Refer to the previous exercise. TIMESCL

(a) Run the ANOVA and report the results.

(b) Use a multiple-comparisons method to compare the three groups. State your conclusions.

(c) The researchers hypothesized that the more fine-grained the time unit presented to a participant, the smaller the reported interval would be. To assess this, they fit a least-squares regression line using the group labels 1, 2, and 3 as the predictor variable. They found the slope ($b = 27.8$) to be large and positive and thus concluded the data supported their hypothesis. Do you think this is an appropriate way to assess their hypothesis? Explain your answer.

9.66 Writing contrasts. You've been asked to help some administrators analyze survey data on textbook expenditures collected at a large public university. Let μ_1, μ_2, μ_3, and μ_4 represent the population mean expenditures on textbooks for first-, second-, third-, and fourth-year students, respectively.

(a) Because third and fourth years take higher-level courses, which might use more expensive textbooks, the administrators want to compare the average expenditure of first and second years with the average of third and fourth years. Write a contrast that expresses this comparison.

(b) Write a contrast for comparing first years with second years.

(c) Write a contrast for comparing third years with fourth years.

CASE 9.1 **9.67 Additional ANOVA for the moral strategy example.** Refer to Case 9.1 (page 461) for a description of the study. In addition to rating the likelihood to continue to purchase products, each participant was asked to judge the CEO's degree of immorality. This was done by answering a couple questions on a 0 to 7 scale where the higher the score, the stronger the immorality. MORAL1

(a) Use numerical and graphical methods to describe the three populations.

(b) Examine the conditions necessary for ANOVA. Summarize your findings.

(c) Run the ANOVA and report the results.

CASE 9.1 **9.68 Additional ANOVA for the moral strategy example, continued.** Refer to the previous exercise. MORAL1

(a) Use a multiple-comparisons method to compare the three groups. State your conclusions.

(b) Refer to the results in Exercise 9.46 (page 496) and part (a). The researchers hypothesized that moral decoupling would allow a participant to view the behavior as immoral yet still be likely to purchase products. Does this group appear to be the only one with this behavior? Generate a numerical or graphical summary that helps explain your answer to this question.

9.69 Organic foods and morals? Organic foods are often marketed using moral terms such as "honesty" and "purity." Is this just a marketing strategy, or is there a conceptual link between organic food and morality? In one experiment, 62 undergraduates were randomly assigned to one of three food conditions (organic, comfort, and control).[12] First, each participant was given a packet of four food types from the assigned condition and told to rate the desirability of each food on a 7-point scale. Then, each was presented with a list of six moral transgressions and asked to rate each on a 7-point scale ranging from 1 = not at all morally wrong to 7 = very morally wrong. The average of these six scores was used as the response. ORGANIC

(a) Make a table giving the sample size, mean, and standard deviation for each group. Is it reasonable to pool the variances?

(b) Generate a histogram for each of the groups. Can we feel confident that the sample means are approximately Normal? Explain your answer.

9.70 Organic foods and morals, continued. Refer to the previous exercise. ORGANIC

(a) Analyze the scores using analysis of variance. Report the test statistic, degrees of freedom, and P-value.

(b) Assess the assumptions necessary for inference by examining the residuals. Summarize your findings.

(c) Compare the groups using the least-significant differences method.

(d) A higher score is associated with a harsher moral judgment. Using the results from parts (a) and (c), write a short summary of your conclusions.

9.71 Organic foods and friendly behavior? Refer to Exercise 9.69 for the design of the experiment. After rating the moral transgressions, the participants were told "that another professor from another department is also conducting research and really needs volunteers." They were told that they would not receive compensation or course credit for their help and then were asked to write the number of minutes (out of 30) that they would be willing to volunteer. This sort of question is often used to measure a person's prosocial behavior.

(a) Figure 9.19 contains the Minitab output for the analysis of this response variable. Write a one-paragraph summary of your conclusions.

(b) Figure 9.20 contains a residual plot and a Normal quantile plot of the residuals. Are there any concerns regarding the assumptions necessary for inference? Explain your answer.

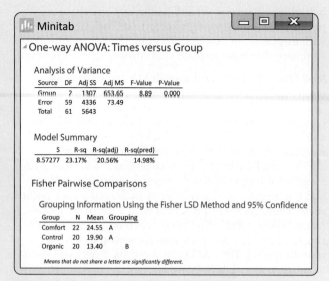

FIGURE 9.19 Minitab output comparing prosocial behavior across three treatment groups, for Exercise 9.71.

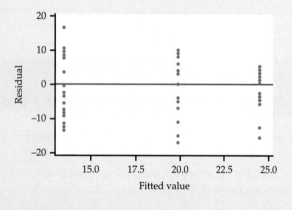

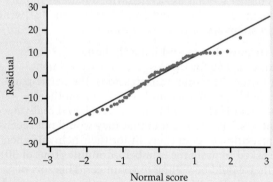

FIGURE 9.20 Residual plot and Normal quantile plot, for Exercise 9.71.

9.72 Examining the residuals. In Example 9.13 (page 476), we were presented with ANOVA output and drew conclusions from the F test without checking whether the data approximately satisfy the necessary assumptions. Let's do that now. TIDE

(a) Fit the ANOVA model and report the test statistic, degrees of freedom, and P-value. Make sure it matches the results presented in Example 9.13.

(b) Use the residuals to check Normality. Are there any outliers or skewness that concern you?

(c) Plot the residuals versus condition. Does the spread in the residuals look reasonably constant? This is a visual way to assess whether we can assume a common standard deviation.

9.73 Zapping memory? DVR recording of television shows is becoming more and more popular. When viewing these shows later, it is common to fast-forward through the commercials. Should advertisers be concerned about this practice? To investigate, a pair of researchers focused on the speed of fast-forward and whether it affects brand name recall.[13]

They recruited 113 participants from a large undergraduate marketing class and randomly assigned them to one of four groups. All groups watched a half-hour episode of the British television show *The Office* interspersed with three breaks of four 30-second commercials. The commercial breaks were shown either at real time, or fast-forwarded at 300%, 1800%, or 3000%. These speeds were chosen to represent common speeds of a DVR. At the end of the episode, each participant was asked to recall as many brands advertised as possible. RECALL

(a) Make a table giving the sample size, mean, and standard deviation for each group.

(b) Is it reasonable to pool the variances? Explain your answer.

(c) Generate a histogram for each of the groups. The data are not Normal because they are counts ranging between 0 and 12. Can we feel confident that the sample means will be approximately Normal? Defend your answer.

9.74 Zapping memory, continued. Refer to the previous exercise. RECALL

(a) Analyze the number of brands recalled using one-way ANOVA. Report the test statistic, degrees of freedom, and P-value.

(b) Even though you assessed the model assumptions in the previous exercise, let's check the assumptions again by examining the residuals. Summarize your findings.

(c) Compare the groups using the least-significant differences method.

(d) Using the results from parts (a), (b), and (c), write a short summary of your conclusions.

9.75 Financial incentives for weight loss. The use of financial incentives has shown promise in promoting weight loss and healthy behaviors. In one study, 104 employees of the Children's Hospital of Philadelphia, with BMIs of 30 to 40 kilograms per square meter (kg/m^2), were each randomly assigned to one of three weight-loss programs.[14] Participants in the control program were provided a link to weight-control information. Participants in the individual-incentive program received this link but were also told that $100 would be given to them each time they met or exceeded their target monthly weight loss. Finally, participants in the group-incentive program received similar information and financial incentives as the individual-incentive program, but were also told that they were placed in secret groups of five and at the end of each four-week period, those in their group who met their goals throughout the period would equally split an additional $500. The study ran for 24 weeks and the total change in weight (in pounds) was recorded. LOSS

(a) Make a table giving the sample size, mean, and standard deviation for each group.

(b) Is it reasonable to pool the variances? Explain your answer.

(c) Generate a histogram for each of the programs. Can we feel confident that the sample means are approximately Normal? Defend your answer.

9.76 Financial incentives for weight loss, continued. Refer to the previous exercise. LOSS

(a) Analyze the change in weight using analysis of variance. Report the test statistic, degrees of freedom, P-value, and your conclusions.

(b) Even though you assessed the model assumptions in the previous exercise, let's check the assumptions again by examining the residuals. Summarize your findings.

(c) Compare the groups using the least-significant differences method.

(d) Using the results from parts (a), (b), and (c), write a short summary of your conclusions.

9.77 Changing the response variable. Refer to the previous two exercises, where we compared three weight-loss programs using change in weight measured in pounds. Suppose that you decide to instead make the comparison using change in weight measured in kilograms. LOSS

(a) Convert the weight loss from pounds to kilograms by dividing each response by 2.2.

(b) Analyze these new weight changes using analysis of variance. Compare the test statistic, degrees of freedom, and P-value you obtain here with those reported in part (a) of the previous exercise. Summarize what you find.

9.78 Does sleep deprivation affect your work? Sleep deprivation experienced by physicians during residency training and the possible negative consequences are of concern to many in the health care community. One study of 33 resident anesthesiologists compared their changes from baseline in reaction times on four tasks.[15] Under baseline conditions, the physicians reported getting an average of 7.04 hours of sleep. While on duty, however, the average was 1.66 hours. For each of the tasks, the researchers reported a statistically significant increase in the reaction time when the residents were working in a state of sleep deprivation.

(a) If each task is analyzed separately as the researchers did in their report, what is the appropriate statistical method to use? Explain your answer.

(b) Is it appropriate to use a one-way ANOVA with $I = 4$ to analyze these data? Explain why or why not.

9.79 Promotions and the expected price of a product. If a supermarket product is frequently offered at a reduced price, do customers expect the price of the product to be lower in the future? This question was examined by researchers in a study conducted on students enrolled in an introductory management course at a large midwestern university. For 10 weeks, 160 subjects read weekly ads for the same product. Students were randomly assigned to read one, three, five, or seven ads featuring price promotions during the 10-week period. They were then asked to estimate what the product's price would be the following week.[16] Table 9.1 gives the data. PPROMO

TABLE 9.1 Price promotion data

Number of promotions	Expected price (dollars)
1	3.78 3.82 4.18 4.46 4.31 4.56 4.36 4.54 3.89 4.13 3.97 4.38 3.98 3.91
	4.34 4.24 4.22 4.32 3.96 4.73 3.62 4.27 4.79 4.58 4.46 4.18 4.40 4.36
	4.37 4.23 4.06 3.86 4.26 4.33 4.10 3.94 3.97 4.60 4.50 4.00
3	4.12 3.91 3.96 4.22 3.88 4.14 4.17 4.07 4.16 4.12 3.84 4.01 4.42 4.01
	3.84 3.95 4.26 3.95 4.30 4.33 4.17 3.97 4.32 3.87 3.91 4.21 3.86 4.14
	3.93 4.08 4.07 4.08 3.95 3.92 4.36 4.05 3.96 4.29 3.60 4.11
5	3.32 3.86 4.15 3.65 3.71 3.78 3.93 3.73 3.71 4.10 3.69 3.83 3.58 4.08
	3.99 3.72 4.41 4.12 3.73 3.56 3.25 3.76 3.56 3.48 3.47 3.58 3.76 3.57
	3.87 3.92 3.39 3.54 3.86 3.77 4.37 3.77 3.81 3.71 3.58 3.69
7	3.45 3.64 3.37 3.27 3.58 4.01 3.67 3.74 3.50 3.60 3.97 3.57 3.50 3.81
	3.55 3.08 3.78 3.86 3.29 3.77 3.25 3.07 3.21 3.55 3.23 2.97 3.86 3.14
	3.43 3.84 3.65 3.45 3.73 3.12 3.82 3.70 3.46 3.73 3.79 3.94

(a) Make a Normal quantile plot for the data in each of the four treatment groups. Summarize the information in the plots and draw a conclusion regarding the Normality of these data.

(b) Summarize the data with a table containing the sample size, mean, and standard deviation for each group.

(c) Is the assumption of equal standard deviations reasonable here? Explain why or why not.

(d) Carry out a one-way ANOVA. Give the hypotheses, the test statistic with its degrees of freedom, and the P-value. Summarize your conclusion.

9.80 Compare the means. Refer to the previous exercise. Use the Bonferroni or another multiple-comparisons procedure to compare the group means. Summarize the results and support your conclusions with a graph of the means.

CASE 9.1 **9.81 Considering a transformation.** In Example 9.8 (page 470), we compared the likelihood to purchase among three groups. We performed ANOVA, even though the data were non-Normal with possible nonconstant variance, because of the robustness of the procedure. For this exercise, let's consider a transformation. MORAL

(a) We have data that must be between 0 and 100. This kind of constraint can result in skewed distributions and unequal variances in a similar fashion to the binomial distribution as p moves away from 0.5 toward 0 or 1. For data like these, there is a special transformation, the arcsine square root transformation, that often is helpful. Construct this new response variable

$$\sin^{-1}(\sqrt{x_{ij}/100})$$

(b) Construct histograms of this response variable for each population. Compare the distributions of the transformed variable with those in Figure 9.3 (page 463). Does the spread appear more similar? Do the data also look more Normal?

(c) Perform ANOVA on the transformed variable. Do the results vary much from those in Figure 9.7 (page 470)?

CASE 9.1 **9.82 Comparing confidence intervals.** Refer to the previous exercise. MORAL

(a) Construct the simultaneous confidence interval for the average difference in purchase likelihood between the moral decoupling group and the control group.

(b) Construct the simultaneous confidence interval for the average difference in transformed purchase likelihood between the moral decoupling group and the control group.

(c) We can't directly compare the two intervals because they are on a different scale. Back-transform the upper and lower endpoints of your confidence interval in part (b). This is done by taking the sine of each endpoint, squaring them, and then multiplying by 100.

(d) Now compare the confidence intervals in parts (a) and (c). Write a summary paragraph explaining which interval you prefer.

9.83 Power for the weight-loss study. You are planning another study of financial incentives for a weight-loss study similar to that described in Exercise 9.75. The standard deviations given in that exercise range from 9.08 to 11.50. To perform power calculations, assume that the standard deviation is $\sigma = 11.50$. You have three groups, each with n subjects, and you would like to reject the ANOVA H_0 when the alternative $\mu_1 = -1.0$, $\mu_2 = -8.0$, and $\mu_3 = -4.0$ is true. Use software to make a table of powers against this alternative (similar to the table in Example 9.28, page 495) for the following numbers in each group: $n = 35, 45, 55, 65$, and 75. What sample size would you choose for your study?

9.84 Same power? Repeat the previous exercise for the alternative $\mu_1 = -2.0$, $\mu_2 = -9.0$, and $\mu_3 = -5.0$. Why are the results the same?

9.85 Planning another organic foods study. Suppose that you are planning a new organic foods study using the same moral outcome variable as described in Exercise 9.69 (page 499). Your study will randomly choose shoppers from a large local grocery store.

(a) Explain how you would select the shoppers to participate in your study.

(b) Use the data from Exercise 9.69 to perform power calculations to determine sample sizes for your study.

(c) Write a report that could be understood by someone with a limited background in statistics and that describes your proposed study and why you think it is likely that you will obtain interesting results.

9.86 The noncentral F distribution. The noncentral F distribution is defined by three parameters: the numerator and denominator degrees of freedom and a **noncentrality parameter λ.** When all n_i are equal, $\bar{\mu}$ is the ordinary average of the μ_i and

$$\lambda = \frac{n\sum(\mu_i - \bar{\mu})}{\sigma^2}$$

Large λ points to an alternative far away from H_0 that is more likely to be detected by the F test. Let's use this information and revisit the two alternatives of Exercises 9.51 and 9.52 (pages 496–497).

(a) What are the numerator and denominator degrees of freedom for these two exercises, given $n = 10$?

(b) Using the information in the exercises and $n = 10$, compute λ for the two alternatives.

(c) Using these values of λ, which of these two alternatives is more likely to be detected? Does this agree with your answer in Exercise 9.52?

9.87 The effect of an outlier. Refer to the weight-loss study described in Exercise 9.75 (page 500). LOSS

(a) Suppose that when entering the data into the computer, you accidentally entered the first observation as 53 pounds rather than 5.3 pounds. Run the ANOVA with this incorrect observation, and record the F statistic, the estimate of the within-group variance s_p^2, and the estimated treatment means.

(b) Alternatively, suppose that when entering the data into the computer, you accidentally entered Observation #101 as 79.4 pounds rather than 19.4 pounds. Run the ANOVA with this incorrect observation, and record the same information requested in part (a).

(c) Compare the results of each of these two cases with the results obtained with the correct data set. What happens to the within-group variance s_p^2? Do the estimated treatment means move closer or further apart? What effect do these changes have on the F test?

(d) What do these two cases illustrate about the effects of an outlier in an ANOVA? Write a one-paragraph summary.

(e) Explain why a table of means and standard deviations for each of the three treatments would help you to detect an incorrect observation.

9.88 Changing units and ANOVA. Refer to Exercise 9.77 (page 501). Suggest a general conclusion about what happens to the F test statistic, degrees of freedom, P-value, and conclusion when you perform ANOVA on data after changing the units through a linear transformation $y = ax + b$, where a and b are chosen constants. In Exercise 9.77, the constants were $a = 1/2.2$ and $b = 0$.

9.89 Smartphones and sleep. Many studies have shown the relationship between the use of electronic media and sleep disturbances. In one recent study, researchers followed adolescents over a two-year period. A total of 591 adolescents were divided into owners (had a smartphone the entire two-year period), new owners (obtained a smartphone during the two-year period), and nonowners. Here are the results for sleeping hours on a school day taken at the beginning and end of the two-year period.[17]

Group	n	Baseline $\bar{x}$	s	Year 2 $\bar{x}$	s
Owners	383	7.81	1.96	7.28	1.76
New owners	153	8.23	1.48	7.54	1.11
Nonowners	55	8.61	0.89	8.00	1.48

(a) Plot the means with time on the x axis and connect the means associated with a group with a line. Does it look like the average decrease in sleep time over the two years is the same for each group? Explain your answer.

(b) The researchers compared baseline averages and Year 2 averages separately using ANOVA. Do you see any concerns with either of these model fits? If so, discuss your concerns.

(c) One might consider fitting a one-way ANOVA using all six combinations of group and time. Is that reasonable here? If not, which model conditions are not satisfied in this case?

9.90 Pooling variances, continued. Refer to Exercise 9.59 (page 497). Based on our rule of thumb (page 467), we consider it reasonable to use the assumption of equal standard deviations in our analysis. However, when sample sizes vary substantially, we need to use caution. As demonstrated in Exercise 9.59, the pooled standard error is closer to the standard deviation of the third group than any of the other three standard deviations. Assuming these sample standard deviations are close to the population standard deviations, explain the impact of using the pooled standard error on the coverage of the simultaneous confidence intervals between means. In particular, would you expect the coverage of the interval for the difference between the first and second groups to be larger or smaller than $(1 - \alpha)100\%$? Explain your answer.

CHAPTER 10

Inference for Proportions

Introduction

Frequently, data on *categorical variables,* expressed as proportions or percents, are used to make business decisions. For example,

- IKEA is planning next year's ad campaign and wants to know what proportion of its customers next year will be college students.

- Whole Foods is considering expanding either its fresh produce or fresh meats section and would like to know what percent of its customers are vegetarians.

- To project the earnings of *Cars 4*, Pixar would like to know what proportion of people who watched *Cars 3* also watched *Cars 2.*

When we record categorical variables, such as these, our data consist of *counts* of the numbers of observations for each value of the variable. Frequently, we reexpress these counts as *proportions* (the count divided by the total number of observations) or *percents* (100 times proportion). The parameter of interest in this setting is the *population proportion.*

Just as in the case of inference about population means, we may be concerned with a single population or with comparing two populations. Inference about one or two proportions is very similar to inference about means. In particular, inference for both means and proportions is based on sampling distributions that are approximately Normal.

We begin with a detailed study of inference about a single population proportion. Section 10.2 concerns methods for comparing two proportions.

CHAPTER OUTLINE

10.1 Inference for a Single Proportion

10.2 Comparing Two Proportions

parameter, p. 295

Similar to our comparison of two means, we include methods for both comparing proportions from two independent populations.

10.1 Inference for a Single Proportion

When you complete this section, you will be able to:

- Identify the sample size, the count, and the sample proportion for a single sample.
- Calculate the standard error of a sample proportion and the margin of error.
- Construct the large-sample confidence interval for a single proportion.
- Use the large-sample significance test to test a null hypothesis about a population proportion.
- Find the sample size needed for a desired margin of error.
- Find the sample size needed for a significance test.

CASE 10.1 Trends in the Workplace The Pew Research Center created the American Trends Panel (ATP) in 2014 to monitor opinions of the U.S. population.[1] In one study, the ATP was asked about the impact of various workplace trends. A total of 4573 U.S. adults responded. When asked about the outsourcing of jobs to other countries, 30% said that this trend has hurt their job or career. ■

For problems involving a single proportion, we will use n for the sample size and X for the count of the outcome of interest. Often, we will use the terms "success" and "failure" for the two possible outcomes. When we do this, X is the number of successes.

We would like to know the proportion of U.S. adults who would respond Yes to the question about the outsourcing of jobs hurting their job or career. This population proportion is the *parameter* of interest. The *statistic* used to estimate this unknown parameter is the sample proportion, a random variable. The sample proportion is $\hat{p} = X/n$.

population proportion, sample proportion p. 295

EXAMPLE 10.1

CASE 10.1 Data for the Outsourcing of Jobs The sample size is the number of ATP members who responded to the Pew survey question, $n = 4573$. A report on the survey tells us that 30% of the respondents said that outsourcing of jobs to other countries has hurt their job or career. Thus, the sample proportion is $\hat{p} = 0.30$. We can calculate the count X from the information given; it is the sample size times the proportion responding Yes, $X = n\hat{p} = 4573(0.3) = 1372$. ■

EXAMPLE 10.2

CASE 10.1 Estimating the Proportion of U.S. Adults The sample proportion $\hat{p}$ in Example 10.1 is a discrete random variable. It can take the values 0, 1/4573, 2/4573, ..., 4573/4573. Because this panel was put together to represent the U.S. population, we can use $\hat{p}$ to estimate the population proportion. For our particular sample, the estimate of the proportion of U.S. adults who would answer Yes is

$$\hat{p} = \frac{1372}{4573} = 0.30$$ ■

10.1 Inference for a Single Proportion

APPLY YOUR KNOWLEDGE

10.1 Community banks. The American Bankers Association Community Bank Insurance Survey for 2017 had responses from 123 banks. Of these, 64 were community banks, defined to be banks with assets of $1 billion or less.[2]

(a) What is the sample size n for this survey?

(b) What is the count X? Describe the count in a short sentence.

(c) Find the sample proportion $\hat{p}$.

10.2 Choosing a credit card. In Example 1.6 (page 9), we looked at data from a survey of 1659 adults that asked about their reasons for choosing a credit card. Receiving cash back for their purchases was the most important reason for 680 of those surveyed.

(a) What is the sample size n for this survey?

(b) What is the count X? Describe the count in a short sentence.

(c) Find the sample proportion $\hat{p}$.

distribution of $\hat{p}$, p. 329

In many cases, a probability model for $\hat{p}$ can be based on the binomial distributions for counts. In Chapter 6, we described this situation as the *binomial setting* and described the sampling distribution of $\hat{p}$. If the sample size n is very small, we can base tests and confidence intervals for p on the discrete distribution of $\hat{p}$. We will focus on situations where the sample size is sufficiently large that we can approximate the distribution of $\hat{p}$ by a Normal distribution.

SAMPLING DISTRIBUTION OF A SAMPLE PROPORTION

Choose an SRS of size n from a large population that contains population proportion p of "successes." Let X be the count of successes in the sample, and let $\hat{p}$ be the **sample proportion** of successes,

$$\hat{p} = \frac{X}{n}$$

Then:

- For large sample sizes, the distribution of $\hat{p}$ is **approximately Normal.**
- The **mean** of the distribution of $\hat{p}$ is p.
- The **standard deviation** of $\hat{p}$ is

$$\sqrt{\frac{p(1-p)}{n}}$$

Figure 10.1 summarizes these facts in a form that recalls the idea of sampling distributions. Our inference procedures are based on this Normal approximation. These procedures are similar to those for inference about the mean of a Normal distribution (page 345). We will see, however, that there are a few extra details involved, caused by the added difficulty of approximating the discrete distribution of $\hat{p}$ by a continuous Normal distribution.

Large-sample confidence interval for a single proportion

We use the sample proportion $\hat{p} = X/n$ to estimate population proportion p. Notice that $\sqrt{p(1-p)/n}$, the standard deviation of $\hat{p}$, depends upon the unknown parameter p. In our calculations, we estimate it by replacing the population parameter p with the sample estimate $\hat{p}$. Therefore, the standard error of $\hat{p}$ is $SE_{\hat{p}} = \sqrt{\hat{p}(1-\hat{p})/n}$. This quantity is our estimate of the standard deviation of $\hat{p}$.

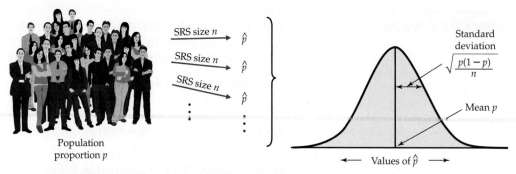

FIGURE 10.1 Draw a large SRS from a population in which the proportion p are successes. The sampling distribution of the sample proportion $\hat{p}$ of successes has approximately a Normal distribution.

68–95–99.7 rule, p. 45

If the sample size is large, the distribution of $\hat{p}$ will be approximately Normal with mean p and standard deviation $SE_{\hat{p}}$. It follows that $\hat{p}$ will be within two standard deviations ($2SE_{\hat{p}}$) of the unknown parameter p about 95% of the time. This is how we use the Normal approximation to construct the large-sample confidence interval for p. Here are the details.

CONFIDENCE INTERVAL FOR A POPULATION PROPORTION

Choose an SRS of size n from a large population with unknown proportion p of successes. The **sample proportion** is

$$\hat{p} = \frac{X}{n}$$

The **standard error of $\hat{p}$** is

$$SE_{\hat{p}} = \sqrt{\frac{\hat{p}(1-\hat{p})}{n}}$$

and the **margin of error** for confidence level C is

$$m = z^* SE_{\hat{p}}$$

where z^* is the value for the standard Normal density curve with area C between $-z^*$ and z^*. The **large-sample level C confidence interval** for p is

$$\hat{p} \pm m$$

This interval is approximately correct when the number of successes and the number of failures are both at least 10.

EXAMPLE 10.3

CASE 10.1 **Confidence Interval for the Outsourcing of Jobs** The sample survey in Example 10.1 found that 1372 of a sample of 4573 U.S. adults reported that outsourcing of jobs to other countries has hurt their job or career. The sample proportion is

$$\hat{p} = \frac{X}{n} = \frac{1372}{4573} = 0.30002$$

The standard error is

$$SE_{\hat{p}} = \sqrt{\frac{\hat{p}(1-\hat{p})}{n}} = \sqrt{\frac{0.30002(1-0.30002)}{4573}} = 0.0067768$$

The z critical value for 95% confidence is $z^* = 1.96$, so the margin of error is
$$m = 1.96\,\text{SE}_{\hat{p}} = (1.96)(0.0067768) = 0.013283$$
The confidence interval is
$$\hat{p} \pm m = 0.300 \pm 0.013$$

We are 95% confident that between 28.7% and 31.3% of U.S. adults would report outsourcing of jobs to other countries has hurt their job or career. ∎

In performing these calculations, we have kept a large number of digits for our intermediate calculations. However, when reporting the results, we prefer to use rounded values, for example, "30.0% with a margin of error of 1.3%." *You should always focus on what is important. Reporting extra digits that are not needed can divert attention from the main point of your summary.* There is no additional information to be gained by reporting $\hat{p} = 0.30002$ with a margin of error of 0.013283. Do you think it would be better to report 30% with a 1% margin of error?

Remember that the margin of error in any confidence interval includes only random sampling error. If people do not respond honestly to the questions asked, for example, your estimate is likely to miss by more than the margin of error. Similarly, response bias can also be present. In our outsourcing example, the response variable measures whether the adults said that they thought outsourcing hurt their job or career, not whether outsourcing did, in fact, hurt their job or career.

Because the calculations for statistical inference for a single proportion are relatively straightforward, we often do them with a calculator or in a spreadsheet. Figure 10.2 gives output from JMP and Minitab for the data in Example 10.1. There are alternatives to the Normal approximation method that we have presented that are used by some software packages. JMP uses one of these, called the score method, as a default, but provides options for selecting different methods. In general, the alternatives will give very similar results, particularly for large sample sizes.

As usual, the outputs report more digits than are useful. When you use software, be sure to think about how many digits are meaningful for your purposes. Do not clutter your report with information that is not meaningful.

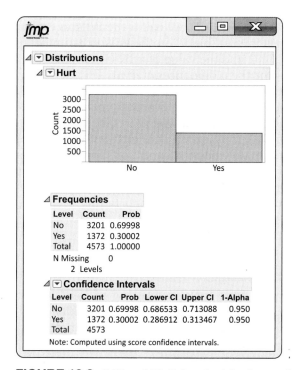
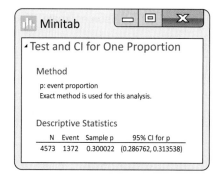

FIGURE 10.2 JMP and Minitab output for the confidence interval in Exercise 10.3.

APPLY YOUR KNOWLEDGE

10.3 Community banks. Refer to Exercise 10.1 (page 507).

(a) Find $SE_{\hat{p}}$, the standard error of $\hat{p}$. Explain the meaning of the standard error in simple terms.

(b) Give the 95% confidence interval for p in the form of estimate plus or minus the margin of error.

(c) Give the confidence interval as an interval of percents.

10.4 Choosing a credit card. Refer to Exercise 10.2 (page 507).

(a) Find $SE_{\hat{p}}$, the standard error of $\hat{p}$.

(b) Give the 95% confidence interval for p in the form of estimate plus or minus the margin of error.

(c) Give the confidence interval as an interval of percents.

> **BEYOND THE BASICS**
>
> **Plus four confidence interval for a single proportion**
>
> Suppose we have a sample where the count is zero ($X = 0$). Then, because $\hat{p} = 0$, the standard error and the margin of error will both be 0. The confidence interval for any confidence level would be the single point 0. Clearly, a confidence interval based on the large-sample Normal approximation does not make sense in this situation.
>
> Both computer studies and careful mathematics show that we can do better by moving the sample proportion $\hat{p}$ away from 0 and 1.[3] There are several ways to do this. Here is a simple adjustment that works very well in practice.
>
> The adjustment is based on the following idea: act as if we have four additional observations, two of which are successes and two of which are failures. The new sample size is $n + 4$ and the count of successes is $X + 2$. Because this estimate was first suggested by Edwin Bidwell Wilson in 1927, we call it the **Wilson estimate.** To compute a confidence interval based on the Wilson estimate, first replace the value of X by $X + 2$ and the value of n by $n + 4$. Then use these values in the formulas for the confidence interval.
>
> In Example 10.1, we had $X = 1372$ and $n = 4573$. To apply the "plus four" approach, we use the z procedure with $X = 1374$ and $n = 4577$. You can use this interval when the sample size is at least 10 ($n \geq 10$).
>
> In general, the large sample interval will agree pretty well with the Wilson estimate when the conditions for the application of the large sample method are met (the number of successes and failures are both at least 10). The Wilson interval is most useful when these conditions are not met and the sample proportion is close to zero or one.

Wilson estimate

APPLY YOUR KNOWLEDGE

10.5 Use plus four for outsourcing of jobs. Refer to Example 10.3 (page 508). Compute the plus four 95% confidence interval, and compare this interval with the one given in that example.

10.6 New-product sales. Yesterday, your top salesperson called on twelve customers and obtained orders for your new product from all 12. Suppose that it is reasonable to view these eight customers as a random sample of all customers.

(a) Give the plus four estimate of the proportion of her customers who would buy the new product. Notice that we don't estimate that all customers will buy, even though all eight in the sample did.

(b) Give the margin of error and the confidence interval for 95% confidence. (You may see that the upper endpoint of the confidence interval is greater than 1. In that case, take the upper endpoint to be 1.)

(c) Do the results apply to all the people on your sales force? Explain why or why not.

10.7 Construct an example. Make up an example where the large-sample method and the plus four method give very different intervals. Do not use a case where either $X = 0$ or $X = 1$.

Significance test for a single proportion

We know that the sample proportion $\hat{p} = X/n$ is approximately Normal, with mean $\mu_{\hat{p}} = p$ and standard deviation $\sigma_{\hat{p}} = \sqrt{p(1-p)/n}$. To construct confidence intervals, we use the standard error $SE_{\hat{p}}$ in place of $\sigma_{\hat{p}}$ because the standard deviation depends upon the unknown parameter p. When performing a significance test, however, the null hypothesis specifies a value for p, which we will call p_0. This means that when we test $H_0: p = p_0$, we substitute p_0 for p in the expressions for $\mu_{\hat{p}}$ and $\sigma_{\hat{p}}$. Here are the details.

z SIGNIFICANCE TEST FOR A POPULATION PROPORTION

Choose an SRS of size n from a large population with unknown proportion p of successes. To test the hypothesis $H_0: p = p_0$, compute the **z statistic**

$$z = \frac{\hat{p} - p_0}{\sqrt{\dfrac{p_0(1-p_0)}{n}}}$$

In terms of a standard Normal random variable Z, the P-value for a test of H_0 against

$H_a: p > p_0$ is $P(Z \geq z)$

$H_a: p < p_0$ is $P(Z \leq z)$

$H_a: p \neq p_0$ is $2P(Z \geq |z|)$

These P-values are approximately correct when the expected number of successes np_0 and the expected number of failures $n(1 - p_0)$ are both at least 10.

EXAMPLE 10.4

Comparing Two Sunblock Lotions Your company produces a sunblock lotion designed to protect the skin from both UVA and UVB exposure from the sun. You hire a company to compare your product with the product sold by your major competitor. The testing company exposes skin on the backs of a sample of 20 people to UVA and UVB rays and measures the protection provided by each product. For 13 of the subjects, your product provided better protection, while for the other seven subjects, your competitor's product provided better protection. Do you have evidence to support a commercial claiming that your product provides superior UVA and UVB protection? For the data we have $n = 20$ subjects and $X = 13$ successes. To answer the claim question, we test

$$H_0: p = 0.5$$
$$H_a: p \neq 0.5$$

The expected numbers of successes (your product provides better protection) and failures (your competitor's product provides better protection) are $20 \times 0.5 = 10$ and $20 \times 0.5 = 10$. Both are at least 10, so we can use the z test. The sample proportion is

$$\hat{p} = \frac{X}{n} = \frac{13}{20} = 0.65$$

The test statistic is

$$z = \frac{\hat{p} - p_0}{\sqrt{\frac{p_0(1-p_0)}{n}}} = \frac{0.65 - 0.5}{\sqrt{\frac{(0.5)(0.5)}{20}}} = 1.34$$

From Table A, we find $P(Z \leq 1.34) = 0.9099$, so the probability in the upper tail is $1 - 0.9099 = 0.0901$. The P-value is the area in both tails, $P = 2 \times 0.0901 = 0.1802$. JMP and Minitab outputs for the analysis appear in Figure 10.3. Note that JMP reports P-values using two different methods, "Likelihood Ratio" and "Pearson." The Minitab output reports the z significance test. Although the P-values are different, all three methods lead to the same conclusion. We report that the sunblock testing data are compatible with the hypothesis of no difference between your product and your competitor's ($\hat{p} = 0.65$, $z = 1.34$, $P = 0.18$). The data do not provide your company with enough evidence to support the advertising claim. This conclusion, of course, throws your marketing department into a panic. Was the study adequately powered? We discuss sample size considerations next. ∎

Note that we used a two-sided hypothesis test when we compared the two sunblock lotions in Example 10.4. In settings like this, we must start with the view that either product could be better if we want to prove a claim of superiority. Thinking or hoping that your product is superior cannot be used to justify a one-sided test.

FIGURE 10.3 JMP and Minitab output for the significance test in Example 10.4.

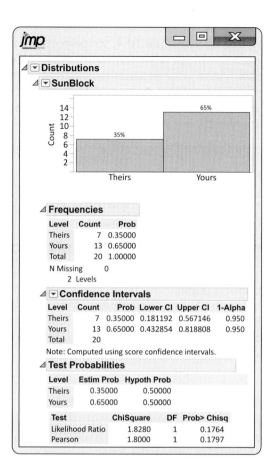
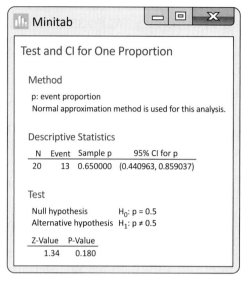

APPLY YOUR KNOWLEDGE

10.8 Draw a picture. Draw a picture of a standard Normal curve and shade the tail areas to illustrate the calculation of the P-value for Example 10.4.

10.9 What does the confidence interval tell us? Inspect the outputs in Figure 10.3 and report the confidence interval for the percent of people who would get better sun protection from your product than from your competitor's. Be sure to convert from proportions to percents and round appropriately. Interpret the confidence interval and compare this way of analyzing data with the significance test.

10.10 The effect of X on conclusions. In Example 10.4, suppose that your product provided better UVA and UVB protection for 15 of the 20 subjects. Perform the significance test and summarize the results.

10.11 The effect of n on conclusions. In Example 10.4, consider what would have happened if you had paid for 60 subjects to be tested, with 65% having better UVA and UVB protection with your product.

(a) Perform the significance test and summarize the results.

(b) Compare these results with those found in Example 10.4, and write a short summary of the effect of the sample size on these significance tests.

In Example 10.4, we treated an outcome as a success whenever your product provided better sun protection. Would we get the same results if we defined success as an outcome where your competitor's product was superior? In this setting, the null hypothesis is still $H_0: p = 0.5$. You will find that the z test statistic is unchanged except for its sign and that the P-value for the two-sided alternative remains the same.

APPLY YOUR KNOWLEDGE

10.12 Yes or no? In Example 10.4, we performed a significance test to compare your sunblock with your competitor's. Success was defined as the outcome where your product provided better protection. Now, take the viewpoint of your competitor, and define success as the outcome where your competitor's product provides better protection. In other words, n remains the 20, but X is now 7.

(a) Verify that the P-value of the two-sided significance test remains the same.

(b) Find the 95% confidence interval for this setting, and compare it with the intervals in Figure 10.3 that used $X = 13$. Write a short summary.

Choosing a sample size for a confidence interval

sample size for a desired m, p. 349

In Chapter 7, we showed how to choose the sample size n to obtain a confidence interval for a population mean with specified margin of error m. Because we are also using the Normal distribution for inference about a population proportion in this chapter, sample size selection proceeds in much the same way.

Recall that the margin of error of the large-sample confidence interval for a population proportion is

$$m = z^* \text{SE}_{\hat{p}} = z^* \sqrt{\frac{\hat{p}(1-\hat{p})}{n}}$$

Choosing a confidence level C fixes the critical value z^*. The margin of error also depends on the value of $\hat{p}$ and the sample size n. Because we don't know the value of $\hat{p}$ until we gather the data, we must guess a value to use in the calculations. We will call the guessed value p^*. Here are two ways to get p^*:

• Use a sample estimate obtained from a pilot study or from similar studies done earlier.

• Use $p^* = 0.5$. Because the margin of error is largest when $\hat{p} = 0.5$, this choice gives a sample size that is somewhat larger than we really need for the confidence level we choose. It is a safe choice no matter what the data later show.

Once we have chosen p^* and the margin of error m that we want, we can find the n we need to achieve this margin of error. Here is the result.

SAMPLE SIZE FOR DESIRED MARGIN OF ERROR

The level C confidence interval for a proportion p will have a margin of error approximately equal to a specified value m when the sample size is

$$n = \left(\frac{z^*}{m}\right)^2 p^*(1 - p^*)$$

Here, z^* is the critical value for confidence C, and p^* is a guessed value for the proportion of successes in the future sample.

The margin of error will be less than or equal to m if p^* is chosen to be 0.5. The sample size required is then given by

$$n = \left(\frac{z^*}{2m}\right)^2$$

Note that to use the confidence interval, which is based on the Normal approximation, we still require that the number of successes and the number of failures if the sample be at least 10.

The value of n obtained by this method is not particularly sensitive to the choice of p^* as long as p^* is not too far from 0.5. However, if your actual sample turns out to have $\hat{p}$ smaller than about 0.3 or larger than about 0.7, the sample size based on $p^* = 0.5$ may be much larger than needed.

EXAMPLE 10.5

Planning a Sample of Customers Your company has received complaints about its customer support service. You intend to hire a consulting company to carry out a sample survey of customers. Before contacting the consultant, you want some idea of the sample size you will have to pay for. One critical question is the degree of satisfaction with your customer service, measured on a 5-point scale. You want to estimate the proportion p of your customers who are satisfied (i.e., who choose either "satisfied" or "very satisfied," the two highest levels on the 5-point scale).

You want to estimate p with 95% confidence and a margin of error less than or equal to 3%, or 0.03. For planning purposes, you are willing to use $p^* = 0.5$. To find the sample size required, we use the formula

$$n = \left(\frac{z^*}{2m}\right)^2 = \left[\frac{1.96}{(2)(0.03)}\right]^2 = 1067.1$$

We round up to get $n = 1068$. (Always round up. Rounding down would give a margin of error slightly greater than 0.03.)

Similarly, for a 2.5% margin of error, we have (after rounding up)

$$n = \left[\frac{1.96}{(2)(0.025)}\right]^2 = 1537$$

and for a 2% margin of error,

$$n = \left[\frac{1.96}{(2)(0.02)}\right]^2 = 2401 \blacksquare$$

News reports frequently describe the results of surveys with sample sizes between 1000 and 1500 and a margin of error of about 3%. These surveys generally use sampling procedures more complicated than simple random sampling, so the calculation of confidence intervals is more involved than what we have studied in this section. The calculations in Example 10.5 nonetheless show, in principle, how such surveys are planned.

In practice, many factors influence the choice of a sample size. Case 10.2 illustrates one set of factors.

CASE 10.2

Marketing Christmas Trees An association of Christmas tree growers in Indiana sponsored a sample survey of Indiana households to help improve the marketing of Christmas trees.[4] The researchers decided to use a telephone survey and estimated that each telephone interview would take about two minutes. Nine students trained in agribusiness marketing were to make the phone calls between 1:00 P.M. and 8:00 P.M. on a Sunday. After discussing problems related to people not being at home or being unwilling to answer the questions, the survey team proposed a sample size of 500. Several of the questions asked demographic information about the household. The key questions of interest had responses of Yes or No, for example, "Did you have a Christmas tree last year?" The primary purpose of the survey was to estimate various sample proportions for Indiana households. An important issue in designing the survey was, therefore, whether the proposed sample size of $n = 500$ would be adequate to provide the sponsors of the survey with the information they required. $\blacksquare$

To address this question, we calculate the margins of error of 95% confidence intervals for various values of $\hat{p}$.

EXAMPLE 10.6

Margins of Error In the Christmas tree market survey, the margin of error of a 95% confidence interval for any value of $\hat{p}$ and $n = 500$ is

$$m = z^* \text{SE}_{\hat{p}}$$

$$= 1.96 \sqrt{\frac{\hat{p}(1-\hat{p})}{500}}$$

The results for various values of $\hat{p}$ are

$\hat{p}$	m	$\hat{p}$	m
0.05	0.019	0.60	0.043
0.10	0.026	0.70	0.040
0.20	0.035	0.80	0.035
0.30	0.040	0.90	0.026
0.40	0.043	0.95	0.019
0.50	0.044		

The survey team judged these margins of error to be acceptable and used a sample size of 500 in their survey. ■

The table in Example 10.6 illustrates two points. First, the margins of error for $\hat{p} = 0.05$ and $\hat{p} = 0.95$ are the same. In fact, the margins of error will always be the same for $\hat{p}$ and $1 - \hat{p}$. This is a direct consequence of the form of the confidence interval. Second, the margin of error varies only between 0.040 and 0.044 as $\hat{p}$ varies from 0.3 to 0.7, and the margin of error is greatest when $\hat{p} = 0.5$, as we claimed earlier. It is true in general that the margin of error will vary relatively little for values of $\hat{p}$ between 0.3 and 0.7. Therefore, when planning a study, it is not necessary to have a very precise guess for p. If $p^* = 0.5$ is used and the observed $\hat{p}$ is between 0.3 and 0.7, the actual interval will be a little shorter than needed, but the difference will be quite small.

APPLY YOUR KNOWLEDGE

10.13 Is there interest in a new product? One of your employees has suggested that your company develop a new product. You decide to take a random sample of your customers and ask whether there is interest in the new product. The response is on a 1 to 5 scale, with 1 indicating "definitely would not purchase"; 2, "probably would not purchase"; 3, "not sure"; 4, "probably would purchase"; and 5, "definitely would purchase." For an initial analysis, you will record the responses 1, 2, and 3 as No and 4 and 5 as Yes. What sample size would you use if you wanted the 95% margin of error to be 0.14 or less?

10.14 More information is needed. Refer to the previous exercise. Suppose that, after reviewing the results of the previous survey, you proceeded with preliminary development of the product. Now you are at the stage where you need to decide whether to make a major investment to produce and market the product. You will use another random sample of your customers, but now you want the margin of error to be smaller. What sample size would you use if you wanted the 95% margin of error to be 0.03 or less?

Choosing a sample size for a significance test

power, p. 441

In Chapter 8, we introduced the idea of power for a significance test. This idea applies to the significance test for a proportion that we studied in this section. There are some differences in the details, but the basic ideas are the same. Fortunately, software can take care of the details, so we just need to concentrate on choosing the appropriate input and interpreting the output.

To find the required sample size, we need to specify

- the value of p_0 in the null hypothesis H_0: $p = p_0$

- the alternative hypothesis, two-sided (H_a: $p \neq p_0$) or one-sided (H_a: $p > p_0$ or H_a: $p < p_0$)

- a value of p for the alternative hypothesis

- the significance level (α, the probability of rejecting the null hypothesis when it is true); usually we choose 5% ($\alpha = 0.05$) for significance level

- the desired power (probability of rejecting the null hypothesis when it is false); usually we choose 80% (0.80) for power

EXAMPLE 10.7

Sample Size for Comparing Two Sunblock Lotions In Example 10.4, we performed the significance test for comparing two sunblock lotions in a setting where each subject used the two lotions and the product that provided better protection was recorded. Although your product performed better 13 times in 20 trials, the value of $\hat{p} = 13/20 = 0.65$ was not sufficiently far from the null hypothesized value of $p_0 = 0.5$ for us to reject the H_0 ($P = 0.18$). Let's suppose that the true percent of the time that your lotion would perform better is $p_0 = 0.65$ and we plan to test the null hypothesis H_0: $p = 0.5$ versus the two-sided alternative H_a: $p \neq 0.5$ at the 5% level.

What sample size n should we choose if we want to have an 80% chance of rejecting H_0? Outputs from JMP and Minitab are given in Figure 10.4. JMP indicates that $n = 89$ should be used, while Minitab suggests $n = 85$. The difference is due to the slightly different test methods that can be used in this setting. ∎

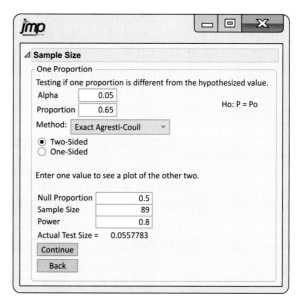

FIGURE 10.4 JMP and Minitab output for the sample size calculation in Example 10.7.

FIGURE 10.4 Continued

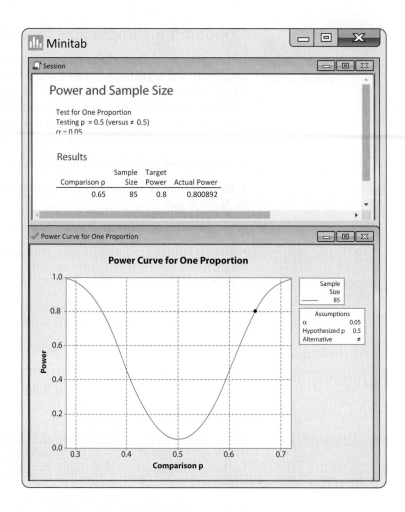

Note that Minitab provides a graph as a function of the value of the proportion for the alternative hypothesis. Similar plots can be produced by JMP. In some situations, you might want to specify the sample size n and have software compute the power. This option is also available in JMP, Minitab, and other software.

APPLY YOUR KNOWLEDGE

10.15 Compute the sample size for a different alternative. Refer to Example 10.7. Use software to find the sample size needed for a two-sided test of the null hypothesis that $p = 0.5$ versus the two-sided alternative with $\alpha = 0.05$ and 80% power if the alternative is $p = 0.6$.

10.16 Compute the power for a given sample size. Consider the setting in Example 10.7. You have a budget that will allow you to test 120 subjects. Use software to find the power of the test for this value of n.

SECTION 10.1 SUMMARY

- Inference about a population proportion is based on an SRS of size n. When n is large, the distribution of the **sample proportion** $\hat{p} = X/n$ is approximately Normal with mean p and standard deviation $\sqrt{p(1-p)/n}$.

- The estimated standard deviation of the distribution of $\hat{p}$ is the **standard error of $\hat{p}$**

$$\mathrm{SE}_{\hat{p}} = \sqrt{\frac{\hat{p}(1-\hat{p})}{n}}$$

- The **margin of error** for confidence level C is
$$m = z^* \text{SE}_{\hat{p}}$$
where z^* is the value for the standard Normal density curve with area C between $-z^*$ and z^*.

- The **large-sample level C confidence interval** for p is
$$\hat{p} \pm m$$
We recommend using this method when the number of successes and the number of failures are both at least 10.

- Tests of $H_0: p = p_0$ are based on the **z statistic**
$$z = \frac{\hat{p} - p_0}{\sqrt{\frac{p_0(1-p_0)}{n}}}$$
with P-values calculated from the $N(0,1)$ distribution. Use this test when the expected number of successes np_0 and the expected number of failures $n(1-p_0)$ are both at least 10.

- The **sample size** required to obtain a confidence interval of approximate margin of error m for a population proportion is found from
$$n = \left(\frac{z^*}{m}\right)^2 p^*(1-p^*)$$
where p^* is a guessed value for the proportion, and z^* is the standard Normal critical value for the desired level of confidence. To ensure that the margin of error of the interval is less than or equal to m no matter what $\hat{p}$ may be, use
$$n = \left(\frac{z^*}{2m}\right)^2$$

- Power calculations for significance tests on a single population proportion are easily performed using software. In addition to the characteristics of the significance test (null and alternative hypotheses, significance level), you can specify the power and determine the sample size or specify the sample size and determine the power.

SECTION 10.1 EXERCISES

For Exercises 10.1 and 10.2, see page 507; for 10.3 and 10.4, see page 510; for 10.5 to 10.7, see pages 510–511; for 10.8 to 10.11, see page 513; for 10.12, see page 514; for 10.13 and 10.14, see page 516; and for 10.15 and 10.16, see page 518.

10.17 What's wrong? Explain what is wrong with each of the following.

(a) A large-sample confidence interval for an unknown proportion p is $\hat{p}$ plus or minus its standard error.

(b) You can use a significance test to evaluate the hypothesis $H_0: \hat{p} = 0.4$ versus the one-sided alternative.

(c) The large-sample confidence interval for a population proportion always has 95% confidence.

10.18 What's wrong? Explain what is wrong with each of the following.

(a) A student project used a confidence interval to describe the results in a final report. The confidence level was negative 99%.

(b) The margin of error for a confidence interval used for an opinion poll takes into account the fact that some of the questions were poorly worded.

(c) If the P-value for a significance test is 0.6, we can conclude that the null hypothesis is more likely to be true than false.

10.19 Draw some pictures. Consider the binomial setting with $n = 50$ and $p = 0.4$.

(a) The sample proportion $\hat{p}$ will have a distribution that is approximately Normal. Give the mean and the standard deviation of this Normal distribution.

(b) Draw a sketch of this Normal distribution. Mark the location of the mean.

(c) Find a value d^* for which the probability is 95% that $\hat{p}$ will be between $p \pm d^*$. Mark these two values on your sketch.

10.20 Smartphones and purchases. A research study asked 4024 smartphone users about how they used their phones. In response to a question about purchases, 2057 reported that they purchased an item after using their smartphone to search for information about the item.

(a) What is the sample size n for this survey?

(b) In this setting, describe the population proportion p in a short sentence.

(c) What is the count X? Describe the count in a short sentence.

(d) Find the sample proportion $\hat{p}$.

(e) Find $SE_{\hat{p}}$, the standard error of $\hat{p}$.

(f) Give the 95% confidence interval for p in the form of estimate plus or minus the margin of error.

(g) Give the confidence interval as an interval of percents.

10.21 Working while sick. A survey commissioned by the Southern Cross Healthcare Group reported that 46% of New Zealanders turn up for work despite being sick.[5] Assume that the sample size is 235.

(a) What number of survey respondents reported that they turn up for work despite being sick? You will need to round your answer. Why?

(b) Find a 95% confidence interval for the proportion of New Zealanders who report that they turn up for work despite being sick.

(c) Convert the estimate and your confidence interval to percents.

(d) Discuss reasons the estimate might be biased.

10.22 Nonconforming switches. Suppose that an SRS of size 200 is taken from a large shipment of switches. Each switch is tested to determine whether it conforms with the shipment contract specifications. Twenty-one nonconforming switches are found in the SRS.

(a) Find the sample proportion.

(b) What is the margin of error for 95% confidence?

(c) Find the 95% confidence interval for the proportion of nonconforming switches in the large shipment from which the SRS was selected.

10.23 Significance test for nonconforming switches. Refer to the previous exercise. You want to test the null hypothesis that the true proportion of nonconforming switches is 7% versus the alternative that it is greater than 7%.

(a) State the null and alternative hypotheses for performing a significance test in this setting.

(b) Find the test statistic.

(c) Find the P-value.

(d) Write a short summary of your conclusion.

10.24 Customer preferences for your new product. A sample of 60 potential customers was asked to use your new product and the product of the leading competitor. After one week, they were asked to indicate which product they preferred. In the sample, 40 potential customers said that they preferred your product.

(a) Find the sample proportion.

(b) What is the margin of error for 95% confidence?

(c) Find the 95% confidence interval for the proportion of potential customers who prefer your product.

10.25 How many potential customers should you sample? Refer to the previous exercise. If you want the 95% margin of error to be 0.07 or less, what would you choose for a sample size? Explain how you calculated your answer and show your work.

10.26 How much influence do social media have on purchasing decisions? A Gallup poll asked this question of 18,525 U.S. adults aged 18 and older.[6] The response "No influence at all" was given by 62% of the respondents. Find a 99% confidence for the true proportion of U.S. adults who would choose "No influence at all" as their response.

10.27 Canadian Millennials use new technologies. A survey of 1700 Canadian Millennials asked about using new technologies to discover music and bands that they like.[7] Of these, 1037 reported that they have used new technologies to discover music and bands that they like.

(a) What proportion of the Canadian Millennials in the sample have used new technologies to discover music and bands that they like?

(b) Find the 95% margin of error for the estimate.

(c) Compute the 95% confidence interval for the population proportion.

(d) Write a short paragraph explaining the meaning of the confidence interval.

(e) Do you prefer to report the sample proportion with the margin of error or the confidence interval? Give reasons for your answer.

10.28 Country food and Inuits. Country food includes seal, caribou, whale, duck, fish, and berries and is an important part of the diet of the aboriginal people called Inuits, who inhabit Inuit Nunaat, the northern region of what is now called Canada. A survey of Inuits in Inuit Nunaat reported that 3274 out of 5000 respondents said that at least half of the meat and fish they eat is country food.[8] Find the sample proportion and a 95% confidence interval for the population proportion of Inuits whose diet is at least half country food.

10.29 Mathematician tosses coin 10,000 times! The South African mathematician John Kerrich, while a prisoner of war during World War II, tossed a coin 10,000 times and obtained 5067 heads.

(a) Give a 95% confidence interval to see what probabilities of heads are roughly consistent with Kerrich's result.

(b) Is this significant evidence at the 5% level that the probability that Kerrich's coin comes up heads is not 0.5?

10.30 "Guitar Hero" and "Rock Band." An electronic survey of 7061 game players of "Guitar Hero" and "Rock Band" reported that 67% of players who do not currently play a musical instrument said that they are likely to begin playing a real musical instrument in the next two years.[9] The reports describing the survey do not give the number of respondents who do not currently play a musical instrument.

(a) Explain why it is important to know the number of respondents who do not currently play a musical instrument.

(b) Assume that half of the respondents do not currently play a musical instrument. Find the count of players who said that they are likely to begin playing a real musical instrument in the next two years.

(c) Give a 99% confidence interval for the population proportion who would say that they are likely to begin playing a real musical instrument in the next two years.

(d) The survey collected data from two separate consumer panels. There were 3300 respondents from the LightSpeed consumer panel and the others were from Guitar Center's proprietary consumer panel. Comment on the sampling procedure used for this survey and how it would influence your interpretation of the findings.

10.31 "Guitar Hero" and "Rock Band." Refer to the previous exercise.

(a) How would the result that you reported in part (c) of the previous exercise change if only 25% of the respondents said that they did not currently play a musical instrument?

(b) Do the same calculations for a case in which the percent is 75%.

(c) The main conclusion of the survey that appeared in many news stories was that 67% of players of "Guitar Hero" and "Rock Band" who do not currently play a musical instrument said that they are likely to begin playing a real musical instrument in the next two years. What can you conclude about the effect of the three scenarios—part (b) in the previous exercise and parts (a) and (b) in this exercise—on the margin of error for the main result?

10.32 How would the confidence interval change? Refer to Exercise 10.28. Would a 90% confidence interval be wider or narrower than the one that you found in that exercise? Verify your results by computing the interval.

10.33 How would the confidence interval change? Refer to Exercise 10.28. Would a 99% confidence interval be wider or narrower than the one that you found in that exercise? Verify your results by computing the interval.

10.34 Can we use the z test? In each of the following cases, is the sample large enough to permit safe use of the z test? (The population is very large.)

(a) $n = 50$ and $H_0: p = 0.4$.

(b) $n = 50$ and $H_0: p = 0.1$.

(c) $n = 500$ and $H_0: p = 0.04$.

(d) $n = 19$ and $H_0: p = 0.5$.

10.35 Shipping the orders on time. As part of a quality improvement program, your mail-order company is studying the process of filling customer orders. According to company standards, an order is shipped on time if it is sent within two working days of the time it is received. You select an SRS of 200 of the 8000 orders received in the past month for an audit. The audit reveals that 185 of these orders were shipped on time. Find a 95% confidence interval for the true proportion of the month's orders that were shipped on time.

10.36 Instant versus fresh-brewed coffee. A matched pairs experiment compares the taste of instant coffee with fresh-brewed coffee. Each subject tastes two unmarked cups of coffee, one of each type, in random order and states which he or she prefers. Of the 40 subjects who participate in the study, 14 prefer the instant coffee and the other 26 prefer fresh-brewed. Take p to be the proportion of the population that prefers fresh-brewed coffee.

(a) Find a 95% confidence interval for p.

(b) Test the claim that a majority of people prefer the taste of fresh-brewed coffee. Report the z statistic and its P-value. Is your result significant at the 5% level? What is your practical conclusion?

CASE 10.2 **10.37 Checking the demographics of a sample.** Of the 500 households that responded to the Christmas tree marketing survey, 38% were from rural areas (including small towns), and the other 62% were from urban areas (including suburbs). How well does the rural versus urban split in the sample correspond to the demographics of the state of Indiana as a whole? According to the census, 36% of Indiana households are in rural areas, and the remaining 64% are in urban areas. Let p be the proportion of rural respondents. Set up hypotheses about p_0 and perform a test of significance to examine how well the sample represents the state in regard to rural versus urban residents. Summarize your results.

CASE 10.2 **10.38 More on demographics.** In the previous exercise, we arbitrarily chose to state the hypotheses in terms of the proportion of rural respondents. We could as easily have used the proportion of *urban* respondents.

(a) Write hypotheses in terms of the proportion of urban residents to examine how well the sample represents the state in regard to rural versus urban residents.

(b) Perform the test of significance and summarize the results.

(c) Compare your results with the results of the previous exercise. Summarize and generalize your conclusion.

10.39 High-income households on a mailing list. Land's Beginning sells merchandise through the mail. It is considering buying a list of addresses from a magazine. The magazine claims that at least 40% of its subscribers have household incomes in excess of $150,000. Land's Beginning would like to estimate the proportion of high-income people on the list. Verifying income is difficult, but another company offers this service. Land's Beginning will pay to verify the incomes of an SRS of people on the magazine's list. They would like the margin of error of the 95% confidence interval for the proportion to be 0.06 or less. Use the guessed value $p^* = 0.40$ to find the required sample size.

10.40 Change the specs. Refer to the previous exercise. For each of the following variations on the design specifications, state whether the required sample size will be larger, smaller, or the same as that found in Exercise 10.39.

(a) Use a 99% confidence interval.

(b) Change the allowable margin of error to 0.03.

(c) Use a planning value of $p^* = 0.35$.

(d) Use a different company to do the income verification.

10.41 Be an entrepreneur. A student organization wants to start a nightclub for students under the age of 21. To assess support for this proposal, the organization will select an SRS of students and ask each respondent if he or she would patronize this type of establishment. About 75% of the student body are expected to respond favorably.

(a) What sample size is required to obtain a 95% confidence interval with an approximate margin of error of 0.04?

(b) Suppose that 65% of the sample responds favorably. Calculate the margin of error of the 95% confidence interval. For the sample size, use the value that you computed in part (a).

10.42 Are the customers dissatisfied? A cell phone manufacturer would like to know what proportion of its customers are dissatisfied with the service received from their local distributor. The customer relations department will survey a random sample of customers and compute a 95% confidence interval for the proportion that are dissatisfied. From past studies, the department believes that this proportion will be about 0.10.

(a) Find the sample size needed if the margin of error of the confidence interval is to be about 0.03.

(b) Suppose 14% of the sample say that they are dissatisfied. What is the margin of error of the 99% confidence interval?

10.43 Increase student fees? You have been asked to survey students at a large college to determine the proportion that favor an increase in student fees to support an expansion of the student newspaper. Each student will be asked whether he or she is in favor of the proposed increase. Using records provided by the registrar, you can select a random sample of students from the college. After careful consideration of your resources, you decide that it is reasonable to conduct a study with a sample of 250 students.

(a) Construct a table of the margins of error for 95% confidence when $\hat{p}$ takes the values 0.1, 0.2, 0.3, 0.4, 0.5, 0.6, 0.7, 0.8, and 0.9.

(b) Make a graph of margin of error versus the value of $\hat{p}$.

10.44 Justify the cost of the survey. A former editor of the student newspaper agrees to underwrite the study in the previous exercise because she believes the results will demonstrate that most students support an increase in fees. She is willing to provide funds for a sample size of 500. Write a short summary for your benefactor of why the increased sample size will provide better results.

10.45 Are the customers dissatisfied? Refer to Exercise 10.42, where you computed the sample size based on the width of a confidence interval. Now we will use the same setting to determine the sample size based on a significance test. You want to test the null hypothesis that the population proportion is 0.10 using a two-sided test with $\alpha = 0.05$ and 80% power. Use 0.20 as the proportion for the alternative. What sample size would you recommend? Note that you need to specify an alternative hypothesis to answer this question.

10.46 Nonconforming switches. Refer to Exercises 10.22 and 10.23, where you found a confidence interval and performed a significance test for nonconforming switches. Find the sample size needed for testing the null hypothesis that the population proportion is 0.07 versus the one-sided alternative that the population proportion is greater than 0.07. Use $\alpha = 0.05$, 80% power, and 0.14 as the alternative for your calculations.

10.2 Comparing Two Proportions

When you complete this section, you will be able to:

- Identify the counts and sample sizes for a comparison between two proportions, compute the sample proportions, and find their difference.
- Use the large-sample method to find the confidence interval for a difference between two proportions and interpret the confidence interval.
- Use the large-sample method to perform a significance test for comparing two proportions and interpret the results.
- Determine when the large sample method is appropriate for either a confidence interval or a significance test.
- Find the sample size needed for a desired margin of error for the difference in proportions.
- Find the sample size needed for a significance test comparing two proportions.

Because comparative studies are so common, we now discuss how to compare the proportions of two groups (such as men and women) that have some characteristic. The setting and notation are very similar to the material in Section 8.2, where we compared two means. We call the two groups being compared Population 1 and Population 2 and the two population proportions of "successes" p_1 and p_2. The data consist of two independent SRSs. The sample sizes are n_1 for Population 1 and n_2 for Population 2. The proportion of successes in each sample estimates the corresponding population proportion. Here is the notation we will use in this section:

Population	Population proportion	Sample size	Count of successes	Sample proportion
1	p_1	n_1	X_1	$\hat{p}_1 = X_1/n_1$
2	p_2	n_2	X_2	$\hat{p}_2 = X_2/n_2$

To compare the two unknown population proportions, start with the difference between the two sample proportions,

$$D = \hat{p}_1 - \hat{p}_2$$

When both sample sizes are sufficiently large, the sampling distribution of the difference D is approximately Normal. This is because each of the sample proportions is approximately Normal and the difference between two Normal random variables is Normal. What are the mean and the standard deviation of D? Each of the two $\hat{p}$'s has the mean and standard deviation given in the box on page 507. Because the two samples are independent, the two $\hat{p}$'s are also independent. We can apply the rules for means and variances of sums of random variables. Here is the result, which is summarized in Figure 10.5. You are asked to go through the details in Exercise 10.49.

rules for means and variances, p. 236, 240–241

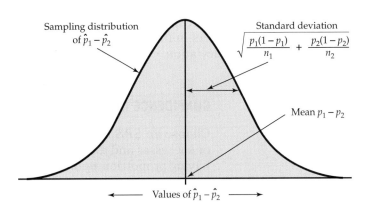

FIGURE 10.5 The sampling distribution of the difference between two sample proportions is approximately Normal. The mean and standard deviation are found from the two population proportions of successes, p_1 and p_2.

SAMPLING DISTRIBUTION OF $\hat{p}_1 - \hat{p}_2$

Choose independent SRSs of sizes n_1 and n_2 from two populations with proportions p_1 and p_2 of successes. Let $D = \hat{p}_1 - \hat{p}_2$ be the difference between the two sample proportions of successes. Then

- As both sample sizes increase, the sampling distribution of D becomes **approximately Normal.**
- The **mean** of the sampling distribution is $p_1 - p_2$.
- The **standard deviation** of the sampling distribution is

$$\sigma_D = \sqrt{\frac{p_1(1-p_1)}{n_1} + \frac{p_2(1-p_2)}{n_2}}$$

APPLY YOUR KNOWLEDGE

10.47 Rules for means and variances. Suppose $p_1 = 0.4$, $n_1 = 30$, $p_2 = 0.6$, and $n_2 = 40$. Find the mean and the standard deviation of the sampling distribution of $\hat{p}_1 - \hat{p}_2$.

10.48 Effect of the sample sizes. Suppose $p_1 = 0.4$, $n_1 = 120$, $p_2 = 0.6$, and $n_2 = 160$.

(a) Find the mean and the standard deviation of the sampling distribution of $\hat{p}_1 - \hat{p}_2$.

(b) The sample sizes here are four times as large as those in the previous exercise, while the population proportions are the same. Compare the mean and standard deviation for this exercise with those that you found in the previous exercise. What is the effect of multiplying the sample sizes by 4?

10.49 Rules for means and variances. It is quite easy to verify the mean and standard deviation of the difference D.

(a) What are the means and standard deviations of the two sample proportions $\hat{p}_1$ and $\hat{p}_2$? (Look at the box on page 524 if you need to review this.)

(b) Use the addition rule for means of random variables (page 524): what is the mean of $D = \hat{p}_1 - \hat{p}_2$?

(c) The two samples are independent. Use the addition rule for variances of random variables (page 524) to find the variance of D.

Large-sample confidence intervals for a difference in proportions

The large-sample estimate of the difference in two proportions $p_1 - p_2$ is the corresponding difference in sample proportions $\hat{p}_1 - \hat{p}_2$. To obtain a confidence interval for the difference, we once again replace the unknown parameters in the standard deviation by estimates to obtain an estimated standard deviation, or standard error. Here is the confidence interval we want.

CONFIDENCE INTERVAL FOR COMPARING TWO PROPORTIONS

Choose an SRS of size n_1 from a large population having proportion p_1 of successes and an independent SRS of size n_2 from another population having proportion p_2 of successes.

The large-sample estimate of the difference in proportions is

$$D = \hat{p}_1 - \hat{p}_2 = \frac{X_1}{n_1} - \frac{X_2}{n_2}$$

The **standard error of the difference** is

$$SE_D = \sqrt{\frac{\hat{p}_1(1-\hat{p}_1)}{n_1} + \frac{\hat{p}_2(1-\hat{p}_2)}{n_2}}$$

and the **margin of error for confidence level C** is

$$m = z^* SE_D$$

where z^* is the value for the standard Normal density curve with area C between $-z^*$ and z^*. The **large-sample level C confidence interval** for $p_1 - p_2$ is

$$(\hat{p}_1 - \hat{p}_2) \pm m$$

This method is approximately correct when the number of successes and the number of failures in each of the samples are at least 10.

Social Media in the Supply Chain In addition to traditional marketing strategies, marketing through social media has become an increasingly important component of the supply chain. This is particularly true for relatively small companies that do not have large marketing budgets. One study of Austrian food and beverage companies compared the use of audio/video sharing through social media by large and small companies.[10] Companies were classified as small or large based on whether their annual sales were greater than or less than 135 million euros. We use company size as the explanatory variable. It is categorical with two possible values. Media is the response variable with values Yes for the companies who use audio/visual sharing on social media in their supply chain, and No for those that do not.

Here is a summary of the data. We let X denote the count of the number of companies that use audio/visual sharing. ■

Size	n	X	$\hat{p} = X/n$
1 (small companies)	178	150	0.8427
2 (large companies)	52	27	0.5192

The study in Case 10.3 suggests that smaller companies are more likely to use audio/visual sharing through social media than are large companies. Let's explore this possibility using a confidence interval.

EXAMPLE 10.8

Small Companies versus Large Companies First, we find the estimate of the difference:

$$D = \hat{p}_1 - \hat{p}_2 = \frac{X_1}{n_1} - \frac{X_2}{n_2} = 0.8427 - 0.5192 = 0.3235$$

Next, we calculate the standard error:

$$SE_D = \sqrt{\frac{0.8427(1-0.8427)}{178} + \frac{0.5192(1-0.5192)}{52}} = 0.07447$$

For 95% confidence, we use $z^* = 1.96$, so the margin of error is
$$m = z^* \text{SE}_D = (1.96)(0.07447) = 0.1460$$
The large-sample 95% confidence interval is
$$D \pm m = 0.3235 \pm 0.1460 = (0.18, 0.47)$$

With 95% confidence we can say that the difference in the proportions is between 0.18 and 0.47. Alternatively, we can report that the percent usage of audio/visual sharing through social media by smaller companies is about 32% higher than the percent for large companies, with a 95% margin of error of 15%. ∎

JMP and Minitab for Example 10.8 appear in Figure 10.6. Note that JMP uses a slightly different approximation. Other statistical packages provide output that is similar.

In surveys such as this, small companies and large companies typically are not sampled separately. The respondents to a single sample of companies are classified after the fact as small or large. The sample sizes are then random and reflect the characteristics of the population sampled. Two-sample significance tests and confidence intervals are still approximately correct in this situation, even though the two sample sizes were not fixed in advance.

In Example 10.8, we chose small companies to be the first population. Had we chosen large companies as the first population, the estimate of the difference would be negative (−0.3235). Because it is easier to discuss positive numbers, we generally choose the first population to be the one with the higher proportion. The choice does not affect the substance of the analysis.

FIGURE 10.6 JMP and Minitab outputs for Example 10.8.

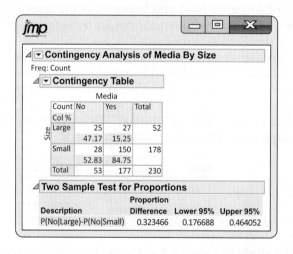
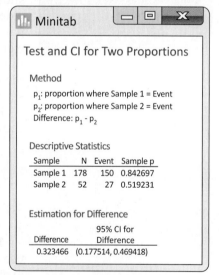

APPLY YOUR KNOWLEDGE

10.50 Sex and commercial preference. A study was designed to compare two energy drink commercials. Each participant was shown the commercials in random order and was asked to select the better one. Commercial A was selected by 85 out of 120 women and 68 out of 130 men. Give an estimate of the difference in sex proportions that favored Commercial A. Also construct a large-sample 95% confidence interval for this difference.

10.51 Sex and commercial preference, revisited. Refer to Exercise 10.50. Construct a 95% confidence interval for the difference in proportions that favor Commercial B. Explain how you could have obtained these results from the calculations you did in Exercise 10.50.

BEYOND THE BASICS

Plus four confidence intervals for a difference in proportions

Just as in the case of estimating a single proportion, a small modification of the sample proportions greatly improves the behavior of the confidence interval when the sample proportions are near 0 or 1.[11] The confidence intervals will be approximately the same as the large-sample confidence intervals when the criteria using those intervals are satisfied. When the criteria are not met, the plus four intervals will still be valid when both sample sizes are at least five.

As before, we add two successes and two failures to the actual data. Here, however, we divide them equally between the two samples. That is, *add one success and one failure to each sample*. Note that this adds 2 to both n_1 and n_2. We then perform the calculations for the large-sample procedure with the modified data. As in the case of a single sample, we use the term **Wilson estimate** for the estimated difference in proportions produced in this way.

Wilson estimates

For example, in Example 10.8, we observed $X_1 = 150$, $n_1 = 178$, $X_2 = 27$, and $n_2 = 52$. For the plus four procedure, we would use $X_1 = 151$, $n_1 = 180$, $X_2 = 28$, and $n_2 = 54$.

APPLY YOUR KNOWLEDGE

10.52 Social media and the supply chain using plus four. Refer to Example 10.8 (page 525), where we computed a 95% confidence interval for the difference in the proportions of small companies and large companies that use audio/visual sharing through social media as part of their supply chain. Redo the computations using the plus four method, and compare your results with those obtained in Example 10.8.

10.53 Social media and the supply chain using plus four. Refer to the previous exercise and to Example 10.8. Suppose that the sample sizes were smaller, but that the proportions remained approximately the same. Specifically, assume that 17 out of 20 small companies used social media and 13 out of 25 large companies used social media. Compute the plus four interval for 95% confidence. Then, compute the corresponding z interval and compare the results.

10.54 Sex and commercial preference. Refer to Exercises 10.50 and 10.51, where you analyzed data about sex and the preference for one of two commercials. The study also asked the same subjects to give a preference for two other commercials, C and D. Suppose that 92 women preferred Commercial C and that 125 men preferred Commercial C.

(a) The large-sample confidence interval for comparing two proportions should not be used for these data. Why?

(b) Compute the plus four confidence interval for the difference in proportions.

Significance tests

Although we prefer to compare two proportions by giving a confidence interval for the difference between the two population proportions, it is sometimes useful to test the null hypothesis that the two population proportions are the same (H_0: $p_1 = p_2$ or H_0: $p_1 - p_2 = 0$).

As we've done for other tests, we standardize $D = \hat{p}_1 - \hat{p}_2$ by subtracting its mean $p_1 - p_2$ and then dividing by its standard deviation

$$\sigma_D = \sqrt{\frac{p_1(1-p_1)}{n_1} + \frac{p_2(1-p_2)}{n_2}}$$

If n_1 and n_2 are large, this standardized difference is approximately $N(0,1)$.

When constructing a confidence interval, we use sample estimates in place of the unknown population proportions p_1 and p_2 in the expression for σ_D. Although this approach would lead to a valid significance test, we follow the more common practice of replacing the unknown σ_D with an estimate that takes into account the null hypothesis that $p_1 = p_2$. If these two proportions are equal, we can view all the data as coming from a single population. Let p denote the common value of p_1 and p_2. The standard deviation of $D = \hat{p}_1 - \hat{p}_2$ is then

$$\sigma_{Dp} = \sqrt{\frac{p(1-p)}{n_1} + \frac{p(1-p)}{n_2}}$$

$$= \sqrt{p(1-p)\left(\frac{1}{n_1} + \frac{1}{n_2}\right)}$$

The subscript on σ_{Dp} reminds us that this is the standard deviation under the special condition that the two populations share a common proportion p of successes.

We estimate the common value of p by the overall proportion of successes in the two samples:

$$\hat{p} = \frac{\text{number of successes in both samples}}{\text{number of observations in both samples}} = \frac{X_1 + X_2}{n_1 + n_2}$$

pooled estimate of p This estimate of p is called the **pooled estimate** because it combines, or pools, the information from two independent samples.

To estimate the standard deviation of D, substitute $\hat{p}$ for p in the expression for σ_{Dp}. The result is a standard error for D under the condition that the null hypothesis H_0: $p_1 = p_2$ is true. The test statistic uses this standard error to standardize the difference between the two sample proportions.

SIGNIFICANCE TESTS FOR COMPARING TWO PROPORTIONS

Choose an SRS of size n_1 from a large population having proportion p_1 of successes and an independent SRS of size n_2 from another population having proportion p_2 of successes. To test the hypothesis

$$H_0: p_1 = p_2$$

compute the **z statistic**

$$z = \frac{\hat{p}_1 - \hat{p}_2}{SE_{Dp}}$$

where the **pooled standard error** is

$$SE_{Dp} = \sqrt{\hat{p}(1-\hat{p})\left(\frac{1}{n_1} + \frac{1}{n_2}\right)}$$

based on the **pooled estimate** of the common proportion of successes

$$\hat{p} = \frac{X_1 + X_2}{n_1 + n_2}$$

In terms of a standard Normal random variable Z, the P-value for a test of H_0 against

$H_a: p_1 > p_2$ is $P(Z \geq z)$

$H_a: p_1 < p_2$ is $P(Z \leq z)$

$H_a: p_1 \neq p_2$ is $2P(Z \geq |z|)$

This P-value is approximately correct when the number of successes and the number of failures in each of the samples are at least five.

EXAMPLE 10.9

CASE 10.3 **Social Media in the Supply Chain** Example 10.8 (page 525) analyzes data on the use of audio/visual sharing through social media by small and large companies. Are the proportions of social media users the same for the two types of companies? Here is the data summary:

Size	n	X	$\hat{p} = X/n$
1 (small companies)	178	150	0.8427
2 (large companies)	52	27	0.5192

The sample proportions are certainly quite different, but we need a significance test to verify that the difference is too large to easily result from the role of chance in choosing the sample. Formally, we compare the proportions of social media users in the two populations (small companies and large companies) by testing the hypotheses

$$H_0: p_1 = p_2$$
$$H_a: p_1 \neq p_2$$

The pooled estimate of the common value of p is

$$\hat{p} = \frac{150 + 27}{178 + 52} = \frac{177}{230} = 0.7696$$

This is just the proportion of social media users in the entire sample.

First, we compute the standard error

$$SE_{Dp} = \sqrt{(0.7696)(1 - 0.7696)\left(\frac{1}{178} + \frac{1}{52}\right)} = 0.0664$$

and then we use this in the calculation of the test statistic

$$z = \frac{\hat{p}_1 - \hat{p}_2}{SE_{Dp}} = \frac{0.8427 - 0.5192}{0.0664} = 4.87$$

The difference in the sample proportions is almost five standard deviations away from zero. The P-value is $2P(Z \geq 4.87)$. In Table A, the largest entry we have is $z = 3.49$ with $P(Z \leq 3.49) = 0.9998$. So, $P(Z > 3.49) = 1 - 0.9998 = 0.0002$. Therefore, we can conclude that $P < 2 \times 0.0002 = 0.0004$. Our report: 84% of small companies use audio/visual sharing through social media versus 52% of large companies; the difference is statistically significant ($z = 4.87$, $P < 0.0004$). ∎

Figure 10.7 gives the JMP and Minitab outputs for Example 10.9. Carefully examine the output to find all the important pieces that you would need to report the results of the analysis and to draw a conclusion. Note that the slight differences in results are due to the use of different approximations.

Some experts would expect the usage of social media would be greater for small companies than for large companies because small companies do not have the resources for large, expensive marketing efforts. These experts might choose the one-sided alternative $H_a: p_1 > p_2$. Because the z statistic is positive, the P-value would be half of the value obtained for the two-sided test. With a z statistic so large, this distinction is of no practical importance.

FIGURE 10.7 JMP and Minitab outputs for Example 10.9.

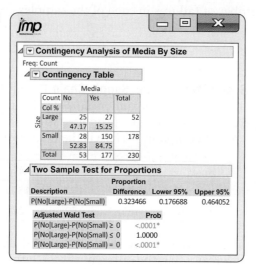
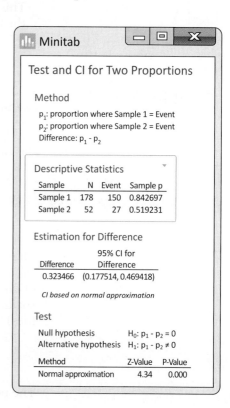

APPLY YOUR KNOWLEDGE

10.55 Sex and commercial preference. Refer to Exercise 10.50 (page 526), which compared women and men with regard to their preference for one of two commercials.

(a) State appropriate null and alternative hypotheses for this setting. Give a justification for your choice.

(b) Use the data given in Exercise 10.50 (page 526) to perform a two-sided significance test. Give the test statistic and the P-value.

(c) Summarize the results of your significance test.

10.56 What about preference for Commercial B? Refer to Exercise 10.51, where we changed the roles of the two commercials in our analysis. Answer the questions given in the previous exercise for the data altered in this way. Describe the results of the change.

Choosing a sample size for two sample proportions

In Section 10.1, we studied methods for determining the sample size using two settings. First, we used the margin of error for a confidence interval for a single proportion as the criterion for choosing n (page 507). Second, we used the power of the significance test for a single proportion as the determining factor (page 511). We follow the same approach here for comparing two proportions.

Using the margin of error

Recall that the large-sample estimate of the difference in proportions is

$$D = \hat{p}_1 - \hat{p}_2 = \frac{X_1}{n_1} - \frac{X_2}{n_2}$$

and its standard error is

$$SE_D = \sqrt{\frac{\hat{p}_1(1-\hat{p}_1)}{n_1} + \frac{\hat{p}_2(1-\hat{p}_2)}{n_2}}$$

The margin of error for confidence level C is

$$m = z^* SE_D$$

where z^* is the value for the standard Normal density curve with area C between $-z^*$ and z^*.

SAMPLE SIZE FOR DESIRED MARGIN OF ERROR

The level C confidence interval for a difference in two proportions will have a margin of error approximately equal to a specified value m when the sample size for each of the two proportions is

$$n = \left(\frac{z^*}{m}\right)^2 (p_1^*(1-p_1^*) + p_2^*(1-p_2^*))$$

Here, z^* is the critical value for confidence C, and p_1^* and p_2^* are guessed values for p_1 and p_2, the proportions of successes in the population.

The margin of error will be less than or equal to m if p_1^* and p_2^* are chosen to be 0.5. The common sample size required is then given by

$$n = \left(\frac{1}{2}\right)\left(\frac{z^*}{m}\right)^2$$

Note that to use the confidence interval, which is based on the Normal approximation, we still require that the number of successes and the number of failures in each of the samples be at least 10.

EXAMPLE 10.10

Confidence Interval–Based Sample Sizes for Preferences of Women and Men
Consider the setting in Exercise 10.50, where we compared the preferences of women and men for two commercials. Suppose we want to do a study in which we perform a similar comparison using a 95% confidence interval that will have a margin of error of 0.2 or less. What should we choose for our sample size? Using $m = 0.2$ and z^* in our formula, we have

$$n = \left(\frac{1}{2}\right)\left(\frac{z^*}{m}\right)^2 = \left(\frac{1}{2}\right)\left(\frac{1.96}{0.2}\right)^2 = 48.02$$

We would include 48 women and 48 men in our study. ∎

Note that we have rounded the calculated value, 48.02, down because it is very close to 48. In this case our margin of error could be slightly larger than 0.2. If we always round up, we will guarantee that our margin of error will always be less than or equal to the value of m that we used in our calculations when we assume that the two proportions are 0.05.

APPLY YOUR KNOWLEDGE

10.57 What would the margin of error be? Consider the setting in Example 10.10.

(a) Compute the margins of error for $n_1 = 48$ and $n_2 = 48$ for each of the following scenarios: $p_1 = 0.4$, $p_2 = 0.5$; $p_1 = 0.7$, $p_2 = 0.6$; and $p_1 = 0.3$, $p_2 = 0.6$.

(b) If you think that any of these scenarios is likely to fit your study, should you reconsider your choice of $n_1 = 48$ and $n_2 = 48$? Explain your answer.

Use the power of the significance test

When we studied using power to compute the sample size needed for a significance test for a single proportion, we used software. We will do the same for the significance test for comparing two proportions.

Some software allows us to consider significance tests that are a little more general than the version we studied in this section. Specifically, we used the null hypothesis $H_0: p_1 = p_2$, which we can rewrite as $H_0: p_1 - p_2 = 0$. The generalization allows us to consider the null hypothesis $H_0: p_1 - p_2 = \Delta_0$, where Δ_0 can be different from zero. To compute the sample size for the significance test, we will need to specify $\Delta_0 = 0$.

Here is a summary of the inputs needed for software to perform the calculations:

- the value of Δ_0 in the null hypothesis $H_0: p_1 - p_2 = \Delta_0$

- the alternative hypothesis, two-sided ($H_a: \Delta_0 \neq 0$) or one-sided ($H_a: \Delta_0 > 0$ or $H_a: \Delta_0 < 0$)

- values for p_1 and p_2 in the alternative hypothesis

- the significance level (α, the probability of rejecting the null hypothesis when it is true); usually we choose 5% ($\alpha = 0.05$) for the significance level

- power (probability of rejecting the null hypothesis when it is false); usually we choose 80% (0.80) for power

EXAMPLE 10.11

Sample Sizes for Preferences of Women and Men Refer to Example 10.10, where we used the margin of error to find the sample sizes for comparing the preferences of women and men for two commercials. Let's find the sample sizes required for a significance test that the two proportions who prefer Commercial A are equal ($\Delta_0 = 0$) using a two-sided alternative with $p_1 = 0.65$ and $p_2 = 0.4$, $\alpha = 0.05$, and 80% (0.80) power. Outputs from JMP and Minitab are given in Figure 10.8. The JMP output indicates that we need $n_1 = 60$ women and $n_2 = 60$ men for our study. Using a slightly different calculation, Minitab reports that 62 subjects of each sex are needed. ∎

FIGURE 10.8 JMP and Minitab outputs for Example 10.11.

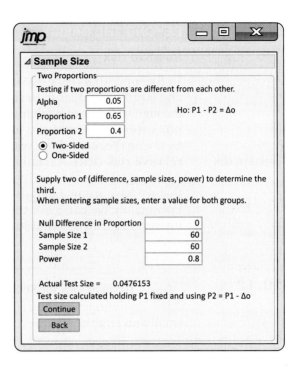

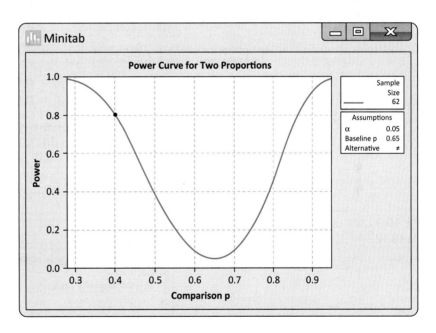

Note that the Minitab output gives the power curve for different alternatives. All of these have $p_1 = 0.65$, while p_2, which Minitab calls the "Comparison p," varies from about 0.3 to 0.9. We see that the power is essentially 100% (1) at these extremes. It is 0.05, the significance level, at $p_2 = 0.65$, which corresponds to the null hypothesis.

APPLY YOUR KNOWLEDGE

10.58 Find the sample sizes. Consider the setting in Example 10.11. Change p_1 to 0.75 and p_2 to 0.35. Find the required sample sizes.

BEYOND THE BASICS

Relative risk

In Example 10.8 (page 525), we compared the proportions of small and large companies with respect to their use of audio/visual sharing through social media, giving a confidence interval for the *difference* of proportions. Alternatively, we might choose to make this comparison by giving the *ratio* of the two proportions. This ratio is often called the **relative risk** (RR). A relative risk of 1 means that the proportions $\hat{p}_1$ and $\hat{p}_2$ are equal. Confidence intervals for relative risk apply the principles that we have studied, but the details are somewhat complicated. Fortunately, we can leave the details to software and concentrate on interpreting and communicating the results.

EXAMPLE 10.12

Relative Risk for Social Media in the Supply Chain The following table summarizes the data on the proportions of social media use for small and large companies:

Size	n	X	$\hat{p} = X/n$
1 (small companies)	178	150	0.8427
2 (large companies)	52	27	0.5192

The relative risk for this sample is

$$\text{RR} = \frac{\hat{p}_1}{\hat{p}_2} = \frac{0.8427}{0.5192} = 1.62$$

Confidence intervals for the relative risk entire population of food and beverage companies are based on this sample relative risk. Figure 10.9 gives output from JMP. Our summary: small companies are about 1.62 times as likely to use audio/visual sharing through social media as part of their supply chain as large companies are; the 95% confidence interval is (1.24, 2.12). ■

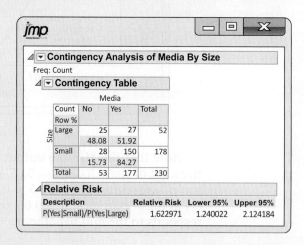

FIGURE 10.9 JMP output for Example 10.12.

In Example 10.12, the confidence interval is clearly not symmetric about the estimate; that is, 1.62 is much closer to 1.24 than it is to 2.12. This is true, in general, for confidence intervals for relative risk.

Relative risk, comparing proportions by a ratio rather than by a difference, is particularly useful when the proportions are small. This way of describing results is often used for epidemiology and medical studies.

SECTION 10.2 SUMMARY

- The **estimate of the difference in two population proportions** is

$$D = \hat{p}_1 - \hat{p}_2$$

where

$$\hat{p}_1 = \frac{X_1}{n_1} \quad \text{and} \quad \hat{p}_2 = \frac{X_2}{n_2}$$

- The **standard error of the difference** is

$$\text{SE}_D = \sqrt{\frac{\hat{p}_1(1-\hat{p}_1)}{n_1} + \frac{\hat{p}_2(1-\hat{p}_2)}{n_2}}$$

and the **margin of error for confidence level C** is

$$m = z^* \text{SE}_D$$

where z^* is the value for the standard Normal density curve with area C between $-z^*$ and z^*.

- The **z large-sample level C confidence interval** for the difference in two proportions $p_1 - p_2$ is

$$(\hat{p}_1 - \hat{p}_2) \pm m$$

We recommend using this method when the number of successes and the number of failures in both samples are at least 10.

- Significance tests of H_0: $p_1 = p_2$ use the **z statistic**

$$z = \frac{\hat{p}_1 - \hat{p}_2}{\text{SE}_{Dp}}$$

which has the $N(0, 1)$ distribution under the null hypothesis. In this statistic,

$$\text{SE}_{Dp} = \sqrt{\hat{p}(1-\hat{p})\left(\frac{1}{n_1} + \frac{1}{n_2}\right)}$$

where $\hat{p}$ is the **pooled estimate** of the common value of p_1 and p_2,

$$\hat{p} = \frac{X_1 + X_2}{n_1 + n_2}$$

We recommend using this test when the number of successes and the number of failures in each of the samples are at least 5.

- The **sample sizes** for each of the two proportions needed for a specified value m of the margin of error for the difference in two proportions are

$$n = \left(\frac{z^*}{m}\right)^2 \left(p_1^*(1-p_1^*) + p_2^*(1-p_2^*)\right)$$

Here z^* is the critical value for confidence C, and p_1^* and p_2^* are guessed values for p_1 and p_2, the proportions of successes in the future sample.

- The margin of error will be less than or equal to m if p_1^* and p_2^* are chosen to be 0.5. The common **sample size** required is then given by

$$n = \left(\frac{1}{2}\right)\left(\frac{z^*}{m}\right)^2$$

SECTION 10.2 EXERCISES

For Exercises 10.47 to 10.49, see page 524; for 10.50 and 10.51, see page 526; for 10.52 to 10.54, see page 527; for 10.55 and 10.56, see page 530; for 10.57, see page 532; and for 10.58, see page 533.

10.59 To tip or not to tip. A study of tipping behaviors examined the relationship between the color of the shirt worn by the server and whether the customer left a tip.[12] There were 418 male customers in the study; 40 of the 69 who were served by a server wearing a red shirt left a tip. Of the 349 who were served by a server wearing a different colored shirt, 130 left a tip.

(a) What is the explanatory variable for this setting? Explain your answer.

(b) What is the response variable for this setting? Explain your answer.

(c) What are the parameters for this study? Explain your answer.

10.60 Confidence interval for tipping. Refer to the previous exercise.

(a) Find the proportion of tippers for the red-shirted servers and the proportion of tippers for the servers with other colored shirts.

(b) Find a 95% confidence interval for the difference in proportions.

(c) Write a short paragraph summarizing your results.

10.61 Significance test for tipping. Refer to the previous two exercises.

(a) Give a null hypothesis for this setting in terms of the parameters. Explain the meaning of the null hypothesis in simple terms.

(b) Give an alternative hypothesis for this setting in terms of the parameters. Explain the meaning of the alternative hypothesis in simple terms. Give a reason for your choice of this particular alternative hypothesis.

(c) Are the conditions satisfied for the use of the significance test based on the Normal distribution? Explain your answer.

10.62 Significance test details for tipping. Refer to the previous exercise.

(a) What is the distribution of the test statistic if the null hypothesis is true?

(b) Find the test statistic.

(c) Find the P-value.

(d) Use a sketch of a Normal distribution to explain the interpretation of the P-value that you found in part (c).

(e) Write a brief summary of the results of your significance test. Include enough details so that someone reading your summary could reproduce all your results.

10.63 Draw a picture. Suppose that there are two binomial populations. For the first, the true proportion of successes is 0.4; for the second, it is 0.5. Consider taking independent samples from these populations, 50 from the first and 40 from the second.

(a) Find the mean and the standard deviation of the distribution of $\hat{p}_1 - \hat{p}_2$.

(b) This distribution is approximately Normal. Sketch this Normal distribution, and mark the location of the mean.

(c) Find a value d for which the probability is 0.95 that the difference in sample proportions is within $\pm d$. Mark these values on your sketch.

10.64 What's wrong? For each of the following, explain what is wrong and why.

(a) A z statistic is used to test the null hypothesis that $p_1 \neq p_2$.

(b) If two sample sizes are equal, then the sample proportions are equal.

(c) A 95% confidence interval for the difference in two proportions includes errors due to poorly worded questions.

10.65 College student summer employment. Suppose that 84% of college men and 86% of college women were employed last summer. A sample survey interviews SRSs of 200 college men and 200 college women. The two samples are independent.

(a) What is the approximate distribution of the proportion $\hat{p}_W$ of women who worked last summer? What is the approximate distribution of the proportion $\hat{p}_M$ of men who worked?

(b) The survey wants to compare men and women. What is the approximate distribution of the difference in the proportions who worked, $\hat{p}_M - \hat{p}_W$?

10.66 A corporate liability trial. A major court case on liability for contamination of groundwater took place in the town of Woburn, Massachusetts. A town well in Woburn was contaminated by industrial chemicals. During the period that residents drank water from this well, there were 16 birth defects among 414 births. In years when the contaminated well was shut off and water was supplied from other wells, there were three birth defects among 228 births. The plaintiffs suing the firms responsible for the contamination claimed that these data show that the rate of birth defects was higher when the contaminated well was in use.[13] How statistically significant is the evidence? Be sure to state what assumptions your analysis requires and to what extent these assumptions seem reasonable in this case.

CASE 10.2 **10.67 Natural versus artificial Christmas trees.** In the Christmas tree survey introduced in Case 10.2 (page 515), respondents who had a tree during the holiday season were asked whether the tree was natural or artificial. Respondents were also asked if they lived in an urban area or in a rural area. Of the 421 households displaying a Christmas tree, 160 lived in rural areas and 261 were urban residents. The tree growers want to know if there is a difference in preference for natural trees versus artificial trees between urban and rural households. Here are the data:

Population	n	X(natural)
1 (rural)	160	64
2 (urban)	261	89

(a) Give the null and alternative hypotheses that are appropriate for this problem, assuming we have no prior information suggesting that one population would have a higher preference than the other.

(b) Test the null hypothesis. Give the test statistic and the P-value, and summarize the results.

(c) Give a 90% confidence interval for the difference in proportions.

10.68 Summer employment of college students. A university financial aid office polled an SRS of undergraduate students to study their summer employment. Not all students were employed the previous summer. Here are the results for men and women:

	Men	Women
Employed	622	533
Not employed	58	82
Total	680	615

(a) Is there evidence that the proportion of male students employed during the summer differs from the proportion of female students who were employed? State H_0 and H_a, compute the test statistic, and give the P-value.

(b) Give a 95% confidence interval for the difference between the proportions of male and female students who were employed during the summer. Does the difference seem practically important to you?

10.69 Effect of the sample size. Refer to the previous exercise. Similar results from a smaller number of students may not have the same statistical significance. Specifically, suppose that 124 of 136 men surveyed were employed and 106 of 122 women surveyed were employed. The sample proportions are essentially the same as in the earlier exercise.

(a) Compute the z statistic for these data, and report the P-value. What do you conclude?

(b) Compare the results of this significance test with your results in Exercise 10.69. What do you observe about the effect of the sample size on the results of these significance tests?

10.70 Find the sample size. Consider testing the null hypothesis that two proportions are equal versus the two-sided alternative with $\alpha = 0.05$, 90% power, and equal sample sizes in the two groups.

(a) For each of the following situations, find the required sample size: (i) $p_1 = 0.1$ and $p_2 = 0.2$, (ii) $p_1 = 0.2$ and $p_2 = 0.3$, (iii) $p_1 = 0.3$ and $p_2 = 0.4$, (iv) $p_1 = 0.4$ and $p_2 = 0.5$, (v) $p_1 = 0.5$ and $p_2 = 0.6$, (vi) $p_1 = 0.6$ and $p_2 = 0.7$, (vii) $p_1 = 0.7$ and $p_2 = 0.8$, and (viii) $p_1 = 0.8$ and $p_2 = 0.9$.

(b) Write a short summary describing your results.

CHAPTER 10 REVIEW EXERCISES

10.71 The Internet of Things. The Internet of Things (IoT) refers to connecting computers, phones, and many other types of devices so that they can communicate and interact with each other.[14] A Pew Internet study asked a panel of 1606 experts whether they thought that the IoT would have "widespread and beneficial effects on the everyday lives of the public by 2025." Eighty-three percent of the panel gave a positive response.[15]

(a) How many of the experts responded Yes to the question? Show your work.

(b) Describe the population proportion for this setting.

(c) What is the sample proportion? Show your work.

(d) What is the standard error for the sample proportion? Show your work.

(e) What is the margin of error for 95% confidence? Show your work.

(f) Give the 95% confidence interval for the population proportion.

10.72 A new Pew study of the Internet of Things. Refer to the previous exercise. Suppose Pew would like to do a new study next year to see if expert opinion has changed since the original study was performed. Assume that a new panel of 1606 experts would be asked the same question.

(a) Using 95% confidence, compute the margin of error for the difference in proportions between the two studies for each of the following possible values of the sample proportion for the new study: (i) 0.73, (ii) 0.78, (iii) 0.82, (iv) 0.84, (v) 0.88, and (vi) 0.93.

(b) Summarize your results with a graph.

(c) Write a short summary describing what you have found in this exercise.

10.73 Find the power. Refer to the previous exercise. Consider performing a significance test to compare the population proportions for the two studies. Use $\alpha = 0.05$ and a two-sided alternative. Assume that the proportion in the first study is 0.83.

(a) Find the power of the significance test for each of the following population proportions in the new study: (i) 0.73, (ii) 0.78, (iii) 0.82, (iv) 0.84, (v) 0.88, and (vi) 0.93.

(b) Display your results with a graph.

(c) Write a short summary describing what you have found in this exercise.

10.74 Worker absences and the bottom line. A survey of 1286 companies found that 32% of them did not measure how worker absences affect their company's bottom line.[16]

(a) How many of the companies responded that they do measure how worker absences affect their company's bottom line? Show your work.

(b) Describe the population proportion for this setting.

(c) What is the sample proportion? Show your work.

(d) What is the standard error for the sample proportion? Show your work.

(e) What is the margin of error for 95% confidence? Show your work.

(f) Give the 95% confidence interval for the population proportion.

10.75 The new worker absence study. Refer to the previous exercise. Suppose you would like to do a new study next year to see if there has been a change in the percent of companies that do not measure how worker absences affect their company's bottom line. Assume that a new sample of 1286 companies will be used for the new study.

(a) Compute the 95% margin of error of the difference in proportions between the two studies for each of the following possible values of the sample proportion for the new study: (i) 0.32, (ii) 0.37, (iii) 0.42, and (iv) 0.52.

(b) Summarize your results with a graph.

(c) Write a short summary describing what you have found in this exercise.

10.76 Find the power. Refer to the previous exercise. Consider performing a significance test to compare the population proportions for the two studies. Use $\alpha = 0.05$ and a one-sided alternative. Assume that the proportion for the first study is 0.32.

(a) Find the power of the significance test for each of the following population proportions in the new study: (i) 0.32, (ii) 0.37, (iii) 0.42, and (iv) 0.52.

(b) Display your results with a graph.

(c) Write a short summary describing what you have found in this exercise.

10.77 Worker absences and the bottom line. Refer to Exercises 10.74 through 10.76. Suppose that the companies participating in the new studies are the same as the companies in the original study. Would your answers to any of the parts of Exercises 10.75 and 10.76 change? Explain your answer.

10.78 Effect of the Fox News app. A survey that sampled smartphone users quarterly compared the proportions of smartphone users who visited the Fox News website before and after the introduction of a Fox News app. A report of the survey stated that 17.6% of smartphone users visited the Fox News website before the introduction of the app versus 18.5% of users after the app was introduced.[17] Assume that the sample sizes were 5600 for each condition.

(a) What is the explanatory variable for this study?

(b) What is the response variable for this study?

(c) Give a 95% confidence interval for the difference in the proportions.

10.79 A significance test for the Fox News app. Refer to the previous exercise.

(a) State an appropriate null hypothesis for this setting.

(b) Give an alternative hypothesis for this setting. Explain the meaning of the alternative hypothesis in simple terms, and explain why you chose this particular alternative hypothesis.

(c) Are the conditions satisfied for the use of the significance test based on the Normal distribution?

10.80 Perform the significance test for the Fox News app. Refer to the previous exercise.

(a) What is the test statistic?

(b) What is the distribution of the test statistic if the null hypothesis is true?

(c) Find the P-value.

(d) Use a sketch of the Normal distribution to explain the interpretation of the P-value that you calculated in part (c).

(e) Write a brief summary of the results of your significance test. Include enough detail so that someone reading your summary could reproduce all your results.

10.81 Power for a similar significance test. Refer to Exercises 10.78 through 10.80. Suppose you were planning a similar study for a different app. Assume that the population proportions are the same as the sample proportion in the Fox News study. The numbers of smartphone users will be the same for the before and after groups. Assume 80% power with a test using $\alpha = 0.05$. Find the number of users needed for each group.

10.82 What would the margin of error be? Refer to the previous exercise. Using the sample sizes for the two groups that you found there, what would you expect the

95% margin of error to be for the estimated difference between the two proportions? For your calculations, assume that the sample proportions would be the same as given for the original setting in Exercise 10.78.

10.83 The parrot effect: how to increase your tips. An experiment examined the relationship between tips and server behavior in a restaurant.[18] In one condition, the server repeated the customer's order word for word, while in the other condition, the orders were not repeated. Tips were received in 47 of the 60 trials under the repeat condition and in 31 of the 60 trials under the no-repeat condition.

(a) Find the sample proportions and compute a 95% confidence interval for the difference in population proportions.

(b) Use a significance test to compare the two conditions. Summarize the results.

10.84 The parrot effect: how to increase your tips. Refer to the previous exercise.

(a) The study was performed in a restaurant in the Netherlands. Two waitresses performed the tasks. How do these facts relate to the type of conclusions that can be drawn from this study? Do you think that the parrot effect would apply in other countries?

(b) Design a study to test the parrot effect in a setting that is familiar to you. Be sure to include complete details about how the study will be conducted and how you will analyze the results.

10.85 Does the new process give a better product? Eight percent of the products produced by an industrial process over the past several months fail to conform to the specifications. The company modifies the process in an attempt to reduce the rate of nonconformities. In a trial run, the modified process produces 13 nonconforming items out of a total of 200 produced. Do these results demonstrate that the modification is effective? Support your conclusion with a clear statement of your assumptions and the results of your statistical calculations.

10.86 How much is the improvement? In the setting of the previous exercise, give a 95% confidence interval for the proportion of nonconforming items for the modified process. Then, taking $p_0 = 0.08$ to be the old proportion and p the proportion for the modified process, give a 95% confidence interval for $p - p_0$.

10.87 Choosing sample sizes. For a single proportion, the margin of error of a confidence interval is largest for any given sample size n and confidence level C when $\hat{p} = 0.5$. This has led us to use $p^* = 0.5$ for planning purposes. A similar result is true for a two-sample problem. The margin of error of the confidence interval for the difference between two proportions is largest when $\hat{p}_1 = \hat{p}_2 = 0.5$. Use these conservative values in the following calculations, and assume that the sample sizes n_1 and n_2 have the common value n. Calculate the margins of error of the 95% confidence intervals for the difference in two proportions for the following choices of n: 25, 50, 100, 220, and 500. Present the results in a table and with a graph. Summarize your conclusions.

10.88 Choosing sample sizes, continued. As the previous exercise noted, using the guessed value 0.5 for both $\hat{p}_1$ and $\hat{p}_2$ gives a conservative margin of error in confidence intervals for the difference between two population proportions. You are planning a survey and will calculate a 95% confidence interval for the difference in two proportions when the data are collected. You would like the margin of error of the interval to be less than or equal to 0.08. You will use the same sample size n for both populations.

(a) How large a value of n is needed?

(b) Give a general formula for n in terms of the desired margin of error m and the critical value z^*.

10.89 Unequal sample sizes. You are planning a survey in which a 95% confidence interval for the difference between two proportions will present the results. You will use the conservative guessed value 0.5 for $\hat{p}_1$ and $\hat{p}_2$ in your planning. You would like the margin of error of the confidence interval to be less than or equal to 0.12. It is very difficult to sample from the first population, so it will be impossible for you to obtain more than 25 observations from this population. Taking $n_1 = 25$, can you find a value of n_2 that will guarantee the desired margin of error? If so, report the value; if not, explain why not.

10.90 Students change their majors. In a random sample of 980 students from a large public university, it was found that 454 of the students changed majors during their college years.

(a) Give a 99% confidence interval for the proportion of students at this university who change majors.

(b) Express your results from part (a) in terms of the *percent* of students who change majors.

(c) University officials are more interested in the *number* of students who change majors than in the proportion. The university has 30,000 undergraduate students. Convert your confidence interval in part (a) to a confidence interval for the number of students who change majors during their college years.

10.91 Statistics and the law. *Casteneda v. Partida* is an important court case in which statistical methods were used as part of a legal argument. When reviewing this case, the Supreme Court used the phrase "two or three standard deviations" as a criterion for statistical significance. This Supreme Court review has served as the basis for many subsequent applications of statistical methods in legal settings. (The two or three standard deviations referred to by the Court are values of the z statistic and correspond to P-values of approximately 0.05 and 0.0026.) In *Casteneda* the plaintiffs alleged that the method for selecting juries in a county in Texas was biased against Mexican Americans.[19] For the period of time at issue, there were 181,535 persons eligible for jury duty, of whom 143,611 were Mexican Americans. Of the 870 people selected for jury duty, 339 were Mexican Americans.

(a) What proportion of eligible jurors were Mexican Americans? Let this value be p_0.

(b) Let p be the probability that a randomly selected juror is a Mexican American. The null hypothesis to be tested is $H_0: p = p_0$. Find the value of $\hat{p}$ for this problem, compute the z statistic, and find the P-value. What do you conclude? (A finding of statistical significance in this circumstance does not constitute proof of discrimination. It can be used, however, to establish a prima facie case. The burden of proof then shifts to the defense.)

(c) We can reformulate this exercise as a two-sample problem. Here, we wish to compare the proportion of Mexican Americans among those selected as jurors with the proportion of Mexican Americans among those not selected as jurors. Let p_1 be the probability that a randomly selected juror is a Mexican American, and let p_2 be the probability that a randomly selected nonjuror is a Mexican American. Find the z statistic and its P-value. How do your answers compare with your results in part (b)?

10.92 The future of gamification as a marketing tool. Gamification is an interactive design that includes rewards such as points, payments, and gifts. A Pew survey of 1021 technology stakeholders and critics was conducted to predict the future of gamification. A report on the survey said that 42% of those surveyed thought that there would be no major increases in gamification by 2020. On the other hand, 53% said that they believed there would be significant advances in the adoption and use of gamification by 2020.[20] Analyze these data using the methods that you learned in this chapter, and write a short report summarizing your work.

10.93 Where do you get your news? A report produced by the Pew Research Center's Project for Excellence in Journalism summarized the results of a survey on how people get their news. Of the 2342 people in the survey who own a desktop or laptop, 1639 reported that they get their news from the desktop or laptop.[21]

(a) Identify the sample size and the count.

(b) Find the sample proportion and its standard error.

(c) Find and interpret the 95% confidence interval for the population proportion.

(d) Are the guidelines for use of the large-sample confidence interval satisfied? Explain your answer.

10.94 Should you bet on Punxsutawney Phil? There is a gathering every year on February 2 at Gobbler's Knob in Punxsutawney, Pennsylvania. A groundhog, always named Phil, is the center of attraction. If Phil sees his shadow when he emerges from his burrow, tradition says that there will be six more weeks of winter. If he does not see his shadow, spring has arrived. How well has Phil predicted the arrival of spring for several years? The National Oceanic and Atmospheric Administration has collected data for the 25 years from 1988 to 2012. For each year, whether Phil saw his shadow is recorded. This is compared with the February temperature for that year, classified as above or below normal. For 18 of the 25 years, Phil saw his shadow, and for six of these years, the temperature was below normal. For the years when Phil did not see his shadow, two of these years had temperatures below normal.[22] Analyze the data, and write a report on how well Phil predicts whether winter is over.

Dean Drobot/Shutterstock

Inference for Categorical Data

Introduction

In Chapter 8 we studied methods for analyzing random variables that were means from one and two populations. Chapter 9 extended these analyses to the case of more than two means using ANOVA. In Chapter 10 we studied methods analyzing categorical random variables with two possible values from one and from two populations. There are times in business, however, when we want to compare categorical variables from more than two or compare categorical variables with more than two outcomes. For example, in the first section of this chapter we extend our analysis methods to handle these situations. In these settings, statistical inference requires a new family of probability distributions, the chi-square distributions.

- IKEA is planning next year's ad campaign and has classified its customers by sex and age, expressed as five age ranges. They would like to study the relationship between these two categorical variables to target their ads in an effective way.

- Whole Foods is considering expanding either its fresh produce or fresh meats section and classified its customers by the frequency they visit Whole Foods—once a week or less, twice a week, three or more times per week—and whether they purchased fresh produce, fresh meat, or neither during their most recent visit.

- Is there a relationship between the flexibility of a company and its competitiveness? A sample of companies is classified into three levels of flexibility and three levels of competitiveness. A team of researchers wants to explore the relationship between these two categorical variables.

CHAPTER 11

CHAPTER OUTLINE

11.1 Inference for Two-Way Tables

11.2 Goodness of Fit

This material also connects with our study of relationships between pairs of variables in Chapters 2 and 4. In particular, the graphical and numerical summaries of Section 2.5 are helpful visualization tools for the inference we study here.

Finally, in a second section we study a single population version of inference for a categorical variable. This is similar to the inference for the mean of a single population that we studied in Section 8.1.

Frequently, data on *categorical variables*, expressed as proportions or percents, are used to make business decisions. For example,

11.1 Inference for Two-Way Tables

When you complete this section, you will be able to:

- Find the joint distribution, the marginal distributions, and the conditional distributions for a two-way table of counts.
- Identify these quantities from software output. Choose appropriate conditional distributions to describe relationships in a two-way table.
- Compute expected counts, the chi-square statistic, and its degrees of freedom.
- Perform a chi-square test using software and use the *P*-value to draw a conclusion.
- For a 2 × 2 table, explain the relationship between the chi-square test and the large-sample test for comparing two proportions.

When we studied inference for two populations in Section 10.2, we recorded the number of observations in each group (n) and the count of those that are "successes" (X).

EXAMPLE 11.1

Sex and Commercial Preference In Exercise 10.50 (page 526) we examined a study where subjects were asked to express a preference for one of two commercials, A and B. The proportions of subjects preferring Commercial A were compared for women and men. Here is the data summary.

Sex	n	X	$\hat{p}$
Women	120	85	$85/120 = 0.78$
Men	130	68	$68/130 = 0.52$

The table gives the number n of women and the number of men in the study. The count X is the number of subjects who preferred Commercial A. To compare women with men, we calculated sample proportions from these counts. Those are shown in the third column. ∎

Two-way tables

Let's now consider summarizing these data in a different way. Rather than recording just the counts of successes (the number of subjects who prefer Commercial A) for women and men, we record counts for both outcomes in a two-way table **two-way table** where the columns of the table represent populations (women, men) and the rows represent the outcomes (prefers A, prefers B).

EXAMPLE 11.2

Sex and Commercial Preference Here is the two-way table of counts, classifying women and men by their commercial preference:

Commercial preferred	Sex		
	Women	Men	Total
A	85	68	153
B	35	62	97
Total	120	130	250

marginal totals

We've included an extra row and column that contain the totals. We call these **marginal totals** because they appear in the margins of our two-way table. Check that this table simply rearranges the information in Exercise 10.50. ■

explanatory and response variable, p. 64
scatterplot, p. 66

Because we are interested in how sex influences commercial preference, we view sex as an explanatory variable and commercial preference as a response variable. This is why we put sex in the columns (like the x axis in a scatterplot) and commercial preference in the rows (like the y axis in a scatterplot).

Be sure that you understand how this table is obtained from the table in Example 11.1. *Most errors in the use of categorical data methods come from a misunderstanding of how these tables are constructed.*

We call this particular two-way table a 2×2 table because there are two rows (A and B for the preferred commercial) and two columns (Women and Men). The advantage of two-way tables is that they can present data for variables having more than two categories. Suppose, for example, that we asked the subjects which of three commercials (A, B, and C) they preferred. The response variable would then have 3 levels, so our table would be 3×2, with three rows and two columns.

marginal distribution, p. 106
conditional distribution, p. 108

Given a two-way table, we can describe the data in a variety of ways. In Section 2.5 and Section 4.3, we discussed these in detail. In reviewing this material, pay particular attention to *marginal distributions,* and *conditional distributions.* Here, we advance from describing data to inference in the setting of two-way tables. Our data are counts of observations, classified according to two categorical variables. The question of interest is whether there is a relation between the row variable and the column variable. For example, is there a relation between sex and commercial preference? In Exercise 10.50 (page 526) we compared the proportions of women and men who preferred Commercial A. We now think about these data from a slightly different point of view: is there a relationship between sex and commercial preference?

We illustrate this progression to inference for two-way tables with data that form a 2×3 table. Keep in mind, however, that the methodology applies to all two-way tables.

CASE 11.1 **Are Flexible Companies More Competitive?** A study designed to address this question examined characteristics of 61 food and beverage companies. Each company was asked to describe its own level of competitiveness and level of flexibility.[1]

Options for competitiveness were "High," "Medium," and "Low." No companies chose the third option, so this categorical variable has two levels. They were given four options for flexibility, but again one option, "No flexibility," was not chosen. Here are the characterizations of the other three options:

FLXCOM

1. Adaptive flexibility: responds to issues eventually

2. Parallel flexibility: identifies issues and responds to them
3. Preemptive flexibility: anticipates issues and responds before they develop into a problem

We can think of this categorical variable to be ordered and measuring the degree of flexibility, with adaptive being the least flexible, followed by parallel, and then preemptive. ■

To start our analysis of the relationship between competitiveness and flexibility we will organize the data in a two-way table. The following example gives the details.

EXAMPLE 11.3

FLXCOM

The Two-Way Table Two categorical variables were measured for each company. Each company was classified according to competitiveness—"High" or "Medium"—and flexibility—"Adaptive," "Parallel," or "Preemptive." The study author described a theory where more flexibility could lead to more competitiveness. Therefore, we treat flexibility as the explanatory variable here and make it the column variable. Here is the 2×3 table of counts with the marginal totals:

	Flexibility			
Competitiveness	Adaptive	Parallel	Preemptive	Total
Medium	12	21	3	36
High	2	15	8	25
Total	14	36	11	61

■

The variable Competitiveness has two outcomes and the variable Flexibility has three outcomes. Each combination of outcomes for the two categorical variables defines a **cell**. A two-way table with r rows and c columns contains rc cells. The 2×3 table in Example 11.3 has six cells.

The entries in the two-way table of Example 11.3 are the observed, or sample, counts of companies in each category. For example, there were 12 adaptive companies that were medium competitiveness and two adaptive companies that were high competitiveness. The table includes the marginal totals, calculated by summing over the rows or columns. The grand total, $n = 61$, is the sum of the row totals and is also the sum of the column totals. It is the total number of companies in the study.

In this study, we have data on two variables for a single sample of 61 companies. The same table might also have arisen from two separate samples, one from medium competitive companies and the other from highly competitive companies. Fortunately, the same inference applies in both cases. When we studied relationships between quantitative variables in Chapter 2, we noted that not all relationships involve an explanatory variable and a response variable. The same is true for categorical variables that we study here. Two-way tables can be used to display the relationship between any two categorical variables.

APPLY YOUR KNOWLEDGE

CASE 10.3 **11.1 Small companies versus large companies.** In Case 10.3 (page 525) we analyzed data from a study which compared the use of social media in the supply chain by small and large companies. The study included 178 small companies and 52 large companies. Among small companies, 150 used audio/visual sharing on social media in their supply chain. The corresponding number for large companies was 27.

(a) For these data, should one of these categorical variables be the explanatory variable and the other be the response variable? If yes, specify the explanatory variable. Regardless, give a reason for your answer. Display these data using an $r \times c$ table.

(b) How many cells will the $r \times c$ table have?

(c) Add the marginal totals to your table.

11.2 A reduction in force. A human resources manager wants to assess the impact of a planned reduction in force (RIF) on employees over the age of 40. Various laws state that discrimination against this group is illegal. For this reason we use age as a categorical variable with two possible values: 40 or less and over 40.) The company has 838 employees over 40 and 645 who are 40 years of age or less. The current plan for the RIF will terminate 130 employees: 94 who are over 40, and 36 who are 40 or less. Display these data in a two-way table. (Be careful. Remember that each employee should be counted in exactly one cell.)

Describing relations in two-way tables

To describe relations between categorical variables, we compute and compare percents. Section 2.5 (page 105) discusses methods for describing relationships in two-way tables. You should review that material now if you have not already studied it.

Analysis of two-way tables in practice uses statistical software to carry out the arithmetic. We use output from some typical software packages for the data of Case 11.1 to describe inference for two-way tables.

The count in each cell of a two-way table can be viewed as a percent of the grand total, of the row total, or of the column total. In the first case, we are describing the *joint distribution* of the two variables; in the other two cases, we are examining the *conditional distributions*.

joint and conditional probability, p. 201

We learned many of the ideas related to conditional distributions when we studied conditional probability in Section 4.3. When analyzing data, you should use the context of the problem to decide which percents are most appropriate. *Software usually prints out all three, but usually, not all are important for a specific problem.*

EXAMPLE 11.4

FLXCOM

CASE 11.1 **Software Output** Figure 11.1 shows the output from JMP and Minitab for the data of Case 11.1. We named the variables Competitiveness and Flexibility. The two-way table appears in the outputs in expanded form. Each cell contains five entries. They appear in different orders or with different labels, but the two outputs contain the same information. The count is the first cell entry in both outputs. The row and column totals appear in the margins, just as in Example 11.3. The cell count as a percent of the row total is labeled as "Row %," or "% of Row." The row % for the cell with the count for High Competitiveness and Preemptive Flexibility is 8/25, or 32%. Similarly, the cell count as a percent of the column total is also given. For this cell the column percent is 72.73%.

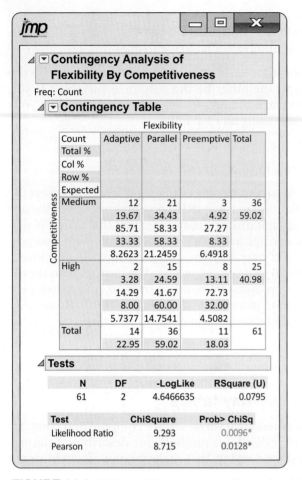

FIGURE 11.1 JMP and Minitab outputs, Example 11.4.

Another entry is the cell count divided by the total number of observations (the joint distribution). Often, this is not very useful and tends to clutter up the output. We discuss the last entry, "Expected count," and other parts of the output in Examples 11.5 and 11.6. ∎

In Case 11.1, we are interested in comparing competitiveness for the three levels of flexibility. We examine the column percents to make this comparison. Here they are, rounded from the output for clarity:

	Column percents for flexibility		
	Flexibility		
Competitiveness	Adaptive	Parallel	Preemptive
Medium	86%	58%	27%
High	14%	42%	73%
Total	100%	100%	100%

The "Total" row reminds us that the sum of the column percents is 100% for each level of flexibility.

APPLY YOUR KNOWLEDGE

11.3 Read the output. Look at Figure 11.1. What percent of companies are highly competitive? What percent of highly competitive companies are classified as parallel for flexibility?

11.4 Read the output. Look at Figure 11.1. What type of flexibility characterizes the largest percent of companies? What is this percent?

EXAMPLE 11.5

FLXCOM

mosaic plot, p. 110

CASE 11.1 **Graphical Displays** Figure 11.2 is a bar chart from Minitab that displays the percent of highly competitive companies for each level of flexibility. It shows a clear pattern: as we move from adaptive flexibility to parallel flexibility, to preemptive flexibility, the proportion of highly competitive companies increases from 14% to 42%, to 73%. The mosaic plot from JMP in Figure 11.3 displays the joint distribution, the distribution of competitiveness for the three levels of flexibility, and the marginal distributions. Which graphical display do you prefer for this example? ■

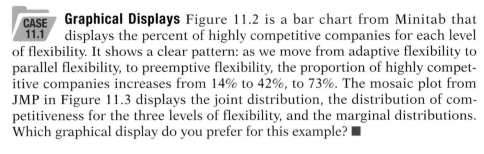

FIGURE 11.2 Bar graph from Minitab displaying the relationship between competitiveness and flexibility, Example 11.5.

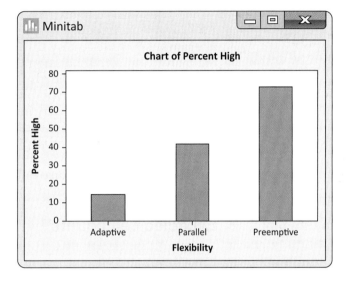

FIGURE 11.3 Mosaic plot from JMP displaying the relationship between competitiveness and flexibility, Example 11.5.

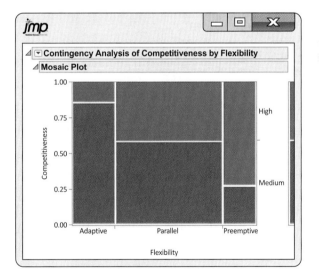

APPLY YOUR KNOWLEDGE

11.5 Sex and commercial preference. Refer to Exercise 11.1 (page 544), where you created a 2×2 table of counts for the commercial preferences of women and men. Make a graphical display of the data. Give reasons for the choices of what information to include in your plot.

11.6 A reduction in force. Refer to Exercise 11.2 (page 545), where you summarized data regarding a reduction in force. Make a graphical display of the data. Give reasons for the choices of what information to include in your plot.

The null hypothesis: no association

The differences in the percents of highly competitive companies among the three types of flexibility vary from 14% to 73%. A statistical test tells us whether this variation can be plausibly attributed to chance. Specifically, if there is no association between competitiveness and flexibility, how likely is it that a sample would show differences as large or larger than those displayed in Figures 11.2 and 11.3?

The null hypothesis H_0 of interest in a two-way table is this: there is *no association* between the row variable and the column variable. For Case 11.1, this null hypothesis says that competitiveness and flexibility are not related. The alternative hypothesis H_a is that there *is* an association between these two variables but the alternative H_a does not specify any particular direction for the association. For $r \times c$ tables in general, the alternative includes many different possibilities. Because it includes all the many kinds of association that are possible, we cannot describe H_a as either one-sided or two-sided.

In our example, the hypothesis H_0 that there is no association between competitiveness and flexibility is equivalent to the statement that the distributions of the competitiveness variable are the same for companies in the three categories of flexibility. For $r \times c$ tables like that in Example 11.3, there are c distributions for the row variable, one for each population. The null hypothesis then says that the c distributions of the row variable are identical. The alternative hypothesis is that the distributions are not all the same.

Expected cell counts

expected cell counts

To test the null hypothesis in $r \times c$ tables, we compare the observed cell counts with **expected cell counts** calculated under the assumption that the null hypothesis is true. Our test statistic is a numerical measure of the distance between the observed and expected cell counts.

EXAMPLE 11.6

FLXCOM

Expected Counts from Software The expected counts for Case 11.1 appear in the computer outputs shown in Figure 11.1. For example, the expected count for the parallel flexibility and highly competitive cell is 14.75.

How is this expected count obtained? Look at the percents in the right margin of the table in Figure 11.1. We see that 40.98% of all companies are highly competitive. If the null hypothesis of no relation between competitiveness and flexibility is true, we expect this overall percent to apply to all levels of flexibility. For our example, we expect 40.98% of the companies that use parallel flexibility to be highly competitive. There are 36 companies that use parallel flexibility, so the expected count is 40.98% of 36, or 14.75. The other expected counts are calculated in the same way. ■

The reasoning of Example 11.6 leads to a simple formula for calculating expected cell counts. To compute the expected count for highly successful companies that use parallel flexibility, we multiplied the proportion of highly competitive companies (25/61) by the number of companies that use parallel flexibility (36). From Figure 11.1, we see that the numbers 25 and 36 are the row and column totals for the cell of interest and that 61 is n, the total number of observations for the table. The expected cell count is, therefore, the product of the row and column totals divided by the table total.

EXPECTED CELL COUNTS

The **expected count** in any cell of a two-way table when the null hypothesis of no association is true is

$$\text{expected count} = \frac{\text{row total} \times \text{column total}}{n}$$

APPLY YOUR KNOWLEDGE

11.7 Expected counts. We want to calculate the expected count of companies that use adaptive flexibility and are highly competitive. From Figure 11.1, how many companies use adaptive flexibility? What proportion of all companies are highly competitive? Explain in words why, if there is no association between flexibility and competitiveness, the expected count we want is the product of these two numbers. Verify that the formula gives the same answer.

11.8 An alternative view. Refer to Figure 11.1. Verify that you can obtain the expected count for the highly competitive by adaptive flexibility cell by multiplying the number of highly competitive companies by the percent of companies that use adaptive flexibility. Explain your calculations in words.

The chi-square test

To test the H_0 that there is no association between the row and column classifications, we use a statistic that compares the entire set of observed cell counts with the entire set of expected counts. First, take the difference between each observed count and its corresponding expected count, and then square these values so that they are all 0 or positive. A large squared difference means less if it comes from a cell that we think will have a large count, so divide each squared difference by the expected count, a kind of standardization. Finally, sum over all cells. The result is called the *chi-square statistic* X^2. The chi-square statistic was invented by the English statistician Karl Pearson (1857–1936) in 1900, for purposes slightly different from ours. It is the oldest inference procedure still used in its original form. With the work of Pearson and his contemporaries at the beginning of the twentieth century, statistics first emerged as a separate discipline.

standardized observation, p. 47

CHI-SQUARE STATISTIC

The **chi-square statistic** is a measure of how much the observed cell counts in a two-way table diverge from the expected cell counts. The recipe for the statistic is

$$X^2 = \sum \frac{(\text{observed count} - \text{expected count})^2}{\text{expected count}}$$

where "observed" represents an observed sample count, "expected" represents the expected count for the same cell, and the sum is over all cells in the table.

If the expected counts and the observed counts are very different, a large value of X^2 will result. Therefore, large values of X^2 provide evidence against the null hypothesis. To obtain a P-value for the test, we need the sampling

FIGURE 11.4 (a) The $\chi^2(2)$ density curve; (b) the $\chi^2(4)$ density curve.

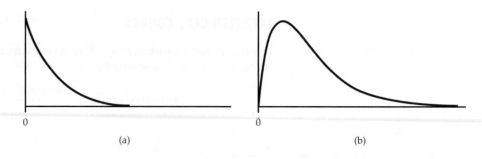

chi-square distribution

distribution of X^2 under the assumption that H_0 (no association between the row and column variables) is true. We once again use an approximation, related to the Normal approximations that we employed in Chapter 10. The result is a new distribution, the **chi-square distribution,** which we denote by χ^2 (χ is the lowercase form of the Greek letter chi).

Like the t distributions, the χ^2 distributions form a family described by a single parameter, the degrees of freedom, k. We use $\chi^2(k)$ to indicate a particular member of this family. Figure 11.4 displays the density curves of the $\chi^2(2)$ and $\chi^2(4)$ distributions. As the figure suggests, χ^2 distributions take only positive values and are skewed to the right. Software gives P-values for significance tests, and Table F in the back of the book gives upper critical values for the χ^2 distributions.

CHI-SQUARE TEST FOR TWO-WAY TABLES

The null hypothesis H_0 is that there is no association between the row and column variables in a two-way table. The alternative is that these variables are related.

If H_0 is true, the chi-square statistic X^2 has approximately a χ^2 distribution with $(r-1)(c-1)$ degrees of freedom.

The P-value for the chi-square test is

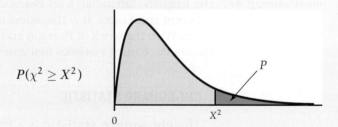

$P(\chi^2 \geq X^2)$

where χ^2 is a random variable having the $\chi^2(k)$ distribution with $k = (r-1)(c-1)$.

The chi-square test always uses the upper tail of the χ^2 distribution because any deviation from the null hypothesis makes the statistic larger. The approximation of the distribution of X^2 by χ^2 becomes more accurate as the cell counts increase. Moreover, it is more accurate for tables larger than 2×2.

For tables larger than 2×2, we use this approximation whenever the average of the expected counts is 5 or more and the smallest expected count is 1 or more. For 2×2 tables, we require that all four expected cell counts be 5 or more.[2] When the data are not suitable for the chi-square approximation to be

Fisher Exact Test

useful, other methods are available. These are provided in the output of many statistical software programs. A test based on the hypergeometric distribution (see Exercise 5.74 on page 270), called the **Fisher Exact Test,** is the usual choice for 2×2 tables when the counts are small.

EXAMPLE 11.7

FLXCOM

CASE 11.1 **Are Flexible Companies More Competitive?** The results of the chi-square significance test that we described appear in the lower portion of the computer outputs in Figure 11.1 (page 546) for the flexibility and competitiveness example. They are labeled "Pearson." The outputs also give an alternative significance test called the likelihood ratio test. The results here are very similar.

Because all the expected cell counts are moderately large, the χ^2 distribution provides accurate P-values. We see that $X^2 = 8.715$ and $k = 2$. Examine the outputs and find the P-value in each output. The rounded value is $P = 0.01$. As a check, we verify that the degrees of freedom are correct for a 2×3 table:

$$k = (r-1)(c-1) = (2-1)(3-1) = 2$$

The chi-square test confirms that the data contain clear evidence against the null hypothesis that there is no relationship between competitiveness and flexibility. Under H_0, the chance of obtaining a value of X^2 greater than or equal to the calculated value of 8.715 is small—less than one time in 100. ∎

The test does not tell us what kind of relationship is present. *You should always accompany a chi-square test with percents and figures such as those in Figures 11.1, 11.2, and 11.3 along with a description of the nature of the relationship.*

randomized comparative experiment, p. 147

The observational study of Case 11.1 cannot tell us whether being flexible is a *cause* of being highly competitive. The association may be explained by confounding with other variables that have not been measured. A randomized comparative experiment that assigns companies to the three types of competitiveness would settle the issue of causation. As is often the case, however, an experiment like this isn't practical.

What is the practical value of our analysis? If we can argue that the effects of possible confounding variables are not very important and that the relationship makes sense based on our understanding of how businesses work, we could use the results for prediction. If a company would take steps to improve its flexibility, we predict that the company would be more competitive. There is much uncertainty in this prediction, but the potential benefit from making a change may be worth the risk.

APPLY YOUR KNOWLEDGE

11.9 Degrees of freedom. A chi-square significance test is performed to examine the association between two categorical variables in a 6×4 table. What are the degrees of freedom associated with the test statistic?

11.10 The P-value. A test for association gives $X^2 = 13.67$ with $k = 6$. How would you report the P-value for this problem? Use software or Table F in the back of the book. Illustrate your solution with a sketch.

The chi-square test and the z test

We began this chapter by converting a "compare two proportions" setting (Example 11.1) into a 2×2 table. We now have two ways to test the hypothesis of equality of two population proportions: the chi-square test and the two-sample test from Section 10.2. In fact, *these tests always give exactly the same result* because the chi-square statistic is equal to the square of the z statistic, and $\chi^2(1)$ critical values are equal to the squares of the corresponding $N(0,1)$ critical values. Exercise 11.11 asks you to verify this for Example 11.1. The advantage of the z test is that we can test either one-sided or two-sided alternatives and add confidence intervals to the significance test. The chi-square test always tests the two-sided alternative for a 2×2 table. The advantage of the chi-square test is that it is much more general: we can compare more than two population proportions or, more generally yet, ask about relations in two-way tables of any size.

APPLY YOUR KNOWLEDGE

11.11 Social media in the supply chain. Sample proportions from Example 11.1 and the two-way table in Example 11.2 (page 545) report the same information in different ways. We saw in Example 10.9 (page 529) that the z statistic for the hypothesis of equal population proportions is $z = 4.87$ with $P < 0.0004$.

(a) Find the chi-square statistic X^2 for this two-way table and verify that it is equal (up to roundoff error) to z^2.

(b) Verify that the 0.001 critical value for chi-square with $k = 1$ (Table F) is the square of the 0.0005 critical value for the standard Normal distribution (Table A). The 0.0005 critical value corresponds to a P-value of 0.001 for the two-sided z test.

(c) Explain carefully why the two hypotheses

H_0: $p_1 = p_2$ (z test)

H_0: no relation between company size and social media use   (X^2 test)

say the same thing about the population.

Models for two-way tables

The chi-square test for the presence of a relationship between the two directions in a two-way table is valid for data produced from several different study designs. The precise statement of the null hypothesis "no relationship" in terms of population parameters is different for different designs. We now describe two of these settings in detail. *An essential requirement is that each experimental unit or subject is counted only once in the data table.*

Comparing several populations: the first model

Case 2.2 (wine sales in three environments, page 105) is an example of *separate and independent random samples* from each of c populations. The c columns of the two-way table represent the populations. There is a single categorical response variable, wine type. The r rows of the table correspond to the values of the response variable.

We know that the z test for comparing the two proportions of successes and the chi-square test for the 2×2 table are equivalent. The $r \times c$ table allows us to compare more than two populations or more than two categories of response, or both. In this setting, the null hypothesis "no relationship between column variable and row variable" becomes

H_0: The distribution of the response variable is the same in all c populations.

Because the response variable is categorical, its distribution consists of just the probabilities of its r values. The null hypothesis says that these probabilities (or population proportions) are the same in all c populations.

EXAMPLE 11.8

 Music and Wine Sales In the market research study of Case 2.2, we compare three populations:

Population 1: bottles of wine sold when no music is playing
Population 2: bottles of wine sold when French music is playing
Population 3: bottles of wine sold when Italian music is playing

We have three samples, of sizes 84, 75, and 84, a separate sample from each population. The null hypothesis for the chi-square test is

H_0: The proportions of each wine type sold are the same in all three populations.

The parameters of the model are the proportions of the three types of wine that would be sold in each of the three environments. There are three proportions (for French wine, Italian wine, and other wine) for each environment. ∎

More generally, if we take an independent SRS from each of c populations and classify each outcome into one of r categories, we have an $r \times c$ table of population proportions. There are c different sets of proportions to be compared. There are c groups of subjects, and a single categorical variable with r possible values is measured for each individual.

A careful analysis of the music and wine data suggests that there is a relationship between the music being played and the wine sales. In particular, when French music is playing, more French wine is sold than when Italian music or no music is being played. A similar effect is seen for Italian music and Italian wine. These relationships make sense. This suggests that we can predict, for example, that if we play French music, we will increase the sales of French wine. Would you recommend this strategy as a way to increase sales of French wine?

> **MODEL FOR COMPARING SEVERAL POPULATIONS USING TWO-WAY TABLES**
>
> Select independent SRSs from each of c populations, of sizes $n_1, n_2, \ldots, n_c$. Classify each individual in a sample according to a categorical response variable with r possible values. There are c different probability distributions, one for each population.
>
> The null hypothesis is that the distributions of the response variable are the same in all c populations. The alternative hypothesis says that these c distributions are not all the same.

Testing independence: the second model

A second model for which our analysis of $r \times c$ tables is valid is illustrated by the competitiveness and flexibility study of Case 11.1. There, a *single* sample of size n from a *single* population was classified according to two categorical random variables.

EXAMPLE 11.9

 Competitiveness and Flexibility The single population studied is

Population: food and beverage companies

The researchers had a sample of 61 companies. They measured two categorical variables for each company:

Column variable: flexibility (adaptive, parallel, or preemptive)
Row variable: competitive (medium or high)

The null hypothesis for the chi-square test is

H_0: The row variable and the column variable are independent.

The parameters of the model are the probabilities for each of the six possible combinations of values of the row and column variables. If the null hypothesis is true, the multiplication rule for independent events says that these can be found as the products of outcome probabilities for each variable alone. ∎

multiplication rule for independent events, p. 191

More generally, take an SRS from a single population and record the values of two categorical variables, one with r possible values and the other with c possible values. The data are summarized by recording the number of individuals for each possible combination of outcomes for the two random variables. This gives an $r \times c$ table of counts. Each of these $r \times c$ possible outcomes has its own probability. The probabilities give the joint distribution of the two categorical variables.

Each of the two categorical random variables has a distribution. These are the marginal distributions because they are the sums of the population proportions in the rows and columns.

marginal distributions, p. 106

The null hypothesis "no relationship" now states that the row and column variables are independent. The multiplication rule for independent events tells us that the joint probabilities are the products of the marginal probabilities.

EXAMPLE 11.10

FLXCOM

The Joint Distribution and the Two Marginal Distributions The joint probability distribution gives a probability for each of the six cells in our 3×2 table of "Flexibility" and "Competitive." The marginal distribution for "Flexibility" gives probabilities for adaptive, parallel, and preemptive, the three possible categories of flexibility. The marginal distribution for "Competitive" gives probabilities for medium and high, the two possible types of competitiveness.

Independence between "Flexibility" and "Competitive" implies that the joint distribution can be obtained by multiplying the appropriate terms from the two marginal distributions. For example, the probability that a company is adaptive (flexibility) *and* medium (competitive) is equal to the probability that it is adaptive (flexibility) *times* the probability it is medium (competitive). The hypothesis that "Flexibility" and "Competitive" are independent says that the multiplication rule applies to *all* outcomes. ∎

MODEL FOR EXAMINING INDEPENDENCE IN TWO-WAY TABLES

Select an SRS of size n from a population. Measure two categorical variables for each individual.

The null hypothesis is that the row and column variables are independent. The alternative hypothesis is that the row and column variables are dependent.

BEYOND THE BASICS

Meta-analysis

Policymakers wanting to make decisions based on research are sometimes faced with the problem of summarizing the results of many studies. These studies may show effects of different magnitudes, some

meta-analysis

highly significant and some not significant. What *overall conclusion* can we draw? **Meta-analysis** is a collection of statistical techniques designed to combine information from different but similar studies. Each individual study must be examined with care to ensure that its design and data quality are adequate. The basic idea is to compute a measure of the effect of interest for each study. These are then combined, usually by taking some sort of weighted average, to produce a summary measure for all of the studies. Of course, a confidence interval for the summary is included in the results. Here is an example.

EXAMPLE 11.11

Vitamin A Saves Lives of Young Children Vitamin A is often given to young children in developing countries to prevent night blindness. It was observed that children receiving vitamin A appear to have reduced death rates. To investigate the possible relationship between vitamin A supplementation and death, a large field trial with more than 25,000 children was undertaken in Aceh Province of Indonesia. About half of the children were given large doses of vitamin A, and the other half were controls. The researchers reported a 34% reduction in mortality (deaths) for the treated children who were one to six years old compared with the controls. Several additional studies were then undertaken. Most of the results confirmed the association: treatment of young children in developing countries with vitamin A reduces the death rate, but the size of the effect varied quite a bit.

How can we use the results of these studies to guide policy decisions? To address this question, a meta-analysis was performed on data from eight studies.[3] Although the designs varied, each study provided a two-way table of counts. Here is the table for the study conducted in Aceh Province. A total of $n = 25{,}200$ children were enrolled in the study. Approximately half received vitamin A supplements. One year after the start of the study, the number of children who had died was determined.

	Vitamin A	Control
Dead	101	130
Alive	12,890	12,079
Total	12,991	12,209

relative risk

The summary measure chosen was the **relative risk**, the ratio formed by dividing the proportion of children who died in the vitamin A group by the proportion of children who died in the control group. For Aceh, the proportion who died in the vitamin A group was

$$\frac{101}{12{,}991} = 0.00777$$

or 7.7 per thousand. For the control group, the proportion who died was

$$\frac{130}{12{,}209} = 0.01065$$

or 10.6 per thousand. The relative risk is therefore

$$\frac{0.00777}{0.01065} = 0.73$$

Relative risk less than 1 means that the vitamin A group has the lower mortality rate.

The relative risks for the eight studies were

0.73   0.50   0.94   0.71   0.70   1.04   0.74   0.80

A meta-analysis combined these eight results to produce a relative risk estimate of 0.77 with a 95% confidence interval of (0.68, 0.88). That is, vitamin A supplementation reduced the mortality rate to 77% of its value in an untreated group. The confidence interval does not include 1, so we can reject the null hypothesis of no effect (a relative risk of 1). The researchers examined many variations of this meta-analysis, such as using different weights and leaving out one study at a time. These variations had little effect on the final estimate. ■

After these findings were published, large-scale programs to distribute high-potency vitamin A supplements were started. These programs have saved hundreds of thousands of lives since the meta-analysis was conducted and the arguments and uncertainties were resolved.

SECTION 11.1 SUMMARY

- Relate the concepts of **joint distribution, marginal distribution,** and **conditional distribution** to the counts in a two-way table.

- The **null hypothesis** for $r \times c$ tables of count data is that there is no relationship between the row variable and the column variable.

- **Expected cell counts** under the null hypothesis are computed using the formula

$$\text{expected count} = \frac{\text{row total} \times \text{column total}}{n}$$

- The null hypothesis is tested by the **chi-square statistic,** which compares the observed counts with the expected counts:

$$X^2 = \sum \frac{(\text{observed} - \text{expected})^2}{\text{expected}}$$

- Under the null hypothesis, X^2 has approximately the **chi-square distribution** with $(r-1)(c-1)$ degrees of freedom. The P-value for the test is

$$P(\chi^2 \geq X^2)$$

where χ^2 is a random variable having the $\chi^2(k)$ distribution with $k = (r-1)(c-1)$.

- The chi-square approximation is adequate for practical use when the average expected cell count is 5 or greater and all individual expected counts are 1 or greater, except in the case of 2×2 tables. All four expected counts in a 2×2 table should be 5 or greater.

- To analyze a two-way table, first **compute percents or proportions** that describe the relationship between the row and column variables. Then calculate **expected counts,** the **chi-square statistic,** and the **P-value.**

SECTION 11.1 EXERCISES

For Exercises 11.1 and 11.2, see pages 544–545; for 11.3 and 11.4, see page 546; for 11.5 and 11.6, see page 547; for 11.7 and 11.8, see page 549; for 11.9 and 11.10, see page 551; and for 11.11, see page 552.

11.12 To tip or not to tip. A study of tipping behaviors examined the relationship between the color of the shirt worn by the server and whether the customer left a tip.[4] Here are the data for 418 male customers who participated in the study. TIPMALE

Tip	Shirt color					
	Black	White	Red	Yellow	Blue	Green
Yes	22	25	40	31	25	27
No	49	43	29	41	42	43

(a) Use numerical summaries to describe the data. Give a justification for the summaries that you choose.

(b) State appropriate null and alternative hypotheses for this setting.

(c) Give the results of the significance test for these data. Be sure to include the test statistic, the degrees of freedom, and the P-value.

(d) Make a mosaic plot if you have the needed software.

(e) Write a short summary of what you have found, including your conclusion.

11.13 To tip or not to tip: women customers. Refer to the previous exercise. Here are the data for the 304 female customers who participated in the study. TIPFEM

Tip	Shirt color					
	Black	White	Red	Yellow	Blue	Green
Yes	18	16	15	19	16	18
No	33	32	38	31	31	37

Using the questions for the previous exercise as a guide, analyze these data and compare the results with those you found for the male customers.

11.14 Evaluating the price and math anxiety. Subjects in a study were asked to arrange for the rental of two tents, each for two weeks. They were offered two options for the price: (A) $40 per day per tent with a discount of $50 per tent per week, or (B) $40 per day per tent with a discount of 20%. The subjects were classified by their level of math anxiety as Low, Moderate, or High.[5] The percents of subjects choosing the higher-priced option that is easier to compute (A) were 15%, 20%, and 45% for the low, medium, and high math anxiety groups, respectively. Assume that there are 40 subjects in each of these groups.

(a) Give the two-way table of counts for this study.

(b) Use numerical summaries to describe the data. Give a justification for the summaries that you choose.

(c) State appropriate null and alternative hypotheses for this setting.

(d) Give the results of the significance test for these data. Be sure to include the test statistic, the degrees of freedom, and the P-value.

(e) Write a short summary of what you have found, including your conclusion.

11.15 Education and the workforce. The educational profile of the workforce has changed as millennials move into the workforce while older workers retire. A PEW research study examined the relationship between generations and educational status.[6] Individuals were classified by birth date ranges: Millennial, 1981–1996; Gen X, 1965–1980; Boomer, 1946–1964; and Silent, before 1946. The following table gives the percents of women who have earned a bachelor's degree while they were young for each generation:

	Generation			
	Millennial	Gen X	Boomer	Silent
Percent college	36	28	20	9

Assume that these percents are based on samples of size 800 for each generation.

(a) Create a two-way table of counts based on the categorical variables generation and college degree (yes or no).

(b) Summarize the data numerically and graphically.

(c) Perform a significance test to assess the relationship between the two categorical variables. Be sure to give the test statistic, the degrees of freedom, the P-value, and your conclusion.

(d) Using your results from parts (a), (b), and (c) write a short report summarizing what you have found.

11.16 Analyze the data for men. Refer to the previous exercise. Here are the corresponding data for men:

	Generation			
	Millennial	Gen X	Boomer	Silent
Percent college	29	24	22	15

Answer the questions from the previous exercise using these data.

11.17 Compare the women with the men. Refer to the previous two exercises. For each generation, compare the proportions of college graduates for women and men. Use the questions from Exercise 11.15 as a guide for each of your four analyses. Then summarize what you have found based on these four analyses.

11.18 Nonresponse in a survey. A business school conducted a survey of companies in its state. It mailed a questionnaire to 220 small companies, 220 medium-sized companies, and 220 large companies. The rate of nonresponse is important in deciding how reliable survey results are. Here are the data on response to this survey. NRESP

	Small	Medium	Large
Response	139	70	46
No response	81	130	174
Total	220	220	220

Note that you answered parts (a) through (c) of this exercise if you completed Exercise 2.112 (page 116).

(a) What was the overall percent of nonresponse?

(b) Describe how nonresponse is related to the size of the business. (Use percents to make your statements precise.)

(c) Draw a bar graph to compare the nonresponse percents for the three size categories.

(d) State and test an appropriate null hypothesis for these data.

11.19 Hiring practices. A company has been accused of age discrimination in hiring for operator positions. Lawyers for both sides look at data on applicants for the past three years. They compare hiring rates for applicants younger than 40 years and those 40 years or older. HIRING

Age	Hired	Not hired
Younger than 40	84	1140
40 or older	4	178

Note that you answered parts (a) through (d) of this exercise if you completed Exercise 2.111 (page 116).

(a) Find the two conditional distributions of hired/not hired: one for applicants who are less than 40 years old and one for applicants who are not less than 40 years old.

(b) Based on your calculations, make a graph to show the differences in distribution for the two age categories.

(c) Describe the company's hiring record in words. Does the company appear to discriminate on the basis of age?

(d) What lurking variables might be involved here?

(e) Use a significance test to determine whether the data indicate that there is a relationship between age and whether an applicant is hired.

11.20 Obesity and health. Recent studies have shown that earlier reports underestimated the health risks associated with being overweight. The error was due to overlooking lurking variables. In particular, smoking tends both to reduce weight and to lead to earlier death. Note that you answered part (a) of this exercise if you completed Exercise 2.117 (page 117).

(a) Illustrate Simpson's paradox by a simplified version of this situation. That is, make up tables of overweight (yes or no) by early death (yes or no) by smoker (yes or no) such that

- Overweight smokers and overweight nonsmokers both tend to die earlier than those not overweight.

- But when smokers and nonsmokers are combined into a two-way table of overweight by early death, persons who are not overweight tend to die earlier.

(b) Perform significance tests for the combined data set and for the smokers and nonsmokers separately. If all P-values are not less than 0.05, redo your tables so that all results are statistically significant at this level.

11.21 Discrimination? Wabash Tech has two professional schools, business and law. Here are two-way tables of applicants to both schools, categorized by sex and admission decision. (Although these data are made up, similar situations occur in reality.) DISCR

	Business			Law	
	Admit	Deny		Admit	Deny
Male	480	120	Male	10	90
Female	180	20	Female	100	200

Note that you answered parts (a) through (d) of this exercise if you completed Exercise 2.116 (page 116).

(a) Make a two-way table of sex by admission decision for the two professional schools together by summing entries in these tables.

(b) From the two-way table, calculate the percent of male applicants who are admitted and the percent of female applicants who are admitted. Wabash admits a higher percent of male applicants.

(c) Now compute separately the percents of male and female applicants admitted by the business school and by the law school. Each school admits a higher percent of female applicants.

(d) This is Simpson's paradox: both schools admit a higher percent of the women who apply, but overall Wabash admits a lower percent of female applicants than of male applicants. Explain carefully, as if speaking to a skeptical reporter, how it can happen that Wabash appears to favor males when each school individually favors females.

(e) Use the data summary that you prepared in part (a) to test the null hypothesis that there is no relationship between sex and whether an applicant is admitted to a professional school at Wabash Tech.

(f) Test the same null hypothesis using the business school data only.

(g) Do the same for the law school data.

(h) Compare the results for the two schools.

11.2 Goodness of Fit

When you complete this section, you will be able to:

- Compute expected counts given a sample size and the probabilities specified by a null hypothesis for a chi-square goodness-of-fit test.
- Find the chi-square test statistic, its degrees of freedom, and the P-value.
- Interpret the results of a chi-square goodness-of-fit significance test.

In the first section of this chapter, we discussed the use of the chi-square test to compare distributions of categorical variables from c populations. We now consider a slight variation on this scenario in which we compare the distribution of a sample from one population with a hypothesized distribution. The ideas are very similar to those we used in Chapter 5 to assess the compatibility of data with binomial and Poisson models (pages 260 and 265). Here is an example that illustrates the basic ideas.

EXAMPLE 11.12

DEMO

The Demographics of Your Customers Your company's products are marketed to a population that consists primarily of young customers. The following table gives the age distribution of your customers for the past several years.

12 to 17 years	18 to 24 years	25 years and over
35	50	15

Last year your marketing strategy changed by using the Internet to a greater extent and relying less on television advertising. You would like to know if this change has impacted the age distribution of your customers. You select a random sample of 300 current customers to receive a survey that asks about their age and other demographics. To encourage their participation, each customer who completes the survey receives a discount of 50% on a future purchase with a maximum discount of $100.00. A total of 273 customers participated in the survey. Here is the age distribution for your sample:

12 to 17 years	18 to 24 years	25 years and over
78	158	37

Let's see how well the age distribution in your sample of current customers matches the age distribution of customers from the past several years. We start with the 12–17 years age group. Historically, 35% of the customers are 12 to 17 years old. If age distribution has not changed, we expect 35% of our sample to be in this age range. With 273 customers in our sample, we calculate

expected count for ages 12 to 17 = (0.35)273 = 95.55

Here are the expected counts for all age groups:

12 to 17 years	18 to 24 years	25 years and over
95.55	136.50	40.95

APPLY YOUR KNOWLEDGE

11.22 Calculate the expected counts. Refer to Example 11.12. Verify that the expected counts for the age groups 18 to 24 years and 25 years and over agree with the table above. DEMO

11.23 Calculate the expected counts In the setting of Example 11.12, find the expected counts if the number of current customers in the sample was 100.

As we saw with the expected counts in the analysis of two-way tables in Section 11.1, we do not really expect the observed counts to be *exactly* equal to the expected counts. Different samples under the same conditions would give different counts. We expect the average of these counts to be equal to the expected counts when the null hypothesis is true. How close do we think the counts and the expected counts should be?

We can think of our table of observed counts in Example 11.12 as a one-way table with three cells, each with a count of the number of subjects sampled from a particular state. Our question of interest is translated into a null hypothesis that says that the observed proportions of current customers can be viewed as random samples from the population of customers from the past several years. The alternative hypothesis is that the age distribution of customers has shifted in some way.

Our analysis of these data is very similar to the analyses of two-way tables that we studied in Section 11.1. We have already computed the expected counts. We now construct a chi-square statistic that measures how far the observed counts are from the expected counts. Here is a summary of the procedure:

THE CHI-SQUARE GOODNESS-OF-FIT TEST

Data for n observations of a categorical variable with k possible outcomes are summarized as observed counts, $n_1, n_2, \ldots, n_k$, in k cells. The null hypothesis specifies probabilities $p_1, p_2, \ldots, p_k$ for the possible outcomes. The alternative hypothesis says that the true probabilities of the possible outcomes are not the probabilities specified in the null hypothesis.

For each cell, multiply the total number of observations n by the specified probability to determine the expected counts:

$$\text{expected count} = np_i$$

The **chi-square statistic** measures how much the observed cell counts differ from the expected cell counts. The formula for the statistic is

$$X^2 = \sum \frac{(\text{observed count} - \text{expected count})^2}{\text{expected count}}$$

The degrees of freedom are $k-1$, and P-values are computed from the chi-square distribution.

Use this procedure when the expected counts are all 5 or more.

EXAMPLE 11.13

DEMO

The Goodness-of-Fit Test for the Customer Survey For the customers aged 12 to 17, the observed count is 78. In Example 11.12, we calculated the expected count, 95.55. The contribution to the chi-square statistic for this age group is

$$\frac{(\text{observed count} - \text{expected count})^2}{\text{expected count}} = \frac{(78 - 95.55)^2}{95.55} = 3.22$$

We use the same approach to find the contributions to the chi-square statistic for the other two age groups. The expected counts are all at least 5, so we can proceed with the significance test.

The sum of these three values is the chi-square statistic,

$$X^2 = 6.99$$

The degrees of freedom are the number of cells minus 1: $k = 3 - 1 = 2$. We calculate the P-value using Table F or software. From Table F, we can determine $P < 0.05$. We conclude that the observed counts are not compatible with the hypothesized proportions. The data provide evidence that our age distribution of the customers has changed. ■

APPLY YOUR KNOWLEDGE

11.24 Compute the chi-square statistic. For each of the other two age groups, compute the contribution to the chi-square statistic using the method illustrated for the customers who are 12 to 17 years old in Example 11.13. Use the expected counts that you calculated in Exercise 11.13 for these calculations. Show that the sum of these values is the chi-square statistic. DEMO

EXAMPLE 11.14

The Goodness-of-Fit Test from Software Software output from Minitab for this problem is given in Figure 11.5. The P-value is 0.030. ■

DEMO

For tables of counts, the contribution to chi-square for a cell is defined as

$$\text{residual} = \frac{\text{observed count} - \text{expected count}}{\sqrt{\text{expected count}}}$$

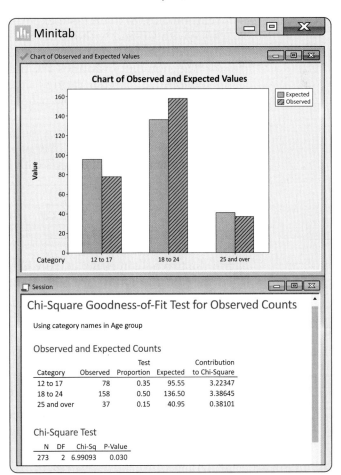

FIGURE 11.5 Minitab output, Example 11.14.

residuals for regression, p. 89

These residuals are very much like the residuals that we studied for regression analysis. Note that the chi-square statistic is the sum of the squares of these residuals. These values are given in the column labeled "Contribution to Chi-Square" in Figure 11.5. If we reject the null hypothesis (with a large value or the chi-square statistic), the residuals tell us which cells are contributing the most to the rejection.

Some software packages do not provide routines for computing the chi-square goodness-of-fit test.[7] However, there is a very simple trick that can be used to produce the results from software that can analyze two-way tables. Make a two-way table in which the first column contains k cells with the observed counts. Add a second column with counts that correspond *exactly* to the probabilities specified by the null hypothesis, with a very large number of observations. Then perform the chi-square significance test for two-way tables.

APPLY YOUR KNOWLEDGE

11.25 Distribution of M&M colors. M&M Mars Company has varied the mix of colors for M&M'S plain chocolate candies over the years. These changes in color blends are the result of consumer preference tests. Most recently, the color distribution is reported to be 13% brown, 14% yellow, 13% red, 20% orange, 24% blue, and 16% green.[8] You open up a 14-ounce bag of M&M'S and find 61 brown, 59 yellow, 49 red, 77 orange, 141 blue, and 88 green. Use a goodness-of-fit test to examine how well this bag fits the percents stated by the M&M Mars Company. MM

EXAMPLE 11.15

The Sign Test as a Goodness-of-Fit Test In Exercise 8.14 (page 412), you used a sign test to examine the effect of cold plasma (CP) technology on the taste of fruit juice. In this example, we modify the results slightly to illustrate this relationship. Sixteen taste experts rated, on a scale of 0 to 100, the taste of juice with CP and without CP. Six experts rated the juice with CP higher and 10 experts rated the juice without CP higher.

The sign test examines the null hypothesis that the experts are equally likely to give a higher rating to the juice with the CP than the juice without CP. With $n = 16$, the sample proportion is $\hat{p} = 6/16$ and the null hypothesis is $H_0: p = 0.5$.

To look at these data from the viewpoint of goodness of fit, we think of the data as two counts: the number of experts with higher ratings for the juice with CP and the number of experts with higher scores for the juice without CP.

Counts		
With CP	Without CP	Total
6	10	16

If the two outcomes are equally likely, the expected counts are both 8 (16×0.5). The expected counts are both greater than 5, so we can proceed with the significance test.

The test statistic is

$$X^2 = \frac{(6-8)^2}{8} + \frac{(10-8)^2}{8}$$
$$= 0.5 + 0.5$$
$$= 1.0$$

We have $k = 2$, so the degrees of freedom are 1. From Table F we conclude that $P > 0.25$. We do not have evidence to conclude that CP has an effect on the taste of the juice. ∎

APPLY YOUR KNOWLEDGE

11.26 Is the coin fair? In Exercise 5.130 (page 290) we learned that the South African statistician John Kerrich tossed a coin 10,000 times while imprisoned by the Germans during World War II. The coin came up heads 5067 times.

(a) Formulate the question about whether the coin was fair as a goodness-of-fit hypothesis.

(b) Perform the chi-square significance test and write a short summary of the results.

SECTION 11.2 SUMMARY

- The **chi-square goodness-of-fit test** examines the null hypothesis that the probabilities of the k possible outcomes for a categorical variable are equal to a particular set of values, $p_1, p_2, \ldots, p_k$. The data for the test are the observed counts in the k cells, $n_1, n_2, \ldots, n_k$.

- **Expected cell counts** under the null hypothesis are

$$\text{expected count} = np_i$$

where n is the total number of observations.

- The **chi-square statistic** measures how much the observed cell counts differ from the expected cell counts. The formula for the statistic is

$$X^2 = \sum \frac{(\text{observed count} - \text{expected count})^2}{\text{expected count}}$$

The degrees of freedom are $k-1$, and P-values are computed from the chi-square distribution. Use this procedure when the expected counts are all 5 or more.

SECTION 11.2 EXERCISES

For Exercises 11.22 and 11.23, see page 560; for 11.24, see page 561; for 11.25, see page 562; and for 11.26, see page 563.

11.27 Goodness of fit to a standard Normal distribution. Computer software generated 1000 random numbers that should look as if they are from the standard Normal distribution. They are categorized into five groups: (1) less than or equal to −0.80, (2) greater than −0.80 and less than or equal to −0.25, (3) greater than −0.25 and less than or equal to 0.25, (4) greater than 0.25 and less than or equal to 0.80, and (5) greater than 0.80. The counts in the five groups are 212, 185, 188, 195, and 220, respectively. Find the probabilities for these five intervals using Table A. Then compute the expected number for each interval for a sample of 1000. Finally, perform the goodness-of-fit test and summarize your results.

11.28 More on the goodness of fit to a standard Normal distribution. Refer to the previous exercise.

(a) Use software to generate your own sample of 900 uniform random variables on the interval from 0 to 1, and perform the goodness-of-fit test using the intervals from the previous exercise.

(b) Choose a different set of intervals than the ones used in the previous exercise. Rerun the goodness-of-fit test.

(c) Compare the results you found in parts (a) and (b). Which intervals would you recommend?

11.29 Goodness of fit to the uniform distribution. Computer software generated 1000 random numbers that should look as if they are from the uniform distribution on the interval 0 to 1 (see page 273). They are categorized into four groups: (1) less than or equal to 0.25, (2) greater than 0.25 and less than or equal to 0.50, (3) greater than 0.50 and less than or equal to 0.75, and (4) greater than 0.75. The counts in the four groups are 231, 252, 270, and 247, respectively.

(a) The probabilities for these four intervals are all the same. What is this probability?

(b) Compute the expected number for each interval for a sample of 1000.

(c) Finally, perform the goodness-of-fit test and summarize your results.

11.30 More on goodness of fit to the uniform distribution. Refer to the previous exercise.

(a) Use software to generate your own sample of 900 uniform random variables on the interval from 0 to 1, and perform the goodness-of-fit test using the intervals from the previous exercise.

(b) Choose a different set of intervals than the ones used in the previous exercise. Rerun the goodness-of-fit test.

(c) Compare the results you found in parts (a) and (b). Which intervals would you recommend?

CHAPTER 11 REVIEW EXERCISES

11.31 What's wrong? Explain what is wrong with each of the following.

(a) Observed cell counts are computed under the assumption that the null hypothesis is true.

(b) A chi-square test was used to test the null hypothesis that there is an association between two categorical variables.

(c) The P-value for a chi-square significance test was 1.05.

11.32 Plot the test statistic and the P-values. Here is a 2 × 2 two-way table of counts. The two categorical variables are U and V, and the possible values for each of these variables are 0 and 1. Notice that the second row depends upon a quantity that we call a. For this exercise, you will examine how the test statistic and its corresponding P-value depend upon this quantity. Notice that the row sums are both 100.

	V	
U	0	1
0	60	60
1	60 + a	60 − a

(a) Consider setting a equal to zero. Find the percent of zeros for the variable V when $U = 0$. Do the same for the case where $U = 1$. With this choice of a, the data match the null hypothesis as closely as possible. Explain why.

(b) Consider the tables where the values of a are equal to 0, 5, 10, 15, 20, and 25. For each of these scenarios, find the percent of zeros for V when $U = 1$. Notice that this percent does not vary with a for $U = 0$.

(c) Compute the test statistic and P-value for testing the null hypothesis that there is no association between the row and column variables for each of the values of a given in part (b).

(d) Plot the values of the X^2 test statistic versus the percent of zeros for V when $U = 1$. Do the same for the P-values. Summarize what you have learned from this exercise in a short paragraph.

11.33 Plot the test statistic and the P-values. Here is a 2 × 2 two-way table of counts. The two categorical variables are U and V, and the possible values for each of these variables are 0 and 1. COUNTS

	U	
V	0	1
0	6	6
1	7	3

(a) Find the percent of zeros for V when $U = 0$. Do the same for the case where $U = 1$. Find the value of the test statistic and its P-value.

(b) Now multiply all of the counts in the table by 2. Verify that the percent of zeros for V when $U = 0$ and the percent of zeros for the V when $U = 1$ do not change. Find the value of the test statistic and its P-value for this table.

(c) Answer part (b) for tables where all counts are multiplied by 4, 6, and 8. Summarize all your results graphically, and write a short paragraph describing what you have learned from this exercise.

11.34 Trends in broadband market. The Pew Internet and American Life Project collects data about the impact of the Internet on various aspects of American life.[9] One set of surveys has tracked the use of broadband in homes over a period of several years.[10] Here are some data on the percent of homes that access the Internet using broadband:

Date of survey	2001	2005	2009	2013	2018
Homes with broadband	6%	33%	63%	70%	89%

Assume a sample size of 2250 for each survey.

(a) Display the data in a two-way table of counts.

(b) Test the null hypothesis that the proportion of homes that access the Internet using broadband has not changed over this period of time. Report your test statistic with degrees of freedom and the P-value. What do you conclude?

11.35 Trends for adults aged 18 to 29 years. Refer to the previous exercise. The same report gives the same data for individuals in different age ranges. Here are the data for adults aged 18 to 29 years:

Date of survey	2001	2005	2009	2013	2018
Percent of individuals	72%	83%	92%	97%	98%

Assume a sample size of 540 for each percent.

(a) Answer the questions given in the previous exercise for these data.

(b) Use your analysis from this exercise and the previous one to write a short report summarizing the changes in broadband access that have occurred over this period of time. In what ways are the patterns for households and individuals aged 18 to 29 similar? In what ways are they different?

11.36 How robust are the conclusions? Refer to Exercise 11.34 on the use of broadband to access the Internet. In that exercise, the percents were read from a graph, and we assumed that the sample size was 2250 for all the surveys. Investigate the robustness of your conclusions in Exercise 11.34 against the use of 2250 as the sample size for all surveys and to roundoff and slight errors in reading the graph. Assume that the actual sample sizes ranged from 2200 to 2600. Assume also that the percents reported are all accurate to within ±2%. In other words, if the reported percent is 33%, then we can assume that the actual survey percent is between 31% and 35%. Reanalyze the data using at least five scenarios that vary the percents and the sample sizes within the assumed ranges. Summarize your results in a report, paying particular attention to the consequences for your conclusions in Exercise 11.34.

11.37 Find the P-value. For each of the following situations give the degrees of freedom and an appropriate bound on the P-value (give the exact value if you have software available) for the X^2 statistic for testing the null hypothesis of no association between the row and column variables.

(a) A 2×4 table with $X^2 = 11.24$.

(b) A 3×2 table with $X^2 = 11.24$.

(c) A 5×2 table with $X^2 = 11.24$.

(d) A 2×3 table with $X^2 = 11.24$.

11.38 Health care fraud. Most errors in billing insurance providers for health care services involve honest mistakes by patients, physicians, or others involved in the health care system. However, fraud is a serious problem. The National Health Care Anti-Fraud Association estimates that approximately tens of billions of dollars are lost to health care fraud each year.[11] When fraud is suspected, an audit of randomly selected billings is often conducted. The selected claims are then reviewed by experts and each claim is classified as allowed or not allowed. The distributions of the amounts of claims are frequently highly skewed, with a large number of small claims and small number of large claims. Simple random sampling would likely be overwhelmed by small claims and would tend to miss the large claims, so stratification is often used. See the section on stratified sampling in Chapter 3 (page 136). Here are data from an audit that used three strata based on the sizes of the claims (small, medium, and large).[12] BERRORS

Stratum	Sampled claims	Number not allowed
Small	59	7
Medium	20	4
Large	6	1

(a) Construct the 3×2 table of counts for these data and include the marginal totals.

(b) Find the percent of claims that were not allowed in each of the three strata.

(c) State an appropriate null hypothesis to be tested for these data.

(d) Perform the significance test and report your test statistic with degrees of freedom and the P-value. State your conclusion.

11.39 Population estimates. Refer to the previous exercise. One reason to do an audit such as this is to estimate the number of claims that would not be allowed if all claims in a population were examined by experts. We have estimates of the proportions of such claims from each stratum based on our sample. With our simple random sampling of claims from each stratum, we have unbiased estimates of the corresponding population proportion for each stratum. Therefore, if we take the sample proportions and multiply by the population sizes, we would have the estimates that we need. Here are the population sizes for the three strata:

Stratum	Claims in strata
Small	3142
Medium	236
Large	51

(a) For each stratum, estimate the total number of claims that would not be allowed if all claims in the stratum had been audited.

(b) (Optional) Give margins of error for your estimates. (*Hint:* You first need to find standard errors for your sample estimates; see Chapter 10, page 508. Then you need to use the rules for means given in Chapter 6, page 311, to find the standard errors for the population estimates. Finally, you need to multiply by z^* to determine the margins of error.)

11.40 Construct a table. Construct a 4×2 table of counts where there is no apparent association between the row and column variables.

11.41 Jury selection. Exercise 10.91 (page 539) concerns *Casteneda v. Partida,* the case in which the Supreme Court decision used the phrase "two or three standard deviations" as a criterion for statistical significance. There were 181,535 persons eligible for jury duty, of whom 143,611 were Mexican Americans. Of the 870 people selected for jury duty, 339 were Mexican Americans. We are interested in finding out if there is an association between being a Mexican American and being selected as a juror. Formulate this problem using a two-way table of counts. Construct the 2×2 table using the variables "Mexican American or not" and "juror or not." Find the X^2 statistic and its P-value. Square the z statistic that you obtained in Exercise 10.91 and verify that the result is equal to the X^2 statistic.

11.42 Which model? This exercise concerns the material in Section 11.1 on models for two-way tables. Look at Exercises 11.18, 11.34, 11.38, and 11.43. For each exercise, state whether you are comparing several populations based on separate samples from each population (the first model for two-way tables) or testing independence between two categorical variables based on a single sample (the second model).

11.43 A reduction in force. In economic downturns or to improve their competitiveness, corporations may undertake a "reduction in force" (RIF), in which substantial numbers of employees are laid off. Federal and state laws require that employees be treated equally regardless of their age. In particular, employees over the age of 40 years are a "protected class." Many allegations of discrimination focus on comparing employees over 40 with their younger coworkers. Here are the data for a recent RIF. RIF1

	Over 40	
Released	No	Yes
Yes	8	43
No	503	765

(a) Complete this two-way table by adding marginal and table totals. What percent of each employee age group (over 40 or not) were laid off? Does there appear to be a relationship between age and being laid off?

(b) Perform the chi-square test. Give the test statistic, the degrees of freedom, the P-value, and your conclusion.

11.44 Employee performance appraisal. A major issue that arises in RIFs like that in the previous exercise is the extent to which employees in various groups are similar. If, for example, employees over 40 receive generally lower performance ratings than younger workers, that might explain why more older employees were laid off. We have data on the last performance appraisal. The possible values are "partially meets expectations," "fully meets expectations," "usually exceeds expectations," and "continually exceeds expectations." Because there were very few employees who partially met expectations, we combine the first two categories. Here are the data. RIF2

	Over 40	
Performance appraisal	No	Yes
Partially or fully meets expectations	86	234
Usually exceeds expectations	352	494
Continually exceeds expectations	64	36

Note that the total number of employees in this table is less than the number in the previous exercise because some employees do not have a performance appraisal. Analyze the data. Do the older employees appear to have lower performance evaluations?

11.45 Titanic! In 1912, the luxury liner *Titanic,* on its first voyage, struck an iceberg and sank. Some passengers got off the ship in lifeboats, but many died. Think of the *Titanic* disaster as an experiment in how the people of that time behaved when faced with death in a situation where only some can escape. The passengers are a sample from the population of their peers. Here is information about who lived and who died, by sex and economic status.[13] (The data leave out a few passengers whose economic status is unknown.) TITANIC

Men			Women		
Status	Died	Survived	Status	Died	Survived
Highest	111	61	Highest	6	126
Middle	150	22	Middle	13	90
Lowest	419	85	Lowest	107	101
Total	680	168	Total	126	317

(a) Compare the percents of men and of women who died. Is there strong evidence that a higher proportion of men die in such situations? Why do you think this happened?

(b) Look only at the women. Describe how the three economic classes differ in the percent of women who died. Are these differences statistically significant?

(c) Now look only at the men and answer the same questions.

11.46 Goodness of fit to a binomial distribution. Recall from Chapter 5 (page 249) that the binomial distribution, $B(n, p)$, is a model for the distribution of the count, X, of successes in a sample of size n where the probability of success is p. Use statistical software to generate 500 observations from the $B(3, 0.4)$ distribution and perform the goodness-of-fit test for these data.

11.47 Goodness of fit to a Poisson distribution. Recall from Chapter 5 (page 261) that the Poisson distribution, is a model for the distribution of the count, X, of successes in the Poisson setting. The parameter of the Poisson distribution is μ, the mean number of successes. Use statistical software to generate 500 observations from the Poisson distribution with $\mu = 0.9$. Group the observations into three categories: $X = 0, X = 1, X = 2$, and $X \geq 3$. Perform the goodness of fit for the data categorized in this way.

11.48 Alternative goodness of fit to a Poisson distribution. Refer to the previous exercise. Reanalyze the data using three categories: $X = 0, X = 1, X \geq 2$. Compare your results with what you found in the previous exercise.

11.49 Suspicious results? An instructor who assigned an exercise similar to the one described in the previous exercise received homework from a student who reported a P-value of 0.999. The instructor suspected that the student did not use the computer for the assignment but just made up some numbers for the homework. Why was the instructor suspicious? How would this scenario change if there were 2000 students in the class?

Feverpitch/Deposit Photos

CHAPTER 12

Inference for Regression

Introduction

One of the most common uses of statistical methods in business and economics is to predict, or forecast, a response based on one or several explanatory (predictor) variables. In predictive analytics, these forecasts are then used by companies to make decisions. Here are some examples:

- Lime uses the day of the week, hour of the day, and current weather forecast to predict scooter- and bike-sharing demand around a city. This information is incorporated into the company's nightly redistribution strategy.

- Amazon wants to describe the relationship between dollars spent in its Digital Music department and dollars spent in its Online Grocery department by 18- to 25-year-olds this past year. This information will be used to determine a new advertising strategy.

- Panera Bread, when looking for a new store location, develops a model to predict profitability using the amount of traffic near the store, the proximity to competitive restaurants, and the average income level of the neighborhood.

Prediction is most straightforward when there is a straight-line relationship between a quantitative response variable y and a single quantitative explanatory variable x. This is **simple linear regression**, the topic of this chapter. In Chapter 13, we will consider the more common setting involving more than one explanatory (predictor) variable. Because both settings share many of the same ideas, we introduce inference for regression under the simple setting.

CHAPTER OUTLINE

12.1 Inference about the Regression Model

12.2 Using the Regression Line

12.3 Some Details of Regression Inference

simple linear regression

least-squares line, p. 83

In Chapter 2, we saw that the least-squares line can be used to predict y for a given value of x. Now we consider the use of significance tests and confidence intervals in this setting. To do this, we will think of the least-squares line, $b_0 + b_1 x$, as an estimate of a regression line for the population—just as in Chapter 8, where we viewed the sample mean $\bar{x}$ as the estimate of the population mean μ, and in Chapter 10, where we viewed the sample proportion $\hat{p}$ as the estimate for the population proportion p.

parameters and statistics, p. 295

We write the population regression line as $\beta_0 + \beta_1 x$. The numbers β_0 and β_1 are *parameters* that describe this population line. The numbers b_0 and b_1 are *statistics* calculated by fitting a line to a sample. The fitted intercept b_0 estimates the intercept of the population line β_0, and the fitted slope b_1 estimates the slope of the population line β_1.

Our discussion begins with an overview of the simple linear regression model and inference about the slope β_1 and the intercept β_0. Because regression lines are most often used for prediction, we then consider inference about either the mean response or an individual future observation on y for a given value of the explanatory variable x. We conclude the chapter with more of the computational details, including the use of analysis of variance (ANOVA). If you plan to read Chapter 13 on regression involving more than one explanatory variable, these details will be very useful.

ANOVA, p. 458

12.1 Inference about the Regression Model

When you complete this section, you will be able to:

- Describe the simple linear regression model in terms of a population regression line and the distribution of deviations of the response variable y from this line.
- Use linear regression output from statistical software to find the least-squares regression line and estimated regression standard deviation.
- Use plots of the residuals to visually check the assumptions of the simple linear regression model.
- Construct and interpret a confidence interval for the population intercept and for the population slope.
- Perform a significance test for the population intercept and for the population slope and summarize the results.

the one-way ANOVA model, p. 465

Simple linear regression studies the relationship between a quantitative response variable y and a quantitative explanatory variable x. We expect that different values of x will be associated with different mean responses for y. We encountered a situation similar to this in Chapter 9, when we considered the possibility that different treatment groups had different mean responses.

Figure 12.1 illustrates the statistical model from Chapter 9 for comparing the items per hour entered by three groups of financial clerks using new

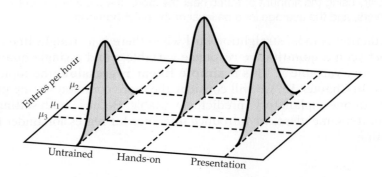

FIGURE 12.1 The statistical model for comparing the responses to three treatments. The responses vary within each treatment group according to a Normal distribution. The mean may be different in the three treatment groups.

data entry software. Group 1 received no training, Group 2 received one hour of hands-on training, and Group 3 attended an hour-long presentation describing the entry process. Entries per hour is the response variable y. Treatment (or type of training) is the explanatory variable. The model has two important parts:

- The mean entries per hour may be different in the three populations. These means are μ_1, μ_2 and μ_3 in Figure 12.1.

- Individual entries per hour vary within each population according to a Normal distribution. The three Normal curves in Figure 12.1 describe these responses. These Normal distributions have the same spread, indicating that the population standard deviations are assumed to be equal.

Statistical model for simple linear regression

In linear regression, the explanatory variable x is quantitative and can have many different values. Imagine, for example, giving different lengths x of hands-on training to different groups of clerks. We can think of these groups as belonging to **subpopulations,** one for each possible value of x. Each subpopulation consists of all individuals in the population having the same value of x. If we gave $x = 1$ hour of training to some subjects, $x = 2$ hours of training to some others, and $x = 4$ hours of training to some others, these three groups of subjects would be considered samples from the corresponding three subpopulations.

The statistical model for simple linear regression assumes that, for each value of x (or subpopulation), the response variable y is Normally distributed with a mean that depends on x. We use μ_y to represent these means. In general, the means μ_y can change as x changes according to any sort of pattern. In simple linear regression, we assume that the means all lie on a line when plotted against x.

To summarize, this model has two important parts:

- The mean entries per hour μ_y changes as the number of training hours x changes and these means all lie on a straight line; that is, $\mu_y = \beta_0 + \beta_1 x$.

- Individual entries per hour y for subjects with the same amount of training x vary according to a Normal distribution. This variation, measured by the standard deviation σ, is the same for all values of x.

Figure 12.2 illustrates this statistical model. The line describes how the mean response μ_y changes with x; it is called the **population regression line.** The three Normal curves show how the response y will vary for three different values of the explanatory variable x. Each curve is centered at its mean response μ_y. All three curves have the same spread, measured by their common standard deviation σ.

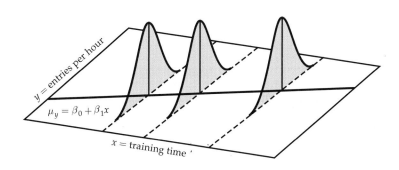

FIGURE 12.2 The statistical model for linear regression. The responses vary within each subpopulation according to a Normal distribution. The mean response is a straight-line function of the explanatory variable.

From data analysis to inference

The data for a simple linear regression problem are the n pairs of (x, y) observations. The model takes each x to be a fixed known quantity, like the hours of training that a clerk receives.[1] The response y for a given x is a Normal random variable. Our regression model describes the mean and standard deviation of this random variable.

We will use Case 12.1 to explain the fundamentals of simple linear regression. In practice, regression calculations are always done by software, so we rely on computer output for the arithmetic. Later in the chapter, we show formulas for doing the calculations. These formulas are useful in understanding analysis of variance (see Section 12.3) and multiple regression (see Chapter 13).

CASE 12.1 **The Relationship between Income and Education for Entrepreneurs**
Numerous studies have shown that better-educated employees have higher incomes. Is this also true for entrepreneurs? Does more years of formal education translate into higher income? We know about the extremely successful entrepreneurs, such as Oprah Winfrey and her amazing rags-to-riches story. Cases like this, however, are anecdotal and most likely not representative of the population of entrepreneurs. One study explored this question using the National Longitudinal Survey of Youth (NLSY), which followed a large group of individuals aged 14 to 22 for roughly 10 years.[2] The researchers studied both employees and entrepreneurs, but we just focus on entrepreneurs here.

The researchers defined *entrepreneurs* as those individuals who were self-employed or who were the owner/director of an incorporated business. For each of these individuals, they recorded the education level and income. The education level (Educ) was defined as the years of completed schooling prior to starting the business. The income level (Inc) was the average annual total earnings since starting the business.

smoothed curve, p. 69

We consider a random sample of 100 entrepreneurs. Figure 12.3 is a scatterplot of the data with a fitted smoothed curve to help us visualize the relationship. The explanatory variable x is the entrepreneur's education level. The response variable y is the income level. ■

Let's briefly review some of the ideas from Chapter 2 regarding least-squares regression. We always start with a plot of the data, as in Figure 12.3,

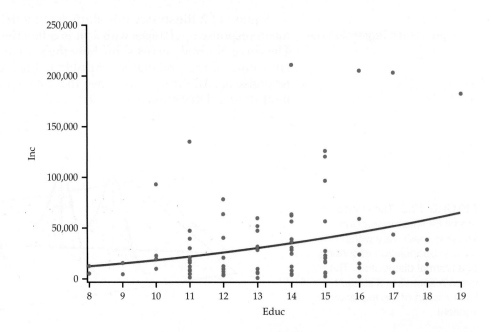

FIGURE 12.3 Scatterplot, with smoothed curve, of average annual income versus years of education for a sample of 100 entrepreneurs.

FIGURE 12.4 Scatterplot, with smoothed curve (black) and regression line (red), of log average annual income versus years of education for a sample of 100 entrepreneurs. The smoothed curve is almost the same as the least-squares regression line.

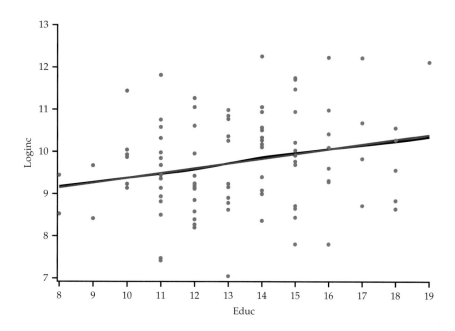

to verify that the relationship is approximately linear with no outliers. *There is no point in fitting a linear model if the relationship does not, at least approximately, appear linear.* For the data of Case 12.1, the smoothed curve looks roughly linear but the distributions of incomes about it are skewed to the right. At each education level, there are many small incomes and just a few very large incomes. It also looks like the smoothed curve is being pulled toward those very large incomes, suggesting those observations could be influential.

influential observations, p. 95

A common remedy for a skewed variable such as income is to consider transforming it prior to fitting a model. Here, the researchers considered the natural logarithm of income (Loginc). Figure 12.4 is a scatterplot of Loginc versus Educ with a fitted curve and the least-squares regression line. The smoothed curve nearly overlaps the fitted line, suggesting a very linear association. In addition, the observations in the y direction are more equally dispersed above and below this fitted line than with the curve in Figure 12.3. Lastly, those four very large incomes no longer appear to be influential. Given these results, we continue our discussion of least-squares regression using the transformed y data.

log transformation, p. 70

EXAMPLE 12.1

CASE 12.1

ENTRE

Prediction of Loginc from Educ The fitted line in Figure 12.4 is the least-squares regression line for predicting y (log income) from x (years of formal schooling). The equation of this line is

$$\hat{y} = 8.2546 + 0.1126x$$

or

$$\text{predicted Loginc} = 8.2546 + 0.1126 \times \text{Educ}$$

We can use the least-squares regression equation to find the predicted log income corresponding to a given education level. The difference between the observed value and the predicted value is the residual. For example, Entrepreneur 4 has 15 years of formal schooling and a log income of $y = 10.2274$. The predicted log income of this person is

residuals, p. 90

$$\hat{y} = 8.2546 + (0.1126)(15) = 9.9436$$

so the residual is

$$y - \hat{y} = 10.2274 - 9.9436 = 0.2838 \ \blacksquare$$

Recall that the least-squares line is the line that minimizes the sum of the squares of the residuals. The least-squares regression line also always passes through the point $(\bar{x}, \bar{y})$. These are helpful facts to remember when considering the fit of this line to a data set. You can also use the *Correlation and Regression* applet, introduced in Chapter 2, to visually explore residuals and the properties of the least-squares line.

In Section 2.2 (page 74), we discussed the correlation as a measure of linear association between two quantitative variables. In Section 2.3, we learned to interpret the square of the correlation as the fraction of the variation in y that is explained by x in a simple linear regression.

interpretation of r^2, p. 88

EXAMPLE 12.2

CASE 12.1 **Correlation between Loginc and Educ** For Case 12.1, the correlation between log income and education level is $r = 0.2394$. Because the squared correlation $r^2 = 0.0573$, indicating that the change in Loginc along the regression line as Educ increases explains only 5.7% of the variation. The remaining 94.3% is due to other differences among these entrepreneurs. The entrepreneurs in this sample live in different parts of the United States; some are single and others are married, and some may have had a difficult upbringing. All of these factors could be associated with income and, therefore, add to the variability if they are not included in the model. ∎

APPLY YOUR KNOWLEDGE

CASE 12.1 **12.1 Predict Loginc.** In Case 12.1, Entrepreneur 12 has Educ = 13 years and a log income of $y = 10.7649$. Using the least-squares regression equation in Example 12.1, find the predicted Loginc and the residual for this individual.

12.2 Draw the fitted line. Suppose you fit 10 pairs of (x, y) data using least squares. Draw the fitted line if $\bar{x} = 5$, $\bar{y} = 4$, and the residual for the pair $(3, 4)$ is 1.

Having reviewed the basics of least-squares regression, we are now ready to discuss inference for regression. To do this:

- We regard the 100 entrepreneurs for whom we have data as a simple random sample from the population of all entrepreneurs in the United States.

- We use the regression line calculated from this sample as a basis for inference about the population. For example, for a given level of education, we want not just a prediction, but a prediction with a margin of error and a level of confidence for the log income of any entrepreneur in the United States.

Our statistical model assumes that the responses y are Normally distributed with a mean μ_y that depends upon x in a linear way. Specifically, the population regression line

$$\mu_y = \beta_0 + \beta_1 x$$

describes the relationship between the mean log income μ_y and the number of years of formal education x in the population. The slope β_1 is the average change in log income for each additional year of education. It turns out that a change in natural logs is a good approximation for the percent change [see Example 14.11 (page 698) for more details]. Thus, another way to view β_1 in

extrapolation, p. 100

this setting is as the average percent change in income for an additional year of education. The intercept β_0 is the mean log income when an entrepreneur has $x = 0$ years of formal education. This parameter, by itself, is not interesting in this example because zero years of education is very unusual. The value $x = 0$ is also well outside the data's range.

Because the means μ_y lie on the line $\mu_y = \beta_0 + \beta_1 x$, they are all determined by β_0 and β_1. Thus, once we have estimates of β_0 and β_1, the linear relationship determines the estimates of μ_y for all values of x. Linear regression allows us to do inference not only for those subpopulations for which we have data, but also for those subpopulations corresponding to x's not present in the data. These x-values can be both within and outside the range of observed x's. *Use extreme caution when predicting outside the range of the observed x's, because there is no assurance that the same linear relationship between μ_y and x holds.*

We cannot observe the population regression line because the observed responses y vary about their means. In Figure 12.4, we see the least-squares regression line that describes the overall pattern of the data, along with the scatter of individual points about this line. The statistical model for linear regression makes the same distinction, as shown in Figure 12.2 with the line and three Normal curves. The population regression line describes the on-the-average relationship, whereas the Normal curves describe the variability in y for each value of x.

As we did in Chapter 9, we can think of this regression model as being of the form

DATA = FIT + RESIDUAL, p. 464

$$\text{DATA} = \text{FIT} + \text{RESIDUAL}$$

The FIT part of the model consists of the subpopulation means, given by the expression $\beta_0 + \beta_1 x$. The RESIDUAL part represents deviations of the data from the line of population means.

The model assumes that these deviations are Normally distributed with standard deviation σ. We use ε (the lowercase Greek letter epsilon) to stand for the RESIDUAL part of the statistical model. A response y is the sum of its mean and a chance deviation ε from the mean. The deviations ε represent "noise"—that is variations in y due to other causes that prevent the observed (x,y)-values from forming a perfectly straight line.

SIMPLE LINEAR REGRESSION MODEL

Given n observations of the explanatory variable x and the response variable y,

$$(x_1, y_1), (x_2, y_2), \ldots, (x_n, y_n)$$

The **statistical model for simple linear regression** states that the observed response y_i when the explanatory variable takes the value x_i is

$$y_i = \beta_0 + \beta_1 x_i + \varepsilon_i$$

Here, $\mu_y = \beta_0 + \beta_1 x_i$ is the mean response when $x = x_i$. The deviations ε_i are independent and Normally distributed with mean 0 and standard deviation σ.

The parameters of the model are β_0, β_1, and σ.

Use of a simple linear regression model can be justified in a wide variety of circumstances. Sometimes, we observe the values of two variables, and we formulate a model with one of these as the response variable and the other as the explanatory variable. This is the setting for Case 12.1, where the response variable is log income (Loginc) and the explanatory variable is the number of years of formal education (Educ). In other settings, the values of the explanatory variable are chosen by the persons designing the study. The scenario illustrated by Figure 12.2 is an example. Here, the explanatory variable is training time, which is set at a few carefully selected values. The response variable is the number of entries per hour.

APPLY YOUR KNOWLEDGE

12.3 Understanding a linear regression model. Consider a linear regression model for the number of financial entries per hour with $\mu_y = 56.82 + 2.4x$ and standard deviation $\sigma = 4.4$. The explanatory variable x is the number of hours of hands-on training.

(a) What is the slope of the population regression line?

(b) Explain clearly what this slope says about the change in the mean of y for an additional hour of training.

(c) What is the intercept of the population regression line?

(d) Explain clearly what this intercept says about the mean number of entries per hour.

12.4 Understanding a linear regression model, continued. Refer to the previous exercise.

(a) What is the subpopulation mean when $x = 3$ hours?

(b) What is the subpopulation distribution when $x = 3$ hours?

(c) Between what two values would approximately 95% of the observed responses y fall when $x = 3$ hours?

For the simple linear regression model to be valid, one essential assumption is that the relationship between the means of the response variable for the different values of the explanatory variable is approximately linear. This is the FIT part of the model. Another essential assumption concerns the RESIDUAL part of the model. The assumption states that the deviations are an SRS from a Normal distribution with mean zero and standard deviation σ. If the data are collected through some sort of random sampling, the SRS assumption is often easy to justify. This is the case in our two scenarios, in which both variables are observed in a random sample from a population or the response variable is measured at several predetermined values of the explanatory variable that were randomly assigned to clerks.

In many other settings, particularly in business applications, we analyze all of the data available and there is no random sampling. Here, we often justify the use of inference for simple linear regression by viewing the data as coming from some sort of process. Here is one example.

EXAMPLE 12.3

Profits and Foot Traffic Panera Bread wants to select the location for a new store. To help with this decision, company managers use information from all the current stores to determine the relationship between profits and foot traffic outside the establishment. The regression model they use says that

$$\text{Profits} = \beta_0 + \beta_1 \times \text{Foot Traffic} + \varepsilon$$

The slope β_1 is, as usual, a rate of change: it is the expected increase in annual profits associated with each additional person walking by the store. The intercept β_0 is needed to describe the line but has no interpretive importance because no stores have zero foot traffic. Nevertheless, foot traffic does not completely determine profit. The ε term in the model accounts for differences among individual stores with the same foot traffic. A store's proximity to other restaurants, for example, could be important but is not included in the FIT part of the model. In Chapter 13, we consider moving variables like this out of the RESIDUAL part of the model by allowing for more than one explanatory variable in the FIT part. ■

APPLY YOUR KNOWLEDGE

12.5 U.S. versus overseas stock returns. Returns on common stocks in the United States and overseas appear to be growing more closely correlated as various countries' economies become more interdependent. Suppose that the following population regression line connects the total annual returns (in percent) on two indexes of stock prices:

$$\text{Mean overseas return} = -0.3 + 0.12 \times \text{U.S. Return}$$

(a) What is β_0 in this line? What does this number say about overseas returns when the U.S. market is flat (0% return)?

(b) What is β_1 in this line? What does this number say about the relationship between U.S. and overseas returns?

(c) We know that overseas returns will vary in years that have the same return on U.S. common stocks. Write the regression model based on the population regression line given in the problem statement. What part of this model allows overseas returns to vary when U.S. returns remain the same?

12.6 Fixed and variable costs. In some mass-production settings, there is a linear relationship between the number x of units of a product in a production run and the total cost y of making these x units.

(a) Write a population regression model to describe this relationship.

(b) The fixed cost is the component of total cost that does not change as x increases. Which parameter in your model is the fixed cost?

(c) Which parameter in your model shows how total cost changes as more units are produced? Do you expect this number to be greater than 0 or less than 0? Explain your answer.

(d) Actual data from several production runs will not fall directly on a straight line. What term in your model allows variation among runs of the same size x?

Estimating the regression parameters

The method of least squares presented in Chapter 2 fits the least-squares line to summarize the relationship between the observed values of an explanatory variable and a response variable. Now we want to use this line as a basis for inference about a population from which our observations are a sample. In this setting, the slope b_1 and intercept b_0 of the least-squares line

$$\hat{y} = b_0 + b_1 x$$

estimate the slope β_1 and the intercept β_0 of the population regression line, respectively.

This inference should be done only when the statistical model for regression is reasonable. Model checks are needed and some judgment is required. Because many of these checks rely on the residuals, let's briefly review the methods introduced in Chapter 2 for fitting the linear regression model to data and then discuss the model checks.

Using the formulas from Chapter 2, the slope of the least-squares line is

$$b_1 = r \frac{s_y}{s_x}$$

and the intercept is

$$b_0 = \bar{y} - b_1 \bar{x}$$

correlation, p. 75

Here, r is the correlation between the observed values of y and x, s_y is the standard deviation of the sample of y's, and s_x is the standard deviation of the sample of x's. Notice that if the estimated slope is 0, so is the correlation, and vice versa. We discuss this connection in more depth later in this section.

The remaining parameter to be estimated is σ, which measures the variation of y about the population regression line. More precisely, σ is the standard deviation of the Normal distribution of the deviations ε_i in the regression model. We don't observe these ε_i, so how can we estimate σ?

residuals, p. 90

Recall that the vertical deviations of the points in a scatterplot from the fitted regression line are the residuals. We use e_i for the residual of the ith observation:

$$e_i = \text{Observed Response} - \text{Predicted Response}$$
$$= y_i - \hat{y}_i$$
$$= y_i - b_0 - b_1 x_i$$

The residuals e_i are the observable quantities that correspond to the unobservable model deviations ε_i. The e_i sum to 0, and the ε_i come from a population with mean 0. Because we do not observe the ε_i, we use the residuals to estimate σ and check the model assumptions of the ε_i.

To estimate σ, we work first with the variance and take the square root to obtain the standard deviation. For simple linear regression, the estimate of σ^2 is the average squared residual

$$s^2 = \frac{1}{n-2} \sum e_i^2$$
$$= \frac{1}{n-2} \sum (y_i - \hat{y}_i)^2$$

We average by dividing the sum by $n - 2$ so as to make s^2 an unbiased estimator of σ^2. We subtract 2 from n because we're using the data to also estimate β_0 and β_1. In addition, it turns out that when any $n - 2$ residuals are known, we can find the other two residuals.

The quantity $n - 2$ is the degrees of freedom of s^2. The estimate of the **regression standard deviation** σ is given by

regression standard deviation σ

$$s = \sqrt{s^2}$$

We call s the *regression standard error*.

ESTIMATING THE REGRESSION PARAMETERS

In the simple linear regression setting, we use the **slope b_1** and **intercept b_0** of the least-squares regression line to estimate the slope β_1 and intercept β_0 of the population regression line, respectively.

The standard deviation σ in the model is estimated by the **regression standard error**

$$s = \sqrt{\frac{1}{n-2} \sum (y_i - \hat{y}_i)^2}$$

In practice, we use software to calculate b_1, b_0, and s from the (x, y) pairs of data. Here are the results for the income example of Case 12.1.

EXAMPLE 12.4

ENTRE

Reading Simple Regression Output Figure 12.5 displays Excel output for the regression of log income (Loginc) on years of education (Educ) for our sample of 100 entrepreneurs in the United States. In this output, we find the correlation $r = 0.2394$ and the squared correlation that we used in Example 12.2, along with the intercept and slope of the least-squares line. The regression standard error s is labeled simply "Standard Error."

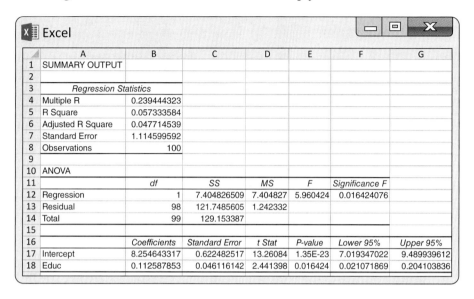

FIGURE 12.5 Excel output for the regression of log average income on years of education, for Example 12.4.

The three parameter estimates are

$$b_0 = 8.254643317 \quad b_1 = 0.112587853 \quad s = 1.114599592$$

After rounding, the fitted regression line is

$$\hat{y} = 8.2546 + 0.1126x$$

As usual, we ignore the parts of the output that we do not yet need. We will return to the output for additional information later.

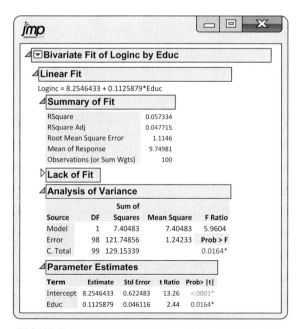
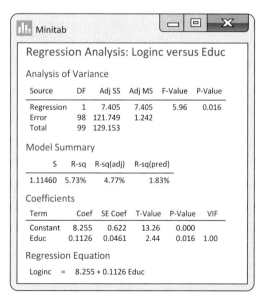

FIGURE 12.6 JMP, Minitab, and R outputs for the regression of log average income on years of education. The data are the same as in Figure 12.5.

FIGURE 12.6 Continued

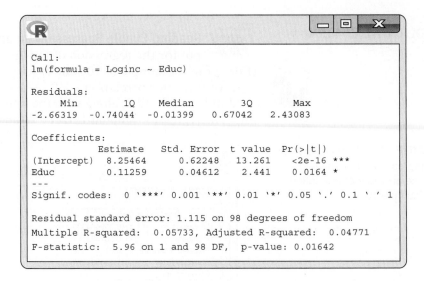

Figure 12.6 shows the regression output from three other software packages. Although the formats differ, you should be able to find the results you need. Once you know what to look for, you can understand statistical output from almost any software. ■

APPLY YOUR KNOWLEDGE

12.7 Research and development spending. The National Science Foundation collects data on research and development spending by universities and colleges in the United States.[3] Here are the data for the years 2012–2015: NSF

Year	2012	2013	2014	2015
Spending (billions of dollars)	65.9	67.1	67.3	68.7

(a) Create a scatterplot that shows the increase in research and development spending over time. Does the pattern suggest that the spending is increasing linearly over time? Explain your answer.

(b) Find the equation of the least-squares regression line for predicting spending from year. Add this line to your scatterplot.

(c) For each of the four years, find the residual. Use these residuals to calculate the regression standard error s. (Do these calculations with a calculator or spreadsheet.)

(d) Write the regression model for this setting. What are your estimates of the unknown parameters in this model?

(e) Use your least-squares equation to predict research and development spending for the year 2016. The actual spending for that year was $72.0 billion. Add this point to your plot and comment on how well the model predicted the actual outcome.

(*Comment:* These are *time series data*. Simple regression is often a good fit to time series data over a limited span of time. See Chapter 14 for methods designed specifically for use with time series.)

Conditions for regression inference

You can fit a least-squares line to any set of explanatory-response data when both variables are quantitative. The simple linear regression model, which is the basis for inference, imposes several conditions on this fit. *We should always verify these conditions before proceeding to inference.* There is no point in trying to do statistical inference if we cannot trust the results.

The conditions concern the population, but we can observe only our sample. Thus, in doing inference, we act as if **the sample is an SRS from the population.** For the study described in Case 12.1, the researchers used a national survey. Participants were chosen to be a representative sample of the United States, so we can treat this sample as an SRS. *The potential for bias should always be considered, especially when the sample includes volunteers.*

outliers and influential observations, p. 95

The next condition is that **there is a linear relationship in the population,** described by the population regression line. We can't observe the population line, so we check this condition by asking if the sample data show a roughly linear pattern in a scatterplot. We also check for any outliers or influential observations that could affect the least-squares fit.

The model also says that **the standard deviation of the responses about the population line is the same for all values of the explanatory variable.** In practice, this means the spread in the observations above and below the least-squares line should be roughly the same as x varies.

residual plots, p. 91

Plotting the residuals against the explanatory variable or against the predicted values is a helpful and frequently used visual aid to check both of these conditions. This technique is often better than creating a scatterplot because a residual plot magnifies any patterns that exist. The residual plot in Figure 12.7 for the data of Case 12.1 looks satisfactory. There is no obvious pattern in the residuals versus x, no data points seem out of the ordinary, and the residuals appear equally dispersed throughout the range of the explanatory variable.

Normal quantile plot, p. 53

The final condition is that **the response varies Normally about the population regression line.** If that is the case, we expect the residuals e_i to also be Normally distributed.[4] A Normal quantile plot or histogram of the residuals is commonly used to check this condition. For the data of Case 12.1, a Normal quantile plot of the residuals (Figure 12.8) shows no serious deviations

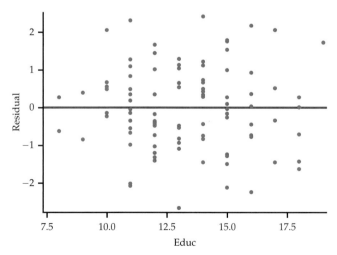

FIGURE 12.7 Plot of the regression residuals against the explanatory variable for the annual income data.

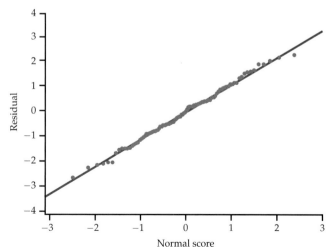

FIGURE 12.8 Normal quantile plot of the regression residuals for the average annual income data.

from a Normal distribution. The data give us no reason to doubt the simple linear regression model, so we proceed to inference.

Notice that Normality of the distributions of the response and explanatory variables is not required. The Normality condition applies to the distribution of the model deviations, which we assess using the residuals. For the entrepreneur problem, we transformed y to get a more linear relationship and residuals that are more Normal with constant variance. The fact that the distribution of the transformed y approaches Normality is purely a coincidence.

While not the case here, sometimes x is not a fixed known quantity but rather is measured with error. Even if all the conditions for linear regression are satisfied, *this regression model is not appropriate if the error in measuring x is large relative to the spread of the x's*. If this is a concern, seek expert advice, as more advanced inference methods are needed.

LINEAR REGRESSION MODEL CONDITIONS

To use the least-squares line as a basis for inference about a population, each of the following conditions should be approximately met:

- The sample is an SRS from the population.
- There a linear relationship between x and y.
- The standard deviation of the responses y about the population regression line is the same for all x.
- The model deviations are Normally distributed.

Confidence intervals and significance tests

Chapter 8 presented confidence intervals and significance tests for means and differences in means. In each case, inference rested on the standard errors of estimates and on t distributions. Inference for the slope and intercept in linear regression is similar in principle. For example, the t^* confidence intervals have the form

$$\text{estimate} \pm t^* \text{SE}_{\text{estimate}}$$

where t^* is a critical value of a t distribution. It is the formulas for the estimate and standard error that are different.

Confidence intervals and tests for the slope and intercept are based on the sampling distributions of the estimates b_1 and b_0. Here are some important facts about these sampling distributions when the simple linear regression model is true:

- Both b_1 and b_0 have Normal distributions.
- The mean of b_1 is β_1 and the mean of b_0 is β_0. That is, the slope and intercept of the fitted line are unbiased estimators of the slope and intercept of the population regression line.

unbiased estimator, p. 300

- The standard deviations of b_1 and b_0 are multiples of the regression standard deviation σ. (We give details later.)

central limit theorem, p. 313

Normality of b_1 and b_0 is a consequence of Normality of the individual deviations ε_i in the regression model. If the ε_i are not Normal, a general form of the central limit theorem tells us that the distributions of b_1 and b_0 will be approximately Normal when we have a large sample. On the one hand, this means **regression inference is robust against moderate lack of Normality.** *On the other hand, outliers and influential observations can invalidate the results of inference for regression.*

Because b_1 and b_0 have Normal sampling distributions, standardizing these estimates gives standard Normal z statistics. The standard deviations of these estimates are multiples of σ. Because we do not know σ, we estimate it by s, the regression standard error. When we do this, we get t distributions with degrees of freedom $n - 2$, the degrees of freedom of s. We give formulas for the standard errors SE_{b_1} and SE_{b_0} in Section 12.3. For now, we concentrate on the basic ideas and let software do the calculations.

INFERENCE FOR THE REGRESSION SLOPE

A **level C confidence interval** for the slope β_1 of the population regression line is

$$b_1 \pm t^* \text{SE}_{b_1}$$

In this expression, t^* is the value for the $t(n - 2)$ density curve with area C between $-t^*$ and t^*. The **margin of error** is $m = t^* \text{SE}_{b_1}$.

To test the hypothesis $H_0: \beta_1 = \beta_1^*$, compute the ***t* statistic**

$$t = \frac{b_1 - \beta_1^*}{\text{SE}_{b_1}}$$

Most software provides the test of the hypothesis $H_0: \beta_1 = 0$. In that case, the t statistic reduces to

$$t = \frac{b_1}{\text{SE}_{b_1}}$$

The **degrees of freedom** are $n - 2$. In terms of a random variable T having the $t(n - 2)$ distribution, the P-value for a test of H_0 against

$H_a: \beta_1 > \beta_1^*$ is $P(T \geq t)$

$H_a: \beta_1 < \beta_1^*$ is $P(T \leq t)$

$H_a: \beta_1 \neq \beta_1^*$ is $2P(T \geq |t|)$

Formulas for confidence intervals and significance tests for the intercept β_0 are exactly the same, replacing b_1 and SE_{b_1} by b_0 and its standard error SE_{b_0}, respectively. *Although computer outputs may include a test of $H_0: \beta_0 = 0$, this information often has little practical value.* From the equation for the population regression line, $\mu_y = \beta_0 + \beta_1 x$, we see that β_0 is the mean response corresponding to $x = 0$. In many situations, this subpopulation does not exist or is not interesting. That is the case for Case 12.1, but Exercises 12.5 and 12.6 (page 577) are two settings where this information is meaningful.

The test of $H_0: \beta_1 = 0$ is always quite useful. When we substitute $\beta_1 = 0$ in the model, the x term drops out and we are left with

$$\mu_y = \beta_0$$

This model says that the mean of y does not vary with x. In other words, all the y's come from a single population with mean β_0, which we would estimate by $\bar{y}$ and then perform inference using the methods of Section 8.1. The hypothesis H_0: $\beta_1 = 0$, therefore, says that there is no straight-line relationship between y and x and that linear regression of y on x is of no value for predicting y.

EXAMPLE 12.5

ENTRE

Does Loginc Increase with Educ? The Excel regression output in Figure 12.5 (page 579) for the entrepreneur problem contains the information needed for inference about the regression coefficients. You can see that the slope of the least-squares line is $b_1 = 0.1126$ and the standard error of this statistic is $\text{SE}_{b_1} = 0.0461$.

Given that the response y is on the log scale, this slope also approximates the percent change in the original variable for a unit change in x. In this case, one extra year of education is associated with an increase in income of approximately 11.3%.

A 95% confidence interval for the slope β_1 of the regression line in the population of all entrepreneurs in the United States is

$$b_1 \pm t^* \text{SE}_{b_1} = 0.1126 \pm (1.984)(0.0461)$$
$$= 0.1126 \pm 0.0915$$
$$= 0.0211 \text{ to } 0.2041$$

This interval contains only positive values, suggesting an increase in Loginc for an additional year of schooling. In terms of percent change, we are 95% confident that the average increase in income for one additional year of education is between 2.1% and 20.4%.

The t statistic and P-value for the test of H_0: $\beta_1 = 0$ against the two-sided alternative H_a: $\beta_1 \neq 0$ appear in the columns labeled "t Stat" and "P-value." The t statistic for the significance of the regression is

$$t = \frac{b_1}{\text{SE}_{b_1}} = \frac{0.1126}{0.0461} = 2.44$$

and the P-value for the two-sided alternative is 0.0164. If we expected beforehand that income rises with education, our alternative hypothesis would be one-sided, H_a: $\beta_1 > 0$. The P-value for this H_a is one-half the two-sided value given by Excel; that is, $P = 0.0082$. In both cases, there is strong evidence that the mean log income level increases as education increases.

The t distribution for this problem has $n - 2 = 98$ degrees of freedom. Table D has no row for 98 degrees of freedom. In Excel, the critical value and P-value can be obtained by using the functions = T.INV(0.975, 98) and = T.DIST.2T(2.44, 98), respectively. If you do not have access to software, we suggest taking a conservative approach and using the next *lower* degrees of freedom in Table D (80 degrees of freedom). This makes our interval a bit wider than we actually need for 95% confidence and the P-value a bit larger. ∎

conservative, p. 421

elasticity

In this example, we can discuss percent change in income for a unit change in education because the response variable y is on the log scale and x is not. In business and economics, we often encounter models in which both variables are on the log scale. In these cases, the slope approximates the percent change in y for a 1% change in x. This relationship is known as **elasticity,** a very important concept in economic theory.

APPLY YOUR KNOWLEDGE

Treasury bills and inflation. *When inflation is high, lenders require higher interest rates to make up for the loss of purchasing power of their money while it is loaned out. Table 12.1 displays the return for six-month Treasury bills (annualized) and the rate of inflation as measured by the change in the government's Consumer Price Index in the same year.[5] An inflation rate of 5% means that the same set of goods and services costs 5% more. The data cover 60 years, from 1958 to 2017. Figure 12.9 is a scatterplot of these data. Figure 12.10 shows Excel regression output for predicting T-bill return from inflation rate. Exercises 12.8 through 12.10 ask you to use this information.* INFLAT

12.8 Look at the data. Give a brief description of the form, direction, and strength of the relationship between the inflation rate and the return on Treasury bills. What is the equation of the least-squares regression line for predicting T-bill return?

12.9 Is there a relationship? What are the slope b_1 of the fitted line and its standard error? Use these numbers to test by hand the hypothesis that there is no straight-line relationship between inflation rate and T-bill return against the alternative that the return on T-bills increases as the rate of inflation increases. State the hypotheses, give both the t statistic and its degrees of freedom, and use Table D to approximate the P-value. Then compare your results with those given by Excel. (Excel's P-value rounded to 2.40E-10 is shorthand for 0.00000000024. We would report this as "< 0.0001.")

TABLE 12.1 Return on Treasury bills and rate of inflation

Year	T-bill percent	Inflation percent	Year	T-bill percent	Inflation percent	Year	T-bill percent	Inflation percent
1958	3.01	1.76	1978	7.58	9.02	1998	4.83	1.61
1959	3.81	1.73	1979	10.04	13.20	1999	4.75	2.68
1960	3.20	1.36	1980	11.32	12.50	2000	5.90	3.39
1961	2.59	0.67	1981	13.81	8.92	2001	3.34	1.55
1962	2.90	1.33	1982	11.06	3.83	2002	1.68	2.38
1963	3.26	1.64	1983	8.74	3.79	2003	1.05	1.88
1964	3.68	0.97	1984	9.78	3.95	2004	1.58	3.26
1965	4.05	1.92	1985	7.65	3.80	2005	3.39	3.42
1966	5.06	3.46	1986	6.02	1.10	2006	4.81	2.54
1967	4.61	3.04	1987	6.03	4.43	2007	4.44	4.08
1968	5.47	4.72	1988	6.91	4.42	2008	1.62	0.09
1969	6.86	6.20	1989	8.03	4.65	2009	0.28	2.73
1970	6.51	5.57	1990	7.46	6.11	2010	0.20	1.50
1971	4.52	3.27	1991	5.44	3.06	2011	0.10	2.96
1972	4.47	3.41	1992	3.54	2.90	2012	0.13	1.74
1973	7.20	8.71	1993	3.12	2.75	2013	0.09	1.50
1974	7.95	12.34	1994	4.64	2.67	2014	0.06	0.76
1975	6.10	6.94	1995	5.56	2.54	2015	0.16	0.73
1976	5.26	4.86	1996	5.08	3.32	2016	0.46	2.07
1977	5.52	6.70	1997	5.18	1.70	2017	1.05	2.11

FIGURE 12.9 Scatterplot of the percent return on Treasury bills against the rate of inflation the same year, for Exercises 12.8 to 12.10.

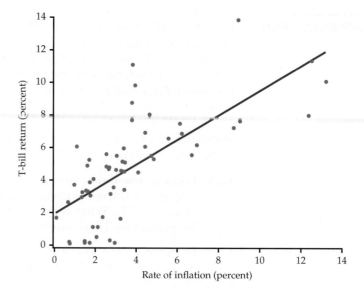

FIGURE 12.10 Excel output for the regression of the percent return on Treasury bills against the rate of inflation the same year, for Exercises 12.8 to 12.10.

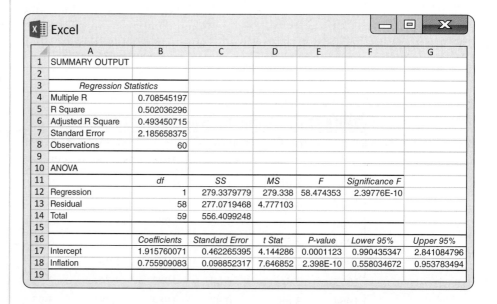

12.10 Estimating the slope. Using Excel's values for b_1 and its standard error, find a 95% confidence interval for the slope β_1 of the population regression line. Compare your result with Excel's 95% confidence interval. What does the confidence interval tell you about the change in the T-bill return rate for a 1% increase in the inflation rate?

The word "regression"

To "regress" means to go backward. Why are statistical methods for predicting a response from an explanatory variable called "regression"? Sir Francis Galton (1822–1911) was the first to apply regression to biological and psychological data. He looked at examples such as the heights of children versus the heights of their parents. He found that the taller-than-average parents tended to have children who were also taller than average, but not as tall as their parents. Galton called this fact "regression toward mediocrity," and the name came to be applied to the statistical method. Galton also invented the correlation coefficient r and named it "correlation."

Why are the children of tall parents shorter on the average than their parents? The parents are tall in part because of their genes. But they are also tall in part by chance. Looking at tall parents selects those in whom chance produced height. Their children inherit their genes, but not necessarily their good luck. As a group, the children are taller than average (genes), but their heights vary by chance about the average, some upward and some downward. The children, unlike the parents, were not selected because they were tall and thus, on average, are shorter. A similar argument can be used to describe why children of short parents tend to be taller than their parents.

Here's another example. Students who score at the top on the first exam in a course are likely to do less well on the second exam. Does this show that they stopped studying? No—they scored high in part because they knew the material but also in part because they were lucky. On the second exam, they may still know the material but be less lucky. As a group, they will still do better than average but not as well as they did on the first exam. The students at the bottom on the first exam will tend to move up on the second exam, for the same reason.

regression fallacy

The **regression fallacy** is the assertion that *regression toward the mean* shows that there is some systematic effect at work: students with top scores now work less hard, or managers of last year's best-performing mutual funds lose their touch this year, or heights get less variable with each passing generation as tall parents have shorter children and short parents have taller children. The Nobel economist Milton Friedman says, "I suspect that the regression fallacy is the most common fallacy in the statistical analysis of economic data."[6] Beware.

APPLY YOUR KNOWLEDGE

12.11 Hot funds? Explain carefully to a naive investor why the mutual funds that had the highest returns this year will, as a group, probably do less well relative to other funds next year.

12.12 Mediocrity triumphant? In the early 1930s, a man named Horace Secrist wrote a book titled *The Triumph of Mediocrity in Business*. Secrist found that businesses that did unusually well or unusually poorly in one year tended to be nearer the average in profitability at a later year. Why is it a fallacy to say that this fact demonstrates an overall movement toward "mediocrity"?

Inference about correlation

The correlation between log income and level of education for the 100 entrepreneurs is $r = 0.2394$. This value appears in the Excel output in Figure 12.5 (page 579), where it is labeled "Multiple R."[7] We might expect a positive correlation between these two measures in the population of all entrepreneurs in the United States. Is the sample result convincing evidence that this is true?

population correlation ρ

This question concerns a new population parameter, the **population correlation**. This is the correlation between the log income and level of education when we measure these variables for every member of the population. We call the population correlation ρ, the Greek letter rho. To assess the evidence that $\rho > 0$ in the population, we must test the hypotheses

$$H_0: \rho = 0$$
$$H_a: \rho > 0$$

It is natural to base the test on the sample correlation $r = 0.2394$. Indeed, most computer packages with routines to calculate sample correlations provide the result of this significance test. We can also use regression software by exploiting the close link between correlation and the regression slope. The population correlation ρ is zero, positive, or negative exactly when the slope β_1 of the population regression line is zero, positive, or negative, respectively. In fact, the t statistic for testing H_0: $\beta_1 = 0$ also tests H_0: $\rho = 0$. What is more, this t statistic can be written in terms of the sample correlation r.

TEST FOR ZERO POPULATION CORRELATION

To test the hypothesis H_0: $\rho = 0$, either use the t statistic for the regression slope or compute this statistic from the sample correlation r:

$$t = \frac{r\sqrt{n-2}}{\sqrt{1-r^2}}$$

This t statistic has $n - 2$ degrees of freedom.

EXAMPLE 12.6

ENTRE

Correlation between Loginc and Educ The sample correlation between Loginc and Educ is $r = 0.2394$ for a sample of size $n = 100$. Figure 12.11 contains Minitab output for this correlation calculation. Minitab calls this a Pearson correlation to distinguish it from other kinds of correlations it can calculate. The P-value for a two-sided test of H_0: $\rho = 0$ is 0.016 and the P-value for our one-sided alternative is 0.008.

We can also get this result from the Excel output in Figure 12.5 (page 579). In the "Educ" line, notice that $t = 2.441$ with two-sided P-value 0.0164. Thus, $P = 0.00082$ for our one-sided alternative.

Finally, we can calculate t directly from r as follows:

$$t = \frac{r\sqrt{n-2}}{\sqrt{1-r^2}}$$

$$= \frac{0.2394\sqrt{100-2}}{\sqrt{1-(0.2394)^2}}$$

$$= \frac{2.3699}{0.9709} = 2.441$$

If we are not using software, we can compare $t = 2.441$ with critical values from the t table (Table D) with 80 (largest row less than or equal to $n - 2 = 98$) degrees of freedom. ∎

The alternative formula for the test statistic is convenient because it uses only the sample correlation r and the sample size n. Remember that correlation, unlike regression, does not require a distinction between the explanatory and response variables. For variables x and y, there are two regressions (y on x and x on y) but just one correlation. Both regressions produce the same t statistic.

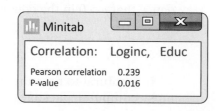

FIGURE 12.11 Minitab output for the correlation between log average income and years of education, for Example 12.6.

The distinction between the regression setting and correlation is important only for understanding the conditions under which the test for zero population correlation makes sense. In the regression model, we take the values of the explanatory variable x as given. The values of the response y are Normal random variables, with means that are a straight-line function of x. In the model for testing correlation, we think of the setting where we obtain a random sample from a population and measure both x and y. Both are assumed to be Normal random variables. In fact, they are taken to be **jointly Normal.** This implies that the conditional distribution of y for each possible value of x is Normal, just as in the regression model.

jointly Normal

APPLY YOUR KNOWLEDGE

12.13 T-bills and inflation. We expect the interest rates on Treasury bills to rise when the rate of inflation rises and to fall when inflation falls. That is, we expect a positive correlation between the return on T-bills and the inflation rate.

(a) Find the sample correlation r for the 60 years in Table 12.1 in the Excel output in Figure 12.10 (page 586).

(b) From r, calculate the t statistic for testing correlation. What are its degrees of freedom? Use Table D to give an approximate P-value. Compare your result with the P-value from part (a).

(c) Verify that your t for correlation calculated in part (b) has the same value as the t for slope in the Excel output.

CASE 12.1 **12.14 Two regressions.** We have regressed Loginc on Educ, with the results appearing in Figures 12.5 and 12.6. Use software to regress Educ on Loginc for the same data. ENTRE

(a) What is the equation of the least-squares line for predicting years of education from log income? Is it a different line than the regression line in Figure 12.4? To answer this question, plot two points for each equation and draw a line connecting them.

(b) Verify that the two lines cross at the mean values of the two variables. That is, substitute the mean Educ into the line in Figure 12.5, and show that the predicted log income equals the mean of Loginc of the 100 subjects. Then substitute the mean Loginc into your new line, and show that the predicted years of education equals the mean Educ for the entrepreneurs.

(c) Verify that the two regressions give the same value of the t statistic for testing the hypothesis of zero population slope. You could use either regression to test the hypothesis of zero population correlation.

SECTION 12.1 SUMMARY

- **Least-squares regression** fits a straight line to data to predict a quantitative response variable y from a quantitative explanatory variable x. Inference about regression requires additional conditions.

- The **simple linear regression model** says that a **population regression line** $\mu_y = \beta_0 + \beta_1 x$ describes how the mean response in an entire population varies as x changes. The observed response y for any x has a Normal distribution with a mean given by the population regression line and with the same standard deviation σ for any value of x.

- The **parameters** of the simple linear regression model are the intercept β_0, the slope β_1, and the regression standard deviation σ. The **slope** b_1 and intercept b_0 of the least-squares line estimate the slope β_1 and intercept β_0 of the population regression line, respectively.

- The parameter σ is estimated by the **regression standard error**

$$s = \sqrt{\frac{1}{n-2}\sum(y_i - \hat{y}_i)^2}$$

where the differences between the observed and predicted responses are the **residuals**

$$e_i = y_i - \hat{y}_i$$

- Prior to inference, always examine the residuals for Normality, constant variance, and any other remaining patterns in the data. **Plots of the residuals** are commonly used as part of this examination.

- The regression standard error s has $n - 2$ **degrees of freedom.** Inference about β_0 and β_1 uses t distributions with $n - 2$ degrees of freedom.

- **Confidence intervals for the slope** of the population regression line have the form $b_1 \pm t^* SE_{b_1}$. In practice, you will use software to find the slope b_1 of the least-squares line and its standard error SE_{b_1}.

- To test the hypothesis that the population slope is zero, use the t **statistic** $t = b_1 / SE_{b_1}$, also given by software. This null hypothesis says that straight-line dependence on x has no value for predicting y.

- The t test for zero population slope also tests the null hypothesis that the **population correlation** is zero. This t statistic can be expressed in terms of the sample correlation, $t = r\sqrt{n-2}/\sqrt{1-r^2}$.

SECTION 12.1 EXERCISES

For Exercises 12.1 and 12.2, see page 574; for 12.3 and 12.4, see page 576; for 12.5 and 12.6, see page 577; for 12.7, see page 580; for 12.8 to 12.10, see pages 585–586; for 12.11 and 12.12, see page 587; and for 12.13 and 12.14, see page 589.

12.15 Assessment value versus sales price. Real estate is typically assessed annually for property tax purposes. This assessed value, however, is not necessarily the same as the fair market value of the property. Table 12.2 lists the sales price and assessed value for an SRS of 35 properties recently sold in a midwestern county.[8] Both variables are measured in thousands of dollars. HSALES

(a) What proportion have a selling price greater than the assessed value? Do you think this proportion is a good estimate for the larger population of all homes recently sold? Explain your answer.

(b) Make a scatterplot with assessed value on the horizontal axis. Briefly describe the relationship between assessed value and selling price.

(c) Based on the scatterplot, there are two properties with very large assessed values. Do you think it is more appropriate to consider all 35 properties for linear regression analysis or to just consider the 33 properties? Explain your decision.

(d) Report the least-squares regression line for predicting selling price from assessed value using all 35 properties. What is the regression standard error?

(e) Now remove the two properties with the highest assessments and refit the model. Report the least-squares regression line and regression standard error.

(f) Compare the two sets of results. Describe how these large x values impact the results.

12.16 Assessment value versus sales price, continued. Refer to the previous exercise. Let's consider linear regression analysis using all 35 properties. HSALES

(a) Obtain the residuals and plot them versus assessed value. Is there anything unusual to report? Describe the reasoning behind your answer.

(b) Do the residuals appear to be approximately Normal? Describe how you assessed this.

(c) Do you think all the conditions for inference are approximately met? Explain your answer.

(d) Construct a 95% confidence interval for the intercept and slope, and summarize the results.

12.1 Inference about the Regression Model

TABLE 12.2 Sales price and assessed value (in thousands of $) of 35 homes in a midwestern county

Property	Sales price	Assessed value	Property	Sales price	Assessed value	Property	Sales price	Assessed value
1	116.9	94.9	13	200.0	205.6	25	200.0	200.6
2	161.0	160.0	14	146.6	152.9	26	162.5	92.3
3	202.0	233.3	15	215.0	167.4	27	256.8	251.0
4	300.0	255.1	16	125.0	139.3	28	286.0	184.3
5	137.5	123.9	17	139.9	128.2	29	90.0	102.0
6	178.0	157.4	18	238.0	198.2	30	284.3	272.4
7	350.0	395.5	19	120.9	93.4	31	229.9	217.0
8	150.9	126.8	20	142.5	92.3	32	235.0	199.7
9	122.5	109.7	21	282.2	257.6	33	419.0	335.8
10	270.5	241.9	22	279.0	243.5	34	149.0	209.8
11	267.5	254.4	23	110.0	109.2	35	255.4	258.1
12	174.9	135.0	24	130.0	125.1			

12.17 Are the assessment value and sales price different? Refer to the previous two exercises. HSALES

(a) Again create the scatterplot with assessed value on the horizontal axis. If, on average, sales price and the assessed value are the same, the population regression line should be $y = x$. Draw this line on your scatterplot and compare it to the least squares line.

(b) Explain why we cannot simply test $H_0: \beta_1 = 1$ versus the two-sided alternative to assess if the least-squares line is different from $y = x$.

(c) Use methods from Chapter 8 to test the hypothesis that, on average, the sales price equals the assessed value.

12.18 Are female CEOs older? A pair of researchers looked at the age and sex of large sample of CEOs.[9] To investigate the relationship between these two variables, they fit a regression model with age as the response variable and sex as the explanatory variable. The explanatory variable was coded $x = 0$ for males and $x = 1$ for females. The resulting least-squares regression line was

$$\hat{y} = 55.643 - 2.205x$$

(a) What is the expected age for a male CEO ($x = 0$)?

(b) What is the expected age for a female CEO ($x = 1$)?

(c) What is the difference in the expected age of female and male CEOs?

(d) Relate your answers to parts (a) and (c) to the least-squares estimates b_0 and b_1.

(e) The t statistic for testing $H_0: \beta_1 = 0$ was reported as -6.474. Based on this result, what can you conclude about the average ages of female and male CEOs?

(f) To compare the average age of male and female CEOs, the researchers could have instead performed a two-sample t test (Chapter 8). Will this regression approach provide the same result? Explain your answer.

TABLE 12.3 In-state tuition and fees (in dollars) for 33 public universities

School	2013	2017	School	2013	2017	School	2013	2017
Penn State	16,992	18,436	Ohio State	10,037	10,591	Texas	9790	10,136
Pittsburgh	17,100	19,080	Virginia	12,458	16,781	Nebraska	8075	8901
Michigan	13,142	14,826	California–Davis	13,902	14,382	Iowa	8061	8964
Rutgers	13,499	14,638	California–Berkeley	12,864	13,928	Colorado	10,529	12,086
Michigan State	12,908	14,460	California–Irvine	13,149	15,516	Iowa State	7726	8636
Maryland	9161	10,399	Purdue	9992	9992	North Carolina	8340	9005
Illinois	14,750	15,868	California–San Diego	13,302	14,028	Kansas	10,107	10,824
Minnesota	13,618	14,417	Oregon	9763	11,571	Arizona	10,391	11,877
Missouri	10,104	9787	Wisconsin	10,403	10,533	Florida	6263	6381
Buffalo	7022	7976	Washington	12,397	10,974	Georgia Tech	10,650	12,418
Indiana	10,209	10,533	UCLA	12,696	13,749	Texas A&M	8506	10,403

12.19 Public university tuition: 2013 versus 2017.
Table 12.3 shows the in-state undergraduate tuition in 2013 and 2017 for 33 public universities.[10] TUIT

(a) Plot the data with the 2013 tuition on the x axis and describe the relationship. Are there any outliers or unusual values? Does a linear relationship between the tuition in 2013 and 2017 seem reasonable?

(b) Fit the simple linear regression model and give the least-squares regression line and regression standard error.

(c) Obtain the residuals and plot them versus the 2013 tuition amount. Describe anything unusual in the plot.

(d) Do the residuals appear to be approximately Normal? Explain.

(e) Remove any unusual observations and repeat parts (b)–(d).

(f) Compare the two sets of least-squares results. Describe any impact these unusual observations have on the results.

12.20 More on public university tuition. Refer to the previous exercise. Use all 33 observations for this exercise. TUIT

(a) Give the null and alternative hypotheses for examining if there is a linear relationship between 2013 and 2017 tuition amounts.

(b) Write down the test statistic and P-value for the hypotheses stated in part (a). State your conclusions.

(c) Construct a 95% confidence interval for the slope. What does this interval tell you about the annual percent increase in tuition between 2013 and 2017?

(d) The tuition at CashCow U was $9200 in 2013. What is the predicted tuition in 2017?

(e) The tuition at Moneypit U was $18,895 in 2013. What is the predicted tuition in 2017?

(f) Discuss the appropriateness of using the fitted equation to predict tuition for each of these universities.

12.21 The timing of initial public offerings.
Initial public offerings (IPOs) have tended to group together in time and in sector of business. Some researchers hypothesize this clustering is due to managers either speeding up or delaying the IPO process in hopes of taking advantage of a "hot" market, which will provide the firm with high initial valuations of its stock.[11] The researchers collected information on 196 public offerings listed on the Warsaw Stock Exchange over a six-year period. For each IPO, they obtained the length of the IPO offering period (the time between the approval of the prospectus and the IPO date) and three market return rates. The first rate was for the period between the date the prospectus was approved and the "expected" IPO date. The second rate was for the period 90 days prior to the "expected" IPO date. The last rate was between the approval date and 90 days after the "expected" IPO date. The "expected" IPO date was the median length of the 196 IPO periods. They regressed the length of the offering period (in days) against each of the three rates of return. Here are the results:

Period	b_0	b_1	P-value	r
1	48.018	−129.391	0.0008	−0.238
2	49.478	114.785	<0.0001	0.414
3	47.613	−41.646	0.0463	−0.143

(a) What does this table tell you about the relationship between the IPO offering period and the three market return rates?

(b) The researchers argue that since the strongest correlation is for the second period and the weakest correlation is for the third period, there is evidence supporting their hypothesis. Do you agree with this conclusion? Explain your answer.

12.22 The relationship between log income and education level for employees. Recall Case 12.1 (page 572). The researchers also looked at the relationship between education and log income for employees. An employee was defined as a person whose main employment status is a salaried job. Based on a sample of 100 employees: EMPL

(a) Construct a scatterplot of log income versus education. Describe the relationship between the two variables. Is a linear relationship reasonable? Explain your answer.

(b) Report the least-squares regression line.

(c) Obtain the residuals and use them to assess the assumptions needed for inference.

(d) In Example 12.5 (page 584), we constructed a 95% confidence interval for the slope of the entrepreneur population; it was (0.0208 to 0.2044). Construct a 95% confidence interval for the slope of the employee population.

(e) Compare the two confidence intervals. Do you think there is a difference in the two slopes? Explain your answer.

12.23 Incentive pay and job performance. In the National Football League (NFL), incentive bonuses now account for roughly 25% of player compensation.[12] Does tying a player's salary to performance bonuses result in better individual or team success on the field? Focusing on linebackers, let's look at the relationship between a player's end-of-the-year production rating and the percent of his salary devoted to incentive payments in that same year. PERPLAY

(a) Use numerical and graphical methods to describe the two variables and summarize your results.

(b) Neither variable is Normally distributed. Does that necessarily pose a problem for performing linear regression? Explain.

(c) Construct a scatterplot of the data and describe the relationship. Are there any outliers or unusual values? Does a linear relationship between the percent of salary from incentive payments and player rating seem reasonable? Is it a very strong relationship? Explain.

(d) Run the simple linear regression and give the least-squares regression line.

(e) Obtain the residuals and assess whether the assumptions for the linear regression analysis are reasonable. Include all plots and numerical summaries that you used to make this assessment.

12.24 Incentive pay and job performance, continued. Refer to the previous exercise. PERPLAY

(a) Now run the simple linear regression for the variables square root of rating and percent of salary from incentive payments.

(b) Obtain the residuals and assess whether the assumptions for the linear regression analysis are reasonable. Include all plots and numerical summaries that you used to make this assessment.

(c) Construct a 95% confidence interval for the square root increase in rating given a 1% increase in the percent of salary from incentive payments.

(d) Consider the values 0%, 20%, 40%, 60%, and 80% salary from incentives. Compute the predicted rating for this model and for the one in Exercise 12.23. For the model in this exercise, you will need to square the predicted value to get back to the original units.

(e) Plot the predicted values versus the percents, and connect those values from the same model. For which regions of percent do the predicted values from the two models vary the most?

(f) Based on your comparison of the regression models (both predicted values and residuals), which model do you prefer? Explain.

12.25 Predicting public university tuition: 2008 versus 2017. Refer to Exercise 12.19. The data file also includes the in-state undergraduate tuition for the year 2008. TUIT

(a) Plot the data with the 2008 tuition on the x axis, then describe the relationship. Are there any outliers or unusual values? Does a linear relationship between the tuition in 2008 and 2017 seem reasonable?

(b) Fit the simple linear regression model and give the least-squares regression line and regression standard error.

(c) Obtain the residuals and plot them versus the 2008 tuition amount. Describe anything unusual in the plot.

(d) Do the residuals appear to be approximately Normal? Explain.

12.26 Compare the analyses. In Exercises 12.19 and 12.25, you used two different explanatory variables to predict university tuition in 2017. Summarize the two analyses and compare the results. If you had to choose between the two, which explanatory variable would you choose? Give reasons for your answers.

Age and income. *The data file for the following exercises contains the age and income of a random sample of 5712 men between the ages of 25 and 65 who have a bachelor's degree but no higher degree. Figure 12.12 is a scatterplot of these data. Figure 12.13 displays Excel output for regressing income on age. The line in the scatterplot is the least-squares regression line. Exercises 12.27 through 12.29 ask you to interpret this information.* INAGE

12.27 Looking at age and income. The scatterplot in Figure 12.12 has a distinctive form.

(a) Age is recorded as of the last birthday. How does this explain the vertical stacks of incomes in the scatterplot?

(b) Give some reasons that older men in this population might earn more than younger men. Give some reasons that younger men might earn more than older men. What do the data show about the relationship between age and income in the sample? Is the relationship very strong?

(c) What is the equation of the least-squares line for predicting income from age? What specifically does the slope of this line tell us?

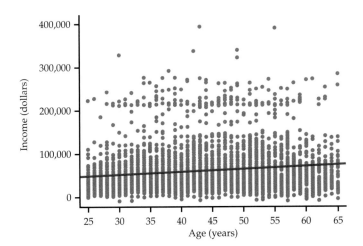

FIGURE 12.12 Scatterplot of income against age for a random sample of 5712 men aged 25 to 65, for Exercises 12.27 to 12.29.

FIGURE 12.13 Excel output for the regression of income on age, for Exercises 12.27 to 12.29.

	A	B	C	D	E	F	G
1	SUMMARY OUTPUT						
2							
3	*Regression Statistics*						
4	Multiple R	0.18774782					
5	R Square	0.03524924					
6	Adjusted R Square	0.03508029					
7	Standard Error	47620.2923					
8	Observations	5712					
9							
10	ANOVA						
11		*df*	*SS*	*MS*	*F*	*Significance F*	
12	Regression	1	4.73102E+11	4.73102E+11	208.62713	1.79127E-46	
13	Residual	5710	1.29485E+13	2267692234			
14	Total	5711	1.34216E+13				
15							
16		*Coefficients*	*Standard Error*	*t Stat*	*P-value*	*Lower 95%*	*Upper 95%*
17	Intercept	24874.3745	2637.419757	9.431329401	5.749E-21	19704.03079	30044.7182
18	Age	892.113523	61.7639029	14.44393054	1.791E-46	771.0328323	1013.194214

12.28 Income increases with age. We see that older men do, on average, earn more than younger men, but the increase is not very rapid. (Note that the regression line describes many men of different ages—data on the same men over time might show a different pattern.)

(a) We know even without looking at the Excel output that there is highly significant evidence that the slope of the population regression line is greater than 0. Why do we know this?

(b) Excel gives a 95% confidence interval for the slope of the population regression line. What is this interval?

(c) Give a 99% confidence interval for the slope of the population regression line.

12.29 Was inference justified? You see from Figure 12.12 that the incomes of men at each age are (as expected) not Normal but right-skewed.

(a) How is this apparent on the plot?

(b) Nonetheless, your confidence interval in the previous exercise will be quite accurate even though it is based on Normal distributions. Why?

12.30 Regression to the mean? Suppose a large population of test takers take the GMAT. You fear some cheating may have occurred so you ask those people who scored in the top 10% to take the exam again.

(a) If their scores, on average, decrease, is this evidence that there was cheating? Explain your answer.

(b) If these same people were asked to take the test a third time, would you expect their scores to decline even further? Explain your answer.

12.31 T-bills and inflation. Exercises 12.8 through 12.10 interpret the part of the Excel output in Figure 12.10 (page 586) that concerns the slope—that is, the rate at which T-bill returns increase as the rate of inflation increases. Use this output to answer questions about the intercept.

(a) The intercept β_0 in the regression model is meaningful in this example. Explain what β_0 represents. Why should we expect β_0 to be greater than 0?

(b) What values does Excel give for the estimated intercept b_0 and its standard error SE_{b_0}?

(c) Is there good evidence that β_0 is greater than 0?

(d) Write the formula for a 95% confidence interval for β_0. Verify that the hand calculation (using the Excel values for b_0 and SE_{b_0}) agrees approximately with the output in Figure 12.10.

12.32 Is the correlation significant? Two studies looked at the relationship between customer-relationship management (CRM) implementation and organizational structure. One study reported a correlation of $r = 0.33$ based on a sample of size $n = 25$. The second study reported a correlation of $r = 0.22$ based on a sample of size $n = 62$. For each, test the null hypothesis that the population correlation $\rho = 0$ against the one-sided alternative $\rho > 0$. Are the results significant at the 5% level? What conclusions would you draw based on both studies?

12.33 Correlation between the prevalences of adult binge drinking and underage drinking. A group of researchers compiled data on the prevalence of adult binge drinking and the prevalence of underage drinking in 42 states.[13] A correlation of 0.32 was reported.

(a) Test the null hypothesis that the population correlation $\rho = 0$ against the alternative $\rho > 0$. Are the results significant at the 5% level?

(b) Explain this correlation in terms of the direction of the association and the percent of variability in the prevalence of underage drinking that is explained by the prevalence of adult binge drinking.

TABLE 12.4			Net new money (millions of $) flowing into stock and bond mutual funds					
Year	Stocks	Bonds	Year	Stocks	Bonds	Year	Stocks	Bonds
1984	4336	13,058	1996	216,937	3141	2008	−215,757	30,039
1985	6643	63,127	1997	227,106	29,166	2009	2013	371,123
1986	20,386	102,618	1998	156,875	74,656	2010	−24,385	232,351
1987	19,231	6797	1999	187,565	−4767	2011	−129,363	117,734
1988	−14,948	−4488	2000	315,705	−50,115	2012	−152,678	306,256
1989	6774	−1226	2001	33,483	88,463	2013	159,481	−70,771
1990	12,915	6833	2002	−29,310	141,865	2014	25,458	43,600
1991	39,888	59,258	2003	144,077	32,750	2015	−75,620	−25,270
1992	78,983	70,989	2004	171,945	−15,102	2016	−258,030	106,897
1993	127,260	72,169	2005	123,938	25,294	2017	−159,640	260,162
1994	114,525	−61,362	2006	147,804	59,448			
1995	124,392	−5922	2007	73,307	110,609			

(c) The researchers collected information from 42 of 50 states so almost all the data available was used in the analysis. Provide an argument for the use of statistical inference in this setting.

12.34 Stocks and bonds. How is the flow of investors' money into stock mutual funds related to the flow of money into bond mutual funds? Table 12.4 shows the net new money flowing into stock and bond mutual funds in the years 1984 to 2017, in millions of dollars.[14] "Net" means that funds flowing out are subtracted from those flowing in. If more money leaves than arrives, the net flow will be negative. FLOW

(a) Make a scatterplot with cash flow into stock funds as the explanatory variable. Find the least-squares line for predicting net bond investments from net stock investments. What do the data suggest?

(b) Is there statistically significant evidence of some straight-line relationship between the flows of cash into bond funds and stock funds? (State the hypotheses, give a test statistic and its P-value, and state your conclusion.)

(c) Generate a plot of the residuals versus year. Describe any unusual patterns you see in this plot.

(d) Given the 2008 financial crisis and its lingering effects, remove the data for the years after 2007 and refit the remaining years. Is there statistically significant evidence of a straight-line relationship?

(e) Compare the least-squares regression lines and regression standard errors using all the years and using only the years before 2008.

(f) How would you report these results in a paper? In other words, how would you handle the difference in relationship before and after 2008?

12.35 Size and selling price of houses. Table 12.5 describes a random sample of 30 houses sold in a

TABLE 12.5		Selling price and size of homes			
Price ($1000)	Size (sq ft)	Price ($1000)	Size (sq ft)	Price ($1000)	Size (sq ft)
268	1897	142	1329	83	1378
131	1157	107	1040	125	1668
112	1024	110	951	60	1248
112	935	187	1628	85	1229
122	1236	94	816	117	1308
128	1248	99	1060	57	892
158	1620	78	800	110	1981
135	1124	56	492	127	1098
146	1248	70	792	119	1858
126	1139	54	980	172	2010

Midwest city during a recent year.[15] In this exercise, we examine the relationship between size and price. HSIZE

(a) Plot the selling price versus the number of square feet. Describe the pattern. Does r^2 suggest that size is quite helpful for predicting selling price?

(b) Do a linear regression analysis. Give the least-squares line and the results of the significance test for the slope. What does your test tell you about the relationship between house size and selling price?

12.36 Are inflows into stocks and bonds correlated? Is the correlation between net flow of money into stock mutual funds and into bond mutual funds significantly different from 0? Use the regression analysis you did in Exercise 12.34, part (b), to answer this question with no additional calculations. FLOW

12.37 Do larger houses have higher prices? We expect that there is a positive correlation between the sizes of houses in the same market and their selling prices. HSIZE

(a) Use the data in Table 12.5 to test this hypothesis. (State the hypotheses, find the sample correlation r and the t statistic based on it, and give an approximate P-value and your conclusion.)

(b) How do your results in part (a) compare to the test of the slope in Exercise 12.35, part (b)?

(c) To what extent do you think that these results would apply to other regions in the United States?

12.38 Highway MPG and CO_2 emissions. Let's investigate the relationship between highway miles per gallon (MPGHwy) and carbon dioxide emissions (CO2 Emissions) for cars that use premium gasoline as reported by Natural Resources Canada.[16] PREM

(a) Make a scatterplot of the data and describe the pattern.

(b) Plot MPGHwy versus the logarithm of CO_2 emissions. Are these points closer to a straight line?

(c) Regress MPGHwy by the logarithm of CO_2 emissions. Give a 95% confidence interval for the slope of the population regression line. Describe what this interval tells you in terms of percent change in CO_2 emissions for every one mile increase in highway miles per gallon.

12.39 Influence? Your scatterplot in Exercise 12.35 shows one house whose selling price is quite high for its size. Rerun the analysis without this outlier. Does this one house influence r^2, the location of the least-squares line, or the t statistic for the slope in a way that would change your conclusions? HSIZE

12.40 Correlation between the observed and predicted y's. Using your choice of software, fit the data of Case 12.1 and obtain the log income predicted values $\hat{y}$. Then compute the correlation between these predicted values $\hat{y}$ and log income values y and compare it to the correlation between x and y that is reported in Example 12.2 (page 574). Describe what you find. ENTRE

12.2 Using the Regression Line

When you complete this section, you will be able to:

- Construct and interpret a confidence interval for a mean response when $x = x^*$.
- Construct and interpret a prediction interval for a future observation when $x = x^*$.
- Identify when a prediction interval should be constructed instead of a confidence interval.

Predictive analytics involves the use of various techniques from statistics to make predictions that help support decision making. One of the most common reasons to fit a line to data is to predict the response to a particular value of the explanatory, or predictor, variable. The method is simple: just substitute the value of x into the equation of the line. For example, the least-squares line for predicting log income of entrepreneurs from their years of education (Case 12.1) is

$$\hat{y} = 8.2546 + 0.1126x$$

For an Educ of 16, our least-squares regression equation gives the prediction

$$\hat{y} = 8.2546 + (0.1126)(16) = 10.0562$$

Confidence and prediction intervals

In terms of inference, there are two different uses of this prediction. First, we can estimate the *mean* log income in the subpopulation of entrepreneurs with 16 years of education. Second, we can predict the log income of *one individual entrepreneur* with 16 years of education.

For each use, the actual prediction is the same, $\hat{y} = 10.0562$. *It is the margin of error that is different.* Individual entrepreneurs with 16 years of education don't all have the same log income. Thus, we need a larger margin of error when predicting an individual's log income than when estimating the mean log income of all entrepreneurs who have 16 years of education.

To emphasize the distinction between predicting a single outcome and estimating the mean of all outcomes in the subpopulation, we use different terms for the two resulting intervals.

- To estimate the *mean* response, we use a *confidence interval*. This is an ordinary confidence interval for the parameter

$$\mu_y = \beta_0 + \beta_1 x^*$$

The regression model says that μ_y is the mean of responses y when $x = x^*$. It is a fixed number whose value we don't know because we don't know β_0 and β_1.

prediction interval, p. 383

- To estimate an *individual* response y, we use a *prediction interval*. A prediction interval estimates a single random response y rather than a parameter like μ_y. Even if we know β_0 and β_1, the response y is not a fixed number. The model says that y varies Normally with a mean that depends on x.

Fortunately, the meaning of a prediction interval is very much like the meaning of a confidence interval. A 95% prediction interval, like a 95% confidence interval, is right 95% of the time in repeated use. Consider doing the following many times:

1. Draw a sample of n observations (x, y) and one additional observation (x^*, y).

2. Calculate the 95% prediction interval for y when $x = x^*$ using the n observations.

Being right means that the y value in the additional observation will be in the calculated interval 95% of the time.

Each interval has the usual form

$$\hat{y} \pm t^* \text{SE}$$

where $t^*\text{SE}$ is the margin of error. The main distinction is that because it is more difficult to predict a single observation (random variable) than the mean of a subpopulation (fixed value), the margin of error for the prediction interval is wider than the margin of error for the confidence interval. Formulas for computing these quantities are given in Section 12.3. For now, we rely on software to do the arithmetic.

CONFIDENCE AND PREDICTION INTERVALS FOR REGRESSION RESPONSE

A level C **confidence interval for the mean response** μ_y when x takes the value x^* is

$$\hat{y} \pm t^* \text{SE}_{\hat{\mu}}$$

Here, $\text{SE}_{\hat{\mu}}$ is the standard error for estimating a mean response. The **margin of error** is $m = t^* \text{SE}_{\hat{\mu}}$.

A level C **prediction interval for a single observation** on y when x takes the value x^* is

$$\hat{y} \pm t^* \text{SE}_{\hat{y}}$$

Here, $\text{SE}_{\hat{y}}$ is the standard error for predicting an individual response and the **margin of error** is $m = t^*\text{SE}_{\hat{y}}$.

The standard error $\text{SE}_{\hat{y}}$ is larger than the standard error $\text{SE}_{\hat{\mu}}$.

In both cases, t^* is the value for the $t(n-2)$ density curve with area C between $-t^*$ and t^*.

Before moving on to the examples, it is important to note that predicting an individual response is an exception to the general fact that regression inference is robust against lack of Normality. *The prediction interval relies on Normality of individual observations, not just on the approximate Normality of statistics like the slope b_1 and intercept b_0 of the least-squares line.* In practice, this means that we should regard prediction intervals as rough approximations.

EXAMPLE 12.7

ENTRE

Predicting Loginc from Educ Jacob Brown is an entrepreneur with Educ = 16 years of education. We don't know his log income, but we can use the data on other entrepreneurs to predict it.

Statistical software usually allows prediction of the response for each x-value in the data and also for new values of x. Here is the output from the prediction option in the Minitab regression command for $x^* = 16$ when we ask for 95% intervals:

Fit	SE Fit	95% CI	95% PI
10.0560	0.167802	(9.72305, 10.3890)	(7.81924, 12.2929)

The "Fit" entry gives the predicted log income, 10.0560. This agrees with our hand calculation within the rounding error. Minitab gives both 95% intervals; you must then choose which one you want. We are predicting a single response, so the prediction interval "95% PI" is the right choice. We are 95% confident that Jacob's log income lies between 7.81924 and 12.2929. This is a wide range because the data are widely scattered about the least-squares line. The 95% confidence interval for the mean log income of all entrepreneurs with EDUC = 16, given as "95% CI," is much narrower. ■

Note that Minitab reports only one of the two standard errors—the standard error for estimating the mean response, $\text{SE}_{\hat{\mu}} = 0.1678$. A graph will help us to understand the difference between the two types of intervals.

EXAMPLE 12.8

ENTRE

Comparing the Two Intervals Figure 12.14 displays the data, the least-squares line, and both intervals. The confidence interval for the mean is the solid red vertical line. The prediction interval for Jacob's individual log income level is the dashed black vertical line. You can see that the prediction interval is much wider and that it matches the vertical spread of entrepreneurs' log incomes about the regression line. ■

Some software packages will graph the intervals for all values of the explanatory variable within the range of the data. With this type of display, we can see the difference between the two types of intervals across the range of x.

FIGURE 12.14 Confidence interval for mean log income (solid red) and prediction interval for individual log income (dashed black) for an entrepreneur with 16 years of education. Both intervals are centered on the predicted value from the least-squares line, which is $\hat{y} = 10.056$ for $x^* = 16$.

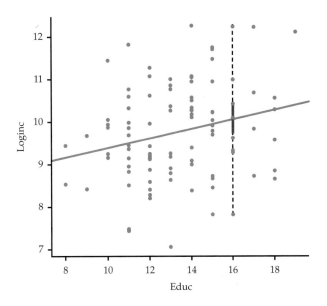

EXAMPLE 12.9

CASE 12.1

Graphing the Confidence Intervals The confidence intervals for the log income data are graphed in Figure 12.15. For each value of Educ, we see the estimated value on the solid line and the confidence limits on the dashed curves. ■

FIGURE 12.15 95% confidence intervals for mean response for the annual income data, for Example 12.9.

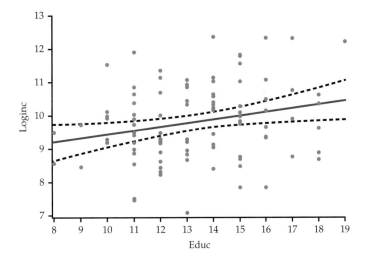

Notice that the intervals get wider as the values of Educ move away from the mean of this variable. This phenomenon reflects the fact that we have less information for estimating means that correspond to extreme values of the explanatory variable.

EXAMPLE 12.10

CASE 12.1

Graphing the Prediction Intervals The prediction intervals for the log income data are graphed in Figure 12.16. As with the confidence intervals, we see the predicted values on the solid line and the prediction limits on the dashed curves. ■

It is much easier to see the curvature of the confidence limits in Figure 12.15 than the curvature of the prediction limits in Figure 12.16. One reason

FIGURE 12.16 95% prediction intervals for individual response for the annual income data, for Example 12.10.

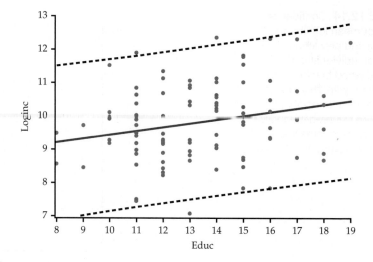

for this is that the prediction intervals in Figure 12.16 are dominated by the entrepreneur-to-entrepreneur variation. On the one hand, because the prediction intervals are concerned with individual predictions, they contain a very large proportion of the observations. On the other hand, the confidence intervals are designed to contain mean values and are not concerned with individual observations.

Because we transformed income to better fit the model conditions for simple linear regression, our predictions are on the log scale (in log dollars). To **back-transform** get a prediction on the original scale (dollars), we can **back-transform** the predictions from our linear model. The inverse of the logarithm function is the exponential function. Thus, we could naively estimate the average income by exponentiating the log dollars estimate and exponentiating the endpoints of the confidence interval to get an approximate 95% confidence interval for the average income. For example, the average income for entrepreneurs with $x = 16$ years of education is $e^{10.0560} = \$23{,}295$ and the 95% confidence interval is (\$16,698, \$32,500).

These calculations, however, provide an estimate and confidence interval for the median income rather than the mean. In fact, the regression model **logNormal distribution** implies that income has a **logNormal distribution.** An estimate and confidence interval for the mean income can still be constructed from the regression model estimates but the calculations are more complicated. We suggest seeking expert advice when a situation like this arises.

APPLY YOUR KNOWLEDGE

12.41 Predicting the mean Loginc. In Example 12.7, software predicts the mean log income of entrepreneurs with 16 years of education to be $\hat{y} = 10.0560$. We also see that the standard error of this estimated mean is $SE_{\hat{\mu}} = 0.167802$. These results come from data on 100 entrepreneurs.

(a) Use these facts and $t^* = 1.984$ to verify by hand Minitab's 95% confidence interval for the mean log income when Educ = 16.

(b) Use the same information to construct a 90% confidence interval for the mean log income when Educ = 16. Make sure to specify what degrees of freedom are used to obtain t^*.

12.42 Predicting the return on Treasury bills. Table 12.1 (page 585) gives data on the rate of inflation and the percent return on Treasury bills for 60 years. Figures 12.9 and 12.10 analyze these data. You think that next year's inflation rate will be 2.25%. Figure 12.17 displays part of the

FIGURE 12.17 Minitab output for the regression of the percent return on Treasury bills against the rate of inflation in the same year, for Exercise 12.42. The output includes predictions of the T-bill return when the inflation rate is 2.25%.

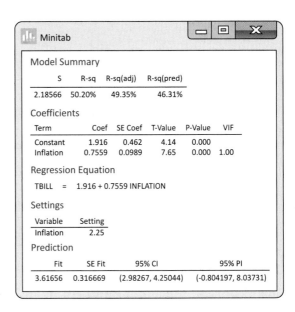

Minitab regression output, including predicted values for $x^* = 2.25$. The basic output agrees with the Excel results in Figure 12.10.

(a) Verify the predicted value $\hat{y} = 3.617$ from the equation of the least-squares line.

(b) What is your 95% interval for predicting next year's return on Treasury bills?

BEYOND THE BASICS

Nonlinear regression

The simple linear regression model assumes that the relationship between the response variable and the explanatory variable can be summarized with a straight line. When the relationship is not linear, we can sometimes transform one or both of the variables so that the relationship becomes linear. Case 12.1 is an example in which the relationship of $\log y$ with x is linear. In other circumstances, we use **nonlinear models** that directly express a curved relationship using parameters that are not just intercepts and slopes.

Here is a typical example of a model that involves parameters β_0 and β_1 in a nonlinear way:

$$y_i = \beta_0 x_i^{\beta_1} + \varepsilon_i$$

This nonlinear model still has the form

$$\text{DATA} = \text{FIT} + \text{RESIDUAL}$$

The FIT term describes how the mean response μ_y depends on x. Figure 12.18 shows the form of the mean response for several values of β_1 when $\beta_0 = 1$. Choosing $\beta_1 = 1$ produces a straight line, but other values of β_1 result in a variety of curved relationships.

We cannot write simple formulas for the estimates of the parameters β_0 and β_1, but software can calculate both estimates and approximate standard errors for the estimates. If the deviations ε_i follow a Normal distribution, we can do inference both on the model parameters and for prediction. The details become more complex, but the ideas remain the same as those we have studied.

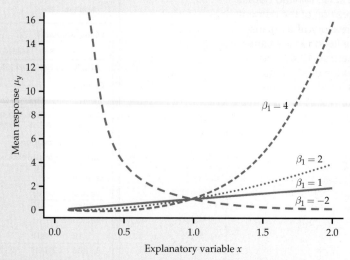

FIGURE 12.18 The nonlinear model $\mu_y = \beta_0 x^{\beta_1}$ includes these and other relationships between the explanatory variable x and the mean response.

SECTION 12.2 SUMMARY

- The **estimated mean response** for the subpopulation corresponding to the value x^* of the explanatory variable is found by substituting $x = x^*$ in the equation of the least-squares regression line:

$$\text{estimated mean response} = \hat{y} = b_0 + b_1 x^*$$

- The **predicted value of the response** y for a single observation from the subpopulation corresponding to the value x^* of the explanatory variable is found in exactly the same way:

$$\text{predicted individual response} = \hat{y} = b_0 + b_1 x^*$$

- **Confidence intervals for the mean response** μ_y when x has the value x^* have the form

$$\hat{y} \pm t^* \text{SE}_{\hat{\mu}}$$

- **Prediction intervals** for an individual response y have a similar form with a larger standard error:

$$\hat{y} \pm t^* \text{SE}_{\hat{y}}$$

In both cases, t^* is the value for the $t(n-2)$ density curve with area C between $-t^*$ and t^*. Software often gives these intervals. The standard error $\text{SE}_{\hat{y}}$ for an individual response is larger than the standard error $\text{SE}_{\hat{\mu}}$ for a mean response because it must account for the variation of individual responses around their mean.

SECTION 12.2 EXERCISES

For Exercises 12.41 and 12.42, see pages 600–601.
Many of the following exercises require use of software that will calculate the intervals required for predicting mean response and individual response.

12.43 More on public university tuition. Refer to Exercises 12.19 and 12.20 (page 592). TUIT

(a) The tuition at CashCow U was $9200 in 2013. Find the 95% prediction interval for its tuition in 2017.

(b) The tuition at Moneypit U was $18,895 in 2013. Find the 95% prediction interval for its tuition in 2017.

(c) Compare the widths of these two intervals. Which is wider and why?

12.44 More on assessment value versus sales price. Refer to Exercise 12.17 (page 591). Suppose we're interested in determining whether the population regression line differs from $y = x$. We'll look at this three ways. HSALES

(a) Construct a 95% confidence interval for each property in the data set. If the model $y = x$ is reasonable, then the assessed value used to predict the sales price should be in the interval. Is this true for all (x, y) pairs?

(b) The model $y = x$ means $\beta_0 = 0$ and $\beta_1 = 1$. Test each of these hypotheses. Is there enough evidence to reject either of them?

(c) Recall that not rejecting H_0 does not imply H_0 is true. A test of "equivalence" would be a more appropriate method to assess similarity. Suppose that, for the slope, a difference within ±0.05% is considered not different. Construct a 90% confidence interval for the slope and see if it falls entirely within the interval (0.95, 1.05). If it does, we would conclude that the slope is not different from 1. What is your conclusion using this method?

12.45 Predicting 2017 tuition from 2013 tuition. Refer to Exercise 12.19 (page 592). TUIT

(a) Find a 95% confidence interval for the mean tuition amount corresponding to a 2013 tuition of $10,403.

(b) Find a 95% prediction interval for a future response corresponding to a 2013 tuition of $10,403.

(c) Write a short paragraph interpreting the meaning of the intervals in terms of public universities.

(d) Do you think that these results can be applied to private universities? Explain why or why not.

12.46 Predicting 2017 tuition from 2008 tuition. Refer to Exercise 12.25 (page 593). TUIT

(a) Find a 95% confidence interval for the mean tuition amount corresponding to a 2008 tuition of $7568.

(b) Find a 95% prediction interval for a future response corresponding to a 2008 tuition of $7568.

(c) Write a short paragraph interpreting the meaning of the intervals in terms of public universities.

(d) Do you think that these results can be applied to private universities? Explain why or why not.

12.47 Compare the estimates. Case 20 in Table 12.3 (Wisconsin) has a 2008 tuition of $7568 and a 2013 tuition of $10,403. A predicted 2017 tuition amount based on 2013 tuition was computed in Exercise 12.45, while one based on the 2008 tuition was computed in Exercise 12.46. Compare these two estimates and explain why they differ. Use the idea of a prediction interval to interpret these results.

12.48 Is the price right? Refer to Exercise 12.35 (page 595), where the relationship between the size of a home and its selling price is examined. HSIZE

(a) Suppose that you have a client who is thinking about purchasing a home in this area that is 1750 square feet in size. The asking price is $180,000. What advice would you give this client?

(b) Answer the same question for a client who is looking at a 1300-square-foot home that is selling for $110,000.

12.49 Predicting income from age. Figures 12.12 and 12.13 (pages 593 and 594) analyze data on the age and income of 5712 men between the ages of 25 and 65. Here is Minitab output predicting the income for ages 30, 40, 50, and 60 years:

Prediction

Fit	SE Fit	95% CI	95% PI
51638	948	(49780, 53496)	(−41735, 145010)
60559	637	(59311, 61807)	(−32803, 153921)
69480	822	(67870, 71091)	(−23888, 162848)
78401	1307	(75840, 80963)	(−14988, 171790)

(a) Use the regression line from Figure 12.12 to verify the "Fit" for age 30 years.

(b) Report the 95% confidence interval for the income of all 30-year-old men.

(c) Joseph is 30 years old. You don't know his income, so give a 95% prediction interval based on his age alone. How useful do you think this interval is?

12.50 Predict what? The two 95% intervals for the income of 30-year-olds given in Exercise 12.49 are very different. Explain briefly to someone who knows no statistics why the second interval is so much wider than the first. Start by looking at 30-year-olds in Figure 12.12.

12.51 Predicting income from age, continued. Use the computer outputs in Figure 12.13 and Exercise 12.49 to give a 90% confidence interval for the mean income of all 40-year-old men.

12.52 T-bills and inflation. Figure 12.17 (page 601) gives part of a regression analysis of the data in Table 12.1 relating the return on Treasury bills to the rate of inflation. The output includes prediction of the T-bill return when the inflation rate is 2.25%.

(a) Use the output to give a 90% confidence interval for the mean return on T-bills in all years having 2.25% inflation.

(b) You think that next year's inflation rate will be 2.25%. It isn't possible, without complicated arithmetic, to give a 90% prediction interval for next year's T-bill return based on the output displayed. Why not?

12.53 Two confidence intervals. The data used for Exercise 12.49 include 195 men who are 30 years old. The mean income of these men is $\bar{y} = \$49{,}880$ and the standard deviation of these 195 incomes is $s_y = \$38{,}250$.

(a) Use the one-sample t procedure to give a 95% confidence interval for the mean income μ_y of 30-year-old men.

(b) Why is this interval different from the 95% confidence interval for μ_y in the regression output? (*Hint:* Which data are used by each method?)

12.54 Size and selling price of houses. Table 12.5 (page 595) gives data on the size in square feet of a random sample of houses sold in a Midwest county along with their selling prices. HSIZE

(a) Find the mean size $\bar{x}$ of these houses and also their mean selling price $\bar{y}$. Give the equation of the least-squares regression line for predicting price from size, and use it to predict the selling price of a house of mean size. (You knew the answer, right?)

(b) Zoey and Aiden are selling a house in this Midwest county whose size is equal to the mean of this sample.

Give an interval that predicts the price they will receive with 95% confidence.

(c) Compare the prediction interval you used in part (b) to the prediction interval for a new observation from a $N(\mu,\sigma)$ population described in Chapter 7 (page 385).

12.3 Some Details of Regression Inference

When you complete this section, you will be able to

- Use ANOVA table output to perform the ANOVA F test and draw appropriate conclusions regarding H_0: $\beta_1 = 0$.
- Use ANOVA table output to compute the square of the sample correlation and provide an interpretation of it in terms of explained variation.
- Perform, using a calculator or spreadsheet, inference in simple linear regression when software is not available.
- Distinguish the formulas for the standard error that we use for a confidence interval for the mean response and the standard error that we use for a prediction interval when $x = x^*$.

We have assumed that you will use software to handle regression in practice. If you do, it is much more important to understand what the standard error of the slope SE_{b_1} means than it is to know the formula your software uses to find its numerical value. For that reason, we have not yet given formulas for the standard errors. We have also not explained the block of output from software that is labeled "ANOVA" or "Analysis of Variance." This section addresses both of these omissions.

Standard errors

In this section, we give the formulas for all the standard errors we have encountered, for two reasons. First, you may want to see how these formulas can be obtained from facts you already know. The second reason is more practical: some software (in particular, spreadsheet programs) does not automate inference for prediction. Fortunately, almost all software does the hard work of calculating the regression standard error s. With s in hand, the rest is straightforward—but only if you know the details.

Tests and confidence intervals for the slope of a population regression line start with the slope b_1 of the least-squares line and with its standard error SE_{b_1}. If you are willing to skip some messy algebra, it is easy to see where SE_{b_1} and the similar standard error SE_{b_0} of the intercept come from.

1. The regression model takes the explanatory values x_i to be fixed numbers and the response values y_i to be independent random variables all having the same standard deviation σ.

2. The least-squares slope is $b_1 = rs_y/s_x$. Here is the first bit of messy algebra that we skip: it is possible to write the slope b_1 as a linear function of the responses, $b_1 = \sum a_i y_i$. The coefficients a_i depend on the x_i, so they are fixed numbers, too.

rules for variances, pp. 240–241

3. We can find the variance of b_1 by applying the rule for the variance of a sum of independent random variables; it is just $\sigma^2 \sum a_i^2$. A second piece of messy algebra shows that this simplifies to

$$\sigma_{b_1}^2 = \frac{\sigma^2}{\sum (x_i - \bar{x})^2}$$

The standard deviation σ about the population regression line is, of course, not known. If we estimate it by the regression standard error s based on the residuals from the least-squares line, we get the standard error of b_1. Here are the results for both the slope and the intercept.

STANDARD ERRORS FOR SLOPE AND INTERCEPT

The standard error of the slope b_1 of the least-squares regression line is

$$\text{SE}_{b_1} = \frac{s}{\sqrt{\sum(x_i - \bar{x})^2}}$$

The standard error of the intercept b_0 is

$$\text{SE}_{b_0} = s\sqrt{\frac{1}{n} + \frac{\bar{x}^2}{\sum(x_i - \bar{x})^2}}$$

The critical fact is that both standard errors are multiples of the regression standard error s. In a similar manner, accepting the results of yet more messy algebra, we get the standard errors for the two uses of the regression line that we have studied.

STANDARD ERRORS FOR TWO USES OF THE REGRESSION LINE

The standard error for estimating the mean response when the explanatory variable x takes the value x^* is

$$\text{SE}_{\hat{\mu}} = s\sqrt{\frac{1}{n} + \frac{(x^* - \bar{x})^2}{\sum(x_i - \bar{x})^2}}$$

The standard error for predicting an individual response when $x = x^*$ is

$$\text{SE}_{\hat{y}} = s\sqrt{1 + \frac{1}{n} + \frac{(x^* - \bar{x})^2}{\sum(x_i - \bar{x})^2}}$$

$$= \sqrt{\text{SE}_{\hat{\mu}}^2 + s^2}$$

Once again, both standard errors are multiples of s. The only difference between the two prediction standard errors is the extra 1 under the square root sign in the standard error for predicting an individual response. This added term reflects the additional variation in individual responses, just as it did in Section 7.4 (page 385) when we considered predicting a new observation from a $N(\mu, \sigma)$ population. It also implies that $\text{SE}_{\hat{y}}$ is always greater than $\text{SE}_{\hat{\mu}}$.

EXAMPLE 12.11

Prediction Intervals from a Spreadsheet In Example 12.7, we used statistical software to predict the log income of Jacob, who has Educ = 16 years of education. Suppose that we have only the Excel spreadsheet. The prediction interval then requires some additional work.

Step 1. From the Excel output in Figure 12.5 (page 579), we know that $s = 1.1146$. Excel can also find the mean and variance of the Educ x for the 100 entrepreneurs. They are $\bar{x} = 13.28$ and $s_x^2 = 5.901$.

Step 2. We need the value of $\sum(x_i - \bar{x})^2$. Recalling the definition of the variance, we see that this is just

$$\sum(x_i - \bar{x})^2 = (n-1)s_x^2$$
$$= (99)(5.901) = 584.2$$

Step 3. The standard error for predicting Jacob's log income from his years of education, $x^* = 16$, is

$$\text{SE}_{\hat{y}} = s\sqrt{1 + \frac{1}{n} + \frac{(x^* - \bar{x})^2}{\sum(x_i - \bar{x})^2}}$$

$$= 1.1146\sqrt{1 + \frac{1}{100} + \frac{(16 - 13.28)^2}{584.2}}$$

$$= 1.1146\sqrt{1 + \frac{1}{100} + \frac{7.3984}{584.2}}$$

$$= (1.1146)(1.01127) = 1.12716$$

Step 4. We predict Jacob's log income from the least-squares line (Figure 12.5 again):

$$\hat{y} = 8.2546 + (0.1126)(16) = 10.0562$$

This agrees with the "Fit" from software in Example 12.7. The 95% prediction interval requires the 95% critical value for $t(98)$. Using Excel, the function = T.INV(0.975, 98) gives $t^* = 1.984$. The interval is

$$\hat{y} \pm t^* \text{SE}_{\hat{y}} = 10.0562 \pm (1.984)(1.12716)$$
$$= 10.0562 \pm 2.2363$$
$$= 7.8199 \text{ to } 12.2925$$

This agrees with the software result in Example 12.7, with a small difference due to roundoff. ■

The formulas for the standard errors of mean estimation and predicting an individual response show us one more thing about prediction. They both contain the term $(x^* - \bar{x})^2$, the squared distance of the value x^* for which we want to do prediction from the mean $\bar{x}$ of the x-values in our data. We see that prediction is most accurate (smallest margin of error) at the mean and grows less accurate as we move away from the mean of the explanatory variable. *If you know the values of x for which you want to do prediction, try to collect data centered near these values.*

APPLY YOUR KNOWLEDGE

12.55 T-bills and inflation. Figure 12.10 (page 586) gives the Excel output for regressing the annual return on Treasury bills on the annual rate of inflation. The data appear in Table 12.1 (page 585). Starting with the regression standard error $s = 2.1857$ from the output and the variance of the inflation rates in Table 12.1 (use your calculator), find the standard error of the regression slope SE_{b_1}. Check your result against the Excel output. INFLAT

12.56 Predicting T-bill return. Figure 12.17 (page 601) uses statistical software to predict the return on Treasury bills in a year when the inflation rate is 2.25%. Let's do this calculation without specialized software. Figure 12.10 contains Excel regression output. Use a calculator or software to find the variance s_x^2 of the annual inflation rates in Table 12.1 (page 585). From this information, find the 95% prediction interval for the T-bill return. Check your result against the software output in Figure 12.17. INFLAT

Analysis of variance for regression

Software output for regression problems, such as those in Figures 12.5, 12.6, and 12.10, reports values under the heading of "ANOVA" or "Analysis of Variance." Analysis of variance (ANOVA) is the term for statistical analyses that break down the variation in data into separate pieces that correspond to different sources of variation. We used it in Chapter 9 to compare several population means by breaking down the total variation, expressed by sums of squares, into the variation among groups and the variation within groups. In the regression setting, the observed variation in the responses y_i also comes from two sources:

- As the explanatory variable x moves, it pulls the response with it along the regression line. In Figure 12.4, for example, entrepreneurs with 15 years of education generally have higher log incomes than those entrepreneurs with 9 years of education. The least-squares line drawn on the scatterplot describes this tie between x and y.

- When x is held fixed, y still varies because not all individuals who share a common x have the same response y. There are several entrepreneurs with 11 years of education, and their log income values are scattered above and below the least-squares line.

We discussed these sources of variation in Chapter 2, where the main point was that the squared correlation r^2 is the proportion of the total variation in the responses that comes from the first source, the straight-line tie between x and y. Analysis of variance for regression expresses these two sources of variation in algebraic form so that we can calculate the breakdown of overall variation into two parts. Skipping quite a bit of messy algebra, the **analysis of variance equation** that always holds is

total variation in y	=	variation along the line	+	variation about the line
SST	=	SSR	+	SSE
$\sum(y_i - \bar{y})^2$	=	$\sum(\hat{y}_i - \bar{y})^2$	+	$\sum(y_i - \hat{y}_i)^2$

This breakdown is commonly summarized in the form of an ANOVA table:

Source	Degrees of freedom	Sum of squares	Mean square	F
Regression	DFR = 1	SSR = $\sum(\hat{y}_i - \bar{y})^2$	MSR = SSR/DFR	MSR/MSE
Residual	DFE = $n - 2$	SSE = $\sum(y_i - \hat{y}_i)^2$	MSE = SSE/DFE	
Total	DFT = $n - 1$	SST = $\sum(y_i - \bar{y})^2$		

The "total variation in y," which we label SST, is expressed by the sum of the squares of the deviations $y_i - \bar{y}$. Similar to Chapter 9, it is just $n - 1$ times the variance of the responses. The "variation along the line," labeled SSR, has the same form but is the variation among the *predicted* responses $\hat{y}_i$. The predicted responses lie on the least-squares regression line—they show how y moves in response to x. The more y moves in response to x, the larger this term will be. The "variation about the line," labeled SSE, is the sum of squares of the *residuals* $y_i - \hat{y}_i$. It measures the size of the scatter of the observed responses above and below the line. If all the responses fell exactly on a straight line, the residuals would all be 0. In such a case, there would be no variation about the line (SSE = 0) and the total variation would equal the variation along the line (SST = SSR). The other extreme is when $b_1 = 0$. In that case, $\hat{y}_i = \bar{y}$ so SSR = 0 and SST = SSE.

EXAMPLE 12.12

ANOVA for Entrepreneur Income Study Figure 12.19 repeats Figure 12.5; it shows the Excel output for the regression of log income on years of education (Case 12.1). The three terms in the analysis of variance equation appear under the "SS" heading, reflecting the fact that each of the three terms is a sum of squared quantities. You can read the output as follows:

$$\text{SST} = \text{SSR} + \text{SSE}$$
$$129.1534 = 7.4048 + 121.7486$$

FIGURE 12.19 Excel output for the regression of log annual income on years of education, for Examples 12.12 and 12.13. We now concentrate on the analysis of variance part of the output.

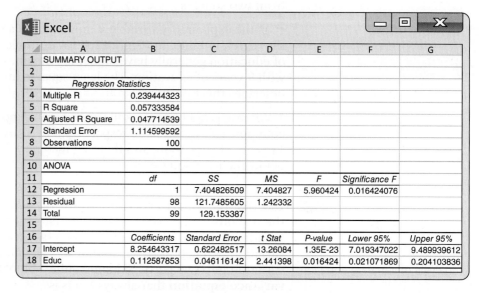

Excel uses the names "Total," "Regression," and "Residual" for these three sources of variation. Other software (see Figure 12.6, page 579) may use other row names.

The proportion of variation in log incomes explained by regressing on years of education is

$$r^2 = \frac{\text{SSR}}{\text{SST}}$$
$$= \frac{7.4048}{129.1534} = 0.0573$$

This agrees with the "R Square" value in the output. Only about 6% of the variation in log incomes is explained by the linear relationship between log income and years of education. The rest is variation in log incomes among entrepreneurs with the same level of education. ■

degrees of freedom, p. 33

The remaining columns of the ANOVA table are similar to those described in Chapter 9. For the degrees of freedom column, the total degrees of freedom (DFT) are $n-1=99$ (the degrees of freedom for the variance of $n = 100$ observations). We know that the degrees of freedom for the residuals (DFE) and for t statistics in simple linear regression are $n-2$. Therefore, it is no surprise that the degrees of freedom for the residual sum of squares are also $n-2 = 98$. That leaves just 1 degree of freedom for regression (DFR), because degrees of freedom in ANOVA are added as follows:

$$\text{DFT} = \text{DFR} + \text{DFE}$$
$$n - 1 = 1 + n - 2$$

mean squares, p. 472

The next column reports the mean squares, obtained by dividing each sum of squares by its degrees of freedom. The total mean square (not given in the output) is just the variance of the responses y_i. The residual mean square (MSE) is the square of our old friend, the regression standard error:

$$\text{MSE} = \frac{\text{SSE}}{\text{DFE}}$$
$$= \frac{\sum(y_i - \hat{y}_i)^2}{n-2}$$
$$= s^2$$

This is why the JMP output in Figure 12.6 labels s as the "Root Mean Square Error."

The next column is named "F" and is the ratio of the two calculated mean squares:

$$F = \frac{\text{MSR}}{\text{MSE}}$$

ANOVA F statistic, p. 473

F distribution, p. 473

This ANOVA F statistic provides a different way to test for the overall significance of the regression. Its P-value is displayed in the last column named "Significance F." This is computed from an F distribution with 1 and $n - 2 = 98$ degrees of freedom. Table E in the back of the book contains the F critical values to compare the ANOVA F statistic against.

Recall that if regression on x has no value for predicting y, we expect the slope of the population regression line to be close to zero. That is, the null hypothesis of "no linear relationship" is H_0: $\beta_1 = 0$. To test H_0, we standardize the slope of the least-squares line to get a t statistic. The ANOVA approach starts instead with sums of squares. If regression on x has no value for predicting y, we expect the SSR to be only a small part of the SST, most of which will be made up of the SSE. That, in turn, means we expect F to be small.

For simple linear regression, the ANOVA F statistic always equals the square of the t statistic for testing H_0: $\beta_1 = 0$. That is, the two tests amount to the same thing. Let's verify that relationship using Case 12.1.

EXAMPLE 12.13

CASE 12.1 **ANOVA for Entrepreneur Income Study, Continued** The Excel output in Figure 12.19 contains the values for the analysis of variance equation for sums of squares and also the corresponding degrees of freedom. The residual mean square is

$$\text{MSE} = \frac{\text{SSE}}{\text{DFE}}$$
$$= \frac{121.7486}{98} = 1.2423$$

The square root of the residual MS is $\sqrt{1.2423} = 1.1146$. This is the regression standard error s, as claimed. The ANOVA F statistic is

$$F = \frac{\text{MSR}}{\text{MSE}}$$
$$= \frac{7.4048}{1.2423} = 5.9604$$

The square root of F is $\sqrt{5.9604} = 2.441$. Sure enough, this is the value of the t statistic for testing the significance of the regression, which also appears in the Excel output. The P-value for F, $P = 0.0164$, is the same as the two-sided P-value for t. ∎

We have now explained almost all the results that appear in a typical regression output such as Figure 12.19. ANOVA shows exactly what r^2 means in regression. Aside from this, ANOVA seems redundant; it repeats in less clear form information that is found elsewhere in the output. This is true in simple linear regression, but ANOVA comes into its own in *multiple regression*, the topic of the next chapter.

APPLY YOUR KNOWLEDGE

T-bills and inflation. *Figure 12.10 (page 586) gives Excel output for the regression of the rate of return on Treasury bills against the rate of inflation during the same year. Exercises 12.57 through 12.59 use this output.*

12.57 A significant relationship? The output reports *two* tests of the null hypothesis that regressing on inflation does *not* help to explain the return on T-bills. State the hypotheses carefully, give the two test statistics, show how they are related, and give the common P-value.

12.58 The ANOVA table. Use the numerical results in the Excel output to verify each of these relationships.

(a) The ANOVA equation for sums of squares.

(b) How to obtain the total degrees of freedom and the residual degrees of freedom from the number of observations.

(c) How to obtain each mean square from a sum of squares and its degrees of freedom.

(d) How to obtain the F statistic from the mean squares.

12.59 ANOVA by-products.

(a) The output gives $r^2 = 0.5020$. How can you obtain this from the ANOVA table?

(b) The output gives the regression standard error as $s = 2.1857$. How can you obtain this from the ANOVA table?

SECTION 12.3 SUMMARY

- The **analysis of variance (ANOVA) equation** for simple linear regression expresses the total variation in the responses as the sum of two sources: the linear relationship of y with x and the residual variation in responses for the same x. The equation is expressed in terms of **sums of squares.**

- Each sum of squares has a **degrees of freedom.** A sum of squares divided by its degrees of freedom is a **mean square.** The residual mean square is the square of the regression standard error.

- The **ANOVA table** gives the degrees of freedom, sums of squares, and mean squares for total, regression, and residual variation. The **ANOVA F statistic** is the ratio $F =$ Regression MS/Residual MS. In simple linear regression, F is the square of the t statistic for the hypothesis that regression on x does not help explain y.

- The **square of the sample correlation** can be expressed as

$$r^2 = \frac{\text{Regression SS}}{\text{Total SS}} = \frac{\text{SSR}}{\text{SST}}$$

and is interpreted as the proportion of the variability in the response variable y that is explained by the explanatory variable x in the linear regression.

SECTION 12.3 EXERCISES

For Exercises 12.55 and 12.56, see page 606; and for 12.57 to 12.59, see page 610.

12.60 What's wrong? For each of the following statements, explain what is wrong and why.

(a) In simple linear regression, the standard error for a future observation is s, the measure of spread about the regression line.

(b) In an ANOVA table, SSE is the sum of the deviations.

(c) There is a close connection between the correlation r and the intercept of the regression line.

(d) The squared correlation r^2 is equal to MSR/MST.

12.61 What's wrong? For each of the following statements, explain what is wrong and why.

(a) In simple linear regression, the null hypothesis of the ANOVA F test is $H_0: \beta_0 = 0$.

(b) In an ANOVA table, the mean squares add; in other words, MST = MSR + MSE.

(c) The smaller the P-value for the ANOVA F test, the greater the explanatory power of the model.

(d) The total degrees of freedom in an ANOVA table are equal to the number of observations n.

U.S. versus overseas stock returns. *How are returns on common stocks in overseas markets related to returns in U.S. markets? Consider measuring U.S. returns by the annual rate of return on the Standard & Poor's 500 stock index and overseas returns by the annual rate of return on the Morgan Stanley Europe, Australasia, Far East (EAFE) index. Both are recorded in percents. Here is part of the Minitab output for regressing the EAFE returns on the S&P 500 returns for the 29 years 1989 to 2017.*

The regression equation is
Eafe = −3.11 + 0.820 S&P

Analysis of Variance

Source	DF	SS	MS	F
Regression		5821.3		
Residual Error				
Total	28	10454.7		

Exercises 12.62 through 12.66 use this output. EAFE

12.62 The ANOVA table. Complete the analysis of variance table by filling in the "Residual Error" row and the other missing items in the DF, MS, and F columns.

12.63 s and r^2. What are the values of the regression standard error s and the squared correlation r^2?

12.64 Estimating the standard error of the slope. The standard deviation of the S&P 500 returns for these years is 17.58%. From this and your work in the previous exercise, find the standard error for the least-squares slope b_1. Give a 90% confidence interval for the slope β_1 of the population regression line.

12.65 Inference for the intercept? The mean of the S&P 500 returns for these years is 12.01. From this and information from the previous exercises, find the standard error for the least-squares intercept b_0. Use this to construct a 95% confidence interval. Finally, explain why the intercept β_0 is meaningful in this example.

12.66 Predicting the return for a future year. Suppose the S&P annual return for a future year is 0%. Using the information from the previous four exercises, construct the appropriate 95% interval. Also, explain why this interval is or is not the same interval constructed in Exercise 12.65.

Gross domestic product per capita and net savings. *The gross domestic product (GDP) measures the aggregate amount of good and services produced in an economy. Growing GDP is a primary focus of policymakers. A random sample of 38 emerging economies (countries) was taken to assess whether there was a positive linear relationship between GDP per capita and a country's adjusted net savings.[17] Adjusted net savings is defined as a country's net savings plus expenditures on education minus depletion of a country's air, minerals, and forests. It is reported as a percent of the gross national income. Figure 12.20 contains JMP output for the regression of the logarithm of GDP per capita (Lgdpc) on adjusted net savings (Sav). Exercises 12.67 through 12.74 concern this analysis. You can take it as given that an examination of the data shows no serious violations of the conditions required for regression inference.*

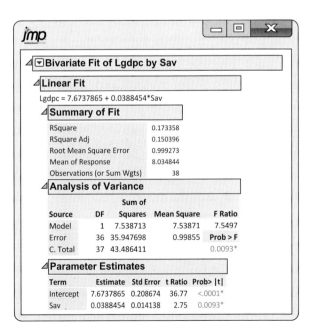

FIGURE 12.20 JMP output for the regression of GDP per capita of 38 emerging economies on the country's adjusted net savings, for Exercises 12.67 to 12.74.

12.67 Significance in two senses.

(a) Provide an explanation for using the logarithm of GDP per capita as the response variable rather than GDP per capita.

(b) Is there good evidence that adjusted net savings helps explain Lgdpc? (State the hypotheses, give a test statistic and P-value, and state a conclusion.)

(c) What percent of the variation in Lgdpc among these economies is explained by a regression on adjusted net savings?

(d) Use your findings in parts (a) and (b) as the basis for a short description of the distinction between statistical significance and practical significance.

12.68 Estimating the slope. Explain clearly what the slope β_1 of the population regression line tells us in this setting. Give a 95% confidence interval for this slope.

12.69 Predicting Lgdpc. An additional calculation shows that the variance of the adjusted net savings for these 38 countries is $s_x^2 = 135.03$. JMP labels the regression standard error s as "Root Mean Square Error" and the sample mean of the responses $\bar{y}$ as "Mean of Response." Starting from these facts, give a 95% confidence interval for the mean Lgdpc for all countries with an adjusted net savings $x = 15.0$. [Hint: The least-squares regression line always goes through $(\bar{x}, \bar{y})$.]

12.70 Predicting Lgdpc. Will a 95% prediction interval for a country's Lgdpc when $x = 15.0$ be wider or narrower than the confidence interval found in the previous exercise? Explain why we should expect this result. Then give the 95% prediction interval.

12.71 F versus t. How do the ANOVA F statistic and its P-value relate to the t statistic for the slope and its P-value? Identify these results on the output and verify their relationship (up to roundoff error).

12.72 The regression standard error. How can you obtain s from the ANOVA table? Do this, and verify that your result agrees with the value that JMP reports for s.

12.73 Squared correlation. JMP gives the squared correlation r^2 as "RSquare." How can you obtain r^2 from the ANOVA table? Do this, and verify that your result agrees with the RSquare reported by JMP.

12.74 Correlation. The regression in Figure 12.20 takes adjusted net savings as explaining Lgdpc. As an alternative, we could take Lgdpc as explaining adjusted net savings. We would then reverse the roles of the variables, regressing Sav on Lgdpc. Both regressions lead to the same conclusions about the correlation between Lgdpc and Sav. What is this correlation r? Is there good evidence that it is positive?

CHAPTER 12 REVIEW EXERCISES

12.75 What's wrong? For each of the following statements, explain what is wrong and why.

(a) The slope describes the change in x for a unit change in y.

(b) The population regression line is $y = b_0 + b_1 x$.

(c) A 95% confidence interval for the mean response is the same width regardless of x.

(d) The residual for the ith observation is $\hat{y}_i - y_i$

12.76 What's wrong? For each of the following statements, explain what is wrong and why.

(a) The parameters of the simple linear regression model are b_0, b_1, and s.

(b) To test $H_0: b_1 = 0$, you would use a t test.

(c) For any value of the explanatory variable x, the confidence interval for the mean response will be wider than the prediction interval for a future observation.

(d) The least-squares line is the line that maximizes the sum of the squares of the residuals.

12.77 Interpreting a residual plot. Figure 12.21 shows four plots of residuals versus x. For each plot, comment on the regression model conditions necessary for inference. Which plots suggest a reasonable fit to the linear regression model? What actions might you take to remedy the problems in each of the other plots?

12.78 Are the results consistent? A researcher surveyed $n = 214$ hotel managers to assess the relationship between customer-relationship management (CRM) and organizational culture.[18] Each variable was an average of more than twenty-five 5-point Likert survey responses and, therefore, was treated as a quantitative variable. The researcher reports a sample correlation of $r = 0.74$ and an ANOVA F statistic of 60.35 for a simple linear regression of CRM on organizational culture. Using the relationship between testing $H_0: \rho = 0$ and testing $H_0: \beta_1 = 0$, show that these two results are not consistent. (Hint: It's far more likely that there was a typo and $r = 0.47$.)

12.79 College debt versus adjusted in-state costs. Kiplinger's "Best Values in Public Colleges" provides a ranking of U.S. public colleges based on a combination of various measures of academics and affordability.[19] We'll consider a random collection of 40 colleges from Kiplinger's 2018 report and focus on the average debt in dollars at graduation (AveDebt) and the in-state cost per year after need-based aid (InCostAid). BESTVAL

(a) A scatterplot of these two variables with a smoothed curve is shown in Figure 12.22. Describe the relationship. Are there any possible outliers or unusual values? Does a linear relationship between InCostAid and AveDebt seem reasonable?

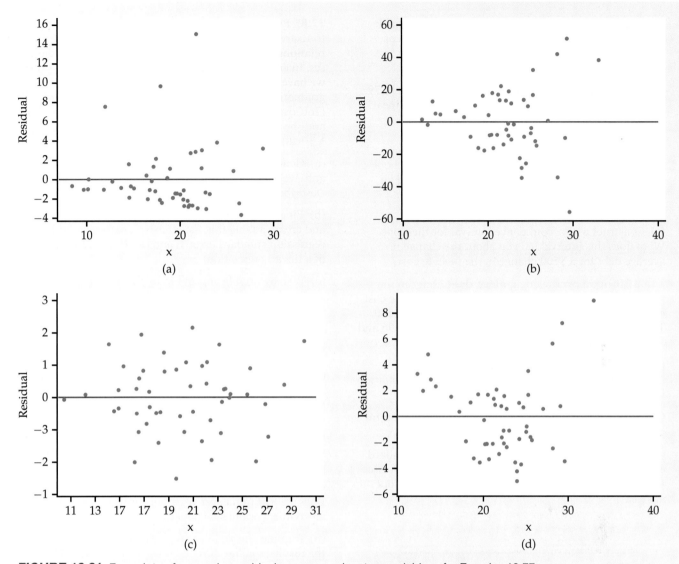

FIGURE 12.21 Four plots of regression residuals versus explanatory variable x, for Exercise 12.77.

(b) Based on the scatterplot, approximately how much does the average debt change for an additional $1000 of annual cost?

(c) The University of North Carolina at Chapel Hill is a school with an adjusted in-state cost of $4843. Discuss the appropriateness of using this data set to predict the average debt at graduation for students attending this school.

12.80 Can we consider this an SRS? Refer to the previous exercise. The report states that Kiplinger's rankings focus on traditional four-year public colleges with broad-based curricula and on-campus housing. Each year, the researchers start with more than 500 schools and then narrow the list down to roughly 120 based on academic quality before ranking them. The data set in the previous exercise is an SRS from Kiplinger's published list of 100 schools. When investigating the relationship between the average debt and the in-state

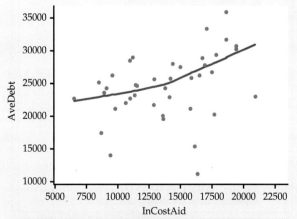

FIGURE 12.22 Scatterplot of average debt (in dollars) at graduation versus the in-state cost per year (in dollars) after need-based aid, for Exercise 12.79.

cost after adjusting for need-based aid, is it reasonable to consider this to be an SRS from the population of interest? Write a short paragraph explaining your answer.

12.81 Predicting college debt. Refer to Exercise 12.79. Figure 12.23 contains JMP output for the simple linear regression of AveDebt on InCostAid. BESTVAL

(a) State the least-squares regression line.

(b) The University of California at Irvine is one school in this sample. It has an in-state cost of $9621 and an average debt at graduation of $20,466. What is the residual?

(c) Construct a 95% confidence interval for the slope. What does this interval tell you about the change in average debt for a $500 change in the in-state cost?

12.82 More on predicting college debt. Refer to the previous exercise. Purdue University has an in-state cost of $7788 and an average debt at graduation of $27,530. Texas A&M University has an in-state cost of $11,396 and an average debt at graduation of $24,072. BESTVAL

(a) Using your answer to part (a) of the previous exercise, what is the predicted average debt at graduation for a student attending Purdue University?

(b) What is the predicted average debt at graduation for a student attending Texas A&M University?

(c) Without doing any calculations, would the standard error for the estimated average debt be larger for Purdue University or the Texas A&M University? Explain your answer.

12.83 Predicting college debt: Other measures. Refer to Exercise 12.79. Let's now look at AveDebt and its relationship with all six measures available in the data set. In addition to the in-state cost after aid (InCostAid), we have the admittance rate (Admit), the four-year graduation rate (Grad4Rate), the in-state cost before aid (TotCostIn), the out-of-state cost before aid (TotCostOut), and the out-of-state cost after aid (OutCostAid). BESTVAL

(a) Generate scatterplots of each explanatory variable and AveDebt. Do all these relationships look linear? Describe what you see.

(b) Fit each of the explanatory variables separately and create a table that lists the explanatory variable, regression standard error s, and the P-value for the test of a linear association.

(c) Which variable appears to be the best single explanatory variable of average debt at graduation? Explain your answer.

12.84 Yearly number of tornadoes. The Storm Prediction Center of the National Oceanic and Atmospheric Administration maintains a database of tornadoes, floods, and other weather phenomena. Table 12.6 summarizes the annual number of tornadoes in the United States between 1953 and 2017.[20] (Note: These are time series data with a very weak correlation, so simple linear regression is reasonable here. See Chapter 14 for methods designed specifically for use with time series.) TWISTER

(a) Make a plot of the total number of tornadoes by year. Does a linear trend over years appear reasonable? Are there any outliers or unusual patterns? Explain your answer.

(b) Run the simple linear regression and summarize the results, making sure to construct a 95% confidence interval for the average annual increase in the number of tornadoes.

(c) Obtain the residuals and plot them versus year. Is there anything unusual in the plot?

(d) Are the residuals Normal? Justify your answer.

(e) The number of tornadoes in 2004 is much larger than expected under this linear model. Remove this observation and rerun the simple linear regression. Compare these results with the results in part (b). Do you think this observation should be considered an outlier and removed? Explain your answer.

12.85 Plot indicates model assumptions. Construct a plot with data and a regression line that fits the simple linear regression model framework. Then construct another plot that has the same slope and intercept but a much smaller value of the regression standard error s.

12.86 Significance tests and confidence intervals. The significance test for the slope in a simple linear regression gave a value $t = 2.08$ with 18 degrees of freedom. Would the 95% confidence interval for the slope include the value zero? Give a reason for your answer.

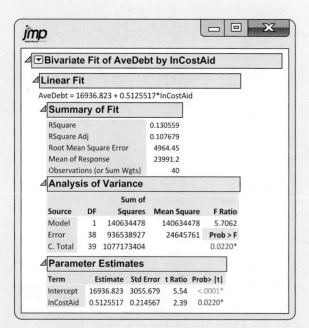

FIGURE 12.23 JMP output for the regression of average debt (in dollars) at graduation on the in-state cost (in dollars) per year, for Exercise 12.81.

TABLE 12.6 Annual number of tornadoes in the United States between 1953 and 2017

Year	Number of tornadoes	Year	Number of tornadoes	Year	Number of tornadoes	Year	Number of tornadoes
1953	422	1970	653	1987	656	2004	1817
1954	550	1971	889	1988	702	2005	1265
1955	593	1972	741	1989	856	2006	1103
1956	504	1973	1102	1990	1133	2007	1096
1957	858	1974	945	1991	1132	2008	1692
1958	564	1975	919	1992	1297	2009	1156
1959	604	1976	834	1993	1173	2010	1282
1960	616	1977	852	1994	1082	2011	1692
1961	697	1978	789	1995	1235	2012	936
1962	657	1979	855	1996	1173	2013	891
1963	463	1980	866	1997	1148	2014	881
1964	704	1981	782	1998	1424	2015	1183
1965	897	1982	1047	1999	1339	2016	985
1966	585	1983	931	2000	1075	2017	1406
1967	926	1984	907	2001	1215		
1968	660	1985	684	2002	934		
1969	608	1986	765	2003	1374		

12.87 Predicting college debt: One last measure. Refer to Exercises 12.79, 12.81, and 12.83. Given the in-state cost of attending a college prior to and after aid, another measure of potential college debt is the average amount of need-based aid. Create this new variable by subtracting these two costs, and investigate its relationship with average debt at graduation. Write a short paragraph summarizing your findings. BESTVAL

12.88 Brand equity and sales. Brand equity is one of the most important assets of a business. It includes brand loyalty, brand awareness, perceived quality, and brand image. One study examined the relationship between brand equity and sales using simple linear regression analysis.[21] The correlation between brand equity and sales was reported to be 0.757, with a significance level of 0.001.

(a) Explain in simple language the meaning of these results.

(b) The study examined quick-service restaurants in Korea and was based on 394 usable surveys from a total of 950 that were distributed to shoppers at a mall. Write a short narrative commenting on the design of the study and how well you think the results would apply to other settings.

12.89 Hotel sizes and numbers of employees. A human resources study of hotels collected data on the hotel size, measured by number of rooms, and the number of employees for 14 hotels in Canada.[22] Here are the data: HOTSIZE

Employees	Rooms	Employees	Rooms
1200	1388	275	424
180	348	105	240
350	294	435	601
250	413	585	1590
415	346	560	380
139	353	166	297
121	191	228	108

(a) To what extent can the number of employees be predicted by the size of the hotel? Plot the data and summarize the relationship.

(b) Is this the type of relationship that you would expect to see before examining the data? Explain why or why not.

(c) Calculate the least-squares regression line and add it to the plot.

(d) Give the results of the significance test for the regression slope with your conclusion.

(e) Find a 95% confidence interval for the slope.

12.90 How can we use the results? Refer to the previous exercise.

(a) If one hotel had 100 more rooms than another hotel, how many additional employees would you expect the first hotel to have? HOTSIZE

(b) Give a 95% confidence interval for your answer in part (a).

(c) The study collected these data from 14 hotels in Toronto. Discuss how well you think the results can be generalized to other hotels in Toronto, to hotels in Canada, and to hotels in other countries.

12.91 Check the outliers. The plot that you generated in Exercise 12.89 has two observations that appear to be outliers. HOTSIZE

(a) Identify these points on a plot of the data.

(b) Rerun the analysis with the other 12 hotels, and summarize the effect of the two possible outliers on the results that you gave in Exercise 12.89.

12.92 Selling a large house. Among the houses for which we have data in Table 12.5 (page 595), just four have floor areas of 1800 square feet or more. Give a 90% confidence interval for the mean selling price of houses with floor areas of 1800 square feet or more. HSIZE

12.93 Agricultural productivity. Few sectors of the economy have increased their productivity as rapidly as agriculture. Let's describe this increase. Productivity is defined as output per unit input. "Total factor productivity" (TFP) takes all inputs (labor, capital, fuels, and so on) into account. The data set AGPROD contains TFP for the years 1948–2015.[23] The TFP entries are index numbers; that is, they give each year's TFP as a percent of the value for 1948. AGPROD

(a) Plot TFP against year. It appears that around 1980, the rate of increase in TFP changed. How is this apparent from the plot? What was the nature of the change?

(b) Regress TFP on year using only the data for the years 1948–1980. Add the least-squares line to your scatterplot. The line makes the finding in part (a) clearer.

(c) Give a 95% confidence interval for the annual rate of change in TFP during the period 1948–1980.

(d) Regress TFP on year for the years 1981–2015. Add this line to your plot. Give a 95% confidence interval for the annual rate of improvement in TFP during these years.

(e) Write a brief report on trends in U.S. farm productivity since 1948, making use of your analysis in parts (a) through (d).

12.94 CEO pay and gross profits. Starting in 2018, publicly traded companies must disclose their workers' median pay and the compensation ratio between a worker and the company's CEO. Does this ratio say something about the performance of the company? CNBC collected this ratio and the gross profits per employee from a variety of companies.[24] CNBC

(a) Generate a scatterplot of the gross profit per employee (Profit) versus the CEO pay ratio (Ratio). Describe the relationship.

(b) To compensate for the severe right skewness of both variables, take the logarithm of each variable. Generate a scatterplot and describe the relationship between these transformed variables.

(c) Fit a simple linear regression for log Profits versus log Ratio.

(d) Examine the residuals. Are the model conditions approximately satisfied? Explain your answer.

(e) Construct a 95% confidence interval for β_1 and interpret the result in terms of a percent change in y for a percent change in x.

PhotoSky/Shutterstock

CHAPTER 13

Multiple Regression

Introduction

In Chapters 2 and 12, we studied methods for inference in the setting of a linear relationship between a quantitative response variable y and a *single* quantitative explanatory variable x. In this chapter, we look at situations in which *several explanatory variables,* both quantitative and categorical, work together to explain or predict a single response variable. Here are some examples. Can you identify the response and explanatory variables?

- As part of their R&D, Under Armour investigates elite marathon runners' body temperature in relation to the outside temperature and humidity, their age and resting heart rate, and several new concept versions of HeatGear® apparel.

- Disney Media and Advertising Lab wants a model that will predict a viewer's emotional arousal (via skin conductance) based on different themed advertisements, the volume of the advertisement's background music, whether it was viewed on an HD or standard-definition television, and the sex and age of the viewer.

- Coldwell Banker wants to develop an app for its realtors that predicts the selling price of a house based on house characteristics such as size, number of bedrooms, number of bathrooms, and school district.

In all these examples, the information gleaned from the analysis is used by the company to aid in its decision making, whether it be to advance product development or to better communicate with customers. This is an underlying theme of predictive analytics and why multiple regression is one of its most valuable and fundamental tools.

CHAPTER OUTLINE

13.1 Data Analysis for Multiple Regression

13.2 Inference for Multiple Regression

13.3 Multiple Regression Model Building

To continue our study of inference for regression, we build on the descriptive tools we learned in Chapter 2 and the basics of regression inference from Chapter 12. Many of these tools and ideas carry directly over to the multiple regression setting. For example, we will continue to use scatterplots and correlation to study pairs of variables. We will also continue to use least-squares regression to obtain model parameter estimates.

The presence of several explanatory variables, however, which may assist or substitute for each other in predicting the response, leads to many new ideas—so many that there are textbooks and courses devoted primarily to multiple regression. This chapter does not try to cover everything. Instead, we discuss key methods of data analysis and inference for multiple regression and then introduce some of these additional concepts through the analysis of several case studies.

13.1 Data Analysis for Multiple Regression

When you complete this section, you will be able to:

- Describe a multiple regression model in terms of a response variable, the explanatory variables, and the number of cases.
- Compute the predicted response from a linear model involving multiple explanatory variables.
- Perform preliminary data analysis for multiple regression.
- Fit a multiple regression model using statistical software and obtain the residuals to check model conditions prior to inference.
- Interpret statistical software regression output to obtain the multiple regression equation and regression standard error.

Before jumping directly into data analysis and inference, let's first explore the use of a linear model with five variables. The goal here is to become comfortable with linear models that involve more than one explanatory (or predictor) variable, something we have not yet seen in the textbook.

Using a linear model with multiple variables

Allocation of space or other resource within a business organization is often done using quantitative methods. Characteristics for a subunit of the organization are determined, and then a mathematical formula is used to compute the required needs. Space allocation models are frequently used to plan for a company's upsizing or operational restructuring.

EXAMPLE 13.1

A Space Allocation Model A university uses a linear model with multiple predictors to determine the office space needs in square feet (ft^2) for each department.[1] The formula allocates 210 ft^2 for the department head (Head), 160 ft^2 for each faculty member (Fac), 160 ft^2 for each manager (Mgr), 150 ft^2 for each administrator and lecturer (Lect), 65 ft^2 for each postdoctorate and graduate assistant (Grad), and 120 ft^2 for each clerical and service worker (Clsv). These allocations were not obtained through fitting the model to data, but rather a university committee determined these allocations using information on the numbers of each employee type and space availability in the buildings on campus. ∎

The Chemistry Department in this university has 1 department head, 45.25 faculty, 15.50 managers, 41.52 lecturers, 411.88 graduate assistants, and 25.24 clerical and service workers. Note that fractions of people are possible in these calculations because individuals may have appointments in more than one department. For example, a person with an even split between two departments would be counted as 0.50 in each.

EXAMPLE 13.2

Office Space Needs for the Chemistry Department Let's calculate the office space needs for the Chemistry Department based on these personnel numbers. We start with 210 ft^2 for the department head. We have 45.25 faculty, each needing 160 ft^2. Therefore, the total office space needed for faculty is 45.25×160 ft^2, which is 7240 ft^2. We do the same type of calculation for each personnel category and then sum the results.

Here are the calculations in a table:

Category	Number of employees	Square footage per employee	Employees × square footage
Head	1.00	210	210.0
Fac	45.25	160	7,240.0
Mgr	15.50	160	2,480.0
Lect	41.52	150	6,228.0
Grad	411.88	65	26,772.2
Clsv	25.24	120	3028.8
Total			45,959.0

These calculations use a set of explanatory variables—Head, Fac, Mgr, Lect, Grad, and Clsv—to find the office space needs for the Chemistry Department. Given values of these variables for any other department in the university, we can perform the same calculations to find its office space needs. We organized our calculations for the Chemistry Department in the preceding table. Another way to organize calculations of this type is to give a formula.

EXAMPLE 13.3

The Office Space Needs Formula Let's assume that each department has exactly one head. So the first term in our equation will be the space need for this position, 210 ft^2. To this, we add the space needs for the faculty, 160 ft^2 for each, or 160Fac. Similarly, we add the number of square feet for each category of personnel times the number of employees in the category. The result is the office space needs predicted by the space model. Here is the formula:

PredSpace = 210 + 160Fac + 160Mgr + 150Lect + 65Grad + 120Clsv

The formula combines information from the explanatory variables and computes the office space needs for any department. This prediction generally will not match the actual space being used by a department. The difference between the value predicted by the model and the actual space being used is of interest to the people who assign space to departments.

EXAMPLE 13.4

Compare Predicted Space with Actual Space The Chemistry Department currently uses 50,075 ft² of space. On the other hand, the model predicts a space need of 45,959 ft². The difference between these two quantities is a residual:

$$\text{residual} = \text{ActualSpace} - \text{PredSpace}$$
$$= 50{,}075 - 45{,}959$$
$$= 4116$$

According to the university space needs model, the Chemistry Department has about 4116 ft² more office space than it needs.

Because of this, the university director of space management is considering giving some of this excess space to a department that has actual space less than what the model predicts. Of course, the Chemistry Department does not think that it has excess space. In negotiations with the space management office, the department will explain that it needs all the current space and that its needs are not fully captured by the model. ∎

APPLY YOUR KNOWLEDGE

13.1 Changing the formula. Suppose that the university committee revised the space allocations for postgraduate and graduate students to 75 ft² and for the department head to 225 ft². Revise the formula in Example 13.3 to reflect these new allocations.

13.2 Needs of the Chemistry Department. Refer to the previous exercise.

(a) Use this new formula to predict the space needs of the Chemistry Department.

(b) Find the residual under this new formula and explain what it means in a few sentences.

These space allocation examples illustrate two key ideas that we need for multiple regression. First, we have several explanatory variables that are combined in a linear prediction equation. Second, residuals remain the differences between the actual values and the predicted values.

We now illustrate the techniques of multiple linear regression, including some new ideas, through a series of case studies. In all examples, we use software to do the calculations.

CASE 13.1 **The Inclusive Development Index (IDI)** In 2017, the World Economic Forum introduced a new metric that measures the impact a country's economic policy and growth have on all its citizens. This metric resulted from concerns over widening income gaps and political polarization. Table 13.1 shows this metric and two characteristics of a country's economic policy that should contribute positively to the IDI for 15 advanced economies.[2] Let's consider fitting a multiple linear regression model to see how well these two characteristics predict IDI. Here, IDI represents the response variable, and employment (% of population) and median per

capita daily income ($ US) are the two explanatory variables. The variable Country labels each row of values. ∎

IDI

TABLE 13.1	Countries with advanced economies: IDI, percent employment, and median income		
Country	IDI	Employment (% of population)	Median daily income ($ US)
Australia	5.36	60.9	44.4
Belgium	5.14	49.1	43.8
Canada	5.06	60.8	49.2
Czech Rep.	5.09	57.0	24.3
Estonia	4.74	57.6	22.1
Iceland	6.07	71.1	43.4
Ireland	5.44	54.9	38.0
Israel	4.51	60.3	25.8
Italy	4.31	42.7	34.3
Japan	4.53	57.2	34.8
Rep. of Korea	5.09	58.6	34.2
Netherlands	5.61	59.7	43.3
Norway	6.08	61.7	63.8
Portugal	3.97	51.8	21.2
United Kingdom	4.89	59.6	39.4

Data for multiple regression

The data for a simple linear regression problem consist of pairs of an explanatory variable x and a response variable y. We use n for the number of pairs. The major difference in the data for multiple regression is that we have more than one explanatory variable. Because of this, we use the term "case" instead of "pair."

case, p. 2

EXAMPLE 13.5

Data for the IDI Study In Case 13.1, the cases are the 15 countries. Each case consists of a value for a response variable (IDI) and values for the two explanatory variables (employment and median daily income.). ∎

In general, we have data on n cases and we use k for the number of explanatory variables. Data are often entered into spreadsheets and computer regression programs in a format where each row describes a case and each column corresponds to a different variable.

EXAMPLE 13.6

Spreadsheet Data for IDI Study In Case 13.1, there are 15 countries; employment (Employ) and median per capita daily income (Medinc) are the explanatory variables. Therefore, $n = 15$ and $k = 2$. Figure 13.1 shows the part of an Excel spreadsheet with the first 10 cases. ∎

FIGURE 13.1 First 10 cases of data in an Excel spreadsheet for Example 13.6.

	A	B	C	D
1	Country	IDI	Employ	Medinc
2	Australia	5.36	60.9	44.4
3	Belgium	5.14	49.1	43.8
4	Canada	5.06	60.8	49.2
5	CzechRep	5.09	57.0	24.3
6	Estonia	4.74	57.6	22.1
7	Iceland	6.07	71.1	43.4
8	Ireland	5.44	54.9	38.0
9	Israel	4.51	60.3	25.8
10	Italy	4.31	42.7	34.3

APPLY YOUR KNOWLEDGE

13.3 Assets, interest-bearing deposits, and equity capital. The Federal Deposit Insurance Corporation (FDIC) supplies data for insured commercial banks, by state or other area.[3] In Table 13.2, the cases are the

TABLE 13.2 Insured commercial banks by state or other area

State or area	Assets	Deposits ($ billions)	Equity ($ billions)	State or area	Assets	Deposits ($ billions)	Equity ($ billions)
Alabama	261.8	143.0	34.4	Montana	34.4	19.9	4.1
Alaska	6.4	2.8	0.8	Nebraska	74.1	47.4	7.8
Arizona	24.4	12.0	2.7	Nevada	220.1	179.3	17.5
Arkansas	97.7	64.9	13.5	New Hampshire	11.0	7.6	1.2
California	756.7	380.9	92.1	New Jersey	148.6	92.9	17.2
Colorado	63.0	42.0	6.3	New Mexico	13.3	8.0	1.3
Connecticut	111.2	67.2	12.8	New York	1004.2	603.3	120.4
Delaware	1067.3	608.3	145.8	North Carolina	2033.5	1057.2	242.6
District of Columbia	1.2	0.7	0.1	North Dakota	28.8	19.5	2.9
Florida	196.0	124.9	20.3	Ohio	3077.0	1655.1	314.0
Georgia	312.3	193.6	36.4	Oklahoma	114.9	68.0	12.3
Guam	2.4	1.6	0.2	Oregon	31.3	17.0	4.9
Hawaii	53.4	32.0	5.2	Pennsylvania	247.0	159.2	29.9
Idaho	6.2	3.5	0.7	Rhode Island	132.7	72.1	18.2
Illinois	494.4	287.8	55.8	South Carolina	36.1	22.9	4.9
Indiana	94.9	60.6	11.1	South Dakota	3193.8	1779.8	316.7
Iowa	83.7	55.8	9.1	Tennessee	130.6	81.0	16.3
Kansas	70.3	45.2	8.1	Texas	499.1	249.9	55.5
Kentucky	59.7	37.9	6.7	Utah	654.3	448.3	78.3
Louisiana	76.7	46.1	9.2	Vermont	4.9	3.1	0.5
Maine	26.7	17.2	3.0	Virginia	731.7	472.5	92.4
Maryland	40.6	24.2	4.9	Washington	77.1	44.5	9.8
Massachusetts	390.2	244.0	39.7	West Virginia	29.6	18.2	3.6
Michigan	79.2	44.4	8.8	Wisconsin	112.9	70.6	13.5
Minnesota	73.5	47.1	7.9	Wyoming	8.3	5.8	0.9
Mississippi	94.7	55.3	10.9	Puerto Rico	62.4	72.1	8.5
Missouri	159.0	100.1	16.2				

50 states, the District of Columbia, Guam, and Puerto Rico. Bank assets, interest-bearing deposits, and equity capital are given in billions of dollars. We are interested in describing how assets are explained by total interest-bearing deposits and total equity capital.

BANKS

(a) What is the response variable?

(b) What are the explanatory variables?

(c) What is k, the number of explanatory variables?

(d) What is n, the number of cases?

(e) Is there a label variable? If yes, identify it.

13.4 **Are company profits related to the CEO's profile?** As part of a study, data from 55 *Fortune* 500 companies were obtained.[4] Based on these data, the researchers described the relationship between a company's annual profits and the age and facial width-to-height ratio of its CEO.

(a) What is the response variable?

(b) What is n, the number of cases?

(c) What is k, the number of explanatory variables?

(d) What are the explanatory variables?

Preliminary data analysis for multiple regression

As with any statistical analysis, we begin our multiple regression analysis with a careful examination of the data. We look first at each variable separately, then at relationships among the variables. In both situations, we continue our practice of combining plots and numerical descriptions.

EXAMPLE 13.7

IDI

CASE 13.1 **Describing IDI, Employment, and Income** Figure 13.2 shows descriptive statistics for each of these variables. Figure 13.3 presents histograms. The distributions are not highly skewed, nor are there any obvious outliers. Thus, we have no concerns about influential values at the individual variable level. ∎

FIGURE 13.2 Descriptive statistics for Example 13.7.

Minitab

Descriptive Statistics: IDI, Employ, Medinc

Statistics

Variable	N	Mean	SE Mean	StDev	Minimum	Q1	Median	Q3	Maximum
IDI	15	5.059	0.155	0.602	3.970	4.530	5.090	5.440	6.080
Employ	15	57.53	1.65	6.39	42.70	54.90	58.60	60.80	71.10
Medinc	15	37.47	2.95	11.43	21.20	25.80	38.00	43.80	63.80

FIGURE 13.3 Histograms for Example 13.7.

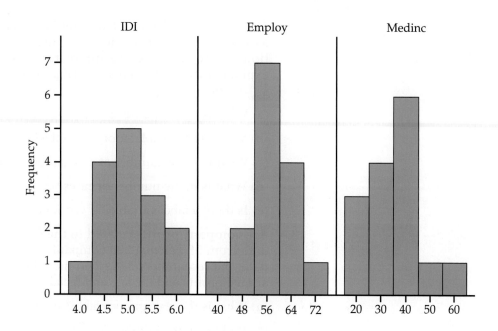

Later in this chapter, we will describe a statistical model that is the basis for inference in multiple regression. *This model does not require Normality for the distributions of the response or explanatory variables*. The Normality assumption applies to the distribution of the model deviations, as was the case for inference in simple linear regression. We look at the distribution of each variable to be used in a multiple regression to determine if there are any unusual patterns or values that may be important when performing our regression analysis.

APPLY YOUR KNOWLEDGE

13.5 Is there a problem? Refer to Exercise 13.4 (page 623). The 55 firms in the sample represented a range of industries, including retail and computer manufacturing. Suppose this resulted in the response variable, annual profits, having a bimodal distribution. Considering that this distribution is not Normal, will this necessarily be a problem for inference in multiple regression? Explain your answer.

13.6 Look at the data. Refer to Exercise 13.3 (page 622). Examine the distribution for Assets, Deposits, and Equity. That is, use graphs to display the distribution of each variable. Based on your examination, how would you describe the data? Are there any cases with values that you consider to be outliers or unusual in any way? Explain your answer. BANKS

Now that we know something about the distributions of the individual variables, we next look at the relations between pairs of variables.

EXAMPLE 13.8

CASE 13.1 **IDI, Employment, and Income in Pairs** With three variables, we also have three pairs of variables to examine. Figure 13.4 gives the three correlations, and Figure 13.5 displays the corresponding scatterplots. We use a smoothed curve (page 69) to help us see the overall pattern of each scatterplot. ∎

FIGURE 13.4 Correlations for Example 13.8.

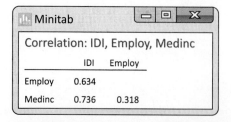

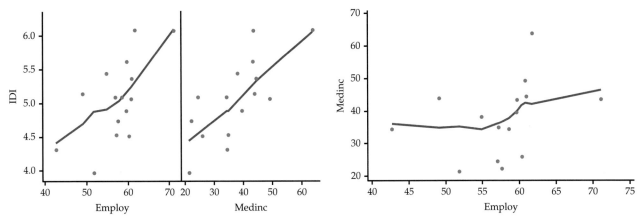

FIGURE 13.5 Scatterplots of pairs of variables for Example 13.8.

Both Employ and Medinc have reasonably strong positive correlations with IDI, suggesting these variables may be useful in explaining IDI. In addition, Employ and Medinc are weakly positively correlated ($r = 0.318$). *Because we will use both variables to explain IDI, we would be concerned if this correlation were high*, something we will later term multicollinearity. Two highly correlated explanatory variables contain about the same information, so both together may explain IDI only a little better than either alone.

The plots are more revealing. The relationship between IDI and each of the two explanatory variables appears reasonably linear and there are no unusual observations. The relationship between Medinc and Employ is relatively flat, suggesting little shared information between these variables. Because each variable is linearly related with IDI, this suggests that some of the variation in IDI that is unexplained by one explanatory variable is explained by the other.

APPLY YOUR KNOWLEDGE

13.7 Examining the pairs of relationships. Refer to Exercise 13.3 (page 622). Examine the relationship between each pair of variables. That is, compute correlations and construct scatterplots. Based on these summaries, describe these relationships. Are there any states or other areas that you consider unusual in any way? Explain your answer. BANKS

13.8 Try logs. Refer to the previous exercise. Because of strong right-skewness, the data set also contains the logarithms of each variable. Find the correlations and generate scatterplots for each pair of transformed variables. Describe the relationships. Are there any states or other areas that you consider unusual in any way? Explain your answer. BANKS

Estimating the multiple regression coefficients

least-squares regression, p. 81

Simple linear regression with a response variable y and one explanatory variable x begins by using the least-squares idea to fit a straight line $\hat{y} = b_0 + b_1 x$ to data on the two variables. Although we now have k explanatory variables, the principle is the same: we use the least-squares idea to fit a linear function

$$\hat{y} = b_0 + b_1 x_1 + b_2 x_2 + \cdots + b_k x_k$$

to the data.

We use a subscript i to distinguish different cases. For the ith case, the predicted response is

$$\hat{y}_i = b_0 + b_1 x_{i1} + b_2 x_{i2} + \cdots + b_k x_{ik}$$

residual, p. 90

As usual, the residual is the difference between the observed value of the response variable and the value predicted by the model. For the ith case, the residual is

$$e_i = y_i - \hat{y}_i$$

The method of least squares chooses the b's that make the sum of squares of the residuals as small as possible. In other words, the *least-squares estimates* are the values that minimize the quantity

$$\sum (y_i - \hat{y}_i)^2$$

As with simple linear regression, it is possible to give formulas for the least-squares estimates. Because the formulas for the multiple regression coefficients are complicated and hand calculation is out of the question, we are content to understand the least-squares principle and to let software do the computations.

multiple regression equation

We call the resulting least-squares equation the **multiple regression equation.** It allows us to predict the response variable for any set of explanatory variables, not just those associated with the cases. Just as in simple regression, however, caution must be taken when considering a set of explanatory variable values that are outside the range of the fitted data.

extrapolation, p. 100

EXAMPLE 13.9

IDI

CASE 13.1 **Predicting IDI from Employment and Income** Our examination of the explanatory and response variables separately and then in pairs did not reveal any severely skewed distributions with outliers or potential influential observations. Outputs for the multiple regression analysis from Excel, JMP, and Minitab are given in Figure 13.6. Notice that the number of digits provided varies with the software used. Rounding the results to three decimal places gives the least-squares equation

$$\widehat{\text{IDI}} = 1.477 + 0.042 \text{Employ} + 0.031 \text{Medinc} \blacksquare$$

FIGURE 13.6 Excel, JMP, and Minitab output for Example 13.9.

Excel

	A	B	C	D	E	F	G
1	SUMMARY OUTPUT						
2							
3	*Regression Statistics*						
4	Multiple R	0.84816337					
5	R Square	0.719381102					
6	Adjusted R Square	0.672611285					
7	Standard Error	0.344357917					
8	Observations	15					
9							
10	ANOVA						
11		df	SS	MS	F	Significance F	
12	Regression	2	3.647904833	1.82395	15.38131	0.000488317	
13	Residual	12	1.4229885	0.11858			
14	Total	14	5.070893333				
15							
16		Coefficients	Standard Error	t Stat	P-value	Lower 95%	Upper 95%
17	Intercept	1.477309284	0.833754479	1.77188	0.101782	-0.33928567	3.293904239
18	Employ	0.041888183	0.015177035	2.75997	0.01728	0.008820265	0.074956102
19	Medinc	0.031282666	0.008487773	3.68562	0.003117	0.012789397	0.049775935

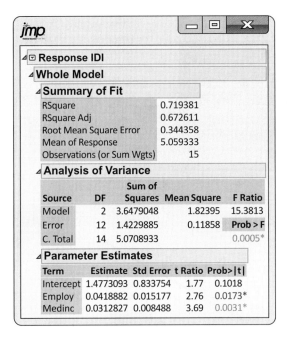
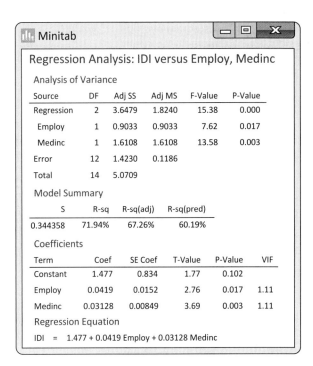

FIGURE 13.6 Continued

APPLY YOUR KNOWLEDGE

13.9 **Predicting bank assets.** Refer to Exercise 13.3 (page 622). Run the regression to predict assets using deposits and equity capital. Give the least-squares regression equation. BANKS

13.10 **Regression after transforming.** Refer to the previous exercise. In Exercise 13.8 (page 625), we considered the log transformation for all variables. Run the regression using the log-transformed variables and report the least-squares equation. BANKS

Regression residuals

The residuals are the errors in predicting the sample responses from the multiple regression equation. Recall that the residuals are the differences between the observed and predicted values of the response variable.

$$e = \text{observed response} - \text{predicted response}$$
$$= y - \hat{y}$$

linear regression model conditions, p. 582

As with simple linear regression, the residuals of a multiple regression sum to zero, and we use them to estimate σ and to check the linear regression model conditions.

The best way to examine them is to use plots. We first examine the distribution of the residuals to see if the residuals appear to be approximately Normal.

EXAMPLE 13.10

IDI

Distribution of the Residuals Figure 13.7 is a histogram of the residuals with a Normal distribution overlay. The distribution of residuals is slightly skewed but does not have any outliers. The Normal quantile plot in Figure 13.8 is reasonably linear. Given the small sample size, these plots are not extremely out of the ordinary. Similar to simple linear regression, inference is robust against moderate lack of Normality, so we're just looking for obvious violations. There do not appear to be any here. ∎

FIGURE 13.7 Histogram of residuals for Example 13.10.

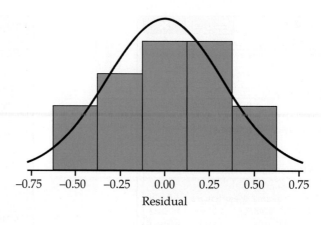

FIGURE 13.8 Normal quantile plot of residuals for Example 13.10.

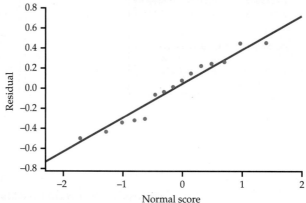

Another important aspect of examining the residuals is to plot them against each explanatory variable. Sometimes, we can detect unusual patterns, such as increasing variance or a curved relationship, when we examine the data in this way.

EXAMPLE 13.11

IDI

CASE 13.1 **Residual Plots** The residuals are plotted versus Employ and versus Medinc in Figure 13.9. In both cases, the residuals appear reasonably randomly scattered above and below zero with constant spread. The smoothed curves suggest a slight curvature in the pattern of residuals but not to the point of considering further analysis. These curves are likely showing more trend in the data than there really is given the small sample size.

Figure 13.10 is the scatterplot of the residuals versus the predicted values. This plot is especially helpful in assessing constant variance. In this case, the spread of the residuals around zero is constant over the entire range of fitted values. There is no evidence of the variance changing with the mean. ∎

FIGURE 13.9 Plot of residuals versus each of the two explanatory variables for Example 13.11.

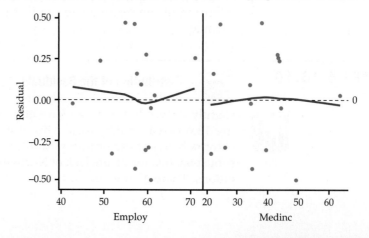

FIGURE 13.10 Plot of residuals versus the predicted values for Example 13.11.

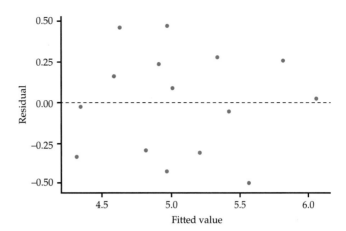

APPLY YOUR KNOWLEDGE

13.11 Examine the residuals. In Exercise 13.9 (page 627), you ran a multiple regression using the bank data. Obtain the residuals and fitted values from this regression and plot the residuals versus each of the explanatory variables and the fitted values. Also, examine the Normality of the residuals using a histogram or Normal quantile plot. Summarize your conclusions. BANKS

13.12 Examine the combined effect of several states. The states of Delaware, New York, Utah, and Virginia have far fewer assets than is predicted by the regression equation. Delete these observations and rerun the multiple regression. Describe how the regression coefficients change. BANKS

13.13 Residuals for the log analysis. In Exercise 13.10 (page 627), you carried out multiple regression using the logarithms of all the variables in Exercise 13.3. Obtain the residuals and fitted values from this regression and examine the residuals as you did in Exercise 13.11. Summarize your conclusions and compare your plots with the plots for the original variables. BANKS

13.14 Examine the effect of Puerto Rico. For the logarithm-transformed data, Puerto Rico has far fewer assets than is predicted by the regression equation. Delete Puerto Rico from the data set and rerun the multiple regression using the transformed data. Describe how the regression coefficients change. BANKS

The regression standard error

Just as the sample standard deviation measures the variability of observations about their mean, we can quantify the variability of the response variable about the predicted values obtained from the multiple regression equation. As in the case of simple linear regression, we first calculate a variance using the squared residuals:

$$s^2 = \frac{1}{n-k-1} \sum e_i^2$$

The quantity $n - k - 1$ is the degrees of freedom associated with s^2. The number of degrees of freedom equals the sample size n minus $(k + 1)$, the number of coefficients b_i in the multiple regression model. In the simple linear regression case, there is just one explanatory variable, so $k = 1$ and the number of degrees of freedom for s^2 is $n - 2$. The regression standard error s is the square root of the sum of squares of residuals divided by the number of degrees of freedom:

regression standard error, p. 578

$$s = \sqrt{s^2}$$

APPLY YOUR KNOWLEDGE

CASE 13.1 **13.15 Reading software output.** Regression software usually reports both s^2 and the regression standard error s. For the IDI data of Case 13.1, the approximate values are $s^2 = 0.119$ and $s = 0.344$. Locate s^2 and s in each of the three outputs in Figure 13.6 (pages 626–627). Give the unrounded values from each output. What name does each software give to s?

CASE 13.1 **13.16 Compare the variability.** Figure 13.2 (page 623) gives the standard deviation s_y of the IDI for the 15 countries. What is this value? The regression standard error s from Figure 13.6 (pages 626–627) also measures the variability of IDI, this time after taking into account the effect of Employ and Medinc. Explain briefly why we expect s to be smaller than s_y. One way to describe how well multiple regression explains the response variable y is to compare s with s_y.

Case 13.1 uses data on the IDI, employment, and income of 15 countries with advanced economies. These countries were an SRS from the population of countries with advanced economies. Inference, as opposed to data analysis, draws conclusions about a population or process from which our data are a sample. Inference is most easily understood when we have an SRS from a clearly defined population. Whether inference from a multiple regression model not based on a random sample is trustworthy is a matter for judgment.

Applications of statistics in business settings frequently involve data that are not random samples. We often justify inference by saying that we are studying an underlying process that generates the data. For example, in salary-discrimination studies, data are collected on all employees in a particular group. The salaries of these current employees reflect the process by which the company sets salaries. Multiple regression builds a model of this process, and inference tells us whether gender or age has a statistically significant effect in the context of this model. Would the BANKS data of Exercise 13.3 also be an example of a sample that is not random?

SECTION 13.1 SUMMARY

- **Data for multiple linear regression** consist of the values of a response variable y and k explanatory variables $x_1, x_2, \ldots, x_k$ for n cases. We organize the data and enter them into software in the form

		Variables			
Case	y	x_1	x_2	$\ldots$	x_k
1	y_1	x_{11}	x_{12}	$\ldots$	x_{1k}
2	y_2	x_{21}	x_{22}	$\ldots$	x_{2k}
$\vdots$					
n	y_n	x_{n1}	x_{n2}	$\ldots$	x_{nk}

- **Data analysis for multiple regression** starts with an examination of the distribution of each of the variables and scatterplots to display the relations between the variables.

- The **multiple regression equation** predicts the response variable by a linear relationship with all the explanatory variables:

$$\hat{y} = b_0 + b_1 x_1 + b_2 x_2 + \cdots + b_k x_k$$

- The (k + 1) coefficients b_i in the multiple regression equation are estimated using the **principle of least squares.**

- The **residuals** for multiple linear regression are

$$e_i = y_i - \hat{y}_i$$

- Examine the **distribution of the residuals** and plot them against each of the explanatory variables and the predicted values to check linear regression model conditions and to identify unusual patterns and outliers.

- The variability of the responses about the multiple regression equation is measured by the **regression standard error s,** where s is the square root of

$$s^2 = \frac{\sum e_i^2}{n - k - 1}$$

SECTION 13.1 EXERCISES

For Exercises 13.1 and 13.2, see page 620; for 13.3 and 13.4, see page 622; for 13.5 and 13.6, see page 624; for 13.7 and 13.8, see page 625; for 13.9 and 13.10, see page 627; for 13.11 to 13.14, see page 629; and for 13.15 and 13.16, see page 630.

13.17 Describing a multiple regression. As part of a study on financial risk-taking, data from 128 employees of a large international bank were obtained through a survey. The researchers were interested in the effect of professional identity on the share invested in risky assets after adjusting for the socioeconomic characteristics age, sex, education level, risk literacy, and nationality.[5]

(a) What is the response variable?

(b) What is n, the number of cases?

(c) What is k, the number of explanatory variables?

(d) What are the explanatory variables?

13.18 Understanding the fitted regression line. The fitted regression equation for a multiple regression is

$$\hat{y} = 1.8 + 2.2x_1 - 1.3x_2$$

(a) If $x_1 = 5$ and $x_2 = 2$, what is the predicted value of y?

(b) For the answer to part (a) to be valid, is it necessary that the values $x_1 = 5$ and $x_2 = 2$ correspond to a case in the data set? Explain why or why not.

(c) If you hold x_2 at a fixed value, what is the effect of an increase of two units in x_1 on the predicted value of y?

13.19 Predicting the price of tablets: individual variables. Suppose your company needs to buy some tablets. To help in the purchasing decision, you decide to develop a model to predict the selling price. You decide to obtain price and product characteristic information on 36 tablets from *Consumer Reports*.[6] The characteristics are screen size, battery life, weight (pounds), ease of use, display, and versatility. The latter three are scored on a 1 to 5 scale. TABLTS

(a) Make a table giving the mean, median, and standard deviation of each variable.

(b) Use histograms to make graphical summaries of each distribution.

(c) Describe these distributions. Are there any unusual observations that may affect a multiple regression? Explain your answer.

(d) The screen size distribution appears bimodal. Is this lack of Normality necessarily a problem? Explain your answer.

13.20 Predicting the price of tablets: pairs of variables. Refer to the tablet data described in the previous exercise. TABLTS

(a) Examine the relationship between each pair of variables using correlation and a scatterplot.

(b) Which characteristic is most strongly correlated with price? Is any pair of characteristics strongly correlated?

(c) Summarize the relationships. Are there any unusual or outlying cases?

13.21 Predicting the price of tablets: multiple regression equation. Refer to the tablet data described in Exercise 13.19. TABLTS

(a) Run a multiple regression to predict price using the six product characteristics. Give the equation for predicted price.

(b) What is the value of the regression standard error s? Verify that this value is the square root of the sum of squares of residuals divided by the degrees of freedom for the residuals.

(c) Obtain the residuals and use graphical summaries to describe the distribution.

13.22 Predicting the price of a tablet. Refer to the previous exercise. TABLTS

(a) Explain how you could use the residuals to help determine which tablets are a good deal.

(b) What is the predicted price for the second tablet? The characteristics are Size = 9.7, Battery = 8.9, Weight = 0.9, Ease = 5, Display = 5, and Versatility = 4.

(c) The stated price for this tablet is $550. Is the predicted price above or below the stated price? Should you consider buying it? Explain your answer.

(d) *Consumer Reports* names Tablet 8 as a "Best Buy." Based on your regression model, do you agree with this assessment? What tablets would you recommend?

13.23 Predicting college debt using multiple regression. In Exercise 12.83 (page 614), you looked at AvgDebt and its relationship with each of the six measures available in the data. Let's now consider a multiple regression using a subset of these measures. BESTVAL

(a) Run a multiple regression to predict average debt at graduation using InCostAid, OutCostAid, Admit, and Grad4Rate.

(b) Give the equation for predicted average debt at graduation as well as the regression standard error s.

(c) Compare this regression standard error s with the value found in Figure 12.23 (page 614). Does it appear this multiple regression model is an improvement over the simple linear regression that just used InCostAid? Explain your answer.

13.24 Predicting college debt using multiple regression, continued. In Exercise 12.82 (page 614), we predicted the average debt at graduation, for students attending Purdue University and Texas A&M, using just InCostAid as a predictor. BESTVAL

(a) Using the equation from part (b) of the previous exercise, what is the predicted debt at graduation for a student attending Purdue University? Its values are InCostAid = 7788; OutCostAid = 26,590; Admit = 56; and Grad4Rate = 49.

(b) What is the predicted debt at graduation for a student attending Texas A&M University? Its values are InCostAid = 11,396; OutCostAid = 38,147; Admit = 66; and Grad4Rate = 52.

(c) Write a short summary comparing these predicted values to the ones using simple linear regresson in Exercise 12.82.

13.25 Predicting market share. You've been asked to develop a regression model to predict the market share of limited-service restaurant chains. You've been supplied with this year's market share (based on current U.S. sales) along with the number of franchises, number of company-owned stores, increase in the total number of franchises and stores from the previous year, and annual sales ($ million) from three years earlier.[7] FFOOD

(a) Make a table giving the mean, the standard deviation, and the five-number summary for each of these variables.

(b) Use histograms to make graphical summaries of the five distributions.

(c) Describe the distributions. Are there any unusual observations?

13.26 Data analysis: pairs of variables. Refer to the previous exercise. FFOOD

(a) Plot market share versus each of the explanatory variables.

(b) Summarize these relationships. Are there any influential observations?

(c) Find the correlation between each pair of variables. Are there any correlations between explanatory variables that are a concern? Explain your answer.

13.27 Multiple regression equation. Refer to the fast-food data in Exercise 13.25. Run a multiple regression to predict market share using all four explanatory variables. FFOOD

(a) Give the equation for predicted market share.

(b) What is the value of the regression standard error s?

13.28 Checking residuals. Refer to the fast-food data in Exercise 13.25. Find the residuals for the multiple regression used to predict market share based on the three explanatory variables. FFOOD

(a) Give a graphical summary of the distribution of the residuals. Are there any outliers in this distribution?

(b) Plot the residuals versus each of the explanatory vairables. Describe each plot and any unusual cases.

Your analyses in Exercises 13.25 through 13.28 point to possibly two or three chains, McDonald's, Starbucks, and Chick-fil-A, as unusual in several respects. How influential are these restaurants? The following four exercises have you investigate.

13.29 Rerun Exercises 13.27 and 13.28 without the data for McDonald's. Assess the model conditions and report the regression equation and regression standard error. FFOOD

13.30 Rerun Exercises 13.27 and 13.28 without the data for McDonald's and Starbucks. Assess the model conditions and report the regression equation and regression standard error. FFOOD

13.31 Rerun Exercises 13.27 and 13.28 without the data for McDonald's, Starbucks, and Chick-fil-A. Assess the model conditions and report the regression equation and regression standard error. FFOOD

13.32 Based on your findings in the previous three exercises and your regression equation from Exercise 13.27, write a short paragraph explaining the impact these cases have on the results as well as the model you would recommend for use in predicting future market share.

13.33 Predicting retail sales. Daily sales at a secondhand shop are recorded over a 25-day period.[8] The daily gross sales and total number of items sold are broken down into items paid by check, cash, and credit card. The owners expect that the daily numbers of cash items, check items, and credit card items sold will accurately predict gross sales. RETAIL

(a) Describe the distribution of each of these four variables using both graphical and numerical summaries. Briefly summarize what you find and note any unusual observations.

(b) Use plots and correlations to describe the relationships between each pair of variables. Summarize your results.

(c) Run a multiple regression and give the least-squares equation.

(d) Analyze the residuals from this multiple regression. Are there any patterns of interest?

(e) One of the owners is troubled by the equation because the intercept is not zero (i.e., no items sold should result in $0 gross sales). Explain to this owner why this isn't a problem.

13.34 CEO architectural firm billings. A summary of firms engaged in commercial architecture in the Indianapolis, Indiana, area provides firm characteristics, including total annual billing in the current year, total annual billing in the previous year, the number of architects, the number of engineers, and the number of staff employed in the firm.[9] Consider developing a model to predict current total billing using the other four variables. ARCH

(a) Using numerical and graphical summaries, describe the distribution of current and past year total billing and the number of architects, engineers, and staff.

(b) For each of the 10 pairs of variables, use graphical and numerical summaries to describe the relationship.

(c) Carry out a multiple regression. Report the fitted regression equation and the value of the regression standard error s.

(d) Analyze the residuals from the multiple regression. Are there any concerns?

(e) A firm did not report its current total billing but had $1 million in billing last year and employs three architects, one engineer, and 17 staff members. What is the predicted total billing for this firm?

13.2 Inference for Multiple Regression

When you complete this section, you will be able to:

- Describe the multiple linear regression model in terms of the population regression equation and the distribution of deviations of the response variable y from this line.
- Use statistical software to obtain individual regression coefficient estimates, their t tests, and the multiple regression ANOVA table.
- Use the ANOVA table to perform the ANOVA F test, determine the squared multiple correlation R^2, and draw appropriate conclusions about the regression model.
- Interpret a confidence interval and a significance test for an individual regression coefficient.
- Explain the differences between the ANOVA F test and the t tests for individual coefficients.
- Construct and interpret a confidence interval for a mean response and a prediction interval for a set of explanatory variables.

To move from using multiple regression for data analysis to inference in the multiple regression setting, we need to make some assumptions about our data. These assumptions are summarized in the form of a statistical model. As with all the models that we have studied, we do not require that the model be exactly correct. We only require that it be approximately true and that the data do not severely violate the assumptions.

Recall that the *simple linear regression model* assumes that the mean of the response variable y depends on the explanatory variable x according to a linear equation

simple linear regression model, p. 575

$$\mu_y = \beta_0 + \beta_1 x$$

For any fixed value of x, the response y varies Normally around this mean and has a standard deviation σ that is the same for all values of x.

In the *multiple regression* setting, the response variable y depends on not one but k explanatory variables, which we denote by $x_1, x_2, \ldots, x_k$. The mean response is a linear function of these explanatory variables:

$$\mu_y = \beta_0 + \beta_1 x_1 + \beta_2 x_2 + \cdots + \beta_k x_k$$

population regression equation

Similar to simple linear regression, this expression is the **population regression equation,** and the observed y's vary about their means given by this equation.

Just as we did in simple linear regression, we can think of this model in terms of subpopulations of responses. The only difference is that each subpopulation now corresponds to a particular set of values for *all* the explanatory variables $x_1, x_2, \ldots, x_k$. The observed y's in each subpopulation are still assumed to vary Normally with a mean given by the population regression equation and standard deviation σ that is the same in all subpopulations.

subpopulations, p. 571

Multiple linear regression model

To form the multiple regression model, we combine the population regression equation with assumptions about the form of the *variation* of the observations about their mean. We again think of the model in the form

DATA = FIT + RESIDUAL

DATA = FIT + RESIDUAL, p. 464

The FIT part of the model consists of the subpopulation mean μ_y. The RESIDUAL part represents the variation of the response y around its subpopulation mean. That is, the model is

$$y = \mu_y + \varepsilon$$

The symbol ε represents the deviation of an individual observation from its subpopulation mean. We assume that these deviations are Normally distributed with mean 0 and an unknown standard deviation σ that does not depend on the values of the x variables.

MULTIPLE LINEAR REGRESSION MODEL

The **statistical model for multiple linear regression** is

$$y_i = \beta_0 + \beta_1 x_{i1} + \beta_2 x_{i2} + \cdots + \beta_k x_{ik} + \varepsilon_i$$

for $i = 1, 2, \ldots, n$.

The **mean response** μ_y is a linear function of the explanatory variables:

$$\mu_y = \beta_0 + \beta_1 x_1 + \beta_2 x_2 + \cdots + \beta_k x_k$$

The **deviations** ε_i are independent and Normally distributed with mean 0 and standard deviation σ. That is, they are an SRS from the $N(0, \sigma)$ distribution.

The parameters of the model are $\beta_0, \beta_1, \beta_2, \ldots, \beta_k$, and σ.

The assumption that the subpopulation means are related to the regression coefficients β by the equation

$$\mu_y = \beta_0 + \beta_1 x_1 + \beta_2 x_2 + \cdots + \beta_k x_k$$

implies that we can estimate all subpopulation means from estimates of the β's. To the extent that this equation is accurate, we have a useful tool for describing how the mean of y varies with any collection of x's.

Predicting Movie Revenue The Internet Movie Database Pro (IMDbPro) provides movie industry information on both movies and television shows. Can information available soon after a movie's release be used to predict total U.S. box office revenue? To investigate this, let's consider a sample of 44 movies released in the first half of 2013 to guarantee they are no longer in the theaters.[10] The response variable is a movie's total U.S. box office revenue (USRevenue). Among the explanatory variables are the movie's opening-weekend revenue (Opening), how many theaters the movie was in for the opening weekend (Theaters), and a score summarizing the movie critics' reviews (Review). This score runs between 1 and 100, with a higher score associated with better overall reviews. All dollar amounts are measured in millions of U.S. dollars. ■

MOVIE

APPLY YOUR KNOWLEDGE

CASE 13.2 **13.35 Look at the data.** Examine the data for total U.S. revenue, opening-weekend revenue, number of opening-weekend theaters, and review score. That is, use graphs to display the distribution of each variable and the relationships between pairs of variables. Based on your examination, how would you describe the data? Are there any movies you consider to be outliers or unusual in any way? Explain your answer. MOVIE

CASE 13.2 **13.36 Look at the transformed data.** Because total U.S. box office revenue and opening-weekend revenue are skewed to the right, take the natural log of these variables and repeat the analysis performed in the previous exercise. MOVIE

EXAMPLE 13.12

CASE 13.2 **A Model for Predicting Movie Revenue** We want to investigate if a linear model that includes a movie's opening-weekend revenue, opening-weekend theater count, and review score can forecast total U.S. box office revenue. We will use the logarithm of total U.S. revenue and opening-weekend revenue in our model (see Exercise 13.36). This transformation is common with economic and financial data as it reduces the possibility of influential observations. We will use the variable names LUSRevenue and LOpening to reflect that the natural logarithm is being used.

This multiple regression model has $k = 3$ explanatory variables: $x_1 =$ LOpening, $x_2 =$ Theaters, and $x_3 =$ Review. Each particular combination of log opening-weekend revenue, opening-weekend theater count, and review score defines a particular subpopulation. Our response variable y is the log U.S. box office revenue.

The multiple regression model for the subpopulation mean of log U.S. box office revenue is

$$\mu_{\text{LUSRevenue}} = \beta_0 + \beta_1 \text{LOpening} + \beta_2 \text{Theaters} + \beta_3 \text{Review}$$

For movies that earn $43.7 million in 3050 theaters their first weekend and score a 58 on reviews, the model gives the subpopulation mean log U.S. box office revenue as

$$\mu_{\text{LUSRevenue}} = \beta_0 + \beta_1 \log(43.7) + \beta_2 3050 + \beta_3 58 \ \blacksquare$$

Estimating the parameters of the model

To estimate the mean U.S. box office revenue in Example 13.12, we must estimate the coefficients β_0, β_1, β_2, and β_3. Inference requires that we also estimate the variability of the responses about their means, represented in the model by the standard deviation σ.

In any multiple regression model, the parameters to be estimated from the data are $\beta_0, \beta_1, \ldots, \beta_k$, and σ. We estimate these parameters by applying least-squares multiple regression as described in Section 13.1 (pages 625–626).

That is, we view the coefficients b_j in the multiple regression equation

$$\hat{y} = b_0 + b_1 x_1 + b_2 x_2 + \cdots + b_k x_k$$

as estimates of the population parameters β_j. The observed variability of the responses about this fitted model is measured by the variance

$$s^2 = \frac{1}{n-k-1} \sum e_i^2$$

and the regression standard error

$$s = \sqrt{s^2}$$

In the model, the parameters σ^2 and σ measure the variability of the responses about the population regression equation. It is natural to estimate σ^2 by s^2 and σ by s.

ESTIMATING THE REGRESSION PARAMETERS

In the multiple linear regression setting, we use the **method of least-squares regression** to estimate the population regression parameters.

The standard deviation σ in the model is estimated by the **regression standard error**

$$s = \sqrt{\frac{1}{n-k-1} \sum (y_i - \hat{y}_i)^2}$$

Inference about the regression coefficients

Confidence intervals and significance tests for each of the regression coefficients β_j have the same form as in simple linear regression. The standard errors of the b's have more complicated formulas, but all are again multiples of s. Statistical software does the calculations.

CONFIDENCE INTERVALS AND SIGNIFICANCE TESTS FOR β_j

A **level C confidence interval** for β_j is

$$b_j \pm t^* \mathrm{SE}_{b_j}$$

where SE_{b_j} is the standard error of b_j and t^* is the value for the $t(n-k-1)$ density curve with area C between $-t^*$ and t^*. The **margin of error** is $m = t^* \mathrm{SE}_{b_j}$.

To test the hypothesis $H_0: \beta_j = \beta_j^*$, compute the **t statistic**

$$t = \frac{b_j - \beta_j^*}{\mathrm{SE}_{b_j}}$$

Most software packages provide the test of the hypothesis $H_0: \beta_j = 0$. In that case, the t statistic reduces to

$$t = \frac{b_j}{\mathrm{SE}_{b_j}}$$

The **degrees of freedom** are $n - k - 1$. In terms of a random variable T having the $t(n - k - 1)$ distribution, the P-value for a test of H_0 against

$H_a: \beta_j > \beta_j^*$ is $P(T \geq t)$

$H_a: \beta_j < \beta_j^*$ is $P(T \leq t)$

$H_a: \beta_j \neq \beta_j^*$ is $2P(T \geq |t|)$

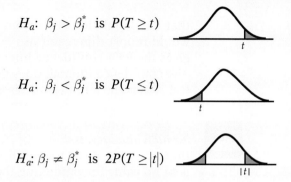

EXAMPLE 13.13

MOVIE

CASE 13.2 **Predicting U.S. Box Office Revenue** In Example 13.12, there are $k = 3$ explanatory variables, and we have data on $n = 44$ movies. The degrees of freedom for multiple regression are therefore

$$n - k - 1 = 44 - 3 - 1 = 40$$

Statistical software output for this fitted model provides many details of the model's fit and the significance of the independent variables. Figure 13.11 shows multiple regression outputs from Excel and JMP. Rounding the results to four decimal places, the regression equation is

$$\widehat{\text{LUSRevenue}} = 0.4867 + 0.9748\text{LOpening} + 0.0001\text{Theaters} + 0.0043\text{Review}$$

and that the regression standard error is $s = 0.1848$.

The outputs present the t statistic for each regression coefficient and its two-sided P-value. For example, the t statistic for the coefficient of LOpening

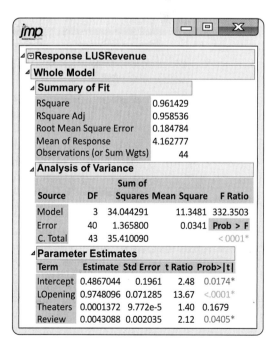

FIGURE 13.11 Multiple regression outputs from Excel and JMP for Examples 13.13, 13.14, and 13.15.

is 13.67 with a very small P-value. The data give strong evidence against the null hypothesis

$$H_0: \beta_1 = 0$$

that the population coefficient for opening-weekend log revenue is zero. We would report this result as $t = 13.67$, df $= 40$, $P < 0.0001$. The software also gives the 95% confidence interval for the coefficient β_1. It is (0.8307, 1.1189). The confidence interval does not include 0, consistent with the fact that the test rejects the null hypothesis at the 5% significance level. ∎

Be very careful in your interpretation of the t tests and confidence intervals for individual regression coefficients. In *simple linear regression*, the model says that $\mu_y = \beta_0 + \beta_1 x$. The null hypothesis $H_0: \beta_1 = 0$ says that regression on x is of no value for predicting the response y, or alternatively, that there is no straight-line relationship between x and y. The corresponding hypothesis for the *multiple regression* model $\mu_y = \beta_0 + \beta_1 x_1 + \beta_2 x_2 + \beta_3 x_3$ of Example 13.13 says that x_1 is of no value for predicting y, *given that x_2 and x_3 are also in the model*. That's a very important difference.

The output in Figure 13.11 shows, for example, that the P-value for opening-weekend theater count is $P = 0.1679$. We can conclude that the number of theaters does not help predict U.S. box office revenue, *given that opening-weekend revenue and review score are used for prediction*. This does *not* mean that the opening-weekend theater count alone is not a good predictor of U.S. box office revenue. In fact, in Exercise 13.36 (page 635), you show there is a strong positive relationship between the number of theaters and the log of total U.S. revenue.

The conclusions of inference about any one explanatory variable in multiple regression depend on what other explanatory variables are also in the model. This is a basic principle for understanding multiple regression. The t tests in Example 13.13 show that the opening-weekend theater count does not significantly aid prediction of the U.S. box office revenue *if the log of opening-weekend revenue and review score are also in the model*. On the other hand, the log of opening-weekend revenue is highly significant *even when the opening-weekend theater count and review score are also in the model*.

The interpretation of a confidence interval for an individual coefficient also depends on the other variables in the model, but in this case only if they remain constant. For example, the 95% confidence interval for LOpening implies that, *given the number of theaters and review score do not change*, a 1% increase in the opening-weekend revenue results in an expected percent increase in total U.S. box office revenue somewhere between 0.831% and 1.119%. While keeping review score constant may be reasonable, it may not make sense to keep the number of theaters fixed. The number of theaters and opening-weekend revenue are positively correlated, so an increase in opening revenue suggests a larger number of theaters.

elasticity, p. 584

APPLY YOUR KNOWLEDGE

CASE 13.2 **13.37 A simpler model.** In the multiple regression analysis using all three variables, Theaters appears to be the least helpful (given that the other two explanatory variables are in the model). Do a new analysis using only the log of the opening-weekend revenue and review score. Give the estimated regression equation for this analysis and compare it with the analysis using all three explanatory variables. Summarize the inference results for the coefficients. Explain carefully to someone who knows no statistics why the conclusions about review score here and in Figure 13.11 differ. 📊 MOVIE

CASE 13.2 **13.38 Reading software outputs.** Carefully examine the outputs from the two software packages given in Figure 13.11. Make a table giving the estimated regression coefficient for the review score (Review), its standard error, the t statistic with degrees of freedom, and the P-value as reported by each of the packages. What do you conclude about this coefficient?

Inference about prediction

Inference about the regression coefficients in simple and multiple regression look similar, but there are important differences in interpretation. That is not the case with inference about prediction where the process and interpretation are much the same. We may wish to give a confidence interval for the mean response for some specific set of values of the explanatory variables. Or we may want a prediction interval for an individual response for the same set of values.

confidence and prediction intervals for regression response, p. 597

The distinction between predicting a mean and individual response is exactly as in simple regression. The prediction interval is again wider because it must allow for the variation of individual responses about the mean. In most software, the commands for prediction inference are the same for multiple and simple regression. The details of the arithmetic performed by the software are, of course, more complicated for multiple regression, but this does not affect interpretation of the output.

What about changes in the model, which we saw can greatly influence inference about the regression coefficients? It is often the case that different models give similar predictions. We expect, for example, that the predictions of the log U.S. box office revenue from opening-weekend revenue and review score will be about the same as predictions based on opening-weekend revenue, opening-weekend theater count, and review score.

Because of this, when prediction is the key goal of a multiple regression, it is common to search for a model that predicts well but does not contain unnecessary predictors. Some refer to this as following the KISS principle.[11] In Section 13.3, we discuss some procedures that can be used for this type of search.

APPLY YOUR KNOWLEDGE

CASE 13.2 **13.39 Predicting U.S. movie revenue.** The movie *Despicable Me 2* was also released during the first half of 2013. It was shown in 3997 theaters, grossing $83.5 million during the first weekend. Its review score was 62. Use software to construct the following. MOVIE

(a) A 95% prediction interval based on the model with all three explanatory variables.

(b) A 95% prediction interval based on the model using only opening-weekend revenue and review score.

(c) Compare the two intervals. Do the models give similar predictions and standard errors?

ANOVA table for multiple regression

The basic ideas of the regression ANOVA table are the same in simple and multiple regression. ANOVA expresses variation in the form of sums of squares. It breaks the total variation into two parts: the sum of squares explained by the regression equation and the sum of squares of the residuals. The ANOVA table has the same form in simple and multiple regression except for the degrees of freedom, which reflect the number k of explanatory variables. Here is the ANOVA table for multiple regression:

Source	Degrees of freedom	Sum of squares	Mean square	F
Regression	DFR = k	SSR = $\sum(\hat{y}_i - \bar{y})^2$	MSR = SSR/DFR	MSR/MSE
Residual	DFE = $n - k - 1$	SSE = $\sum(y_i - \hat{y}_i)^2$	MSE = SSE/DFE	
Total	DFT = $n - 1$	SST = $\sum(y_i - \bar{y})^2$		

The brief notation in the table uses, for example, MSE for the residual mean square. This is common notation; the "E" stands for "error." Of course, no error has been made. "Error" in this context is just a synonym for "residual."

ANOVA for regression, p. 607

The degrees of freedom and sums of squares add, just as in simple regression:

$$SST = SSR + SSE$$

$$DFT = DFR + DFE$$

The estimate of the variance σ^2 for our model is again given by the MSE in the ANOVA table. That is, $s^2 = $ MSE.

The ratio MSR/MSE is again the statistic for the ANOVA F test. In simple linear regression, the F test from the ANOVA table is equivalent to the two-sided t test of the hypothesis that the slope of the regression line is 0. *These two tests are not equivalent in multiple regression.*

In the multiple regression setting, the null hypothesis for the F test states that *all* the regression coefficients (with the exception of the intercept) are 0. One way to write this is

$$H_0\colon \beta_1 = 0 \text{ and } \beta_2 = 0 \text{ and } \cdots \text{ and } \beta_k = 0$$

A shorter way to express this hypothesis is

$$H_0\colon \beta_1 = \beta_2 = \cdots = \beta_k = 0$$

The alternative hypothesis is

$$H_a\colon \text{ at least one of the } \beta_j \text{ is not 0}$$

The null hypothesis says that none of the explanatory variables helps explain the response, at least when used in the form expressed by the multiple regression equation. The alternative states that at least one of them is linearly related to the response.

This test provides an overall assessment of the model to explain the response. The individual t tests assess the importance of a single variable given the presence of the other variables in the model. *While looking at the set of individual t tests to assess overall model significance may be tempting, it is not recommended because it leads to more frequent incorrect conclusions.* The F test also better handles situations when there are two or more highly correlated explanatory variables.

As in simple linear regression, large values of F give evidence against H_0. When H_0 is true, F has the $F(k, n - k - 1)$ distribution. The degrees of freedom for the F distribution are those associated with the regression and residual terms in the ANOVA table. When software is not available, Table E in the back of the book gives upper p critical values of the F distributions for $p = 0.10, 0.05, 0.025, 0.01,$ and 0.001.

ANALYSIS OF VARIANCE F TEST

In the multiple regression model, the hypothesis

$$H_0: \beta_1 = \beta_2 = \cdots = \beta_k = 0$$

versus

H_a: at least one of these coefficients is not zero

is tested by the analysis of variance F statistic

$$F = \frac{\text{MSR}}{\text{MSE}}$$

The P-value is the probability that a random variable having the $F(k, n-k-1)$ distribution is greater than or equal to the calculated value of the F statistic.

EXAMPLE 13.14

MOVIE

F Test for Movie Revenue Model Example 13.13 (page 637) gives the results of multiple regression analysis for predicting U.S. box office revenue. The F statistic is 332.35. The degrees of freedom appear in the ANOVA table. They are 3 and 40. The software packages (see Figure 13.11) report the P-value in different forms: Excel, 9.73E-11 and JMP, < 0.0001. We would report the results as follows: a movie's opening-weekend revenue, opening-weekend theater count, and review score contain information that can be used to predict the movie's total U.S. box office revenue ($F = 332.32$, df = 3 and 40, $P < 0.0001$). ∎

A significant F test does not tell us which explanatory variables explain the response. It simply allows us to conclude that at least one of the coefficients is not zero. We may want to refine the model by eliminating some variables that do not appear to be useful (KISS principle). On the other hand, if we fail to reject the null hypothesis, we have found no evidence that *any* of the coefficients are not zero. In this case, there is little point in attempting to refine the model.

APPLY YOUR KNOWLEDGE

CASE 13.2 **13.40 F test for the model without Theaters.** Rerun the multiple regression using the movie's opening-weekend log revenue and review score to predict the log of U.S. box office revenue. Report the F statistic, the associated degrees of freedom, and the P-value. How do these differ from the corresponding values for the model with the three explanatory variables? What do you conclude? MOVIE

Squared multiple correlation R^2

For simple linear regression, the square of the sample correlation r^2 can be written as the ratio of SSR to SST. We interpret r^2 as the proportion of variation in y explained by linear regression on x. A similar statistic is important in multiple regression.

THE SQUARED MULTIPLE REGRESSION CORRELATION

The statistic

$$R^2 = \frac{\text{SSR}}{\text{SST}} = \frac{\sum(\hat{y}_i - \bar{y})^2}{\sum(y_i - \bar{y})^2}$$

is the proportion of the variation of the response variable y that is explained by the explanatory variables $x_1, x_2, \ldots, x_k$ in a multiple linear regression.

coefficient of determination, p. 475

multiple regression correlation coefficient

In one-way ANOVA, this statistic is called the coefficient of determination and that name is sometimes used here. Often, it is multiplied by 100 and expressed as a percent. The square root of R^2, called the **multiple regression correlation coefficient,** is the correlation between the observations y_i and the predicted values $\hat{y}_i$ (recall Exercise 12.40, page 596). Some software provides a scatterplot of this relationship to help visualize the predictive strength of the model.

EXAMPLE 13.15

CASE 13.2 R^2 **for Movie Revenue Model** Figure 13.11 (page 637) gives the results of the multiple regression analysis to predict U.S. box office revenue. The value of the R^2 statistic is 0.9614, or 96.14%. Be sure that you can find this statistic in the outputs. We conclude that slightly more than 96% of the variation in U.S. box office log revenues can be explained by the movies' opening-weekend log revenues, opening-weekend theater counts, and review scores. Figure 13.12 is a plot of the y_i versus the predicted values $\hat{y}_i$. We can see that the model's predictive strength is very strong. These predicted values are a function of the x's so this plot is similar to that of plotting y by x in simple linear regression to assess the strength of the association. ∎

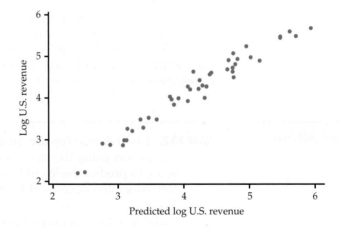

FIGURE 13.12 Scatterplot of the log U.S. revenues versus their predicted values for Example 13.15.

The F statistic for the multiple regression of U.S. box office revenue on opening-weekend revenue, opening-weekend theater count, and review score is highly significant, $P < 0.0001$. There is strong evidence of a relationship among these three variables and eventual box office revenue. The squared multiple correlation tells us that these variables in this multiple regression model explain about 96% of the variability in box office log revenues. The other 4% is represented by the RESIDUAL term in our model and is due to differences among the movies that are not measured by these three variables. For example, these differences might be explained by the movie's rating,

13.2 Inference for Multiple Regression

the genre of the movie, and whether the movie is a sequel. These variables are all categorical, and their inclusion in a multiple regression model is discussed in the next section.

APPLY YOUR KNOWLEDGE

CASE 13.2 **13.41 R^2 for different models.** Use each of the following sets of explanatory variables to predict U.S. box office revenue: (a) LOpening, Theaters; (b) LOpening, Review; (c) Theaters, Review; (d) LOpening; (e) Theaters; (f) Review. Make a table giving the model and the value of R^2 for each. Summarize what you have found. MOVIE

Inference for a collection of regression coefficients

We have studied two different types of significance tests for multiple regression. The ANOVA F test examines the hypothesis that the coefficients for *all* the explanatory variables are zero. On the other hand, we used t tests to examine *individual* coefficients. (For simple linear regression with one explanatory variable, these are two different ways to examine the same question.)

Often, we are interested in an intermediate setting: does a specific set of explanatory variables contribute to explaining the response, given that another set of explanatory variables is already in the model? We formulate such a question as follows: start with the multiple regression model that contains *all the explanatory variables* and test the hypothesis that *a specific set of the coefficients* are all zero. When the specific set involves more than one explanatory variable, we need to consider an F test rather than a set of individual parameter t tests.

F TEST FOR A COLLECTION OF REGRESSION COEFFICIENTS

In the multiple regression model with k explanatory variables, the hypothesis

H_0: q specific explanatory variables all have zero coefficients

versus the hypothesis

H_a: at least one of these q coefficients is not zero

is tested by an F statistic. The degrees of freedom are q and $n - k - 1$. The P-value is the probability that a random variable having the $F(q, n - k - 1)$ distribution is greater than or equal to the calculated value of the F statistic.

Some software allows you to directly state and test hypotheses of this form. Here is a way to find the F statistic by fitting two regression models and looking at the change in R^2.

1. Regress y on all k explanatory variables. Read the R^2-value from the output and call it R_1^2.

2. Then regress y on just the $k - q$ variables that remain after removing the q variables in H_0 from the model. Again read the R^2-value and call it R_2^2. This will be smaller than R_1^2 because removing variables can only decrease R^2.

3. The test statistic is

$$F = \left(\frac{n - k - 1}{q}\right)\left(\frac{R_1^2 - R_2^2}{1 - R_1^2}\right)$$

which has an F distribution with q and $n - k - 1$ degrees of freedom when H_0 is true.

EXAMPLE 13.16

MOVIE

Do Opening-Weekend Theater Count and Review Score Add Predictive Ability? In the multiple regression analysis using all three explanatory variables, we can consider theater count and review score to represent initial non-monetary support of the movie. A question we might ask is

> Do these two non-monetary variables help predict movie revenue, given that the monetary one, opening-weekend revenue, is included?

The same question in another form is

> If we start with a model containing all three variables, does removing theater count and review score reduce our ability to predict revenue?

The first regression run includes $k = 3$ explanatory variables: LOpening, Theaters, and Review. The R^2 for this model is $R_1^2 = 0.9614$.

Now remove the $q = 2$ variables Theaters and Review and redo the regression with just LOpening as the explanatory variable. For this model, we get $R_2^2 = 0.9566$.

The test statistic is

$$F = \left(\frac{n-k-1}{q}\right)\left(\frac{R_1^2 - R_2^2}{1 - R_1^2}\right)$$

$$= \left(\frac{44-3-1}{2}\right)\left(\frac{0.9614 - 0.9566}{1 - 0.9614}\right) = 2.49$$

The degrees of freedom are $q = 2$ and $n - k - 1 = 44 - 3 - 1 = 40$.

This value is slightly larger than the smallest entry in Table E with 2 and 40 degrees of freedom. Thus, $0.05 < P < 0.10$. Software gives $P = 0.0927$. Opening-weekend theater count and review score do not contribute significantly to explaining U.S. box office log revenue when opening weekend log revenue is already in the model. When added, R^2 only improves by 0.48%, from 95.66% to 96.14%. ∎

The hypothesis test in Example 13.16 asks about the coefficients of Theaters and Review in a model that also contains LOpening as an explanatory variable. If we start with a different model, we may get a different answer. For example, we would not be surprised to find that Theaters and Review help explain movie revenue in a model with only these two explanatory variables. *Individual regression coefficients, their standard errors, and significance tests are meaningful only when interpreted in the context of the other explanatory variables in the model.*

APPLY YOUR KNOWLEDGE

CASE 13.2 13.42 Are Theaters and Review useful predictors of LUSRevenue? Run the multiple regression to predict movie log revenue using all three predictors. Then run the model using only Theaters and Review.

 MOVIE

(a) The R^2 for the second model is 0.7811. Does your work confirm this?

(b) Make a table giving the Theaters and Review coefficients and their standard errors, t statistics, and P-values for both models. Explain carefully how your assessment of the value of these two predictors of movie revenue depends on whether opening-weekend log revenue is in the model.

CASE 13.2 **13.43 Is LOpening helpful when Theaters and Review are available?** We saw that Theaters and Review are not useful in a model that contains LOpening. Now, let's examine the other version of this question. Does LOpening help explain USRevenue in a model that contains Theaters and Review? Run the models with all three predictors and with only Theaters and Review. Compare the values of R^2. Perform the F test and give its degrees of freedom and P-value. Carefully state a conclusion about the usefulness of the predictor LOpening when Theaters and Review are available. Also compare this F-test and P-value with the t test for the coefficient of LOpening in Example 13.13 (page 637). MOVIE

SECTION 13.2 SUMMARY

- The statistical model for **multiple linear regression** with response variable y and k explanatory variables $x_1, x_2, \ldots, x_k$ is

$$y_i = \beta_0 + \beta_1 x_{i1} + \beta_2 x_{i2} + \cdots + \beta_k x_{ik} + \varepsilon_i$$

where $i = 1, 2, \ldots, n$. The deviations ε_i are independent Normal random variables with mean 0 and a common standard deviation σ. The **parameters** of the model are $\beta_0, \beta_1, \beta_2, \ldots, \beta_k$, and σ.

- The β's are estimated by the coefficients $b_0, b_1, b_2, \ldots, b_k$ of the multiple regression equation fitted to the data by **the method of least squares**. The parameter σ is estimated by the **regression standard error**

$$s = \sqrt{\text{MSE}} = \sqrt{\frac{\sum e_i^2}{n-k-1}}$$

where the e_i are the **residuals**,

$$e_i = y_i - \hat{y}_i$$

- A **level C confidence interval for the regression coefficient** β_j is

$$b_j \pm t^* \text{SE}_{b_j}$$

where t^* is the value for the $t(n-k-1)$ density curve with area C between $-t^*$ and t^*. The **margin of error** is $m = t^* \text{SE}_{b_j}$.

- Tests of the hypothesis $H_0: \beta_j = 0$ are based on the **individual t statistic:**

$$t = \frac{b_j}{\text{SE}_{b_j}}$$

and the $t(n-k-1)$ distribution.

- The estimate b_j of β_j and the test and confidence interval for β_j are all based on a specific multiple linear regression model. The results of all these procedures change if other explanatory variables are added to or deleted from the model.

- The **ANOVA table** for a multiple linear regression gives the degrees of freedom, sum of squares, and mean squares for the regression and residual sources of variation. The **ANOVA F statistic** is the ratio MSR/MSE and is used to test the null hypothesis

$$H_0: \beta_1 = \beta_2 = \cdots = \beta_k = 0$$

- If H_0 is true, this statistic has the $F(k, n-k-1)$ distribution.

- The **squared multiple correlation** is given by the expression

$$R^2 = \frac{\text{SSR}}{\text{SST}}$$

and is interpreted as the proportion of the variability in the response variable y that is explained by the explanatory variables $x_1, x_2, \ldots, x_k$ in the multiple linear regression.

- The null hypothesis that a **collection of q explanatory variables** all have coefficients equal to zero is tested by an **F statistic** with q degrees of freedom in the numerator and $n - k - 1$ degrees of freedom in the denominator. This statistic can be computed from the squared multiple correlations for the model with all the explanatory variables included (R_1^2) and the model with the q variables deleted (R_2^2):

$$F = \left(\frac{n-k-1}{q}\right)\left(\frac{R_1^2 - R_2^2}{1 - R_1^2}\right)$$

SECTION 13.2 EXERCISES

For Exercise 13.35, see page 635; for 13.36 and 13.37, see pages 635–637; for 13.38 and 13.39, see page 639; for 13.40, see page 641; for 13.41, see page 643; and for 13.42 and 13.43, see pages 644–645.

13.44 Confidence interval for a regression coefficient. In each of the following settings, give a 95% confidence interval for the coefficient of x_1 and interpret what it tells us.

(a) $n = 28$, $\hat{y} = 8.1 + 10.3x_1 + 4.2x_2$, $SE_{b_1} = 5.0$.

(b) $n = 33$, $\hat{y} = 2.1 + 2.8x_1 + 3.6x_2$, $SE_{b_1} = 1.0$.

(c) $n = 28$, $\hat{y} = 1.8 + 4.7x_1 + 2.4x_2 + 3.7x_3$, $SE_{b_1} = 4.5$.

(d) $n = 33$, $\hat{y} = 6.3 + 10.3x_1 + 4.2x_2 + 2.1x_3$, $SE_{b_1} = 2.8$.

13.45 Significance test for a regression coefficient. For each of the settings in the previous exercise, test the null hypothesis that the coefficient of x_1 is zero versus the two-sided alternative.

13.46 What's wrong? In each of the following situations, explain what is wrong and why.

(a) One of the assumptions for multiple regression is that the distribution of each explanatory variable is Normal.

(b) A small P-value for the ANOVA F test implies the model has very strong explanatory power.

(c) All explanatory variables that are significantly correlated with the response variable will have a statistically significant regression coefficient in the multiple regression model.

13.47 What's wrong? In each of the following situations, explain what is wrong and why.

(a) The multiple correlation gives the proportion of the variation in the response variable that is explained by the explanatory variables.

(b) In a multiple regression with a sample size of 35 and four explanatory variables, the test statistic for the null hypothesis $H_0: b_2 = 0$ is a t statistic that follows the $t(30)$ distribution when the null hypothesis is true.

(c) A small P-value for the ANOVA F test implies that all explanatory variables are statistically different from zero.

13.48 Inference basics. You run a multiple regression with 64 cases and three explanatory variables.

(a) What are the degrees of freedom for the F statistic for testing the null hypothesis that all three of the regression coefficients for the explanatory variables are zero?

(b) Software output gives MSE = 37.3. What is the estimate of the standard deviation σ of the model?

(c) The output gives the estimate of the regression coefficient for the first explanatory variable as 1.05 with a standard error of 0.45. Find a 95% confidence interval for the true value of this coefficient.

(d) Test the null hypothesis that the regression coefficient for the first explanatory variable is zero. Give the test statistic, the degrees of freedom, the P-value, and your conclusion.

13.49 Inference basics. You run a multiple regression with 23 cases and four explanatory variables. The ANOVA table includes the sums of squares SSR = 96 and SSE = 153.

(a) Find the F statistic for testing the null hypothesis that the regression coefficients for the four explanatory variables are all zero. Carry out the significance test and report the results.

(b) What is the value of R^2 for this model? Explain what this number tells us.

13.50 Burnout in the workplace. Burnout in the workplace has been a growing problem in recent decades. To better understand the issues, several researchers studied the relationship between the three dimensions of burnout and the five domains of quality of life.[12] The following table summarizes a multiple linear regression fit to a sample of 202 French Canadian workers. The response variable is emotional exhaustion, where a higher score means more exhaustion. The five explanatory variables are such that a high score means the worker is more satisfied with this domain of his or her life.

Explanatory variables	b	s(b)
Physical Health	−1.45	0.17
Psychological Health	−0.82	0.27
Environment	0.27	0.20
Social Relationships	−0.10	0.13
Spirituality	0.39	0.15
R^2		0.18

(a) The reported F statistic is 8.63. What degrees of freedom should be used to compute the P-value?

(b) Do these variables help predict emotional exhaustion? Compute the P-value for the overall F test and state your conclusion.

(c) Using the parameter estimates and standard errors, perform the individual parameter t tests.

(d) Interpret those coefficients that are statistically significant.

13.51 Predicting Taco Bell's market share. Recall Exercises 13.25 through 13.32 (page 632) on predicting market share. Let's consider the model fit to the data that does not include McDonald's. FFOOD

(a) Predict the market share for Taco Bell in three years. Currently, Taco Bell has $9790.15 million in sales, 5799 franchises, 647 company units, and an increase of 168 franchises and units over last year.

(b) Construct a 95% prediction interval and explain what this interval tells you.

13.52 Predicting the price of a tablet. In Exercise 13.22 (page 632), you predicted the price of a tablet with characteristics Size = 9.7, Battery = 8.9, Weight = 0.9, Ease = 5, Display = 5, and Versatility = 4. Construct a 95% prediction interval for this tablet and use it to assess its $550 asking price. TABLTS

CASE 13.2 13.53 Checking the model assumptions. Statistical inference requires us to make some assumptions about our data. These should always be checked prior to drawing conclusions. For brevity, we did not discuss this assessment for the movie revenue data of Section 13.2, so let's do it here. MOVIE

(a) Obtain the residuals for the multiple regression in Example 13.13, and construct a histogram and Normal quantile plot. Do the residuals appear approximately Normal? Explain your answer.

(b) Plot the residuals versus LOpening. Comment on anything unusual in the plot.

(c) Repeat part (b) using the explanatory variables Review and Theaters on the x axis.

(d) Repeat part (b) using the predicted value on the x axis.

(e) Summarize your overall findings from these summaries. Are the model assumptions reasonably satisfied? Explain your answer.

CASE 13.2 13.54 Effect of a potential outlier. Refer to the previous exercise. MOVIE

(a) There is one movie that has a much larger total U.S. box office revenue than predicted. Which is it, and how much more revenue did it obtain compared with that predicted?

(b) Remove this movie and redo the multiple regression. Make a table giving the regression coefficients and their standard errors, t statistics, and P-values.

(c) Compare these results with those presented in Example 13.11. How does the removal of this outlying movie affect the estimated model?

(d) Obtain the residuals from this reduced data set and graphically examine their distribution. Do the residuals appear approximately Normal? Is there constant variance? Explain your answer.

13.55 Compensation model for men's basketball head coaches. Some researchers collected information on $n = 36$ Division I head basketball coaches to determine significant predictors of compensation.[13] The following table summarizes the results of this analysis.

Explanatory variables	b	P-value
Constant	−1,138,240.3	0.641
Tenure	−8153.1	0.884
Career Win %	2,805,207.5	0.485
Tournament Appearances	128,873.4	0.030
Ticket Sales	0.275	0.002
Institutional Support	0.124	0.800
Guaranteed Revenue	−0.872	0.517
Contributions	0.116	0.409
In-kind Contributions	0.658	0.785
Media Rights	0.156	0.211
NCAA Distributions	0.112	0.497
Conference Distributions	0.133	0.286
Concession	−2.2	0.026
Royalties	−0.217	0.350
Camp Revenue	3.2	0.007
Restricted Endowments	−1.7	0.486
Other	−0.581	0.630

(a) The overall F statistic is 4.16. What are the degrees of freedom and P-value of this statistic?

(b) Which of the explanatory variables significantly aid prediction given the presence of all the explanatory variables?

(c) The intercept and coefficient for Tenure are both very large negative values. For someone unfamiliar with regression, this can be bothersome, as salaries for head coaches are in the millions of dollars. Explain to this person why negative estimates are not a concern in this case.

TABLE 13.3	Regression coefficients and t statistics for Exercise 13.57	
Variable	b	t
Intercept	15.47	
Loan size (in dollars)	−0.0015	10.30
Length of loan (in months)	−0.906	4.20
Percent down payment	−0.522	8.35
Cosigner (0 = no, 1 = yes)	−0.009	3.02
Unsecured loan (0 = no, 1 = yes)	0.034	2.19
Total payments (borrower's monthly installment debt)	0.100	1.37
Total income (borrower's total monthly income)	−0.170	2.37
Bad credit report (0 = no, 1 = yes)	0.012	1.99
Young borrower (0 = older than 25, 1 = 25 or younger)	0.027	2.85
Male borrower (0 = female, 1 = male)	−0.001	0.89
Married (0 = no, 1 = yes)	−0.023	1.91
Own home (0 = no, 1 = yes)	−0.011	2.73
Years at current address	−0.124	4.21

13.56 Bank auto loans. Banks charge different interest rates for different loans. A random sample of 2229 loans made by banks for the purchase of new automobiles was studied to identify variables that explain the interest rate charged. A multiple regression was run with interest rate as the response variable and 13 explanatory variables.[14]

(a) The F statistic reported is 71.34. State the null and alternative hypotheses for this statistic. Give the degrees of freedom and the P-value for this test. What do you conclude?

(b) The value of R^2 is 0.297. What percent of the variation in interest rates is explained by the 13 explanatory variables?

13.57 Bank auto loans, continued. Table 13.3 gives the coefficients for the fitted model and the individual t statistic for each explanatory variable in the study described in the previous exercise. The t-values are given without the sign, assuming that all tests are two-sided.

(a) State the null and alternative hypotheses tested by an individual t statistic. What are the degrees of freedom for these t statistics? What values of t will lead to rejection of the null hypothesis at the 5% level?

(b) Which of the explanatory variables have coefficients that are significantly different from zero in this model? Explain carefully what you conclude when an individual t statistic is not significant.

(c) The signs of many of the coefficients are what we might expect before looking at the data. For example, the negative coefficient for loan size means that larger loans get a smaller interest rate. This is reasonable. Examine the signs of each of the statistically significant coefficients and give a short explanation of what they tell us.

13.58 Auto dealer loans. The previous two exercises describe auto loans made directly by a bank. The researchers also looked at 5664 loans made indirectly—that is, through an auto dealer. They again used multiple regression to predict the interest rate using the same set of 13 explanatory variables.

(a) The F statistic reported is 27.97. State the null and alternative hypotheses for this statistic. Give the degrees of freedom and the P-value for this test. What do you conclude?

(b) The value of R^2 is 0.141. What percent of the variation in interest rates is explained by the 13 explanatory variables? Compare this value with the percent explained for direct loans in Exercise 13.57.

13.59 Auto dealer loans, continued. Table 13.4 gives the estimated regression coefficient and individual t statistic for each explanatory variable in the setting of the previous exercise. The t-values are given without the sign, assuming that all tests are two-sided.

(a) What are the degrees of freedom of any individual t statistic for this model? What values of t are significant at the 5% level? Explain carefully what significance tells us about an explanatory variable.

(b) Which of the explanatory variables have coefficients that are significantly different from zero in this model?

(c) The signs of many of these coefficients are what we might expect before looking at the data. For example, the negative coefficient for loan size means that larger loans

TABLE 13.4 Regression coefficients and *t* statistics for Exercise 13.59

Variable	b	t
Intercept	15.89	
Loan size (in dollars)	−0.0029	17.40
Length of loan (in months)	−1.098	5.63
Percent down payment	−0.308	4.92
Cosigner (0 = no, 1 = yes)	−0.001	1.41
Unsecured loan (0 = no, 1 = yes)	0.028	2.83
Total payments (borrower's monthly installment debt)	−0.513	1.37
Total income (borrower's total monthly income)	0.078	0.75
Bad credit report (0 = no, 1 = yes)	0.039	1.76
Young borrower (0 = older than 25, 1 = 25 or younger)	−0.036	1.33
Male borrower (0 = female, 1 = male)	−0.179	1.03
Married (0 = no, 1 = yes)	−0.043	1.61
Own home (0 = no, 1 = yes)	−0.047	1.59
Years at current address	−0.086	1.73

get a smaller interest rate. This is reasonable. Examine the signs of each of the statistically significant coefficients and give a short explanation of what they tell us.

13.60 Direct versus indirect loans. The previous four exercises describe a study of loans for buying new cars. The authors conclude that banks take higher risks with indirect loans because they do not take into account borrower characteristics when setting the loan rate. Explain how the results of the multiple regressions lead to this conclusion.

13.61 Canada's Small Business Financing Program. The Canada Small Business Financing Program (CSBFP) seeks to increase the availability of loans for establishing and improving small businesses. A survey was performed to better understand the experiences of small businesses when seeking loans and the extent to which they are aware of and satisfied with the CSBFP.[15] A total of 1050 survey interviews were completed. To understand the drivers of perceived fairness of CSBFP terms and conditions, a multiple regression was undertaken. The response variable was the subject's perceived fairness scored on a 5-point scale, where 1 means "very unfair" and 5 means "very fair." The 15 explanatory variables included characteristics of the survey participant (gender, francophone, loan history, previous CSBFP borrower) and characteristics of his or her small business (type, location, size).

(a) What are the degrees of freedom for the *F* statistic of the model that contains all the predictors?

(b) The report states that the *P*-value for the overall *F* test is *P* = 0.005 and that the complete set of predictors has an R^2 of 0.031. Explain to a statistical novice how the *F* test can be highly significant but with a very low R^2.

(c) The report also states that only two of the explanatory variables were found significant at the 0.05 level. Suppose the model with just an indicator of previous CSBFP participation and an indicator that the business is in transportation and warehousing explained 2.5% of the variation in the response variable. Test the hypothesis that the other 13 predictors do not help predict fairness when these two predictors are already in the model.

13.62 Compensation and human capital. A study of bank branch manager compensation collected data on the salaries of 82 managers at branches of a large eastern U.S. bank.[16] Multiple regression models were used to predict how much these branch managers were paid. The researchers examined two sets of explanatory variables. The first set included variables that measured characteristics of the branch and the position of the branch manager. These were number of branch employees, a variable constructed to represent how much competition the branch faced, market share, return on assets, an efficiency ranking, and the rank of the manager. A second set of variables was called human capital variables and measured characteristics of the manager. These were experience in industry, gender, years of schooling, and age. For the multiple regression using all the explanatory variables, the value of R^2 was 0.77. When the human capital variables were deleted, R^2 fell to 0.06. Test the null hypothesis that the coefficients for the human capital variables are all zero in the model that includes all the explanatory variables. Give the test statistic with its degrees of freedom and *P*-value, and give a short summary of your conclusion in nontechnical language.

13.3 Multiple Regression Model Building

When you complete this section, you will be able to:

- Fit a curved relationship between y and x through the use of a polynomial model.
- Describe what multicollinearity means and the issues that may arise because of it.
- Use indicator variables in order to include categorical variables in multiple regression.
- Include interaction terms in a regression model when the effect of one explanatory variable depends on the value of another explanatory variable.
- Implement variable selection methods to choose a best model.

Often, we have many explanatory variables, and our goal is to use these to explain the variation in the response variable. A model using just a few of the variables often predicts about as well as the model using all the explanatory variables. We may also find that the reciprocal of a variable is a better choice than the variable itself or that including the square of an explanatory variable improves prediction. How can we find a good model? That is the **model building** issue. A complete discussion would be quite lengthy, so we must be content with illustrating some of the basic ideas with a case study.

model building

HOMES

CASE 13.3 **Prices of Homes** People wanting to buy a home can find information on the Internet about homes for sale in their community. We work with online data for homes for sale in Lafayette and West Lafayette, Indiana.[17] The response variable is Price, the asking price of a home. The online data contain the following explanatory variables: (a) SqFt, the number of square feet for the home; (b) BedRooms, the number of bedrooms; (c) Baths, the number of bathrooms; (d) Garage, the number of cars that can fit in the garage; and (e) Zip, the postal zip code for the address. There are 504 homes in the data set. ∎

The analysis starts with a study of the variables involved. Here is a short summary of this work.

Price, as we expect, has a right-skewed distribution. The mean (in thousands of dollars) is $158 and the median is $130. There is one high outlier at $830, which we delete as unusual in this location. Remember that a skewed distribution for Price does not itself violate the conditions for multiple regression. The model requires that the *residuals* from the fitted regression equation be approximately Normal. We will have to examine how well this condition is satisfied when we build our regression model.

SqFt, the number of square feet for the home, is a quantitative variable that we expect to strongly influence Price. **BedRooms** ranges from one to five. The website uses 5 for all homes with five or more bedrooms. The data contain just one home with one bedroom. **Baths** includes both full baths (with showers or bathtubs) and half baths (which lack bathing facilities). Typical values are 1, 1.5, 2, and 2.5. **Garage** has values of 0, 1, 2, and 3. The website uses the value 3 when three or more vehicles can fit in the garage. There are 50 homes that can fit three or more vehicles into their garage (or possibly garages).

Zip describes location, traditionally the most important explanatory variable for house prices, but Zip is a quite crude description because a single zip code covers a broad area. All of the postal zip codes in this community have 4790 as the first four digits. The fifth digit is coded as the variable Zip. The possible values are 1, 4, 5, 6, and 9. There is only one home with zip code 47901.

FIGURE 13.13 Scatterplot matrix of the case study variables in zip code 47904.

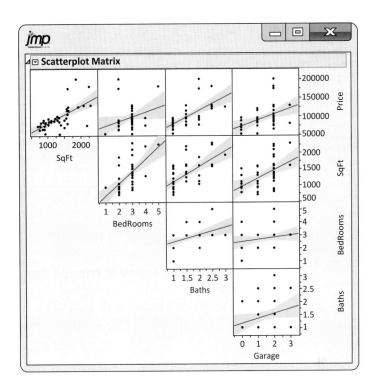

scatterplot matrix

We first look at the houses in each zip code separately. We start by examining only the homes in zip code 47904, corresponding to Zip = 4. Figure 13.13 displays a **scatterplot matrix** of the 44 houses for sale in 47904. These plots combine all the pairwise scatterplots into a matrix for easy viewing. Most statistical packages provide the option to create these plots. JMP also includes the option to include in each panel the simple least-squares fit with confidence intervals.

The first row of the matrix reveals that each explanatory variable is positively correlated with Price. This is not unexpected. We also see some moderately strong correlations among the explanatory variables. This, again, is not surprising. For example, we'd expect a house with three bedrooms to be larger (more square feet) than a house with two bedrooms.

We also see in the scatterplots some unusual cases. Most homes for sale in this area are moderately priced. There are, however, a few houses with prices that are somewhat high relative to the others. Similarly, some values for SqFt are relatively high. Because we do not want our analysis to be overly influenced by these homes, we exclude any home with Price greater than $150,000 and any home with SqFt greater than 1800 ft². Seven homes were excluded by these criteria, leaving 37 homes for analysis.

EXAMPLE 13.17

HOM04

CASE 13.3 **Price and Square Feet** Let's begin our model building by examining the relationship between Price and SqFt. Figure 13.14 displays the relationship between SqFt and Price. This is similar to the upper-left panel in Figure 13.13 except it no longer includes those seven houses and we have replaced the least-squares line with a "smooth" fit. The relationship is approximately linear but curves up somewhat for the higher-priced homes. ■

Because the relationship is approximately linear, let's start by examining the simple linear regression of Price on SqFt.

FIGURE 13.14 Plot of price versus square feet for Example 13.17.

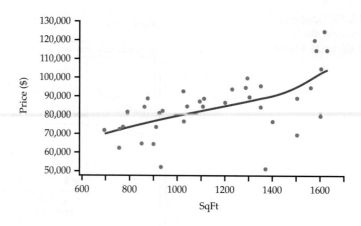

EXAMPLE 13.18

HOM04

CASE 13.3 **Regression of Price on Square Feet** Figure 13.15 gives the regression output. The number of degrees of freedom in the "Corrected Total" line in the ANOVA table is 36. This is correct for the $n = 37$ homes that remain after we excluded seven of the original 44. The fitted model is

$$\widehat{\text{Price}} = 45{,}298 + 34.32\text{SqFt}$$

The coefficient for SqFt is statistically significant ($t = 4.57$, df $= 35$, $P < 0.0001$). Each additional square foot of area raises selling prices by $34.32 on the average. From the R^2, we see that 37.3% of the variation in the home prices is explained by a linear relationship with square feet. We hope that multiple regression will allow us to improve on this first attempt to explain selling price. ■

FIGURE 13.15 Linear regression output for predicting price using square feet for Example 13.18.

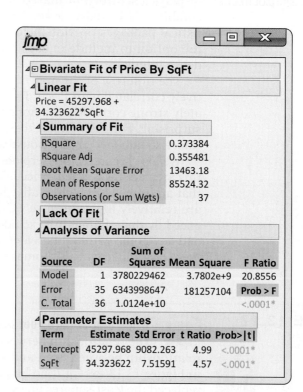

APPLY YOUR KNOWLEDGE

CASE 13.3 **13.63 Distributions.** Make histograms of the prices and of the square feet for the 44 homes in Table 13.5. Do the seven homes excluded in Example 13.17 appear unusual for this location? HOM04

TABLE 13.5 Homes for sale in zip code 47904

Id	Price ($ thousands)	SqFt	Bed Rooms	Baths	Garage	Id	Price ($ thousands)	SqFt	Bed Rooms	Baths	Garage
01	52,900	932	1	1.0	0	23	75,000	2188	4	1.5	2
02	62,900	760	2	1.0	0	24	76,900	1400	3	1.5	2
03	64,900	900	2	1.0	0	25	81,900	796	2	1.0	2
04	69,900	1504	3	1.0	0	26	84,500	864	2	1.0	2
05	76,900	1030	3	2.0	0	27	84,900	1350	3	1.0	2
06	87,900	1092	3	1.0	0	28	89,600	1504	3	1.0	2
07	94,900	1288	4	2.0	0	29	87,000	1200	2	1.0	2
08	52,000	1370	3	1.0	1	30	89,000	876	2	1.0	2
09	72,500	698	2	1.0	1	31	89,000	1112	3	2.0	2
10	72,900	766	2	1.0	1	32	93,900	1230	3	1.5	2
11	73,900	777	2	1.0	1	33	96,000	1350	3	1.5	2
12	73,900	912	2	1.0	1	34	99,900	1292	3	2.0	2
13	81,500	925	3	1.0	1	35	104,900	1600	3	1.5	2
14	82,900	941	2	1.0	1	36	114,900	1630	3	1.5	2
15	84,900	1108	3	1.5	1	37	124,900	1620	3	2.5	2
16	84,900	1040	2	1.0	1	38	124,900	1923	3	3.0	2
17	89,900	1300	3	2.0	1	39	129,000	2090	3	1.5	2
18	92,800	1026	3	1.0	1	40	173,900	1608	2	2.0	2
19	94,900	1560	3	1.0	1	41	179,900	2250	5	2.5	2
20	114,900	1581	3	1.5	1	42	199,500	1855	2	2.0	2
21	119,900	1576	3	2.5	1	43	80,000	1600	3	1.0	3
22	65,000	853	3	1.0	2	44	129,000	2296	3	2.5	3

CASE 13.3 **13.64 Plot the residuals.** Obtain the residuals from the simple linear regression in Example 13.18 and plot them versus SqFt. Describe the plot. Does it suggest that the relationship might be curved? HOM04

CASE 13.3 **13.65 Predicted values.** Use the simple linear regression equation to obtain the predicted price for a home that has 1250 ft^2. Do the same for a home that has 1750 ft^2. HOM04

Models for curved relationships

Figure 13.14 (page 652) suggests that the relationship between SqFt and Price may be slightly curved. One simple kind of curved relationship is a quadratic function. To model a quadratic function with multiple regression, create a new variable that is the square of the explanatory variable and include it in the regression model. There are now $k = 2$ explanatory variables, x and x^2. The model is

$$y = \beta_0 + \beta_1 x + \beta_2 x^2 + \varepsilon$$

with the usual conditions on the ε.

EXAMPLE 13.19

Quadratic Regression of Price on Square Feet To predict price using a quadratic function of square feet, first create a new variable by squaring each value of SqFt. Call this variable SqFt2. Figure 13.16 displays the output for multiple regression of Price on SqFt and SqFt2. The fitted model is

$$\widehat{\text{Price}} = 81{,}273 - 30.14\text{SqFt} + 0.0271\text{SqFt2}$$

This model explains 38.6% of the variation in Price, little more than the 37.3% explained by simple linear regression of Price on SqFt. The coefficient of SqFt2 is not significant ($t = 0.84$, df $= 34$, $P = 0.41$). That is, the squared term does not significantly improve the fit when the SqFt term is present. We conclude that adding SqFt2 to our model is not helpful. ∎

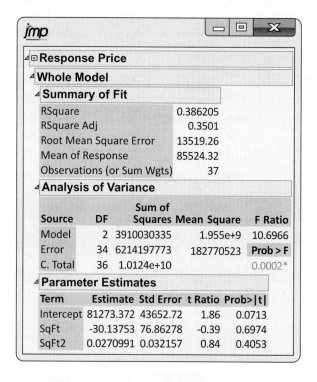

FIGURE 13.16 Quadratic regression output for predicting price using square feet for Example 13.19.

The output in Figure 13.16 is a good example of the need for care in interpreting multiple regression. The individual t tests for *both* SqFt and SqFt2 are not significant. Yet, the overall F test for the null hypothesis that both coefficients are zero *is* significant ($F = 10.70$, df $= 2$ and 34, $P = 0.0002$). To resolve this apparent contradiction, remember that a t test assesses the contribution of a single variable, *given that the other variables are present in the model*. Once either SqFt or SqFt2 is present, the other contributes very little. This is a consequence of the fact that these two variables are highly correlated. This phenomenon is called **collinearity** or **multicollinearity**. In extreme cases, collinearity can cause numerical instabilities, and the results of the regression calculations can become very imprecise. Collinearity can exist between seemingly unrelated variables and can be hard to detect in models with many explanatory variables. Some statistical software packages will calculate a **variance inflation factor (VIF)** value for each explanatory variable in a model. VIF values greater than 10 are generally considered an indication that severe collinearity exists among the explanatory variables in a model. Exercise 13.86 (page 668) explores the calculation and use of VIF values. In this particular case, we could dispense with either SqFt or SqFt2, but the F

test tells us that we cannot drop both of them. It is natural to keep SqFt and drop its square, SqFt2.

Multiple regression can fit a *polynomial* model of any degree:

$$y = \beta_0 + \beta_1 x + \beta_2 x^2 + \cdots + \beta_k x^k + \varepsilon$$

In general, we include all powers up to the highest power in the model. A relationship that curves first up and then down, for example, might be described by a cubic model with explanatory variables x, x^2, and x^3. Other transformations of an explanatory variable, such as the square root and the logarithm, can also be used to model curved relationships.

APPLY YOUR KNOWLEDGE

CASE 13.3 **13.66 The relationship between SqFt and SqFt2.** Using the data set for Example 13.19, plot SqFt2 versus SqFt. Describe the relationship. We know that it is not linear, but is it approximately linear? What is the correlation between SqFt and SqFt2? The plot and correlation demonstrate that these variables are collinear and explain why neither of them contributes much to a multiple regression once the other is present. HOM04

CASE 13.3 **13.67 Predicted values.** Use the quadratic regression equation in Example 13.19 to predict the price of a home that has 1250 ft^2. Do the same for a home that has 1750 ft^2. Compare these predictions with the ones from an analysis that uses only SqFt as an explanatory variable (e.g., Exercise 13.65). HOM04

Models with categorical explanatory variables

Although adding the square of SqFt failed to improve our model significantly, Figure 13.14 (page 652) does suggest that the price rises a bit more steeply for larger homes. Perhaps some of these homes have other desirable characteristics that increase the price. Let's examine another explanatory variable.

EXAMPLE 13.20

CASE 13.3 **Price and the Number of Bedrooms** Figure 13.17 gives a plot of Price versus BedRooms. We see that there appears to be a curved relationship. However, all but two of the homes have either two or three bedrooms. One home has one bedroom and another has four. These two cases are why the relationship appears to be curved. To avoid this situation, we group the

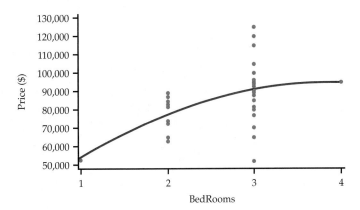

FIGURE 13.17 Plot of price versus the number of bedrooms for Example 13.20.

four-bedroom home with those that have three bedrooms (BedRooms = 3) and the one-bedroom home with the homes that have two bedrooms.

The price of the four-bedroom home is in the middle of the distribution of the prices for the three-bedroom homes. On the other hand, the one-bedroom home has the lowest price of all the homes in the data set. This observation may require special attention later. ∎

"Number of bedrooms" is now a *categorical variable* that places homes in two groups: one/two bedrooms and three/four bedrooms. Software often allows you to simply declare that a variable is categorical. Then the values for the two groups don't matter. We could use the values 2 and 3 for the two groups. If you work directly with the variable, however, it is better to indicate whether the home has three or more bedrooms. We will take the "number of bedrooms" categorical variable to be Bed3 = 1 if the home has three or more bedrooms and Bed3 = 0 if it does not. Bed3 is called an *indicator (or dummy) variable*.

INDICATOR VARIABLES

An indicator (or dummy) variable is a variable with the values 0 and 1. To use a categorical variable that has I possible values in a multiple regression, create $K = I - 1$ indicator variables to use as the explanatory variables. This can be done in many different ways. Here is one common choice:

$$X1 = \begin{cases} 1 & \text{if the categorical variable has the first value} \\ 0 & \text{otherwise} \end{cases}$$

$$X2 = \begin{cases} 1 & \text{if the categorical variable has the second value} \\ 0 & \text{otherwise} \end{cases}$$

$$\vdots$$

$$XK = \begin{cases} 1 & \text{if the categorical variable has the next to last value} \\ 0 & \text{otherwise} \end{cases}$$

We need only $I - 1$ variables to code I different values because the last value is identified by "all $I - 1$ indicator variables are 0."

EXAMPLE 13.21

HOM04

Price and the Number of Bedrooms Figure 13.18 displays the output for the regression of Price on the indicator variable Bed3. This model explains 19% of the variation in price. This is about one-half of the 37.3% explained by SqFt, but it suggests that Bed3 may be a useful explanatory variable.

The fitted equation is

$$\widehat{\text{Price}} = 75,700 + 15,146 \text{Bed3}$$

The coefficient for Bed3 is significantly different from 0 ($t = 2.88$, df = 35, $P = 0.0068$). This coefficient is the slope of the least-squares line. That is, it is the increase in the average price when Bed3 increases by 1. The indicator variable Bed3 has only two values, so we can clarify the interpretation.

The predicted price for homes with two or fewer bedrooms (Bed3 = 0) is

$$\widehat{\text{Price}} = 75,700 + 15,146(0) = 75,700$$

FIGURE 13.18 Output for predicting price using whether there are three or more bedrooms for Example 13.21.

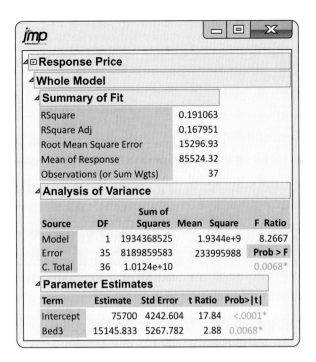

That is, the intercept 75,700 is the mean price for homes with Bed3 = 0. The predicted price for homes with three or more bedrooms (Bed3 = 1) is

$$\widehat{\text{Price}} = 75{,}700 + 15{,}146(1) = 90{,}846$$

That is, the slope 15,146 says that homes with three or more bedrooms are priced $15,146 higher on the average than homes with two or fewer bedrooms. When we regress on a single indicator variable, both intercept and slope have simple interpretations. ■

Example 13.21 shows that regression on one indicator variable essentially models the means of the two groups. A regression model for a categorical variable with I possible values requires $I-1$ indicator variables and therefore I regression coefficients (including the intercept) to model the I category means (this is like the one-way ANOVA model). Indicator variables can also be used in models with other variables to allow for different regression lines for different groups. Exercises 13.80 and 13.81 (page 667) explore the use of an indicator variable in a model with another quantitative variable for this purpose.

Here is an example of a categorical explanatory variable with four possible values.

EXAMPLE 13.22

CASE 13.3

HOM04

Price and the Number of Bathrooms The homes in our data set have 1, 1.5, 2, or 2.5 bathrooms. Figure 13.19 gives a plot of price versus the number of bathrooms with a "smooth" fit. The relationship does not appear to be linear, so we start by treating the number of bathrooms as a categorical variable. We require three indicator variables for the four values. To use the homes that have one bath as the basis for comparisons, we let this correspond to "all indicator variables equal to 0." The indicator variables are

$$\text{Baths15} = \begin{cases} 1 & \text{if the home has 1.5 baths} \\ 0 & \text{otherwise} \end{cases}$$

$$\text{Baths2} = \begin{cases} 1 & \text{if the home has 2 baths} \\ 0 & \text{otherwise} \end{cases}$$

$$\text{Baths25} = \begin{cases} 1 & \text{if the home has 2.5 baths} \\ 0 & \text{otherwise} \end{cases}$$

Multiple regression using these three explanatory variables gives the output in Figure 13.20. The overall model is statistically significant ($F = 12.59$, df = 3 and 33, $P < 0.0001$), and it explains 53.4% of the variation in price. This is somewhat more than the 37.3% explained by square feet.

The fitted model is

$$\widehat{\text{Price}} = 77{,}504 + 20{,}553\,\text{Baths15} + 12{,}616\,\text{Baths2} + 44{,}896\,\text{Baths25}$$

The coefficient of each of the indicator variables is positive and statistically significant, indicating a higher mean price than a house with one bathroom. ∎

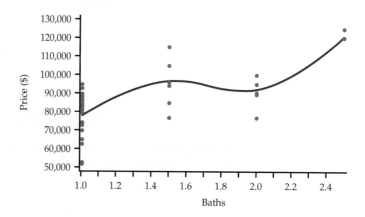

FIGURE 13.19 Plot of price versus the number of bathrooms for Example 13.22.

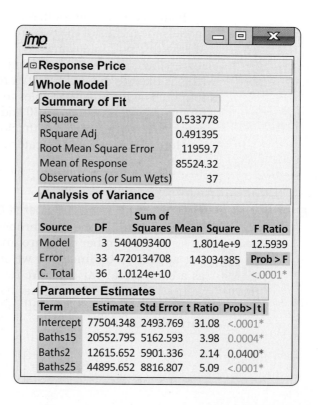

FIGURE 13.20 Output for predicting price using indicator variables for the number of bathrooms for Example 13.22.

APPLY YOUR KNOWLEDGE

CASE 13.3 13.68 Find the means. Using the data set for Example 13.21, find the mean price for the homes that have three or more bedrooms and the mean price for those that do not.

(a) Compare these sample means with the predicted values given in Example 13.21.

(b) What is the difference between the mean price of the homes in the sample that have three or more bedrooms and the mean price of those that do not? Verify that this difference is the coefficient for the indicator variable Bed3 in the regression in Example 13.21. HOM04

CASE 13.3 13.69 Compare the means. Regression on a single indicator variable compares the mean responses in two groups. It is, in fact, equivalent to the pooled t test for comparing two means (Chapter 8, page 429). Use the pooled t test to compare the mean price of the homes that have three or more bedrooms with the mean price of those that do not. Verify that the test statistic, degrees of freedom, and P-value agree with the t test for the coefficient of Bed3 in Example 13.21. HOM04

CASE 13.3 13.70 Modeling the means. Following the pattern in Example 13.22, use the output in Figure 13.20 to write the equations for the predicted mean price for the following:

(a) Homes with 1 bathroom.

(b) Homes with 1.5 bathrooms.

(c) Homes with 2 bathrooms.

(d) Homes with 2.5 bathrooms.

(e) How can we interpret the coefficient for one of the indicator variables, say Baths2, in language understandable to house shoppers?

More elaborate models

We now suspect that a model that uses square feet, number of bedrooms, and number of bathrooms may explain price reasonably well. Before examining such a model, we use an insight based on careful data analysis to improve the treatment of number of baths.

Figure 13.19 reminds us that the data describe only two homes with 2.5 bathrooms. Our model with three indicator variables fits a mean for these two observations. The pattern of Figure 13.19 reveals an interesting feature: adding a half bath to either one or two baths raises the predicted price by a similar, quite substantial, amount. If we use this information to construct a model, we can avoid the problem of fitting one parameter to just two houses.

EXAMPLE 13.23

HOM04

An Alternative Bath Model Starting again with one-bath homes as the base (all indicator variables 0), let B2 be an indicator variable for an extra full bath and let Bh be an indicator variable for an extra half bath. Thus, a home with two baths has Bh = 0 and B2 = 1. A home with 2.5 baths has Bh = 1 and B2 = 1. Regressing Price on Bh and B2 gives the output in Figure 13.21.

FIGURE 13.21 Output for predicting price using the alternative coding for bathrooms, for Example 13.23.

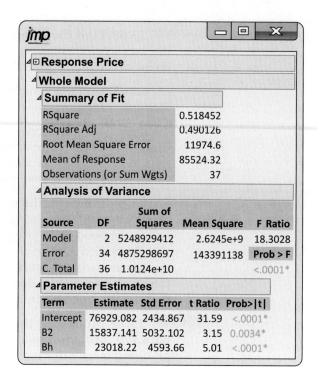

The overall model is statistically significant ($F = 18.30$, df = 2 and 34, $P < 0.0001$), and it explains 51.8% of the variation in price. This compares favorably with the 53.4% explained by the model with three indicator variables for bathrooms. The fitted model is

$$\widehat{\text{Price}} = 76{,}929 + 15{,}837\text{B2} + 23{,}018\text{Bh}$$

That is, an extra full bath adds $15,837 to the mean price and an extra half bath adds $23,018. The t statistics show that both regression coefficients are significantly different from zero ($t = 3.15$, df = 34, $P = 0.0034$; and $t = 5.01$, df = 34, $P < 0.0001$). ∎

So far we have learned that the price of a home is related to the number of square feet, whether there are three or more bedrooms, whether there is an additional full bathroom, and whether there is an additional half bathroom. Let's try a model including all these explanatory variables.

EXAMPLE 13.24

HOM04

Square Feet, Bedrooms, and Bathrooms Figure 13.22 gives the output for predicting price using SqFt, Bed3, B2, and Bh. The overall model is statistically significant ($F = 11.48$, df = 4 and 32, $P < 0.0001$), and it explains 58.9% of the variation in price.

The individual t for Bed3 is not statistically significant ($t = -0.97$, df = 32, $P = 0.3374$). That is, in a model that contains square feet and information about the bathrooms, there is no additional information in the number of bedrooms that is useful for predicting price. This happens because the explanatory variables are related to each other: houses with more bedrooms tend to also have more square feet and more baths.

FIGURE 13.22 Output for predicting price using square feet, bedroom, and bathroom information for Example 13.24.

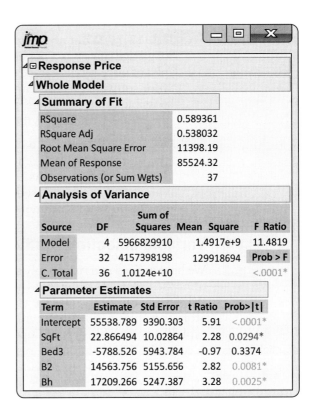

Therefore, we redo the regression without Bed3. The output appears in Figure 13.23. The value of R^2 has decreased slightly to 57.7%, but now all the coefficients for the explanatory variables are statistically significant. The fitted regression equation is

$$\widehat{\text{Price}} = 59{,}268 + 16.78\text{SqFt} + 13{,}161\text{B2} + 16{,}859\text{Bh} \blacksquare$$

FIGURE 13.23 Output for predicting price using square feet and bathroom information for Example 13.24.

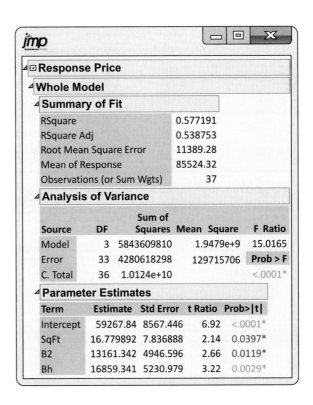

APPLY YOUR KNOWLEDGE

CASE 13.3 **13.71 What about garages?** We have not yet examined the number of garage spaces as a possible explanatory variable for price. Make a scatterplot of price versus garage spaces. Describe the pattern. Use a "smooth" fit if your software has this capability. Otherwise, find the mean price for each possible value of Garage, plot the means on your scatterplot, and connect the means with lines. Is the relationship approximately linear? HOM04

CASE 13.3 **13.72 The home with three garages.** There is only one home with three garage spaces. We might either place this house in the Garage = 2 group or remove it as unusual. Either decision leaves Garage with values 0, 1, and 2. Based on your plot in the previous exercise, which choice do you recommend?

Variable selection methods

We have arrived at a reasonably satisfactory model for predicting the asking price of houses. But it is clear that there are many other models we might consider. We have not used the Garage variable, for example. What is more, the explanatory variables can *interact* with each other. This means that the effect of one explanatory variable depends upon the value of another explanatory variable. We account for this situation in a regression model by including **interaction terms**.

interaction terms

The simplest way to construct an interaction term is to multiply the two explanatory variables together. Thus, if we wanted to allow the effect of an additional half bath to depend upon whether there is an additional full bath, we would create a new explanatory variable by taking the product B2 × Bh. Interaction terms that are the product of an indicator variable and another variable in the model can be used to allow for different slopes for different groups. Exercise 13.80 (page 667) explores the use of such an interaction term in a model.

Considering interactions increases the number of possible explanatory variables. If we start with the five explanatory variables SqFt, Bed3, B2, Bh, and Garage, there are 10 interactions between pairs of variables. That is, there are now 15 possible explanatory variables in all. From 15 explanatory variables, it is possible to build 32,767 different models for predicting price. Clearly, we need to automate the process of examining all possible models.

Modern regression software offers *variable selection methods* that examine all possible multiple regression models for a given set of exploratory variables. The software then presents us a list of the top models based on some selection criteria. Available criteria include R^2, adjusted R^2, the regression standard error s, AIC, and BIC. The latter four criteria balance fit against model simplicity in different ways and are appropriate for comparing models with different numbers of explanatory variables. R^2 should be used only when comparing linear models with the same number of explanatory variables.

EXAMPLE 13.25

 Predicting Asking Price Software tells us that the highest available R^2 increases as we increase the number of explanatory variables as follows:

 HOM04

Variables	R^2
1	0.44
2	0.57
3	0.62
4	0.66
5	0.72
⋮	
13	0.77

Because of collinearity problems, models with 14 or all 15 explanatory variables cannot be used. The highest possible R^2 is 77%, using 13 explanatory variables. ∎

There are only 37 houses in the data set. A model with too many explanatory variables will fit the accidental wiggles in the prices of these specific houses and may do a poor job of predicting prices for other houses. This is called **overfitting**.

overfitting

There is no formula for choosing the "best" multiple regression model. We tend to prefer smaller models to larger models because they avoid overfitting. However, too small a model may not fit the data or predict well. Model selection criteria such as the regression standard error s or AIC balance these two goals but in different ways. Using one of these is preferred to R^2 when trying to determine a best model because R^2 does not account for overfitting.

We also want a model that makes intuitive sense. For example, the best single predictor ($R^2 = 0.44$) is SqFtBh, the interaction between square feet and having an extra half bath. In general, we would not use a model that has interaction terms unless the explanatory variables involved in the interaction are also included. The variable selection output does suggest that we include SqFtBh and, therefore, SqFt and Bh.

EXAMPLE 13.26

HOM04

One More Model Figure 13.24 gives the output for the multiple regression of Price on SqFt, Bh, and the interaction between these variables. This model explains 55.7% of the variation in price. The coefficients for SqFt and Bh are significant at the 10% level but not at the 5% level ($t = 1.82$, df $= 33$, $P = 0.0777$; and $t = -1.79$, df $= 33$, $P = 0.0825$). However, the interaction is significantly different from zero ($t = 2.29$, df $= 33$, $P = 0.0284$).

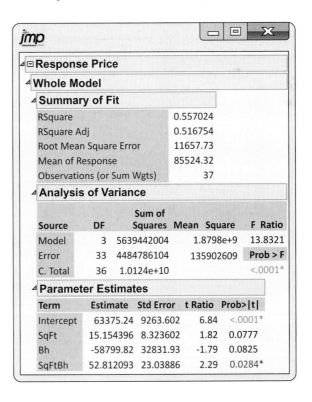

FIGURE 13.24 Output for predicting price using square feet, extra half bathroom, and the interaction for Example 13.26.

Because the interaction is significant, we keep the terms SqFt and Bh in the model even though their P-values do not quite pass the 0.05 standard. The fitted regression equation is

$$\widehat{\text{Price}} = 63{,}375 + 15.15\text{SqFt} - 58{,}800\text{Bh} + 52.81\text{SqFtBh} \quad \blacksquare$$

The negative coefficient for Bh seems odd. We expect an extra half bath to increase price. A plot will help us to understand what this model says about the prices of homes.

EXAMPLE 13.27

CASE 13.3 **Interpretation of the Fit** Figure 13.25 plots Price against SqFt, using different plot symbols to show the values of the categorical variable Bh. Homes without an extra half bath are plotted as blue circles, and homes with an extra half bath are plotted as green squares. Look carefully: none of the smaller homes have an extra half bath.

The lines on the plot graph the model from Example 13.26. The presence of the categorical variable Bh and the interaction Bh × SqFt account for the *two* lines. That is, the model expresses the fact that the relationship between price and square feet depends on whether there is an additional half bath. ∎

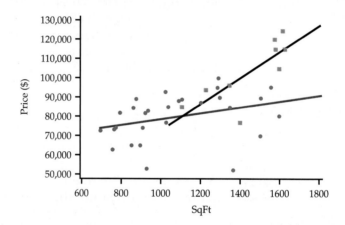

FIGURE 13.25 Plot of the data and model for price predicted by square feet, extra half bathroom, and the interaction for Example 13.27.

We can determine the equations of the two lines from the fitted regression equation. Homes that lack an extra half bath have $Bh = 0$. When we set $Bh = 0$ in the regression equation

$$\widehat{Price} = 63{,}375 + 15.15\,SqFt - 58{,}800\,Bh + 52.81\,SqFtBh$$

we get

$$\widehat{Price} = 63{,}375 + 15.15\,SqFt$$

For homes that have an extra half bath, $Bh = 1$ and the regression equation becomes

$$\widehat{Price} = 63{,}375 + 15.15\,SqFt - 58{,}800 + 52.81\,SqFt$$
$$= (63{,}375 - 58{,}800) + (15.15 + 52.81)SqFt$$
$$= 4575 + 67.96\,SqFt$$

In Figure 13.25, we graphed this line starting slightly below the price of the least-expensive home with an extra half bath. These two equations tell us that an additional square foot increases asking price by $15.15 for homes without an extra half bath and by $67.96 for homes that do have an extra half bath.

APPLY YOUR KNOWLEDGE

CASE 13.3 **13.73 Comparing some predicted values.** Refer to the model in Example 13.26 and consider two homes, both with 1500 ft². Suppose the first has an extra half bath and the second does not. Find the predicted price for each home and then find the difference.

CASE 13.3 **13.74 Suppose the homes are larger.** Refer to the previous exercise. Consider two additional homes, both with 1700 ft², one with an extra half bath and one without. Find the predicted prices and the difference. How does this difference compare with the difference you obtained in the previous exercise? Explain what you have found.

CASE 13.3 **13.75 How about the smaller homes?** Would it make sense to do the same calculations as in the previous two exercises for homes that have 800 ft^2? Explain why or why not.

CASE 13.3 **13.76 Residuals.** Once we have chosen a model, we must examine the residuals for violations of the conditions of the multiple regression model. Examine the residuals from the model in Example 13.26. HOM04

(a) Plot the residuals against SqFt. Do the residuals show a random scatter, or is there some systematic pattern?

(b) The residual plot shows three somewhat low residuals, between −$20,000 and −$40,000. Which homes are these? Is there anything unusual about these homes?

(c) Make a Normal quantile plot or histogram of the residuals. Is the distribution roughly Normal?

BEYOND THE BASICS

Regression trees

In multiple linear regression, we seek to determine a single population regression equation that holds over the entire data range. As we saw in Case 13.3, building this single regression equation can be quite complicated, especially when explanatory variables interact and relationships between the response and explanatory variables are nonlinear.

regression trees

A conceptually simple yet powerful alternative that can handle nonlinear relationships and interactions is **regression trees.** This approach splits, or partitions, the data space into smaller and smaller regions such that these smaller regions can be fit using a simple model like the mean. These regions, or **leaves,** are just like our subpopulations in regression. The difference is that there is no longer an equation directly linking these means to the explanatory variables.

leaves

Figure 13.26 contains the start of a regression tree for predicting house price using all 504 homes in Case 13.3 and the explanatory variables SqFt, BedRooms, Baths, Garage, and Zip. The figure shows the tree after three splits. The data were first split using SqFt < 2360. This node of 413 cases was then split using SqFt < 1674. Finally the node with SqFt < 1674 (251 cases) was split using Baths < 1.5. The Difference statistic reported in the nodes that split represent the difference in the means of the resulting nodes. For example, in the node with SqFT < 1674,

$$\text{difference} = 114{,}552.05 - 80{,}065 = 34{,}487.05$$

These splits result in a model with four leaves (i.e., four predicted values) that has $R^2 = 62.6\%$ and pooled standard error $s_p = \$53{,}248$.

pooled standard error, p. 467

These partitions are determined by searching each explanatory variable within each node for a split value that minimizes the sum of squared error. The LogWorth statistic in the nodes that split is related to the magnitude of this change. You can see that change is getting smaller and smaller with each split.

Fortunately, there are fast, reliable algorithms to do this searching for us. Once we have the tree, the predicted values are obtained by simply moving down the tree based on the observed explanatory variables. The mean value in the resulting leaf serves as the estimate. In Figure 13.26, for example, we see that for cases where Baths = 1 and SqFt < 1674, the predicted value is $80,065.

FIGURE 13.26 JMP output of the regression tree for predicting price after three splits of the entire Case 13.3 data set.

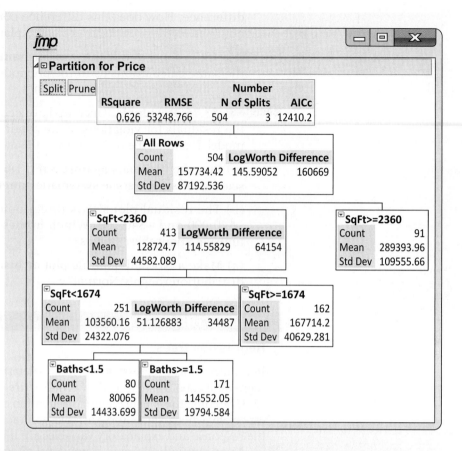

cross-validation

validation sample

Given the ease of splitting, it is not difficult to create overcomplex trees that do not generalize well. In practice, it is common to use **cross-validation** to avoid this overfitting. The goal of cross-validation is to assess how well a model predicts new data that were not used in the estimation. This is done by dividing the data into training and testing samples. The model is built using the training data and assessed using the testing sample. Common proportions for this split are 70% training and 30% testing. In the case of regression trees, there is often a third sample, known as the **validation sample,** that is used to determine when to stop splitting.

SECTION 13.3 SUMMARY

- Start the model building process by performing **data analysis** for multiple linear regression. Examine the distribution of the variables and the form of relationships among them.

- Note any **categorical explanatory variables** that have very few cases for some values. If it is reasonable, combine these values with other values. If not, delete these cases and examine them separately.

- For **curved relationships,** consider transformations or additional explanatory variables that will account for the curvature. Sometimes adding a **quadratic term** improves the fit.

- Examine the possibility that the effect of one explanatory variable depends upon the value of another explanatory variable. **Interactions** can be used to model this situation.

- Examine software outputs from modern **variable selection methods** to see what models of each subset size have the highest R^2.

SECTION 13.3 EXERCISES

For Exercises 13.63 to 13.65, see pages 652–653; for 13.66 and 13.67, see page 655; for 13.68 to 13.70, see page 659; for 13.71 and 13.72, see page 662; and for 13.73 to 13.76, see pages 664–665.

13.77 Quadratic models. Sketch each of the following quadratic equations for values of x between 0 and 5. Then describe the relationship between μ_y and x in your own words.

(a) $\mu_y = 10 + x + x^2$.

(b) $\mu_y = 10 - x + x^2$.

(c) $\mu_y = 10 + x - x^2$.

(d) $\mu_y = 10 - x - x^2$.

13.78 Models with indicator variables. Suppose that x is an indicator variable with the value 0 for Group A and 1 for Group B. The following equations describe relationships between the value of μ_y and membership in Group A or B. For each equation, give the value of the mean response μ_y for Group A and for Group B.

(a) $\mu_y = 5 + 10x$.

(b) $\mu_y = 10 + 5x$.

(c) $\mu_y = 15 - 5x$.

13.79 Differences in means. Verify that the coefficient of x in each part of the previous exercise is equal to the mean for Group B minus the mean for Group A. Do you think that this will be true in general? Explain your answer.

13.80 Models with interactions. Suppose that x_1 is an indicator variable with the value 0 for Group A and 1 for Group B, and x_2 is a quantitative variable. Each of the following models describes a relationship between μ_y and the explanatory variables x_1 and x_2. For each model, substitute the value 0 for x_1, and write the resulting equation for μ_y in terms of x_2 for Group A. Then substitute $x_1 = 1$ to obtain the equation for Group B, and sketch the two equations on the same graph. Describe in words the difference in the relationship for the two groups.

(a) $\mu_y = 20 + 10x_1 + 3x_2 + 2x_1x_2$.

(b) $\mu_y = 20 + 10x_1 + 3x_2 - 2x_1x_2$.

(c) $\mu_y = 20 - 10x_1 - 3x_2 + 2x_1x_2$.

13.81 Differences in slopes and intercepts. Refer to the previous exercise. Verify that the coefficient of x_1x_2 is equal to the slope for Group B minus the slope for Group A in each of these cases. Also, verify that the coefficient of x_1 is equal to the intercept for Group B minus the intercept for Group A in each of these cases. Do you think these two results will be true in general? Explain your answer.

13.82 Write the model. For each of the following situations write a model for μ_y of the form

$$\mu_y = \beta_0 + \beta_1 x_1 + \beta_2 x_2 + \cdots + \beta_k x_k$$

where k is the number of explanatory variables. Be sure to give the value of k and, if necessary, explain how each of the x's is coded.

(a) A model where the explanatory variable is a categorical variable with four possible values. (*Note:* This model is similar to our one-way ANOVA model.)

(b) A model where there are three explanatory variables. One of these is categorical with two possible values; another is categorical with three possible values. Include a term that would model an interaction of the first categorical variable and the third (quantitative) explanatory variable.

(c) A cubic regression where terms up to and including the third power of an explanatory variable are included in the model.

13.83 Brain injuries in Canadian football players. Multiple concussions have been shown to be associated with neurodegenerative diseases. In one study, the brain volumes of the left hippocampus of 53 retired Canadian football players, 25 age- and education-matched controls, and controls from the Centre for Aging and Neuroscience database were compared.[18] Using two indicator variables ($C_1 = 1, C_2 = 0$ if matched control and $C_1 = 0, C_2 = 1$ if retired football player), the researchers compared three simple linear regressions by fitting the model

$$\text{volume} = \beta_0 + \beta_1 C_1 + \beta_2 C_2 + \beta_3 \text{Age} + \beta_4 C_1 \text{Age} + \beta_5 C_2 \text{Age} + \varepsilon$$

The following table summarizes the results.

Explanatory variable	b	P-value
Intercept	−0.046	0.350
C_1	0.412	0.047
C_2	0.119	0.347
Age	−0.484	<0.001
C_1Age	0.169	0.528
C_2Age	−0.351	0.017

(a) Using what you've learned from Exercises 13.80 and 13.81, report the three linear regressions of volume versus age.

(b) Using the t tests for the individual coefficients, summarize what this model tells you about the left hippocampus volume and age across these three groups.

13.84 CEO pay and gross profits. In Exercise 12.94 (page 616), you assessed the relationship between the logarithm of a CEO's pay ratio and the logarithm of the company's gross profit per employee. These companies, however, are divided up into different industries. Is the relationship the same for each of these industries? The data set CNBC1 contains the centered values for log(RATIO) and log(PROFIT) plus two indicator variables and their interactions with the centered log(RATIO). Use them to compare the linear relationships and write a short paragraph of your findings. CNBC1

CASE 13.2 **13.85 Predicting movie revenue.** A plot of theater count versus box office revenue suggests that the relationship may be slightly curved. MOVIE

(a) Examine this question by running a regression to predict the box office revenue using the theater count and the square of the theater count. Report the relevant test statistic with its degrees of freedom and P-value, and summarize your conclusion.

(b) Now view this analysis in the framework of testing a hypothesis about a collection of regression coefficients, which you studied in Section 13.2 (page 643). The first model includes theater count and the square of theater count, while the second includes only theater count. Run both regressions and find the value of R^2 for each. Find the F statistic for comparing the models based on the difference in the values of R^2. Carry out the test and report your conclusion.

(c) Verify that the square of the t statistic that you found in part (a) for testing the coefficient of the quadratic term is equal to the F statistic that you found in part (b).

CASE 13.2 **13.86 Assessing collinearity in the movie revenue model.** Many software packages will calculate VIF values for each explanatory variable. In this exercise, you calculate the VIF values using several multiple regressions, and then use them to see if there is collinearity among the movie explanatory variables. MOVIE

(a) Use statistical software to estimate the multiple regression model for predicting Review based on LOpening and Theaters. Calculate the VIF value for Review using R^2 from this model and the formula

$$\text{VIF} = \frac{1}{1-R^2}$$

(b) Use statistical software to estimate the multiple regression model for predicting LOpening based on Theaters and Review. Calculate the VIF value for LOpening using R^2 from this model and the formula from part (a).

(c) Use statistical software to estimate the multiple regression model for predicting Theaters based on LOpening and Review. Calculate the VIF value for Theaters using R^2 from this model and the formula from part (a).

(d) Do any of the calculated VIF values indicate severe collinearity among the explanatory variables? Explain your response.

CASE 13.2 **13.87 Predicting movie revenue, continued.** Refer to Exercise 13.85. Although a quadratic relationship between total U.S. revenue and theater count provides a better fit than the linear model, it does not make sense that box office revenue would again increase for very low budgeted movies (unless you are the Syfy Channel). An alternative approach is to describe the relationship using a transformation of y. For the movie prediction model, we considered the log of y. MOVIE

(a) Generate a scatterplot of LUSRevenue versus Theaters. Does there are appear to be a linear association?

(b) Fit the model with theater count and obtain the fitted values. Backtransform these fitted values to the orginal scale by computing $e^{\hat{y}}$.

(c) Compare these results with the fitted values you obtained from the quadratic fit of Exercise 13.85. Which model do you prefer? Explain your answer.

CASE 13.2 **13.88 Predicting movie revenue: model selection.** Refer to the data set on movie revenue in Case 13.2 (page 635). In addition to the movie's opening-weekend log revenue, opening-weekend theater count, and its review, the data set also includes a column named Sequel. Sequel is 1 if the corresponding movie is a sequel, and Sequel is 0 if the movie is not a sequel. Assuming opening-weekend log revenue (LOpening) is in the model, there are eight possible regression models. For example, one model just includes LOpening; another model includes LOpening and Theaters; and another model includes LOpening, Sequel, and Theaters. Run these eight regressions and make a table giving the regression coefficients, the value of R^2, and the value of s for each regression. (If an explanatory variable is not included in a particular regression, enter a value 0 for its coefficient in the table.) Mark coefficients that are statistically significant at the 5% level with an asterisk (*). Summarize your results and state which model you prefer. MOVIE

CASE 13.2 **13.89 Effect of an outlier.** In Exercise 13.54 (page 647), we identified a movie that had much higher revenue than predicted. Remove this movie and repeat the previous exercise. Does the removal of this movie change which model you prefer? MOVIE

CHAPTER 13 REVIEW EXERCISES

13.90 Discrimination at work? A survey of 457 engineers in Canada was performed to identify the relationship of race, language proficiency, and location of training in finding work in the engineering field. In addition, each participant completed the Workplace Prejudice and Discrimination Inventory (WPDI), which is designed to measure perceptions of prejudice on the job, primarily due to race or ethnicity. The score of the WPDI ranged from 16 to 112, with higher scores indicating more perceived discrimination. The following table summarizes two multiple regression models used to predict an engineer's WPDI score. The first explanatory variable indicates whether the engineer was foreign trained ($x = 1$) or locally trained ($x = 0$). The next set of seven variables indicate race and the last six are demographic variables.

	Model 1		Model 2	
Explanatory variables	b	$s(b)$	b	$s(b)$
Foreign trained	0.55	0.21	0.58	0.22
Chinese			0.06	0.24
South Asian			−0.06	0.19
Black			−0.03	0.52
Other Asian			−0.38	0.34
Latin American			0.20	0.46
Arab			0.56	0.44
Other (not white)			0.05	0.38
Mechanical	−0.19	0.25	−0.16	0.25
Other (not electrical)	−0.14	0.20	−0.13	0.21
Masters/PhD	0.32	0.18	0.37	0.18
30–39 years old	−0.03	0.22	−0.06	0.22
40 or older	0.32	0.25	0.25	0.26
Female	−0.02	0.19	−0.05	0.19
R^2	0.10		0.11	

(a) The F statistics for these two models are 7.12 and 3.90, respectively. What are the degrees of freedom and P-value of each statistic?

(b) The F statistics for the multiple regressions are highly significant, but the R^2 are relatively low. Explain to a statistical novice how this can occur.

(c) Do foreign-trained engineers perceive more discrimination than do those trained locally? To address this, test if the first coefficient in each model is equal to zero. Summarize your results.

(d) Use the results of the two models to test whether the collection of explanatory variables on race contribute to explaining the response, given the variables on foreign training and demographics are already in the model.

CASE 10.1 **13.91 Education and income.** Recall Case 10.1 (page 506), which looked at the relationship between an entrepreneur's log income and level of education. In addition to the level of education, the entrepreneur's age and a measure of his or her perceived control of the environment (locus of control) was also obtained. The larger the locus of control, the more in control one feels. ENTRE1

(a) Write the model that you would use for a multiple regression to predict log income from education, locus of control, and age.

(b) What are the parameters of your model?

(c) Run the multiple regression and give the estimates of the model parameters.

(d) Find the residuals and examine their distribution. Summarize what you find.

(e) Plot the residuals versus each of the explanatory variables. Describe the plots. Does your analysis suggest that the model assumptions may not be reasonable for this problem?

CASE 10.1 **13.92 Education and income, continued.** Refer to the previous exercise. Provided the data meet the requirements of the multiple regression model, we can now perform inference. ENTRE1

(a) Test the hypothesis that the coefficients for education, locus of control, and age are all zero. Give the test statistic with degrees of freedom and the P-value. What do you conclude?

(b) What is the value of R^2 for this model and data? Interpret what this numeric summary means to someone unfamiliar with it.

(c) Give the results of the hypothesis test for the coefficient for education. Include the test statistic, degrees of freedom, and the P-value. Do the same for the other two variables. Summarize your conclusions from these three tests.

13.93 Compare regression coefficients. Again refer to Exercise 13.91. ENTRE1

(a) In Example 10.5 (page 515), parameter estimates for the model that included just EDUC were obtained. Compare those parameter estimates with the ones obtained from the full model that also includes age and locus of control. Describe any changes.

(b) Consider a 36-year-old entrepreneur with 12 years of education and a locus of control of −0.25. Compare the predicted log incomes based on the full model and the model that includes only education level.

(c) In Example 12.12 (page 608), we computed r^2 for the model that included only education level. It was 0.0573.

Use r^2 and R^2 to test whether age and locus of control together are helpful predictors, given Educ is already in the model.

13.94 Business-to-business (B2B) marketing. A group of researchers were interested in determining the likelihood that a business currently purchasing office supplies via a catalog would switch to purchasing from the website of the same supplier. To do this, they performed an online survey using the business clients of a large Australian-based stationery provider with both a catalog and a Web-based business.[19] Results from 1809 firms, all currently purchasing via the catalog, were obtained. The following table summarizes the regression model.

Variable	b	t
Staff interpersonal contact with catalog	−0.08	3.34
Trust of supplier	0.11	4.66
Web benefits (access and accuracy)	0.08	3.92
Previous Web purchases	0.18	8.20
Previous Web information search	0.08	3.47
Key catalog benefits (staff, speed, security)	−0.08	3.96
Web benefits (speed and ease of use)	0.36	3.97
Problems with Web ordering and delivery	−0.06	2.65

(a) The F statistic is reported to be 78.15. What degrees of freedom are associated with this statistic?

(b) This F statistic can be expressed in terms of R^2 as

$$F = \left(\frac{n-k-1}{k}\right)\left(\frac{R^2}{1-R^2}\right)$$

Use this relationship to determine R^2.

(c) **The coefficients listed in the table are standardized coefficients.** These are obtained when each variable is standardized (subtract its mean, divide by its standard deviation) prior to fitting the regression model. These coefficients then represent the change in standard deviations of y for a one standard deviation change in x. This typically allows one to determine which independent variables have the greatest effect on the dependent variable. Using this idea, what are the top two variables in this analysis?

Exercises 13.95 through 13.98 use the PPROMO data set shown in Table 13.6.

13.95 Discount promotions at a supermarket. How does the frequency that a supermarket product is promoted at a discount affect the price that customers expect to pay for the product? Does the percent reduction also affect this expectation? These questions were examined by researchers in a study that used 160 subjects. The treatment conditions corresponded to the number of promotions (one, three, five, or seven) that were described during a 10-week period and the percent that the product was discounted (10%, 20%, 30%, and 40%). Ten students were randomly assigned to each of the 4 × 4 = 16 treatments.[20] PPROMO

TABLE 13.6 Expected price data

Number of promotions	Percent discount	Expected price ($)									
1	40	4.10	4.50	4.47	4.42	4.56	4.69	4.42	4.17	4.31	4.59
1	30	3.57	3.77	3.90	4.49	4.00	4.66	4.48	4.64	4.31	4.43
1	20	4.94	4.59	4.58	4.48	4.55	4.53	4.59	4.66	4.73	5.24
1	10	5.19	4.88	4.78	4.89	4.69	4.96	5.00	4.93	5.10	4.78
3	40	4.07	4.13	4.25	4.23	4.57	4.33	4.17	4.47	4.60	4.02
3	30	4.20	3.94	4.20	3.88	4.35	3.99	4.01	4.22	3.70	4.48
3	20	4.88	4.80	4.46	4.73	3.96	4.42	4.30	4.68	4.45	4.56
3	10	4.90	5.15	4.68	4.98	4.66	4.46	4.70	4.37	4.69	4.97
5	40	3.89	4.18	3.82	4.09	3.94	4.41	4.14	4.15	4.06	3.90
5	30	3.90	3.77	3.86	4.10	4.10	3.81	3.97	3.67	4.05	3.67
5	20	4.11	4.35	4.17	4.11	4.02	4.41	4.48	3.76	4.66	4.44
5	10	4.31	4.36	4.75	4.62	3.74	4.34	4.52	4.37	4.40	4.52
7	40	3.56	3.91	4.05	3.91	4.11	3.61	3.72	3.69	3.79	3.45
7	30	3.45	4.06	3.35	3.67	3.74	3.80	3.90	4.08	3.52	4.03
7	20	3.89	4.45	3.80	4.15	4.41	3.75	3.98	4.07	4.21	4.23
7	10	4.04	4.22	4.39	3.89	4.26	4.41	4.39	4.52	3.87	4.70

(a) Plot the expected price versus the number of promotions. Do the same for expected price versus discount. Summarize the results.

(b) These data come from a designed experiment with an equal number of observations for each promotion by discount combination. Find the means and standard deviations for expected price for each of these combinations. Describe any patterns that are evident in these summaries.

(c) Using your summaries from part (b), make a plot of the mean expected price versus the number of promotions for the 10% discount condition. Connect these means with straight lines. On the same plot, add the means for the other discount conditions. Summarize the major features of this plot.

13.96 Run the multiple regression. Refer to the previous exercise. Run a multiple regression using promotions and discount to predict expected price. Write a summary of your results. PPROMO

13.97 Residuals and other models. Refer to the previous exercise. Analyze the residuals from your analysis, and investigate the possibility of using quadratic and interaction terms as predictors. Write a report recommending a final model for this problem with a justification for your recommendation. PPROMO

13.98 Can we generalize the results? The subjects in this experiment were college students at a large Midwest university who were enrolled in an introductory management course. They received the information about the promotions during a 10-week period during their course. Do you think that these facts about the data would influence how you would interpret and generalize the results? Write a summary of your ideas regarding this issue. PPROMO

13.99 Determinants of innovation capability. A study of 367 Australian small/medium enterprise (SME) firms looked at the relationship between perceived innovation marketing capability and two marketing support capabilities, market orientation and management capability. All three variables were measured on the same scale such that a higher score implies a more positive perception.[21] Given the relatively large sample size, the researchers grouped the firms into three size categories (micro, small, and medium) and analyzed each separately. The following table summarizes the results.

Explanatory variable	Micro $n=108$		Small $n=173$		Medium $n=86$	
	b	$s(b)$	b	$s(b)$	b	$s(b)$
Market orientation	0.69	0.08	0.47	0.06	0.37	0.12
Management capability	0.14	0.08	0.39	0.06	0.38	0.12
F statistic	87.6		117.7		37.2	

(a) For each firm size, test if these two explanatory variables together are helpful in predicting the perceived level of innovation capability. Make sure to specify degrees of freedom.

(b) Using the table, test if each explanatory variable is a helpful predictor given that the other variable is already in the model.

(c) Using the table and your results to parts (a) and (b), summarize the relationship between innovation capability and the two explanatory variables. Are there any differences in this relationship across different-sized firms?

13.100 Are separate analyses needed? Refer to the previous exercise. Suppose you wanted to generate a similar table but have it based on results from only one multiple regression rather than on three.

(a) Describe what additional explanatory variables you would need to include in your regression model and write out the model.

(b) In the actual table, the importance (b coefficient) of marketing orientation appears to decrease as the firm size increases. Based on your model in part (a), describe an F test to see if marketing coefficient is different across the three firm sizes.

(c) Explain why this F test is more appropriate compared to using t tests to compare each pair of coefficients.

13.101 Game-day spending. Game-day spending (ticket sales and food and beverage purchases) is critical for the sustainability of many professional sports teams. In the National Hockey League (NHL), nearly half the franchises generate more than two-thirds of their annual income from game-day spending. Understanding and possibly predicting this spending would allow teams to respond with appropriate marketing and pricing strategies. To investigate this possibility, a group of researchers looked at data from one NHL team over a three-season period ($n = 123$ home games).[22] The following table summarizes the multiple regression used to predict ticket sales.

Explanatory variables	b	t
Constant	12,493.47	12.13
Division	−788.74	−2.01
Nonconference	−474.83	−1.04
November	−1800.81	−2.65
December	−559.24	−0.82
January	−925.56	−1.54
February	−35.59	−0.05
March	−131.62	−0.21
Weekend	2992.75	8.48
Night	1460.31	2.13
Promotion	2162.45	5.65
Season 2	−754.56	−1.85
Season 3	−779.81	−1.84

(a) Which of the explanatory variables significantly aid prediction in the presence of all the explanatory variables? Show your work.

(b) The overall F statistics was 13.59. What are the degrees of freedom and P-value of this statistic?

(c) The value of R^2 is 0.52. What percent of the variance in ticket sales is explained by these explanatory variables?

(d) The constant predicts the number of tickets sold for a nondivisional conference game with no promotions, played during the day during the week in October during Season 1. What is the predicted number of tickets sold for a divisional conference game with no promotions played on a weekend evening in March during Season 3?

(e) Would a 95% confidence interval for the mean response or a 95% prediction interval be more appropriate to include with your answer to part (d)? Explain your reasoning.

13.102 Correlations may not be a good way to screen for multiple regression predictors. We use a constructed data set in this problem to illustrate this point. DSETA

(a) Find the correlations between the response variable Y and each of the explanatory variables X_1 and X_2. Plot the data and run the two simple linear regressions to verify that no evidence of a relationship is found by this approach. Some researchers would conclude at this point that there is no point in further exploring the possibility that X_1 and X_2 could be useful in predicting Y.

(b) Analyze the data using X_1 and X_2 in a multiple regression to predict Y. The fit is quite good. Summarize the results of this analysis.

(c) What do you conclude about an analytical strategy that first looks at one candidate predictor at a time and selects from these candidates for a multiple regression based on some threshold level of significance?

13.103 The multiple regression results do not tell the whole story. We use a constructed data set in this problem to illustrate this point. DSETB

(a) Run the multiple regression using X_1 and X_2 to predict Y. The F test and the significance tests for the coefficients of the explanatory variables fail to reach the 5% level of significance. Summarize these results.

(b) Now run the two simple linear regressions using each of the explanatory variables in separate analyses. The coefficients of the explanatory variables are statistically significant at the 5% level in each of these analyses. Verify these conclusions with plots and correlations.

(c) What do you conclude about an analytical strategy that looks only at multiple regression results?

Exercises 13.104 through 13.110 use the CROPS data file, which contains the U.S. yield (bushels/acre) of corn and soybeans from 1957 to 2018.[23]

13.104 Corn yield varies over time. Run the simple linear regression using year to predict corn yield. CROPS

(a) Summarize the results of your analysis, including the significance test results for the slope and R^2 for this model.

(b) Analyze the residuals with a Normal quantile plot. Is there any indication in the plot that the residuals are not Normal?

(c) Plot the residuals versus soybean yield. Does the plot indicate that soybean yield might be useful in a multiple linear regression with year to predict corn yield? Explain your answer.

13.105 Can soybean yield predict corn yield? Run the simple linear regression using soybean yield to predict corn yield. CROPS

(a) Summarize the results of your analysis, including the significance test results for the slope and R^2 for this model.

(b) Analyze the residuals with a Normal quantile plot. Is there any indication in the plot that the residuals are not Normal?

(c) Plot the residuals versus year. Does the plot indicate that year might be useful in a multiple linear regression with soybean yield to predict corn yield? Explain your answer.

13.106 Use both predictors. From the previous two exercises, we conclude that year *and* soybean yield may be useful together in a model for predicting corn yield. Run this multiple regression. CROPS

(a) Explain the results of the ANOVA F test. Give the null and alternative hypotheses, the test statistic with degrees of freedom, and the P-value. What do you conclude?

(b) What percent of the variation in corn yield is explained by these two variables? Compare it with the percent explained in the simple linear regression models of the previous two exercises.

(c) Give the fitted model. Why do the coefficients for year and soybean yield differ from those in the previous two exercises?

(d) Summarize the significance test results for the regression coefficients for year and soybean yield.

(e) Give a 95% confidence interval for each of these coefficients.

(f) Plot the residuals versus year and versus soybean yield. What do you conclude?

(g) There is one case that is not predicted well with this model. What year is it? Remove this case and refit the model. Compare the estimated parameters with the results from part (c). Does this case appear to be influential? Explain your answer.

13.107 Try a quadratic. We need a new variable to model the curved relation that we see between corn yield and year in the residual plot of the last exercise. Let year2 = (year − 1985)^2. (When adding a squared term to a multiple regression model, we sometimes subtract the mean of the variable being squared before squaring. This eliminates the correlation between the linear and quadratic terms in the model and thereby reduces collinearity.) CROPS

(a) Run the multiple linear regression using year, year2, and soybean yield to predict corn yield. Give the fitted regression equation.

(b) Give the null and alternative hypotheses for the ANOVA F test. Report the results of this test, giving the test statistic, degrees of freedom, P-value, and conclusion.

(c) What percent of the variation in corn yield is explained by this multiple regression? Compare this with the model in the previous exercise.

(d) Summarize the results of the significance tests for the individual regression coefficients.

(e) Analyze the residuals and summarize your conclusions.

13.108 Compare models. Run the model to predict corn yield using year and the squared term year2 defined in the previous exercise. CROPS

(a) Summarize the significance test results.

(b) The coefficient for year2 is not statistically significant in this run, but it was highly significant in the model analyzed in the previous exercise. Explain how this can happen.

(c) Obtain the fitted values for each year in the data set, and use these to sketch the curve on a plot of the data. Plot the least-squares line on this graph for comparison. Describe the differences between the two regression functions. For what years do they give very similar fitted values? For what years are the differences between the two relatively large?

13.109 Do a prediction. Use the simple linear regression model with corn yield as the response variable and year as the explanatory variable to predict the corn yield for the year 2014, and give the 95% prediction interval. Also, use the multiple regression model where year and year2 are both explanatory variables to find another predicted value with the 95% interval. Explain why these two predicted values are so different. The actual yield for 2014 was 167.4 bushels per acre. How well did your models predict this value? CROPS

13.110 Predict the yield for another year. Repeat the previous exercise doing the prediction for 2020. Compare the results of this exercise with the previous one. Also explain why the predicted values are beginning to differ more substantially. CROPS

13.111 Predicting U.S. movie revenue. Refer to Case 13.2 (page 635). The data set MOVIE contains several other explanatory variables that are available at the time of release that we did not consider in the examples and exercises. These include

- Budget: The amount designated to the making of the movie.

- Minutes: The length of the movie in minutes.

- Ratings: The IMDb rating of the movie based on members opinions.

- Sequel: A variable indicating if the movie is a sequel or not.

Using these explanatory variables and Opening, Budget, and Theaters, determine the best model for predicting U.S. revenue. MOVIE

13.112 Price-fixing litigation. Multiple regression is sometimes used in litigation. In the case of *Cargill, Inc. v. Hardin*, the prosecution charged that the cash price of wheat was manipulated in violation of the Commodity Exchange Act. In a statistical study conducted for this case, a multiple regression model was constructed to predict the price of wheat using three supply-and-demand explanatory variables.[24] Data for 14 years were used to construct the regression equation, and a prediction for the suspect period was computed from this equation. The value of R^2 was 0.989.

(a) The fitted model gave the predicted value $2.136 with standard error $0.013. Express the prediction as an interval. (The degrees of freedom were large for this analysis, so use 100 as the df to determine t^*.)

(b) The actual price for the period in question was $2.13. The judge decided that the analysis provided evidence that the price was not artificially depressed, and the opinion was sustained by the court of appeals. Write a short summary of the results of the analysis that relate to the decision and explain why you agree or disagree with it.

13.113 Predicting CO_2 emissions. The data set CO2MPG contains an SRS of 200 passenger vehicles sold in Canada in 2014. There appears to be a quadratic relationship between CO_2 emissions and mile per gallon highway (MPGHwy). CO2MPG

(a) Create two new centered variables MPG = MPGHwy − 35 and MPG2 = MPG × MPG and fit a quadratic regression for each fuel type (FuelType). Create a table of parameter estimates and comment on the similarities and differences in the coefficients across fuel types.

(b) Create three indicator variables for fuel type, three interaction variables between MPG and each of the indicators, and three interaction variables between MPG2 and each the indicator variables. Fit this model to the entire data set. Use the estimated coefficients to construct the quadratic equation for each of the fuel types. How do they compare to the equations in part (a)?

13.114 Prices of homes. Consider the data set used for Case 13.3 (page 650). This data set includes information for several other zip codes. Pick a different zip code and analyze the data. Compare your results with what we found for zip code 47904 in Section 13.3. HOMES

Bruce Leighty/Getty Images

CHAPTER 14

Time Series Forecasting

Introduction

Many businesses and government agencies routinely track data over time. Quarterly sales figures, annual health benefits costs, monthly product demand, weekly production, daily stock prices, and hourly number of restaurant customers are all examples of *time series* data.

- Hilton Worldwide is one of the largest hospitality groups with brands that include Hilton, DoubleTree, and Embassy Suites. Hilton Worldwide makes annual forecasts of occupancy rates and revenue per available room. Favorable forecasts lead to decisions to add new rooms worldwide and benefit the company in terms of investor relations.

- Kimberly-Clark, whose leading brands include Kleenex and Huggies, utilizes sales data to generate forecasts that trigger shipments to stores. As a result of improved forecasting, Kimberly-Clark has reduced its cash conversion cycle, cut its total supply chain expenses, and increased gross margins.

- Federal and state agencies use time series techniques known as smoothing to highlight medium to long-term movements in hundreds of economic indicators: for example, the unemployment rate, unemployment insurance claims, housing starts, and the prime interest rate.

- Financial analysts use time series methods to model stock prices and to model the volatility of stock returns.

CHAPTER OUTLINE

14.1 Assessing Time Series Behavior

14.2 Random Walks

14.3 Basic Smoothing Models

14.4 Regression-Based Forecasting Models

675

Overview of Time Series Forecasting

> **TIME SERIES**
>
> Measurements of a variable, commonly taken at regular intervals over time, form a **time series.**

Large and small companies alike depend on forecasts of numerous time series variables to guide their business decisions and plans. In the short term, forecasting is typically used to predict demand for products or services. Demand forecasting helps in operational decisions such as establishing daily or weekly production levels or staffing. For the longer term, businesses use forecasting to make investment decisions such as determining capacity or deciding where to locate facilities. It is not uncommon for companies to provide forecasts of key quantities in their annual reports; for example, a 2020 annual report may contain forecasts for how the company will perform in 2021.

Whether the forecasts are short or long term, the first step in making reasonable future predictions is to gain an understanding of the time behavior of the key variables. We handle time series data with the same approach used in earlier chapters:

- Plot the data, then add numerical summaries.
- Identify overall patterns and deviations from those patterns.
- When the overall pattern is quite regular, use a compact mathematical model to describe it.

In this chapter, we focus on the plots and calculations that are most helpful when describing time series data. We learn to identify patterns common to time series data as well as the models that are commonly used to describe those patterns. These time-series models are fundamental tools in any predictive analytics toolkit. To recognize when patterns exist in time series data, it is useful to understand the meaning of a time series that lacks a pattern. We explore such a time series in the next section.

14.1 Assessing Time Series Behavior

When you complete this section, you will be able to:

- Use a time plot to describe the behavior of a time series.
- Explain the characteristics and nature of a stationary process versus a nonstationary process.
- Explain the characteristics and nature of a random process versus a nonrandom process.
- Perform a runs test for randomness and summarize the results.
- Interpret an autocorrelation function (ACF) as a check for randomness.
- Construct a prediction interval for a future observation of a time series demonstrating random behavior associated with a Normal distribution.

When possible, data analysis should always begin with a plot. What we choose as an initial plot can be critical. Consider the case of a manufactured part with a dimensional specification of 5 millimeters (mm) and tolerances of ±0.003.

FIGURE 14.1 Histogram of a sample of the dimensions (millimeters) of 50 manufactured parts, with tolerance limits indicated.

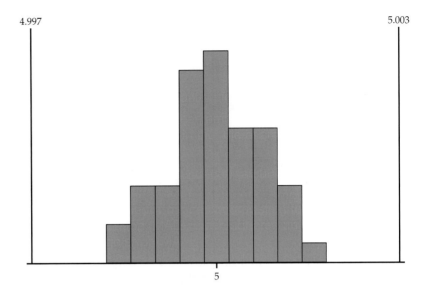

This means that any part with dimension less than 4.997 mm or greater than 5.003 mm is considered defective. Suppose 50 parts were sampled from the manufacturing process, one drawn each minute for 50 minutes. It is tempting to start the data analysis with a histogram plot such as Figure 14.1. The tolerance limits have been added to the plot. The histogram shows a fairly symmetric distribution centered near the specification of 5 mm. Relative to the tolerance limits, the histogram suggests there is little chance of producing a defective item. But before handing out a quality award to manufacturing, let us recall a *time plot*, which was in introduced in Chapter 1.

TIME PLOT

A **time plot** of a variable is a plot of each observation against the time at which it was measured. Time is marked on the horizontal scale of the plot, and the variable you are studying is marked on the vertical scale. Connecting sequential data points by lines helps emphasize any change over time.

We have previously seen an example of a time plot in Figure 1.12 (page 21). There we saw that T-bill interest rates show a series in which successive observations are close together in value, and as a result, the series exhibits a "meandering" or "snakelike" type of movement. Figure 14.2 is a time plot of the 50 manufactured parts plotted in order of manufacture and exhibits a general upward trend over time.

Compared to the histogram, this graph gives a dramatically different conclusion about the manufacturing process. The dimension is growing in size over time. In fact, if the pattern continues, we would expect the process to produce many defective parts. For this process, the histogram is misleading. In general, a histogram is used to describe a *single* population, but the population is changing over time here. The moral of the story is clear. *The first step in the analysis of time series data should always be the construction of a time plot.*

Time series can exhibit a variety of patterns over time. We explore many of these patterns in this chapter and learn how to exploit them for forecasting purposes. But before we do so, let us explore a special type of process that is patternless.

FIGURE 14.2 Time plot of the 50 manufactured parts in order of manufacture along with tolerance limits.

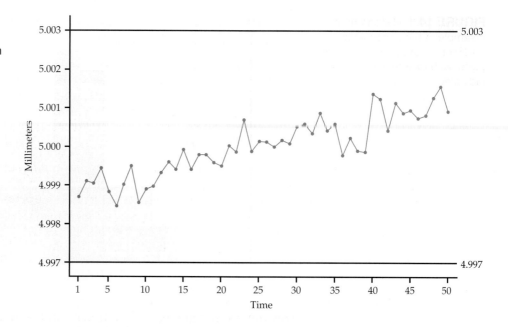

ADIDAS

Adidas Stock Price Returns Whether it is by smartphone, laptop, or tablet, investors around the world are habitually checking the ups and downs of individual stock prices or indices (like the Dow Jones or S&P 500). Even the most casual investor knows that stock prices constantly change over time. If the changes are large enough, investors can make a fortune or suffer a great loss.

CASE 14.1

Figure 14.3 plots the weekly stock price returns of Adidas stock (traded in the Frankfurt Stock Exchange) as a percent from the beginning of January 2016 through the first week of July 2018.[1] To add perspective, a horizontal line at the average has been superimposed on the plot. This line is often referred to as a *centerline*. The centerline serves a couple of purposes. First, it provides a tangible feel for the level of the process data. Second, it serves as a reference line to aid our eyes in determining whether systematic patterns exist in the data. There are three distinct features of the returns data that can be singled out:

1. The general (overall) level of the series stays constant. The centerline at the average of 0.664881% is a good description of the general level of the series from beginning to end.

FIGURE 14.3 Time plot of weekly returns of Adidas stock (January 2016 through July 2018).

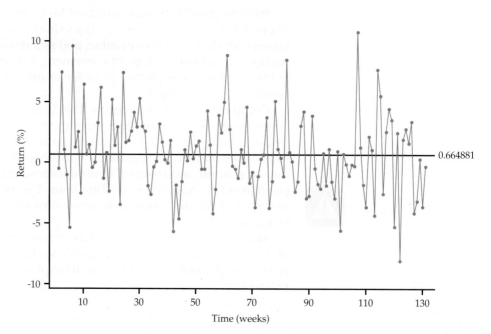

2. The scattering of the observations around the centerline—variation of the data—appears about the same throughout the series.

3. Sometimes, consecutive returns bounce from one side of the average to the other side, while at other times, they form short "strings" remaining on one side of the average. However, overall, there is no persistent pattern in the sequence of consecutive observations. The fact that a return is above or below the average seems to provide no insight as to whether future returns will fall above or below the average. ∎

stationary process

random process

Features 1 and 2—constant general level and variation—define a **stationary process.** If, in addition, the sequence of observations is unpredictable (feature 3), we say that the process is a **random process.** Thus, a random process is a stationary process. However, as we will see by illustration, a stationary process is not necessarily a random process.

Random process model

Moving forward, it will be convenient to use the subscript t to represent the time period. As an example, for a time series y, y_1 represents the value of the series at period 1 and, likewise, y_2 represents the value of the series at period 2. Basically, t is a time index for the periods. Periods are regular time intervals such as hours, shifts, days, weeks, months, quarters, years, and so on.

A random process, like the Adidas returns series of Case 14.1, can be represented as follows:

$$y_t = \mu + \varepsilon_t$$

The deviations ε_t represent "noise" that prevents us from observing the value of μ. These deviations are independent, with mean 0 and standard deviation σ. This representation of y observations varying randomly around an underlying mean μ due to random deviations was first introduced in Section 7.4. The only distinction here is that the random deviations occur over time as reflected by the subscript t.

random deviation, p. 384

The more technical statistical jargon for random is independently and identically distributed or, simply, **iid.** Independence suggests that the individual process observations taken at different time points are unrelated. Thus, knowing that certain observations are high or low is not suggestive that other observations (past or future) are high or low. The other phrase of "identically distributed" means that the underlying distribution of process outcomes is the same for each time period. Random or iid processes do not have to be associated with the Normal distribution. The requirement for iid behavior is that the distribution, whatever it may be, remains the same over time. In the important case of a process being iid and Normal, it can be abbreviated as *iidn*.

iid

Nonrandom processes

Now that we have experienced a time plot of a random process, let us return to processes, like the manufacturing one in Figure 14.2, that clearly are not exhibiting themselves as random. There are many patterns that are nonrandom and they are shown in Figure 14.4. In attempting to characterize the nature of a nonrandom process, we need to be on the lookout for one or more of the following violations of a random process:

Violation 1: The overall level of the process does not remain constant over time.

Violation 2: There is a persisting pattern in the process. As an informal test, if a centerline does not seem to serve as a reasonable prediction

FIGURE 14.4 Time plots of nonrandom processes.

for any individual observation in the future, then this is suggestive of a persisting pattern.

Violation 3: The scattering of points relative to the general movement of the series is not constant.

If any one of these violations occur, we say that the process is "not random" or "nonrandom." These common phrases/terms do not suggest that the process is not subject to random variation. But rather, they suggest a process behavior other than *pure* random variation which is constant in mean and variance over time. The more technical jargon for a nonrandom process is non-iid.

For the series shown in the first row of Figure 14.4, both violations 1 and 2 apply. In both cases, the overall process level is not constant but rather it is systematically moving away (upward or downward) as time passes. Unlike a random process, a horizontal centerline would not be a useful guide for predicting future behavior of this process; instead, future observations are expected to be farther away from the centerline. Such a nonrandom process

trend is said to exhibit a **trend**. *Do not confuse the term "trending" to mean upward as suggested in the context of news or social media topics that are gaining popularity.* The nonconstant overall level of a trend process implies that it is a

nonstationary process **nonstationary process.** Notice that the variation around the systematic trend is similarly dispersed, which implies that violation 3 is not at work here.

meandering The plots in the second row of Figure 14.4 are characterized as being **meandering.** The term "meandering" is meant to suggest the image of a winding river; the metaphor is not perfect since a river can loop backward while a time series cannot. The appearance of this image results because successive observations tend to be close together in value. With this in mind, for either plot, ask yourself "Given the values of past observations, is the centerline a reasonable prediction for the next observation in the future?" The answer is definitely "no." Instead, a more reasonable prediction for the next observation would be a value close to the most recently observed value of the time series. Thus, both of these meandering series are nonrandom in terms of violation 2.

For each of the processes shown in the second row of Figure 14.4, the variation around the meandering pattern remains about the same from the beginning to the end of the time plot. The primary difference between the two processes is in the long-run implications. Namely, the general level of the first (left graph) process "averages out" to some constant level, while the general level of the second (right graph) process is not constant in that the series appears to be "running" away. Hence, the first nonrandom process is stationary, while the second nonrandom process is nonstationary.

Consider now the bottom left plot of Figure 14.4. We see a process that tends to bounce from one side of the center line to the other. Given the oscillating pattern of the data, we would predict the next observation in the future to be above the centerline if the most recent observation is below the centerline, and vice-versa. As such, we would designate the process as nonrandom based on violation 2. The process is stationary since it appears that the average level remains constant at the center line along with the fact that the scattering of points around the center line seems fairly constant.

Finally, the bottom right plot reflects a process for which dispersion around the center line is steadily increasing. Note that neither violation 1 nor violation 2 is at hand since the overall process level is keeping steady and the successive observations appear unpredictable; the systematic pattern is related to the variation, not the level of the process. The best guess for the future outcomes is at the centerline but with increasing uncertainty as one projects farther into the future. The nonconstant variation indicates that this process can also be designated as nonstationary.

With Figure 14.4, we have illustrated only a handful of nonrandom processes. There are, of course, many more possible nonrandom scenarios. Some other examples can be created simply by combining two or more scenarios found in Figure 14.4. For instance, it is quite possible to have a trend process with steadily increasing variation around the trend line as time evolves. An important type of nonrandom behavior not illustrated in Figure 14.4 is **seasonality**. Seasonality is commonly due to the effects of weather (e.g., energy consumption, agricultural seasons, winter vs. summer clothing spending) and due to calendar- or schedule-related factors such as the timing of holidays (e.g., Christmas spending), of school breaks, and of accounting periods. Seasonality is not restricted to monthly or quarterly time intervals; for example, a transit system or a restaurant observes seasonal patterns depending on the day or the hour of the day.

Let us now turn to an example of a time series exhibiting a combination of nonrandom effects.

AMAZON

Amazon Sales Once just an online bookseller, Amazon has been continually expanding its business from providing a wide selection of consumer goods for online shoppers to providing services and technology for businesses to build their own online operations. Figure 14.5 displays a time plot of Amazon's quarterly sales (in millions of dollars), beginning with the first quarter of 2010 and ending with the second quarter of 2018.[2] To help in our interpretation, the four quarters of each year are numbered. Breaking down the plot, we find:

- Sales are steadily trending upward with time at an increasing rate.
- Sales show strong seasonality, with the fourth quarter being the strongest.

FIGURE 14.5 Time plot of Amazon quarterly sales (first quarter 2010 through second quarter 2018).

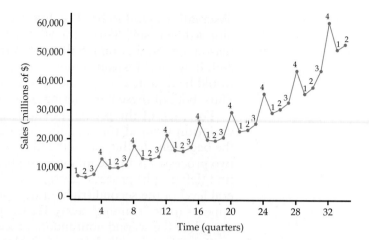

- The variation of sales appears to increase with time. This is most evident with the fourth quarter in that the amount of increase from the third quarter to the fourth quarter increases over time. As a confirmation of this increase, below are the sales changes by year from the third to the fourth quarter.

Year	Third- to fourth-quarter change
2010	5,388
2011	6,555
2012	7,462
2013	8,495
2014	8,749
2015	10,389
2016	11,027
2017	16,709

 The strong fourth-quarter seasonality here is clearly related to the holiday shopping season from Thanksgiving to Christmas. *However, be aware that companies can follow different fiscal periods. For example, the first quarter of Apple, Inc., spans the holiday shopping season.* ∎

Our visual inspection of the Amazon sales series reveals a series influenced by both trend and seasonal components. In this chapter, we learn how to use the regression methods of Chapters 12 and 13 to estimate each of the components embedded in the time series with the goal of building predictive models for the time series. In the process of estimating the trend and seasonal components of the Amazon series, we later reveal that the series is also influenced by a meandering pattern.

In summary, we have seen that a time plot is a fundamental tool for establishing a basic impression of underlying process behavior. With this said, before proceeding to the modeling of a time series, let us first add two more tools to the data analysis toolkit for assessing the behavior of a time series.

Runs test

One simple numerical check for randomness of a time series is a runs test. As a first step, the runs test classifies each observation as being above (+) or below (−) some reference value such as the sample mean. Based on these classifications, a run is a string of consecutive pluses or minuses. To illustrate, suppose that we observe the following sequence of 10 observations:

5 6 7 8 13 14 15 16 17 19

What do you notice about the sequence? It is distinctly nonrandom because the observations increase in value. Assigning + and − symbols relative to the sample mean of 12, we have the following:

$$- - - - + + + + +$$

In this sequence, there are two runs: − − − − and + + + + +. The trending values result in runs of longer length which, in turn, leave us with a very small number of runs. Imagine now a very different nonrandom sequence of 10 observations in which consecutive observations oscillate from one side of the mean to the other side. In such a case, there would be 10 runs, each of length one (either + or −).

These extreme examples bring out the essence of the runs test. Namely, if we observe too few runs or too many runs, then we should suspect that the process is not random. In hypothesis-testing framework, we are considering the following competing hypotheses:

H_0: Observations arise from a random process.
H_a: Observations arise from a nonrandom process.

We utilize the following facts about the number of runs to test the hypothesis of a random process.

RUNS TEST FOR RANDOMNESS

For a sequence of n observations, let n_A be the number of observations above the mean, and let n_B be the number of observations below or equal to the mean. If the underlying process generating the observations is random, then the number of runs statistic R has mean

$$\mu_R = \frac{2 n_A n_B + n}{n}$$

and standard deviation

$$\sigma_R = \sqrt{\frac{2 n_A n_B (2 n_A n_B - n)}{n^2 (n-1)}}$$

The **runs test** rejects the hypothesis of a random process when the observed number of runs R is far from its mean. For sequences of at least 10 observations, the runs statistic is well approximated by the Normal distribution.

EXAMPLE 14.1

CASE 14.1 **Runs Test and Stock Price Returns** In Figure 14.3 (page 678), we observed $n = 131$ weekly returns of Adidas stock. For this series, consider subtracting off the sample mean (0.664881) from each observation and denoting whether the resulting difference is positive (+) or negative (−). Figure 14.6 shows the resulting sequence of pluses and minuses with the first four runs identified. Going through the whole sequence, we find 65 runs, and we also find $n_A = 64$ observations above the sample mean and $n_B = 67$ observations below or equal to the sample mean. Given these counts, we can

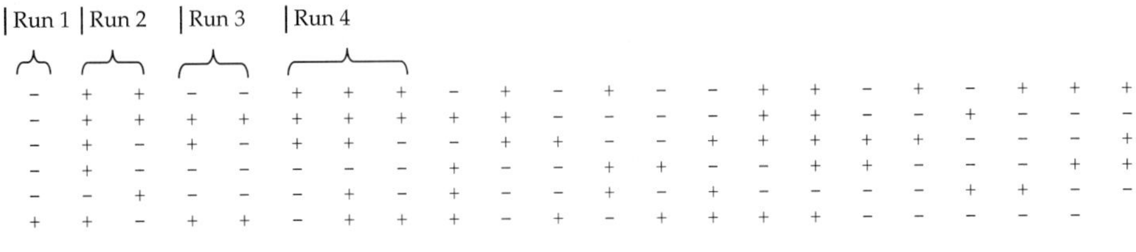

FIGURE 14.6 Counting runs in the Adidas returns series.

compute the mean and standard deviation for the number of runs under the null hypothesis of a random process. The mean is

$$\mu_R = \frac{2n_A n_B + n}{n}$$

$$= \frac{2(64)(67) + 131}{131} = 66.47$$

and the standard deviation is

$$\sigma_R = \sqrt{\frac{2n_A n_B (2n_A n_B - n)}{n^2(n-1)}}$$

$$= \sqrt{\frac{2(64)(67)[2(64)(67) - 131]}{131^2(131-1)}} = \sqrt{32.464} = 5.698$$

The 65 observed runs deviates by 1.47 from the expected number (mean) of runs, or by 0.258 standard deviations as obtained by computing the following test statistic:

$$z = \frac{R - \mu_R}{\sigma_R} = \frac{65 - 66.47}{5.698} = -0.258$$

The results can be summarized with a P-value. Because evidence against randomness is associated with either too many or too few runs, using Table A, we have a two-sided test with a P-value of

$$P = 2P(Z \geq |-0.258|) = 0.795$$

The large P-value indicates that there is very little evidence to reject the hypothesis of a random process. ∎

Figure 14.7 shows Minitab and JMP output for the runs test; JMP runs test is an easily obtained add-in. Except for a slight difference in the P-value due to rounding, we find all the values counted or computed in Example 14.1 to be the same as those found in the two software outputs.

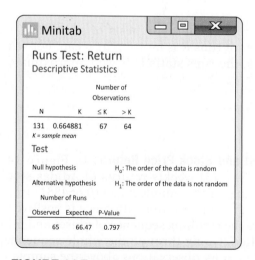
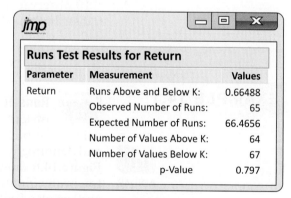

FIGURE 14.7 Minitab and JMP runs test output for the Adidas returns series.

EXAMPLE 14.2

Runs Test and Amazon Sales Figure 14.8 shows R's runs test output (using the TSA package) applied to the Amazon sales series. Here, we see that there are only six observed runs versus the expected number of 17.47059. For the R output, n1 represents the number of observations below

FIGURE 14.8 R runs test output for the Amazon sales series.

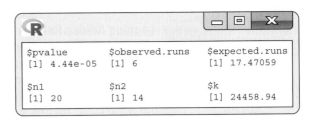

or equal to the mean and n2 represents the number of observations greater than the mean. The reported P-value of 4.44×10^{-5} shows that the evidence is extremely strong against the null hypothesis of randomness. ∎

Although it is convenient to have numerical guidance with the runs test, this and other numerical checks should not be a replacement for your visual examination of the data series. For example, the runs test does not consider the extent of how far a given observation deviates above or below the mean. Thus, the runs test will fail to signal anything unusual with a process showing unpredictability in the sequence of observations but having nonconstant variation. Or, an insignificant runs count might be the result of offsetting nonrandom behavior such as oscillation for the first half and meandering for the second half of the series.

APPLY YOUR KNOWLEDGE

CASE 14.1 **14.1 Adidas prices.** In Case 14.1 and Example 14.1, we found the weekly returns of Adidas stock to be consistent with a random process. What can we say about the time series of the weekly closing prices? ADIDAS

(a) Using software, construct a time plot of weekly prices. Is the series consistent with a random process? If not, explain the nature of the inconsistency.

(b) The sample mean of the prices is $\bar{y} = 154.336$. Count the number of runs in the series. Is your count consistent with your conclusion in part (a)?

14.2 Spam process. Most email servers keep inboxes clean by automatically moving incoming mail that is determined to be spam to a "Junk" folder. Here are the weekly counts of the number of emails moved to a Junk folder for 10 consecutive weeks:

$$11 \quad 6 \quad 17 \quad 11 \quad 10 \quad 11 \quad 3 \quad 8 \quad 11 \quad 13$$

Answer the following without the aid of software.

(a) Find the observed and expected number of runs.

(b) Determine the P-value of the hypothesis test for randomness.

Autocorrelation function

The runs test serves as a simple numerical check for the question of randomness. However, most statistical software provide a more comprehensive check of randomness based on computing the correlations among the observations over time.

The idea is that if a process is exhibiting some sort of persistent pattern over time, then the observations will potentially reflect some sort of association relative to prior observations. To check whether associations exist, we need a way of relating observations made at different time periods. We illustrate this by example.

EXAMPLE 14.3

ADIDAS

lagging

Lagging Adidas Returns Consider again the 131 weekly Adidas returns of Case 14.1. By using the established system of notation, the time series can be denoted as Return_t, where $t = 1, 2, \ldots, 131$. To compare observations with prior observations, we create new variables that take on earlier values of the original variable. This is accomplished by a process known as **lagging.**

Lagging a variable means we create another variable by shifting the data so that original observations are lined up with prior observations from a certain number of periods back. Here, for example, are the return data (rounded to the hundredths place) along with lagged variables going one and two periods back:

t	Return_t	Return_{t-1}	Return_{t-2}
1	−0.57	*	*
2	7.39	−0.57	*
3	1.01	7.39	−0.57
4	−1.09	1.01	7.39
5	−5.41	−1.09	1.01
⋮	⋮	⋮	⋮
129	0.24	−3.21	−4.18
130	−3.68	0.24	−3.21
131	−0.36	−3.68	0.24

lag variable

The variable Return_{t-1} is called a **lag one variable** and the variable Return_{t-2} is called a **lag two variable.** For any given time period, notice that the lag one variable takes on the value of the immediately preceding observation, while the lag two variable takes on the value of the observation from two periods back. The symbol * denotes a missing value. For example, there is a missing value for the lag one variable in period one because there is no available observation prior to it. ∎

Once lagged variables are created, we can more easily investigate the potential presence of associations between observations over time. For example, Figure 14.9(a) shows a scatterplot of the original observations Return_t against the lag one observations Return_{t-1}. Similarly, Figure 14.9(b) shows Return_t against Return_{t-2}. The scatterplots show no visual evidence of relationships between observations one and two lags apart. The lack of associations between observations over time is consistent with a random process.

Going beyond the visual inspection of the scatterplots, we can compute the sample correlation between Return_t and Return_{t-1} and also the sample correlation between Return_t and Return_{t-2}. Because we are not looking to compute the correlation between two arbitrary variables but rather a correlation computed from observations taken from a single time series, the computed correlation is commonly referred to as an **autocorrelation** (*self*-correlation).

autocorrelation

AUTOCORRELATION

The correlation between successive values y_{t-1} and y_t of a time series is called **lag one autocorrelation.** The correlation between values two periods apart is called **lag two autocorrelation.** In general, the correlation between values k periods apart is called **lag k autocorrelation.**

14.1 Assessing Time Series Behavior

FIGURE 14.9 (a) Scatterplot of returns versus lag one returns and (b) scatterplot of returns versus lag two returns.

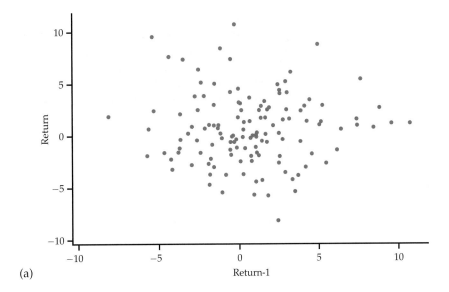

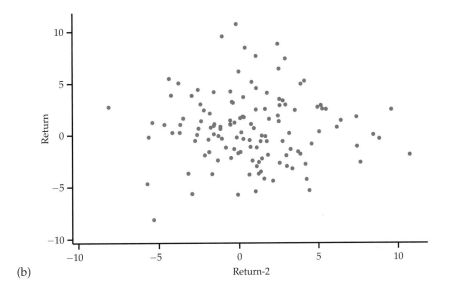

EXAMPLE 14.4

ADIDAS

correlation test, p. 588

CASE 14.1 **Lag One and Lag Two Autocorrelations for Adidas Returns** Figure 14.10 shows the sample lag one and lag two autocorrelations, computed by Minitab. The output also provides the P-values for testing the null hypothesis $H_0: \rho = 0$ that the population correlation is 0. We find for each autocorrelation, the associated P-value is not significant at the 5% level of significance. Thus, for either lag, the evidence is not strong enough to reject the null hypothesis of underlying correlation of 0. ∎

FIGURE 14.10 Minitab output of lag one and lag two correlations for Adidas returns data.

688 Chapter 14 Time Series Forecasting

autocorrelation function (ACF)

Examples 14.3 and 14.4 explored the first two lags of the Adidas returns series. We could have gone further back to create a third lag, a fourth lag, and so on. Lagging a variable and then computing the autocorrelations for different lags can be a cumbersome task. Fortunately, there is a convenient alternative offered by statistical software. In particular, software will compute the autocorrelations for several lags in one shot and then will plot the autocorrelations as a bar graph. This resulting graph is known as an **autocorrelation function (ACF)**.

EXAMPLE 14.5

ADIDAS

ACF for Adidas Returns ACF output from JMP and Minitab is given in Figure 14.11. Notice that JMP shows an autocorrelation value of 1 at lag 0. The autocorrelation at lag 0 is always 1 since this is simply a correlation of the series with itself. The lag 0 autocorrelation can be ignored. Indeed, we see that Minitab does not report it. *Be careful to note that if software reports the lag 0 autocorrelation, this simply measures the correlation of the series with itself, which is always equal to a value of 1.*

From either software output, notice that the first and second autocorrelations are plotted negative. From the JMP output, we see that the first two autocorrelations are reported to be -0.0037 and -0.0198, which are basically the values shown in Figure 14.10.[3]

What we also find with the ACF is a band of lines superimposed symmetrically around the 0 correlation value. These lines serve as a test of the significance of the autocorrelations. If a process is truly random, the underlying process autocorrelation at any lag k is theoretically 0. However, as we have learned throughout the book, sample statistics (in this case, sample autocorrelations) are subject to sampling variation.

The lines seen on the ACF attempt to incorporate sampling variability of the autocorrelation statistic. In particular, if the underlying autocorrelation at a particular lag k is truly 0, then with repeated samples from the process, we would expect 95% of the lag k sample autocorrelations to fall within the band limits. Therefore, for any given lag, the limits are an implementation of a 5% level significance test of the null hypothesis that the underlying process autocorrelation is 0. The computation of the band limits varies slightly by software. We see from Figure 14.11 that none of the sample autocorrelations breach the limits. The ACF leaves us with no suspicion against randomness, which confirms both our initial visual inspection of the time series with Figure 14.3 (page 678) and the results of the runs test in Example 14.1 (page 683). ■

A word of caution is required when using the ACF significance limits shown in Figure 14.11. *The problem is that although the probability of false rejection is 0.05 when testing any one autocorrelation, the chances are collectively higher than 0.05 that at least one of several sample autocorrelations will fall outside the limits when the process is truly random.* There are alternative tests (e.g., the Ljung-Box Q test seen in the JMP ACF output) that have been designed to simultaneously test more than one autocorrelation and to control the false rejection rate. However, these tests are topics of a more advanced treatment of time series analysis than done here. Short of advanced testing procedures, background knowledge of the time series under investigation with common sense go a long way to help minimize the possibility of overreacting. If one sample autocorrelation falls outside the ACF limits at some "oddball" lag but all the remaining autocorrelations are insignificant, there is probably good reason to resist the temptation to reject randomness.

FIGURE 14.11 JMP and Minitab autocorrelation (ACF) output for the Adidas returns series.

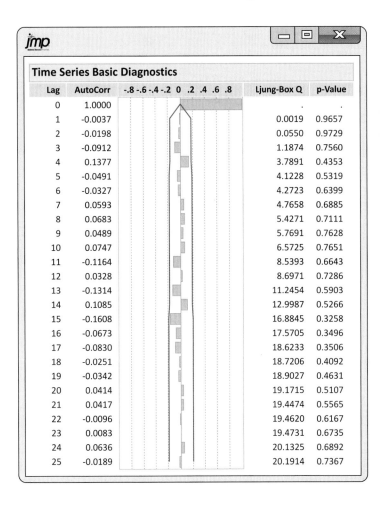

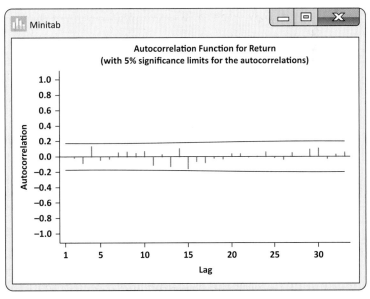

EXAMPLE 14.6

AMAZON

CASE 14.2 **ACF for Amazon Sales** Figure 14.12 displays the ACF output from R software for the Amazon sales series. In contrast to the ACF output of Figure 14.11, the ACF of Figure 14.12 leaves us with a very different impression about the time series in question. We find that many of sample autocorrelations go beyond the significance limits. Furthermore, the sample autocorrelations show a decay-like pattern. The sample autocorrelations for

FIGURE 14.12 R output of the ACF for the Amazon sales series.

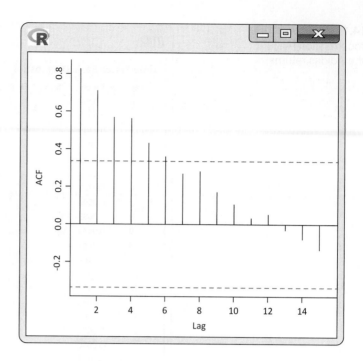

a random process are expected to "swim" between the limits with no pattern as seen in the Adidas returns ACF of Figure 14.11. In the end, the ACF for the Amazon sales series is providing us with strong evidence that the underlying process generating the observations is not random. ■

APPLY YOUR KNOWLEDGE

CASE 14.1 **14.3 Adidas prices.** Continue the study of Adidas closing prices from Exercise 14.1. ADIDAS

(a) Using software, create a lag one variable for prices. Make a scatterplot of prices plotted against the lag one variable. Describe what you see.

(b) Find the correlation between prices and their first lag. Test the null hypothesis that the underlying lag one correlation ρ is 0. What is the P-value? What do you conclude?

CASE 14.1 **14.4 Adidas prices.** Continue with the study of the Adidas price series. ADIDAS

(a) Obtain the ACF for the price series. How many autocorrelations are beyond the ACF significance limits?

(b) Aside from autocorrelations falling beyond significance limits, what else do you see in the ACF that gives evidence against randomness?

Forecasts of a random process

forecast

Our ultimate goal is to make a **forecast** of future values. For nonrandom patterned processes, the forecasts exploit the nature of the pattern. We explore the modeling and forecasting of these processes in the sections to follow. Here, we will focus on a forecast of a random process.

Given the random behavior of the weekly Adidas returns, an intuitive forecast for future values would simply be the sample mean, which is superimposed on Figure 14.3 (page 678). As introduced in Section 7.4, a single best guess of a new outcome of y is called a point prediction. It was noted (page 384) that the choice of $\hat{y} = \bar{y}$ is grounded in the least-squares criterion of minimizing the sum of squared deviations between the y values and the predicted $\hat{y}$ values.

point prediction, p. 384

Often, we are interested in going beyond a single-valued guess of a future observation to report a range of likely future observations. An interval for the prediction of an individual observation is known as a prediction interval. Establishing a prediction interval depends on the distribution of the individual observations. For a sample of n independent observations from a Normally distributed population having unknown mean μ and known standard deviation σ, a level C prediction interval for new observations of y is

prediction interval, p. 385

$$\bar{y} \pm z^{*}\sigma\sqrt{1 + \frac{1}{n}}$$

When σ is unknown, we replace it with the sample standard deviation s and use the interval

$$\bar{y} \pm t^{*}s\sqrt{1 + \frac{1}{n}}$$

where $s\sqrt{1 + \frac{1}{n}}$ is the standard error of prediction and t^{*} is the value for the $t(n-1)$ density curve with area C between $-t^{*}$ and t^{*}.

EXAMPLE 14.7

ADIDAS

CASE 14.1 **Prediction Interval for Adidas Returns** Figure 14.13 provides a histogram of the weekly Adidas returns data ($n = 131$). We can also supplement the histogram with a Normal quantile plot (not shown). We see that the data are fairly compatible with the Normal distribution. It is worth noting that stock price returns are often found to follow a symmetric distribution that is more similar to a t distribution than the Normal (i.e., more probability in the tails).

FIGURE 14.13 Histogram of Adidas returns.

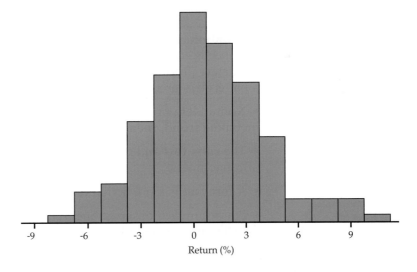

For our illustration, however, the Normal distribution is sufficient. The sample mean of the returns is $\bar{y} = 0.664881\%$. The sample standard deviation can be found to be $s = 3.33639\%$. For $C = 95\%$ and degrees of freedom $n - 1 = 130$, we can use Excel's formula of "=T.INV.2T(0.05,130)" to find $t^{*} = 1.97838$. The 95% prediction interval for future observations is

$$\bar{y} \pm t^{*}s\sqrt{1 + \frac{1}{n}} = 0.664881 \pm 1.97838(3.33639)\sqrt{1 + \frac{1}{131}}$$
$$= (-5.9609\%, \ 7.2907\%)$$

Figure 14.14 shows the JMP produced 95% prediction limits. ■

FIGURE 14.14 JMP output of 95% prediction interval limits for Adidas returns, Example 14.7.

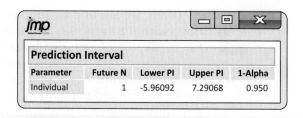

Example 14.7 describes the precise construction of the 95% prediction interval. However, given $n = 131$ is fairly large, the factor of $1/n$ contributes very little to the standard error of prediction. In addition, t^* is close to the value of 2. As such, we can reasonably approximate the prediction interval using the 68–95–99.7 rule:

68–95–99.7 rule, p. 45

$$\bar{y} \pm 2s = 0.664881 \pm 2(3.33639) = (-6.0079\%, 7.3377\%)$$

In our study of the Adidas returns, we have found that the returns behave randomly over time and are approximately Normal. As noted earlier (page 679), we can refer to such a process as approximate iidn. Thus, the prediction interval construction presented is applicable for iidn (or approximately so) processes.

This brings up a point that should be emphasized. Randomness and Normality are *distinct* concepts. A process can be random and Normal, but it can also be random and not Normal. *It is a common mistake to believe that randomness implies Normality.*

SECTION 14.1 SUMMARY

- Data collected at regular time intervals form a **time series.**

- Analysis of time series data should always begin with a **time plot.**

- A **stationary process** is a process for which the general mean and variance is constant over time. A **random process** is a special case of a stationary process. A random process is distinguished by the fact that it is a patternless process of independent observations.

- Nonrandom processes can exhibit a variety of systematic behaviors including **trend, meandering,** and **seasonality.**

- The **runs test** and the **autocorrelation function (ACF)** are statistical tests of randomness.

- A **forecast** is a prediction of a future value of the time series.

- For a sample n observations from a random process following the Normal distribution having unknown mean μ and unknown standard deviation σ, a level C **prediction interval** for a future observation of y is

$$\bar{y} \pm t^* s \sqrt{1 + \frac{1}{n}}$$

where t^* is the value for the $t(n-1)$ density curve with area C between $-t^*$ and t^*.

SECTION 14.1 EXERCISES

For Exercises 14.1 and 14.2, see page 685; and for 14.3 and 14.4, see page 690.

14.5 All same values. Suppose 50 observations in a data sequence are all equal to 20.

(a) What are the values of n_A and n_B?

(b) What would be the observed number of runs?

(c) Explain why software is unable to compute the P-value for this situation.

14.6 Annual inflation rate. Here are the annual inflation rates for the United States for the years 2004 through 2017:

Year	2004	2005	2006	2007	2008	2009	2010
Percent	2.7	3.4	3.2	2.8	3.8	−0.4	1.6

Year	2011	2012	2013	2014	2015	2016	2017
Percent	3.2	2.1	1.5	0.76	0.73	2.07	2.11

(a) Manually determine the values of n_A and n_B, and the number of runs around the sample mean.

(b) Assuming the underlying process is random, what is the expected number of runs?

(c) Using Table A, determine the P-value of the hypothesis test of randomness.

14.7 Runs test output. Here is the runs test output for a time series. Refer to Example 14.2 (page 684) for the meaning of the notations of n1 and n2.

```
$pvalue       $observed.runs    $expected.runs
[1] ???       [1] 41            [1] 38.33333

$n1           $n2               $k
[1] 35        [1] 40            [1] 106.4281
```

(a) What are the values of n_A and n_B?

(b) How many observations are in the data series?

(c) What is the sample mean of time series observations?

(d) Determine the missing P-value.

14.8 Randomness versus distribution. A Normal random process is defined as a process that generates independent observations that are well described by the Normal distribution. Consider daily data on the average waiting time (minutes) for patients at a health clinic over the course of 50 consecutive workdays. CLINIC

(a) Use statistical software to make a time plot of these data. From your visual inspection of the plot, what do you conclude about the behavior of the process?

(b) Obtain an ACF for the series. What do you conclude?

(c) Obtain a histogram and a Normal quantile plot of these data. What do these plots suggest?

(d) Based on what you learned from parts (a), (b), and (c), summarize the overall nature of the clinic waiting-time process.

14.9 Toronto Raptors point spreads. Consider the consecutive game point spreads of the NBA team Toronto Raptors for the 2017–2018 regular season.[4] RAPTORS

(a) Use statistical software to make a time plot of these data. From your visual inspection of the plot, what do you conclude about the behavior of the process?

(b) Obtain an ACF for the series. What do you conclude?

(c) Obtain a histogram and a Normal quantile plot of these data. What do these plots suggest?

(d) What is the standard error of prediction? Show computations.

(e) Assuming that the general process behavior remains the same into postseason (playoff) games, what is the 95% prediction interval for the point spread of the next game in the future? Show the computations.

14.2 Random Walks

When you complete this section, you will be able to:

- Distinguish the behavior of a random walk process from a random process.
- Identify a random walk with and without drift.
- Apply the differencing operation to break down a random walk to a random process.
- Perform a test for drift of a random walk.
- Make forecasts for a random walk with or without drift.
- Use the logarithm transformation to approximate period-to-period percent changes (returns) and to stabilize the variance of a time series.
- Distinguish the nature of a deterministic trend versus a stochastic trend.

In the previous section, we explored a random process, which serves as a good starting point in the study of time series. For random processes, the point prediction is typically chosen to be simply the sample mean $\bar{y}$. However, most time series we encounter, like the Amazon sales series, will not be random.

random walk

In such cases, we will want to apply statistical methods that adapt to the underlying patterns in the time series.

But before we move on to such methods, we consider a very special non-random process called a **random walk.** We give it special attention because it serves as a model for many economics and finance applications, especially price data from financial and commodity markets. A random walk process is written as follows:

$$y_t = \delta + y_{t-1} + \varepsilon_t$$

Notice that the above model is very different than the one given for the random process (page 679), as there is the term of y_{t-1} found on the right side of the equation. If the y variable represents a stock price, then the random walk model says that the observed price for period t is equal to a constant δ plus the previous period's price plus a random deviation. As defined earlier, these deviations are independent, with mean 0 and standard deviation σ.

Random walks exhibit curious behavior as can be seen with our next example.

EXAMPLE 14.8

Simulating Random Walks Consider a simple version of the random walk with $\delta = 0$:

$$y_t = y_{t-1} + \varepsilon_t$$

This model implies that the observed value of y at period t is simply the observed value of y at period $t - 1$ plus a random deviation. Let us assume for our study that the error deviations follow the standard Normal distribution, that is, $N(0,1)$. Furthermore, we start the time series at $y_1 = 0$.

Because the error deviation distribution is symmetrically centered on 0, there is a 50% chance of a deviation being positive or negative. This implies that there is a 50% chance that the y variable will increase or decrease from one period to the next. In that light, how might the random walk process evolve over time? It would seem the equal chance of y going up or down from period to period will result in random walk observations hovering close to the initial y value of 0.

Figure 14.15 shows three simulation runs of our random walk setting for 500 consecutive periods. Notice first that the simulated series look remarkably like stock price charts. In one case (blue line), the random walk series

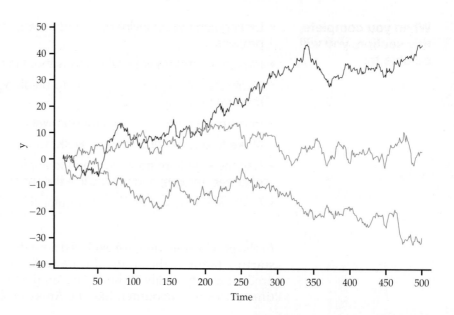

FIGURE 14.15 Three simulated random walk series for Example 14.8.

takes a nose dive downward with no indication of returning back. The second case (maroon line), on the other hand, takes off with the observations steadily increasing. Finally, in the third case (green line), the series wanders around the 0 value. With the first two cases, it would be *incorrect* to conclude that the processes will continue to trend in the same direction. Moving into the future from $t = 500$, the three random walks can move in any direction, as they did starting from the first period. ∎

It should be clear from Example 14.8 that a random walk is anything but random. Don't be confused by the word "random" in "random walk." A random walk process is a highly nonrandom process. Furthermore, given there is no guarantee that a random walk will stay around any centerline value, random walks are nonstationary processes. But, as it turns out, a simple technique allows us to remove the nonrandom component and unlock the random component embedded in the process. We can gain insights about a random walk process by rearranging terms:

$$y_t = \delta + y_{t-1} + \varepsilon_t$$
$$y_t - y_{t-1} = \delta + \varepsilon_t$$

Recognize that $y_t - y_{t-1}$ represents the *change* in y from one period to the next. In the area of time series, period-to-period changes of a time series are known as **first differences.** For convenience, write as $d_t = y_t - y_{t-1}$. This gives us the following relationship:

$$d_t = \delta + \varepsilon_t$$

The above is simply the model for a random process (page 679). Thus, a random walk is a process for which the first differences are random with the underlying mean of the differences being δ. Accordingly, the least-squares estimate of δ is simply $\bar{d}$.

RANDOM WALK

The model for a **random walk** process is given by

$$y_t = \delta + y_{t-1} + \varepsilon_t$$

where the parameter δ is the mean first difference (period-to-period change) and the deviations ε_t are independent, with mean 0 and standard deviation σ.

If the mean change δ in the random walk is zero, then the random walk is said to have no **drift.** With no drift, there is no tendency of the random walk to go in any particular direction, as was seen with the random walk model simulated in Example 14.8. If δ is not 0, then the random walk is said to have drift. Random walks with $\delta > 0$ will tend to drift upward, while random walks with $\delta < 0$ will tend to drift downward.

EXAMPLE 14.9

HONDA1

Honda Prices Figure 14.16 plots the weekly closing stock prices of Honda Motor Co. from the beginning of January 2015 through mid-July 2018.[5] We can see that the prices are meandering around (i.e., wavelike movements), much like the simulated series of Figure 14.15. The price series is clearly a nonrandom process. Now consider the first differences of the prices shown in Figure 14.17. The average price change is $\bar{d} = \$0.0152$. The price changes are demonstrating random behavior around the average. A runs test and ACF would confirm this conclusion. Thus, the Honda price series is well described by the random walk model. ∎

FIGURE 14.16 Weekly closing prices of Honda stock (January 2015 through mid-July 2018).

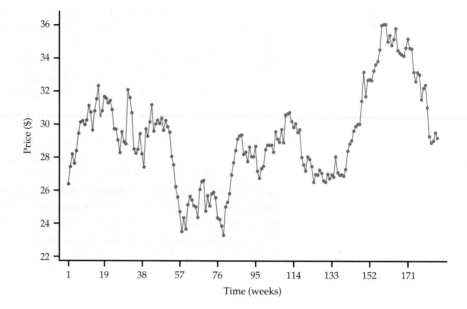

FIGURE 14.17 Weekly price changes (first differences) of Honda stock (January 2015 through mid-July 2018).

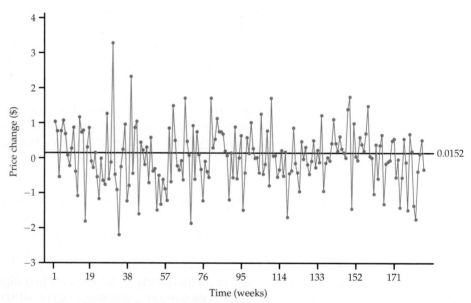

To make a forecast of a future price under a random walk process, consider the model but shift the subscript one period into the future:

$$y_{t+1} = \delta + y_t + \varepsilon_{t+1}$$

The parameter δ represents the mean price change. For the Honda series, the sample mean of the weekly price changes is $0.0152, or 1.5 cents. However, a standard one-sample t test of the null hypothesis $\delta = 0$ results in a P-value of 0.805; thus, there is not sufficient evidence to reject the null hypothesis. For all practical purposes, we can view the drift parameter δ as 0. This implies that we are proceeding with a random walk model with no drift. The term ε_{t+1} represents the *future* random deviation, which is unknown. Because the random deviations have mean 0, our "best" guess for ε_{t+1} is simply 0. In the end, the forecast equation for predicting Honda's closing price for time $t+1$ is given by:

$$\hat{y}_{t+1} = y_t$$

In other words, the best forecast of the next period's price is simply the current period's price. Using the current observation as the forecast of the next period is known as a **naïve forecast**.

EXAMPLE 14.10

Honda Prices Given that the data suggest that the Honda weekly price series of Example 14.9 can be viewed for practical purposes as a random walk with no drift, we use the naïve forecasting model to predict the next period's price. There are 186 observations in the price series, with the last closing price being $y_{186} = \$29.19$. The naïve forecast for next week's price is

$$\hat{y}_{187} = y_{186} = \$29.19$$ ∎

one-step ahead forecast

Example 14.10 illustrated a **one-step ahead forecast.** A two-step ahead forecast would be a prediction of the time series two periods into the future. For the naïve forecast model, the two-step ahead forecast would be the same as the one-step ahead forecast. In the case of the Honda price series, the two-step ahead forecast of period 188 made at period 186 is simply $29.19.

If the price data suggest the need to incorporate drift, then we would use the average price change $\bar{d}$ as the estimate of δ. Substituting this estimate into the random walk model and recognizing that our best guess for a future random deviation is 0, the one-step ahead forecast is

$$\hat{y}_{t+1} = y_t + \bar{d}$$

A two-step ahead forecast would call for adding another $\bar{d}$ to the one-step ahead forecast. By adding $\bar{d}$ for each additional step into the future, we have a general forecasting equation for random walks.

RANDOM WALK FORECASTS

In general, for a random walk process, the *l*-**step ahead forecast** of y_{t+l} made at time t is

$$\hat{y}_{t+l} = y_t + l\bar{d}$$

where $\bar{d}$ is the sample mean of the first differences (period-to-period changes). In the case where $\bar{d}$ is chosen to be 0, the general forecasting equation becomes a naïve forecast.

How the *l*-step ahead forecasts evolve pushing into the future depends on the value of $\bar{d}$. When $\bar{d}$ is taken to be 0, the forecasts project out horizontally ("flat") from the last observed value y_t. However, for a positive $\bar{d}$, the *l*-step ahead forecasts will trend upward, while for a negative $\bar{d}$, the *l*-step ahead forecasts will trend downward.

APPLY YOUR KNOWLEDGE

14.10 Honda prices. In Example 14.10, a naïve forecast was made for prediction of the next week's closing price. For this exercise, use $\bar{d} = \$0.0152$ in the forecasting of future closing prices. This value of $\bar{d}$ is reported in Example 14.9.

(a) What is the forecast for the closing price one week into the future?

(b) What is the forecast for the closing price two weeks into the future?

(c) Based on the values of the last observed price and of $\bar{d}$, write the equation for the *l*-step ahead forecast.

14.11 St. Louis financial stress index. Several of the banks in the Federal Reserve system have developed indices to serve as barometers of the health of the U.S. financial system. For example, the Federal Reserve Bank of St. Louis constructed an index known as the St. Louis Financial

Stress Index (STLFSI). The STLFSI is designed to track distress in the U.S. financial system on a weekly basis. An index reading of zero implies that financial conditions are normal, while a positive reading indicates that the financial system is experiencing some degree of stress. Consider data on weekly STLFSI readings from January 1, 2016, to July 27, 2018.[6] STLFSI

(a) Test the randomness of the STLFSI series using all approaches (time plot, runs test, and ACF) available with your software. What do you conclude?

(b) Obtain the first differences for the STLFSI series and test them for randomness. What do you conclude?

(c) Would you conclude that the STLFSI series behaves as a random walk? Explain.

(d) Find the P-value for a two-sided test of $\delta = 0$, where δ is the mean difference of the index values. State your conclusion using the $\alpha = 0.05$ significance level.

(e) Based on your conclusion of part (d), is a naïve forecast justified for prediction of the stress index value one week into the future? Explain. Provide the naïve forecast of the stress index value.

Price changes versus returns

Our discussion of the random walk model centered around the ups and downs of prices. Earlier, in Section 14.1, we looked at the ups and downs of Adidas prices in terms of percents. Percent changes in prices are referred to as returns. The advantage of using returns over price changes is made evident with our next example.

EXAMPLE 14.11

APPLE

Apple Prices In Example 14.9 (page 695), we looked at weekly closing prices over an approximate 3.5-year horizon. Consider now the weekly closing prices of Apple stock over a longer time horizon, namely, from the beginning of January 2010 through the first week of December 2018.[7] The price series is shown in Figure 14.18(a). We can see that the prices started out at around 20 and have increased to around 170. Visually, there appears to be a strong tendency for the series to drift upward, suggesting that this series might be modeled as a random walk with drift.

Figure 14.18(b) shows the corresponding price changes. The price changes appear random over time which is suggestive of random walk behavior of the price series. However, a salient point of the graph is that the variation of the price changes is increasing substantially with time. ∎

The phenomenon we are witnessing with the increasing variability of price changes is common with economic time series. In particular, many economic time series change as a rate or percent. Suppose that prices change on average by 1%, whether up or down. This would mean that if the price level is $10, the expected price change would be $0.10. But if the price level is at $100, then the expected price change would be $1.00, which is 10 times greater than when the price level was $10.

We can compute returns using the standard formula:

$$\text{Return}_t = \frac{y_t - y_{t-1}}{y_{t-1}}$$

In finance, there can be some ambiguity with the use of the term "return." It can be defined as the change in price (or value), which, in the formula, is given by the numerator term. But it is also common practice to have return represent the change in value as a percent. When return is calculated as a percent, we

FIGURE 14.18 (a) Weekly closing prices of Apple stock (January 2010 through beginning December 2018) and (b) corresponding weekly price changes (first differences).

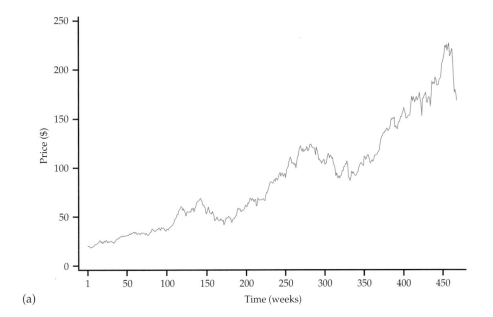

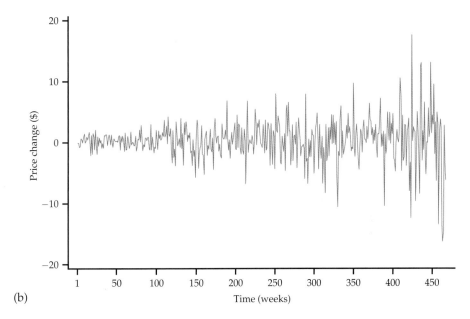

could refer to the computed quantity as a "rate of return." We, however, will go with the standard practice of simply calling the computed values "returns."

For example, the closing price for Apple on November 30, 2018 was 174.72, while the closing price for a week later, on December 7, 2018, was 168.49. This means the return was

$$\text{Return} = \frac{168.49 - 174.72}{174.72} = -0.0357 \text{ or } -3.57\%$$

Consider now the following calculation using the natural log:

$$\log(168.49) - \log(174.72) = -0.0363 \text{ or } -3.63\%$$

The closeness of these values is not accidental. It turns out that the difference of natural logged values is approximately a percent change with the approximation being nearly exact if the percent change is small. We have seen the benefits of the log transformation in many examples. Given its benefits, along with its nice interpretation, it is no wonder that we find that so many data analyses in economics and finance are routinely done in logged units.

EXAMPLE 14.12

APPLE

Apple Returns by Means of Log Transformation To approximate the percent changes of Apple prices, we first take the log of prices. The logged prices are shown in Figure 14.19(a). We now take the differences of these logged values. The differences can be either left in decimal form or multiplied by 100 to be in the form of a percent. Figure 14.19(b) shows the differences of logged Apple prices left in decimal form. Compare this time plot with the price changes of Figure 14.18(b). We can clearly see that the returns approximated by the log transformation have much more constant variance. The vertical axis is directly interpretable. For example, we find that weekly returns have had decreases and increases greater than 10%. ■

FIGURE 14.19 (a) Weekly logged closing prices of Apple stock (January 2010 through beginning December 2018) and (b) corresponding weekly logged price changes (first differences).

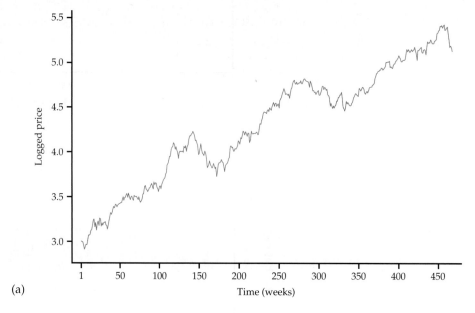
(a)

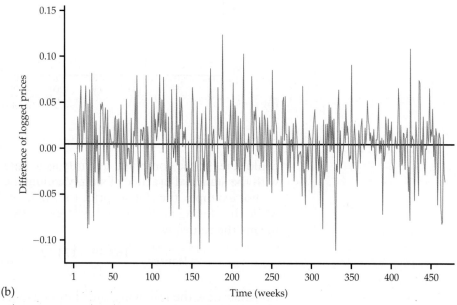
(b)

The stabilizing effect of the log transformation on the variance of a time series can be quite useful in simplifying modeling efforts. Take a look back at Figure 14.5 (page 682), which shows the time plot of Amazon sales. You should not be surprised that our eventual modeling of the Amazon sales series will involve the log transformation to help stabilize its increasing variance over time.

EXAMPLE 14.13

Prediction of Apple Prices Example 14.12 studied the approximate weekly returns of Apple prices. How might we use the return data to make a prediction of next week's Apple price? The sample mean of the returns shown in Figure 14.19(b) is $\bar{r} = 0.00455773$. Given that the variance of the return data over time is stable, we can use the standard one-sample t test to conduct a two-sided test of $\delta = 0$. This results in a P-value of 0.006. This P-value suggests evidence of a nonzero underlying mean return, which implies a random walk process with a positive drift. Our fitted equation is then:

$$\widehat{\log(y_t)} = 0.00455773 + \log(y_{t-1})$$

There are 467 observations in the price series shown in Figure 14.18(a), with the last closing price being $y_{467} = \$168.49$. There are a couple of ways to obtain the prediction of next week's price (one-step ahead forecast). One way is to apply the l-step ahead forecast equation (page 697) to the logged prices with $l = 1$ and $\bar{d} = \bar{r}$ as shown below:

$$\begin{aligned}\widehat{\log(y_{468})} &= \log(y_{467}) + (1)\bar{r} \\ &= \log(168.49) + (1)(0.00455773) \\ &= 5.13143\end{aligned}$$

At this stage, we untransform the log prediction value by applying the exponential function.

$$e^{5.13143} = \$169.26$$

By realizing that $\bar{r}$ is the estimated weekly return, we can equivalently obtain the prediction of next week's closing price by taking the current price plus the estimated dollar return from that price:

$$\begin{aligned}\hat{y}_{468} &= y_{467} + \bar{r}y_{467} \\ &= (1 + \bar{r})y_{467} \\ &= (1 + 0.00455773)168.49 \\ &= \$169.26\end{aligned}$$

Note that in all the above calculations it is important to use the decimal form of the average return, namely, 0.00455773. It would be *incorrect* to use the average return as a percent value, namely, 0.455773%. ∎

APPLY YOUR KNOWLEDGE

14.12 Computing returns. On August 2, 2018, the S&P 500 closed at 2827.22. On August 3, 2018, the S&P closed at 1931.51.

(a) Compute the percent change in the S&P from August 2 to August 3—that is, the daily return as a percent.

(b) Approximate the daily return as a percent using logs. Compare your answer with part (a).

14.13 Apple price forecasts. In Example 14.13, a one-step ahead forecast of the Apple closing price was calculated based on the estimated average return.

(a) Using the estimated average return value, what would be the two-step and three-step ahead forecasts for Apple closing prices?

(b) Based on the values of the last observed price and of $\bar{r}$, write the equation for the l-step ahead forecast.

Deterministic and stochastic trends

The time plot of Apple logged price series seen in Figure 14.19(a) shows a series having a clear tendency to move upward. In the end, we found that this series is well modeled by a random walk with drift. In Section 14.1, we spoke generically about a nonrandom process having trend as being a process that systematically moves away in some direction (upward or downward) as time passes. The top two time series displayed in Figure 14.4 (page 680) showed examples of trending series. With this in mind, it would seem that a random walk with drift can be characterized as a trending series. Indeed, a random walk with drift can be said to be a process with trend. However, it is worth noting an important classification of trends. Broadly speaking, trends can be classified as being either deterministic or stochastic.

deterministic trend

Simply stated, a **deterministic trend** is an explicit function of the time index ($t = 1, 2, 3, \ldots$) and the nature of its movement remains constant. A popular example of a process with a deterministic trend component is a linear trend process:

$$y_t = \beta_0 + \beta_1 t + \varepsilon_t$$

The deterministic trend component is $\beta_0 + \beta_1 t$. For any given time period t, we observe a y_t value, which has been pushed above or below the trend line by the random deviation ε_t. The random deviation at time t affects only the y_t. Thus, the impact of any random deviation is on only one time period. As a result, we would simply observe a time series with random scattering around a straight line. The path of the trend remains constant and is predictable. Deterministic trends are not restricted to being linear. Below is a quadratic trend process:

$$y_t = \beta_0 + \beta_1 t + \beta_2 t^2 + \varepsilon_t$$

We can even have exponential trends. With all deterministic trends (linear or curved), the resulting series is *nonstationary* in that the mean level is not constant as the series progressively moves in a certain direction. However, once the trend component is modeled and removed from the data, the residual series will behave as a stationary process. For this reason, deterministic trend

trend stationary

processes are referred to as being **trend stationary.** In Section 14.4, we will show how the regression method can be used to model deterministic trends.

stochastic trend

A process with a **stochastic trend** is also a nonstationary process. Unlike a deterministic trend, a stochastic trend is not an explicit function of t. Let us take a look again at the model for a random walk with drift,

$$y_t = \delta + y_{t-1} + \varepsilon_t$$

Like the linear trend process there is a random deviation ε_t affecting the value of y_t. However, the component of $(\delta + y_{t-1})$ is not a function of t, but rather, it is related to the *value* of a previous observation. This distinction has a profound impact on how the process will evolve over time. To get a sense of things, suppose we are sitting at time t and just observed y_t. Let us now consider where the process will be l periods into the future:

$$\begin{aligned} y_{t+l} &= \delta + y_{t+l-1} + \varepsilon_{t+l} \\ &= \delta + (\delta + y_{t+l-2} + \varepsilon_{t+l-1}) + \varepsilon_{t+l} \\ &\vdots \\ &= y_t + l\delta + (\varepsilon_{t+1} + \varepsilon_{t+2} \cdots + \varepsilon_{t+l}) \end{aligned}$$

Moving l periods into the future, the $l\delta$ term will push the process that amount away from the current position of y_t which gives the trend-like movement. The direction of that push depends on the sign of δ. But notice that the evolution of the process is also influenced by the cumulative effect of *all* past random deviations, that is, the effect of any given random deviation carries over into future observations. In contrast, remember that for a deterministic trend process, the random deviation for any given period is temporary in that it affects only the y observation of that period. This cumulative effect of random deviations is at the heart of the movement of a random walk without drift. Indeed, we have seen in Figure 14.15 (page 694) that the paths of random walks without drift are unpredictable meandering paths. Adding a drift term simply "nudges" these paths to move in a particular direction.

EXAMPLE 14.14

Simulating Deterministic and Stochastic Trends To visualize the nature of the paths of a deterministic trend process versus a stochastic trend process, consider two simple models:

$$\text{Deterministic trend: } y_t = 2t + \varepsilon_t$$
$$\text{Stochastic trend: } y_t = 2 + y_{t-1} + \varepsilon_t$$

The random deviation term (with standard deviation σ) is controlled to be the same for both scenarios. To simplify the comparison, we will initiate the stochastic trend process to a fixed constant value of $y_0 = 0$. By doing so, notice that for period 1, both processes are given by $y_1 = 2 + \varepsilon_t$. Since the mean of the random deviation term is 0, the mean of y for period 1 is $\mu_y = 2$. Moving forward in time, it is easy to show that at time t, the mean of y is the same for each process and is given by $\mu_y = 2t$.

If the means are the same for each process at any given time period, then how do these processes differ? It turns out that the standard deviation of y at time t for our deterministic trend process is $\sigma_y = \sigma$, while for the stochastic trend it is $\sigma_y = \sqrt{t}\sigma$. Thus, the variability of the y observations around the mean at any given time period remains *constant* for the deterministic trend process, while the variability *grows* as t increases for the stochastic trend process.

Figure 14.20 shows 24 simulation runs for over 100 periods: 12 runs (in maroon) are associated with the deterministic trend process and 12 runs (in blue) are associated with the stochastic trend process. Notice that all the deterministic trend series consistently stay near the straight line $y = 2t$. Furthermore, the variability of observations around this straight line is constant throughout time. Like the deterministic trend series, the stochastic trend series have tendencies to move upward. However, the paths of the stochastic trend series are far from predictable as they meander farther and farther away from the paths of the deterministic trend series.

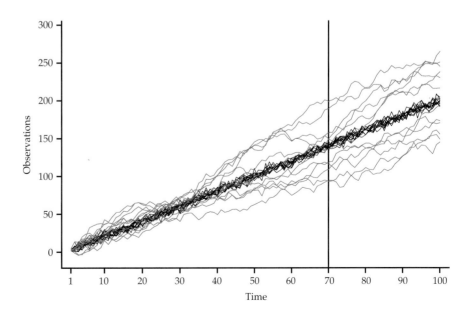

FIGURE 14.20 Simulated deterministic and stochastic trend processes.

On the graph, a reference line is placed at $t = 70$. Theoretically, as noted above, for both processes, the mean of y at this time period would be $\mu_y = 2t = 2(70) = 140$. Indeed, as can be visually seen, if we were to average the y observations on the blue lines versus the maroon lines, the averages would be about the same value at around 140. However, the y observations for the stochastic trend series have much greater variation compared to the y observations on the stochastic trend series. This variation of the observations from the stochastic trend series can be seen to increase with time. ∎

Unlike deterministic trends, stochastic trends cannot be modeled with a regression model based on the time index t. As we have seen with the random walk with drift process, the capturing of the nonrandom component is done by taking the first differences of the series. The first differences transform the original series to a random process which is a *stationary* process. At this point, the differenced series is simply fit with the sample mean of the differences ($\bar{d}$). Herein lies the key strategy for modeling stochastic trends. Namely, begin by taking the first differences of the series. Check if the differenced series is stationary in terms of the overall level of the process, known as mean stationary. Remember that a stationary process is not necessarily a random process. If the process is mean stationary, then find an appropriate model for the differenced series. If the first differences are not mean stationary, then take differences of the differences (known as second differences) to see if stationarity is then obtained. If so, then the second differences are appropriately modeled. Theoretically, the differencing can continue but, in practice, the first difference is enough in most cases.

How we fit the differenced series that has been made stationary depends on the nature of the stationarity. In the case of a random walk with drift, the first differences follow the simplest of stationary processes, namely, a random process. In the cases where the stationary process is not random, we may be able to use regression to model the process. That is our focus in Section 14.4. However, when one partakes in using the differencing technique to "tame" stochastic trends, a class of models known as ARIMA models (also known as Box-Jenkins models) is typically considered. ARIMA stands for AutoRegressive-Integrated-Moving-Average. The terms are explained as follows:

1. "AutoRegressive" refers to the use of past values of the series as predictor variables.

2. "Integrated" refers to the degree of differencing needed to obtain stationarity in the mean level.

3. "Moving Average" refers to the use of past random deviations as predictor variables. Since the values of random deviations are unknown, the approach basically uses past residual values from a preliminary autoregressive fit as predictor variables. As note of caution, "moving average" with an ARIMA model has *nothing* to do with the smoothing technique of moving averages discussed in Section 14.3.

ARIMA modeling is beyond the scope of this textbook. Knowledge of ARIMA modeling should be part of any sophisticated forecaster's toolkit. Without denying the usefulness of the ARIMA approach, it is reassuring that the techniques presented in this chapter effectively model many real-world time series.

BEYOND THE BASICS

Dickey-Fuller tests

We have learned that random walks can be analyzed with the simple approach of differencing. For our purposes, if we find that differences behave as a random process, then we deem the random walk to be a satisfactory model for the original series. However, one can conduct a more formal test of whether the process is a random walk.

The random walk model is a special case of what is known as an autoregressive lag one or AR(1) model given as follows:

$$y_t = \delta + \phi y_{t-1} + \varepsilon_t$$

When $\phi = 1$, we have the random walk with drift model. An important fact is that if $|\phi| < 1$, then the AR(1) process is stationary. Remember that for a stationary process, the overall level of the process remains constant over time. A random walk (with or without drift) is nonstationary. This is due to the fact that $\phi = 1$ and, thus, its absolute value is not less than 1. In technical time series jargon, a random walk is said to be a unit root process.

There are a number of ways to estimate ϕ as explained in the time series literature. One way, as shown in Section 14.4, is to use the regression method. The idea is simply to treat y_t as the response variable and y_{t-1} as the explanatory (predictor) variable. In terms of the notation of Chapter 12, we can write the AR(1) model as:

$$y_t = \beta_0 + \beta_1 y_{t-1} + \varepsilon_t$$

Refer now to Example 14.13 (page 701), where we found that the differencing approach led to the relation of

$$\widehat{\log(y_t)} = 0.00455773 + \log(y_{t-1})$$

Suppose we were to fit an AR(1) model by means of regression. Figure 14.21 shows the output. The resulting regression model is given by:

$$\widehat{\log(y_t)} = 0.0222579 + 0.9958784 \log(y_{t-1})$$

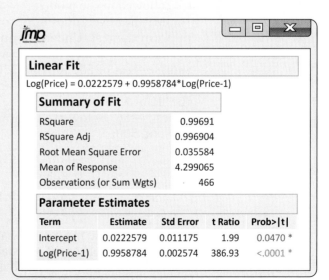

FIGURE 14.21 Regression output for logged closing price of Apple stock regressed on its lag.

inference for the regression slope, p. 583

Dickey-Fuller tests

We find that the estimate of β_1 is $b_1 = 0.9958784$, which is very close to a random walk value of $\beta_1 = 1$. To see if we have a random walk, we might consider a formal test of $\beta_1 = 1$. In Section 12.1, we learned that the test of $H_0: \beta_1 = 1$ would involve the computation of the following t statistic:

$$t = \frac{b_1 - 1}{SE_{b_1}}$$

For standard applications, under the null hypothesis, we have learned that the above statistic follows the t distribution with $(n-2)$ degrees of freedom. However, in the case of an AR(1) model, the test statistic does *not* follow the standard t distribution under the null hypothesis of $\beta_1 = 1$. The problem stems from the fact that the null hypothesis is associated with a nonstationary process. However, two statisticians (Dave Dickey and Wayne Fuller) developed a collection of tests (known as **Dickey-Fuller tests** for unit roots) which allow us to test $\beta_1 = 1$ against a variety of alternatives.

One Dickey-Fuller test considers the following competing hypotheses:

$$H_0: y_t = \beta_0 + y_{t-1} + \varepsilon_t, \quad \beta_1 = 1$$
$$H_a: y_t = \beta_0 + \beta_1 y_{t-1} + \varepsilon_t, \quad \beta_1 < 1$$

The test can be thought of as testing the competing hypotheses of nonstationarity versus stationarity. The alternative is one-sided. As it turns out, the test statistic (denoted by $t_{\beta_1=1}$) is computed no differently than the t statistic from a standard regression. Referring to Figure 14.21, the test statistic is

$$t_{\beta_1=1} = \frac{b_1 - 1}{SE_{b_1}} = \frac{0.9958784 - 1}{0.002574} = -1.601$$

JMP conveniently computes the above test statistic without the need to obtain a regression and then do the manual computation. In JMP's time series platform, for this series, JMP automatically reports "Single Mean ADF = -1.601375." The abbreviation "ADF" stands for Augmented Dickey Fuller and "Single Mean" implies that there is an intercept term (β_0) allowed in the model.

JMP, however, does not provide a P-value or critical values to perform the test. Under the null hypothesis, the test statistic follows a Dickey-Fuller distribution. A table of one-sided critical values (for sample sizes of 25, 50, 100, 250, 500, and infinity) is available in the time series literature.[8] This table can also be reproduced in R using the "fUnitRoots" package. The sample size for the Apple returns series is 466. Below are the lower 5% critical values for n equal to 250 and 500:

```
qadf(0.05,N = 250,trend = "c")
-2.88
qadf(0.05,N = 500,trend = "c")
-2.86
```

The critical value for $n = 466$ can be interpolated between the values. In the end, the observed test statistic value of -1.601 is not below the critical value which implies that we do not reject the null hypothesis of $\beta_1 = 1$ at the 5% significance level. Thus, the hypothesis that the logged Apple series follows a random walk is not rejected. If R package "akima" is installed with the "fUnitRoots" package, the P-value for any sample size greater than or equal to 25 can be approximated from the Dickey-Fuller distribution as seen below:

```
padf(-1.601,N = 466,trend = "c")
0.462243
```

The *P*-value of 0.462243 is quite large. Thus, the data fail to provide evidence against the null hypothesis of $\beta_1 = 1$. What if we had naïvely used the *t* distribution for computing the *P*-value? We would have found a *P*-value of 0.055 which edges on suggesting rejection of the null hypothesis; that is, the standard *t* test is leaning toward the conclusion that the process is not a random walk and the process is stationary. This substantial difference in *P*-values clearly demonstrates the need for these specialized unit root tests. We have shown only one version of the Dickey-Fuller testing procedure; there are many other versions. In addition, there is a popular alternative battery of tests known as Phillips-Perron tests. There are many good references in the time series and econometrics literature for all these tests.[9]

SECTION 14.2 SUMMARY

- The **first differences** are the period-to-period changes of a time series.

- A **random walk** is a distinctly nonrandom process for which its first differences are random.

- A **random walk without drift** is associated with an underlying zero mean change between successive observations, while a **random walk with drift** is associated with an underlying nonzero mean change between successive observations.

- For a random walk without drift, the best forecast is a **naïve forecast.**

- For a random walk with drift, the *l*-step ahead forecast is given by

$$\hat{y}_{t+l} = y_t + l\bar{d}$$

where $\bar{d}$ is the sample mean of the first differences.

- The difference between two values in **natural logarithmic** units is a good approximation for percent change between the values in original units when the percent is small to moderate in value.

- A **deterministic trend** is an explicit function of the time index *t* while a **stochastic trend** is not. Trending processes, deterministic or stochastic, are nonstationary processes. Deterministic trend processes follow predictable paths while stochastic trend processes take on unpredictable meandering paths.

SECTION 14.2 EXERCISES

For Exercises 14.10 and 14.11, see pages 697–698; for 14.12 and 14.13, see page 701.

If asked to test for randomness of series, use all approaches (time plot, runs test, and ACF) available with your software.

14.14 Consumer sentiment index. Each month, the University of Michigan conducts a survey of consumer attitudes concerning both the present situation as well as expectations regarding economic conditions. The results of the survey are used to construct a Consumer Sentiment Index. The index has been normalized to have a value of 100 for the base year of 1966. Consider the quarterly values for the index from the first quarter of 1990 to the second quarter of 2018.[10]

UMCSENT

(a) Test the randomness of the index series. What do you conclude?

(b) Obtain the first differences for the series and plot these differences over time. Also, test the m for randomness. What do you conclude?

(c) Would you conclude that the consumer index series behaves as a random walk? Explain.

(d) Find the P-value for a two-sided test of $\delta = 0$, where δ is the mean difference of the prices. State your conclusion using the $\alpha = 0.05$ significance level. What does your conclusion imply about the nature of the index series?

(e) Based on your conclusion in part (c), make a forecast for the value of the index for the third quarter of 2018.

14.15 Gold prices. Consider the monthly data of the price of gold ($ per troy ounce) from January 2000 to June 2018.[11] GPRICE

(a) Test the randomness of the index series. What do you conclude?

(b) Obtain the first differences for the series and plot these differences over time. Also, test them for randomness. What do you conclude?

(c) Take the log of the gold prices and then take their first differences. What do these differences approximate?

(d) Plot the differences of the logged prices over time. Describe what you observe with this plot in comparison to the plot from part (b).

(e) Test the log differences for randomness. What do you conclude?

14.16 Gold prices. Continue the previous exercise. GPRICE

(a) Find the P-value for a two-sided test of $\delta = 0$, where δ is the mean difference of the logged prices. State your conclusion using the $\alpha = 0.05$ significance level.

(b) Based on your conclusion in part (a), provide a prediction of the gold price for July 2018.

(c) Obtain a histogram and Normal quantile plot for the log differences. What does the plot suggest?

(d) Assuming the Normal distribution for the log differences, find a 95% prediction interval for the next period's return.

(e) Use the prediction interval limits found in part (d) to obtain a 95% prediction interval for the price of gold for July 2018.

14.17 U.S.-Canadian exchange rates. Consider the daily U.S.-Canadian exchange rates (one Canadian dollar to one U.S. dollar) from the beginning of 2015 through the last week of July 2018 (bank holidays excluded).[12] CANRATE

(a) Test the randomness of the exchange rate series. What do you conclude?

(b) Obtain the first differences for the series and test them for randomness. What do you conclude?

(c) Would you conclude that the exchange rate series behaves as a random walk? Explain.

(d) Find the P-value for a two-sided test of $\delta = 0$, where δ is the mean difference of the prices. State your conclusion using the $\alpha = 0.05$ significance level. What does your conclusion imply about the nature of the exchange rate series?

(e) Based on your conclusion in part (c), make a forecast for the value of the exchange rate for the next week in the future.

14.18 Apple returns. Consider the approximated returns plot in Figure 14.19(b) from Example 14.12 (page 700). APPLE

(a) Obtain a histogram and Normal quantile plot of the returns data. What do you conclude about the distribution of the returns?

(b) Find a 95% prediction interval for return for the next week in the future.

(c) Use the prediction interval limits for the future return found in part (b) to obtain a 95% prediction interval for Apple's closing price for next week.

14.19 U.S.-Canadian exchange rates. Refer to Exercise 14.17. CANRATE

(a) Obtain a histogram and Normal quantile plot of the daily changes in exchange rates. What do you conclude?

(b) Test the daily changes against the null hypothesis that the underlying mean change is 0. What do you conclude? What is the implication of your conclusion in terms of the exchange rate series having drift or not?

(c) Given your conclusion in part (b), what would be your forecast in general for the next day's exchange rate?

(d) Find a 95% prediction interval for daily change in exchange rate for the next period.

(e) Use the prediction interval limits for future daily change in exchange rates found in part (d) to obtain a 95% prediction interval for the exchange rate for the next day in the future.

14.3 Basic Smoothing Models

When you complete this section, you will be able to:

- Perform moving-average forecasting for a chosen span.
- Describe the relationship between amount of smoothing and moving-average span length and explain the role of smoothing for highlighting long-term movements.
- Explain when moving averages are and are not well suited for forecasting.
- Compare the relative performance of competing forecast models using mean absolute deviation (MAD), mean squared arror (MSE), or mean absolute percentage error (MAPE)
- Explain scenarios applicable for an additive seasonal model versus a multiplicative seasonal model.
- Compute seasonal indexes and seasonal ratios based on the moving-average method and use them to produce a deseasonalized time series.
- Perform exponential smoothing forecasting and describe the weighting scheme applied to observations used in the forecast.

In Section 14.1, we learned that the forecasting of a random process involves averaging *all* the observations to obtain an estimate of the underlying mean μ. This is because the mean of a random process does not change over time. As such, a random process is sometimes referred to as a **constant mean process.** However, in practice, it is often the case that the time series will not behave as a purely constant mean process. This section focuses on models for these processes.

constant mean process

Figure 14.22 shows two different time series of length 100 where, in each case, a forecast based on the average of all past observations is problematic. In Figure 14.22(a), there is an abrupt shift in the process to a higher mean level. It is clear that a forecast based on the average of all 100 observations to predict period 101 is not a suitable choice. In Figure 14.22(b), the series does not have an abrupt shift but rather is slowly meandering around some horizontal value. Again, the average of all 100 observations to predict period 101 does not seem to be a reasonable choice. As an alternative strategy, we might consider using only the most recent observations if we suspect the process is prone to shifts and/or meandering. This is illustrated in Figure 14.22, where the average of the most recent four observations are used.

In general, by averaging observations, we are attempting to wash out the random fluctuations in the time series to get a better sense of the evolution of the process. In the area of forecasting, the idea of reducing or canceling the obscuring effect of random variation is known as **smoothing.** There are varying amounts of smoothing that can be applied to a time series with the average of all past observations being the most extreme. But as we see from Figure 14.22, too much smoothing can be counterproductive in that it not only washes out the random fluctuations but also washes out the underlying movements of the two series.

smoothing

In this section, we will explore two basic smoothing methods (moving averages and simple exponential smoothing) that are often used by practitioners. For certain scenarios, these methods can provide accurate short-term forecasts. They are also useful to gain a general feel for the longer-term movements in a time series.

FIGURE 14.22 One-step ahead forecast of period 101 using overall average versus average of last four observations for two scenarios: (a) series with a mean shift and (b) meandering series.

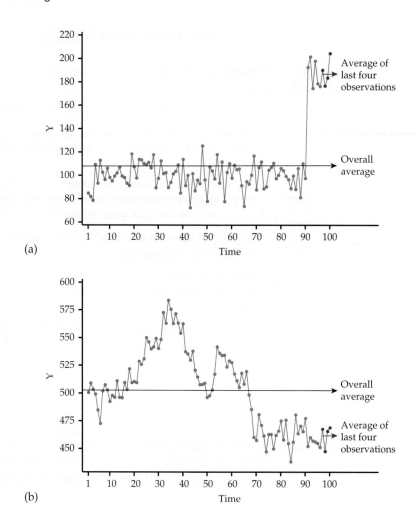

Moving-average models

Perhaps the most common method used in practice to smooth out short-term fluctuations is the *moving-average* model. A moving average can be thought of as a rolling average in that the average of the last several values of the time series is used to forecast the next value.

MOVING-AVERAGE FORECAST MODEL

At time t, the **moving-average forecast model** uses the average of the last k values of the time series as the one-step ahead forecast of y_{t+1}. The equation is

$$\hat{y}_{t+1} = \frac{y_t + y_{t-1} + \cdots + y_{t-k+1}}{k}$$

The number of preceding values included in the moving average is called the **span** of the moving average. More generally, the average of the last k observations serves as the l-step ahead forecast of y_{t+l} made at time t.

Some care should be taken in choosing the span k for a moving-average forecast model. As a general rule, larger spans smooth the time series more than smaller spans by averaging many ups and downs in each calculation. Smaller spans tend to follow the ups and downs of the time series. Note that for extreme case of $k = 1$, the moving-average forecast is equivalent to the naïve forecast (page 696) associated with a random walk model with no drift. More generally, the moving-average forecast model considers more than just the last observation.

LAKE

CASE 14.3

Lake Michigan Water Levels The water levels of the Great Lakes have received much attention in the media. As the world's largest single source of freshwater, more than 40 million people in the United States and Canada depend on the lakes for drinking water. The economies that greatly depend on the lakes include agriculture, shipping, hydroelectric power, fishing, recreation, and water-intensive industries such as steel making and paper and pulp production.

In the early 1990s, there was concern that the water levels were too high, with the risk of damaging shoreline properties. The fear that the lake would engulf shoreline properties was so high that cities like Chicago considered building protection systems that would have cost billions of dollars. Then in the first decade of the twenty-first century, there were great concerns about the water levels being lower than historical averages! The lower levels caused wetlands along the shores to dry up, threatening the reproductive cycle of numerous fish species and ultimately the fishing industry. The shipping and steel industries were also feeling the effects in that freighters had to carry lighter loads to avoid running aground in shallower harbors.

Figure 14.23 displays the average annual water levels (in feet) of Lake Michigan from 1909 through 2017 taken at the Calumet Harbor, Illinois, monitoring station.[13] Calumet Harbor is of major commercial importance because it is the primary link between the Great Lakes, inland waterways, and foreign ports. We see that the lake levels meander about the overall mean level of 578.771 feet. Each time the series drifts away from the overall mean level (above or below), it reverts back toward the mean. There does not seem to be any strong evidence for a long-term trend. ∎

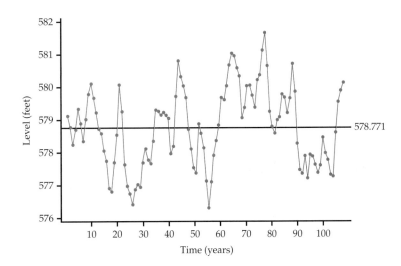

FIGURE 14.23 Time plot of average annual water levels (in feet) of Lake Michigan (1909 through 2017).

From Figure 14.23, we see that the lake levels at the end of the series have drifted above the overall mean level. If we wish to predict the lake level one year into the future, it would be preferable to base the forecast on the most recent lake levels and not on the overall average of all the 109 observations. The idea to use an average on a short span for a meandering series is similarly seen with Figure 14.22(b). Let's now explore the moving-average method with different spans.

EXAMPLE 14.15

LAKE

CASE 14.3 **Smoothing and Forecasting Lake Michigan Water Levels** Figure 14.24 displays moving-average one-step ahead forecasts based on a span of three years and moving averages based on a span of 15 years.

The 15-year moving averages are much more smoothed out than the three-year moving averages. The 15-year moving averages provide a long-term perspective of the cyclic movements of the lake levels. However, for this series, the 15-year moving-average model does not seem to be a good choice for short-term forecasting. Because the 15-year moving averages are "anchored" so many years into the past, this model tends to lag behind when the series shifts in another direction. In contrast, the three-year moving averages are better able to follow the larger ups and downs while smoothing the smaller changes in the time series.

The lake series has 109 observations ending with 2017. Here is the computation for the three-year moving-average forecast of the lake level for 2018:

$$\hat{y}_{110} = \frac{y_{109} + y_{108} + y_{107}}{3} = \frac{1739.561}{3} = 579.854 \text{ ft} \blacksquare$$

The previous example demonstrated how moving averages might be used to not only smooth out the time series but ultimately to make forecasts into the future. In our next example, the primary focus of the moving averages is to highlight the long-term movement of the series.

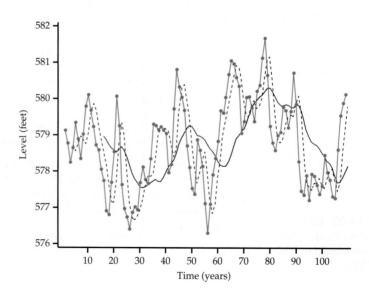

FIGURE 14.24 Time plot of annual lake levels (blue) with 3-year (black) and 15-year (maroon) moving-average forecasts superimposed.

EXAMPLE 14.16

ESALES

E-Commerce Retail Sales Figure 14.25 displays quarterly e-commerce retail sales as a percent of total U.S. retail sales. The series begins with the first quarter of 2013 and ends with the first quarter of 2018.[14] We can see the seasonal regularity to the series: the second quarter's percent tends to be the lowest; then there is a progressive rise in percents going into the third and fourth quarters, followed by a decline in the first quarter. When dealing with seasonal data, it is generally recommended that the length of the season be used for the value of k. In doing so, the average is based on the full cycle of the seasons, which effectively takes out the seasonality component of the data. Superimposed on the series are the moving-average forecasts based on a span of $k = 4$. Notice that the seasonal pattern in the time series is not present in the moving averages. The moving averages are a smoothed-out version of the original time series, reflecting only the general trending in the series, which is upward. ■

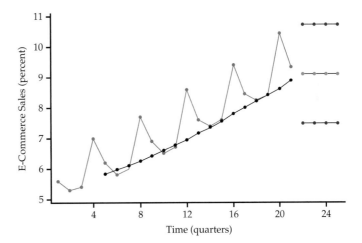

FIGURE 14.25 Time plot of quarterly e-commerce retail sales as a percent of total U.S. retail sales (first quarter 2013 through first quarter 2018) along with moving-average forecasts based on a span of $k = 4$ and prediction limits.

By smoothing out both random and seasonal fluctuations, the moving-average method is helping isolate and highlight the trend movement in the series. This reason alone is why the moving-average method is so popular with business practitioners and economists. Indeed, the moving-average method is the smoothing technique most commonly used by government departments/agencies worldwide for a variety of economic series to summarize general changes in the direction of the economy.

However, the basic moving-average method is not designed for making forecasts in the presence of trends. The problem is that the moving averages are derived from past observations, and the process is trending away from those observations. So, the moving averages are always lagging behind. Figure 14.25 also shows software-generated moving-average forecasts and 95% prediction limits projected into the future. Notice that the moving-average model with its horizontal forecast projection into the future makes no accommodation for the upward trend. The horizontal forecast projection is reflective of an earlier noted fact (page 710) that l-step ahead forecast of y_{t+l} made at time t remains the average of the last k observations regardless of the value of l.

It should be noted that the moving-average model presented here is the most basic version of a wider class of moving-average models. There are adaptations to the basic moving-average model that allow for forecasting in the presence of trend. However, these adaptations are beyond the scope of our presentation. We will instead deal with trend fitting in Section 14.4 using the familiar regression model.

APPLY YOUR KNOWLEDGE

14.20 Railroad dwell time. Government agencies and logistics organizations routinely track performance measures in the transportation sector. In the railroad industry, the most commonly monitored measures per time period include the number of cars operated, average speed, and terminal dwell time for major carriers. One such major carrier is BSNF Railway Company. BSNF operates one of the largest freight railroad networks in North America. Consider nine consecutive weeks of data on the average dwell time (hours) for BSNF-operated trains at the Denver, Colorado, terminal from the week ending June 1, 2018, to the week ending July 27, 2018.[15]

Week	Units dwell time (hrs)
1	29.3
2	24.4
3	24.2
4	26.9
5	27.6
6	30.9
7	27.1
8	28.7
9	25.9

(a) Hand calculate one-step ahead forecasts for weeks 2 through 10 based on a 1-week moving average.

(b) Hand calculate one-step ahead forecasts for weeks 3 through 10 based on a 2-week moving average.

(c) Hand calculate one-step ahead forecasts for weeks 4 through 10 based on a 3-week moving average.

Forecasting accuracy

In Example 14.15 (page 712), the lake level data series was smoothed with moving averages based on k equal to 3 and 15. Simply based on a visual inspection of the smoothed series displayed in Figure 14.24 (page 712), we argued that $k = 3$ would be the better choice for short-terms forecasts such as one-step ahead forecasting.

When comparing competing forecast methods, a primary concern is the relative accuracy of the methods. In this section, we introduce some basic measures of forecast accuracy that are commonly used for model comparison and model choice. Ultimately, how well a forecasting method does is reflected in the **forecast errors**, which are also referred to as prediction errors. A forecast error is simply the difference between the actual value (y) and its forecast ($\hat{y}$). In particular, the forecast error for time period t is given by

$$e_t = y_t - \hat{y}_t$$

As a general principle, smaller forecast errors in magnitude are more desirable. To make a judgment on a forecasting model, we don't simply look at one forecast error but rather a collection of forecast errors in aggregate. One measure of forecasting accuracy is **mean absolute deviation (MAD)**. As the name suggests, it is a measure of the average size of the forecast errors. For a given set of forecast errors (e_t),

$$\text{MAD} = \frac{\sum |e_t|}{m}$$

forecast errors

mean absolute deviation (MAD)

14.3 Basic Smoothing Models

mean squared error (MSE)

where m is the number of forecast errors considered. Another measure is **mean squared error (MSE),** which is a measure of the average size of the forecast errors in squared units:

$$\text{MSE} = \frac{\sum e_t^2}{m}$$

Be careful not to confuse the computation of the above forecast-based MSE with the regression-based MSE presented in Chapters 12 and 13. The two types of MSE are similar in that the numerator is the sum of squared errors (forecast errors or residuals). However, the denominators are different. For the forecast-based MSE, the denominator is simply the number of forecast errors used in the sum, while for regression-based MSE, the denominator is the degrees of freedom for the residuals. To avoid confusion, some software packages (e.g., Minitab) refer to the forecast-based MSE as mean squared deviation (MSD).

Imagine a forecast being off by 5 units from the actual value that equals 50 versus a forecast being off by 5 units from the actual value that equals 100. From a magnitude perspective, the forecast errors are viewed no differently. However, in the first case, the forecast is off by 10%, while in the second case, the forecast is off by only 5%. So, to put the forecast errors in perspective, it can be useful to measure the errors in terms of percents. One such measure is

mean absolute percent error

mean absolute percentage error (MAPE):

$$\text{MAPE} = \frac{\sum \left(\frac{|e_t|}{y_t} \right)}{m} \times 100\%$$

EXAMPLE 14.17

IURATE

Insured Unemployment Rate in Texas Unemployment insurance programs are administered at the state level and provide assistance to unemployed individuals who are looking for work. In addition to tracking the general unemployment rate, state and federal governments monitor the insured unemployment rate, which is the number of people currently receiving unemployment insurance as a percent of the labor force. This labor-related measure is considered useful because there is an observed and verifiable action (filing for unemployment insurance) associated with being among the insured unemployed.

Figure 14.26 displays the weekly insured unemployment rate for 20 weeks for the state of Texas from February to June 2018.[16] Let's consider applying a two-week ($k = 2$) and a three-week ($k = 3$) moving average to the series for one-step ahead forecasts. The one-step ahead forecasts for each moving averages are shown in Figure 14.26. The first forecast that the three-week moving-average model can make is for week 4 and is found to be

$$\hat{y}_4 = \frac{y_1 + y_2 + y_3}{3} = \frac{1.11 + 1.11 + 1.06}{3} = 1.093$$

In comparison, the forecast for week 4 based on the two-week moving average is

$$\hat{y}_4 = \frac{y_2 + y_3}{3} = \frac{1.11 + 1.06}{2} = 1.085$$

Figure 14.26 shows all the one-step ahead forecasts up to week 20. The forecast errors for each moving-average model are also shown. There are 18 forecast errors for the two-week moving-average model and 17 forecast errors for the three-week moving-average model. For comparative purposes, we will use only the weeks that have forecasts from both models and, thus, the

FIGURE 14.26 Excel spreadsheet showing weekly insured unemployment rate in Texas (February 2018 through June 2018) along with moving-average forecasts and forecast errors.

	A	B	C	D	E	F
1	Week	Rate	Forecast (k = 2)	Error (k = 2)	Forecast (k = 3)	Error (k = 3)
2	1	1.11				
3	2	1.11				
4	3	1.06	1.110	−0.050		
5	4	1.04	1.085	−0.045	1.093	−0.053
6	5	1.10	1.050	0.050	1.070	0.030
7	6	1.03	1.070	−0.040	1.067	−0.037
8	7	1.05	1.065	−0.015	1.057	−0.007
9	8	0.99	1.040	−0.050	1.060	−0.070
10	9	1.05	1.020	0.030	1.023	0.027
11	10	0.99	1.020	−0.030	1.030	−0.040
12	11	1.02	1.020	0.000	1.010	0.010
13	12	0.95	1.005	−0.055	1.020	−0.070
14	13	1.01	0.985	0.025	0.987	0.023
15	14	0.97	0.980	−0.010	0.993	−0.023
16	15	1.01	0.990	0.020	0.977	0.033
17	16	0.98	0.990	−0.010	0.997	−0.017
18	17	1.05	0.995	0.055	0.987	0.063
19	18	1.05	1.015	0.035	1.013	0.037
20	19	1.08	1.050	0.030	1.027	0.053
21	20	1.03	1.065	−0.035	1.060	−0.030

number of forecast errors m is 17. For the three-week moving-average model, the calculations of MAD, MSE, and MAPE are as follows:

$$\text{MAD} = \frac{|-0.053| + |0.030| + \cdots + |-0.030|}{17} = 0.0367$$

$$\text{MSE} = \frac{(-0.053)^2 + (0.030)^2 + \cdots + (-0.030)^2}{17} = 0.0017$$

$$\text{MAPE} = \frac{\left(\frac{|-0.053|}{1.04}\right) + \left(\frac{|0.030|}{1.10}\right) + \cdots + \left(\frac{|-0.030|}{1.03}\right)}{17} \times 100 = 3.595\%$$

Similar calculations can be made for the two-week moving average. Below is a summary of the forecasting accuracy values:

Measure	k = 2	k = 3
MAD	0.0315	0.0367
MSE	0.0013	0.0017
MAPE	3.069%	3.595%

From above, we find that the two-week moving-average model slightly outperforms the three-week moving-average model on all accuracy measures. ∎

In Example 14.17, we found for the particular set of data that the two-week moving-average model had better forecast accuracy than the three-week moving-average model. However, the series we studied was fairly short. In general, we would need a longer series to arrive at a more confident verdict of relative performance. In addition, it is always good practice to continue monitoring a model's performance moving into the future with newly observed data.

APPLY YOUR KNOWLEDGE

14.21 Software reporting. In Example 14.17, the accuracy measures of MAD, MSE, and MAPE were computed based on values shown in Figure 14.26. If we were to have software compute MAD with $k = 2$ using the 20 weekly rate values shown in Figure 14.26, the reported MAD value would be 0.0325, not the value of 0.0315 shown in Example 14.17. Explain why this is the case and show the details of computation resulting in the value of 0.0325.

14.22 Insured unemployment rate in Texas. Refer to Example 14.17 and the data shown in Figure 14.26.

(a) Provide the one-step forecasts for weeks 2 through 21 based on a naïve forecast (page 696).

(b) Compute the forecast errors for weeks 4 through 20.

(c) Compute MAD, MSE, and MAPE based on the forecast errors found in part (b).

(d) Based on the accuracy measures from part (c), how do the naïve forecasts compare with the two- and three-week moving-average forecasts?

Moving average and seasonal indexes

In Example 14.16 (page 713), we applied the moving-average method to a quarterly e-commerce series that showed strong seasonality and trending. The moving averages isolated the general trend. But, as Figure 14.25 (page 713) illustrated, we should not rely on moving averages for predicting future observations of a trending series. However, by comparing the seasonal observations relative to the general trend, we have a means of estimating the seasonal component.

The approach utilizes the moving averages to provide a baseline for the general level of the series *not* to provide forecasts. Instead of projecting the moving averages into the future, we use them as a summary of the past. Consider, for example, the first computed moving average on quarterly data:

$$\frac{y_1 + y_2 + y_3 + y_4}{4}$$

If we were to use this average as a forecast, then it would forecast period 5. Look at the first plotted moving-average value shown in Figure 14.25 (page 713). Notice that it has been placed at period 5. Instead of using the moving average as forecast for the future, consider the average as representing the *past* level of the time series. A minor difficulty can arise when using the moving average to estimate the past level of the time series. Because the moving average was based on periods 1, 2, 3, and 4, it is technically centered on period $t = 2.5$. This is problematic because we will want to compare the observations that fall on the whole-number time periods with the level of the series to estimate the seasonality component. The solution to this difficulty can be recognized by considering the next moving average in the quarterly series:

$$\frac{y_2 + y_3 + y_4 + y_5}{4}$$

The preceding moving average represents the level of the series at $t = 3.5$. We now have one moving average representing $t = 2.5$ and another one representing $t = 3.5$. By taking the average of these two moving averages, we now have a new average that will be centered on $t = 3$. This average is referred to as **centered moving average (CMA).**

centered moving average

The second step of averaging the averages is necessary *only* when the span of seasons is even, as with semiannual, quarterly, or monthly data. If the span of seasons is odd, then the initial moving average is the centered moving average. As an example, if we had data for the seven days of the week, then we would take a moving average of span 7. The average of the first seven periods will center on $t = 4$. Let's now illustrate the computation of centered moving averages for series with odd and even spans of seasons.

EXAMPLE 14.18

CTA

CTA Commuters The Chicago Transit Authority (CTA) rapid rail system is well known as the "L" (abbreviation for "elevated"). It is the third busiest system after the New York City Subway and the Washington Metro. Consider the daily count of commuters going through a particular station. The count is based on how many commuters went through all the turnstiles at the station. In particular, the data are daily for the downtown station of Lake/State from October 16, 2017 (Monday), to November 5, 2018 (Sunday).[17] Figure 14.27 displays the time series. There is the strong seasonality with few commuters on the weekends and a hint that Mondays are associated with slightly fewer commuters than the other weekdays. The series also shows a horizontal behavior in the sense that this pattern appears to be consistently above or below some underlying level that is flat over time.

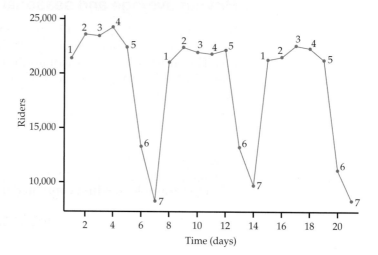

FIGURE 14.27 Time plot of daily Chicago Transit Authority commuters going through Lake/State station (October 16, 2017 through November 5, 2018).

Excel is a convenient way to manually compute the moving averages. Figure 14.28 shows a screenshot of the Excel computations for the CTA data. The first moving average of 19,540.57 is computed as the average of periods 1 through 7. In terms of the Excel spreadsheet shown, enter

$$= \text{AVERAGE}(C2:C8)$$

in cell D5 and copy the formula down to cell D19. As a result, we find that the second moving average of 19,488.86 is computed as the average of periods 2 through 8. Since the number of seasons of 7 is odd, the moving averages shown in Figure 14.28 are centered properly on the respective time periods. For example, the first moving average is properly sitting on $t = 4$, and the second moving average is properly sitting on $t = 5$. Thus, in Figure 14.28, we labeled the moving averages as centered moving averages, or CMA. ∎

Now that we have an appropriate baseline estimate for the level of the process historically, we can compute the seasonal component. How we compute the seasonal component depends on the form of the seasonal effect. Seasonal effects, and their corresponding models, are typically classified as **additive** or **multiplicative**. An additive seasonal model views the seasonal movements as the underlying level of the series *plus* a seasonal component:

additive and multiplicative seasonal effects

$$y = \text{LEVEL} + \text{SEASON}$$

In contrast, a multiplicative seasonal model views the seasonal movements as the underlying level of the series *times* a seasonal component:

$$y = \text{LEVEL} \times \text{SEASON}$$

FIGURE 14.28 Excel spreadsheet used in the computation of centered moving averages and seasonal ratios for Chicago Transit Authority commuter series.

	A	B	C	D	E	F	G
1	Day	t	Commuters	CMA	Observed Ratios	Estimated Ratios	Adjusted Series
2	M	1	21410			1.122	19081.996
3	T	2	23608			1.171	20160.547
4	W	3	23470			1.196	19623.746
5	Th	4	24272	19540.57	1.242	1.204	20159.468
6	F	5	22429	19488.86	1.151	1.161	19318.691
7	Sat	6	13315	19318.29	0.689	0.696	19130.747
8	Sun	7	8280	19102.14	0.433	0.474	17468.354
9	M	8	21048	18749.86	1.123	1.122	18759.358
10	T	9	22414	18712.71	1.198	1.171	19140.905
11	W	10	21957	18700.14	1.174	1.196	18358.696
12	Th	11	21806	18904.57	1.153	1.204	18111.296
13	F	12	22169	18936.71	1.171	1.161	19094.746
14	Sat	13	13227	18811.71	0.703	0.696	19004.31
15	Sun	14	9711	18900.43	0.514	0.474	20487.342
16	M	15	21273	18976.29	1.121	1.122	18959.893
17	T	16	21539	18846.29	1.143	1.171	18393.681
18	W	17	22578	18548.86	1.217	1.196	18877.926
19	Th	18	22337	18350.86	1.217	1.204	18552.326
20	F	19	21259			1.161	18310.939
21	Sat	20	11145			0.696	16012.931
22	Sun	21	8325			0.474	17563.291

seasonal indexes

One approach to capture the seasonal component is to compute **seasonal indexes** (or indices). The idea behind seasonal indexes can be seen by rearranging the level-plus-seasonal model and level-times-seasonal model, respectively.

$$\text{SEASON} = y - \text{LEVEL}$$

$$\text{SEASON} = \frac{y}{\text{LEVEL}}$$

These indexes say that we can isolate the seasonal component by either subtracting or dividing the data by the estimated level of the series. In the case of the multiplicative model, the seasonal index is commonly referred as a **seasonal ratio.**

seasonal ratio

A natural question is "Which seasonal model should we choose for a given time series?" An additive seasonal model is an appropriate choice if the magnitude of the seasonal fluctuations does not vary with the level of the series. A multiplicative seasonal model is an appropriate choice when the seasonal fluctuations are *proportional* to the general level of the series. In such a case, if the general level of the series changes over time, then the amount of the fluctuation for a given season will change even though the proportion remains constant.

Figure 14.29 displays a time plot of monthly total U.S. retail sales (in millions of dollars) from January 2015 to April 2018. The overall level of the series is trending up over time. Relative to the trend, the increase or decrease in sales for any given month is fairly constant from year to year. An additive seasonal model would be an appropriate choice for this series.

For a multiplicative seasonal model example, look back at the time plot of Amazon's quarterly sales shown in Figure 14.5 (page 682) along with the corresponding table of changes from third-quarter to fourth-quarter sales. We clearly see that the seasonal fluctuations increase with the increasing general level of the series. Given that the magnitude of seasonal fluctuation varies with the level of the series, a multiplicative seasonal model would be an appropriate choice.

The U.S. retail series and the Amazon series provide us with examples where the choice between an additive and a multiplicative season model is clear. But there are situations where we can safely choose either version.

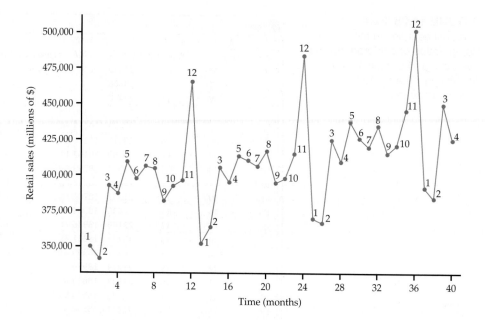

FIGURE 14.29 Time plot of monthly U.S. retail sales (January 2015 through April 2018).

Consider, for example, the CTA commuter series of Example 14.18 and displayed in Figure 14.27 (page 718). The daily fluctuations are fairly constant from week to week. This would suggest the use of an additive seasonal model. However, the level of the series remains nearly constant over time. Applying seasonal ratios to a series with a constant level will result in the increases and decreases being constant over time. This implies that a multiplicative seasonal model is also justified for the CTA commuter series. In the next example, we will use the CTA commuter series to first illustrate the computation of seasonal ratios for a multiplicative seasonal model. We will also provide an example of the computation of seasonal indexes for an additive seasonal model.

EXAMPLE 14.19

CTA

CTA Commuters and Seasonal Ratios In Example 14.18, we used the moving averages to estimate the level of time series, which is basically constant over time. For each of the estimated levels given by the centered moving averages, we calculate the ratio of actual number of commuters divided by the centered moving average. For the spreadsheet shown in Figure 14.28, we would enter

$$= C5/D5$$

in cell E5 and then copy the formula down to cell E19. The resulting ratios are shown in column E of the Excel spreadsheet.

Because we have more than one seasonal ratio observation for a given day, we average these ratios by day. That is, we compute the average of all the Monday ratios, then the average of all the Tuesday ratios, and so on. These averages become our seasonal ratio estimates. The following table displays the seasonal ratios that result. ■

Day	Seasonal ratio
Monday	1.122
Tuesday	1.171
Wednesday	1.196
Thursday	1.204
Friday	1.161
Saturday	0.696
Sunday	0.474

The seasonal ratios are snapshots of the typical ups and downs over the course of a week. Final estimated seasonal ratios should technically average to 1. The above ratios have an average of 1.003429, which is sufficiently close to 1 for practical purposes. We could further adjust them so that they indeed average to exactly 1 by dividing each ratio by 1.003429.

If you think of the ratios in terms of percents, then the ratios show how each day compares to the average level for all seven days of a given week. For example, Monday's ratio is 1.122, or 112.2%, indicating that the average number of commuters for Monday is estimated to be 12.2% above the weekly average. Saturday's ratio is 0.696, or 69.6%, indicating that the average number of commuters for Saturday is typically 30.4% below the weekly average. Notice how the seasonal ratios mimic the pattern that is repeated over the different days in Figure 14.27 (page 718). With seasonal ratios in hand, the next step is to seasonally adjust the series to allow estimation of the underlying level of the series.

EXAMPLE 14.20

CTA

deseasonalized series

Seasonally Adjusted CTA Commuter Series By again rearranging the level-times-seasonal model (page 718), we obtain

$$\text{LEVEL} = \frac{y}{\text{SEASON}}$$

This ratio tells us how to seasonally adjust a series. In particular, by dividing the original series by the seasonal component, we are then left with the level component with no seasonality. The resulting adjusted series is often called a **deseasonalized series.** Dividing each observation of the original series by its respective seasonal ratio found in Example 14.19 results in the adjusted values shown in column G of the Excel spreadsheet shown in Figure 14.28 (page 719). The deseasonalized series is plotted along with the original series in Figure 14.30. The deseasonalized series does not show the regular ups and downs of the original unadjusted data. The deseasonalized series basically resembles a random series around some underlying average. We estimate the underlying average by taking the average of seasonally adjusted observations which can be found to be 18,791.1, as shown in Figure 14.30. Our predictive model is then

$$\hat{y}_t = 18{,}791.1 \times \text{SR}$$

where SR is the seasonality ratio for the appropriate day. ∎

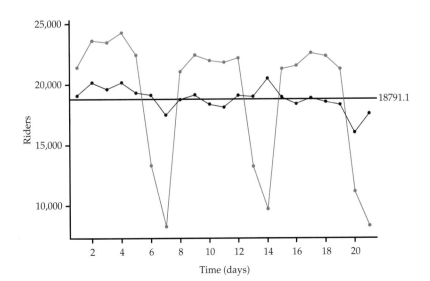

FIGURE 14.30
Deseasonalized Chicago Transit Authority commuter series along with mean fit.

EXAMPLE 14.21

ESALES

E-Commerce and Centered Moving Averages Consider again the quarterly e-commerce retail sales as a percent of total U.S. retail sales from Example 14.16 (page 713). Figure 14.31 shows a screenshot of the Excel computations for this series. The first moving average of 5.825 is computed as the average of the first four periods. This moving average is centered on period $t = 2.5$, while the second moving average of 5.975 is centered on period $t = 3.5$. To orient the average values on the whole-number periods, we need to average the moving averages to find the centered moving averages. The first centered moving average is then

$$\text{CMA} = \frac{5.825 + 5.975}{2} = 5.9$$

This value can be seen in the spreadsheet and was obtained by entering

$$= \text{AVERAGE(D3:D4)}$$

in cell E4. This first centered moving average is properly sitting on $t = 3$. The remaining centered moving averages can be found by copying the formula in cell E4 down to cell E20. The last moving average found in cell D20 involves periods 18 through 21, and it represents $t = 19.5$. The next-to-last moving average in cell D19 represents $t = 18.5$. The average of these two averages gives the last shown centered moving average of 9.001 (cell D20), which represents $t = 19$.

FIGURE 14.31 Excel spreadsheet used in the computation of centered moving averages and seasonal indexes for e-commerce series.

	A	B	C	D	E	F	G	H
1	Quarter	t	Precent of Sales	Moving Average	CMA	Observed Indexes	Estimated Indexes	Adjusted Series
2	1	1	5.60				-0.0178	5.6178
3	2	2	5.30	5.825			-0.5201	5.8201
4	3	3	5.40	5.975	5.900	-0.500	-0.5207	5.9207
5	4	4	7.00	6.100	6.038	0.962	1.0587	5.9413
6	1	5	6.20	6.250	6.175	0.025	-0.0178	6.2178
7	2	6	5.80	6.425	6.338	-0.538	-0.5201	6.3201
8	3	7	6.00	6.600	6.513	-0.513	-0.5207	6.5207
9	4	8	7.70	6.775	6.688	1.012	1.0587	6.6413
10	1	9	6.90	6.950	6.863	0.037	-0.0178	6.9178
11	2	10	6.50	7.175	7.063	-0.563	-0.5201	7.0201
12	3	11	6.70	7.350	7.263	-0.563	-0.5207	7.2207
13	4	12	8.60	7.575	7.463	1.137	1.0587	7.5413
14	1	13	7.60	7.800	7.688	-0.088	-0.0178	7.6178
15	2	14	7.40	8.000	7.900	-0.500	-0.5201	7.9201
16	3	15	7.60	8.210	8.105	-0.505	-0.5207	8.1207
17	4	16	9.40	8.420	8.315	1.085	1.0587	8.3413
18	1	17	8.44	8.628	8.524	-0.084	-0.0178	8.4578
19	2	18	8.24	8.888	8.758	-0.518	-0.5201	8.7601
20	3	19	8.43	9.113	9.001	-0.571	-0.5207	8.9507
21	4	20	10.44				1.0587	9.3813
22	1	21	9.34				-0.0178	9.3578

Figure 14.32 shows the centered moving averages plotted on the e-commerce series. Similar to Figure 14.25 (page 713), the averages show the general trend. However, unlike Figure 14.25, these averages are shifted back in time to provide an appropriate estimate of where the level of the process *was* as opposed to an incorrect forecast of where the process might be. ■

FIGURE 14.32 Centered moving averages (maroon) superimposed on e-commerce series (blue).

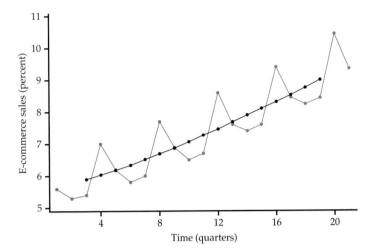

Now with the baseline of the level of the time series with the centered moving average, we can estimate the seasonal departures away (above and below) from this baseline. In Example 14.19 (page 720), we considered a multiplicative seasonal model and, thus, estimated these departures as percents, as summarized by seasonal ratios. Let's now consider estimation of the seasonal departures from an additive seasonal model perspective.

EXAMPLE 14.22

E-Commerce Series and Seasonal Indexes The e-commerce series displayed in Figure 14.32 shows the seasonal fluctuations as remaining constant, although the general level of the series is trending up. For this reason, an additive seasonal model is the suitable choice. For each of the estimated levels given by the centered moving averages shown in Figure 14.31, we calculate the indexes as the observed values minus the centered moving averages. For the spreadsheet shown in Figure 14.31, we would enter

$$= C4 - E4$$

in cell F4 and then copy the formula down to cell F20.

We compute the average of all the quarter 1 index values, the average of all the quarter 2 index values, and so on to arrive at the following seasonal index estimates:

Quarter	Seasonal index
1	−0.0275
2	−0.5298
3	−0.5304
4	1.0490

For an additive seasonal model, final estimated seasonal index values should technically average to 0. The above values have an average of −0.009675. Even though this average value is close to 0, we will make adjustments to the index values because the magnitudes of the data series are relatively small. To adjust the index values, we simply subtract the average value of −0.009675 from each index value. The adjusted seasonal index estimates are:

Quarter	Seasonal index
1	−0.0178
2	−0.5201
3	−0.5207
4	1.0587

∎

With the seasonal component estimated with the seasonal indexes, we are ready to deseasonalize the original series so that the level of the series can be highlighted. As illustrated in Example 14.20 (page 721), for a multiplicative seasonal model, the process of deseasonalizing the series involves dividing out the seasonal component. For an additive seasonal model, it stands to reason that the process of deseasonalizing involves subtracting out the seasonal component.

EXAMPLE 14.23

ESALES

Seasonally Adjusted E-Commerce Series By rearranging the level-plus-seasonal model (page 718), we obtain

$$\text{LEVEL} = y - \text{SEASON}$$

This suggests that seasonal adjustment is accomplished by subtracting the seasonal component from the original series. Subtracting the corresponding seasonal index value found in Example 14.22 from each observation of the original series results in the adjusted values shown in column H of the Excel spreadsheet shown in Figure 14.31.

Figure 14.33 displays the deseasonalized series. This adjusted series shows a clear upward trend. At this stage, we need to estimate the trend in the adjusted series. The approach we will take is to use regression to estimate the trend component. This will be illustrated in Section 14.4. Once we have the trend estimated, forecasts are made off the trend model and then adjusted by the seasonal indexes. We will complete the modeling and forecasting of this series in Example 14.31 (page 740). ∎

FIGURE 14.33
Deseasonalized e-commerce series.

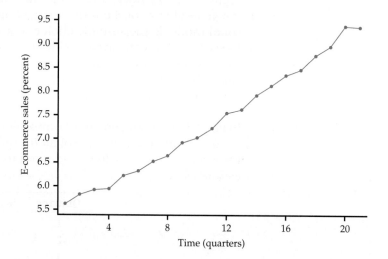

Many economic time series are routinely adjusted by season to make the overall trend, if it exists, in the numbers more apparent. Depending on the economic series, government agencies may release one or the other or both versions of the series, so be careful to notice whether the data have been seasonally adjusted when using government sources or when reading government press releases.

APPLY YOUR KNOWLEDGE

14.23 Seasonality ratios and predictions. Example 14.20 (page 721) gave the predictive model of the CTA commuter application. Based on this model, provide predictions for the number of commuters for each day of the week.

14.24 Seasonality indexes for CTA commuter series. Example 14.19 (page 720) showed the computation of the seasonal ratios for the CTA commuter application. For this exercise, consider the modeling of the time series by means of an additive seasonal model with seasonal indexes. CTA

(a) Using the centered moving averages shown in Figure 14.28 (page 719), determine the seasonal index values. The final seasonal index values should be adjusted so that their average is 0.

(b) Use the estimated seasonal indexes from part (a) to determine the values of the deseasonalized series. Find the average of the deseasonalized series.

(c) Based on the seasonal indexes from part (a) and the average found in part (b), provide predictions of the number of commuters for each day of the week.

14.25 Seasonality ratios and e-commerce series. In Example 14.22 (page 723), seasonal indexes were estimated for the e-commerce series. These seasonal indexes were used to produce the deseasonalized series shown in Figure 14.33. Assuming that the trend continues, explain how the nature of future forecasts will generally differ over time if seasonal indexes are used versus if seasonal ratios are used.

Exponential smoothing models

Moving-average forecast models appeal to our intuition. Using the average of several of the most recent data values to forecast the next value of the time series is easy to understand conceptually. However, two criticisms can be made against moving-average models. First, our forecast for the next time period ignores all but the last k observations in our data set. If you have 100 observations and use a span of $k = 5$, your forecast will not use 95% of your data! Second, the data values used in our forecast are all weighted equally. In many settings, the current value of a time series depends more on the most recent value and less on past values. We may improve our forecasts if we give the most recent values greater "weight" in our forecast calculation. *Exponential smoothing* models address both of these criticisms.

Several variations exist on the basic exponential smoothing model. We will look at the details of the *simple exponential smoothing model*, which we will refer to as, simply, the *exponential smoothing model*. More complex variations exist to handle time series with specific features, but the details of these models are beyond the scope of this chapter.

> **EXPONENTIAL SMOOTHING MODEL**
>
> The **exponential smoothing model** uses a weighted average of the observed value y_t and the forecasted value $\hat{y}_t$ as the forecast for time period $t + 1$. The forecasting equation is
>
> $$\hat{y}_{t+1} = wy_t + (1-w)\hat{y}_t$$
>
> The weight w is called the **smoothing constant** for the exponential smoothing model and is traditionally assumed have a range of $0 \leq w \leq 1$.

It should be noted that in the time series literature, there is no consistency in the assumption about the range of w in terms of inclusion or exclusion of the values of 0 and 1 as choices for w. Even with software there are differences. Excel allows w to be 1 but not 0. JMP and Minitab allow w to be 0 or 1. At the end of this section, we will note that statistical software can even allow for a wider range for w.

Choosing the smoothing constant w in the exponential smoothing model is similar to choosing the span k in the moving-average model—both relate directly to the smoothness of the model. Smaller values of w correspond to greater smoothing of movements in the time series. Larger values of w put most of the weight on the most recent observed value, so the forecasts tend to be close to the most recent movement in the series. Some series are better suited for a larger w, while others are better suited for a smaller w. Let's explore a couple of examples to gain insights.

EXAMPLE 14.24

PMI

PMI Series On the first business day of each month, the Institute for Supply Management (ISM) issues the *ISM Manufacturing Report on Business*. This report is viewed by many economists, business leaders, and government agencies as an important short-term economic barometer of the manufacturing sector of the economy. In the report, one of the economic indicators reported is PMI.

PMI is a composite index based on a survey of purchasing managers at roughly 300 U.S. manufacturing firms. The components of the index relate to new orders, employment, order backlogs, supplier deliveries, inventories, and prices. The index is scaled as a percent—a reading above 50% indicates that the manufacturing economy is generally expanding, while a reading below 50% indicates that it is generally declining.

Figure 14.34 displays the monthly PMI values from January 2010 through May 2018.[18] Also displayed are the forecasts from the exponential smoothing with $w = 0.1$ and $w = 0.7$.

The PMI series shows a meandering movement due to successive observations tending to be close to each other. With successive observations being close, a larger smoothing constant should work better for tracking and forecasting the series. Indeed, we can see from Figure 14.34 that the forecasts based on $w = 0.7$ track much closer to the PMI series than the forecasts based on $w = 0.1$. With a smoothing constant of 0.1, the model smooths out much of the movement (both short term and long term) in the series. As a result, the forecasts react slowly to momentum shifts in the series. ∎

When successive observations do not show a persistence to be close to each other, a larger value for w is no longer a preferable choice, as we can see with the next example.

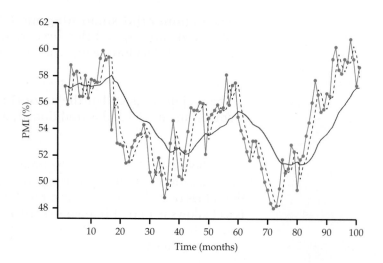

FIGURE 14.34 Monthly PMI index series (January 2010 to May 2018) with two exponential smoothing models: with $w = 0.1$ (maroon) and with $w = 0.7$ (black).

EXAMPLE 14.25

ADIDAS

CASE 14.1 **Adidas Returns** In Case 14.1 (page 678), we first studied the weekly returns of Adidas stock with a time plot. The time plot along with the runs test (Example 14.1, page 683) and ACF (Example 14.5, page 688) showed that the returns series is consistent with a random series.

Figure 14.35(a) shows the last 15 observations of the series with forecasts based on $w = 0.7$. In this case, each one-step ahead forecast $\hat{y}_{t+1}$ is close in value to the most recent observation y_t. Because the returns bounce around randomly, these forecasts also bounce around in the same way and are often considerably off from the actual observations they are attempting to predict. Looking closely at Figure 14.35(a), we see that when a return is higher than the average, the forecast for the next period's return is also higher than the average. But, with randomness, the next period's return can easily be below the average, resulting in the forecast being considerably off mark. Based on a similar argument, a return that is below average can result in a forecast for the next period being considerably off. In other words, a larger w is giving unnecessary weight to random movements, which, by their very nature, have no predictive value.

Figure 14.35(b) shows the forecasts based on a smaller smoothing constant, $w = 0.1$. With a small smoothing constant, we get a more smoothed-out forecast curve, less reactive to the up-and-down, short-term random movements of the returns. As we progressively make the smoothing constant smaller, the forecast curve converges to a horizontal line at the overall average of the observations. ∎

FIGURE 14.35 Adidas returns (blue) with two exponential smoothing models (maroon): (a) $w = 0.7$ and (b) $w = 0.1$.

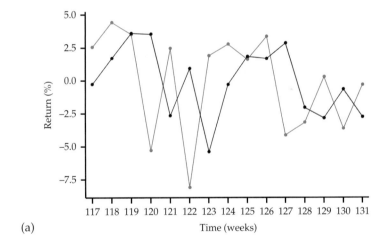

(a)

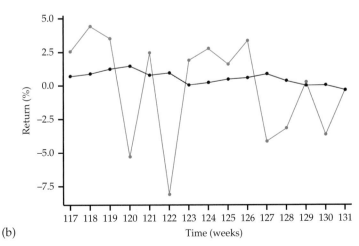

(b)

A little algebra is needed to see how the exponential smoothing model utilizes the data differently than the moving-average model. We start with the forecasting equation for the exponential smoothing model, imagine forecasting the value of the time series for the time period $t+1$, and progressively substitute backward in time:

$$\begin{aligned}\hat{y}_{t+1} &= wy_t + (1-w)\hat{y}_t \\ &= wy_t + (1-w)[wy_{t-1} + (1-w)\hat{y}_{t-1}] \\ &= wy_t + w(1-w)y_{t-1} + (1-w)^2\hat{y}_{t-1} \\ &= wy_t + w(1-w)y_{t-1} + (1-w)^2[wy_{t-2} + (1-w)\hat{y}_{t-2}] \\ &= wy_t + w(1-w)y_{t-1} + w(1-w)^2 y_{t-2} + (1-w)^3\hat{y}_{t-2} \\ &\vdots \\ &= wy_t + w(1-w)y_{t-1} + w(1-w)^2 y_{t-2} + \cdots + w(1-w)^{t-1}y_1 + (1-w)^t \hat{y}_1\end{aligned}$$

This alternative version of the forecast equation shows exactly how our forecast depends on the values of the time series. Notice that the calculation of $\hat{y}_{t+1}$ can be tracked all the way back to the initial forecast $\hat{y}_1$. It is common practice to initiate the exponential smoothing model by using the actual value of the first time period y_1 as the forecast for the first time period $\hat{y}_1$. Excel and JMP indeed initiate the exponential smoothing model in this manner. However, Minitab's default is to use the average of the first six observations as the initial forecast. This default can be changed so that the initial forecast is the first observation.

We also learn from the backtracking that the forecast for y_{t+1} uses *all* available values of the time series $y_1, y_2, \ldots, y_t$, not just the most recent k-values like a moving-average model would. The weights on the observations decrease exponentially in value by a factor of $(1-w)$ as you read the equation from left to right down to the y_1. This factor of $(1-w)$ is known as the **damping factor**.

damping factor

While the second version of our forecasting equation reveals some important properties of the exponential smoothing model, it is easier to use the first version of the equation for calculating forecasts.

EXAMPLE 14.26

Forecasting PMI Consider forecasting June 2018 PMI using an exponential smoothing model with $w = 0.7$. The PMI series ended on $t = 101$. This means that the forecasting of June 2018 is forecasting period 102:

$$\hat{y}_{102} = 0.7y_{101} + (1 - 0.7)\hat{y}_{101}$$

We need the forecasted value $\hat{y}_{101}$ to finish our calculation. However, to calculate $\hat{y}_{101}$, we will need the forecasted value of $\hat{y}_{100}$! In fact, continuing to apply this logic, we need the values of all past forecasts before we can calculate $\hat{y}_{102}$. We calculate the first few forecasts here and leave the remaining calculations for software. Taking $\hat{y}_1$ to be y_1, the calculations begin as follows:

$$\begin{aligned}\hat{y}_2 &= 0.7 \times y_1 + (1-0.7) \times \hat{y}_1 \\ &= (0.7)(57.2) + (0.3)(57.2) \\ &= 57.2\end{aligned}$$

$$\begin{aligned}\hat{y}_3 &= 0.7 \times y_2 + (1-0.7) \times \hat{y}_2 \\ &= (0.7)(55.8) + (0.3)(57.2) \\ &= 56.22\end{aligned}$$

$$\begin{aligned}\hat{y}_4 &= 0.7 \times y_3 + (1-0.7) \times \hat{y}_3 \\ &= (0.7)(58.8) + (0.3)(56.22) \\ &= 58.026\end{aligned}$$

Software continues our calculations to arrive at a forecast for y_{101} of 57.986. We use this value to complete our forecast calculation for y_{102}:

$$\hat{y}_{102} = 0.7 \times y_{101} + (1 - 0.7) \times \hat{y}_{101}$$
$$= (0.7)(58.7) + (0.3)(57.986)$$
$$= 58.486 \quad \blacksquare$$

With the forecasted value for June 2018 from Example 14.26, the forecast for July requires only one calculation:

$$\hat{y}_{103} = (0.7)(y_{102}) + (0.3)(58.486)$$

Once we observe the actual PMI for June 2018 (y_{102}), we can enter that value into the preceding forecast equation. Updating forecasts from exponential smoothing models only requires that we keep track of last period's forecast and last period's observed value. In contrast, moving-average models require that we keep track of the last k observed values of the time series.

In Examples 14.24 (page 726) and 14.25 (page 727), we illustrated situations in which either larger or smaller smoothing constants are preferable. However, our discussions revolved around only values of 0.1 and 0.7 for w. These choices are quite arbitrary and were used only for illustrative purposes. In practice, we want to fine-tune the smoothing constant with the hope of obtaining tighter forecasts.

One approach is to try different values of w along with a measure of forecast accuracy and find the value of w that does best. For example, we can search for the value of w that minimizes the accuracy measures of MAD, MSE, or MAPE introduced earlier (pages 714–715). Spreadsheets can be utilized to do the computations on different possible values of w to help us hone in on a reasonable choice of w. As an alternative, some statistical software packages (such as JMP and Minitab) provide an option to estimate the "optimal" smoothing constant. It turns out that the exponential smoothing forecast equation is the optimal (in terms of minimal MSE) equation for a very special case within a larger class of ARIMA models (page 704).[19] This special ARIMA model has a certain parameter that is directly related to the smoothing constant w. So, by estimating the parameter, we have an estimate for w.

EXAMPLE 14.27

PMI

Optimal Smoothing Constant for PMI Series Figure 14.36 shows output from JMP and Minitab for the estimation of the optimal smoothing constant. JMP reports the value as 0.8659, and Minitab reports the value as 0.8643. The slight difference in these values is because each software package uses different procedures for estimating ARIMA parameters. ∎

When we defined the exponential smoothing model (page 725), we stated that the smoothing constant is *traditionally* assumed to fall in the range of $0 \leq w \leq 1$. A smoothing constant value in this range has intuitive appeal as being a weighted average of the observation and forecast in which the weights are positive percents adding up to 100%. For example, a w of 0.40 implies 40% weight on the observation value and 60% weight on the forecast value.

However, it turns out that when the exponential smoothing model is viewed from the perspective of being connected to an ARIMA model, then the smoothing constant can possibly take on a value greater than 1. When finding an optimal value for w, Minitab allows w to fall in the range of $0 \leq w < 2$. JMP provides the user with a number of options in its estimation of the smoothing constant: (1) constrained to the range of $0 \leq w \leq 1$, (2) constrained to the range of $0 \leq w < 2$, or (3) allowing w to be any possible value with no constraints. Without going into the technical reasons, it can be shown that as long as w is

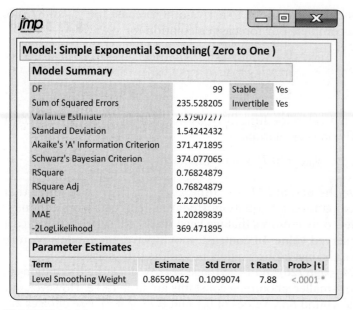
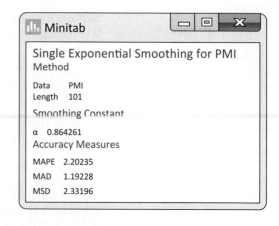

FIGURE 14.36 JMP and Minitab output summarizing the optimal exponential smoothing constant.

in the range of $0 \leq w < 2$, then the forecasting model is considered stable. If software estimates w to be greater than 1, then there are a couple of options. The first option is to go ahead and use it in the exponential smoothing model. There is nothing technically wrong with doing so. However, there are practitioners who are uncomfortable with a smoothing constant greater than 1 and prefer to have the traditional constraint. In such a case, the next best option in terms of minimizing MSE is to choose a value of 1 for w.

Similar to the moving-average model, the simple exponential smoothing model is best suited for forecasting time series with no strong trend. It is also not designed to track seasonal movements. Variations on the simple exponential smoothing model have been developed to handle time series with a trend (double exponential smoothing and Holt's exponential smoothing), with seasonality (seasonal exponential smoothing), and with both trend and seasonality (Winters exponential smoothing). Your software may offer one or more of these smoothing models.

APPLY YOUR KNOWLEDGE

14.26 Domestic average airfare. The Bureau of Transportation Statistics conducts a quarterly survey to monitor domestic and international airfares.[20] Here are the average airfares (inflation-adjusted to 2018 dollars) for U.S. domestic flights for 2004 through 2018:

Year	Airfare	Year	Airfare	Year	Airfare
2004	$402.26	2009	$359.91	2014	$411.67
2005	$391.50	2010	$383.48	2015	$395.70
2006	$405.48	2011	$402.45	2016	$361.94
2007	$390.30	2012	$406.13	2017	$352.85
2008	$400.27	2013	$407.99	2018	$346.49

(a) Hand calculate forecasts for the time series using an exponential smoothing model with $w = 0.2$. Provide a forecast for the average U.S. domestic airfare for 2019.

(b) Hand calculate forecasts for the time series using an exponential smoothing model with $w = 0.8$. Provide a forecast for the average U.S. domestic airfare for 2019.

(c) Write the forecast equation for the 2020 average U.S. domestic airfare based on the exponential smoothing model with $w = 0.2$.

14.27 Domestic average airfare. Refer to the subsection on forecasting accuracy (pages 714–715) for a presentation of the forecast accuracy measures MAD, MSE, and MAPE.

(a) Based on the forecast values for years 2004 through 2018 calculated in part (a) of Exercise 14.26, calculate MAD, MSE, and MAPE.

(b) Based on the forecast values for years 2004 through 2018 calculated in part (b) of Exercise 14.26, calculate MAD, MSE, and MAPE.

SECTION 14.3 SUMMARY

- **Moving-average forecast models** use the average of the last k observed values to forecast the next period's value. The number k is called the **span** of the moving average. Larger values of k result in a smoother model.

- A **forecast error** for time t is given by

$$e_t = y_t - \hat{y}_t$$

where y_t is the observed value of the time series at time t and $\hat{y}_t$ is its forecast.

- Three of the most common measures of forecast accuracy are **mean absolute deviation (MAD), mean squared error (MSE),** and **mean absolute percent error (MAPE).** These measures are computed as follows

$$\text{MAD} = \frac{\sum |e_t|}{m}$$

$$\text{MSE} = \frac{\sum e_t^2}{m}$$

$$\text{MAPE} = \frac{\sum \left(\frac{|e_t|}{y_t} \right)}{m} \times 100\%$$

- Seasonal effects are typically classified as **additive** or **multiplicative.**

- By appropriately centering the moving-average computations, **seasonal indexes** or **seasonal ratios** can be computed to capture seasonal effects. For additive seasonal models, **seasonally adjusted** or **deseasonalized** data can be obtained by subtracting the seasonal indexes from the original data. For multiplicative seasonal models, seasonal adjustment is made by dividing the original data by the seasonal ratios. Government agencies typically release seasonally adjusted data for economic time series.

- The forecast equation for the **simple exponential smoothing model** is a weighted average of last period's observed value and last period's forecasted value. The degree of smoothing is determined by the choice of a smoothing constant w. Some practitioners restrict w to the range of $0 \leq w \leq 1$, but some software allows for a range of $0 \leq w < 2$.

SECTION 14.3 EXERCISES

For Exercise 14.20, see page 714; for 14.21 and 14.22, see pages 716–717; for 14.23 to 14.25, see pages 724–725; and for 14.26 and 14.27, see pages 730–731.

14.28 It's exponential. Exponential smoothing models are so named because the weights

$$w, w(1-w), w(1-w)^2, w(1-w)^3, \ldots$$

decrease in value exponentially. Assume that $t = 100$ and we are making a forecast for period 101. For this forecast, consider the weights for observations $y_{100}, y_{99}, y_{98}, y_{97}, y_{96}$, and y_{95}, respectively. Use software to do the calculations.

(a) Calculate the weights for a smoothing constant of $w = 0.1$.

(b) Calculate the weights for a smoothing constant of $w = 0.5$.

(c) Calculate the weights for a smoothing constant of $w = 0.9$.

(d) Plot each set of weights from parts (a), (b), and (c). The weight values should be measured on the vertical axis, while the horizontal axis can simply be numbered $1, 2, \ldots, 5, 6$ for the 6 coefficients from each part. Be sure to use a different plotting symbol and/or color to distinguish the three sets of weights and connect the points for each set. Also, label the plot so that it is clear which curve corresponds to each value of w that is used.

(e) Describe each curve in part (d). Which curve puts more weight on the most recent value of the time series and little weight on past observations when calculating a forecast? Which curve shows the weights to be closer to each other in value for the observations in question?

14.29 It's exponential. In the previous exercise, you explored the behavior of the exponential smoothing model weights when the smoothing constant is between 0 and 1. We noted in the section that software can actually report an optimal smoothing constant greater than 1. Suppose that software reports an optimal w of 1.3.

(a) What is the value of the damping factor?

(b) Starting with the current observation y_t and going back to y_{t-7}, calculate the weights for each of these observations.

(c) Describe how the weights behave moving back in time. What values are some of these weights taking on that would not be the case if w is constrained not to exceed 1?

14.30 Business visitors to Australia. Consider data on the monthly number of international business visitors to Australia from January 2014 through December 2017.[21] VISIT

(a) Using an Excel spreadsheet, compute the seasonal ratios. The final seasonal ratio values should be adjusted so that their average is 1.

(b) Produce and plot the deseasonalized series. What are your impressions of this plot?

(c) Find the average of the deseasonalized series and provide predictions of the number of business visitors for January, February, and March 2018.

14.31 Business visitors to Australia. As with the previous exercise, consider data on the monthly number of international business visitors to Australia from January 2014 through December 2017. VISIT

(a) Using an Excel spreadsheet, compute the seasonal indexes. The final seasonal index values should be adjusted so that their average is 0.

(b) Produce and plot the deseasonalized series. What are your impressions of this plot?

(c) Find the average of the deseasonalized series and provide predictions of the number of business visitors for January, February, and March 2018.

14.32 Philadelphia property violations. Consider data on the number of monthly violations issued by the City of Philadelphia Department of Licenses and Inspections as related to the code on appearance and maintenance of property and vacant lots.[22] Examples of violations include unsafe property structures, graffiti, and high weeds on vacant lots. The time series starts on January 2014 and ends on July 2018. PROPVIO

(a) Using an Excel spreadsheet, compute the seasonal ratios. The final seasonal ratio values should be adjusted so that their average is 1.

(b) Produce and plot the deseasonalized series. What are your impressions of this plot?

(c) Find the average of the deseasonalized series and provide predictions of the number of violations for August, September, and October 2018.

14.33 Moving averages and linear trend. The moving-average model provides reasonable predictions only under certain scenarios. Consider monthly seasonally adjusted Consumer Price Index (CPI) data, starting with January 1990 and ending in July 2014.[23] CPI

(a) Make a time plot of the CPI series. Describe its movement over time.

(b) Using software, calculate moving-average forecasts for spans of $k = 50$ and 100. Superimpose the moving averages (on a single time plot) on a plot of the original time series.

(c) As the span increases, what do you observe about the plot of the moving averages?

(d) At the stock market analysis website **stockcharts.com**, it is stated that moving averages "are best suited for trend identification and trend following purposes, not for prediction." Explain whether your results from part (a) are consistent with this claim.

14.34 Philadelphia property violations. Refer to Exercise 14.32 and consider again data on the number of monthly property violations issued in Philadelphia. PROPVIO

(a) Using an Excel spreadsheet, compute the seasonal indexes. The final seasonal index values should be adjusted so that their average is 0.

(b) Produce and plot the deseasonalized series. What are your impressions of this plot?

(c) Find the average of the deseasonalized series and provide predictions of the number of violations for August, September, and October 2018.

14.35 First period weight. On page 728, the exponential smoothing equation was expanded to show the weights on past observations and on the initial forecast ($\hat{y}_1$).

(a) As most software implement, take the initial forecast to be y_1. Referring to the expanded equation, use algebra to determine in the most simplified form the weight applied to y_1.

(b) Suppose $w = 0.2$ and $t = 10$. Based on part (a), what is the weight on y_1 in the forecast of period 11? Show calculations.

14.36 What's wrong? Refer to Example 14.26 (page 728) in which the monthly PMI series was forecasted using an exponential smoothing model with $w = 0.7$. Open the provided file in Excel. Click the "Data" tab and then select the "Data Analysis" option. You will then find the Data Analysis pop-up box. Scroll down and select "Exponential Smoothing." In the Input Range box enter "$B$1:$B$102" and make sure the Labels box has a check mark in it. Enter 0.7 in the Damping factor box and enter "$D$2" in the Output Range box. Click OK to find the one-step ahead forecasts in column C. PMI

(a) What is the value of the forecast for February 2010, that is, period 2 ($\hat{y}_2$)? Based on this value, what is Excel in general taking as the initial forecast for the first time period?

(b) What does Excel report as the forecasts for period 3 ($\hat{y}_3$) and period 101 ($\hat{y}_{101}$)?

(c) In Example 14.26, what values are reported for $\hat{y}_3$ and $\hat{y}_{101}$? Compare these values with the values from part (b). Are they the same? If not, explain why.

(d) Redo the Excel exponential smoothing option so that similar results as those reported in Example 14.26 are obtained. Report Excel's forecast for period 101 to five decimal places (hundred thousandths place).

14.37 Exponential smoothing for insured unemployment rate. Consider the monthly insured unemployment rate in Texas time series from Example 14.17 (page 715). IURATE

(a) Calculate and plot (on a single time plot) exponential smoothing model forecast values using smoothing constants of $w = 0.1$ and 0.9.

(b) Comment on the smoothness of each exponential smoothing model in part (a).

(c) Based on the forecasts calculated in part (a), calculate MAD, MSE, and MAPE.

(d) The series ended on June 2018. For each model in part (a), calculate the forecasts for the insured unemployment rate of July 2018.

14.38 Exponential smoothing for railroad dwell time. Consider the weekly average dwell time (hours) for BSNF trains at the Denver, Colorado, terminal from Exercise 14.20 (page 714).

(a) Calculate and plot (on a single time plot) exponential smoothing model forecast values using smoothing constants of $w = 0.2$ and 0.8.

(b) Comment on the smoothness of each exponential smoothing model in part (a).

(c) Based on the forecasts calculated in part (a), calculate MAD, MSE, and MAPE.

(d) For each model in part (a), calculate the forecast for average dwell time for week 10.

14.39 Exponential smoothing for insured unemployment rate. Continue analysis of the monthly insured unemployment rate time series considered in Exercise 14.37. IURATE

(a) Use statistical software to determine the optimal smoothing constant w. Results may vary slightly between software packages.

(b) The series ended with the hires rate of June 2018. Based on the reported optimal w, calculate the forecast for the insured unemployment rate of July 2018.

14.40 Exponential smoothing for railroad dwell time. Continue with the analysis of the time series of the weekly railroad average dwell time considered in Exercise 14.38.

(a) Use statistical software to determine the optimal smoothing constant w. Results may vary slightly between software packages.

(b) Based on the reported optimal w, calculate the forecast for average dwell time for week 10.

14.41 Exponential smoothing forecast equation. We have learned that the exponential smoothing forecast equation is written as

$$\hat{y}_{t+1} = wy_t + (1-w)\hat{y}_t$$

(a) Show that the equation can be written as

$$\hat{y}_{t+1} = \hat{y}_t + we_t$$

where e_t is the forecast error for period t.

(b) Explain in words how the reexpressed equation can be interpreted.

CASE 14.3 14.42 Unconstrained optimal w. Consider the annual lake levels of Lake Michigan that were modeled with moving averages in Example 14.15 (page 712).

(a) Use statistical software to determine the optimal smoothing constant w. For JMP users, select "Unconstrained" from the Constraints options pull-down box. Results may vary slightly between software packages.

(b) Save the forecasts (predictions) to the data table of the software. Based on the reported optimal w and the forecast for the year 2017 found in the data table, hand compute the forecast of the lake level for the year 2018.

14.4 Regression-Based Forecasting Models

When you complete this section, you will be able to:
- Use regression to model deterministic linear and quadratic trends.
- Use seasonal indicators in a regression model to model additive seasonal behavior.
- Explain why the logarithmic transformation converts a multiplicative seasonal scenario to an additive seasonal scenario.
- Model an exponential trend by fitting a regression-based trend model to logged transformed observations.
- Interpret a partial autocorrelation function (PACF) for the identification of lagged variables to be used as potential predictor variables.
- Use regression to model purely autoregressive behaviors and to model autoregressive behavior in combination with trend and/or seasonal behaviors.

A time plot along with randomness checks (e.g., runs test and ACF) provide us with basic tools to assess whether a time series is a random process. If random, our "modeling" of the series trivializes to using the sample mean as our forecast equation. However, if it is nonrandom, then the challenge is to model the systematic patterns. In the special case of random walks (Section 14.2), modeling involves only the simple technique of differencing to capture the systematic pattern. In Section 14.3, we explored basic smoothing methods that are based on some form of averaging of past observations to track the series and to produce forecasts. In this section, we learn how the regression methods of Chapters 12 and 13 can be used to capture a variety of systematic patterns.

Modeling deterministic trends

A trend is a steady movement of the time series in a particular direction. Trends are frequently encountered and are of practical importance. A trend might reflect growth in company sales. Continual process improvement efforts might cause a process, such as defect levels, to steadily decline.

In Section 14.2 (page 702), we classified trends as being deterministic or stochastic. As noted earlier, deterministic trends are functions of the time index t, have predictable paths, and are easily estimated using regression. Stochastic trend processes, such as a random walk with drift, take on quite unpredictable paths and are modeled using the differencing operation. In this section, we will focus on modeling deterministic trends. Let's first explore the fitting of a linear trend.

EXAMPLE 14.28

CABLE

Monthly Cable Sales Consider the monthly sales data for festoon cable manufactured by a global distributor of electric wire and cable taken over a three-year horizon starting with January.[24] A festoon cable is a flat cable used in overhead material-handling equipment such as cranes, hoists, and overhead automation systems. The sales data are specifically the total number of linear feet (in thousands of feet) sold in a given month. In the time plot of Figure 14.37, the upward trend is clearly visible.

FIGURE 14.37 Time plot of monthly sales of festoon cable.

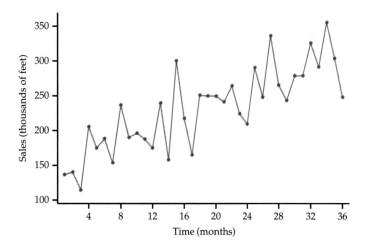

FIGURE 14.38 Excel trend regression output for festoon cables series.

If we ignore the line connections made between successive observations, the plot of Figure 14.37 can be viewed as a scatterplot of the response variable (sales) against the explanatory variable (Time), which is simply the indexed numbers 1, 2,..., 36. With this perspective, we can use simple linear regression (Chapter 12) to estimate the upward trend. From the Excel regression output of Figure 14.38, we find the estimated line to be

$$\widehat{\text{Sales}}_t = 147.186 + 4.533t$$

where t is a time index of the number of months elapsed since the first month of the time series; that is, $t = 1$ corresponds to first month (January), $t = 2$ corresponds to second month (February), and so on. The equation of the line provides a mathematical model for the observed trend in sales. The estimated slope of 4.533 indicates that festoon cable sales increased an average of 4.533 thousand feet per month. Conventional regression inference can be performed. For example, in Exercise 14.43, we ask you to test the null hypothesis that the trend coefficient is zero. R^2 is 65.1%, meaning that 65.1% of the original variation in the sales series has been explained by the trend fit. Figure 14.39 is the time plot of sales series with the estimated trend line superimposed. ∎

Conditions for regression inference, p. 581

In Chapters 12 and 13, we discussed how plots of the residuals can be used to check whether certain conditions of the regression model are being met. When using regression for time series data, we need to recognize that the residuals are themselves time ordered. If we are successful in modeling

FIGURE 14.39 Trend line fitted to monthly sales of festoon cable.

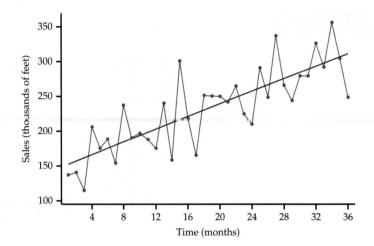

the systematic component(s) of a time series, then the residuals will behave as a random process. Figure 14.40 shows a time plot of the residuals. The plot indicates that the residuals are consistent with randomness, which, in turn, implies that the linear trend is a good fit for the time series. Additionally, a histogram and Normal quantile plot (not shown) indicate that residuals are compatible with the Normal distribution.

FIGURE 14.40 Time plot of the residuals from the linear trend fit for festoon cable series.

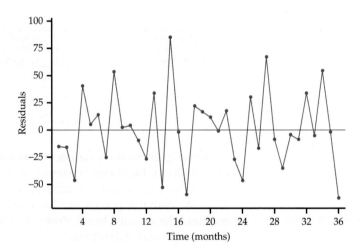

In Example 14.28, we created a column of the time index values (1, 2, ..., 36) to represent our x variable, and then we ran a regression of sales on the time index. It turns out that some software packages, including Excel and Minitab, provide a direct trend-fitting option, saving us from having to create the time index variable and then separately running the regression procedure. We use this convenient option in Example 14.29.

In Chapter 13, we used regression techniques to fit a variety of models to data. For example, we used a polynomial model to predict the prices of houses in Example 13.19 (page 654). Trends in time series data may also be best described by a curve like a polynomial. Often, curved trends represent growth in a series that is more rapid than that given by a linear trend model. The next two examples consider two different economic time series associated with the Chinese economy. Considering China's rapid economic growth, it is not surprising to find nonlinear trending in many China-related economic series.

 Fitting curved relationships, p. 654

EXAMPLE 14.29

UWAGE

Urban Wage Rate Consider annual data on the average wage (in yuan) of employed people living in urban settings in China ranging from 2000 to 2016.[25] Figure 14.41 shows an Excel produced time plot of the annual wage rates along with a fitted linear trend created using Excel's trendline option with charts. Notice that Excel reports the estimated linear trend model as

$$y = 3698.1x - 936.13$$

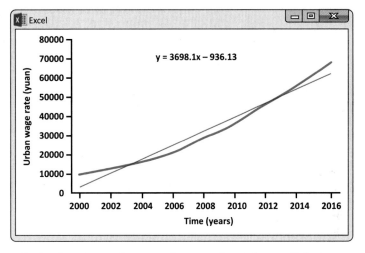

FIGURE 14.41 Excel output of a linear trend fit for the annual average urban wage rate series (2000 through 2016).

Even though the horizontal axis takes on the values of the years (2000 to 2016), the x shown in the Excel equation does not represent the values of 2000 to 2016. *It would be incorrect to put year values (e.g., 2016) into the Excel reported equation.* Instead, x is what we call t in Example 14.28 which, in this case, takes on the values 1, 2, ..., 17. Using the format of Example 14.28, the estimated linear trend model would be written as

$$\hat{y}_t = -936.13 + 3698.1t$$

From Figure 14.41, it is clear that the linear trend model fit is systematically off. The linear model is not able to capture the curvature in the series. One possible approach to fitting curved trends is to introduce the square of the time index to the model:

$$\hat{y}_t = b_0 + b_1 t + b_2 t^2$$

quadratic trend model

Such a fitted model is known as a **quadratic trend model.** Figure 14.42 shows the best-fitting quadratic trend model, again produced by Excel. In comparison to the linear trend model, the quadratic trend model fits the data series remarkably well. ∎

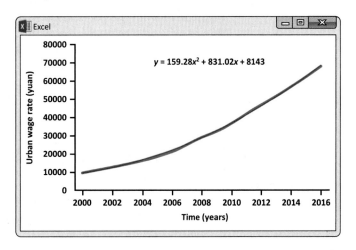

FIGURE 14.42 Excel output of a quadratic trend fit for the annual average urban wage rate series (2000 through 2016).

Quadratic trends can be quite flexible in adapting to a variety of curvatures in practice but there can be nonlinear patterns that challenge the quadratic model, as seen with our next example.

EXAMPLE 14.30

INVEST

Total Investment in China Consider Figure 14.43, which shows the best-fitting quadratic trend model (produced by Minitab's Trend Analysis platform) to a time plot of annual total investment in fixed assets (in 100 million yuan) in China from 1990 to 2016.[26] What we see is an increasing rate of growth in investment with time. However, the quadratic model is systematically off of the actual series for several years.

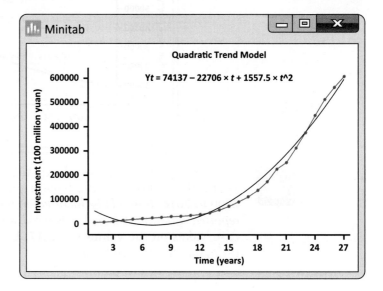

FIGURE 14.43 Minitab output of a quadratic trend fit for the annual total investment in China series (1990 through 2016).

exponential trend model

An alternative model that is often well suited to a rapidly growing (or decaying) data series is an **exponential trend model.** Figure 14.44 shows the investment series data with an exponential trend produced by Minitab. This model does a much nicer job than the quadratic model in tracking the growth in the series. Minitab reports the estimated exponential trend to be

$$\hat{y}_t = 5014.9(1.20148^t)$$

The estimated model has a direct interpretation of the growth rate. In particular, as the time index t increases by one unit, we multiply the estimate for the total investment from the previous year by 1.20148. This implies that the yearly growth rate in the total investment is estimated to be 20.148%. However, it appears that actual growth might be slowing down in that the exponential fit is overpredicting by a bit in years 2015 and 2016. As more data come in, it will be important to monitor the forecasts relative to actual observations. When the forecasts are consistently off, this is a critical signal to reestimate the trend model.

Figure 14.45 displays the exponential trend fit output from Excel. As noted in Example 14.29, Excel's variable label of x is the same as the time index t. The mathematical constant e seen in the Excel report was introduced in Chapter 5 with the Poisson distribution on page 262. Aside from rounding errors, we see here that the reported models are the same in the end:

$$\hat{y}_t = 5014.89972 e^{0.18356 t}$$
$$= 5014.89972 (e^{0.18356})^t$$
$$= 5014.89972 (1.20149^t) \blacksquare$$

FIGURE 14.44 Minitab output of an exponential trend fit for the annual total investment in China series (1990 through 2016).

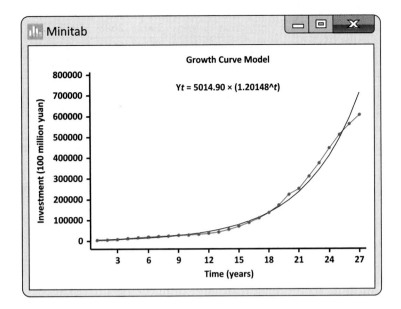

FIGURE 14.45 Excel output of an exponential trend fit for the annual total investment in China series (1990 through 2016).

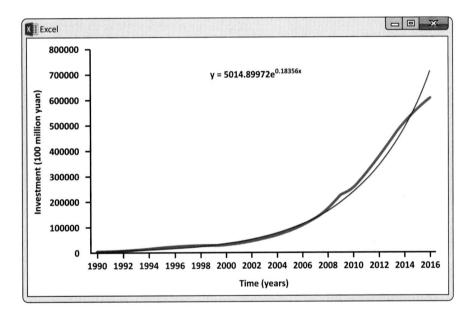

In the above example, it would seem that standard regression is not applicable to modeling exponential trends because the two software-reported equations do not have the format of an intercept *plus* terms of coefficients times predictor variables. However, you will soon learn that standard regression can be used to model exponential trends by utilizing the logarithmic transformation.

APPLY YOUR KNOWLEDGE

14.43 Monthly cable sales. Refer to the computer output for the linear trend fit of Example 14.28 (page 734) shown in Figure 14.38 (page 735).

(a) Test the null hypothesis that the regression coefficient for the trend term is zero. Give the test statistic, P-value, and your conclusion. What is the implication of your conclusion?

(b) The series ended at $t = 36$. Provide forecasts for festoon cable sales for periods $t = 37$ and $t = 38$.

14.44 Percent of Brazilian Internet users. Following are data on the time series of the percent of Brazilians using the Internet for 10 consecutive years ending in 2016.[27]

Year	2007	2008	2009	2010	2011	2012	2013	2014	2015	2016
Percent (%)	30.88	33.83	39.22	40.65	45.69	48.56	51.04	54.55	58.33	60.87

We usually want a time series longer than just 10 time periods, but this short series will give you a chance to do some calculations by hand. Hand calculations take the mystery out of what computers do so quickly and efficiently for us.

(a) Make a time plot of these data.

(b) Using the following summary information, refer to the slope and intercept formulas given on page 83 to calculate the least-squares regression line for predicting the percent of Brazilians using the Internet. The variable t simply takes on the values $1, 2, 3, \ldots, 10$ in time order.

Variable	Mean	Standard deviation	Correlation
t	5.5	3.02765	0.998
Percent users	46.362	10.15602	

(c) Sketch the least-squares line on your time plot from part (a). Does the linear model appear to fit these data well?

(d) Interpret the slope in the context of the application.

14.45 Total investment in China. Refer to the exponential trend fit reported in Example 14.30 for the annual total investment in fixed assets (in 100 million yuan).

(a) The series ended on year 2016. Provide a forecast for total investment for year 2017.

(b) In light of the discussion of Example 14.30, what concern might you have with the forecast of part (a)?

14.46 Urban wage rate. Refer to the quadratic trend fit shown in Figure 14.42 for the annual average wage (in yuan) of employed people in urban settings in China. The series ended on year 2016. Provide forecasts for the average wage rates for years 2017 and 2018.

Modeling seasonality

In Section 14.3, we explored the use of moving averages for estimating the seasonal component. With the estimated seasonal component, we are then able to deseasonalize the series to highlight the nature of the general level of the series. When a deseasonalized series exhibits flat level behavior as seen with the CTA commuter series shown in Figure 14.30 (page 721), we can simply model the level with the mean. However, if the deseasonalized series exhibits a trending level as seen with the e-commerce series shown in Figure 14.33 (page 724), we can no longer fit the level behavior with the mean. But, as we have learned, regression can be used to model trends.

EXAMPLE 14.31 **Trend Fit of Seasonally Adjusted E-Commerce Series** Consider again the deseasonalized e-commerce series obtained in Example 14.23 and displayed in Figure 14.33 (page 724). The trending of the deseasonalized series appears linear. We can come up with a trendline equation using regression, as seen in

FIGURE 14.46
Deseasonalized e-commerce series along with linear trend fit.

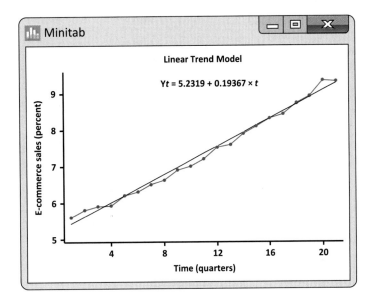

Figure 14.46. Using the final seasonal index values reported in Example 14.22 (page 723), the trend-and-season predictive model is then

$$\hat{y}_t = (5.2319 + 0.19367t) + \text{SI}$$

where SI is the seasonal index for the appropriate quarter corresponding to the value of t. ∎

The above example combines two distinct approaches (moving averages and regression) to arrive at the final seasonal model. Consider now the modeling of seasonality exclusively within a regression framework.

EXAMPLE 14.32

OFFICE

Monthly Sales for Office Supply and Stationery Stores The Census Bureau tracks a variety of retail and service sales using the Monthly Retail Trade Survey.[28] Consider, in particular, monthly sales (in millions of dollars) from January 2014 through April 2018 for office supply and stationery stores.

Figure 14.47 plots monthly sales with the months labeled "1" for January, "2" for February, and so on. The plot reveals interesting characteristics of sales for this sector:

• Sales are decreasing over time. The decrease is reflected in the superimposed trendline fit. This is a good opportunity to underscore that trending doesn't necessarily imply upward movement.

• A distinct pattern repeats itself every 12 months: there is a moderate peak with December and January sales; sales progressively decrease in the spring and early summer months; and, finally, there is a dramatic increase in August sales followed by a September drop. ∎

indicator variables, p. 656

A trend line fitted to the data in Figure 14.47 ignores the seasonal variation in the sales time series. Because of this, using a trend model to forecast sales for, say, August 2018 will undoubtedly result in a gross underestimate because the line underestimates sales for all the Augusts in the data set. We need to take the month-to-month pattern into account if we wish to accurately forecast sales in a specific month. To do so within a regression framework, we make use of indicator variables introduced in Chapter 13.

FIGURE 14.47 Trend line fitted to monthly office supply and stationery sales (January 2014 to April 2018).

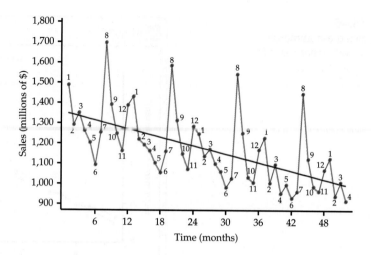

EXAMPLE 14.33

OFFICE

Monthly Sales for Office Supply and Stationery Stores We need only $I - 1$ indicator variables to represent I different categories. Viewing each month as a category, we begin by creating $12 - 1 = 11$ indicator variables.

$$Jan = \begin{cases} 1 & \text{if the month is January} \\ 0 & \text{otherwise} \end{cases}$$

$$Feb = \begin{cases} 1 & \text{if the month is February} \\ 0 & \text{otherwise} \end{cases}$$

$$\vdots$$

$$Nov = \begin{cases} 1 & \text{if the month is November} \\ 0 & \text{otherwise} \end{cases}$$

December data are indicated when all 11 indicator variables are 0.

We can extend the trend model with these indicator variables. The new model captures the linear trend along with the seasonal pattern in the time series.

$$Sales_t = \beta_0 + \beta_1 t + \beta_2 Jan + \cdots + \beta_{12} Nov + \varepsilon_t$$

Fitting this multiple regression model to our data, we get the regression output shown in Figure 14.48. The improved model fit is reflected in the R^2-values for the two models: $R^2 = 98.2\%$ for the trend-and-season model and $R^2 = 34.4\%$ for the trend-only model. The significance of seasonal variables can be collectively tested with an F test introduced in Chapter 13. The heart of the test is based on the change in R^2. In our case, we find the seasonal variables significantly contribute to the prediction of sales ($F = 131.95$ and $P = 7.41 \times 10^{-27}$). ∎

F test for collection of regression coefficients, p. 643

From a hypothesis-testing perspective, the implication of the F test noted in the above example is that we want to hold on to at least one of the explanatory variables. However, from Figure 14.48, we see that there is an insignificant P-value for the September indicator variable. We could attempt to remove explanatory variables with the guidance of t statistics and with support of an F test for collection of regression coefficients. Alternatively, we can consider variable selection methods noted in Chapter 13. With 13 possible explanatory variables (trend plus 12 monthly indicator variables), there are 8192 different possible models for predicting sales. As mentioned in Chapter 13, statistical software can examine all possible multiple regression models, allowing the user to select models based on criteria that balance fit against model simplicity. Two such criteria noted in Chapter 13 are AIC and BIC.

variable selection methods, p. 662

FIGURE 14.48 Trend-and-season model fitted to monthly office supply and stationery sales, using all 11 indicator variables along with trend variable.

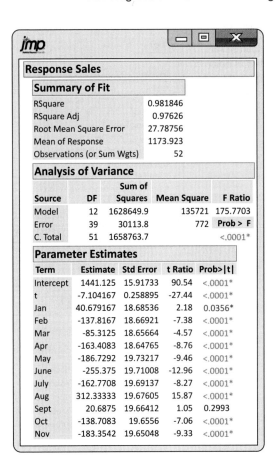

EXAMPLE 14.34

OFFICE

Variable Selection Using the All Possible Models option in JMP, the output shown in Figure 14.49(a) is reported. Preferable models are associated with *smaller* AIC (or BIC) values. The combination with the lowest AIC and BIC amongst the reported models has nine explanatory variables (combination in bold). The regression output for the AIC-selected model is shown in Figure 14.49(b). Comparing outputs, we find that the adjusted R^2 value for the AIC-selected model is slightly *larger* than for the model based on using the trend variable along with all of the allowable seasonal indicator variables.

For the AIC-selected model, the months of April, May, July, and November are not included. These excluded months are associated with the values of 0 for all the monthly indicator variables found in the model. Adjusting for the trend term, the model views the sales for the excluded months to be no different than each other. Inspecting the time plot of Figure 14.47 with particular attention to the labels of "4" (April), "5" (May), "7" (July), and "11" (November), it indeed appears that these months have little separation from each other. ∎

The excluded months of April, May, July, and November can be viewed as a baseline for interpreting the coefficients of the included months. Each of the included months has an estimated trend line a certain amount below (June) or above (January, February, March, August, September, October, and December) the baseline group depending on the magnitude of the month's coefficient.

When using indicator variables to incorporate seasonality into a trend model as we did in Examples 14.33 and 14.34, we *add* the indicator variables to the trend model. This implies that these models are additive seasonal models as introduced in Section 14.3 (page 718). As discussed earlier, seasonal

FIGURE 14.49 (a) JMP variable selection output with lowest AIC combination highlighted in bold and (b) regression output of the variable selected trend-and-seasonal model.

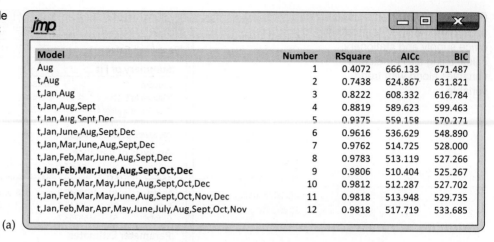

Model	Number	RSquare	AICc	BIC
Aug	1	0.4072	666.133	671.487
t,Aug	2	0.7438	624.867	631.821
t,Jan,Aug	3	0.8222	608.332	616.784
t,Jan,Aug,Sept	4	0.8819	589.623	599.463
t,Jan,Aug,Sept,Dec	5	0.9375	559.158	570.271
t,Jan,June,Aug,Sept,Dec	6	0.9616	536.629	548.890
t,Jan,Mar,June,Aug,Sept,Dec	7	0.9762	514.725	528.000
t,Jan,Feb,Mar,June,Aug,Sept,Dec	8	0.9783	513.119	527.266
t,Jan,Feb,Mar,June,Aug,Sept,Oct,Dec	**9**	**0.9806**	**510.404**	**525.267**
t,Jan,Feb,Mar,May,June,Aug,Sept,Oct,Dec	10	0.9812	512.287	527.702
t,Jan,Feb,Mar,May,June,Aug,Sept,Oct,Nov,Dec	11	0.9818	513.948	529.735
t,Jan,Feb,Mar,Apr,May,June,July,Aug,Sept,Oct,Nov	12	0.9818	517.719	533.685

(a)

Response Sales

Summary of Fit

RSquare	0.980605
RSquare Adj	0.976449
Root Mean Square Error	27.67629
Mean of Response	1173.923
Observations (or Sum Wgts)	52

Analysis of Variance

Source	DF	Sum of Squares	Mean Square	F Ratio
Model	9	1626592.7	180733	235.9503
Error	42	32171.0	766	Prob > F
C. Total	51	1658763.7		<.0001 *

Parameter Estimates

Term	Estimate	Std Error	t Ratio	Prob>|t|
Intercept	1267.4636	9.532091	132.97	<.0001 *
t	-7.095717	0.256814	-27.63	<.0001 *
Jan	214.12932	14.08452	15.20	<.0001 *
Feb	35.625041	14.08052	2.53	0.0152 *
Mar	88.120758	14.08121	6.26	<.0001 *
June	-81.91639	15.3921	-5.32	<.0001 *
Aug	485.77504	15.38051	31.58	<.0001 *
Sept	194.12076	15.38114	12.62	<.0001 *
Oct	34.716475	15.38605	2.26	0.0293 *
Dec	173.40791	15.40873	11.25	<.0001 *

(b)

additive models are appropriate when the magnitude of the seasonal fluctuations does not vary with the level of the series, as appears to be the case with the monthly sales for office supply and stationery stores. However, if seasonal fluctuations vary proportionally with the level of the series, then consideration needs to be given to multiplicative seasonal models. Such seasonal behavior is seen with the Amazon sales series (see Figure 14.5, page 682), for which the magnitude of the seasonal fluctuation for a given season increases as the sales series grows. As illustrated with Example 14.19 (page 720), one approach to the modeling of multiplicative season effects is based on the use of seasonal ratios.

The question is how might we approach the building of multiplicative seasonal models within a regression framework? To answer this question, it helps to write again the basic construct of a multiplicative seasonal model, namely,

$$y = \text{LEVEL} \times \text{SEASON}$$

Recognizing that the logarithm of a product is the sum of logarithms, consider the result of applying the logarithm to the above product:

$$\log(y) = \log(\text{LEVEL} \times \text{SEASON}) = \log(\text{LEVEL}) + \log(\text{SEASON})$$

We can see that the logarithm changes the modeling of level and seasonality from a multiplicative relationship to an additive relationship. When the level component is following an exponential trend $y = ae^{bt}$, we can gain another insight from the application of the natural logarithm:

$$\log(y) = \log(ae^{bt}) = \log(a) + \log(e^{bt}) = \log(a) + bt$$

This breakdown implies that fitting an exponential trend model can be accomplished by fitting a simple linear trend model with $\log(y)$ as the response variable. Because the logarithm can simplify the seasonal and trend-fitting process, it is commonly used for time series model building.

EXAMPLE 14.35

AMAZON

Fitting and Forecasting Amazon Sales In Case 14.2 (page 681), we observed that the time series was growing at an increasing rate, which leads us to consider an exponential trend model. Additionally, we saw that the seasonal variation increases with the growth of sales. This indicates that an additive seasonal model is probably not the best choice.

It seems that this data series is a good candidate for logarithmic transformation. Figure 14.50 displays the sales series in logged units. Compare this time plot with Figure 14.5 (page 682). In logged units, the trend is now nearly linear. Focusing on changes from third quarter to fourth, the seasonal fluctuations appear to decrease a bit over the years but then the 2017 change is seen to be greater than the 2016 change. So, there has been some stabilization of the seasonal fluctuations with the transformation. All in all, it seems reasonable to model logged sales as a linear trend with additive seasonal indicator variables:

$$\log(\text{Sales}_t) = \beta_0 + \beta_1 t + \beta_2 Q1 + \beta_3 Q2 + \beta_4 Q3 + \varepsilon_t$$

where Q1, Q2, and Q3 are indicator variables for quarters 1, 2, and 3, respectively. Statistical software calculates the fitted model to be

$$\log(\widehat{\text{Sales}}_t) = 9.1991 + 0.0571t - 0.3072Q1 - 0.3629Q2 - 0.3375Q3$$

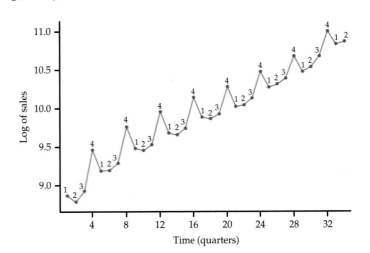

FIGURE 14.50 Time plot of Amazon quarterly sales in logged units.

To make predictions of sales in the original units, millions of dollars, we can first predict sales in logged units using the preceding fitted equation. The series ended with the second quarter of 2018 and with $t = 34$. The next period would then be the third quarter of 2018 with $t = 35$, which means a forecast of

$$\log(\widehat{\text{Sales}}_{35}) = 9.1991 + 0.0571t - 0.3072Q1 - 0.3629Q2 - 0.3375Q3$$
$$= 9.1991 + 0.0571(35) - 0.3072(0) - 0.3629(0) - 0.3375(1)$$
$$= 10.8601$$

At this stage, we untransform the log prediction value by applying the exponential function, which on your calculator is the e^x key.

$$e^{10.8601} = 52057.28$$

mean response, p. 575

However, it should be noted that by exponentiating the predicted value of $\log(y)$, the untransformed value provides an estimate of the median response at the given values of the predictor variables. Recall that standard regression provides an estimate for the mean response at given values of the predictor variables. If prediction of the mean is desired, then an adjustment factor must be applied to the untranformed prediction.[29]

regression standard error, p. 578

The adjustment is accomplished by multiplying the untransformed prediction by $e^{s^2/2}$, where s is the regression standard error from the log-based regression. From the regression output for the fitting of logged sales, we find that $s = 0.0680333$. The predicted mean sales is then:

$$\widehat{\text{Sales}}_{35} = 52057.289 e^{s^2/2}$$
$$= 52057.28 e^{0.0680333^2/2}$$
$$= 52057.28(1.0023)$$
$$= 52177.01$$

In this application, the regression standard error is small, which results in only a slight adjustment to the initial untransformed value. The forecast is in millions of dollars. So third-quarter sales are forecasted to be about $52.177 billion. ∎

Figure 14.51 displays the original sales series with the trend-and-season model predictions using the mean adjustment described in the preceding example. The fitted values not only follow the trend nicely, but also do a fairly good job of adapting to the changes in seasonal variation.

Be aware that the adjustment made in Example 14.35 to the untransformed log prediction is *not* automatically done by most software. For example, consider the exponential trend-fitted models reported by Minitab and

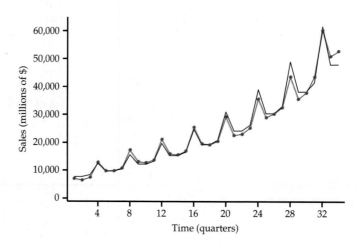

FIGURE 14.51 Trend-and-season model fitted to Amazon sales.

Excel in Example 14.30 (page 738) for the total investment in fixed assets series. These models are simply the result of exponentiating a linear trend fit of log(y) with no adjustments made. Technically, this implies that the reported exponential trend models from software are estimating the median response in original units. However, as we saw in Example 14.35, the differences in predicted values with or without the adjustment are relatively small when the log fit has a small regression standard error, as is also the case in the application of Example 14.30.

Finally, it is important to note that *no* adjustments are required when untransforming a prediction interval. To obtain a prediction interval for a response in original units, we first obtain the prediction interval using the regression for the transformed data and then simply untransform back the lower and upper endpoints of the prediction interval. Hence, if the prediction interval for log(y) is (a,b), then the prediction interval for y in original units is (e^a, e^b).

APPLY YOUR KNOWLEDGE

14.47 Monthly office supply and stationery stores sales. Consider the monthly office supply and stationery stores sales series discussed in Examples 14.32–14.34. Using the trend-and-season fitted model shown in 14.49(b) (page 744), provide forecasts for the eight remaining months of 2018.

14.48 E-commerce retail sales. Consider the time series on the quarterly e-commerce retail sales as a percent of total U.S. retail sales first introduced in Example 14.16 (page 713). Additive seasonal indexes were subsequently estimated for this series in Example 14.22 (page 723) to then produce a deseasonalized series as illustrated in Example 14.23 (page 724). Finally, in Example 14.31 (page 740), the deseasonalized series was fit with a linear trend line to produce a forecast equation based on seasonal indexes as shown in Example 14.31. In this exercise, approach the modeling of the series exclusively using regression. In particular, create indicator variables for the first, second, and third quarters along with a trend index (t). Call these indicator variables Q1, Q2, and Q3. ESALES

(a) Use statistical software to fit a multiple regression model of trend plus the three indicator variables. Write the estimated trend-and-season model.

(b) You will notice that all quarterly indicators in the model have negative coefficient values. (*Hint:* the baseline trend model for quarter 4 is when all the indicator variables have value 0.)

(c) The series ended on the first quarter of 2018. Produce a table of forecasts for the remaining quarters of 2018 based on the seasonal index model of Example 14.31 and the regression model of part (a). Are the forecasts close in value? If so, explain why.

14.49 Total investment in China. In Example 14.30 (page 738), an exponential trend model was fit to the time series of annual total investment in fixed assets in China. Based on the fitted model provided in the example, what is the fitted model for total investment in logged units? This problem should be done without the use of computer software. Only a calculator is needed.

Residual checking

In our discussions (page 736) following the trend fit of cable sales in Example 14.28, we noted that if the residuals from a regression applied to time series data are consistent with a random process, then the estimated model has captured well the systematic components of the series.

Because the cable series seems to be influenced only by trend effect, the residuals from the trend-only model are consistent with a random process, as seen in Figure 14.40 (page 736). In contrast, fitting a trend-only model to a time series influenced by both trend and seasonality will result in residuals that show seasonal variation. Similarly, fitting a seasonal-only model to a time series influenced by both trend and seasonality will result in residuals that show trend behavior.

Modeling the trend and seasonal components correctly does not, however, guarantee that the residuals will be random. What trend and seasonal models do not capture is the influence of past values of time series observations on current values of the same series. A simple form of this dependence on past values is known as *autocorrelation*. In Section 14.1, we introduced the ACF, which provides us with correlation estimates between observations separated by certain numbers of periods apart known as the lags. Let's see what insights the ACF can provide us in our modeling of the Amazon sales series.

EXAMPLE 14.36

CASE 14.2

Residual Analysis of the Amazon Sales Model Figure 14.52 plots the residuals from the trend-and-season regression model for logged Amazon sales given in Example 14.35. Notice the pattern of long runs of positive residuals and long runs of negative residuals. The residual series appears to "snake" or "meander" around the mean line. This appearance is due to the fact that successive observations tend to be close to each other. A precise technical term for this behavior is **positive autocorrelation.**

positive autocorrelation

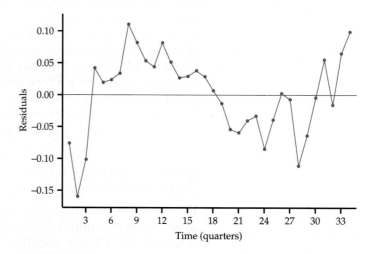

FIGURE 14.52 Time plot of residuals from the trend-and-season model for logged Amazon sales.

Our visual impression of nonrandomness is confirmed by the ACF of the residuals shown in Figure 14.53. We see that the lag 1 autocorrelation stands out well beyond the significance limits. Thereafter, the remaining autocorrelations show a "block-like" pattern within the ACF significance limits which is further indication of a persisting nonrandomness to the residual series. ∎

Durbin-Watson statistic

When residuals are correlated, it is often due to a dependency between successive residual values as seen in Example 14.36. Because of this fact, there is a test statistic known as the **Durbin-Watson statistic** that focuses on testing of the lag 1 autocorrelation of the residuals. Most software packages provide the option of reporting the Durbin-Watson statistic with the regression output. Some software packages provide an associated P-value, while many do not. Without a reported P-value, the precise interpretation entails the use of a specialized table of critical values—that we shall not discuss here. The Durbin-Watson statistic provides no information about autocorrelations at higher lags. If using the Durbin-Watson statistic, view it as only a quick and preliminary check for possible problems in the residuals. We recommend

FIGURE 14.53 Minitab ACF of residuals from the trend-and-season model for logged Amazon sales.

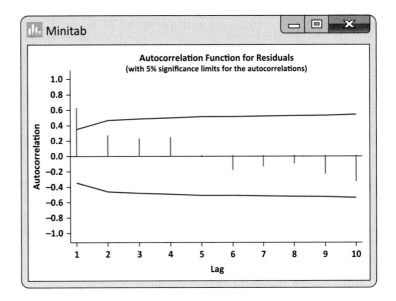

the use of the ACF as a more comprehensive check of the randomness of the residuals over different lags.

Autocorrelation in the residuals indicates an opportunity to improve the model fit. There are several approaches for dealing with autocorrelation in a time series. In the next section, we explore using past values of the time series as explanatory variables added to the model.

APPLY YOUR KNOWLEDGE

14.50 Monthly office supply and stationery stores sales. Refer to the AIC-selected model of Example 14.34 (page 743) for the monthly office supply and stationery stores sales series. Fit the same model shown in Figure 14.49(b) (page 744) and obtain the residuals. OFFICE

(a) Make a time plot of the residuals. Do the residuals suggest an adequacy with the trend-and-seasonal model fit? Explain.

(b) Obtain an ACF for the residual series. What do you conclude? What is the implication about the adequacy of the trend-and-seasonal model fit?

(c) Obtain a histogram and Normal quantile plot for residuals. What do you conclude?

Modeling with lagged variables

With the examples so far, we have used a time period variable and/or seasonal indicator variables as the predictor variables. If such explanatory variables are sufficient in modeling the patterns in the time series, then the residuals will be consistent with a random process. As we saw for the Amazon sales series in Example 14.36, residuals can still exhibit nonrandomness, even after trend and seasonal components are incorporated in the model. Our analysis revealed that the residuals showed a meandering behavior.

In Section 14.2, we learned that certain strongly meandering series, known as random walks, can be modeled by a simple approach, namely, differencing the series, leaving us to work with a random series. As noted earlier, differencing is an important technique used in the modeling of ARIMA models (page 704). With this said, differencing is not necessarily the appropriate tool for analyzing all meandering series. In fact, differencing can sometimes be counterproductive and inadvisable. This leads us to examples where the use of regression with the use of *past* values of the time series is a useful course of action.

> ### LAG VARIABLE
>
> A **lag variable** is a variable based on the past values of the time series.

Recall that we introduced the idea of a lag variable in our development of the ACF. Notationally, if y_t represents the time series in question, then lag variables are given by $y_{t-1}, y_{t-2}, \ldots$, where y_{t-1} is called a *lag one variable*, y_{t-2} is called a *lag two variable* and so on. Lag variables can be added to the regression model as explanatory variables. The regression coefficients for the lag variables represent multiples of past values for the prediction of future values.

autoregressive

Sometimes, the best explanatory variables are simply past values of the response variable. **Autoregressive** time series models take advantage of the linear relationship between successive values of a time series to predict future values of the series. The simplest of autoregressive models is one that uses only the previous period's value as a predictor.

> ### FIRST-ORDER AUTOREGRESSIVE MODEL
>
> A **first-order autoregressive model** specifies a linear relationship between successive values of the time series. The shorthand for this model is AR(1), and it is given by
>
> $$y_t = \beta_0 + \beta_1 y_{t-1} + \varepsilon_t$$

The AR(1) model was mentioned in the "Beyond the Basics" section (page 705) in that the random walk model is a special case of the AR(1) model with $\beta_1 = 1$. This allowed us to present a special class of tests for random walks known as unit root tests. The AR(1) model can be expanded to include more lag terms. For example, a time series that is dependent on lag one and lag two variables is referred to as an AR(2) model. In general, an AR(p) model is built on the explanatory variables of $y_{t-1}, y_{t-2}, \ldots, y_{t-p}$. Autoregressive models constitute a subset of general ARIMA models; ARIMA models were briefly described earlier (page 704).

EXAMPLE 14.37

NFLX

stationary process, p. 679

Daily Trading Volume of Netflix For many investors and stock analysts, trading volume (quantity of shares that change owners) is viewed as an important metric for showing level of interest in a stock.[30] As such, there is particular interest in studying the movements of trading volume over time for making stock trading decisions. For our purposes, we ask whether trading volumes are random or if they exhibit some form of regularity. Figure 14.54 is a time plot of daily trading volumes for Netflix stock from January 2 to July 31, 2018, with the mean trading volume indicated. It appears that there are "strings" of observations either all below or all above the mean. With these strings, we see that if an observation is below (above) the mean, then the next observation quite often will also be below (above) the mean. But, notice that as the observations tend to meander away from the mean level, they "revert" back to the mean level. The constancy of the general level is a property of a stationary process. The ACF of the volumes shown in Figure 14.55 confirms that the series exhibits nonrandom behavior. ∎

Notice from Figure 14.54 that the more extreme volumes pull in one direction, namely, upward. If we were to continue our analysis of the volumes, we would find the distribution of residuals to be right-skewed. This situation is

FIGURE 14.54 Time plot of daily trading volume of Netflix stock (January 2 to July 31, 2018).

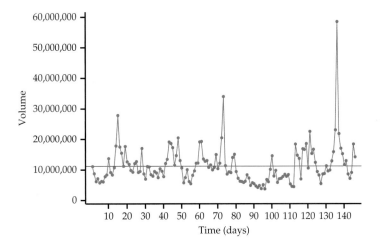

FIGURE 14.55 Minitab ACF of daily trading volume of Netflix stock.

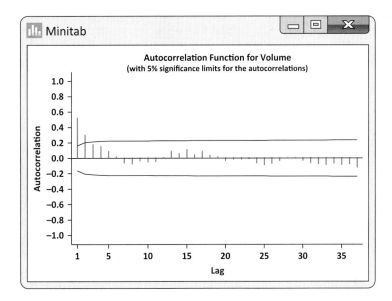

similar to the situation of Case 12.1, in which we needed to take the log transformation of income due to its skewness (page 572). We do the same with volumes and transform them with the natural log.

In pursuing the use of lag variables as predictor variables, the natural question is how many lag variables should we use? Looking at the ACF in Figure 14.55, we see that the first two autocorrelations are significant. Does that imply that we should entertain a multiple regression on two lag variables? Imagine that there is a strong positive lag one autocorrelation. This would result in adjacent observations being close to each other. In turn, observations two periods apart will also tend to be close to each other, resulting in a positive lag two autocorrelation. So, the lag two autocorrelation is reflecting the effects of the lag one autocorrelation. What we need to know is the correlation between observations two apart after we have adjusted for the effects of the lag one autocorrelation. Similarly, we would want the correlation between observations three apart after we have adjusted for the effects of the lag one and lag two autocorrelations. Such an adjusted correlation is known as partial autocorrelation. Akin to the ACF, a plot of the partial autocorrelations against the lags is called a **partial autocorrelation function (PACF).**

partial autocorrelation function (PACF)

In Figure 14.56, JMP-produced ACF and PACF for the logged volumes are presented side by side. As with the ACF, the PACF superimposes significance limits based on the 5% level of significance. The PACF is showing only the lag

FIGURE 14.56 JMP ACF and PACF of logged daily trading volume of Netflix stock.

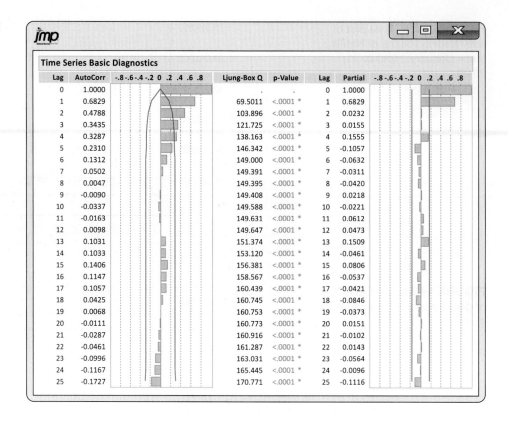

one correlation to be significant; remember to ignore the correlation at lag 0, which is always 1. This suggests that we need only the lag one variable for our model. In other words, we should pursue the fitting of an AR(1) model.

There is another aspect of Figure 14.56 that suggests the pursuit of an AR(1) model. In general, a stationary AR(p) process is associated with an ACF with exponentially decaying autocorrelations, while its PACF will show significant partial autocorrelations up to lag p followed by a sharp drop (cutoff) in their values. Indeed, from Figure 14.56, the ACF is showing decaying autocorrelations, and the PACF shows a spike at lag one and then a sharp drop in the partial autocorrelation values thereafter. This is a classic signature of a stationary AR(1) process.

As we have done with the trend and seasonal applications, we will estimate the AR(1) model using regression. Recall that the method of least squares is used to estimate the regression parameters. In more technical language, these estimates are called ordinary least squares (OLS) estimates. For small samples, OLS estimates of autoregressive models—like the AR(1)—have a discernible bias. However, it is a reassuring fact that as the sample size increases, this bias dissipates to zero. As such, for many applications (including ours), OLS estimators provide perfectly satisfactory results. With this said, there are processes within the general class of ARIMA models for which we cannot directly apply OLS to estimate model parameters. Advanced treatments of time series modeling will necessarily include exposure to alternatives to OLS estimators.

EXAMPLE 14.38

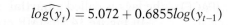

NFLX

Fitting Netflix Trading Volumes Figure 14.57 displays regression output for the estimated AR(1) fitted model of logged volumes. In summary, the fitted model is given by

$$\widehat{log(y_t)} = 5.072 + 0.6855 log(y_{t-1})$$

From the R^2 value, the fitted model explains about 47% of the variation in the logged volume series. The P-value for the two-sided test of H_0: $\beta_1 = 0$ is reported to

FIGURE 14.57 Estimated AR(1) model for logged Netflix volumes.

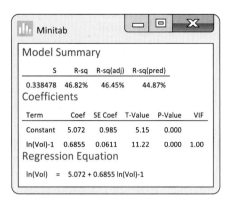

be 0.000. This indicates strong evidence of a relationship between the logged volume variable and its lag. This significant result is consistent with what we observed with the PACF of Figure 14.56. (*Note:* unlike the testing of $\beta_1 = 1$ (page 706), the testing of $\beta_1 = 0$ can be conducted with standard regression output.)

To check the adequacy of the AR(1) fitted model, we turn our attention to the time plot of residuals in Figure 14.58(a). The residuals show a mix of short runs and some oscillation with no strong tendency toward one or the other behavior. In other words, the residuals resemble a random series. The ACF of residuals shown in Figure 14.58(b) confirm that the residuals are indeed consistent with randomness.

The log volume series ends with $log(y_{146}) = 16.7203$, which is slightly greater than the overall mean of 16.121. The forecast for the next-day log volume is

$$\widehat{log(y_{147})} = 5.072 + (0.6855)(16.7203) = 16.5338$$

Notice that the forecast of 16.5338 is slightly less than the observed value of 16.7203. The forecast reflects the tendency of the series to revert back to the overall mean. ■

Our forecast of the next day's log volume was based on an *observed* data value—today's log volume, which is a known value in our time series. If we wish to forecast two days into the future, we have to base our forecast on an *estimated* value because the next day's value is not a known value in our time series.

FIGURE 14.58 (a) Time plot of residuals from AR(1) fit and (b) Minitab ACF of residuals from AR(1) fit.

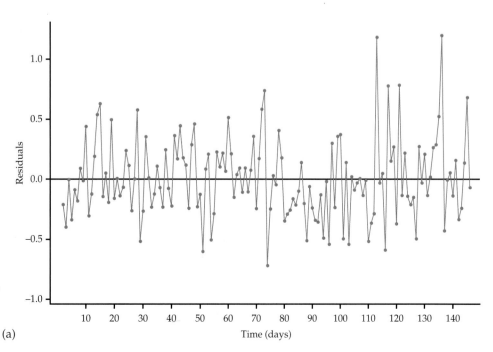

(a)

FIGURE 14.58 Continued

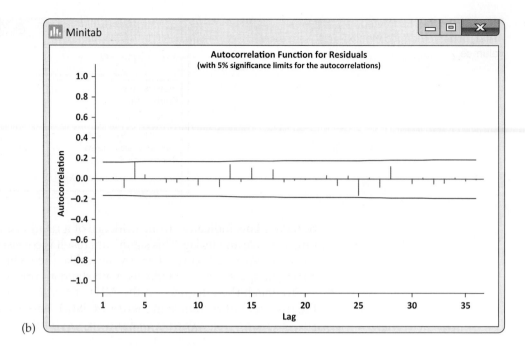

(b)

EXAMPLE 14.39

Forecasting Two Periods Ahead Because our time series ends with $\log(y_{146})$, the value of $\log(y_{147})$ is not known. In its place, we will use the value of $\widehat{\log(y_{147})}$ calculated in Example 14.38:

$$\widehat{\log(y_{148})} = 5.072 + 0.6855\widehat{\log(y_{147})}$$
$$= 5.072 + (0.6855)(16.5338)$$
$$= 16.4059$$

The forecast two periods into the future shows further reversion back to the overall mean. This forecast trajectory is intuitive for a process meandering above and below some horizontal mean level. In the short term, future observations are expected to be close to recent observations. However, in the long run, it is not possible to know if the process has meandered above or below the overall mean. Thus, the best guess in the long run is simply the overall mean. ∎

The process shown in Example 14.39 can be repeated to produce forecasts even further into the future.

In Chapters 12 and 13, we calculated prediction intervals for the response in our regression model. The regression procedures of these chapters can be used to construct prediction intervals for one-step ahead forecasts based on an AR(1) model fit. A difficulty arises when seeking two-step ahead or more prediction intervals. The one-step ahead forecast $\widehat{\log(y_{147})}$ depends on our estimated values of β_0 and β_1 and the *known* value of $\log(y_{146})$. However, our forecast of $\log(y_{148})$ depends on estimated values of β_0, β_1, *and* $\log(y_{147})$. The additional uncertainty involved in estimating $\log(y_{147})$ makes the two-ahead prediction interval wider than the one-step ahead prediction interval. However, as the long-run forecasts settle down to the overall mean, the width of the prediction interval stabilizes to a constant width, mimicking prediction intervals for random processes that are based on the overall mean as a forecast.

Standard regression procedures do not incorporate the extra uncertainty associated with using estimates of future values in the regression model. In this situation, you want to use time series routines in statistical software to calculate the appropriate prediction intervals. Using JMP's AR(1) model-fitting

FIGURE 14.59 Time plot of logged Netflix volumes with forecasts and prediction limits.

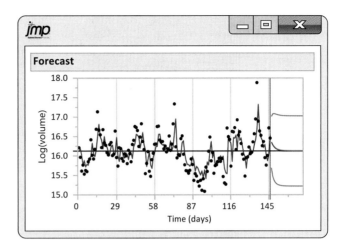

routine, Figure 14.59 displays forecasts along with prediction limits for several days into the future. Notice that the forecasts do indeed converge to the overall mean line. Also, it can be seen that the width of the prediction interval initially increases and then converges to a constant width.

EXAMPLE 14.40

LAKE

CASE 14.3 **Modeling Lake Michigan Water Levels** Consider again the annual average water level of Lake Michigan studied in Example 14.15 (page 712). Refer to Figure 14.23 (page 711) to see the meandering behavior of the lake levels with tendency to revert back after drifting away (upward or downward). There does not seem to be any strong evidence for a long-term trend. Given the meandering behavior of the lake levels with no obvious trending, the series seems to be a good candidate for modeling with lag variables only.

Figure 14.60 shows JMP-produced side-by-side ACF and PACF output. Notice that the autocorrelations of the ACF are exponentially decaying while

FIGURE 14.60 JMP ACF and PACF of lake levels.

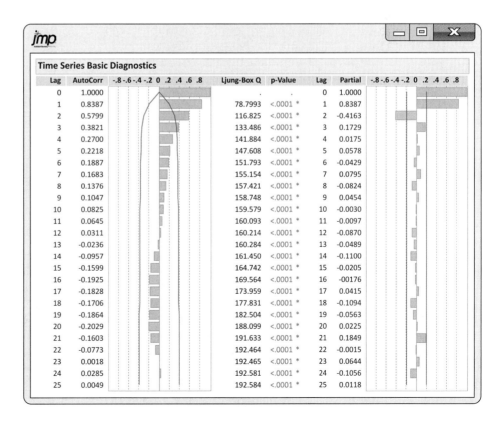

partial autocorrelations for lags one and two of the PACF standout. These patterns are consistent with an AR(2) process. Accordingly, we create a lag one and lag two variable to estimate an AR(2) model. We find the following estimated model:

$$\hat{y}_t = 127.7 + 1.2163 y_{t-1} - 0.4369 y_{t-2}$$

The residuals from this fit can be found to exhibit random behavior, indicating that the AR(2) is a good model for the data. Figure 14.61 shows that the AR(2) model does a nice job tracking the water levels over time. ∎

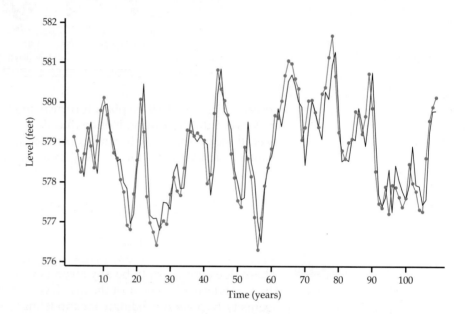

FIGURE 14.61 AR(2) fitted model (maroon) superimposed on annual lake levels (blue).

The Netflix and Lake Michigan applications gave us opportunities to model series influenced only by autoregressive effects, one as an AR(1) model and the other as an AR(2) model. These autoregressive models are based exclusively on lag variables as predictor variables. In some cases, we need to include lag variables along with other predictor variables to capture the systematic effects.

EXAMPLE 14.41

AMAZON

CASE 14.2 **Fitting and Forecasting Amazon Sales** Example 14.36 (page 748) showed us that the trend-and-season model given in Example 14.35 (page 745) fails to capture the lingering autocorrelation effect between successive observations. One strategy is to add a lag variable to the trend-and-seasonal model. In other words, we need to consider a model for logged sales that combines all the effects:

$$\log(\text{Sales}_t) = \beta_0 + \beta_1 t + \beta_2 Q1 + \beta_3 Q2 + \beta_4 Q3 + \beta_5 \log(\text{Sales}_{t-1}) + \varepsilon_t$$

From software, we find the estimated model to be:

$$\widehat{\log(\text{Sales}_t)} = 3.2097 + 0.0182t - 0.5272 Q1 - 0.3836 Q2 - 0.3119 Q3 + 0.6804 \log(\text{Sales}_{t-1})$$

Figure 14.62 shows the ACF for the residuals from the above fit. Compare this ACF with the trend-and-seasonal residuals ACF of Figure 14.53 (page 749). Other than a slight breach of the significance limits at lag 4, we can see that the lag effects have been basically captured and thus result in a better fitting model.

In terms of forecasting, the series ends with second-quarter 2018 sales of $52,886 (in millions). The ending quarter is the 34th period in the series, so our notation is $y_{34} = 52{,}886$.

FIGURE 14.62 Minitab ACF of residuals for trend-and-season-lag model for logged Amazon sales.

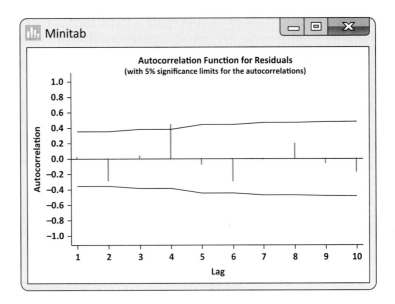

First, use the model to forecast the logarithm of third-quarter 2018 sales:

$$\log(\widehat{Sales}_{35}) = 3.2097 + 0.0182(35) - 0.5272(0) - 0.3836(0) - 0.3119(1)$$
$$+ 0.6804 \log(Sales_{34})$$
$$= 3.5348 + 0.6804 \log(52886)$$
$$= 3.5348 + (0.6804)(10.87589)$$
$$= 10.93476$$

We can now untransform the log predicted value:

$$e^{10.93476} = 56092.65$$

Recall from the discussion of Example 14.35 (page 745) that the preceding fitted value provides a prediction of median sales. If we want a prediction of mean sales, we need to obtain the regression standard error from our log fit. From the regression output for the fitting of logged sales, we find $s = 0.051032$. The predicted mean sales is then:

$$\widehat{Sales}_{35} = 56092.65 e^{s^2/2}$$
$$= 56092.65 e^{0.051032^2/2}$$
$$= 56092.65(1.0013)$$
$$= 56165.57 \quad \blacksquare$$

In Example 14.35 (page 745), the exponential trend-and-season model predicted sales to be $52.177 billion. Our inclusion of the lag one term in the model adjusted the forecast up by $3.989 billion.

APPLY YOUR KNOWLEDGE

14.51 PMI series. Refer to the monthly PMI series discussed in Example 14.24 (page 726). PMI

(a) Test the randomness of the PMI series with an ACF. What do you conclude?

(b) Create a lag one variable of PMI and plot PMI versus its lag. Does the plot suggest the use of the lag variable as a possible predictor variable? Explain.

(c) Obtain a PACF for the PMI index series. What does the PACF suggest as a possible model to fit to the data?

14.52 PMI series, continued. Let y_t denote the observed value of PMI for time period t. PMI

(a) Use software to fit a simple linear regression model, using y_t as the response variable and y_{t-1} as the explanatory variable. Record the estimated regression equation.

(b) Test the residuals for randomness with a time plot and an ACF. Does it appear that the AR(1) accounts for the systematic movements in the PMI series? Explain.

(c) The PMI series ends on May 2018. Use the fitted AR(1) model from part (a) to obtain PMI forecasts for the next three months. What do you notice about the forecast values? In what way are they similar to the forecast values shown in Figure 14.59 (page 755)?

BEYOND THE BASICS

ARCH models

The primary focus of the time series methods presented in this chapter is to track and predict the mean level of a series. Our modeling proceeded on the basis that the variance of the series is unchanging over time. Financial and macroeconomic time series often reflect time-changing variance. In Figures 14.5 (page 682) and 14.18(b) (page 699), we observed steadily increasing variance. For those applications, we applied the logarithmic transformation to the time series. The transformation was used not to model the variance per se but rather to stabilize it before modeling the mean level. In finance applications, variability is often referred to as volatility and its direct modeling plays an important role in the management of financial risk.

EXAMPLE 14.42

HONDA2

Honda Returns Consider the weekly closing prices of Honda Motor Co. over a time horizon of more than three decades through mid-July 2018.[31] Figure 14.63 shows the weekly returns as approximated by the first differences of the logged prices. The returns appear to be a random process around an average of essentially 0. However, a close inspection of the returns reveals periods of increased or decreased variance (volatility). More specifically, there appears to be dependence in the level of variability in the sense that a volatile period tends to be followed by another volatile period. Similarly, a relatively quiet period tends to be followed by another quiet period. Such behavior in a financial market is referred to as "volatility clustering." ∎

To model changing variance, an autoregressive conditionally heteroscedastic (ARCH) model was proposed by Nobel Prize economist Robert Engle. The simplest ARCH model is a first-order ARCH model or ARCH(1) model. ARCH(1) models the returns process as follows:

$$r_t = \sigma_t a_t$$
$$\sigma_t^2 = \alpha_0 + \alpha_1 r_{t-1}^2$$

where r_t is the return at time t and the random deviations a_t represent noise following the standard Normal distribution (i.e., $\mu_a = 0$ and

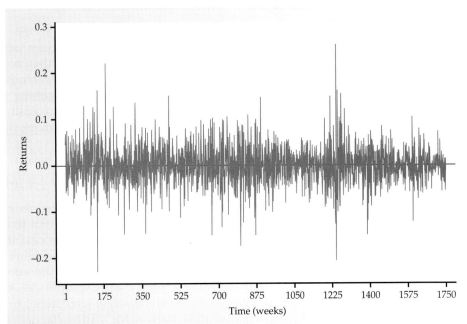

FIGURE 14.63 Approximated returns of Honda stock (January 1985 through mid-July 2018).

$\sigma_a = 1$). We use the notation of a_t for the noise process here as opposed to our earlier use of ε_t (page 679) because a_t represents a special case of a noise process in that its standard deviation is equal to 1. With this setup, the returns have a mean of 0 but their variance is allowed to change as implied by the subscript t found with σ_t and σ_t^2. To ensure that σ_t^2 is nonnegative, the parameters of α_0 and α_1 are restricted to be nonnegative.

The equations don't have the immediate appearance of an autoregressive model as suggested in the name of the model. The "trick" is to square the first equation and switch around the second equation:

$$r_t^2 = \sigma_t^2 a_t^2$$
$$\alpha_0 + \alpha_1 r_{t-1}^2 = \sigma_t^2$$

Subtracting the two equations gives us

$$r_t^2 - (\alpha_0 + \alpha_1 r_{t-1}^2) = \sigma_t^2 a_t^2 - \sigma_t^2$$

We can now express r_t^2 in the following manner:

$$\begin{aligned} r_t^2 &= \alpha_0 + \alpha_1 r_{t-1}^2 + \sigma_t^2 a_t^2 - \sigma_t^2 \\ &= \alpha_0 + \alpha_1 r_{t-1}^2 + \sigma^2(a_t^2 - 1) \\ &= \alpha_0 + \alpha_1 r_{t-1}^2 + w_t \end{aligned}$$

We now see the appearance of an AR(1) model in that there is a linear relationship between successive squared return values! The w_t term represents a zero-mean noise process. However, this noise process does not follow the Normal distribution. Instead, it is related to the chi-square (χ^2) distribution.

EXAMPLE 14.43

HONDA2

ARCH(1) Fit Since in Example 14.42 the average return is essentially equal to zero, we will immediately square the return values shown in Figure 14.63. If the average is significantly different than zero, the modeling of the described ARCH(1) would call for subtracting the average from each of the return values and *then* squaring the resulting values. Even though more sophisticated estimation methods are typically used for estimating the ARCH model, a standard regression can provide us with a satisfactory fit. Figure 14.64 shows the regression output for the squared values regressed on their lags. Based on the output, we have

$$\hat{r}_t^2 = 0.0013527 + 0.2619145 r_{t-1}^2$$

A pertinent question is whether the estimated slope coefficient of α_1 ($\hat{\alpha}_1 = 0.2619145$) is significantly different than 0. As noted just prior to this example, the underlying error process for the ARCH model is not Normal. For small sizes, this non-Normality would undermine standard regression inference. However, the sample size here is nearly 1750 observations. For such a large sample size, the t statistic for testing the slope equaling 0 can be safely interpreted in the standard manner. In particular, the t statistic for testing $H_0: \alpha_1 = 0$ is 11.34, and the associated P-value is less than 0.0001, implying there is strong evidence that successive squared returns are related. Thus, there is evidence of the volatility clustering in the original return series.

The primary use of an ARCH model is to predict future variances. Returning to the initial setup of the ARCH model, we have the following estimated equation:

$$\hat{\sigma}_t^2 = 0.0013257 + 0.2619145 r_{t-1}^2$$

As an example, the return series ends with $r_{1750} = -0.010239$. The forecast for the variance of next week's return in the future is

$$\hat{\sigma}_{1751}^2 = 0.0013257 + 0.2619145(-0.010239)^2 = 0.0013532$$

The sample variance of *all* the observed returns can be found to be $s^2 = 0.0017941$, which is larger than next week's predicted variance. This prediction of lower variance (volatility) is consistent with the time plot of the returns in Figure 14.63. Namely, the end of the series is showing a cluster of lower volatility in the returns relative to the overall variance of the return series. ∎

ARCH models can be expanded to include more lags of the squared returns than just the first lag. There is also a generalization of the ARCH model known as a GARCH model that allows for variance at any given time to be modeled as a function of past squared returns and also of past variances of the process.

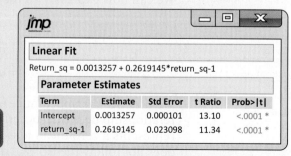

FIGURE 14.64 Regression estimate of ARCH(1) model.

SECTION 14.4 SUMMARY

• Regression methods can be used to model the deterministic trend and **seasonal variation** in a time series.

• **Indicator variables** can be used to model the seasons in a time series. When indicator variables are used in a regression applied to untransformed data, the indicator variables capture **additive** seasonal effects.

• Transformations, such as the logarithm, can simplify the regression modeling process for trend and seasonal fitting. Seasonal indicator variables used in the modeling of logged data capture **multiplicative** seasonal effects.

• Examine the **residuals** from a regression-based time series model to see if there is any evidence of systematic patterns, such as **autocorrelation,** that are not adequately captured by the model.

• The **first-order autoregressive model AR(1)** is appropriate when successive values of a time series are linearly related. An AR(1) model can be estimated by regressing the time series on a **lag one variable.** In some cases, more lags can be added to the model to improve the fit.

• The **partial autocorrelations function (PACF)** shows adjusted autocorrelations that help in determining how many lags to add to the model.

• Lag variables can be added to trend and seasonal models to capture autocorrelation effects that are not captured by trend and seasonal indicator variables.

SECTION 14.4 EXERCISES

For Exercises 14.43 to 14.46, see pages 739–740; for 14.47 to 14.49, see page 747; for 14.50, see page 749; and for 14.51 and 14.52, see pages 757–758.

14.53 AT&T DSL business. AT&T provides customers Internet service by means of using standard copper or phone lines for connection (DSL) and by means of dedicated fiber-optic connection (UVerse). Consider a time series of the quarterly number of AT&T customers (in thousands) who have DSL connections.[32] The series begins with the third quarter of 2015 and ends with the second quarter of 2018. DSL

(a) Make a time plot of these data. Describe the overall movement of the data series.

(b) Does there appear to be seasonality?

(c) Fit the time series with a linear trend model. Report the estimated model.

(d) Superimpose the fitted linear trend model on the data series. Does the fit seem adequate? If not, explain why not.

14.54 AT&T DSL business. Continue the previous exercise. Enter the data into a column of Excel and produce a line chart (Insert Tab → Charts). DSL

(a) Right-click on the line graph, choose "Add Trendline," and choose "Polynomial, Order 2" to fit a quadratic trend model. Make sure to put a check mark in "Display equation on chart." Report the fitted equation. Discuss in what way the quadratic fit is an improvement over the linear fit of Exercise 14.53.

(b) Now, double-click on the line graph to return to the trendline options. This time choose "Exponential." Report the fitted equation. Discuss the exponential fit in comparison to the linear fit of Exercise 14.53 and to the quadratic fit of part (a).

(c) The exponential trend fit from Excel is in the form ae^{bt} which is algebraically equal to $a(e^b)^t$ which can be written as ar^t, where $r = e^b$. Compute the value of r with a calculator and provide an interpretation of this value in the context of the application.

(d) Return again to the trendline options and for both the quadratic and exponential fits, project the forecasts 10 periods into the future by entering "10" in the "Forward" forecast dialog box. What are the paths of the forecast projections for each model? Explain which fitted trend model provides a more plausible projection into the future.

14.55 Hotel occupancy rate. A fundamental measure of the well-being of the hotel industry is occupancy rate. Consider monthly data on the average hotel occupancy rate in the United States, beginning in January 2014 and ending in June 2018.[33] HOTEL

(a) Use statistical software to make a time plot of these data.

(b) Does this time series exhibit a trend? If so, describe the trend.

(c) Do these data exhibit a regular, repeating pattern (seasonal variation)? If so, describe the repeating pattern.

14.56 Hotel occupancy rate. Continue the previous exercise. HOTEL

(a) Make indicator variables for the months of the year, and fit a linear trend along with 11 indicators to the data series. Report your estimated model.

(b) Does the regression output suggest the presence of a trend? Explain how you reach your conclusion.

(c) Check the residuals from the trend-seasonal model for randomness with a time plot and an ACF (if available with software). What do you conclude?

(d) Based on the fit, make a forecast for occupancy rate for July 2018.

14.57 Hourly earnings. Consider monthly data on the average hourly earnings of production and nonsupervisory employees in private U.S. industry, beginning in January 2014 and ending in July 2018.[34] EARN

(a) Make a time plot of these data. Describe the overall movement of the data series.

(b) Does this time series exhibit a trend? If so, describe the trend.

(c) Do these data exhibit a regular, repeating pattern (seasonal variation)? If so, describe the repeating pattern.

14.58 Hourly earnings. Continue the previous exercise. EARN

(a) Make indicator variables for the months of the year, and fit a linear trend along with 11 indicators to the data series. Report your estimated model.

(b) Does the regression output suggest the presence of seasonality? Explain how you reach your conclusion.

(c) Based on the regression output, which variables appear not to be contributing significantly to the fit? Rerun the regression without those variables and report your estimated model.

(d) Based on the fit from part (c), make a forecast for average hourly earnings for August 2018.

(e) *Optional:* If your software has variable selection guided by AIC or BIC, find and report the model with the smallest AIC (or BIC). Does this model differ from the one fit in part (c)? If so, use this model to forecast average hourly earnings for August 2018 and compare your results with the forecast of part (c).

14.59 Runs test and autocorrelation. Suppose that you have three different time series, S1, S2, and S3. The observed number of runs for series S1 is significantly less than the expected number of runs. The observed number of runs for series S2 is significantly greater than the expected number of runs. The observed number of runs for series S3 is not significantly different from the expected number of runs. If you were to plot observations of a given series against its first lag, explain what you would likely see.

14.60 Monthly retail sales. Figure 14.29 (page 720) shows the monthly total U.S. retail sales (in millions of dollars) of general merchandise stores beginning in January 2015 and ending in April 2015. GRETAIL

(a) Make indicator variables for the months of the year, and fit a linear trend along with 11 indicators to the data series. Report your estimated model.

(b) Check the residuals from the trend-seasonal model for randomness with a time plot and an ACF (if available with software). What do you conclude?

(c) Provide a forecast for sales for May 2018.

14.61 A more compact model. Suppose that you fit a trend-and-seasonal model to a time series of quarterly sales and you find the following:

$$\hat{y}_t = 300 + 20t - 15Q_1 - 15Q_2 - 15Q_3$$

Reexpress this fitted model in a more compact form using only one indicator variable.

14.62 MLB batting average. Consider data on the annual average batting average of all Major League Baseball (MLB) teams in a given year. The data series begins with 1960 and ends with 2018.[35] MLB

(a) Make a time plot of the batting average series. Describe any important features of the time series.

(b) Create a lag one variable and plot the batting average of a given year against its lag. Does the plot suggest the use of the lag variable as a possible predictor variable? Explain.

(c) If your software has the option, obtain a PACF for the batting average series. What does the PACF suggest as a possible model to fit to the data?

CASE 14.2 **14.63 Amazon sales.** In Example 14.41 (page 756), the one-step ahead forecast was calculated. Using the model of Example 14.41, determine the two-step ahead forecast in original dollar units.

14.64 MLB batting average. Continue the analysis of MLB batting averages from Exercise 14.62. MLB

(a) Use software to fit a simple linear regression model, using y_t as the response variable and y_{t-1} as the explanatory variable. Recall this is referred to as an AR(1) model (page 750). Report the estimated regression equation.

(b) Check the residuals from the AR(1) for randomness with a time plot and an ACF (if available with software). What do you conclude?

(c) Use the fitted AR(1) model from part (a) to obtain forecasts for batting averages in 2019 and 2020.

14.65 Time index. Refer to Example 14.29 (page 737) in which trend models (linear and quadratic) were fitted to the urban wage rate series for the years of 2000 to 2016. Figure 14.41 (page 737) shows the Excel-reported linear trend model equation to be $y = 3698.1x - 936.1$. As we noted in the example, the x in the Excel equation does not represent the years of 2000 to 2016, but it is the time index t which, in this case, ranges from 1 to 17. In this exercise, you will explore the use of an index based on the actual year as opposed to the standard time index t. To reduce the effects of rounding error, use the following linear trend model on t:

$$\hat{y}_t = -936.125 + 3698.105392 t$$

Now, consider the variable *Year* taking on the values of 2000, 2001, ..., 2016. UWAGE

(a) As an equation, express the variable t as a function of *Year*.

(b) Substitute the expression for t from part (a) into the linear trend model. Simplify the linear trend model into form of $a + b$ *Year*. What are the values of a and b? How has the use of *Year* in place of t changed the linear trend equation?

(c) Use software to estimate and report the linear trend model based on *Year*. Compare the software-reported trend model with the algebraically derived model of part (b).

(d) Predict the wage rate for 2017 using the trend model based on t and based on *Year*. Confirm that the predictions are equal.

CHAPTER 14 REVIEW EXERCISES

14.66 Just use last month's figures! Working with the financial analysts at your company, you discover that when it comes to forecasting various time series, they often just use last period's value as the forecast for the current period. As noted in the chapter, this is known as a naïve forecast (page 696).

(a) If you could pick the estimates of β_0 and β_1 in the AR(1) model, could you pick values such that the AR(1) forecast equation would provide the same forecasts your company's analysts use? If so, specify the values that accomplish this.

(b) What span k in a moving-average forecast model would provide the same forecasts your company's analysts use?

(c) What smoothing constant w in a simple exponential smoothing model would provide the same forecasts your company's analysts use?

(d) Under what circumstances is the naïve forecast that your company's analysts are using the most appropriate option? Explain your response.

14.67 Forecast accuracy measures. Refer to Example 14.17 (page 715) in which MAD, MSE, and MAPE were calculated for two-week and three-week moving-average models. The measures were computed from the 17 weeks' worth of forecast errors (weeks 4 through 20) shown in Figure 14.26 (page 716). Looking at the computation of MAD and MSE in Example 14.17, we see that the numerator of MAD is the sum of forecast errors in absolute value, while the numerator of MSE is the sum of forecast errors squared. A common mistake is to believe that the square root of the numerator of MSE equals the numerator of MAD.

(a) For the three-week moving-average model of Example 14.17 with reported MSE of 0.0017, compute the square root of the numerator of this MSE.

(b) For the reported MAD of 0.0367, find the value of the numerator of MAD. Compare this value with the value found in part (a).

(c) MAD is in original units while MSE is in squared units. As such, some software report the square root of MSE, which is called root mean square error (RMSE). Calculate RMSE for the two-week and three-week moving-average models of Example 14.17.

14.68 MLB batting average. Consider the annual average batting average series (1960 to 2018) from Exercise 14.62. MLB

(a) Set up an Excel spreadsheet to calculate forecasts based on four-year moving averages. Provide a forecast for batting average in 2019.

(b) Compute the forecast errors for years 1964 to 2018 and calculate MAD, MSE, and MAPE.

14.69 MLB batting average. Continue the analysis of the annual average batting average series (1960 to 2018). MLB

(a) Set up an Excel spreadsheet to calculate forecasts for the time series using an exponential smoothing model with $w = 0.8$. Provide a forecast for the batting average in 2019.

(b) Based on the forecasts calculated for years 1960 to 2018 in part (a), calculate MAD, MSE, and MAPE.

(c) Based on the forecasts calculated for years 1980 to 2016, calculate MAD, MSE, and MAPE. Which smoothing constant value provided better forecasting accuracy?

(d) Use statistical software to determine the optimal smoothing constant w. Results may vary slightly between software.

(e) Based on the reported optimal w, calculate the forecast for the batting average in 2019.

14.70 Los Angeles public library program attendees. Consider monthly data on the number of program attendees at Los Angeles public libraries from July 2013 through July 2017.[36] Library programs include speaker series, homework help, computer training, and storytelling for children. LIBRARY

(a) Using an Excel spreadsheet, compute the seasonal indexes. The final seasonal index values should be adjusted so that their average is 0.

(b) Produce and plot the deseasonalized series. What are your impressions of this plot?

(c) Find the average of the deseasonalized series and provide predictions of the number of program attendees for August, September, and October 2017.

14.71 Los Angeles public library program attendees. Consider data from the previous exercise. LIBRARY

(a) Using an Excel spreadsheet, compute the seasonal ratios. The final seasonal ratio values should be adjusted so that their average is 1.

(b) Produce and plot the deseasonalized series. What are your impressions of this plot?

(c) Find the average of the deseasonalized series and provide predictions of the number of program attendees for August, September, and October 2017.

14.72 CO emissions. Carbon monoxide (CO) is a colorless, odorless gas that is highly toxic. A major source of CO is motor vehicle emissions. Consider data on average CO emissions (grams per mile) for gasoline passenger cars by production year for years 2000 to 2018.[37] Gasoline-electric hybrids, but not electric cars, are included in the estimates. CO

(a) Make a time plot of these data. Describe the overall movement of the data series.

(b) Apply the natural logarithm to the series and make a time plot of the logged data. Describe the pattern of the logged series. What does this pattern suggest about the type of trend model that will fit the original data well? Refer to the discussion in the chapter in your explanation.

14.73 United Airlines monthly price changes. Consider a time series on the monthly changes in the closing prices of United Airlines stock for the year 2017. UNITED

(a) The sample mean of the price changes is $\bar{y} = -0.220833$. Find the observed and expected number of runs.

(b) Determine the P-value of the hypothesis test for randomness. Carry out the test using $\alpha = 0.05$ and summarize the results.

(c) Obtain an ACF for the series. What do you conclude? Is your conclusion consistent with that of part (b)?

14.74 CO emissions. Continue the analysis of annual average CO emissions of gasoline passenger cars from Exercise 14.72. CO

(a) Fit the CO series with a linear trend model. Report the estimated model.

(b) Fit the logged CO series with a linear trend model. Report the estimated model.

(c) Use the fitted model from part (a) to make predictions for CO emissions for each of the years (2000–2018).

(d) Use the fitted model from part (b) to make predictions for logged CO emissions for each of the years (2000–2018). Using the exponential function, convert the logged predictions back to original units. Do *not* apply any further adjustments to the converted values as illustrated with the Amazon application.

(e) Compute the forecast errors for the predictions from parts (c) and (d). *Note:* use the predictions in original units from part (d), not the logged predictions.

(f) Calculate MAD, MSE, and MAPE for the two competing models. Which model performs better based on these accuracy measures?

14.75 United Airlines monthly price changes. Continue the analysis of monthly changes in the closing prices of United stock for 2017 from Exercise 14.73. Moving into 2018, below are the next seven months of price changes:

Month	Jan	Feb	Mar	Apr	May	Jun	Jul
Price change	−0.03	1.68	−1.93	2.05	0.14	10.67	1.32

(a) The sample mean of the price changes for 2017 is $\bar{y} = -0.220833$. Use this sample mean as the forecast for all the monthly price changes in 2018. With this forecast, compute forecast errors for the seven months of 2018.

(b) The monthly price change for December 2017 was 0.42. The naïve forecast for January 2018 is 0.42. Moving forward month by month, find the naïve forecasts and corresponding forecast errors for the seven months of 2018.

(c) Calculate MAD, MSE, and MAPE for the two competing set of forecasts. Which forecast scheme performs better? Provide an intuitive explanation as to why one scheme outperformed the other.

14.76 Annual average temperature in New York City. From climate observing sites throughout the United States, the U.S. government keeps track of various meteorological elements (e.g., temperature, precipitation, and snow fall) at these locations over time. Consider data on the average annual temperature (1900–2017) from the New York City station located in Central Park.[38] NYCTEMP

(a) Make a time plot of the temperature series. Describe the overall movement of the series.

(b) Fit the temperature series with a linear trend model. Report the estimated model.

(c) Check the residuals from the trend model for randomness with a time plot and an ACF (if available with software). What do you conclude?

(d) Test the null hypothesis that the regression coefficient for the trend term is zero. Give the test statistic, P-value, and your conclusion.

(e) Interpret the slope estimate in the context of the application.

14.77 NFL offense. In the National Football League (NFL), many argue that rule changes over the years are favoring offenses. Consider a time series of the average number of points scored by teams in the NFL per regular season from 1990 through 2017.[39] NFLPTS

(a) Make a time plot. Is there evidence that the average number of points scored by teams is trending in one direction? Describe the general movement of the series.

(b) Fit a trend model to the data and report the estimated model. Test the null hypothesis that the regression coefficient for the trend term is zero. Give the test statistic, P-value, and your conclusion.

(c) Check the residuals from the trend model for randomness with a time plot and an ACF (if available with software). What do you conclude?

(d) Based on the trend fit, forecast the average number of points scored by team for the 2018, 2019, and 2020 seasons.

14.78 Annual average temperature in Los Angeles. In Exercise 14.76, consideration was given to the average annual temperature for New York City. In this exercise, consider the average annual temperature (1909–2017) from the Los Angeles downtown station located at the USC campus.[40] LATEMP

(a) Make a time plot of the temperature series. Describe the overall movement of the series.

(b) Fit a trend model to the data and report the estimated model.

(c) Check the residuals from the trend model for randomness with a time plot and an ACF (if available with software). What do you conclude?

(d) Create a lag one variable of temperature and fit a regression model on a trend variable and the lag one variable. Report the estimated model.

(e) Check the residuals from the trend and lag model for randomness with a time plot and an ACF (if available with software). What do you conclude?

(f) Based on the fitted trend-lag model, provide forecasts for the next three years into the future. What direction are the forecasts taking?

(g) Expand out your forecast horizon. Make forecasts for the next 25 years into the future. Calculations are more easily done with a spreadsheet. Plot the forecasts. Discuss the trajectory of the forecasts. Explain why this trajectory makes sense in light of the general movement of the series and in light of the ending observed temperature for 2017.

14.79 U.S. poverty rate. Consider a time series on the annual poverty rate of U.S. residents aged 18 to 64 from 1980 through 2016.[41] POVERTY

(a) Make a time plot. Describe the movement of the data over time.

(b) Obtain the first differences for the series and test them for randomness. What do you conclude?

(c) Would you conclude that the poverty rate series behaves as a random walk? Explain.

14.80 U.S. poverty rate. Continue the analysis of the annual U.S. poverty rate. POVERTY

(a) Set up an Excel spreadsheet to calculate forecasts for the time series using an exponential smoothing model with $w = 0.2$. Provide a forecast for the poverty rate in 2017.

(b) Set up an Excel spreadsheet to calculate forecasts for the time series using an exponential smoothing model with $w = 0.8$. Provide a forecast for the poverty rate in 2017.

(c) Based on the forecasts calculated for years 1980 to 2016 in part (a), calculate MAD, MSE, and MAPE.

(d) Based on the forecasts calculated for years 1980 to 2016, calculate MAD, MSE, and MAPE. Which smoothing constant value provided better forecasting accuracy?

14.81 Annual precipitation. Global temperatures are increasing. Lake Michigan water levels meander up and down (see Figure 14.23, page 711). Do all environmental processes exhibit time series patterns? Consider a time series of the annual precipitation (inches) in Washington, DC from 1885 through 2017.[42] PRECIP

(a) Make a time plot of the precipitation series.

(b) Obtain an ACF for the series. What do you conclude?

(c) Using software, perform a runs test on the series. What do you conclude?

14.82 Annual precipitation. Continue the analysis of the annual precipitation time series. PRECIP

(a) Make a histogram and Normal quantile plot of the precipitation data. What do you conclude from these plots?

(b) What is the standard error of prediction? Show the computations.

(c) What is a 90% prediction interval for the annual precipitation for 2018?

14.83 Guatemala population density. Consider a time series on the annual population density (number of people per square kilometer) in Guatemala from 1961 through

2017.[43] Enter the data into a column of Excel and produce a line chart (Insert Tab → Charts). DENSITY

(a) From the Excel time plot, describe the movement of the data over time.

(b) Since population growth is often exponential, double-click on the line graph to return to the trendline options. This time choose "Exponential." Report the fitted equation.

(c) Now, consider a quadratic trend fit. Right-click on the line graph, choose "Add Trendline," and choose "Polynomial, Order 2" to fit a quadratic trend model. Make sure to put a check mark in "Display equation on chart." Report the fitted equation. From a visual perspective, which trend model (exponential or quadratic) seems to better track the series?

(c) Obtain the forecast errors for each trend model fit and calculate MAD, MSE, and MAPE. Which model has better forecast accuracy measures?

14.84 OPEC basket prices. In 2005, Organization of the Petroleum Exporting Countries (OPEC) introduced a weighted average of prices for petroleum blends produced by OPEC members. OPEC uses the basket price to monitor world oil market conditions. Consider data on the daily basket price from the beginning of January 1, 2016, to August 9, 2018.[44] OPEC

(a) Make a time plot of the price series. Describe any important features of the time series.

(b) Obtain the first differences for the series and test them for randomness using an ACF. What do you conclude?

(c) Would you conclude that the price series behaves as a random walk? Explain.

14.85 Exponential smoothing for OPEC basket prices. Continue the analysis of OPEC basket prices. OPEC

(a) Use statistical software to determine the optimal smoothing constant w. Does this optimal w fall in the traditional range for w? Explain.

(b) To see if the exponential smoothing model with optimal smoothing constant captures the pattern in the OPEC series, check the forecast errors for randomness. In Minitab's exponential smoothing platform, save the forecast errors by clicking the "Storage" button and then selecting "Residuals." From there, obtain an ACF of the residuals. With JMP's exponential smoothing fit, you can simply scroll down and open the grey arrow associated with "Residuals" to then find an ACF of the residuals. From the ACF, what do you conclude about the model fit?

(c) Save the forecasts (predictions) to the data table of the software. Based on the reported optimal w and the forecast for the last day (August 9, 2018) found in the data table, hand compute the forecast of the next daily OPEC price.

14.86 Toronto Raptors point spreads. Refer to Exercise 14.9. In that exercise, the time series of the consecutive regular season game point spreads for the NBA team Toronto Raptors was studied. The results of the exercise show that the series appears to resemble a random series. Without any other information, the point spread series is simply fit with a horizontal line such as the mean. However, by introducing other information, we position ourselves to explain some portion of the previously unexplained variability. Let us consider information of where the game was played, namely, home or away. RAPTORS

(a) In the data file, there is information on where the game was played. Create an indicator variable called Home that is "1" for home games and "0" otherwise. Fit a regression model of point spreads (response variable) on the predictor variable of Home. Report the regression model.

(b) Is there significant evidence that game location is associated with point spreads? State the hypothesis, give the test statistic and P-value, and state your conclusion.

(c) Check the residuals from the trend model for randomness with a time plot and an ACF (if available with software). What do you conclude?

(d) Assuming that the general process behavior remains the same into postseason (playoff) games, what is the predicted point spread for a home game versus an away game?

14.87 Los Angeles public library program attendees. Consider the monthly number of program attendees at Los Angeles public libraries time series from Exercises 14.70 and 14.71. LIBRARY

(a) Make indicator variables for the months of January through November, and fit a linear trend along with 11 indicators to the data series. Report your estimated model.

(b) Check the residuals from the trend-seasonal model for randomness with a time plot and an ACF (if available with software). What do you conclude?

(c) Create a lag one variable for number of attendees and rerun regression on the 11 indicator variables, the trend variable, and the lag one variable. Based on the regression output, which variables appear to not be contributing significantly to the fit? Rerun the regression without those variables and report the estimated model.

(d) Check the residuals from your second-run regression of part (c) for randomness with a time plot and an ACF (if available with software). What do you conclude?

(e) The series ended on July 2017. Based on the second-run regression of part (c), forecast the number of attendees for August, September, and October 2017.

(f) *Optional:* If your software has variable selection guided by AIC or BIC, find and report the model with the smallest AIC (or BIC). Does this model differ from the one fit in part (c)?

14.88 Philadelphia parking violations. Consider data on the number of monthly parking violations issued by the City of Philadelphia from January 2014 through December 2017.[45] PARKING

(a) Does the seasonal variation appear to be additive or multiplicative in nature? Justify your answer.

(b) Make indicator variables for the months of January through November, and fit a linear trend along with 11 indicators to the data series. Report your estimated model.

(c) Check the residuals from the trend-seasonal model for randomness with a time plot and an ACF (if available with software). What do you conclude?

(d) Based on the regression output in part (b), which variables appear to not be contributing significantly to the fit? Rerun the regression without those variables and report the estimated model.

(e) Based on the fit from part (d), make forecasts for number of parking violations for January, February, and March 2018.

(f) *Optional:* If your software has variable selection guided by AIC or BIC, find and report the model with the smallest AIC (or BIC). Does this model differ from the one fit in part (d)? If so, use this model to forecast the number of parking violations for January, February, and March 2018. Compare these forecasts with those of part (e).

14.89 Philadelphia parking violations. Consider the number of monthly parking violations from the previous exercise. PARKING

(a) Using an Excel spreadsheet, compute the seasonal indexes. The final seasonal index values should be adjusted so that their average is 0.

(b) Produce and plot the deseasonalized series. What are your impressions of this plot?

(c) Fit a trend model and report the P-value of the trend coefficient. Is there enough evidence of the presence of a trend?

(d) Given your conclusion of part (c) along with the seasonal indexes, forecast the number of parking violations for January, February, and March 2018.

14.90 Daily trading volume of Netflix. Refer to Example 14.38 (page 752) in which a forecast of logged trading volume of Netflix stock for period 147 was made.

(a) Exponentiate the log prediction value back to original units. What is the interpretation of the estimate?

(b) Refer to Example 14.35 (page 745) and make a similar adjustment to the untranformed value of part (a). (*Note*: You will need to refer to the regression output given in Figure 14.57, page 753.) What is the interpretation of the adjusted estimate?

Tables

TABLE A: Standard Normal Probabilities

TABLE B: Random Digits

TABLE C: Binomial Probabilities

TABLE D: t Distribution Critical Values

TABLE E: F Distribution Critical Values

TABLE F: χ^2 Distribution Critical Values

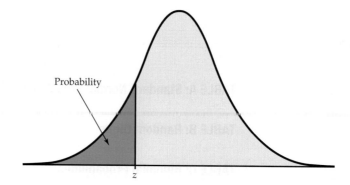

Table entry for z is the area under the standard Normal curve to the left of z.

TABLE A Standard Normal Probabilities

z	.00	.01	.02	.03	.04	.05	.06	.07	.08	.09
−3.4	.0003	.0003	.0003	.0003	.0003	.0003	.0003	.0003	.0003	.0002
−3.3	.0005	.0005	.0005	.0004	.0004	.0004	.0004	.0004	.0004	.0003
−3.2	.0007	.0007	.0006	.0006	.0006	.0006	.0006	.0005	.0005	.0005
−3.1	.0010	.0009	.0009	.0009	.0008	.0008	.0008	.0008	.0007	.0007
−3.0	.0013	.0013	.0013	.0012	.0012	.0011	.0011	.0011	.0010	.0010
−2.9	.0019	.0018	.0018	.0017	.0016	.0016	.0015	.0015	.0014	.0014
−2.8	.0026	.0025	.0024	.0023	.0023	.0022	.0021	.0021	.0020	.0019
−2.7	.0035	.0034	.0033	.0032	.0031	.0030	.0029	.0028	.0027	.0026
−2.6	.0047	.0045	.0044	.0043	.0041	.0040	.0039	.0038	.0037	.0036
−2.5	.0062	.0060	.0059	.0057	.0055	.0054	.0052	.0051	.0049	.0048
−2.4	.0082	.0080	.0078	.0075	.0073	.0071	.0069	.0068	.0066	.0064
−2.3	.0107	.0104	.0102	.0099	.0096	.0094	.0091	.0089	.0087	.0084
−2.2	.0139	.0136	.0132	.0129	.0125	.0122	.0119	.0116	.0113	.0110
−2.1	.0179	.0174	.0170	.0166	.0162	.0158	.0154	.0150	.0146	.0143
−2.0	.0228	.0222	.0217	.0212	.0207	.0202	.0197	.0192	.0188	.0183
−1.9	.0287	.0281	.0274	.0268	.0262	.0256	.0250	.0244	.0239	.0233
−1.8	.0359	.0351	.0344	.0336	.0329	.0322	.0314	.0307	.0301	.0294
−1.7	.0446	.0436	.0427	.0418	.0409	.0401	.0392	.0384	.0375	.0367
−1.6	.0548	.0537	.0526	.0516	.0505	.0495	.0485	.0475	.0465	.0455
−1.5	.0668	.0655	.0643	.0630	.0618	.0606	.0594	.0582	.0571	.0559
−1.4	.0808	.0793	.0778	.0764	.0749	.0735	.0721	.0708	.0694	.0681
−1.3	.0968	.0951	.0934	.0918	.0901	.0885	.0869	.0853	.0838	.0823
−1.2	.1151	.1131	.1112	.1093	.1075	.1056	.1038	.1020	.1003	.0985
−1.1	.1357	.1335	.1314	.1292	.1271	.1251	.1230	.1210	.1190	.1170
−1.0	.1587	.1562	.1539	.1515	.1492	.1469	.1446	.1423	.1401	.1379
−0.9	.1841	.1814	.1788	.1762	.1736	.1711	.1685	.1660	.1635	.1611
−0.8	.2119	.2090	.2061	.2033	.2005	.1977	.1949	.1922	.1894	.1867
−0.7	.2420	.2389	.2358	.2327	.2296	.2266	.2236	.2206	.2177	.2148
−0.6	.2743	.2709	.2676	.2643	.2611	.2578	.2546	.2514	.2483	.2451
−0.5	.3085	.3050	.3015	.2981	.2946	.2912	.2877	.2843	.2810	.2776
−0.4	.3446	.3409	.3372	.3336	.3300	.3264	.3228	.3192	.3156	.3121
−0.3	.3821	.3783	.3745	.3707	.3669	.3632	.3594	.3557	.3520	.3483
−0.2	.4207	.4168	.4129	.4090	.4052	.4013	.3974	.3936	.3897	.3859
−0.1	.4602	.4562	.4522	.4483	.4443	.4404	.4364	.4325	.4286	.4247
−0.0	.5000	.4960	.4920	.4880	.4840	.4801	.4761	.4721	.4681	.4641

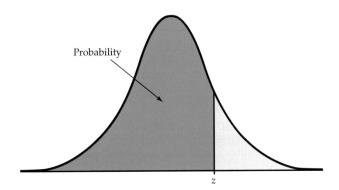

Table entry for z is the area under the standard Normal curve to the left of z.

TABLE A Standard Normal Probabilities (continued)

z	.00	.01	.02	.03	.04	.05	.06	.07	.08	.09
0.0	.5000	.5040	.5080	.5120	.5160	.5199	.5239	.5279	.5319	.5359
0.1	.5398	.5438	.5478	.5517	.5557	.5596	.5636	.5675	.5714	.5753
0.2	.5793	.5832	.5871	.5910	.5948	.5987	.6026	.6064	.6103	.6141
0.3	.6179	.6217	.6255	.6293	.6331	.6368	.6406	.6443	.6480	.6517
0.4	.6554	.6591	.6628	.6664	.6700	.6736	.6772	.6808	.6844	.6879
0.5	.6915	.6950	.6985	.7019	.7054	.7088	.7123	.7157	.7190	.7224
0.6	.7257	.7291	.7324	.7357	.7389	.7422	.7454	.7486	.7517	.7549
0.7	.7580	.7611	.7642	.7673	.7704	.7734	.7764	.7794	.7823	.7852
0.8	.7881	.7910	.7939	.7967	.7995	.8023	.8051	.8078	.8106	.8133
0.9	.8159	.8186	.8212	.8238	.8264	.8289	.8315	.8340	.8365	.8389
1.0	.8413	.8438	.8461	.8485	.8508	.8531	.8554	.8577	.8599	.8621
1.1	.8643	.8665	.8686	.8708	.8729	.8749	.8770	.8790	.8810	.8830
1.2	.8849	.8869	.8888	.8907	.8925	.8944	.8962	.8980	.8997	.9015
1.3	.9032	.9049	.9066	.9082	.9099	.9115	.9131	.9147	.9162	.9177
1.4	.9192	.9207	.9222	.9236	.9251	.9265	.9279	.9292	.9306	.9319
1.5	.9332	.9345	.9357	.9370	.9382	.9394	.9406	.9418	.9429	.9441
1.6	.9452	.9463	.9474	.9484	.9495	.9505	.9515	.9525	.9535	.9545
1.7	.9554	.9564	.9573	.9582	.9591	.9599	.9608	.9616	.9625	.9633
1.8	.9641	.9649	.9656	.9664	.9671	.9678	.9686	.9693	.9699	.9706
1.9	.9713	.9719	.9726	.9732	.9738	.9744	.9750	.9756	.9761	.9767
2.0	.9772	.9778	.9783	.9788	.9793	.9798	.9803	.9808	.9812	.9817
2.1	.9821	.9826	.9830	.9834	.9838	.9842	.9846	.9850	.9854	.9857
2.2	.9861	.9864	.9868	.9871	.9875	.9878	.9881	.9884	.9887	.9890
2.3	.9893	.9896	.9898	.9901	.9904	.9906	.9909	.9911	.9913	.9916
2.4	.9918	.9920	.9922	.9925	.9927	.9929	.9931	.9932	.9934	.9936
2.5	.9938	.9940	.9941	.9943	.9945	.9946	.9948	.9949	.9951	.9952
2.6	.9953	.9955	.9956	.9957	.9959	.9960	.9961	.9962	.9963	.9964
2.7	.9965	.9966	.9967	.9968	.9969	.9970	.9971	.9972	.9973	.9974
2.8	.9974	.9975	.9976	.9977	.9977	.9978	.9979	.9979	.9980	.9981
2.9	.9981	.9982	.9982	.9983	.9984	.9984	.9985	.9985	.9986	.9986
3.0	.9987	.9987	.9987	.9988	.9988	.9989	.9989	.9989	.9990	.9990
3.1	.9990	.9991	.9991	.9991	.9992	.9992	.9992	.9992	.9993	.9993
3.2	.9993	.9993	.9994	.9994	.9994	.9994	.9994	.9995	.9995	.9995
3.3	.9995	.9995	.9995	.9996	.9996	.9996	.9996	.9996	.9996	.9997
3.4	.9997	.9997	.9997	.9997	.9997	.9997	.9997	.9997	.9997	.9998

TABLE B Random Digits

Line								
101	19223	95034	05756	28713	96409	12531	42544	82853
102	73676	47150	99400	01927	27754	42648	82425	36290
103	45467	71709	77558	00095	32863	29485	82226	90056
104	52711	38889	93074	60227	40011	85848	48767	52573
105	95592	94007	69971	91481	60779	53791	17297	59335
106	68417	35013	15529	72765	85089	57067	50211	47487
107	82739	57890	20807	47511	81676	55300	94383	14893
108	60940	72024	17868	24943	61790	90656	87964	18883
109	36009	19365	15412	39638	85453	46816	83485	41979
110	38448	48789	18338	24697	39364	42006	76688	08708
111	81486	69487	60513	09297	00412	71238	27649	39950
112	59636	88804	04634	71197	19352	73089	84898	45785
113	62568	70206	40325	03699	71080	22553	11486	11776
114	45149	32992	75730	66280	03819	56202	02938	70915
115	61041	77684	94322	24709	73698	14526	31893	32592
116	14459	26056	31424	80371	65103	62253	50490	61181
117	38167	98532	62183	70632	23417	26185	41448	75532
118	73190	32533	04470	29669	84407	90785	65956	86382
119	95857	07118	87664	92099	58806	66979	98624	84826
120	35476	55972	39421	65850	04266	35435	43742	11937
121	71487	09984	29077	14863	61683	47052	62224	51025
122	13873	81598	95052	90908	73592	75186	87136	95761
123	54580	81507	27102	56027	55892	33063	41842	81868
124	71035	09001	43367	49497	72719	96758	27611	91596
125	96746	12149	37823	71868	18442	35119	62103	39244
126	96927	19931	36089	74192	77567	88741	48409	41903
127	43909	99477	25330	64359	40085	16925	85117	36071
128	15689	14227	06565	14374	13352	49367	81982	87209
129	36759	58984	68288	22913	18638	54303	00795	08727
130	69051	64817	87174	09517	84534	06489	87201	97245
131	05007	16632	81194	14873	04197	85576	45195	96565
132	68732	55259	84292	08796	43165	93739	31685	97150
133	45740	41807	65561	33302	07051	93623	18132	09547
134	27816	78416	18329	21337	35213	37741	04312	68508
135	66925	55658	39100	78458	11206	19876	87151	31260
136	08421	44753	77377	28744	75592	08563	79140	92454
137	53645	66812	61421	47836	12609	15373	98481	14592
138	66831	68908	40772	21558	47781	33586	79177	06928
139	55588	99404	70708	41098	43563	56934	48394	51719
140	12975	13258	13048	45144	72321	81940	00360	02428
141	96767	35964	23822	96012	94591	65194	50842	53372
142	72829	50232	97892	63408	77919	44575	24870	04178
143	88565	42628	17797	49376	61762	16953	88604	12724
144	62964	88145	83083	69453	46109	59505	69680	00900
145	19687	12633	57857	95806	09931	02150	43163	58636
146	37609	59057	66967	83401	60705	02384	90597	93600
147	54973	86278	88737	74351	47500	84552	19909	67181
148	00694	05977	19664	65441	20903	62371	22725	53340
149	71546	05233	53946	68743	72460	27601	45403	88692
150	07511	88915	41267	16853	84569	79367	32337	03316

TABLE B	Random Digits (continued)							
Line								
151	03802	29341	29264	80198	12371	13121	54969	43912
152	77320	35030	77519	41109	98296	18984	60869	12349
153	07886	56866	39648	69290	03600	05376	58958	22720
154	87065	74133	21117	70595	22791	67306	28420	52067
155	42090	09628	54035	93879	98441	04606	27381	82637
156	55494	67690	88131	81800	11188	28552	25752	21953
157	16698	30406	96587	65985	07165	50148	16201	86792
158	16297	07626	68683	45335	34377	72941	41764	77038
159	22897	17467	17638	70043	36243	13008	83993	22869
160	98163	45944	34210	64158	76971	27689	82926	75957
161	43400	25831	06283	22138	16043	15706	73345	26238
162	97341	46254	88153	62336	21112	35574	99271	45297
163	64578	67197	28310	90341	37531	63890	52630	76315
164	11022	79124	49525	63078	17229	32165	01343	21394
165	81232	43939	23840	05995	84589	06788	76358	26622
166	36843	84798	51167	44728	20554	55538	27647	32708
167	84329	80081	69516	78934	14293	92478	16479	26974
168	27788	85789	41592	74472	96773	27090	24954	41474
169	99224	00850	43737	75202	44753	63236	14260	73686
170	38075	73239	52555	46342	13365	02182	30443	53229
171	87368	49451	55771	48343	51236	18522	73670	23212
172	40512	00681	44282	47178	08139	78693	34715	75606
173	81636	57578	54286	27216	58758	80358	84115	84568
174	26411	94292	06340	97762	37033	85968	94165	46514
175	80011	09937	57195	33906	94831	10056	42211	65491
176	92813	87503	63494	71379	76550	45984	05481	50830
177	70348	72871	63419	57363	29685	43090	18763	31714
178	24005	52114	26224	39078	80798	15220	43186	00976
179	85063	55810	10470	08029	30025	29734	61181	72090
180	11532	73186	92541	06915	72954	10167	12142	26492
181	59618	03914	05208	84088	20426	39004	84582	87317
182	92965	50837	39921	84661	82514	81899	24565	60874
183	85116	27684	14597	85747	01596	25889	41998	15635
184	15106	10411	90221	49377	44369	28185	80959	76355
185	03638	31589	07871	25792	85823	55400	56026	12193
186	97971	48932	45792	63993	95635	28753	46069	84635
187	49345	18305	76213	82390	77412	97401	50650	71755
188	87370	88099	89695	87633	76987	85503	26257	51736
189	88296	95670	74932	65317	93848	43988	47597	83044
190	79485	92200	99401	54473	34336	82786	05457	60343
191	40830	24979	23333	37619	56227	95941	59494	86539
192	32006	76302	81221	00693	95197	75044	46596	11628
193	37569	85187	44692	50706	53161	69027	88389	60313
194	56680	79003	23361	67094	15019	63261	24543	52884
195	05172	08100	22316	54495	60005	29532	18433	18057
196	74782	27005	03894	98038	20627	40307	47317	92759
197	85288	93264	61409	03404	09649	55937	60843	66167
198	68309	12060	14762	58002	03716	81968	57934	32624
199	26461	88346	52430	60906	74216	96263	69296	90107
200	42672	67680	42376	95023	82744	03971	96560	55148

TABLE C Binomial Probabilities

Entry is $P(X = k) = \binom{n}{k} p^k (1-p)^{n-k}$

						p				
n	k	.01	.02	.03	.04	.05	.06	.07	.08	.09
2	0	.9801	.9604	.9409	.9216	.9025	.8836	.8649	.8464	.8281
	1	.0198	.0392	.0582	.0768	.0950	.1128	.1302	.1472	.1638
	2	.0001	.0004	.0009	.0016	.0025	.0036	.0049	.0064	.0081
3	0	.9703	.9412	.9127	.8847	.8574	.8306	.8044	.7787	.7536
	1	.0294	.0576	.0847	.1106	.1354	.1590	.1816	.2031	.2236
	2	.0003	.0012	.0026	.0046	.0071	.0102	.0137	.0177	.0221
	3				.0001	.0001	.0002	.0003	.0005	.0007
4	0	.9606	.9224	.8853	.8493	.8145	.7807	.7481	.7164	.6857
	1	.0388	.0753	.1095	.1416	.1715	.1993	.2252	.2492	.2713
	2	.0006	.0023	.0051	.0088	.0135	.0191	.0254	.0325	.0402
	3			.0001	.0002	.0005	.0008	.0013	.0019	.0027
	4									.0001
5	0	.9510	.9039	.8587	.8154	.7738	.7339	.6957	.6591	.6240
	1	.0480	.0922	.1328	.1699	.2036	.2342	.2618	.2866	.3086
	2	.0010	.0038	.0082	.0142	.0214	.0299	.0394	.0498	.0610
	3		.0001	.0003	.0006	.0011	.0019	.0030	.0043	.0060
	4						.0001	.0001	.0002	.0003
	5									
6	0	.9415	.8858	.8330	.7828	.7351	.6899	.6470	.6064	.5679
	1	.0571	.1085	.1546	.1957	.2321	.2642	.2922	.3164	.3370
	2	.0014	.0055	.0120	.0204	.0305	.0422	.0550	.0688	.0833
	3		.0002	.0005	.0011	.0021	.0036	.0055	.0080	.0110
	4					.0001	.0002	.0003	.0005	.0008
	5									
	6									
7	0	.9321	.8681	.8080	.7514	.6983	.6485	.6017	.5578	.5168
	1	.0659	.1240	.1749	.2192	.2573	.2897	.3170	.3396	.3578
	2	.0020	.0076	.0162	.0274	.0406	.0555	.0716	.0886	.1061
	3		.0003	.0008	.0019	.0036	.0059	.0090	.0128	.0175
	4				.0001	.0002	.0004	.0007	.0011	.0017
	5								.0001	.0001
	6									
	7									
8	0	.9227	.8508	.7837	.7214	.6634	.6096	.5596	.5132	.4703
	1	.0746	.1389	.1939	.2405	.2793	.3113	.3370	.3570	.3721
	2	.0026	.0099	.0210	.0351	.0515	.0695	.0888	.1087	.1288
	3	.0001	.0004	.0013	.0029	.0054	.0089	.0134	.0189	.0255
	4			.0001	.0002	.0004	.0007	.0013	.0021	.0031
	5							.0001	.0001	.0002
	6									
	7									
	8									

TABLE C — Binomial Probabilities (continued)

Entry is $P(X = k) = \binom{n}{k} p^k (1-p)^{n-k}$

n	k	p=.10	.15	.20	.25	.30	.35	.40	.45	.50
2	0	.8100	.7225	.6400	.5625	.4900	.4225	.3600	.3025	.2500
	1	.1800	.2550	.3200	.3750	.4200	.4550	.4800	.4950	.5000
	2	.0100	.0225	.0400	.0625	.0900	.1225	.1600	.2025	.2500
3	0	.7290	.6141	.5120	.4219	.3430	.2746	.2160	.1664	.1250
	1	.2430	.3251	.3840	.4219	.4410	.4436	.4320	.4084	.3750
	2	.0270	.0574	.0960	.1406	.1890	.2389	.2880	.3341	.3750
	3	.0010	.0034	.0080	.0156	.0270	.0429	.0640	.0911	.1250
4	0	.6561	.5220	.4096	.3164	.2401	.1785	.1296	.0915	.0625
	1	.2916	.3685	.4096	.4219	.4116	.3845	.3456	.2995	.2500
	2	.0486	.0975	.1536	.2109	.2646	.3105	.3456	.3675	.3750
	3	.0036	.0115	.0256	.0469	.0756	.1115	.1536	.2005	.2500
	4	.0001	.0005	.0016	.0039	.0081	.0150	.0256	.0410	.0625
5	0	.5905	.4437	.3277	.2373	.1681	.1160	.0778	.0503	.0313
	1	.3280	.3915	.4096	.3955	.3602	.3124	.2592	.2059	.1563
	2	.0729	.1382	.2048	.2637	.3087	.3364	.3456	.3369	.3125
	3	.0081	.0244	.0512	.0879	.1323	.1811	.2304	.2757	.3125
	4	.0004	.0022	.0064	.0146	.0284	.0488	.0768	.1128	.1562
	5		.0001	.0003	.0010	.0024	.0053	.0102	.0185	.0312
6	0	.5314	.3771	.2621	.1780	.1176	.0754	.0467	.0277	.0156
	1	.3543	.3993	.3932	.3560	.3025	.2437	.1866	.1359	.0938
	2	.0984	.1762	.2458	.2966	.3241	.3280	.3110	.2780	.2344
	3	.0146	.0415	.0819	.1318	.1852	.2355	.2765	.3032	.3125
	4	.0012	.0055	.0154	.0330	.0595	.0951	.1382	.1861	.2344
	5	.0001	.0004	.0015	.0044	.0102	.0205	.0369	.0609	.0937
	6			.0001	.0002	.0007	.0018	.0041	.0083	.0156
7	0	.4783	.3206	.2097	.1335	.0824	.0490	.0280	.0152	.0078
	1	.3720	.3960	.3670	.3115	.2471	.1848	.1306	.0872	.0547
	2	.1240	.2097	.2753	.3115	.3177	.2985	.2613	.2140	.1641
	3	.0230	.0617	.1147	.1730	.2269	.2679	.2903	.2918	.2734
	4	.0026	.0109	.0287	.0577	.0972	.1442	.1935	.2388	.2734
	5	.0002	.0012	.0043	.0115	.0250	.0466	.0774	.1172	.1641
	6		.0001	.0004	.0013	.0036	.0084	.0172	.0320	.0547
	7				.0001	.0002	.0006	.0016	.0037	.0078
8	0	.4305	.2725	.1678	.1001	.0576	.0319	.0168	.0084	.0039
	1	.3826	.3847	.3355	.2670	.1977	.1373	.0896	.0548	.0313
	2	.1488	.2376	.2936	.3115	.2965	.2587	.2090	.1569	.1094
	3	.0331	.0839	.1468	.2076	.2541	.2786	.2787	.2568	.2188
	4	.0046	.0185	.0459	.0865	.1361	.1875	.2322	.2627	.2734
	5	.0004	.0026	.0092	.0231	.0467	.0808	.1239	.1719	.2188
	6		.0002	.0011	.0038	.0100	.0217	.0413	.0703	.1094
	7			.0001	.0004	.0012	.0033	.0079	.0164	.0312
	8					.0001	.0002	.0007	.0017	.0039

(*Continued*)

TABLE C Binomial Probabilities (continued)

Entry is $P(X = k) = \binom{n}{k} p^k (1-p)^{n-k}$

n	k	p=.01	.02	.03	.04	.05	.06	.07	.08	.09
9	0	.9135	.8337	.7602	.6925	.6302	.5730	.5204	.4722	.4279
	1	.0830	.1531	.2116	.2597	.2985	.3292	.3525	.3695	.3809
	2	.0034	.0125	.0262	.0433	.0629	.0840	.1061	.1285	.1507
	3	.0001	.0006	.0019	.0042	.0077	.0125	.0186	.0261	.0348
	4			.0001	.0003	.0006	.0012	.0021	.0034	.0052
	5						.0001	.0002	.0003	.0005
	6									
	7									
	8									
	9									
10	0	.9044	.8171	.7374	.6648	.5987	.5386	.4840	.4344	.3894
	1	.0914	.1667	.2281	.2770	.3151	.3438	.3643	.3777	.3851
	2	.0042	.0153	.0317	.0519	.0746	.0988	.1234	.1478	.1714
	3	.0001	.0008	.0026	.0058	.0105	.0168	.0248	.0343	.0452
	4			.0001	.0004	.0010	.0019	.0033	.0052	.0078
	5					.0001	.0001	.0003	.0005	.0009
	6									.0001
	7									
	8									
	9									
	10									
12	0	.8864	.7847	.6938	.6127	.5404	.4759	.4186	.3677	.3225
	1	.1074	.1922	.2575	.3064	.3413	.3645	.3781	.3837	.3827
	2	.0060	.0216	.0438	.0702	.0988	.1280	.1565	.1835	.2082
	3	.0002	.0015	.0045	.0098	.0173	.0272	.0393	.0532	.0686
	4		.0001	.0003	.0009	.0021	.0039	.0067	.0104	.0153
	5				.0001	.0002	.0004	.0008	.0014	.0024
	6							.0001	.0001	.0003
	7									
	8									
	9									
	10									
	11									
	12									
15	0	.8601	.7386	.6333	.5421	.4633	.3953	.3367	.2863	.2430
	1	.1303	.2261	.2938	.3388	.3658	.3785	.3801	.3734	.3605
	2	.0092	.0323	.0636	.0988	.1348	.1691	.2003	.2273	.2496
	3	.0004	.0029	.0085	.0178	.0307	.0468	.0653	.0857	.1070
	4		.0002	.0008	.0022	.0049	.0090	.0148	.0223	.0317
	5			.0001	.0002	.0006	.0013	.0024	.0043	.0069
	6						.0001	.0003	.0006	.0011
	7								.0001	.0001
	8									
	9									
	10									
	11									
	12									
	13									
	14									
	15									

TABLE C Binomial Probabilities (continued)

Entry is $P(X = k) = \binom{n}{k} p^k (1-p)^{n-k}$

n	k	p=.10	.15	.20	.25	.30	.35	.40	.45	.50
9	0	.3874	.2316	.1342	.0751	.0404	.0207	.0101	.0046	.0020
	1	.3874	.3679	.3020	.2253	.1556	.1004	.0605	.0339	.0176
	2	.1722	.2597	.3020	.3003	.2668	.2162	.1612	.1110	.0703
	3	.0446	.1069	.1762	.2336	.2668	.2716	.2508	.2119	.1641
	4	.0074	.0283	.0661	.1168	.1715	.2194	.2508	.2600	.2461
	5	.0008	.0050	.0165	.0389	.0735	.1181	.1672	.2128	.2461
	6	.0001	.0006	.0028	.0087	.0210	.0424	.0743	.1160	.1641
	7			.0003	.0012	.0039	.0098	.0212	.0407	.0703
	8				.0001	.0004	.0013	.0035	.0083	.0176
	9						.0001	.0003	.0008	.0020
10	0	.3487	.1969	.1074	.0563	.0282	.0135	.0060	.0025	.0010
	1	.3874	.3474	.2684	.1877	.1211	.0725	.0403	.0207	.0098
	2	.1937	.2759	.3020	.2816	.2335	.1757	.1209	.0763	.0439
	3	.0574	.1298	.2013	.2503	.2668	.2522	.2150	.1665	.1172
	4	.0112	.0401	.0881	.1460	.2001	.2377	.2508	.2384	.2051
	5	.0015	.0085	.0264	.0584	.1029	.1536	.2007	.2340	.2461
	6	.0001	.0012	.0055	.0162	.0368	.0689	.1115	.1596	.2051
	7		.0001	.0008	.0031	.0090	.0212	.0425	.0746	.1172
	8			.0001	.0004	.0014	.0043	.0106	.0229	.0439
	9					.0001	.0005	.0016	.0042	.0098
	10							.0001	.0003	.0010
12	0	.2824	.1422	.0687	.0317	.0138	.0057	.0022	.0008	.0002
	1	.3766	.3012	.2062	.1267	.0712	.0368	.0174	.0075	.0029
	2	.2301	.2924	.2835	.2323	.1678	.1088	.0639	.0339	.0161
	3	.0852	.1720	.2362	.2581	.2397	.1954	.1419	.0923	.0537
	4	.0213	.0683	.1329	.1936	.2311	.2367	.2128	.1700	.1208
	5	.0038	.0193	.0532	.1032	.1585	.2039	.2270	.2225	.1934
	6	.0005	.0040	.0155	.0401	.0792	.1281	.1766	.2124	.2256
	7		.0006	.0033	.0115	.0291	.0591	.1009	.1489	.1934
	8		.0001	.0005	.0024	.0078	.0199	.0420	.0762	.1208
	9			.0001	.0004	.0015	.0048	.0125	.0277	.0537
	10					.0002	.0008	.0025	.0068	.0161
	11						.0001	.0003	.0010	.0029
	12								.0001	.0002
15	0	.2059	.0874	.0352	.0134	.0047	.0016	.0005	.0001	.0000
	1	.3432	.2312	.1319	.0668	.0305	.0126	.0047	.0016	.0005
	2	.2669	.2856	.2309	.1559	.0916	.0476	.0219	.0090	.0032
	3	.1285	.2184	.2501	.2252	.1700	.1110	.0634	.0318	.0139
	4	.0428	.1156	.1876	.2252	.2186	.1792	.1268	.0780	.0417
	5	.0105	.0449	.1032	.1651	.2061	.2123	.1859	.1404	.0916
	6	.0019	.0132	.0430	.0917	.1472	.1906	.2066	.1914	.1527
	7	.0003	.0030	.0138	.0393	.0811	.1319	.1771	.2013	.1964
	8		.0005	.0035	.0131	.0348	.0710	.1181	.1647	.1964
	9		.0001	.0007	.0034	.0116	.0298	.0612	.1048	.1527
	10			.0001	.0007	.0030	.0096	.0245	.0515	.0916
	11				.0001	.0006	.0024	.0074	.0191	.0417
	12					.0001	.0004	.0016	.0052	.0139
	13						.0001	.0003	.0010	.0032
	14								.0001	.0005
	15									

(Continued)

TABLE C — Binomial Probabilities (continued)

n	k	p=.01	.02	.03	.04	.05	.06	.07	.08	.09
20	0	.8179	.6676	.5438	.4420	.3585	.2901	.2342	.1887	.1516
	1	.1652	.2725	.3364	.3683	.3774	.3703	.3526	.3282	.3000
	2	.0159	.0528	.0988	.1458	.1887	.2246	.2521	.2711	.2818
	3	.0010	.0065	.0183	.0364	.0596	.0860	.1139	.1414	.1672
	4		.0006	.0024	.0065	.0133	.0233	.0364	.0523	.0703
	5			.0002	.0009	.0022	.0048	.0088	.0145	.0222
	6				.0001	.0003	.0008	.0017	.0032	.0055
	7						.0001	.0002	.0005	.0011
	8								.0001	.0002
	9									
	10									
	11									
	12									
	13									
	14									
	15									
	16									
	17									
	18									
	19									
	20									

n	k	p=.10	.15	.20	.25	.30	.35	.40	.45	.50
20	0	.1216	.0388	.0115	.0032	.0008	.0002	.0000	.0000	.0000
	1	.2702	.1368	.0576	.0211	.0068	.0020	.0005	.0001	.0000
	2	.2852	.2293	.1369	.0669	.0278	.0100	.0031	.0008	.0002
	3	.1901	.2428	.2054	.1339	.0716	.0323	.0123	.0040	.0011
	4	.0898	.1821	.2182	.1897	.1304	.0738	.0350	.0139	.0046
	5	.0319	.1028	.1746	.2023	.1789	.1272	.0746	.0365	.0148
	6	.0089	.0454	.1091	.1686	.1916	.1712	.1244	.0746	.0370
	7	.0020	.0160	.0545	.1124	.1643	.1844	.1659	.1221	.0739
	8	.0004	.0046	.0222	.0609	.1144	.1614	.1797	.1623	.1201
	9	.0001	.0011	.0074	.0271	.0654	.1158	.1597	.1771	.1602
	10		.0002	.0020	.0099	.0308	.0686	.1171	.1593	.1762
	11			.0005	.0030	.0120	.0336	.0710	.1185	.1602
	12			.0001	.0008	.0039	.0136	.0355	.0727	.1201
	13				.0002	.0010	.0045	.0146	.0366	.0739
	14					.0002	.0012	.0049	.0150	.0370
	15						.0003	.0013	.0049	.0148
	16							.0003	.0013	.0046
	17								.0002	.0011
	18									.0002
	19									
	20									

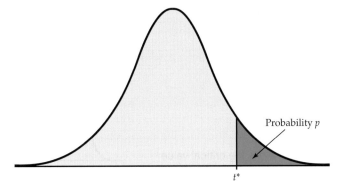

Table entry for p and C is the critical value t^* with probability p lying to its right and probability C lying between $-t^*$ and t^*.

TABLE D *t* **Distribution Critical Values**

df	\.25	\.20	\.15	\.10	\.05	\.025	\.02	\.01	\.005	\.0025	\.001	\.0005
1	1.000	1.376	1.963	3.078	6.314	12.71	15.89	31.82	63.66	127.3	318.3	636.6
2	0.816	1.061	1.386	1.886	2.920	4.303	4.849	6.965	9.925	14.09	22.33	31.60
3	0.765	0.978	1.250	1.638	2.353	3.182	3.482	4.541	5.841	7.453	10.21	12.92
4	0.741	0.941	1.190	1.533	2.132	2.776	2.999	3.747	4.604	5.598	7.173	8.610
5	0.727	0.920	1.156	1.476	2.015	2.571	2.757	3.365	4.032	4.773	5.893	6.869
6	0.718	0.906	1.134	1.440	1.943	2.447	2.612	3.143	3.707	4.317	5.208	5.959
7	0.711	0.896	1.119	1.415	1.895	2.365	2.517	2.998	3.499	4.029	4.785	5.408
8	0.706	0.889	1.108	1.397	1.860	2.306	2.449	2.896	3.355	3.833	4.501	5.041
9	0.703	0.883	1.100	1.383	1.833	2.262	2.398	2.821	3.250	3.690	4.297	4.781
10	0.700	0.879	1.093	1.372	1.812	2.228	2.359	2.764	3.169	3.581	4.144	4.587
11	0.697	0.876	1.088	1.363	1.796	2.201	2.328	2.718	3.106	3.497	4.025	4.437
12	0.695	0.873	1.083	1.356	1.782	2.179	2.303	2.681	3.055	3.428	3.930	4.318
13	0.694	0.870	1.079	1.350	1.771	2.160	2.282	2.650	3.012	3.372	3.852	4.221
14	0.692	0.868	1.076	1.345	1.761	2.145	2.264	2.624	2.977	3.326	3.787	4.140
15	0.691	0.866	1.074	1.341	1.753	2.131	2.249	2.602	2.947	3.286	3.733	4.073
16	0.690	0.865	1.071	1.337	1.746	2.120	2.235	2.583	2.921	3.252	3.686	4.015
17	0.689	0.863	1.069	1.333	1.740	2.110	2.224	2.567	2.898	3.222	3.646	3.965
18	0.688	0.862	1.067	1.330	1.734	2.101	2.214	2.552	2.878	3.197	3.611	3.922
19	0.688	0.861	1.066	1.328	1.729	2.093	2.205	2.539	2.861	3.174	3.579	3.883
20	0.687	0.860	1.064	1.325	1.725	2.086	2.197	2.528	2.845	3.153	3.552	3.850
21	0.686	0.859	1.063	1.323	1.721	2.080	2.189	2.518	2.831	3.135	3.527	3.819
22	0.686	0.858	1.061	1.321	1.717	2.074	2.183	2.508	2.819	3.119	3.505	3.792
23	0.685	0.858	1.060	1.319	1.714	2.069	2.177	2.500	2.807	3.104	3.485	3.768
24	0.685	0.857	1.059	1.318	1.711	2.064	2.172	2.492	2.797	3.091	3.467	3.745
25	0.684	0.856	1.058	1.316	1.708	2.060	2.167	2.485	2.787	3.078	3.450	3.725
26	0.684	0.856	1.058	1.315	1.706	2.056	2.162	2.479	2.779	3.067	3.435	3.707
27	0.684	0.855	1.057	1.314	1.703	2.052	2.158	2.473	2.771	3.057	3.421	3.690
28	0.683	0.855	1.056	1.313	1.701	2.048	2.154	2.467	2.763	3.047	3.408	3.674
29	0.683	0.854	1.055	1.311	1.699	2.045	2.150	2.462	2.756	3.038	3.396	3.659
30	0.683	0.854	1.055	1.310	1.697	2.042	2.147	2.457	2.750	3.030	3.385	3.646
40	0.681	0.851	1.050	1.303	1.684	2.021	2.123	2.423	2.704	2.971	3.307	3.551
50	0.679	0.849	1.047	1.299	1.676	2.009	2.109	2.403	2.678	2.937	3.261	3.496
60	0.679	0.848	1.045	1.296	1.671	2.000	2.099	2.390	2.660	2.915	3.232	3.460
80	0.678	0.846	1.043	1.292	1.664	1.990	2.088	2.374	2.639	2.887	3.195	3.416
100	0.677	0.845	1.042	1.290	1.660	1.984	2.081	2.364	2.626	2.871	3.174	3.390
1000	0.675	0.842	1.037	1.282	1.646	1.962	2.056	2.330	2.581	2.813	3.098	3.300
z^*	0.674	0.841	1.036	1.282	1.645	1.960	2.054	2.326	2.576	2.807	3.091	3.291
	50%	60%	70%	80%	90%	95%	96%	98%	99%	99.5%	99.8%	99.9%
	\multicolumn{12}{c}{Confidence level C}

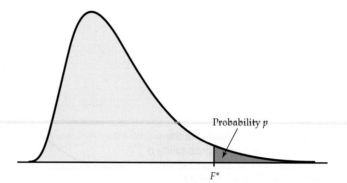

Table entry for p is the critical value F^* with probability p lying to its right.

TABLE E   F Distribution Critical Values

			Degrees of freedom in the numerator								
		p	1	2	3	4	5	6	7	8	9
Degrees of freedom in the denominator	1	.100	39.86	49.50	53.59	55.83	57.24	58.20	58.91	59.44	59.86
		.050	161.45	199.50	215.71	224.58	230.16	233.99	236.77	238.88	240.54
		.025	647.79	799.50	864.16	899.58	921.85	937.11	948.22	956.66	963.28
		.010	4052.18	4999.50	5403.35	5624.58	5763.65	5858.99	5928.36	5981.07	6022.47
		.001	405284	500000	540379	562500	576405	585937	592873	598144	602284
	2	.100	8.53	9.00	9.16	9.24	9.29	9.33	9.35	9.37	9.38
		.050	18.51	19.00	19.16	19.25	19.30	19.33	19.35	19.37	19.38
		.025	38.51	39.00	39.17	39.25	39.30	39.33	39.36	39.37	39.39
		.010	98.50	99.00	99.17	99.25	99.30	99.33	99.36	99.37	99.39
		.001	998.50	999.00	999.17	999.25	999.30	999.33	999.36	999.37	999.39
	3	.100	5.54	5.46	5.39	5.34	5.31	5.28	5.27	5.25	5.24
		.050	10.13	9.55	9.28	9.12	9.01	8.94	8.89	8.85	8.81
		.025	17.44	16.04	15.44	15.10	14.88	14.73	14.62	14.54	14.47
		.010	34.12	30.82	29.46	28.71	28.24	27.91	27.67	27.49	27.35
		.001	167.03	148.50	141.11	137.10	134.58	132.85	131.58	130.62	129.86
	4	.100	4.54	4.32	4.19	4.11	4.05	4.01	3.98	3.95	3.94
		.050	7.71	6.94	6.59	6.39	6.26	6.16	6.09	6.04	6.00
		.025	12.22	10.65	9.98	9.60	9.36	9.20	9.07	8.98	8.90
		.010	21.20	18.00	16.69	15.98	15.52	15.21	14.98	14.80	14.66
		.001	74.14	61.25	56.18	53.44	51.71	50.53	49.66	49.00	48.47
	5	.100	4.06	3.78	3.62	3.52	3.45	3.40	3.37	3.34	3.32
		.050	6.61	5.79	5.41	5.19	5.05	4.95	4.88	4.82	4.77
		.025	10.01	8.43	7.76	7.39	7.15	6.98	6.85	6.76	6.68
		.010	16.26	13.27	12.06	11.39	10.97	10.67	10.46	10.29	10.16
		.001	47.18	37.12	33.20	31.09	29.75	28.83	28.16	27.65	27.24
	6	.100	3.78	3.46	3.29	3.18	3.11	3.05	3.01	2.98	2.96
		.050	5.99	5.14	4.76	4.53	4.39	4.28	4.21	4.15	4.10
		.025	8.81	7.26	6.60	6.23	5.99	5.82	5.70	5.60	5.52
		.010	13.75	10.92	9.78	9.15	8.75	8.47	8.26	8.10	7.98
		.001	35.51	27.00	23.70	21.92	20.80	20.03	19.46	19.03	18.69
	7	.100	3.59	3.26	3.07	2.96	2.88	2.83	2.78	2.75	2.72
		.050	5.59	4.74	4.35	4.12	3.97	3.87	3.79	3.73	3.68
		.025	8.07	6.54	5.89	5.52	5.29	5.12	4.99	4.90	4.82
		.010	12.25	9.55	8.45	7.85	7.46	7.19	6.99	6.84	6.72
		.001	29.25	21.69	18.77	17.20	16.21	15.52	15.02	14.63	14.33

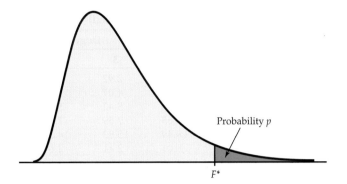

Table entry for p is the critical value F^* with probability p lying to its right.

TABLE E	*F* Distribution Critical Values (continued)										
	Degrees of freedom in the numerator										
	10	12	15	20	25	30	40	50	60	120	1000
	60.19	60.71	61.22	61.74	62.05	62.26	62.53	62.69	62.79	63.06	63.30
	241.88	243.91	245.95	248.01	249.26	250.10	251.14	251.77	252.20	253.25	254.19
	968.63	976.71	984.87	993.10	998.08	1001.41	1005.60	1008.12	1009.80	1014.02	1017.75
	6055.85	6106.32	6157.28	6208.73	6239.83	6260.65	6286.78	6302.52	6313.03	6339.39	6362.68
	605621	610668	615764	620908	624017	626099	628712	630285	631337	633972	636301
	9.39	9.41	9.42	9.44	9.45	9.46	9.47	9.47	9.47	9.48	9.49
	19.40	19.41	19.43	19.45	19.46	19.46	19.47	19.48	19.48	19.49	19.49
	39.40	39.41	39.43	39.45	39.46	39.46	39.47	39.48	39.48	39.49	39.50
	99.40	99.42	99.43	99.45	99.46	99.47	99.47	99.48	99.48	99.49	99.50
	999.40	999.42	999.43	999.45	999.46	999.47	999.47	999.48	999.48	999.49	999.50
	5.23	5.22	5.20	5.18	5.17	5.17	5.16	5.15	5.15	5.14	5.13
	8.79	8.74	8.70	8.66	8.63	8.62	8.59	8.58	8.57	8.55	8.53
	14.42	14.34	14.25	14.17	14.12	14.08	14.04	14.01	13.99	13.95	13.91
	27.23	27.05	26.87	26.69	26.58	26.50	26.41	26.35	26.32	26.22	26.14
	129.25	128.32	127.37	126.42	125.84	125.45	124.96	124.66	124.47	123.97	123.53
	3.92	3.90	3.87	3.84	3.83	3.82	3.80	3.80	3.79	3.78	3.76
	5.96	5.91	5.86	5.80	5.77	5.75	5.72	5.70	5.69	5.66	5.63
	8.84	8.75	8.66	8.56	8.50	8.46	8.41	8.38	8.36	8.31	8.26
	14.55	14.37	14.20	14.02	13.91	13.84	13.75	13.69	13.65	13.56	13.47
	48.05	47.41	46.76	46.10	45.70	45.43	45.09	44.88	44.75	44.40	44.09
	3.30	3.27	3.24	3.21	3.19	3.17	3.16	3.15	3.14	3.12	3.11
	4.74	4.68	4.62	4.56	4.52	4.50	4.46	4.44	4.43	4.40	4.37
	6.62	6.52	6.43	6.33	6.27	6.23	6.18	6.14	6.12	6.07	6.02
	10.05	9.89	9.72	9.55	9.45	9.38	9.29	9.24	9.20	9.11	9.03
	26.92	26.42	25.91	25.39	25.08	24.87	24.60	24.44	24.33	24.06	23.82
	2.94	2.90	2.87	2.84	2.81	2.80	2.78	2.77	2.76	2.74	2.72
	4.06	4.00	3.94	3.87	3.83	3.81	3.77	3.75	3.74	3.70	3.67
	5.46	5.37	5.27	5.17	5.11	5.07	5.01	4.98	4.96	4.90	4.86
	7.87	7.72	7.56	7.40	7.30	7.23	7.14	7.09	7.06	6.97	6.89
	18.41	17.99	17.56	17.12	16.85	16.67	16.44	16.31	16.21	15.98	15.77
	2.70	2.67	2.63	2.59	2.57	2.56	2.54	2.52	2.51	2.49	2.47
	3.64	3.57	3.51	3.44	3.40	3.38	3.34	3.32	3.30	3.27	3.23
	4.76	4.67	4.57	4.47	4.40	4.36	4.31	4.28	4.25	4.20	4.15
	6.62	6.47	6.31	6.16	6.06	5.99	5.91	5.86	5.82	5.74	5.66
	14.08	13.71	13.32	12.93	12.69	12.53	12.33	12.20	12.12	11.91	11.72

(*Continued*)

TABLE E F Distribution Critical Values (continued)

		Degrees of freedom in the numerator								
	p	1	2	3	4	5	6	7	8	9
8	.100	3.46	3.11	2.92	2.81	2.73	2.67	2.62	2.59	2.56
	.050	5.32	4.46	4.07	3.84	3.69	3.58	3.50	3.44	3.39
	.025	7.57	6.06	5.42	5.05	4.82	4.65	4.53	4.43	4.36
	.010	11.26	8.65	7.59	7.01	6.63	6.37	6.18	6.03	5.91
	.001	25.41	18.49	15.83	14.39	13.48	12.86	12.40	12.05	11.77
9	.100	3.36	3.01	2.81	2.69	2.61	2.55	2.51	2.47	2.44
	.050	5.12	4.26	3.86	3.63	3.48	3.37	3.29	3.23	3.18
	.025	7.21	5.71	5.08	4.72	4.48	4.32	4.20	4.10	4.03
	.010	10.56	8.02	6.99	6.42	6.06	5.80	5.61	5.47	5.35
	.001	22.86	16.39	13.90	12.56	11.71	11.13	10.70	10.37	10.11
10	.100	3.29	2.92	2.73	2.61	2.52	2.46	2.41	2.38	2.35
	.050	4.96	4.10	3.71	3.48	3.33	3.22	3.14	3.07	3.02
	.025	6.94	5.46	4.83	4.47	4.24	4.07	3.95	3.85	3.78
	.010	10.04	7.56	6.55	5.99	5.64	5.39	5.20	5.06	4.94
	.001	21.04	14.91	12.55	11.28	10.48	9.93	9.52	9.20	8.96
11	.100	3.23	2.86	2.66	2.54	2.45	2.39	2.34	2.30	2.27
	.050	4.84	3.98	3.59	3.36	3.20	3.09	3.01	2.95	2.90
	.025	6.72	5.26	4.63	4.28	4.04	3.88	3.76	3.66	3.59
	.010	9.65	7.21	6.22	5.67	5.32	5.07	4.89	4.74	4.63
	.001	19.69	13.81	11.56	10.35	9.58	9.05	8.66	8.35	8.12
12	.100	3.18	2.81	2.61	2.48	2.39	2.33	2.28	2.24	2.21
	.050	4.75	3.89	3.49	3.26	3.11	3.00	2.91	2.85	2.80
	.025	6.55	5.10	4.47	4.12	3.89	3.73	3.61	3.51	3.44
	.010	9.33	6.93	5.95	5.41	5.06	4.82	4.64	4.50	4.39
	.001	18.64	12.97	10.80	9.63	8.89	8.38	8.00	7.71	7.48
13	.100	3.14	2.76	2.56	2.43	2.35	2.28	2.23	2.20	2.16
	.050	4.67	3.81	3.41	3.18	3.03	2.92	2.83	2.77	2.71
	.025	6.41	4.97	4.35	4.00	3.77	3.60	3.48	3.39	3.31
	.010	9.07	6.70	5.74	5.21	4.86	4.62	4.44	4.30	4.19
	.001	17.82	12.31	10.21	9.07	8.35	7.86	7.49	7.21	6.98
14	.100	3.10	2.73	2.52	2.39	2.31	2.24	2.19	2.15	2.12
	.050	4.60	3.74	3.34	3.11	2.96	2.85	2.76	2.70	2.65
	.025	6.30	4.86	4.24	3.89	3.66	3.50	3.38	3.29	3.21
	.010	8.86	6.51	5.56	5.04	4.69	4.46	4.28	4.14	4.03
	.001	17.14	11.78	9.73	8.62	7.92	7.44	7.08	6.80	6.58
15	.100	3.07	2.70	2.49	2.36	2.27	2.21	2.16	2.12	2.09
	.050	4.54	3.68	3.29	3.06	2.90	2.79	2.71	2.64	2.59
	.025	6.20	4.77	4.15	3.80	3.58	3.41	3.29	3.20	3.12
	.010	8.68	6.36	5.42	4.89	4.56	4.32	4.14	4.00	3.89
	.001	16.59	11.34	9.34	8.25	7.57	7.09	6.74	6.47	6.26
16	.100	3.05	2.67	2.46	2.33	2.24	2.18	2.13	2.09	2.06
	.050	4.49	3.63	3.24	3.01	2.85	2.74	2.66	2.59	2.54
	.025	6.12	4.69	4.08	3.73	3.50	3.34	3.22	3.12	3.05
	.010	8.53	6.23	5.29	4.77	4.44	4.20	4.03	3.89	3.78
	.001	16.12	10.97	9.01	7.94	7.27	6.80	6.46	6.19	5.98
17	.100	3.03	2.64	2.44	2.31	2.22	2.15	2.10	2.06	2.03
	.050	4.45	3.59	3.20	2.96	2.81	2.70	2.61	2.55	2.49
	.025	6.04	4.62	4.01	3.66	3.44	3.28	3.16	3.06	2.98
	.010	8.40	6.11	5.19	4.67	4.34	4.10	3.93	3.79	3.68
	.001	15.72	10.66	8.73	7.68	7.02	6.56	6.22	5.96	5.75

Degrees of freedom in the denominator

TABLE E F Distribution Critical Values (continued)

			Degrees of freedom in the numerator							
10	12	15	20	25	30	40	50	60	120	1000
2.54	2.50	2.46	2.42	2.40	2.38	2.36	2.35	2.34	2.32	2.30
3.35	3.28	3.22	3.15	3.11	3.08	3.04	3.02	3.01	2.97	2.93
4.30	4.20	4.10	4.00	3.94	3.89	3.84	3.81	3.78	3.73	3.68
5.81	5.67	5.52	5.36	5.26	5.20	5.12	5.07	5.03	4.95	4.87
11.54	11.19	10.84	10.48	10.26	10.11	9.92	9.80	9.73	9.53	9.36
2.42	2.38	2.34	2.30	2.27	2.25	2.23	2.22	2.21	2.18	2.16
3.14	3.07	3.01	2.94	2.89	2.86	2.83	2.80	2.79	2.75	2.71
3.96	3.87	3.77	3.67	3.60	3.56	3.51	3.47	3.45	3.39	3.34
5.26	5.11	4.96	4.81	4.71	4.65	4.57	4.52	4.48	4.40	4.32
9.89	9.57	9.24	8.90	8.69	8.55	8.37	8.26	8.19	8.00	7.84
2.32	2.28	2.24	2.20	2.17	2.16	2.13	2.12	2.11	2.08	2.06
2.98	2.91	2.85	2.77	2.73	2.70	2.66	2.64	2.62	2.58	2.54
3.72	3.62	3.52	3.42	3.35	3.31	3.26	3.22	3.20	3.14	3.09
4.85	4.71	4.56	4.41	4.31	4.25	4.17	4.12	4.08	4.00	3.92
8.75	8.45	8.13	7.80	7.60	7.47	7.30	7.19	7.12	6.94	6.78
2.25	2.21	2.17	2.12	2.10	2.08	2.05	2.04	2.03	2.00	1.98
2.85	2.79	2.72	2.65	2.60	2.57	2.53	2.51	2.49	2.45	2.41
3.53	3.43	3.33	3.23	3.16	3.12	3.06	3.03	3.00	2.94	2.89
4.54	4.40	4.25	4.10	4.01	3.94	3.86	3.81	3.78	3.69	3.61
7.92	7.63	7.32	7.01	6.81	6.68	6.52	6.42	6.35	6.18	6.02
2.19	2.15	2.10	2.06	2.03	2.01	1.99	1.97	1.96	1.93	1.91
2.75	2.69	2.62	2.54	2.50	2.47	2.43	2.40	2.38	2.34	2.30
3.37	3.28	3.18	3.07	3.01	2.96	2.91	2.87	2.85	2.79	2.73
4.30	4.16	4.01	3.86	3.76	3.70	3.62	3.57	3.54	3.45	3.37
7.29	7.00	6.71	6.40	6.22	6.09	5.93	5.83	5.76	5.59	5.44
2.14	2.10	2.05	2.01	1.98	1.96	1.93	1.92	1.90	1.88	1.85
2.67	2.60	2.53	2.46	2.41	2.38	2.34	2.31	2.30	2.25	2.21
3.25	3.15	3.05	2.95	2.88	2.84	2.78	2.74	2.72	2.66	2.60
4.10	3.96	3.82	3.66	3.57	3.51	3.43	3.38	3.34	3.25	3.18
6.80	6.52	6.23	5.93	5.75	5.63	5.47	5.37	5.30	5.14	4.99
2.10	2.05	2.01	1.96	1.93	1.91	1.89	1.87	1.86	1.83	1.80
2.60	2.53	2.46	2.39	2.34	2.31	2.27	2.24	2.22	2.18	2.14
3.15	3.05	2.95	2.84	2.78	2.73	2.67	2.64	2.61	2.55	2.50
3.94	3.80	3.66	3.51	3.41	3.35	3.27	3.22	3.18	3.09	3.02
6.40	6.13	5.85	5.56	5.38	5.25	5.10	5.00	4.94	4.77	4.62
2.06	2.02	1.97	1.92	1.89	1.87	1.85	1.83	1.82	1.79	1.76
2.54	2.48	2.40	2.33	2.28	2.25	2.20	2.18	2.16	2.11	2.07
3.06	2.96	2.86	2.76	2.69	2.64	2.59	2.55	2.52	2.46	2.40
3.80	3.67	3.52	3.37	3.28	3.21	3.13	3.08	3.05	2.96	2.88
6.08	5.81	5.54	5.25	5.07	4.95	4.80	4.70	4.64	4.47	4.33
2.03	1.99	1.94	1.89	1.86	1.84	1.81	1.79	1.78	1.75	1.72
2.49	2.42	2.35	2.28	2.23	2.19	2.15	2.12	2.11	2.06	2.02
2.99	2.89	2.79	2.68	2.61	2.57	2.51	2.47	2.45	2.38	2.32
3.69	3.55	3.41	3.26	3.16	3.10	3.02	2.97	2.93	2.84	2.76
5.81	5.55	5.27	4.99	4.82	4.70	4.54	4.45	4.39	4.23	4.08
2.00	1.96	1.91	1.86	1.83	1.81	1.78	1.76	1.75	1.72	1.69
2.45	2.38	2.31	2.23	2.18	2.15	2.10	2.08	2.06	2.01	1.97
2.92	2.82	2.72	2.62	2.55	2.50	2.44	2.41	2.38	2.32	2.26
3.59	3.46	3.31	3.16	3.07	3.00	2.92	2.87	2.83	2.75	2.66
5.58	5.32	5.05	4.78	4.60	4.48	4.33	4.24	4.18	4.02	3.87

(Continued)

TABLE E F Distribution Critical Values (continued)

| | | \multicolumn{9}{c}{Degrees of freedom in the numerator} |
	p	1	2	3	4	5	6	7	8	9
18	.100	3.01	2.62	2.42	2.29	2.20	2.13	2.08	2.04	2.00
	.050	4.41	3.55	3.16	2.93	2.77	2.66	2.58	2.51	2.46
	.025	5.98	4.56	3.95	3.61	3.38	3.22	3.10	3.01	2.93
	.010	8.29	6.01	5.09	4.58	4.25	4.01	3.84	3.71	3.60
	.001	15.38	10.39	8.49	7.46	6.81	6.35	6.02	5.76	5.56
19	.100	2.99	2.61	2.40	2.27	2.18	2.11	2.06	2.02	1.98
	.050	4.38	3.52	3.13	2.90	2.74	2.63	2.54	2.48	2.42
	.025	5.92	4.51	3.90	3.56	3.33	3.17	3.05	2.96	2.88
	.010	8.18	5.93	5.01	4.50	4.17	3.94	3.77	3.63	3.52
	.001	15.08	10.16	8.28	7.27	6.62	6.18	5.85	5.59	5.39
20	.100	2.97	2.59	2.38	2.25	2.16	2.09	2.04	2.00	1.96
	.050	4.35	3.49	3.10	2.87	2.71	2.60	2.51	2.45	2.39
	.025	5.87	4.46	3.86	3.51	3.29	3.13	3.01	2.91	2.84
	.010	8.10	5.85	4.94	4.43	4.10	3.87	3.70	3.56	3.46
	.001	14.82	9.95	8.10	7.10	6.46	6.02	5.69	5.44	5.24
21	.100	2.96	2.57	2.36	2.23	2.14	2.08	2.02	1.98	1.95
	.050	4.32	3.47	3.07	2.84	2.68	2.57	2.49	2.42	2.37
	.025	5.83	4.42	3.82	3.48	3.25	3.09	2.97	2.87	2.80
	.010	8.02	5.78	4.87	4.37	4.04	3.81	3.64	3.51	3.40
	.001	14.59	9.77	7.94	6.95	6.32	5.88	5.56	5.31	5.11
22	.100	2.95	2.56	2.35	2.22	2.13	2.06	2.01	1.97	1.93
	.050	4.30	3.44	3.05	2.82	2.66	2.55	2.46	2.40	2.34
	.025	5.79	4.38	3.78	3.44	3.22	3.05	2.93	2.84	2.76
	.010	7.95	5.72	4.82	4.31	3.99	3.76	3.59	3.45	3.35
	.001	14.38	9.61	7.80	6.81	6.19	5.76	5.44	5.19	4.99
23	.100	2.94	2.55	2.34	2.21	2.11	2.05	1.99	1.95	1.92
	.050	4.28	3.42	3.03	2.80	2.64	2.53	2.44	2.37	2.32
	.025	5.75	4.35	3.75	3.41	3.18	3.02	2.90	2.81	2.73
	.010	7.88	5.66	4.76	4.26	3.94	3.71	3.54	3.41	3.30
	.001	14.20	9.47	7.67	6.70	6.08	5.65	5.33	5.09	4.89
24	.100	2.93	2.54	2.33	2.19	2.10	2.04	1.98	1.94	1.91
	.050	4.26	3.40	3.01	2.78	2.62	2.51	2.42	2.36	2.30
	.025	5.72	4.32	3.72	3.38	3.15	2.99	2.87	2.78	2.70
	.010	7.82	5.61	4.72	4.22	3.90	3.67	3.50	3.36	3.26
	.001	14.03	9.34	7.55	6.59	5.98	5.55	5.23	4.99	4.80
25	.100	2.92	2.53	2.32	2.18	2.09	2.02	1.97	1.93	1.89
	.050	4.24	3.39	2.99	2.76	2.60	2.49	2.40	2.34	2.28
	.025	5.69	4.29	3.69	3.35	3.13	2.97	2.85	2.75	2.68
	.010	7.77	5.57	4.68	4.18	3.85	3.63	3.46	3.32	3.22
	.001	13.88	9.22	7.45	6.49	5.89	5.46	5.15	4.91	4.71
26	.100	2.91	2.52	2.31	2.17	2.08	2.01	1.96	1.92	1.88
	.050	4.23	3.37	2.98	2.74	2.59	2.47	2.39	2.32	2.27
	.025	5.66	4.27	3.67	3.33	3.10	2.94	2.82	2.73	2.65
	.010	7.72	5.53	4.64	4.14	3.82	3.59	3.42	3.29	3.18
	.001	13.74	9.12	7.36	6.41	5.80	5.38	5.07	4.83	4.64
27	.100	2.90	2.51	2.30	2.17	2.07	2.00	1.95	1.91	1.87
	.050	4.21	3.35	2.96	2.73	2.57	2.46	2.37	2.31	2.25
	.025	5.63	4.24	3.65	3.31	3.08	2.92	2.80	2.71	2.63
	.010	7.68	5.49	4.60	4.11	3.78	3.56	3.39	3.26	3.15
	.001	13.61	9.02	7.27	6.33	5.73	5.31	5.00	4.76	4.57

Degrees of freedom in the denominator

TABLE E F Distribution Critical Values (continued)

			Degrees of freedom in the numerator							
10	12	15	20	25	30	40	50	60	120	1000
1.98	1.93	1.89	1.84	1.80	1.78	1.75	1.74	1.72	1.69	1.66
2.41	2.34	2.27	2.19	2.14	2.11	2.06	2.04	2.02	1.97	1.92
2.87	2.77	2.67	2.56	2.49	2.44	2.38	2.35	2.32	2.26	2.20
3.51	3.37	3.23	3.08	2.98	2.92	2.84	2.78	2.75	2.66	2.58
5.39	5.13	4.87	4.59	4.42	4.30	4.15	4.06	4.00	3.84	3.69
1.96	1.91	1.86	1.81	1.78	1.76	1.73	1.71	1.70	1.67	1.64
2.38	2.31	2.23	2.16	2.11	2.07	2.03	2.00	1.98	1.93	1.88
2.82	2.72	2.62	2.51	2.44	2.39	2.33	2.30	2.27	2.20	2.14
3.43	3.30	3.15	3.00	2.91	2.84	2.76	2.71	2.67	2.58	2.50
5.22	4.97	4.70	4.43	4.26	4.14	3.99	3.90	3.84	3.68	3.53
1.94	1.89	1.84	1.79	1.76	1.74	1.71	1.69	1.68	1.64	1.61
2.35	2.28	2.20	2.12	2.07	2.04	1.99	1.97	1.95	1.90	1.85
2.77	2.68	2.57	2.46	2.40	2.35	2.29	2.25	2.22	2.16	2.09
3.37	3.23	3.09	2.94	2.84	2.78	2.69	2.64	2.61	2.52	2.43
5.08	4.82	4.56	4.29	4.12	4.00	3.86	3.77	3.70	3.54	3.40
1.92	1.87	1.83	1.78	1.74	1.72	1.69	1.67	1.66	1.62	1.59
2.32	2.25	2.18	2.10	2.05	2.01	1.96	1.94	1.92	1.87	1.82
2.73	2.64	2.53	2.42	2.36	2.31	2.25	2.21	2.18	2.11	2.05
3.31	3.17	3.03	2.88	2.79	2.72	2.64	2.58	2.55	2.46	2.37
4.95	4.70	4.44	4.17	4.00	3.88	3.74	3.64	3.58	3.42	3.28
1.90	1.86	1.81	1.76	1.73	1.70	1.67	1.65	1.64	1.60	1.57
2.30	2.23	2.15	2.07	2.02	1.98	1.94	1.91	1.89	1.84	1.79
2.70	2.60	2.50	2.39	2.32	2.27	2.21	2.17	2.14	2.08	2.01
3.26	3.12	2.98	2.83	2.73	2.67	2.58	2.53	2.50	2.40	2.32
4.83	4.58	4.33	4.06	3.89	3.78	3.63	3.54	3.48	3.32	3.17
1.89	1.84	1.80	1.74	1.71	1.69	1.66	1.64	1.62	1.59	1.55
2.27	2.20	2.13	2.05	2.00	1.96	1.91	1.88	1.86	1.81	1.76
2.67	2.57	2.47	2.36	2.29	2.24	2.18	2.14	2.11	2.04	1.98
3.21	3.07	2.93	2.78	2.69	2.62	2.54	2.48	2.45	2.35	2.27
4.73	4.48	4.23	3.96	3.79	3.68	3.53	3.44	3.38	3.22	3.08
1.88	1.83	1.78	1.73	1.70	1.67	1.64	1.62	1.61	1.57	1.54
2.25	2.18	2.11	2.03	1.97	1.94	1.89	1.86	1.84	1.79	1.74
2.64	2.54	2.44	2.33	2.26	2.21	2.15	2.11	2.08	2.01	1.94
3.17	3.03	2.89	2.74	2.64	2.58	2.49	2.44	2.40	2.31	2.22
4.64	4.39	4.14	3.87	3.71	3.59	3.45	3.36	3.29	3.14	2.99
1.87	1.82	1.77	1.72	1.68	1.66	1.63	1.61	1.59	1.56	1.52
2.24	2.16	2.09	2.01	1.96	1.92	1.87	1.84	1.82	1.77	1.72
2.61	2.51	2.41	2.30	2.23	2.18	2.12	2.08	2.05	1.98	1.91
3.13	2.99	2.85	2.70	2.60	2.54	2.45	2.40	2.36	2.27	2.18
4.56	4.31	4.06	3.79	3.63	3.52	3.37	3.28	3.22	3.06	2.91
1.86	1.81	1.76	1.71	1.67	1.65	1.61	1.59	1.58	1.54	1.51
2.22	2.15	2.07	1.99	1.94	1.90	1.85	1.82	1.80	1.75	1.70
2.59	2.49	2.39	2.28	2.21	2.16	2.09	2.05	2.03	1.95	1.89
3.09	2.96	2.81	2.66	2.57	2.50	2.42	2.36	2.33	2.23	2.14
4.48	4.24	3.99	3.72	3.56	3.44	3.30	3.21	3.15	2.99	2.84
1.85	1.80	1.75	1.70	1.66	1.64	1.60	1.58	1.57	1.53	1.50
2.20	2.13	2.06	1.97	1.92	1.88	1.84	1.81	1.79	1.73	1.68
2.57	2.47	2.36	2.25	2.18	2.13	2.07	2.03	2.00	1.93	1.86
3.06	2.93	2.78	2.63	2.54	2.47	2.38	2.33	2.29	2.20	2.11
4.41	4.17	3.92	3.66	3.49	3.38	3.23	3.14	3.08	2.92	2.78

(Continued)

TABLE E F Distribution Critical Values (continued)

			Degrees of freedom in the numerator								
		p	1	2	3	4	5	6	7	8	9
Degrees of freedom in the denominator	28	.100	2.89	2.50	2.29	2.16	2.06	2.00	1.94	1.90	1.87
		.050	4.20	3.34	2.95	2.71	2.56	2.45	2.36	2.29	2.24
		.025	5.61	4.22	3.63	3.29	3.06	2.90	2.78	2.69	2.61
		.010	7.64	5.45	4.57	4.07	3.75	3.53	3.36	3.23	3.12
		.001	13.50	8.93	7.19	6.25	5.66	5.24	4.93	4.69	4.50
	29	.100	2.89	2.50	2.28	2.15	2.06	1.99	1.93	1.89	1.86
		.050	4.18	3.33	2.93	2.70	2.55	2.43	2.35	2.28	2.22
		.025	5.59	4.20	3.61	3.27	3.04	2.88	2.76	2.67	2.59
		.010	7.60	5.42	4.54	4.04	3.73	3.50	3.33	3.20	3.09
		.001	13.39	8.85	7.12	6.19	5.59	5.18	4.87	4.64	4.45
	30	.100	2.88	2.49	2.28	2.14	2.05	1.98	1.93	1.88	1.85
		.050	4.17	3.32	2.92	2.69	2.53	2.42	2.33	2.27	2.21
		.025	5.57	4.18	3.59	3.25	3.03	2.87	2.75	2.65	2.57
		.010	7.56	5.39	4.51	4.02	3.70	3.47	3.30	3.17	3.07
		.001	13.29	8.77	7.05	6.12	5.53	5.12	4.82	4.58	4.39
	40	.100	2.84	2.44	2.23	2.09	2.00	1.93	1.87	1.83	1.79
		.050	4.08	3.23	2.84	2.61	2.45	2.34	2.25	2.18	2.12
		.025	5.42	4.05	3.46	3.13	2.90	2.74	2.62	2.53	2.45
		.010	7.31	5.18	4.31	3.83	3.51	3.29	3.12	2.99	2.89
		.001	12.61	8.25	6.59	5.70	5.13	4.73	4.44	4.21	4.02
	50	.100	2.81	2.41	2.20	2.06	1.97	1.90	1.84	1.80	1.76
		.050	4.03	3.18	2.79	2.56	2.40	2.29	2.20	2.13	2.07
		.025	5.34	3.97	3.39	3.05	2.83	2.67	2.55	2.46	2.38
		.010	7.17	5.06	4.20	3.72	3.41	3.19	3.02	2.89	2.78
		.001	12.22	7.96	6.34	5.46	4.90	4.51	4.22	4.00	3.82
	60	.100	2.79	2.39	2.18	2.04	1.95	1.87	1.82	1.77	1.74
		.050	4.00	3.15	2.76	2.53	2.37	2.25	2.17	2.10	2.04
		.025	5.29	3.93	3.34	3.01	2.79	2.63	2.51	2.41	2.33
		.010	7.08	4.98	4.13	3.65	3.34	3.12	2.95	2.82	2.72
		.001	11.97	7.77	6.17	5.31	4.76	4.37	4.09	3.86	3.69
	100	.100	2.76	2.36	2.14	2.00	1.91	1.83	1.78	1.73	1.69
		.050	3.94	3.09	2.70	2.46	2.31	2.19	2.10	2.03	1.97
		.025	5.18	3.83	3.25	2.92	2.70	2.54	2.42	2.32	2.24
		.010	6.90	4.82	3.98	3.51	3.21	2.99	2.82	2.69	2.59
		.001	11.50	7.41	5.86	5.02	4.48	4.11	3.83	3.61	3.44
	200	.100	2.73	2.33	2.11	1.97	1.88	1.80	1.75	1.70	1.66
		.050	3.89	3.04	2.65	2.42	2.26	2.14	2.06	1.98	1.93
		.025	5.10	3.76	3.18	2.85	2.63	2.47	2.35	2.26	2.18
		.010	6.76	4.71	3.88	3.41	3.11	2.89	2.73	2.60	2.50
		.001	11.15	7.15	5.63	4.81	4.29	3.92	3.65	3.43	3.26
	1000	.100	2.71	2.31	2.09	1.95	1.85	1.78	1.72	1.68	1.64
		.050	3.85	3.00	2.61	2.38	2.22	2.11	2.02	1.95	1.89
		.025	5.04	3.70	3.13	2.80	2.58	2.42	2.30	2.20	2.13
		.010	6.66	4.63	3.80	3.34	3.04	2.82	2.66	2.53	2.43
		.001	10.89	6.96	5.46	4.65	4.14	3.78	3.51	3.30	3.13

TABLE E   F Distribution Critical Values (continued)

Degrees of freedom in the numerator											
10	12	15	20	25	30	40	50	60	120	1000
1.84	1.79	1.74	1.69	1.65	1.63	1.59	1.57	1.56	1.52	1.48
2.19	2.12	2.04	1.96	1.91	1.87	1.82	1.79	1.77	1.71	1.66
2.55	2.45	2.34	2.23	2.16	2.11	2.05	2.01	1.98	1.91	1.84
3.03	2.90	2.75	2.60	2.51	2.44	2.35	2.30	2.26	2.17	2.08
4.35	4.11	3.86	3.60	3.43	3.32	3.18	3.09	3.02	2.86	2.72
1.83	1.78	1.73	1.68	1.64	1.62	1.58	1.56	1.55	1.51	1.47
2.18	2.10	2.03	1.94	1.89	1.85	1.81	1.77	1.75	1.70	1.65
2.53	2.43	2.32	2.21	2.14	2.09	2.03	1.99	1.96	1.89	1.82
3.00	2.87	2.73	2.57	2.48	2.41	2.33	2.27	2.23	2.14	2.05
4.29	4.05	3.80	3.54	3.38	3.27	3.12	3.03	2.97	2.81	2.66
1.82	1.77	1.72	1.67	1.63	1.61	1.57	1.55	1.54	1.50	1.46
2.16	2.09	2.01	1.93	1.88	1.84	1.79	1.76	1.74	1.68	1.63
2.51	2.41	2.31	2.20	2.12	2.07	2.01	1.97	1.94	1.87	1.80
2.98	2.84	2.70	2.55	2.45	2.39	2.30	2.25	2.21	2.11	2.02
4.24	4.00	3.75	3.49	3.33	3.22	3.07	2.98	2.92	2.76	2.61
1.76	1.71	1.66	1.61	1.57	1.54	1.51	1.48	1.47	1.42	1.38
2.08	2.00	1.92	1.84	1.78	1.74	1.69	1.66	1.64	1.58	1.52
2.39	2.29	2.18	2.07	1.99	1.94	1.88	1.83	1.80	1.72	1.65
2.80	2.66	2.52	2.37	2.27	2.20	2.11	2.06	2.02	1.92	1.82
3.87	3.64	3.40	3.14	2.98	2.87	2.73	2.64	2.57	2.41	2.25
1.73	1.68	1.63	1.57	1.53	1.50	1.46	1.44	1.42	1.38	1.33
2.03	1.95	1.87	1.78	1.73	1.69	1.63	1.60	1.58	1.51	1.45
2.32	2.22	2.11	1.99	1.92	1.87	1.80	1.75	1.72	1.64	1.56
2.70	2.56	2.42	2.27	2.17	2.10	2.01	1.95	1.91	1.80	1.70
3.67	3.44	3.20	2.95	2.79	2.68	2.53	2.44	2.38	2.21	2.05
1.71	1.66	1.60	1.54	1.50	1.48	1.44	1.41	1.40	1.35	1.30
1.99	1.92	1.84	1.75	1.69	1.65	1.59	1.56	1.53	1.47	1.40
2.27	2.17	2.06	1.94	1.87	1.82	1.74	1.70	1.67	1.58	1.49
2.63	2.50	2.35	2.20	2.10	2.03	1.94	1.88	1.84	1.73	1.62
3.54	3.32	3.08	2.83	2.67	2.55	2.41	2.32	2.25	2.08	1.92
1.66	1.61	1.56	1.49	1.45	1.42	1.38	1.35	1.34	1.28	1.22
1.93	1.85	1.77	1.68	1.62	1.57	1.52	1.48	1.45	1.38	1.30
2.18	2.08	1.97	1.85	1.77	1.71	1.64	1.59	1.56	1.46	1.36
2.50	2.37	2.22	2.07	1.97	1.89	1.80	1.74	1.69	1.57	1.45
3.30	3.07	2.84	2.59	2.43	2.32	2.17	2.08	2.01	1.83	1.64
1.63	1.58	1.52	1.46	1.41	1.38	1.34	1.31	1.29	1.23	1.16
1.88	1.80	1.72	1.62	1.56	1.52	1.46	1.41	1.39	1.30	1.21
2.11	2.01	1.90	1.78	1.70	1.64	1.56	1.51	1.47	1.37	1.25
2.41	2.27	2.13	1.97	1.87	1.79	1.69	1.63	1.58	1.45	1.30
3.12	2.90	2.67	2.42	2.26	2.15	2.00	1.90	1.83	1.64	1.43
1.61	1.55	1.49	1.43	1.38	1.35	1.30	1.27	1.25	1.18	1.08
1.84	1.76	1.68	1.58	1.52	1.47	1.41	1.36	1.33	1.24	1.11
2.06	1.96	1.85	1.72	1.64	1.58	1.50	1.45	1.41	1.29	1.13
2.34	2.20	2.06	1.90	1.79	1.72	1.61	1.54	1.50	1.35	1.16
2.99	2.77	2.54	2.30	2.14	2.02	1.87	1.77	1.69	1.49	1.22

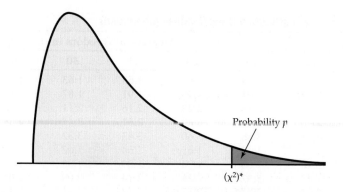

Table entry for p is the critical value $(\chi^2)^*$ with probability p lying to its right.

TABLE F χ^2 Distribution Critical Values

df	.25	.20	.15	.10	.05	.025	.02	.01	.005	.0025	.001	.0005
1	1.32	1.64	2.07	2.71	3.84	5.02	5.41	6.63	7.88	9.14	10.83	12.12
2	2.77	3.22	3.79	4.61	5.99	7.38	7.82	9.21	10.60	11.98	13.82	15.20
3	4.11	4.64	5.32	6.25	7.81	9.35	9.84	11.34	12.84	14.32	16.27	17.73
4	5.39	5.99	6.74	7.78	9.49	11.14	11.67	13.28	14.86	16.42	18.47	20.00
5	6.63	7.29	8.12	9.24	11.07	12.83	13.39	15.09	16.75	18.39	20.51	22.11
6	7.84	8.56	9.45	10.64	12.59	14.45	15.03	16.81	18.55	20.25	22.46	24.10
7	9.04	9.80	10.75	12.02	14.07	16.01	16.62	18.48	20.28	22.04	24.32	26.02
8	10.22	11.03	12.03	13.36	15.51	17.53	18.17	20.09	21.95	23.77	26.12	27.87
9	11.39	12.24	13.29	14.68	16.92	19.02	19.68	21.67	23.59	25.46	27.88	29.67
10	12.55	13.44	14.53	15.99	18.31	20.48	21.16	23.21	25.19	27.11	29.59	31.42
11	13.70	14.63	15.77	17.28	19.68	21.92	22.62	24.72	26.76	28.73	31.26	33.14
12	14.85	15.81	16.99	18.55	21.03	23.34	24.05	26.22	28.30	30.32	32.91	34.82
13	15.98	16.98	18.20	19.81	22.36	24.74	25.47	27.69	29.82	31.88	34.53	36.48
14	17.12	18.15	19.41	21.06	23.68	26.12	26.87	29.14	31.32	33.43	36.12	38.11
15	18.25	19.31	20.60	22.31	25.00	27.49	28.26	30.58	32.80	34.95	37.70	39.72
16	19.37	20.47	21.79	23.54	26.30	28.85	29.63	32.00	34.27	36.46	39.25	41.31
17	20.49	21.61	22.98	24.77	27.59	30.19	31.00	33.41	35.72	37.95	40.79	42.88
18	21.60	22.76	24.16	25.99	28.87	31.53	32.35	34.81	37.16	39.42	42.31	44.43
19	22.72	23.90	25.33	27.20	30.14	32.85	33.69	36.19	38.58	40.88	43.82	45.97
20	23.83	25.04	26.50	28.41	31.41	34.17	35.02	37.57	40.00	42.34	45.31	47.50
21	24.93	26.17	27.66	29.62	32.67	35.48	36.34	38.93	41.40	43.78	46.80	49.01
22	26.04	27.30	28.82	30.81	33.92	36.78	37.66	40.29	42.80	45.20	48.27	50.51
23	27.14	28.43	29.98	32.01	35.17	38.08	38.97	41.64	44.18	46.62	49.73	52.00
24	28.24	29.55	31.13	33.20	36.42	39.36	40.27	42.98	45.56	48.03	51.18	53.48
25	29.34	30.68	32.28	34.38	37.65	40.65	41.57	44.31	46.93	49.44	52.62	54.95
26	30.43	31.79	33.43	35.56	38.89	41.92	42.86	45.64	48.29	50.83	54.05	56.41
27	31.53	32.91	34.57	36.74	40.11	43.19	44.14	46.96	49.64	52.22	55.48	57.86
28	32.62	34.03	35.71	37.92	41.34	44.46	45.42	48.28	50.99	53.59	56.89	59.30
29	33.71	35.14	36.85	39.09	42.56	45.72	46.69	49.59	52.34	54.97	58.30	60.73
30	34.80	36.25	37.99	40.26	43.77	46.98	47.96	50.89	53.67	56.33	59.70	62.16
40	45.62	47.27	49.24	51.81	55.76	59.34	60.44	63.69	66.77	69.70	73.40	76.09
50	56.33	58.16	60.35	63.17	67.50	71.42	72.61	76.15	79.49	82.66	86.66	89.56
60	66.98	68.97	71.34	74.40	79.08	83.30	84.58	88.38	91.95	95.34	99.61	102.69
80	88.13	90.41	93.11	96.58	101.88	106.63	108.07	112.33	116.32	120.10	124.84	128.26
100	109.14	111.67	114.66	118.50	124.34	129.56	131.14	135.81	140.17	144.29	149.45	153.17

Answers to Odd-Numbered Exercises

Chapter 1: Examining Distributions

1.1 Europe; 121

1.3 Who: The cases are Initial Coin Offerings (ICOs); there are nine cases. What: There are four variables—ID, Name, Location, and amount in dollars. The purpose of the data is to list ICOs that have raised more than $100 million.

1.5 **(a)** If you were interested in attending a large college, you would want to know the number of graduates. **(b)** If you were interested in making sure you graduate, you would want to know the graduation rate.

1.7 **(a)** The cases are employees. **(b)** Employee identification number—label, last name—label, first name—label, middle initial—label, department—categorical, number of years—quantitative, salary—quantitative, education—categorical, age—quantitative. **(c)** Sample data would vary.

	A	B	C	D	E	F	G	H	I
1	EIN	Last	First	Middle	Department	Years	Salary	Education	Age
2	001	Marley	Bob	M	Sales	4	45000	some college	34
3	002	Fisher	Margeret	A	Sales	8	54000	college degree	37
4	003	Marin	Jane	E	Admin	2	39000	high school	25

1.9 **(a)** Categorical. **(b)** Quantitative. **(c)** Categorical. **(d)** Categorical. **(e)** Categorical. **(f)** Categorical. **(g)** Categorical. For all quantitative variables, numerical summaries would be meaningful; for categorical variables, numerical summaries are *not* meaningful.

1.11 Answers will vary. Examples include (1) How many hours per week do you study?—quantitative, unit of measurement is hours. (2) How many nights per week do you usually study?—quantitative, nights. (3) Do you usually study alone or with others?—categorical, people. (4) Do you feel like you study too much?—enough, not enough categorical.

1.13 **(a)** The states are the cases. **(b)** The name of the state is the label variable. **(c)** Number of students from the state who attend college—quantitative, number of students who attend college in their home state—quantitative. **(d)** Answers will vary. This would tell you which states have large percents of students that like to stay "at home" versus small percents, which indicate students' preference to leave home to attend college.

1.15 Answers may vary. Both tell the same story, so it is down to personal preference.

1.17 The Cost Centers would include Parts and materials, Manufacturing equipment, Salaries, Maintenance, and Office lease. We need to include Office lease even though it gives more than 80%, because otherwise we would only have the top 75% according to the data. So, to get the other 5%, we need to put Office lease in, giving us 82.12% total.

1.19 **(b)** Most people will prefer the Pareto because it emphasizes the largest categories.

1.21 Answers will vary. One solution is to have the highest range include 100, so $90 < \text{score} \leq 100$, $80 < \text{score} \leq 90$, etc.

1.23 Answers will vary. One example is shown.

```
0 | 3 4
1 | 1 3 6 7 9
2 | 2 3 5 5 7 8
3 | 1 1 2 5 6
4 | 1 5 7
5 | 4 5
6 | 8
7 | 2
```

1.25 **(a)** Histogram would be the best to show; the variable is quantitative. **(b)** Pareto chart would be the best to prioritize those characteristics that they liked best; pie chart might also be suitable. **(c)** Stemplot would be best because it is a small data set; histogram might also be suitable. **(d)** Pie chart is likely best in this situation to divide all the customers into groups from the whole; Pareto chart or bar graph might also be suitable.

1.27 **(b)** Chrome has by far the largest percent of market share, followed by Internet Explorer and Firefox. Other browsers have relatively very little market share.

1.29 **(b)** India and the United States have the most Facebook users in the sample of countries; however, India has 1 million more Facebook users than the United States. In the same sample of countries, the United Kingdom has the smallest number of Facebook users.

S-1

1.31 The distribution is fairly symmetric. The center is around 15.0% and the range is between 9.4% and 19.1%.

1.33 $\bar{X} = 27.32$.

1.35 $\bar{X} = 196.575$.

1.37 $M = 103.5$.

1.39 (a)

```
0 | 7
1 |
2 |
3 | 9
4 | 2 9
5 | 5 8
6 | 0 2 4 6
7 | 5
8 |
9 | 9
```

(b) The distribution is fairly symmetric with a slight skew to the left. Japan stands out with an unusually low growth rate of 0.7% and China stands out with an unusually high growth rate of 9.9%. (c) The mean growth rate is 5.55. Because the distribution is slightly skewed to the left, the median is a better measure of the center than the mean. (d) The median growth rate is 5.65. The median is a good measure of the center.

1.41 The time is right-skewed, with a long right tail. The mean is much higher than the median because of the skew. Answers will vary on preference.

1.43 Without Venezuela: $s = 15.7$; with Venezuela: $s = 44.9$.

1.45 (a) $\bar{X} = 196.6$, $s = 342$. (b) Min = 1, $Q1 = 54.5$, $M = 103.5$, $Q3 = 200$, Max = 2631. (c) The five-number summary is a better summary because the distribution is heavily skewed and has potential outliers.

1.47 (a) $M = 27.35$, $Q1 = 6.45$, $Q3 = 192.2$. (c) $M = 19.75$, $Q1 = 4.7$, $Q3 = 72.4$. These measures are less sensitive to outliers, except for $Q3$ which is noticeably different. This difference can be explained by the fact that the original data set was highly skewed rightward.

1.49 (b) Mozambique, Guinea, Bhutan, Liberia, Libya, Sierra Leone, and Timor-Leste represent extreme negative values in the distribution of trade balance, based on the IQR rule. (c) $\bar{X} = -0.023$, $s = 0.069$, $M = -0.024$, $Q1 = -0.056$, $Q3 = 0.017$. The distribution and numerical summaries are almost identical before and after the outliers are removed. (d) Overall, the distribution is very symmetrical, so that if some countries export a lot, there are other countries that import just as much. The mean and median trade balances are very close to 0. The outliers had almost no effect on the distribution or numerical summaries. Essentially, the outliers form longer tails on the curve.

1.51 Answers will vary.

1.53 (b) $\bar{X} = 19.48$, $s = 22.48$, $M = 11.35$, $Q1 = 7.90$, $Q3 = 24.05$. (c) The distribution is strongly right-skewed, with several brands far more valuable than most others. This is shown in the numerical summaries, with 75% of brand values less than $Q3 = 24.05$. Additionally, the median brand value is only 11.35. The mean value is 19.48, substantially higher than the median, again indicating the skew. Thus, brands like Apple and those listed in the problem dwarf the competition.

1.55 (a) $\bar{X} = 122.92$. (b) $M = 102.5$. (c) The data are right-skewed, which pulls the mean, making it higher than the median. The median is preferred.

1.57 (a) $IQR = 62$. (b) London is the outlier. (c) The data are right-skewed. (d) The data are right-skewed

(e)
```
0 | 2 4 6 6 7 9
1 | 1 1 1 4 9
2 |
3 |
4 | 2
```

(f) Answers will vary.

1.59 (a) With outliers included ($n = 175$), $\bar{X} = .053$, $M = 0.049$. After deleting the 16 outliers ($n = 159$), $\bar{X} = 0.05$, $M = 0.049$. The median remains the same. The mean changes slightly. (b) With outliers included ($n = 175$), $s = 0.014$, $Q1 = 0.045$, $Q3 = 0.058$. After deleting the outliers ($n = 159$), $s = 0.008$, $Q1 = 0.044$, $Q3 = 0.056$. The standard deviation changes substantially. The quartiles change slightly. (c) Answers will vary.

1.61 (a) Min = 9.4, $Q1 = 14.1$, $M = 14.9$, $Q3 = 15.8$, Max = 19.1. (b) $IQR = 15.8 - 14.1 = 1.7$, $Q1 - 1.5 \times IQR = 11.6$. So, Alaska with 9.4%, Texas with 11.5%, and Utah with 10% are low outliers. $Q3 + 1.5 \times IQR = 18.4$. So, Florida with 19.1% is a high outlier.

1.63 Applet, answers will vary.

1.65 The means and standard deviations are the same. $\bar{X} = 7.5$, $s = 2.03$. The stemplots (rounded to one decimal) show very different distributions. Data A are strongly left-skewed with a couple possible low outliers; Data B are equally distributed between 5 and 9 but has one high outlier at 12.5.

Data A:

```
3 | 1
4 | 7
5 |
6 | 1
7 | 3
8 | 1 1 7 8
9 | 1 1 3
```

Data B:

```
 5 | 3 6 8
 6 | 6 9
 7 | 0 7 9
 8 | 5 8
 9 |
10 |
11 |
12 | 5
```

1.67 (a) Picking the same number for all four observations results in a standard deviation of 0. (b) Picking 10, 10, 20, and 20 results in the largest standard deviation = 5.77. (c) For part (a), you may pick any number as long as all observations are the same. For part (b), only one choice provides the largest standard deviation.

1.69 The 5% trimmed mean is 16.04. The original mean was 19.48. The 5% trimmed mean is influenced by the large outliers as the original mean.

1.71 Answers will vary. Verify that the density curve is symmetric.

1.73 (a) The mean is at point C; the median is at point B. (b) The mean and median are both at point A. (c) The mean is at point A; the median is at point B.

1.75 (a) 16%. (b) 58.6 and 69.4. (c) 2.5%.

1.77 According to the rule, 99.7% of students will fall between $\mu \pm 3\sigma$. Therefore, 99.7% of students have scores between 155 and 329.

1.79 For $X = 200$, $Z = \dfrac{200 - 242}{29} = -1.45$; the proportion less than 200 is the area to the left, which is 0.0735. For the proportion greater than or equal to 200, we calculate $1 - 0.0735 = 0.9265$.

1.81 To get the top 25% of students, we need to solve for the 75th percentile. The corresponding Z is 0.67. So $X = 242 + 29(0.67) = 261.43$.

1.83 (a) The points fall below the 45° line; they form a straight line at first but then, near the right side, begin to increase steeply, indicating the right-skew. (b) It is likely part of a very long tail as it aligns perfectly with the curvature; see the Normal quantile plot.

1.85 (a) The area under the density curve is always equal to 1. (b) A density curve is a mathematical model for the distribution of a quantitative variable. (c) The density curve for a distribution that is skewed to the right can have more than one peak.

1.87 (c) The distribution shifts along the horizontal axis without changing its shape.

1.91 (a) According to the rule, 68% of women speak between 5232 and 23,362 words per day; 95% of women speak between −3833 and 32,427 words per day; and 99.7% of women speak between −12,898 and 41,492 words per day. (b) This rule doesn't seem to fit because you can't speak a negative number of words per day; therefore, the percents don't make sense. (c) According to the rule, 68% of men speak between 5004 and 23,116 words per day; 95% of men speak between −4052 and 32,172 words per day; and 99.7% of men speak between −13,108 and 41,228 words per day. Similar to the women, the rule doesn't seem to fit because, again, you can't speak a negative number of words per day, so the percents don't make sense. (d) The data do show that the mean number of words spoken per day is higher for women than for men, but it is very close; in addition, the standard deviations are nearly the same, making the intervals very close. To put it in perspective, the women only speak about 1%–2% more words per day on average. That small of a difference could be due just to chance.

1.93 (a) Using $Z = (X - \mu)/\sigma$ or $Z = (X - 72)/10$, the standardized values are: −0.5, 0.7, 1.4, 0.5, −0.6, 1.8, 1.3, −2.8, 0, 1.1. (b) For the top 15% we need the 85th percentile; Table A gives $Z = 1.04$. (c) Only two scores have standardized values bigger than 1.04, the 94 and the 98.

1.95 Answers will vary. The wider curve has a standard deviation of about 0.4. The narrower curve has a standard deviation about 0.2.

1.97 (a) Using the 95% rule, we go 2σ, or 32, from the mean in each direction. $\mu - 2\sigma = 266 - 32 = 234$. $\mu + 2\sigma = 266 + 32 = 298$. So 95% of pregnancies last between 234 and 298 days. (b) The shortest 2.5% of pregnancies fall 2 standard deviations below the mean, or $\mu - 2\sigma = 266 - 32 = 234$. So the shortest pregnancies are 234 days or less.

1.99 (a) 0.0202. (b) 0.9798. (c) 0.0456. (d) $0.9554 - 0.0202 = 0.9352$.

1.101 (a) For 200, $Z = \dfrac{200 - 250}{25} = -2$; the area to the left of this is 0.0228. For 175, $Z = \dfrac{175 - 250}{25} = -3$;

the area to the left of this is 0.0013. Subtracting gives 0.0228 − 0.0013 = 0.0215. So the proportion between 175 and 200 is 0.0215. Answers will vary due to rounding and use of technology versus the normal tables. **(b)** To solve, we need either the 10th or 90th percentile. 250 − x corresponds with the 10th percentile → $Z = \frac{(250 + x) - 250}{25} = -1.28$. 250 + x corresponds with the 90th percentile → $Z = 1.28$. So, $1.28 = \frac{(250 + x) - 250}{25}$; solving gives $x = 32$.

1.103 **(a)** The area to the left of the first quartile is 25%. The corresponding Z is −0.67. The area to the left of the third quartile is 75%. The corresponding Z is 0.67. **(b)** For the first quartile, $X = \mu + \sigma(Z) = 266 + 16(-0.67) = 255.28$. For the third quartile, $X = \mu + \sigma(Z) = 266 + 16(0.67) = 276.72$.

1.105 Answers will vary.

1.107 **(b)** The distribution is right-skewed; this is also shown in the Normal quantile plot. **(c)** Answers will vary.

1.109 Answers will vary.

1.111 **(a)** Population values did not change that much from 2011 to 2017. Ontario and Quebec have the largest populations, followed by Alberta and British Columbia. For the population over age 65, most regions have similar percents of over 65 except the three territories that are most northern; similar to the overall population, these numbers didn't change much between 2011 and 2017. For percent of population over 65, four areas dominate: Ontario, Quebec, British Columbia, and Alberta. Again, the numbers did not change drastically between 2011 and 2017. **(b)** Answers will vary. Most marketing techniques for targeting seniors should mention the four areas with the highest population over age 65: Ontario, Quebec, British Columbia, and Alberta.

1.113 Answers will vary. The most popular industry among the top 100 brands is Technology, followed by Financial Services, Consumer Packaged Goods, Automotive, and Luxury.

1.115 **(a)** Brand—categorical, brewery—categorical, percent alcohol—quantitative, calories per 12 ounces—quantitative, carbohydrates in grams—quantitative. **(b)** Brand is the label. **(c)** A case is a domestic brand of beer; there are 175 cases.

1.117 Many of the brands of beer come from the MillerCoors brewery, followed by Flying Dog Brewery, Sierra Nevada, Anheuser Busch, and Budweiser.

1.119 $\bar{X} = 1242.20$, $s = 2555.30$, $M = 236.5$, $Q1 = 80$, $Q3 = 675$. For 2007, the distribution is also strongly right-skewed and has similar numerical summaries. The leading countries in 2007 were United States, Canada, South Asia, and Spain.

1.121 1, c—currently there are more females in college than males. 2, b—there should be more right-handed students than left-handed. 3, d—height should be Normally distributed. 4, a—should be right-skewed, because some students will study much more than others.

1.123 Gender and automobile preference are categorical. Age and household income are quantitative.

Chapter 2: Examining Relationships

2.1 **(a)** The cases are employees. **(b)** The label could be the employee's name or ID. **(c)** Answers will vary. **(d)** The explanatory variable is how much sleep they get; the response is how effectively they work.

2.3 State is the label; all other variables are quantitative.

2.7 The skew is gone from both distributions. Both are close to symmetrical, with a single peak in the middle and roughly bell shaped.

2.9 **(a)** A strong negative linear relationship. **(b)** A weak positive relationship that is not linear. **(c)** A more complicated relationship. Answers will vary; below is an example of two distinct populations with separate relationships plotted together. **(d)** No apparent relationship.

2.11 We expect the relationship between 2017 and 1997 to be weaker because the time difference is larger. The data for year 1997 would be the explanatory variable; the data for year 2017 would be the response. We would expect the 1997 data to explain, and possibly cause, changes in the 2017 data. The form is roughly linear; the direction is positive; the strength is moderate. Low- and middle-income countries are the outlier.

2.13 Percent alcohol is somewhat right-skewed. Carbohydrates, shown below, is fairly symmetric.

Variable	Obs	Mean	Std. Dev.	Min	Max
alcoholabv	175	0.0527143	0.0144146	0.004	0.115
carbohydrates	160	12.0625	4.93525	1.9	32.1

2.15 **(b)** Two out of three territories have smaller percents of the population over 65 than any of the provinces. Additionally, the same two territories have larger percents of the population under 15 than any of the provinces.

2.17 **(a)** We would expect time to explain the count, so time should be on the x axis. **(b)** As time increases, the count goes down. **(c)** The form is curved; the direction is negative; the strength is very strong. **(d)** The first data point at time 1 is somewhat of an outlier because it doesn't line up as well as the other times do. **(e)** A curve might fit the data better than a simple linear trend.

2.19 **(a)** 2008 data should explain the 2018 data. **(c)** There are 167 points; some of the data for 2008 are missing. **(d)** The form is somewhat linear; the direction is positive; the strength is moderate. **(e)** Suriname is an outlier for both 2008 and 2018. **(f)** The relationship is somewhat linear, though there are observations that don't follow the linear trend well.

2.21 There is a negative relationship between MPG and CO_2 emissions; better MPG is associated with lower CO_2 emissions. The relationship, however, is not linear but curved. There also seem to be two distinct lines or groups. This relationship is very similar to what we found in Example 2.7 when using highway MPG, with the patterns seen in the plot nearly identical to the form we saw in Example 2.7.

2.23 $r = 0.9729$.

2.25 **(b)** The relationship between x and y is very strong but it is not linear; it has a curved relationship or parabola. **(c)** $r = 0$. **(d)** The correlation is only good for measuring the strength of a linear relationship.

2.27 **(a)** $r = 0.9825$. **(b)** Yes, there is a very strong linear relationship between the 2007 and 2017 data.

2.29 **(a)** Yes, the relationship between time and count is quite strong because the data points form a nice curve. **(b)** $r = -0.964$. **(c)** The correlation is not a good numerical summary because the data form a curve, not a line; a transformation is needed to get a better relationship.

2.31 **(a)** $r = 0.9051$. **(b)** Yes, the relationship between percent alcohol and calories is quite linear, so the correlation gives a good numerical summary of the relationship.

2.33 **(b)** $r = -0.9113$. **(c)** No, although the relationship is mostly linear, there is an outlier, Nunavut, with a high percent of under 15 and a very low percent over 65.

2.35 **(a)** $r = 0.9797$. **(b)** The correlation went up from 0.9729 before taking the logs to 0.9797 thereafter. Though the correlation went up a little bit, the log didn't help much with the explanation of the data.

2.37 **(b)** There is a strong positive relationship between city MPG and highway MPG. **(c)** 0.9230. **(d)** A large and positive correlation value indicates that there is a strong positive relationship.

2.39 Applet; answers will vary.

2.41 The magazine report is wrong because they are interpreting a correlation close to 0 as a negative association rather than no association. Answers will vary. "A new study shows no linear association (relationship) between how much a company pays its CEO and how well their stock performs."

2.43 **(a)** A correlation is reserved for quantitative data; because color is categorical, it cannot have any correlation. **(b)** A correlation can never exceed 1, which indicates a perfect linear relationship. **(c)** The correlation is not dependent on order and remains the same between two variables regardless of order.

2.45 There are 8 (one is just barely above the line) positive prediction errors and 7 negative prediction errors.

2.47 **(a)** $b_1 = 0.039$. $b_0 = 3.61$. **(c)** and **(d)**

Country	Prediction	Prediction Error
Australia	5.3416	0.02
Belgium	5.3182	−0.18
Canada	5.5288	−0.47
CzechRep	4.5577	0.53
Estonia	4.4719	0.27
Iceland	5.3026	0.77
Ireland	5.092	0.35
Israel	4.6162	−0.11
Italy	4.9477	−0.64
Japan	4.9672	−0.44
KoreaRep	4.9438	0.15
Netherlands	5.2987	0.31
Norway	6.0982	−0.02
Portugal	4.4368	−0.47
United Kingdom	5.1466	−0.26

2.49 Applet; answers will vary.

2.51 The residuals sum to −0.04. This is due to rounding error.

2.53 The lines are very similar, with and without Texas. Texas is not an influential observation.

2.55 **(b)** For $x = 12$, $y = 800{,}000 + 9000(12) = \$908{,}000$. **(c)** $y = 800{,}000 + 12{,}000x$.

2.57 **(b)** Plan B gets cheaper after 130 minutes.

2.59 **(a)** $\hat{y} = 178.19 + 1.06x$. **(b)** For $x = 740$, $\hat{y} = 178.19 + 1.06 = 178.19 + 1.06(740) = 962.59$. **(c)** Residual $= y - \hat{y} = 465 - 962.59 = -497.59$. The prediction using the 2007 data is much better than it is in the 1997 data, which produces a fairly large residual.

2.61 **(a)** $\hat{y} = 6.59306 - 0.26062x$. **(b)** For $x = 1$, $\hat{y} = 6.59306 - 0.26062(1) = 6.3324$. For $x = 3$, $\hat{y} = 6.59306 - 0.26062(3) = 5.8112$. For $x = 5$, $\hat{y} = 6.59306 - 0.26062(5) = 5.29$. For $x = 7$, $\hat{y} = 6.59306 - 0.26062(7) = 4.7687$. **(c)** For $x = 1$, residual $= y - \hat{y} = 6.3596 - 6.3324 = 0.0271$. For $x = 3$, residual $= y - \hat{y} = 5.7589 - 5.8112 = -0.0523$. For $x = 5$, residual $= y - \hat{y} = 5.3132 - 5.29 = 0.0232$. For $x = 7$, residual $= y - \hat{y} = 4.7707 - 4.7687 = 0.0020$. **(d)** Taking logs of count does not improve the performance of the model. It is still not linear.

2.63 **(a)** The vehicles with high city MPG don't follow the regression line; rather, they have a much lower highway MPG than the regression line would predict. **(b)** For the vehicles with high city MPG, all of the residuals are negative, creating a curve in the plot, suggesting a possible transformation is necessary. **(c)** The residual is negative for this vehicle; its predicted value is much larger. **(d)** Prius C, Prius V, RAV4 Hybrid AWD.

2.65 **(a)** $\hat{y} = 5.77 + 2858.20x$.

2.67 **(b)** The residual plot looks fairly random, although there is one potential low outlier with a very small percent alcohol. **(c)** There is nothing unusual about the location of New Belgium Fat Tire; it is right in the middle of the plot among many other similar brands of beer.

2.69 **(a)** The correlation is 0.9999, so the calibration does not need to be repeated. **(b)** $\hat{y} = 1.65709 + 0.1133x$. For $x = 500$, $\hat{y} = 1.65709 + 0.1133(500) = 58.30709$. Because the relationship is so strong, $r = 0.9999$, we would expect our predicted absorbance to be very accurate.

2.71 **(a)** There seems to be a weak positive linear relationship between y and x, but with one extreme outlier with a very high x-value. **(b)** $\hat{y} = -8.07 + 1.93x$. **(d)** $r^2 = 0.8343$ or 83.43%. **(e)** Although the x variable accounts for 83.43% of the variation in y, there is a very high outlier for x, which pulls the regression line unnaturally. This is seen in the jump of the R-square value from 27% up to 83%, indicating that this observation is very influential in the analysis. This is also demonstrated by the systematic pattern in both the scatterplot and the residual plot, with most of the data points forming a line except for the outlier.

2.73 Applet; answers will vary.

2.75 $r = \text{sqrt}(0.36) = 0.6$, but the value should be negative because the relationship was negative as described, so $r = -0.6$.

2.79 No, the condition of the patient is a possible lurking variable because larger hospitals likely have more resources to treat more seriously injured or ill patients, which also explains why they stay longer.

2.81 **(a)** A lurking variable can be categorical. **(b)** It is actually impossible for all the residuals to be positive. Even if many of the residuals are positive, this tells us nothing regarding the relationship (positive or negative) of the variables. We need to look at the slope to determine if the relationship is positive or negative. **(c)** Whether the relationship is negative or positive does not tell us anything as to whether there is causation.

2.83 If we predicted just next year's sales, the prediction would probably be reasonable, assuming there were no major changes in the company. However, we should not try to predict sales 5 years from now; there is no guarantee the sales will still follow a straight line, as there could be significant changes in the company in the next 5 years.

2.85 Answers will vary. A simple example is a married man who may be willing to work more (for various reasons), which raises his income.

2.87 Answers will vary. One plausible explanation is the opposite relationship; being heavy causes the switch to artificial sweeteners instead of sugar.

2.89 Answers will vary. Parents' education or income, socioeconomic status, etc.

2.91

Music	Frequency	Percent
French	75	30.86
Italian	84	34.57
None	84	34.57

2.93 (a)

	Table of FieldOfStudy by Country							
FieldOfStudy	Canada	France	Germany	Italy	Japan	UK	US	Total
SocBusLaw	64	153	66	125	259	152	878	1697
SciMathEng	35	111	66	80	136	128	355	911
ArtsHum	27	74	33	42	123	105	397	801
Educ	20	45	18	16	39	14	167	319
Other	30	289	35	58	97	76	272	857
Total	176	672	218	321	654	475	2069	4585

(b)

Country	Frequency	Percent
Canada	176	3.84
France	672	14.66
Germany	218	4.75
Italy	321	7
Japan	654	14.26
UK	475	10.36
US	2069	45.13

(c)

FieldOfStudy	Frequency	Percent
ArtsHum	801	17.47
Educ	319	6.96
Other	857	18.69
SciMathEng	911	19.87
SocBusLaw	1697	37.01

2.95 (a)

Wine	Frequency
French	30
Italian	19
Other	35
Total	84

(b)

Wine	Frequency	Percent
French	30	35.71
Italian	19	22.62
Other	35	41.67

(d) Yes, the percent went up from 13.1% to 22.62%.

2.97 (a) **Conditional Distribution of FieldOfStudy Given Country**

	Table of FieldOfStudy by Country						
FieldOfStudy	Canada	France	Germany	Italy	Japan	UK	US
SocBusLaw	36.36	22.77	30.28	38.9	39.6	32	42.4
SciMathEng	19.89	16.52	30.28	24.9	20.8	27	17.2
ArtsHum	15.34	11.01	15.14	13.1	18.81	22.1	19.2
Educ	11.36	6.7	8.26	4.98	5.96	2.95	8.07
Other	17.05	43.01	16.06	18.1	14.83	16	13.2
Total	100	100	100	100	100	100	100

(c) The graph shows that most countries have similarities in their distributions of students among fields of study. Most countries have the most students in social sciences, business, and law, followed in second by science, math, and engineering. France, however, is unique as it has a huge percent in other, much more than the other countries shown. Also, the UK has an extremely low percent in education.

2.99 (a) The first approach looks at the distribution of fields of studies within each country, suggesting the popularity of fields among each country. The second approach is somewhat biased or misleading because of different population sizes, so bigger countries will generally have more students in each field of study. (b) Answers will vary. Most students will likely find the first approach more meaningful. (c) Answers will vary. The first approach shows popularity of fields for each country; the second approach shows which countries have the largest number of students in each field of study.

2.101 (a) For patients in poor condition, $57/1500 = 0.038$ or 3.8% of Hospital A's patients died; $8/200 = 0.04$ or 4% of Hospital B's patients died. (b) For patients in good condition, while $6/600 = 0.01$ or 1% of Hospital A's patients died, $8/600 = 0.0133$ or 1.33% of Hospital B's patients died. (c) The percent of deaths for both conditions is lower for Hospital A, so recommend Hospital A. (d) Because Hospital A had so many more patients in poor condition (1500) compared to good condition patients (600), its overall percent is mostly representing poor-condition patients, who have a high death rate. Similarly, Hospital B had very few patients in poor condition (200) compared to good condition patients (600), so its overall percent is mostly representing good-condition patients, who have a low death rate, making their overall percent lower.

2.103 (a) 18 to 29: 91%, 30 to 49: 85%, 50 to 64: 68%, 65 and over: 40%. (b) 18 to 29: 9%, 30 to 49: 15%, 50 to 64: 32%, 65 and over: 60%.

2.105 There were 21,140 students total; 20,032 agree and 1108 disagree; 11,358 female and 9782 male. 96% of females and 93% of males agreed that trust and honesty are essential. A slightly higher percent of females than males said that trust and honesty are essential, but overall the results were quite similar for both males and females.

Conditional Distribution of Trust-Essential Given Gender		
TrustEssential	Male	Female
Yes	93.00	96.28
No	7.00	3.72
Total	100	100

2.107 (a) For those who get enough sleep, $151/266 = 0.568$ or 56.8% are high exercisers and $115/266 = 0.432$ or 43.2% are low exercisers. (b) For those who don't get enough sleep, $148/390 = 0.379$ or 37.9% are high exercisers and $242/390 = 0.621$ or 62.1% are low exercisers. (c) Those who get enough sleep are more likely to be high exercisers than those who don't get enough sleep. (d) Answers will vary.

2.109 (a) For age 15 to 19: 89.7% are full-time and 10.3% are part-time. For age 20 to 24: 81.82% are full-time and 18.18% are part-time. For age 25 to 34: 50.06% are full-time and 49.94% are part-time. For age 35 and over: 27.15% are full-time and 72.85% are part-time. (d) Students aged 15–24 are much more likely to be full-time, while students aged 35 and over are more likely to be part-time. Students aged 25–34 are about equally likely to be full- or part-time students. (e) Because there are only 2 categories for status, if we are given the percent of full-time students, the percent of part-time students must be 100% minus the percent for full-time. (f) Both are valid descriptions; it mostly depends on what condition the student(s) you are interested in is. If we are interested in a particular age group, the current analysis likely has more meaning, whereas if we are interested in a particular status, the previous analysis has more meaning.

2.111 (a) For younger than 40: 6.7% were hired, 93.3% were not. For 40 or older: 2.2% were hired, 97.8% were not. (c) The percent of hired is greater for the younger than 40 group; the company looks like it is discriminating. (d) Education could be different among groups, making them more or less qualified.

2.113 (a) 27,792 never married; 65,099 married; 11,362 widowed; 13,749 divorced. (b) Joint distribution and marginal distributions

Percent	Never-Married	Married	Widowed	Divorced	Total
18To24	10.26	1.84	0.02	0.14	12.3
25To39	8.03	15.44	0.15	2.12	25.7
40To64	4.43	29.68	2.09	7.35	43.5
65AndOver	0.83	8.21	7.37	2.04	18.5
Total	23.55	55.17	9.63	11.65	100

(c)

Conditional Distribution of Age Given Marital Status				
Percent	Never-Married	Married	Widowed	Divorced
18To24	43.58	3.33	0.2	1.19
25To39	34.08	27.99	1.56	18.18
40To64	18.8	53.8	21.68	63.09
65AndOver	3.54	14.88	76.56	17.54
Total	100	100	100	100

Conditional Distribution of Marital Status Given Age					
Percent	Never-Married	Married	Widowed	Divorced	Total
18To24	83.7	15	0.16	1.13	100
25To39	31.19	60	0.58	8.23	100
40To64	10.17	68.16	4.79	16.88	100
65AndOver	4.52	44.48	39.93	11.07	100

(d) Not shown. (e) More than half of women are married; of that group, age 40 to 64 is the most common, followed by 25 to 39. Almost 25% never married, but most of that group is represented by younger age groups. Widowed and divorced have relatively small percents across the board, though the 65 and over group is most likely to be widowed and the 40 to 64 group is most likely to be divorced.

2.115 33,748 never married; 64,438 married; 2968 widowed; 9964 divorced. More than half of men are married; of that group, age 40 to 64 is the most common, followed by 25 to 39 and 65 and over. More than 30% never married, very few of whom are 65 and over. Fewer than 3% of men are widowed, and the vast majority are 65 and over. About 9% are divorced, two-thirds of whom are in the 40 to 64 age group.

Joint and Marginal Distributions					
Percent	Never-Married	Married	Widowed	Divorced	Total
18To24	12.16	1.12	0.01	0.06	13.3
25To39	11.42	14.43	0.07	1.61	27.5
40To64	6.18	31.18	0.68	5.98	44
65AndOver	0.62	11.26	1.91	1.32	15.1
Total	30.37	57.99	2.67	8.97	100

Conditional Distribution Given Marital Status				
Percent	Never-Married	Married	Widowed	Divorced
18To24	40.03	1.93	0.2	0.63
25To39	37.59	24.88	2.63	17.96
40To64	20.35	53.77	25.61	66.71
65AndOver	2.03	19.42	71.56	14.69
Total	100	100	100	100

Conditional Distribution given Age					
Percent	Never-Married	Married	Widowed	Divorced	Total
18To24	91.14	8.4	0.04	0.43	100
25To39	41.48	52.41	0.26	5.85	100
40To64	14.04	70.82	1.55	13.59	100
65AndOver	4.08	74.55	12.65	8.72	100

2.117 Answers will vary. One example is shown below.

Smokers	Overweight	Not
Early Death	30	260
No	110	1400

Non Smokers	Overweight	Not
Early Death	100	100
No	1000	1200

Combined	Overweight	Not
Early Death	130	360
No	1110	2600

2.119 (a) The log 2007 data (explanatory) should explain the log 2017 data (response). (b) There is a strong positive linear relationship. (c) $\hat{y} = -17.7 + 1.08x$. (f) Answers will vary.

2.121 (a) There is a weak positive linear relationship. There is one high X outlier and one high Y outlier. (b) $\hat{y} = 109.82 + 0.12635x$. (c) $\hat{y} = 109.82 + 0.12635(160) = 130.0361$. (d) Residual = $107 - 130.0361 = -23.0361$. (e) R-square = 10.27%.

2.123 (a) There is little to no relationship. There are four or five potential X outliers with much larger production values than the rest of the observations. (b) $\hat{y} = 112.25678 + 0.11496x$. (c) $\hat{y} = 112.25678 + 0.11496(125) = 126.6262$. (d) Residual = $136 - 126.6262 = 9.3738$. (e) R-square = 2.29%. This R-square is also quite small, suggesting sales is not strongly related to production. There don't seem to be any strong relationships between any of the variables in this data.

2.125 (a) The residual plot shows that the data are not linear; a curve would provide a much better fit. (b) By zooming in on the residuals, the residual plot emphasizes the curve that could potentially be overlooked in the scatterplot.

2.127 **(a)** The regression line is $\hat{y} = 41.25263 + 3.93308x$. So the prediction is $\hat{y} = 41.25263 + 3.93308(25) = \$139{,}579.63$. **(b)** The regression line is $\hat{y} = 3.86751 + 0.04832x$. So the prediction is $\hat{y} = 3.86751 + 0.04832(25) = 5.07551$, or $\$160{,}053.80$. **(c)** The log prediction is better because the data are curved. **(d)** Even if r^2 is high, this doesn't mean a linear fit is appropriate. If the data follow a curve, a transformation is needed and should give an even higher r^2. **(e)** Graphs can show you trends that numerical summaries cannot.

2.129 **(a)** $\hat{y} = 2990.4 + 0.99216x$. **(b)** The residual plot shows one Y outlier, which is a low value.

2.131 Graduation rates can be different based on the difficulty of programs and/or how good the incoming students are. The residual, or difference between actual graduation rate and predicted graduation rate, is better because it shows if a program is doing better or worse than what is expected given the other variables regarding the incoming students.

2.133 Answers will vary.

2.135 **(a)** Higher amps means a bigger motor and more weight. **(b)** As amps increase, so does weight. **(c)** $\hat{y} = 5.8 + 0.4x$; $r^2 = 45.71\%$. **(d)** For every 1 amp increase, weight increases by 0.4 pounds. **(e)** 2.5 amps. **(f)** Yes, there is a slight curvature in the residual plot, suggesting that at the highest amp level, weights may go down somewhat.

2.137 A correlation measures the strength of a linear relationship; that is to say, the relationship between Fund A and Fund B is consistent along a straight line. It doesn't mean they have to change by the same amount or a slope of 1. So as long as Fund A moves 20% and Fund B moves 10% consistently, up or down, you will still remain on the same regression line, and they will remain perfectly correlated.

2.139 **(b)** The strength decreases with length until 9 inches, then levels off. There are no outliers. **(c)** The line does not adequately describe the relationship because the relationship changes after length 9 inches. **(d)** The two lines adequately explain the data. Ask the wood expert what happens at 10 inches.

2.141 **(a)** Smokers 76.12%, nonsmokers 68.58%.

	Smoker		
Outcome	Yes	No	Total
Dead	139	230	369
Alive	443	502	945
Total	582	732	1314

(b) Age 18 to 44 alive: smokers 93.4%, nonsmokers 96.18%. Age 45 to 64 alive: smokers 68.16%, nonsmokers 73.87%. Age 65 and over alive: smokers 14.29%, nonsmokers 14.51%. **(c)** The percents of smokers are 45.86% (18 to 44), 55.18% (45 to 64), 20.25% (65 and over). This confirms the authors' explanation.

Chapter 3: Producing Data

3.1 This is anecdotal evidence; the preference of a group of friends likely would not generalize to the entire college.

3.3 This is anecdotal evidence; the preference of Samantha likely would not generalize to most young people.

3.5 Answers will vary.

3.7 Observational study, because they are just observing which brand has the most market share.

3.9 This is an experiment; there is a treatment group that is randomly assigned.

3.11 This is anecdotal. You have just noticed that the tuna can seems lacking and there are no real data to support the claim. If you wanted to collect data to test your theory, you could measure the contents of a random sample.

3.13 **(a)** This is an experiment. Groups were chosen to receive extra milk or not. **(b)** The control group data alone are observational, because there was no treatment imposed of additional milk. **(c)** TBBMC is the response. Age is not the explanatory variable; it functions as a control variable. The treatment of "additional milk" is the explanatory variable.

3.15 **(a)** Experimental data were collected. The response variable is the severity of the cold and the treatment is the pill. **(b)** This is an experiment because a treatment was imposed and responses were recorded.

3.17 The population is all forest owners in this region. The sample is the 772 to whom the survey was sent. The response rate is $348/772 = 45.1\%$.

3.19 **(a)** The population consists of all manufacturers in the United States. **(b)** The population consists of a subsidized housing facility with 320 apartments.

3.21 Answers will vary.

3.23 Using line 121, the junior associates chosen are 14, 21, 25, 29. Continuing from line 121, the senior associates chosen are 07, 09. Names may vary based on number assignments. Junior associates chosen using the method I stated in the comment would be Rollins, Knapp, Xie, Fletcher, and Bell. Seniors chosen are Traylor and Eastern. Here is my Excel printout.

	A	B	C
1	Associate	Name	
2	Junior	Abrams	00-02
3	Junior	Armstrong	03-05
4	Junior	Bell	06-08
5	Junior	Boyer	09-11
6	Junior	Cantrell	12-14
7	Junior	Colver	15-17
8	Junior	Dawson	18-20
9	Junior	Dowell	21-23
10	Junior	Enos	24-26
11	Junior	Fletcher	27-29
12	Junior	Gastineau	30-32
13	Junior	Haas	33-35
14	Junior	Heathcote	36-38
15	Junior	Howe	39-41
16	Junior	Johnson	42-44
17	Junior	Kim	45-47
18	Junior	Knapp	48-50
19	Junior	Lewicki	51-54
20	Junior	Marks	55-57
21	Junior	Mills	58-60
22	Junior	Perkins	61-62
23	Junior	Perez	63-65
24	Junior	Reed	66-68
25	Junior	Rollins	69-71
26	Junior	Seeger	72-74
27	Junior	Sloan	75-77
28	Junior	Swihart	78-80
29	Junior	Turner	81-83
30	Junior	Xie	84-86
31	Junior	Zhang	87-89
32	Senior	Cline	00-09
33	Senior	Hardin	10-19
34	Senior	Boudreaux	20-29
35	Senior	Kiesler	30-39
36	Senior	Eastern	40-49
37	Senior	Oden	50-59
38	Senior	Eifert	60-69
39	Senior	Traylor	70-79
40	Senior	Haarms	80-89
41	Senior	Whilby	90-99

3.25 (a) Households not listed in the telephone directory are omitted. They may not own a phone or choose to have their number unlisted. (b) Random digit dialing would include those that are unlisted but own a phone.

3.27 (a) Material from the first chapter is likely not representative of the reading level for the entire book. Answers will vary to do the randomization correctly. One example is to randomly select pages and evaluate the reading level of these. (b) Students who attend a particular class at 7:30 A.M. are certainly not representative of all students. A list of all students needs to be used in which you randomly selected 50 to participate. (c) Taking subjects from the top or bottom of an alphabetized list gives different chances of being selected than taking subjects based on the first letter of their name and is not equally likely. Using number assignments and random digits is an appropriate selection process.

3.29 The population is the number of text messages received in the next 100 days. (b) The sample is the number of text messages received in the next 10 days.

3.31 The complexes selected are 08, 12, 17, 23, 25. Complexes may vary based on number assignments. My solution differs significantly, so I post a screenshot.

	A	B	C
1	Apartment	Random number	In the sample
2	Ashley Oaks	00-02	
3	Bay Pointe	03-05	
4	Beau Jardin	06-08	
5	Bluffs	09-11	
6	Brandon Place	12-14	
7	Briarwood	15-17	
8	Brownstone	18-20	
9	Burberry	21-23	
10	Cambridge	24-26	5
11	Chauncey Village	27-29	
12	Country Squire	30-32	
13	Country View	33-35	
14	Country Villa	36-38	
15	Crestview	39-41	
16	Del-Lynn	42-44	1
17	Fairington	45-47	
18	Fairway Knolls	48-50	
19	Fowler	51-54	
20	Franklin Park	55-57	
21	Georgetown	58-60	
22	Greenacres	61-62	
23	Lahr house	63-65	
24	Mayfair Village	66-68	
25	Nobb Hill	69-71	
26	Pemberly Courts	72-74	
27	Peppermill	75-77	4
28	Pheasant Run	78-80	
29	Richfield	81-83	
30	Sagamore Ridge	84-86	
31	Salem Courthouse	87-89	
32	Village Manor	90-92	2
33	Waterford Court	93-95	3
34	Williamsburg	96-98	

3.33 Answers will vary.

3.35 If the same set of random digits is used, then the process is no longer random.

3.37 Call the initial number chosen the "leading" number. All numbers form groups based off their leading number. Certainly, each leading number has an equal chance of being selected. Then, for each subsequent number, it is automatically chosen if its leading number is chosen. Which means all numbers (i.e., all groups) have an equal chance to be chosen. It is not a simple random sample because certain numbers can't be chosen together; so not all samples are possible, only the preformed groups based off leading numbers.

3.39 Answers will vary based on selection process.

3.41 (a) The first sentence is slanted toward yes because it suggests a possible link between cell phones and brain cancer. (b) The question is slanted toward agreeing because it starts "Do you agree..." and also contains a possible advantage of "reduce administrative costs." (c) This is slanted toward a favorable response by the phrases "escalating environmental degradation" and "incipient resource depletion."

3.43 It is an experiment because the instructor has assigned the students to one of two instructional methods (paper-based or web-based). The experimental units are students. The treatments are type of instruction, paper-based and web-based. The response variable is the change in standardized drawing test (pre versus post). The factor is the instructional method, with levels paper-based and web-based. The results would likely not be widely generalizable because the students were all in the same course but could probably be generalized to the same or similar courses in computer graphics technology.

3.45 For those students who volunteered, they could have attributes that lead them to volunteer and also do better on the final. For example, they might be more willing to work hard or might enjoy studying, which may make them more willing to agree to use the new software and also do better on the final.

3.47

Factor A	
Instruction Type	
Paper-based	Web-based
1	2

Random assignment, followed by two groups of students (25 students in each). Then one group receives paper-based and the other web-based.

3.49

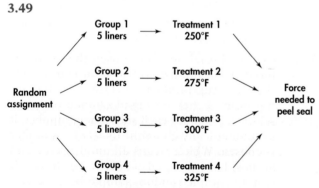

Answers will vary if using software. Using labels 01–20 and line 130, the assignments are:

Group 1: 05, 15, 17, 18, 19
Group 2: 14, 09, 06, 16, 20
Group 3: 01, 02, 05, 10, 03
Group 4: 04, 08, 07, 12, 13

3.51 "A significant difference" means that the difference found between the sexes is unlikely to have occurred by chance alone and that sex is likely a contributor to the difference found in earnings. "No significant difference" means that the difference between black and white students is small enough that it is likely due to just chance. Whichever group happens to have more or less earnings, the difference is not due to race.

3.53 Because the experimenter measured their anxiety and also taught the group how to meditate, the experimenter could biasedly rate the group that meditated lower or higher in anxiety based on their expectation of whether the meditation would help. Separate from the experimenter's possible bias, the subjects themselves could behave more or less anxiously during the final evaluation based on their interaction with the experimenter during the meditation instruction, which could also bias the results.

3.55 In a completely randomized design: 10 students each are randomly assigned to two groups, then one group is randomly assigned the software that highlights trends, the other receives the regular software, and at the end you compare the money made by the two groups. In a matched pairs design: each student uses both types of software in random order for half the time, then the difference between the money made with and without the trend highlights is compared.

3.57 (a) More than 5 subjects (even all 10) could end up in the same group. (b) Batches of rats could be similar, so randomization should be done to ensure that each rat is treated separately, or that batches of rats are divided randomly among treatments equally. (c) Not all subjects have an equal chance to be in each group (first name is not random).

3.59 Answers will vary. Possibly blocking variables include anything related to their prior experience with teamwork (those who are already on a team versus those who are not, etc.).

3.61 Answers will vary.

3.63 Applet; answers will vary.

3.65 Design (a) will produce more trustworthy data because it is completely randomized. Design (b) would be a true matched design, and hence better than design (a) if the same women who do not receive pollen randomly received it at a different time.

3.67 (a) Experiment; the subjects are assigned a treatment. (b) The explanatory variable is the price history shown (with and without promotions); the response is the price they would expect to pay.

3.69 (a) The 120 children are the subjects. (b) The factor is the sets of beverages. The levels are the three sets. The response variable is the child's choice (milk or fruit drink.)

(c)

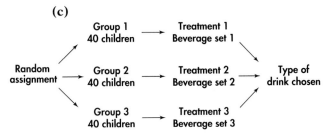

(d) Assign numbers 001–120 to the children. Using line 145, the subjects chosen are 060, 102, 050, 005, and 006.

3.71 (a)

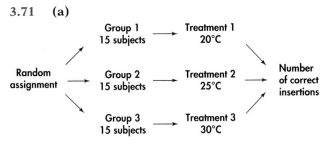

(b) Each worker performs the task three times, once at each temperature, for 30 minutes, and the number of insertions is recorded. The order of temperature assigned is randomized so 15 workers first perform it at 20°C, 25°C, and 30°C. The difference in number of insertions between the three temperatures is the response.

3.73 Answers will vary. **(a)** Minimal. **(b)** Minimal. **(c)** Not minimal.

3.75 Answers will vary.

3.77 Answers will vary.

3.79 Answers will vary.

3.81 It is unethical to say nothing about the identifying codes if respondents expect anonymity.

3.83 This is not anonymous because the survey was done in the person's home. Confidentiality is possible if the individual's name and other identifying information are removed before the results are publicized.

3.85 (a) The subjects should be told what kind of questions will be asked and how long it will take. **(b)** Answers will vary. **(c)** Revealing the sponsor could bias the poll, especially if the respondent doesn't like or agree with the sponsor. However, the sponsor should be announced once the results are made public; that way people can know the motivation for the study and judge whether it was done appropriately, etc.

3.87 Answers will vary. However, participation should be voluntary and not coerced. Psychology 002 is ok if the students are given alternative assignments to earn similar credits.

3.89 It is ethical if all the procedures to safeguard subjects are followed.

3.91 Yes, if the IRB grants approval for deception in the study.

3.93 It is an experiment; even though not random, the researcher imposes treatment. The explanatory variables are the cases of soft drinks. The response variable is the price paid.

3.95 Answers will vary.

3.97 This is not an experiment because there is no treatment imposed. The explanatory variable is different price promotions. The response variable is the subject's preference.

3.99 (a)

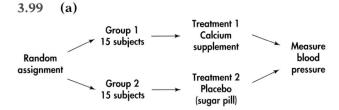

(b) Using line 136, the subjects chosen for group 1 are 08, 14, 20, 09, 24, 12, 11, 16, 22, 15, 13, 17, 28, 04, 10. The rest go in group 2.

3.101 (a) Answers will vary. The placebo effect could be at work because no control group was used. Also, job conditions could have changed drastically over the past month. **(b)** If the experimenter is biased toward meditation either way, the evaluation likely is not objective. **(c)** Answers will vary. Important factors include using a control group that doesn't receive the meditation and making sure the evaluator doesn't know which employees meditated.

3.103 Experiment because a treatment is imposed (tasting the two pizzas).

3.105 (a) Answers will vary. **(b)** During the experiment, the lights may help because they are new and catch the attention of drivers, but over time if they become standard, then they will be less noticeable.

3.107 (a) Label the students 0001, 0002, . . . , 2640. **(b)** Answers will vary.

3.109 (a) The population is all students at your school. **(b)** Answers will vary. A stratified sample is likely a good idea to make sure certain groups are included: majors, departments, schools, gender, etc. **(c)** Answers will vary. Some possible problems would be undercoverage, nonresponse, response bias.

3.111 Answers will vary. Experiments are more difficult to conduct because they impose treatments; they control lurking variables but may lack realism. Observational studies are usually easier to perform but do not control lurking variables and may have biases depending on what is studied; they also cannot show causation.

3.113 Answers will vary. The factors are time of day and zip code (used or not). The treatments would be all the different combinations of time of day and zip code. Lurking variables and solutions will also vary. To handle day of the week, you could mail equal numbers of letters each day.

3.115 Answers will vary.

3.117 Subjects are more likely to be truthful in a CASI survey than in a CAPI survey. For something negative like drug use, the CASI survey will likely produce a higher percent admitting to drug use.

3.119 Answers will vary.

3.121 (a) Software will select a simple random sample for you. (b) Randomized comparative experiments may be superior to matched pair designs and block designs. (c) Multistage samples can be used for assigning smaller groups within the population in stages.

Chapter 4: Probability: The Study of Randomness

4.1 (a) 1/9. (b) 1/9. (c) 8/9.

4.3 Answers will vary. One example would be a business agreement: either they make a deal or it falls through.

4.5 (a) If the dice lands on a 1, the victim knows the perpetrator; otherwise, not. (b) If each dice lands on a 1, the victim knows the perpetrator; otherwise, not. (c) If the number generated is 1, the victim knows the perpetrator; otherwise, not.

4.7 (a) 0.105, 21/200. Yes 0.1 is close to 0.105; the next 200 should also give a similar proportion as well as the combined proportion if 400 repetitions, because the numbers are random. (b) Answers will vary but should be close to 0.1.

4.9 (a) The price changes over time appear quite random and equally likely to be either positive or negative. (b) There doesn't appear to be any relationship between the price change and the lag price change. Yes, the series behaves as a series of independent trials because there is no discernible pattern in the plot. (c) Answers will vary.

4.11 Applet; answers will vary.

4.13 (a) 0. (b) 1. (c) 0.01. (d) 0.6.

4.15 (a) Theoretically, it is possible to never roll a 6 because, each time, there is a chance the die will not be a 6. $S = \{1, 2, 3, \ldots\}$. (b) Theoretically, no matter how many tweets the student made, he or she could make 1 additional tweet, which makes the outcomes infinite. $S = \{0, 1, 2, \ldots\}$.

4.17 By Rule 4: $1 - (0.08 + 0.04 + 0.05) = 0.83$. This is slightly easier than using Rule 3 by taking the 5 colors included and adding their probabilities because, here, we can just add the 3 colors that are not included and use the complement rule.

4.19 By the disjoint rule: $0.187 + 0.267 = 0.454$. By the complement rule: $1 - 0.454 = 0.546$.

4.21 (a) $0.436 + 0.426 + 0.045 + 0.040 + 0.034 + 0.019 = 1$. (b) $0.436 + 0.426 = 0.862$.

4.23 $0.079 + 0.602 = 0.681$.

4.25 (a) The rank of one student should not affect the other students' rank as they likely attended different high schools. (b) $(0.41)(0.41) = 0.1681$. (c) $(0.41)(0.01) = 0.0041$.

4.27 (a) $P(\text{all are truthful}) = (0.7561)^5 = 0.2471$. (b) $P(\text{at least one misrepresented}) = 1 - 0.2471 = 0.7529$.

4.29 (a) $P(\text{string will remain bright}) = P(\text{no lights fail}) = (1 - 0.02)^{20} = 0.6676$. (b) The probability increases with fewer lights; the probability decreases with more lights.

4.31 (a) For small business: $0.00 + 0.00 + 0.07 = 0.07$. For big business: $0.01 + 0.04 + 0.36 = 0.41$. (b) For small business: $1 - 0.07 = 0.93$. For big business: $1 - 0.41 = 0.59$.

4.33 (a) $0.32 + 0.35 + 0.13 + 0.07 + 0.13 = 1$. (b) $0.35 + 0.32 = 0.67$.

4.35 (a) For Canada: $1 - (0.124 + 0.014 + 0.055 + 0.011) = 0.796$. For United States: $1 - (0.052 + 0.006 + 0.028 + 0.012) = 0.901$. (b) For Canada: $0.124 + 0.014 + 0.055 = 0.193$. For United States: $0.052 + 0.006 + 0.028 = 0.086$. Canada is doing much better than the United States regarding sustainable transportation.

4.37 (a) $1 - 0.92 = 0.08$. (b) $0.92 + 0.01 = 0.93$. (c) $1 - 0.93 = 0.07$.

4.39 (a) 0.12, because the probabilities should add up to 1. (b) $0.20 + 0.16 + 0.14 = 0.5$.

4.41 (a) Legitimate. (b) Not legitimate; probabilities add to more than 1. (c) Not legitimate; probabilities add to less than 1.

4.43 (a) Not equally likely. (b) Equally likely: there are equal number of kings and twos in the deck. (c) Equally likely: theoretically, a person should be equally likely to turn right or left. (d) Not equally likely.

4.45 (a) $(0.09)^6 = 0.0000005$. (b) $(1 - 0.09)^6 = 0.5679$. (c) $6(0.09)(0.91)^5 = 0.337$.

4.47 (a) $(0.65)^3 = 0.2746$. (b) Since the years are independent, the probability remains the same,

$(1 - 0.65) = 0.35$. **(c)** P(two rises or two falls) $=$ P(two rises) $+ P$(two falls) $= (0.65)^2 + (0.35)^2 = 0.545$.

4.49 **(a)** The statement is corrected by replacing *disjoint* with *independent*. Disjoint events cannot both occur. **(b)** A probability can never be bigger than 1. Also, the probability of both A and B happening is $(0.6)(0.5)$ if A and B are independent. **(c)** A probability can never be negative. The probability of the complement of A is 1 minus the probability of A or $P(A^C) = 1 - P(A)$.

4.51 $1 - (1 - 0.02)^5 = 0.0961$.

4.53 Because $(A$ and $B)$ and $(A$ and $B^C)$ are disjoint, then $P(A)$ can be written as $P(A) = P(A$ and $B) + P(A$ and $B^C)$; subtraction gives $P(A$ and $B^C) = P(A) - P(A$ and $B)$. Because A and B are independent, applying the multiplication rule gives $P(A$ and $B^C) = P(A) - P(A)P(B) = P(A)(1 - P(B)) = P(A)P(B^C)$; thus, A and B^C are independent.

4.55 P(sum is even or sum is more than 8) $= P$(sum is even) $+ P$(sum is more than 8) $- P$(sum is even *and* sum is more than 8) $= 18/36 + 10/36 - 4/36 = 2/3 = 0.6666$.

4.57 **(a)** $0.52637 + 0.07314 = 0.59951$. **(b)** $0.39012 + 0.52637 = 0.91649$.

4.59 **(a)**

	A_1	A_2	Probability
B_1	0.15	0.39	0.54
B_2	0.22	0.24	0.46
Probability	0.37	0.63	1

(b)

	A_1	A_2	
B_1	30	78	108
B_2	44	48	92
	74	126	200

(c) $P(B_1|A_1) = 0.15/0.37 = 0.405$; $P(A_1|B_1) = 0.15/0.54 = 0.278$.

4.61 P(dollar falls and supplier demands renegotiation) $= P$(dollar falls)P(supplier demands renegotiation|dollar falls) $= (0.4)(0.8) = 0.32$.

4.63 The multiplication rule was used to calculate $P(DE)$.

4.65 **(a)** 0.00511 represents the joint probability. **(b)** $0.00073/0.007 = 0.10$.

4.67 $P(A$ or B or $C) = 0.3 + 0.2 + 0.1 = 0.6$. **(c)** $P(A$ or B or $C)^c = 1 - 0.6 = 0.4$.

4.69 **(a)** $P(B|A) = P(A$ and $B)/P(A) = 0.4/0.6 = 0.67$.

4.71 **(a)**

Outcome	Men	Women
Four-year institution	0.324	0.396
Two-year institution	0.1316	0.1484

(b) P(four-year institution|woman) $= P$(four-year institution and woman)/P(woman) $= 0.396/(0.396 + 0.1484) = 0.7274$.

4.73 **(a)** The tree diagrams are different because each branch on the tree is a conditional probability given the previous branches. So by rearranging the order of the branches, we are necessarily changing the probabilities (assuming the events are not independent). **(b)** Using the multiplication rule, $P(A$ and $B) = P(B|A) \times P(A) = P(A|B) \times P(B)$.

4.75 $P(A|B) = P(A$ and $B)/P(B) = 0.082/0.261 = 0.3142$. If A and B are independent, then $P(A|B) = P(A)$ but because $P(A) = 0.138$, A and B are not independent.

4.77 **(a)** The vehicle is a light truck is the event A^C, $P(A^C) = 0.69$. **(b)** The vehicle is an imported car is the event $(A$ and $B)$, $P(A$ and $B) = 0.08$.

4.79 **(a)** P(Male and Enthusiastic). **(b)** No, that is a joint probability.

4.81 P(HI) $= 0.07314$, P(Mortgage) $= 0.48093$, P(Mortgage and HI) $= 0.05419$; independent if P(Mortgage and HI) $= P$(Mortgage)P(HI).

4.83 $P(O|D) = P(O$ and $D)/P(D) = (0.08)(0.4)/0.0356 = 0.8989$. Given that the customer defaults on the loan, there is an 89.89% chance that the customer has overdrawn an account. If the chance that this customer will overdraw his account is 25%, $P(O|D) = 0.8163$.

4.85 $P(A$ and B and $C) = P(C|A$ and $B)P(A$ and $B) = (0.56)(0.006461) = 0.00361816$.

4.87 **(a)** P(undergraduate and score at least 600) $= P$(score at least 600|undergraduate)P(undergraduate) $= (0.5)(0.4) = 0.2$. P(graduate and score at least 600) $= P$(score at least 600|graduate)P(graduate) $= (0.7)(0.6) = 0.42$. **(b)** P(score at least 600) $= P$(score at least 600 and undergraduate) $+ P$(score at least 600 and graduate) $= (0.5)(0.4) + (0.7)(0.6) = 0.62$.

4.89 P(undergraduate|score at least 600) $= P$(undergraduate and score at least 600)/P(score at least 600) $= 0.2/0.62 = 0.3226$ or 32.26%.

4.91 Yes. $P(A)P(B) = (0.6)(0.5) = 0.3 = P(A$ and $B)$.

4.93 **(a)** $P(D|L) = P(D$ and $L)/P(L) = [(0.03)(0.88)]/[(0.03)(0.88) + (0.97)(0.40)] = 0.0637$. **(b)** Since only 6.37% of those who are late (and denied) will actually default, 93.63% are expected to not default. So out of 100, we would expect 93 or 94 not to default. **(c)** The manager's policy seems unreasonable. Because only 3% of customers default, and while it's true that percent more than doubles to 6.37% if they are late, the vast majority of these customers still do not default, and it is bad business to deny them credit.

4.95 **(a)** $P(S1) = 423/995 = 0.4251$. $P(S2) = 367/995 = 0.3688$. $P(S3) = 205/995 = 0.2060$. **(b)** $P(S3|D) =$

$P(D|S3)/P(S3)/[P(D|S1)/P(S1) + P(D|S2)/P(S2) + P(D|S3)/P(S3)] = (0.006)(0.2060)/[(0.004)(0.4251) + (0.003)(0.3688) + (0.006)(0.2060)] = 0.3057$.

4.97 (a) If A and B are disjoint, then $P(A \text{ or } B) = 1.2$, which is not possible. (b) The smallest possible value is 0.2; otherwise the probabilities will add to more than 1. The largest possible value is 0.4, which is the entire probability for event B. (c) For independent events, $P(A \text{ or } C) = P(A) + P(C) - P(A)P(C) = 0.8 + 0.1 - (0.8)(0.1) = 0.82$.

4.99 (a) $P(A) = 1/36$. $P(B) = 15/36$. (b) $P(A) = 1/36$. $P(B) = 15/36$. (c) $P(A) = 10/36$. $P(B) = 6/36$. (d) $P(A) = 10/36$. $P(B) = 6/36$.

4.101 (a) All the probabilities are between 0 and 1 and sum to 1. (b) P(tasters agree) $= 0.03 + 0.07 + 0.25 + 0.20 + 0.06 = 0.61$. (c) P(Taster 1 rates higher than 3) $= 0.00 + 0.02 + 0.05 + 0.20 + 0.02 + 0.00 + 0.01 + 0.01 + 0.02 + 0.06 = 0.39$. P(Taster 2 rates higher than 3) $= 0.00 + 0.02 + 0.05 + 0.20 + 0.02 + 0.00 + 0.01 + 0.01 + 0.02 + 0.06 = 0.39$.

4.103 P(Taster 2's rating is higher than 3 | Taster 1's rating is 3) $= P$(Taster 2's rating is higher than 3 and Taster 1's rating is 3)$/P$(Taster 1's rating is 3) $= (0.05 + 0.01)/(0.01 + 0.05 + 0.25 + 0.05 + 0.01) = 0.1622$.

4.105 To avoid double counting, the joint probabilities, which are represented by the areas that overlap, must be subtracted to accurately calculate $P(A \text{ or } B \text{ or } C)$.

Chapter 5: Random Variables and Probability Distributions

5.1 (a) $0.09 + 0.36 + 0.35 + 0.13 + 0.05 + 0.02 = 1$. It is a legitimate discrete distribution. (b) The American household owns at least one car. Using the complement rule, $P(X \geq 1) = 1 - P(X = 0) = 1 - 0.09 = 0.91$. (c) $P(X > 2) = 0.13 + 0.05 + 0.02 = 0.20$ or 20%.

5.3 (a) 0.44. (b) 0. (c) 0.7. (d) 1. (e) 0.

5.5

x	1	2	3	4	5	6
$P(X = x)$	0.05	0.05	0.13	0.26	0.36	0.15

5.7 (a) $P(X \leq 3) = P(X = 1) + P(X = 2) + P(X = 3) = 0.05 + 0.05 + 0.13 = 0.23$. (b) $P(X = 4 \text{ or } X = 5) = 0.26 + 0.36 = 0.62$. (c) $P(X = 8) = 0$.

5.9 (a) $0.27 + 0.33 + 0.16 + 0.14 + 0.06 + 0.03 + 0.01 = 1$. It is a legitimate probability distribution. (b) $P(X \geq 5) = P(X = 5) + P(X = 6) + P(X = 7) = 0.06 + 0.03 + 0.01 = 0.10$. (c) $P(X > 5) = P(X = 6) + P(X = 7) = 0.03 + 0.01 = 0.04$.

(d) $P(2 < X \leq 4) = P(X = 3) + P(X = 4) = 0.16 + 0.14 = 0.30$. (e) $P(X \neq 1) = 1 - P(X = 1) = 1 - 0.27 = 0.73$. (f) $P(X > 2) = P(X = 3) + P(X = 4) + P(X = 5) + P(X = 6) + P(X = 7) = 0.16 + 0.14 + 0.06 + 0.03 + 0.01 = 0.40$. Much easier to use the complement $P(X > 2) = 1 - P(X \leq 2) = 1 - 0.60$.

5.11 (a) Time is continuous. (b) Hits are discrete (you can count them). (c) Yearly income is discrete (you can count money).

5.13 (b) $P(X \geq 1) = 1 - P(X = 0) = 1 - 0.1 = 0.9$. (c) "At most, two nonword errors." $P(X \leq 2) = P(X = 0) + P(X = 1) + P(X = 2) = 0.1 + 0.3 + 0.3 = 0.7$, $P(X < 2) = P(X = 0) + P(X = 1) = 0.1 + 0.3 = 0.4$.

5.15 The possible values of X are $0 and $5, each with probability 0.5. The mean is $\$0(0.5) + \$5(0.5) = \$2.50$.

5.17 $\mu_y = 28 - 2(6) = 16$.

5.19 (a) $\mu_X = 1(0.3) + 2(0.5) + 3(0.2) = 1.9$. $\mu_Y = 1(0.1) + 2(0.15) + 3(0.4) + 4(0.3) + 5(0.5) = 3.05$. (b) $\mu_{8000X} = 8000\mu_X = 8000(1.9) = \$15,200$. $\mu_{30000Y} = 30000\mu_Y = 30000(3.05) = \$91,500$. (c) $\mu_{X+Y} = 4.95$. $\mu_{8000X+30000Y} = 8000\mu_X + 30000\mu_Y = \$106,700$.

5.21 (a) $\sigma_X^2 = (1 - 1.9)^2(0.3) + (2 - 1.9)^2(0.5) + (3 - 1.9)^2(0.3) = 0.49$, $\sigma_X = 0.7$. (b) $\sigma_Y^2 = (1 - 3.05)^2(0.1) + (2 - 3.05)^2(0.15) + (3 - 3.05)^2(0.4) + (4 - 3.05)^2(0.3) + (5 - 3.05)^2(0.05) = 1.0475$. $\sigma_Y = 1.02347$.

5.23 $\mu_{X-Y} = \$100$, $\sigma_{X-Y}^2 = \sigma_X^2 + \sigma_Y^2 - 2\rho\sigma_X\sigma_Y = (100)^2 + (80)^2 - 2(0.4)(100)(80) = \$100$. When two variables have a positive correlation, if one variable increases, the other will also increase. If one variable decreases, the other will also decrease. This results in the average differences being smaller than if they were independent. With independence, if one variable increases, the other could increase or decrease.

5.25 $\mu_R = (0.2)(0.351) + (0.8)(0.66) = 0.5982$. $\sigma_R^2 = (0.2^2)(4.195^2) + (0.8^2)(4.287^2) + (2)(0.449)(0.2 \times 4.195)(0.8 \times 4.287) = 15.05$. $\sigma_R = 3.879$.

5.27 $\mu_X = (-1)(0.1) + (0)(0.4) + (1)(0.3) + (2)(0.2) = 0.6$; $\sigma_X^2 = (-1 - 0.6)^2(0.1) + (0 - 0.6)^2(0.4) + (1 - 0.6)^2(0.3) + (2 - 0.6)^2(0.2) = 0.84$; $\sigma = 0.9165$.

5.29 (a) $65 + 2(0.5)(4)(7) = 93$; 9.6.
(b) $65 - 2(0.5)(4)(7) = 37$; 6.1.
(c) $505 - 2(2)(3)(0.5)(4)(7) = 337$; 18.36.

5.31 (a) $S = \{HHH, HHT, HTH, THH, TTH, THT, HTT, TTT\}$.

(b)

Value of D	−3	−1	1	3
Probability	1/8	3/8	3/8	1/8

(c) $\mu_D = 0$. $\sigma_D = 1.732$.

5.33 $\sigma^2 = (0 - 0.1538)^2(0.8507) + (1 - 0.1538)^2 (0.1448) + (2 - 0.1538)^2(0.0045) = 0.1391$, $\sigma = 0.373$.

5.35 Answers will vary. The average value of the 30,000 values should be close to the theoretical mean of \$0.50 as in Example 5.3.

5.37 (a) $\sigma_{X+Y}^2 = \sigma_X^2 + \sigma_Y^2 + 2\rho\sigma_X\sigma_Y = \sigma_{0.7X}^2 + \sigma_{0.3Y}^2 + 2\rho\sigma_{0.7X}\sigma_{0.3Y} = (0.7)^2(4)^2 + (0.3)^2(2)^2 + 2(-1)(0.7)(4)(0.3)(2) = (0.49)(16) + (0.09)(4) + 2(-1)(0.7)(4)(0.3)(2) = 7.84 + 0.36 + -3.36 = 4.84$. $\sigma_{X+Y} = 2.2$. The portfolio does not have zero risk. (b) $\sigma_{X+Y}^2 = \sigma_X^2 + \sigma_Y^2 + 2\rho\sigma_X\sigma_Y = \sigma_{0.5X}^2 + \sigma_{0.5Y}^2 + 2\rho\sigma_{0.5X}\sigma_{0.5Y} = (0.5)^2(4)^2 + (0.5)^2(2)^2 + 2(-1)(0.5)(4)(0.5)(2) = (0.25)(16) + (0.25)(4) + 2(-1)(0.5)(4)(0.5)(2) = 4 + 1 + -4 = 1$. The portfolio does not have zero risk. (c) $\sigma_{X+Y}^2 = \sigma_X^2 + \sigma_Y^2 + 2\rho\sigma_X\sigma_Y = \sigma_{0.5X}^2 + \sigma_{0.5Y}^2 + 2\rho\sigma_{0.5X}\sigma_{0.5Y} = (0.5)^2(4)^2 + (0.5)^2(4)^2 + 2(-1)(0.5)(4)(0.5)(4) = (0.25)(16) + (0.25)(16) + 2(-1)(0.5)(4)(0.5)(4) = 4 + 4 + -8 = 0$. The portfolio does have zero risk. (d) The online quote is not a universally true statement. "Perfectly negatively correlated investments *with equal variance and equal mix...*"

5.39 (a) $\mu_X = 540(0.1) + 545(0.25) + 550(0.3) + 555(0.25) + 560(0.1) = 550°C$. $\sigma_X^2 = (540 - 550)^2(0.1) + (545 - 550)^2(0.25) + (550 - 550)^2(0.3) + (555 - 550)^2(0.25) + (560 - 550)^2(0.1) = 32.5$. $\sigma_X = 5.7°C$. (b) $\mu_{X-550} = \mu_X - 550 = 0°C$. $\sigma_{X-550} = 5.7°C$. (c) $\mu_Y = \frac{9}{5}\mu_X + 32 = \frac{9}{5}(550) + 32 = 1022°F$. $\sigma_Y^2 = \left(\frac{9}{5}\right)^2 \sigma_X^2 = \left(\frac{9}{5}\right)^2 (32.5) = 105.3$. $\sigma_Y = 10.26°F$.

5.41 30/70, 5.44%.

X	Y	σ^2	σ
0	1	32.21	5.68
0.1	0.9	30.54	5.53
0.2	0.8	29.69	5.45
0.3	0.7	29.64	5.44
0.4	0.6	30.40	5.51
0.5	0.5	31.96	5.65
0.6	0.4	34.33	5.86
0.7	0.3	37.51	6.12
0.8	0.2	41.50	6.44
0.9	0.1	46.30	6.80
1	0	51.90	7.20

5.43 $X \sim B(20, 0.5)$.

5.45 X is not binomial; there is not a fixed number of trials n.

5.47 (a) 0.1780; 0.0376. (b) 0.1780; 0.0376. (c) They are the same.

5.49 (a) $P(X \leq 11) = 0.01$. (b) Smaller. (c) The claim is likely false.

5.51 (a) $S = \{0, 1, 2, 3, 4, 5\}$. (b) $P(X = 2) = \binom{5}{2}(0.2439)^2 1 - 0.2439^3 = 0.2571$. (c) $P(X \geq 1) = 1 - P(X = 0) = 1 - \binom{5}{0}(0.2439)^0(1 - 0.2439)^5 = 0.75$.

5.53 (a) $\mu_X = np = 20(0.8) = 16$. (b) $\sigma_X = \sqrt{np(1-p)} = \sqrt{20(0.8)(1-0.8)} = 1.789$. (c) For $p = 0.9$, $\sigma_X = \sqrt{20(0.9)(1-0.9)} = 1.342$. For $p = 0.99$, $\sigma_X = \sqrt{20(0.99)(1-0.99)} = 0.445$. As p gets closer to 1, the standard deviation gets smaller.

5.55 $P(X \geq 1) = 1 - P(X = 0) = 1 - 0.4066 = 0.5934$.

5.57 (a) 0.95. (b) 0.647.

5.59 (a) Each flip is independent, and prior tosses have no impact on the outcome of a new toss. (b) Each flip is independent, and prior tosses have no impact on the outcome of a new toss. (c) p is a parameter for the binomial, not $\hat{p}$. (d) There is no fixed number of trials n.

5.61 (a) $B(200, p)$, where p is the probability that a student says he or she usually feels irritable in the morning. (b) This is not binomial; there is not a fixed n. (c) $B(500, 1/12)$. (d) This is not binomial because separate cards are not independent.

5.63 (a) $n = 5$, $p = 0.56$. (b) $S = \{0, 1, 2, 3, 4, 5\}$. (c)

Value	Prob
0	0.016492
1	0.104947
2	0.267137
3	0.339993
4	0.216359
5	0.055073

(d) $\mu_X = 5(0.56) = 2.8$. $\sigma_X = \sqrt{np(1-p)} = \sqrt{5(0.56)(1-0.56)} = 1.11$.

5.65 (a) 0.38. (b) 0.42. (c) 0.29.

5.67 (a) Answers will vary—the events are not disjoint, etc. (b) $X \sim B(5, 0.1)$. $P(X \geq 1) = 1 - P(X = 0) = 1 - (0.9)^5 = 1 - 0.5905 = 0.4095$. (c) $X \sim B(40, 0.1)$. $\mu_X = np = 40(0.1) = 4$.

5.69 (a) $\mu_X = np = 1500(0.13) = 195$. $\sigma_X = \sqrt{np(1-p)} = \sqrt{1500(0.13)(1-0.13)} = 13.025$. (b) 0.0002. (c) The small probability computed in part (b) indicates that the sample with 150 or fewer blacks is not representative of the population.

5.71 **(a)** Your vote will matter if the other 10 committee members are split 5 and 5 on a vote. So using $X \sim B(10, 0.5)$, we want $P(X = 5) = 0.2461$. **(b)** Your vote will matter if the other 522 town voters are split 261 and 261 on the vote. So using $X \sim B(522, 0.5)$, we want $P(X = 261)$. Using the Normal approximation, $X \sim N(261, 11.42)$, $P(X = 261) = P(260.5 < X < 261.5) = P(-0.04 < Z < 0.04) = 0.5160 - 0.4840 = 0.0320$ using continuity correction (0.0350 from software).

5.73 Y has possible values $1, 2, 3, \ldots$, etc. $P(Y = k) = (5/6)^{k-1}(1/6)$, because we must have $k - 1$ failures before the success on the kth trial.

5.75

N	$P(X = 10)$
1000	0.110902468
10000	0.107348923
100000	0.106997648
1000000	0.106962588
10000000	0.106959083

The probability decreases as N increases.

5.77 Poisson with $\mu = 2.768$. **(a)** $P(X = 0) = 0.0628$. **(b)** $P(X \geq 3) = 1 - P(X \leq 2) = 0.5229$.

5.79 **(a)** Poisson with $\mu = 48.7$. Using Excel, $P(X \geq 50) = 1 - P(X \leq 49) = 1 - 0.5550 = 0.4450$. Or using the Normal approximation, we have $P(X \geq 50) = P\left(Z \geq \frac{50 - 48.7}{\sqrt{48.7}}\right) = P(Z \geq 0.19) = 0.4247$. **(b)** $\sigma_X = \sqrt{48.7} = 6.9785$. $\sigma_X = \sqrt{97.4} = 9.8691$. **(c)** Poisson with $\mu = 97.4$. Using Excel, $P(X \geq 100) = 1 - P(X \leq 99) = 1 - 0.5905 = 0.4095$. Or using the Normal approximation, we have $P(X \geq 100) = P\left(Z \geq \frac{100 - 97.4}{\sqrt{97.4}}\right) = P(Z \geq 0.26) = 0.3974$.

5.81 **(a)** Poisson with $\mu = 56$ per day. **(b)** 7.483. **(c)** $P(X \geq 50) = 0.8061$ (0.7881 using Normal).

5.83 $P(X = 0) = 0.5971$. $1532 \times 0.5971 = 914.7 \sim 915$. $1099 - 915 = 184$.

5.85 **(a)** $P(X = 0) = e^{-\mu}\mu^0/0! = e^{-\mu}(1)/1 = e^{-\mu}$. **(b)** $P(X = k) = e^{-\mu}\mu^k/k! = e^{-\mu}\mu^{k-1}(\mu)/[k(k-1)!] = (\mu/k)(e^{-\mu}\mu^{k-1}/(k-1)!) = (\mu/k) P(X = k-1)$. **(c)** $e^{-3} = 0.0498$. **(d)** Because $P(X = 1) = \frac{\mu}{1} P(X = 0)$, we multiply $P(X = 0)$ by $3/1 = 3$; because $P(X = 2) = \frac{\mu}{2} P(X = 1)$, we multiply $P(X = 1)$ by $3/2 = 1.5$.

5.87 **(a)** 0. **(b)** 0.69. Convert seconds to minutes 28 seconds = 0.47 minutes. So you miss the train if X is less than 3.47. **(c)** 0:2:30.

5.89 **(a)** $P(X < 30) = P(Z < -1.43) = 0.0766$. **(b)** $P(Z > 0.71) = 0.2375$. **(c)** $P(-2.14 < Z < 1.43) = 0.9074$.

5.91 **(a)** $X = 68$. **(b)** $X = 142$.

5.93 **(a)** 120; 6.93. **(b)** Yes. **(c)** 0.002. This is a low probability and the 70% successes could have been due to chance.

5.95 EXPON.DIST(1,0.14,1).

5.97 **(a)** True if X is a discrete random variable. **(b)** True for discrete random variables, but the area under the density curve is the probability of a continuous random variable. **(c)** The uniform distribution is symmetric. **(d)** The uniform distribution does not have a density curve that lies above the horizontal axis for all values of X.

5.99 **(a)** 1; 2.3. **(b)** 0.81. **(c)** 0.75. **(d)** 0. (The probability statement is not written correctly.) **(e)** $X = 4.2$. **(f)** $X = -2.2$.

5.101 **(a)** 0.1056. **(b)** 0.1056. **(c)** 0.1866. **(d)** 264. **(e)** 165.

5.103 **(a)** 154; 5.95. **(b)** (142.1, 165.9). **(c)** 170 falls in the upper 2.5% of the Normal distribution so students at this institution like hybrid instruction more than students do nationally.

5.105 **(a)** Yes, $np = 27.3 > 10$ and $n(1 - p) = 72.7 > 10$. **(b)** 0.679 (using the Normal tables). **(c)** 0.721.

5.107 **(a)** 466.5; 18.07. **(b)** $P(X \leq 450) = P(Z \leq 0.91) = 0.1814$. **(c)** $P(X \leq 450.5) = P(Z \leq 0.89) = 0.1867$.

5.109 (4424.9, 4695.1).

5.111 **(a)** $P(X \leq 2) = 0.16$. **(b)** $P(5 \leq X \leq 15) = 0.38$.

5.113 **(a)** 16.25. **(b)** EXP.DIST(16.25,0.1417,1).

5.115 **(a)** 40%. **(b)** The mean of $X = (0.4)(-6) + (0.6)(7) = 1.8$.

5.117 **(a)** The possible values and probabilities for Y are shown the table.

Value of Y	15	195
Probability	0.8	0.2

(b) $\mu_Y = (0.8)(15) + (0.2)(195) = 51$. $\sigma_Y^2 = (15 - 51)^2(0.8) + (195 - 51)^2(0.2) = 5184$, $\sigma_Y = 72$. **(c)** There are no rules for a quadratic function of a random variable; we must use the definitions.

5.119 1/18.

5.121 **(a)** P(at least one bag every day) = P(at least one bag)^{30} = $(1 - P(\text{no bags}))^{30}$ = $(1 - 0.202)^{30}$ = 0.0011484. **(b)** "At least one day will demand 0 bags." **(c)** P(365 days without a demand for 12 bags) = P(not a demand for 12 bags)^{365} = $(1 - P(\text{demand for 12 bags}))^{365}$ = $(1 - 0.003)^{365}$ = 0.3340.

5.123 **(a)** P(man does not die in the next five years) = $1 - P$(man dies within the next five years) = $1 - (0.00039 + 0.00044 + 0.00051 + 0.00057 + 0.00060) = 0.99749$. **(b)** $\mu = 875(0.99749) - 99,825(0.00039) - 99,650(0.00044) - $

99,475(0.00051) − 99,300(0.00057) − 99,125(0.00060) = \$623.22.

5.125 (a) $0.301 + 0.176 + 0.125 = 0.602$.
(b) $X \sim B(10, 0.602)$. $P(X \geq 4) = 1 - P(X \leq 3) = 1 - 0.0533 = 0.9467$. (c) $P(X \geq 620) = 0.1290$.
(d) $X \sim N(6.02, 1.548)$. $P(X \geq 4) = P(X \geq 3.5) = P\left(Z \geq \dfrac{3.5 - 6.02}{1.548}\right) = P(Z \geq -1.63) = 0.9484$.

5.127 (a) $m = 575$. Excel command is BINOM.DIST (575, 1000, 0.602, TRUE) = 0.04381.

5.129 (a) $\mu_X = np = 20(0.25) = 5$. (b) $P(X \geq 10) = 0.013864417$. (c) $P(X \geq 550) = 0.0057$ (0.0052 using the Normal approximation).

5.131 $P(X \geq 1) = 1 - P(X = 0) = 1 - 0.9666 = 0.0334$.

5.133 The Poisson distribution is not appropriate because the rate is not constant and increases between the midnight and 6 A.M. period.

5.135 Poisson with $\mu = 15$. $P(X \geq 19) = 1 - P(X \leq 18) = 0.1805$.

5.137 $P(8 \leq X \leq 10) = P\left(\dfrac{8-9}{0.0724} < Z < \dfrac{10-9}{0.0724}\right) = P(-13.8 \leq Z \leq 13.8) \approx 1$. The probability is essentially 1.

5.139 (a) Mean arrival per minute is $16.25/10 = 1.625$. (b) $\lambda = 1/\mu = 1/1.625 = 0.615$. (c) 0.615. (d) The standard deviation = the mean = 0.615 minutes per customer. (e) =EXPON.DIST(0.5, 0.615, TRUE) = 0.265. (f) $P(X > 1) = 1 - P(X \leq 1) = $ 1-EXPON.DIST(1, 0.615, TRUE) = 0.54.

Chapter 6: Sampling Distributions

6.1 The statistic is the proportion of respondents in the sample of 1008 employees that directly experienced abusive conduct at work; the population is workers; the population parameter is the proportion of the entire population that directly experienced abusive conduct at work.

6.3 (a) 0.7. (b) The histogram for sample size $n = 20$. (c) The larger sample will have 95% of the sample proportions lie between 0.6 and 0.8; the smaller sample will have 95% of the sample proportions lie between 0.5 and 0.9.

6.5 (a) Graduates that found a job within 6 weeks; 100 graduates of university's business school; 82%. (b) The average number of times a student has visited the career services office; 32,585 currently enrolled college students; 1.9. (c) Understatement of sales due to shipments made but not recorded as sales; 200 shipping documents; 2%.

6.7 (a) bias = $90 - 100 = -10$. (b) -4.08. (c) It is a downward biased.

6.9 The sample size is unknown.

6.11 (a) Answers will vary. (b) Answers will vary. (c) Answers will vary.

6.13 A sample size of 107 is large enough to give an appropriate estimate.

6.15 (a) The population is the number of steps per day taken by users of the *Azumio Argus* app; for a sample of 400,000 users, the statistic is the average number of steps taken and the standard deviation. (b) The shape of the population distribution is right-skewed.

6.17 $\mu_{\bar{X}} = 420$, $\sigma_{\bar{X}} = 1$. Larger sample gives the same mean but a smaller standard deviation.

6.19 (a) $m_1 = \pm 2\left(\dfrac{84}{\sqrt{200}}\right)$. (b) $m_2 = \pm 2\left(\dfrac{84}{\sqrt{50}}\right)$. (c) m_2 is greater than m_1 by a factor of 2; the margin of error increases as the sample size gets smaller.

6.21 Applet; answers will vary.

6.23 (a) The standard deviation will be $\dfrac{30}{\sqrt{10}}$. (b) A larger sample will result in a smaller standard deviation. (c) Only the standard deviation depends on the sample size n. (d) $\bar{x}$ will fall within $\mu \pm \dfrac{2\sigma}{\sqrt{n}}$.

6.25 Parameter, statistic.

6.27 (a) Defining X as $2B$ might be incorrect if B_1 and B_2 are not independent. (b) The variance of $2B = 0.0004$, which is half of the variance reported in Example 6.12.

6.29 (a) Larger, to decrease the standard deviation. (b) 0.085. (c) $n = 68$.

6.31 (a) $\mu = 69.4$. (b) Software; answers will vary. (c) Software; answers will vary.

6.33 $P(\text{Total} > 4000) = P\left(\bar{X} > \dfrac{4000}{19} = 210.526\right) = P\left(Z > \dfrac{210.526 - 190}{35/\sqrt{19}} = 2.56\right) = 0.0052$.

6.35 (a) $P(X > 72) = P\left(Z > \dfrac{72-69}{2.8} = 1.07\right) = 0.1423$.
(b) $P(\bar{X} > 72) = P\left(Z > \dfrac{72-69}{2.8/\sqrt{2}} = 1.52\right) = 0.0643$.
(c) $P(\bar{X} > 72) = P\left(Z > \dfrac{72-69}{2.8/\sqrt{5}} = 2.40\right) = 0.0082$.
(d) It becomes less likely to have a sample mean greater than 6 ft because as the sample size increases, the standard deviation of the sample mean decreases, which will pull the mean closer to the population mean which is 69 in.

6.37 $n = 300$, $X = 21$, $\hat{p} = 7.1\%$.

6.39 (a) 0.04. (b) Yes, the population is more than 20 times larger than the sample. (c) 0.0113. (d) The mean is different but the standard deviation is the same for accurate documents.

6.41 (a) 0.4794. (b) 0.3. (c) 0.0011.
6.43 (a) 0.075. (b) 0.2 and 0.5. (c) 0.45. (d) No, it is not unusual.
6.45 (a) 0.11. (b) 0.0188. (c) 0.13 and 0.21. (d) Less than the national level.
6.47 (a) It is unclear whether all 80 members of the fraternity are undergraduates. (b) The minimum sample size is 100,000.
6.49 (a) $\mu_{\hat{p}} = p = 0.69$. $\sigma_{\hat{p}} = \sqrt{\frac{p(1-p)}{n}} = \sqrt{\frac{0.69(1-0.69)}{300000}} = 0.000844$. (b) $2\sigma = 0.0017$, so between 0.6883 and 0.6917. (c) No, the actual percents are much more variable than the interval, suggesting that the percent has changed from season to season.
6.51 The variability of the sample proportion is relatively small. Therefore, the sample proportion is reliable estimator for the population proportion. The governor is justified in applying the survey results.
6.53 (a) 3100 employees who are in favor of the proposal. (b) 50 employees who are in favor of the proposal. (c) The statistic is the proportion of employees who are in favor of the proposal; it is calculated by dividing the number of employees who are in favor of the proposal by 50. (d) The population parameter is the proportion of employees who are in favor of the proposal divided by 3100.
6.55 (a) The proportion of medium- and large-sized companies are not equal; biased high. (b) 450,000; 1,010,000.
6.57 79.57; it is smaller than the theoretical value.
6.59 (a) 15,317.30; 28.9. (b) 15,286.7; 62.6. (c) The normal quantile plot shows that the simulated means are normally distributed.
6.61 (a) For mean sample: 300.28; for median sample: 300.34. (b) 8.49. (c) 8.46; it is marginally smaller compared to the value calculated in part (b). (d) 10.27; small bias, small variability.
6.63 (a) $\mu_{\bar{x}} = \mu$; $\sigma_{\bar{x}} = \sigma$. (b) For all n, $\mu_{\bar{x}} = 0$ and $\sigma_{\bar{x}} = 0$. (c) $\mu_{\hat{p}} = 1$ or $\mu_{\hat{p}} = 0$; $\sigma_{\hat{p}} = 0$. (d) 0.5.

Chapter 7: Introduction to Inference

7.1 $\sigma_{\bar{x}} = \frac{\sigma}{\sqrt{n}} = \frac{320}{\sqrt{100}} = \frac{320}{10} = 32$.
7.3 $64.
7.5 Applet; answers will vary.
7.7 Doubled; for $n = 2834, m = 1.96\left(\frac{642.10}{\sqrt{2834}}\right) = 23.64$; for $n = 709, m = 1.96\left(\frac{642.10}{\sqrt{709}}\right) = 47.27$.
7.9 $n = \left(\frac{1.96(11500)}{1000}\right)^2 = 508.05$, so $n = 508$.
7.11 (a) 1933. (b) Answers will vary for reasons of nonresponses; nonresponses are always to be considered a source of bias. It is likely the 40% of those who responded are different from those who didn't respond, so our estimated margin of error is not a good measure of accuracy.
7.15 They will have the same margin of error because the sample sizes are the same, $n = 100$.
7.17 (a) She forgot to divide the standard deviation by $\sqrt{n}$. (b) Inference is about the population mean, not the sample mean. (c) Confidence does not mean probability; furthermore, making probability statements about μ doesn't make sense because it's fixed, not random. (d) The Central Limit Theorem guarantees that the sample mean will be Normally distributed, not the original values. "... the sample mean of alumni ratings will be approximately Normal."
7.19 (a) $(-\infty, \infty)$. This is useless because it gives us no information about what μ is. (b) $\bar{X} \pm 0$. The chance that $\bar{X}$ is exactly μ has 0 probability, so in the long run none of our confidence intervals would cover μ, making our confidence 0%.
7.21 Because there is nonresponse, the accuracy is in question regardless of the small margin of error. There is no guarantee the respondents are similar to the nonrespondents.
7.23 $\bar{X} = 28.8 \frac{\text{miles}}{\text{gallon}} \times \frac{1 \text{ gallon}}{3.785 \text{ liters}} \times \frac{1.609 \text{ kilometers}}{1 \text{ mile}} = 12.24$. The margin of error is 1.21.
7.25 $\bar{X} = \frac{28.8}{3.80} = 7.58$; $m = \frac{2.85}{3.80} = 0.75$.
7.27 (a) No, we are only 95% confident that the interval covers the true index value for the population. (b) We believe the actual index for the population is in this interval with 95% confidence. (c) 0.076. (d) No, the interval only accounts for error due to random sampling.
7.29 = CONFIDENCE.NORM(0.05,973.21,1511) = 49.07; 64.4899.
7.31 $H_0: \mu = 0$; $H_a: \mu < 0$.
7.33 (a) 2.4. (b) 4.8. (c) −2.4. (d) −4.8.
7.35 (a) $Z = -1.58$. (b) 0.1142. No.
7.37 (a) Yes. (b) No. (c) Because $0.034 \leq 0.05$, we reject H_0. Because $0.034 > 0.01$, we do not reject H_0.
7.39 0.0456; 0.0026.
7.41 (a) 2.5. (b) 0.0062. (c) For a two-sided alternative, the P-value would be doubled. 0.0124.
7.43 (a) Since the P-value is less than 0.05, we reject H_0 for a two-sided test. Therefore, 70 is not included

in the 95% confidence interval. **(b)** Since the P-value is greater than 0.01, we fail to reject H_0 for a two-sided test. Therefore, 70 is included in the 99% confidence interval.

7.45 **(a)** -1.5. **(b)** Values beyond ± 2.576. **(c)** Fail to reject the null hypothesis; the absolute value of the z statistic is not greater than the critical value (2.576).

7.47 **(a)** The null hypothesis should contain the = sign, so the hypothesis should be "... the average weekly demand is equal to 100 units." **(b)** The square root is missing; the standard deviation should be $\frac{10}{\sqrt{25}}$. **(c)** The hypotheses are always about the parameter, not the statistic; so it should be $H_0: \mu = 17$.

7.49 **(a)** For $\alpha = 0.05$, we reject H_0 when Z is either bigger than 1.96 or smaller than -1.96. **(b)** You cannot just compare $\bar{X}$ to μ; we need to do a hypothesis test. **(c)** The P-value should be the two tail areas, so it should be $2P(Z > 1.2)$. **(d)** $\sqrt{n}$ is missing from the formula.

7.51 **(a)** $H_0: \mu = 28; H_a: \mu > 28$. **(b)** $H_0: \mu = 4;$ $H_a: \mu \neq 4$. **(c)** $H_0: \mu = 90\%; H_a: \mu < 90\%$.

7.53 **(a)** p_M = percent of males; p_F = percent of females. $H_0: p_M = p_F; H_a: p_M > p_F$. **(b)** μ_A = mean score for Group A; μ_B = mean score for Group B. $H_0: \mu_A = \mu_B; H_a: \mu_A > \mu_B$. **(c)** ρ = correlation between income and the percent of disposable income that is saved. $H_0: \rho = 0; H_a: \rho > 0$.

7.55 **(a)** Answers will vary. $H_0: \mu_A = \mu_B; H_a: \mu_A \neq \mu_B$. **(b)** Do not reject the null hypothesis. There is no evidence that exercise affects how students perform on their final exam in statistics. **(c)** How did they measure exercise—was it observational or experimental, how did they get the sample, etc.?

7.57 **(a)** 0.031. **(b)** 0.969, but since the test statistic is positive, this alternative is not possible. **(c)** 0.062.

7.59 $Z = \dfrac{11.2 - 6.9}{\frac{2.7}{\sqrt{5}}} = 3.56$; P-value ≈ 0 (<0.0002).

The new poems are by another author.

7.61 $Z = \dfrac{158.4 - 160}{\frac{5}{\sqrt{50}}} = -2.26$; P-value $= 0.0238$.

There is evidence that the population mean is not 160 bushels per acre. Yes, the conclusion is still correct for non-Normality because of the Central Limit Theorem.

7.63 **(a)** The P-value $= 0.1676$. No. **(b)** No.

7.65 At $\bar{X} = 0.8$, it is significant at $\alpha = 0.01$. If α is smaller, $\bar{X}$ has to be farther away from μ_0 to be significant.

$\bar{X}$	Significance
0.1	Not significant
0.2	Not significant
0.3	Not significant
0.4	Not significant
0.5	Not significant
0.6	Not significant
0.7	Not significant
0.8	Significant
0.9	Significant
1	Significant

7.67 The test is significant when $\bar{X} = 0.3$. Larger samples are able to detect smaller differences between $\bar{X}$ and μ_0.

$\bar{X}$	Significance
0.1	Not significant
0.2	Not significant
0.3	Significant
0.4	Significant
0.5	Significant
0.6	Significant
0.7	Significant
0.8	Significant
0.9	Significant
1	Significant

7.69 The P-value doubles for each $\bar{X}$ value because our P-value now represents two tail areas.

$\bar{X}$	P-value
0.1	0.7518
0.2	0.5271
0.3	0.3428
0.4	0.2059
0.5	0.1139
0.6	0.0578
0.7	0.0269
0.8	0.0114
0.9	0.0044
1	0.0016

7.71 If the P-value is less than 0.01, it must also be less than 0.05.

7.73 Any number between 2.576 and 2.807 (or -2.576 and -2.807).

7.75 **(a)** $Z = \dfrac{541.4 - 525}{\frac{100}{\sqrt{100}}} = 1.64$; P-value $= 0.0505$.

No. **(b)** $Z = 1.65$; P-value $= 0.0495$. Yes. **(c)** Even though the first result is not statistically significant, we would still consider it of practical significance because it is more than a 15-point improvement.

7.77 If $\alpha = 0.20$, then we would be making a mistake 20% of the time.

7.79 Answers will vary.

7.81 **(a)** If SES had no effect on LSAT, there would still be some small differences due to chance variation. "Statistically insignificant" means that the effect is small enough that it could just be due to this chance. **(b)** If the effect were large, it could be of practical importance even though it wasn't statistically significant; this is especially true if the sample was small.

7.83 **(a)** There seems to be no relationship between x and y. **(b)** $r = 0.07565$. P-value $= 0.8355$. Yes, there is not a significant correlation between x and y. **(c)** The plot is identical. The correlation is the same, $r = 0.07565$. The P-value is smaller (P-value $= 0.7513$). **(d)** The correlation has not changed, but the P-value gets smaller as n increases.

n	R	P-value
10	0.07565	0.8355
20	0.07565	0.7513
30	0.07565	0.6911
40	0.07565	0.6427
50	0.07565	0.6016
60	0.07565	0.5657

(e) $n = 680$. **(f)** Even with no relationship and a very small correlation, a big enough sample size can show statistical significance, warning us to make sure the effect is worth our attention rather than just "trusting" the statistics.

7.85 $\alpha = 0.05/6 = 0.0083$. The three tests with P-values of 0.007, 0.004, and <0.001 are significant.

7.87 **(a)** $Z^* = 2.87$. **(b)** It will get bigger.

7.89 **(a)** $X \sim B(80, 0.05)$. **(b)** $P(X \geq 2) = 0.9139$.

7.90 The power for $n = 20$ is 0.6736. Answers may vary if using software.

7.91 **(a)** 472.56.

(b) $472.56 \pm 1.645(2)\sqrt{1 + \frac{1}{20}} = (469.19, 475.93)$.

7.93 **(a)** The width of the interval decreases; theoretically, the width of the interval can be pushed closer and closer to 0. **(b)** $z \times \sigma\sqrt{1}$. **(c)** $1.96 \times 5\sqrt{1}$.

7.95 **(a)** 227. **(b)** 196.

7.97 **(a)** 75. **(b)** 10.98.

7.99 Applet; answers will vary. At first, the answer may not be close to 95% but eventually with enough repeats, the percent hits should be quite close to 95%.

7.101 $25.21 \pm 1.96 \dfrac{68.65}{\sqrt{464}} = (18.96, 31.46)$.

7.103 $68.94 \pm 1.96 \dfrac{\sigma}{\sqrt{340}} = (67.40, 70.47)$; $\sigma = 14.39$.

7.105 **(a)** 2×10^{-15}. **(b)** The probability in part (a) is smaller than the probability of 1 in 175 million.

7.107 **(a)** The distribution is roughly symmetric.

2	034
3	011246
4	3

(b) (16.01, 44.79). **(c)** $H_0: \mu = 25$; $H_a: \mu > 25$; $Z = (30.4 - 25)/(7/\sqrt{(10)}) = 2.44$; P-value $= 0.0074$. Reject H_0. Mean odor is greater than 25.

7.109 **(a)** The ideal population is all nonprescription medication customers. The actual population consists of those listed in the Indianapolis telephone directory. **(b)** Food stores: (15.22, 22.12), Mass merchandisers: (27.77, 36.99), Pharmacies: (43.68, 53.52). **(c)** Yes, the confidence interval for the pharmacies gives values much higher than in the other two intervals.

7.111 Software may vary; the answer will be (26.06, 34.74) with possible rounding error.

7.113 **(a)** 0.327 is the P-value. **(b)** $2P(Z \leq -0.9839) = 0.327$. **(c)** We fail to reject H_0 since the P-value is greater than 0.05, which means the interval will contain the population mean of 473.

7.115 **(a)** -22.92. **(b)** P-value ≈ 0; the number of runs is different from 58.92.

7.117 **(a)** Agree; two-sided P-value is equal to 2 times the P-value for a one-sided test. **(b)** Disagree; the two-sided P-value is $2P(Z \geq |1.8|)$. **(c)** Agree; half of the normal distribution is equal to 0.5; therefore, a one-sided P-value cannot exceed 0.5. **(d)** Disagree, it is exactly equal to 0.05.

7.119 **(a)** The difference between the groups is so large that we do not believe it is attributed to chance. **(b)** 95% confidence means our results, in the long run, will be correct 95% of the time. **(c)** Not necessarily, because there likely are lurking variables. For example, it is possible that those mothers willing to sign up for the training program are also more actively seeking employment, which could account for the difference.

Chapter 8: Inference for Means

8.1 (a) 42. (b) 15.

8.3 $879 \pm 2.131 \frac{168}{\sqrt{16}} = (789, 969)$.

8.5 $H_0: \mu = 800$; $H_1: \mu \neq 800$; $t = 1.18$; P-value = 0.0795; there is not strong evidence to reject the null hypothesis at the 5% level.

8.7 (a) $H_0: \mu = 1.5$; $H_1: \mu > 1.5$; the market direction is up when the average return is greater than 1.5%. (b) $t = 1.54$ df = 24; $t^*_{0.05} = 1.71$; there is not strong evidence that market direction is up because the t statistic is smaller than the critical value.

8.9 P-value = 0.083.

8.11 90% confidence interval: (−4.16, 2.16); the confidence interval is within the 0.00 ± 4 region; therefore, we can conclude at the 10% level that the two means are equivalent.

8.13 Yes, although the data are strongly skewed, $n \geq 40$ implies we can still use the t procedure.

8.15 (a) As the degrees of freedom increase, the $t(k)$ distribution approaches the $N(0, 1)$ distribution. (b) The margin of error of the sample mean is $t^* \frac{s}{\sqrt{n}}$. (c) $\bar{x}$ should be replaced with μ in the hypotheses. (d) If there are no outliers or skewness in the data, one sample t procedure can be safely used.

8.17 (a) df = $n - 1 = 28 - 1 = 27$. (b) 2.052 and 2.158. (c) 0.025 and 0.02. (d) $0.02 < $ P-value < 0.025. (e) It is significant at the 5% level; it is not significant at the 1% level. (f) From software, P-value = 0.0212.

8.19 (a) df = $n - 1 = 200 - 1 = 199$. (b) $0.0025 < $ P-value < 0.005. (c) From software, P-value = 0.0033.

8.21 (a) The histogram suggests that it is appropriate to use the t procedures in analyzing the data. (b) (25.15, 25.84).

8.23 (a) Because $n \geq 40$, we can still use the t procedure for non-Normal distributions. (b) $H_0: \mu = 0$; $H_a: \mu \neq 0$; $t = \frac{-0.55 - 0}{\frac{2.16}{\sqrt{69}}} = -2.12$; df = 68; $0.02 < $ P-value < 0.04. The data are significant at the 5% level; there is evidence that the mean rating is different from zero. (People do not feel that the trips take the same amount of time.)

8.25 (b) According to the graphs, the distribution of the data appears to be Normal; there appears to be no outliers and the data are fairly symmetric. (c) Since the distribution of the data appears to be Normal, it is appropriate to apply the t procedures in analyzing the data.

8.27 (a) $H_0: \mu = 70$; $H_1: \mu \neq 70$; $P > 0.5$; there is strong evidence that the amount of iodine in a serving is 70.0 micrograms. (b) $H_0: \mu = 70.5$; $H_1: \mu \neq 70.5$; $0.005 < P < 0.001$; there is not strong evidence that the amount of iodine in a serving is 70.0 micrograms. (c) 70 micrograms lies in the 95% confidence interval from the previous exercise; 70.5 micrograms lies outside of the 95% confidence interval from the previous exercise.

8.29 (a) The data are slightly left-skewed but because $n \geq 40$, we can still use the t procedures. (b) $\bar{X} = 35,288.2$; $s = 10,362$; $SE_{\bar{X}} = 1092.3$. The margin of error for 90% is 1817.5. (c) (33,470.7, 37,105.7). (d) $H_0: \mu = 33,000$; $H_a: \mu > 33,000$; $t = 2.091$; df = 89; $0.01 < $ P-value < 0.02 (0.0195 from software). The data are significant at the 5% level, and there is evidence that the mean pounds of product treated in 1 hour is greater than 33,000.

8.31 (a) There are three large outliers, making the data not Normal.

0	599
1	1356778
2	001234566888
3	25799
4	135579
5	359
6	0
7	
8	46
9	4

(b) $\bar{X} = 34.134$; $s = 21.248$; $SE_{\bar{X}} = 3.3597$. (c) From software, the 95% confidence interval is (27.3384, 40.9296).

8.33 (a) A matched pairs t test is appropriate because the question of interest concerns the difference between the two strategies from the S&P 500 Index. (b) There appears to be one outlier in the data; otherwise, the data are reasonably symmetric. (c) Since $n \geq 15$, it is appropriate to analyze the data using the t procedures.

8.35 (a) $H_0: \mu = 10$; $H_1: \mu < 10$. (b) The t procedures can be applied to integer or noninteger data. (c) P-value < 0.0001; since 0.05 is the significance level, there is strong evidence that the null hypothesis can be rejected because the P-value < 0.05.

8.37 The 90% C.I. is (−21.1682, −5.4718). The mean time for right-hand threads is 104.12; for left it is 117.44. The ratio is 88.66%. On an assembly line with an 8-hour period, this amounts to saving more than 54 minutes or almost an

8.39 H_0: $\mu_d = 0$; H_a: $\mu_d \neq 0$. (b) $t = -2.37$; df $= 7$; $0.025 < P$-value < 0.02. The data are significant at the 5% level, and there is enough evidence to show a difference between the yields of these two varieties.

8.41 (a) H_0: median $= 0$; H_a: median > 0 or H_0: $p = 0.5$; H_a: $p > 0.5$. (b) $Z = 2.86$; P-value $= 0.0021$. There is evidence that the median is greater than zero.

8.43 $100 - 94 \pm 2.017 \sqrt{\frac{11^2}{25} + \frac{8^2}{25}} = (0.51, 11.49)$. With the smaller sample sizes, the interval got wider.

8.45 (a) H_0: $\mu_1 = \mu_2$; H_1: $\mu_1 \neq \mu_2$; $t = 2.21$; $0.025 < P$-value < 0.05. The data are significant at the 5% level, and there is enough evidence to show a difference between the two approaches. (b) Because the confidence interval does not contain 0, we reject H_0.

8.47 $t = 3.35$; df $= 113$; the t statistic, 3.35, is very close to the t statistic (3.39) reported in Example 8.13.

8.49 (a) Because 0 is in the interval, we fail to reject the null hypothesis. (b) Generally, a larger sample will result in a smaller margin of error.

8.51 (a) H_0: $\mu_1 = \mu_2$; H_1: $\mu_1 \neq \mu_2$; $t = -2.32$; $0.02 < P$-value < 0.05; fail to reject the null hypothesis at the 1% level. (b) H_0: $\mu_1 = \mu_2$; H_1: $\mu_1 < \mu_2$; $t = -2.32$; $0.01 < P$-value < 0.02; fail to reject the null hypothesis at the 1% level.

8.53 H_0: $\mu_{\text{Brown}} = \mu_{\text{Blue}}$; H_a: $\mu_{\text{Brown}} > \mu_{\text{Blue}}$.
(c) $t = \dfrac{0.55 - (-0.38)}{\sqrt{\dfrac{(1.68)^2}{40} + \dfrac{(1.53)^2}{40}}} = 2.59$. df $= 39$;
$0.005 < P$-value < 0.01. The data show that brown-eyed students appear more trustworthy compared with their blue-eyed counterparts.

8.55 (a) Both distributions are Normally distributed, except the low-intensity class has a low outlier. (b) H_0: $\mu_H = \mu_L$; H_a: $\mu_H \neq \mu_L$; $t = 5.30$; df $= 14$; P-value < 0.001 (< 0.0001 from software). The data are significant at the 5% level, and there is evidence the noise levels are different between the high- and low-intensity fitness classes. (c) Because the low-intensity class has an outlier, the t test is not appropriate. (d) $t = 6.31$; df $= 13$; P-value < 0.001 (< 0.0001 from software). Removing the outlier didn't change the results. (e) Because the outlier is not affecting the results, it is probably okay to report both tests. It would be a good idea to investigate the outlier and see why it had such a low decibel value; if it were drastically different in some way, it might be good to remove it and only report the test without it after mentioning its removal.

8.57 (a) H_0: $\mu_1 = \mu_2$; H_1: $\mu_1 > \mu_2$; $t = 7.34$; P-value < 0.0001; reject the null hypothesis at the 5% level.

8.59 (a) Answers will vary. But there are likely differences about this company's workers that could not be generalized to other workers. (b) (4.37, 5.43). With 95% confidence, the drill and blast workers have between 4.37 and 5.43 more exposure to respirable dust than the outdoor concrete workers. (c) H_0: $\mu_{DB} = \mu_{OC}$; H_a: $\mu_{DB} \neq \mu_{OC}$; $t = 18.47$; df $= 114$; P-value < 0.001. There is significant evidence that the drill and blast workers have more exposure to respirable dust than the outdoor concrete workers. (d) Because $n_1 + n_2 \geq 40$, the t procedures can be used for skewed data.

8.61 (a) Because $n_1 + n_2 \geq 40$, we can use the t procedures on skewed data. (b) H_0: $\mu_{\text{days}} = \mu_{\text{month}}$; H_a: $\mu_{\text{days}} \neq \mu_{\text{month}}$; $t = -2.42$; df $= 89$; $0.01 < P$-value < 0.02. The data are significant at the 5% level, and there is evidence the means of the two groups are different. Those who are told 30/31 days have a smaller expectation interval on average than those who are told 1 month.

8.63 (a) H_0: $\mu_L = \mu_H$; H_a: $\mu_L \neq \mu_H$; $t = 8.23$; df $= 13$; P-value < 0.001 (< 0.0001 from software). The data are significant at both the 5% and 1% levels, and there is evidence the two groups are different in mean ego strength. (b) No, they were all college faculty who volunteered and would not represent all middle-aged men. (c) No, the study was observational; we would need an experiment to show causation.

8.65 (a) $(-1.32, 7.32)$. (b) With 95% confidence, the mean change in sales from last year to this year is between -1.32 and 7.32. Because the interval covers 0 and includes some negative values, it is possible sales have actually decreased.

8.67 (a)

Variable	N	Mean	Std Dev
With_NotTold	48	35.1170833	6.6297309
Without_NotTold	49	40.0048980	6.6554975

(b) Both Normal quantile plots show the two variables are both roughly Normally distributed. (c) H_0: $\mu_W = \mu_{WO}$; H_a: $\mu_W \neq \mu_{WO}$; $t = -3.62$; df $= 47$; P-value < 0.001. The data are significant at the 5% level, and there is evidence the two groups are different in total cost for those with and without feedback among those who were not told they were on a budget. The results are similar to those in Example 7.10; feedback helped reduce spending.

8.69 There could be things that are similar about the next seven employees who need new computers as well as the following seven, which could bias the results (like being from the same office or department).

8.71 When the standard deviations are similar, the Satterthwaite DF are closer to $n_1 + n_2 - 2$. When one standard deviation is much larger, the Satterthwaite DF are closer to the smaller of $n_1 - 1$ and $n_2 - 1$.

n_1	n_2	s_1	s_2	Satterthwaite DF
15	28	4.5026	2.7783	19.865
14	28	2.8517	2.7783	25.502

8.73 (a) df = 4; $t^* = 2.776$. (b) df = 8; $t^* = 2.306$. (c) Because the critical value is smaller for the pooled t test, it is easier to show significance than the unpooled t test.

8.75 (a) The sample size will increase because as the margin of error decreases, the ratio $\left(\frac{z^* s}{m}\right)^2$ increases. (b) The sample size will decrease because the 90% confidence interval requires a relatively smaller z^*; therefore, the ratio $\left(\frac{z^* s}{m}\right)^2$ decreases. (c) The sample will decrease because a smaller s will cause the ratio $\left(\frac{z^* s}{m}\right)^2$ to decrease.

8.77 (a) The sample size will decrease because as the margin of error increases, the ratio $\left(\frac{\sqrt{2} z^* s}{m}\right)^2$ decreases. (b) The sample size will increase because 99% confidence requires a relatively larger z^*; therefore, the ratio $\left(\frac{\sqrt{2} z^* s}{m}\right)^2$ increases. (c) The sample size will decrease because it is equal to $\left(\frac{z^* s}{m}\right)^2$ rather than $\left(\frac{\sqrt{2} z^* s}{m}\right)^2$.

8.79 (a) 0.65. (b) 1. (c) 0.05.

8.81 Decrease; because the difference we are trying to find is smaller, it is harder to detect, so the power will be smaller.

8.83 (a) The sample size would have to be quadrupled to reduce the margin of error by half. (b) The power at $\mu = 15$ is the same as the power at $\mu = 25$. (c) Lowering the significance level from 0.05 to 0.01 will decrease the power. (d) An increase in the sample size does not affect Type I error.

8.85 No, the confidence interval is for the mean monthly rate, not the individual apartment rates.

8.87 (a) $n = \left(\frac{1.96(5.1)}{1}\right)^2 = 99.9$, so $n = 100$. (b) We would need to use the bigger sample to make sure both margin of error conditions are met.

8.89 (a) $t^* = 2.390$. (b) Reject H_0 when $t > 2.390$, or when $\bar{X} > 710.4$. (c) Given $\mu = 300$,
$$P(\bar{X} > 710.4) = P\left(Z > \frac{710.4 - 300}{\frac{1330}{\sqrt{60}}}\right) =$$
$P(Z > 2.39) = 0.008$; this power is sufficient.

8.91 0.542.

8.93 For 70, df = 35 + 35 − 2 = 68, so using df = 60, $\alpha = 0.05$, we get $t^* = 2.000$; $= \dfrac{4.5}{8\sqrt{\frac{1}{35} + \frac{1}{35}}} = 2.353$;
$P(Z > 2.000 - 2.353) = P(Z > -0.353) = 0.638$.
For 90, df = 45 + 45 − 2 = 88, so using df = 80, $\alpha = 0.05$, we get $t^* = 1.990$; $= \dfrac{4.5}{8\sqrt{\frac{1}{45} + \frac{1}{45}}} = 2.668$;
$P(Z > 1.990 - 2.668) = P(Z > -0.668) = 0.748$.
For 110, df = 55 + 55 − 2 = 108, so using df = 100, $\alpha = 0.05$, we get $t^* = 1.984$; $= \dfrac{4.5}{8\sqrt{\frac{1}{55} + \frac{1}{55}}} = 2.950$;
$P(Z > 1.984 - 2.950) = P(Z > -0.950) = 0.829$.

8.95 (a) Using df = 20 + 20 − 2 = 38, so using df = 30 and $\alpha = 0.01$, we get $t^* = 2.75$.
$\delta = \dfrac{|0.5|}{0.7\sqrt{\frac{1}{20} + \frac{1}{20}}} = 2.2588$. $P(Z > 2.750 - 2.2588) = P(Z > 0.49) = 0.3121$ (0.32601 from software using df = 38). (b) Using df = 30 + 30 − 2 = 58, so using df = 50 and $\alpha = 0.05$, we get $t^* = 2.009$. $\delta = \dfrac{|0.5|}{0.7\sqrt{\frac{1}{30} + \frac{1}{30}}} = 2.7664$.
$P(Z > 2.009 - 2.7664) = P(Z > -0.76) = 0.7764$ (0.77432 from software using df = 58).

8.97 The margin of error will decrease; $m = \dfrac{1.96(15,500)}{\sqrt{50}} = 4296.4$. With $n = 25$ the marginal error will increase.

8.99 (a) A split stemplot is shown.

```
0 | 8
1 | 14
1 | 56679
2 | 01111112233344
2 | 5788889
3 | 11223334
3 | 559
4 | 00
4 | 5578
5 | 3
5 | 66
6 | 
6 | 7
```

Because $n \geq 40$, we can use t procedures for skewed data. **(b)** (26.06, 33.18). Yes, the interval is essentially the same as the bootstrap intervals.

8.101 **(a)** P-value $= 0.038$; we reject H_0. **(b)** P-value $= 0.962$; we fail to reject H_0.

8.103 **(a)** A single sample. We only have one group we are interested in. **(b)** Two independent samples. The two groups from different years form independent samples. **(c)** Matched pairs. We would have each customer rate both floor plans and subtract to obtain the differences.

8.105 **(a)** $49.692 - 50.545 \pm 2.262\sqrt{\frac{5.373}{10} + \frac{3.703}{10}} = (-3.01, 1.30)$. (Or $(-2.859, 1.153)$ from software).

(b) $-0.853 \pm 2.262\left(\frac{1.269}{\sqrt{10}}\right) = (-1.761, 0.055)$.

(c) The estimates (centers) are the same, but the margin of error for the two-sample procedure is much larger than for the matched pairs procedure.

8.107 (674.6, 686.1).

8.109 $H_0: \mu_C = \mu_N$; $H_a: \mu_C > \mu_N$;

$t = \dfrac{4.44 - 3.95}{\sqrt{\dfrac{(3.98)^2}{90} + \dfrac{(2.88)^2}{90}}} = 0.95$;

df $= 89$; $0.15 < P$-value < 0.20. There is no significant evidence that those treated rudely were willing to pay more. This is a two-sample one-sided significance test; we assume the data were randomly selected and that the data contain no outliers.

8.111 $H_0: \mu_C = \mu_N$; $H_a: \mu_C > \mu_N$;

$t = \dfrac{2.90 - 2.98}{\sqrt{\dfrac{(3.28)^2}{90} + \dfrac{(3.24)^2}{90}}} = -0.16$; P-value > 0.25.

There is no significant evidence that those treated rudely were willing to pay more. This is a two-sample one-sided significance test; we assume the data were randomly selected and that the data contain no outliers. We found similar results here as in Exercise 8.109, where condescending did not boost sales.

8.113 $H_0: \mu = 4.88$; $H_a: \mu > 4.88$; $t = \dfrac{5.91 - 4.88}{\dfrac{0.57}{\sqrt{148}}} = 21.98$; df $= 147$; P-value < 0.0005. There is evidence that the hotel managers on average score higher in masculinity than the general male population.

8.115 **(a)** There are no outliers, so t procedures can be used for $n \geq 40$.

12	22234
12	5556777888888899
13	01111223344444
13	5556677788
14	113

(b) From software, the 95% confidence interval is (13.00, 13.31).

8.117 **(a)** Matched pairs t test; the hockey players are not drawn from independent samples. **(b)** $H_0: \mu = 0$; $H_0: \mu \neq 0$; $t = 0.58$; df $= 25$; P-value $= 0.564$; there is no change in accuracy across all hockey players. **(c)** Two-sample t test; sexes are drawn from independent samples. **(d)** $H_0: \mu_1 = \mu_2$; $H_0: \mu_1 \neq \mu_2$; $t = 0.503$; df $= 24$; P-value $= 0.619$; the change in psychomotor demand is not different between males and females.

8.119 **(a)** $11.99 \pm 2.365\left(\dfrac{0.828}{\sqrt{8}}\right) = (11.30, 12.68)$. **(b)** The original confidence interval, which was expressed in minutes, could have been divided by 60 to get the confidence interval in part (a).

8.121 The 95% confidence interval is (1.02, 7.66).

8.123 $78.32 \pm 2.064\left(\dfrac{33.3563}{\sqrt{25}}\right) = (64.55, 92.09)$. The average percent of purchases for which the alternate supplier offered a lower price is 64.55% and 92.09%. So, a large percent of the time, the alternate vendor gave better prices, meaning the original supplier's prices are not competitive.

8.125 **(a)** $H_0: p = 0.5$; $H_a: p > 0.5$. **(b)** $X \sim B(20, 0.5)$; $P(X \geq 14) = 0.0577$. This is not significant at the 5% level. There is not strong evidence that the intensive language training improves Spanish listening skills.

8.127 **(a)** $H_0: \mu_S = \mu_W$; $H_a: \mu_S < \mu_W$;

$t = \dfrac{35.12 - 37.32}{\sqrt{\dfrac{(4.31)^2}{750} + \dfrac{(3.83)^2}{412}}} = -8.95$; df $= 411$;

P-value < 0.0005. There is significant evidence that the experienced workers will outperform the students during the first minute. **(b)** Because $n_1 + n_2 \geq 40$, the t procedures can be used for skewed data. **(c)** 29.66 and 44.98. **(d)** The scores for the 1st minute are much lower than the scores for the 15th minute.

8.129 $H_0: p = 0.5$; $H_a: p \neq 0.5$; $Z = 1.67$; $2P(Z \geq 1.67) = 0.095$. There is no strong evidence that the two endowment effects are different.

Chapter 9: One-Way Analysis of Variance

9.1 (a) ANOVA does not test the sample means; it should be the population means. (b) This is not within-group variation; it is between-group variation. (c) A one-way ANOVA can compare two or more populations. (d) You need an experiment to show causation.

9.3 This population is representative of the University of Pennsylvania as long as the participants were randomly chosen; this population might not be representative of the city because students from the University of Pennsylvania, and, hence, participants, may not come from the city; this population might not be representative of the United States but rather universities across the United States.

9.5 $I = 3$; $n_1 = 20$; $n_2 = 20$; $n_3 = 20$; DFG = 2; DFE = 57. The parameters of the model are: μ_1, μ_2, μ_3, and σ.

9.7 (a) Yes, the largest s is less than twice the smallest. (b) $\mu_1 = 60.3$; $\mu_2 = 85.6$; $\mu_3 = 69.7$; $\sigma = 19.07$.

9.9 The Normal quantile plot shows the residuals are Normally distributed.

9.11 MST = SST/DFT = $51,229.418/120 = 426.9118$. The mean is 69.0052; the variance is 426.9118.

9.13 (a) DFG = 4; DFE = 30. (b) Bigger than 2.69. (c) Bigger than 2.56. (d) More observations per group give a larger DFE, which makes the F critical value smaller. Conceptually, more data give more information showing the group differences and giving more evidence the differences we are observing are real and not due to chance. Thus, a larger DFE will give a smaller critical value and smaller P-value.

9.15 (a) Take the square root of the Within Groups MS. (b) s_p can be found by taking the square root of the MSE.

9.17 (a) Response: rating score (1 to 7). The populations are (1) elementary statistics students, (2) health and safety students, and (3) cooperative housing students. $I = 3$, $n_1 = 237$, $n_2 = 147$, $n_3 = 86$, $N = 470$. (b) Response: acceptance rating (1 to 7). The populations are (1) students who attend varsity football or basketball games only, (2) students who also attend other varsity competitions, and (3) students who did not attend any varsity games. $I = 3$, $n_1 = 37$, $n_2 = 18$, $n_3 = 59$, $N = 114$. (c) Response: sales. The populations are sales on days offering (1) a free drink, (2) free chips, (3) a free cookie, and (4) nothing. $I = 4$, $n_1 = 5$, $n_2 = 5$, $n_3 = 5$, $n_4 = 5$, $N = 20$.

9.19 For part (a): (a) DFG = 2; DFE = 467, DFT = 469. (b) $H_0: \mu_1 = \mu_2 = \mu_3$. H_a: not all of the μ_i are equal. (c) $F(2, 469)$. For part (b): (a) DFG = 2, DFE = 111, DFT = 113. (b) $H_0: \mu_1 = \mu_2 = \mu_3$. H_a: not all of the μ_i are equal. (c) $F(2, 111)$. For part (c): (a) DFG = 3, DFE = 16, DFT = 19. (b) $H_0: \mu_1 = \mu_2 = \mu_3 = \mu_4$. H_a: not all of the μ_i are equal. (c) $F(3, 16)$.

9.21 Answers will vary. Most situations don't use random sampling and/or are too specific to be generalizable.

9.23 (a) We are not comparing the variances; we are comparing the means of several populations. (b) This is true for contrasts, not multiple comparisons. Multiple comparisons are used when we have no specific relations among means in mind before looking at the data. (c) Rejecting the null, we conclude that at least one mean is different, not all. (d) This is backward. The F statistics are large when the between-group is much larger than the within-group variation.

9.25 (a) $F = 137/50 = 2.74$; DFG = 3; DFE = 40; $0.05 < P$-value < 0.10. (b) MSG = $36/2 = 18$; MSE = $100/24 = 4.2$; $F = 18/4.2 = 4.29$; DFG = 2; DFE = 24; $0.025 < P$-value < 0.05.

9.27 More variation between the groups makes the F statistic larger and the P-value smaller. This happens because the more variation we have between groups suggests that the differences we are seeing are actually due to actual differences between the groups, not just chance, giving us more evidence of group differences.

9.29 (a) Yes. (b) $s_p^2 = \dfrac{(21-1)(5.66)^2 + (21-1)(5.30)^2 + (21-1)(6.14)^2}{21+21+21-2} = 32.07$; $s_p = 5.66$. (c) Yes, there appears to be a difference in average gains in performance; the estimated common standard deviation is 5.66, which is a value similar to the individually reported standard deviations; this suggests that the variability in the means is not driven by group deviations.

9.31 (a) DFG = 2; DFE = 254; MSG = $183.59/2 = 91.795$; MSE = $2643.53/254 = 10.4076$; $F = 8.82$. (b) $H_0: \mu_1 = \mu_2 = \mu_3$; H_a: not all of the μ_i are equal. (c) $F(2, 254)$; P-value < 0.001. There are significant differences among the groups. (d) $s_p^2 = $ MSE $= 10.4076$; $s_p = 3.226$.

9.33 (a) The normal quantile graph demonstrates that the residuals follow a normal distribution. (b) $\sigma = 19.42$; this value is close to the value reported in Exercise 9.7, which is 19.07. (c) $F = 8.8$; df in numerator = 2; df in denominator = 57; P-value = 0.0005; there are significant differences among the three groups.

9.35 (a) As long as the data are quantitative, we will feel comfortable in applying the ANOVA procedure. (b) Yes, the conditions are better met under this assumption. (c) $F = 5.83$;

df in numerator = 3; df in denominator = 117; P-value = 0.001; there are significant differences among the four groups. **(d)** The results in part (c) are similar to the results reported in part (d) of 9.34.

9.37 A contrast has more power because it is designed to express a specific question; a t test answers a relatively more general question, which makes it a less powerful test.

9.39 $0.0121/0.0040 = 3.025$; $0.2849/0.0950 = 3.00$; $0.6060/0.2020 = 3$; the P-values are nearly identical.

9.41 The power is 0 because the power of a test is the probability that we make the right decision in rejecting the null hypothesis when it is not correct.

9.43 $SE_{c_1} = 6\sqrt{\frac{\left(\frac{1}{4}\right)^2}{8} + \frac{\left(\frac{1}{4}\right)^2}{8} + \frac{\left(\frac{1}{4}\right)^2}{8} + \frac{\left(\frac{1}{4}\right)^2}{8} + \frac{(-1)^2}{8}} = 2.37$.

9.45 $1.2 \pm 2.03(2.37) = (-3.61, 6.01)$.

9.47 **(a)** $1/2\mu_1 + 1/2\mu_2 = 1/2\mu_3 + 1/2\mu_4$. **(b)** $\mu_1 + \mu_3 = \mu_2 + \mu_4$. **(c)** $\mu_1 - \mu_2 = \mu_3 - \mu_4$.

9.49 Answers will vary.

9.51 Answers will vary.

9.53 The ANOVA F test result is equal to t^2 when $I = 2$.

9.55 **(a)** DFG = 5; DFE = 24. 2.10, 2.62, 3.15, 3.90, 5.98. **(c)** $0.025 < P$-value < 0.05. **(d)** Reject H_0; the P-value is smaller than α. **(e)** No, a rejection of the null hypothesis only indicates that at least one mean is different, not all.

9.57 In setting (c) in 9.56, the ANOVA F test result is most likely to be significant because the between-group variation is relatively large.

9.59 **(a)** Yes, the largest s is less than twice the smallest s; $25 < 2(13) = 26$. **(b)** $s_p^2 = 312.1909$. **(c)** $s_p = 17.67$. **(d)** The third group has the largest sample size, and will influence, or weight, the pooled standard deviation more.

9.61 **(a)** Observational. **(b)** Yes, the mean for SC appears higher than the others. **(c)** Yes, the largest s is less than twice the smallest. **(d)** Yes, it is reasonable to use the one-way ANOVA procedure as long as the data are quantitative. **(e)** MSG = SSG/DFG = 543.14/2 = 272.57; MSE = $s_p^2 = 3.136$; F statistic = 86.587; P-value = 0.

9.63 Using the least-significance differences method, each group is significantly different from every other group.

9.65 **(a)** $H_0: \mu_1 = \mu_2 = \mu_3$; H_a: not all of the μ_i are equal; $F = 5.31$, P-value = 0.0067. There are significant interval differences among the three groups. **(b)** The Bonferroni procedure shows that Group 2 is not significantly different from either Group 1 or Group 3; however, Group 3 is significantly different (larger) from Group 1. **(c)** This is not appropriate. The regression assumes that Group 2 (coded as 2) would have twice the effect of Group 1 (coded as 1), Group 3 (coded as 3) would have 3 times the effect of Group 1, etc. This is likely not true.

9.67 **(a)**

		Immorality	
Level of Grp	N	Mean	Std Dev
C	41	6.4512	0.5788
D	43	6.2209	0.8040
R	37	5.6892	1.2767

(b) There is no reason to believe the cases are not independent. Constant variance is violated: the largest s is more than twice the smallest s; $1.277 > 2(0.579) = 1.158$. The sample sizes are large enough that the sample means should be approximately Normally distributed. ANOVA should not be used because the standard deviations are too different to be assumed equal. **(c)** $H_0: \mu_C = \mu_D = \mu_R$; H_a: not all of the μ_i are equal; $F = 7.0$; P-value = 0.0013. There are significant immorality judgment differences among the three groups.

9.69 **(a)** Yes, the largest s is less than twice the smallest s; $0.621669 < 2(0.572914) = 1.146$.

		Score	
Level of Food	N	Mean	Std Dev
Comfort	22	4.8873	0.5729
Control	20	5.0825	0.6217
Organic	20	5.5835	0.5936

(b) While the distributions aren't Normal, there are no outliers or extreme departures from Normality that would invalidate the results. We can likely proceed with the ANOVA.

9.71 **(a)** $H_0: \mu_1 = \mu_2 = \mu_3$; H_a: not all of the μ_i are equal; $F = 8.89$; P-value = 0.000. There are significant differences in the number of minutes that the three groups are willing to volunteer. According to the Tukey multiple comparison, the Comfort group is willing to donate significantly more minutes than the Organic group. In other words, the Comfort group shows more prosocial behavior than the Organic group. The Control group is in the middle, not significantly different from either the Comfort or Organic group in the number of minutes they are willing to donate. **(b)** The residual plot shows a slight decrease in variability, suggesting a possible violation of constant variance. The Normal quantile plot looks fine and shows a roughly Normal distribution.

9.73 (a)

Speed	n	$\bar{x}$	s
1	29	5.14	1.62
2	25	6.68	1.91
3	26	3.31	2.1
4	30	1.67	1.7

(b) Yes, the largest s is less than twice the smallest. (c) The sample means will be approximately normal as the sample size increases.

9.75 (a)

		Loss	
Level of Group	N	Mean	Std Dev
Ctrl	35	−1.0086	11.5007
Grp	34	−10.7853	11.1392
Indiv	35	−3.7086	9.0784

(b) Yes, the largest s is less than twice the smallest s; $11.501 < 2(9.078) = 18.156$. (c) All three distributions are roughly Normal.

9.77 (a) All weight loss values are divided by 2.2. (b) $F = 7.77$; DF = 2 and 101; P-value = 0.0007. The results are identical with the transformed data regarding the test statistic, DF, and P-value. Transforming the response variable by a fixed amount has no effect on the ANOVA results.

9.79 (a) All four Normal quantile plots show roughly Normal distributions with only minor departures from Normality. (b)

		Price	
Level of Promotions	N	Mean	Std Dev
1	40	4.2240	0.2734
3	40	4.0628	0.1742
5	40	3.7590	0.2526
7	40	3.5488	0.2750

(c) Yes, the largest s is less than twice the smallest s; $0.275 < 2(0.174) = 0.348$. (d) $H_0: \mu_1 = \mu_2 = \mu_3 = \mu_4$; H_a: not all of the μ_i are equal; $F = 59.90$; DF = 3, 156; P-value < 0.0001. There are significant price estimate differences among the four groups, which read different numbers of promotions.

9.81 (a)–(b) The distributions of the transformed data are much more Normal than the original likelihood histograms. The spreads are all very similar now between 0.2 and 0.25. (c) $F = 9.56$; P-value = 0.0001. There are significant differences among groups for the transformed data. The results of this ANOVA are quite similar to the results of the ANOVA on the untransformed data.

9.83 Answers will vary; $n = 65$ or 75 would be good choices.

n	DFG	DFE	F^*	Power
35	2	102	3.09	0.609
45	2	132	3.06	0.729
55	2	162	3.05	0.818
65	2	192	3.04	0.881
75	2	222	3.04	0.924

9.85 (a) Answers will vary. (b) For $n = 20$, the power is already 0.924. (c) Answers will vary.

n	DFG	DFE	F^*	Power
10	2	27	3.35	0.621
15	2	42	3.22	0.821
20	2	57	3.16	0.924
25	2	72	3.12	0.97
30	2	87	3.10	0.989

9.87 (a) The results are nearly identical to before: $F = 7.78$; P-value = 0.0007. The estimated variance s_p^2 went from 112.81 to 140.65, but the means got farther apart, as shown in the table below.

		Loss	
Level of Group	N	Mean	Std Dev
Ctrl	35	0.3543	14.6621
Grp	34	−10.7853	11.1392
Indiv	35	−3.7086	9.0784

(b) The results are not as significant: $F = 3.12$; P-value = 0.0485. The estimated variance s_p^2 went from 112.81 to 183.27, but the means got closer together, as shown in the table below.

		Loss	
Level of Group	N	Mean	Std Dev
Ctrl	35	−1.0086	11.5007
Grp	34	−9.0206	18.4317
Indiv	35	−3.7086	9.0784

(c) With the first outlier, the means got farther apart, suggesting more significance, but the estimated variance s_p^2 went from 112.81 to 140.65, suggesting a worse fit, which resulted in a very similar F value and P-value. With the second outlier, the means got closer together, suggesting less significance, and the estimated variance s_p^2 went from 112.81 to 183.27, also suggesting a much worse fit, which resulted in a P-value much less significant than originally and almost not significant. (d) In both cases, the estimate variance s_p^2 got much worse.

Generally, we would expect outliers to make it harder to see significance. But as shown in the first example, if the outlier pulls the means farther apart, this may not be true. **(e)** We can see the incorrect observation because the standard deviation for the group with the outlier becomes much larger than the standard deviations for the other groups.

9.89 **(a)** It appears that the decrease in average sleep time is the sample for each group. **(b)** A two-way ANOVA procedure would be appropriate in this context. **(c)** This would not address the difference in means between the baseline and Year 2.

Chapter 10: Inference for Proportions

10.1 **(a)** $n = 123$. **(b)** $X = 64$; it is the count of banks with assets of $1 billion or less. **(c)** $\hat{p} = \frac{64}{123} = 0.52$.

10.3 **(a)** $SE_{\hat{p}} = \sqrt{\frac{(0.52)(1-0.52)}{123}} = 0.045$; the standard error tells us how much $\hat{p}$ varies. **(b)** 0.52 ± 0.088. **(c)** $(43.2\%, 61\%)$.

10.5 $(0.433, 0.607)$.

10.7 Answers will vary. Anything with a very small n and a very high or low X (close to n or 0).

10.9 The 95% confidence interval is $0.65 \pm 1.96\sqrt{\frac{(0.65)(1-0.65)}{20}} = (44.1\%, 85.9\%)$. With 95% confidence, the percent of people who would get better protection from your product is between 44.1% and 85.9%. The confidence interval gives similar information to the significance test that the percent is not significantly different from 50% because 50% is inside our interval.

10.11 **(a)** $10 \hat{p} = 0.65$; $Z = 2.3$; P-value $= 0.02$; The data show a significant difference between your product and your competitor's. **(b)** As the sample size increases, the Z test statistic increases and the P-value gets smaller, making the data more significant.

10.13 $n = \left(\frac{z^*}{m}\right)^2 p^*(1-p^*) = \left(\frac{1.96}{0.14}\right)^2 0.5(1-0.5) = 49$; $n = 49$.

10.15 From software, $n = 47$.

10.17 **(a)** It should be plus or minus the margin of error. **(b)** Significance tests are applied to two-sided tests. **(c)** The large sample confidence interval can be used for 90%, 95%, and 99%.

10.19 **(a)** $\mu_{\hat{p}} = 0.4$; $\sigma_{\hat{p}} = 0.0693$. **(c)** The values are 0.26 and 0.54.

10.21 **(a)** 109. **(b)** $(0.40, 0.53)$. **(c)** $(40\%, 53\%)$. **(d)** Respondents may not have answered honestly.

10.23 **(a)** $H_0: p = 0.07$; $H_a: p > 0.07$. **(b)** $Z = \frac{0.105 - 0.07}{\sqrt{\frac{(0.07)(1-0.07)}{200}}} = 1.94$ **(c)** P-value $= 0.026$. **(d)** The data show that the proportion of nonconforming switches is different from the assumed 7% at the 5% level of significance.

10.25 Using the estimated proportion of 0.07 for p^*, $n = \left(\frac{z^*}{m}\right)^2 p^*(1-p^*) = \left(\frac{1.96}{0.07}\right)^2 0.67(1-0.67) = 173.34$, so $n = 174$.

10.27 $\hat{p} = 0.61$. **(b)** $m = 1.96\sqrt{\frac{0.61(1-0.61)}{1700}} = 0.0232$. **(c)** $0.61 \pm 0.023 = (0.587, 0.633)$. **(d)** 95% of all random samples will have the mean proportion lie in the interval $(0.587, 0.633)$. **(e)** Answers will vary.

10.29 **(a)** $\hat{p} = \frac{5067}{10000} = 0.5067$; $Z = \frac{0.5067 - 0.5}{\sqrt{\frac{(0.5)(1-0.5)}{10000}}} = 1.34$; P-value $= 0.1802$. The data do not provide evidence that the coin is biased. **(b)** The 95% confidence interval is $0.5067 \pm 1.96\sqrt{\frac{(0.5067)(1-0.5067)}{10000}} = (0.497, 0.516)$.

10.31 **(a)** The 99% confidence interval is $0.67 \pm 2.576\sqrt{\frac{(0.67)(1-0.67)}{1765}} = (0.641, 0.699)$. **(b)** The 99% confidence interval is: $0.67 \pm 2.576\sqrt{\frac{(0.67)(1-0.67)}{5296}} = (0.653, 0.687)$. **(c)** While it's true that if the sample size gets smaller or bigger, the margin of error will go up or down, in all three cases, the sample is quite large and the results don't differ that much, so the main conclusion still holds.

10.33 Wider; the confidence interval is $(0.637, 0.672)$.

10.35 $(0.889, 0.962)$.

10.37 $H_0: p = 0.36$; $H_a: p \neq 0.36$; $\hat{p} = 0.38$; $Z = \frac{0.38 - 0.36}{\sqrt{\frac{(0.36)(1-0.36)}{500}}} = 0.93$; P-value $= 0.3524$. The data do not show a difference between the sample and the state census information. The sample is a reasonable representation of the state with regard to rural versus urban residence.

10.39 $n = \left(\frac{z^*}{m}\right)^2 p^*(1-p^*) = \left(\frac{1.96}{0.06}\right)^2 0.4(1-0.4) = 256.11$; $n = 257$.

10.41 (a) $n = \left(\dfrac{z^*}{m}\right)^2 p^*(1-p^*) = \left(\dfrac{1.96}{0.04}\right)^2 0.75(1-0.25) = 450.19$; $n = 451$. (b) $m = 1.96\sqrt{\dfrac{0.75(1-0.75)}{451}} = 0.04$. (c) $m = 1.96\sqrt{\dfrac{0.65(1-0.65)}{451}} = 0.044$.

10.43 (a)

$\hat{p}$	Margin of error
0.1	0.0372
0.2	0.0496
0.3	0.0568
0.4	0.0607
0.5	0.0620
0.6	0.0607
0.7	0.0568
0.8	0.0496
0.9	0.0372

10.45 Using software, $n = 80$.

10.47 $\mu_{\hat{p}_1-\hat{p}_2} = -0.2$; $\sigma_{\hat{p}_1-\hat{p}_2} = 0.118$.

10.49 (a) $\mu_{\hat{p}_1} = p_1$; $\sigma_{\hat{p}_1} = \sqrt{\dfrac{p_1(1-p_1)}{n_1}}$. $\mu_{\hat{p}_2} = p_2$; $\sigma_{\hat{p}_2} = \sqrt{\dfrac{p_2(1-p_2)}{n_2}}$. (b) $\mu_{\hat{p}_1-\hat{p}_2} = \mu_{\hat{p}_1} - \mu_{\hat{p}_2} = p_1 - p_2$. (c) $\sigma^2_{\hat{p}_1-\hat{p}_2} = \sigma^2_{\hat{p}_1} + (-1)^2\sigma^2_{\hat{p}_2} = \dfrac{p_1(1-p_1)}{n_1} + \dfrac{p_2(1-p_2)}{n_2}$.

10.51 $(-0.304, -0.067)$. We can just reverse the sign of the interval in the previous exercise.

10.53 The plus-four interval is $0.8182 - 0.5185 \pm 1.96\sqrt{\dfrac{0.8182(1-0.8182)}{22} + \dfrac{0.5185(1-0.5185)}{27}} = (0.052, 0.548)$. The z interval is $0.85 - 0.52 \pm 1.96\sqrt{\dfrac{0.85(1-0.85)}{20} + \dfrac{0.52(1-0.52)}{25}} = (0.079, 0.581)$. The intervals are somewhat different when the sample sizes are small.

10.55 (a) H_0: $p_1 = p_2$; H_α: $p_1 \neq p_2$. Not knowing anything about the two commercials, there is no reason to believe men or women will prefer Commercial A more; so the test should be two-sided. (b) $Z = 3.0$; P-value $= 0.003$. (c) The data show evidence of a difference between women and men concerning preference of Commercial A.

10.57 (a) Using $p_1 = 0.4$ and $p_2 = 0.5$, $m = 1.96\sqrt{\dfrac{(0.4)(1-0.4)}{48} + \dfrac{(0.5)(1-0.5)}{48}} = 0.198$; using $p_1 = 0.7$ and $p_2 = 0.6$, $m = 1.96\sqrt{\dfrac{(0.7)(1-0.7)}{48} + \dfrac{(0.6)(1-0.6)}{48}} = 0.190$; using $p_1 = 0.3$ and $p_2 = 0.6$, $m = 1.96\sqrt{\dfrac{(0.3)(1-0.3)}{48} + \dfrac{(0.6)(1-0.6)}{48}} = 0.190$. (b) No, under all three conditions, the margin of error is slightly smaller than the desired 0.2 as given in Example 10.10.

10.59 (a) The explanatory variable is the color of the shirt; it is used to explain whether a tip was given. (b) The response variable is whether a tip was given; it is the interest of the study. (c) p_1 is the true proportion of male customers who would leave a tip for a red-shirt server. p_2 is the true proportion of male customers who would leave a tip for a different-colored-shirt server.

10.61 (a) H_0: $p_1 = p_2$. For male customers, the percent who tip a red-shirted server is the same as the percent who tip a different-colored-shirt server. (b) Answers will vary. H_α: $p_1 > p_2$. For male customers, the percent who tip a red-shirted server is higher than the percent who tip a different-colored-shirt server. Because the color red was singled out, it is logical that we might expect it to have higher percent of tips. (Note: H_α: $p_1 \neq p_2$ is also acceptable with a suitable argument.) (c) Yes, we have at least 10 successes and failures in each sample.

10.63 (a) $\mu_{\hat{p}_1-\hat{p}_2} = -0.1$, $\sigma_{\hat{p}_1-\hat{p}_2} = 0.1051$. (c) -0.31 and 0.11.

10.65 (a) For women, $N(0.86, 0.025)$; for men, $N(0.84, 0.026)$.

10.67 (a) H_0: $p_1 = p_2$; H_α: $p_1 \neq p_2$. (b) $\hat{p}_1 = \dfrac{64}{160} = 0.4$; $\hat{p}_2 = \dfrac{89}{261} = 0.3410$. $\hat{p} = \dfrac{64+89}{160+261} = 0.3634$; $Z = \dfrac{0.4 - 0.3410}{0.3634\sqrt{\dfrac{1}{160} + \dfrac{1}{261}}} = 1.22$; P-value $= 0.2224$. The data do not show a difference in preference for natural trees versus artificial tress between urban and rural households. (c) The 90% confidence interval is $0.4 - 0.3410 \pm 1.645\sqrt{\dfrac{0.4(1-0.4)}{160} + \dfrac{0.3410(1-0.3410)}{261}} = (-0.021, 0.139)$.

10.69 (a) $\hat{p}_1 = \dfrac{124}{136} = 0.9118$; $\hat{p}_2 = \dfrac{106}{122} = 0.8689$; $\hat{p} = \dfrac{124+106}{136+122} = 0.8915$; $Z = \dfrac{0.9118 - 0.8689}{0.8915\sqrt{\dfrac{1}{136} + \dfrac{1}{122}}} = 1.11$;

P-value = 0.2670. The data show no significant difference between the proportions of male and female students employed during the summer. **(b)** The smaller sample sizes are not big enough to detect the practical difference we observed in the previous exercise.

10.71 (a) $X = 1606(0.83) = 1332.98$ or 1333. **(b)** p is the true proportion of all experts who would give a positive response to the question. **(c)** $\hat{p} = 0.83$; it is given in the problem. **(d)** $\sqrt{\dfrac{\hat{p}(1-\hat{p})}{n}} = \sqrt{\dfrac{0.83(1-0.83)}{1606}} = 0.0094$. **(e)** $m = 1.96(0.0094) = 0.018$. **(f)** $0.83 \pm 0.018 = (0.812, 0.848)$.

10.73 (a)

$\hat{p}$	Power
0.77	0.99
0.79	0.82
0.81	0.31
0.83	0.03
0.85	0.34
0.87	0.89
0.89	1

(c) As the difference between the two proportions gets smaller, the power decreases; if the two proportions are quite different, then the power is quite large.

10.75 (a)
0.0350
0.0367
0.0371
0.0374

$\hat{p}$	Margin of error
0.32	0.0361
0.37	0.0367
0.42	0.0371
0.52	0.0374

(c) As the second proportion increases (moves closer to 0.5), the margin of error of the difference in proportions increases somewhat; but note that due to the large sample size the change is not drastic.

10.77 The results in the previous exercises assume independent samples; if the same companies were used, this would not be the case and the results from comparing the two studies would not be valid.

10.79 (a) H_0: $p_1 = p_2$. **(b)** H_a: $p_1 < p_2$. The proportion of smartphone users who visit the Fox News website before the Fox News app was introduced is less than the proportion of smartphone users who visit the Fox News website after the Fox News app was introduced. It is likely that the app helps bring traffic to the Fox News website; hence, the one-sided alternative. (*Note:* H_a: $p_1 \neq p_2$ is also acceptable with a suitable argument.) **(c)** Assuming the samples were random and the samples were independent (i.e., not the same group), then the conditions for the Normal distribution are met because we have at least 10 successes and failures in each sample.

10.81 Using software, we would need 22,580 for each group.

10.83 (a) $\hat{p}_1 = \dfrac{47}{60} = 0.7833$; $\hat{p}_2 = \dfrac{31}{60} = 0.5167$. The 95% confidence interval is $0.7833 - 0.5167 \pm 1.96 \sqrt{\dfrac{0.7833(1-0.7833)}{60} + \dfrac{0.5167(1-0.5167)}{60}} =$ $(0.103, 0.431)$. **(b)** $Z = \dfrac{0.7833 - 0.5167}{0.65\sqrt{\dfrac{1}{60} + \dfrac{1}{60}}} = 3.06$; *P*-value = 0.0022 (two-sided). The data show a significant difference between the proportion of customers who tip when the order is repeated and the proportion of customers who tip when the order is not repeated. The confidence interval showed an increase of between 10.3% and 43.1% in tips when the order is repeated.

10.85 We are assuming items are independent and the 300 produced represent a random sample for the modified process. H_0: $p = 0.08$; H_a: $p < 0.08$; $\hat{p} = \dfrac{13}{200} = 0.065$; $Z = -0.782$; *P*-value = 0.22. The data do not show significant evidence that the proportion of nonconforming items in the modified process is less than 8%; the modification was effective.

10.87 As the sample size increases, the margin of error decreases.

n	Margin of error
20	0.3099
50	0.1960
100	0.1386
220	0.0934
500	0.0620

10.89 It is not possible. The formula for the margin of error is given by
$$m = 1.96\sqrt{\dfrac{0.5(1-0.5)}{25} + \dfrac{0.5(1-0.5)}{n_2}}.$$
Plugging in 0.1 for m gives $n_2 = -33.8$, which is not possible.

10.91 (a) $p_0 = \dfrac{143{,}611}{181{,}535} = 0.791$. (b) $\hat{p} = \dfrac{339}{870} = 0.39$;

$Z = \dfrac{0.39 - 0.791}{\sqrt{\dfrac{0.791(1-0.791)}{870}}} = -29.11$; P-value ≈ 0

(< 0.0002 from Table A). The data show evidence that the proportion of jurors selected who are Mexican American is significantly less than 79.1% (the percent of Mexican Americans in the population). (c) $\hat{p}_1 = \dfrac{339}{143611} = 0.002361$;

$\hat{p}_2 = \dfrac{531}{37924} = 0.014002$; $\hat{p} = \dfrac{339 + 531}{143611 + 37924} = 0.004792$; $Z = \dfrac{0.002361 - 0.014002}{0.004792\sqrt{\dfrac{1}{143611} + \dfrac{1}{37924}}} =$

-29.20; P-value ≈ 0 (<0.0002 from Table A). The answer is nearly identical to part (b).

10.93 (a) $n = 2342$; $X = 1639$. (b) $\hat{p} = \dfrac{1639}{2342} = 0.7$;

$SE_{\hat{p}} = \sqrt{\dfrac{0.7(1-0.7)}{2342}} = 0.0095$. (c) The 95% confidence interval is $0.7 \pm 1.96\sqrt{\dfrac{(0.7)(1-0.7)}{2342}} =$

(0.681, 0.718). With 95% confidence, the proportion of people who get their news from a desktop or laptop is between 68.1% and 71.8%. (d) Yes, we have at least 10 successes and failures in the sample.

Chapter 11: Inference for Categorical Data

11.1 (a) Company size is the explanatory and media is the response because we are interested in how gender influences commercial preference. (b) $r \times c = 2 \times 2$. Gender is the column variable because it is the explanatory variable; commercial preference is the row variable because it is the response variable.

	Media		
Company Size	Yes	No	Total
Small	150	28	178
Large	27	25	52
Total	177	53	230

(b) Four cells.
(c)

	Media		
Company Size	Yes	No	Total
Small	150	28	178
Large	27	25	52
Total	177	53	230

11.3 40.98%; 60%.

11.5 Answers will vary. It seems appropriate to look at commercial preference for each gender.

11.7 14. 40.98%. 22.95% of companies use adaptive flexibility, and of that 22.95%, 40.98% of them should be highly competitive, assuming no association between flexibility and competitiveness. From the formula, $(25)(14)/(61) = 5.74$.

11.9 df $= (6-1)(4-1) = 15$.

11.11 (a) $X^2 = 23.74 \approx z^2 = 23.72$. (b) $(z^*)^2 = 3.291^2 = 10.83 = X^{2*}$. (c) The z test null hypothesis indicates that the proportions are equal or that small and large companies use social media equally; in other words, the size of the company doesn't matter in determining social media use—there is no relation between company size and social media use, the null hypothesis for the X^2 test.

11.13 (a) Answers will vary. The percent that tip for each shirt color are: Black 35.29%, White 33.33%, Red 28.30%, Yellow 38.00%, Blue 34.04%, Green 32.73%. (b) H_0: There is no association between whether a female customer tips and shirt color of the server. H_a: There is an association between whether a female customer tips and shirt color of the server. (c) From software, $X^2 = 1.1913$; df $= 5$; P-value $= 0.9457$ (using Table F, P-value > 0.25). (d) A conditional graph by ShirtColor is shown. (e) The data do not provide evidence of an association between whether the female customer left a tip and shirt color worn by the server.

11.15 (a)

	Generation			
	Millennial	GenX	Boomer	Silent
Yes	288	224	160	72
No	512	576	640	728

(b) Answers will vary. (c) From software, $\chi^2 = 178.7678$; df $= 3$; P-value $= 0$. The data provide evidence of an association between college degree and generation. (d) The results suggest there is an association between college degree and generation.

11.17 More women from the Millennial and GenX groups have college degrees; more men from the Boomer and Silent groups have college degrees.

11.19 (a) For younger than 40: 6.9% were hired, 93.1% were not. For 40 or older: 2.2% were hired, 97.8% were not. (c) The company appears to hire a lot more from the younger than 40 age group than from the 40 or older age group. (d) The qualifications of the candidates is a lurking variable. **(e)** H_0: There is no association between age and whether an applicant is hired. H_a: There is an association between age and whether an applicant is hired. From software, $\chi^2 = 5.8767$; df $= 1$; P-value $= 0.0153$. The data do not provide evidence of an association between age and whether an applicant is hired.

11.21 (a)

Gender	Admit	Deny	Total
Male	490	210	700
Female	280	220	500
Total	770	430	1200

(b) 70% of males and 56% of females overall are admitted. **(c)** For Business: 80% of males and 90% of females are admitted. For Law: 10% of males and 33.33% of females are admitted. **(d)** Because a huge number of males apply to business school and the business school has a higher overall admittance rate (82.5%) than the law school (27.5%), this pulls up the overall percent of males that are admitted to both schools. Similarly, a large number of females apply to law school, but with low admittance rates; this makes the overall percent of females that are admitted to both schools look bad, when within each school females are admitted with a higher percent than males. **(e)** H_0: There is no association between gender and whether an applicant is admitted. H_a: There is an association between gender and whether an applicant is admitted. From software, $X^2 = 24.8626$; df $= 1$; P-value < 0.0001 (using Table F, P-value < 0.0005). The data provide evidence of an association between gender and whether an applicant is admitted. **(f)** H_0: There is no association between gender and whether an applicant is admitted. H_a: There is an association between gender and whether an applicant is admitted. $X^2 = 10.3896$; df $= 1$; P-value $= 0.0013$ (using Table F, $0.001 < P$-value < 0.0025). The data provide evidence of an association between gender and whether an applicant is admitted. **(g)** H_0: There is no association between gender and whether an applicant is admitted. H_a: There is an association between gender and whether an applicant is admitted. $X^2 = 20.4807$; df $= 1$; P-value < 0.0001 (using Table F, P-value < 0.0005). The data provide evidence of an association between gender and whether an applicant is admitted. **(h)** The results for the two schools are similar.

11.23 35; 50; 15.

11.25 $\chi^2 = \dfrac{(61-61.75)^2}{61.75} + \dfrac{(59-66.5)^2}{66.5} + \dfrac{(49-61.75)^2}{61.75} + \dfrac{(77-95)^2}{95} + \dfrac{(141-114)^2}{114} + \dfrac{(88-76)^2}{76} = 15.19$. df $= 5$; P-value $= 0.0096$. The data provide evidence that the bag is different from the percents stated by the M&M Mars Company.

11.27 From Table A, the probabilities are (1) 0.2119. (2) $0.4013 - 0.2119 = 0.1894$. (3) $0.5987 - 0.4013 = 0.1974$. (4) $0.7881 - 0.5987 = 0.1894$. (5) $1 - 0.7881 = 0.2119$. $\chi^2 = \dfrac{(212-211.9)^2}{211.9} + \dfrac{(185-189.4)^2}{189.4} + \dfrac{(188-197.4)^2}{197.4} + \dfrac{(195-189.4)^2}{189.4} + \dfrac{(220-211.9)^2}{211.9} = 1.0251$ df $= 4$; P-value > 0.90 (0.906 from software). The data do not provide evidence that the data are different from a standard Normal distribution.

11.29 (a) The probability for each interval is 0.25. **(b)** The expected number for each interval is 250. $\chi^2 = \dfrac{(231-250)^2}{250} + \dfrac{(252-250)^2}{250} + \dfrac{(270-250)^2}{250} + \dfrac{(247-250)^2}{189.4} = 3.0960$ df $= 3$; P-value > 0.25 (0.377 from software). The data do not provide evidence that the data are different from a Uniform (0, 1) distribution.

11.31 (a) Observed cell counts are computed under the assumption that the *alternative* hypothesis is true. **(b)** The null hypothesis should be that there *is no* association between two categorical variables. **(c)** A P-value cannot be greater than 1.

11.33 (a) $U = 0$: 46%; $U = 1$: 67%; $\chi^2 = 0.9026$; P-value $= 0.3421$.
(b)

Multiplier	χ^2	P-value
×1	0.9026	0.3421
×2	1.8051	0.1791
×4	3.6103	0.0574
×6	5.4154	0.0200
×8	7.2205	0.0072

(c) In relation to sample size, collecting twice as much data that demonstrate the same association doubles the χ^2 value and makes the data more significant. Similarly, collecting four times as much data that portray the same association quadruples the χ^2 value and makes the data even more significant, etc. In addition, as the multiplier increases, the P-value decreases.

11.35 (a)

Broadband	Year					Total
	2001	2005	2009	2013	2018	
Yes	389	448	497	524	529	2387
No	151	92	43	16	11	313
Total	540	540	540	540	540	2700

H_0: There is no association between year and Internet access using broadband. H_a: There is an association between year and Internet access using broadband. From software, $\chi^2 = 251.11$; df = 4; P-value = 0. The data provide evidence of an association between year and Internet access using broadband. Broadband usage has increased steadily from 6% in 2001 to 89% in 2018.

(b) In the previous exercise, broadband usage rapidly increased from 2001 to 2018; in the current exercise, broadband usage exhibited a steady increase from 2001 to 2018.

11.37 (a) df = 3; P-value = 0.01. (b) df = 2; P-value = 0.004. (c) df = 4; P-value = 0.024. (d) df = 2; P-value = 0.004.

11.39 (a) For small: $\frac{7}{59}(3142) = 372.8$, so 373; for medium: $\frac{4}{20}(236) = 47.2$, so 47; for large: $\frac{1}{6}(51) = 8.5$, so 9. (b) For small: $m = 1.96\sqrt{(3142)^2 \frac{\left(\frac{7}{59}\right)\left(1-\frac{7}{59}\right)}{59}} = 259.26$; for medium: $m = 1.96\sqrt{(236)^2 \frac{\left(\frac{4}{20}\right)\left(1-\frac{4}{20}\right)}{20}} = 41.37$; for large: $m = 1.96\sqrt{(51)^2 \frac{\left(\frac{1}{6}\right)\left(1-\frac{7}{6}\right)}{6}} = 15.21$.

11.41 From software, $X^2 = 852.4330$; df = 1; P-value < 0.0001 (using Table F, P-value < 0.0005). $Z^2 = (-29.2)^2 = 852.64 = X^2$ with rounding error.

11.43 (a) 5.32% of those over 40 while only 1.57% of those under 40 were laid off. Yes, it looks like a higher percent of over 40 were laid off than under 40.

11.45 (a) 80.19% of men and 28.44% of women died. From software, $X^2 = 332.2054$; df = 1; P-value < 0.0001 (using Table F, P-value < 0.0005). The data provide evidence that a higher proportion of men died than women. Answers will vary. (b) Among the women, 4.55% of Highest, 12.62% of Middle, and 51.44% of Lowest died. From software, $X^2 = 103.7665$; df = 2; P-value < 0.0001 (using Table F, P-value < 0.0005). The data provide evidence of an association between death and economic status for the women. (c) Among the men, 64.53% of Highest, 87.21% of Middle, and 83.13% of Lowest died. From software, $X^2 = 34.6206$; df = 2; P-value < 0.0001 (using Table F, P-value < 0.0005). The data provide evidence of an association between death and economic status for the men.

11.47 Answers will vary.

11.49 It is highly unlikely to get a P-value so high; the random sample would have to produce an almost perfectly uniform result with an equal number (~100) in each of the five groups. Most random samples will vary slightly, and getting a result so exact would be extremely rare. If the class had 2000 students, it is plausible that one (some) of the 2000 students might get an extremely rare result that provides an unusually high P-value.

Chapter 12: Inference for Regression

12.1 $\hat{y} = 9.7184$. Residual $= y - \hat{y} = 10.7649 - 9.7184 = 1.0465$.

12.3 (a) 2.4. (b) For each unit change in x, y changes by 2.4 on average. (c) 56.82. (d) When the number of hours of hands-on training is zero, the mean number of financial entries per hour is 56.82.

12.5 (a) −0.3; when the U.S. market return is 0%, the overseas returns is predicted to be −0.3. (b) 0.12; for each unit increase in U.S. returns, the mean overseas return will increase by 0.12. (c) MEAN OVERSEAS RETURN = $\beta_0 + \beta_1 \times$ U.S. RETURN + ε. The ε term allows overseas returns to vary when U.S. returns remain the same.

12.7 (a) The upward-sloping least squares line indicates that spending is increasing linearly over time. (b) $\hat{y} = -1664.36 + 0.86x$. (c) −0.06, 0.28, −0.38, 0.16. $s = 0.35496$. (d) SPENDING = $\beta_0 + \beta_1 \times$ YEAR + ε. The estimate for β_0 is −1664.36, the estimate for β_1 is 0.86, and the estimate for ε is 0. (e) $\hat{y} = -1664.36 + 0.86(2016) = 69.4$; the predicted value ($69.4) is smaller than the actual value ($72.0), which indicates that the regression model poorly forecasted spending in 2016.

12.9 $b_1 = 0.7559$. $SE_{b_1} = 0.09885$. $H_0: \beta_1 = 0$, $H_a: \beta_1 \neq 0$. $t = 7.65$, df $= 58$, P-value < 0.001. The results are the same as Excel's.

12.11 The mutual funds that had the highest returns this year were high in part because they did well but also in part because they were lucky; so we might expect them to do well again next year but probably not as well as this year.

12.13 **(a)** $r = 0.7086$. P-value < 0.001. There is a significant positive correlation between T-bills and inflation rate. **(b)** $t = 7.65$, df $= 58$, P-value < 0.001. The results are the same. **(c)** Both values are confirmed to be the same.

12.15 **(a)** 26 out of 35; this might be reasonable because the sellers might expect potential buyers to offer relatively low bids for their homes. **(b)** The relationship between the selling price and assessed values is positive; this positive relationship indicates that the selling price increases in response to an increase in assessed values. **(c)** The answer is subjective; some researchers may choose to leave all 35 observations in the analysis to avoid criticism of manipulating the results; other researchers may choose to exclude the 2 large assessed values to reduce the error variance in their results. **(d)** $\hat{y} = 27.7182 + 0.94404x$. $s = 32.802$. **(e)** $\hat{y} = 28.03822 + 0.93822x$. $s = 29.708$. **(f)** The inclusion of the 2 large assessed values appeared to increase the error variance of the regression; in other words, regression standard error is larger in the regression model with the inclusion of the 2 large assessed values compared with the regression model that excludes these values.

12.17 **(b)** The estimated slope coefficient from the regression model measures the influence of assessed values on sales price; estimates from the regression model cannot be used to test whether the mean value of sales prices and the mean value of assessed values are equal. **(c)** Assuming that the population variances are equal, the pooled two-sample $t = 0.9624$, P-value $= 0.3392$, which indicates that we fail to reject the null hypothesis that the mean sales price is equal to the mean assessed value.

12.19 **(a)** The relationship is linear, upward sloping, and relatively strong. There appear to be no outliers; a linear model seems reasonable. **(b)** $\hat{y} = 43.51 + 1.09x$. $s = 942.48$. **(c)** The plot looks random and scattered, there is nothing unusual in the plot, and the assumptions appear valid. **(d)** The Normal quantile plot below shows the distribution is Normal. **(e)** Washington and Virginia were excluded; $\hat{y} = 73.13 + 1.085x$. $s = 612.04$. **(f)** The two results are nearly identical; the slope coefficient is 1.09 for both results, but the intercept coefficients are different; the regression standard error decreases when Washington and Virginia are excluded.

12.21 **(a)** All three market return rates are significant predictors of the IPO offering period; however, the first two are much stronger predictors than the rate for the third period. **(b)** Answers will vary.

12.23 **(a)** Percent is strongly right-skewed; a lot of players have a small percent of their salary devoted to incentive payments. Rating is also right-skewed. **(b)** Only the residuals need to be Normal; because they are somewhat right-skewed, it could pose a threat to the results. **(c)** The relationship is quite scattered. The direction is positive, but the linear relationship is weak. A large number of observations fall close to 0 percent. **(d)** $\hat{y} = 6.24693 + 0.10634x$. **(e)** The residual plot looks good; no apparent violations. The Normal quantile plot shows the violation of Normality and the right-skew we saw earlier.

12.25 **(a)** The relationship is linear, positive, and strong. There are no outliers; a linear model seems reasonable. **(b)** The regression line is $\hat{y} = 3194.33 + 1.1x$. $s = 1757.4$. **(c)** The plot looks random and scattered, there is nothing unusual in the plot, and the assumptions appear valid. **(d)** The Normal quantile plot below shows the distribution is Normal.

12.27 **(a)** All ages are reported as integers forming the stacks. **(b)** Answers will vary. Older men could have more experience; younger men could have more recent education. The data show that there is no relationship between age and income for men, or a very weak one. **(c)** $\hat{y} = 24,874 + 892.11x$. For each 1 year older a man gets, his predicted income goes up by $892.11.

12.29 **(a)** For the stack at each age, there are very few with large incomes, showing the right-skew. **(b)** Regression inference is robust against moderate lack of Normality, especially given our large sample size.

12.31 **(a)** β_0 is the return on T-bills when there is no inflation. Without inflation, we would expect a positive return on any invested money. **(b)** $b_0 = 1.91576$. $SE_{b_0} = 0.462227$. **(c)** $t = 4.14$, df $= 58$, P-value < 0.0001. There is significant evidence that the intercept is greater than 0. **(d)** Using df $= 58$, $1.91576 \pm 2.001(0.462227)$, which gives the same answer as Excel (with rounding error).

12.33 **(a)** $0.01 < P$-value < 0.02 if we use the T-table; using Excel P-value $= 0.019$, the correlation is significantly greater than 0 at the 5% level. **(b)** States with more adult binge drinking are more likely to have underage drinking. $R^2 = 0.1024$. 10.24% of the variation in underage drinking can be accounted for by the prevalence of adult binge drinking. **(c)** Even though most states were used, it is assumed that sampling took place for each state; thus, we can still infer about the true unknown correlation. (Had we obtained different samples from each state, we would have gotten different results.)

12.35 **(a)** The relationship is linear, positive, and of moderate strength. $R^2 = 0.4306$. 43.06% of the variation in house price can be attributed to house size. House size is fairly useful in determining house price. **(b)** $\hat{y} = 21.39844 + 0.07659x$. $t = 4.60$, df $= 28$, P-value < 0.0001. There is a significant linear relationship between price and size. For each additional square foot, house price increases by $76.59.

12.37 **(a)** $H_0: \beta_1 = 0$, $H_a: \beta_1 > 0$; $r = 0.6562$, $t = 4.60$, df $= 28$, P-value < 0.0005. **(b)** The results are identical. **(c)** The housing markets are likely different in other cities, so these results would not apply to them.

12.39 With the outlier removed: R^2 decreases from 43.06% to 38%, the slope decreases from 0.07659 to 0.05790, and the t-value decreases from 4.60 to 4.07. Overall, these changes appear to be minor, and our conclusions are the same with and without the outlier; therefore, the outlier is deemed to not be influential.

12.41 **(a)** $10.0560 \pm 1.984(0.167802)$, $(9.72305, 10.3890)$. **(b)** df $= 98$, $t^* = 1.660$; $10.0560 \pm 1.660(0.167802)$, $(9.777449, 10.334551)$.

12.43 **(a)** $(8095.7293, 12029.417)$. **(b)** $(18434.193, 22807.22)$. **(c)** Moneypit U is wider because it is farther from the mean for Y2017.

12.45 **(a)** $(11,024, 11,722)$. **(b)** $(9,419, 13,330)$. **(c)** For all public universities with tuition of $10,403 in 2008, the mean tuition in 2013 will be between $11,024 and $11,722 with 95% confidence. For an individual public university with tuition of $10,403 in 2008, the predicted tuition in 2013 will be between $9419 and $13,330 with 95% confidence. **(d)** No, private universities likely are different from public universities.

12.47 Using the 2013 tuition, the prediction interval is $(9419, 13330)$; using the 2008 tuition, the prediction interval is $(7881, 15,170)$. The intervals are different because they are based on two different models using different predictors. That said, the intervals don't differ that much, with the first interval entirely in the second interval. It could be that the 2013 data better predict the 2017 tuition because such data are closer in time to 2017 and thus give a better estimate than the 2008 data.

12.49 **(a)** $\hat{y} = 24{,}874.3745 + 892.1135(30) = 51{,}637.78$. **(b)** $(49{,}780, 53{,}496)$. **(c)** $(-41{,}735, 145{,}010)$. The interval isn't very useful; we could have guessed he was in a similar range without any statistics.

12.51 $(59{,}510, 61{,}603)$.

12.53 **(a)** $49{,}880 \pm 1.984 \dfrac{38{,}250}{\sqrt{195}} = (44{,}446, 55{,}314)$. **(b)** The interval from part (a) includes information from only the 195 men who are 30 years old, while the interval for the mean response in the output uses the data from all 5712 men to give a better estimate for the 30-year-olds.

12.55 $SE_{b_1} = 0.099$.

12.57 $H_0: \beta_1 = 0$ or no linear relationship, $H_a: \beta_1 \neq 0$ or there is a linear relationship. $t = 7.65$, $F = 58.457$, $F = t^2$. P-value < 0.0001.

12.59 $r^2 = $ regressionSS/totalSS $= 279.367/556.4375 = 0.5021$. **(b)** $s = $ SQRT(ResidualMS) $= $ SQRT$(4.778) = 2.1859$.

12.61 **(a)** $H_0: \beta_1 = 0$. **(b)** The sum of squares add up, SSR + SSE = SST. **(c)** The smaller the P-value, the greater the significance of the regression. **(d)** The total degrees of freedom in an ANOVA table are equal to $n - 1$.

12.63 $s = 13.1$. $r^2 = 5821.3/10454.7 = 0.5568$.

12.65 $SE_{b_0} = 2.96$. The interval is $(-9.21, 2.99)$. The intercept is meaningful because it tells us what the EAFE is when the S&P is 0, meaning no return in U.S. markets.

12.67 **(a)** Using the natural logarithm of GDP per capita remedies the skewness of the GDP per capita data. **(b)** Yes, there is statistical evidence suggesting that net savings is a predictor of the logarithm of GDP per capita. $H_0: \beta_1 = 0$, $H_a: \beta_1 = 0$; $t = 2.75$; P-value $= 0.0093$; The t test statistic and associated P-value indicate that the effect of net savings is statistically different from 0, or that the null hypothesis can be rejected. **(c)** 17.34%. **(d)** Statistical significance refers to whether an effect exists, whereas practical significance refers to the magnitude of the effect.

12.69 $SE_{\hat{\mu}} = 0.99927\sqrt{\dfrac{1}{38} + \dfrac{(15 - 9.2)^2}{37(135.03)}} = 0.182$; $8.256 \pm 2.028(0.182) = 7.9$ to 8.6.

12.71 $F = t^2$; $7.6 = (2.75)^2$; the P-value for the F statistic is equal to the P-value for the t statistic.

12.73 $r^2 = \dfrac{SSR}{SST} = \dfrac{7.54}{43.49} = 0.1734$.

12.75 **(a)** y and x are reversed; the slope describes the change in y for a unit change in x. **(b)** The population regression line uses parameters $y = \beta_0 + \beta_1 x + \varepsilon$. **(c)** This is incorrect; the width of the interval widens the farther x is from $\bar{x}$. **(d)** The correct equation is $y - \hat{y}$.

12.77 The residual plot in Panel A shows that most of the residuals are skewed below the value of 0; in Panel B, the residuals appear to be skewed to the left; in Panel C, the residuals appear to be equally dispersed throughout the range of x; in Panel D, the residuals do not appear to be equally dispersed throughout the range of x. The residual plot in Panel C suggests a reasonable fit to the linear regression model. For the other plots, a remedy would be to transform the dependent variable and/or explanatory variable by taking the natural logarithm.

12.79 **(a)** The data are weakly linear and positive. There appears to be one unusually small AveDebt observation, which may be an outlier. A linear model seems appropriate. **(b)** Answers will vary because it is difficult to tell from the scatterplot. The actual value is around $513. **(c)** Because the in-state cost at University of North Carolina–Chapel Hills falls outside the range for our data set; it would be extrapolation and likely yield an incorrect prediction.

12.81 **(a)** $\hat{y} = 16{,}937 + 0.51255x$. **(b)** $\hat{y} = 16{,}937 + 0.51255(9621) = 21{,}868$. Residual $= y - \hat{y} = 20{,}466 - 21{,}868 = -1402$. **(c)** The interval is (0.078, 0.947). For each $500 of in-state cost, we expect an average debt of between $39 and $474 at graduation with 95% confidence.

12.83 **(a)** There is a linear relationship exhibited between each of the explanatory variables and AveDebt. **(b)** Grad4Rate is the single best explanatory variable; it has the most significant P-value and the smallest model standard deviation s. **(c)** Grad is the single best explanatory variable; it has the most significant P-value and the smallest model standard deviation s.

12.85 Answers will vary. The plot with smaller s should have the data points closer to the line.

12.87 The effect of need-based aid on average debt is negative; the estimated regression slope is -0.0457, indicating that for every $1000 of need-based aid, average debt goes down by $45.7, but it is not significant. $H_0: \beta_1 = 0$, $H_a: \beta_1 \neq 0$. $t = -0.26$, df $= 38$, P-value $= 0.795$. There is no significant linear relationship between average debt and need-based aid.

12.89 **(a)** The relationship is linear, upward sloping, and relatively strong. **(b)** Yes, a bigger hotel requires more employees to take care of both the rooms and the guests, so we would expect a positive slope. **(c)** $\hat{y} = 101.98 + 0.5136x$. **(d)** $H_0: \beta_1 = 0$, $H_a: \beta_1 \neq 0$. $t = 4.29$, df $= 12$, P-value $= 0.0011$. There is a significant linear relationship between the number of employees and the number of rooms for hotels. **(e)** (0.25259, 0.77459).

12.91 **(a)** They are very large hotels with more than 1000 rooms. **(b)** The analysis changes drastically. $t = 1.89$, df $= 10$, P-value $= 0.0880$. The data are no longer significant at the 5% level with these outliers removed. The previous results likely are not valid, as the significance seen was just due to these two outliers.

12.93 **(a)** The rise from 1948 to 1980 is fairly consistent, then shifts and increases at a much faster rate from 1981 to 2011. **(b)** The least squares line using data for the years 1948–1980 is linear and upward sloping. **(c)** (1.55206, 1.78825). **(d)** (2.84583, 3.35647). **(e)** Overall, there has been rapid growth in agricultural productivity since 1948. But there was a significant shift in that productivity around 1980. Before that, the growth, on average, was between 155% and 179% each year; after 1980, the growth was between 293% and 353% each year. These results are at 95% confidence.

Chapter 13: Multiple Regression

13.1 PredSpace $= 225 + 160$Fac $+ 160$Mgr $+ 150$Lect $+ 75$Grad $+ 120$Clsv.

13.3 **(a)** Assets. **(b)** Total interest-bearing deposits and total equity capital. **(c)** $k = 2$. **(d)** $n = 53$. **(e)** The label variable is State or area.

13.5 No. Multiple regression does not require the response to be Normally distributed, only the residuals need to be Normally distributed.

13.7 Correlations are 0.9990, 0.9942, and 0.9923. All the plots show a strong relationship between the two variables, but it is not quite linear because there is some curvature in each of the plots. Additionally, four states stick out on all plots: Delaware, North Carolina, Ohio, and South Dakota.

13.9 $\hat{y} = -17.63188 + 1.362408 X_1 + 2.40093 X_2$.

13.11 The three residual plots show the curvature problem we noted in the previous exercise; the Normal quantile plot shows that the residuals are not Normally distributed because of several outliers on either tail.

13.13 The residual plots are much better for the log-transformed data; there is no longer any curvature in the plots. The Normal quantile plot shows the data are much better and roughly Normally distributed.

13.15 For Excel: $s^2 = 0.118582375$; $s = 0.344357917$; s is called the Standard Error. for Minitab: $s^2 = 0.118582$; $s = 0.344358$; s is called S. for JMP: $s^2 = 0.118582$; $s = 0.344358$; s is called the Root Mean Square Error.

13.17 **(a)** The share invested in risky assets. **(b)** 128. **(c)** 6. **(d)** Professional identity, age, sex, education level, risk literacy, and nationality.

13.19 (a)

Variable	Mean	Median	Std Dev
Price	697.5	590	329.4
Size	10.76	10.05	1.317
Battery	10.46	10.6	1.556
Weight	1.292	1	0.4674
Ease	4.917	5	0.2803
Display	4.694	5	0.5248
Versatility	3.25	3	0.4392

(c) Price is a right-tail only distribution, skewed to the right. Size has a bimodal distribution. Battery is nearly Normal. Ease and Versatility all only have two different values (while Display only has 3 values) even though they were rated on a 1 to 5 scale. There aren't really any unusual observations that might affect the regression analysis. **(d)** No, we do not make any assumption on the distribution of explanatory variables, so this is perfectly fine.

13.21 (a) $\hat{y} = 626.609 + 179.441\text{Size} + 17.233\text{Battery} - 149.259\text{Weight} - 459.600\text{Ease} + 104.831\text{Display} - 24.741\text{Versatility}$. **(b)** $s = 190.46$. **(c)** A Normal quantile plot shows a potential outlier.

13.23 (a)–(b) $\hat{y} = -26.223 + 0.554\text{InCostAid} - 0.200\text{OutCostAid} + 151.896\text{Admit} + 261.361\text{Grad4Rate}$; $s = 3655.4$. **(c)** Yes, the multiple regression model is an improvement as evidenced by the smaller regression standard error.

13.25 (a)

Variable	Mean	Std Dev	Minimum	Lower quartile	Median	Upper quartile	Maximum
Share	4.744	4.218	1.8	2.31	3.875	5.15	19.47
Franchises	6003	6484	0	2062	4730	7214	27205
Company	1022	1764	0	129	478.5	933	7338
Sales	8.244	7.855	3.2	4.1	5.65	8.55	35.4
Burger	0.3125	0.4787	0	0	0	1	1

(c) McDonald's is an outlier for Share and Sales, Subway is an outlier for Franchises, and Starbucks is an outlier for Company. Otherwise, it is hard to tell the distributions of the other restaurants since they are being squished on the histograms because of the outliers. Also, Burger has only two possible values.

13.27 (a) $\hat{y} = 0.7255544 - 0.0000877\text{Franchises} - 0.0001526\text{Company} + 0.5842836\text{Sales} - 0.3717163\text{Burger}$. **(b)** $s = 0.45045$.

13.29 $\hat{y} = 0.1589813 - 0.0001527\text{Franchises} - 0.0003321\text{Company} + 0.7630378\text{Sales} - 0.4888361\text{Burger}$; $s = 0.35075$.

13.31 $\hat{y} = 0.522583 - 0.000151\text{Franchises} - 0.0006792\text{Company} + 0.7267082\text{Sales} - 0.4855795\text{Burger}$; $s = 0.29496$.

13.33 (a) All four variables are somewhat right-skewed. There is a potential outlier for gross sales.

Variable	Mean	Std Dev	Minimum	Lower quartile	Median	Upper quartile	Maximum
Gross_Sales	320.30	180.09	92.3	225.5	263.29	394.28	890.5
Cash_Items	20.52	11.80	5	11	19	26	55
Check_Items	20.04	14.07	3	13	15	26	57
Credit_Cards_Items	7.68	7.98	0	2	5	11	28

(b) All three explanatory variables look linearly related with gross sales, but each scatterplot has a few semi-outlying observations that could be potentially influential. From the correlation matrix, we can see that both cash items and check items have quite strong linear relationships with gross sales, but they also have some correlation between them.

	Gross_Sales	Cash_Items	Check_Items	Credit_Cards_Items
Gross_Sales	1	0.81696	0.82106	0.45794
Cash_Items	0.81696	1	0.51642	0.35271
Check_Items	0.82106	0.51642	1	0.17645
Credit_Cards_Items	0.45794	0.35271	0.17645	1

(c) $\hat{y} = 0.34126 + 7.10034\text{CashItems} + 6.98713\text{CheckItems} + 4.45787\text{CreditCardItems}$. **(d)** The Normal quantile plot shows a roughly Normal distribution with no outliers. The three residual plots all look pretty good (random) but show a couple of semi-outlying observations we identified earlier. **(e)** The intercept is not significantly different from 0; P-value $= 0.9887$.

13.35 U.S. revenue and opening weekend revenue are right-skewed. The scatterplots show that there is some linear relationship between the explanatory variables as well. The scatterplots show no outliers.

13.37 $\hat{y} = 0.6977223 + 1.061375\text{Ln(opening)} + 0.0033747\text{Review}$. The slope for Ln(opening) increased from 0.98 to 1.1, while the slope for review decreased from 0.004 to 0.003. An increase in opening weekend revenue is associated with an increase in U.S. revenue; an increase in review is associated with an increase in U.S. revenue; the slope for review decreased due to the omission of theaters and including the natural logarithm of opening weekend revenue.

13.39 **(a)** $(-25.5145, 158.2006)$. **(b)** $(-28.0983, 154.2399)$. **(c)** The intervals are quite close; the first model gives a little wider interval because it has a little more standard error for prediction.

13.41 The best model is the model with LOpening and Review with an R^2 of 0.9595. LOpening and Theaters is the second best model with an R^2 of 0.9571.

Model	R-square
Revenue, LOpening, Theaters	0.9571
Revenue, LOpening, Review	0.9595
Revenue, Theaters, Review	0.7811
Revenue, LOpening	0.9566
Revenue, Theaters	0.7088
Revenue, Review	0.1124

13.43 The R^2 value for all three predictors is 0.9585. The R^2 value for just Theaters and Review is 0.7188. $F = 2.53$; df $= 2$ and 40; P-value < 0.1. Opening does contribute significantly to explaining U.S. box office revenue when Review and Theaters are already in the model. The conclusion is the same as the t test.

13.45 **(a)** The coefficient is not statistically different from 0. **(b)** The coefficient is statistically different from 0. **(c)** The coefficient is not statistically different from 0. **(d)** The coefficient is statistically different from 0.

13.47 **(a)** This is true for the *squared* multiple correlation. **(b)** We should not have a slope in the hypotheses; it should be a parameter, $H_0: \beta_2 = 0$. **(c)** A significant F test implies that *at least one* explanatory variable is statistically different from zero, not all.

13.49 MSR $= 96/4$. MSE $= 153/18$; $F = $ MSR/MSE $= 2.82$. Using df of 4 and 18, $0.05 < P$-value < 0.10, the data are not significant at the 5% level exact value is 0.0558; and there is not enough evidence to say that at least one of the slopes is not zero. **(b)** $R^2 = 96/(96+253) = 0.39$; 39% of the variation in the response variable is explained by all the explanatory variables.

13.51 **(a)** $\hat{y} = 0.1589813 - 0.0001527(5799) - 0.0003321(647) + 0.7630378(9.8) - 0.4888361(0) = 6.54$. **(b)** $(5.54, 7.53)$.

13.53 **(a)** The residuals are right-skewed and not Normally distributed. **(b)** There is one outlier in the residual plot against LOpening. **(c)** The residual plots against each explanatory variable exhibit one outlier. **(d)** The residual plot against the predicted values exhibits one outlier. **(e)** The model assumptions are not reasonably satisfied; the residuals are right-skewed, and there are several outliers in the data set that are potentially influencing the regression analysis.

13.55 **(a)** df of 16 and 19; P-value $= 0.002$. **(b)** Tournament appearances, Ticket Sales, Concession, and Camp Revenue. **(c)** The coefficients for these variables are not statistically different from 0.

13.57 **(a)** $H_0: \beta_i = 0$; $H_a: \beta_i \neq 0$; df $= 2215$; $|t| > 1.961$. **(b)** Loan size, Length of loan, Percent down, Cosigner, Unsecured loan, Total income, Bad credit report, Young borrower, Own home, and Years at current address are significant. Those that aren't significant only mean that the particular variable is not useful after the other variables are considered included in the model already. **(c)** Having a larger loan size gives a smaller interest rate. Having a longer loan gives a smaller interest rate. Having a larger percent down payment gives a smaller interest rate. Having a cosigner gives a smaller interest rate. Having an unsecured loan gives a larger interest rate. Having larger total income gives a smaller interest rate. Having a bad credit report gives a larger interest rate. Being a young borrower gives a larger interest rate. Owning a home gives a smaller interest rate. More years at current address gives a smaller interest rate.

13.59 **(a)** df $= 5650$; $|t| > 1.960$. Any variable that is significant tells us that the particular variable is useful in predicting the response after the other variables are considered included in the model already. **(b)** Only Loan size, Length of loan, Percent down, and Unsecured loan are significant. **(c)** Having a larger loan size gives a smaller interest rate. Having a longer loan gives a smaller interest rate. Having a larger percent down payment gives a smaller interest rate. Having an unsecured loan gives a larger interest rate.

13.61 (a) df are 15 and 1034. (b) The F test is significant, meaning the model is good at predicting the response variable, but there is still a lot of variance that is unexplained (a lot of scatter around our current regression line) because R^2 is small. This small R^2 just means there are other potential predictors that may also help us, in addition to our current predictors, to account for this remaining scatter or variation in the response. (c) $F = (1034/13)[(0.031 - 0.025)/(1 - 0.031)] = 0.4925$; df are 13 and 1034; P-value > 0.10. The added 13 variables do not contribute significantly in explaining the response when these two predictors are already in the model.

13.63 The seven houses all appear in the tails of the distributions helping to form the right-skew, but they are not outliers.

13.65 88,198; 105,358.

13.67 85,941.75; 111,521.75. Both predicted values are higher using the quadratic equation instead of the linear equation.

13.69 $t = 2.88$; df $= 35$; P-value $= 0.0068$.

13.71 Other than the one house with three garage spaces, the plot is quite linear between 0 and 2 spaces.

13.73 The difference is $106,515 - $86,100 = 20,415$.

13.75 No, it would not make sense. As we saw in the interaction plot, homes less than 800 square feet don't have an extra half bath.

13.77 (a) The relationship is curved; as x increases, μ_Y also increases; but at larger values of x, μ_Y increases more rapidly. (b) The relationship is curved; as x increases, μ_Y decreases at first but then starts to increase slowly; but at larger values of x, μ_Y increases more rapidly. (c) The relationship is curved; as x increases, μ_Y increases at first but then starts to decrease slowly; but at larger values of x, μ_Y decreases more rapidly. (d) The relationship is curved; as x increases, μ_Y decreases, but at larger values of x, μ_Y decreases more rapidly.

13.79 (a) $15 - 5 = 10$, which is the slope for x. (b) $15 - 10 = 5$, which is the slope for x. (c) $10 - 15 = -5$, which is the slope for x. Yes, it is true in general as long as x is an indicator variable with values 0 and 1.

13.81 (a) $5 - 3 = 2$, which is the coefficient of $x_1 x_2$; $30 - 20 = 10$, which is the coefficient of x_1. (b) $1 - 3 = -2$, which is the coefficient of $x_1 x_2$; $30 - 20 = 10$, which is the coefficient of x_1. (c) $-1 - (-3) = 2$, which is the coefficient of $x_1 x_2$. $30 - 20 = 10$, which is the coefficient of x_1. These results will be true in general as long as x_2 is an indicator variable with values 0 and 1.

13.83 (a) For age, $\hat{y} = -0.046 - 0.484 \text{age}$; for age of matched control, $\hat{y} = 0.366 - 0.315 \text{age}$; for age of retired football player, $\hat{y} = 0.073 - 0.835 \text{age}$. (b) Education matched controls, age, and age for retired football players have statistically significant impacts on volume.

13.85 (a) $t = -0.2298$; df $= 41$; P-value $= 0.058$. The coefficient for Theaters is statistically different from 0 at the 10% level. (b) $R^2 = 0.6491$; $R^2 = 0.5826$; $F = 7.78$. (c) $t^2 = 2.79^2 = 7.8$.

13.87 (c) Answers will vary.

13.89 Model 8 is the preferred model.

			Regression coefficients				
Model	R-square	s	Intercept	LOpening	Review	Theaters	Sequel
1	0.9549	0.17359	0.787556	1.081097*	0	0	0
2	0.9663	0.1723	0.7149826	1.068429*	0.0023277	0	0
3	0.9679	0.17011	0.517095	0.9872508*	0.0032197	0.0001285	0
4	0.9656	0.17397	0.6830437	1.034196*	0	0.000079	0
5	0.9676	0.1711	0.6224187	1.038033*	0	0.0001013	−0.105529
6	0.9670	0.17256	0.7161768	1.083415*	0.0016348	0	−0.0692719
7	0.9664	0.17118	0.7604304	1.096005*	0	0	−0.0920791
8	0.9688	0.16998	0.5109617	1.000405*	0.0025043	0.0001333	−0.0748798

13.91 (a) $\text{LOGINC} = \beta_0 + \beta_1 \text{EDUC} + \beta_2 \text{LOC} + \beta_3 \text{AGE} + \varepsilon$. (b) The parameters are $\beta_0, \beta_1, \beta_2, \beta_3$, and σ. (c) $b_0 = 7.44972$; $b_1 = 0.08542$; $b_2 = 0.22743$; $b_3 = 0.03825$. (d) The Normal quantile plot shows the residuals are Normally distributed. (e) All three residual plots look good (random). Both linearity and constant variance are valid.

13.93 (a) For the model with EDUC: $b_0 = 8.2546$; $b_1 = 0.1126$. For the model with all three: $b_0 = 7.4497$; $b_1 = 0.08542$. With just Education, the intercept was larger and the effect size of each year of education was larger, 0.1126, on LogIncome. Once we account for both Locus of control and Age, the intercept isn't quite as large, and the effect size of each year of education goes down to 0.08542. **(b)** For the full model: $\hat{y} = 9.7948$. For the EDUC model: $\hat{y} = 9.6057$. The predictions don't seem too different unless we undo the log transformation; the predicted incomes are \$17,940.21 and \$14,849.18, which seems like a substantial difference. **(c)** $R^2 = 0.1273$; $F = (96/2)(0.1273 - 0.0573)/(1 - 0.1273) = 3.85$; df are 2 and 96; $0.01 < P\text{-value} < 0.025$. Locus of control and Age are helpful predictors in explaining LogIncome when Education is already in the model.

13.95 (a) As the number of promotions increases, the expected price goes down. For discount, the expected prices for 10% and 20% seem similar, as do the expected prices for 30% and 40%, which are lower than the expected prices for 10% and 20%. **(b)–(c)** The drop of expected price is fairly consistent with an increase in promotions. Similarly, the drop in price is fairly consistent with increase in percent discount; however, the 40% discount consistently yields higher expected prices than when the 30% discount is used.

Promotions	Discount	Mean	Std Dev
1	10	4.92	0.1520234
	20	4.689	0.2330689
	30	4.225	0.3856092
	40	4.423	0.1847551
3	10	4.756	0.2429083
	20	4.524	0.2707274
	30	4.097	0.2346179
	40	4.284	0.2040261
5	10	4.393	0.2685372
	20	4.251	0.2648459
	30	3.89	0.1628906
	40	4.058	0.1759924
7	10	4.269	0.2699156
	20	4.094	0.2407488
	30	3.76	0.2617887
	40	3.78	0.2143725

13.97 The Normal quantile plot shows a slight left-skew in the residuals. The residual plot for promotions looks good (random). The residual plot for Discount shows a slight curve and suggests a possible quadratic model. Investigating a quadratic term for discount and possible interaction terms shows that none of the interaction terms test significant. After removing these, the quadratic term for discount is significant ($t = 4.26$; $P\text{-value} < 0.0001$). The equation becomes $\hat{y} = 5.54049 - 0.10164\text{Promotions} - 0.05969\text{Discount} + 0.000845625\text{Discount}^2$. This model has an $R^2 = 0.6114$, which is somewhat better than the 56.62% for the model without the quadratic term. It is possible to leave out the quadratic term to simplify interpretation; otherwise, the model with this term seems to be best in terms of prediction.

13.99 (a) For Micro: $F = 87.6$; df are 2 and 105; $P\text{-value} < 0.001$. For Small: $F = 117.7$; df are 2 and 170; $P\text{-value} < 0.001$. For Medium: $F = 37.2$; df are 2 and 83; $P\text{-value} < 0.001$. The two explanatory variables are helpful in predicting the perceived level of innovation capability for each firm size. **(b)** For Micro: $t = 8.625$; df $= 105$; $P\text{-value} < 0.001$; and Market Orientation is significant. $t = 1.75$; df $= 105$; $0.05 < P\text{-value} < 0.010$; and Management Capability is not significant. For Small: $t = 7.83$; df $= 170$; $P\text{-value} < 0.001$; and Market Orientation is significant. $t = 6.5$; df $= 170$; $P\text{-value} < 0.001$; and Management Capability is significant. For Medium: $t = 3.08$; df $= 83$; $0.002 < P\text{-value} < 0.005$; and Market Orientation is significant. $t = 3.17$; df $= 83$; $0.002 < P\text{-value} < 0.005$; and Management Capability is significant. **(c)** For all three sizes, the overall model was very significant. However, for the Micro size, the Management Capability was not needed and was not significant, given that Market Orientation is in the model. For the other two sizes, Small and Medium, both variables tested significant at the 5% level, and both were useful in predicting perceived level of innovation capability.

13.101 (a) Using $\alpha = 0.05$ and df $= 123 - 12 - 1 = 110$ (use 100), for significance we need $|t| > 1.984$. So Division, November, Weekend, Night, and Promotion are all significant in the presence of all the other explanatory variables. **(b)** df $= 12$ and 110; $P\text{-value} < 0.001$. **(c)** 52%. **(d)** $12{,}493.47 - 788.74 + 2992.75 + 1460.31 - 131.62 - 779.81 = 15{,}246.36$. **(e)** Because we don't expect the same setting for very many games, the mean response interval doesn't make sense; so a prediction interval is more appropriate to represent this particular game and its specific settings.

13.103 (a) The multiple regression equation is $\hat{y} = 0.90602 + 0.02709x_1 + 0.21144x_2$. $F = 3.15$; $P\text{-value} = 0.0588$. Likewise, neither predictor tests significant when added last: $x_1: t = 0.86$; $P\text{-value} = 0.3991$. $x_2: t = 1.07$; $P\text{-value} = 0.2919$. The data do not show a significant multiple linear regression between y and the predictors x_1 and x_2.

(b) For Y and X_1: $r = 0.39319$. For Y and X_2: $r = 0.40896$. Both $X_1(F = 5.12$; P-value $= 0.0316)$ and $X_2(F = 5.62$; P-value $= 0.0248)$ are significant in predicting Y in a simple linear regression. **(c)** An insignificant multiple regression F test doesn't necessarily imply that all predictors are not useful; we should explore other strategies and/or tests to verify that none of the predictors are useful in different models/settings. In this case, x_1 and x_2 are highly correlated ($r = 0.7033$) and their t tests will likely be insignificant when they are used in the same model together.

13.105 $\hat{y} = -31.59464 + 4.268859$SoyBeanYield; there is a significant simple linear regression between corn yield and soybean yield; soybean yield can significantly predict corn yield. $R^2 = 0.9321$. **(b)** The Normal quantile plot shows that the residuals are more Normal but have a slight right-skew. **(c)** There is somewhat of a relationship between the residuals and year, suggesting that it might be useful in the model with soybean yield to predict corn yield.

13.107 (a) $\hat{y} = -1255.255 + 0.6369891$Year $- 0.0130097$Year2 $+ 3.139897$SoyBeanYield. **(b)** $H_0: \beta_1 = \beta_2 = \beta_3 = 0$; H_a: at least one $\beta_i \neq 0$; $F = 471.14$; df are 3 and 58; P-value $= 0$. There is a significant multiple linear regression between corn yield and the predictors Year, Year2, and SoyBeanYield. Together, the predictors can significantly predict corn yield. **(c)** $R^2 = 96.06\%$, higher than 95.22%. **(d)** For Year: $H_0: \beta_1 = 0$; $H_a: \beta_1 \neq 0$; $t = 3.69$; df $= 58$; P-value $= 0.001$. Year is significant in predicting corn yield in a model already containing Year2 and SoyBeanYield. For Year2: $H_0: \beta_2 = 0$; $H_a: \beta_2 \neq 0$; $t = -3.50$; df $= 58$, P-value $= 0.001$. Year2 is significant in predicting corn yield in a model already containing Year and SoyBeanYield. For SoyBeanYield: $H_0: \beta_3 = 0$; $H_a: \beta_3 \neq 0$; $t = 7.72$; df $= 58$; P-value $= 0$. SoyBeanYield is significant in predicting corn yield in a model already containing Year and Year2. **(e)** The Normal quantile plot shows a roughly Normal distribution; there is one observation with a fairly high residual. The residual plots all look good (random); the residual plot for Year is much better and doesn't have the rising and falling that the previous plot had. Overall, the model fit is much better using the quadratic term for Year than without.

13.109 For Year alone, the 95% prediction interval is (142.1, 184.0). For Year and Year2, the 95% prediction interval is (142.0, 185.0). The two predicted values are different because we are near the edge of the data for Year, and as we saw in the previous exercise, this will cause the greatest differences using the quadratic term.

13.111 Answers will vary.

13.113 (a) Types X and Z are very similar and show very few differences in all of the coefficients. Types D and E are very different. Type E has a much smaller slope for MPG than all the other types, and the MPG2 effect is quite large—more than double all the rest. Type D also has a slightly smaller slope for MPG than X, but it has an extremely small slope for MPG2.

Type	Regression coefficients		
	Intercept	mpg	mpg2
D	267.3823	−5.42585	0.04619
E	160.84557	−3.89582	0.30631
X	235.16637	−7.18033	0.12751
Z	243.75987	−7.88188	0.13832

(b) Answers will vary depending on how the indicator variables were created. Setting Z has the default type ($X_1 = X_2 = X_3 = 0$), the parameter estimates are in the table shown. So the estimates for the intercept, MPG, and MPG2 will match type Z's estimates exactly. To recoup the others, we just set $X_1 = 1$ and $X_2 = X_3 = 0$ for Type D, etc., yielding an intercept of $243.75987 + 23.62243 = 267.3823$, a slope for MPG of $-7.88188 + 2.45603 = -5.42585$, a slope for MPG2 of $0.13832 - 0.09214 = 0.04619$, etc. This yields the same equations as part (a).

Parameter	Estimate
intercept	243.75987
X1	23.62243
X2	−82.91430
X3	−8.59350
mpg	−7.88188
MPGX1	2.45603
MPGX2	3.98607
MPGX3	0.70155
mpg2	0.13832
MPG2X1	−0.09214
MPG2X2	0.16798
MPG2X3	−0.01081

Chapter 14: Time Series Forecasting

14.1 **(a)** The series is not random; it is increasing over time. **(b)** There are only two runs. Since there should be many more runs than just two, the process isn't random. This is consistent with part (a).

14.3 **(a)** The price and lag one price are linearly related. **(b)** $r = 0.9873$. $H_0: \rho = 0$; $H_a: \rho \neq 0$.

P-value < 0.0001. The underlying lag one correlation is significantly different from 0. The price and lag one price are significantly correlated.

14.5 (a) $N_A = 0$; $N_B = 50$. (b) 1. (c) The standard deviation for runs is 0.

14.7 (a) $N_A = 40$; $N_B = 8$. (b) $n = 81$. (c) 106.4281. (d) P-value $= 0.0821$.

14.9 (a) The plot looks fairly random. (b) The ACF shows that the process is random. (c) The histogram and Normal quantile plot show the data are right-skewed. (d) $\bar{y} \pm t^* s \sqrt{1 + \frac{1}{n}} = 7.78 \pm 1.99(11.86)\sqrt{1 + \frac{1}{82}} = (-15.97, 31.53)$.
(e) $\bar{y} \pm t^* s \sqrt{1 + \frac{1}{n}} = 7.78 \pm 1.99(11.86)\sqrt{1 + \frac{1}{82}} = (-15.97, 31.53)$.

14.11 (a) Runs test: $Z = -10.78$; P-value $= 0$; the runs test, time plot, and ACF suggest the STLFSI series is not random. (b) For the first differences, ACF show they are random. (c) The STLFSI series does behave like a random walk because the first differences appear to be random. (d) P-value $= 0.4512$; do not reject the null hypothesis. (e) Since the first differences appear as random, it is justified to use the naïve forecast; 0.009.

14.13 (a) 170.03; 170.79. (b) $\widehat{y_{T+l}} = (1 + l\bar{r})y_T$.

14.15 (a) Runs test: $Z = -13.99$; P-value $= 0$; the runs test and the ACF suggest the price of gold series is not random. (b) For the first differences of index, the ACF show they are random. (c) These differences approximate the percent change in the price of gold. (d) The first differences of the log price data show much more constant variance than the original differences. (e) For the first differences of the log of price, the ACF show they are random.

14.17 (a) The runs test: $Z = 26.8$; P-value $= 0$; the runs test and the ACF suggest the exchange rate is not random. (b) For the first differences of rate, the ACF show they are random. (c) Yes, because the exchange rates are not random but their first differences are random. (d) P-value $= 0.5306$; cannot reject the null hypothesis. (e) 2.60.

14.19 (a) The histogram and Normal quantile plot show a Normal distribution with one high potential outlier. (b) P-value $= 0.5306$; the underlying mean is not significantly different from 0. We have no evidence of a drift in the exchange rate series. (c) The best forecast is a naïve forecast, or today's rate. (d) $(-1.3, 1.4)$. (e) $(1.33, 1.37)$.

14.21 The values are different because in Example 14.7, MAD was calculated as $\frac{|-0.045| + \cdots + |-0.035|}{17} = 0.0315$; using the 20 weekly rate values, MAD is calculated as $\frac{|-0.05| + |-0.045| + \cdots + |-0.035|}{18} = 0.0325$.

14.23

Monday	21,083.61
Tuesday	22,004.38
Wednesday	22,474.16
Thursday	22,624.48
Friday	21,816.47
Saturday	13,078.61
Sunday	8,906.981

14.25 The nature of future forecasts will vary because the averages of the deseasonalized series, whether computed using the seasonal index or seasonal ratio, will be different.

14.27 (a) MAD $= 16.54$; MSE $= 465.04$; MAPE $= 4.46$. (b) MAD $= 14.24$; MSE $= 317.77$; MAPE $= 3.64$.

14.29 (a) $1 - w = 1 - 1.3 = -0.3$.
(b)

Obs	Weight
$t - 1$	-0.390
$t - 2$	0.117
$t - 3$	-0.035
$t - 4$	0.011
$t - 5$	-0.003
$t - 6$	0.001
$t - 7$	-0.390

(c) With a negative damping factor, the effect of the weights is negative and positive for every other observation.

14.31 (a)

Jan	$-10,599.2$
Feb	$5,759.144$
Mar	$8,188.31$
Apr	$-2,314.47$
May	$5,064.699$
Jun	$-2,847.8$
Jul	$-2,833.91$
Aug	791.0881
Sep	$-1,563.08$
Oct	$5,667.477$
Nov	$12,493.87$
Dec	$-17,806.1$

(b) The deseasonalized series does not exhibit a trend. **(c)** Average: 52,302.89; January: 41,703.69; February: 58,062.03; March: 60,491.20.

14.33 **(a)** The CPI series is steadily increasing over time. **(b)** and **(c)** The plot gets smoother with a larger span. **(d)** The results are consistent with the quote. Both moving average predictions would grossly underestimate the CPI values. However, they do show the general pattern, or upward trend, of the CPI data.

14.35 **(a)** $(1-w)^t[w(1-w)^{-1}+1]$.
(b) $(1-0.2)^{10}[0.2(1-0.2)^{-1}+1] = 0.13$.

14.37 **(a) (b)** With a relatively smaller smoothing constant, we have a more smoothed out forecast. **(c)** For $w = 0.2$: 0.0354, 0.002, 0.0350; for $w = 0.9$: 0.041, 0.002, 0.040. **(d)** 1.03; 1.02.

14.39 **(a)** Answers will vary. **(b)** Answers will vary.

14.41 **(a)** $\hat{y}_{t+1} = wy_t + (1-w)\hat{y}_t = \hat{y}_t + w(y_t - \hat{y}_t) = \hat{y}_t + we_t$. **(b)** The future forecast is the sum of the weighted current value and weighted residual.

14.43 **(a)** $H_0: \beta_1 = 0$; $H_a: \beta_1 \neq 0$; $t = 7.97$; P-value = 2.74599E-09. The data show a significant non-zero trend term. **(b)** $\hat{y} = 147.186 + 4.533(37) = 314.907$; $\hat{y} = 147.186 + 4.533(38) = 319.44$.

14.45 **(a)** 855,707.9. **(b)** Actual growth might be slowing down in that the exponential fit is overpredicting by a bit in years 2015 and 2016; therefore, the predicted growth in 2017 might be overstated.

14.47

May	891
June	802
July	877
August	1356
September	1057
October	891
November	849
December	1015

14.49 $\log(y_t) = \log(5014.9) + \log(1.20148)\,t$.

14.51 **(a)** The ACF suggests that PMI is not a random series. **(b)** The lag variable is a good predictor; there is a strong positive relationship between PMI and its lag. **(c)** The PACF suggests that the fitting of an AR(1) model is appropriate.

14.53 **(a)** The number of wireline voice connections has been steadily decreasing over this time period. **(b)** There is no seasonal pattern. **(c)** $\hat{y} = 2120.667 - 124.0769t$. **(d)** The data appear to curve rather than follow a straight line.

14.55 **(a)** and **(b)** Overall, occupancy rates are going up slightly. **(c)** There is a seasonal trend every year, with rates peaking in June and July and lows during December and January.

14.57 **(a)** Earning is increasing over time. **(b)** Yes, there is a trend; earning is increasing over time. **(c)** It is hard to discern a seasonal pattern, though earning seems to be somewhat stable at times and then starts to climb rapidly during other times.

14.59 S1 would show a strong positive relationship with its first lag, S2 would show a strong negative relationship with its first lag, and S3 would show no particular relationship with its first lag.

14.61 $\hat{y} = 300 + 20t + 15Q_4$.

14.63 $Predlog(y_{36}) = 3.2097 + 0.0182(36) - 0.5272(0) - 0.3836(0) - 0.3119(0) + 0.6804 \log(56,165.57) = 11.31$. Undoing the transformation yields $e^{10.46589} = 81,633.91$. $PredSales = 81,633.91 (1.0013) = \$81,740.03$.

14.65 **(a)** $t = $ year $- 1999$.
(b) $\hat{y} = -739,448.8 + 3698.105392\,year$.
(c) $\hat{y} = -73,9449 + 3698.105392\,year$.
(d) $\hat{y} = -73,9448.8 + 3698.105392(2018) = 65,630$; $\hat{y} = -936.125 + 3698.105(18) = 65,630$.

14.67 **(a)** 0.17. **(b)** 0.62; this value is different from the square root of the numerator of MSE. **(c)** Two-week RMSE = 0.036; three-week RMSE = 0.041.

14.69 **(a)** 0.249. **(b)** 0.0039; 0.000023; 1.57%. **(c)** 0.0117; 0.000015; 1.16%. **(d)** Answers will vary. **(e)** Answers will vary.

14.71 **(a)**

Jan	1.56
Feb	0.91
Mar	0.90
Apr	1.05
May	0.80
Jun	0.75
Jul	0.73
Aug	0.90
Sep	1.04
Oct	0.97
Nov	0.82
Dec	1.54

(b) The deseasonalized series does not exhibit a trending level. **(c)** Average: 32,300.49; August: 29,070.44; September: 33,592.51; October: 31,331.48.

14.73 **(a)** 5; 6.8. **(b)** P-value = 0.25; the process appears random. **(c)** The series appears to be random.

14.75

(a)	(b)
0.190833	0.42
1.900833	−0.45
−1.70917	2.13
2.270833	−4.06
0.360833	6.11
10.89083	−5.97
1.540833	16.64

(c) For part (a): 2.695; 18.977; 269.488%; for part (b): 5.111; 53.037; 627.902%. The forecast using the sample mean performs better since the naïve forecast ignores past values the forecasted variable.

14.77 (a) Points exhibits an upward trend. (b) $\hat{y} = 19.36349 - 0.1256705t$; $t = 8.13$; P-value $= 0$; the trend term is statistically different from 0. (c) The residuals appear to be random. (d) 23.0; 23.1; 23.3.

14.79 (a) The poverty rate exhibits increases and decreases over time; recently, the poverty rate has fallen. (b) The first differences of the poverty rate are not random. (c) The first differences in a random walk are random; because the first differences for poverty are not random, it is unlikely the poverty series behaves like a random walk.

14.81 (a) and (b) The series appears to be random. (c) P-value $= 0.38$; the process is random.

14.83 (a) The annual population density in Guatemala is linear over time. (b) $\hat{y} = 41.304e^{-0.0241t}$. (c) $\hat{y} = 40.554 + 0.9655t + 0.019t^2$. (d) For exponential: 83.29; 8418.53; 89.4%; for polynomial: 212.65; 60986.55; 219.75.

14.85 Answers will vary.

14.87 (a) $\hat{y} = 22{,}801.11 + 31.16218t - 160.4122Jan + 4244.676Feb + 9619.263Mar + 8740.601Apr + 4177.189May + 26{,}055.78June + 26{,}305.64July + 5507.899Aug + 4654.487Sep + 12{,}369.57Oct + 3628.973Nov + 2436.912Dec$. (b) The residuals appear to be random. (c) The lag one variable, January, May, August, September, November are not significant; $\hat{y} = 18{,}569.16 + 0.2361117y_{t-1} + 3757.162Feb + 8115.461Mar + 5991.6Apr + 24{,}639.32June + 19{,}042.92July + 10{,}649.99Oct$. (d) The residuals appear to be random. (e) August: 18,580.97; September: 18,581.20; October: 29,231.43. (f) Answers will vary.

14.89 (a)

Jan	−10,739
Feb	−7,811.79
Mar	7,157.182
Apr	7,538.414
May	3,140.318
Jun	3,469.273
Jul	−8,259.98
Aug	2,672.029
Sep	−3,478.46
Oct	17,331.79
Nov	−6,137.15
Dec	−5,617.59

(b) The deseasonalized series exhibits an upward trend. (c) $\hat{y} = 12{,}1362.1 + 515.7103t$; P-value $= 0$; there is statistical evidence for the presence of a trend. (d) January: 135,892.87; February: 139,335.82; March: 154,820.51.

Index

Aggregation, 113
Alternative hypothesis. *See* Hypothesis, alternative
Analysis of variance (ANOVA)
    one-way, 457–480
    regression, 607–610, 639–640, 645
    two-way, 458, Chapter 16
    verifying the conditions for, 469–470
Analysis of variance table
    one-way, 470–472, 475, 480
    regression, 607–610, 639–640, 645
    two-way, 16-14–16-15, 16-21
Anecdotal data, 124, 129
Anonymity, 164
Applet
    Central Limit Theorem, 317–318
    Confidence Intervals, 343–344, 390
    Correlation and Regression, 80, 89, 95, 98, 574
    Distribution of the One-Sample t Statistic, 409
    Law of Large Numbers, 181, 233
    Mean and Median, 27, 39
    Normal Approximation to Binomial, 278
    Normal Curve, 49
    One-Variable Statistical Calculator, 14
    One-Way ANOVA, 473, 481, 497
    Probability, 178, 181
    Simple Random Sample, 134, 142, 154, 159
    Statistical Significance, 375–376
ARCH models, 758–760
ARIMA model, 704, 729, 749
Assignable cause, 15-8
Association, 68, 72, 102
Autocorrelation, 686, 748
Autocorrelation function, 685–690, 692
Autoregressive model, 750–756, 761
Available data, 125, 129

Back-transform, 600
Bar graph, 9, 22, 107, 114, 547
Bayes, Thomas, 209
Bayes's rule, 208–211, 213
Behavioral and social experiments, 167–168
Benford's law, 187–189, 222–223, 231
Bias, 132. *See also* Unbiased estimator
    in a sample, 132, 138, 139, 141
    in an experiment, 147, 158
    of an estimator, 300–302, 305
Big data, 103, 133, 378, 380, 392
Binomial coefficient, 257, 267
Binomial distribution. *See* Distribution, binomial
Binomial setting, 250, 267, 18-1
Block, 156–157, 158

Block design. *See* Experiment, block design
Bonferroni method, 382–383, 490–493
Bootstrap, 298, 412–413, 451
Boxplot, 30, 36
    modified, 30, 32
    side-by-side, 30, 32, 460
Buffon, Count, 178–179, 190

Capability, 15-38–15-40
Capability indices, 15-41–15-45
Capture-recapture sampling, 140
Case, 2, 6, 621, 630
Categorical variable. *See* Variable, categorical
Causation, 102, 104
Cause-and-effect diagram, 15-6
Cell, 105, 544, 16-3
Census, 127, 129
Center of a distribution, 25–27, 36
Centered moving average, 717
Central limit theorem, 312–321, 328, 334–335, 340, 344, 346–347
Chi-square distribution. *See* Distribution, chi-square
Chi-square statistic, 549, 556, 18-11
    and the z statistic, 551–552, 18-11
Classes, 13
Clinical trial, 165, 168
Coefficient, 85
Coefficient of determination, 475
Coin tossing, 177–184, 220–221, 234, 246–247, 250–251, 268–269
Collinearity, 654, 668
Column variable. *See* Variable, row and column
Common cause, 15-8–15-10, 15-16
Comparative experiment. *See* Experiment, comparative
Complement of an event. *See* Event, complement of
Conditional distribution. *See* Distribution, conditional
Conditional probability. *See* Probability, conditional
Confidence interval, 340–352
    and two-sided tests, 368–370
    behavior, 348–349
    bootstrap, 412–413, 451
    simultaneous, 492
    t for contrast, 486
    t for difference of means, 421–422, 432
        pooled, 429, 433
    t for matched pairs, 407–408
    t for mean response in regression, 597, 602
    t for multiple regression coefficients, 636, 645
    t for one mean, 398, 414
    t for regression slope, 583, 590

Confidence interval (*continued*)
   z for difference of proportions, 524–525, 535
   z for one mean, 344–348, 352
   z for one proportion, 507–508, 519
Confidence level, 343, 352
Confidentiality, 163–164, 168
Confounding, 144, 158
Conservative, 421
Consumer Expenditure Survey, 416
Continuity correction, 280–281, 285–286, 289–290
Contrasts, 483–489
Control, statistical, 15-7, 15-34, 15-38
Control chart, 15-6–15-9
   c chart, 15-51–15-53
   I chart, 15-28–15-31, 15-34
   MR chart, 15-28–15-31, 15-34
   p chart, 15-46–15-51, 15-53
   R chart, 15-13–15-21, 15-34
   s chart, 15-22–15-23, 15-34
   three-sigma, 15-12, 15-34
   $\bar{x}$ chart, 15-13–15-23, 15-34
Control chart constant, 15-13–15-14, 15-23–15-24
Control group, 147, 153
Convenience sample, 132
Correlation, 75–78, 102, 104
   and regression, 88–89, 96
   between random variables, 240–243
   cautions about, 99–104
   inference for, 587–589, 590
   population, 587, 590
   squared, 88, 96, 607, 610
   squared multiple, 641–642, 645
Count, 249–250, 259, 267–268, 325, 331
   distribution of, 250–252, 261–265, 267–268, 277–282
Countably infinite, 221, 226, 261
Critical value
   of chi-square distribution, 550, Table F
   of F distribution, 474, Table E
   of standard Normal distribution, Table A
   of t distribution, 398, Table D
Critical value approach, 371, 373
Cross-validation, 666
Cumulative probability. *See* Probability, cumulative
Cumulative proportion, 11, 48
Current Population Survey, 137, 139, 164
Curved relationships, 71, 72, 92, 601–602, 653–655

Data, 2–7
Data Warehouse, 128
Database, 128
Deciles, 60
Decision, relation to inference, 445–449
Degrees of freedom
   approximation for, 421, 432
   for a collection of explanatory variables, 643, 646
   for one-way ANOVA, 473–474
   for two-way ANOVA, 16-7, 16-14
   of chi-square distribution, 550, 556
   of chi-square test, 550, 556

   of correlation t, 588
   of F distribution, 473
   of noncentral F distribution, 502
   of one-sample t, 397, 414
   of pooled two-sample t, 428–429, 433
   of regression ANOVA, 607–608, 640, 645
   of regression s, 578, 590, 629
   of regression t, 583, 590, 636–637, 645
   of the variance, 33, 301–302
   of two-sample t, 421, 432
Deming, W. Edwards, 15-2, 15-36
Density curve, 40–41, 57, 224–227, 231, 273–275, 282–284
   skewed, 42–43, 57
   symmetric, 41–43, 57
Density estimation, 56
Design of an experiment, 143–158
Direction, of a relationship, 68, 72.
   See also Correlation
Disjoint events. *See* Event, disjoint
Distribution, 8, 22
   binomial, 249–255, 267
      and logistic regression, Chapter 18
      formula, 256–258, 267
      Normal approximation, 277–281, 285
      Poisson approximation, 264–265, 268
      use in the sign test, 411
   categorical variable, 8, 22
   chi-square, 550, 556
   conditional, 108, 114, 543
   exponential, 282–285
   F, 473
   hypergeometric, 252, 270–271
   joint, 545, 547, 554
   jointly Normal, 589
   logNormal, 600
   marginal, 106, 114, 543
   noncentral F, 493, 502
   noncentral t, 443, 449
   Normal, 44, 48–52, 57, 275–277, 284
      standard, 47, 58, 275
   Poisson, 261–262, 268
      Normal approximation, 281–282, 285
   probability. *See* Probability
   quantitative variable, 13, 22
   sampling. *See* Sampling distribution
   skewed, 19, 22
   symmetric, 19, 22
   t, 396–398, 414
   uniform, 43, 224–226, 231, 273–274, 284
Distribution-free procedure. *See* Nonparametric procedure

Effect size, 442
Elasticity, 584, 638
Empirical, 177
Environmental Protection Agency (EPA), 415–416
Equivalence testing, 407–408
Estimator, 299–303, 305, 311–312, 321, 325–326, 331

Ethics, 161–168
Excel, 2, 4, 86, 135, 150, 180–182, 247–248, 253, 263, 271, 275–276, 282–284, 286–287, 306, 323, 332, 355, 363, 390–391, 404, 427, 477, 498, 579, 594, 608, 622, 626, 637, 691, 716, 718–722, 724–725, 728, 732–733, 735–739, 747, 761, 763–767, 16-20
Expected cell count, 548–549, 556, 560, 563
Expected value, 230. *See also* Mean, of a random variable
Experiment, 127, 129, 459
   behavioral and social science, 167–168
   block design, 156–158
   cautions about, 154, 158
   comparative, 146–147, 158
   completely randomized, 148
   design of, 143–158
   matched pairs design, 156, 158
   principles, 152, 158
   randomized comparative, 147, 152
Experimental units, 144, 158
Explanatory variable. *See* Variable, explanatory
Exploratory data analysis, 8, 22
Exponential smoothing model, 725–730
Extrapolation, 99–100, 103
Event, 184, 194
   complement of, 185,–186, 194, 212–213
   disjoint, 185–186, 192, 194, 198–199, 213
   empty, 199
   independent, 191–192, 194, 211–213
   intersection, 205, 212–213
   mutually exclusive, 185, 194
   union, 198–199, 205, 212–213

F distribution. *See* Distribution, F
F test. *See* Significance test, F
Factor, experimental, 144, 158, 458, 16-2
Factorial, 257, 267
False-positive, 381
First difference, 695, 704, 707
Fisher, Sir R. A., 377, 449, 473
Five-number summary, 30, 36
Flowchart, 15-4–15-5
Forecast, 690, 692, 697
Forecast errors, 714
Form, of a relationship, 68, 72
Frequency. *See* Count
Friedman, Milton, 587

Galton, Sir Francis, 587
General Social Survey, 127
Goodness of fit, 559–563
Gosset, William S., 397
Grand mean, 15-12

Histogram, 13, 22
Hot hand, 234
Hypothesis, 355–358, 365, 372
   alternative, 358, 372
   null, 358, 372
   one-sided, 359, 372
   two-sided, 359, 372
Hypothesis testing. *See* Significance test

Independence
   in two-way tables, 554
   of events, 191–192, 194, 211–213
   of observations, 679, 692
   of random variables, 240–241, 246
   of trials, 179
Inference, statistical. *See* Statistical inference
Influential observations, 93, 95, 96
Informed consent, 163, 168
Institutional Review Boards (IRB), 162, 168
Instrument, 5
Interaction, 145, 16-8–16-12, 16-13
   terms in multiple regression, 662
Intercept of a line, 84, 96
Interquartile range, 31
Intervention, 127

JMP, 86, 111, 180–181, 212, 247, 260, 266, 272, 306, 323, 332, 363, 377, 387, 391, 404, 427, 445, 470, 477, 479, 484, 488, 494, 509, 513, 517, 526, 530, 533, 546, 579, 611, 614, 627, 637, 651, 666, 684, 688–689, 691–692, 706, 725, 728–730, 733, 743–744, 751–752, 754–755, 766, 15-17, 15-26, 15-33, 16-18, 16-20, 17-7–17-10, 17-12–17-13, 17-19–17-20, 17-23, 17-28–17-29, 18-12, 18-15, 18-19
Joint probability table, 201

Kerrich, John, 178–179, 190, 290
Kruskal-Wallis test, Chapter 17

Label, 2, 6
Lag variable, 686, 750
Lagging, 686
Large numbers, law of, 231–234
Leaf
   in a regression tree, 665
   in a stemplot, 17
Least-significant difference (LSD) method, 490
Least squares, 83, 87, 95, 384, 388, 625–626, 752
Least-squares line, 83, 95, 570
Level of a factor, 144, 465, 16-2
Linear relationship, 68, 72
Log transformation, 69, 72, 453, 467, 573, 700, 745–747, 751
Logistic regression. *See* Regression, logistic
Logit, 18-4, 18-9
Lurking variable. *See* Variable, lurking

Machine learning, 128
Main effects, 16-5, 16-7–16-12, 16-13
Major League Baseball (MLB), 206–208
Margin of error, 303, 305
   for a difference in two means, 421, 432
   for a difference in two proportions, 525, 535
   for a single mean, 342–352, 398, 414

Margin of error (*continued*)
    for a single proportion, 508, 519
    for regression mean, 598
    for regression response, 597–598
    for regression slope, 583, 636, 645
Marginal means, 16-8, 16-12
Matched pairs design. *See also* Experiments, matched pairs design
    inference for, 406–409, 411–412
Mean, 25, 36
    of a random variable, 219
    of binomial distribution, 259, 267
    of density curve, 42, 57, 231
    of difference of sample means, 420
    of difference of sample proportions, 524
    of exponential distribution, 282, 285
    of Normal distribution, 44–58, 275, 284
    of Poisson distribution, 262, 268
    of uniform distribution, 274, 284
    population, 232, 295
    rules for, 234–236, 247
    sample mean, 311, 321
    sample proportion, 326, 331, 507
Mean absolute deviation, 714, 731
Mean absolute percentage error, 715, 731
Mean response, estimated, 596–600
Mean square, 472
    in forecasting, 715, 731
    in multiple linear regression, 640
    in one-way ANOVA, 472, 480
    in simple linear regression, 608–609
    in two-way ANOVA, 16-14
Meandering, 680
Median, 26, 36
    inference for, 411–412
    of density curve, 42, 57
Meta-analysis, 128, 554–555
Minimum variance portfolio, 244
Minitab, 85, 180–181, 247, 253, 267, 272, 287, 306, 323, 332, 363, 391, 404, 427, 439, 443, 477, 494, 500, 509, 513, 518, 526, 530, 533, 546, 579, 588, 601, 623, 624, 627, 684, 687–689, 715, 725, 728–730, 736, 738–739, 741, 749, 751, 753–754, 757, 766, 15-17, 15-25, 15-30–15-31, 15-33, 16-20, 17-9–17-10, 17-19, 17-21, 17-23, 17-28–17-29, 18-12, 18-15, 18-16, 18-23
Model, mathematical, 40
Model building, 650
Mosaic plot, 110, 114, 192, 212, 547
Moving average model, 710–713
Moving range, 15-28
Multicollinearity. *See* Collinearity
Multiple comparisons, 489–492
Multiple regression. *See* Regression, multiple

Naïve forecast, 696–697, 711
National Association of Colleges and Employers (NACE), 451
National Collegiate Athletic Association (NCAA), 18-24
National Football League (NFL), 205–206, 592

National Hockey League (NHL), 671
National Longitudinal Survey of Youth (NLSY), 572
National Oceanic and Atmospheric Administration (NOAA), 614
National Science Foundation (NSF), 580
Natural Resources Canada, 596
Neyman, Jerzy, 449
Noncentrality parameter, 502
Nonlinear regression. *See* Regression, nonlinear
Nonparametric procedure, 410, Chapter 17
Nonresponse, 138, 141
Nonsense correlation, 102
Nonstationary process, 680
Normal distribution. *See* Distribution, Normal
Normal probability plot. *See* Normal quantile plot
Normal quantile plot, 53–54, 58
Normal score, 53
Null hypothesis. *See* Hypothesis, null

Observational study, 128, 129
Odds, 18-3
Odds ratio, 18-8, 18-9
Office of the Superintendent of Bankruptcy Canada (OSB), 415
Outliers, 19, 68, 72, 95
    $1.5 \times IQR$ rule, 31
Out-of-control signal, 15-17, 15-19–15-21, 15-34, 15-49–15-50, 15-52–15-53
Overdispersion, 261, 266
Overfitting, 663

Parameter, 295–296, 304
Pareto chart, 11, 22, 15-5
Partial autocorrelation function, 751, 761
Pattern of a distribution, 18–19, 22
Pearson, Egon, 449
Pearson, Karl, 178–179, 190
Percent, 8, 10
Percentile, 29
Pie chart, 10, 22
Placebo, 146
Placebo effect, 146
Poisson setting, 262, 268
Pooled estimator
    of population proportion, 528, 535
    of variance in ANOVA, 467, 472, 16-6
    of variance in two samples, 428, 433
Population, 130, 141, 294
Population regression equation, 571, 574, 589, 634
Portfolio analysis, 235
Power, 441–443, 449
    and sample size, 442
    and Type II error, 448
    for one-way ANOVA, 493–495
    for a single proportion, 517–518
    for two proportions, 531–533
    increasing, 442
    of *t* test
        one-sample, 443–444
        two-sample, 444–445

Power curve, 443, 494, 518, 533
Prediction, 84, 95, 384–388
Prediction interval, 385–388, 597–598, 602, 691–692, 747, 754–755
Predictive analytics, 8, 84, 383, 569, 596, 617, 676
Probability, 178–179
   conditional, 201–205, 212
   cumulative, 253, 263, 276, 283
   equally likely outcomes, 189–190
   finite sample space, 187
   joint, 201, 212
   marginal, 200, 213
   posterior, 210, 213
   prior, 210, 213
Probability distribution, 221, 226–227
   mean of, 219, 245
   standard deviation of, 239, 241, 245–246
   variance of, 238–241, 245–246
Probability histogram, 222, 227
Probability model, 182, 194
Probability rules, 185, 194, 213
   addition, 185, 194, 198–199, 213
   complement, 185, 194, 213
   multiplication, 191, 194, 203, 213
Probability sample. *See* Sample, probability
Probit regression, 18-8
Process, 15-2–15-3, 15-9
Proportion
   population, 295, 507
   sample, 295, 507
P-value, 362, 372

Quantitative variable. *See* Variable, quantitative
Quartiles, 28–29, 36
   of a density curve, 41
Queuing theory, 283

R, 180–181, 247, 253, 263, 306, 323, 332, 355, 363, 391–392, 580, 684–685, 689–690, 706–707, 17-8–17-9, 17-19, 17-21, 17-23, 17-29
Random deviations, 384, 388, 679, 694, 702–703, 758
Random digit dialing, 140
Random digits, 134, 141
Random phenomenon, 178–179
Random process, 679, 692
Random sample. *See* Sample
Random variable, 220–226
   continuous, 226–227
   discrete, 221–224, 226–227
   mean of, 230–231, 245–246
   standard deviation of, 238–239, 245
   variance of, 238–239, 245–246
Random walk, 694–695, 707
Randomize, how to, 149
Randomized comparative experiment.
      See Experiment, randomized comparative
Randomness, 176–179
Range statistic, 15-12

Rate, 5, 7
Rational subgroup, 15-10–15-11
REDCap, 128
Regression
   and correlation, 87–88, 96
   cautions about, 99–103
   conditions for, 581–582, 627–628
   equation, 83, 95–96, 626
   interaction terms, 662–664
   least-squares, 83, 95
   line, 81, 95
   logistic, Chapter 18
   model building, 650–664
   multicollinearity, 654
   multiple, 617–668
   nonlinear, 70
   polynomial, 655
   quadratic, 653–654, 666
   simple linear, Chapter 12
   standardized coefficients, 670
   variable selection methods, 662–663, 742–744
   variance inflation factor (VIF), 654, 668
   with categorical explanatory variables, 655–658
Regression fallacy, 587
Regression trees, 665–666
Rejection region, 371, 373
Relative risk, 534, 555
Replication, in experiments, 152, 168
Resample, 413
Residuals, 89–90, 96, 466, 573–574, 590, 626, 631, 702, 704, 715, 735, 747–749, 756, 761
   distribution, 93
   plots, 91, 500, 581, 612–613, 628–629, 736, 748–749, 753–754, 757
Resistant measure, 26, 27, 30, 31, 34, 37
Response bias, 139, 141, 346
Response rate, 131
Response variable. *See* Variable, response
Risk, investment, 35
Risk pooling, 243
Robustness, 409–410, 414, 424, 451, 480, 583
Rounding, 17
Row variable. *See* Variable, row and column
Run, in coin tossing, 233
Run chart, 15-5
Runs rules for control charts, 15-17, 15-36
Runs test for randomness, 682–685, 692

Sample, 130, 141, 295
   cautions about, 138
   convenience, 132
   design of, 131, 141
   frame, 131
   multistage, 137, 141
   probability, 136, 141
   simple random, 133, 141
   stratified, 136, 141
   survey, 127, 129
   systematic, 142–143

Sample size, choosing
    confidence interval for a difference in means, 440–441, 449
    confidence interval for a difference in proportions, 531, 535
    confidence interval for a mean, 349, 352, 438–440, 449
    confidence interval for a proportion, 514, 519
    for a desired power, 441–443, 449
    for one-way ANOVA, 493–495
    significance test for a difference in means, 444–445, 449
    significance test for a difference in proportions, 532, 536
    significance test for a mean, 443–444, 449
    significance test for a proportion, 517, 519
Sample space, 182, 194
    finite, 187
Sampling, 127, 130–141
Sampling distribution, 296–298, 304
    of difference of means, 420–421
    of difference of two proportions, 524
    of one sample t statistic, 396–397, 409
    of regression estimators, 582–583
    of sample mean, 312–313, 321
    of sample proportion, 329, 331, 507, 518
Sampling frame, 131
Sampling variability, 296
SAS, 440, 441, 491
Satterthwaite approximation, 421, 432, 437
Scatterplot, 66, 72, 15-6
Scatterplot matrix, 651
Scatterplot smoothing, 69, 572–573, 624–625, 18-7–18-8
Seasonal effects
    additive, 718, 731, 743–744, 761
    multiplicative, 718, 731, 744–745, 761
Seasonal ratio, 719–720, 731
Seasonality, 681–682, 692, 717–724, 731, 740–747, 756–757
Seasonally adjusted, 721, 724, 731
Secrist, Horace, 587
Shape of a distribution, 19–20
Shewhart, Walter, 15-2, 15-10
Sign test. *See* Significance test, sign test
Significance, statistical, 153, 364, 372
    and practical significance, 378–379
    and Type I error, 448
Significance level, 364, 376–378
Significance test, 355–373
    chi-square for goodness of fit, 560, 563
    chi-square for multiple logistic regression, 18-19–18-20
    chi-square for two-way table, 550, 556
    common practice, 448
    equivalence, 407–408
    F test for a collection of regression coefficients, 643, 646
    F test for one-way ANOVA, 473–474
    F test in multiple regression, 640–641, 645

    F test in regression, 609, 610
    F tests for two-way ANOVA, 16-15, 16-21
    Fisher exact test, 551
    multiple comparison procedure, 490
    runs test for randomness, 382–385
    sign test, 411–412, 562
    t test for contrast, 486
    t test for correlation, 588, 590
    t test for matched pairs, 406–407
    t test for multiple regression coefficients, 636–637, 645
    t test for one mean, 400–403
    t test for regression slope, 583
    t test for two means, 423–424
        pooled, 428–429
    using, 376–381
    z test for logistic regression coefficient, 18-11, 18-17, 18-19–18-20
    z test for one mean, 366–368
    z test for one proportion, 511, 519
    z test for two means, 420, 432
    z test for two proportions, 528–529, 535
Simple linear regression. *See* Regression, simple linear
Simple random sample. *See* Sample, simple random
Simpson's paradox, 111–112, 114
Simulation, 297
Six sigma quality, 290, 15-45
68–95–99.7 rule, 45, 57
Skewed distribution. *See* Distribution, skewed
Slope, 84, 95
Small numbers, law of, 233–234
Smoothing. *See* Scatterplot smoothing
Special cause, 15-8
Spread of a distribution, 28, 33, 36
Spreadsheet, 4. *See also* Excel
SPSS, 404, 427
Standard deviation, 33, 36
    of binomial distribution, 259, 268
    of density curve, 43, 57
    of difference of sample means, 420
    of difference of sample proportions, 524
    of Normal distribution, 44, 60
    of Poisson distribution, 262, 268
    of prediction error, 385
    of random variable, 239, 245
    of sample mean, 311, 321
    of sample proportion, 326, 331, 507
    pooling, 428–429, 433, 467
    rules for pooling, 467
Standard error, 396
    for regression prediction, 605
    of a contrast, 486
    of a difference of sample proportions, 525, 528
    of a sample mean, 396, 414
    of a sample proportion, 508, 518
    of logistic regression parameter, 18-11, 18-16
    of mean regression response, 605
    of regression intercept and slope, 605

pooled, 429, 433, 467, 665, 16-6
regression, 578, 590, 629, 631, 636, 645
Standard Normal distribution. *See* Distribution, Normal
Standard Normal Standardized observation, 47, 57
Stationary process, 679, 692
Statistic, 295, 304
Statistical inference, 294–305, 338–339
Statistical process control, 15-7–15-9
Statistical significance. *See* Significance, statistical
Stem-and-leaf plot. *See* Stemplot
Stemplot, 16, 22
back-to-back, 18
rounding, 17
splitting stems, 18
Strata, 136
Strength, of a relationship, 68, 72. *See also* Correlation
Subgroup, 15-10
Subjects, experimental, 144, 158
Subpopulation, 571, 634
Sums of squares, 471
for two-way ANOVA, 16-7
in multiple linear regression, 639–640
in one-way ANOVA, 471–472
in simple linear regression, 607–608
Survey, 127, 130–141
Symmetric distribution. *See* Distribution, symmetric

t confidence intervals. *See* Confidence interval, t
t distribution. *See* Distribution, t
t significance tests. *See* Significance test, t
Table
Table A Standard Normal Probabilities, 50–52
Table B Random digits, 134, 141
Table C Binomial probabilities, 254
Table D t distribution critical values, 398
Table E F critical values, 474
Table F χ^2 distribution critical values, 550
Test of significance. *See* Significance test
Test statistic, 360–361, 366, 367, 564, 588, 18-11, 18-17. *See also* Significance test
Testing hypotheses. *See* Significance test
Text, as data, 128
Three-way table, 113
Time plot, 20, 22, 677
Time series, 675–761
Total probability, law of, 208, 213
Training data set, 128
Transformation
arcsine square root, 502
linear, 503
log, 69–70, 72, 453, 573, 700, 745–747, 751
power, 15-32–15-33
square, 483
square root, 16-24
Treatment, experimental, 127, 129, 144, 147, 158
Tree diagram, 206–207

Trend, 680, 692
deterministic, 702–704, 707, 734–739
exponential, 738–739
linear, 702, 734–737
quadratic, 702, 737–738
stochastic, 702–704, 707
Tuskegee Study, 165
Two-sample problems, 419
Two-way table, 105–114, 200, 542–543
counts to probabilities, 200–201
data analysis for, 105–113
hypothesis, 548
inference for, 542–554
models for, 552–554
Type I and II error, 446–449

Unbiased estimator, 300
Uncountably infinite, 224
Undercoverage, 138, 141
Unit, of measurement, 3
United States Department of Agriculture (USDA), 426, 453

Validation data set, 128, 666
Value, 2, 6
Variability of a statistic, 300
Variable, 2, 6
categorical, 3, 7, 71
dependent, 64
explanatory, 64, 72
independent, 64
indicator, 656–657
lurking, 100–101, 103
nominal, 5
ordered, 5
quantitative, 3, 7
response, 64, 72
row and column, 105, 114
Variable selection methods, 662–663, 742
Variance, 33, 36. *See also* Standard deviation
of random variable, 238–239, 245–246
rules for, 240–241, 246
Variance inflation factor, 654
Variation
between-group, 461, 471, 481
within-group, 461, 471, 481
Venn diagram, 185
Volatility, 235
Voluntary response, 132, 141

Wilcoxon rank sum test, Chapter 17
Wilson estimate, 510, 527
Wording questions, 139, 141

z confidence intervals. *See* Confidence interval, z
z-score, 47, 57
z significance tests. *See* Significance test, z
Zero inflation, 266